Foundations
for
Advanced Mathematics

About the Cover Design

Pictured on the cover is a piece of a computer-generated image of the Mandelbrot Set. This set belongs to a family of shapes known as **fractals**, which have the special properties:

[1] **self-similarity** [for example, the resemblance in shape between each floret of a stalk of broccoli and the original stalk on which it appears] and

[2] a **fractional number of dimensions** [as opposed to the usual *whole* number of dimensions of a figure in 2-space or 3-space].

With the advent of computer technology, interest in Fractal Geometry, a branch of mathematics first discussed in the early part of the 20th century, was reawakened. The fractal set shown was named after the contemporary mathematician who was the first to glimpse its intricate contours. Awed by the wondrous detail, Mandelbrot set out to discover "entirely new shapes that nobody has seen anywhere."

Ann Xavier Gantert has been involved in mathematics education at all levels of instruction, from the primary grades through college. She is known for her contributions to the development of an integrated approach to the teaching of mathematics, and has coauthored a series of textbooks reflecting this philosophy.

Howard Brenner supervises and trains teachers of secondary mathematics, and has taught mathematics at the secondary and college levels. He has a special interest in preparing students for mathematics competitions, is active in professional organizations, and has served as the editor of a professional journal.

This text has been reviewed and endorsed by:

Teacher's Edition

Foundations for
Advanced Mathematics

Ann Xavier Gantert

Department of Mathematics
Nazareth Academy
Rochester, New York

Howard Brenner

Assistant Principal, Mathematics
Fort Hamilton High School
New York City

When ordering this book, please specify: *either* **N 566 T** *or*
FOUNDATIONS FOR ADVANCED MATHEMATICS
TEACHER'S EDITION

Dedicated to serving

AMSCO
SCHOOL PUBLICATIONS, INC.
315 HUDSON STREET, NEW YORK, N.Y. 10013

our nation's youth

ISBN 0-87720-284-2

PRINTED IN THE UNITED STATES OF AMERICA

1 2 3 4 5 6 7 8 9 10   99 98 97 96 95 94 93 92

Acknowledgements

Cover Design & Text Design
Electronic PrePress Production

K A. W. KINGSTON PUBLISHING, INC.
4515 S. McClintock Dr. Tempe, AZ

Illustrations and Line Art

EDWIN C. BUFFUM
CONSTANCE F. KIRWIN
DONNA L. FREEMAN

Contributed to Exercises for Challenge   KEITH CONRAD

Reviewed Chapter 13   LEONARD MALEY

Assisted in Preparing Solutions Manual   PHYLLIS BRENNER

Photo Credits

Stock • Boston

Photo	Page	Photographer
Ramses II, Egypt	1	ROBERT CAPUTO
Washington Monument	126	MARTIN ROGERS
Hydroponics	162	LIANE ENKELIS
Generators	242	NUBAR ALEXANIAN
Truck Fleet	286	MICHAEL DWYER
School Orchestra	347	BOB DAEMMRICH
Surveyors	414	ROBERT RATHE
Dish Antenna	564	PETER MENZEL
St. Louis Arch	581	FRANK SITEMAN
3 Mile Island	584	LIONEL DELEVINGNE
Pharmaceutical Lab	592	STACY PICK

Tony Stone Worldwide

Photo	Page	Photographer
U. S. Capitol Building	48	ED PRITCHARD
Archeologic Remains	208	DAVID AUSTEN

St. Louis Science Center

Photo	Page	Photographer
Planetarium Building	586	COURTESY PHOTO

Dedication

To Patricia   To Klabe
A. X. G.   H. B.

About *This Book*

Intention

Foundations for Advanced Mathematics is intended to help students develop the skills and self-confidence necessary to study Calculus, Discrete Mathematics, Statistics, or other college-level courses in higher mathematics.

As a prerequisite, students should have completed Geometry and Algebra II (either in a traditional or integrated approach), but thorough mastery of these subjects is not expected. This text, by starting each topic at ground-zero, reaches out to students at their actual level of understanding and offers the support to bring them to a more advanced level of accomplishment.

Students who will not pursue further studies in mathematics will find here a rich overview and intertwining of topics, to provide a sense of the strength and pervasiveness of the mathematical disciplines.

Topics and Presentation

The topics selected for inclusion were chosen from the compendium of topics generally accepted for study at this level, with emphasis on those areas recommended for increased attention by the Standards of the National Council of Teachers of Mathematics, a document recognized as the guide for the future of mathematics education throughout the United States.

- Problem solving is used as a motivation for and an outcome of the study of each topic.

- Unifying themes, such as the concepts of function and transformation, serve to integrate and connect topics.

- Real-world applications connect topics to a variety of disciplines, exposing students to the usefulness and power of mathematics.

- Technology is introduced to enhance understanding, encourage exploration, and allow the use of real data.

To achieve an appropriate level of mathematical rigor, both intuitive and formal approaches are employed. A variety of proofs, offered in different styles, are included in the text—direct proof, indirect proof, analytic proof, proof by mathematical induction. Additional proofs are reserved for the Teacher's Edition, thereby allowing the teacher to further customize the level of rigor.

Features of the Student Text

To assist students and teachers, this well-organized, comprehensive text includes:

- exposition written for student readability
- color highlights, graphics, and variations in typography for emphasis and clarity

Graphs of Inequalities

To graph an inequality that is bounded by an ellipse or hyperbola, work with the related equality.

For example, every point on the ellipse $\frac{x^2}{9} + \frac{y^2}{4} = 1$ satisfies the equation of the ellipse. The remaining points in the plane are either inside the ellipse, satisfying the inequality $\frac{x^2}{9} + \frac{y^2}{4} < 1$, or outside the ellipse, satisfying the inequality $\frac{x^2}{9} + \frac{y^2}{4} > 1$.

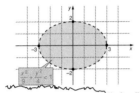 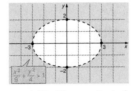

an ellipse on a calculator, it is necessary to enter the

- a calculator strand that teaches the underlying mathematics, using scientific and graphing calculators for discovery and computation

On the calculator, enter the two parts of the equation of the ellipse.

Under Y_1, enter:

2 [2nd] [√] [( ] 1 [−] [X|T] [x²] [÷] 9 [ )]

Under Y_2, enter:

[(−)] 2 [2nd] [√] [( ] 1 [−] [X|T] [x²] [÷] 9 [ )]

Set the range for x from −3.5 to 3.5 and for y from −3 to 3.

The draw function on the calculator can be used to display the solution sets of the inequalities related to the ellipse $\frac{x^2}{9} + \frac{y^2}{4} = 1$. Press [2nd] [DRAW] to display the menu. Then highlight **7:Shade(** and press [ENTER]. The instruction that appears on the screen must be completed by expressions for the lower boundary and upper boundary of the curve.

- detailed examples with step-by-step guidance

its appear

1	−5	6	7		$\underline{4}$
1					

Step 1
Write just the coefficients of the terms of the dividend and the opposite of the constant term of the divisor. Bring down the leading coefficient.

1	−5	6	7		$\underline{4}$
	4				
1	−1				

Step 2
Multiply the leading coefficient by the opposite of the constant term (1 × 4). Place the product under the second coefficient, −5, and add.

1	−5	6	7		$\underline{4}$
	4	−4			
1	−1	2			

Step 3
Multiply the sum from the previous step by the opposite of the constant term (−1 × 4). Place the product under the next coefficient, 6, and add.

1	−5	6	7		$\underline{4}$
	4	−4	8		
1	−1	2	15		

Step 4
Repeat the previous step until the last coefficient is used.

in the final line of the computation, 1 −1 2 15,

Example 2

Use synthetic division to find $P(-4)$ when $P(x) = 2x^4 - 13x^2 + x - 7$.

Solution

Rewrite the polynomial as $2x^4 + 0x^3 - 13x^2 + x - 7$ so that the coefficient of each power of x is represented. Synthetic division can be used to find $P(-4)$, since the Remainder Theorem states that $P(-4)$ is the remainder when $P(x)$ is divided by $x + 4$.

$$
\begin{array}{r|rrrrr}
-4 & 2 & 0 & -13 & 1 & -7 \\
 & & -8 & 32 & -76 & 300 \\
\hline
 & 2 & -8 & 19 & -75 & 293
\end{array}
$$

List the coefficients, write −4 as the synthetic divisor, and perform the synthetic division.

Answer: $P(-4) = 293$

- ample exercises that are graded in difficulty and varied in nature

Exercises 2.4 *(continued)*

47. The area of an equilateral triangle is $3\sqrt{2}$ in.². If the length of one side is $(\sqrt[4]{x+2})$ in., find x. (Use the formula $A = s^2 \frac{\sqrt{3}}{4}$.)

48. The area of a rectangle is $(\sqrt{x+5})$ ft.² and the length of the diagonal is $(\sqrt{x-18})$ ft. If one side of the rectangle is 3 ft., find the length of the other side.

49. If r, s, and t are real numbers greater than 0, prove that if $\sqrt{x+r} = t - \sqrt{x+s}$ has a real solution for x, then $t^2 + s \geq r$.

50. The formula for the surface area of a sphere is $s = 4\pi r^2$. If a spherical balloon is being filled with air so that its surface area is increasing at a rate of (6π) cm²/min. and the current radius is $(\frac{1}{2}\sqrt{10})$ cm, find the radius after 1 minute.

A graphing calculator or a computer program designed to graph functions entered by the user can often help to solve an equation. For example, the solution to the equation $\sqrt{x^2+7} = 10 - \sqrt{39-x}$ can be found by drawing the graphs of $y = \sqrt{x^2+7}$ and $y = 10 - \sqrt{39-x}$ and finding the x-coordinate of the point of intersection.

Begin by selecting the range of values for x and y. Enter a large range in order to determine the part of the graph at which the point or points of intersection are located. For example, enter values of x and y from −20 to 20, a scale of 1 for both x and y, and a resolution of 1.

[RANGE] [(−)] 20 [ENTER] 20 [ENTER] 1 [ENTER]

[(−)] 20 [ENTER] 20 [ENTER] 1 [ENTER] 1 [ENTER]

Now enter the equations as y_1 and y_2.

[Y=] [2nd] [√] [( ] [X|T] [x²] [+] 7 [ )] [ENTER]

10 [−] [2nd] [√] [( ] 39 [−] [X|T] [ )] [GRAPH]

Note that there are two points at which the graphs intersect. Move the cursor to the intersection at the right, press [ZOOM] and select **2:Zoom In**. The section about the point of intersection will be enlarged. Press [TRACE] and then move the cursor as close as possible to the intersection. The values are $x = 3$, $y = 4$. Repeat the use of [ZOOM] and [TRACE] to find the coordinates of the other point of intersection. For this point, the values, to the nearest tenth, are $x = -2.4$, $y = 3.6$. The two x values, 3 and −2.4, are the solutions of the equation.

51. [calc] E is the midpoint of side $\overline{AB}$ of rectangle $ABCD$ and F is a point on side $\overline{BC}$. If $AD = 9$, $BF = 5$, and $DE + EF = 28$, find AB.

to the nearest

Each chapter of the text contains:

- **Chapter Summary and Review**

  The important concepts developed in each chapter are summarized in outline form. A set of review exercises reinforces understanding.

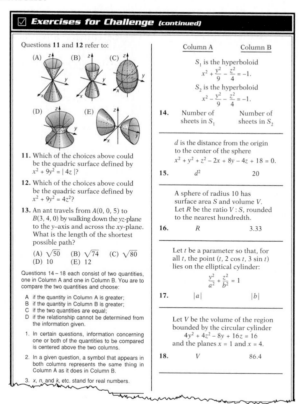

- **Exercises for Challenge**

  Additional nonroutine situations are presented in the format of college entrance and achievement examinations.

Answers to the odd-numbered exercises are bound into the student text. Also available are a separate *Answer Key* containing the answers to all of the exercises, and a *Solutions Manual* with detailed solutions.

Features of the Teacher's Edition

To make this Teacher's Edition useful and comfortable, it contains:

- the complete student text, with exposition in full-size type
- answers to all of the exercises, placed in context—with additional answer pages inserted as needed

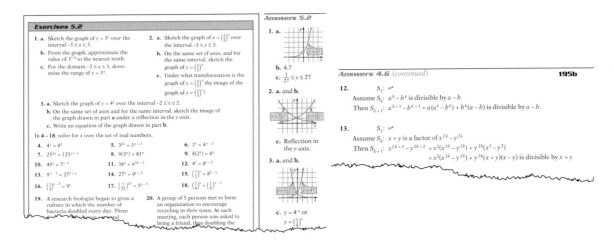

The extensive teacher commentary preceding each chapter includes:
- a detailed listing of the contents of each section of the chapter, followed by an overview

Chapter Contents

2.1 First-Degree Equations and Inequalities
algebraic equations or inequalities; domain of a variable; root of an equation or inequality; solution set; algorithm; examples of various first-degree equations or inequalities, including those with rational expressions; values not in the domain of a variable; literal equations

2.2 Using an Open Sentence to Solve a Problem
steps for an algebraic solution to a word problem; problems where the product of two quantities equals a third, such as coin, motion, work, investment, and mixture problems; average rate problems

2.3 Quadratic Equations and Inequalities
standard form of a quadratic equation; values of a, b, and c; solution by factoring; completion of the square; quadratic formula; nature of the roots using the discriminant; solution of quadratic inequalities by factoring; word problems leading to quadratic equations and inequalities

2.4 Additional Techniques for Solving Equations
solution set obtained by raising each side of an equation to a power; radical equations; extraneous solutions; changing equations to quadratic form; word problems involving radicals and powers

2.5 Solving Equations of Degree Greater Than 2
degree of a polynomial equation; Remainder Theorem; Factor Theorem; factors of the polynomial in a polynomial equation

Suggestions for enrichment include:
- supplemental topics and computer programs

The inequality $\frac{x^2}{9} + \frac{y^2}{4} < 1$ represents the region enclosed by the ellipse $\frac{x^2}{9} + \frac{y^2}{4} = 1$. The area of this region can be approximated by using probability, first approximating the area in the first quadrant. If a point in the region enclosed by the rectangle $ABCD$ is chosen at random, the probability that that point lies in the region enclosed by the ellipse is:

$$\frac{\text{Area of the ellipse}}{\text{Area of } ABCD}$$

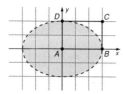

This value can be found empirically by generating the coordinates of T random points in the region enclosed by the rectangle $ABCD$ and counting the number, S, that lie in the interior of the ellipse. The probability that a point lies in the region enclosed by the ellipse is $\frac{S}{T}$. Therefore:

$$\frac{\text{Area of the ellipse}}{\text{Area of } ABCD} = \frac{S}{T}$$

$$\text{Area of the ellipse} = \frac{S}{T} \times \text{Area of } ABCD$$

This method of finding an area is called the *Monte Carlo Method*.

The following computer program can be used to generate the coordinates of 100 points, test the coordinates in the inequality that defines the area enclosed by the ellipse, and count the number of points that satisfy that inequality. The pairs are chosen for that portion of the ellipse in the first quadrant by generating x-coordinates from 0 to 3 and y-coordinates from 0 to 2. Since the area of $ABCD$ is 6, the area of one-quarter of the ellipse is $6\left(\frac{S}{100}\right)$, and the area of the entire region enclosed by the ellipse is $4(6)\left(\frac{S}{100}\right)$ or $24\left(\frac{S}{T}\right)$.

```
100 REM  THIS PROGRAM WILL
          GENERATE RANDOM NUMBERS,
110 REM  TEST THE PAIR AS A SOLUTION
          OF AN INEQUALITY,
120 REM  AND USE A PROPORTION TO
          FIND THE AREA UNDER A
          CURVE.
130 S = 0
140 FOR I = 1 TO 100
150 X = RND(1) * 3
160 Y = RND(1) * 2
170 IF X^2 / 9 + Y^2 / 4 < 1 THEN
          S = S + 1
180 NEXT I
190 A = 24 * S / 100
200 PRINT "A = ";A
210 END
```

Run the program several times and find the average of the areas generated. Compare this average with the theoretical area given by the formula $A = \pi a b$.

The method described above can be used to approximate the area bounded above by

- activities to promote understanding and discovery

Activity (*10.4*) _____

Directions:
    Discoveries:
        What If?

- proofs that further customize the level of mathematical rigor

You may wish to introduce some vector proofs of familiar geometric theorems. Begin by asking students to study the following vector proof. Preliminary to the proof are two definitions and a theorem, which refer to the diagram shown.

■ **Definition:** *Quadrilateral ABCD is a parallelogram if* $\overrightarrow{AB} = \overrightarrow{DC}$. (*Ask students to explain.*)

■ **Definition:** *If M is the midpoint of* $\overrightarrow{AC}$, *then* $\overrightarrow{AM} = \overrightarrow{MC}$.

□ **Theorem:** *Any vector can be expressed uniquely as the sum of scalar multiples of two given nonzero vectors, as long as the two vectors do not have the same direction.*

Prove: The diagonals of a parallelogram bisect each other.

In parallelogram $ABCD$, point M is the intersection of the diagonals. Therefore, $\overrightarrow{AM} = s\overrightarrow{AC}$, $\overrightarrow{DM} = t\overrightarrow{DB}$, $\overrightarrow{MC} = \overrightarrow{AC} - s\overrightarrow{AC}$, and $\overrightarrow{MB} = \overrightarrow{DB} - t\overrightarrow{DB}$.

By vector addition: $\overrightarrow{AD} + t\overrightarrow{DB} = \overrightarrow{AM}$ and $\overrightarrow{MB} + \overrightarrow{BC} = \overrightarrow{MC}$

By substitution: $\overrightarrow{AD} + t\overrightarrow{DB} = s\overrightarrow{AC}$ and $(\overrightarrow{DB} - t\overrightarrow{DB}) + \overrightarrow{BC} = (\overrightarrow{AC} - s\overrightarrow{AC})$

Since $ABCD$ is a parallelogram, $\overrightarrow{AD} = \overrightarrow{BC}$. Therefore, $\overrightarrow{BC} + t\overrightarrow{DB} = s\overrightarrow{AC}$ and $\overrightarrow{BC} = s\overrightarrow{AC} - t\overrightarrow{DB}$

$\overrightarrow{AC} - t\overrightarrow{DB} = \overrightarrow{AC} - s\overrightarrow{AC}$ or $(1 - t)\overrightarrow{DB} + s\overrightarrow{AC} = t\overrightarrow{DB} + (1 - s)\overrightarrow{AC}$

- for each chapter:

Suggested Test Items

    Bonus   **Answers to Suggested Test Items**

Course Planning Guide

This text can be used for a number of one-year or one-semester courses, as shown in the following proposed course outlines. Note that the suggested amount of instructional time is given both in days, assuming a typical 40 – 50 minute high school period, and in hours, assuming 45 – 60 hours for a typical one-semester college course.

One-Year Pre-Calculus for Students Who Have Studied Trigonometry

	Chapter	Days	Hours
1	Real and Complex Numbers	4 – 6	3 – 4
	quick review 1.1 - 1.5; cover 1.6 - 1.7		
2	Equations and Inequalities	13 – 16	10 – 13
	quick review 2.1; cover 2.2 - 2.7		
3	Functions and Their Graphs	18 – 20	14 – 16
4	Sequences, Series, and Induction	12 – 15	10 – 12
5	Exponents and Logarithms	10 – 12	9 – 10
6	Vectors	15 – 18	13 – 14
7	Matrices	8 – 10	6 – 8
	omit 7.4 - 7.6		
8	The Trigonometric Functions	4 – 6	3 – 4
	omit 8.1 - 8.4		
9	Applications of Trigonometry	12 – 15	10 – 13
10	The Conic Sections	10 – 12	8 – 10
11	Polar Coordinates	12 – 15	10 – 12
	Total	**118 – 145**	**96 – 116**

One-Year Pre-Calculus for Students Who Have Not Studied Trigonometry

		Days	Hours
1	Real and Complex Numbers	10 – 12	8 – 10
2	Equations and Inequalities	15 – 18	12 – 14
3	Functions and Their Graphs	18 – 20	14 – 16
4	Sequences, Series, and Induction	12 – 15	10 – 12
5	Exponents and Logarithms	10 – 12	8 – 10
6	Vectors	15 – 18	12 – 14
8	The Trigonometric Functions	16 – 20	13 – 16
9	Applications of Trigonometry	16 – 20	13 – 16
10	The Conic Sections	8 – 10	6 – 8
	Total	**120 – 145**	**96 – 116**

One-Year Liberal Arts Course

		Days	Hours
2	Equations and Inequalities	15 – 18	12 – 14
3	Functions and Their Graphs	18 – 20	14 – 16
4	Sequences, Series, and Induction	12 – 15	10 – 12
5	Exponents and Logarithms	10 – 12	8 – 10
6	Vectors	15 – 18	12 – 14
7	Matrices	8 – 10	6 – 8
	omit 7.4 - 7.6		
10	The Conic Sections	10 – 12	8 – 10
11	Polar Coordinates	10 – 12	8 – 10
	omit 11.5 - 11.6		
12	Curved Surfaces	9 – 12	7 – 9
13	Statistics	13 – 16	10 – 13
	Total	**120 – 145**	**95 – 116**

One-Semester Course in Pre-Calculus

	Chapter	Days	Hours
1	Real and Complex Numbers *quick review 1.1 – 1.5; cover 1.6 – 1.7*	4 – 6	3 – 4
2	Equations and Inequalities *quick review 2.1; cover 2.2 – 2.7*	13 – 16	10 – 13
3	Functions and Their Graphs	16 – 20	13 – 16
4	Sequences, Series, and Induction *omit 4.6*	10 – 12	8 – 10
5	Exponents and Logarithms *omit 5.1 – 5.2*	6 – 8	5 – 6
10	The Conic Sections	10 – 12	8 – 10
	Total	**59 – 74**	**47 – 59**

One-Semester Course in Discrete Mathematics

	Chapter	Days	Hours
3	Functions and Their Graphs *quick review 3.1 – 3.2; cover 3.3 – 3.7*	13 – 16	10 – 13
4	Sequences, Series, and Induction *omit 4.1 – 4.4*	6 – 9	5 – 7
6	Vectors	14 – 17	11 – 14
7	Matrices	13 – 16	10 – 13
13	Statistics	14 – 17	11 – 14
	Total	**60 – 75**	**47 – 61**

One-Semester Course in Advanced Algebra

	Chapter	Days	Hours
1	Real and Complex Numbers *quick review 1.1 – 1.3; cover 1.4 – 1.7*	5 – 8	4 – 6
2	Equations and Inequalities	16 – 18	13 – 14
4	Sequences, Series, and Induction	12 – 19	10 – 15
5	Exponents and Logarithms	10 – 12	8 – 10
6	Vectors	16 – 18	13 – 14
	Total	**59 – 75**	**48 – 59**

One-Semester Course in Analysis

	Chapter	Days	Hours
1	Real and Complex Numbers	10 – 12	8 – 10
2	Equations and Inequalities	15 – 17	12 – 14
3	Functions and Their Graphs	15 – 17	12 – 14
5	Exponents and Logarithms	10 – 12	8 – 10
10	The Conic Sections	10 – 12	8 – 10
	Total	**60 – 70**	**48 – 58**

One-Semester Course in Analytic Geometry

	Chapter	Days	Hours
3	Functions and Their Graphs	15 – 20	12 – 16
6	Vectors	15 – 18	13 – 14
10	The Conic Sections	10 – 12	8 – 10
11	Polar Coordinates	12 – 15	10 – 12
12	Curved Surfaces	8 – 10	6 – 8
	Total	**60 – 75**	**49 – 60**

Table of *Contents*

Chapter 9   *Applications of Trigonometry*

Chapter 10   *The Conic Sections*

Chapter 11   *Polar Coordinates*

Chapter 12 — Curved Surfaces

Chapter 13 — Statistics

Appendix — Tables

Teacher's Chapter *1*

Real and Complex Numbers

Chapter Contents

At the beginning of any course, it is necessary to review some basic concepts in order to establish the foundations on which the course will be built. However, review alone does not take full advantage of the enthusiam present at the beginning of a new year. This first chapter utilizes enough new approaches to the material being reviewed to challenge creativity and to maintain interest.

1.1   Natural Numbers and Integers

Each year, for many years, students have re-examined the properties of the set of natural numbers, the set of integers, and equality as an introduction to their work in mathematics. Each time, the level of maturity they bring to the review should increase their awareness of the importance of the principles outlined. Each successive set of numbers builds on the previously defined set, except the first set, the natural numbers, which remains undefined.

Begin by asking for a reply to a young child who wants to know, "What is three?" Most students will reply by pointing out a set of three objects: three fingers, three pencils, three books, etc. The discussion should highlight the difficulty of explaining a concept that is not based on previously-defined terms.

This first section establishes the natural numbers as an undefined set and the way in which the number line allows visualization of some of the properties assigned to that set. Contrast the importance of the order of the set of natural numbers with the arbitrary order of the alphabet.

The first four of Peano's Postulates are the foundations of addition. Compare these postulates, which enable us to prove that addition is always possible, to the names and symbols for the natural numbers together with the addition facts memorized as a child, which enable us to identify $3 + (2 + 7)$ as the number 12. The fifth postulate, often referred to as the *induction postulate*, is not used in this chapter but is presented here for completeness.

It is the basis of the Principle of Mathematical Induction that students will study in Chapter 4.

The limitations on the set of natural numbers motivate the introduction of 0 and the negative numbers. The operations on the integers are then defined in terms of related operations on the natural numbers.

1.2   Factors and Multiples of the Integers

The theorems of this section are basic to the proofs of some important theorems that will be presented in later chapters. For example, the Divisibility of a Product Theorem is needed to prove the Rational Roots Theorem for the solution of higher-degree equations. The proofs are not rigorous proofs but are sufficiently convincing for students at this level.

Two methods for finding the greatest common factor are given. The use of prime factorization is often simpler for numbers whose prime factors are easily obtained. The Euclidean Algorithm is demonstrated with an example. Since the remainders in this sequence of divisions form a decreasing set of natural numbers, the process must end in a finite number of steps because of the existence of a smallest natural number. The repetitive process makes it possible to perform the computations by means of a simple BASIC program.

Computer Activity (1.2)

```
100  REM THIS PROGRAM WILL FIND THE
110  REM GREATEST COMMON FACTOR OF
120  REM TWO INTEGERS.
130  PRINT "TO FIND THE GREATEST
              COMMON FACTOR OF"
140  PRINT "TWO INTEGERS, ENTER
              THE INTEGERS"
150  PRINT "SEPARATED BY A COMMA."
160  INPUT A,B
170  PRINT
180  LET A1 = A
190  LET B1 = B
200  IF A > B THEN GOSUB 330
210  LET R = B - INT (B / A) * A
220  IF R = 0 THEN 260
230  LET B = A
240  LET A = R
```

```
250   GOTO 210
260   PRINT "THE GREATEST COMMON
              FACTOR OF"
270   PRINT A1; " AND "; B1; " IS "; A
280   PRINT
290   PRINT "DO YOU WANT TO DO
              ANOTHER?(Y OR N)"
300   INPUT R$
310   IF R$ = "Y" OR R$ = "YES" THEN 130
320   GOTO 370
330   LET T = A
340   LET A = B
350   LET B = T
360   RETURN
370   END
```

1.3   The Real Numbers

This section uses the decimal representations of real numbers to distinguish rational from irrational numbers.

Ask if it is possible to predict the length of the block of digits that repeat for a given rational number. Students should be aware that the repeating decimal for $\frac{a}{b}$ can have no more than $b - 1$ digits in the block of repeating digits.

Students should recognize the difference between a decimal in which no repetition is apparent in the digits displayed, such as 0.142857..., which may be rational or irrational, and one in which a pattern ensures that repetition cannot occur, such as 0.122333444455555.... Emphasize that no matter how many nonrepeating digits have been determined for the decimal representation of a number, no finite number of nonrepeating digits can establish that a number is irrational. The proof that a number is irrational requires reasoning that is independent of a finite portion of the decimal expansion.

1.4   Properties of the Real Numbers

After reviewing the field properties of the set of real numbers, this section uses *positive* as an undefined term to define *greater than*. It can be proved as a theorem that a positive number is greater than 0.

Theorem: If a is positive, $a > 0$.
  Proof: $a = 0 + a$; therefore, a is the positive number required by the definition to conclude that $a > 0$.

Theorem: If a is positive, then $-a < 0$.
  Proof: $0 = -a + a$; therefore, a is the positive number required by the definition to conclude that $0 > -a$. But $0 > -a$ implies $-a < 0$.

These proofs seem so simple that some students fail to appreciate that anything has been proved. Elicit a description of what has happened, mentioning that as the study of mathematical structure progresses, assumptions that were accepted informally in earlier studies can now be proven.

1.5   Complex Numbers

Some students wonder if the set of complex numbers is a subset of some larger set. Establish that the set of complex numbers completes the development of the number system commonly used today.

The text demonstrates the addition of two complex numbers as a vector addition in the complex plane.

Note: The text uses overbar notation: $\overline{AB}$ for line segment AB, $\overrightarrow{AB}$ for a ray, $\overleftrightarrow{AB}$ for a line, and $\vec{AB}$ for a vector. Without an overbar, AB represents the distance between A and B.

Activity (1.5) _____

Directions

This activity demonstrates the product $(2 + i)(-1 + 3i)$. By the distributive property, $(2 + i)(-1 + 3i) = 2(-1 + 3i) + i(-1 + 3i)$.

1. Draw the real and imaginary axes on a sheet of graph paper.
2. Graph $-1 + 3i$ on the complex plane as A.

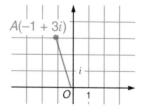

3. $2(-1 + 3i) = -2 + 6i$

Multiply $-1 + 3i$ by 2 by finding point A', the image of point A under a dilation D_2. Point A' is on $\overrightarrow{OA}$ at a distance from the origin equal to $2 \cdot OA$.

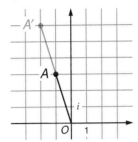

4. $i(-1 + 3i) = -i + 3i^2 = -i - 3 = -3 - i$

Multiply $-1 + 3i$ by i by finding point B, the image of A under a counterclockwise rotation of 90° about the origin. $\angle AOB$ is a right angle, and $OA = OB$.

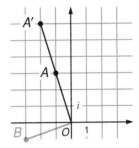

5. Use vector addition of $\overrightarrow{OA'}$ and $\overrightarrow{OB}$ to find point C. Point C is the graph of $-5 + 5i$, the product of $2 + i$ and $-1 + 3i$.

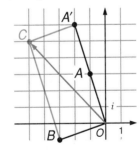

To show graphically that multiplication of complex numbers is commutative, repeat Steps 1 – 5 using $-1(2 + i) + 3i(2 + i)$. To multiply by a negative number, reflect over the origin. To multiply by $3i$, first rotate 90° (to multiply by i), then use the dilation D_3.

Choose other pairs of complex numbers and perform the multiplication graphically. Compare the result with the product obtained algebraically.

Observations

1. Multiplication by a positive constant is equivalent to a dilation.

2. Multiplication by a negative constant is equivalent to a dilation and a reflection over the origin.

3. Multiplication by i is equivalent to a counterclockwise rotation of 90° about the origin.

1.6 Polynomials

This section reviews familiar operations with polynomials, emphasizing their similarity to the integers. Like the integers, the polynomials are closed, commutative, and associative under addition and multiplication. There is an identity polynomial for addition, 0, and for multiplication, 1. Every polynomial has an inverse for addition, but only polynomials of degree 0 have inverses for multiplication.

1.7 Factoring Polynomials

After reviewing familiar procedures for factoring over the set of integers, this section introduces factoring over the set of complex numbers.

Students may ask about factoring trinomials with leading coefficients other than 1. Trinomials such as $3x^2 - x + 1$ can be factored over the set of complex numbers using either of the following methods.

Method 1

$$3x^2 - x + 1 = 3\left(x^2 - \tfrac{1}{3}x + \tfrac{1}{3}\right)$$

$$= 3\left(x^2 - \tfrac{1}{3}x + \tfrac{1}{36} + \tfrac{11}{36}\right)$$

$$= 3\left[\left(x - \tfrac{1}{6}\right)^2 - \left(\tfrac{\sqrt{11}}{6}i\right)^2\right]$$

$$= 3\left(x - \tfrac{1}{6} + \tfrac{\sqrt{11}}{6}i\right)\left(x - \tfrac{1}{6} - \tfrac{\sqrt{11}}{6}i\right)$$

Method 2

$3x^2 - x + 1 = \frac{1}{3}(9x^2 - 3x + 3)$

$$= \frac{1}{3}\left(9x^2 - 3x + \frac{1}{4} + \frac{11}{4}\right)$$

$$= \frac{1}{3}\left[\left(3x - \frac{1}{2}\right)^2 - \left(\frac{\sqrt{11}}{2}i\right)^2\right]$$

$$= \frac{1}{3}\left(3x - \frac{1}{2} + \frac{\sqrt{11}}{2}i\right)\left(3x - \frac{1}{2} - \frac{\sqrt{11}}{2}i\right)$$

You may choose not to include factoring trinomials over the complex numbers using the method of completing the square and expressing the result as the difference of two squares (as shown in the text on page 37). If so, omit Exercises 31, 32, 34, and 36.

Suggested Test Items

In **1 – 5**, list the symbols for all of the sets to which the number belongs. Let N = {Natural Numbers}, W = {Whole Numbers}, Z = {Integers}, Q = {Rational Numbers}, R = {Real Numbers}, and C = {Complex Numbers}.

1. 7　　　　**2.** $\sqrt{7}$　　　　**3.** $\frac{3}{5}$　　　　**4.** -5　　　　**5.** $0.1\overline{5}$

In **6 – 9**, write the additive inverse of the given number if it exists in the given set.

6. $4 \in N$　　　　**7.** $4 \in Z$　　　　**8.** $\sqrt{7} \in R$　　　　**9.** $-1 + i \in C$

In **10 – 13**, write the multiplicative inverse of the given number if it exists in the given set.

10. $4 \in N$　　　　**11.** $4 \in Q$　　　　**12.** $\sqrt{2} \in R$　　　　**13.** $-1 + i \in C$

14. Write $\frac{7}{12}$ as a repeating decimal.　　　**15.** Write $0.2\overline{6}$ as the ratio of integers.

In **16 – 21**, perform the indicated operation and write the result in $a + bi$ form.

16. $(-1 + i) + (1 - i)$　　　**17.** $(-1 + i)(1 - i)$　　　**18.** $(2 - 3i)(-2i)$

19. $1 \div (3 - 4i)$　　　**20.** $(5 - 3i)^2$　　　**21.** $(1 + 7i) \div (1 + i)$

In **22 – 24**, perform the indicated operation.

22. $(4x - x^2) + (5 + 2x^2)$　　　**23.** $(x^3 - 27) \div (x - 3)$　　　**24.** $(5x^2 - 2x)(x - 1)$

25. Find the quotient and remainder when $(3x^3 - 5x^2 + 3)$ is divided by $(x^2 - 2)$.

In **26 – 33**, write the expression in simplest form.

26. $2\sqrt{28} - \sqrt{63}$　　　**27.** $\dfrac{-6 + 2\sqrt{45}}{3}$　　　**28.** $8\sqrt{3}(2\sqrt{12})$　　　**29.** $\dfrac{1}{x} - \dfrac{1}{x - 3}$

30. $\dfrac{1}{x^2 + 8x + 12} + \dfrac{1}{x^2 - 4}$　　　　　**31.** $\dfrac{x^2 - x}{x^2 - 1}(x + 1)$

32. $\dfrac{2}{x^2 - 9} + \dfrac{1}{3x + 9}$　　　　　**33.** $\dfrac{x^2 - 5x}{5x + 25} \div \dfrac{x^2 - 10x + 25}{x^2 - 25}$

In **34 – 39**, factor over the set of integers.

34. $12x^3 - 3x$ **35.** $8x^2 - 2x - 3$ **36.** $x^4 - 5x^2 + 4$

37. $36x^2 - 9$ **38.** $8x^3 - 1$ **39.** $20x^5 - 5x$

In **40 – 45**, factor over the set of complex numbers.

40. $x^2 + 4$ **41.** $x^2 - 7$ **42.** $x^4 - 25$

43. $x^3 - 8x^2 + 2x - 16$ **44.** $x^2 - 4x + 2$ **45.** $x^4 + 4x^2 - 1$

46. Find the solution set of $12 - 3x > 6$.

47. Find the greatest common factor of 72 and 120.

48. 🖩 Find the GCF of 2,405 and 910.

Bonus

On day 1, a librarian put 50 new books, which had been numbered 1 through 50, out on a special display shelf. On day 2, every second book was checked out. On day 3, every book with a number divisible by 3 was checked out if it was still on the shelf, or returned if it had been out. On day 4, every book with a number divisible by 4 was checked out if it was on the shelf, or returned if it had been out. This pattern continued for fifty days. At the end of day 50, which books were on the shelf?

Answers to Suggested Test Items

1. N, W, Z, Q, R, C **2.** R, C

3. Q, R, C **4.** Z, Q, R, C

5. Q, R, C **6.** Does not exist

7. -4 **8.** $-\sqrt{7}$

9. $1 - i$ **10.** Does not exist

11. $\frac{1}{4}$ **12.** $\frac{1}{\sqrt{2}}$ or $\frac{\sqrt{2}}{2}$

13. $\frac{1}{-1 + i}$ or $\frac{-1 - i}{2} = -\frac{1}{2} - \frac{1}{2}i$

14. $0.58\overline{3}$ **15.** $\frac{4}{15}$

16. $0 + 0i$ **17.** $0 + 2i$

18. $-6 - 4i$ **19.** $\frac{3}{25} + \frac{4}{25}i$

20. $16 - 30i$ **21.** $4 + 3i$

22. $x^2 + 4x + 5$ **23.** $x^2 + 3x + 9$

24. $5x^3 - 7x^2 + 2x$

25. Quotient $3x - 5$; Remainder $6x - 7$

26. $\sqrt{7}$ **27.** $-2 + 2\sqrt{5}$

28. 96 **29.** $-\dfrac{3}{x(x - 3)}$

30. $\dfrac{2}{(x + 6)(x - 2)}$ **31.** x

32. $\dfrac{1}{3(x - 3)}$ **33.** $\dfrac{x}{5}$

34. $3x(2x + 1)(2x - 1)$

35. $(2x + 1)(4x - 3)$

36. $(x + 2)(x - 2)(x + 1)(x - 1)$

37. $9(2x + 1)(2x - 1)$

38. $(2x - 1)(4x^2 + 2x + 1)$

39. $5x(2x^2 + 1)(2x^2 - 1)$

40. $(x + 2i)(x - 2i)$

41. $(x + \sqrt{7})(x - \sqrt{7})$

42. $(x + i\sqrt{5})(x - i\sqrt{5})(x + \sqrt{5})(x - \sqrt{5})$

43. $(x + i\sqrt{2})(x - i\sqrt{2})(x - 8)$

44. $(x - 2 + \sqrt{2})(x - 2 - \sqrt{2})$

45. $(x^2 + 2 + \sqrt{5})(x^2 + 2 - \sqrt{5})$

46. $\{x \mid x < 2\}$

47. 24

48. 65

Bonus

All the books numbered 1, 4, 9, 16, 25, 36, and 49; i.e., all the books with numbers that are squares.

To determine whether a book will be in the library or out, consider how many factors are in the book's number. Since no books were taken out on Day 1, do not count 1 as a factor.

If 1 is not counted, most numbers have an odd number of factors. Books with such numbers will be out at the end of Day 50.

Example: $12 = 1 \cdot 12 = 2 \cdot 6 = 3 \cdot 4$

There are 5 factors, namely 2, 3, 4, 6, and 12. The book will be out on Day 2, in on Day 3, out on Day 4, in on Day 6, out on Day 12, and will not come in again.

Again not counting 1, only perfect squares have an even number of factors, since one factor repeats.

Example: $16 = 1 \cdot 16 = 2 \cdot 8 = 4 \cdot 4$

There are 4 factors, namely 2, 4, 8, and 16. The book will be out on Day 2, in on Day 4, out on Day 8, in on Day 16, and will not go out again.

Book number 1 has remained on the shelf. The other books on the shelf at the end of Day 50 are those numbered 4, 9, 16, 25, 36, and 49.

Chapter 1

Real and Complex Numbers

*T*he records of early civilizations contain an almost universal use of a single line or stroke to represent the number one and the repetition of that symbol to represent two, three, and so on. The characters used to represent numbers have developed through the ages, and as symbolism developed, so has the understanding of numbers as a basic tool of mathematics.

Chapter Table of Contents

1.1   *Natural Numbers and Integers*

In the history of humankind, counting was the first mathematical task. In the numerical structure of mathematics, the numbers that answer the question "How many?" are the most basic.

The *natural numbers*, $N = \{1, 2, 3, 4, ...\}$, are also called the *counting numbers*. The process of counting highlights one of the fundamental properties of the natural numbers: for each natural number there is a next, greater natural number, or *successor*. With this concept, the Italian mathematician Peano demonstrated in 1890 that the entire natural number system could be derived from 5 basic postulates.

Postulates of the Natural Numbers

- There exists a natural number 1.
- Every natural number has a successor. The successor of *a* is *a* + 1 and the antecedent (preceding number) of *a* + 1 is *a*.
- The number 1 has no antecedent.
- If two successors are equal, then their antecedents are equal. If *a* + 1 = *b* + 1, then *a* = *b*.
- Let *S* be a set of natural numbers for which the following are true:
  1 ∈ *S*
  If *S* contains *a*, then *S* contains *a* + 1.
  Then *S* is the set of natural numbers.

The act of counting, or the idea of a *unique successor*, suggests a way of representing the natural numbers on a ray. Let the endpoint of a ray pointing to the right represent the number 1. Then represent each number by a point to the right of its antecedent so that the distances between the points that represent successive numbers are equal. The point that represents a number is called the *graph* of that number.

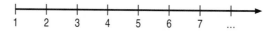

In the set of natural numbers, addition is a *binary operation*—that is, addition assigns to every pair of natural numbers a unique number that is the sum. The existence of a successor for every natural number allows us to add natural numbers if the associative property of addition is postulated, as shown in the following example.

Example 1

    Add 5 + 3.

Solution

$5 + 3 = 5 + (2 + 1)$	3 is the successor of 2.
$= 5 + ((1 + 1) + 1)$	2 is the successor of 1.
$= ((5 + 1) + 1) + 1$	Associative property
$= (6 + 1) + 1$	The successor of 5 is 6.
$= 7 + 1$	The successor of 6 is 7.
$= 8$	The successor of 7 is 8.

The natural numbers are sufficient to answer questions such as "How many days are in a week?" But in order to answer "How many elephants are walking on the ceiling of this room?" you need another number, 0. The set of natural numbers together with 0 forms the set of *whole numbers*, $W = \{0, 1, 2, 3, 4, ...\}$.

If 0 is defined as the number such that $0 + 1 = 1$, it can be shown, assuming the associative property of addition, that for any natural number a, $a + 0 = 0 + a = a$. This statement is called the *identity property of addition*.

In the set of whole numbers, subtraction is defined in terms of addition:

■ **Definition of Subtraction:** *$a - b = c$ if and only if there exists a whole number c such that $a = b + c$.*

Thus, for example: $12 - 7 = 5$ because $12 = 7 + 5$, but $7 - 12$ is not defined in the set of whole numbers since there is no whole number that when added to 12 will equal 7.

Since subtraction is not always possible in the set of whole numbers, the number system must be expanded further. For each natural number a, an *opposite*, $-a$, is defined such that $a + (-a) = 0$. This last statement is the *inverse property of addition*.

You can think of the natural numbers as the same set of numbers as the positive integers. Then the opposites of the positive integers are the negative integers. The negative integers, 0, and the positive integers form the set of *integers*, Z.

    $Z = \{..., -2, -1, 0, 1, 2, ...\}$

To graph the set of integers on a horizontal number line, choose any two points on the line. Let the point that is to the left be the graph of 0 and the point to the right be the graph of 1.

When the distance from 0 to 1 is used as the unit of measure, the positive integers can be located just as the natural numbers were, and the negative integers can be located to the left of 0 so that each positive integer and its opposite are equidistant from 0.

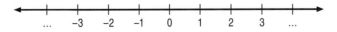

Addition in the set of integers is defined in terms of the addition in the set of whole numbers.

■ *Definition of Addition:* If a and b are whole numbers and $-a$ and $-b$ are the opposites of a and b, then

$$a + (-b) = \begin{cases} a - b \text{ if } a - b \in W \\ -(b - a) \text{ if } a - b \notin W \end{cases}$$

> $a - b \in W$ means $a - b$ is a member of the set of whole numbers.
>
> $a - b \notin W$ means $a - b$ is not a member of the set of whole numbers.

$$-a + (-b) = -(a + b)$$

In the set of integers, Z, addition is a binary operation.

Example 2

  a. $7 + (-3) = 7 - 3 = 4$, since $7 - 3$ is a whole number.

  b. $3 + (-7) = -(7 - 3) = -4$, since $3 - 7$ is not a whole number.

  c. $-7 + (-3) = -(7 + 3) = -10$

Two numbers are equal if and only if they have the same graph on the number line.

Recall the following equality postulates for the numbers of any set S.

Equality Postulates

For all a, b, and $c \in S$:
- Reflexive Property of Equality: $a = a$
- Symmetric Property of Equality: If $a = b$, then $b = a$.
- Transitive Property of Equality: If $a = b$ and $b = c$, then $a = c$.
- Addition Property of Equality: If $a = b$, then $a + c = b + c$.
- Multiplication Property of Equality: If $a = b$, then $ac = bc$.

These properties of equality, together with the commutative and associative properties of addition, enable you to use the definition of subtraction given for whole numbers to subtract any two integers and to solve an equation of the form $x + a = b$.

Example 3

Prove that if $x + a = b$, then $x = b + (-a)$.

Solution

$x + a = b$	Given
$x + a + (-a) = b + (-a)$	Addition property of equality
$x + 0 = b + (-a)$	Inverse property of addition
$x = b + (-a)$	Identity property of addition

Example 4

Prove that $5 - (-3) = 8$.

Solution

Let: $5 - (-3) = c$

$5 = -3 + c$	Definition of subtraction
$3 + 5 = 3 + (-3) + c$	Addition property of equality
$8 = 0 + c$	Inverse property of addition
$8 = c$	Identity property of addition

Therefore, $5 - (-3) = 8$.

1. Let $a - b = x$.
Therefore, by the definition of subtraction, $a = b + x$.

$a = b + x$	
$-b + a = -b + b + x$	*Addition Property of Equality*
$-b + a = 0 + x$	*Inverse Property of Addition*
$-b + a = x$	*Identity Property of Addition*
$-b + a = a - b$	*Transitive Property of Equality*
$a - b = -b + a$	*Symmetric Property of Equality*
$a - b = a + (-b)$	*Commutative Property of Addition*

2. 9

3. 9

4. −9

5. −9

6. −13

7. −13

8. −3

9. −13

The set of integers is a commutative group for addition because the five properties of a commutative group hold.

The Five Properties of a Commutative Group

For all $a, b, c \in Z$:

1.	$a + b \in Z$	Closure
2.	$a + (b + c) = (a + b) + c$	Associative
3.	There exists an integer 0 such that for every a, $a + 0 = 0 + a = a$.	Identity
4.	For every a, there exists $-a \in Z$ such that $a + (-a) = (-a) + a = 0$.	Inverse
5.	$a + b = b + a$	Commutative

Exercises 1.1

1. Use the definition of subtraction and the addition property of equality to prove that in the set of integers, $a - b = a + (-b)$.

In **2 – 9**, find the sum or difference in Z.

2. $15 - 6$ 3. $15 + (-6)$ 4. $6 + (-15)$ 5. $6 - 15$

6. $-5 + (-8)$ 7. $-5 - 8$ 8. $-8 + 5$ 9. $-8 - 5$

In **10 – 15**, name the property or definition that justifies the statement.

10. If $2 + 3 = 5$, then $5 = 2 + 3$.   11. $5 = 5$

12. If $7 - 5 = 2$, then $7 = 5 + 2$.   13. If $a = 8$, then $3a = 3(8)$.

14. If $a = 2 + 1$ and $2 + 1 = 3$, then $a = 3$.   15. If $x + 2 = 7$, then $x + 2 + (-2) = 7 + (-2)$.

16. **a.** In the set of integers, prove that if $a + 1 = b + 1$ then $a = b$.

 b. Is this proof valid in the set of natural numbers? Explain.

In **17 – 20**, state the postulate of the natural numbers that justifies the statement.

17. There is a first natural number.   18. There is no last natural number.

19. If the successor of b equals the successor of the successor of a, then $b = a + 1$.   20. The set of natural numbers is not the empty set.

Answers 1.1

10. Symmetric Property of Equality

11. Reflexive Property of Equality

12. Definition of Subtraction

13. Multiplication Property of Equality

14. Transitive Property of Equality

15. Addition Property of Equality

---

16. a.

$a + 1 = b + 1$	*Given*
$a + 1 + (-1) = b + 1 + (-1)$	*Addition Property of Equality*
$a + 0 = b + 0$	*Inverse Property of Addition*
$a = b$	*Identity Property of Addition*

b. The proof is not valid because there are no additive inverses and no additive identity element in the set of natural numbers. However, the statement is valid because if two successors are equal, then their antecedents are equal.

17. The number 1 has no antecedent.

18. Every natural number has a successor.

19. If two successors are equal, then their antecedents are equal.

20. There exists a natural number 1.

1.2 Factors and Multiples of the Integers

Prime and Composite Numbers

The product of any two integers is an integer. Thus, for all integers a and b, ab is an integer. The numbers a and b are *factors* of ab. At first, the discussion will be limited to the positive integers.

Every positive integer c can be written as $1 \cdot c$. Therefore, every positive integer greater than 1 has at least two factors, 1 and itself. If c has exactly two factors, 1 and c, then c is *prime*. For example, the numbers 2, 3, 5, 7, and 11 are the first five primes.

If c has more than two factors, then c is *composite*. Since $4 = 1 \cdot 4$ and $4 = 2 \cdot 2$, 4 has three factors: 1, 2, and 4. Since $6 = 1 \cdot 6$ and $6 = 2 \cdot 3$, 6 has 4 factors: 1, 2, 3, and 6. Therefore, 4 and 6 are composite numbers.

Disregarding order, every natural number can be written as the product of primes in exactly one way. For example:

$$12 = 2 \cdot 6 = 2 \cdot 2 \cdot 3 = 2^2 \cdot 3 \qquad\qquad 12 = 3 \cdot 4 = 3 \cdot 2^2$$

The prime factorization of a positive integer can be used to determine all of its factors.

Example 1

At the end of a game show, the host picks a number of people from the audience and gives each an equal number of ten-dollar bills. If he has 24 such bills to give away, how many people could he pick?

Solution

The numbers of people he could pick are the factors of 24. Since the prime factorization of 24, $2^3 \cdot 3$, contains 2^3, each factor of 24 must contain one of four possible powers of 2: $2^0, 2^1, 2^2$, and 2^3. Since the prime factorization of 24 contains 3 (or 3^1), each factor of 24 must contain one of two possible powers of 3: 3^0 and 3^1.

Recall the *Counting Principle*: If one activity can occur in any of m ways and, following that, a second activity can occur in any of n ways, then both activities can occur in the given order in mn ways.

Since there are four ways of selecting factors that have powers of 2, and two ways of selecting factors that have powers of 3, there are then $4 \cdot 2$, or 8, factors of 24. A diagram is a useful problem-solving technique here. A *tree diagram* can be used to determine the factors.

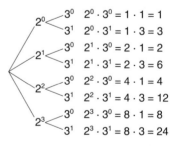

Therefore, the host could pick 1, 2, 3, 4, 6, 8, 12 or 24 people.

The previous example demonstrates the following theorem.

☐ **Number of Factors Theorem:** *If the prime factorization of a natural number a is* $a = p_1^{b_1} p_2^{b_2} \ldots p_n^{b_n}$, *where* $p_1, p_2, \ldots, p_n$ *are the prime factors, then the number of factors of a is* $(b_1 + 1)(b_2 + 1) \cdots (b_n + 1)$.

Note that in the example, the prime factorization of 24 is $2^3 \cdot 3^1$ and there are $(3 + 1)(1 + 1) = 4(2) = 8$ factors of 24.

Divisibility

Division in the set of integers is defined in terms of multiplication. The formal definition is stated as follows:

■ **Definition of Division:** *b divides a if and only if there exists an integer c such that a = bc. If c exists, then a ÷ b = c.*

Example 2

Can Uncle Zeke divide 10,101 shares of stock equally among his 37 nieces and nephews?

Solution

Does 37 divide 10,101 equally?
Since $10{,}101 = 37 \cdot 273$, then $10{,}101 \div 37 = 273$.
Therefore, each niece and nephew would get 273 shares of stock.

Example 3

When motor oil was on sale, Mr. Ellis purchased a number of quarts for himself and his neighbor. Later, his charge statement showed that the total amount of the purchase before tax was $16.91, but it did not show the cost of one quart. Mr. Ellis, who has forgotten the sale price of the motor oil, needs to ask his neighbor for payment for 9 quarts. Can the cost of a quart of oil be determined?

Solution

Use a guess and check approach to the problem with the help of a calculator (since the factors of 1,691 are not obvious).

Store 1,691 in the memory so that it will not be necessary to enter that number for each trial. Since 1,691 is odd, it can have only odd factors. Begin by trying 11 since you know that he bought more than 9 quarts.

Enter: 1691 | STO▸ | | ÷ | 11 | = |

Enter: | RCL | | ÷ | 13 | = |

Continue dividing by odd numbers until the quotient is an integer.

Enter: | RCL | | ÷ | 19 | = |

Display: ⟦   89. ⟧

Since it is unlikely that the motor oil cost $0.19, assume that the cost of the oil is $0.89 and that Mr. Ellis purchased 19 quarts, 9 for his neighbor and 10 for himself.

The definition of divisibility, closure of the set of integers under addition, and the distributive property of multiplication over addition are needed to prove the following theorem.

☐ **Divisibility of a Sum Theorem:** *If b divides c, and b divides d, then b divides c + d.*

Proof

b divides $c \rightarrow c = bx_1$ for some integer x_1

b divides $d \rightarrow d = bx_2$ for some integer x_2

Therefore, $c + d = bx_1 + bx_2$

$$= b(x_1 + x_2)$$

Since the sum of two integers is an integer, $x_1 + x_2$ is an integer, and b divides $c + d$.

Example 4

How many integers between 1 and 100 inclusive are divisible by 2, or by 3, or by both 2 and 3?

Solution

Use the strategy of a simpler related problem.

Find the number of integers from 1 to 6 that satisfy the given conditions. There are 4 such integers: 2, 3, 4, 6.

Now separate the integers into blocks of 6. In each block, the integers of the forms $6k + 2$, $6k + 3$, $6k + 4$, and $6k + 6$ are divisible by 2, or by 3, or by both 2 and 3 (the divisibility of a sum theorem).

Since $6(16) = 96$, the integers from 1 to 96 can be separated into 16 blocks. Of the remaining integers, 98, 99, and 100 satisfy the given conditions.

Therefore, there are $4(16) + 3$, or 67, integers from 1 to 100 that are divisible by 2, or by 3, or by both 2 and 3.

If b does not divide a, then the largest integer, q, can be found such that $a = bq + r$, where r and $b - r$ are positive integers. In the expression $bq + r$, q is the quotient and r is the remainder.

Example 5

If I have 80 cents, what is the largest number of 25-cent stamps that I can buy?

Solution

Since $80 = 25 \cdot 3 + 5$, I can buy 3 stamps and will have 5 cents left over.

Greatest Common Factor

The *greatest common factor* (GCF) of two integers a and b is the largest integer that divides both a and b. Every common factor of a and b is a factor of the GCF of a and b.

To find the GCF of two integers, look at their prime factorizations. The GCF is the product of the smallest power of all of the primes that are factors of both numbers.

Example 6

Find the GCF of 84 and 112.

Solution

$$84 = 2^2 \cdot 3 \cdot 7$$
$$112 = 2^4 \cdot 7$$

Therefore, the GCF of 84 and 112 is $2^2 \cdot 7 = 28$.

Any factor of 28 is a common factor of 84 and 112. Therefore, $\pm 1, \pm 2,$ $\pm 4, \pm 7, \pm 14,$ and ± 28 are all common factors of 84 and 112, with 28 being the greatest common factor. To symbolize that the greatest common factor of 84 and 112 is 28, write $(84, 112) = 28$.

Although it is always possible to find the GCF of two integers by using their prime factorizations, this method can sometimes be a lengthy process of trial and error. Another method of finding the GCF of two integers is the Euclidean Algorithm.

Example 7

Find the GCF of 56 and 124 using the Euclidean Algorithm.

Solution

Step 1
Using the smaller number as the divisor, write the larger number in terms of a quotient and remainder.

$$124 = 2 \cdot 56 + 12$$

Step 2
Using the remainder as the new divisor, write the previous divisor in terms of a quotient and remainder.

$$56 = 4 \cdot 12 + 8$$

Step 3
Repeat Step 2 until a remainder of 0 is obtained.
The last nonzero remainder, 4, is the GCF.

$$12 = 1 \cdot 8 + 4$$
$$8 = 2 \cdot 4 + 0$$

Proof

$$124 = 2 \cdot 56 + 12 \text{ or } 124 - 2 \cdot 56 = 12$$

If a number is a divisor of 124 and a divisor of 56, then it is a divisor of $124 - 2 \cdot 56$ and, therefore, of 12. All the divisors of both 124 and 56 are divisors of 12.

$$56 = 4 \cdot 12 + 8 \text{ or } 56 - 4 \cdot 12 = 8$$

If a number is a divisor of 56 and a divisor of 12, then it is a divisor of $56 - 4 \cdot 12$ and, therefore, of 8. All the divisors of both 56 and 12 are divisors of 8.

$$12 = 1 \cdot 8 + 4 \text{ or } 12 - 1 \cdot 8 = 4$$

If a number is a divisor of 12 and a divisor of 8, then it is a divisor of $12 - 1 \cdot 8$ and, therefore, of 4. All the divisors of both 12 and 8 are divisors of 4.

$$8 = 2 \cdot 4$$

Since 4 divides 8, 4 is the GCF of 4 and 8. Therefore, by working backward through the sequence, it can be shown that 4 is the GCF of 8 and 12, of 12 and 56, and of 56 and 124. Therefore, (56, 124) = 4.

Relatively-Prime Integers

Since 1 is a factor of every number, every pair of numbers has 1 as a common factor. If two numbers have no common factor other than 1, then 1 is the GCF. Two integers are *relatively prime* if their GCF is 1.

Example 8

Show that 98 and 165 are relatively prime.

Solution

$$165 = 1 \cdot 98 + 67$$
$$98 = 1 \cdot 67 + 31$$
$$67 = 2 \cdot 31 + 5$$
$$31 = 6 \cdot 5 + 1$$
$$5 = 5 \cdot 1 + 0$$

The GCF of 98 and 165 is 1. Therefore, 98 and 165 are relatively prime.

☐ **Divisibility of a Product Theorem:** *If a divides bc, and a and b are relatively prime, then a divides c.*

Proof

If a divides bc, then there exists an integer x such that $ax = bc$. These two equal numbers, ax and bc, have the same prime factorization. Therefore, all the prime factors of a are prime factors of bc. But a and b have no common prime factors. Therefore, all the prime factors of a are factors of c, and a divides c.

Least Common Multiple

If a and b are integers and a is a factor of b, then b is a multiple of a. For example, since $111 = 3 \cdot 37$, then 37 is a factor of 111, and 111 is a multiple of 37.

The *least common multiple* (LCM) of two integers a and b is the smallest positive integer that is a multiple of both a and b. Every multiple of a and b is a multiple of the LCM.

To find the LCM of two integers, look at their prime factorizations. The LCM is the product of the largest power of each of the primes that are factors of one or both numbers.

Example 9 _____

Find the LCM of 90 and 440.

Solution

$$90 = 2 \cdot 3^2 \cdot 5$$
$$440 = 2^3 \cdot 5 \cdot 11$$

Therefore, the LCM of 90 and 440 is $2^3 \cdot 3^2 \cdot 5 \cdot 11 = 3{,}960$.

Note that the GCF of 90 and 440 is $2 \cdot 5$, or 10. The LCM of 90 and 440 is their product divided by their GCF.

$$\frac{90(440)}{10} = 3{,}960 \text{ or } 90(440) = 10(3{,}960)$$

If a and b are integers, then $ab =$ (GCF of a and b)(LCM of a and b).

Example 10 _____

Eastbound trains leave the Main Street Station every 24 minutes and northbound trains leave every 42 minutes. If both an eastbound and a northbound train left at 6 o'clock, when is the next time that two trains will leave together?

Solution

The LCM of 24 and 42 will determine how often the train times will coincide.

$$24 = 2^3 \cdot 3$$
$$42 = 2 \cdot 3 \cdot 7$$

Therefore, the LCM is $2^3 \cdot 3 \cdot 7 = 168$.

Since $168 = 2 \cdot 60 + 48$, the trains will leave together every 2 hours and 48 minutes. The next time that they will leave together after 6 o'clock is at 8:48.

Example 11 _____

Find the GCF and the LCM of 120 and 76.

Solution

Method 1 (Prime Factorization)

$$120 = 2^3 \cdot 3 \cdot 5$$
$$76 = 2^2 \cdot 19$$

Therefore, the GCF is $2^2 = 4$, and the LCM is $2^3 \cdot 3 \cdot 5 \cdot 19 = 2{,}280$. This is verified by the fact that $120 \cdot 76 = 4 \cdot 2{,}280$.

Method 2 (Euclidean Algorithm)

$$120 = 1 \cdot 76 + 44$$
$$76 = 1 \cdot 44 + 32$$
$$44 = 1 \cdot 32 + 12$$
$$32 = 2 \cdot 12 + 8$$
$$12 = 1 \cdot 8 + 4$$
$$8 = 2 \cdot 4 + 0$$

Therefore, the GCF is 4, and the LCM is $\frac{120 \cdot 76}{4} = 2{,}280$.

1. $2^3 \cdot 3^2 \cdot 5$

2. $2^2 \cdot 3 \cdot 13$

3. $3^2 \cdot 5^2 \cdot 7$

4. $3^3 \cdot 7^2 \cdot 11$

5. $5 \cdot 7 \cdot 13^2$

6. 6

7. 29

8. 4

9. 1

10. 702

11. 342

12. 650

13. 630

14. 12 different ways:
    1 row of 60
    2 rows of 30
    3 rows of 20
    4 rows of 15
    5 rows of 12
    6 rows of 10
    10 rows of 6
    12 rows of 5
    15 rows of 4
    20 rows of 3
    30 rows of 2
    60 rows of 1

15. 8 or 10

16. a. 12   b. 24

Example 12

A ticket for a bus tour costs $21 if purchased at least a week in advance and $25 if purchased less than a week before the trip. After one tour had departed, the business manager found that the revenue for the trip was $1,500 but that no record had been kept of the number of tickets sold. Is it possible to determine the number of tickets sold?

Solution

The number of tickets sold can be determined by finding the solutions, in integers, of the equation $21x + 25y = 1{,}500$, where x represents the number of tickets sold at least a week in advance and y represents the number of tickets sold less than a week in advance.

Apply the strategy of using a simpler related problem. That is, solve the equation $21x + 25y = 1$ and then multiply the solutions by 1,500.

To solve the simpler equation, apply the strategy of making a list. Use a calculator to find multiples of 21 and 25 that differ by 1. Entering a constant is helpful here. Some calculators have a constant key, labeled ⎣ K ⎦.

Enter	Display
21 ⎣ × ⎦ ⎣ K ⎦ 1 ⎣ = ⎦	21.
2 ⎣ = ⎦	42.
3 ⎣ = ⎦	63.
⋮	

Clear the calculator and perform a similar sequence using 25.

The following table results.

Examine the entries in the table and locate entries that differ by 1. The numbers 126 and 125 are the required values.

	Multiples of 21	Multiples of 25
1	21	25
2	42	50
3	63	75
4	84	100
5	105	125
6	126	150

Since $126 - 125 = 21(6) + 25(-5) = 1$, a possible solution for the equation $21x + 25y = 1$ is $x = 6$ and $y = -5$. Therefore, a possible solution for $21x + 25y = 1{,}500$ is $x = 1{,}500(6) = 9{,}000$ and $y = 1{,}500(-5) = -7{,}500$.

---

17. Let a be a positive integer.
    $$a = p_1^{b_1} \cdot p_2^{b_2} \cdots p_n^{b_n}$$
    $$a^2 = p_1^{2b_1} \cdot p_2^{2b_2} \cdots p_n^{2b_n}$$
    The number of factors of $a^2 = (2b_1 + 1)(2b_2 + 1) \cdots (2b_n + 1)$. Since each factor $(2b_i + 1)$ is odd, the number of factors of a^2 is the product of n odd numbers and is therefore odd.

18. 11,390,625   19. 46,656

20. a. 2, 4, 5   b. 2, 4, 5
    c. $(n + 3)^2 - 1 = n^2 + 6n + 9 - 1$
    $$= (n^2 - 1) + 6n + 9$$
    3 divides $6n + 9$. If 3 divides $n^2 - 1$, then by the Divisibility of a Sum Theorem, 3 divides the sum $(n^2 - 1) + (6n + 9)$ and therefore, $(n + 3)^2 - 1$.

21. a. Q 6; R 448   b. Q 2; R 435

22. (2, 24), (5, 19), (8, 14), (11, 9), (14, 4)

$$21(9,000) + 25(-7,500) = 1,500$$

Although this is a solution in integers, it is not an acceptable solution to the problem since the values of x and y must be positive integers. However, this solution can be used to obtain a solution in positive integers if one exists.

The equation will remain unchanged if $21(-25k) + 25(21k)$, the equivalent of 0, is added to the left member of the equation.

$$21(9,000) + 21(-25k) + 25(-7,500) + 25(21k) = 1,500$$
$$21(9,000 - 25k) + 25(-7,500 + 21k) = 1,500$$

For any integral value of k, $x = 9,000 - 25k$ and $y = -7,500 + 21k$ is a solution. For a solution in positive integers, find the values of k for which $9,000 - 25k$ and $-7,500 + 21k$ are both positive.

$$9,000 - 25k > 0 \qquad \text{and} \qquad -7,500 + 21k > 0$$
$$-25k > -9,000 \qquad\qquad 21k > 7,500$$
$$k < 360 \qquad\qquad\qquad k > 357.14286$$

Therefore, $k = 358$ and $k = 359$ provide solutions.

For $k = 358$, $x = 9,000 - 25(358) = 50$ and $y = -7,500 + 21(358) = 18$

For $k = 359$, $x = 9,000 - 25(359) = 25$ and $y = -7,500 + 21(359) = 39$

The total number of bus tour tickets sold was either $50 + 18 = 68$ or $25 + 39 = 64$.

An equation of the form $ax + by = c$, where a, b, and c are integers and the solution must be in integers, is called a *Diophantine equation*. A Diophantine equation has solutions in integers if c is a multiple of the greatest common factor of a and b. However, a solution in *positive* integers is not always possible.

Exercises 1.2

In **1 – 5**, find the prime factorization of the number.

1. 360   **2.** 156   **3.** 1,575   **4.** 14,553   **5.** 5,915

In **6 – 9**, find the GCF of the pair of numbers.

6. 54, 78   **7.** 87, 145   **8.** 12, 196   **9.** 24, 385

In **10 – 13**, find the LCM of the pair of numbers.

10. 54, 78   **11.** 38, 171   **12.** 325, 130   **13.** 18, 35

---

23. Let the three-digit number be $100h + 10t + u$.

$$100h + 10t + u = 99h + h + 9t + t + u$$
$$= (h + t + u) + (99h + 9t)$$

$99h + 9t$ is divisible by 9. If $(h + t + u)$ is divisible by 9, then by the Divisibility of a Sum Theorem, $(h + t + u) + (99h + 9t)$ is divisible by 9.

24. Let the three-digit number be $100h + 10t + u$.

$$100h + 10t + u = 99h + h + 11t - t + u$$
$$= (h + u - t) + (99h + 11t)$$

If $h + u = t$, then $h + u - t = 0$.
Therefore, $100h + 10t + u = 0 + 99h + 11t$, which is divisible by 11.

27. Let $a = 2, b = 3$, and $c = 5$.

2 does not divide 3, and 2 does not divide 5, but 2 does divide the sum $(3 + 5)$ or 8. This counterexample disproves the given statement.

28. If a divides b, then $b = ax$. If a does not divide c, then $c = ay + r$, where a does not divide r.

$b + c = a(x + y) + r$

$\dfrac{b + c}{a} = (x + y) + \dfrac{r}{a}$

Since $(x + y)$ is an integer, then $(x + y) + \dfrac{r}{a}$ is a mixed number, and a does not divide $b + c$.

29. Divide both sides by 3.

$3x + 4y = \dfrac{1}{3}$

If x and y are integers, then $3x + 4y$ is an integer, and cannot be equal to $\dfrac{1}{3}$.

Exercises 1.2 *(continued)*

14. In a lecture hall, there are 60 chairs to be arranged in rows. In how many different ways could the chairs be placed if there is to be an equal number of chairs in each row?

15. An auditorium contains 200 seats. In each row, there are the same number of seats. The number of seats in each row is more than 6 and less than 15. How many seats could there be in each row?

16. Find the smallest positive integer:
   a. with exactly 6 factors
   b. with exactly 8 factors

17. Show that every perfect-square integer has an odd number of factors.

18. Find the smallest integer divisible by 15 that is both a square and a cube.

19. Find the smallest integer divisible by 72 that is both a square and a cube.

20. a. Find 3 values of n such that 3 divides $n^2 - 1$.
   b. Find 3 values of n such that 3 divides $(n + 3)^2 - 1$.
   c. Prove that if 3 divides $n^2 - 1$, then 3 divides $(n + 3)^2 - 1$.

21. Find the quotient and remainder in the set of integers when a is divided by b.
   a. $a = 3{,}784, b = 556$
   b. $a = 1{,}397, b = 481$

22. Find all pairs of positive integers x and y such that $5x + 3y = 82$.

23. Prove that if the sum of the digits of a three-digit number is divisible by 9, then the number is divisible by 9. (A three-digit number can be written as $100h + 10t + u$).

24. Prove that if the sum of the first and last digits of a three-digit number equals the middle digit, then the number is divisible by 11.

25. Prove that if the sum of the first and last digits of a three-digit number equals 11 plus the middle digit, then the number is divisible by 11.

26. A *palindrome* is a word, phrase, or number that is the same when read backward or forward. Prove that there are no four-digit palindromes that are primes.

27. Prove or disprove that if a does not divide b, and a does not divide c, then a does not divide $b + c$.

28. Prove or disprove that if a divides b, and a does not divide c, then a does not divide $b + c$.

29. Show that $9x + 12y = 1$ has no solution in the set of integers.

---

25. Let the three-digit number be $100h + 10t + u$.
$100h + 10t + u = 99h + h + 11t - t + u$
$= (h + u - t) + (99h + 11t)$
If $h + u = 11 + t$, then $h + u - t = 11$.
Therefore, $100h + 10t + u = 11 + 99h + 11t$, which is divisible by 11.

26. Let the four-digit palindrome be $1{,}000a + 100b + 10b + a$.
$1{,}000a + 100b + 10b + a = 1{,}001a + 110b$
$= 11 \cdot 91a + 11 \cdot 10b$
$= 11(91a + 10b)$
Therefore, every four-digit palindrome is divisible by 11.

1.3 The Real Numbers

The Rational Numbers

The quotient of two integers is not always an integer. However, the indicated ratio, or quotient, of two integers can be located as a point on the number line, and represents an important idea in the application of numbers to everyday situations. For example, it is possible to have $\frac{5}{8}$ of a piece of lumber or $\frac{5}{8}$ of a length of fabric, by dividing the length of the piece into eight equal segments and cutting at the endpoint of the fifth of those segments.

In the same way, it is possible to find the graph of $\frac{5}{8}$ on the number line by separating the segment from 0 to 1 into eight equal segments and marking the point at the right of the fifth segment. A number such as $\frac{5}{8}$ is called a *rational number*.

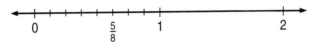

■ **Definition of a Rational Number:** *A rational number is a number that can be expressed as the ratio $\frac{a}{b}$, where a and b are integers and $b \neq 0$.*

Since the ratio $\frac{a}{b}$ can also be expressed as $a \div b$, the decimal representation for $\frac{a}{b}$ can be found using division.

Example 1

Find the decimal representations of $\frac{5}{12}$ and $\frac{3}{22}$.

Solution

$$12\overline{)5.00000...} = 0.41666...$$

$$22\overline{)3.0000000...} = 0.1363636...$$

You can see that in the division by 12, every step after the third will repeat the third step. In the division by 22, every two steps after the third will repeat the second and third steps. The resulting decimals are called *repeating*, or *periodic, decimals*. The repetition is often indicated by placing a bar over the digit or digits that repeat. For example, $\frac{5}{12} = 0.41\overline{6}$ and $\frac{3}{22} = 0.1\overline{36}$.

Every rational number can be written as a repeating decimal. To find the repeating decimal for the rational number $\frac{a}{b}$, divide a by b. After each step in the division, there are b possible remainders: 0, 1, 2, ..., $b - 1$. If the remainder is 0, then the decimal is a *terminating decimal*, that is, a repeating decimal with repeating zeros. For example:

$$\tfrac{1}{8} = 0.125 = 0.125000... = 0.125\overline{0}$$

If the remainder is not 0, there are $b - 1$ other possibilities. If there has not been a repetition before the $(b - 1)$ step, all the possible remainders have already been used, and the next remainder must either be 0 or a repetition of one of the previous remainders. At that point, the decimal must terminate or repeat. For example:

$$\frac{4}{33} = 0.121212\cdots = 0.\overline{12} \qquad\qquad \frac{1}{7} = 0.\overline{142857} \qquad\qquad \frac{1}{4} = 0.250\overline{0}$$

Example 2

Write the following as a ratio of integers.

    a. $1.\overline{189}$ **b.** $0.15\overline{6}$

Solution

Suppose the repetition occurs in blocks of n digits.

a. $1.\overline{189}$ has a repetition that occurs in a block of 3 digits.

    Let: $x = 1.\overline{189}$

$$1{,}000x = 1{,}189.\overline{189} \qquad \text{Multiply each side by } 10^3.$$

$$-\quad x = \qquad 1.\overline{189} \qquad \text{Subtract the first equation from the second.}$$

$$\overline{999x = 1{,}188}$$

$$x = \frac{1{,}188}{999} \qquad \text{Divide by 999.}$$

$$x = \frac{44}{37} \qquad \text{Express the fraction in simplest form.}$$

b. $0.15\overline{6}$ has a repetition that occurs in a block of 1 digit.

    Let: $y = 0.15\overline{6}$

$$10\,y = 1.56\overline{6} \qquad \text{Multiply each side by } 10^1.$$

$$-\quad y = 0.15\overline{6} \qquad \text{Subtract the first equation from the second.}$$

$$\overline{9y = 1.41}$$

$$y = \frac{1.41}{9} \qquad \text{Divide by 9.}$$

$$y = \frac{141}{900} = \frac{47}{300} \qquad \text{Express the fraction in simplest form.}$$

The procedure above suggests that every repeating decimal represents a rational number.

The Irrational Numbers

There are points on the number line that represent numbers that are not rational numbers. Every point on the number line that is not the graph of a rational number is the graph of an *irrational number*.

The area of a floor tile is often 1 square foot. Consider two such tiles that are cut along a diagonal and placed as shown. The area of the resulting square is 2 square feet. The length of one side of that square is $\sqrt{2}$ feet. It is not possible to write $\sqrt{2}$ as the ratio of two integers; that is, $\sqrt{2}$ is not a rational number.

The graph of $\sqrt{2}$ can be located on a number line. Construct a square with one vertex at 0 and an adjacent vertex at 1. With the center of a circle at 0 and the length of the diagonal as the radius, draw an arc that intersects the positive ray of the number line. This point of intersection is the graph of $\sqrt{2}$.

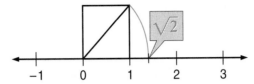

An indirect argument can be used to prove that $\sqrt{2}$ is not rational. Such an argument uses the fact that the square of an integer has an even number of prime factors. This is true since every prime factor of a occurs twice in the prime factorization of a^2. For example:

$$30 = 2 \cdot 3 \cdot 5 \text{ and } 30^2 = 2^2 \cdot 3^2 \cdot 5^2$$
$$12 = 2^2 \cdot 3 \text{ and } 12^2 = (2^2)^2 \cdot 3^2 = 2^4 \cdot 3^2$$

☐ **Theorem:** $\sqrt{2}$ *is not rational.*

Proof

Assume that $\sqrt{2}$ is rational.

Therefore, $\sqrt{2} = \dfrac{a}{b}, b \neq 0.$

$$2 = \frac{a^2}{b^2} \qquad \text{Square each side of the equation.}$$
$$2b^2 = a^2 \qquad \text{Multiply each side by } b^2.$$

The prime factorization of $2b^2$ has the factor 2 as well as the factors of b^2. Therefore, it has an odd number of prime factors. The prime factorization of a^2 has an even number of prime factors.

But there cannot be two different factorizations for the same number, a^2 and $2b^2$. Since there is a contradiction, the assumption must be false. Therefore, $\sqrt{2}$ is not rational.

The union of the set of rational numbers and the set of irrational numbers is the set of *real numbers*.

Every real number can be written as a decimal with an infinite number of digits to the right of the decimal point. Since a repeating decimal always represents a rational number, an irrational number has a nonrepeating, nonterminating decimal representation.

For example:

0.1010010001... is a nonrepeating, nonterminating decimal.

$\pi = 3.1415926...$ is a nonrepeating, nonterminating decimal.

In the first example, the changing pattern prevents the decimal from repeating. In the second case, the lack of a repeating pattern in the seven decimal places shown (or in the hundreds of decimal places that have been found using computer technology) does not constitute a proof. Other evidence proves that π is irrational and, therefore, cannot have a repeating decimal.

Every point on the number line is the graph of a real number and every real number has a graph on the number line. There is a one-to-one correspondence between the real numbers and the points on the number line.

Operations with Radicals

Irrational numbers are also expressed as the nth root of a real number.

The square root of a is one of the two equal factors of a. If $x \cdot x = a$, then x is one of the square roots of a. Since $(\sqrt{5})(\sqrt{5}) = 5$ and $(-\sqrt{5})(-\sqrt{5}) = 5$, there are two square roots of 5, a positive and a negative one.

In general, every positive real number has two square roots, a positive one called the *principal square root*, and a negative one.

The cube root of a is one of the three equal factors of a. If $x \cdot x \cdot x = a$, then x is the cube root of a, or $x = \sqrt[3]{a}$. Every nonzero real number has three cube roots. One of these is a real number, the *principal cube root*. You will study the other cube roots in a later chapter.

In general, for a natural number n, the nth root of a is one of the n equal factors of a. If the index n is even, every positive number has two real nth roots, a positive one and a negative one. The positive one is the principal nth root. If n is odd, every real number has one real root, the principal nth root.

If $a > 0$ and $b > 0$:

$$\sqrt[n]{a} \cdot \sqrt[n]{b} = \sqrt[n]{ab} \qquad \text{and} \qquad \sqrt[n]{ab} = \sqrt[n]{a} \cdot \sqrt[n]{b}$$

$$\frac{\sqrt[n]{a}}{\sqrt[n]{b}} = \sqrt[n]{\frac{a}{b}} \qquad \text{and} \qquad \sqrt[n]{\frac{a}{b}} = \frac{\sqrt[n]{a}}{\sqrt[n]{b}}$$

These principles, together with the commutative, associative, and distributive properties, determine how we compute with irrational numbers expressed as radicals.

Example 3

a. $\sqrt{40} = \sqrt{4} \cdot \sqrt{10} = 2\sqrt{10}$

To simplify, factor out the largest perfect square.

b. $\sqrt[3]{24} = \sqrt[3]{8} \cdot \sqrt[3]{3} = 2\sqrt[3]{3}$

Factor out the largest perfect cube.

c. $\sqrt{3} \cdot \sqrt{15} = \sqrt{45}$
$= \sqrt{9} \cdot \sqrt{5} = 3\sqrt{5}$

To multiply, radicals must have the same index. Multiply radicands. Simplify.

d. $\dfrac{\sqrt[4]{2,592}}{\sqrt[4]{162}} = \sqrt[4]{16} = 2$

To divide, radicals must have the same index. Divide radicands. Simplify.

e. $\sqrt{12} + \sqrt{3} = \sqrt{4} \cdot \sqrt{3} + \sqrt{3}$
$= 2\sqrt{3} + \sqrt{3}$
$= \sqrt{3}\,(2 + 1) = 3\sqrt{3}$

To add, radicals must have the same index and the same radicand.

A radical is said to be in simplest form when the radicand is the smallest possible positive integer. To simplify a square root radical with a fractional radicand, first express the radicand as an equivalent fraction with a denominator that is a perfect square. Then factor the radicand into the product of a perfect-square fraction and an integer.

Example 4 _____

a. $\sqrt{\frac{9}{8}} = \sqrt{\frac{18}{16}} = \sqrt{\frac{9}{16} \cdot 2} = \sqrt{\frac{9}{16}} \cdot \sqrt{2} = \frac{3}{4}\sqrt{2}$

b. $\sqrt{.2} = \sqrt{.20} = \sqrt{.04 \cdot 5} = \sqrt{.04} \cdot \sqrt{5} = .2\sqrt{5}$

c. $\sqrt{\frac{3}{2}} + \sqrt{\frac{2}{3}} = \sqrt{\frac{6}{4}} + \sqrt{\frac{6}{9}}$

$= \sqrt{\frac{1}{4} \cdot 6} + \sqrt{\frac{1}{9} \cdot 6}$

$= \sqrt{\frac{1}{4}} \cdot \sqrt{6} + \sqrt{\frac{1}{9}} \cdot \sqrt{6}$

$= \frac{1}{2}\sqrt{6} + \frac{1}{3}\sqrt{6} = \frac{5}{6}\sqrt{6}$

An approximate rational value for an irrational number expressed in radical form can be found by using a calculator. Almost all calculators have a key labeled $\boxed{\sqrt{x}}$ or $\boxed{\sqrt{\ }}$ to find the square root of a number.

For example, to find a rational approximation for $\sqrt{76}$:

Enter: 76 $\boxed{\sqrt{x}}$

Display: $\boxed{8.7177978}$

On some calculators, you must use the square-root key first.

Enter: $\boxed{\sqrt{\ }}$ 76 $\boxed{=}$

To find an approximation for radicals with a higher index, use a scientific calculator that has a power key, $\boxed{y^x}$ or $\boxed{\wedge}$. Two methods are possible.

Example 5 _____

Find $\sqrt[3]{17}$.

Solution

Method 1   Write the root as a power with a fractional exponent:

$\sqrt[3]{17} = (17)^{\frac{1}{3}}$

Enter: 17 $\boxed{y^x}$ $\boxed{(}$ 1 $\boxed{\div}$ 3 $\boxed{)}$ $\boxed{=}$ or

Enter: 17 $\boxed{\wedge}$ $\boxed{(}$ 1 $\boxed{\div}$ 3 $\boxed{)}$ $\boxed{=}$

Display: $\boxed{2.5712816}$

1. $0.75\overline{0}$

2. $0.58\overline{3}$

3. $1.\overline{3}$

4. $0.38\overline{8}$

5. $-3.\overline{238095}$

6. $\frac{1}{8}$

7. $\frac{113}{900}$

8. $\frac{5}{33}$

9. $\frac{50}{27}$

10. $\frac{129}{110}$

Method 2   Since finding the nth root and raising to the nth power are inverse operations, use: INV y^x

Enter: 17 INV y^x 3 =

Display: $\boxed{2.5712816}$

Exercises 1.3

In **1 – 5**, write the rational number as a repeating decimal.

1. $\frac{3}{4}$ **2.** $\frac{7}{12}$ **3.** $\frac{4}{3}$ **4.** $\frac{7}{18}$ **5.** $-3\frac{5}{21}$

In **6 – 10**, write the decimal as a ratio of two integers.

6. $0.125\overline{0}$ **7.** $0.12\overline{5}$ **8.** $0.\overline{15}$ **9.** $1.\overline{851}$ **10.** $1.1\overline{72}$

11. Write an infinite decimal that is an irrational number.

12. Prove that $\sqrt{3}$ is irrational.

 In **13 – 22**, find the first 7 digits of the nonrepeating, nonterminating, decimal representation of the given irrational number.

13. $\sqrt{2}$ **14.** $\sqrt{3}$ **15.** $\sqrt{5}$ **16.** $\sqrt{17}$ **17.** $\sqrt{\frac{1}{2}}$

18. $\sqrt[3]{6}$ **19.** $\sqrt[4]{12}$ **20.** $\sqrt[3]{.04}$ **21.** $\sqrt[5]{1.8}$ **22.** $\sqrt[4]{\frac{3}{2}}$

23. Write an irrational number that is between 0.23 and 0.24.

24. Identify each of the following as rational or irrational.

   a. The length of the diagonal of a square if the length of one side is s centimeters, and s is a natural number.

   b. The probability of a sum of 6 when two fair dice are tossed.

   c. The amount for which a check is written.

   d. The circumference of a circle if the length of the radius in meters is r, and r is a rational number.

   e. The perimeter of a square when a side of the square measures $\sqrt{a}$ cm and a is a prime number.

   f. The area of a square when a side of the square measures $\sqrt{a}$ cm and a is a prime number.

11. 0.1234567891011121314...
(Answers will vary.)

12. Assume that $\sqrt{3}$ is rational. Then $\sqrt{3}$ can be written as the ratio of integers. Let a and b be integers such that:

$$\sqrt{3} = \frac{a}{b}$$
$$3 = \frac{a^2}{b^2}$$
$$3b^2 = a^2$$

Since a^2 and b^2 have an even number of prime factors, the left member of the last equation has an odd number of prime factors and the right member of the equation has an even number of prime factors. But each side of the equation must represent the same number and therefore have the same number of prime factors. Since our assumption led to a false statement, it must be false. Therefore, $\sqrt{3}$ is not rational, or $\sqrt{3}$ is irrational.

Exercises 1.3 *(continued)*

In **25 – 54**, perform the indicated operations and express the results in simplest radical form.

25. $\sqrt{2} \cdot \sqrt{6}$

26. $\sqrt{8} \cdot \sqrt{12}$

27. $\sqrt{5} \cdot \sqrt{50}$

28. $5\sqrt{6}\,(2\sqrt{15})$

29. $8\sqrt{5}\,(\sqrt{5} + 2)$

30. $7\sqrt{10}\,(3 - 2\sqrt{5})$

31. $(\sqrt{2} + 1)(\sqrt{2} - 1)$

32. $(6 - 2\sqrt{3})(6 + 2\sqrt{3})$

33. $(\sqrt{5} - 1)^2$

34. $(2\sqrt{3} + 5)^2$

35. $(1 + \sqrt{5})(3 + 2\sqrt{5})$

36. $(8 + 5\sqrt{2})(3 - 2\sqrt{2})$

37. $\sqrt{18} + \sqrt{50}$

38. $\sqrt{24} - \sqrt{6}$

39. $\sqrt{200} + 3\sqrt{2}$

40. $2\sqrt{8} + 5\sqrt{50}$

41. $6\sqrt{12} - \sqrt{48}$

42. $\sqrt{2} \cdot \sqrt{12} + \sqrt{54}$

43. $\sqrt[3]{4} \cdot \sqrt[3]{4}$

44. $\sqrt[5]{4} \cdot \sqrt[5]{24}$

45. $\sqrt[3]{16} \cdot \sqrt[3]{54}$

46. $\sqrt{3} \cdot \sqrt{\tfrac{1}{6}}$

47. $\sqrt{\tfrac{3}{4}} \cdot \sqrt{\tfrac{1}{2}}$

48. $\sqrt{\tfrac{1}{2}} + \sqrt{\tfrac{1}{8}}$

49. $\dfrac{\sqrt{8}}{\sqrt{18}}$

50. $\dfrac{\sqrt{40}}{2}$

51. $\dfrac{\sqrt{200}}{5}$

52. $\dfrac{8 + \sqrt{60}}{2}$

53. $\dfrac{-6 + \sqrt{20}}{\tfrac{1}{2}}$

54. $\dfrac{\sqrt{20} + \sqrt{40}}{\sqrt{5}}$

55. a. Write $0.\overline{1}$ as a ratio of integers.

   b. Use the value of $0.\overline{1}$ to predict the values of $0.\overline{2}$, $0.\overline{3}$, ..., $0.\overline{9}$.

   c. Use the fractional values of $0.\overline{2}$ and $0.2\overline{0}$ to find the value of $0.0\overline{2}$.

   d. What if the repeating decimal has a block of 2 digits, such as $0.\overline{02}$ or $0.\overline{34}$?

25. $2\sqrt{3}$

26. $4\sqrt{6}$

27. $5\sqrt{10}$

28. $30\sqrt{10}$

29. $40 + 16\sqrt{5}$

30. $21\sqrt{10} - 70\sqrt{2}$

31. 1

32. 24

33. $6 - 2\sqrt{5}$

34. $37 + 20\sqrt{3}$

35. $13 + 5\sqrt{5}$

36. $4 - \sqrt{2}$

37. $8\sqrt{2}$

38. $\sqrt{6}$

39. $13\sqrt{2}$

40. $29\sqrt{2}$

41. $8\sqrt{3}$

42. $5\sqrt{6}$

43. $2\sqrt[3]{2}$

44. $2\sqrt[5]{3}$

45. $6\sqrt[3]{4}$

46. $\tfrac{1}{2}\sqrt{2}$

47. $\tfrac{1}{4}\sqrt{6}$

48. $\tfrac{3}{4}\sqrt{2}$

49. $\tfrac{2}{3}$

50. $\sqrt{10}$

51. $2\sqrt{2}$

52. $4 + \sqrt{15}$

53. $-12 + 4\sqrt{5}$

54. $2 + 2\sqrt{2}$

13. 1.414213...

14. 1.732050...

15. 2.236067...

16. 4.123105...

17. 0.7071067...

18. 1.817120...

19. 1.861209...

20. 0.3419951...

21. 1.124746...

22. 1.106681...

23. 0.231011011101111...
(Answers will vary.)

24. a. irrational   **b.** rational
c. rational   **d.** irrational
e. irrational   **f.** rational

55. a. $0.\overline{1} = \tfrac{1}{9}$

$0.\overline{a} = a(0.\overline{1}) = a\left(\tfrac{1}{9}\right)$

b. $0.\overline{2} = 2(0.\overline{1}) = 2\left(\tfrac{1}{9}\right) = \tfrac{2}{9}$

$0.\overline{9} = 9(0.\overline{1}) = 9\left(\tfrac{1}{9}\right) = 1.\overline{0}$

Note that $0.\overline{9}$ and $1.\overline{0}$ are both repeating decimals for the integer 1.

c. $0.0\overline{2} = 0.\overline{2} - 0.20 = \tfrac{2}{9} - \tfrac{1}{5} = \tfrac{1}{45}$.

d. The denominator is 99.
$0.\overline{02} = \tfrac{2}{99}; \; 0.\overline{34} = \tfrac{34}{99}$

1.4 Properties of the Real Numbers

The Field Properties

The set of real numbers under the operations of addition and multiplication forms a *field*. Let R = {real numbers}, with a, b, and $c \in R$.

1. The set of real numbers is closed under addition.
$$a + b \in R$$

2. Addition is associative in the set of real numbers.
$$a + (b + c) = (a + b) + c$$

3. There exists a unique real number 0 that is the identity element for addition.
$$a + 0 = 0 + a = a$$

4. Every real number, a, has an additive inverse, $-a$.
$$a + (-a) = (-a) + a = 0$$

5. Addition is commutative in the set of real numbers.
$$a + b = b + a$$

6. The set of real numbers is closed under multiplication.
$$a \cdot b \in R$$

7. Multiplication is associative in the set of real numbers.
$$a \cdot (b \cdot c) = (a \cdot b) \cdot c$$

8. There exists a unique real number 1 that is the identity element for multiplication.
$$a \cdot 1 = 1 \cdot a = a$$

9. Every nonzero real number, a, has a multiplicative inverse, a^{-1}.
$$a \cdot a^{-1} = a^{-1} \cdot a = 1$$

10. Multiplication is commutative in the set of real numbers.
$$a \cdot b = b \cdot a$$

11. Multiplication is distributive over addition in the set of real numbers.
$$a \cdot (b + c) = a \cdot b + a \cdot c$$

Order Properties of the Real Numbers

The graphs of the real numbers on the number line suggest the order property of the real numbers.

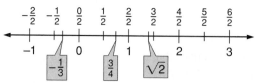

☐ **Trichotomy Property:** *For each pair of real numbers a and b, exactly one of the following is true: $a < b$, $a = b$, or $a > b$.*

- If two real numbers have the same graph on the number line, then they are equal. For example, $3 = \frac{6}{2}$.

- If the graph of a is to the right of the graph of b, then $a > b$. For example, $3 > 1$.

- If the graph of a is to the left of the graph of b, then $a < b$. For example, $-2 < 0$.

Since moving to the right on the number line can be associated with the addition of a positive number, this idea gives rise to the following definition.

■ **Definition:** *$a > b$ if and only if there exists a positive real number x such that $a = b + x$. $b < a$ if and only if $a > b$.*

Since inequality is defined in terms of equality, the properties of equality that were previously accepted as postulates (that is, without proof) can now be used to prove the properties of inequality.

☐ **Transitive Property of Inequality:** *If $a > b$ and $b > c$, then $a > c$.*

> *Proof*
>
> $a > b \rightarrow a = b + x$ for a positive x.
> $b > c \rightarrow b = c + y$ for a positive y.
>
> $a = (c + y) + x$ Substitute $c + y$ for b.
> $a = c + (y + x)$ Associative property
>
> Since the sum of two positive numbers is positive, $y + x$ is a positive number and $a > c$.

☐ **Addition Property of Inequality:** *If $a > b$, then $a + c > b + c$.*

> *Proof*
>
> $a > b \rightarrow a = b + x$ for a positive x.
>
> $a + c = (b + x) + c$ Add c to both sides.
> $a + c = (b + c) + x$ Associative and commutative properties
>
> Thus, $a + c > b + c$.

Example 1

    a. $5 > -2 \rightarrow 5 + 3 > -2 + 3$, or $8 > 1$ Adding 3.

    b. $-1 > -7 \rightarrow -1 + (-2) > -7 + (-2)$, or $-3 > -9$ Adding -2.

☐ **Multiplication Property of Inequality:** *If $a > b$ and $c > 0$, then $ac > bc$. If $a > b$ and $c < 0$, then $ac < bc$.*

> *Proof*
>
> If $a > b$ and $c > 0$, then $ac > bc$.
> $a > b \rightarrow a = b + x$ for a positive x.
>
> $ac = (b + x)c$ Multiply by c.
> $ac = bc + xc$ Distributive property
>
> Since the product of two positive numbers is positive, xc is positive. Therefore, $ac > bc$.

1. Commutative Property of Addition

2. Distributive Property of Multiplication over Addition

3. Multiplication Property of Inequality

4. Additive Identity Property

5. Associative Property of Addition

6. Commutative Property of Multiplication

7. Commutative Property of Addition

8. Associative Property of Multiplication

9. Reflexive Property of Equality and Identity Property of Multiplication

10. Inverse Property of Addition

11. $x > -2$

12. $x < 0.5$

13. $x > 5$

14. $x > 2$

15. $x < 1$

16. $x < 0$

17. The cost of a can of soup is less than $0.56\frac{2}{3}$. When buying one can of soup, the customer may have to pay up to $.57.

18. more than 1.2 ounces but less than 1.6 ounces

19. 4 books

20. more than 2 books

If $a > b$ and $c < 0$, then $ac < bc$.
$a > b \rightarrow a = b + x$ for a positive x.

$$a + (-x) = b + x + (-x) \quad \text{Add } (-x).$$
$$a + (-x) = b \quad \text{Additive inverse}$$
$$(a + (-x))c = bc \quad \text{Multiply by } c.$$
$$ac + (-x)c = bc \quad \text{Distributive property}$$

Since x is positive, $-x$ is negative. It was given that $c < 0$, that is, c is negative. The product of two negative numbers, $(-x)c$, is positive. Therefore, $bc > ac$ or $ac < bc$.

Thus, it is proven that when each side of an inequality is multiplied by a negative number, the sense of the inequality is reversed.

Since division by c is the same as multiplication by $\frac{1}{c}$, it is also true that when each side of an inequality is divided by a negative number, the sense of the inequality is reversed.

Example 2 _____

a.　$5 > -2 \rightarrow 3(5) > 3(-2)$, or $15 > -6$　　Multiplying by 3.

b.　$-1 > -7 \rightarrow -2(-1) < -2(-7)$, or $2 < 14$　　Multiplying by -2.

The set of real numbers with the operations of addition and multiplication forms an *ordered field*, which means that the system satisfies the eleven properties of a field and the four properties of order: trichotomy, transitive, addition, and multiplication.

Example 3 _____

Name the field property used.

a.　$5(203) = 5(200) + 5(3)$　　a. Distributive property

b.　$\frac{3}{4} \cdot \frac{2}{2} = \frac{6}{8} \rightarrow \frac{3}{4} = \frac{6}{8}$　　b. Identity property for multiplication $\left(\frac{2}{2} = 1\right)$

c.　$3 + a + (-a) = 3 + 0$　　c. Additive inverse property

Example 4 _____

Solve for x:　$5 - 2x > 11$

Solution

$$5 - 2x > 11$$
$$5 + (-5) - 2x > 11 + (-5) \quad \text{Add } -5.$$
$$-2x > 6$$
$$\frac{-2x}{-2} < \frac{6}{-2} \quad \text{Divide by } -2, \text{ reversing the inequality.}$$
$$x < -3$$

Example 5 _____

Mario left the house with only a $10 bill in his wallet. He must buy 8 gallons of gas and have at least $1.48 left to buy a loaf of bread. What is the most he can spend for each gallon of gas and still have enough money left to buy the bread?

21. Assume that $a < b$.

$$a < b \qquad\qquad a < b$$
$$a + a < a + b \qquad a + b < b + b$$
$$2a < a + b \qquad\qquad a + b < 2b$$
$$a < \frac{a + b}{2} \qquad\qquad \frac{a + b}{2} < b$$

Therefore, $a < \dfrac{a + b}{2} < b$.

Solution

Let x = the cost of the gas. After Mario buys 8 gallons of gas, the amount of money left must be greater than or equal to $1.48, the cost of the bread.

$$10.00 - 8x \geq 1.48$$
$$-8x \geq -8.52$$
$$x \leq 1.065 \qquad \text{Order reverses.}$$

Therefore, Mario can spend no more than $1.065 per gallon.

Exercises 1.4

In **1 – 10**, name the property of the real numbers that is illustrated.

1. $-3 + 7 = 7 + (-3)$

2. $\frac{1}{2}\left(4\frac{1}{2}\right) = \frac{1}{2}(4) + \frac{1}{2}\left(\frac{1}{2}\right) = 2\frac{1}{4}$

3. If $-2x > -6$, then $x < 3$.

4. $x + 0 = x$

5. $13 + (27 + 52) = (13 + 27) + 52$

6. $9\left(17 \cdot \frac{1}{3}\right) = 9\left(\frac{1}{3} \cdot 17\right)$

7. $18 + (17 + 12) = 18 + (12 + 17)$

8. $9\left(\frac{1}{3} \cdot 17\right) = \left(9 \cdot \frac{1}{3}\right)17$

9. If $\frac{1}{3} \cdot \frac{5}{5} = \frac{5}{15}$, then $\frac{5}{15} = \frac{1}{3}$.

10. $x + 8 + (-8) = x + 0$

In **11 – 16**, solve the inequality.

11. $x + 2 > 0$

12. $6x < 3$

13. $3x - 7 > 8$

14. $3 - x < x - 1$

15. $9 > 7x + 2$

16. $x + 1 > 2x + 1$

17. From the cost of 3 cans of soup, the cashier deducted the value of a 20-cent coupon. The final cost of the soup was less than $1.50. What is the possible range for the cost of a can of soup?

18. After using 10 slices of a 1-pound loaf of bread, Grace found that the remainder of the loaf weighed less than 4 ounces. What is the possible range for the weight of a slice of bread?

19. A mail-order house offers paperback books for $4.50 each, plus $1.50 for postage and handling on every order. What is the largest number of books that can be purchased for less than $24.00?

20. The books in Exercise 19 may be purchased from a local bookstore for $5.25 each. How many books must be purchased before the mail order cost is less than the local bookstore cost?

21. Prove that $\frac{a + b}{2}$ is between a and b. Hint: Assume that $a < b$. First prove that $a < \frac{a + b}{2}$.

22. Prove that for the natural numbers a, b, c, and d, $\frac{a}{b} < \frac{c}{d}$ if and only if $ad < bc$.

23. Prove that if $a > b$ and $b > 0$, then $\frac{1}{a} < \frac{1}{b}$.

24. **a.** Show by counterexample that $x < x^2$ is not always true.
b. For what real numbers is $x < x^2$ true?
c. For what real numbers is $x > x^2$ true?
d. For what real numbers is $x = x^2$ true?

22. Since $a, b, c,$ and d are natural numbers, they are positive.

If $\quad ad < bc$

$$\frac{1}{bd}(ad) < \frac{1}{bd}(bc)$$

$$\frac{a}{b} < \frac{c}{d}$$

If $\quad \frac{a}{b} < \frac{c}{d}$

$$bd\left(\frac{a}{b}\right) < bd\left(\frac{c}{d}\right)$$

$$ad < bc$$

23. If $a > b$ and $b > 0$, then $a > 0$.

$$a > b$$
$$\frac{1}{ab}(a) > \frac{1}{ab}(b)$$
$$\frac{1}{b} > \frac{1}{a}$$

or $\quad \dfrac{1}{a} < \dfrac{1}{b}$

24. a. If $x = 0.1$, $x^2 = (0.1)^2 = 0.01$; $0.1 < (0.1)^2$ is false.

b. For $x < 0$ or $x > 1$, $x < x^2$ is true.

c. For $0 < x < 1$, $x > x^2$ is true.

d. For $x = 0$ or $x = 1$, $x = x^2$ is true.

1.5 Complex Numbers

It would seem that when the set of numbers that can be put into one-to-one correspondence with the points of the real number line was determined, the development of the number system would be finished. However, with the real numbers, it is still not possible to solve all equations of the form $x^2 = c$, where c is a constant.

If $c > 0$, the solutions of $x^2 = c$ are the real numbers $\pm\sqrt{c}$, and if $c = 0$, the solution is the real number 0. But if $c < 0$, then $x^2 = c$ has no solution in the set of real numbers. Therefore, we define a number i, whose square is –1.

$$i^2 = -1 \text{ and } i = \sqrt{-1}$$

Since the square of every nonzero real number is positive, i is not a real number. Thus, the number i is called the *imaginary unit* and can be used to write any number that is the square root of a negative number.

Example 1 _____

    a. $\sqrt{-25} = \sqrt{25}\sqrt{-1} = 5i$

    b. $\sqrt{-9} \cdot \sqrt{-4} = \sqrt{9}\sqrt{-1} \cdot \sqrt{4}\sqrt{-1} = 3i \cdot 2i = 6i^2 = -6$

A *complex number* is a number of the form $a + bi$, where a and b are real numbers and i is the imaginary unit.

If $b = 0$, the complex number $a + 0i$ is a real number. If $b \neq 0$, the complex number $a + bi$ is an *imaginary number*. If $a = 0$ and $b \neq 0$, the imaginary number $0 + bi$ is a *pure imaginary number*.

Two complex numbers $a + bi$ and $c + di$ are equal if and only if $a = c$ and $b = d$.

The set of complex numbers, C, is the union of the set of real numbers and the set of imaginary numbers.

Example 2 _____

    Write $\sqrt{50} + \sqrt{-100}$ in $a + bi$ form.

Solution

    $\sqrt{50} + \sqrt{-100} = \sqrt{25}\sqrt{2} + \sqrt{100}\sqrt{-1} = 5\sqrt{2} + 10i$

The Complex Number Plane

To graph the set of complex numbers, the intersection of two number lines, a real number line and a pure imaginary number line, is used. The real number line is drawn horizontally and the pure imaginary number line is drawn vertically.

The point of intersection of the two lines, 0 on the real number line and $0i$ on the pure imaginary number line, is the graph of the complex number $0 + 0i$.

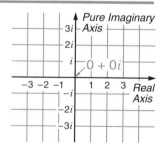

The graph of the complex number $3 + 4i$ is located in the same way that you would locate the graph of the real number pair $(3, 4)$ on the coordinate plane. The graph is the intersection of the vertical line through the real number 3 and the horizontal line through the pure imaginary number $4i$.

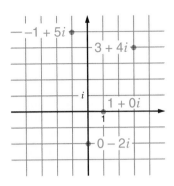

The graphs of the complex numbers $-1 + 5i$, $0 - 2i$, and $1 + 0i$ are also shown in the figure.

Note that the real number $1 + 0i$ is on the real axis and the pure imaginary number $0 - 2i$ is on the pure imaginary axis.

Addition of Complex Numbers

To add two complex numbers, use the commutative and distributive properties as you would use them to find the sum of two binomials.

For example, $(3 + 7i) + (-2 + i) = (3 - 2) + (7 + 1)i = 1 + 8i$.

Addition of two complex numbers can be represented graphically.

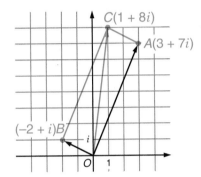

In the figure, point A is the graph of $3 + 7i$ and point B is the graph of $-2 + i$. Locate point C such that $OACB$ is a parallelogram. Point C is the graph of $1 + 8i$, the sum of $3 + 7i$ and $-2 + i$.

This sum can be thought of in terms of vector addition. The sum of $\overrightarrow{OA}$ and $\overrightarrow{OB}$ is $\overrightarrow{OC}$, the diagonal of the parallelogram determined by $\overrightarrow{OA}$ and $\overrightarrow{OB}$.

Since $0 + 0i = 0$, the number $0 + 0i$ plays the same role in the set of complex numbers that 0 plays in the set of real numbers.

$$(a + bi) + (0 + 0i) = (0 + 0i) + (a + bi) = a + bi$$

The identity for addition in the set of complex numbers is $0 + 0i$.

The additive inverses of the real numbers a and b are used to form the additive inverse of $a + bi$.

$$(a + bi) + (-a - bi) = 0 + 0i$$

Every complex number $a + bi$ has an additive inverse $(-a - bi)$.

Subtraction of Complex Numbers

Subtraction in the set of complex numbers is defined in the same way that it is defined in the set of real numbers, $x - y = x + (-y)$.

Example 3

$$(3 + 7i) - (-2 + i) = (3 + 7i) + (2 - i)$$
$$= (3 + 2) + (7 - 1)i$$
$$= 5 + 6i$$

To represent the subtraction of $-2 + i$ from $3 + 7i$ graphically, first locate point A, the graph of $3 + 7i$, and point B, the graph of $-2 + i$. Then find point B', the graph of the additive inverse of $-2 + i$, namely $2 - i$.

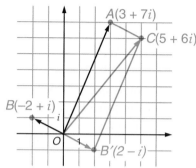

Note that the graph of the additive inverse of point B is the reflection of point B over the origin or the image of point B under a rotation of 180° about the origin.

The graph of $(3 + 7i) - (-2 + i)$ is the point C, $5 + 6i$, the vertex of the parallelogram $OACB'$ determined by $\overline{OA}$ and $\overline{OB'}$.

Multiplication of Complex Numbers

To multiply two complex numbers, use the commutative and distributive properties as you would use them to find the product of two binomials. Then, to write the product in $a + bi$ form, substitute $i^2 = -1$, and simplify.

Example 4

$$(2 + 3i)(-2 + i) = (2 + 3i)(-2) + (2 + 3i)(i)$$
$$= -4 - 6i + 2i + 3i^2$$
$$= -4 - 6i + 2i + 3(-1) \qquad i^2 = -1$$
$$= (-4 - 3) + (-6 + 2)i$$
$$= -7 - 4i$$

Example 5

$$(2 + 3i)(2 - 3i) = (2 + 3i)(2) + (2 + 3i)(-3i)$$
$$= 4 + 6i - 6i - 9i^2$$
$$= 4 + 6i - 6i - 9(-1)$$
$$= (4 + 9) + (6 - 6)i$$
$$= 13 + 0i = 13$$

Note that the product of two imaginary numbers may be either imaginary or real. The complex numbers $a + bi$ and $a - bi$ are called *complex conjugates* and their product is the real number $a^2 + b^2$.

$$(a + bi)(a - bi) = a^2 - b^2i^2 = a^2 + b^2$$

Observe that $(a + bi)$ and $(a - bi)$ are factors of $a^2 + b^2$.

Since $1 + 0i = 1$, the number $1 + 0i$ plays the same role in the set of complex numbers that 1 plays in the set of real numbers.

$$(a + bi)(1 + 0i) = (1 + 0i)(a + bi) = a + bi$$

The identity for multiplication in the set of complex numbers is $1 + 0i$.

Division of Complex Numbers

The fact that the product of complex conjugates is a real number can be used to show that the quotient of complex numbers is a complex number.

$$(2 + 3i) \div (-2 + i) = \frac{2 + 3i}{-2 + i}$$

To write this quotient in the form of a complex number, $a + bi$, multiply the numerator and denominator by the complex conjugate of the denominator. The resulting fraction will have a real number in the denominator.

$$\frac{2 + 3i}{-2 + i} \cdot \frac{-2 - i}{-2 - i} = \frac{-4 - 6i - 2i - 3i^2}{4 - i^2}$$

$$= \frac{-4 - 8i + 3}{4 + 1} = \frac{-1 - 8i}{5} = \frac{-1}{5} + \frac{-8}{5}i$$

The procedure for dividing complex numbers can be used to demonstrate that the number $\dfrac{1}{a + bi}$ can also be written in standard complex form.

$$\frac{1}{a + bi} \cdot \frac{a - bi}{a - bi} = \frac{a - bi}{a^2 - b^2 i^2} = \frac{a - bi}{a^2 + b^2} = \frac{a}{a^2 + b^2} - \frac{b}{a^2 + b^2}i$$

Every nonzero complex number $a + bi$ has a multiplicative inverse $\dfrac{1}{a + bi}$.

The Complex Number Field

The set of complex numbers under the operations of addition and multiplication forms a field. In contrast to the real numbers, the complex numbers are not an ordered field because it is not possible to say that one complex number is greater than another.

Example 6

Express the roots of $x^2 + 20 = 0$ in terms of i.

Solution

$$x^2 + 20 = 0$$

$$x^2 = -20 \qquad \text{Add } -20 \text{ to each side.}$$

$$x = \pm \sqrt{-20} \qquad \text{Take the square root of each side.}$$

$$= \pm \sqrt{4}\sqrt{5}\sqrt{-1}$$

$$= \pm 2i\sqrt{5} \qquad \text{Express the result in terms of } i.$$

Check each solution.

If $x = 2i\sqrt{5}$:

$$x^2 + 20 = 0$$

$$(2i\sqrt{5})^2 + 20 \overset{?}{=} 0$$

$$4i^2(5) + 20 \overset{?}{=} 0$$

$$4(-1)(5) + 20 \overset{?}{=} 0$$

$$-20 + 20 = 0 \quad ✔$$

If $x = -2i\sqrt{5}$:

$$x^2 + 20 = 0$$

$$(-2i\sqrt{5})^2 + 20 \overset{?}{=} 0$$

$$4i^2(5) + 20 \overset{?}{=} 0$$

$$4(-1)(5) + 20 \overset{?}{=} 0$$

$$-20 + 20 = 0 \quad ✔$$

The roots of $x^2 + 20 = 0$ are $\pm 2i\sqrt{5}$.

1. $2i$

2. $6i$

3. $11i$

4. $2i\sqrt{10}$

5. $2i\sqrt{2}$

6. $-4i$

7. $-5i\sqrt{3}$

8. $-i$

9. i

10. $-i$

11. $1 + 5i$

12. $3 - i$

13. $-8 + 4i$

14. $\frac{2}{5} - \frac{4}{5}i$

15. $3 + 2i$

16. $-1 + 2i$

17. $8 + 0i$

18. $10 + 0i$

19. $3 - 5i$

20. $8 - 6i$

21. $-\frac{3}{2} - \frac{7}{2}i$

22. $0 - \frac{1}{2}i$

23. $-1 + 2i$;

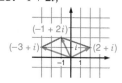

24. $5 + 0i$;

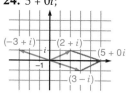

Example 7 _____

Express $i^7 + i^{10}$ in $a + bi$ form.

Solution

$$i^7 = (i^2)^3 \cdot i \qquad\qquad i^{10} = (i^2)^5$$
$$= (-1)^3 \cdot i \qquad\qquad\quad = (-1)^5$$
$$= -i \qquad\qquad\qquad\quad = -1$$

Therefore, $i^7 + i^{10} = -1 - i$.

Note the cyclic nature of the powers of i:

$$i^1 = i \qquad\qquad\qquad i^5 = i^4(i) = i$$
$$i^2 = -1 \qquad\qquad\quad i^6 = i^4(i^2) = -1$$
$$i^3 = i^2(i) = -i \qquad i^7 = i^4(i^3) = -i$$
$$i^4 = i^2(i^2) = 1 \qquad i^8 = i^4(i^4) = 1$$

Thus, the value of any power of i that is greater than 4 must be one of $i, -1, -i,$ or 1.

Symbolically, if $n > 4$ and $n = 4k + r$ where $0 < r \le 4$:

$$i^n = i^{4k+r} = (i^4)^k(i^r) = 1(i^r) = i^r$$

Exercises 1.5

In **1 – 10**, write the expression in terms of i.

1. $\sqrt{-4}$ **2.** $\sqrt{-36}$ **3.** $\sqrt{-121}$ **4.** $\sqrt{-40}$ **5.** $\sqrt{-8}$

6. $-\sqrt{-16}$ **7.** $-\sqrt{-75}$ **8.** i^3 **9.** i^9 **10.** $-i^{21}$

In **11 – 22**, perform the indicated operations and write the result in $a + bi$ form.

11. $(2 + \sqrt{-4}) + (\sqrt{-9} - 1)$ **12.** $(2 + \sqrt{-4}) - (\sqrt{-9} - 1)$

13. $(2 + \sqrt{-4})(\sqrt{-9} - 1)$ **14.** $(2 + \sqrt{-4}) \div (\sqrt{-9} - 1)$

15. $(4 - i) + (-1 + 3i)$ **16.** $i(2 + i)$

17. $(3 + i) - (-5 + i)$ **18.** $(3 + i)(3 - i)$

19. $(8 - 2i) \div (1 + i)$ **20.** $(-3 + i)^2$

21. $\dfrac{5 + 2i}{-1 + i}$ **22.** $\dfrac{i(2 + i)}{-4 - 2i}$

In **23 – 24**, find the sum or difference graphically.

23. $(2 + i) + (-3 + i)$ **24.** $(2 + i) - (-3 + i)$

25. Prove that if $3 + i$ is a root of $ax^2 + bx + c = 0$, then $3 - i$ is also a root.

---

25. If $3 + i$ is a root of $ax^2 + bx + c = 0$, then

$$a(3 + i)^2 + b(3 + i) + c = 0$$
$$a(9 + 6i + i^2) + b(3 + i) + c = 0$$
$$(8a + 3b + c) + (6a + b)i = 0 + 0i$$

Therefore, $8a + 3b + c = 0$ and $6a + b = 0$.

$(3 - i)$ is a root if

$$a(3 - i)^2 + b(3 - i) + c = 0$$
$$a(9 - 6i + i^2) + b(3 - i) + c = 0$$
$$(8a + 3b + c) + (-6a - b)i = 0 + 0i$$

which is true if $8a + 3b + c = 0$ and $-6a - b = 0$.

But since $(3 + i)$ is a root, $8a + 3b + c = 0$ and $6a + b = 0$ or $-6a - b = 0$.

1.6 Polynomials

A *monomial* is a constant, a variable, or the product of constants and variables. A *polynomial* is the sum of monomials.

A polynomial in x is an algebraic expression of the form

$$a_n x^n + a_{n-1} x^{n-1} + a_{n-2} x^{n-2} + \cdots + a_1 x + a_0$$

The a's represent constants, n is a whole number, and $a_n \neq 0$ if $n \neq 0$.

Since the variable in a polynomial represents a number, the polynomial represents a number from the largest closed set of which the coefficients and the value of the variable are elements.

Example 1

Consider the polynomial $3x^3 - 5x + 3$, which is a polynomial over the set of integers since all the coefficients are integers.

If $x = 2$, an integer,
$$3x^3 - 5x + 3 = 3(2)^3 - 5(2) + 3$$
$$= 24 - 10 + 3$$
$$= 17, \text{ an integer.}$$

If $x = \sqrt{2}$, a real number,
$$3x^3 - 5x + 3 = 3(\sqrt{2})^3 - 5(\sqrt{2}) + 3$$
$$= 6\sqrt{2} - 5\sqrt{2} + 3$$
$$= \sqrt{2} + 3, \text{ a real number.}$$

If $x = 3i$, a complex number,
$$3x^3 - 5x + 3 = 3(3i)^3 - 5(3i) + 3$$
$$= 81i^3 - 15i + 3$$
$$= 81(-i) - 15i + 3$$
$$= 3 - 96i, \text{ a complex number.}$$

Since a polynomial represents a number from some subset of the complex numbers, those properties that hold for all complex numbers can be applied to the set of polynomials.

- Addition and multiplication are commutative in the set of polynomials.
  For example:
    a. $2x^2 + 5x = 5x + 2x^2$
    b. $(x - 2)(x^2 + x) = (x^2 + x)(x - 2)$

- Addition and multiplication are associative in the set of polynomials.
  For example:
    a. $(2x^2 + 5x) + 3x = 2x^2 + (5x + 3x)$
    b. $[x^3(x - 2)](x + 2) = x^3[(x - 2)(x + 2)]$

- Multiplication is distributive over addition in the set of polynomials. For example:
$$2x^2(5x^2 - 2x) = 2x^2(5x^2) + 2x^2(-2x)$$

Using these properties of polynomials, it can be shown that the set of polynomials is closed under addition, subtraction, and multiplication.

Example 2

a. The sum of two polynomials is a polynomial.
$$\begin{aligned}(3x^2 + 7x) + (3 - 12x + x^2) &= (3x^2 + x^2) + (7x - 12x) + 3 \\ &= x^2(3 + 1) + x(7 - 12) + 3 \\ &= 4x^2 - 5x + 3\end{aligned}$$

b. The difference of two polynomials is a polynomial.
$$\begin{aligned}(3x^2 + 7x) - (3 - 12x + x^2) &= (3x^2 + 7x) + (-3 + 12x - x^2) \\ &= (3x^2 - x^2) + (7x + 12x) - 3 \\ &= x^2(3 - 1) + x(7 + 12) - 3 \\ &= 2x^2 + 19x - 3\end{aligned}$$

c. The product of two polynomials is a polynomial.
$$\begin{aligned}(3x - 2)(x^2 - 5x) &= (3x - 2)(x^2) + (3x - 2)(-5x) \\ &= 3x^3 - 2x^2 - 15x^2 + 10x \\ &= 3x^3 - 17x^2 + 10x\end{aligned}$$

Like the integers, the set of polynomials is not closed under division. To divide by a monomial, use the distributive property of division over addition.

Example 3

$$\frac{3x^2 - 12x + 2}{3x} = \frac{3x^2}{3x} - \frac{12x}{3x} + \frac{2}{3x} = x - 4 + \frac{2}{3x}$$

A monomial may include only the product of constants and variables. Therefore, $\frac{2}{3x}$ is not a monomial and $x - 4 + \frac{2}{3x}$ is not a polynomial.

To divide a polynomial by a polynomial, use the algorithm that is used to divide by a natural number of two or more digits.

Example 4

Divide $6x^2 - 7x + 3$ by $2x - 1$.

Solution

$$
\begin{array}{r}
3x - 2 \\
2x - 1 \overline{)\, 6x^2 - 7x + 3} \\
6x^2 - 3x \\
\hline
-4x + 3 \\
-4x + 2 \\
\hline
1
\end{array}
$$

Step 1   Divide $6x^2$ by $2x$.
Step 2   Multiply $3x(2x - 1)$.
Step 3   Subtract $6x^2 - 3x$ from $6x^2 - 7x$.
Step 4   Bring down the next monomial term, $+3$.
 Repeat the sequence of operations in Steps 1–4, as often as necessary.

The process ends when the degree of the remainder is smaller than the degree of the divisor.
$$\frac{6x^2 - 7x + 3}{2x - 1} = 3x - 2 + \frac{1}{2x - 1}, \text{ or } 6x^2 - 7x + 3 = (2x - 1)(3x - 2) + 1$$

Example 5

The distance that a car travels in x hours is $(3x^2 - x)$ miles. Find the time that it would take the car to travel $\left(3x^2 - \frac{1}{3}\right)$ miles at the same rate.

Solution

Solve for rate and for time: distance = (rate)(time)

$$\text{rate} = \frac{\text{distance}}{\text{time}} \qquad\qquad \text{time} = \frac{\text{distance}}{\text{rate}}$$

$$\text{rate} = \frac{3x^2 - x}{x} \qquad\qquad \text{time} = \frac{3x^2 - \frac{1}{3}}{3x - 1}$$

$$= \frac{3x^2}{x} - \frac{x}{x}$$

$$= 3x - 1$$

$$\begin{array}{r} x + \frac{1}{3} \\ 3x - 1 \overline{\smash{)}3x^2 + 0x - \frac{1}{3}} \\ \underline{3x^2 - x} \\ x - \frac{1}{3} \\ \underline{x - \frac{1}{3}} \\ 0 \end{array}$$

It would take $\left(x + \frac{1}{3}\right)$ hours.

An expression that is the quotient of two polynomials is a *rational expression*. To write the sum of two rational expressions as a single rational expression, it is necessary to find a common denominator. It is always possible to use the product of the denominators as a common denominator, although this may not be the *least* common denominator.

Example 6

$$\frac{x+5}{3} + \frac{2x}{x-1} = \frac{x+5}{3} \cdot \frac{x-1}{x-1} + \frac{2x}{x-1} \cdot \frac{3}{3}$$

$$= \frac{x^2 + 4x - 5}{3x - 3} + \frac{6x}{3x - 3}$$

$$= \frac{x^2 + 10x - 5}{3x - 3}$$

Exercises 1.6

In **1 – 42**, perform the indicated operations and express the result in simplest form.

1. $7x^3 - 4x + 5x - x^3$ **2.** $3x^4 - (12x^5 - 2x^4)$ **3.** $(3x^2 - 7x) + (2x - x^2)$

4. $x^5 - x^2(x^2 - x^3)$ **5.** $(4x - 3)(3x - 1)$ **6.** $(4x - 3) - (3x - 1)$

7. $(4x - 3) - 5(3x - 1)$ **8.** $(4x^3 - 8) - (x^3 - 8)$ **9.** $8x(x^2 - 2x) - 3x(x^2 + 5x)$

1. $6x^3 + x$

2. $-12x^5 + 5x^4$

3. $2x^2 - 5x$

4. $2x^5 - x^4$

5. $12x^2 - 13x + 3$

6. $x - 2$

7. $-11x + 2$

8. $3x^3$

9. $5x^3 - 31x^2$

10. $\frac{1}{6}x^2 + 2x + 6$

11. $3x^2 + x - 2$

12. $2x^2 - 2x - 2$

13. $x - 10 + \frac{27}{x + 2}$

14. $3x^2 - 5 + \frac{4}{x^2}$

15. $3x - 5.5 + \frac{3.5}{2x + 1}$

16. $x^2 + x + 1$

17. $\frac{x^4}{10} - 40$

18. x^4

19. $2x^3 - 4x^2 + 3$

20. $x^3 - 1$

21. $2x^4 - 3x^3 - 3x^2 + 4x - 1$

22. $\frac{4x}{x^2 - 4}$

23. $\frac{3x + 5}{2(x + 3)}$

24. $\dfrac{2x + 1}{x(x + 1)}$

25. $\dfrac{-x^2 + 4x + 3}{x^2 + x}$

26. $\dfrac{-x}{x^2 + 6x + 8}$

27. $\dfrac{x^2 + x - 1}{x^2 + x}$

28. $\dfrac{4x}{3x^2 + 2x - 1}$

29. $\dfrac{4}{x^2 + 4x}$

30. $\dfrac{x^3 - 4x}{2x^2 - 4}$

31. $\dfrac{-x^2}{5x + 25}$

32. $\dfrac{4}{x^4 - 4}$

33. $\dfrac{6x}{x^2 - 25}$

34. $\dfrac{5x - 12}{x^2 - 5x + 6}$

35. $\dfrac{5x^2 - 15x}{2x^2 - 3x - 2}$

36. $\dfrac{21x^3 + 3x^2}{9x^2 - 1}$

37. $\dfrac{x}{x - 5}$

38. $\dfrac{2x^2}{x^2 - 1}$

39. $\dfrac{x^2 + x + 1}{x + 1}$

40. $\dfrac{9x^2 - 6x + 2}{3x - 1}$

41. $\dfrac{x^2}{x - 1}$

42. $\dfrac{3x^2}{x - 1}$

43. $(9x^2 - 15x + 10)$ dollars

44. $(5a^2 + 23a - 10)$ miles

Exercises 1.6 (continued)

10. $\left(\dfrac{1}{2}x + 3\right)\left(\dfrac{1}{3}x + 2\right)$

11. $(3x^2 - 1) + (x - 1)$

12. $(2x^2 - 4x + 1) + (2x - 3)$

13. $(x^2 - 8x + 7) \div (x + 2)$

14. $(12x^4 - 20x^2 + 16) \div 4x^2$

15. $(6x^2 - 8x - 2) \div (2x + 1)$

16. $(x^3 - 1) \div (x - 1)$

17. $(0.2x^2 - 4)(0.5x^2 + 10)$

18. $\left(\dfrac{1}{2}x^4 + \dfrac{1}{3}x^2\right) + \left(\dfrac{1}{2}x^4 - \dfrac{1}{3}x^2\right)$

19. $(6x^4 - 12x^3 + 9x) \div 3x$

20. $(x - 1)(x^2 + x + 1)$

21. $(2x - 1)(x^3 - x^2 - 2x + 1)$

22. $\dfrac{2}{x + 2} + \dfrac{2}{x - 2}$

23. $\dfrac{1}{2} + \dfrac{x + 1}{x + 3}$

24. $\dfrac{1}{x} + \dfrac{1}{x + 1}$

25. $\dfrac{3}{x} - \dfrac{x - 1}{x + 1}$

26. $\dfrac{1}{x + 2} - \dfrac{2}{x + 4}$

27. $\dfrac{x - 1}{x} + \dfrac{1}{x + 1}$

28. $\dfrac{1}{x + 1} + \dfrac{1}{3x - 1}$

29. $\dfrac{1}{x} - \dfrac{1}{x + 4}$

30. $\dfrac{x}{2} - \dfrac{x}{x^2 - 2}$

31. $\dfrac{x}{x + 5} - \dfrac{x}{5}$

32. $\dfrac{1}{x^2 - 2} - \dfrac{1}{x^2 + 2}$

33. $\dfrac{3}{x - 5} + \dfrac{3}{x + 5}$

34. $\dfrac{3}{x - 3} + \dfrac{2}{x - 2}$

35. $\dfrac{7x}{2x + 1} - \dfrac{x}{x - 2}$

36. $\dfrac{5x^2}{3x - 1} + \dfrac{2x^2}{3x + 1}$

37. $1 + \dfrac{5}{x - 5}$

38. $2 + \dfrac{2}{x^2 - 1}$

39. $x + \dfrac{1}{x + 1}$

40. $3x - 1 + \dfrac{1}{3x - 1}$

41. $x + 1 + \dfrac{1}{x - 1}$

42. $3x + \dfrac{3x}{x - 1}$

In **43 – 48**, express the result as a polynomial or as the sum of a polynomial and a rational expression.

43. Payroll deductions of $(7x - 4x^2)$ dollars and $(5x^2 - 3)$ dollars are made weekly from Melissa's base salary of $(10x^2 - 8x + 7)$ dollars. What is her take-home pay?

44. Find the distance that a train can travel in $(a + 5)$ hours at $(5a - 2)$ miles per hour.

45. Find the rate at which a boy is walking if he walks $(x^2 - 1)$ miles in $(x - 2)$ hours.

46. Find the time it will take a plane to fly $(b^3 - 2b^2)$ miles at $(b - 3)$ miles per hour.

47. If an electronic scanner can read $(x^2 - 7x + 12)$ characters in $(x - 4)$ seconds, how many characters can it read in 1 second?

48. Metro Theatre grossed $(x^2 + 11x + 30)$ dollars for Tuesday's performance. If the price per ticket was fixed at $(5 + x)$ dollars, how many tickets were sold?

49. a. The area of a rectangle is $(x^2 + 3x - 7)$ square feet and the width of the rectangle is $(x - 1)$ feet. Express the perimeter of the rectangle in terms of x.

b. Is the expression that represents the perimeter a polynomial?

c. If $x = 3$, find the area and perimeter of the rectangle.

---

45. $\left(x + 2 + \dfrac{3}{x - 2}\right)$ mph

46. $\left(b^2 + b + 3 + \dfrac{9}{b - 3}\right)$ hours

47. $(x - 3)$ characters

48. $(x + 6)$ tickets

49. a. $\left(4x + 6 - \dfrac{6}{x - 1}\right)$ feet

b. no

c. Area = 11 square feet
Perimeter = 15 feet

1.7 Factoring Polynomials

Many of the principles that were discussed concerning the divisibility of integers apply to polynomials.

If $P(x)$, $Q(x)$, and $S(x)$ represent polynomials in x, and $P(x) \cdot Q(x) = S(x)$, then:

- $P(x)$ divides $S(x)$, and $Q(x)$ divides $S(x)$.
- $P(x)$ is a factor of $S(x)$, and $Q(x)$ is a factor of $S(x)$.
- $S(x)$ is a multiple of $P(x)$, and $S(x)$ is a multiple of $Q(x)$.

To factor a polynomial, find two or more polynomials whose product is the given polynomial. Factoring often consists of examining the product and working backward on the process of multiplication.

Common Monomial Factors

The common monomial factor of a polynomial is found by inspection. Take the smallest power of the variable that is common to all of the terms, and multiply by the GCF of the coefficients. For example:

$$12x^4 + 9x^2 = 3x^2(4x^2 + 3)$$

In the example, the common monomial factor is the product of the GCF, 3, and the smallest common power of x, x^2. The binomial factor is a prime polynomial over the set of integers since its coefficients are relatively prime and it has no polynomial factor of a smaller degree. Finding the greatest common monomial factor, if it is not 1, is usually the first step in factoring a polynomial.

Binomial Factors

After factoring out the common monomial factor, examine the polynomial factor to see if it can be factored into binomial factors.

A general trinomial can be factored by looking for factors of the first and last terms and working backward on the steps of the multiplication of two binomials with all possible combinations of these factors. The examples that follow illustrate.

$$x^2 + 4x + 3$$
$$(x + 3)(x + 1)$$
$$+3x$$
$$+1x$$
$$+4x$$

$$3x^2 + 2x - 8$$
$$(3x - 4)(x + 2)$$
$$-4x$$
$$+6x$$
$$+2x$$

$$9x^4 - 6x^2 + 1$$
$$(3x^2 - 1)(3x^2 - 1)$$
$$-3x^2$$
$$-3x^2$$
$$-6x^2$$

Two special cases are of particular importance.

- The difference of two squares:

$$a^2 - b^2 = (a + b)(a - b)$$

- The perfect-square trinomial:

$$a^2 + 2ab + b^2 = (a + b)(a + b) = (a + b)^2$$

$$a^2 - 2ab + b^2 = (a - b)(a - b) = (a - b)^2$$

The following examples illustrate.

    a. $x^2 - 64 = (x + 8)(x - 8)$

    b. $x^2 + 12x + 36 = (x + 6)(x + 6) = (x + 6)^2$

    c. $4x^2 - 20x + 25 = (2x - 5)(2x - 5) = (2x - 5)^2$

Common Binomial Factor

A polynomial can sometimes be factored by identifying a common binomial factor. Compare the roles of y and $(x - 2)$ in the following examples.

    a. $3xy - 5y = y(3x - 5)$

    b. $3x(x - 2) - 5(x - 2) = (x - 2)(3x - 5)$

Sometimes it is possible to factor a polynomial of four terms by first considering the common monomial factor of each pair of terms of the given polynomial. For example:

$$x^3 - 3x^2 + 2x - 6 = x^2(x - 3) + 2(x - 3)$$
$$= (x - 3)(x^2 + 2)$$

Note: The factors $(x - 3)$ and $(x^2 + 2)$ are prime polynomials over the integers.

Sum and Difference of Two Cubes

You are familiar with the factors of the difference of two squares. It is possible to show by division that $(a - b)$ is a factor of $a^3 - b^3$, the difference of two cubes.

$$
\require{enclose}
\begin{array}{r}
a^2 + ab + b^2 \\[-3pt]
a - b \enclose{longdiv}{a^3 + 0a^2b + 0ab^2 - b^3} \\
\end{array}
$$

$$
\begin{array}{r}
a^3 - a^2b \\ \hline
a^2b + 0ab^2 \\
a^2b - ab^2 \\ \hline
ab^2 - b^3 \\
ab^2 - b^3 \\ \hline
0
\end{array}
$$

Therefore, $a^3 - b^3 = (a - b)(a^2 + ab + b^2)$.

In the same way, it can be shown that $a^3 + b^3 = (a + b)(a^2 - ab + b^2)$. For example:

    a. $x^3 - 27 = (x - 3)(x^2 + 3x + 9)$ **b.** $8x^3 - 1 = (2x - 1)(4x^2 + 2x + 1)$

    c. $x^3 + 27 = (x + 3)(x^2 - 3x + 9)$ **d.** $64x^3 + 125 = (4x + 5)(16x^2 - 20x + 25)$

Factoring a Polynomial Over the Complex Numbers

When $x^4 - x^2 - 2$ is factored over the set of integers, the prime polynomial factors are $(x^2 - 2)(x^2 + 1)$. However, these factors are not prime polynomials over the set of complex numbers. The special techniques that you have learned, such as the factorization of the difference of two squares, make it possible to factor these polynomial factors over the set of complex numbers.

$$x^2 - 2 = x^2 - (\sqrt{2})^2 = (x + \sqrt{2})(x - \sqrt{2})$$

$$x^2 + 1 = x^2 - (-1) = x^2 - i^2 = (x + i)(x - i)$$

Therefore, when factoring $x^4 - x^2 - 2$ over the set of complex numbers:

$$x^4 - x^2 - 2 = (x^2 - 2)(x^2 + 1)$$
$$= (x + \sqrt{2})(x - \sqrt{2})(x + i)(x - i)$$

In general, for $a > 0$:

$$x^2 - a = x^2 - (\sqrt{a})^2$$
$$= (x + \sqrt{a})(x - \sqrt{a})$$

$$x^2 + a = x^2 - (-a)$$
$$= x^2 - (i\sqrt{a})^2$$
$$= (x + i\sqrt{a})(x - i\sqrt{a})$$

The trinomial $x^2 + 6x + 10$ cannot be factored over the set of integers, but it can be factored over the set of complex numbers. To do this, first express the trinomial as the difference of two squares by finding the constant necessary to write a perfect-square trinomial whose first two terms are $x^2 + 6x$.

Recall that $x^2 + 2ax + a^2 = (x + a)^2$. The constant term, a^2, is the square of one-half the coefficient of x. To write the perfect-square trinomial $x^2 + 6x + a^2$, let $a = \frac{1}{2}(6)$, or 3, and thus $a^2 = 9$.

Therefore,

$$x^2 + 6x + 9 = (x + 3)^2$$

and

$$x^2 + 6x + 10 = x^2 + 6x + 9 + 1$$
$$= (x + 3)^2 - (-1)$$
$$= (x + 3)^2 - i^2 \qquad \text{Difference of two squares}$$
$$= (x + 3 + i)(x + 3 - i)$$

Example 1

Factor the polynomial $x^3 - 8$ over the set of:

a. integers **b.** complex numbers

1. $(x + 10)(x - 10)$

2. $(3x + 1)(3x - 1)$

3. $(x + 9)^2$

4. $(x + 6)^2$

5. $(2x - 1)(x + 5)$

6. $(3x + 1)(2x - 3)$

7. $(4x - 3)(2x - 1)$

8. $3x(3x - 4)$

9. $5(x^2 - 3x + 1)$

10. $5x^2(2x + 1)(2x - 1)$

11. $2x(3x - 2)(2x + 1)$

12. $3x(3x^2 - 1)(2x^2 + 1)$

13. $(x - 5)(x^2 + 5x + 25)$

Solution

a. Since $x^3 - 8$ is the difference of two cubes,
$x^3 - 8 = (x - 2)(x^2 + 2x + 4)$.

b. The trinomial factor $x^2 + 2x + 4$ can be factored further over the set of complex numbers by writing it as the difference of two squares.

Since $x^2 + 2x + 1 = (x + 1)^2$:

$$x^2 + 2x + 4 = x^2 + 2x + 1 + 3$$
$$= (x + 1)^2 - (-3)$$
$$= (x + 1)^2 - (i\sqrt{3})^2$$
$$= (x + 1 + i\sqrt{3})(x + 1 - i\sqrt{3})$$

Answer: $x^3 - 8 = (x - 2)(x + 1 + i\sqrt{3})(x + 1 - i\sqrt{3})$

Operations with Rational Expressions

In order to write the sum $\dfrac{1}{x^2 - 1} + \dfrac{1}{2x + 2}$ as a single rational expression, it is necessary to first find a common denominator. One possible common denominator is $(x^2 - 1)(2x + 2)$. To find a simpler common denominator, look at the prime factorization of the given denominators.

$$x^2 - 1 = (x - 1)(x + 1) \qquad 2x + 2 = 2(x + 1)$$

The least common multiple of the two given denominators is $2(x - 1)(x + 1)$. Therefore,

$$\frac{1}{x^2 - 1} + \frac{1}{2x + 2} = \frac{1}{(x - 1)(x + 1)} \cdot \frac{2}{2} + \frac{1}{2(x + 1)} \cdot \frac{x - 1}{x - 1}$$

$$= \frac{2}{2(x - 1)(x + 1)} + \frac{x - 1}{2(x + 1)(x - 1)}$$

$$= \frac{x + 1}{2(x - 1)(x + 1)}$$

This result has the factor $(x + 1)$ in the numerator and in the denominator. By dividing the numerator and the denominator by $(x + 1)$, the fraction will be in lowest terms. Note the restriction on the domain of x.

$$\frac{x + 1}{2(x - 1)(x + 1)} = \frac{\overset{1}{\cancel{x + 1}}}{2(x - 1)\underset{1}{\cancel{(x + 1)}}} = \frac{1}{2(x - 1)} \quad \text{where } x \neq \pm 1$$

Example 2

Express $\left(\dfrac{1}{x} + \dfrac{1}{x^2 - x} \right) \div \dfrac{2}{x^2 - 1}$ in simplest form.

Solution

Step 1   Add the fractions in the parentheses.

Step 2   Divide by multiplying by the reciprocal of the divisor.

14. $3(2x - 1)(4x^2 + 2x + 1)$

15. $x^2(2x^2 - 5)(x^2 + 2)$

16. $5(x + 1)(x^2 - x + 1)(3x^3 - 1)$

17. $(3x - 1)(3x + 1)(9x^2 + 1)$

18. $3x^2(2x - 1)(2x + 1)$

19. $3(2x^2 - 1)(2x^2 + 1)$

20. $6x(2x - 1)(2x + 1)$

21. $(x + 1)(x - 3)$

22. $(x - 12)(x^2 - 2)$

23. $a(x - 1)(ax^3 - 1)$

24. $(x - 5)(x - 1)(x^2 + x + 1)$

25. $(x + \sqrt{5})(x - \sqrt{5})$

26. $(x + 2\sqrt{2})(x - 2\sqrt{2})$

27. $4x(x + \sqrt{3})(x - \sqrt{3})$

28. $(x + 3i)(x - 3i)$

29. $7(x^2 - i^2) = 7(x + i)(x - i)$

Step 3 Reduce the result to lowest terms by canceling any common factors in the numerator and denominator.

$$\left(\frac{1}{x} + \frac{1}{x^2 - x}\right) \div \frac{2}{x^2 - 1} = \left(\frac{1}{x} \cdot \frac{x-1}{x-1} + \frac{1}{x(x-1)}\right) \div \frac{2}{x^2 - 1}$$

$$= \left(\frac{x-1}{x(x-1)} + \frac{1}{x(x-1)}\right) \div \frac{2}{(x-1)(x+1)}$$

$$= \frac{x}{x(x-1)} \cdot \frac{(x-1)(x+1)}{2}$$

$$= \frac{(x+1)}{2} \quad \text{where } x \neq 0, \pm 1$$

Exercises 1.7

In **1 – 24**, factor into prime polynomials over the set of integers.

1. $x^2 - 100$ **2.** $9x^2 - 1$ **3.** $x^2 + 18x + 81$

4. $x^2 + 12x + 36$ **5.** $2x^2 + 9x - 5$ **6.** $6x^2 - 7x - 3$

7. $8x^2 - 10x + 3$ **8.** $9x^2 - 12x$ **9.** $5x^2 - 15x + 5$

10. $20x^4 - 5x^2$ **11.** $12x^3 - 2x^2 - 4x$ **12.** $18x^5 + 3x^3 - 3x$

13. $x^3 - 125$ **14.** $24x^3 - 3$ **15.** $2x^6 - x^4 - 10x^2$

16. $15x^6 + 10x^3 - 5$ **17.** $81x^4 - 1$ **18.** $12x^4 - 3x^2$

19. $12x^4 - 3$ **20.** $24x^3 - 6x$ **21.** $x(x+1) - 3(x+1)$

22. $x^3 - 12x^2 - 2x + 24$ **23.** $a^2x^4 - a^2x^3 - ax + a$ **24.** $x^4 - 5x^3 - x + 5$

In **25 – 36**, factor over the set of complex numbers.

25. $x^2 - 5$ **26.** $x^2 - 8$ **27.** $4x^3 - 12x$ **28.** $x^2 + 9$

29. $7x^2 + 7$ **30.** $2x^4 - 8$ **31.** $x^2 + 4x + 5$ **32.** $x^2 - 10x + 29$

33. $x^2 + x - \frac{3}{4}$ **34.** $x^2 + 3x + \frac{13}{4}$ **35.** $81x^4 - 1$ **36.** $x^3 - 125$

In **37 – 45**, perform the operations. Express the result in simplest form.

37. $\dfrac{x}{x+3} - \dfrac{18}{x^2 - 9}$ **38.** $\dfrac{x}{2x+4} - \dfrac{2}{x^2 + 2x}$ **39.** $\dfrac{1}{2x} + \dfrac{1}{x^2 - 2x}$

40. $\dfrac{4x^2 - 9}{4x + 6} \div (6x - 9)$ **41.** $\dfrac{1}{4x} - \dfrac{1}{x^2 + 4x}$ **42.** $\dfrac{3x}{2}\left(\dfrac{1}{3x} - \dfrac{1}{5x}\right)$

43. $\left(x + 1 + \dfrac{1}{x-1}\right) \div \dfrac{2x}{x-1}$ **44.** $\dfrac{4}{x^2 - 1}\left(\dfrac{x+2}{4} + \dfrac{1}{4x}\right)$ **45.** $\dfrac{x}{2x-1}\left(\dfrac{2x+7}{2} - \dfrac{2}{x}\right)$

46. The area of a rectangle is represented by $x^3 - x$. What are the possible expressions for the dimensions of the rectangle if for all integral values of x, the dimensions are integers?

42. $\dfrac{1}{5}$

43. $\dfrac{x}{2}$

44. $\dfrac{x+1}{x(x-1)}$

45. $\dfrac{x+4}{2}$

46. 1 by $x^3 - x$,
x by $x^2 - 1$,
$x^2 + x$ by $x - 1$,
or $x + 1$ by $x^2 - x$.

30. $2(x + i\sqrt{2})(x - i\sqrt{2})(x + \sqrt{2})(x - \sqrt{2})$

31. $(x + 2 + i)(x + 2 - i)$

32. $(x - 5 + 2i)(x - 5 - 2i)$

33. $\left(x + \frac{3}{2}\right)\left(x - \frac{1}{2}\right)$

34. $\left(x + \frac{3}{2} + i\right)\left(x + \frac{3}{2} - i\right)$

35. $(3x + i)(3x - i)(3x + 1)(3x - 1)$

36. $(x - 5)\left(x + \frac{5}{2} + \frac{5\sqrt{3}}{2}i\right)\left(x + \frac{5}{2} - \frac{5\sqrt{3}}{2}i\right)$

37. $\dfrac{x - 6}{x - 3}$

38. $\dfrac{x - 2}{2x}$

39. $\dfrac{1}{2(x - 2)}$

40. $\dfrac{1}{6}$

41. $\dfrac{1}{4(x + 4)}$

Sets of Numbers

N = Natural Numbers	Z = Integers	Q = Rational Numbers
W = Whole Numbers	Z^+ = Positive Integers	R = Real Numbers
	Z^- = Negative Integers	C = Complex Numbers

The natural numbers are the basic building blocks of the other sets of numbers. You can think of each set of numbers as the result of combining a known set with a new set.

Postulates of the Natural Numbers

- There exists a natural number 1.
- Every natural number has a successor. The successor of a is $a + 1$ and the antecedent (preceding number) of $a + 1$ is a.
- The number 1 has no antecedent.
- If two successors are equal, then their antecedents are equal.
    If $a + 1 = b + 1$, then $a = b$.
- Let S be a set of natural numbers such that $1 \in S$, and if S contains a, then S contains $a + 1$. Then, S is the set of natural numbers.

Equality Postulates

For all a, b, and $c \in C$:
- Reflexive Property of Equality: $a = a$.
- Symmetric Property of Equality: If $a = b$, then $b = a$.
- Transitive Property of Equality: If $a = b$ and $b = c$, then $a = c$.
- Addition Property of Equality: If $a = b$, then $a + c = b + c$.
- Multiplication Property of Equality: If $a = b$, then $ac = bc$.

- Subtraction can be defined in terms of addition in those sets of numbers in which each element has an additive inverse:

  $a - b = a + (-b)$

  Subtraction is not closed in the set of whole numbers.

- Division can be defined in terms of multiplication in those sets of numbers in which each nonzero number has a multiplicative inverse:

  $a \div b = a \cdot \dfrac{1}{b}$

  Division is not closed in the set of integers or in the set of whole numbers.

- In the set of integers, if b divides a, then there exists an integer c such that $a = bc$. If $a = bc$, a is a multiple of b and of c, and b and c are factors of a.

- A prime number is a natural number that has exactly two factors. A composite number is a natural number that has more than two factors.

  Every integer can be written as the product of primes in exactly one way, disregarding the order of the primes.

Theorems

- **Number of Factors Theorem:** If the prime factorization of a natural number a is $a = p_1^{b_1} p_2^{b_2} \ldots p_n^{b_n}$, then a has $(b_1 + 1)(b_2 + 1) \ldots (b_n + 1)$ factors.

- **Divisibility of a Sum Theorem:** If a divides b and a divides c, then a divides $b + c$.

- **Divisibility of a Product Theorem:** If a divides bc and a and b are relatively prime, then a divides c.

- The greatest common factor (GCF) of two or more integers is the largest integer that divides each of the given integers.

- The least common multiple (LCM) of two or more integers is the smallest positive integer that is a multiple of each of the given integers.

- A rational number is a number that can be expressed as the ratio $\dfrac{a}{b}$ where a and b are integers and $b \neq 0$. Every rational number can be written as a repeating decimal. Every repeating decimal represents a rational number.

- An irrational number is a real number that cannot be expressed as the ratio of integers. The decimal representation of an irrational number does not terminate or repeat.

- The real numbers can be put into one-to-one correspondence with the points on the number line. The point that corresponds to a real number a is called the graph of a.

- $a > b$ if and only if there exists a positive number x such that $a = b + x$.
- $b < a$ if and only if $a > b$.
- The Trichotomy Property: For each pair of real numbers, a and b, exactly one of the following is true: $a < b$, $a = b$, or $a > b$.
- The Transitive Property of Inequality: If $a > b$ and $b > c$, then $a > c$.
- The Addition Property of Inequality: If $a > b$, then $a + c > b + c$.
- The Multiplication Property of Inequality: If $a > b$ and $c > 0$, then $ac > bc$.
  If $a > b$ and $c < 0$, then $ac < bc$.
- The set of real numbers with the operations of addition and multiplication forms an ordered field.
- $i^2 = -1$ and $i = \sqrt{-1}$. A pure imaginary number is a number of the form bi, where $b \neq 0$. A complex number is a number of the form $a + bi$, where a and b are real numbers. The graph of a complex number is a point on the complex number plane determined by a horizontal real axis and a vertical imaginary axis.

The set of complex numbers under the operations
of addition and multiplication forms a field.

- The set is closed under addition.
- Addition is associative.
- There is an identity element for addition.
- There is an additive inverse for every element.
- Addition is commutative.

- The set is closed under multiplication.
- Multiplication is associative.
- There is an identity element for multiplication.
- There is a multiplicative inverse for every nonzero element.
- Multiplication is commutative.

- Multiplication is distributive over addition.

- A polynomial in x is an algebraic expression of the form
  $a_n x^n + a_{n-1} x^{n-1} + a_{n-2} x^{n-2} + \cdots + a_1 x + a_0$, where the a's are constants, n is a whole number, and $a_n \neq 0$ if $n \neq 0$. The degree of the polynomial is n.
- A polynomial may be factored over the set of integers by looking for the following characteristics:

Common monomial factor	Difference of two squares
Perfect-square trinomial	Binomial factors of a general trinomial
Common binomial factor	Sum and difference of two cubes

- The difference of two squares and the perfect-square trinomial can be used to factor a polynomial over the set of complex numbers.

Chapter 1 Review Exercises

In **1 – 12**, list the symbols for all of the sets to which the number belongs.
Let N = {Natural Numbers}, W = {Whole Numbers}, Z = {Integers},
Q = {Rational Numbers}, R = {Real Numbers}, and C = {Complex Numbers}.

1. 1   **2.** 0   **3.** -4 **4.** $-\frac{3}{4}$

5. $\sqrt{3}$ **6.** i **7.** π **8.** $3i - 1$

9. $\sqrt{2} - i$ **10.** $0.010010001...$ **11.** $-\sqrt{5}$ **12.** $\sqrt{-8}$

13. Every natural number is an integer, but every integer is not necessarily a natural number. Write a short sentence justifying this statement.

In **14 – 17**, write the additive inverse of the given number if it exists in the given set.

14. $1 \in N$ **15.** $-4 \in Z$ **16.** $\sqrt{3} \in R$ **17.** $3 - i \in C$

In **18 – 25**, write the multiplicative inverse of the given number if it exists in the given set.

18. $1 \in N$ **19.** $-4 \in Z$ **20.** $-4 \in Q$ **21.** $\sqrt{3} \in R$

22. $0 \in Q$ **23.** $0 \in R$ **24.** $\frac{5}{8} \in Q$ **25.** $3 - i \in C$

26. Write $\frac{5}{37}$ as a repeating decimal.   **27.** Write $0.3\overline{18}$ as the ratio of integers.

28. Write $0.\overline{9}$ as the ratio of integers. Does your answer surprise you? Why or why not? Write $0.\overline{3}$ and $0.\overline{6}$ as the ratio of integers. Add these two values together, both in fraction and decimal form.

In **29 – 34**, perform the indicated operation and write the result in $a + bi$ form.

29. $(5 - 2i) + (-8 + i)$ **30.** $(5 - 2i)(-8 + i)$ **31.** $(2 - 3i)(i)$

32. $1 \div (5 - 12i)$ **33.** $(3 - 4i)^2$ **34.** $2i^2 + 3i^9$

In **35 – 37**, perform the indicated operation.

35. $(x^2 + 5x) - (5 - x^2)$ **36.** $(x^3 - 125) \div (x - 5)$ **37.** $(2x - 3)^3$

38. Find the quotient and remainder when $(2x^4 - 7x^2 + 8)$ is divided by $(x^2 - 3)$.

39. If $(x + 1)$ spotted owls are observed in an area of $(x^3 + 1)$ acres in the Pacific Northwest, how many owls could be expected to live in a forest of $(x^2 + 4)$ acres?

Review Answers

1. W, N, Z, Q, R, C
2. W, Z, Q, R, C
3. Z, Q, R, C
4. Q, R, C
5. R, C
6. C
7. R, C
8. C
9. C
10. R, C
11. R, C
12. C
13. If an integer ≤ 0, it is not a natural number.
14. Does not exist
15. 4
16. $-\sqrt{3}$
17. $-3 + i$
18. 1
19. Does not exist
20. $-\frac{1}{4}$
21. $\frac{\sqrt{3}}{3}$
22. Does not exist
23. Does not exist
24. $\frac{8}{5}$
25. $\frac{3}{10} + \frac{1}{10}i$
26. $0.\overline{135}$

27. $\frac{7}{22}$

28. $\begin{aligned} 0.\overline{3} &= \frac{1}{3} \\ + 0.\overline{6} &= \frac{2}{3} \\ \hline 0.\overline{9} &= \frac{3}{3} = 1 \end{aligned}$ Not surprising; the infinitely repeating decimal 0.999... approaches 1 as a limit. (Answers will vary.)

29. $-3 - i$
30. $-38 + 21i$
31. $3 + 2i$

32. $\frac{5}{169} + \frac{12}{169}i$
33. $-7 - 24i$
34. $-2 + 3i$
35. $2x^2 + 5x - 5$
36. $x^2 + 5x + 25$
37. $8x^3 - 36x^2 + 54x - 27$

40. As a mail order supervisor, you have $(x^3 - 4x^2 + 7x + 5)$ orders to fill and you divide them equally among $(x - 1)$ employees. How many orders remain unassigned?

41. In today's mail, you received a check for $(x^4 - 3x^3 + 12x^2 - 7x + 4)$ dollars for a truckload of aluminum cans that your church group collected for recycling. If the price of aluminum is $(x^2 + 2)$ dollars per ton, how many tons of aluminum did your group turn in for recycling?

In **42 – 44**, factor over the set of integers.

42. $3x^3 - 12x$

43. $12x^2 - 10x - 8$

44. $x^6 + 7x^3 - 8$

In **45 – 47**, factor over the set of complex numbers.

45. $x^4 - 16$

46. $x^5 - 2x^3 + x^2 - 2$

47. $x^2 - 4x + 2$

In **48 – 51**, solve for x and write the solution set as a subset of the given set of numbers.

48. $7x - 5 = 12$; Integers

49. $x^2 + 4 = 0$; Complex Numbers

50. $3x + 2 < 8$; Real Numbers

51. $6 - 2x > 0$; Whole Numbers

In **52 – 59**, perform the indicated operations, expressing the result in simplest form.

52. $\sqrt{5}(\sqrt{35} + \sqrt{140})$

53. $(1 + \sqrt{5})^2$

54. $\dfrac{3 + \sqrt{12}}{\sqrt{3}}$

55. $\dfrac{2}{5 - 2\sqrt{6}}$

56. $\dfrac{1}{x^2 - 4x} - \dfrac{1}{4x - 16}$

57. $\dfrac{x^2 - 25}{x^2 - 10x + 25} \div (2x + 10)$

58. $\dfrac{1}{x^2}\left(\dfrac{x}{x - 1} + \dfrac{x}{x + 1}\right)$

59. $\dfrac{x(x - 1) + (x - 1)^2}{x^2 - 1}$

60. Every day at school, Nicole purchases fruit juice for lunch and pays for it with the exact change consisting of 4 different coins. Today she determined that she had spent $34.03 on fruit juice since the school year began. How much does the fruit juice cost and how many days has she been buying it at school?

61. Joan went to her closet this morning and found 3 different blouses to wear. If she has 4 skirts to match and 2 pairs of shoes, how many different combinations are possible for her to wear to school today?

62. Jack has $28 left from his paycheck this month. He decides to buy his girlfriend roses for her birthday and to go to a movie and have pizza with his friends. If roses are $2.25 each, how many can he buy his girlfriend and still have at least $15 left for a movie and pizza?

63. Selected compact discs and audio cassettes are on sale: $10 each for the discs and $7 each for the cassettes. For $87, how many of each can be purchased? (Disregard tax.)

38. $2x^2 - 1$, R = 5

39. $\left(\dfrac{x^2 + 4}{x^2 - x + 1}\right)$ owls

40. 9 orders

41. $\left(x^2 - 3x + 10 - \dfrac{x + 16}{x^2 + 2}\right)$ tons

42. $3x(x + 2)(x - 2)$

43. $2(2x + 1)(3x - 4)$

44. $(x - 1)(x^2 + x + 1)(x + 2)(x^2 - 2x + 4)$

45. $(x - 2)(x + 2)(x - 2i)(x + 2i)$

46. $(x - \sqrt{2})(x + \sqrt{2})(x + 1)\left(x - \dfrac{1}{2} - \dfrac{i\sqrt{3}}{2}\right)\left(x - \dfrac{1}{2} + \dfrac{i\sqrt{3}}{2}\right)$

47. $(x - 2 - \sqrt{2})(x - 2 + \sqrt{2})$

48. ø

49. $\{-2i, 2i\}$

50. $\{x \mid x < 2\}$

51. $\{0, 1, 2\}$

52. $15\sqrt{7}$

53. $6 + 2\sqrt{5}$

54. $\sqrt{3} + 2$

55. $10 + 4\sqrt{6}$

56. $\dfrac{-1}{4x}$

57. $\dfrac{1}{2(x - 5)}$

58. $\dfrac{2}{(x - 1)(x + 1)}$

64. 🖩 A calculator can be used for the computation when applying the Euclidean Algorithm for finding the greatest common factor of two integers a and b. As you work with specific numbers, record the values of a, b, q, and r for each step.

1. Enter: a ÷ b =
   Record the integral part of the quotient as q.

2. Enter: a − q × b =
   Record the answer as r.

3. Replace a by b, and b by r.

   Repeat Steps 1 through 3 until $r = 0$.
   The last nonzero value of r is the greatest common factor.

Find the greatest common factor and the least common multiple of 204 and 36.

65. a. Given any two integers a and b, where $a < b$, let the GCF of a and b be f, and the LCM of a and b be m. Write an inequality showing the order relationship of a, b, f, and m.

   b. Prove that if $f = a$, then $m = b$.

66. In the postulates for the natural numbers, it is stated that the number 1 has no antecedent. If this restriction were removed, what set of numbers would be generated?

67. If b does not divide a, then the largest integer, q, can be found such that $a = bq + r$, where r and $b - r$ are positive integers. Explain why r and $b - r$ must be positive.

In **68 – 69**, give a counterexample to disprove the given statement.

68. The set of irrational numbers is closed under multiplication.

69. The set of imaginary numbers is closed under addition.

70. Find the quotient and remainder when $ix + 1$ is divided by $x - i$.

71. Find the quotient when $x^2 + ix + 2$ is divided by $x - i$. Is the quotient a polynomial over the set of complex numbers?

72. Prove that the set of complex numbers is not an ordered field. *Hint:* Assume that it is and use the Trichotomy Property with $a = i$ and $b = 0$.

73. a. Locate the number $4 + 5i$ on the complex plane.

   b. Find the distance from the origin to $4 + 5i$.

   c. Derive a general formula for the distance from the origin to $a + bi$ of the complex plane.

---

59. $\dfrac{2x - 1}{x + 1}$

60. 41 cents; 83 days

61. 24 combinations

62. 5 roses

63. 8 discs and 1 cassette or 1 disc and 11 cassettes

64. GCF = 12
LCM = 612

65. a. $f \le a < b \le m$

   b. $\dfrac{ab}{f} = m$

   Substitute $f = a$.

   $\dfrac{ab}{a} = b = m$

66. set of integers

67. By definition, in dividing a by b, we seek the largest integer q such that $bq \le a$. If $bq < a$, then $a = bq + r$, for $r > 0$.

If $r < 0$, there is no largest value of q; for example, in dividing 25 by 8,
$25 = 8 \cdot 4 + (-7)$
$= 8 \cdot 6 + (-23)$,
etc.

If $b - r < 0$, then $r > b$, and we do not have the largest possible q; for example,
$25 = 8 \cdot 1 + 17$.

---

68. $\sqrt{2} \cdot \sqrt{2} = 2$

69. $2i + (-2i) = 0i = 0$

70. quotient $= i$; R $= 0$

71. quotient $= x + 2i$; yes

72. Let $a = i, b = 0$. Suppose $a < b$ or $a = b$ or $a > b$:
If $a < b$; then $i < 0$; $i \cdot i > 0 \cdot i, i^2 > 0, -1 > 0$. False
If $a = b$; then $i = 0$. False
If $a > b$; then $i > 0, i^2 > 0, -1 > 0$. False

73. a.

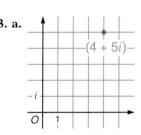

$(4 + 5i)$

b. $\sqrt{41}$ **c.** $d = \sqrt{a^2 + b^2}$

1. 16

2. 26

3. 23

4. 63.7%

5. 1

6. E

7. D

8. C

9. E

10. E

11. D

12. D

13. B

☑ Exercises for Challenge

1. What is the smallest positive integer that has exactly 5 distinct divisors?

2. The measures of the sides of a rectangle are integers and its area is 12 square units. What is the largest possible perimeter of the rectangle?

3. My age is less than 30 years. When my age is divided by 2 the remainder is 1, when it is divided by 3 the remainder is 2, and when it is divided by 5 the remainder is 3. How old am I?

4. A square is inscribed in a circle. Express to the nearest tenth of a percent, the ratio of the area of the square to the area of the circle.

5. What is the remainder when 3^{10} is divided by 11?

6. If a is divisible by 20 and b is divisible by 36, then of the following numbers, which is the largest by which ab must be divisible?
(A) 4   (B) 16   (C) 80
(D) 180   (E) 720

7. The integer n is divisible by 101,101. All of the following must be factors of n *except* for:
(A) 7   (B) 11   (C) 13
(D) 17   (E) 77

8. The positive integer m is a factor of the positive integer n. Which of the following is *never* true?
(A) n is a factor of m.
(B) m is a factor of n + 1.
(C) $m > n$.   (D) m is even.
(E) m is odd.

9. If (a, b) is a solution of $7x + 9y = 1$, then a solution of $14x + 18y = -2$ must be
(A) (b, a) (B) $(-a, b)$
(C) $(a, -b)$ (D) $(-2a, -2b)$
(E) $(-a, -b)$

10. The least common multiple of a and b is $2b$. Which of the following must be true?
(A) a is even and b is odd.
(B) Both a and b are even.
(C) b is even and a is odd.
(D) Both a and b are odd.
(E) a is even and b could be even or odd.

11. Which of the following is a square root of i?
(A) $\frac{1}{2} + \frac{1}{2}i$ (B) $-\frac{1}{\sqrt{2}} + \frac{1}{2}i$
(C) $\frac{1}{2} + \frac{1}{\sqrt{2}}i$ (D) $\frac{1}{\sqrt{2}} + \frac{1}{\sqrt{2}}i$
(E) $\frac{1}{\sqrt{2}} - \frac{1}{\sqrt{2}}i$

12. $\dfrac{(x^3 + 1)(x^4 - 1)}{(x^2 + 1)(x - 1)}$ is equal to
(A) $x^4 + x^3 + x^2 + x + 1$
(B) $x^4 + x^3 + 2x^2 + x + 1$
(C) $x^4 + x^3 + x^2 - 1$
(D) $x^4 + x^3 + x + 1$
(E) $x^4 + x^3 - x + 1$

13. On a number line, the graph of which of the following rational numbers is closest to the graph of $\sqrt{2}$?
(A) $\frac{14,142}{10,000}$ (B) $\frac{577}{408}$ (C) $\frac{239}{169}$
(D) $\frac{238}{169}$ (E) $\frac{99}{70}$

14. Which of the following is *never* true?
- (A) The sum of a rational number and an irrational number is a rational number.
- (B) The product of two irrational numbers is an irrational number.
- (C) The sum of two irrational numbers is a rational number.
- (D) The square root of an irrational number is an irrational number.
- (E) The square root of a rational number is a rational number.

15. If 12 divides a and 15 divides b then 60 must divide which of the following?
(A) $a + b$ (B) $5a + b$ (C) $a + 4b$
(D) $5a + 4b$ (E) $a - b$

16. Which of the following is true about the polynomial $2x^2 - 5x - 1$?

I   The polynomial is a prime polynomial over the set of integers.

II   The polynomial has the factors $2\left(x - \frac{5 + \sqrt{33}}{4}\right)\left(x - \frac{5 - \sqrt{33}}{4}\right)$ over the set of real numbers.

III   The polynomial has the factors $2\left(x + \frac{5 + \sqrt{33}}{4}\right)\left(x + \frac{5 - \sqrt{33}}{4}\right)$ over the set of real numbers.

(A) I only   (B) II only
(C) III only   (D) I and II only
(E) I and III only

17. Which of the following is true?

I $\sqrt{-3} \cdot \sqrt{-5} = \sqrt{15}$

II $\sqrt{-3} \cdot \sqrt{-5} = -\sqrt{15}$

III $\sqrt{-3} \cdot \sqrt{-5} = i\sqrt{15}$

(A) I only   (B) II only
(C) III only   (D) I and II only
(E) II and III only

Questions 18 – 22 each consist of two quantities, one in Column A and one in Column B. You are to compare the two quantities and choose:

A if the quantity in Column A is greater;
B if the quantity in Column B is greater;
C if the two quantities are equal;
D if the relationship cannot be determined from the information given.

1. In certain questions, information concerning one or both of the quantities to be compared is centered above the two columns.

2. In a given question, a symbol that appears in both columns represents the same thing in Column A as it does in Column B.

3. x, n, and k, etc. stand for real numbers.

Column A	Column B

p and q are consecutive primes and $p < q < 30$

18. $q - p$ 8

$a \le \frac{26}{15}$ and $b \ge \sqrt{3}$

19. a b

$a < b < 0$

20. $\dfrac{1}{a}$ $\dfrac{1}{b}$

$a < -1$ and $b < 0$

21. $\dfrac{ab}{a + 1}$ b

The integer a has 5 prime factors and the integer b has 3 prime factors.

22.   The number of   8
prime factors of ab

14. A

15. D

16. D

17. B

18. B

19. D

20. A

21. B

22. D

Teacher's Chapter 2

Equations and Inequalities

Chapter Contents

This chapter reviews the solution of linear and quadratic equations and inequalities, and how these are used in problem solving. Emphasis is placed on the postulates of the real number system in order to prepare students for subsequent chapters. Higher-order equations and inequalities are introduced later in the chapter, along with a number of important theorems from the theory of equations. These include the Remainder Theorem, Factor Theorem, Rational Root Theorem, and Descartes' Rule of Signs. Finally, the approximation of irrational roots is discussed.

2.1 First-Degree Equations and Inequalities

The purpose of this section is to build on what students have already learned about solving linear equations and inequalities. The procedures for finding solutions are more rigorously examined, using examples that go beyond earlier requirements.

Since a linear equation of the form $ax + b = c$ has the general solution $x = \frac{c - b}{a}$ with $a \neq 0$, a computer program can be used to solve.

Computer Activity (2.1)

```
10  REM SOLVING LINEAR EQUATIONS
        OF THE FORM AX + B = C
20  REM BY ENTERING VALUES OF A,
        B, AND C
25  PRINT
30  PRINT "ENTER VALUES FOR A,
        B, AND C SEPARATED BY
        COMMAS."
40  INPUT A, B, C
50  IF A = 0 THEN GOTO 120
60  X = (C - B)/A
70  PRINT "X = "; X
80  PRINT "DO YOU WISH TO CONTINUE
        (Y OR N)?"
90  INPUT A$
100 IF A$ = "N" THEN GOTO 140
110 GOTO 30
120 PRINT "A CANNOT BE ZERO."
130 GOTO 25
140 END
```

2.2 Using an Open Sentence to Solve a Problem

Emphasize the importance of the domain in solving equations and inequalities and the assumed domain when solving word problems. For example, elicit that a problem involving the lengths of the sides of a rectangle has an assumed domain of positive real numbers for its solution set, while a problem involving a number of coins has an assumed domain of natural numbers.

Emphasize, also, the importance of identifying a reasonable answer.

The classic four-step approach to solving a problem is reviewed and detailed for an algebraic solution. Problems involving a relationship in the form $A \times B = C$ are used as examples. Students should be encouraged to present their analysis of a problem, discussing alternate representations and solutions. You may wish to present average speed and average price per gallon of fuel (see Exercises 27, 28, and 29) as harmonic means of the numbers being averaged.

A number of situations in science, industry, and investing exist where components must be mixed to make a blend whose overall quality is directly proportional to the percentage of the better-quality component. Following, are two examples of this concept representing two investments.

Example 1

$1,000 @ 6% =	$ 60.00
$4,000 @ 8% =	$320.00
Total: $5,000	$380.00

(Rate of return: 7.6%)

If the amount of money placed in the higher-yielding investment is dropped from $\frac{4}{5}$ of the total to $\frac{2}{5}$ of the total, then

$3,000 @ 6% =	$180.00
$2,000 @ 8% =	$160.00
Total: $5,000	$340.00

(Rate of return: 6.8%)

Students will discover that in this situation the range of possible returns is from 6% to 8%, depending on how the $5,000 is divided between the two investments. If $\frac{4}{5}$ of the money is invested at the higher rate of return, then the total income from the investments will be an amount equal to the number that corresponds to the point that is $\frac{4}{5}$ of the way across the interval between 6% and 8% (6% + $\frac{4}{5}$(2%) = 7.6%). Similarly, when $\frac{2}{5}$ of the money is invested in the higher-yielding investment, the net result is only $\frac{2}{5}$ of the way across the interval (6% + $\frac{2}{5}$(2%) = 6.8%).

The second example of this concept involves the computation of octane rating for gasoline. One way in which octane ratings are computed is to prepare a standard fuel composed of only heptane and isooctane. The octane rating for this fuel is simply the percentage of the fuel that is isooctane. Actual gasolines are then compared to these standard fuels. For example, if an actual blend of gasoline is rated at 87 octane, it has the same antiknock properties of a standard fuel that is 87% isooctane.

Example 2

If a car requires 91 octane gasoline, how many liters of 94 octane gas and how many liters of 86 octane gasoline must be placed in the tank if 30 liters are needed to fill the tank?

Answer: 18.75 liters of 94 octane
11.25 liters of 86 octane

2.3 Quadratic Equations and Inequalities

The methods for solving quadratic equations by factoring, by completion of the square, and by the quadratic formula are reviewed.

When the discriminant is reviewed, emphasize the fact that the rules stated for determining the nature of the roots apply only when the coefficients of the quadratic equation are rational numbers. For equations with irrational coefficients, $b^2 - 4ac$ does not describe the nature of the roots. For example,

for the equation $x^2 - \sqrt{5}x - 1 = 0$, the value of $b^2 - 4ac = 9$, yet the roots of the equation are irrational.

When solving problems that lead to quadratic equations, students must be especially alert to the reasonableness of an answer. Since every quadratic equation has two solutions, each of these must be evaluated in terms of the original problem to judge if it is a possible answer to the problem.

Quadratic inequalities are solved by considering the possible combinations of positive and negative factors that can be solutions of the given inequality. An alternate method displays the signs of the factors for all real numbers on the number line. This latter method is particularly useful when the number of possible combinations of positive and negative factors is large. For example, $(x - a)(x - b)(x - c) > 0$ is true when 0 or 2 factors are negative. Therefore, there are four possible cases; all are positive, $(x - a)$ is positive and the others are negative, $(x - b)$ is positive and the others are negative, and $(x - c)$ is positive and the others are negative.

2.4 Additional Techniques for Solving Equations

The discussion of solving equations is expanded to radical equations, after stating the principle that if both sides of an equation are raised to a power, the solution set of the original equation is a subset of the solution set of the new equation. Stress that checking for extraneous roots is essential to the solution process.

You may wish to show additional examples of equations that, when squared, have solution sets different from those of the original equations. For example, the equations $2 + \sqrt{x + 4} = x$ and $2 - \sqrt{x + 4} = x$ result in the same equation when squared, although the original equations are different. Both original equations have solutions that are also solutions of $x^2 - 5x = 0$, the square of each equation.

However, both 0 and 5 are roots of $x^2 - 5x = 0$, while 0 is the only root of $2 - \sqrt{x + 4} = x$ and 5 is the only root of $2 + \sqrt{x + 4} = x$.

The solution of a quadratic equation, reviewed in the previous section, plays an important role in the techniques of this section, and will continue to be important in subsequent sections of this chapter.

2.5   Solving Equations of Degree Greater than 2

Ask students to compare the field properties of the set of integers with these same properties of the set of polynomials over the integers, over the rational numbers, or over the real numbers. Note particularly that, except for the polynomials of degree 0, polynomials do not have multiplicative inverses. Like the integers, polynomials can be considered to be prime or composite. (Ask students how they would define a prime polynomial over the set of integers, or over the set of real numbers.) These similarities allow us to consider the prime polynomial factors of a polynomial and the remainder when a polynomial is divided by a polynomial of smaller degree.

Students often have difficulty with the conclusion of the Remainder Theorem, the remainder is $P(a)$. Emphasize that the theorem is stating that two numbers are equal; the value of the remainder after $P(x)$ is divided by $x - a$ and the value of the polynomial when x is replaced by a.

The proof of the Factor Theorem, though very simple, can be elusive for students who do not readily see the relationship between a remainder of 0 when dividing, and the quotient and divisor as factors.

Students who have access to a graphing calculator or to a computer graphing program can benefit from seeing the graph of a polynomial when they are finding the zeros of the polynomial algebraically.

The following computer program can be used to find specific values of a function.

Computer Activity (2.5) _____

```
100  REM EVALUATES A POLYNOMIAL
         FUNCTION F(X) FOR VALUES OF X
110  REM POLYNOMIAL MAY BE OF
         DEGREE 25 OR LESS
130  DIM A(25)
140  FOR I = 1 TO 25
150  A(I) = 0
160  NEXT I
170  PRINT "WHAT IS THE DEGREE OF
         THE POLYNOMIAL?"
180  INPUT N
190  PRINT "ENTER THE COEFFICIENTS,
         ONE AT A TIME,"
200  PRINT "STARTING WITH THE
         LEADING COEFFICIENT."
210  FOR I = N TO 0 STEP -1
220  PRINT "A("; I; ") = ";
230  INPUT A(I)
240  NEXT I
250  PRINT "ENTER A VALUE OF X"
260  INPUT X
270  Y = 0
280  FOR I = N TO 0 STEP -1
290  Y = Y * X + A(I)
295  PRINT Y
300  NEXT I
310  PRINT "X = "; X, "Y = "; Y
320  PRINT
330  PRINT "ENTER ANOTHER VALUE OF
         X OR TYPE 999 TO END"
350  INPUT X
360  IF X > < 999 THEN 270
370  END
```

2.6   More About Polynomial Open Sentences

Synthetic division is introduced in this section. It is a basic skill for the rest of the chapter and students should be able to do synthetic division quickly and accurately. The short form of the synthetic division table should be emphasized so that repeated trials can be done efficiently.

The Rational Root Theorem is introduced and a "proof by example" is demonstrated for a cubic equation. The critical step in this proof (if a divides $-4b^3$, and a and b are relatively prime, then a divides -4) uses the Divisibility of a Product Theorem from Chapter 1.

This should be sufficient to convince students that the theorem works for any polynomial. The following proof could also be presented.

If $x = \dfrac{p}{q}$ is a zero of

$$f(x) = a_n x^n + a_{n-1} x^{n-1} + \cdots + a_0,$$

then $qx - p$ is a factor of $f(x)$ and the other factor is $b_{n-1} x^{n-1} + b_{n-2} x^{n-2} + \cdots + b_0$.
Then:

$$f(x) = (qx - p)(b_{n-1} x^{n-1} + b_{n-2} x^{n-2} + \cdots + b_0)$$

$$= q b_{n-1} x^n + q b_{n-2} x^{n-1} + \cdots - p b_0$$

Equating coefficients of like powers of x, $q b_{n-1} = a_n$ and $-p b_0 = a_0$. Therefore, q is a factor of a_n and p is a factor of a_0.

2.7 Approximate Roots of Polynomial Equations

Since irrational roots are more common than rational roots when solving higher-degree equations, it is useful to have methods to find rational approximations. Graphs are used as a tool in explaining the significance of a sign change in the function, but curve sketching is not a primary objective at this point. Descartes' Rule of Signs is stated, but not proven.

So that students can have the benefit of working with a variety of polynomials, calculators and computers can play an important role. A computer program that plots graphs or a graphing calculator are helpful for estimating where a change in sign occurs in the function. Another approach would be to use a computer program to enable students to evaluate a polynomial function over a particular range for values that change by a specified interval. For example, a polynomial function $P(x)$ can be evaluated from $x = 1$ to $x = 3$ using an interval of 0.5.

Computer Activity (2.7)

```
100  REM EVALUATES A POLYNOMIAL FUNC-
         TION F(X) OVER A GIVEN RANGE
110  REM WITH A SPECIFIED INTERVAL.
120  REM POLYNOMIAL MAY BE OF
         DEGREE 25 OR LESS
130  DIM A(25)
140  FOR I = 1 TO 25
150  A(I) = 0
160  NEXT I
170  PRINT "WHAT IS THE DEGREE OF
         THE POLYNOMIAL?"
180  INPUT N
190  PRINT "ENTER THE COEFFICIENTS,
         ONE AT A TIME,"
200  PRINT "STARTING WITH THE
         LEADING COEFFICIENT."
210  FOR I = N TO 0 STEP -1
220  PRINT "A("; I; ") = ";
230  INPUT A(I)
240  NEXT I
250  PRINT "ENTER THE SMALLEST VALUE
         OF X TO BE USED"
260  INPUT S
270  PRINT "ENTER THE LARGEST VALUE
         OF X TO BE USED"
280  INPUT L
281  PRINT "ENTER THE INTERVAL TO
         BE USED"
282  INPUT T
290  FOR K = S TO L STEP T
300  Y = 0
310  FOR I = N TO 0 STEP -1
320  Y = Y * K + A(I)
330  NEXT I
340  PRINT "X = "; K, "Y = "; Y
350  NEXT K
360  PRINT "ENTER ANOTHER RANGE OF
         VALUES (TYPE Y OR N)."
370  INPUT A$
380  IF A$ = "Y" THEN 250
390  END
```

The examples in this section were chosen so that trial values between $P(-3)$ and $P(3)$ are sufficient to answer the questions, but students should be aware that, for many polynomials, it is necessary to try values beyond this domain before finding changes in the sign of the value of the function. For example, $x^3 + 2x^2 - 85x - 350 = 0$ or $x^2 + 1.2x - 21.32 = 0$ will not show a sign change in the interval $[-3, 3]$. This may also be illustrated by sketching some graphs of polynomials where the sign changes occur further out on the x-axis.

Suggested Test Items

1. Find the solution set:
   $$-3[4 - 3(x + 1)] = 6(2x - 1) - 93$$

2. ▨ Find x to the nearest hundredth:
   $$1.26x - 0.175 = 6.2x - 1.614$$

3. Solve for x.
   a. $\dfrac{3}{2x - 5} = \dfrac{2}{x + 1}$

   b. $\dfrac{5x}{x + 4} - 3 = -\dfrac{20}{x + 4}$

   c. $\dfrac{x^2 + 7}{x + 3} + 1 > \dfrac{x^2 - 5}{x + 3}$

   d. $\dfrac{6}{x - 2} + \dfrac{11}{3x - 6} = \dfrac{19}{x + 14}$

4. Describe the solution set of the equation, where n is a real number.
   a. $\dfrac{8 - n}{8 - n} = 0$
   b. $\dfrac{8 - n}{8 - n} = 1$
   c. $8n = 8n + 8$
   d. $\dfrac{8n}{n} = 1$

5. The measure of one of two complementary angles is 3 degrees less than $\frac{1}{2}$ the other. Find the measure of each of the angles.

6. Solve each literal equation for the variable indicated.
   a. $s = \frac{1}{5}P$, for P
   b. $L = 2W + 5$, for W
   c. $s = \frac{1}{2}gt^2$, for g
   d. $\dfrac{1}{x} + \dfrac{1}{y} = 1$, for y

7. Howard Gant received a $20,000 inheritance, which he invested in two bonds. The first bond yields 9% and the second yields 8%. If his total yearly income from both bonds is 8.75% of the $20,000, how much was invested in each bond?

8. Maria Hernandez spent $3,000 for fuel in each of three consecutive years. If the cost per gallon for each year was $.80, $1.10, and $1.25, respectively, what was the average price per gallon over the three-year period? Give your answer to the nearest cent.

9. Wilson Simms wishes to strengthen a 10% mixture of acid to a 30% mixture. How much pure acid should he add to 7 liters of the original solution?

10. Ms. Willis won an election over her opponent by getting 300 votes. If 10 people had changed their votes, she would have lost the election by 4 votes. How many people voted in the election?

11. Solve for x by factoring:
    $$8x^2 + 14x < -3$$

12. Solve by completing the square:
    $$3y^2 - y + 2 = 0$$

13. Solve by using the quadratic formula:
    $$\sqrt{2}x^2 - 3x + \sqrt{2} = 0$$

14. Evaluate the discriminant to determine the nature of the roots of $3x^2 - 0.5x - 0.7 = 0$.

15. ▨ Estimate the roots:
    $$(2 + \sqrt{3})x^2 - 3\sqrt{2}x - 2 = 0$$
    Answer to the nearest hundredth.

16. Find all the possible pairs of integers that have a sum of 10 and the sum of whose squares is 148.

17. A picture with area of 70 cm² is contained in a frame that is 1.5 cm wide. If the difference between the length and the width of the frame is 3 cm, find the outer dimensions of the frame.

18. The lengths of the legs of a right triangle are represented by x and $x + 4$, and the length of the hypotenuse is represented by $x + 10$. Find the length of each side of the triangle to the nearest tenth.

19. Solve for x: $x = \sqrt{48 - 2x}$

20. Solve for x: $\sqrt{2x + 5} = \sqrt{x + 2} + 1$

21. Solve for n: $n^4 + 5n^2 - 36 = 0$

22. Solve for y: $6(y + 1)^2 - 5(y + 1) - 50 = 0$

23. Find the solution set of each equation.
 a. $x(x - 3)(x - 2)(x + \sqrt{2}) = 0$
 b. $x^3 - x^2 - 4x - 6 = 0$
 c. $x^3 - 5x^2 + 17x - 13 = 0$

24. Use synthetic division:
$$\frac{8x^3 + 8x^2 + 4x + 1}{8x + 4}$$

25. If $2i$ is a root of $x^4 + 6x^2 + 8 = 0$, find the other roots.

26. When $5x^3 - 6x^2 - 73x - 3$ is divided by $x - 4$, find the remainder.

27. Show that $x - r$, where r is a real number, cannot be a factor of $4x^4 + 2x^2 + 2$.

28. If $P(x) = x^3 + 2x^2 - 85x - 350$ and $P(10) = 0$, name one linear factor of $P(x)$.

29. If -2 is a root of $x^3 - kx + 1 = 0$, find k.

30. Use synthetic division to determine if $-i$ is a root of $x^3 - ix^2 + 3x + 5i = 0$.
 Note: $i = \sqrt{-1}$

31. Write the set of possible rational roots for $f(x) = 14x^3 + 3x^2 + 73x - 2 = 0$.

32. When $x^{99} + 2x^{48} - 2$ is divided by $x - 1$, find the remainder.

33. A box with an open top is to be constructed from a piece of cardboard that is 8 cm on each side by cutting congruent squares from each corner and folding up the sides. Find 2 values of the length of a side of a square that will make the volume of the box 32 cm³. Estimate the irrational value to the nearest tenth.

34. Approximate the real zeros of each function to the nearest tenth.
 a. $f(x) = x^3 - 4x^2 - 5x + 10$
 b. $f(x) = x^3 + 5x - 12$

35. Find the number of possible positive real zeros and the number of possible negative real zeros for each polynomial function.
 a. $P(x) = x^3 + 2x^2 - 3x + 5$
 b. $P(x) = 2x^4 - 3x^3 - x^2 + 7x - 1$

36. Find an upper bound and a lower bound for the zeros of each polynomial function.
 a. $P(x) = 8x^3 - 12x^2 + 2x - 1$
 b. $P(x) = x^4 - 5x - 1$

Bonus

Approximate the solutions of $x^4 - 16x^2 + 4 \geq 0$ to the nearest tenth.

Answers to Suggested Test Items

1. $\{32\}$ **2.** $x = 0.29$

3. **a.** $x = 13$
    b. no solution
    c. $x < -15$ or $x > -3$
    d. $x = 18\frac{4}{7}$

4. **a.** empty set
    b. $\{n \mid n \in R, n \neq 8\}$
    c. empty set
    d. empty set

5. The angles measure $62°$ and $28°$.

6. **a.** $P = 5s$ **b.** $W = \dfrac{L-5}{2}$
    c. $g = \dfrac{2s}{t^2}$ **d.** $y = \dfrac{x}{x-1}$

7. $15,000 @ 9\%$
    $5,000 @ 8\%$ **8.** \$1.01

9. 2 liters of acid   **10.** 584 people

11. $-\frac{3}{2} < x < -\frac{1}{4}$ **12.** $y = \dfrac{1 \pm i\sqrt{23}}{6}$

13. $x = \sqrt{2}, \; x = \dfrac{\sqrt{2}}{2}$

14. $d = 8.65$
    real, irrational, and unequal

15. $1.50, -0.36$ **16.** $(12, -2)$

17. The frame is $10 \text{ cm} \times 13 \text{ cm}$.

18. $17.0, 21.0, 27.0$ **19.** $x = 6$

20. $x = \pm 2$ **21.** $\{\pm 3i, \pm 2\}$

22. $\left\{-\frac{7}{2}, \frac{7}{3}\right\}$

23. **a.** $\{0, 3, 2, -\sqrt{2}\}$
    b. $\{-1 \pm i, 3\}$
    c. $\{1, 2 \pm 3i\}$

24. $x^2 + \frac{1}{2}x + \frac{1}{4}$ **25.** $-2i, \pm i\sqrt{2}$

26. The remainder is -71.

27. $P(x) = 4x^4 + 2x^2 + 2$
    $P(r) = 4r^4 + 2r^2 + 2$
    Since $r^4 \geq 0$ and $r^2 \geq 0$,
    $P(r) \neq 0$ for any value of r.
    Therefore, r is not a zero, and
    $x - r$ is not a factor of $P(x)$.

28. $x - 10$ **29.** $k = \frac{7}{2}$

30. $-i$ is not a root.

31. $\left\{\pm 1, \pm \frac{1}{2}, \pm \frac{1}{7}, \pm \frac{1}{14}, \pm 2, \pm \frac{2}{7}\right\}$

32. 1   **33.** $0.8 \text{ cm}, 2 \text{ cm}$

34. **a.** $-1.8, 1.2, 4.6$
    b. 1.6

35. **a.** 2 or 0 positive real zeros
      1 negative real zero
    b. 1 or 3 positive real zeros
      1 negative real zero

36. **a.** An upper bound is 2.
      A lower bound is 0.
    b. An upper bound is 2.
      A lower bound is -1.

Bonus $x^4 - 16x^2 + 4 \geq 0$

Let $x^2 = y$ and solve $y^2 - 16y + 4 = 0$.

$$y = \frac{16 \pm \sqrt{240}}{2} = 8 \pm 7.746$$

$$x^2 = 8 + 7.746 = 15.746 \qquad x^2 = 8 - 7.746 = 0.254$$
$$x = \pm\sqrt{15.746} = \pm 4.0 \qquad x = \pm\sqrt{0.254} = \pm 0.5$$

$(x + 4.0)(x - 4.0)(x + 0.5)(x - 0.5) \geq 0$

The product is positive when 0, 2, or 4 factors are positive.

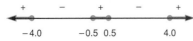

Answer: $\{x \mid x \leq -4.0 \text{ or } -0.5 \leq x \leq 0.5 \text{ or } x \geq 4.0\}$

Chapter 2

Equations and Inequalities

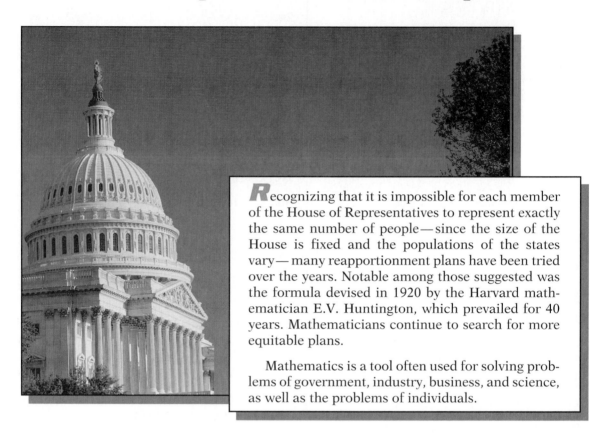

*R*ecognizing that it is impossible for each member of the House of Representatives to represent exactly the same number of people—since the size of the House is fixed and the populations of the states vary—many reapportionment plans have been tried over the years. Notable among those suggested was the formula devised in 1920 by the Harvard mathematician E.V. Huntington, which prevailed for 40 years. Mathematicians continue to search for more equitable plans.

Mathematics is a tool often used for solving problems of government, industry, business, and science, as well as the problems of individuals.

Chapter Table of Contents

2.1 First-Degree Equations and Inequalities

An algebraic equation or inequality in one variable is a mathematical sentence stating that of two algebraic expressions, at least one of which is in terms of the variable, the first is either less than, equal to, or greater than the second. In most cases considered here, the domain, or set of permissible values of the variable, is the set of all real numbers for which both sides of the equation or inequality are defined.

The following are examples of first-degree open sentences in x, that is, x is to the first power.

$$x + 10 = 15 \qquad 3x < 12 \qquad 2x = x + x \qquad x = x + 1 \qquad \frac{3}{x - 2} = \frac{2}{x}$$

In the first four sentences, both sides of the sentence are defined for every real number. In the last sentence, however, if $x = 2$ the left side is undefined and if $x = 0$ the right side is undefined. Therefore, the domain for this sentence is the set of real numbers with 0 and 2 excluded, $\{x \in R \text{ and } x \neq 0, 2\}$.

If the variable of an equation or inequality is replaced by one of the values from the domain, then either a true or a false statement results. A value that results in a true statement is called a *solution*, or a *root*, of the equation or inequality.

For the equation $x + 10 = 15$, there is exactly one value that makes the statement true. Therefore, the *solution set* of the equation is $\{5\}$.

The inequality $3x < 12$ is true for all values of x less than 4. The solution set is $\{x \mid x < 4\}$.

The equation $2x = x + x$ is an example of an *identity*, an equation that is true for all values in the domain of the variable.

The equation $x = x + 1$ is false for all real numbers. Therefore, the solution set is the empty set, written as $\{\ \}$ or $\emptyset$.

A simple open sentence can often be solved by inspection, but generally it is necessary to use some *algorithm*, or set of rules, to find the solution set.

Although different methods are used for different types of equations and inequalities, the algorithms usually involve using the addition and multiplication properties of equality and inequality to form *equivalent open sentences*, that is, equations and inequalities with the same solution sets as the originals.

A solution can be checked by substituting it for the variable in the original open sentence and determining if the resulting statement is true.

Example 1

Solve the equation $3 - (x + 4) = 2(x - 5)$.

Solution

$$3 - (x + 4) = 2(x - 5)$$

$3 - x - 4 = 2x - 10$	Use the distributive property.
$-1 - x = 2x - 10$	Combine like terms.
$-1 = 3x - 10$	Add x to each side.
$9 = 3x$	Add 10 to each side.
$3 = x$	Divide each side by 3.

Check

$$3 - (x + 4) = 2(x - 5)$$
$$3 - (3 + 4) \overset{?}{=} 2(3 - 5)$$
$$3 - 7 \overset{?}{=} 2(-2)$$
$$-4 = -4$$

Answer: $x = 3$

Note that in a check, each side is evaluated independently; the method of solution is not repeated. Recall the *order of operations* for evaluating arithmetic expressions. First perform the operations within parentheses, then multiply and divide from left to right, and finally, add and subtract from left to right.

To solve an equation involving rational expressions, multiply each side of the equation by the least common denominator (LCD) of the fractions.

When checking a possible solution, be sure to consider whether that possibility is in the domain of the variable.

Example 2 _____

Find the solution: $\dfrac{3}{x - 2} = \dfrac{2}{x}$

Solution

$x(x - 2)\left(\dfrac{3}{x - 2}\right) = x(x - 2)\left(\dfrac{2}{x}\right)$	LCD $= x(x - 2)$
$3x = 2(x - 2)$	Cancel common factors.
$3x = 2x - 4$	
$3x + (-2x) = 2x - 4 + (-2x)$	
$x = -4$	

The domain of x for this equation is the set of all real numbers except 2 and 0. Since -4 is not one of the excluded values, it is a solution of the equation. This can be verified by checking.

When solving an inequality involving rational expressions, two cases must be considered, depending on whether the least common denominator of the fractions is positive or negative.

Example 3 _____

Solve the inequality: $\dfrac{6x}{x - 4} + 3 < \dfrac{6}{x - 4}$

Solution

Case I: $\quad\quad x - 4 > 0$	*Case II:* $\quad\quad x - 4 < 0$
$x > 4$	$x < 4$
$\dfrac{6x}{x-4} + 3 < \dfrac{6}{x-4}$	$\dfrac{6x}{x-4} + 3 < \dfrac{6}{x-4}$
$\dfrac{6x}{x-4}(x-4) + 3(x-4) < \dfrac{6}{x-4}(x-4)$	$\dfrac{6x}{x-4}(x-4) + 3(x-4) > \dfrac{6}{x-4}(x-4)$
$6x + 3x - 12 < 6$	$6x + 3x - 12 > 6$
$9x < 18$	$9x > 18$
$x < 2$	$x > 2$
There are no values of x such that $x > 4$ and $x < 2$	
Solution for *Case I*: { }	Solution for *Case II*: $2 < x < 4$

The solution set for the inequality is the union of the solution sets for the two cases.

Answer: The solution set for the inequality is $\{x \mid 2 < x < 4\}$.

Example 4 _____

Solve: $\dfrac{2}{x+3} - \dfrac{1}{x-2} = \dfrac{-10}{x^2 + x - 6}$

Solution

Factor the denominator on the right side, and multiply each side by the LCD of the denominators, $(x+3)(x-2)$.

$$\frac{2}{x+3} - \frac{1}{x-2} = \frac{-10}{x^2+x-6}$$

$$\frac{2}{x+3} - \frac{1}{x-2} = \frac{-10}{(x+3)(x-2)}$$

$$(x+3)(x-2)\left(\frac{2}{x+3}\right) - (x+3)(x-2)\left(\frac{1}{x-2}\right) = (x+3)(x-2)\left(\frac{-10}{(x+3)(x-2)}\right)$$

$$2(x-2) - (x+3) = -10$$

$$2x - 4 - x - 3 = -10$$

$$x - 7 = -10$$

$$x - 7 + 7 = -10 + 7$$

$$x = -3$$

The domain of x for this equation is the set of reals excluding -3 and 2.

Answer: The solution set is ø.

Literal Equations

A formula such as that used to find the area of a trapezoid, $A = \frac{1}{2}h(a + b)$, where h is the height, and a and b are the lengths of the bases, is called a *literal equation*.

It is sometimes useful to solve a literal equation for one variable in terms of the others. To do this, use the properties of equality to rewrite the equation as an equivalent equation with the desired variable alone on one side. Example 5 shows how the properties of equality are used to manipulate the variables A, h, and b in the same way they are used on the constants 108, 9, and 8.

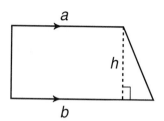

Example 5

Solve for a:

a. $108 = \frac{1}{2}(9)(a + 8)$

b. $A = \frac{1}{2}h(a + b)$

Solution

a.
$$108 = \frac{1}{2}(9)(a + 8)$$
$$216 = 9(a + 8)$$
$$216 = 9a + 72$$
$$216 - 72 = 9a$$
$$\frac{216 - 72}{9} = a$$

b.
$$A = \frac{1}{2}h(a + b)$$
$$2A = h(a + b)$$
$$2A = ha + hb$$
$$2A - hb = ha$$
$$\frac{2A - hb}{h} = a$$

The solution in part **b** expresses the value of a in terms of A, h, and b. Note that the solution in part **a** could be obtained from the solution in part **b** by letting $A = 108$, $h = 9$, and $b = 8$.

Example 6

The equation $mv = Ft + mv_0$ is used in physics to express the velocity at time t in terms of applied force F, mass m, and initial velocity v_0. Solve this equation for m.

Solution

$$mv = Ft + mv_0$$

$$mv - mv_0 = Ft \qquad \text{Gather the terms containing } m \text{ on one side.}$$

$$m(v - v_0) = Ft \qquad \text{Factor out } m.$$

$$m = \frac{Ft}{v - v_0} \qquad \text{Divide by the coefficient of } m.$$

Exercises 2.1

In **1 – 8**, find the solution set for the equation or inequality.

1. $3x + 8 = 17$

2. $-16 < y + 8$

3. $4r - 10 > 12 - r$

4. $w - 10 = 4w + 4$

5. $3(x - 2) < 5(x - 3)$

6. $8(y + 1) = 4(y - 3)$

7. $3(9 + 4a) + 2a = 7(2 - 5a)$

8. $-2[7 + 5(8x - 10)] = -20(3x - 6) - 14$

 In **9 – 10**, solve, rounding the answer to the nearest tenth.

9. $80.3x - 1.29 = 1.07x + 38.325$

10. $1.22(x - 7.1) > 14.8(2.1x - 4)$

In **11 – 22**, find the solution of the equation or inequality.

11. $\dfrac{5x}{4} + \dfrac{1}{2} = x - \dfrac{1}{2}$

12. $\dfrac{y}{8} - \dfrac{y}{5} > 10$

13. $\dfrac{10}{2x} - \dfrac{2}{x} < \dfrac{3}{2}$

14. $\dfrac{1}{2x - 7} = \dfrac{2}{5x + 8}$

15. $\dfrac{4}{3x - 2} - \dfrac{1}{9x^2 - 4} = \dfrac{7}{3x + 2}$

16. $\dfrac{1}{x - 2} + \dfrac{2}{2 - x} < 10$

17. $\dfrac{4}{2y^2 + y - 6} = \dfrac{3}{y^2 - y - 6}$

18. $\dfrac{1}{x - 2} - \dfrac{3}{2 - x} = \dfrac{6}{x}$

19. $\dfrac{a + 1}{a - 1} - \dfrac{a - 1}{a + 1} = \dfrac{6}{a^2 - 1}$

20. $\dfrac{c}{2c^2 - 5c + 2} = \dfrac{c + 1}{2c^2 + c - 1}$

21. $\dfrac{x - 1}{x^2 + x} + \dfrac{3x - 1}{2x^2 + x} = \dfrac{5x}{2x^2 + 3x + 1}$

22. $3 + \dfrac{5}{2x + 2} = \dfrac{3x^2 + 2}{x^2 - 1}$

In **23 – 26**, find the solution set.

23. $2x = 4x - 2x$

24. $\dfrac{2x}{x} = 1$

25. $\dfrac{3 - x}{3 - x} = 1$

26. $2x + 1 > 2x$

In **27 – 36**, solve for the variable indicated. In problem and solution, assume no denominator is equal to 0.

27. $E = IR$; for R

28. $A = \frac{1}{2}a(b + c)$; for a

29. $C = \frac{5}{9}(F - 32)$; for F

30. $F = \frac{9}{5}C + 32$; for C

31. $R = \dfrac{K\ell}{d^2}$; for K

32. $s = 2a + ab$; for a

33. $\dfrac{1}{x} = \dfrac{1}{y} + \dfrac{1}{z}$; for y

34. $s = \dfrac{ra_n - a}{r - 1}$; for r

35. $E = I\left(R + \dfrac{r}{n}\right)$; for n

36. $\dfrac{1}{x} = \dfrac{1}{y - x} + \dfrac{2}{y + x}$; for x

Answers 2.1

1. $\{3\}$

2. $\{y \mid y > -24\}$

3. $\left\{r \mid r > \frac{22}{5}\right\}$

4. $\left\{-\frac{14}{3}\right\}$

5. $\left\{x \mid x > \frac{9}{2}\right\}$

6. $\{-5\}$

7. $\left\{-\frac{13}{49}\right\}$

8. $\{-1\}$

9. $x = 0.5$

10. $x < 1.7$

11. $x = -4$

12. $y < -\dfrac{400}{3}$

13. $x > 2$ or $x < 0$

14. $x = -22$

15. $x = \dfrac{7}{3}$

16. $x < \dfrac{19}{10}$ or $x > 2$

17. $y = -\dfrac{3}{2}$

18. $x = 6$

19. $a = \dfrac{3}{2}$

20. no solution

21. $x = 2$

22. $x = 3$

23. all real values

24. $\{\ \}$

25. $\{x \mid x \neq 3\}$

26. all reals

27. $R = \dfrac{E}{I}$

28. $a = \dfrac{2A}{(b + c)}$

29. $F = \dfrac{9}{5}C + 32$

30. $C = \dfrac{5}{9}(F - 32)$

31. $K = \dfrac{Rd^2}{\ell}$

32. $a = \dfrac{s}{(2 + b)}$

33. $y = \dfrac{xz}{(z - x)}$

34. $r = \dfrac{s - a}{s - a_n}$

35. $n = \dfrac{Ir}{E - IR}$

36. $x = \dfrac{y}{3}$

2.2 Using an Open Sentence to Solve a Problem

The general procedure for problem solving involves a 4-step process:

1. Read   2. Plan   3. Solve   4. Check

To effect an algebraic solution using one variable, consider the following details.

Using an Algebraic Solution for a Problem

1. Read and understand the problem. Select a variable and state what it represents. Express all unknown quantities in terms of the variable.

2. Look for information in the problem or recall familiar formulas that can be used to write an equation or inequality using the variable.

3. Solve the equation or inequality and use the solution set to find the unknown values.

4. Check your answer by verifying that the answer satisfies the conditions of the original problem. State the answer.

In carrying out an algebraic solution, it is also useful to apply other problem-solving strategies, such as drawing a diagram or making a table.

Example 1

If the lengths of two opposite sides of a square are increased by 8 cm and the lengths of the other two sides are decreased by 2 cm, the perimeter of the resulting rectangle is 80 cm. Find the length of the rectangle.

Solution

Step 1   Read, and select a variable.

Let s = the length of the side of the square.

Then $s + 8$ = the length of the resulting rectangle.

And $s - 2$ = the width of the resulting rectangle.

Draw a diagram to visualize the situation.

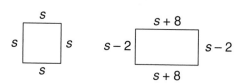

Step 2   Write an equation.

$$P_{\text{rectangle}} = 2\ell + 2w$$
$$80 = 2(s + 8) + 2(s - 2)$$

Step 3   Solve the equation. Find values for the expressions.

$$80 = 2(s + 8) + 2(s - 2)$$
$$80 = 2s + 16 + 2s - 4$$
$$80 = 4s + 12$$
$$68 = 4s$$
$$17 = s$$
$$25 = s + 8$$
$$15 = s - 2$$

Step 4   Check in the original problem.

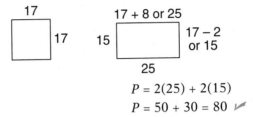

$$P = 2(25) + 2(15)$$
$$P = 50 + 30 = 80 \ ✔$$

Answer: The length of the rectangle is 25 cm.

A number of problems in daily life, in business, and in the physical sciences are based on relationships that can be written in the form $A \times B = C$. Some examples are:

Rate of Travel × Time Traveled = Distance Traveled

Rate of Work × Time Worked = Work Done

Number of Coins × Value of Coin = Cash Value

Number of Units × Price per Unit = Cost

Amount Invested × Rate of Interest = Return on Investment

Volume of Solution × Percent of Concentrate = Volume of Concentrate

Example 2 _____

Mai J. uses only lowfat and skim milk in her Natural Foods Cafe. From her distributor, she buys lowfat milk for $.75 per liter and skim milk for $.80 per liter. In the past week, she paid $66.25 for 85 liters of milk. How many liters of each type of milk did she purchase?

Solution

Step 1   Let x = the number of liters of lowfat milk.
Then $85 - x$ = the number of liters of skim milk.

A table can be used to organize the information.

Type of Milk	Number of Liters	× Price per Liter ($)	= Cost ($)
Lowfat	x	.75	$.75x$
Skim	$85 - x$	.80	$.80(85 - x)$

Step 2   The sum of the costs for each type of milk must equal the total paid, $66.25.

$$.75x + .80(85 - x) = 66.25$$

Step 3 $$75x + 80(85 - x) = 6,625$$
$$75x + 6,800 - 80x = 6,625$$
$$-5x + 6,800 = 6,625$$
$$-5x = -175$$
$$x = 35$$
$$85 - x = 50$$

Step 4   **Check:** 35 liters @ $.75 / liter = $26.25
50 liters @ $.80 / liter = $40.00
$$\overline{}$$
Total cost = $66.25 ✓

Answer: She purchased 35 liters of lowfat milk and 50 liters of skim milk.

When a problem involves an inequality, often the value of the variable is known to be positive. In these circumstances, when multiplying by an algebraic expression that involves the variable, you know whether the multiplier is positive or negative and, therefore, only one case need be considered.

Example 3

A race car did two laps around a 1-mile track. If the average speed for the two laps was less than 115 m.p.h. and the average speed for the first lap was 150 m.p.h., what was the average speed for the second lap?

Solution

Step 1   The average speed for two laps is not the same as the average of the two average speeds. Instead, the average speed for two laps depends on the total distance and the total time.

To find the total time, first find the time for each lap.

Let x = the average speed for the second lap. $x > 0$

Formula: Time = Distance ÷ Rate

First Lap: Time = $1 \div 150 = \dfrac{1}{150}$

Second Lap: Time = $1 \div x = \dfrac{1}{x}$

Step 2   Average Speed for 2 Laps = $\dfrac{\text{Total Distance}}{\text{Total Time}}$

$$115 > \frac{1 + 1}{\frac{1}{150} + \frac{1}{x}}$$

Step 3
$$115\left(\frac{1}{150} + \frac{1}{x}\right) > 2$$

Cross multiply. Since $x > 0$, $\frac{1}{150} + \frac{1}{x} > 0$.

$$\frac{115}{150} + \frac{115}{x} > 2 \qquad \text{Distribute.}$$

$$115x + 115(150) > 2(150)x \qquad LCD = 150x$$

$$115x + 17{,}250 > 300x$$

$$17{,}250 > 185x$$

$$\frac{17{,}250}{185} > x$$

$$x < 93.\overline{243}$$

Step 4   **Check:** Since the average speed for the two laps was less than 115 m.p.h., the high first lap average of 150 m.p.h. has to be balanced by a lower average speed for the second lap. Thus, less than $93.\overline{243}$ m.p.h. is a reasonable answer.

Verify: $115 < \dfrac{1 + 1}{\frac{1}{150} + \frac{1}{93}}$

Note, as previously stated, that 115 is not the numerical average of the two values 150 and $93.\overline{243}$.

Answer: The average speed for the second lap is a positive number less than $93.\overline{243}$ m.p.h.

Example 4

When moving from Florida to New York, Idalia realized that she needed a stronger antifreeze solution in her car. She decided to change it herself by draining out some of the solution and replacing it with pure antifreeze. If the car contains 20 quarts of a solution that is 40% antifreeze and she wishes to increase the strength of the solution to 50% antifreeze, how many quarts should she drain and replace with pure antifreeze?

Solution

An important relationship in the problem is:

$$\begin{array}{ccc} \text{Volume of} \\ \text{Solution} \end{array} \times \begin{array}{c} \text{Percent of} \\ \text{Concentrate} \end{array} = \begin{array}{c} \text{Volume of Concentrate} \\ \text{in Solution} \end{array}$$

Think of the cooling system solution as being in three different states as represented by the diagrams.

x quarts of the 40% solution contains $0.60x$ quarts of water and $0.40x$ quarts of antifreeze.

| 12 qt. of water | $(12 - 0.60x)$ qt. of water | $(12 - 0.60x)$ qt. of water |
| 8 qt. of antifreeze | $(8 - 0.40x)$ qt. of antifreeze | $(8 - 0.40x + x)$ qt. of antifreeze |

The new solution is to be 50% antifreeze:

$$\frac{\text{Number of Quarts of Antifreeze}}{\text{Total Capacity of Cooling System}} = 50\%$$

$$\frac{8 - 0.40x + x}{20} = \frac{1}{2} \qquad \text{\small 50\% = } \tfrac{1}{2}.$$

$$\frac{8 + 0.60x}{20} = \frac{1}{2} \qquad \text{\small Combine like terms in the numerator on the left side.}$$

$$8 + 0.60x = 10 \qquad \text{\small Multiply by 20.}$$

$$0.60x = 2 \qquad \text{\small Subtract 8.}$$

$$60x = 200 \qquad \text{\small Multiply by 100.}$$

$$x = \frac{200}{60} \qquad \text{\small Divide by 60.}$$

$$x = 3\tfrac{1}{3}$$

Answer: $3\tfrac{1}{3}$ quarts of coolant must be removed and replaced by pure antifreeze to obtain a 50% solution.

Exercises 2.2

Solve each problem by using one variable.

1. An old machine in a manufacturing plant has been in use twice as long as a newer one. In 4 years, the sum of the years the two machines have been in use will be 32 years. Find the number of years that the newer machine has been in use.

2. Between the ages of 25 and 40, a man gained enough weight to double his weight at age 25. He then went on a diet and lost 125 pounds, bringing him down to 145 pounds, the normal weight for his height. How much did he weigh when he was 25?

Exercises 2.2 (continued)

3. The lengths of the sides of a triangle are in the ratio 2:5:6. If 1 foot is added to each side of the triangle, the perimeter of the triangle formed is more than 29 feet. Find the length of the shortest side of the original triangle.

4. In degrees, the measure of each base angle of an isosceles triangle is $6\frac{3}{5}^{\circ}$ less than $\frac{1}{5}$ the measure of the vertex angle. Find the measure of each of the angles of the triangle.

5. The degree measure of one of two supplementary angles is 9 less than twice the measure of the other. Find the measure of the smaller of the two angles.

6. The length of a rectangle is 2 cm less than twice its width. If the perimeter is less than 92 cm, find the width of the rectangle.

7. A towel that is three times as long as it is wide shrinks 3 inches in length after being washed, but the width remains the same. If the towel loses a total of 54 square inches in area, what are its original dimensions?

8. A square sheet of rubber is stretched so that its length is increased by 8 inches and its width is decreased by 2 inches. If the stretched piece of rubber has a perimeter of 40 inches, find its original dimensions.

9. A delivery company picked up, from a bookstore, a shipment of packages that had an average weight of 5 pounds and, from a clothing store, a shipment of packages that had an average weight of 2 pounds. If the average weight of the packages in both shipments was 3 pounds, what was the ratio of the number of packages shipped by the bookstore to the number of packages shipped by the clothing store?

10. Mr. Fritz enlarged his square garden by extending one pair of parallel sides by 20 feet each. In doing so, he doubled the amount of fencing that he needed to enclose the garden. What were the dimensions of the enlarged garden?

11. The larger of two numbers is 15 less than 8 times the smaller. If the larger is divided by the smaller, the quotient is 7 and the remainder is 5. Find the numbers.

12. Find two consecutive integers such that the reciprocal of the smaller increased by 4 times the reciprocal of the larger is equal to 11 times the reciprocal of the product of the integers.

13. The lengths of two opposite sides of a square are increased by 4 cm and the lengths of the other sides are reduced to 2 cm less than one-half the original length. The perimeter of the resulting rectangle is 52 cm. Find the length of the sides of the square.

14. The length of one side of a rectangle is 2 ft. greater than the length of the other. If the longer side is doubled and the shorter side is tripled, the resulting rectangle has a perimeter of 58 ft. Find the dimensions of the original rectangle.

Answers 2.2

3. more than 4 ft.
4. 138°, 21°, 21°
5. 63°
6. less than 16 cm
7. 18 in. × 54 in.
8. 7 in. × 7 in.
9. 1:2
10. 10 ft. × 30 ft.
11. 145, 20
12. 2, 3
13. 16 cm
14. 5 ft. × 7 ft.

Exercises 2.2 *(continued)*

15. The longest side of a triangle is 1 m longer than twice the length of the shortest side, and the remaining side is 7 m longer than the shortest side. If the perimeter is 5 times the length of the shortest side, find the length of each side of the triangle.

16. A compound consists of two chemicals, one worth $.64 per ounce and the other $.40 per ounce. If 2 pounds of the compound can be purchased for $17.60, what percent of the compound is the chemical worth $.40 per ounce?

17. A fuel broker received two shipments of gasoline totaling 600 gallons. One shipment cost $1.00 per gallon and the other $1.25 per gallon. If he sells the 600 gallons for $1.10 per gallon, he will receive more than what he paid for the two shipments. How many gallons of the less expensive gasoline did he receive?

18. Mrs. Rich invested $50,000 in two stocks. The first stock showed an appreciation of 8% at the end of the year, but the second lost 2% within the same period. If her total income from the two investments was $3,080, how much money did she originally invest in each of the stocks?

19. How many pounds of coffee worth *c* cents/lb. must be mixed with 20 lb. of coffee worth *d* cents/lb. to make a mixture worth *m* cents/lb.? Answer in terms of *c*, *d*, and *m*.

20. Mr. Jones invests $100,000 in two investments. If the first investment consists of $25,000 at 6% interest, what rate of return must he get on the remaining $75,000 in order to have an average rate of return of at least 8% on his total investment?

21. A dental company produces two amalgams. One contains *a*% silver and the other contains *b*% silver. How many grams of each should be mixed to produce 60 grams of a new alloy that contains *c*% silver? Answer in terms of *a*, *b*, and *c*, and assume that $a < c < b$.

22. A factory makes silicon chips for computers. The management calculates that it costs $.02 to make each chip produced without defects, but it costs $.30 for each chip that turns out defective. If, for one day, the total production cost was $2,560 and 100,000 chips were produced, how many of the chips produced were defective?

23. How many ounces of a 90% solution of disinfectant should be added to 15 oz. of a 10% solution of disinfectant to produce a 30% solution?

24. How many liters of a 10%-acid solution and how many liters of a 6%-acid solution should be mixed to produce 100 liters of a 9%-acid solution?

---

15. 8 m, 15 m, 17 m

16. 37.5%

17. more than 360 gal. @ $1.00

18. $40,800, $9,200

19. $\dfrac{20(m-d)}{c-m}$

20. $8\frac{2}{3}\%$ or more

21. $\dfrac{60(c-b)}{a-b}$ g of *a*% silver

$\dfrac{60(a-c)}{a-b}$ g of *b*% silver

22. 2,000 chips

23. 5 oz.

24. 75 liters of 10% and 25 liters of 6%

25. A manufacturing process requires the use of a 60% solution of acid. Since no solution of that strength is on hand, the foreman decides to evaporate water from a weaker solution. How much water must be evaporated from 32 oz. of a 40% solution of acid to make a 60% solution of acid?

26. A school nurse uses a 3% solution of hydrogen peroxide as an antiseptic. She purchased an 8-ounce bottle of a 5% solution, poured some into another bottle to store it, and replaced what had been poured off with distilled water. How much of the 5% solution should have been replaced with water to obtain a 3% solution?

27. **a.** A race car does two laps around a track that is 2 miles long. If it travels 150 m.p.h. on the first lap and 100 m.p.h. on the second, what is its average speed for the two laps?

    b. Solve the problem again, assuming the track is *m* miles long.

28. On a car trip to a resort, the driver averaged 40 m.p.h. What must the average speed be on the return trip over the same route to achieve an average speed of 45 m.p.h. for the entire trip? Give your answer to the nearest m.p.h.

29. During a fuel crisis, Mr. Jones spent an average of $1.40 per gallon on heating oil. The following year, the price went down, and he could raise the thermostat. If he spent $1,500 for fuel each year, and his average cost for the two-year period was $1.25 per gallon, how much did he pay per gallon during the second year? Give your answer to the nearest tenth of a cent per gallon.

30. In a small town, 150 people voted on a referendum that required a simple majority for passage. Although the referendum passed, if 5 people had changed their vote from *for* to *against*, it would have lost by 2 votes. How many of the voters actually voted for the referendum?

31. A teller had intended to change a $5.00 bill using 36 coins consisting of quarters and nickels. He inadvertently interchanged the number of nickels and quarters, causing an error that resulted in a shortage in his drawer. How much was he short due to the error?

32. If one typist can complete the typing of a manuscript in 50 hours and another typist requires 60 hours to do the job, how long would it take the two typists to complete the job together? *Hint:* Part done in 1 hour × Number of hours = 1 completed job.

33. A landscaper who can redesign a garden in 10 days, works for 3 days and then gets help from another gardener. They work together to complete the job in another 3 days. How long would it have taken the second gardener to do the job alone?

34. A painter works twice as fast as his helper. Together they paint a house in 10 days. How long would it have taken the painter if he had painted the house alone?

Answers 2.2

25. $10\frac{2}{3}$ oz.

26. $3\frac{1}{5}$ oz.

27. **a.** 120 m.p.h.
    b. 120 m.p.h.

28. 51 m.p.h.

29. $1.129

30. 79

31. $.80

32. $27\frac{3}{11}$ hours

33. $7\frac{1}{2}$ days

34. 15 days

2.3 Quadratic Equations and Inequalities

■ **Definition:** *A quadratic equation is an equation that can be written in the form*

$$ax^2 + bx + c = 0$$

where x is the variable, a, b, and c are real numbers, and $a \neq 0$.

The following are examples of quadratic equations with the values of a, b, and c given.

> **Note:** A quadratic equation is in *standard form* when the terms of one side are in descending order, and the other side is 0.

$x^2 - 7x + 1 = 0$	$a = 1$	$b = -7$	$c = 1$
$3x^2 + 12x = 0$	$a = 3$	$b = 12$	$c = 0$
$x^2 - 9 = 0$	$a = 1$	$b = 0$	$c = -9$

Solution by Factoring

Recall that one of the properties of the field of real numbers states: If a and b are real numbers and $ab = 0$, then $a = 0$ or $b = 0$. This property of the real numbers is used to solve a quadratic equation by factoring.

Example 1

Find the solution set of the equation $x^2 = 5x + 50$.

Solution

$$x^2 = 5x + 50$$

$x^2 - 5x - 50 = 0$	Rewrite in standard form.
$(x - 10)(x + 5) = 0$	Factor the left side.
$x - 10 = 0 \quad \mid \quad x + 5 = 0$	Set each factor equal to 0.
$x = 10 \quad \mid \quad x = -5$	Solve each linear equation.

Check Substitute each possibility into the original equation.

$x^2 = 5x + 50$	$x^2 = 5x + 50$
$10^2 \stackrel{?}{=} 5(10) + 50$	$(-5)^2 \stackrel{?}{=} 5(-5) + 50$
$100 \stackrel{?}{=} 50 + 50$	$25 \stackrel{?}{=} -25 + 50$
$100 = 100 \ ✓$	$25 = 25 \ ✓$

Answer: The solution set is $\{-5, 10\}$.

When a quadratic equation has equal roots, the solution is called a *double root*.

Example 2

Find the solution set of $4x^2 + 12x = -9$.

Solution

$$4x^2 + 12x + 9 = 0$$
$$(2x + 3)(2x + 3) = 0$$
$$2x + 3 = 0$$
$$2x = -3$$
$$x = -\frac{3}{2}$$

Answer: The solution set is $\left\{ -\frac{3}{2} \right\}$.

Only quadratic equations with rational roots can be solved by factoring over the set of integers. Others can be solved by factoring over the set of real numbers. For example:

$$x^2 = 5$$
$$x^2 - 5 = 0$$
$$(x + \sqrt{5})(x - \sqrt{5}) = 0$$

$$x + \sqrt{5} = 0 \quad \Big| \quad x - \sqrt{5} = 0$$
$$x = -\sqrt{5} \quad \Big| \quad x = \sqrt{5}$$

The solution set is $\{\pm\sqrt{5}\}$.

The equation could also have been solved by taking the square root of each side.

$$x^2 = 5$$
$$x = \pm\sqrt{5}$$

Since this method of taking the square root of each side can be generalized for any quadratic equation if one side of the equation is a perfect square and the other side is a constant, it can be used to solve quadratic equations of this form even when these quadratics are not factorable in the set of integers.

Example 3

Solve $x^2 + 2x + 1 = 17$.

Solution

$$x^2 + 2x + 1 = 17$$
$$(x + 1)^2 = 17 \qquad \text{Write the left side as a square.}$$
$$x + 1 = \pm\sqrt{17} \qquad \text{Take the square root of each side.}$$

$$x + 1 = \sqrt{17} \quad \Big| \quad x + 1 = -\sqrt{17} \qquad \text{Solve each linear equation.}$$
$$x = \sqrt{17} - 1 \quad \Big| \quad x = -\sqrt{17} - 1$$

Completing the Square

Relying still on taking the square root of each side is the method of *completing the square*. This method involves forming a perfect-square trinomial on one side. The expression $x^2 + 2ax + a^2$ is a perfect-square trinomial, where $2a$ is the coefficient of the x term and a^2 is the constant term.

$$x^2 + 2ax + a^2 = (x + a)^2$$

Note that when the coefficient of x^2 is 1, the constant is equal to the square of one-half the coefficient of x.

Example 4

Solve $x^2 - 10x + 1 = 0$ by completing the square.

Solution

$x^2 - 10x = -1$	Add -1 to each side so that only the terms in x remain on the left.
$x^2 - 10x + \left(\frac{-10}{2}\right)^2 = -1 + \left(\frac{-10}{2}\right)^2$	Add the square of one-half the coefficient of x to each side.
$x^2 - 10x + 25 = 24$	Simplify each side.
$(x - 5)^2 = 24$	Express the left side as a square.
$x - 5 = \pm\sqrt{24}$	Take the square root of each side.

$$x - 5 = \sqrt{24} \quad \bigg| \quad x - 5 = -\sqrt{24}$$
$$x = 2\sqrt{6} + 5 \quad \bigg| \quad x = -2\sqrt{6} + 5$$

Write two equations, using the positive and the negatve square roots of the constant.

Answer: The solution set is $\{5 \pm 2\sqrt{6}\}$.

Example 5

Find the solution set for $4x^2 + 12x - 15 = 0$.

Solution

$4x^2 + 12x - 15 = 0$	
$\dfrac{4x^2}{4} + \dfrac{12x}{4} - \dfrac{15}{4} = \dfrac{0}{4}$	Divide by 4 to get a leading coefficient of 1.
$x^2 + 3x - \dfrac{15}{4} = 0$	
$x^2 + 3x = \dfrac{15}{4}$	Add $\dfrac{15}{4}$.
$x^2 + 3x + \left(\dfrac{3}{2}\right)^2 = \dfrac{15}{4} + \left(\dfrac{3}{2}\right)^2$	Add the square of one-half the coefficient of x.
$\left(x + \dfrac{3}{2}\right)^2 = 6$	Write the left side as a square. Simplify the right side.
$x + \dfrac{3}{2} = \pm\sqrt{6}$	Take the square root of each side.
$x = -\dfrac{3}{2} \pm\sqrt{6}$	Subtract $\dfrac{3}{2}$.

Answer: The solution set is $\{-\frac{3}{2} \pm\sqrt{6}\}$.

The Quadratic Formula

The general quadratic equation $ax^2 + bx + c = 0$ can also be solved by completing the square. The result, called the *quadratic formula*, expresses x in terms of a, b, and c, and can be used to solve any quadratic equation.

$$ax^2 + bx + c = 0$$

$$ax^2 + bx = -c \qquad \text{Subtract } c.$$

$$x^2 + \frac{b}{a}x = -\frac{c}{a} \qquad \text{Divide by } a, \ (a \neq 0).$$

$$x^2 + \frac{b}{a}x + \left(\frac{1}{2} \cdot \frac{b}{a}\right)^2 = -\frac{c}{a} + \left(\frac{1}{2} \cdot \frac{b}{a}\right)^2 \qquad \begin{array}{l}\text{Add the square of one-half the}\\\text{coefficient of } x.\end{array}$$

$$\left(x + \frac{b}{2a}\right)^2 = -\frac{c}{a} + \frac{b^2}{4a^2} \qquad \text{Write the left side as a square.}$$

$$\left(x + \frac{b}{2a}\right)^2 = \frac{b^2 - 4ac}{4a^2} \qquad \text{Simplify the right side.}$$

$$x + \frac{b}{2a} = \frac{\pm\sqrt{b^2 - 4ac}}{2a} \qquad \begin{array}{l}\text{Take the square root of}\\\text{each side.}\end{array}$$

$$x = \frac{-b \pm \sqrt{b^2 - 4ac}}{2a} \qquad \text{Subtract } \frac{b}{2a}.$$

☐ **The Quadratic Formula:** *The solutions of the quadratic equation* $ax^2 + bx + c = 0$ $(a \neq 0)$

are: $x = \dfrac{-b \pm \sqrt{b^2 - 4ac}}{2a}$

Example 6

Use the quadratic formula to solve: $4x^2 + 3x = 5$

Solution

$$4x^2 + 3x - 5 = 0 \qquad \text{Write the equation in standard form.}$$

$$a = 4, \ b = 3, \ c = -5 \qquad \text{Determine the values of } a, \ b, \text{ and } c.$$

$$x = \frac{-b \pm \sqrt{b^2 - 4ac}}{2a} \qquad \text{Write the quadratic formula.}$$

$$= \frac{-3 \pm \sqrt{3^2 - 4(4)(-5)}}{2(4)} \qquad \begin{array}{l}\text{Substitute the values of } a, \ b, \text{ and } c\\\text{into the quadratic formula.}\end{array}$$

$$= \frac{-3 \pm \sqrt{9 + 80}}{8} = \frac{-3 \pm \sqrt{89}}{8} \qquad \text{Simplify.}$$

Answer: The solution set is $\left\{\dfrac{-3 \pm \sqrt{89}}{8}\right\}$.

The Discriminant

Without actually solving the equation, it is often useful to determine the nature of the solutions of a quadratic equation. A number of questions arise regarding the nature of the roots of a quadratic equation. For example:

- Are they equal?
- Are they real or imaginary?
- If they are real, are they rational or irrational?

These questions can be answered for a particular quadratic equation by studying the value of the expression $b^2 - 4ac$. This expression, which appears under the radical sign in the quadratic formula, is called the *discriminant*

Reconsider Example 6 to study how the discriminant, 89, describes the nature of the roots, $\frac{-3 \pm \sqrt{89}}{8}$. Since 89 is positive, the roots are real numbers. Since 89 is not a perfect square, the roots are irrational numbers. Since adding $\sqrt{89}$ is different from subtracting $\sqrt{89}$, the two roots are unequal. Thus, the discriminant establishes the nature of the roots of $4x^2 + 3x = 5$ as real, irrational, and unequal. Note that when a, b, and c are rational, irrational roots occur in pairs; that is, if $p + \sqrt{q}$ is a root, then $p - \sqrt{q}$ is the other root. Imaginary roots also occur in pairs, called *conjugates*, of the form $p \pm qi$.

The information that can be obtained about the roots of a quadratic equation with rational coefficients from the value of the discriminant is summarized as follows.

If a, b, and $c \in$ {rational numbers}

and...	then...
$b^2 - 4ac$ is a nonzero perfect square	the roots are real, rational, and unequal.
$b^2 - 4ac$ is not a perfect square	the roots are real, irrational, and unequal.
$b^2 - 4ac = 0$	the roots are real, rational, and equal. (There is a double root.)
$b^2 - 4ac < 0$	the roots are imaginary and unequal.

Example 7

Determine the nature of the roots: $2x^2 + 5x = -6$

Solution

The given equation in standard form is $2x^2 + 5x + 6 = 0$, with $a = 2$, $b = 5$, and $c = 6$. Since these coefficients are rational, the discriminant can be used to determine the nature of the roots.

$$b^2 - 4ac = 5^2 - 4(2)(6)$$
$$= 25 - 48$$
$$= -23$$

Thus, $b^2 - 4ac < 0$, making the roots imaginary and unequal.

Example 8

Find the dimensions of a rectangle whose area is 40 and whose perimeter is 20.

Solution

The area of the rectangle is the product of the length and width. If one dimension is x, then the other is $\frac{40}{x}$.

The equation for the perimeter is

$$2x + 2\left(\frac{40}{x}\right) = 20$$

$$2x^2 + 80 = 20x \qquad \text{Multiply by } x.$$

$$2x^2 - 20x + 80 = 0 \qquad \text{Standard form.}$$

$$x^2 - 10x + 40 = 0 \qquad \text{Divide by 2.}$$

Use the discriminant to determine the nature of the roots.

$$a = 1, b = -10, c = 40 \qquad b^2 - 4ac = (-10)^2 - 4(1)(40)$$

$$= 100 - 160 = -60$$

Since the discriminant is -60, the roots of the equation are imaginary, and the rectangle cannot exist.

Applications of Quadratic Equations

The remaining examples in this section involve applications requiring the use of a quadratic equation to solve a problem.

The formula $s = -16t^2 + vt + h$ can be used to find s, the distance from the ground (in feet) of a falling object after t seconds if it is thrown from a height of h feet with an initial velocity of v feet per second.

Example 9

An object is thrown vertically downward with an initial velocity of 30 feet per second from a point 200 feet above the ground. Use the equation given above to find, to the nearest second, the time needed for it to reach the ground.

Solution

Since the velocity of an object traveling downward is taken as negative, $v = -30$ ft./sec. Since the object is thrown from 200 feet above the ground, $h = 200$ ft. When the object reaches the ground, $s = 0$.

$$s = -16t^2 + vt + h$$

$$0 = -16t^2 - 30t + 200$$

$$0 = 8t^2 + 15t - 100 \qquad \text{Divide by } -2.$$

$$a = 8, b = 15, \text{ and } c = -100$$

$$t = \frac{-b \pm \sqrt{b^2 - 4ac}}{2a} \qquad\qquad t \approx \frac{-15 \pm 58.5}{16}$$

$$= \frac{-15 \pm \sqrt{(15^2) - 4(8)(-100)}}{2(8)} \qquad \approx \frac{-15 + 58.5}{16} \text{ or } t \approx \frac{-15 - 58.5}{16}$$

$$= \frac{-15 \pm \sqrt{225 + 3,200}}{16} \qquad \approx \frac{43.5}{16} \qquad\qquad \text{Reject the negative value.}$$

$$= \frac{-15 \pm \sqrt{3,425}}{16} \qquad \approx 2.7$$

Answer: To the nearest second, the time needed to reach the ground is 3 seconds.

Problems involving area often require the solution of a quadratic equation.

Example 10

A rectangular grass field 60 m by 100 m is mowed by cutting a strip of uniform width. How wide, to the nearest meter, must the strip at the outer edge be so that one-third of the field's area is mowed?

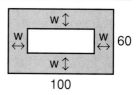

Solution

Let w = the width of the path.
Then $100 - 2w$ = the length of the unmowed area.
And $60 - 2w$ = the width of the unmowed area.

Total area − Unmowed area = Mowed area

$$60(100) - (100 - 2w)(60 - 2w) = \tfrac{1}{3}(60)(100)$$

$$6{,}000 - [6{,}000 - 120w - 200w + 4w^2] = 2{,}000$$

$$320w - 4w^2 = 2{,}000$$

$$4w^2 - 320w + 2{,}000 = 0$$

$$w^2 - 80w + 500 = 0$$

$a = 1, b = -80,$ and $c = 500$

$$w = \frac{-b \pm \sqrt{b^2 - 4ac}}{2a}$$

$$= \frac{-(-80) \pm \sqrt{(-80)^2 - 4(1)(500)}}{2(1)}$$

$$= \frac{80 \pm \sqrt{6{,}400 - 2{,}000}}{2}$$

$$= \frac{80 \pm \sqrt{4{,}400}}{2}$$

$$= \frac{80 \pm \sqrt{400}\sqrt{11}}{2}$$

$$= \frac{80 \pm 20\sqrt{11}}{2}$$

$$= 40 \pm 10\sqrt{11}$$

$w \approx 40 + 10(3.32)$	$w \approx 40 - 10(3.32)$
$\approx 40 + 33.2$	$\approx 40 - 33.2$
≈ 73.2	≈ 6.8

The value on the left, although a solution to the equation, is not a reasonable answer to the problem because it is greater than the width of the field.

Answer: To the nearest meter, the width of the strip is 7 m.

Quadratic Inequalities

To solve a quadratic inequality by factoring, recall the rules for the product of two signed numbers.

- If $a > 0$ and $b > 0$, then $ab > 0$.
- If $a < 0$ and $b < 0$, then $ab > 0$.
- If $a < 0$ and $b > 0$, then $ab < 0$.

Therefore, the product of two factors is positive if both factors are positive or if both factors are negative. The product of two factors is negative if one factor is positive and the other factor is negative.

Example 11

Solve: $x^2 - 3x - 10 > 0$

Solution

$$x^2 - 3x - 10 > 0$$
$$(x - 5)(x + 2) > 0$$

Since the product is positive, the two factors must be both positive or both negative.

$x - 5 > 0$ and $x + 2 > 0$ or	$x - 5 < 0$ and $x + 2 < 0$
$x > 5$ and $\quad x > -2$ or	$x < 5$ and $\quad x < -2$
The values of x that are greater than both 5 and -2 must be greater than 5.	The values of x that are less than both 5 and -2 must be less than -2.
$x > 5$ or	$x < -2$

Answer: The solution set is $\{x \mid x < -2 \text{ or } x > 5\}$.

On a number line, the value of x for which a linear factor is 0 is the boundary between the numbers that make the factor negative and the numbers that make the factor positive. The product of two factors will be positive when both factors have the same sign, zero when one or both factors are zero, and negative when the factors have opposite signs. The solution to Example 11 can be shown on a number line as follows.

Example 12

Solve: $x^2 < 4x + 5$

Solution

$$x^2 < 4x + 5$$
$$x^2 - 4x - 5 < 0$$
$$(x - 5)(x + 1) < 0$$

1. $x = 0, x = -\frac{1}{4}$

2. $x = -\frac{3}{4}, x = \frac{3}{4}$

3. $-4 < t < 7$

4. $w = -3, w = -3$

5. $x < -5$ or $x > \frac{1}{2}$

6. $y < -\frac{3}{2}$ or $y > 11$

7. $x = \frac{3}{2a}, x = -\frac{1}{a}$

8. $y = -a - b, y = b - a$

9. $x = -1, x = -5$

10. $x = 4, x = 0$

11. $x = 0, x = -10$

12. $y = 3, y = 0$

13. $s = 7, s = -1$

14. $x = \frac{3 \pm \sqrt{6}}{3}$

15. $x = \frac{4 \pm \sqrt{10}}{3}$

16. $x = \frac{-\sqrt{2} \pm \sqrt{6}}{2}$

17. $x = \frac{-1 \pm \sqrt{13}}{6}$

18. $x = \frac{1 \pm \sqrt{13}}{6}$

19. $x = \frac{1}{4}, x = -\frac{3}{4}$

20. $y = 1 \pm \sqrt{5}$

21. $x = \frac{17 \pm \sqrt{213}}{38}$

22. $x = \frac{\pm\sqrt{70}}{14}$

23. $x = \frac{\pm\sqrt{35}}{7}$

24. $y = \frac{-\sqrt{2} \pm \sqrt{30}}{2}$

25. $x = 4 \pm \sqrt{7}$

26. $s = \frac{-7 \pm \sqrt{33}}{2}$

27. $x = 1 \pm a\sqrt{2}$

28. $x = 2\sqrt{2}, x = \frac{\sqrt{2}}{2}$

29. $x = -2i, x = -i$

30. $x = 0.8, x = -0.5$

31. $x = 0.29, x = 0.25$

32. $x = 1.68, x = -0.25$

Since the product of two factors is negative, one factor must be positive and the other negative.

$x - 5 > 0$ and $x + 1 < 0$ $\qquad$ or $\qquad$ $x - 5 < 0$ and $x + 1 > 0$

$\quad x > 5$ and $\quad x < -1$ $\qquad$ or $\qquad$ $\quad x < 5$ and $\quad x > -1$

$\qquad$ no values of x $\qquad\qquad\qquad\qquad\qquad -1 < x < 5$

Answer: The solution set is $\{x \mid -1 < x < 5\}$.

Note that, in this example, only the case where $x - 5$ was negative and $x + 1$ was positive provided values of x. Since the value of $x + 1$ must be greater than the value of $x - 5$, then $x + 1$ must be the positive factor and $x - 5$ the negative factor.

Exercises 2.3

In **1 – 8**, solve by factoring.

1. $8x^2 + 2x = 0$

2. $16x^2 - 9 = 0$

3. $t^2 - 3t - 28 < 0$

4. $w^2 + 6w + 9 = 0$

5. $2x^2 + 9x > 5$

6. $19y + 33 < 2y^2$

7. $2a^2x^2 - ax - 3 = 0$
(Solve for x in terms of a.)

8. $(y + a)^2 = b^2$
(Solve for y in terms of a and b.)

In **9 – 16**, solve by completing the square.

9. $x^2 + 6x + 9 = 4$

10. $x^2 - 4x = 0$

11. $x^2 + 10x = 0$

12. $y^2 - 3y = 0$

13. $s^2 = 6s + 7$

14. $3x^2 - 6x + 1 = 0$

15. $3x^2 - 8x + 2 = 0$

16. $x^2 + x\sqrt{2} - 1 = 0$

In **17 – 29**, solve by using the quadratic formula.

17. $3x^2 + x - 1 = 0$

18. $3x^2 - x - 1 = 0$

19. $8x^2 + 4x = \frac{3}{2}$

20. $y^2 - 2y = 4$

21. $19x^2 - 17x + 1 = 0$

22. $14x^2 - 5 = 0$

23. $7x^2 - 5 = 0$

24. $y^2 + y\sqrt{2} = 7$

25. $(x - 3)^2 = 2x$

26. $(s + 2)^2 = -3s$

27. $x^2 - 2x + (1 - 2a^2) = 0$ (Solve for x in terms of a.)

28. $\sqrt{2}x^2 - 5x + \sqrt{8} = 0$

29. $ix^2 - 3x - 2i = 0$ (*Note:* $i = \sqrt{-1}$)

 In **30 – 32**, solve, rounding as indicated.

30. $5.7x^2 - 1.6x - 2.1 = 0$ (to the nearest tenth)

31. $8.06x^2 - 4.41x - 0.6 = 0$ (to the nearest hundredth)

32. $(1 + \sqrt{2})x^2 - 2\sqrt{3}x - 1 = 0$ (to the nearest hundredth)

Exercises 2.3 (continued)

In **33 – 41**, find the nature of the roots by the discriminant. If the discriminant cannot be used (a, b, and c not all rational), solve the equation to determine the nature of the roots.

33. $x^2 - 3x - 4 = 0$

34. $y^2 - 7y = 0$

35. $t^2 + 3 = 0$

36. $16x^2 - 7x = 5$

37. $3x^2 = 0$

38. $x^2 + 0.5x - 0.3 = 0$

39. $s^2 + s\sqrt{5} - 1 = 0$

40. $\dfrac{1}{y+1} - \dfrac{1}{y} - 3 = 0$

41. $x^2 + ix - 1 = 0$

In **42 – 44**, use the formula $s = -16t^2 + vt + h$ to solve the problem.

42. A rock is dropped from the top of an embankment into water that is 256 ft. below.
   a. In how many seconds will the rock hit the water?
   b. How long will it take before the rock is 112 feet above the water?

43. A coin is thrown vertically upward from the top of a table with an initial velocity of 40 ft./sec. How long will it take for the coin to fall back to the table?

44. A coin is thrown down into a lake from a bridge that is 200 ft. above the water. If the initial velocity of the coin is 10 ft./sec., how many seconds pass before the coin travels 150 ft.? Round to the nearest tenth.

45. A page has the dimensions 20 cm by 28 cm. How wide can a uniform margin around the page be in order to allow at least 384 cm² for print?

46. A poster has 234 in.² of printed material, a 1-in. margin at the the top and bottom, and a $1\frac{1}{2}$-in. margin at the sides. What are the dimensions of the poster if the width is $\frac{4}{5}$ of the length?

47. In the center of a circular garden, there is a fountain with a circular base. The radius of the garden is 1 ft. less than twice the radius of the fountain. If the area of the garden is 33 sq. ft., find, to the nearest hundredth:
   a. the radius of the base of the fountain.
   b. The radius of the garden.

48. Use the equation $s = -16t^2 + vt + h$ to find an expression for the time t, in seconds, it takes for an object thrown from a height h feet above the ground with an initial velocity of v ft./sec. to be s ft. from the ground. *Hint:* Solve for t in terms of s, v, and h.

49. A 20-meter-long path connects opposite corners of a rectangular park. A second path, 8 meters long, is the shortest distance from one of the other corners of the park to the 20-meter path. How far from the ends of the longer path do the two paths meet?

50. At a local gasoline station, super unleaded gas costs $.07 a gallon more than regular unleaded. For $10.00, a customer gets 0.4 gallon less super unleaded gas than regular unleaded. To the nearest tenth of a cent, find the cost of super unleaded gasoline.

51. Show that if the roots of the equation $ax^2 + bx + c = 0$ are r_1 and r_2, then:
   a. $r_1 + r_2 = \dfrac{-b}{a}$ **b.** $r_1 r_2 = \dfrac{c}{a}$

Answers 2.3

33. real, rational, unequal

34. real, rational, unequal

35. imaginary, unequal

36. real, irrational, unequal

37. real, rational, equal

38. real, irrational, unequal

39. real, irrational, unequal

40. imaginary, unequal

41. imaginary, unequal

42. a. 4 sec.
   b. 3 sec.

43. 2.5 sec.

44. 2.8 sec.

45. 2 cm or less

46. 16 in. × 20 in.

47. a. 2.12 ft.
   b. 3.24 ft.

48. $t = \dfrac{v \pm \sqrt{v^2 - 64s + 64h}}{32}$

49. 16 m and 4 m

50. $1.358

51. $r_1 = \dfrac{-b + \sqrt{b^2 - 4ac}}{2a}$

$r_2 = \dfrac{-b - \sqrt{b^2 - 4ac}}{2a}$

a. $r_1 + r_2 = \dfrac{-2b}{2a} = -\dfrac{b}{a}$

b. $r_1 r_2 = \dfrac{(-b)^2 - (\sqrt{b^2 - 4ac})^2}{(2a)^2}$

$= \dfrac{b^2 - b^2 + 4ac}{4a^2}$

$= \dfrac{4ac}{4a^2} = \dfrac{c}{a}$

2.4 Additional Techniques for Solving Equations

Raising Each Side of an Equation to a Power

The equation $x = 3$ has only one element in its solution set, $\{3\}$. However, squaring both sides of the equation $x = 3$ gives $x^2 = 9$, which has a solution set with two elements, $\{\pm 3\}$. This demonstrates that when both sides of an equation are raised to an integral power, the resulting equation may not be equivalent to the original equation. However, the solution set of the original equation is a subset of the solution set of the resulting equation. This principle can be stated as follows:

The solution set of an equation is a subset of the solution set of the equation formed by raising each side of the original equation to the nth power, where n is a positive integer.

This principle is used to solve equations that contain radical expressions.

Example 1

Find the solution set of the equation $\sqrt{2x + 7} = x - 4$.

Solution

Since the entire left side of the equation is expressed as a square root, raise each side of the equation to the second power.

$$(\sqrt{2x + 7})^2 = (x - 4)^2$$
$$2x + 7 = x^2 - 8x + 16$$

The resulting quadratic equation, when written in standard form, can be solved by factoring.

$$x^2 - 10x + 9 = 0$$
$$(x - 9)(x - 1) = 0$$
$$x - 9 = 0 \quad | \quad x - 1 = 0$$
$$x = 9 \quad | \quad x = 1$$

According to the principle, the solution set of the original equation, $\sqrt{2x + 7} = x - 4$, is a subset of the solution set of $(\sqrt{2x + 7})^2 = (x - 4)^2$, which is $\{1, 9\}$. Thus, the possible solution sets of the original equation are ø, $\{1\}$, $\{9\}$, and $\{1, 9\}$.

By checking $x = 1$ and $x = 9$ in the original equation, the correct solution set can determined.

$$\text{For } x = 1: \qquad \text{For } x = 9:$$
$$\sqrt{2x + 7} = x - 4 \qquad \sqrt{2x + 7} = x - 4$$
$$\sqrt{2(1) + 7} \overset{?}{=} 1 - 4 \qquad \sqrt{2(9) + 7} \overset{?}{=} 9 - 4$$
$$\sqrt{9} \overset{?}{=} -3 \qquad \sqrt{25} \overset{?}{=} 5$$
$$3 \neq -3 \qquad 5 = 5 \ ✔$$

Since $x = 1$ does not check, it is called an *extraneous solution*

Answer: The solution set is $\{9\}$.

Example 2 _____

Find the solution set of the equation $3\sqrt{3x + 4} - 2 = x + 6$.

Solution

Since squaring immediately would not accomplish the goal of elim-inating the radical expression, first, isolate the term with the radical.

$$3\sqrt{3x + 4} - 2 = x + 6$$
$$3\sqrt{3x + 4} = x + 8 \qquad \text{Isolate the radical term.}$$
$$(3\sqrt{3x + 4})^2 = (x + 8)^2 \qquad \text{Square each side.}$$
$$9(3x + 4) = x^2 + 16x + 64$$
$$27x + 36 = x^2 + 16x + 64$$
$$x^2 - 11x + 28 = 0$$
$$(x - 7)(x - 4) = 0$$
$$x - 7 = 0 \quad | \quad x - 4 = 0$$
$$x = 7 \quad | \quad x = 4$$

The solution set of the equation is a subset of $\{4, 7\}$.

Check

$$3\sqrt{3x + 4} - 2 = x + 6 \qquad\qquad 3\sqrt{3x + 4} - 2 = x + 6$$
$$3\sqrt{3(4) + 4} - 2 \stackrel{?}{=} 4 + 6 \qquad 3\sqrt{3(7) + 4} - 2 \stackrel{?}{=} 7 + 6$$
$$3\sqrt{16} - 2 \stackrel{?}{=} 10 \qquad\qquad 3\sqrt{25} - 2 \stackrel{?}{=} 13$$
$$12 - 2 \stackrel{?}{=} 10 \qquad\qquad\qquad 15 - 2 \stackrel{?}{=} 13$$
$$10 = 10 \;\checkmark \qquad\qquad\qquad 13 = 13 \;\checkmark$$

Answer: The solution set is $\{4, 7\}$.

If an equation of three or more terms contains two radicals, the radicals can only be isolated one at a time. In the following example, it is necessary to square both sides of the equation twice, in order to write an equivalent equation without radicals.

Example 3 _____

Find the solution set of the equation:

$$\sqrt{2x + 11} = \sqrt{5x + 1} - 1$$

Notice:
One radical is already isolated on the left side.

Solution

$$(\sqrt{2x + 11})^2 = (\sqrt{5x + 1} - 1)^2 \qquad \text{Square each side.}$$

$$2x + 11 = 5x + 1 - 2\sqrt{5x + 1} + 1$$

$$2x + 11 = 5x - 2\sqrt{5x + 1} + 2$$

$$9 - 3x = -2\sqrt{5x + 1} \qquad \text{Isolate the remaining radical.}$$

$$(9 - 3x)^2 = (-2\sqrt{5x + 1})^2 \qquad \text{Square each side.}$$

$$81 - 54x + 9x^2 = 4(5x + 1)$$

$$81 - 54x + 9x^2 = 20x + 4$$

$$9x^2 - 74x + 77 = 0$$

$$(9x - 11)(x - 7) = 0$$

$$9x - 11 = 0 \quad \Big| \quad x - 7 = 0$$

$$x = \tfrac{11}{9} \quad \Big| \quad x = 7$$

The solution set is a subset of $\left\{\tfrac{11}{9}, 7\right\}$.

Check
$$\sqrt{2x + 11} = \sqrt{5x + 1} - 1 \qquad\qquad \sqrt{2x + 11} = \sqrt{5x + 1} - 1$$

$$\sqrt{2\left(\tfrac{11}{9}\right) + 11} \overset{?}{=} \sqrt{5\left(\tfrac{11}{9}\right) + 1} - 1 \qquad \sqrt{2(7) + 11} \overset{?}{=} \sqrt{5(7) + 1} - 1$$

$$\sqrt{\tfrac{22}{9} + 11} \overset{?}{=} \sqrt{\tfrac{55}{9} + 1} - 1 \qquad\qquad \sqrt{14 + 11} \overset{?}{=} \sqrt{35 + 1} - 1$$

$$\sqrt{\tfrac{121}{9}} \overset{?}{=} \sqrt{\tfrac{64}{9}} - 1 \qquad\qquad\qquad \sqrt{25} \overset{?}{=} \sqrt{36} - 1$$

$$\tfrac{11}{3} \overset{?}{=} \tfrac{8}{3} - 1 \qquad\qquad\qquad\qquad 5 \overset{?}{=} 6 - 1$$

$$\tfrac{11}{3} \neq \tfrac{5}{3} \qquad\qquad\qquad\qquad\qquad 5 = 5 \ \checkmark$$

Answer: The solution set is {7}.

Converting to Quadratic Form

Some equations of degree higher than 2 can be converted to quadratic form to accomplish a solution.

Example 4

Find the solution set: $x^4 - 2x^2 - 8 = 0$.

Solution

Although the given equation is of the fourth degree, it can be converted to quadratic form because only even powers are present.

$$x^4 - 2x^2 - 8 = 0$$
$$(x^2)^2 - 2x^2 - 8 = 0$$

Let $x^2 = r$. $\qquad r^2 - 2r - 8 = 0$

Solve the equation in r and use the results to obtain the solution set for the original equation in x.

$$r^2 - 2r - 8 = 0$$
$$(r - 4)(r + 2) = 0$$
$$r - 4 = 0 \mid r + 2 = 0$$
$$r = 4 \mid \quad r = -2$$

Replace r by x^2 and solve for x.

$$x^2 = 4 \quad \bigg| \quad x^2 = -2$$
$$x = \pm 2 \quad \bigg| \quad x = \pm i\sqrt{2}$$

Check Substituting each of the 4 values -2, 2, $-i\sqrt{2}$, and $i\sqrt{2}$ for x into the original equation leads to a true statement.

Answer: The solution set of $x^4 - 2x^2 - 8 = 0$ is $\{\pm 2, \pm i\sqrt{2}\}$.

Some radical equations can also be converted to quadratic form. The substitutions used in the conversions are more apparent if the equation is written with fractional exponents. Recall that $\sqrt[m]{x^n} = x^{\frac{n}{m}}$.

Example 5

Find the solution set for $2\sqrt[3]{x^2} - 5\sqrt[3]{x} + 2 = 0$.

Solution

First, rewrite the equation using fractional exponents.
$$2x^{\frac{2}{3}} - 5x^{\frac{1}{3}} + 2 = 0$$

This equation is quadratic in $x^{\frac{1}{3}}$. Let $r = x^{\frac{1}{3}}$, then $r^2 = x^{\frac{2}{3}}$ and the equation can be rewritten as $2r^2 - 5r + 2 = 0$.

Solve the equation in r.
$$2r^2 - 5r + 2 = 0$$
$$(2r - 1)(r - 2) = 0$$
$$2r - 1 = 0 \mid r - 2 = 0$$
$$r = \tfrac{1}{2} \mid \quad r = 2$$

Replace r by $x^{\frac{1}{3}}$ and solve for x.

$$x^{\frac{1}{3}} = \tfrac{1}{2} \qquad \bigg| \qquad x^{\frac{1}{3}} = 2$$
$$\left(x^{\frac{1}{3}}\right)^3 = \left(\tfrac{1}{2}\right)^3 \quad \bigg| \quad \left(x^{\frac{1}{3}}\right)^3 = 2^3$$
$$x = \tfrac{1}{8} \qquad \bigg| \qquad x = 8$$

Note: Since both sides of these equations were raised to a power, the solutions $x = \frac{1}{8}$ and $x = 8$ must be checked in the original equation. Both values do check.

Answer: The solution set for the original equation is $\left\{\frac{1}{8}, 8\right\}$.

1. $\{25\}$

2. $\{\ \}$

3. $\{16\}$

4. $\{25\}$

5. $\{5\}$

6. $\{3\}$

7. $\{32\}$

8. $\left\{-\frac{1}{27}\right\}$

9. $\{\pm 27\}$

10. $\{\ \}$

11. $\left\{\frac{2}{3}\right\}$

12. $\left\{\frac{4}{5}\right\}$

13. $\{-4, -2\}$

14. $\left\{\frac{4}{9}, \frac{1}{9}\right\}$

15. $\left\{\frac{1}{2}\right\}$

16. $\{5\}$

17. $\{3, 2\}$

18. $\{\ \}$

19. $\{2\}$

20. $\left\{-\frac{1}{12}, \frac{8}{3}\right\}$

21. $\{\ \}$

22. $\{8\}$

23. $\left\{\frac{4}{3}\right\}$

24. $\{\ \}$

25. $\{2\}$

26. $\{6\}$

27. $\{1\}$

28. $\{11, 7\}$

29. $\{4\}$

30. $\{0\}$

31. $\left\{\pm\sqrt{w^2 - v^2}\right\}$

32. $\left\{\pm 2i, \pm\sqrt{10}\right\}$

33. $\left\{\pm\frac{\sqrt{3}}{3}, \pm\sqrt{3}\right\}$

34. $\{3, 0\}$

35. $\{-1\}$

36. $\left\{\pm i\frac{\sqrt{3}}{3}, \pm 1\right\}$

37. $\{-10, 7, -2, -1\}$

Exercises 2.4

In **1 – 41**, write the solution set.

1. $5 + \sqrt{x} = 10$

2. $\sqrt{x} + 12 = 8$

3. $\sqrt{x} - 4 = 0$

4. $\sqrt[3]{x + 2} = 3$

5. $\sqrt{6x + 6} = 6$

6. $\sqrt{1 + 5x} + 8 = 12$

7. $2 + \sqrt[4]{\frac{x}{2}} = 4$

8. $\sqrt[3]{\frac{1}{x}} = -3$

9. $\sqrt[3]{x^2} - 3 = 6$

10. $2\sqrt{4x} + 3 = 0$

11. $\sqrt{3x} = \sqrt{2}$

12. $\sqrt{5x + 12} = \sqrt{10x + 8}$

13. $\sqrt{2x + 8} = x + 4$

14. $\sqrt{x} = x + \frac{2}{9}$

15. $x + \frac{1}{2} = \frac{\sqrt{2x} + 3}{4}$

16. $2\sqrt{4 + x} = x + 1$

17. $\sqrt{8x^2 - x - 5} = 3x - 1$

18. $\sqrt{3x} + 1 = -\sqrt{5x + 1}$

19. $\sqrt{4x - 4} + \sqrt{x + 7} = 5$

20. $3\sqrt{x^2 - 2x} = x + \frac{4}{3}$

21. $\sqrt{x - 5} = 1 - \sqrt{x}$

22. $2\sqrt[3]{(3x + 3)^3} = 54$

23. $\sqrt{x} + \sqrt{x - 1} = \frac{1}{\sqrt{x - 1}}$

24. $\sqrt{x + 1} - \frac{2}{\sqrt{x + 6}} = \sqrt{x + 6}$

25. $\sqrt{3x + 1} + \sqrt{x + 5} = 2\sqrt{7}$

26. $\sqrt{2x + 6} - \sqrt{\frac{x}{3}} = \sqrt{x + 2}$

27. $\sqrt{6x - 1} + \sqrt{2x + 3} = 2\sqrt{10x - 5}$

28. $\sqrt{\sqrt{36x + 4} - x} = 3$

29. $\sqrt{x + \sqrt{2x - 4}} = \sqrt{x + 2}$

30. Solve for x: $\sqrt{x + m} = \sqrt{m} + \sqrt{x}, m > 0$

31. Solve for x: $w = \sqrt{x^2 + v^2}$

32. $x^4 - 6x^2 - 40 = 0$

33. $3x^4 - 10x^2 + 3 = 0$

34. $(x + 1)^2 - 5(x + 1) + 4 = 0$

35. $x^{-2} + 2x^{-1} + 1 = 0$

36. $x^{-4} + 2x^{-2} - 3 = 0$

37. $(x^2 + 3x)^2 - 68(x^2 + 3x) - 140 = 0$

38. $(x^2 + x)^2 - 14(x^2 + x) + 24 = 0$

39. $\left(x - \frac{2}{x}\right)^2 - 8\left(x - \frac{2}{x}\right) + 15 = 0$

40. $2x^{\frac{2}{3}} - 3x^{\frac{1}{3}} - 2 = 0$

41. $3^{2x} - 10(3^x) = -9$

42. Find the 4 fourth roots of 16 by solving $x^4 = 16$.

43. The area of a right triangle is 60 in.² and the length of its hypotenuse is 17 in. Find the lengths of the legs.

44. If 12 is added to a number, the square root of the sum is one-quarter of the number. Find the number.

45. The area of a rectangle is $9\sqrt{2}$ cm². If the length of the diagonal is $3\sqrt{3}$ cm, find the dimensions of the rectangle.

46. The length of the longer leg of a right triangle is 1 m less than the length of the hypotenuse and the length of the shorter leg is 1 m more than one-fourth the length of the longer leg. Find the lengths of the sides of the triangle.

---

38. $\{-4, 3, -2, 1\}$

39. $\left\{\frac{5 \pm\sqrt{33}}{2}, \frac{3 \pm\sqrt{17}}{2}\right\}$

40. $\left\{-\frac{1}{8}, 8\right\}$

41. $\{2, 0\}$

42. $x = \pm 2, x = \pm 2i$

43. 8 in. and 15 in.

44. 24

45. 3 cm and $3\sqrt{2}$ cm

46. 7 m, 24 m, 25 m

Exercises 2.4 (continued)

47. The area of an equilateral triangle is $3\sqrt{2}$ in.^{2}. If the length of one side is $(\sqrt[4]{x+2})$ in., find x. (Use the formula $A = s^2 \frac{\sqrt{3}}{4}$.)

48. The area of a rectangle is $(\sqrt{x+5})$ ft.^{2} and the length of the diagonal is $(\sqrt{x-18})$ ft. If one side of the rectangle is 3 ft., find the length of the other side.

49. If r, s, and t are real numbers greater than 0, prove that if $\sqrt{x+r} = t - \sqrt{x+s}$ has a real solution for x, then $t^2 + s \geq r$.

50. The formula for the surface area of a sphere is $s = 4\pi r^2$. If a spherical balloon is being filled with air so that its surface area is increasing at a rate of (6π) cm^2/min. and the current radius is $\left(\frac{1}{2}\sqrt{10}\right)$ cm, find the radius after 1 minute.

A graphing calculator or a computer program designed to graph functions entered by the user can often help to solve an equation. For example, the solution to the equation $\sqrt{x^2+7} = 10 - \sqrt{39-x}$ can be found by drawing the graphs of $y = \sqrt{x^2+7}$ and $y = 10 - \sqrt{39-x}$ and finding the x-coordinate of the point of intersection.

Begin by selecting the range of values for x and y. Enter a large range in order to determine the part of the graph at which the point or points of intersection are located. For example, enter values of x and y from -20 to 20, a scale of 1 for both x and y, and a resolution of 1.

Now enter the equations as y_1 and y_2.

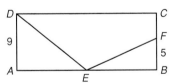

Note that there are two points at which the graphs intersect. Move the cursor to the intersection at the right, press ZOOM and select **2:Zoom In**. The section about the point of intersection will be enlarged. Press TRACE and then move the cursor as close as possible to the intersection. The values are $x = 3$, $y = 4$. Repeat the use of ZOOM and TRACE to find the coordinates of the other point of intersection. For this point, the values, to the nearest tenth, are $x = -2.4$, $y = 3.6$. The two x values, 3 and -2.4, are the solutions of the equation.

51. E is the midpoint of side $\overline{AB}$ of rectangle $ABCD$ and F is a point on side $\overline{BC}$. If $AD = 9$, $BF = 5$, and $DE + EF = 28$, find AB.

52. Use a graph to find, to the nearest tenth, the length of $\overline{AB}$ in Exercise 51 if $DE + EF = 33$.

47. 94 in.

48. 2 ft.

49. Square both sides.

$$x + r = t^2 - 2t\sqrt{x+s} + x + s$$
$$2t\sqrt{x+s} = t^2 + s - r$$

On the left side, $t > 0$ and $x + s \geq 0$.

$\therefore t^2 + s - r \geq 0$ and $t^2 + s \geq r$.

50. 2 cm

51. 24

52. 29.7

2.5 Solving Equations of Degree Greater Than 2

A *polynomial equation* is of the form $P(x) = 0$, where $P(x)$ is a polynomial in x. The *degree* of the polynomial is the highest power of the variable that appears in the polynomial. Generally, the higher the degree of the polynomial in a polynomial equation, the more difficult it is to solve. Techniques for solving first- and second-degree polynomial equations have already been discussed. You have also seen solutions to some higher-degree equations, namely those that can be converted to quadratic form. You will now learn some new techniques and, in some cases, combine these with techniques presented earlier.

Solving Higher-Degree Equations by Factoring

The principle of the complex number system that is used to solve quadratic equations by factoring can be extended. If $a_1, a_2, a_3, \ldots a_n$ are complex numbers and $a_1 \cdot a_2 \cdot a_3 \cdots \cdot a_n = 0$, then at least one of the a's must be 0.

Example 1 _____

Solve $x^3 + x^2 - 6x = 0$.

Solution

In this equation of degree 3, x is a factor of each term.

$$x^3 + x^2 - 6x = 0$$
$$x(x^2 + x - 6) = 0 \qquad \text{Factor out the common monomial } x.$$
$$x\,(x + 3)\,(x - 2) = 0 \qquad \text{Factor the trinomial.}$$
$$x = 0 \,\bigg|\, x + 3 = 0 \,\bigg|\, x - 2 = 0 \qquad \text{Set each factor equal to 0, and solve.}$$
$$ x = -3 \,\bigg|\, x = 2$$

Check Substituting each of the 3 values into the original equation leads to a true statement.

Answer: The solution set is $\{-3, 0, 2\}$.

The roots of the preceding equation $x^3 + x^2 - 6x = 0$ can be found from the graph of the polynomial function $y = x^3 + x^2 - 6x$. Use a calculator to draw the graph of this polynomial function for values of x and of y from -10 to 10 with a scale factor of 1. First press RANGE to determine if these are the values already entered in the calculator. If not, change the given values.

(−) 10 ENTER 10 ENTER 1 ENTER

(−) 10 ENTER 10 ENTER 1 ENTER

1 ENTER

Now enter the polynomial to be graphed.

| Y= | X\|T | ^ | 3 | + | X\|T | ^ | 2 | − | 6 | X\|T | ENTER | GRAPH |

Observe that the graph crosses the x-axis at -3, 0, and 2, the roots of the equation.

Graph each of the following polynomial functions one at a time and find the roots by locating the x-intercepts of the graph. (You may wish to adjust the range of y for a better display.) Then factor each polynomial and check that the roots observed from the graph correspond to the factors.

$$y_1 = x^3 - 2x^2 - 25x + 50 \qquad y_2 = x^3 + 4x^2 - 9x - 36$$
$$y_3 = x^3 - 5x^2 - 14x \qquad y_4 = x^3 - 2x^2 - 11x + 12$$

You should find the following:

	roots	factors
y_1	$-5, 2, 5$	$(x + 5)(x - 2)(x - 5)$
y_2	$-4, -3, 3$	$(x + 4)(x + 3)(x - 3)$
y_3	$-2, 0, 7$	$x(x + 2)(x - 7)$
y_4	$-3, 1, 4$	$(x + 3)(x - 1)(x - 4)$

Example 2

Solve $x^3 - 2x^2 - 2x = 0$.

Solution

$$x^3 - 2x^2 - 2x = 0$$
$$x(x^2 - 2x - 2) = 0 \qquad \text{Factor out the common monomial } x.$$

Since these are the only factors over the set of integers, set each factor equal to 0, and use the quadratic formula to solve the quadratic equation.

$$x = 0 \qquad x^2 - 2x - 2 = 0$$

$$x = \frac{-b \pm \sqrt{b^2 - 4ac}}{2a}$$
$$x = \frac{-(-2) \pm \sqrt{(-2)^2 - 4(1)(-2)}}{2(1)} \qquad a = 1, b = -2, c = -2$$
$$x = \frac{2 \pm \sqrt{12}}{2}$$
$$x = \frac{2 \pm 2\sqrt{3}}{2}$$
$$x = 1 \pm \sqrt{3}$$

Check Substituting each of the 3 values, 0, $1 + \sqrt{3}$, and $1 - \sqrt{3}$, into the original equation leads to a true statement.

Answer: The solution set is $\{0, 1 \pm \sqrt{3}\}$.

Use a graphing calculator or computer program to display the graph of $y = x^3 - 2x^2 - 2x$. Compare the roots found in Example 2 with the x-coordinates of the points at which the graph crosses the x-axis. If you are using the interval from -10 to 10 for the range as you did in Example 1, it may be necessary to use ZOOM to get a more detailed graph.

Enter: `ZOOM` 2

Move the cursor to 1 on the x-axis and press `ENTER`. Observe, on the enlarged graph, that the graph crosses the x-axis between -1 and 0, at 0, and between 2 and 3. Using `TRACE`, you can determine the nonzero values to be about $-.7$ and 2.7, which are the approximate values of $1 - \sqrt{3}$ and $1 + \sqrt{3}$.

The relationship between the roots of a polynomial equation and the intersections of the graph of the corresponding polynomial function with the x-axis will be used to approximate irrational roots of higher-degree equations that cannot be found by factoring or by the quadratic formula.

The Remainder and Factor Theorems

To say that a polynomial is factorable means that division by a factor results in a remainder of 0. Divisors other than factors produce remainders other than 0. For example, since the polynomial $x^2 + 3x - 10$ has the unique factors $(x + 5)(x - 2)$ over the set of integers, division by either $x + 5$ or $x - 2$ results in a remainder of 0, while division by any other polynomial results in a remainder other than 0.

$$
\begin{array}{r}
x - 2 \\
x + 5 \overline{)x^2 + 3x - 10} \\
\underline{x^2 + 5x} \\
-2x - 10 \\
\underline{-2x - 10} \\
0
\end{array}
\qquad
\begin{array}{r}
x + 4 \\
x - 1 \overline{)x^2 + 3x - 10} \\
\underline{x^2 - x} \\
4x - 10 \\
\underline{4x - 4} \\
-6
\end{array}
$$

Since $x + 5$ is a factor of $P(x) = x^2 + 3x - 10$, the value of the polynomial when $x = -5$ is 0.

$$P(x) = x^2 + 3x - 10 = (x + 5)(x - 2)$$
$$P(-5) = (-5)^2 + 3(-5) - 10 = (-5 + 5)(-5 - 2) = 0(-7) = 0$$

Dividing $P(x)$ by $x - 1$, which is not a factor, results in a remainder of -6. Compare the remainder with $P(1)$.

$$P(x) = x^2 + 3x - 10 = (x - 1)(x + 4) + (-6)$$
$$P(1) = 1^2 + 3(1) - 10 = (1 - 1)(1 + 4) + (-6) = 0(5) + (-6) = -6$$

Notice that the remainder when $P(x)$ is divided by $x - 1$ is equal to $P(1)$. This example leads to the following theorem and its proof.

☐ **Remainder Theorem:** *If a polynomial $P(x)$ is divided by $x - a$, the remainder is $P(a)$.*

Proof

Using $P(x)$ for the polynomial, $Q(x)$ for the quotient, and R for the remainder,

$$P(x) = (x - a) \cdot Q(x) + R$$

If $x = a$:
$$P(a) = (a - a) \cdot Q(a) + R$$
$$P(a) = 0 \cdot Q(a) + R$$
$$P(a) = R$$

Example 3

Find the remainder when $3x^5 + 2x^3 + x^2 - 1$ is divided by $x - 2$.

Solution

$$P(x) = 3x^5 + 2x^3 + x^2 - 1$$

By the Remainder Theorem, the remainder when $P(x)$ is divided by $x - 2$ is $P(2)$. Therefore, find $P(2)$.

$$P(2) = 3(2)^5 + 2(2)^3 + (2)^2 - 1$$
$$P(2) = 3(32) + 2(8) + 4 - 1$$
$$P(2) = 96 + 16 + 4 - 1$$
$$P(2) = 115$$

Answer: The remainder is 115.

Example 4

Find the remainder when $x^3 - 1$ is divided by

a. $x - \sqrt{2}$ **b.** $x - 1$

Solution

a. Use the Remainder Theorem to find $P(\sqrt{2})$.

$$P(\sqrt{2}) = (\sqrt{2})^3 - 1$$
$$= 2\sqrt{2} - 1$$

Answer: The remainder is $2\sqrt{2} - 1$.

b. Use the Remainder Theorem to find $P(1)$.

$$P(1) = 1^3 - 1$$
$$= 0$$

Answer: The remainder is 0.

This last result, a remainder of 0, means that $x - 1$ is a factor of $x^3 - 1$. Therefore, this factor can be used to solve the equation $x^3 - 1 = 0$.

$$x^3 - 1 = 0$$
$$(x - 1) \cdot Q(x) = 0$$
$$x - 1 = 0 \;\Big|\; Q(x) = 0$$
$$x = 1$$

Since $Q(x) = 0$, there may also be some other value of x that is a solution of $x^3 - 1 = 0$. This consequence of the Remainder Theorem can be generalized by stating a corollary of the Remainder Theorem, known as the Factor Theorem.

☐ **Factor Theorem:** *$(x - a)$ is a factor of the polynomial $P(x)$ if and only if $P(a) = 0$.*

To prove the theorem, two statements must be proven. First, if $x - a$ is a factor of $P(x)$, then $P(a) = 0$, and second, if $P(a) = 0$, then $x - a$ is a factor of $P(x)$.

1. $s = 1, s = 2, s = 0$

2. $x = -1, x = 2, x = 4$

3. $x = 0, x = \dfrac{9}{4}, x = -2$

4. $y = 0, y = -\dfrac{1}{2}, y = -1$

5. $t = 0, t = 10, t = -6$

6. $x = 0, x = -2,$
$x = -2, x = 1$

7. $x = \dfrac{3}{2}, x = \dfrac{3}{2},$
$x = -\dfrac{1}{4}, x = 0$

8. $x = 0, x = 1, x = 1$

9. $x = 0, x = 1 \pm i$

10. $x = 0, x = \dfrac{9 \pm \sqrt{77}}{2}$

11. $x = 0, x = \pm 3\sqrt{2}$

12. $x = 1, x = 2$

13. $x = -2, x = -\dfrac{5}{3}$

14. $x = \dfrac{3}{2}, x = \dfrac{1}{2}, x = 1$

15. $x = b, x = 0, x = \dfrac{2b}{3}$

16. $x = -a,$
$x = \dfrac{-a \pm \sqrt{a^2 + 4}}{2}$

Proof

1. If $x - a$ is a factor of $P(x)$,
  then $P(x) = (x - a) \cdot Q(x)$
  and $P(a) = (a - a) \cdot Q(a)$
  $P(a) = 0 \cdot Q(a)$
  $P(a) = 0$

2. If $P(a) = 0$,
  then $P(x) = (x - a) \cdot Q(x) + R$
  and $P(a) = (a - a) \cdot Q(a) + R$
  $0 = 0 \cdot Q(a) + R$
  $0 = R$

  Therefore, $P(x) = (x - a)Q(x)$,
  or $x - a$ is a factor of $P(x)$.

■ **Definition:** *If $P(x)$ is a polynomial and if $P(a) = 0$, then a is a root, also called a zero, of the polynomial.*

Example 5

Use the Factor Theorem to determine whether each binomial is a factor of $2x^3 - x^2 - 7x + 2$.

a. $x - 2$

b. $x + 3$

Solution

a. For $x - 2$ to be a factor, $P(2)$ must be 0.

$P(x) = 2x^3 - x^2 - 7x + 2$
$P(2) = 2(2)^3 - (2)^2 - 7(2) + 2$
$P(2) = 2(8) - 4 - 14 + 2$
$P(2) = 16 - 4 - 14 + 2$
$P(2) = 0$

Since $P(2) = 0$, $x - 2$ is a factor of $2x^3 - x^2 - 7x + 2$.

b. Since $x + 3 = x - (-3)$, $x + 3$ is a factor only if $P(-3) = 0$.

$P(x) = 2x^3 - x^2 - 7x + 2$
$P(-3) = 2(-3)^3 - (-3)^2 - 7(-3) + 2$
$P(-3) = 2(-27) - (9) + 21 + 2$
$P(-3) = -54 - 9 + 21 + 2$
$P(-3) = -40$

Since $P(-3) \neq 0$, $x + 3$ is not a factor of $2x^3 - x^2 - 7x + 2$.

Exercises 2.5

In **1 – 16**, find all of the solutions. *Note: a and b are constants.*

1. $s(s - 1)(s - 2) = 0$

2. $(x + 1)(x - 2)(x - 4) = 0$

3. $4x^3 - x^2 - 18x = 0$

4. $2y^3 + 3y^2 + y = 0$

5. $3t(t^2 - 4t - 60) = 0$

6. $4x(x + 2)^2(x - 1) = 0$

7. $(2x - 3)^2(4x + 1)5x = 0$

8. $x^3 - 2x^2 + x = 0$

9. $2x^3 - 4x^2 + 4x = 0$

10. $x^3 - 9x^2 + x = 0$

11. $x^3 - 18x = 0$

12. $2x(x - 1) = 4(x - 1)$

13. $3(x + 2)^2 = x + 2$ *Hint:* Rewrite as $(x + 2)[3(x + 2) - 1] = 0$.

14. $x(2x - 3)^2 = (3 - 2x)$

15. $2ax(x - b)^2 = ax^2(b - x)$

16. $x(x^2 + 2ax + a^2) = (x + a)$

Exercises 2.5 (continued)

In **17 – 24**, find the remainder when the given polynomial $P(x)$ is divided by the given divisor. If the remainder is 0, state one root of the equation $P(x) = 0$.

$P(x)$	Divisor		$P(x)$	Divisor
17. $x^2 + 7x - 1$	$x - 4$	**18.**	$x^4 - 16$	$x - 2$
19. $x^3 + 1$	$x + 1$	**20.**	$x^4 - 16$	$x + 2$
21. $x^2 + 2x - 8$	$x + 4$	**22.**	$2x^3 - 3x^2 + x - 1$	$x + 2$
23. $x^{10} + 2x^5 + 3x^2$	$x + 1$	**24.**	$x^5 - x^3 + 2x^2 - 32$	$x - 2$

In **25 – 26**, find the value of k if the first polynomial is a factor of the second.

25. $x - 1$; $2x^3 - k$ 　　　　　　　**26.** $x - 4$; $x^2 - kx - 12$

In **27 – 28**, find $P(x)$ under the given conditions.

27. divisor is $x - 7$, quotient is $x + 2$, remainder is 8

28. divisor is $x + 2$, quotient is $x^2 - x + 1$, remainder is -7

29. If $x + 1$ divides $x^{10} - k$, find k.

30. If $x - a$ divides $x^{10} - k$, find k in terms of a.

31. a. Demonstrate that $x - a$ is always a factor of $x^n - a^n$.

　　b. For what values of n is $x + a$ a factor of $x^n - a^n$?

 In **32 – 33**, find the remainder for the given division.

32. $7x^4 + 6x^3 - 12x^2 + 1$ is divided by $x - 8$

33. $x^3 - 2.52x^2 + 1.878x - 0.434$ is divided by $x - 0.62$

34. A wholesaler, who sends shipments of assorted merchandise to retailers in cartons that are in the shape of a cube, plans to use smaller cartons that will be easier to handle. One proposed carton is 4 feet high, with length and width each 2 feet shorter than the present carton. Another proposed carton has the same length as the present carton, a width that is 2 feet shorter and a height that is 5 feet shorter than the present carton. If the two proposed cartons have equal volume, what are the dimensions of the carton presently in use?

Answers 2.5

17. R = 43

18. 2

19. –1

20. –2

21. – 4

22. R = –31

23. R = 2

24. 2

25. $k = 2$

26. $k = 1$

27. $P(x) = x^2 - 5x - 6$

28. $P(x) = x^3 + x^2 - x - 5$

29. $k = 1$

30. $k = a^{10}$

31. $P(x) = x^n - a^n$

　a. $P(a) = a^n - a^n = 0$

　　Since $P(a) = 0$, $x - a$ is a factor of $P(x)$.

　b. $P(-a) = (-a)^n - a^n$

　　　　$= a^n - a^n$ if n is even

　　$x + a$ is a factor if n is even.

32. R = 30,977

33. R = 0

34. 8 ft. × 8 ft. × 8 ft.

2.6 More About Polynomial Open Sentences

In the previous section, the solutions of a polynomial equation were found by factoring the polynomial. One way to make the search for factors easier is to look for values of a such that, when the polynomial is divided by $x - a$, the remainder is 0. This method is based on the converse of the Factor Theorem.

The division of $x^3 - 5x^2 + 6x + 7$ by $x - 4$ is shown at the right. Note how often the same number or the same power of the variable occurs. *Synthetic division* is a technique for writing the essential parts of this division without the repetition. The exponent of x in the quotient is always 1 less than the exponent of x in the same column in the dividend. The powers of x are not written since the power is indicated by the column in which its coefficient appears. Note how the coefficients appear in the synthetic division shown below.

$$
\begin{array}{r}
1x^2 - 1x \ + 2 \ \text{R15} \\
x - 4 \overline{)1x^3 - 5x^2 + 6x + 7} \\
\underline{1x^3 - 4x^2} \\
-1x^2 + 6x \\
\underline{-1x^2 + 4x} \\
2x + 7 \\
\underline{2x - 8} \\
+15
\end{array}
$$

1 −5 6 7 $\lfloor$4	*Step 1* Write just the coefficients of the terms of the dividend and the opposite of the constant term of the divisor. Bring down the leading coefficient.
1	

1 −5 6 7 $\lfloor$4 4 1 −1	*Step 2* Multiply the leading coefficient by the opposite of the constant term (1×4). Place the product under the second coefficient, −5, and add.

1 −5 6 7 $\lfloor$4 4 −4 1 −1 2	*Step 3* Multiply the sum from the previous step by the opposite of the constant term (-1×4). Place the product under the next coefficient, 6, and add.

1 −5 6 7 $\lfloor$4 4 −4 8 1 −1 2 15	*Step 4* Repeat the previous step until the last coefficient is used.

The string of numbers in the final line of the computation, 1 −1 2 15, are the coefficients of the quotient and the remainder. The degree of the quotient is one less than the degree of the dividend. The quotient and remainder are formed by writing a polynomial and remainder as follows: 1 −1 2 15 $\rightarrow x^2 - x + 2$ R15.

This is the same quotient and remainder that were obtained previously using long division. Compare the amount of writing necessary to complete the division using synthetic division and using long division.

$$
\begin{array}{rrrr|r}
1 & -5 & 6 & 7 & 4 \\
  & 4 & -4 & 8 & \\
\hline
1 & -1 & 2 & 15 &
\end{array}
$$

$$
\begin{array}{r}
1x^2 - 1x + 2 \ \text{R}15 \\
x - 4\overline{)\,1x^3 - 5x^2 + 6x + 7} \\
\underline{1x^3 - 4x^2} \\
-1x^2 + 6x \\
\underline{-1x^2 + 4x} \\
2x + 7 \\
\underline{2x - 8} \\
+15
\end{array}
$$

The remainder, 15, means that $P(4) = 15$ by the Remainder Theorem. Since $P(4) \neq 0$, $x - 4$ is *not* a factor of $x^3 - 5x^2 + 6x + 7$.

Example 1

Find the quotient and remainder when $2x^3 - x^2 + x - 3$ is divided by $x - 1$.

Solution

$$
\begin{array}{rrrr|r}
2 & -1 & 1 & -3 & 1
\end{array}
$$

List the coefficients and write 1 as the synthetic divisor.

$$
\begin{array}{rrrr|r}
2 & -1 & 1 & -3 & 1 \\
  & 2 & 1 & 2 & \\
\hline
2 & 1 & 2 & -1 &
\end{array}
$$

Perform the synthetic division.

The first three numbers of the last line are the coefficients of the polynomial quotient. The last number is the remainder.

Answer: The quotient is $2x^2 + x + 2$ and the remainder is -1.

Example 2

Use synthetic division to find $P(-4)$ when $P(x) = 2x^4 - 13x^2 + x - 7$.

Solution

Rewrite the polynomial as $2x^4 + 0x^3 - 13x^2 + x - 7$ so that the coefficient of each power of x is represented. Synthetic division can be used to find $P(-4)$, since the Remainder Theorem states that $P(-4)$ is the remainder when $P(x)$ is divided by $x + 4$.

$$
\begin{array}{rrrrr|r}
2 & 0 & -13 & 1 & -7 & -4 \\
  & -8 & 32 & -76 & 300 & \\
\hline
2 & -8 & 19 & -75 & 293 &
\end{array}
$$

List the coefficients, write -4 as the synthetic divisor, and perform the synthetic division.

Answer: $P(-4) = 293$

Example 3

a. Use synthetic division to determine whether $x - 1$ is a factor of $3x^3 + 5x^2 + 2x - 10$.

b. If $x - 1$ is a factor, state a root of the equation $3x^3 + 5x^2 + 2x - 10 = 0$.

Solution

a. Perform the synthetic division with $3x^3 + 5x^2 + 2x - 10$ as the dividend and $x - 1$ as the divisor.

$$
\begin{array}{rrrr|l}
3 & 5 & 2 & -10 & \underline{1} \\
 & 3 & 8 & 10 & \\
\hline
3 & 8 & 10 & 0 &
\end{array}
$$

Answer: Since the remainder is 0, $x - 1$ is a factor of $3x^3 + 5x^2 + 2x - 10$.

b. Since the remainder in part **a** is 0, 1 is a root of the equation $3x^3 + 5x^2 + 2x - 10 = 0$. You can check this by substituting 1 into the equation.

$$3(1)^3 + 5(1)^2 + 2(1) - 10 = 3 + 5 + 2 - 10 = 0$$

The Rational Root Theorem

Until now, the methods used for finding factors of the polynomial $P(x)$ in order to find roots of the equation $P(x) = 0$ have relied on trial and error. If a factor was found, then the resulting quotient was used to find the remaining roots of the equation. For the previous example, after determining that 1 was a root, the 2 remaining roots of the equation $3x^3 + 5x^2 + 2x - 10 = 0$ could have been found.

$$3x^3 + 5x^2 + 2x - 10 = 0$$
$$(x - 1)(3x^2 + 8x + 10) = 0$$

$$x - 1 = 0 \quad \bigg| \quad 3x^2 + 8x + 10 = 0$$

$$x = 1 \quad \bigg| \quad x = \frac{-8 \pm \sqrt{(8)^2 - 4(3)(10)}}{2(3)}$$

$$x = \frac{-8 \pm \sqrt{64 - 120}}{6}$$

$$x = \frac{-8 \pm \sqrt{-56}}{6}$$

$$x = \frac{-8 \pm 2i\sqrt{14}}{6}$$

$$x = \frac{-4 \pm i\sqrt{14}}{3}$$

Therefore, the solution set of $3x^3 + 5x^2 + 2x - 10 = 0$

is $\left\{1, \dfrac{-4 + i\sqrt{14}}{3}, \dfrac{-4 - i\sqrt{14}}{3}\right\}$.

Thus, the important first step in finding a solution of a higher-degree polynomial equation is to find a linear factor if one exists. A study of the factors of a polynomial can help find a linear factor.

Consider a polynomial equation whose coefficients are integers, such as

$$2x^3 - 7x^2 - 5x + 4 = 0.$$

If the rational number $\frac{a}{b}$ (a and b are relatively prime) is a root of the equation, then

$$2\left(\frac{a}{b}\right)^3 - 7\left(\frac{a}{b}\right)^2 - 5\left(\frac{a}{b}\right) + 4 = 0$$

$$2\left(\frac{a^3}{b^3}\right) - 7\left(\frac{a^2}{b^2}\right) - 5\left(\frac{a}{b}\right) + 4 = 0$$

$2a^3 - 7a^2b - 5ab^2 + 4b^3 = 0$ Multiply each side by b^3.

$-7a^2b - 5ab^2 + 4b^3 = -2a^3$ Add the inverse of any term on the left that does not have a factor of b, to each side of the equation.

$b(-7a^2 - 5ab + 4b^2) = -2a^3$ Factor out b from the terms on the left.

Since b is a factor of the left side, it must be a factor of $-2a^3$. Since a and b are relatively prime, b must be a factor of -2. Therefore, b must have one of these values: 1, –1, 2, or –2.

Similarly, if the terms in a are collected on the left side and a is factored out, the result is $a(2a^2 - 7ab - 5b^2) = -4b^3$, which indicates that a divides $-4b^3$. Since a and b are relatively prime, a must be a factor of -4. Therefore, a must have one of these values: 1, –1, 2, –2, 4, or –4. The following tree diagram shows the possible values for a rational root $\frac{a}{b}$ of the polynomial equation $2x^3 - 7x^2 - 5x + 4 = 0$.

a	b	$\dfrac{a}{b}$
±1	±1	$\pm\frac{1}{1} = \pm 1$
	±2	$\pm\frac{1}{2}$
±2	±1	$\pm\frac{2}{1} = \pm 2$
	±2	$\pm\frac{2}{2} = \pm 1$
±4	±1	$\pm\frac{4}{1} = \pm 4$
	±2	$\pm\frac{4}{2} = \pm 2$
$\dfrac{a}{b} \in \left\{ \pm\frac{1}{2},\ \pm 1,\ \pm 2,\ \pm 4 \right\}$		

If there are any rational roots of the equation, they will be in this set.

The following theorem can be proved using a similar argument.

☐ **Rational Root Theorem:** *Let $P(x)$ be a polynomial with integral coefficients and $\frac{a}{b}$ be a rational number with a and b relatively prime. If $\frac{a}{b}$ is a solution of $P(x) = 0$, then a divides the constant term of the polynomial $P(x)$ and b divides the leading coefficient of $P(x)$.*

This theorem can be used to find any rational roots that may exist for the equation $2x^3 - 7x^2 - 5x + 4 = 0$ previously discussed.

Using the set of 8 possible rational roots shown in the tree diagram, use synthetic division to determine if any are actual roots.

Try 1.

$$\begin{array}{rrrr|l} 2 & -7 & -5 & 4 & \underline{1} \\ & 2 & -5 & -10 & \\ \hline 2 & -5 & -10 & -6 & \end{array}$$

Since the remainder is -6, $x = 1$ is not a root.

Try -1.

$$\begin{array}{rrrr|l} 2 & -7 & -5 & 4 & \underline{-1} \\ & -2 & 9 & -4 & \\ \hline 2 & -9 & 4 & 0 & \end{array}$$

Since the remainder is 0, $x = -1$ is a root of the equation. The polynomial quotient, $2x^2 - 9x + 4$, is also a factor of $2x^3 - 7x^2 - 5x + 4$. This polynomial quotient can be used to form an equation of lower degree, called the *depressed equation*.

$$2x^2 - 9x + 4 = 0$$

To find the remaining roots of the original equation, you could continue to check the possible rational roots using synthetic division, or you could solve the depressed equation by factoring or by using the quadratic formula.

$$2x^2 - 9x + 4 = 0$$
$$(2x - 1)(x - 4) = 0$$
$$2x - 1 = 0 \ | \ x - 4 = 0$$
$$x = \tfrac{1}{2} \ \ \ \ \ \ \ x = 4$$

The solution set of $2x^3 - 7x^2 - 5x + 4 = 0$ is then $\left\{-1, \tfrac{1}{2}, 4\right\}$.

Note: Since they are also rational roots of the equation, synthetic division by $\tfrac{1}{2}$ or 4 would also have resulted in a remainder of 0.

The roots of the polynomial equation $2x^3 - 7x^2 - 5x + 4 = 0$ can be used to write the factors of the polynomial and, thus, to solve the inequality $2x^3 - 7x^2 - 5x + 4 > 0$. Because of the number of cases that are necessary to consider all possible combinations of signs for 3 or more factors, the use of the number line to display the signs of the factors is a simpler approach.

As shown above, the roots of $2x^3 - 7x^2 - 5x + 4 = 0$ are -1, $\tfrac{1}{2}$, and 4. Therefore, the factors of the polynomial are $(x + 1)\left(x - \tfrac{1}{2}\right)(x - 4)$. The product of three factors is positive when 2 or 0 factors are negative.

Therefore, the solution set of $2x^3 - 7x^2 - 5x + 4 > 0$ is $\left\{x \mid -1 < x < \tfrac{1}{2} \text{ or } x > 4\right\}$.

Example 4

For the equation $3x^3 + 4x^2 + 2x - 4 = 0$,

a. Find the set of possible rational roots.

b. Find the rational root(s) of the equation.

Solution

a. By the Rational Root Theorem, the possible rational roots are those such that the numerator divides –4 and the denominator divides 3.

A tree diagram helps when writing the set of possible rational roots.

Factors of –4	Factors of 3	Possible rational roots
	± 1	± 1
± 1	± 3	$\pm\frac{1}{3}$
	± 1	± 2
± 2	± 3	$\pm\frac{2}{3}$
	± 1	± 4
± 4	± 3	$\pm\frac{4}{3}$

b. By trying values from this set using synthetic division, it can be shown that one of these, $x = \frac{2}{3}$, is a root.

$$\begin{array}{r|rrrr} \frac{2}{3} & 3 & 4 & 2 & -4 \\ & & 2 & 4 & 4 \\ \hline & 3 & 6 & 6 & 0 \end{array}$$

Each of the remaining values in the set leaves a nonzero remainder. Since the depressed equation, $3x^2 + 6x + 6 = 0$, is a quadratic, it is only necessary to look at the discriminant to determine the nature of the remaining two roots. The discriminant is

$$b^2 - 4ac = 6^2 - 4(3)(6) = -36.$$

Since the discriminant is negative, the remaining two roots are imaginary.

Answer: $x = \frac{2}{3}$ is the only rational root of $3x^3 + 4x^2 + 2x - 4 = 0$.

Example 5

Show that the equation $x^4 - 6x + 3 = 0$ has no rational roots.

Solution

By the Rational Root Theorem, the set of possible rational roots is $\{\pm 1, \pm 3\}$.

Since the x^3 and x^2 terms do not appear in the equation, the coefficient of these terms is 0. Trying each possible root,

```
1  0  0  -6   3 |1          1  0  0  -6   3 |-1
      1  1   1  -5               -1  1  -1   7
   _____        _____
1  1  1  -5  -2             1 -1  1  -7  10


1  0  0  -6   3 |3          1  0   0   -6    3 |-3
      3  9  27  63              -3   9  -27   99
   _____        _____
1  3  9  21  66             1 -3   9  -33  102
```

Since each possible trial yields a nonzero remainder, the equation $x^4 - 6x + 3 = 0$ has no rational roots.

Example 6

Find all roots of the equation $9x^4 - 3x^3 - 14x^2 - 7x - 1 = 0$.

Solution

By the Rational Root Theorem, the set of possible rational roots is $\left\{\pm 1, \pm\frac{1}{3} \pm \frac{1}{9}\right\}$.

Try $-\frac{1}{3}$.

```
9  -3  -14  -7  -1  |-1/3
       -3    2   4    1
   _____
9  -6  -12  -3    0
```

Since the remainder is 0, $-\frac{1}{3}$ is a root.

Try $-\frac{1}{3}$ again.

```
9   -6  -12  -3  |-1/3
         -3    3   3
   _____
9   -9   -9    0
```

Using the depressed equation, $-\frac{1}{3}$ is a double root.

The use of synthetic division with the other possible rational roots shows that the depressed equation $9x^2 - 9x - 9 = 0$ has no rational roots. Irrational or complex roots can be found by using the quadratic formula.

$$9(x^2 - x - 1) = 0$$

$$x = \frac{-(-1) \pm \sqrt{(-1)^2 - 4(1)(-1)}}{2(1)}$$

$$= \frac{1 \pm \sqrt{5}}{2}$$

Answer: The roots are $-\frac{1}{3}, -\frac{1}{3}, \frac{1 + \sqrt{5}}{2}, \frac{1 - \sqrt{5}}{2}$.

A calculator can be used to evaluate a polynomial but it must be understood that if values of the variable that cannot be expressed exactly as a finite decimal are used, the results may not be exact. Evaluation can be done by substituting the value of the variable in the polynomial.

Example 7

Find $P\left(\frac{5}{6}\right)$ when $P(x) = 6x^3 + x^2 - 11x + 5$.

Solution

Enter:

Display: `1.8  -09`

The value, 1.8×10^{-9} or 0.0000000018, is the error introduced when the approximation of $\frac{5}{6}$ is used to evaluate the expression. Some calculators will display 0.

The polynomial can also be evaluated by using the sequence of operations used in synthetic division.

Enter:

Display: `0.`

In this case, the order in which the operations are performed has eliminated any round-off error. The numbers in the display after entering ═ each time can be recorded to obtain the coefficients of the depressed equation.

Exercises 2.6

In **1 – 6**, find the quotient and the remainder if $P(x)$ is divided by $x - a$.

	$P(x)$	$x - a$		$P(x)$	$x - a$
1.	$3x^3 + 2x^2 - 6x - 20$	$x - 2$	**2.**	$2x^3 - x^2 + x - 3$	$x - 1$
3.	$x^4 - 2x^3 + 5x + 2$	$x + 1$	**4.**	$3x^4 - x^2 + x - 1$	$x - 2$
5.	$4x^2 - x^3 + 1$	$x - 4$	**6.**	$x + x^3 - 2x^5$	$x - 2$

In **7 – 13**, use the Remainder Theorem.

7. Find $P(-3)$ if $P(x) = x^3 + x^2 - 4x + 6$.

8. Find $P(-5)$ if $P(x) = x^3 + 3x^2 - 18x - 40$.

9. Find $P\left(\frac{1}{2}\right)$ if $P(x) = 4x^2 + 12x^3 - 1 - 3x$.

10. Find $P\left(\frac{3}{4}\right)$ if $P(x) = 4x^4 - 3x^3 + 8x - 2$.

11. If $P(x) = x^3 - 7x^2 + 36$ and $P(3) = 0$, find one quadratic factor of $x^3 - 7x^2 + 36$.

12. If $x - 2$ is a factor of $x^3 + 6x^2 + kx - 14$, find k.

13. If $P(x)$ is divided by $x - 2$, the quotient is $x^2 - 6x - 10$ and the remainder is -5. Find $P(x)$.

In **14 – 26**, by synthetic division, determine if the second polynomial is a factor of the first polynomial, $P(x)$. If it is, state a root of $P(x) = 0$; otherwise, state the remainder.

14. $x^3 + 3x^2 - 18x + 38;\ x - 4$

15. $x^3 - 2x^2 + 2x - 15;\ x - 3$

16. $4x^3 + 3x^2 + x - 1;\ x + 1$

17. $x^4 + 2x^2 - 7x + 3;\ x - 3$

18. $x^4 - 3x^2 - 4;\ x + 2$

19. $x^4 + 3x^3 + x;\ x + 2$

20. $4x^5 - x^3 + 10;\ x - \frac{1}{2}$

21. $3x^3 - 2x^2 - 18x + 12;\ x - \frac{2}{3}$

22. $x^2 - 2\sqrt{2}x + 2;\ x - \sqrt{2}$

23. $x^3 - \sqrt{3}x^2 - 2x + 2\sqrt{3};\ x - \sqrt{3}$

26. Yes. $x = -2$

27. $\{\pm 1, \pm 3\}$

28. $\{\pm 1, \pm 2, \pm 3, \pm 4, \pm 6, \pm 12\}$

29. $\left\{\pm 1, \pm\frac{1}{3}\right\}$

30. $\left\{\pm 1, \pm\frac{1}{7}\right\}$

31. $\left\{\pm 1, \pm 2, \pm 3, \pm 6, \pm\frac{1}{3}, \pm\frac{2}{3}\right\}$

32. $\left\{\pm 1, \pm\frac{1}{2}, \pm\frac{1}{4}, \pm\frac{1}{8}\right\}$

33. $x = 1, x = -1$

34. $x = \pm 1, x = \pm 3$

35. $-1 < x < \frac{1}{2}$ or $x > \frac{2}{3}$

36. $x = \frac{2}{3}$

37. $x < -1$ or $\frac{1}{5} < x < 1$

38. $-2 < x < \frac{1}{9}$ or $x > 1$

39. $x = \frac{1}{2}, x = -\frac{1}{2}$

40. $x < -\frac{1}{3}$ or $\frac{1}{3} < x < \frac{1}{2}$

41. $x = 0.2$

Answers 2.6

1. $3x^2 + 8x + 10$
$R = 0$

2. $2x^2 + x + 2$
$R = -1$

3. $x^3 - 3x^2 + 3x + 2$
$R = 0$

4. $3x^3 + 6x^2 + 11x + 23$
$R = 45$

5. $-x^2$
$R = 1$

6. $-2x^4 - 4x^3 - 7x^2 - 14x - 27$
$R = -54$

7. $P(-3) = 0$

8. $P(-5) = 0$

9. $P\left(\frac{1}{2}\right) = 0$

10. $P\left(\frac{3}{4}\right) = 4$

11. $x^2 - 4x - 12$

12. $k = -9$

13. $P(x) = x^3 - 8x^2 + 2x + 15$

14. No. $R = 78$

15. Yes. $x = 3$

16. No. $R = -3$

17. No. $R = 81$

18. Yes. $x = -2$

19. No. $R = -10$

20. No. $R = 10$

21. Yes. $x = \frac{2}{3}$

22. Yes. $x = \sqrt{2}$

23. Yes. $x = \sqrt{3}$

24. No. $R = 0.0048$

25. Yes. $x = 0.4$

42. If x is a rational root, then $x \in \{\pm1, \pm3\}$. For these values, $P(x) \neq 0$.

43. If x is a rational root, then $x \in \{\pm1, \pm2\}$. For these values, $P(x) \neq 0$.

44. $x = 1, x = 1 \pm \sqrt{2}$

45. $x = -2, x = \dfrac{2 \pm \sqrt{2}}{2}$

46. $x = \pm4$

47. $x = 0, x = 0, x = 2$

48. $x = \pm i\sqrt{2}$

49. R = 4

50. $k = -18$

51. $x^2 + (-1 + 2i)x$
$\qquad$ R = -2

52. Dividing gives R = 0.

53. Dividing gives R = 0.

54. Dividing gives R = 0.

55. $x^3 - 6x^2 + 11x - 6 = 0$

56. $x^4 - 5x^2 + 6 = 0$

57. $x^3 - 6x^2 + 7x + 4 = 0$

58. $y = \dfrac{1}{2}, y = \dfrac{9 - \sqrt{17}}{4}$

59. a.
$P(a^{-1}) = a^{-4} + 3a^{-3} +$
$\qquad 4a^{-2} + 3a^{-1} + 1,$
for $a \neq 0$
$a^4 \cdot P(a^{-1}) = 1 + 3a +$
$\qquad 4a^2 + 3a^3 + a^4$

The right side is $P(a)$;
since a is a root, $P(a) = 0$.
$a^4 \cdot P(a^{-1}) = 0$
$P(a^{-1}) = 0, \therefore a^{-1}$ is a root.

b. No, only when the coefficients form a symmetric pattern, as in part **a**: 1, 3, 4, 3, 1

c. $3x^2 - 10x + 3 = 0$
(Answers will vary.)

Exercises 2.6 *(continued)*

24. $x^3 - 0.74x^2 - 0.61x + 0.4514$; $x - 0.7$ **25.** $x^3 + 2.6x^2 - 0.6x - 0.24$; $x - 0.4$

26. $x^3 - (4i - 2)x^2 - 8ix$; $x + 2$ *(Note: $i = \sqrt{-1}$)*

In **27 – 32**, write the set of possible rational solutions.

27. $x^3 - 2x^2 - 2x + 3 = 0$ **28.** $x^3 - x^2 + 3x - 12 = 0$ **29.** $3x^3 - x^2 + 2x + 1 = 0$

30. $7x^3 - 6x^2 - 5x - 1 = 0$ **31.** $\frac{1}{2}x^3 - \frac{1}{3}x^2 + \frac{1}{6}x + 1 = 0$ **32.** $\frac{4}{5}x^3 + \frac{1}{2}x^2 + 2x + \frac{1}{10} = 0$

In **33 – 41**, solve the equation or inequality for rational solutions.

33. $x^3 - x^2 - x + 1 = 0$ **34.** $x^4 - 10x^2 + 9 = 0$ **35.** $6x^3 - x^2 - 5x + 2 > 0$

36. $3x^3 - 2x^2 + 3x - 2 = 0$ **37.** $5x^3 - x^2 - 5x + 1 < 0$ **38.** $9x^3 + 8x^2 - 19x + 2 > 0$

39. $8x^4 - 6x^2 + 1 = 0$ **40.** $x^3 - \frac{1}{2}x^2 - \frac{1}{9}x + \frac{1}{18} < 0$ **41.** $x^3 - 0.2x^2 - 3x + 0.6 = 0$

In **42 – 43**, show that the equation has no rational roots.

42. $x^3 + x^2 - 3 = 0$ $\qquad\qquad$ **43.** $x^4 + 3x^2 + 2 = 0$

In **44 – 45**, find all the roots of the equation.

44. $x^3 - 3x^2 + x + 1 = 0$ $\qquad\qquad$ **45.** $4x^3 - 14x + 4 = 0$

In **46 – 48**, if the indicated values are roots, find the remaining roots.

46. $x = 2$ is a root of $x^3 - 2x^2 - 16x + 32 = 0$ **47.** $x = \frac{1}{2}$ is a root of $2x^4 - 5x^3 + 2x^2 = 0$

48. $x = 3$ is a double root of $x^4 - 6x^3 + 11x^2 - 12x + 18 = 0$

49. Find the remainder when $x^{100} + 4x^{50} - 1$ is divided by $x + 1$. **50.** Find the value of k that makes $x - 3$ a factor of $x^3 + x^2 - 6x + k$.

51. Find the quotient and remainder when $x^3 + (-1 + i)x^2 + (2 + i)x - 2$ is divided by $x - i$. ($i = \sqrt{-1}$)

In **52 – 54**, show that the first polynomial is a factor of the second.

52. $x - 2i$; $x^3 - 2ix^2 + x - 2i$ **53.** $x^2 + 9$; $x^3 - x^2 + 9x - 9$ **54.** $x - 3a^2$; $x^4 - 81a^8$

In **55 – 57**, form a polynomial equation with integral coefficients that has the given numbers as roots.

55. 1, 2, 3 **56.** $\sqrt{2}, -\sqrt{2}, \sqrt{3}, -\sqrt{3}$ **57.** $1 + \sqrt{2}, 1 - \sqrt{2}, 4$

58. A box with an open top is constructed from a square piece of cardboard that is 5 units on each side by cutting $y \times y$ squares from each corner. Find two values for y, if the volume of the box is 8 cubic units.

59. a. Show that if $a \neq 0$ is a root of $x^4 + 3x^3 + 4x^2 + 3x + 1 = 0$, then a^{-1} is also a root of the equation.
$\quad$**b.** Is this true of every polynomial equation?
$\quad$**c.** Write another polynomial equation for which it is true.

60. Show that if r_1, r_2, and r_3 are the roots of $a_3x^3 + a_2x^2 + a_1x + a_0 = 0$ where the a's are the coefficients of the polynomial ($a_3 \neq 0$), then:
$\quad$**a.** $r_1 + r_2 + r_3 = \dfrac{-a_2}{a_3}$ $\qquad$ **b.** $r_1 \cdot r_2 \cdot r_3 = \dfrac{-a_0}{a_3}$

---

60. If r_1, r_2, r_3 are roots, then $(x - r_1)(x - r_2)(x - r_3) = 0$.

Multiplying: $x^3 - r_1x^2 - r_2x^2 + r_1r_2x - r_3x^2 + r_1r_3x + r_2r_3x - r_1r_2r_3 = 0$

$x^3 - (r_1 + r_2 + r_3)x^2 + (r_1r_2 + r_1r_3 + r_2r_3)x - r_1r_2r_3 = 0$

Multiply both sides by a_3.

$a_3x^3 - a_3(r_1 + r_2 + r_3)x^2 + a_3(r_1r_2 + r_1r_3 + r_2r_3)x - a_3r_1r_2r_3 = 0$

Equate coefficients with those of like powers in $a_3x^3 + a_2x^2 + a_1x + a_0 = 0$.

a. $a_2 = -a_3(r_1 + r_2 + r_3)$ $\quad r_1 + r_2 + r_3 = -\dfrac{a_2}{a_3}$ **b.** $a_0 = -a_3r_1r_2r_3$ $\quad r_1r_2r_3 = -\dfrac{a_0}{a_3}$

2.7 Approximate Roots of Polynomial Equations

To gain a better understanding of the solutions of an equation, it is useful to study the graph of a related function. Recall that in one dimension, the equation $y = 4$ is represented by a point on a number line.

In the two-dimensional plane, the graph of $y = 4$ is a line. In order to emphasize the fact that this graph represents ordered pairs of values where x takes on all real number values but y is always 4, the function notation $f(x) = 4$ reinforces the idea that, given any value of x, a corresponding value of y, in this case 4, can be found.

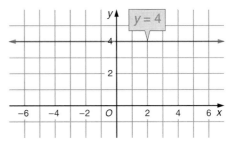

Consider the graph of the polynomial $P(x) = \frac{1}{2}x + 1$, which may be obtained from a table of ordered pairs (x, y) that are solutions of the equation $y = \frac{1}{2}x + 1$. Observe from the graph that the value of the function $P(x) = \frac{1}{2}x + 1$ is 0 when $x = -2$, the value at which the line intersects the x-axis. The values that make a polynomial equal to zero are called the ***zeros of the polynomial***.

x	y
2	2
4	3
-4	-1

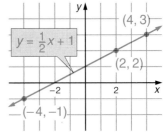

Since the coordinate plane consists only of real number pairs, the x-coordinates of the points that the graph of a function have in common with the x-axis indicate the *real* zeros of the function.

Consider the quadratic function

$$f(x) = x^2 - 2x - 3 \text{ or } y = x^2 - 2x - 3$$

To draw the graph of this function, which is a parabola, choose values of x on both sides of the axis of symmetry of the parabola

$$x = \frac{-b}{2a}$$

$$x = \frac{-(-2)}{2(1)} = 1$$

and find the corresponding values of y.

x	$x^2 - 2x - 3$	y
-2	$(-2)^2 - 2(-2) - 3$	5
-1	$(-1)^2 - 2(-1) - 3$	0
0	$(0)^2 - 2(0) - 3$	-3
1	$(1)^2 - 2(1) - 3$	-4
2	$(2)^2 - 2(2) - 3$	-3
3	$(3)^2 - 2(3) - 3$	0
4	$(4)^2 - 2(4) - 3$	5

After locating the points whose coordinates are given in the table, a smooth curve joining them is drawn. This may be done because the domain of the function is the set of all real numbers. This means that if $f(-2) = 5$ and $f(-1) = 0$, then for every value of x between -2 and -1 there is a corresponding value of y, and for every value of y between 5 and 0, there is a corresponding value of x.

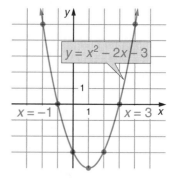

The real zeros of the function $f(x) = x^2 - 2x - 3$ are $x = -1$ and $x = 3$. They appear on the graph as the x-coordinates of the points at which the parabola crosses the x-axis. These are the same real zeros, or solutions, that are obtained when the polynomial equation $x^2 - 2x - 3 = 0$ is solved by factoring.

$$x^2 - 2x - 3 = 0$$
$$(x - 3)(x + 1) = 0$$
$$x - 3 = 0 \ | \ x + 1 = 0$$
$$x = 3 \ \ | \ \ x = -1$$

However, the graph of a quadratic function with imaginary zeros, such as $f(x) = x^2 + 2x + 2$, does not intersect the x-axis. There are no *real* zeros for this function.

x	$x^2 + 2x + 2$	y
-3	$(-3)^2 + 2(-3) + 2$	5
-2	$(-2)^2 + 2(-2) + 2$	2
-1	$(-1)^2 + 2(-1) + 2$	1
0	$(0)^2 + 2(0) + 2$	2
1	$(1)^2 + 2(1) + 2$	5

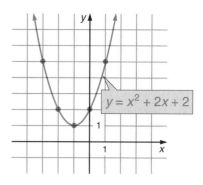

Use the quadratic formula to find the zeros of $x^2 + 2x + 2 = 0$.

$$x = \frac{-2 \pm \sqrt{(2)^2 - 4(1)(2)}}{2(1)}$$
$$= \frac{-2 \pm \sqrt{-4}}{2}$$
$$= \frac{-2 \pm 2i}{2}$$
$$= -1 \pm i$$

The Rational Root Theorem provides a set of possible rational solutions of a polynomial equation, and a trial-and-error process has been used for finding which, if any, of these are actual roots of the equation. In many cases, equations of degree greater than 2 do not have solutions that are rational numbers. The following theorem can be used in these cases to locate irrational roots.

☐ **Location Theorem:** *If the function $y = f(x)$ is a polynomial function such that $f(a)$ and $f(b)$ have opposite signs, then there is at least one real zero of the function between a and b.*

Graphically, the Location Theorem suggests that if the points $(a, f(a))$ and $(b, f(b))$ are on opposite sides of the x-axis, then there must be at least one point between a and b where the graph crosses the x-axis, that is, where $f(x) = 0$.

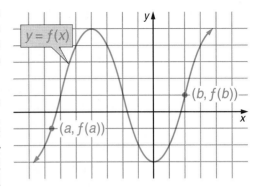

In the figure, $f(b)$ is positive and $f(a)$ is negative. Therefore, by the Location Theorem, the curve must intersect the x-axis in *at least* one point between a and b. In this particular case, there are three zeros between $f(a)$ and $f(b)$. Simply stated, the graph of a polynomial function is a *continuous curve*.

Example 1

Locate, between successive integers, a real zero for the function $f(x) = x^3 - 10x + 2$.

Solution

In order to find a sign change in the value of a function, synthetic division or substitution may be used for integral values of x. (A calculator is helpful when substitution is used.)

Use synthetic division to test values, summarizing the results in a chart.

$$
\begin{array}{rrrr|r}
1 & 0 & -10 & 2 & \underline{-3} \\
  & -3 & 9 & 3 & \\
\hline
1 & -3 & -1 & 5 &
\end{array}
\qquad
\begin{array}{rrrr|r}
1 & 0 & -10 & 2 & \underline{-2} \\
  & -2 & 4 & 12 & \\
\hline
1 & -2 & -6 & 14 &
\end{array}
\qquad
\begin{array}{rrrr|r}
1 & 0 & -10 & 2 & \underline{-1} \\
  & -1 & 1 & 9 & \\
\hline
1 & -1 & -9 & 11 &
\end{array}
$$

$$
\begin{array}{rrrr|r}
1 & 0 & -10 & 2 & \underline{0} \\
  & 0 & 0 & 0 & \\
\hline
1 & 0 & -10 & 2 &
\end{array}
\qquad
\begin{array}{rrrr|r}
1 & 0 & -10 & 2 & \underline{1} \\
  & 1 & 1 & -9 & \\
\hline
1 & 1 & -9 & -7 &
\end{array}
$$

The value of $f(0)$ is 2 and the value of $f(1)$ is –7. The sign of $f(x)$ changes from positive to negative. The Location Theorem insures that there is at least one real zero of the function between $x = 0$ and $x = 1$.

x	$f(x)$
–3	5
–2	14
–1	11
0	2
1	–7

root(s) — sign change

Note: The five trials by synthetic division can be abbreviated further by using a table that eliminates the second line in each division. Compare the following table to the five synthetic divisions.

The first row of the table, for $x = -3$, is obtained by following these steps:

Step 1 Bring down the 1.

Step 2 Multiply 1 by the x-value, -3; add the product to 0.

Step 3 Multiply -3 by the x-value, -3; add the product to -10.

Step 4 Multiply -1 by the x-value, -3; add the product to 2.

Trial Values	Coefficients			
x	1	0	−10	2
−3	1	−3	−1	5
−2	1	−2	−6	14
−1	1	−1	−9	11
0	1	0	−10	2
1	1	1	−9	−7

The steps are repeated for each line, using each x-value.

Example 2

Find a real zero of $f(x) = x^4 + x^3 - 2x^2 + 3x - 4$ to the nearest tenth.

Solution

Since $f(1) = -1$ and $f(2) = 18$, the function has a zero between $x = 1$ and $x = 2$. In order to approximate this zero to the nearest tenth, the values of the function are computed for intervals of one-tenth between 1 and 2. A calculator makes the computation easier, but synthetic division could also be used to find these values. It is only necessary to proceed until a change of sign in $f(x)$ occurs.

x	1	1	−2	3	−4
−2	1	−1	0	3	−10
−1	1	0	−2	5	−9
0	1	1	−2	3	−4
1	1	2	0	3	−1
2	1	3	4	11	18

Since there is a change of sign between $f(1.1)$ and $f(1.2)$, there is a root located between $x = 1.1$ and $x = 1.2$. To determine which value the root is closer to, test $f(1.15)$. Since the sign of $f(1.15)$ is positive, a sign change occurs between $f(1.1)$ and $f(1.15)$, and a real zero is closer to 1.1 than to 1.2.

x	$f(x)$
1.0	−1
1.1	−0.325
1.2	0.522

Answer: One approximate real zero, to the nearest tenth, is $x = 1.1$.

Descartes' Rule of Signs

In the previous two examples, only one of the real zeros of the function was located. However, there may be others. It would be useful to have a way of limiting the number of trials by knowing something about the number of real zeros a function may have. The following theorem, known as Descartes' Rule of Signs, places an upper limit on the number of positive and negative real zeros for a given function.

☐ **Descartes' Rule of Signs:** *The number of positive real zeros of a polynomial P(x), written in standard form, is equal to the number of sign changes in P(x) or the number of sign changes of P(x) less an even number. The number of negative real zeros of a polynomial P(x), written in standard form, is equal to the number of sign changes of P(−x) or the number of sign changes of P(−x) less an even number.*

Example 3

Determine the possible number of positive and negative real roots of $P(x) = 2x^5 - 4x^4 + 3x^2 + 5x - 7$.

Solution

Study $P(x)$ for changes in sign. There is a sign change from the first to the second term, from the second to the third term, and from the fourth to the fifth term. Thus, there are 3 changes in sign in $P(x)$. Write $P(-x)$ to study its sign changes.

$$P(-x) = 2(-x)^5 - 4(-x)^4 + 3(-x)^2 + 5(-x) - 7$$
$$= -2x^5 - 4x^4 + 3x^2 - 5x - 7$$

Note: $P(-x)$ can be formed by simply changing the signs of the co-efficients of the odd powers of x in $P(x)$.

There are sign changes between the second and third terms and between the third and fourth terms. Thus, there are 2 changes in sign in $P(-x)$.

Answer: By Descartes' Rule of Signs, the equation $P(x) = 0$ can have 3 or 1 positive real roots and 2 or 0 negative real roots.

Example 4

Determine the possible number of positive and negative real zeros for $P(x) = x^3 - 5x^2 - 3x - 7$.

Solution

$$P(x) = x^3 - 5x^2 - 3x - 7$$

$P(x)$ has one sign change, between the first and second terms. Therefore, there is 1 positive real zero.

$P(-x) = -x^3 - 5x^2 + 3x - 7$ has two sign changes, between the second and third terms and between the third and fourth terms. Therefore, there are 2 or 0 negative real zeros.

Answer: The equation $P(x) = 0$ can have 1 positive real root, and 2 or 0 negative real roots.

The Upper Bound and Lower Bound Theorems

Study the graph of this function. Notice that there is a zero at $x = 5$ and that no other positive zeros occur. For negative values of x, the function increases continuously, so there are no zeros for negative values of x. It would be useful to be able to determine a value of x, 5 in this case, beyond which no greater real zeros exist and a value of x below which no smaller real zeros exist.

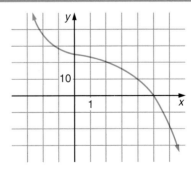

If you consider a polynomial function whose terms are all positive, such as $P(x) = x^3 + 3x^2 + 2x + 7$, then the values of $P(x)$ will be positive for any positive value of x. Therefore, there can be no positive real zeros for this function. The same conclusion can be reached using Descartes' Rule of Signs because there are no sign changes. This principle suggests the following theorem.

☐ **Upper Bound Theorem:** *For a positive number p, if P(x) is divided by x – p and the resulting remainder and the coefficients of the quotient are all positive or are all negative, then the zeros of P(x) have an upper bound at x = p.*

Example 5

Find an upper bound for the zeros of the polynomial function $P(x) = x^3 - 4x^2 - x + 4$.

Solution

Using positive integers for x, form a table of values using a calculator or synthetic division.

The depressed polynomial, when $P(x)$ is divided by $x - 5$, is $x^2 + x + 4$ and the remainder is 24. Since there are no sign changes, $x = 5$ is an upper bound for the real zeros of $P(x)$.

x	1	-4	-1	4
1	1	-3	-4	0
2	1	-2	-5	-6
3	1	-1	-4	-8
4	1	0	-1	0
5	1	1	4	24

The Upper Bound Theorem is based on the fact that for any odd or even power of x, the value of the power of x increases as x increases when x is positive. However, it is not true that the value of a power of x decreases as x decreases when x is negative. For example, if x decreases from -2 to -3, x^4 *increases* from 16 to 81.

One way to address this problem in establishing a lower bound for $y = P(x)$ is to reflect the graph of $y = P(x)$ over the y-axis. The image of any point (x, y) on the graph of $y = P(x)$ when reflected over the y-axis is $(-x, y)$. The reflection of the graph of $y = P(x)$ over the y-axis is the graph of $y = P(-x)$.

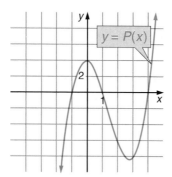
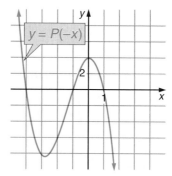

An upper bound of the zeros of $P(-x)$ is the additive inverse of the lower bound of the zeros of $P(x)$. This is summarized in the following theorem.

☐ **Lower Bound Theorem:** *If m is an upper bound for P(–x), then –m is a lower bound for P(x).*

Example 6

Find upper and lower bounds for the zeros of the polynomial function $P(x) = x^3 + 3x^2 - 5x - 13$.

Solution

Using positive integers for x, form a table of values of $P(x)$ using a calculator or synthetic division.

x	1	3	-5	-13
1	1	4	-1	-14
2	1	5	5	-3
3	1	6	13	26

Since there are no sign changes when $x = 3$, 3 is an upper bound for the zeros of $P(x)$. To find a lower bound, first form the polynomial function $P(-x)$.

$P(-x) = -x^3 + 3x^2 + 5x - 13$

Now, using positive integers for x, form a table of values of $P(-x)$ using a calculator or synthetic division.

x	-1	3	5	-13
1	-1	2	7	-6
2	-1	1	7	1
3	-1	0	5	2
4	-1	-1	1	-9
5	-1	-2	-5	-38

From the table, it can be seen that $x = 5$ is an upper bound of $P(-x)$. Therefore, $x = -5$ is a lower bound of $P(x)$.

A graphing calculator can be used to estimate the real zeros of a polynomial function. The Upper and Lower Bound Theorems give an appropriate range for the x-values needed to locate each zero.

For example, to locate the real zeros of the polynomial $P(x) = x^3 + 3x^2 - 5x - 13$ in Example 6, set the range of x from -5, a lower bound, to 3, an upper bound. Note the integers between which the roots are located. Then use TRACE to find approximate values of the roots to the nearest hundredth. Use ZOOM to enlarge the graph around the first root and determine its value. Return to the graph that shows all three roots and use ZOOM to enlarge the graph around the second root and find its value. Repeat the process for the third root.

You should have observed that the roots are between -4 and -3, between -2 and -1, and between 2 and 3. Approximate values of the roots are -3.32, -1.82, and 2.15.

For the following polynomial functions:

$$y_1 = x^3 - 2x^2 - 4x + 6 \qquad y_2 = x^3 - 2x^2 - 4x + 10$$
$$y_3 = x^4 - 2x^3 - 9x^2 + 6x + 18 \qquad y_4 = x^4 - 2x^3 - 9x^2 + 6x + 22$$

a. Determine upper and lower bounds.

b. Use these bounds as the maximum and minimum values of x to graph the functions.

c. Determine the real zeros of the functions to the nearest tenth.

d. Explain why y_2, a 3rd-degree polynomial, has only 1 point of intersection with the x-axis and y_4, a 4th-degree polynomial, has only 2 points of intersection with the x-axis.

Answers 2.7

1. between –3 and –2,
between 0 and 1

2. between –2 and –1,
between 2 and 3

3. at –1,
between 1 and 2,
between 3 and 4

4. between 1 and 2

5. between –2 and –1,
between –1 and 0,
between 1 and 2

6. between –1 and 0,
between 2 and 3

7. between –2 and –1,
between 1 and 2

8. between –2 and –1

9. $P(0)$ is positive

$P\left(\frac{1}{2}\right)$ is negative

$P(1)$ is positive

10. –1.6, 2.6

11. –2.7, 0.7

12. 2.4

13. 0.7

14. –0.8, 0.8

15. –0.7, 0.6

16. –2.2, 0.1

17. –0.6, 1.6

18. $2 < d < 8$

19. positive: 2 or 0
negative: 1

20. positive: 3 or 1
negative: 0

21. positive: 2 or 0
negative: 0

22. positive: 4 or 2 or 0
negative: 0

You should have found the following zeros:

y_1 : –1.87, 1.21, 2.66 y_2 : –2.12

y_3 : –2.33, –.72, 1.25, 3.81 y_4 : 1.91, 3.37

Since the coordinate plane represents ordered pairs of real numbers, the graph shows only real zeros. The functions y_2 and y_4 each have two complex zeros in addition to the real zeros shown on the graph.

Exercises 2.7

In **1 – 8**, locate a zero of the function between two consecutive integers.

1. $f(x) = x^2 + 2x - 2$ **2.** $f(x) = x^2 - x - 3$

3. $f(x) = x^3 - 4x^2 + 5$ **4.** $f(x) = x^3 - 3$

5. $f(x) = x^3 - 3x - 1$ **6.** $f(x) = x^4 - 3x^3 + 2x - 1$

7. $f(x) = 2x^4 - x - 4$ **8.** $f(x) = 4x^3 - 4x + 3$

9. Study the graph of $y = 15x^3 - x^2 - 11x + 3$ shown below.

The zeros located between $x = 0$ and $x = 1$ would not be shown by a sign change, since $P(0)$ and $P(1)$ are both positive. In this case a sign change would occur between $P(0)$ and $P\left(\frac{1}{2}\right)$, and between $P\left(\frac{1}{2}\right)$ and $P(1)$. Verify both of these sign changes.

In **10 – 17**, approximate all real zeros of $f(x)$ to the nearest tenth.

10. $f(x) = x^2 - x - 4$ **11.** $f(x) = x^2 + 2x - 2$

12. $f(x) = x^3 - 2x^2 - 2$ **13.** $f(x) = 2x^3 + 3x^2 - 2$

14. $f(x) = 2x^4 + 3x^2 - 3$ **15.** $f(x) = x^4 - 2x^3 + 2x^2 + x - 1$

16. $f(x) = 2x^4 + 3x^3 + 7x - 1$ **17.** $f(x) = x^4 - x^3 - x - 1$

18. For what values of d will $f(x) = x^3 - x^2 + 2x - d$ have at least one real zero between 1 and 2?

In **19 – 26**, find the possible number of positive real zeros and the possible number of negative real zeros for $P(x)$.

19. $P(x) = x^3 - 3x^2 + 2x + 1$ **20.** $P(x) = x^3 - x^2 + x - 1$

21. $P(x) = x^4 - 3x^3 - 2x + 1$ **22.** $P(x) = x^4 - x^3 + x^2 - x + 1$

23. $P(x) = x^4 + 2x^3 - x^2 + 3x - 1$ **24.** $P(x) = x^8 + 3x^5 + 2x^2 + 7$

25. $P(x) = x^7 + 2x^6 - 4x - 2$ **26.** $P(x) = x^{10} + 3x^7 - 17x^5 + 3x^2 + 1$

23. positive: 3 or 1
negative: 1

24. positive: 0
negative: 2 or 0

25. positive: 1
negative: 2 or 0

26. positive: 2 or 0
negative: 2 or 0

Exercises 2.7 *(continued)*

In **27 – 32**, find an integral upper bound and an integral lower bound
for the zeros of each polynomial function.

27. $P(x) = x^3 + 2x^2 - x + 7$

28. $P(x) = x^4 - 6x + 1$

29. $P(x) = 2x^3 - x^2 + 8x - 1$

30. $P(x) = x^4 + 2x^3 - 2x^2 + x + 3$

31. $P(x) = x^5 - 3x^4 + 1$

32. $P(x) = x^4 - 2x^3 - x^2 + 2x - 3$

33. For $P(x) = x^4 - 2x^3 - x^2 + x - 4$:

   a. Make a table of values of $P(x)$ for integral values of x from
      -3 to 3 using a synthetic division table, and sketch.

   b. Verify by Descartes' Rule of Signs that there is 1 positive
      real root and 1 negative real root. Locate these on the graph.

   c. Use the table of values to identify an upper bound and
      lower bound. Verify your result on your graph.

   d. Estimate each of the real zeros to the nearest tenth.

34. The Upper Bound Theorem may be proved by using the Remainder Theorem.
The beginning of the proof is outlined below. Complete the argument.

    By the Remainder Theorem, $P(x) = (x - b) \cdot Q(x) + P(b)$. By the
    hypothesis of the Upper Bound Theorem, the value of b is posi-
    tive and the values of $P(b)$ and the coefficients of $Q(x)$ are all
    positive or all negative. Then for any value of $x > b$, ...

---

27. upper bound 1
    lower bound -4

28. upper bound 2
    lower bound 0

29. upper bound 1
    lower bound 0

30. upper bound 1
    lower bound -3

31. upper bound 3
    lower bound -1

32. upper bound 2
    lower bound -2

---

33. a.

x	1	-2	-1	1	-4	
-3	1	-5	14	-41	119	
-2	1	-4	7	-13	22	sign change
-1	1	-3	2	-1	-3	
0	1	-2	-1	1	-4	
1	1	-1	-2	-1	-5	
2	1	0	-1	-1	-6	sign change
3	1	1	2	7	17	

b.

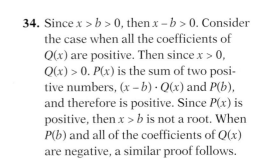

  c. upper bound 3
     lower bound -2

  d. -1.3, 2.5

34. Since $x > b > 0$, then $x - b > 0$. Consider
the case when all the coefficients of
$Q(x)$ are positive. Then since $x > 0$,
$Q(x) > 0$. $P(x)$ is the sum of two positive numbers, $(x - b) \cdot Q(x)$ and $P(b)$,
and therefore is positive. Since $P(x)$ is
positive, then $x > b$ is not a root. When
$P(b)$ and all of the coefficients of $Q(x)$
are negative, a similar proof follows.

Linear Equations and Inequalities

- An algebraic equation or inequality in one variable is a mathematical sentence stating that of two algebraic expressions the first is less than, equal to, or greater than the second. The domain of the variable in the sentence is the set of numbers for which both expressions are defined. Unless otherwise stated, the set of real numbers is the domain for all variables. A solution is a value of the variable that makes the sentence true.

- An identity is an equation that is true for all values of the variable in the domain. When an equation has no solution in the domain, the solution is the empty set, ø. Equivalent equations are equations with the same solution set.

- Equations and inequalities are solved by using properties of the real number system. The solution can be checked by substitution into the original open sentence. When evaluating each side of the sentence in a check, the order of operations is followed.

- To solve an equation containing rational expressions, multiply both sides of the equation by the least common denominator. To solve an inequality containing rational expressions, two cases must be considered: when the LCD is positive and when it is negative.

- Formulas are a form of literal equations. To solve a literal equation, the letter for which the equation is being solved is treated as variable and the other letters are treated as constants.

Algebraic Solutions of Problems

- The four steps for an algebraic solution are:

  Step 1   Use a variable to represent unknown quantities.

  Step 2   Use facts in the problem or familiar formulas to write an equation or inequality.

  Step 3   Solve the equation or inequality and use the solution to obtain the answer to the problem.

  Step 4   Check the answer with the conditions of the original problem. State the answer.

- A number of applications that can be solved by algebraic means make use of relationships having the form $A \times B = C$. These problems concern distances, things of monetary value, and investments and mixtures stated as percents.

Quadratic Equations

- Quadratic equations are of the form $ax^2 + bx + c = 0$ ($a \neq 0$), where a, b, and $c \in R$. There are three methods that can be used to solve a quadratic equation: factoring, completion of the square, and the quadratic formula.

- The discriminant, $b^2 - 4ac$, applicable only when a, b, and c are rational numbers, is used to determine the nature of the roots of a quadratic equation with rational coefficients. The roots are classified as real or imaginary. Real roots may be rational or irrational, and equal or unequal.

Radical Equations

- If each side of an equation is raised to a power, then the solution set of the original equation is a subset of the solution of the resulting equation. Those elements of the solution set of the resulting equation that are *not* roots of the given equation are called extraneous roots.

- To solve an equation containing a radical expression, isolate the radical expression and raise each side to a power that will eliminate the radical. Since extraneous roots may result, it is important to check each solution.

General Polynomial Equations and Inequalities

- The degree of a polynomial is the highest power of the variable.

- In some cases, higher-degree equations can be solved by converting them to equivalent equations in quadratic form.

- Some higher-degree equations can be solved by factoring.

- Remainder Theorem: If a polynomial $P(x)$ is divided by $x - a$, the remainder is $P(a)$.

- Factor Theorem: $(x - a)$ is a factor of the polynomial $P(x)$ if and only if $P(a) = 0$.

- Synthetic division is a shortcut for the algebraic long division of a polynomial by a linear expression.

- For a polynomial equation with integral coefficients, $P(x) = 0$, the Rational Root Theorem states that if $\frac{a}{b}$ is a solution, then a divides the constant term of the polynomial $P(x)$ and b divides the leading coefficient of $P(x)$.

- An inequality of the form $P(x) > 0$ or $P(x) < 0$ can be solved by considering the signs of the factors of $P(x)$.

- The real zeros of a function are the x-coordinates of the intersection points of the graph of the function with the x-axis.

- If a function $y = f(x)$ is a polynomial function such that $f(a)$ and $f(b)$ have opposite signs, then there is at least one real zero between a and b. A calculator or an abbreviated synthetic division table can be used to find a list of values of the function.

1. $\{3\}$

2. $\left\{-\frac{1}{3}\right\}$

3. $\{-2, 3\}$

4. $\{2\}$

5. $\{-1\}$

6. $c = \dfrac{ab}{2(b + a)}$

7. $r = \pm\dfrac{\sqrt{S\pi}}{2\pi}$

8. $\ell = \dfrac{2S - na}{n}$

9. $x = \dfrac{(d - b)y}{(a - c)}$

10. side = 5 cm

11. less than $70,000 @ 4%

12. **a.** $34\frac{2}{7}$ m.p.h.

    b. 60 m.p.h.

13. 4.32 sec.

14. $x = -\frac{1}{4}, x = \frac{3}{2}$

15. $x = 2 \pm \sqrt{2}$

16. **a.** $x = 3 \pm \sqrt{2}$

    b. $x = 2 \pm 3i$

    c. $x = \dfrac{-2 \pm \sqrt{10}}{6}$

    d. $x = \pm 4i$

17. **a.** real, rational, unequal

    b. real, rational, equal

    c. imaginary, unequal

    d. imaginary, unequal

- Descartes' Rule of Signs: The number of positive real zeros of a polynomial function $P(x)$ is equal to the number of sign changes of the coefficients of $P(x)$ or is an even number less than this. The number of negative real zeros of $P(x)$ is equal to the number of sign changes of the coefficients of $P(-x)$ or is an even number less than this.
- The Upper Bound Theorem: If $P(x)$ is divided by $x - p$, with p a positive number, and the resulting quotient and remainder have no sign changes, then the zeros of $P(x)$ have an upper bound at $x = p$.
- The Lower Bound Theorem: If m is an upper bound for $P(-x)$, then $-m$ is a lower bound for $P(x)$.

Chapter 2   Review Exercises

In **1 – 5**, find the solution set.

1. $7x - 4 = 11 + 2x$

2. $5(3x + 2) = 3(2 + x)$

3. $\dfrac{3}{4y - 2} = \dfrac{2}{3y + 1} + \dfrac{1}{10}$

4. $\dfrac{3}{4x^2 - 1} = \dfrac{x - 3}{2x^2 - 5x - 3}$

5. $\dfrac{6}{x - 2} + \dfrac{12}{3x + 9} = \dfrac{x + 1}{3(x + 3)(x - 2)}$

In **6 – 9**, solve the literal equation for the indicated variable.

6. $\dfrac{1}{c} = \dfrac{2}{a} + \dfrac{2}{b}$, for c

7. $S = 4\pi r^2$, for r

8. $S = \dfrac{n}{2}(a + \ell)$, for ℓ

9. $ax + by = cx + dy$, for x

10. In a given square, two opposite sides are increased to 4 cm less than twice their original length, and the other sides are increased to 1 cm more than three times the original length. If the resulting rectangle has an area of 96 cm², find the length of a side of the original square.

11. $100,000 is divided into two investments with respective annual returns of 4% and $7\frac{1}{2}$%. Find the amount invested at the lower rate if the total income at the end of the year is more than $5,050.

12. Mr. Smith goes to work and back by the same route.
    a. Find his average rate of speed for the entire trip if he averages 30 m.p.h. going and 40 m.p.h. returning.
    b. Assume Mr. Smith averages 30 m.p.h. going to work. What speed would he have to average going home if he wants to average 40 m.p.h. for the entire trip?

13. A pebble is dropped from the top of a cliff that is 300 ft. above the water below. How long will it take the pebble to hit the top of a buoy whose top protrudes 2 ft. out of the water? Answer to the nearest hundredth of a second. (Use the formula $s = -16t^2 + vt + h$.)

14. Solve by factoring: $8x^2 - 10x - 3 = 0$.

15. Solve by completing the square: $x^2 - 4x + 2 = 0$.

16. Solve by using the quadratic formula.
    a. $x^2 - 6x + 7 = 0$ **b.** $x^2 - 4x + 13 = 0$
    c. $3x^2 + 2x - \frac{1}{2} = 0$ **d.** $x^2 + 16 = 0$

17. Find the nature of the roots by using the discriminant.
    a. $x^2 - 7x = 0$ **b.** $x^2 + 6x + 9 = 0$
    c. $x^2 + 4x + 13 = 0$ **d.** $x^2 + 10 = 0$

18. A page with printing on it has an area of 88 sq. in. and the printing is surrounded by a uniform margin of 1 in. What are the total dimensions of the page if the length of the printed area is 50% more than its width?

In **19 – 23**, find the solution set.

19. $\sqrt{7x - 6} = 8$

20. $\sqrt{x + 1} = \frac{3\sqrt{5}}{5}$

21. $\sqrt{2x + 6} + \sqrt{x + 2} = 3$

22. $y^4 - 8y^2 - 20 = 0$

23. $(x^2 + 2x)^2 - 7(x^2 + 2x) - 8 = 0$

In **24 – 26**, find all the solutions.

24. $3x^3 - 2x^2 - 4x = 0$ **25.** $x^3 + 15x = 0$ **26.** the four 4th roots of 81

27. Find the remainder when $x^3 + 7x^2 + 5x - 8$ is divided by $x - 2$.

28. If $x - 2$ is a factor of $3x^3 + 2x^2 - k$, find the value of k.

29. Verify that -1 is a root of $x^{15} + 4x^7 + 5 = 0$.

30. Verify that if a is a root of $x^6 + 3x^4 + 6x^2 = 0$, then $-a$ is also a root.

31. Find the quotient and remainder using synthetic division.

    a. $4x^3 + 3x^2 - x + 2$ divided by $x - 1$.   **b.** $x^4 + 2x^2 - x + 3$ divided by $x + 2$.

32. Find $P(4)$ if $P(x) = x^3 - 2x^2 + x - 1$.

33. If $P(1) = 0$ for a polynomial equation $P(x) = 0$, then which of the following must be true?

    (A) x is a factor of $P(x)$.   (B) $x + 1$ is a factor of $P(x)$.
    (C) $x - 1$ is a factor of $P(x)$.   (D) -1 is a root of $P(x) = 0$.

34. Write the set of possible rational solutions for $3x^3 + 5x^2 - x + 8 = 0$.

35. Find the solution set: $2x^3 - 9x^2 - 6x + 5 < 0$

36. If 1 is a root of $x^3 - 7x^2 + 16x - 10 = 0$, find the remaining roots.

37. Form a polynomial equation with integral coefficients whose roots are $3 + 2i$, $3 - 2i$, and 7.

In **38 – 39**, locate a real zero of the function between two consecutive integers.

38. $f(x) = x^2 + x - 4$

39. $f(x) = x^3 - 2x^2 - x + 5$

40. Find, to the nearest tenth, the roots of $f(x) = 9x^3 - 9x^2 - 6x + 3$ between -2 and 2.

41. Find an upper bound for the zeros of $f(x) = x^3 + x^2 - 3x - 1$.

42. For $P(x) = 2x^4 + 6x^3 - 2x^2 + x + 1$, find the number of possible:
    a. positive real zeros
    b. negative real zeros

18. 8 in. × 11 in.

19. $\{10\}$

20. $\left\{\frac{4}{5}\right\}$

21. $\{-1\}$

22. $\{\pm\sqrt{10}, \pm i\sqrt{2}\}$

23. $\{-4, 2, -1\}$

24. $x = 0$, $x = \frac{1 \pm \sqrt{13}}{3}$

25. $x = 0$, $x = \pm i\sqrt{15}$

26. $x = \pm 3$, $x = \pm 3i$

27. R = 38

28. $k = 32$

29. $(-1)^{15} + 4(-1)^7 + 5 \overset{?}{=} 0$
    $-1 + 4(-1) + 5 \overset{?}{=} 0$
                  $0 = 0$

30. Since a is a root, then $P(a) = 0$. Since x occurs only to even powers, then $P(-a) = P(a) = 0$, and $-a$ is a root.

31. a. $4x^2 + 7x + 6$, R = 8
    b. $x^3 - 2x^2 + 6x - 13$, R = 29

32. $P(4) = 35$

33. C

34. $\left\{\pm 1, \pm 2, \pm 4, \pm 8, \pm\frac{1}{3}, \pm\frac{2}{3}, \pm\frac{4}{3}, \pm\frac{8}{3}\right\}$

35. $\{x \mid x < -1$ or $\frac{1}{2} < x < 5\}$

36. $x = 3 \pm i$

37. $x^3 - 13x^2 + 55x - 91 = 0$

38. between -3 and -2
    between 1 and 2

39. between -2 and -1

40. $1.3, -0.7, 0.4$

41. 2

42. a. positive: 2 or 0
    b. negative: 2 or 0

1. 0

2. 2

3. $-\dfrac{1}{2}$

4. 81

5. $(1, 2, 1)$

6. B

7. A

8. C

9. E

10. B

11. C

12. D

13. C

☑ *Exercises for Challenge*

1. For what value(s) of y does
$$\frac{y+1}{y-1} = \frac{y+2}{y-2}?$$

2. For how many different real numbers x does $x^4 = x$?

3. The average of 1 and z equals the average of 1, z, and z^2. Find z when $z < 0$.

4. The degree measures of the acute angles of a right triangle can be represented by x and $\sqrt{x}$. What is the value of x?

5. Let $P(x) = ax^2 + bx + c$. Find the ordered triple (a, b, c) such that $P(-1) = 0$, $P(1) = 4$, and $P(2) = 9$.

6. If $x^2 + 2x = 3$, then $x + 1$ could equal:
(A) -4 \qquad (B) -2 \qquad (C) 0
(D) 1 \qquad\quad (E) 4

7. If $2x + 3 = a$, then $x + a$ equals:
(A) $\frac{3}{2}(a-1)$ (B) $\frac{3}{2}$ \qquad (C) 0
(D) $-\frac{3}{2}$ \qquad (E) $\frac{3}{2}(1-a)$

8. The average of the measures of two angles of a triangle equals the measure of the third angle. What is the sum of the measures of the largest and smallest angles of the triangle?
(A) 60° \qquad (B) 90° \qquad (C) 120°
(D) 145° \qquad (E) 150°

9. The one real zero of $P(x) = 3x^3 - 7x^2 - 7x - 10$ lies between which pair of integers?
(A) -1 and 0 \qquad (B) 0 and 1
(C) 1 and 2 \qquad (D) 2 and 3
(E) 3 and 4

10. The polynomial equation $x^3 + 2x^2 + x + d = 0$ has a real root between 0 and 1 if:
(A) $d < -4$ \qquad (B) $-4 < d < 0$
(C) $d = 0$ \qquad (D) $0 < d < 4$
(E) $d > 4$

11. Rectangle A has width 3 and length 5. The width of rectangle B is equal to the width of rectangle A and the diagonal of rectangle B is 2 units longer than the diagonal of rectangle A. What is the length of rectangle B to the nearest hundredth?
(A) 5.39 \qquad (B) 6.03 \qquad (C) 7.23
(D) 7.59 \qquad (E) 7.64

12. To the nearest tenth of a percent, by what percent must the length of the radius of a circle be increased if the area of the circle is to be doubled?
(A) 14.1% (B) 20.7% (C) 31.4%
(D) 41.4% (E) 57.7%

13. The area of a square is increased by 25%. The length of each side of the square is increased by:
(A) 5% \qquad (B) 6.25%
(C) approximately 11.8%
(D) 25%
(E) approximately 112%

14. Which of the following is the graph of the roots of $x^3 - 1 = 0$ in the complex plane?

(A)   (B)

(C)   (D)

(E)

15. A man who has 2 quarters, 4 dimes, and some nickels finds that he can make change for a dollar in exactly 5 different ways. How many nickels does he have if he uses at least one of each type of coin?

(A) 7   (B) 8   (C) 9
(D) 10   (E) 11

16. A man ran a m.p.h. for 5 minutes and b m.p.h. for 10 minutes. His average speed in miles per hour was:

(A) $\dfrac{a+b}{2}$ (B) $\dfrac{a+b}{15}$

(C) $4(a+b)$ (D) $\dfrac{4}{a+b}$

(E) $\dfrac{a+2b}{3}$

Questions 17 – 21 each consist of two quantities, one in Column A and one in Column B. You are to compare the two quantities and choose:

A if the quantity in Column A is greater;
B if the quantity in Column B is greater;
C if the two quantities are equal;
D if the relationship cannot be determined from the information given.

1. In certain questions, information concerning one or both of the quantities to be compared is centered above the two columns.

2. In a given question, a symbol that appears in both columns represents the same thing in Column A as it does in Column B.

3. x, n, and k, etc. stand for real numbers.

Column A	Column B
	$\|3x - 5\| = 4$

17. x 0

18.   Any root of $x^2 + x - 6 = 0$ Any root of $y^2 + 8y + 16 = 0$

19.   The number of unequal roots of $x^2 - x + 1 = 0$ The number of unequal roots of $(x^2 - x + 1)^2 = 0$

20.   The number of real roots of $x^2 + 5x + c = 0$ when $|c| > 7$ The number of real roots of $x^2 + 5x - c = 0$ when $|c| > 7$

21. 🖩 The number of rational roots of $x^3 - 8x - 3 = 0$ 2

14. A

15. B

16. E

17. A

18. A

19. C

20. D

21. B

Teacher's Chapter *3*

Functions and Their Graphs

Chapter Contents

The concept of function is basic to advanced mathematics and is a unifying theme of the various branches of mathematics. The chapter reviews and extends familiar concepts.

Transformations are presented as a type of function and then line reflections and translations are used to show connections among the graphs of various functions.

Other transformations, applied to graphs of functions and the matrices used to express them, will be studied in later chapters.

3.1 What is a Function?

This section reviews the definition of function and introduces the concept of codomain, of which the range is a subset. The distinction between codomain and range is important in the classification of functions as *onto*. You may wish to introduce the terms *surjection* and *injection* for onto and 1 – 1 functions, respectively, and *bijection* for 1 – 1 correspondence.

Note the words *exactly one* in the definition of function. If f is a function whose domain is A, *every* element of A must be paired with *one* element of the codomain and may not be paired with more than one.

Equations such as $y = \dfrac{1}{x - 2}$ and $y = \sqrt{x - 2}$ do not define functions with domains that are the set of real numbers. In each case, the equation defines a function only on a subset of the reals. These two equations represent the two most common reasons for using a subset of the reals as the domain.

1. The quotient is not a real number when the divisor is 0.

2. The square root of a negative number is not a real number.

Function notation is familiar to students. Call attention to the use of f, g, or any other letter as a convenient name to apply to any function being defined and the use of sin, cos, log, and others, as names that are generally used for specific functions. The names ABS for the absolute value function and INT for the greatest integer function are used in the computer language BASIC, which also allows for user-defined functions.

Computer Activity (3.1) _____

The BASIC language allows the user to define a function using symbolism very similar to the $f(x)$ notation.

```
DEF FNA(X) = 2 * X ^ 2 - 3
```

The symbol DEF tells the computer that a function is being defined. FN followed by some letter identifies the particular function being defined. On some computers, the FN is omitted. The line above, if used in a program, would define the function $A(x) = 2x^2 - 3$.

The following program defines a function, $FNF(x) = -x^2 + 5x$, and will print the second element or function value for any value of x entered by the user.

```
100  REM THE FOLLOWING PROGRAM DEFINES
110  REM A FUNCTION. A FUNCTION VALUE
120  REM IS PRINTED FOR ANY VALUE OF X
130  REM ENTERED BY THE USER.
140  DEF FNF(X) = -1 * X ^ 2 + 5 * X
150  PRINT "ENTER ANY VALUE OF X."
160  INPUT X
170  PRINT
180  PRINT "F("; X; ") = "; FNF(X)
190  END
```

If line 140, which defines the function, is replaced by the following line, the function is not defined for that value of x which makes the denominator equal to 0.

```
140  DEF FNF(X) = 1 / (X - 3)
```

If 3 entered as the value of x, the computer will respond "Division by zero error," or a similar message, since 3 is not in the domain of the defined function.

3.2 The Algebra of Functions

In defining a function as $f(x) = x^2$, the x represents the first element. Any symbol may be used as the first element in defining the same function.

$f(x) = x^2$ defines $\{(x, x^2)\}$

$f(a) = a^2$ defines $\{(a, a^2)\}$

$f(2x - 1) = (2x - 1)^2$ defines $\{(2x - 1, (2x - 1)^2)\}$

$f(-x) = \{(-x, (-x)^2)\}$

Each of these defines the same set of pairs if the domain is the set of real numbers.

An expression such as $f(2x - 1) = (2x - 1)^2$ can be thought of in two ways.

1. If $2x - 1$ is considered to be the first element of a pair of the function, the function maps a number to its square.

2. If x is considered to be the first element of a pair of the function, the function is a composite function that involves two consecutive mappings.

   $x \rightarrow 2x - 1 \rightarrow (2x - 1)^2$

   If $f(x) = x^2$ and $g(x) = 2x - 1$, stress that the composition is written as $(f \circ g)(x)$, since the inner function is applied first.

When forming a function that is the sum (difference, product, or quotient) of two functions, emphasize that the function values (second elements) are operated on. If $(2, 5)$ is an element of f and $(2, 4)$ is an element of g, then $(2, 5 + 4)$ or $(2, 9)$ is an element of $f + g$.

Let $f(x) = x - 3$. Ask questions such as:

Does $f(3) + f(3) = f(6)$?
Does $f(x) + f(x) = f(2x)$?

Does $f(3) + f(3) = 2f(6)$?
Does $f(x) + f(x) = 2f(2x)$?

Does $f(3) + f(3) = 2f(3)$?
Does $f(x) + f(x) = 2f(x)$?

3.3 Line and Point Reflections

For students who have completed three years of Amsco's Integrated Mathematics Series, this is a familiar topic and this section could be omitted or reviewed briefly. If this is a new topic, spend at least a day discussing each type of transformation.

With a few examples, students readily comprehend reflections in the x and y axes. It is important that they also master reflection in the line $y = x$. An understanding of line reflections will enable students to see the relationships between the graph of $f(x)$ and the graphs of $f(-x)$, $-f(x)$ and $f^{-1}(x)$.

Point out that a rotation of $180°$ is equivalent to a point reflection.

3.4 Rotations and Translations

The understanding of translation will enable students to see the relationship between a function that has a graph with its center, or some other critical point, at the origin and a related function with the corresponding point at some general point (h, k).

Activity (3.4)

Directions

1. On the coordinate plane, draw any triangle, ABC.

2. Draw the image of $\triangle ABC$ under the translation $T_{6, 0}$. Let the image of A be A'', of B be B'', and of C be C''.

3. Draw the reflection of $\triangle ABC$ in the vertical line through A. Let the image of A be A', of B be B', of C be C'.

4. Draw the reflection of $\triangle A'B'C'$ over the vertical line that is the perpendicular bisector of AA''.

5. The image drawn in step 4 is $\triangle A''B''C''$.

6. Repeat steps 1 – 5 for a different triangle.

Draw a different triangle and use $T_{-4, 0}$ in step 2. Repeat steps 3 – 5.

Repeat the procedure using $T_{0, 6}$ and horizontal lines.

Discoveries

1. A translation in the horizontal direction is equivalent to the composition of two reflections in vertical lines.

2. The distance between the vertical lines is half of the distance between any point and its image under the translation.

3. A translation in the vertical direction is equivalent to the composition of two reflections in horizontal lines.

4. The distance between the horizontal lines is half of the distance between any point and its image under the translation.

What If?

The translation is $T_{a,\,b}$, with $a, b \neq 0$? (The lines of reflection will be slant lines perpendicular to a line joining any point and its image under the translation.)

3.5   The Inverse of a Function

Only a function that is a 1 – 1 correspondence has an inverse. If f is a function from A to B that is not 1 – 1 or not onto, then f does not have an inverse.

Ask the students to give an example of:

1. a function that is not 1 – 1

Some possible answers are:

   a. $y = x^2$ from R to $[0, \infty)$

   b. $y = |x|$ from R to $[0, \infty)$

   c. An arbitrary set of ordered pairs in which at least two pairs have the same second element.

Show that the result of interchanging the elements of the pairs gives a set that is not a function because at least two pairs have the same first element.

2. a function that is not onto

Some possible answers are:

   a. $y = x^2$ from I^+ to I^+

   b. $y = 3x$ from I to I

   c. An arbitrary set of ordered pairs from A to B in which not all elements of B are used as second elements.

Show that the result of interchanging the elements of the pairs is not a function because not every element of the domain (the codomain of the given function) is paired with an element of the codomain (the domain of the original function).

3.6   Using Related Functions to Graph

Most of the functions that students are asked to graph can be related to the common functions described in Section 3.1, and their graphs can be obtained from the graphs of the common functions by simple transformations. The ability to sketch a graph is an important skill that will be used frequently in advanced mathematics, particularly in the study of calculus.

Activity (3.6)

Directions

Place a sheet of tracing paper or clear plastic over a coordinate grid. Draw a horizontal line. From any point on that line, move up 2 units and right 1 unit and mark a second point. From that second point, move up 2 and right 1 and mark a third point. Draw line ℓ through these three points. This line has a slope of 2 relative to the horizontal line.

1. Place line ℓ so that it passes through the origin, and write its equation. Keeping the horizontal line parallel to the x-axis, move line ℓ up 2 units ($T_{0,\,2}$). Write an equation of the line in its new position.

2. Again place line ℓ so that it passes through the origin. Keeping the horizontal line parallel to the x-axis, move line ℓ down 3 units ($T_{0,\,-3}$). Write an equation of the line in its new position.

3. Again place line ℓ so that it passes through the origin. Keeping the horizontal line parallel to the x-axis, move line ℓ right 2 units ($T_{2, 0}$). Write an equation of the line in its new position.

4. Again place line ℓ so that it passes through the origin. Keeping the horizontal line parallel to the x-axis, move line ℓ left 3 units ($T_{-3, 0}$). Write an equation of the line in its new position.

5. Again place line ℓ so that it passes through the origin. Keeping the horizontal line parallel to the x-axis, move line ℓ up 2 units and left 2 units ($T_{-2, 2}$). Write an equation of the line in its new position.

Discoveries

1. Under $T_{0, 2}$, the image of $y = 2x$ is $y - 2 = 2x$, or $y = 2x + 2$.

2. Under $T_{0, -3}$, the image of $y = 2x$ is $y + 3 = 2x$, or $y = 2x - 3$.

3. Under $T_{2, 0}$, the image of $y = 2x$ is $y = 2(x - 2)$, or $y = 2x - 4$.

4. Under $T_{-3, 0}$, the image of $y = 2x$ is $y = 2(x + 3)$, or $y = 2x + 6$.

5. Under $T_{-2, 2}$, the image of $y = 2x$ is $y - 2 = 2(x + 2)$, or $y = 2x + 6$.

6. For a line, there are infinitely many translations under which a given line is the image.

What If?

The graph of $y = x^2$ were translated? (Each function of the form $y = x^2 + bx + c$ can be graphed by a translation.)

3.7   Rational Functions

Students should consider the nature of a fraction, as a guide to determining the asymptotes of the graph of a rational function. (*Asymptote* is from the Greek for "not falling together.")

If the numerator of a fraction is a constant and the absolute value of the denominator is decreasing (approaching 0), the absolute value of the fraction is increasing (approaching infinity).

If the numerator of a fraction is a constant and the absolute value of the denominator is increasing (approaching infinity), the absolute value of the fraction is decreasing (approaching 0).

If the denominator of a fraction is a constant and the absolute value of the numerator is decreasing (approaching 0), the absolute value of the fraction is decreasing (approaching 0).

If the denominator of a fraction is a constant and the absolute value of the numerator is increasing (approaching infinity), the absolute value of the fraction is increasing (approaching infinity).

If the numerator and denominator of a fraction are both increasing or both decreasing, the value of the fraction may be increasing, decreasing, or remaining constant. When graphing a rational fraction it is usually helpful to express the function values as the sum of a polynomial and a fraction with a constant numerator, if possible.

In **1 – 8**, tell whether or not f is a function. Explain.

1. $f = \{(1, a), (1, b), (1, c)\}$ from $\{1\}$ to $\{a, b, c, d\}$

2. $f = \{(1, a), (2, b), (3, c)\}$ from $\{1, 2, 3, 4\}$ to $\{a, b, c, d\}$

3. $f = \{(1, a), (2, b), (3, c), (4, d)\}$ from $\{1, 2, 3, 4\}$ to $\{a, b, c, d\}$

4. $f(x) = 2x - 1$ from Z to Z **5.** $f(x) = 2x - 1$ from Q to Q **6.** $f(x) = \dfrac{1}{2x - 1}$ from Q to Q

7. $f(x) = \dfrac{1}{2x - 1}$ from R^+ to R^+ **8.** $f(x) = \dfrac{1}{2x - 1}$ from $R/\{\frac{1}{2}\}$ to $R/\{0\}$

In **9 – 13**, let $f(x) = x^3$ and $g(x) = x - 1$. Write an equation for the function.

9. $(f + g)(x)$ **10.** $fg(x)$ **11.** $\dfrac{f}{g}(x)$ **12.** $(f \circ g)(x)$ **13.** $(g \circ f)(x)$

In **14 – 16, a.** What is the largest domain in $R \times R$ for which f is a function? **b.** For what codomain is f onto?

14. $f(x) = x^2 - 1$ **15.** $f(x) = \dfrac{1}{x^2 - 1}$ **16.** $f(x) = \sqrt{x^2 - 4}$

In **17 – 20**, find $f \circ g$ and $g \circ f$.

17. $f = \{(1, 5), (2, 4), (3, 3), (4, 2)\}$ and $g = \{(2, 1), (3, 2), (4, 3), (5, 4)\}$ **18.** $f(x) = 2x + 1$ and $g(x) = \dfrac{1}{x^2}$

19. $f(x) = \sqrt{x + 2}$ and $g(x) = x - 2$ **20.** $f(x) = 3x + 3$ and $g(x) = \frac{1}{3}x - 1$

21. Let $f(x) = \frac{1}{2}x + 3$.
   a. Find $f^{-1}(x)$.   **b.** Sketch the graph of f.   **c.** Sketch the graph of f^{-1}.

22. The graph of f is given at the right. Sketch the graph of:
   a. $f(-x)$ **b.** $-f(x)$
   c. $f(x + 1)$ **d.** $2f(x)$

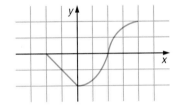

In **23 – 27**, sketch the graph.

23. $y = (x - 2)^2$ **24.** $y = x^2 - 2x$ **25.** $y = 4 - x^2$ **26.** $y = 3 + \sqrt{x + 2}$

27. $y = 3 - \sqrt{x + 2}$

28. Under a line reflection, the image of $A(4, -1)$ is $A'(0, 3)$ and the image of $B(2, -1)$ is $B'(0, 1)$. Find an equation of the line of reflection.

29. When a purchase is made, a tax of 7% of the cost is added to the cost and the result is rounded to the nearest hundredth of a dollar (nearest cent) to obtain the purchase price. Express the amount of money needed to make a purchase as a function of x, the cost of the purchase before tax.

Bonus

Find all possible points B such that $\overline{AC}$ is a leg of isosceles right triangle ABC, where C is the point $(2, -1)$ and A is the point $(4, 2)$.

Answers to Suggested Test Items

1. No, all pairs have the same first element.

2. No, no pair has the first element 4.

3. Yes, each element of the domain is paired to one and only one element of the range.

4. Yes, every integer is paired to exactly one integer.

5. Yes, every rational number is paired to exactly one rational number.

6. No, no pair has the first element $\frac{1}{2}$.

7. No, no pair has the first element $\frac{1}{2}$.

8. Yes, every element of the domain is paired with exactly one element of the range.

9. $(f + g)(x) = x^3 + x - 1$

10. $fg(x) = x^3(x - 1)$ or $fg(x) = x^4 - x^3$

11. $\dfrac{f}{g}(x) = \dfrac{x^3}{x - 1}$

12. $(f \circ g)(x) = (x - 1)^3$

13. $(g \circ f)(x) = x^3 - 1$

14. **a.** R **b.** $\{y \mid y \geq -1\}$

15. **a.** $R/\{1, -1\}$ **b.** $\{y \mid y \leq -1\} \cup \{y > 0\}$

16. **a.** $\{x \mid |x| \geq 2\}$ **b.** nonnegative reals

17. $(f \circ g) = \{(2, 5), (3, 4), (4, 3), (5, 2)\}$
    $(g \circ f) = \{(1, 4), (2, 3), (3, 2), (4, 1)\}$

18. $(f \circ g)(x) = \dfrac{2}{x^2} + 1$
    $(g \circ f)(x) = \dfrac{1}{(2x + 1)^2}$

19. $(f \circ g)(x) = \sqrt{(x - 2) + 2}$
    $\qquad = \sqrt{x}$
    $(g \circ f)(x) = \sqrt{x + 2} - 2$

20. $(f \circ g)(x) = 3\left(\frac{1}{3}x - 1\right) + 3$
    $\qquad = x$
    $(g \circ f)(x) = \frac{1}{3}(3x + 3) - 1$
    $\qquad = x$

21. **a.** $f^{-1}(x) = 2x - 6$

    b.
    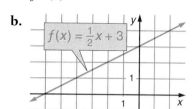
    $f(x) = \frac{1}{2}x + 3$

    c.
    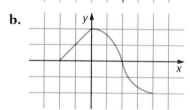
    $f^{-1}(x) = 2x - 6$

22. **a.**
    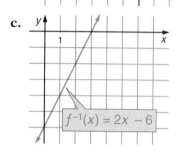

    b.
    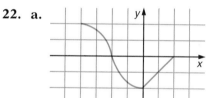

    c.
    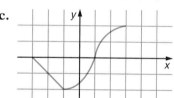

    d.
    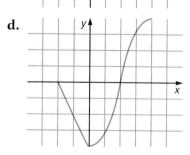

Answers to Suggested Test Items (continued)

23.

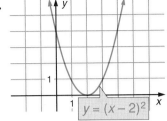

$y = (x - 2)^2$

24.

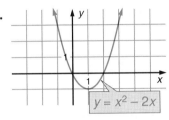

$y = x^2 - 2x$

25.

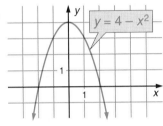

$y = 4 - x^2$

26.

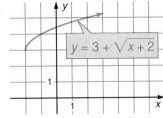

$y = 3 + \sqrt{x + 2}$

27.

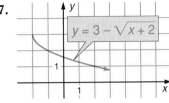

$y = 3 - \sqrt{x + 2}$

28. $y = x - 1$

29. $y = x + \dfrac{[7x + 0.5]}{100}$

Bonus

$(7, 0), (1, 4), (-1, 1), (5, -3)$

    If the right angle is at A, point B is the image of C under a rotation of $90°$ or of $-90°$ about A. Since we know the coordinates of the image of a point under a rotation of $90°$ about the origin, first use the translation $T_{-4, -2}$, thus placing A at the origin, then rotate C $90°$ about the origin, and finally translate the points back to their original positions using $T_{4, 2}$.

$$
\begin{aligned}
T_{4,2} \circ R_{90°} \circ T_{-4, -2}(2, -1) &= T_{4, 2} \circ R_{90°}(-2, -3) \\
&= T_{4, 2}(3, -2) \\
&= (7, 0)
\end{aligned}
$$

$$
\begin{aligned}
T_{4, 2} \circ R_{-90°} \circ T_{-4, -2}(2, -1) &= T_{4, 2} \circ R_{-90°}(-2, -3) \\
&= T_{4, 2}(-3, 2) \\
&= (1, 4)
\end{aligned}
$$

If the right angle is at C, use $T_{-2, 1}$ to translate C to the origin, and rotate point A.

$$
\begin{aligned}
T_{2,-1} \circ R_{90°} \circ T_{-2, 1}(4, 2) &= T_{2, -1} \circ R_{90°}(2, 3) \\
&= T_{2, -1}(-3, 2) \\
&= (-1, 1)
\end{aligned}
$$

$$
\begin{aligned}
T_{2, -1} \circ R_{-90°} \circ T_{-2, 1}(4, 2) &= T_{2, -1} \circ R_{-90°}(2, 3) \\
&= T_{2, -1}(3, -2) \\
&= (5, -3)
\end{aligned}
$$

Chapter 3

Functions and Their Graphs

Rent Me

1 day	$35.00
2 days	$70.00
3 days	$105.00
4 days	$140.00
5 days	$175.00
6 days	$200.00
7 days	$200.00

A car rental agency charges $35 a day to rent a car with unlimited mileage. If a customer keeps a car for more than 5 days, there is a special offer of $200 a week. To help customers estimate the rental cost, the agency displays a chart listing the cost for 1 through 7 days.

Notice that for each value representing the number of days, there is one and only one corresponding value for the cost in dollars. This type of relation is called a *function*.

Functions are used in everyday life to communicate certain kinds of quantitative information.

Chapter Table of Contents

3.1 What is a Function?

The chart on the previous page shows the integers 1 through 7 paired with a subset of the integers through 200. Let $D = \{1, 2, 3, 4, 5, 6, 7\}$ and $C = \{$integers $\leq 200\}$. The set of ordered pairs $\{(1, 35), (2, 70), (3, 105), (4, 140), (5, 175), (6, 200), (7, 200)\}$ is a function whose domain is D and whose codomain is C. Every element of the domain is paired with an element of the codomain, but not every element of the codomain is paired with an element of the domain. The set of second elements of the pairs of the function, $\{35, 70, 105, 140, 175, 200\}$, is a subset of the codomain called the range.

■ **Definition:** A *function* from set A to set B is the set of ordered pairs (x, y) such that every $x \in A$ is paired with exactly one $y \in B$. Set A is the **domain** of the function and set B is the **codomain** of the function. If (x, y) is in f, then we say that f maps x to y, or $f: x \rightarrow y$.

To say that f is a function from A to B is to specify the sets from which the first and second elements of the ordered pairs of f are taken. Note, however, that although *every* element of the domain A must be used as a first element, not every element of the codomain B need be used as a second element. The *range* is the subset of the codomain that contains only the second elements of f.

Example 1 ─────────────────────────────────

If $A = \{a, b, c\}$ and $B = \{p, q, r\}$, explain why each of the following is or is not a function from A to B.

Solution

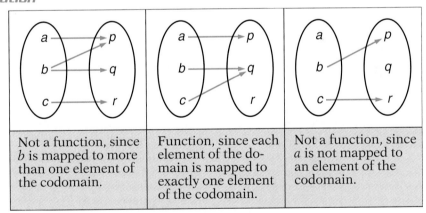

| Not a function, since b is mapped to more than one element of the codomain. | Function, since each element of the domain is mapped to exactly one element of the codomain. | Not a function, since a is not mapped to an element of the codomain. |

Often a function is defined in terms of an equation. For example, every real number x has a square, x^2, that is a nonnegative real number. Therefore, the set of ordered pairs (x, x^2) is a function that maps every real number x to a nonnegative real number x^2. This function can be written as $f = \{(x, y) \mid y = x^2\}$ or as $f(x) = x^2$. The equation $y = x^2$ tells how to find each ordered pair of the function. If $x = -2$, $y = (-2)^2 = 4$. Therefore, $(-2, 4)$ is an ordered pair of the function. Since x can be any real number and y can be any nonnegative real number, the domain of f is the set of real numbers and the range is the set of nonnegative real numbers.

Recall that a polynomial can be defined as:

$$p(x) = a_n x^n + a_{n-1} x^{n-1} + \cdots + a_1 x + a_0$$

Every polynomial function whose second element is a polynomial can have the set of real numbers as its domain. For example, $f(x) = x^2$ is a polynomial function since x^2 is a polynomial.

Not every function that is defined by an equation has the set of real numbers as its domain. It is necessary to exclude from the domain of a function any value of x that would:

1. cause the denominator of any fraction to be 0

2. cause the radicand of any even root to be negative

Example 2

If $f(x) = \sqrt{x - 3}$, find the largest subset of the real numbers that can be:

a. the domain of f **b.** the range of f

Solution

a. Since there is no real number that is the square root of a negative number, the domain must consist of only those values of x for which $x - 3$ is nonnegative, that is:

$$x - 3 \geq 0$$
$$x \geq 3$$

b. The expression $\sqrt{x - 3}$ means the principal or nonnegative square root of $x - 3$. Therefore, the range is the set of nonnegative real numbers.

Answer: **a.** domain of f = {real numbers ≥ 3}
               b. range of f = {real numbers ≥ 0}

Graphs of Functions

On the coordinate plane, every point is the graph of an ordered pair of real numbers, and every ordered pair of real numbers has a graph that is a point of the plane. The set of ordered pairs of real numbers can be represented by $R \times R$ (read: R cross R).

A function in $R \times R$ can be graphed as a subset of the coordinate plane. For example, to draw the graph of $y = x^2$, compute some representative pairs of values, locate the points that are the graphs of the pairs on the coordinate plane, and join the points by a smooth curve.

x	x^2	y
-3	$(-3)^2$	9
-2	$(-2)^2$	4
-1	$(-1)^2$	1
0	$(0)^2$	0
1	$(1)^2$	1
2	$(2)^2$	4
3	$(3)^2$	9

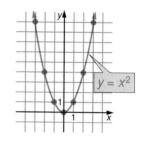

The graph of a function that can be written in the *slope-intercept* form $f(x) = mx + b$ is a straight line. Since the slope of a line, the ratio $\dfrac{\text{change in } y}{\text{change in } x}$, is constant, fewer points are needed to locate the graph. The value of m, the slope, and of b, the y-intercept, can be used.

Example 3

Draw the graph of $\{(x, y) \mid y = 3x - 2\}$.

Solution

Since the line $y = 3x - 2$ is of the form $y = mx + b$, the value of the slope is 3, or $\frac{3}{1}$, meaning that for a change of 3 units in y, there is a corresponding change of 1 unit in x.

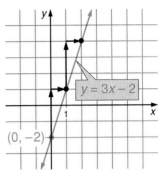

The value of the y-intercept is –2, meaning that the graph crosses the y-axis at the point $(0, -2)$.

To draw the graph, begin at the point $(0, -2)$ and locate a second point by moving up 3 units and right 1 unit. From this second point, again move up 3 units and right 1 unit. The line that is the graph of $y = 3x - 2$ passes through these three points.

Another method for graphing a line is useful when the equation is not written explicitly in the form $y = f(x)$. This method involves working with both the x- and y-intercepts.

Example 4

Draw the graph of $2x - 3y = 6$.

Solution

A point at which a graph crosses a coordinate axis has a coordinate of 0. That is, the line $2x - 3y = 6$ crosses the x-axis at $(3, 0)$ and the y-axis at $(0, -2)$.

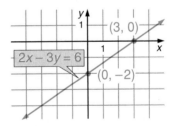

The intercepts of a line can be read from the equation of the line if the equation is rewritten so that the isolated constant term is 1.

$2x - 3y = 6$ The constant term is isolated on the right side.

$\dfrac{2x}{6} - \dfrac{3y}{6} = 1$ Divide by the constant term.

$\dfrac{x}{3} + \dfrac{y}{-2} = 1$ Simplify, expressing the left side as a sum.

Note that, after the equation has been rewritten, the divisor of x is the x-intercept, and the divisor of y is the y-intercept. This example leads to the following general statement.

The equation $\dfrac{x}{a} + \dfrac{y}{b} = 1$ is called the *intercept form* of the equation of a line. The x-intercept is a, and the y-intercept is b. That is, the graph crosses the x-axis at $(a, 0)$, and the y-axis at $(0, b)$.

Functions That Are ONTO

Compare the functions $f = \{(x, y) \mid y = 3x - 2\}$, and $g = \{(x, y) \mid y = x^2\}$, which are functions from R to R, that is, the domain and the codomain are both the set of real numbers. The range of f is the set of all real numbers, the same as the codomain. Thus f is called a function from R onto R. The range of g is the set of nonnegative real numbers, a proper subset of the codomain. Thus g is a function that is not onto.

■ **Definition:** *A function f is* onto *if for every value y in the codomain, there is at least one value x in the domain for which* $f(x) = y$.

A function is *onto* if the range is the same set as the codomain.

A function f is not onto if for at least one y in the codomain, there is no x in the domain for which $f(x) = y$. A function is not onto if its range is a proper subset of the codomain.

Example 5

Each of the diagrams below defines a function from A to B. If $A = \{a, b, c\}$ and $B = \{d, e, f\}$, is the function onto?

Solution

A function from A to B is onto if the codomain is the same set as the range, that is, if every element of the codomain is the image of at least one element of the domain.

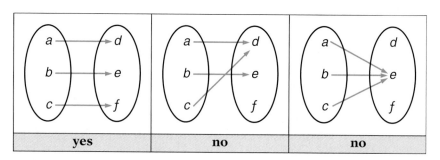

| yes | no | no |

Example 6

For each of the following functions, find the range and determine whether the function is onto. If the function is not onto, give an element of the codomain that is not paired with an element of the domain.

a. $f(x) = x^2$ from the rational numbers to the rational numbers

b. $f(x) = |x| + 1$ from the integers to the positive integers

c. $f(x) = 2x + 1$ from the integers to the integers

d. $f(x) = 2x + 1$ from the rational numbers to the rational numbers

Solution

a. The range is the set of squares of the rational numbers. The function is not onto; the rational number 2 is not in the range.

b. The range is the set of positive integers. The function is onto since every positive integer can be the result of adding 1 to the absolute value of an integer.

c. The range is the set of odd integers. The function is not onto; no even integers are in the range.

d. The range is the set of rational numbers. The function is onto since every rational number can be written as $2x + 1$ when x is a rational number.

Functions That Are ONE-TO-ONE

Compare the function $f = \{(x, y) \mid y = x^2\}$ and the function $g = \{(x, y) \mid y = 3x - 2\}$ again. For each positive real number a^2 in the range of f, there are two real numbers, a and $-a$, in the domain such that $(-a, a^2)$ and (a, a^2) are elements of f. Such a function is referred to as a *many-to-one function*.

For each real number $3a - 2$ in the range of g, there is exactly one real number a in the domain such that $(a, 3a - 2)$ is an element of g. Such a function is called a *one-to-one function*.

■ **Definition:** *A function from set A to set B is said to be one-to-one if no two distinct ordered pairs have the same second element.* Note: One-to-one may also be written as 1 – 1.

If a function is 1 – 1, then no element of the range is paired with more than one element of the domain.

Example 7 _____

For each of the following functions from R to R, determine whether the function is 1 – 1. If it is not 1 – 1, give a counterexample.

Solution

$f(x) = 2x + 5$	$f(x) = \mid x \mid + 1$	$f(x) = x^2 - 2x + 3$
1 – 1	not 1 – 1	not 1 – 1
	(–3, 4) and (3, 4)	(0, 3) and (2, 3)

Some functions, such as $y = 3x - 2$, are both 1 – 1 and onto. Such a function is called a *one-to-one correspondence*. A one-to-one correspondence is a function that is both 1 – 1 and onto.

Some Common Functions in R × R

It is useful to consider some of the frequently-encountered functions in terms of characteristics that have just been discussed. A function is in $R \times R$ if both the domain and the codomain are subsets of R. Thus, if f is a function from A to B, f is in $R \times R$ when A is a subset of R and B is a subset of R.

☐ Constant Function: $f(x) = a$ constant

The range of the constant function is the set whose only element is the constant. It is neither 1 – 1 nor onto.

$f(x) = 3$ is a constant function whose range is {3}.

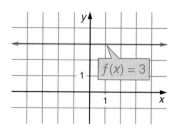

☐ Linear Function: $f(x) = mx + b$

The range of the linear function is the set of real numbers. The function is both 1 – 1 and onto, and, therefore, it is a 1 – 1 correspondence.

$f(x) = \frac{1}{2}x - 1$ is a linear function.

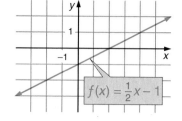

☐ Identity Function: $f(x) = x$

This is a linear function that pairs every real number with itself. Since it is a linear function, it is a 1 – 1 correspondence. A common way of writing the identity function is $y = x$.

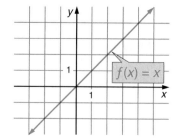

☐ Quadratic Function: $f(x) = ax^2 + bx + c$

The range of the quadratic function is

$$\left\{\text{real numbers} \geq f\left(-\tfrac{b}{2a}\right)\right\} \text{ when } a > 0, \text{ or}$$

$$\left\{\text{real numbers} \leq f\left(-\tfrac{b}{2a}\right)\right\} \text{ when } a < 0.$$

The function is neither 1 – 1 nor onto.

Note: $-\frac{b}{2a}$ is the x-coordinate of the maximum or minimum point of the function.

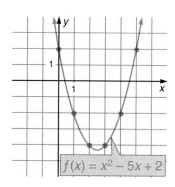

☐ **Polynomial Function:** $f(x) = a_n x^n + a_{n-1} x^{n-1} + \cdots + a_0$, where $a_n \neq 0$.

If n is odd, the range is the set of real numbers and the function is, therefore, onto. The function may or may not be 1 – 1.

If n is even, the range is a proper subset of the real numbers and the function is, therefore, not onto. The function is not 1 – 1.

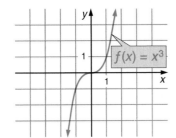

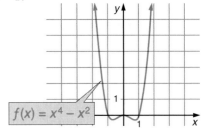

The constant, linear, and quadratic functions are also polynomial functions, of degrees 0, 1, and 2, respectively.

☐ **Absolute-Value Function:** $f(x) = |x|$

$$|x| = \begin{cases} x \text{ if } x \geq 0 \\ -x \text{ if } x < 0 \end{cases}$$

The range is the set of nonnegative real numbers. The function is neither 1 – 1 nor onto.

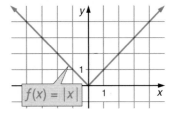

☐ **Step Function:** The graph is a series of steps, for example:

$$f(x) = \begin{cases} 1 \text{ if } x > 0 \\ 0 \text{ if } x = 0 \\ -1 \text{ if } x < 0 \end{cases} \quad \text{or} \quad f(x) = \begin{cases} \dfrac{|x|}{x} \text{ if } x \neq 0 \\ 0 \text{ if } x = 0 \end{cases}$$

The range is the set $\{-1, 0, 1\}$. The function is neither 1 – 1 nor onto. Like the step function, some functions are defined by using different expressions for disjoint intervals of the domain. Such a function is sometimes called a *piecewise* function.

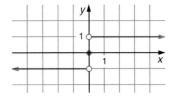

☐ **Greatest-Integer Function:** $f(x) = [x]$

The symbol $[x]$ represents the greatest integer less than or equal to x. The range is the set of integers. The function is neither 1 – 1 nor onto.

The greatest-integer function is another example of a step function.

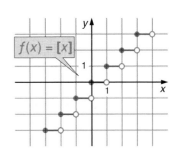

☐ **Square-Root Function:** $f(x) = \sqrt{x}$

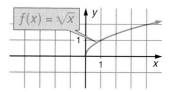
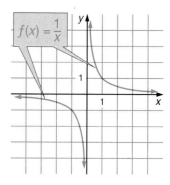

This is not a function from R to R because the largest possible domain for the square-root function is the set of nonnegative real numbers. If the domain and codomain for $f(x) = \sqrt{x}$ are the set of nonnegative real numbers, then the function is 1 – 1 and onto, and, therefore, a 1 – 1 correspondence.

☐ **Reciprocal Function:** $f(x) = \dfrac{1}{x}$

This is also not a function from R to R because the largest possible domain for the reciprocal function is the set of nonzero real numbers. If the domain and codomain for $f(x) = \dfrac{1}{x}$ are the set of nonzero real numbers, then the function is 1 – 1 and onto, and, therefore, a 1 – 1 correspondence.

☐ **Sine Function:** $f(x) = \sin x$

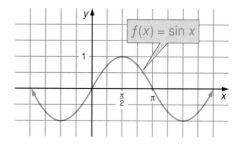

This is a function from R to R. The domain is the set of real numbers and the range is $\{y \mid -1 \le y \le 1\}$. The sine function is neither 1 – 1 nor onto. It is a periodic function with a period of 2π. The sine function is one of a set of functions, called the *trigonometric functions*, that will be investigated in detail in Chapter 8.

☐ **Exponential Function:** $f(x) = b^x$

This is a function from R to R for a constant $b > 0$ and $b \ne 1$. The domain is the set of real numbers and the range is the set of positive real numbers. The exponential function from R to R is 1 – 1 but not onto. The exponential function is a 1 – 1 correspondence from R to R^+ and will be studied together with its inverse function in Chapter 5.

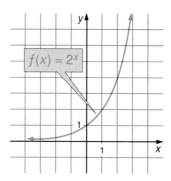
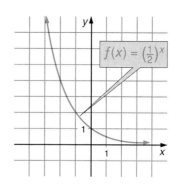

Exercises 3.1

In **1 – 13**, answer these questions:

  a. Is the mapping a function? If not, explain.

  b. If it is a function, is it a 1 – 1 correspondence? Explain.

1. 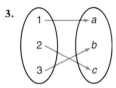   **2.**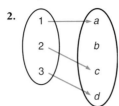

3. 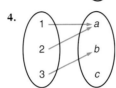   **4.**

5. $f(x) = x^2$ from Z to W **6.** $f(x) = x^2$ from Z to Z **7.** $f(x) = x^2$ from Q to Q

8. $f(x) = 5x - 2$ from Z to Z **9.** $f(x) = 5x - 2$ from R to R **10.** $f(x) = x^2 - 1$ from Z to Z

11. $f(x) = x^2 - 1$ from R to R **12.** $f(x) = \sqrt{x}$ from Z to Z **13.** $f(x) = \sqrt{x}$ from Z^+ to Z^+

In **14 – 32**, graph the function.

14. $y = 4 - x$ **15.** $y = x - 1$ **16.** $3x + 4y = 12$ **17.** $x - 2y = 6$

18. $3y = 12 - 2x$ **19.** $5x = 3y - 15$ **20.** $y = x^2 - 2$ **21.** $y = 2 - x^2$

22. $y = (x - 2)^2$ **23.** $y = \frac{1}{2}x^2 - 4x$ **24.** $y = |x| + 2$ **25.** $y = |x + 2|$

26. $y = [x + 2]$ **27.** $y = [x] + 2$ **28.** $y = 3^x$ **29.** $y = -3^x$

30. $y = 3^{-x}$ **31.** $y = \begin{cases} x \text{ if } x \geq 0 \\ x^2 \text{ if } x < 0 \end{cases}$ **32.** $y = \begin{cases} -x \text{ if } x < 0 \\ 3 \text{ if } x = 0 \\ x^2 \text{ if } x > 0 \end{cases}$

In **33 – 40**, answer these questions:

  a. What is the largest subset of the real numbers that can be the domain and the smallest subset of the real numbers that can be the range if $f(x)$ is a function?

  b. What is $f\left(\frac{5}{4}\right)$?

33. $f(x) = |x - 2|$ **34.** $f(x) = \sqrt{x + 1}$ **35.** $f(x) = \sqrt{3x + 5}$ **36.** $f(x) = \sqrt{x^2 - 1}$

37. $f(x) = \dfrac{|x|}{x}$ **38.** $f(x) = -\dfrac{1}{x}$ **39.** $f(x) = \dfrac{5}{x - 7}$ **40.** $f(x) = \dfrac{1}{\sqrt{x + 2}}$

Answers 3.1

1. a. not a function because some first elements map to more than one second element

2. a. a function

  b. not a 1 – 1 correspondence because the function is not onto (there is no element of the domain that maps to b in the codomain)

3. a. a function

  b. a 1 – 1 correspondence because every element of the domain maps to one and only one element of the codomain and every element of the codomain is the image of an element of the domain

4. a. a function

  b. not a 1 – 1 correspondence because it is not 1 – 1 (contains both $(1, a)$ and $(2, a)$ and not onto (there is no element of the domain that maps to c in the codomain)

5. a. a function

  b. not a 1 – 1 correspondence because it is not 1 – 1 (counterexample: contains both $(1, 1)$ and $(-1, 1)$) and not onto (counterexample: there is no element of the domain that maps to $2 \in W$)

6. a. a function

  b. not a 1 – 1 correspondence because it is not 1 – 1 and not onto. Counterexamples: $(1, 1)$ and $(-1, 1)$ and there are no integers that map to the negative integers

7. a. a function
   b. not a 1 – 1 correspondence because it is not
   1 – 1 and not onto. Counterexamples: (1, 1)
   and (–1, 1) and there are no rational numbers
   that map to the negative rational numbers
   nor to a rational number such as 2

8. a. a function
   b. not a 1 – 1 correspondence because it is not
   onto. Counterexample: there is no element of
   the domain that corresponds to 1 in the
   codomain

9. a. a function
   b. a 1 – 1 correspondence

10. a. a function
   b. not a 1 – 1 correspondence because it is not
   1 – 1 and not onto. Counterexamples: (1, 0)
   and (–1, 0) and there is no element of the
   domain that maps to 2 in the codomain

11. a. a function
   b. not a 1 – 1 correspondence because it is not
   1 – 1 and not onto. Counterexamples: (1, 0)
   and (–1, 0) and there is no element of the
   domain that corresponds to –2 in the
   codomain

12 a. not a function because there are no elements
   of the codomain that are the images of the
   negative integers

13. a. not a function because there are no elements
   of the codomain that are the images of integers that are not perfect squares

14.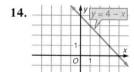

15.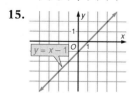

16.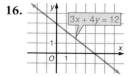

17.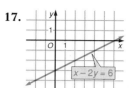

18.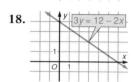

19.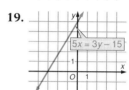

20.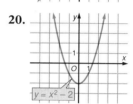

21.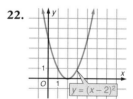

22.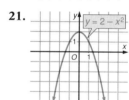

23.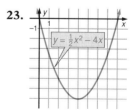

24.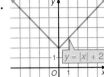

25.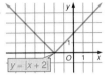

26.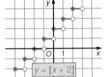

27.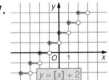

28.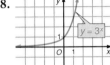

29.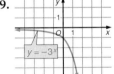

30.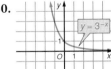

31.

32.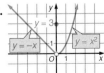

33. **a.** all reals; nonnegative reals

   b. $\frac{3}{4}$

34. **a.** $\{x \mid x \geq -1\}$; $\{y \mid y \geq 0\}$

   b. $\frac{3}{2}$

35. **a.** $\left\{x \mid x \geq \frac{-5}{3}\right\}$; nonnegative reals

   b. $\frac{\sqrt{35}}{2}$

36. **a.** $\{x \mid x \geq 1 \text{ or } x \leq -1\}$; nonnegative reals

   b. $\frac{3}{4}$

37. **a.** nonzero reals; $\{1, -1\}$

   b. 1

38. **a.** nonzero reals; nonzero reals

   b. $-\frac{4}{5}$

39. **a.** $\{x \mid x \neq 7\}$; nonzero reals

   b. $-\frac{20}{23}$

40. **a.** $\{x \mid x > -2\}$; positive reals

   b. $\frac{2\sqrt{13}}{13}$

41. $f(x) = \begin{cases} 35 \text{ if } x \leq 100 \\ 35 + 0.25(x - 100) \\ \quad \text{if } x > 100 \end{cases}$

42. $\lceil x \rceil = \begin{cases} [x] \text{ if } x \text{ is an integer} \\ [x] + 1 \text{ if } x \text{ is not an integer} \end{cases}$

43. **a.** $f(x) = \begin{cases} 3\left[\frac{x}{60}\right] \text{ if } 60 \text{ divides } x \\ 3\left[\frac{x}{60}\right] + 3 \text{ if } 60 \text{ does not divide } x \end{cases}$

   or
   $f(x) = 3\left|\left[\frac{-x}{60}\right]\right|$

   b. $f(x) = 3\left\lceil\frac{x}{60}\right\rceil$

44. Let $F = \{g_1, g_2, g_3, b\}$.

   a. $\{(x, y) \mid y \text{ is the brother of } x\}$ is the set $\{(g_1, b), (g_2, b), (g_3, b)\}$. There is no pair with b as the first element.

   b. In $\{(x, y) \mid y \text{ is the sister of } x\}$, each element of F is paired with more than one second element.

Exercises 3.1 *(continued)*

41. The cost of renting a car is $35 a day with 100 free miles, and an additional charge of $0.25 per mile for each mile over 100. Express the rental cost for one day, $f(x)$, as a function of x, the number of miles that the car is driven.

42. Let $\lceil x \rceil$ represent the smallest integer greater than or equal to x. The function $f = \{(x, y) \mid y = \lceil x \rceil\}$ is often called the *least integer* or *ceiling function*. Use two equations to write $\lceil x \rceil$ in terms of $[x]$, one when x is an integer and the other when x is not an integer.

43. In the local park, paddle boats are rented for $3 per hour or any part of an hour. Express the cost of renting a boat for x minutes, using:
   a. the greatest integer function
   b. the least integer function

44. Let F be the set of children in a family of 3 girls and 1 boy. Explain why each of the following is not a function from F to F:
   a. $\{(x, y) \mid y$ is the brother of $x\}$
   b. $\{(x, y) \mid y$ is the sister of $x\}$

45. Let: $R(x) = [x + 0.5]$
   a. Find $R(3.8)$, $R(2.2)$, $R(4.6)$, and $R(-1.1)$
   b. What is the effect of R?

46. Let: $H(x) = \dfrac{[100x]}{100}$
   a. Find $H(3.8231)$, $H(2.259)$, $H(4.662)$, and $H(-1.1096)$.
   b. What is the effect of H?

47. Write a function that will round x to the nearest tenth.

48. A rectangular garden with one length along the side of a house is to be fenced on the other three sides with 8 m of fencing.
   a. Write the area as a function of the width.   **b.** Find the domain of the function.
   c. What is the largest possible area?   **d.** Find the range of the function.

49. Match the correct function to each graph.

   a.   **b.**   **c.**   **d.**

   (1) $f(x) = 1 - x^3$
   (2) $f(x) = 1 - \sqrt[3]{x}$
   (3) $f(x) = \sqrt[3]{x} - 1$
   (4) $f(x) = x^3 - 1$

45. a. $R(3.8) = 4$
   $R(2.2) = 2$
   $R(4.6) = 5$
   $R(-1.1) = -1$

   b. The function R pairs each x with the integer nearest to x. It could be called a rounding function.

46. a. $H(3.8231) = 3.82$, $H(2.259) = 2.25$, $H(4.662) = 4.66$, $H(-1.1096) = -1.11$

   b. The function H pairs each x with the largest rational number expressed in hundredths that is less than or equal to x.

47. $T(x) = \dfrac{[10x + 0.5]}{10}$

48. a. $y = 8x - 2x^2$ **b.** $0 < x < 4$
   c. 8   **d.** $0 < y \leq 8$

49. a. 2   **b.** 4
   c. 1   **d.** 3

3.2 The Algebra of Functions

When the carpenter who runs the local woodworking shop accepts a job, his helper begins the basic work and the carpenter completes the more detailed, decorative finishing.

Last week they completed five jobs, working the hours on each job shown in the tables. Each of the tables represents a function. Let C be the function that maps each job number to the number of hours that the carpenter worked on that job. Thus, $(3, 8)$ is an element of the function C, or $C(3) = 8$. Let H be the function that maps each job number to the number of hours the helper worked on that job. Thus, $(3, 2)$ is an element of the function H, or $H(3) = 2$. Let T be the function that maps each job number to the total number of hours needed to complete the job.

Carpenter		Helper	
Job #	Hours	Job #	Hours
1	10	1	3
2	3	2	12
3	8	3	2
4	7	4	8
5	12	5	15

Total	
Job #	Hours
1	13
2	15
3	10
4	15
5	27

The table at the left shows the function T, the sum of the functions C and H. For example:

$$T(3) = C(3) + H(3)$$
$$= 8 + 2$$
$$= 10$$

Note that $\{1, 2, 3, 4, 5\}$ is the domain of both C and H, and is, therefore, the domain of T.

This example leads to the following generalization:

If g and h are functions, then $(g + h)(x) = g(x) + h(x)$. The domain of $(g + h)$ is the set of elements common to the domains of g and h, that is, the intersection of the domains of g and h.

Example 1

If $f(x) = x^2$ and $g(x) = 3x$, express $(f + g)(x)$ in terms of x and find $(f + g)(-3)$.

Solution

$$(f + g)(x) = x^2 + 3x$$
$$(f + g)(-3) = (-3)^2 + 3(-3)$$
$$= 9 + (-9)$$
$$= 0$$

This result can be verified by noting that $f(-3) = 9$, $g(-3) = -9$, and, therefore, $f(-3) + g(-3) = 0$.

The difference, product, and quotient of two functions can be defined in a similar way. If g and h are functions, then:

$$(g - h)(x) = g(x) - h(x)$$

$$(gh)(x) = g(x)h(x)$$

$$\left(\frac{g}{h}\right)(x) = \frac{g(x)}{h(x)}, \text{ when } h(x) \neq 0$$

The domain of $g - h$ and gh is the set of elements common to the domains of g and of h, that is, the intersection of the domains of g and h. The domain of $\frac{g}{h}$ is the set of elements common to the domains of g and of h for which $h(x) \neq 0$.

Example 2

Find $f - g$, fg, $\dfrac{f}{g}$, and $\dfrac{g}{f}$ if

$f = \{(0, -1), (1, 0), (2, 3), (3, 8), (4, 15)\}$ and

$g = \{(0, 2), (1, 3), (2, 4), (3, 5)\}$.

Solution

$$f - g = \{(0, -3), (1, -3), (2, -1), (3, 3)\}$$

$$fg = \{(0, -2), (1, 0), (2, 12), (3, 40)\}$$

$$\frac{f}{g} = \left\{\left(0, -\tfrac{1}{2}\right), (1, 0), \left(2, \tfrac{3}{4}\right), \left(3, \tfrac{8}{5}\right)\right\}$$

$$\frac{g}{f} = \left\{(0, -2), \left(2, \tfrac{4}{3}\right), \left(3, \tfrac{5}{8}\right)\right\}$$

It is often useful to consider a function as the sum, difference, product, or quotient of simpler functions.

Example 3

Using simpler functions, write $f(x) = x^3 + x^2$ as:

a. a sum   **b.** a product

Solution

a. Let the simpler functions be $g(x) = x^3$ and $h(x) = x^2$. Then $f(x)$ is already written as a sum:

$$f(x) = g(x) + h(x) = (g + h)(x)$$

b. To express $f(x)$ as a product of two simpler functions, factor.

$$f(x) = x^3 + x^2 = x^2(x + 1)$$

Thus, the simpler functions are $h(x) = x^2$ and $p(x) = x + 1$.

$$f(x) = h(x) \cdot p(x) = (hp)(x)$$

Example 4

If $f(x) = \sqrt{x + 2}$ and $r(x) = x + 2$, what is the largest domain for which $\left(\dfrac{f}{r}\right)(x)$ is a function in $R \times R$?

Solution

The domain of f is the set of all real numbers greater than or equal to –2.

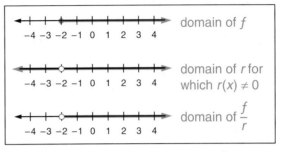

domain of f

domain of r for which $r(x) \neq 0$

domain of $\dfrac{f}{r}$

The domain of r is the set of all real numbers and $r(-2) = 0$.

Since the domain of $\dfrac{f}{r}$ is the intersection of the domain of f and the domain of r for which $r(x) \neq 0$, the domain of $\dfrac{f}{r}$ is the set of real numbers greater than –2.

Composition of Functions

To obtain values for a function such as $h(x) = (x - 2)^2$, it is necessary to carry out two operations: first subtract 2 from x, and then square the result. This function can also be thought of in terms of two mappings:

$$x \rightarrow x - 2 \rightarrow (x - 2)^2$$

Let the two mappings be represented by $f(x) = x - 2$ and $g(x) = x^2$. Therefore, h is the result of first using f to map x to $x - 2$, and then using g to map $x - 2$ to $(x - 2)^2$.

x	$\xrightarrow{f}$	$x - 2$	$\xrightarrow{g}$	$(x - 2)^2$
1	$\longrightarrow$	-1	$\longrightarrow$	$(-1)^2$ or 1
5	$\longrightarrow$	3	$\longrightarrow$	$(3)^2$ or 9
-2	$\longrightarrow$	-4	$\longrightarrow$	$(-4)^2$ or 16
0	$\longrightarrow$	-2	$\longrightarrow$	$(-2)^2$ or 4
x	$\longrightarrow$	$f(x)$	$\longrightarrow$	$g(f(x))$

The function h, which can be written as $h(x) = g(f(x))$ or $(g \circ f)(x)$, is called the *composition* of f and g.

Note that the function g is applied to $f(x)$. Therefore, only the elements of the domain of f that map to elements of the domain of g are elements of the domain of the composition.

If f and g are functions such that the range of f and the domain of g have common elements, then $g \circ f$ is the composition of f and g, written $(g \circ f)(x)$ or $g(f(x))$.

Example 5

Find $g \circ f$ and $f \circ g$ if

$f = \{(0, -1), (1, 0), (2, 1), (3, 2)\}$ and

$g = \{(-1, 1), (0, 0), (1, 1), (2, 4)\}$.

1. $\{(0, 10), (1, 10),$
$(2, 16), (3, 34),$
$(4, 70)\}$

2. $\{(0, 10), (1, 8),$
$(2, 0), (3, -20),$
$(4, -58)\}$

3. $\{(0, 0), (1, 9),$
$(2, 64), (3, 189),$
$(4, 384)\}$

4. $\left\{(1, 9), (2, 1),\right.$
$\left.\left(3, \frac{7}{27}\right), \left(4, \frac{6}{64}\right)\right\}$

5. $\left\{(0, 0), \left(1, \frac{1}{9}\right),\right.$
$(2, 1), \left(3, \frac{27}{7}\right),$
$\left.\left(4, \frac{64}{6}\right)\right\}$

6. a. $x^2 + x + 1$
b. $4x^2 + 2x + 1$
c. $x^4 - x^2 + 1$

7. a. $3x^2 - 12x$
b. $12x^2 + 36x + 15$
c. $3x^4 + 12x^2$

8. a. $x^2 - 6x + 9$
b. $4x^2 + 16x + 16$
c. $x^4 + 6x^2 + 9$

Solution

$$
\begin{array}{ccccccc}
0 & \xrightarrow{f} & -1 & \xrightarrow{g} & 1 & & \\
1 & \longrightarrow & 0 & \dashrightarrow & 0 & & \\
2 & \longrightarrow & 1 & \longrightarrow & 1 & & \\
3 & \longrightarrow & 2 & \longrightarrow & 4 & &
\end{array}
\qquad
\begin{array}{ccccccc}
-1 & \xrightarrow{g} & 1 & \xrightarrow{f} & 0 \\
0 & \longrightarrow & 0 & \longrightarrow & -1 \\
1 & \longrightarrow & 1 & \longrightarrow & 0 \\
2 & \longrightarrow & 4 & &
\end{array}
$$

Note that since the range of g is not the domain of f, only those pairs of each function for which these two sets coincide can be used. The domain of $g \circ f$ is $\{-1, 0, 1\}$.

Answer: $g \circ f = \{(0, 1), (1, 0), (2, 1), (3, 4)\}$
$f \circ g = \{(-1, 0), (0, -1), (1, 0)\}$

Note that, in general, $g \circ f \neq f \circ g$.

Example 6 _____

Express $h(x) = g(f(x))$ as a polynomial in x if $g(x) = x^2 - 3x$ and $f(x) = 2x - 1$.

Solution

$h(x) = g(f(x))$
$h(x) = g(2x - 1)$ Replace $f(x)$.
$h(x) = (2x - 1)^2 - 3(2x - 1)$ In $g(x)$, replace x by $f(x)$.
$h(x) = 4x^2 - 4x + 1 - 6x + 3$
$h(x) = 4x^2 - 10x + 4$

Example 7 _____

If $f(x) = \sqrt{x}$ and $h(x) = 2x - 1$:
a. Write an expression for $(f \circ h)(x)$.
b. Find the domain and range of $(f \circ h)(x)$ in $R \ 3 \ R$.

Solution

a. $(f \circ h)(x) = f(h(x))$
$(f \circ h)(x) = f(2x - 1)$
$(f \circ h)(x) = \sqrt{2x - 1}$

b. Since the domain of f, the set of nonnegative real numbers, must be the range of h, the domain of $f \circ h$ is only those real numbers for which $2x - 1$ is nonnegative.

       $2x - 1 \geq 0$
       $2x \geq 1$
       $x \geq \frac{1}{2}$

Answer:
a. $(f \circ h)(x) = \sqrt{2x - 1}$
b. domain of $f \circ h = \left\{\text{real numbers} \geq \frac{1}{2}\right\}$
    range of $f \circ h = \{\text{real numbers} \geq 0\}$

9. a. $g(x) = x^3$ $h(x) = -3x : f(x) = g(x) + h(x)$
b. $g(x) = x$ $h(x) = x^2 - 3 : f(x) = g(x) \cdot h(x)$

10. a. $g(x) = x^2,$ $h(x) = 5x : f(x) = g(x) + h(x)$
b. $g(x) = x,$ $h(x) = x + 5 : f(x) = g(x) \cdot h(x)$

11. a. $g(x) = 1,$ $h(x) = -\frac{1}{x},$ $x \neq 0 : f(x) = g(x) + h(x)$
b. $g(x) = \frac{1}{x},$ $h(x) = x - 1,$ $x \neq 0 : f(x) = g(x) \cdot h(x)$

12. a. $2x + \frac{1}{x} + 3$ domain : all reals $\neq 0$
b. $2 + \frac{3}{x}$ domain : all reals $\neq 0$
c. $\frac{1}{2x^2 + 3x}$ domain : all reals $\neq 0, -\frac{3}{2}$
d. $\frac{2}{x} + 3$ domain : all reals $\neq 0$
e. $\frac{1}{2x + 3}$ domain : all reals $\neq -\frac{3}{2}$

The domain and codomain of a function are often subsets of the real numbers. The set of real numbers between a and b can be written as the interval from a to b, enclosed by brackets or parentheses to symbolize inclusion or exclusion of endpoints.

$$[a, b] = \{x \mid a \le x \le b\}$$ includes both endpoints a and b

$$(a, b) = \{x \mid a < x < b\}$$ excludes both endpoints a and b

$$(a, b] = \{x \mid a < x \le b\}$$ excludes a, includes b

$$[a, b) = \{x \mid a \le x < b\}$$ includes a, excludes b

When a subset of the real numbers has no upper bound, the symbol for infinity, ∞, is used as the second element of the enclosure.

$$[a, \infty) = \{x \mid a \le x\}$$ includes a, no upper endpoint

$$(a, \infty) = \{x \mid a < x\}$$ excludes a, no upper endpoint

When a subset of the real numbers has no lower bound, the symbol $-\infty$ is used as the first element of the enclosure.

$$(-\infty, b] = \{x \mid x \le b\}$$ no lower endpoint, includes b

$$(-\infty, b) = \{x \mid x < b\}$$ no lower endpoint, excludes b

Examples of the use of intervals for expressing subsets of the real numbers:

- The set of real numbers greater than or equal to 3 but less than 5 is written $[3, 5)$.
- The set of real numbers less than 0 is written $(-\infty, 0)$.
- The domain of $(f \circ h)(x) = \sqrt{2x - 1}$ is the set of real numbers that are greater than or equal to $\frac{1}{2}$, written $[\frac{1}{2}, \infty)$.
- The range of $(f \circ h)(x)$ is the set of nonnegative numbers, written $[0, \infty)$.

Exercises 3.2

In **1 – 5**, let $p = \{(0, 10), (1, 9), (2, 8), (3, 7), (4, 6)\}$ and
$q = \{(0,0), (1, 1), (2, 8), (3, 27), (4, 64)\}$. Find:

1. $p + q$ **2.** $p - q$ **3.** pq **4.** $\dfrac{p}{q}$ **5.** $\dfrac{q}{p}$

In **6 – 8**, given $f(x)$ defined in $R \times R$, write each of the following as a polynomial in x:

a. $f(-x)$ **b.** $f(2x + 1)$ **c.** $f(x^2)$

6. $f(x) = x^2 - x + 1$ **7.** $f(x) = 3x(x + 4)$ **8.** $f(x) = (x + 3)^2$

In **9 – 11**, write the function:
a. as the sum of two functions
b. as the product of two functions

9. $f(x) = x^3 - 3x$ **10.** $f(x) = x(x + 5)$ **11.** $f(x) = \dfrac{x - 1}{x}$

Answers 3.2

13. a. $x^2 + x - 3$
domain : all reals
b. $x^3 - 3x^2$
domain : all reals
c. $\dfrac{x - 3}{x^2}$
domain :
all reals $\ne 0$
d. $x^2 - 6x + 9$
domain : all reals
e. $x^2 - 3$
domain : all reals

14. a. -3
domain : all reals
b. $-x^2 + 3x$
domain: all reals
c. $-1 + \dfrac{3}{x}$
domain :
all reals $\ne 0$
d. $-x + 3$
domain : all reals
e. $-x - 3$
domain : all reals

15. a. $\dfrac{2x - 2}{x(x - 2)}$
domain :
all reals $\ne 0, 2$
b. $\dfrac{1}{x(x - 2)}$
domain :
all reals $\ne 0, 2$
c. $\dfrac{x}{x - 2}$
domain :
all reals $\ne 0, 2$
d. $x - 2$
domain :
all reals $\ne 2$
e. $\dfrac{x}{1 - 2x}$
domain :
all reals $\ne 0, \frac{1}{2}$

16. a. $x^2 + \sqrt{x}$
domain : nonnegative reals
b. $x^2\sqrt{x}$ or $x^{\frac{5}{2}}$
domain : nonnegative reals
c. $\dfrac{\sqrt{x}}{x^2}$ or $\dfrac{1}{x^{\frac{3}{2}}}$
domain : positive reals
d. x domain : nonnegative reals
e. x domain : all reals

17. a. $x^3 + x^2$
domain : reals
b. $x^5 - x^3 + x^2 - 1$
domain : reals
c. $\dfrac{x^2 - x + 1}{x - 1}$
domain : all reals $\ne \pm 1$
d. $x^6 + 2x^3$ domain : reals
e. $x^6 - 3x^4 + 3x^2$ domain : reals

Exercises 3.2 *(continued)*

In **12 – 17**, for the $f(x)$ and $g(x)$ defined in $R \times R$, write each of the following functions and their domains:

a. $(f + g)(x)$ **b.** $(fg)(x)$ **c.** $\left(\dfrac{g}{f}\right)(x)$ **d.** $(f \circ g)(x)$ **e.** $(g \circ f)(x)$

12. $f(x) = 2x + 3$ $g(x) = \dfrac{1}{x}$

13. $f(x) = x^2$ $g(x) = x - 3$

14. $f(x) = -x$ $g(x) = x - 3$

15. $f(x) = \dfrac{1}{x}$ $g(x) = \dfrac{1}{x - 2}$

16. $f(x) = x^2$ $g(x) = \sqrt{x}$

17. $f(x) = x^2 - 1$ $g(x) = x^3 + 1$

In **18 – 22**, if x is an element of the given interval of the real numbers, write an inequality to describe x.

18. $[-2, 8)$ **19.** $(-\infty, 6]$ **20.** $[7, 20]$ **21.** $(0, 5)$ **22.** $[1, \infty)$

23. The tables show the number of hours and rate per hour for 5 jobs completed by a carpenter that were begun by her helper.

Let p be the function that gives the price of each job.

a. Make a table for the function p.

b. Write an expression for $p(x)$ in terms of $c(x)$, $h(x)$, and $r(x)$, the carpenter, helper, and rate functions.

Carpenter		Helper		Rate	
Job #	Hours	Job #	Hours	Job #	Rate/Hr.
1	10	1	3	1	$20
2	3	2	12	2	$15
3	8	3	2	3	$22
4	7	4	8	4	$18
5	12	5	15	5	$18

24. Explain why $f \circ g(x)$ is the empty set in $R \times R$ when $g(x) = -(|x| + 1)$ and $f(x) = \sqrt{x}$.

In **25 – 27**, let $g(x) = x^2$, $h(x) = \sqrt{x}$, and $f(x) = x$. Show by a counterexample that the given function is not equivalent to $f(x)$.

25. $\dfrac{g}{f}(x)$ **26.** $h \circ g(x)$ **27.** $g \circ h\,(x)$

18. $-2 \le x < 8$

19. $x \le 6$

20. $7 \le x \le 20$

21. $0 < x < 5$

22. $x \ge 1$

23. a.

Job #	Price/Job	
1	$20(10 + 3)$	$= \$260$
2	$15(3 + 12)$	$= \$225$
3	$22(8 + 2)$	$= \$220$
4	$18(7 + 8)$	$= \$270$
5	$18(12 + 15)$	$= \$486$

b. $p(x) = r(x) \cdot (c + h)(x)$

24. Range of g is $(-\infty, -1]$ and domain of f is $[0, \infty)$. These two sets have no common elements.

25. $(0, 0)$ in f is not in $\dfrac{g}{f}$

26. $(-1, -1)$ in f is not in $h \circ g$

27. $(-1, -1)$ in f is not in $g \circ h$

3.3 Line and Point Reflections

If you sketch in ink on a sheet of paper and fold the paper before the ink is dry, you will get an image of the sketch so that every point on the sketch corresponds to a point on the image. This is an illustration of the mathematical concept of a transformation of the plane.

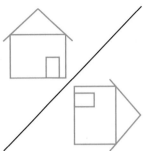

A *transformation* is a function that pairs each point of the plane with exactly one image point of the plane. The domain and codomain of a transformation are the points of the plane, or the set of ordered pairs of real numbers.

A *transformation of the plane* is a one-to-one correspondence between the points of the plane such that each point is mapped to itself or to another point of the plane.

Line Reflections

Definition: *A line reflection in line m maps each point P of the plane to its image P′ such that:*

1. If P is on m, then $P = P'$.

2. If P is not on m, then m is the perpendicular bisector of $\overline{PP'}$.

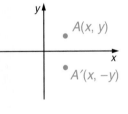

The x-axis, the y-axis, and the line $y = x$ are often used as *reflection lines* in the coordinate plane.

The image of $A(x, y)$ under a reflection in the x-axis is $A'(x, -y)$. To show that this is true, it must be shown that both parts of the definition of a line reflection in m are satisfied.

1. If $A(x, y)$ is on the x-axis, $y = 0$. The image of $A(x, 0)$ is $A(x, 0)$. A is its own image.

2. If A is not on the x-axis, then since $A(x, y)$ and $A'(x, -y)$ have the same first element, they lie on a vertical line. The x-axis is perpendicular to all vertical lines, and, thus, to $\overline{AA'}$. Recall that the midpoint of the segment joining (x_1, y_1) and (x_2, y_2) is $\left(\dfrac{x_1 + x_2}{2}, \dfrac{y_1 + y_2}{2}\right)$, making the midpoint of $\overline{AA'}$ equal to $\left(\dfrac{x + x}{2}, \dfrac{y + (-y)}{2}\right)$, or $(x, 0)$, a point on the x-axis. Thus, the x-axis bisects $\overline{AA'}$.

Therefore, the x-axis is the perpendicular bisector of $\overline{AA'}$.

The following notation is used for the reflection of a point (x, y) in the x-axis.

$$r_{x-\text{axis}}(x, y) = (x, -y)$$

Note the similarity between the reflection notation and the notation for a function f that maps x to y.

$$r_{x-\text{axis}}(x, y) = (x, -y) \qquad f(x) = y$$

$r_{x-\text{axis}}$ corresponds to f, the name of the function.

(x, y) corresponds to x, an element of the domain.

$(x, -y)$ corresponds to y, an element of the codomain.

The image of $A(x, y)$ under a reflection in the y-axis is $A'(-x, y)$. To show that this is true, it must be shown that both parts of the definition of a line reflection in m are satisfied.

1. If $A(x, y)$ is on the y-axis, $x = 0$. The image of $A(0, y)$ is $A(0, y)$. A is its own image.

2. If A is not on the y-axis, then since $A(x, y)$ and $A'(-x, y)$ have the same second element they lie on a horizontal line. The y-axis is perpendicular to all horizontal lines, and, thus, to $\overline{AA'}$. The midpoint of $\overline{AA'}$ is $\left(\dfrac{x + (-x)}{2}, \dfrac{y + y}{2}\right) = (0, y)$, a point on the y-axis. Thus, the y-axis bisects $\overline{AA'}$.

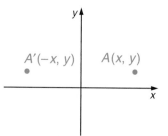

Therefore, the y-axis is the perpendicular bisector of $\overline{AA'}$.

The following notation is used for the reflection of a point (x, y) in the y-axis.

$$r_{y-\text{axis}}(x, y) = (-x, y)$$

The image of the Washington Monument is seen in the water of its pool, called the reflecting pool. The edge of the pool is the line of reflection.

The image of $A(x, y)$ under a reflection in the line $y = x$ is $A'(y, x)$. To show that this is true, it must be shown that both parts of the definition of a line reflection in m are satisfied.

1.   If $A(x, y)$ is on the line $y = x$, then $A(x, y) = A(x, x)$. The image of $A(x, x)$ is $A(x, x)$. A is its own image.

2.   If A is not on the line $y = x$, the image of $A(x, y)$ is $A'(y, x)$. The slope of $\overline{AA'}$ is $\dfrac{y - x}{x - y} = -1$. The slope of $y = x$ is 1. Thus, since the slope of $\overline{AA'}$ is the negative reciprocal of the slope of $y = x$, the line $y = x$ is perpendicular to $\overline{AA'}$. The midpoint of $\overline{AA'}$ is $\left(\dfrac{x + y}{2}, \dfrac{y + x}{2}\right)$, a point on $y = x$. Thus, the line $y = x$ bisects $\overline{AA'}$.

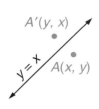

Therefore, the line $y = x$ is the perpendicular bisector of $\overline{AA'}$.

The following notation is used for the reflection of a point (x, y) in the line $y = x$.

$$r_{y = x}(x, y) = (y, x)$$

The following graphs show the images of the point $(-3, 1)$ under the line reflections discussed.

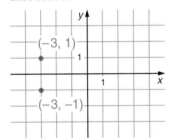
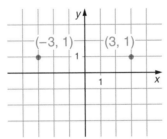
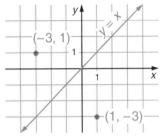

The image of $(-3, 1)$ under a reflection in the x-axis is $(-3, -1)$.

The image of $(-3, 1)$ under a reflection in the y-axis is $(3, 1)$.

The image of $(-3, 1)$ under a reflection in $y = x$ is $(1, -3)$.

Reflections in lines other than the x-axis, y-axis, and the line $y = x$ are also possible.

Example 1

Find the image of $A(2, -4)$ under a reflection in $y = -1$.

Solution

A', the image of $A(2, -4)$, is on a line through A that is perpendicular to $y = -1$. Therefore, $\overline{AA'}$ is a vertical line and A' has the same x-coordinate as A.

Let the coordinates of A' be $(2, y)$. The midpoint of $\overline{AA'}$ is on $y = -1$, that is, the midpoint has y-coordinate -1. Use the midpoint formula to find y.

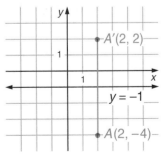

$$\frac{-4 + y}{2} = -1$$
$$-4 + y = -2$$
$$y = 2$$

The image of $A(2, -4)$ is $A'(2, 2)$.

Example 2

Let A' be the image of $A(4, -2)$ under a reflection in the y-axis. Let A'' be the image of A' under a reflection in $y = x$. Find the coordinates of A''.

Solution

$$A'' = r_{y = x}(r_{y\text{-axis}}(4, -2))$$

$$= r_{y = x}(-4, -2)$$

$$= (-2, -4)$$

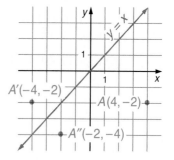

When a second transformation is applied to the image of a first transformation, the result is called the *composition* of the two transformations.

Point Reflections

■ **Definition:** *A* ***point reflection*** *in the point C maps each point P of the plane to its image P' such that:*

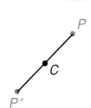

1. Point C is its own image.
2. If $P \neq C$, then C is the midpoint of $\overline{PP'}$.

The most common point reflection in the coordinate plane is a reflection in the origin. The image of (x, y) under a reflection in the origin is $(-x, -y)$. To show that this is true, it must shown that both parts of the definition of a point reflection are satisfied.

1. If $A(x, y) = (0, 0)$, the image of $A(0, 0)$ is $A(0, 0)$. A is its own image.

2. If $A(x, y)$ is not $(0, 0)$ then A' is $(-x, -y)$ and the midpoint of $\overline{AA'}$ is $\left(\dfrac{x + (-x)}{2}, \dfrac{y + (-y)}{2} \right) = (0, 0)$,

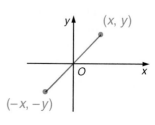

the origin. Therefore, the origin is the midpoint of $\overline{AA'}$.

The following notation is used for a reflection of a point (x, y) in the origin.

$$R_O(x, y) = (-x, -y)$$

For example, the image of $(-3, 1)$ under a reflection in the origin is $(3, -1)$.

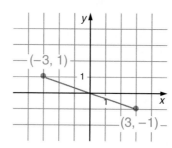

Example 3

Show: $(r_{y\text{-axis}} \circ r_{x\text{-axis}})(x, y) = R_O(x, y)$

Solution

$$(r_{y\text{-axis}} \circ r_{x\text{-axis}})(x, y) = r_{y\text{-axis}}(r_{x\text{-axis}}(x, y))$$
$$= r_{y\text{-axis}}(x, -y)$$
$$= (-x, -y)$$
$$= R_O(x, y)$$

Exercises 3.3

In **1 – 10**, find the image of the point under a reflection in the:

   a. x-axis   **b.** y-axis   **c.** line $y = x$

1. $(1, 3)$ **2.** $(-2, 5)$ **3.** $(3, -2)$ **4.** $(0, -1)$ **5.** $(3, 0)$

6. $(2, 2)$ **7.** $(-3, 3)$ **8.** $(-1, -1)$ **9.** $(0, 0)$ **10.** $(-5, -2)$

In **11 – 20**, find the image of the point under a reflection in the origin.

11. $(1, 3)$ **12.** $(-2, 5)$ **13.** $(3, -2)$ **14.** $(0, -1)$ **15.** $(3, 0)$

16. $(2, 2)$ **17.** $(-3, 3)$ **18.** $(-1, -1)$ **19.** $(0, 0)$ **20.** $(-5, -2)$

21. Find the image of $(-4, 3)$ under a reflection in:

   a. $x = 2$ **b.** $y = 2$ **c.** $(-4, 2)$ **d.** $(1, 3)$ **e.** $(-1, -2)$

In **22 – 25**, given the coordinates of the vertices of $\triangle ABC$, sketch $\triangle ABC$ and $\triangle A'B'C'$, the image of $\triangle ABC$ under a reflection in the:

a. x-axis   **b.** y-axis   **c.** line $y = x$

22. $A(0, 0)$ **23.** $A(0, 0)$ **24.** $A(-1, -3)$ **25.** $A(6, 1)$
    $B(4, 0)$ $B(0, -4)$ $B(-1, -7)$ $B(12, 1)$
    $C(0, 3)$ $C(3, 0)$ $C(-4, -7)$ $C(9, 4)$

26. Show that $(R_O \circ r_{x\text{-axis}})(x, y) = r_{y\text{-axis}}(x, y)$.

27. If the image of (a, b) under a line reflection in $y = 5$ is (a, y), express y in terms of b. *Hint:* If the image of A is A', then the midpoint of $\overline{A A'}$ lies on the line of reflection.

28. If the image of (a, b) under a line reflection:

   a. in $y = c$ is (a, y), express y in terms of b and c

   b. in $x = 2$ is (x, b), express x in terms of a

   c. in $x = c$ is (x, b), express x in terms of a and c

29. Under a line reflection, the image of $A(1, 3)$ is $A'(3, -1)$. Find an equation of the reflection line.

30. Find the image of $(2, -3)$ under a reflection in the line $y = -x$.

---

11. $(-1, -3)$ **12.** $(2, -5)$ **13.** $(-3, 2)$ **14.** $(0, 1)$

15. $(-3, 0)$ **16.** $(-2, -2)$ **17.** $(3, -3)$ **18.** $(1, 1)$

19. $(0, 0)$ **20.** $(5, 2)$

21. a. $(8, 3)$
      b. $(-4, 1)$
      c. $(-4, 1)$
      d. $(6, 3)$
      e. $(2, -7)$

22. 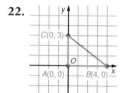   **a.** 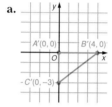   **b.** 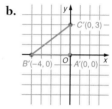   **c.**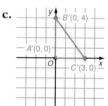

23. 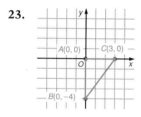   **a.** 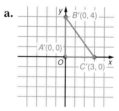   **b.** 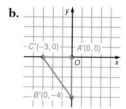   **c.**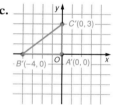

24. 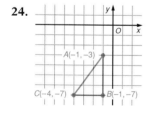   **a.** 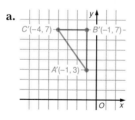   **b.** 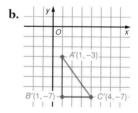   **c.**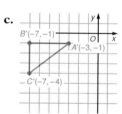

25.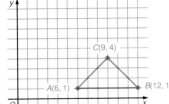

a.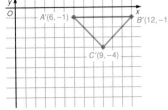

b.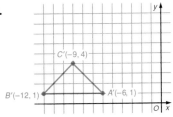

c.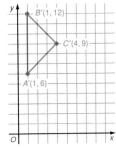

26. $r_{x\text{-axis}}\,(x, y) = (x, -y)$ and $R_O\,(x, y) = (-x, -y)$

$(R_O \circ r_{x\text{-axis}})\,(x, y) = R_O\,(x, -y)$

$= (-x, y)$

$r_{y\text{-axis}}\,(x, y) = (-x, y)$

Therefore, $R_O \circ r_{x\text{-axis}} = r_{y\text{-axis}}$

27. $y = 10 - b$

28. a. $y = 2c - b$
   b. $x = 4 - a$
   c. $x = 2c - a$

29. $y = \dfrac{x}{2}$

30. $(3, -2)$

3.4 Rotations and Translations

■ **Definition:** *A* rotation *of θ about a point C maps each point P of the plane to its image P′ such that:*

    1. Point *C* is its own image.
    2. If $P \neq C$, then, $CP = CP'$ and $m\angle PCP' = \theta$.

If the rotation from *P* to *P′* is counterclockwise, *θ* is positive. If the rotation from *P* to *P′* is clockwise, *θ* is negative. The diagram shows a positive rotation.

The image of *P*(*x*, *y*), under a rotation of 90° about the origin is *P′*(− *y*, *x*).

To show that this is true, it must be shown that both parts of the definition of a rotation about a point are satisfied.

    1. If *P*(*x*, *y*) is at the origin, then *x* = 0 and *y* = 0. The image of *P*(0, 0) is *P*(0, 0). *P* is its own image.

    2. If *P*(*x*, *y*) is not at the origin, then the slope of $\overline{OP}$, which is $\frac{y}{x}$, is the negative reciprocal of the slope of $\overline{OP'}$, which is $\frac{x}{-y}$. Thus, $\overline{OP} \perp \overline{OP'}$ and $m\angle POP' = 90°$.

    By the distance formula,
    $OP = \sqrt{(x - 0)^2 + (y - 0)^2}$ or $\sqrt{x^2 + y^2}$, and
    $OP' = \sqrt{(-y - 0)^2 + (x - 0)^2}$ or $\sqrt{x^2 + y^2}$.
    Thus, $OP = OP'$.

The following notation is used for a 90° rotation of a point (*x*, *y*) about the origin.
$$R_{O,\,90°}(x, y) = (-y, x) \text{ or } R_{90°}(x, y) = (-y, x)$$

Note that when the center of a rotation is the origin, *O* may be omitted from the notation. That is, when no point is given as the center of the rotation, it is understood to be the origin.

For example, the image of (−3, 1) under a rotation of 90° about the origin is (−1, −3).

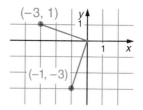

Example 1

Find the image of *P*(*x*, *y*) under the composition of two clockwise rotations of 90° about the origin.

Solution

$$(R_{90°} \circ R_{90°})(x, y) = R_{90°}(-y, x) = (-x, -y)$$

The composition of two rotations of 90° in the same direction about the origin is equivalent to a rotation of 180°. This, in turn, is equivalent to a reflection in the origin.

Translations

■ **Definition:** *A translation maps point P to its image P'
and point Q to its image Q' such that $\overline{PP'} \parallel \overline{QQ'}$ and
PP' = QQ'.*

The image of (x, y) under a translation of a units in
the horizontal direction and b units in the vertical direction is $(x + a, y + b)$. In symbols:

$$T_{a, b}(x, y) = (x + a, y + b)$$

For example, the image of (3, –3) under
$T_{-2, 2}$ is (1, –1).

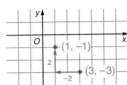

Example 2

If A' is the image of $A(1, 2)$ and B' is the image of $B(3, -1)$ under
$T_{3, -1}$, show that distance is preserved, that is, show $AB = A'B'$.

Solution

$$A' = T_{3, -1}(1, 2) = (4, 1)$$

$$B' = T_{3, -1}(3, -1) = (6, -2)$$

$$AB = \sqrt{(1 - 3)^2 + (2 - (-1))^2} = \sqrt{(-2)^2 + (3)^2} = \sqrt{4 + 9} = \sqrt{13}$$

$$A'B' = \sqrt{(4 - 6)^2 + (1 - (-2))^2} = \sqrt{(-2)^2 + (3)^2} = \sqrt{4 + 9} = \sqrt{13}$$

Therefore, $AB = A'B'$ and distance is preserved.

Exercises 3.4

In **1 – 10**, find the image under a rotation of 90° about the origin.

1. (1, 3)   **2.** (–2, 5)   **3.** (3, –2)   **4.** (0, –1)   **5.** (3, 0)

6. (2, 2)   **7.** (–3, 3)   **8.** (–1, –1)   **9.** (0, 0)   **10.** (–5, –2)

In **11 – 14**, find the coordinates of the image point.

11. $T_{2, 1}(-3, 5)$ **12.** $T_{4, -2}(1, 1)$ **13.** $T_{-1, 0}(2, 4)$ **14.** $T_{-2, 3}(2, -3)$

15. Find the image of (– 4, 3) under a
rotation about the origin of:
  a. 180°   **b.** 270°

16. If under a translation the image of
(3, 1) is (–1, –2), find the image of
(2, –2) under the same translation.

In **17 – 24**, find the image of (4, –3) under the transformation.

17. $r_{x\text{-axis}}$ **18.** $r_{y\text{-axis}}$ **19.** $r_{y = x}$ **20.** $R_{0°}$

21. $R_{90°}$ **22.** $R_{180°}$ **23.** $T_{-2, 2}$ **24.** $T_{-4, 3}$

---

28. Find point V' such that the
wall is the perpendicular
bisector of VV'. The mirror M is placed where
the wall intersects line
GV'. The translation
is a reflection.

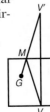

29. Reflect C over the line to C',
and draw line $C'D$.
Place the trans-
former at T, where
line $C'D$ intersects
the power line.
The transformation
is a reflection.

Answers 3.4

1. (–3, 1)
2. (–5, –2)
3. (2, 3)
4. (1, 0)
5. (0, 3)
6. (–2, 2)
7. (–3, –3)
8. (1, –1)
9. (0, 0)
10. (2, –5)
11. (–1, 6)
12. (5, –1)
13. (1, 4)
14. (0, 0)
15. a. (4, –3)
    b. (3, 4)
16. (–2, –5)
17. (4, 3)
18. (–4, –3)
19. (–3, 4)
20. (4, –3)
21. (3, 4)
22. (–4, 3)
23. (2, –1)
24. (0, 0)
25. $(y, -x)$
26. a. (2, 7)
    b. (– 4, –1)
27. (0, 1)

30. Draw a perpendicular from A to point D on the farther bank of the river, and locate A' such that $AA' = CD$, the distance across the river. Place the tunnel at E, where line $A'B$ intersects the bank. Segment FE represents the tunnel. The transformation is a translation.

31. (1) a. $(-2, 4)$
b. $(-3, -1)$
(2) $R_{90°}$

32. (1) a. $(6, 2)$
b. $(15, -1)$
(2) $T_{12, 0}$

33. (1)

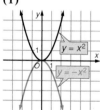

(2) reflection about the x-axis

---

25. If $R_{270°} = R_{180°} \circ R_{90°}$, find $R_{270°}(x, y)$.

26. On the coordinate plane, find the image of $(2, -1)$ under a reflection in the line:
a. $y = 3$ **b.** $x = -1$

27. Under a rotation about C, a point and its image are equidistant from C. Therefore, if A' is the image of A, C lies on the perpendicular bisector of $\overline{AA'}$. Find the center of rotation if the image of $(2, 3)$ is $(-2, 3)$ and the image of $(2, 1)$ is $(0, 3)$.

28. Mrs. Gauger's desk is placed so that she has her back to the office door. She wants to place a mirror on the wall opposite the door so that she can see when someone comes through the doorway. Copy the diagram and show where the mirror should be placed. What transformation is used to locate the position of the mirror?

29. A transformer is to be placed along a power line so that the lines linking the transformer to the homes C and D are the shortest possible. Copy the diagram and indicate where the transformer should be located. What transformation is used to determine the position?

$C \bullet$

$D \bullet$

30. A power line is to be constructed connecting towns A and B. The line must pass through an underground tunnel under the river between the two towns. Because building the tunnel under the river is costly, it will be built perpendicular to the river banks in order to be as short as possible. Find where the underground tunnel should be built in order for the entire line to be as short as possible. What transformation is used to determine the location?

A
$\bullet$

B
$\bullet$

In **31 – 32**: (1) Find the images under the transformations indicated in parts **a** and **b**.
(2) Write a single transformation that is equivalent to the composition in parts **a** and **b**.

31. a. $(r_{y\text{-axis}} \circ r_{y = x})(4, 2)$
b. $(r_{y\text{-axis}} \circ r_{y = x})(-1, 3)$

32. a. $(r_{x = 4} \circ r_{x = -2})(-6, 2)$
b. $(r_{x = 4} \circ r_{x = -2})(3, -1)$

In **33 – 37**: (1) On the same set of axes, draw the graphs of the two given functions.
(2) Describe a transformation under which the graph of the second function is the image of the graph of the first function.

33. $y = x^2$
$y = -x^2$

34. $y = x^2$
$y = x^2 + 2$

35. $y = x^2$
$y = (x + 2)^2$

36. $y = x$
$y = -x$

37. $x = \sqrt{y}$
$y = \sqrt{x}$

---

34. (1)

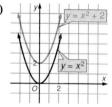

(2) translation
$(x, y) \rightarrow (x, y + 2)$

35. (1)

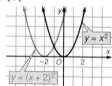

(2) translation
$(x, y) \rightarrow (x - 2, y)$

36. (1)

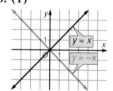

(2) rotation of $90°$ about the origin

37. (1)

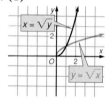

(2) reflection about $y = x$

3.5 The Inverse of a Function

Recall that for the set of real numbers:

0 is the *identity element for addition* since for all $a \in R$, $a + 0 = 0 + a = a$, that is, a remains unchanged when 0 is added. Every real number a has an inverse for addition, $-a$, such that $a + (-a) = -a + a = 0$. Two numbers are *additive inverses* if their sum is the additive identity, 0.

Similarly, 1 is the *identity element for multiplication* since for all $a \in R$, $a \cdot 1 = 1 \cdot a = a$, that is, a remains unchanged when multiplied by 1. Every nonzero real number a has an inverse for multiplication, $\frac{1}{a}$, such that $a\left(\frac{1}{a}\right) = \left(\frac{1}{a}\right)a = 1$.

Two numbers are *multiplicative inverses* if their product is the multiplicative identity, 1.

In the set of functions, the identity function and inverse function for the composition of functions are defined in a similar way.

The Identity Function

Definition: *The identity function is the function I such that $I(x) = x$. I is the function that maps any element of the domain to the same element in the codomain. The domain and codomain of the identity function are the same set, and that set can be any set.*

If $f(x)$ is any function from A to B, then:

$$x \xrightarrow{\ I\ } x \xrightarrow{\ f\ } f(x)$$

$$x \xrightarrow{\ f\ } f(x) \xrightarrow{\ I\ } f(x)$$

Therefore, $f(I(x)) = I(f(x)) = f(x)$.

For example, if $f(x) = x^2 - 2x$ from R to R, then:

$$x \xrightarrow{\ f\ } x^2 - 2x \xrightarrow{\ I\ } x^2 - 2x \quad \text{or} \quad I(f(x)) = f(x)$$

$$x \xrightarrow{\ I\ } x \xrightarrow{\ f\ } x^2 - 2x \qquad \text{or} \quad f(I(x)) = f(x)$$

Inverse Functions

Definition: *The inverse of a function f from A to B is a function f^{-1} from B to A, such that $f^{-1} \circ f = I$ from A to A and $f \circ f^{-1} = I$ from B to B.*

Example 1

Let: $A = \{1, 2, 3\}$ and $B = \{3, 4, 5\}$

$f = \{(1, 3), (2, 4), (3, 5)\}$, a function from A to B
$g = \{(3, 1), (4, 2), (5, 3)\}$, a function from B to A

a. Find $g \circ f$ and $f \circ g$.

b. Draw a conclusion about f and g under composition.

Solution

a. $g \circ f = \{(1, 1), (2, 2), (3, 3)\}$

Note that $g \circ f = I$ from A to A.

$f \circ g = \{(3, 3), (4, 4), (5, 5)\}$

Note that $f \circ g = I$ from B to B.

b. f and g are inverses under composition.

Example 2

If $f(x) = 2x - 4$ and $g(x) = \dfrac{x + 4}{2}$, does $g(x) = f^{-1}(x)$?

Solution

First, find $(f \circ g)(x)$.

$$x \xrightarrow{\;g\;} \frac{x + 4}{2} \xrightarrow{\;f\;} 2\left(\frac{x + 4}{2}\right) - 4 = x$$

Since $f \circ g$ maps x to x, $(f \circ g)(x) = I(x)$.

Then, find $(g \circ f)(x)$.

$$x \xrightarrow{\;f\;} 2x - 4 \xrightarrow{\;g\;} \frac{(2x - 4) + 4}{2} = x$$

Since $g \circ f$ maps x to x, $(g \circ f)(x) = I(x)$.

Therefore, g is the inverse of f.

Answer: Yes, $g(x) = f^{-1}(x)$.

Note that in the above example, f and g are each a $1 - 1$ correspondence from R to R.

Finding the Inverse of a Function

The inverse function f^{-1} of a $1 - 1$ correspondence f can be found by interchanging the coordinates of the ordered pairs of f. For example:

Let: $f = \{(x, y) \mid y = 5 - 2x\}$

Then: $f^{-1} = \{(x, y) \mid x = 5 - 2y\}$

It is customary to write the rule for a function by expressing y in terms of x.

$$x = 5 - 2y$$
$$x - 5 = -2y$$
$$\frac{x - 5}{-2} = \frac{-2y}{-2}$$
$$\frac{5 - x}{2} = y$$

Therefore, $f^{-1} = \left\{(x, y) \mid y = \dfrac{5 - x}{2}\right\}$ or $f^{-1}(x) = \dfrac{5 - x}{2}$.

To verify that the functions are inverses, determine if $(f^{-1} \circ f)(x) = (f \circ f^{-1})(x) = x$.

$$f^{-1} \circ f: \quad x \xrightarrow{\;f\;} 5 - 2x \xrightarrow{\;f^{-1}\;} \frac{5 - (5 - 2x)}{2} = x$$

$$f \circ f^{-1}: \quad x \xrightarrow{\;f^{-1}\;} \frac{5 - x}{2} \xrightarrow{\;f\;} 5 - 2\left(\frac{5 - x}{2}\right) = x$$

Each mapping shows that, for any element x, $f \circ f^{-1}$ and $f^{-1} \circ f$ map x to x. Therefore, $f \circ f^{-1} = f^{-1} \circ f = I$.

Observe from the graphs of these inverse functions, $y = 5 - 2x$ and $y = \dfrac{5 - x}{2}$, that each point on one of the lines has an image point on the other line under a reflection in the line $y = x$.

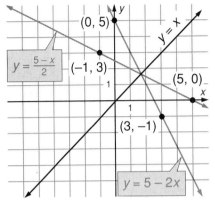

For example, $(3, -1)$ on $y = 5 - 2x$ is the image of $(-1, 3)$ on $y = \dfrac{5 - x}{2}$. Also, $(0, 5)$ is the image of $(5, 0)$.

For each pair (a, b) on one line, there is a corresponding pair (b, a) on the other line. Each line is the image of the other under a reflection in $y = x$.

Example 3

Given: $f(x) = 2x + 6$

a. Find an equation for $f^{-1}(x)$.

b. Graph $f(x)$ and $f^{-1}(x)$ on the same set of axes.

Solution

a.

$y = 2x + 6$	Write $y = f(x)$.
$x = 2y + 6$	Interchange x and y.
$x - 6 = 2y$	Solve for y in terms of x.
$\dfrac{x - 6}{2} = y$	
$\tfrac{1}{2}x - 3 = y$	

Answer: $f^{-1}(x)$ is $y = \tfrac{1}{2}x - 3$.

b. Use the slope-intercept method to graph each line.

$$y = 2x + 6$$
$$\text{slope} = 2$$
$$y\text{-intercept} = 6$$
$$y = \tfrac{1}{2}x - 3$$
$$\text{slope} = \tfrac{1}{2}$$
$$y\text{-intercept} = -3$$

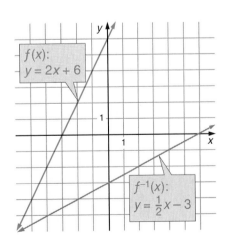

It can be shown that $y = \frac{1}{2}x - 3$ is the image of $y = 2x + 6$ under a reflection in the line $y = x$.

Choose any point on the graph of $f(x)$, such as, $A(-1, 4)$. The corresponding point on $f^{-1}(x)$ is $A'(4, -1)$.

To show that $y = x$ is a line of reflection, show that it is the perpendicular bisector of $\overline{A\,A'}$.

1. $y = x$ passes through the midpoint of $\overline{A\,A'}$.
   Find the midpoint M of $\overline{A\,A'}$.

$$x_M = \frac{-1 + 4}{2} = \frac{3}{2} \qquad y_M = \frac{4 + (-1)}{2} = \frac{3}{2}$$

Since its coordinates are equal, the point $M\left(\frac{3}{2}, \frac{3}{2}\right)$ is on the line $y = x$.

2. The slopes of $y = x$ and $\overline{A\,A'}$ are negative reciprocals.
   Find the slope of each line.

   slope of $\overline{A\,A'}$ is $\frac{4 - (-1)}{-1 - 4} = \frac{5}{-5} = -1$ slope of $y = x$ is 1

   Since their slopes, -1 and 1, are negative reciprocals, the lines are perpendicular.

Thus, $y = x$ is the perpendicular bisector of $\overline{A\,A'}$. Since A can be any point on $f(x)$ and A' the corresponding point on $f^{-1}(x)$, $y = x$ is a line of reflection between the graphs.

Functions Without Inverses

Not every function has an inverse. For example, consider $f = \{(1, 5), (2, 5), (3, 7)\}$. The set of ordered pairs obtained by interchanging the coordinates of the elements of f is $g = \{(5, 1), (5, 2), (7, 3)\}$. Since two pairs of g have the same first coordinate, g is not a function. Therefore, g is not the inverse of f.

Example 4

If $f(x) = x^2$, explain why f from R to R cannot have an inverse function g.

Solution

Recall that a function from set A to set B is the set of ordered pairs (x, y) such that *every* $x \in A$ is paired with *exactly one* $y \in B$. Use counterexamples to show that this definition fails to be satisfied by g.

1. Since there is no real number a such that $(a, -1)$ is a pair in f, there is no pair $(-1, a)$ in g. Therefore, in g, not *every* $x \in R$ is paired with a $y \in R$.

2. Since $(3, 9)$ and $(-3, 9)$ are pairs of f, $(9, 3)$ and $(9, -3)$ are pairs of g. Therefore, in g, there is at least one x that is paired with more than one y.

Restricting the Domain to Allow an Inverse

When a function has no inverse for the given domain and codomain, it is often possible to impose restrictions, thus defining a subset of the function that does have an inverse.

For example, consider the graph of $y = x^2$ and its reflection over the line $y = x$. The reflection is the graph of $y = \pm\sqrt{x}$. Corresponding to every positive real number x, there are two values of y, a positive value and a negative value. However, if the domain of f is restricted to just the positive real numbers and 0, then for every first coordinate there will be no more than one second coordinate. Therefore, the domain of f must be restricted to $[0, \infty)$. Now f is $1-1$. But f must also be onto, that is, the codomain must be the same as the range, $[0, \infty)$.

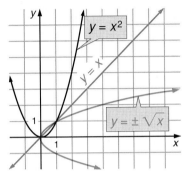 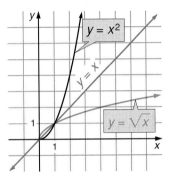

If f is a function from $R^+ \cup \{0\}$ to $R^+ \cup \{0\}$ such that $f(x) = x^2$, then $f^{-1}(x) = \sqrt{x}$.

This is not the only possible restriction. The domain of f could be restricted to $R^- \cup \{0\}$ and the codomain to $R^+ \cup \{0\}$, but the use of the nonnegative real numbers is more common.

Example 5

Restrict the domain of the function illustrated so that the function has an inverse.

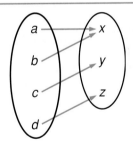

Solution

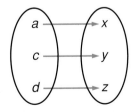   or   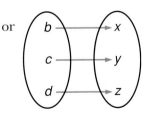

Domain $= \{a, c, d\}$ Domain $= \{b, c, d\}$
Range $= \{x, y, z\}$ Range $= \{x, y, z\}$

Example 6

If $f(x) = \sqrt{9 - x^2}$, then f is a function from S to S where $S = [-3, 3]$, the set of real numbers from -3 to 3 inclusive.

a. Restrict the domain and range of f as necessary in order to form the inverse function, f^{-1}.

b. Write an equation for f^{-1}.

Solution

a. In order to have an inverse, the function must be $1 - 1$ and onto.

In order for f to be $1 - 1$, either the positive or the negative numbers must be eliminated from the domain. Let the restricted domain be $[0, 3]$.

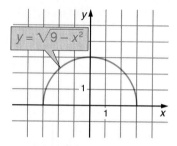

In order for f to be onto, only the range must equal the codomain, that is, the codomain must include only those elements that appear as second coordinates of f.

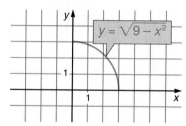

From the graph, it can be seen that the range is the set of nonnegative numbers ≤ 3, or $[0, 3]$.

Answer: The domain and range are each restricted to $[0, 3]$.

b. To write an equation for the inverse, begin with the equation of the function.

$$y = \sqrt{9 - x^2}$$
$$y^2 = 9 - x^2$$
$$x^2 = 9 - y^2 \qquad \text{Interchange } x \text{ and } y.$$
$$x^2 - 9 = -y^2 \qquad \text{Solve for } y.$$
$$9 - x^2 = y^2$$
$$\sqrt{9 - x^2} = y \qquad \text{Use positive values for } y.$$

Answer: An equation for f^{-1} is $y = \sqrt{9 - x^2}$.

Note that f is its own inverse.

Exercises 3.5

In **1 – 4**, write the inverse of the given function.

1. $\{(1, 10), (2, 9), (3, 8), (4, 7)\}$

2. $\{(0, 1), (1, 2), (2, 3), (3, 4)\}$

3. $\{(1, 0), (2, 3), (3, 8), (4, 15)\}$

4. $\{(a, 0), (b, 1), (c, 2), (d, 3)\}$

In **5 – 10**, write an equation for $f^{-1}(x)$.

5. $f(x) = 3x - 1$

6. $f(x) = 2 - x$

7. $f(x) = x - 1$

8. $f(x) = 5x + 5$

9. $f(x) = \frac{1}{3}x - \frac{2}{3}$

10. $f(x) = 1 - \frac{1}{4}x$

11. Let: $f(x) = \sqrt{x + 1}$

    a. Find the largest domain in $R \times R$ for which f is a function.

    b. What is the range of f in $R \times R$?

    c. Find $f^{-1}(x)$, if it exists. What are the domain and range of f^{-1}?

12. If $g(x) = x^3$ defines a function from R to R, find $g^{-1}(x)$.

In **13 – 15**: **a.** Sketch the graph of the given function.
    b. On the same set of axes, sketch the graph of the reflection of the given function over the line $y = x$.
    c. Write an equation of the inverse of the given function.

13. $y = \frac{2}{3}x + 1$

14. $2y + 3x = 12$

15. $y = \sqrt{x}$

In **16 – 17**, restrict the domain of the given function so that the function has an inverse.

16.

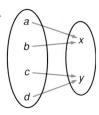

17. $\{(-3, 10), (-2, 5), (-1, 2), (0, 1), (1, 2), (2, 5), (3, 10)\}$

In **18 – 20**, find the largest subsets of the domain and range of the given function for which the function will have an inverse function.

18. $y = -x^2$

19. $y = \sqrt{1 - \frac{x^2}{4}}, x \in [-2, 2]$

20. $y = -x^2 + 2x$

---

1. $\{(10, 1), (9, 2,), (8, 3), (7, 4)\}$

2. $\{(1, 0), (2, 1), (3, 2), (4, 3)\}$

3. $\{(0, 1), (3, 2), (8, 3), (15, 4)\}$

4. $\{(0, a), (1, b), (2, c), (3, d)\}$

5. $f^{-1}(x) = \frac{x + 1}{3}$

6. $f^{-1}(x) = 2 - x$

7. $f^{-1}(x) = x + 1$

8. $f^{-1}(x) = \frac{x - 5}{5}$

9. $f^{-1}(x) = 3x + 2$

10. $f^{-1}(x) = -4x + 4$

11. a. $[-1, \infty)$

    b. $[0, \infty)$

    c. $f^{-1}(x) = x^2 - 1$

    The domain is $[0, \infty)$ and the range is $[-1, \infty)$.

12. $g^{-1}(x) = \sqrt[3]{x}$

13. a. and **b.**

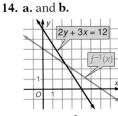

    c. $f^{-1}(x) = \frac{3}{2}x - \frac{3}{2}$

14. a. and **b.**

    c. $f^{-1}(x) = -\frac{2}{3}x + 4$

---

15. a. and **b.**

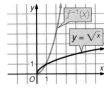

    c. $f^{-1}(x) = x^2$, for $x \geq 0$

16. domain: $\{a, c\}$

17. domain: $\{3, 2, 1, 0\}$

18. domain: $[0, \infty)$ or $(-\infty, 0]$
    range: $(-\infty, 0]$

19. domain: $[-2, 0]$ or $[0, 2]$
    range: $[0, 1]$

20. domain: $[1, \infty)$ or $(-\infty, 1]$
    range: $(-\infty, 1]$

3.6 Using Related Functions to Graph

As mentioned earlier, most functions in $R \times R$ can be graphed by locating some points of the function and drawing a smooth curve through them. This usually requires some awareness of how the function values change as the values of x change, and can be a time-consuming task. However, it is often possible to obtain the graph of a function by applying the strategy of considering the graph of a simpler, related function. The following discussions make use of the graphs of the common functions outlined in Section 3.1.

Reflection in the y-axis

The graph of $f(-x)$ is a reflection in the y-axis of the graph of $f(x)$.

For example, if $f(x) = 2x + 4$, then the graph of $f(-x) = 2(-x) + 4 = -2x + 4$ is the reflection in the y-axis.

To verify, sketch the graphs of $f(x)$ and $f(-x)$ by using intercepts.

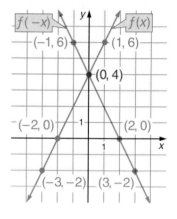

$$f(x) = y = 2x + 4$$
$$-2x + y = 4$$
$$\frac{-2x}{4} + \frac{y}{4} = \frac{4}{4}$$
$$\frac{x}{-2} + \frac{y}{4} = 1$$

$$f(-x) = y = -2x + 4$$
$$2x + y = 4$$
$$\frac{2x}{4} + \frac{y}{4} = \frac{4}{4}$$
$$\frac{x}{2} + \frac{y}{4} = 1$$

The x-intercept is -2.

The y-intercept is 4.

The x-intercept is 2.

The y-intercept is 4.

The coordinates of the points shown on the graph satisfy the definition of reflection in the y-axis.

For an algebraic demonstration, let $g(x) = f(-x)$.

If $(1, 6)$ is a point on $f(x)$, then $(-1, 6)$ is a point on $g(x)$.

$$g(-1) = f(-(-1)) = f(1) = 6$$

If $(-3, -2)$ is a point on $f(x)$, then $(3, -2)$ is a point on $g(x)$.

$$g(3) = f(-(3)) = f(-3) = -2$$

Thus, to graph $y = -2x + 4$, you could sketch the image of the graph of $y = 2x + 4$ under a reflection in the y-axis.

Example 1

Using the graph of $y = x^3$, sketch the graph of $y = -x^3$.

Solution

Let: $f(x) = x^3$

Then: $f(-x) = (-x)^3 = -x^3$

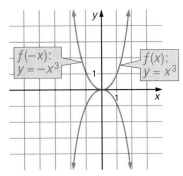

The graph of $y = -x^3$ is the reflection in the y-axis of $y = x^3$.

Use some points to sketch $y = x^3$ whose general shape you should try to remember since it is a commonly occurring curve.

Reflection in the x-axis

The graph of $-f(x)$ is a reflection in the x-axis of the graph of $f(x)$.

For example, if $f(x) = 2x - 1$, then the graph of $-f(x) = -(2x - 1) = -2x + 1$ is the image of the graph of $f(x)$ under a reflection in the x-axis. Sketch the graphs of $f(x)$ and $-f(x)$ by using intercepts.

$$f(x) = y = 2x - 1 \qquad\qquad -f(x) = y = -2x + 1$$

$$-2x + y = -1 \qquad\qquad\qquad 2x + y = 1$$

$$\frac{-2x}{-1} + \frac{y}{-1} = \frac{-1}{-1} \qquad\qquad \frac{2x}{1} + \frac{y}{1} = \frac{1}{1}$$

$$\frac{x}{\frac{1}{2}} + \frac{y}{-1} = 1 \qquad\qquad\qquad \frac{x}{\frac{1}{2}} + \frac{y}{1} = 1$$

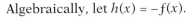

The x-intercept is $\frac{1}{2}$.   The x-intercept is $\frac{1}{2}$.

The y-intercept is -1.   The y-intercept is 1.

Algebraically, let $h(x) = -f(x)$.

If $(2, 3)$ is a point on $f(x)$, then $(2, -3)$ is a point on $h(x)$.

$$h(2) = -f(2) = -3$$

Thus, to graph $y = -(2x - 1)$, you could sketch the image of the graph of $y = 2x - 1$ under a reflection in the x-axis.

Example 2

Using the graph of $y = x^2$, sketch the graph of $y = -x^2$.

Solution

Let: $f(x) = x^2$

Then: $-f(x) = -x^2$

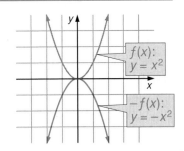

The graph of $y = -x^2$ is the reflection in the x-axis of $y = x^2$.

Translations

Many functions are the images of familiar basic functions under translations. For example, the graph of $y = x^2$ is a parabola whose vertex is at the origin. The image of $y = x^2$ under the translation $T_{1, 2}$ is a parabola since the shape of the curve remains unchanged. The vertex of the image parabola is (1, 2), the image of the origin under $T_{1, 2}$. Consider a horizontal line and a vertical line through (1, 2) to be a new set of axes called the x'-axis and the y'-axis. With reference to this new set of axes, the equation of the image parabola is $y' = (x')^2$. But $x' = x - 1$ and $y' = y - 2$. Therefore, with reference to the x-axis and the y-axis, the equation of the image parabola is:

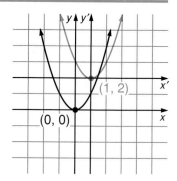

$$y - 2 = (x - 1)^2 \text{ or } y = (x - 1)^2 + 2$$

A graphing calculator, or a computer program designed to graph functions entered by the user, will help you to visualize translations similar to the one described above. Enter each of the following sets of related functions and describe how the graph of the first function of each set is translated to obtain the other functions of that set.

Set 1: $y_1 = 2x$ $y_2 = y_1 - 2$ $y_3 = y_1 - 3$ $y_4 = y_1 - 5$

Set 2: $y_1 = 2x$ $y_2 = y_1 + 2$ $y_3 = y_1 + 3$ $y_4 = y_1 + 5$

Set 3: $y_1 = x^2$ $y_2 = y_1 - 2$ $y_3 = y_1 - 3$ $y_4 = y_1 - 5$

Set 4: $y_1 = x^2$ $y_2 = y_1 + 2$ $y_3 = y_1 + 3$ $y_4 = y_1 + 5$

Set 5: $y_1 = x^2$ $y_2 = (x - 2)^2$ $y_3 = (x - 3)^2$ $y_4 = (x - 5)^2$

Set 6: $y_1 = x^2$ $y_2 = (x + 2)^2$ $y_3 = (x + 3)^2$ $y_4 = (x + 5)^2$

Set 7: $y_1 = x^2$ $y_2 = (x - 2)^2 + 3$ $y_3 = (x - 2)^2 - 3$ $y_4 = (x + 2)^2 + 3$

You should have observed the following relationships:

1. Adding a negative number to the function value moves the graph down.

2. Adding a positive number to the function value moves the graph up.

3. Adding a negative number to x moves the graph to the right.

4. Adding a positive number to x moves the graph to the left.

When using a graphing calculator, it is necessary to enter the subset of the domain and range that you want to display. For example: to graph $y = (x - 25)^2 + 30$, it is necessary to include values of x and y around (25, 30) in order to include the turning point of the parabola; to graph $y = (x + 75)^3 - 20$, it is necessary to enter values around (–75, –20) in order to display the critical portion of the graph.

Thus, in general, the graph of $y - k = f(x - h)$ is the image of $y = f(x)$ under the translation $T_{h, k}$.

Example 3

Using the graph of $y = x^2$, sketch the graph of $y + 1 = (x - 3)^2$.

Solution

Compare $f(x)$: $y = x^2$

and $g(x)$: $y + 1 = (x - 3)^2$

Let $x' = x - 3$ and $y' = y + 1$

$0 = 3 - 3$ $0 = -1 + 1$

Thus, the new origin should be at $(3, -1)$.

Draw the x'-axis and y'-axis with origin at $(3, -1)$ and sketch the familiar curve $y' = (x')^2$, which is the graph of $y + 1 = (x - 3)^2$.

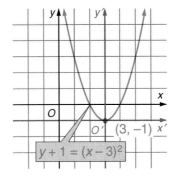

Example 4

Sketch the graphs of $f(x) = x^3$ and $h(x) = (x + 3)^3$.

Solution

$h(x) = f(x + 3)$

Sketch the graph of $h(x)$ by sketching the graph of $f(x)$ under the translation $T_{-3, 0}$, that is, under a translation of 3 units to the left.

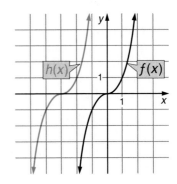

Example 5

Draw the graph of $g(x) = x^2 - 4x + 1$.

Solution

For $g(x) = x^2 - 4x + 1$, complete the square and write the function in terms of the square of a binomial.

Recall that $(x - 2)^2 = x^2 - 4x + 4$. Therefore, the constant term of the trinomial, $g(x)$, must be 4 to complete the square. Begin by adding 3 to each side of the equation.

$$g(x) = x^2 - 4x + 1$$
$$g(x) + 3 = x^2 - 4x + 1 + 3$$
$$g(x) + 3 = x^2 - 4x + 4$$
$$g(x) + 3 = (x - 2)^2$$

Let: $f(x) = x^2$ and $g(x) = y$

$$y + 3 = f(x - 2)$$

The graph of $g(x)$ is the image of $f(x)$ under the translation $T_{2, -3}$.

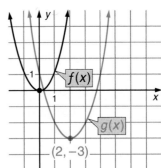

Using Composition of Functions to Graph

Many functions can be graphed by using a composition of two or more transformations.

Example 6

Draw the graph of $y = -x^2 - 2$.

Solution

1. Recognize that $y = x^2$ is the basic curve, and sketch it.

2. Reflect $y = x^2$ in the x-axis, to obtain $y = -x^2$. Call this image h.

3. Find the image of h under $T_{0, -2}$.

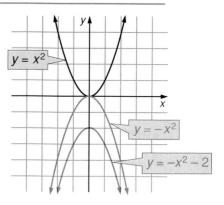

The Graph of $af(x)$

Frequently, the graph of a function can be drawn with reference to the graph of a familiar basic function because the ordinates of points with the same abscissa on the two graphs are related by a constant multiple. For example, if $p(x) = x - 3$ and $q(x) = 2x - 6$, then $q(x) = 2p(x)$. Each point on the graph of $q(x)$ has an ordinate that is twice the ordinate of the point with the same abscissa on the graph of $p(x)$.

For $f(x)$, the function whose graph is shown at the right, the domain is $[-4, 4]$. If $g(x) = 3f(x)$, then each pair $(a, f(a))$ of f can be used to locate a pair $(a, 3f(a))$ of g. The distance from a point in g to the x-axis is three times the distance from a point with the same abscissa in f to the x-axis.

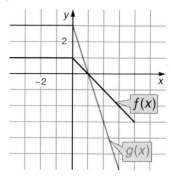

Example 7

Use the graph of $y = x^2$ to sketch the graphs of $y = 2x^2$ and $y = \frac{1}{2}x^2$.

Solution

Let the distance from the x-axis to each point of $y = 2x^2$ be twice the distance of a point of $y = x^2$ from the x-axis.

Let the distance from the x-axis to each point of $y = \frac{1}{2}x^2$ be one-half the distance of a point of $y = x^2$ from the x-axis.

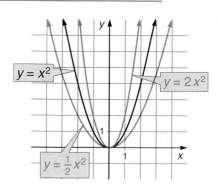

144

The Graph of y = f(ax)

The graphs of two functions can also be related because the abscissas of the points with the same ordinate are related by a constant multiple.

For $f(x)$, the function whose graph is shown below at the left, the domain is $[-6, 6]$. If $g(x) = f(3x)$, then for each ordered pair $(a, f(a))$ in f, there is an ordered pair $(a, f(3a))$, in g. For example, if $(-3, 6)$ is in f, then $(-1, 6)$ is in g; if $(6, 0)$ is in f, then $(2, 0)$ is in g. Note that g is the result of compressing f in the horizontal direction by a factor of $\frac{1}{3}$. If the domain of f is $[-6, 6]$, then the domain of g is $[-2, 2]$.

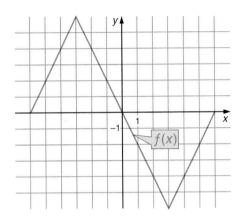
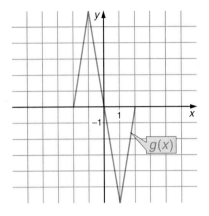

Example 8

Use the graph of $y = x^2$ to sketch the graphs of $y = (2x)^2$ and $y = \left(\frac{1}{2}x\right)^2$.

Solution

If $(1, 1)$ and $(-1, 1)$ are points of $y = x^2$, then $\left(\frac{1}{2}, 1\right)$ and $\left(-\frac{1}{2}, 1\right)$ are points of $y = (2x)^2$, and $(2, 1)$ and $(-2, 1)$ are points of $y = \left(\frac{1}{2}x\right)^2$.

If $(2, 4)$ and $(-2, 4)$ are points of $y = x^2$, then $(1, 4)$ and $(-1, 4)$ are points of $y = (2x)^2$, and $(4, 4)$ and $(-4, 4)$ are points of $y = \left(\frac{1}{2}x\right)^2$.

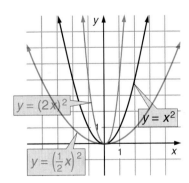

A graphing calculator, or a computer program, can be used to visualize graphs whose ordinates or abscissas are related by a constant multiple. Enter each of the following sets of functions and describe for each set how the graphs of y_2, y_3, and y_4 are related to the graph of y_1.

Set 1: $y_1 = x^2$ $y_2 = 2y_1$ $y_3 = \frac{1}{2}y_1$ $y_4 = -y_1$

Set 2: $y_1 = |x|$ $y_2 = 2y_1$ $y_3 = \frac{1}{2}y_1$ $y_4 = -y_1$

Set 3: $y_1 = x^3$ $y_2 = 2y_1$ $y_3 = \frac{1}{2}y_1$ $y_4 = -y_1$

Set 4: $y_1 = \sin x$ $y_2 = 2y_1$ $y_3 = \frac{1}{2}y_1$ $y_4 = -y_1$

Set 5: $y_1 = x^2$ $y_2 = (2x)^2$ $y_3 = \left(\frac{1}{2}x\right)^2$ $y_4 = (-x)^2$

Set 6: $y_1 = |x|$ $y_2 = |2x|$ $y_3 = \left|\frac{1}{2}x\right|$ $y_4 = |-x|$

Set 7: $y_1 = x^3$ $y_2 = (2x)^3$ $y_3 = \left(\frac{1}{2}x\right)^3$ $y_4 = (-x)^3$

Set 8: $y_1 = \sin x$ $y_2 = \sin 2x$ $y_3 = \sin \frac{1}{2}x$ $y_4 = \sin(-x)$

You should have observed the following relationships:

1. Multiplying a function value by a number greater than 1 stretches the graph in the vertical direction.

2. Multiplying a function value by a positive number less than 1 shrinks the graph in the vertical direction.

3. The graph of $-f(x)$ is the reflection of $f(x)$ in the x-axis.

4. The graph of $f(ax)$ is the graph of $f(x)$:
   (a) compressed horizontally by a factor of $\dfrac{1}{a}$ if $a > 1$, or
   (b) stretched horizontally by a factor of $\dfrac{1}{a}$ if $0 < a < 1$.

5. The graph of $f(-x)$ is a reflection of the graph of $f(x)$ in the y-axis.

Let $g(x) = 4x^2$ or $g(x) = (2x)^2$. Then the graph of $f(x) = x^2$ can be thought of as being stretched in the vertical direction by a factor of 4, since $g(x) = 4f(x)$, or $f(x)$ can be thought of as being compressed in the horizontal direction by a factor of $\frac{1}{2}$, since $g(x) = f(2x)$.

In general, if (x, y) is an ordered pair of the function $f(x)$, then (x, ay) is an ordered pair of the function $af(x)$ and $\left(\dfrac{1}{a}x, y\right)$ is a pair of the function $f(ax)$. The graph of $af(x)$ can be thought of as the result of stretching or compressing the graph of $f(x)$ in the vertical direction

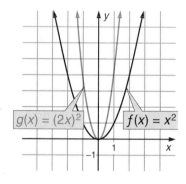

The Graph of the Sum of Two Functions

As some graphs are related by multiples of the ordinates, others are related by the sum of the ordinates. The graphs of $g(x)$ and $h(x)$ are shown at the right. If $f(x) = g(x) + h(x)$, then in order to find the graph of $f(x)$, add $g(x)$ and $h(x)$ for any x in the domain of both $g(x)$ and $h(x)$.

Note: When $g(x) = 0$, then $f(x) = h(x)$ and
when $h(x) = 0$, then $f(x) = g(x)$.

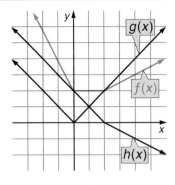

Example 9

Sketch the graph of $f(x) = x^2 + 2x$.

Solution

Sketch the graphs of $h(x) = x^2$ and $g(x) = 2x$ on the same set of axes. Add the y-coordinates of points with the same x-coordinate.

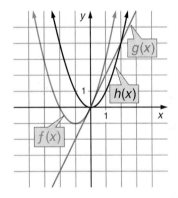

Note that:
at $x = -1$, $h(-1) = 1$, $g(-1) = -2$, and
$f(-1) = 1 + (-2) = -1$;
at $x = -2$, $h(-2) = 4$, $g(-2) = -4$, and
$f(-2) = 4 + (-4) = 0$.

Example 10

Sketch the graph of $f(x) = x + [x]$.

Solution

Sketch the graphs of $h(x) = x$ and $g(x) = [x]$ on the same set of axes. To add the corresponding y-values, note:

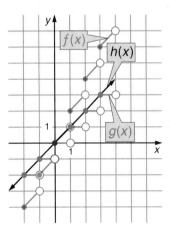

In the interval $[0, 1)$, $g(x) = 0$.
Therefore, $f(x) = h(x)$.

In the interval $[1, 2)$, $g(x) = 1$.
Therefore, $f(x)$ is one unit above $h(x)$.

In the interval $[2, 3)$, $g(x) = 2$.
Therefore, $f(x)$ is two units above $h(x)$.

This pattern is used to draw the graph.

1.

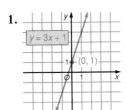

$y = 3x + 1$

$(0, 1)$

2.

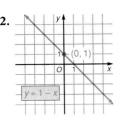

$(0, 1)$

$y = 1 - x$

3.

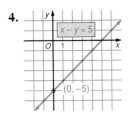

$y = \frac{3}{2}x - 2$

$(0, -2)$

4.

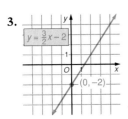

$x - y = 5$

$(0, -5)$

5.

$(0, 4)$

$2x + 3y = 12$

6.

$(0, 5)$

$\frac{x}{5} + \frac{y}{5} = 1$

7.

$(0, 3)$

$(-1, 0)$

$\frac{x}{} - \frac{y}{} = $

Exercises 3.6

In **1 – 5**, sketch the graph using the slope and y-intercept.

1. $y = 3x + 1$ **2.** $y = 1 - x$ **3.** $y = \frac{3}{2}x - 2$ **4.** $x - y = 5$ **5.** $2x + 3y = 12$

In **6 – 10**, write the equation in intercept form and use the x- and y-intercepts to sketch the graph.

6. $x + y = 5$ **7.** $y = 3x + 3$ **8.** $y = 4 - x$ **9.** $y = \frac{3}{4}x - 3$ **10.** $4x - y = 2$

In **11 – 16**, sketch: **a.** $f(x)$ **b.** $-f(x)$ **c.** $f(-x)$

11. $f(x) = 2x - 4$ **12.** $f(x) = 3 - 2x$ **13.** $f(x) = x^2$

14. $f(x) = x^3$ **15.** $f(x) = \sqrt{x}$ **16.** $f(x) = |x|$

In **17 – 18**, given the graph of $f(x)$, sketch: **a.** $-f(x)$ **b.** $f(-x)$

17.

18.

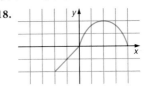

In **19 – 22**, sketch the graph using a translation of $y = x^2$.

19. $y = (x - 2)^2$ **20.** $y + 1 = (x - 1)^2$ **21.** $y = x^2 + 6x + 7$ **22.** $y = x^2 - 2x - 2$

In **23 – 26**, sketch the graph using a translation of $y = |x|$.

23. $y = |x - 1|$ **24.** $y - 3 = |x + 2|$ **25.** $y = |x| + 5$ **26.** $y = 4 + |x - 3|$

27. Translate the graph of $y = \sqrt{x}$ to sketch the graph of:

a. $y = \sqrt{x + 2}$ **b.** $y = 3 + \sqrt{x - 1}$

In **28 – 34**, use translations and reflections to sketch the graph.

28. $y = -x^2 + 2$ **29.** $y = -(x + 2)^2$ **30.** $y = -x^2 + 4x$ **31.** $y = (1 - x)^2$

32. $y = \sqrt{4 - x}$ **33.** $y = (x - 1)^3 + 2$ **34.** $y = -x^3 - 1$

Hint: Sketch $f(x) = (x - 1)^2$.

In **35 – 43**, let $f(x) = x^2$.

a. Express the given function in terms of $f(x)$.

b. Graph the function.

35. $y = (x - 2)^2$ **36.** $y = x^2 - 2$ **37.** $y = 2 - x^2$

38. $y = x^2 - 8x + 10$ **39.** $y = -x^2 - 2x$ **40.** $y = \frac{1}{3}x^2$

41. $y = \left(\frac{1}{3}x\right)^2$ **42.** $y = -\frac{1}{3}x^2$ **43.** $y = -\left(\frac{1}{3}x\right)^2$

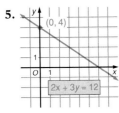

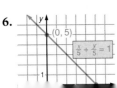

8.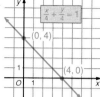

9.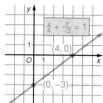

10.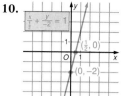

11. a.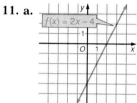

b.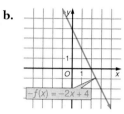

c.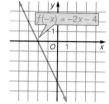

12. a.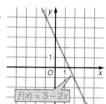

b.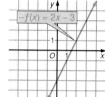

c.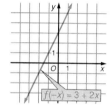

13. a.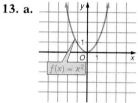

b.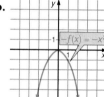

c.

14. a.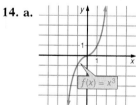

b.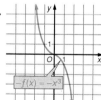

c.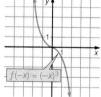

15. a.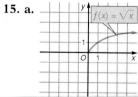

b.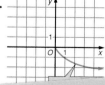

c.

16. a.
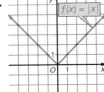

b.

c.

17. a.

b.
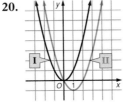

18. a.

b.
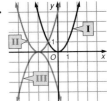

19.
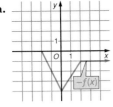

I $y = x^2$ **II** $y = (x - 2)^2$

20.

I $y = x^2$
II $y = (x - 1)^2 - 1$

21.
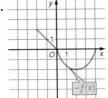

I $y = x^2$
II $y = (x + 3)^2 - 2$

22.
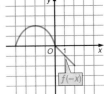

I $y = x^2$
II $y = (x - 1)^2 - 3$

23.
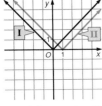

I $y = |x|$ **II** $y = |x - 1|$

24.
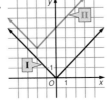

I $y = |x|$
II $y - 3 = |x + 2|$

25.
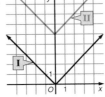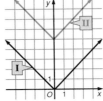

I $y = |x|$
II $y = |x| + 5$

26.
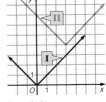

I $y = |x|$
II $y = 4 + |x - 3|$

27. a.
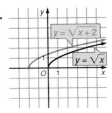

b.
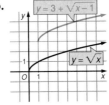

28.

I $y = x^2$
II $y = -x^2$
III $y = -x^2 + 2$

29.

I $y = x^2$
II $y = (x + 2)^2$
III $y = -(x + 2)^2$

30. $y = -x^2 + 4x$
$= -(x-2)^2 + 4$

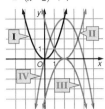 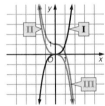

I $y = x^2$ **II** $y = (x-2)^2$
III $y = -(x-2)^2$
IV $y = -(x-2)^2 + 4$

31.

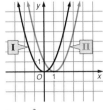

I $y = x^2$
II $y = (1-x)^2$

32.

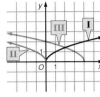

I $y = \sqrt{x}$ **II** $y = \sqrt{-x}$
III $y = \sqrt{4-x}$

33.

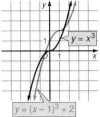

$y = (x-1)^3 + 2$

34.

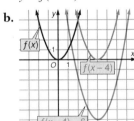

I $y = x^3$ **II** $y = -x^3$
III $y = -x^3 + 1$

35. a. $y = f(x-2)$

b.

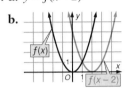

36. a. $y = f(x) - 2$

b.

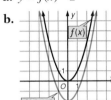

37. a. $y = 2 - f(x)$

b.

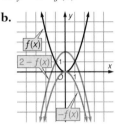

38. a. $y = f(x-4) - 6$

b.

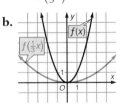

39. a. $y = -f(x+1) + 1$

b.

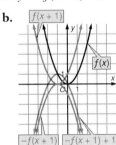

40. a. $y = \frac{1}{3}f(x)$

b.

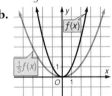

41. a. $y = f\left(\frac{1}{3}x\right)$

b.

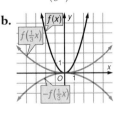

42. a. $y = -\frac{1}{3}f(x)$

b.

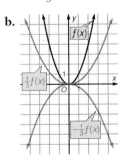

43. a. $y = -f\left(\frac{1}{3}x\right)$

b.

44.

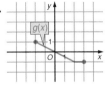

45.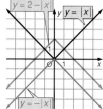

46.

47.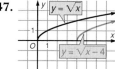

48.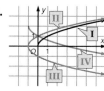

I $y = \sqrt{x}$
II $y = \sqrt{x + 1}$
III $y = -\sqrt{x + 1}$
IV $y = 2 - \sqrt{x + 1}$

49.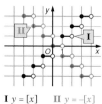

I $y = [x]$ **II** $y = -[x]$

50.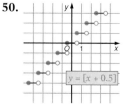

51.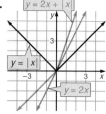

52.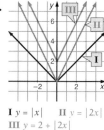

I $y = |x|$ **II** $y = |2x|$
III $y = 2 + |2x|$

53. a. $f(x) = x^4 - 3x^2$
$f(-x) = (-x)^4 - 3(-x)^2$
$= x^4 - 3x^2$
$f(x) = f(-x)$

b. $g(x) = x^2 - 3x$
$g(-x) = (-x)^2 - 3(-x)$
$= x^2 + 3x$
$g(x) \neq g(-x)$

54. a. $h(x) = x^3 - x$
$h(-x) = (-x)^3 - (-x)$
$= -x^3 + x$
$= -(x^3 - x)$
$-h(x) = h(-x)$

b. $g(x) = x^3 + x^2$
$g(-x) = (-x)^3 + (-x)^2$
$= -x^3 + x^2$
$-g(x) = -x^3 - x^2$
$-g(x) \neq g(-x)$

55. even

56. neither

57. even

58. neither

59. neither

60. odd

61.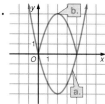

62.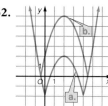

63.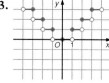

64.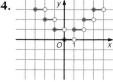

Exercises 3.6 *(continued)*

44. The graph shows the function $f(x)$ whose domain is $[-2, 3]$.

Sketch the graph of g with a domain of $[-2, 3]$, if $g(x) = f\left(\frac{1}{2}x\right)$.

In **45 – 52**, sketch the graph.

45. $y = 2 - |x|$ **46.** $y = |x + 3|$ **47.** $y = \sqrt{x - 4}$ **48.** $y = 2 - \sqrt{x + 1}$

49. $y = -[x]$ **50.** $y = [x + 0.5]$ **51.** $y = 2x + |x|$ **52.** $y = 2 + |2x|$

53. An *even function* is a function f such that $f(x) = f(-x)$. An even function is symmetric with respect to the y-axis, that is, if $(a, b) \in f$, then $(-a, b) \in f$.

   a. Show that $f(x) = x^4 - 3x^2$ is an even function.

   b. Show that $g(x) = x^2 - 3x$ is not an even function.

54. An *odd function* is a function such that $-f(x) = f(-x)$. An odd function is symmetric with respect to the origin, that is, if $(a, b) \in f$, then $(-a, -b) \in f$.

   a. Show that $h(x) = x^3 - x$ is an odd function.

   b. Show that $g(x) = x^3 + x^2$ is not an odd function.

In **55 – 60**, identify the given function as even, odd, or neither.

55. $y = x^2$ **56.** $y = x + 1$ **57.** $y = |x|$

58. $y = \sqrt{x + 1}$ **59.** $y = |x + 1|$ **60.** $y = x(x^2 - 1)$

61. Sketch the graph of:

   a. $p(x) = x^2 - 4x$

   b. $q(x) = |x^2 - 4x|$

62. Sketch the graph of:

   a. $y = |x^2 - 4x| - 2$

   b. $y = |x^2 - 4x - 2|$

In **63 – 68**, sketch the graph of the function described by a piecewise definition.

63. $y = \begin{cases} [x] \text{ if } x \geq 0 \\ [-x] \text{ if } x < 0 \end{cases}$

64. $y = \begin{cases} [x] \text{ if } x \geq 0 \\ -[x] \text{ if } x < 0 \end{cases}$

65. $y = \begin{cases} x^2 \text{ if } x \geq 0 \\ -x^2 \text{ if } x < 0 \end{cases}$

66. $y = \begin{cases} 3 \text{ if } x > 0 \\ 0 \text{ if } x = 0 \\ -3 \text{ if } x < 0 \end{cases}$

67. $y = \begin{cases} -x \text{ if } x \geq 0 \\ x \text{ if } x < 0 \end{cases}$

68. $y = \begin{cases} 0 \text{ if } x \leq 0 \\ 10 \text{ if } 0 < x \leq 5 \\ 2x \text{ if } x > 5 \end{cases}$

65.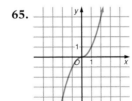

66.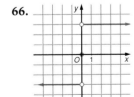

67.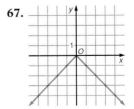

68.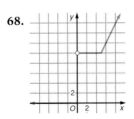

3.7 Rational Functions

If you must travel 20 miles to get to work every morning, the time that it takes to get there varies ***inversely*** as the average rate at which you make the trip. In general, for a constant distance, time varies inversely as the rate of speed. If x and y represent two quantities that vary inversely, then the relationship of x and y can be expressed by the formula $xy = c$ or $y = \frac{c}{x}$, where c is a nonzero constant. The function defined by $y = \frac{c}{x}$ is an example of a *rational function*.

■ **Definition:** *A rational function is the ratio of two polynomial functions.*

If $f(x) = \frac{g(x)}{h(x)}$, *then the domain of f is the set of elements common to the domains of g and of h for which $h(x) \neq 0$.*

Consider the rational function $y = \frac{1}{x}$, whose numerator is a constant function (which is a polynomial of degree 0) and whose denominator is the identity function (which is a polynomial of degree 1).

The domain and range of f are each the set of nonzero real numbers, written $R/\{0\}$. Note that 0 is excluded from the domain because $\frac{1}{x}$ is undefined at $x = 0$, and from the range because the value of $\frac{1}{x}$ can never be 0.

The graph of $y = \frac{1}{x}$ has two branches, one in the first quadrant when x and y are both positive and one in the third quadrant when x and y are both negative.

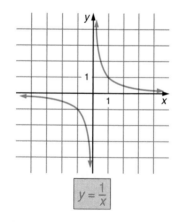

As $|x|$ increases, $\left|\frac{1}{x}\right|$ decreases and approaches 0. As the positive values of x increase, the values of y approach 0 from above, and as the negative values of x decrease, the values of y approach 0 from below. As $|x|$ approaches infinity, the graph approaches, but does not have a point in common with, the x-axis. The x-axis (the line whose equation is $y = 0$) is a *horizontal asymptote* of the graph of $y = \frac{1}{x}$.

Similarly, as $|x|$ decreases, $\left|\frac{1}{x}\right|$ increases. For any positive number B, no matter how large, it is possible to find a positive value for x for which $\frac{1}{x}$ is greater than B and a negative value of x for which $\frac{1}{x}$ is less than $-B$. In other words, as $|x|$ approaches 0, $\left|\frac{1}{x}\right|$ approaches infinity. As $|x|$ approaches 0, the graph of $y = \frac{1}{x}$ approaches, but does not have a point in common with, the y-axis. The y-axis (the line whose equation is $x = 0$) is a *vertical asymptote* of the graph of $y = \frac{1}{x}$.

To formulate a general definition for asymptote, the following notation is convenient.

Let $x \to a$, where a is a real number, mean that x is approaching a. That is, the values of x are getting closer to a but $x \neq a$.

Let $x \to \infty$ mean that x is approaching infinity. That is, the values of x are increasing such that for any large number, the value of x is greater than that number.

Let $x \to -\infty$ mean that the values of x, a negative number, are decreasing such that for any negative number, the value of x is less than that number.

■ **Definition:** *A line* $y = a$ *is a* *horizontal asymptote* *of* f *if, as* $x \to \infty$ *or* $x \to -\infty$, $f(x) \to a$.

A line $x = a$ *is a* *vertical asymptote* *of* f *if, as* $x \to a$, $f(x) \to \infty$ *or* $f(x) \to -\infty$.

Example 1

Sketch the graph of $g(x) = \dfrac{1}{x-1}$, noting asymptotes of the curve and the domain of g.

Solution

Recall that the graph of $y - k = f(x - h)$ is the image of $y = f(x)$ under the translation $T_{h,\,k}$. Since $g(x)$ is the image of $f(x) = \dfrac{1}{x}$ under $T_{1,\,0}$, the graph of g can be drawn by translating the graph of f 1 unit to the right.

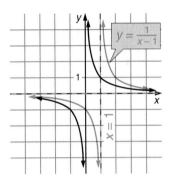

The x-axis is a horizontal asymptote and the line $x = 1$, the image of the y-axis under the translation, is a vertical asymptote.

The domain of g is the set of all real numbers except 1 (written $R/\{1\}$), because $x = 1$ would make the denominator equal to 0.

Example 2

Sketch the graph of $f(x) = \dfrac{3x + 7}{x + 2}$ and write equations of its asymptotes.

Solution

Dividing the numerator by the denominator gives the following:

$$y = 3 + \frac{1}{x + 2} \quad \longleftarrow \quad \begin{array}{r} 3 \\ x + 2 \overline{)\, 3x + 7} \\ \underline{3x + 6} \\ 1 \end{array}$$

$$y - 3 = \frac{1}{x + 2}$$

The graph of f can be drawn by drawing the image of $y = \dfrac{1}{x}$ under $T_{-2,\,3}$. Since the asymptotes of $y = \dfrac{1}{x}$ are the x-axis and the y-axis, the asymptotes of $y - 3 = \dfrac{1}{x + 2}$ are the images of the x-axis and y-axis, that is, the lines $y = 3$ and $x = -2$.

The domain is $R\,/\{-2\}$ and the range is $R\,/\{3\}$.

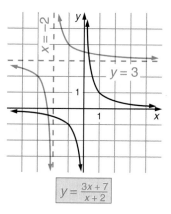

$$y = \frac{3x + 7}{x + 2}$$

Example 3

Sketch the graph of $g(x) = -\dfrac{1}{x}$.

Solution

If $h(x) = \dfrac{1}{x}$, then $g(x) = -h(x)$. Therefore, g is the reflection of h in the x-axis.

The graph is symmetric with respect to the line $y = x$ and $g(x)$ is its own inverse under composition of functions, that is, $g(x) = g^{-1}(x)$. The graph of $g(x)$ is also the reflection of the graph of $h(x)$ under a reflection in the y-axis. The asymptotes of g are the x-axis and the y-axis.

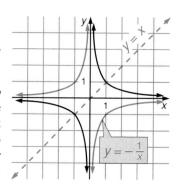

$$y = -\frac{1}{x}$$

In the example above, 0 was excluded from the domain, and $x = 0$ was a vertical asymptote. Also, 0 was excluded from the range, and $y = 0$ was a horizontal asymptote. However, it is not always the case that the values of x and y excluded from the domain and range define the vertical and horizontal asymptotes.

Example 4

Sketch the graph of $f(x) = \dfrac{x^2 - 1}{x + 1}$.

Solution

When $x = -1$, $f(x)$ is undefined. Therefore, the domain of f is $R\,/\{-1\}$.

If $x \neq -1$:

$$\frac{x^2 - 1}{x + 1} = \frac{(x + 1)(x - 1)}{x + 1} = x - 1$$

Since $f(x) = x - 1$, for all values of x except $x = -1$, as $x \to -1$, $f(x) \to -2$, a finite value. The graph of f is the same as the graph of $y = x - 1$ except for the point $(-1, -2)$, which is not an element of f.

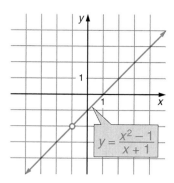

$$y = \frac{x^2 - 1}{x + 1}$$

Example 5

Sketch the graph of $f(x) = \dfrac{1}{x^2}$.

Solution

Note that $f(x) = f(-x)$, since $\dfrac{1}{(-x)^2} = \dfrac{1}{x^2}$. Therefore, the graph is symmetric with respect to the y-axis.

As $|x| \to \infty, y \to 0$.

As $x \to 0, y \to \infty$.

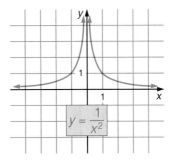

Therefore, the x-axis and the y-axis are the horizontal and vertical asymptotes. The domain is $R/\{0\}$ and the range is R^+.

Example 6

Sketch the graph of $y - 3 = \dfrac{1}{(x + 1)^2}$.

Solution

The graph of $y - 3 = \dfrac{1}{(x + 1)^2}$ is the image of $y = \dfrac{1}{x^2}$ under $T_{-1, 3}$. The domain is $R/\{-1\}$ and the range is $R/\{3\}$. The vertical asymptote is $x = -1$ and the horizontal asymptote is $y = 3$.

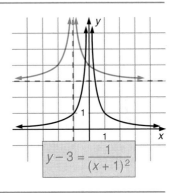

Example 7

Sketch the graph of $f(x) = \dfrac{x^3 - 1}{x}$.

Solution

$$f(x) = \frac{x^3}{x} - \frac{1}{x}$$

$$f(x) = x^2 - \frac{1}{x}$$

Draw the graphs of $y = x^2$ and $y = \dfrac{1}{x}$. To find the graph of $x^2 - \dfrac{1}{x}$, subtract the value of $\dfrac{1}{x}$ from the value of x^2 for any x in the domain of both functions. Consider the values of x in intervals.

When x is negative:

$\dfrac{1}{x}$ is negative and $f(x)$ is x^2 increased by $\left|\dfrac{1}{x}\right|$. This portion of the graph of f is above the graph of $y = x^2$.

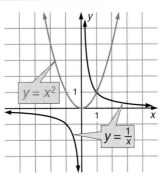

1.

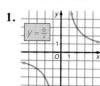

$x = 0, y = 0$

2.

$x = 0, y = 0$

3.

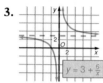

$x = 0, y = 3$

4.

$x = 0, y = 1$

5.

$x = 0, y = 0$

6.

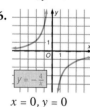

$x = 0, y = 0$

When $0 < x < 1$:

The graph of $y = \frac{1}{x}$ is above the graph of $y = x^2$. Therefore, $x^2 - \frac{1}{x}$ is negative and this portion of the graph of f is below the x-axis.

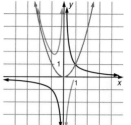

When $x = 1$:

The graphs of $y = \frac{1}{x}$ and $y = x^2$ intersect. Therefore, $f(1) = 0$ and the graph of f has a point on the x-axis.

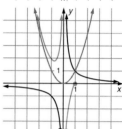

When $x > 1$:

The graph of $y = \frac{1}{x}$ is positive but below the graph of $y = x^2$. Therefore, $x^2 - \frac{1}{x}$ is positive but less than x^2. Thus, this portion of graph of f is above the x-axis but below the graph of $y = x^2$.

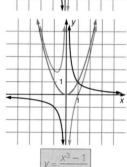

$y = \frac{x^3 - 1}{x}$

Exercises 3.7

In **1 – 12**, sketch the graph and write equations of any asymptotes.

1. $y = \frac{6}{x}$ **2.** $y = -\frac{6}{x}$ **3.** $y = 3 + \frac{6}{x}$ **4.** $y = \frac{x+1}{x}$

5. $y = \frac{4}{x}$ **6.** $y = -\frac{4}{x}$ **7.** $y = \frac{4}{x} - 1$ **8.** $y = \frac{2x+1}{x+2}$

9. $y = \frac{4x-3}{x-1}$ **10.** $y = \frac{1-x^2}{x^2}$ **11.** $y = \frac{2x^2-2}{x^2}$ **12.** $y = \frac{x^3+1}{x}$

13. a. Sketch the graph of $y = \frac{x^2-4}{x-2}$.

    b. As $x \to 2$, does y approach either ∞ or $-\infty$?

    c. Is $x = 2$ an asymptote of the graph? Explain.

7.

$x = 0, y = -1$

8.

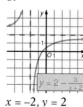

$x = -2, y = 2$

9.

$x = 1, y = 4$

10.

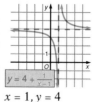

$x = 0, y = -1$

11.

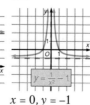

$x = 0, y = 2$

12.

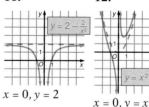

$x = 0, y = x^2$

13. a.

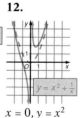

b. no
c. no;
$y = x + 2$
for $x \neq 2$

Chapter 3   *Summary and Review*

Defining a Function

- A function from set A to set B is a set of ordered pairs (x, y) such that every $x \in A$ is paired with exactly one $y \in B$. Set A is the domain of the function and set B is the codomain of the function. The range of the function is the set of all elements of the codomain that are images of the elements of the domain.

- On the coordinate plane, every point is the graph of an ordered pair of real numbers, and every ordered pair of real numbers has a graph that is a point of the plane. The set of ordered pairs of real numbers can be represented by $R \times R$.

  A function in $R \times R$ can be graphed as a subset of the coordinate plane.

- A function f is onto if for every value y in the codomain, there is at least one value x in the domain for which $f(x) = y$.

  A function from A to B is onto if the codomain is the same set as the range.

- A function from set A to set B is one-to-one $(1 - 1)$ if no two distinct ordered pairs have the same second element.

- A one-to-one correspondence is a function that is both $1 - 1$ and onto.

Some Common Functions in *R* × *R*

- Constant Function: $f(x) = a$

- Linear Function: $f(x) = mx + b$

  The equation $y = mx + b$ is the slope-intercept form of an equation of a line, where m represents the slope and b the y-intercept.

  The equation $\dfrac{x}{a} + \dfrac{y}{b} = 1$ is the intercept form of an equation of a line. The x-intercept is a and the y-intercept is b.

- Identity Function: $f(x) = x$

- Quadratic Function: $f(x) = ax^2 + bx + c$

- Polynomial Function: $f(x) = a_n x^n + a_{n-1} x^{n-1} + \ldots + a_0$

- Absolute-Value Function: $f(x) = |x|$ or $|x| = \begin{cases} x \text{ if } x \geq 0 \\ -x \text{ if } x < 0 \end{cases}$

- Step Function: The graph is a series of steps, for example:

$$f(x) = \begin{cases} 1 \text{ if } x > 0 \\ 0 \text{ if } x = 0 \\ -1 \text{ if } x < 0 \end{cases} \quad \text{or} \quad f(x) = \begin{cases} \dfrac{|x|}{x} \text{ if } x \neq 0 \\ 0 \text{ if } x = 0 \end{cases}$$

- Greatest-Integer Function: $f(x) = [x]$
- Square-Root Function: $f(x) = \sqrt{x}$
- Reciprocal Function: $f(x) = \dfrac{1}{x}$
- Sine Function: $f(x) = \sin x$
- Exponential Function: $f(x) = b^x$

The Algebra of Functions

- If g and h are functions, then:
  $$(g + h)(x) = g(x) + h(x)$$
  $$(g - h)(x) = g(x) - h(x)$$
  $$(gh)(x) = g(x)h(x)$$
  $$\left(\frac{g}{h}\right)(x) = \frac{g(x)}{h(x)} \text{ if } h(x) \neq 0$$

  The domain of $g + h$, $g - h$ and gh is the set of elements common to the domains of g and of h. The domain of $\dfrac{g}{h}$ is the set of elements common to the domains of g and of h for which $h \neq 0$.

- If f and g are functions such that the range of f is the domain of g, then $g \circ f$ is the composition of f and g such that $(g \circ f)(x) = g(f(x))$.
  In general, $g \circ f$ is different from $f \circ g$.

- Subsets of the set of real numbers can be designated by writing the endpoints within brackets or parentheses:

  $[a, b] = \{x \mid a \leq x \leq b\}$ $[a, \infty) = \{x \mid a \leq x\}$

  $(a, b) = \{x \mid a < x < b\}$ $(a, \infty) = \{x \mid a < x\}$

  $(a, b] = \{x \mid a < x \leq b\}$ $(-\infty, b] = \{x \mid x \leq b\}$

  $[a, b) = \{x \mid a \leq x < b\}$ $(-\infty, b) = \{x \mid x < b\}$

Transformations

- A transformation of the plane is a one-to-one correspondence between the points of the plane such that each point is mapped to itself or to another point of the plane.

  A transformation is a function that pairs each point of the plane with exactly one image point of the plane. The domain and codomain of the transformation are the points of the plane, or the set of ordered pairs of real numbers in the coordinate plane.

- A line reflection in line m maps each point P of the plane to its image P' such that:
  1. If P is on m, then $P = P'$.
  2. If P is not on m, then m is the perpendicular bisector of $\overline{PP'}$.

- The image of $A(x, y)$ under a reflection in the x-axis is $A'(x, -y)$,
   or $r_{x\text{-axis}}(x, y) = (x, -y)$.
   The image of $A(x, y)$ under a reflection in the y-axis is $A'(-x, y)$,
   or $r_{y\text{-axis}}(x, y) = (-x, y)$.
   The image of $A(x, y)$ under a reflection in the line $y = x$ is $A'(y, x)$,
   or $r_{y = x}(x, y) = (y, x)$.

- A reflection in the point C maps each point P of the plane to its image P' such that:
   1. Point C is its own image.
   2. If $P \neq C$, then C is the midpoint of $\overline{PP'}$.

- The image of $A(x, y)$ under a reflection in the origin is $A'(-x, -y)$,
   or $R_O(x, y) = (-x, -y)$.

- A rotation of θ degrees about a point C maps each point P of the plane to its image P' such that:
   1. Point C is its own image.
   2. If $P \neq C$, then $CP = CP'$ and m$\angle PCP' = \theta°$.

- If the rotation from P to P' is counterclockwise, θ is positive.
   If the rotation from P to P' is clockwise, θ is negative.

- The image of $P(x, y)$ under a rotation of $90°$ about the origin is $P'(-y, x)$,
   or $R_{O, 90°}(x, y) = R_{90°}(x, y) = (-y, x)$.

- A translation maps point P to its image P' and point Q to its image Q' such that
   $\overline{PP'} \parallel \overline{QQ'}$ and $PP' = QQ'$.

- The image of $A(x, y)$ under a translation of a units in the horizontal direction and b units in the vertical direction is $A'(x + a, y + b)$,
   or $T_{a, b}(x, y) = (x + a, y + b)$.

The Inverse of a Function

- The identity function is the function I such that $I(x) = x$. I is the function that maps any element of the domain to the same element in the codomain. The domain and codomain of the identity function are the same set.

- The inverse of a function f from A to B is a function f^{-1} from B to A, such that $f^{-1} \circ f = I$ from A to A and $f \circ f^{-1} = I$ from B to B.

   The inverse function of a $1 - 1$ correspondence f can be found by interchanging the coordinates of the ordered-pair elements of f. If the function is an infinite set of ordered pairs defined by an equation, rewrite the equation by interchanging x and y. Then, solve for y.

- When a function has no inverse for the given domain and codomain, it is often possible to restrict the domain and codomain to define a subset of the function that does have an inverse.

1. 6 from the domain maps to 2 different elements of the codomain.

2. 6 from the domain maps to no element of the codomain.

3. 0 from the domain maps to no element of the codomain.

4. no; it is not onto. There is no element of the domain that maps to a.

5. no; it is not 1 – 1. Every integer maps to 0.

6. yes; it is 1 – 1 and onto. Every real number maps to one and only one real number and every real number is the image of a real number.

7. no; it is not onto. The real numbers less than –1 are not the images of real numbers.

8. $(f + g)(x)$
   $= x^2 + x - 1$

9. $(f - g)(x)$
   $= x^2 - x + 1$

10. $(fg)(x) = x^2(x - 1)$
    $= x^3 - x^2$

11. $(f \circ g)(x) = f(x - 1)$
    $= (x - 1)^2$
    $= x^2 - 2x + 1$

12. $(g \circ f)(x) = g(x^2)$
    $= x^2 - 1$

13. $f(2x + 1) = (2x + 1)^2$
    $= 4x^2 + 4x + 1$

14. $2g(x) = 2(x - 1)$
    $= 2x - 2$

15. $g(2x) = 2x - 1$

Graphing Functions

- The graph of $f(-x)$ is a reflection in the y-axis of the graph of $f(x)$. The graph of $-f(x)$ is a reflection in the x-axis of the graph of $f(x)$.

- The graph of $y - k = f(x - h)$ is the image of $y = f(x)$ under the translation $T_{h, k}$.

- The graph of $y = f(x)$ can be used to sketch the graph of $y = af(x)$, where a is a constant and $a > 0$, by stretching or shrinking the graph in the vertical direction. The ordinate of each point of $y = af(x)$ is a times the ordinate of the point of $y = f(x)$ having the same abscissa.

- The graph of $y = f(x)$ can be used to sketch the graph of $y = f(ax)$, where a is a constant and $a > 0$, by stretching or shrinking the graph in the horizontal direction. The abscissa of each point of $y = f(ax)$ is $\frac{1}{a}$ times the abscissa of the point of $y = f(x)$ having the same ordinate.

- The graph of $h(x) = f(x) + g(x)$ can be drawn by adding the ordinates of the points of f and g that have the same abscissa.

Rational Functions

- A rational function is the ratio of two polynomial functions. If $f(x) = \frac{g(x)}{h(x)}$, then the domain of f is the set of elements common to the domains of f and of h for which $h(x) \neq 0$.

- $x \to a$, where a is a real number, means that x is approaching a, that is, the values of x are getting closer to a but $x \neq a$.

  $x \to \infty$ means that x is approaching infinity, that is, the values of x are increasing such that for any large number, the value of x can be greater than that number.

  $x \to -\infty$ means that the values of x, a negative number, are decreasing such that for any negative number, the value of x can be less than that number.

- A line $y = a$ is a horizontal asymptote of f if as $x \to \infty$ or $x \to -\infty$, $f(x) \to a$. A line $x = a$ is a vertical asymptote of f if as $x \to a$, $f(x) \to \infty$ or $f(x) \to -\infty$.

Chapter 3   Review Exercises

In **1 – 3**, tell why the mapping is not a function.

1.

2.

3. $f(x) = \frac{1}{x}$ from R to R

16.
$g(x) = \frac{x^2}{x + 1}$ and
$h(x) = \frac{1}{x + 1}$
Alternate solution:
$g(x) = x - 1$ and
$h(x) = \frac{2}{x + 1}$

17.

$f(x) = \frac{4}{3}x - 9$

18.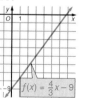

$g(x) = x^2$

$f(x) = g(x - 4) + 1$

19.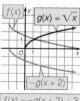

$g(x) = \sqrt{x}$

$-g(x + 2)$

$f(x) = -g(x + 2) + 3$

In **4 – 7**, is the function a 1 – 1 correspondence? Explain.

4.

5. $f = \{(x, y) \mid y = x - [x]\}$ from R to $[0, 1)$

6. $f = \{(x, y) \mid y = 1 - x\}$ from R to R **7.** $f = \{(x, y) \mid y = x^2 - 1\}$ from R to R

In **8 – 15**, let $f(x) = x^2$ and $g(x) = x - 1$. Write the given expression as a polynomial in x.

8. $(f + g)(x)$ **9.** $(f - g)(x)$ **10.** $(fg)(x)$ **11.** $(f \circ g)(x)$

12. $(g \circ f)(x)$ **13.** $f(2x + 1)$ **14.** $2g(x)$ **15.** $g(2x)$

16. If $(g + h)(x) = \dfrac{x^2 + 1}{x + 1}$, find $g(x)$ and $h(x)$. (There are many possibilities.)

In **17 – 19**, sketch the graph of f in $R \times R$.

17. $f(x) = \dfrac{4x}{3} - 9$ **18.** $f(x) = (x - 4)^2 + 1$ **19.** $f(x) = 3 - \sqrt{x + 2}$

20. a. Sketch the graph of $f(x) = (x - 1)^2$.
b. Sketch the image of f under a reflection in the y-axis, and then write its equation.
c. Sketch the image of f under a reflection in the x-axis, and then write its equation.

21. a. Sketch the graph of $y = \sqrt{x - 1}$.
b. On the same set of axes, sketch the graph of $y = x$.
c. On the same set of axes, sketch the graph of $y = x + \sqrt{x - 1}$.

22. The diagram shows the graph of a function f with domain $[-3, 3]$.
a. If the codomain is $[-4, 4]$, is f an onto function?
b. Is f a 1 – 1 function?   **c.** Is f a 1 – 1 correspondence?
Sketch:
d. $-f(x)$ **e.** $f(-x)$ **f.** $2f(x)$ **g.** $f(2x)$
h. the image of f under a reflection in the line $y = x$.
i. the image of f under a rotation of 90° about the origin.

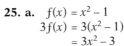

23. a. Sketch the graph of $f(x) = \sqrt{x + 1}$.
b. Sketch g, the reflection of f in the line $y = x$.
c. Is g a function?
d. Is g the inverse of f?

24. a. Sketch the graph of $f(x) = \dfrac{2x}{x - 2}$.
b. What are the domain and range of f?
c. Write equations of the asymptotes of f.

25. If $f(x) = x^2 - 1$, find: **a.** $3f(x)$ **b.** $f(3x)$

26. Write a piecewise function that expresses the cost of renting a car for 1 day if the cost is \$35 a day with 100 free miles plus \$.12 a mile for each mile over 100.

27. Write a piecewise function that expresses the cost of renting a cabin for one week if the charge is \$350 per week for 2 persons plus \$20 for each additional person.

20.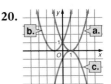

a. $f(x) = (x - 1)^2$
b.
$f(-x) = (-x - 1)^2 = (x + 1)^2$
c. $-f(x) = -(x - 1)^2$

21.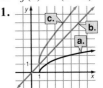

a. $y = \sqrt{x - 1}$ **b.** $y = x$
c. $y = x + \sqrt{x - 1}$

22. a. not onto
b. no; all of the numbers in the interval $[-3, -1]$ map to 1.
c. no; it is neither 1 – 1 nor onto
d.
e.
f.
g.
h.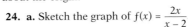
i.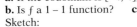

23. a. and **b.**

c. yes
d. yes

24. a.
$f(x) = \dfrac{2x}{x - 2} = 2 + \dfrac{4}{x - 2}$

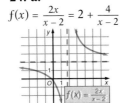

b. $R/\{2\}$; $R/\{2\}$
c. $x = 2$, $y = 2$

25. a. $f(x) = x^2 - 1$
$3f(x) = 3(x^2 - 1)$
$= 3x^2 - 3$
b. $f(x) = x^2 - 1$
$f(3x) = (3x)^2 - 1$
$= 9x^2 - 1$

26. $f(x) = \begin{cases} 0.12(x - 100) + 35, \, x > 100 \\ 35, \, x \leq 100 \end{cases}$

27. $f(x) = \begin{cases} 350 + 20(x - 2), \, x \geq 2 \\ 350, \, x < 2 \end{cases}$

1. $(1, \sqrt{3})$

2. $\{-1, 0\}$

3. $a = 3$

4. 0.768

5. B

6. A

7. C

8. A

9. B

10. C

☑ *Exercises for Challenge*

1. What are the coordinates of the image of the point (2, 0) under a counterclockwise rotation of 60° about the origin?

2. The domain of $f(x) = [x] + [-x]$ is the set of real numbers. What is the range? ($[x]$ = the largest integer less than or equal to x.)

3. If the domain of $f(x) = x^2 - 6x + 5$ is restricted to $(-\infty, a]$, what is the largest value of a for which f is a one-to-one function?

4. 🖩 Find to the nearest thousandth the sum of the areas of the four rectangles in the diagram.

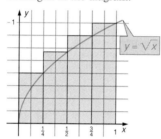

Questions 5 – 9 each consist of two quantities, one in Column A and one in Column B. You are to compare the two quantities and choose:

A if the quantity in Column A is greater;
B if the quantity in Column B is greater;
C if the two quantities are equal;
D if the relationship cannot be determined from the information given.

1. In certain questions, information concerning one or both of the quantities to be compared is centered above the two columns.

2. In a given question, a symbol that appears in both columns represents the same thing in Column A as it does in Column B.

3. x, n, and k, etc. stand for real numbers.

Column A	Column B
Set S has 3 elements and set T has 2 elements.	
5. The number of possible functions from S to T	The number of possible functions from T to S
Set S has 3 elements and set T has 2 elements.	
6. The number of possible functions from S onto T	The number of possible 1–1 functions from S to T
7. The smallest value of $y = x^2 + 6x + 10$	The largest value of $y = -x^2 + 4x - 3$
8. The number of vertical asymptotes of $xy = 3$	The number of vertical asymptotes of $y = \dfrac{x-2}{x^3 - x - 6}$

$$f(x) = 1 + \frac{1}{\sqrt{x}}$$

9. 🖩 $f(f(f(2)))$	$f(f(f(f(2))))$

In **10 – 16**, choose the correct answer.

10. If $f(x) = \left[\dfrac{\left[\frac{x}{2}\right]}{3} \right]$, then $f(-7)$ equals

(A) -4 (B) -3 (C) -2
(D) -1 (E) 0

11. The graph of $y = f(x)$ is shown below.

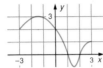

The graph of $y = f(|x|)$ is

(A)

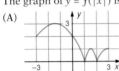

(B)

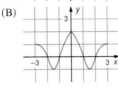

(C)

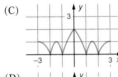

(D)

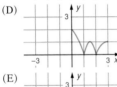

(E)

12. 🖩 If $f(x) = x^3 + 3$ and $g(x) = 2x - 1$, then $(f \circ g)(2.3) - (g \circ f)(2.3)$ equals

(A) 0   (B) 20.322 (C) 24.508
(D) 32.667 (E) 35.489

13. Let $g(x) = \begin{cases} x^2 \text{ if } x \geq 0 \\ -x^2 \text{ if } x < 0 \end{cases}$ from R to R.

Which of the following is the function inverse of g?

(A) $g^{-1}(x) = \begin{cases} \sqrt{x} \text{ if } x \geq 0 \\ -\sqrt{x} \text{ if } x < 0 \end{cases}$

(B) $g^{-1}(x) = \begin{cases} \sqrt{x} \text{ if } x \geq 0 \\ \sqrt{-x} \text{ if } x < 0 \end{cases}$

(C) $g^{-1}(x) = \begin{cases} \sqrt{x} \text{ if } x \geq 0 \\ \sqrt{|-x|} \text{ if } x < 0 \end{cases}$

(D) $g^{-1}(x) = \begin{cases} \sqrt{x} \text{ if } x \geq 0 \\ -\sqrt{-x} \text{ if } x < 0 \end{cases}$

(E) $g^{-1}(x) = \begin{cases} \sqrt{x} \text{ if } x \geq 0 \\ |\sqrt{-x}| \text{ if } x < 0 \end{cases}$

14. Let f, g, and h be functions from R to R. Which of the following could be false?

I $((f + g) \cdot h)(x) = (f \cdot h)(x) + (g \cdot h)(x)$
II $(f \circ (g + h))(x) = (f \circ g)(x) + (f \circ h)(x)$
III $((f + g) \circ h)(x) = (f \circ h)(x) + (g \circ h)(x)$

(A) I only   (B) II only (C) III only
(D) I and II only   (E) II and III only

15. Let f and g be functions from R onto R. Which of the following must be onto?

(A) $f + g$ (B) $f - g$ (C) $f \cdot g$
(D) $\dfrac{f}{g}$ (E) $f \circ g$

16. Let f be a function from R to R that has an inverse f^{-1}. For all a and b in R, $f(a + b) = f(a) \cdot f(b)$. Which of the following is true for f^{-1}?

(A) $f^{-1}(a + b) = f^{-1}(a) \cdot f^{-1}(b)$
(B) $f^{-1}(a + b) = f^{-1}(a) + f^{-1}(b)$
(C) $f^{-1}(ab) = f^{-1}(a) + f^{-1}(b)$
(D) $f^{-1}\left(\dfrac{a}{b}\right) = f^{-1}(a) - f^{-1}(b)$
(E) No rule can be determined.

11. B

12. B

13. D

14. B

15. E

16. C and D

Teacher's Chapter 4

Sequences, Series, and Induction

Chapter Contents

A general introduction to sequence, which includes the definition in terms of function, is followed by an in-depth treatment of arithmetic and geometric sequences and series. Throughout, the mathematics of finance is used as a theme for applications.

The principle of mathematical induction is presented, and inductive proofs are offered for statements that use the natural numbers, many of which are in the form of a series.

Finally, attention is directed to work with the binomial theorem, whose proof depends on the inductive principle.

Pattern recognition plays an important role in this chapter. The calculator makes it possible to emphasize relationships, and in addition, students have the opportunity to build power in algebraic skills.

4.1 Number Sequences and Series

Since a sequence can be defined as a function whose domain is the set of positive integers, a convenient notation uses subscripts to associate each term of the sequence with its domain value or position. While the definitions in this section are explicit, recursive definitions are introduced later, in the work with arithmetic and geometric sequences and series.

Students should be comfortable with a variety of sequences before sigma notation is introduced. Work with factorials and exponents, emphasized in the exercises, is needed throughout the chapter. Mention that 0! is defined as 1 for consistency with the other definitions. Later, more definitions include the 0th case.

4.2 Arithmetic Sequences

Arithmetic sequences provide a good starting point to familiarize students with the basics: first and last terms, sum, and general term.

The concept of a recursive definition is fundamental to computer programming. Encourage students who have had some programming experience to recall examples.

Connecting the concept of average to an arithmetic sequence makes the term *arithmetic mean* significant. Insertion of k arithmetic means might be compared to dividing the difference between two numbers into $(k + 1)$ equal intervals.

4.3 Geometric Sequences

You may wish to spend some time with the construction of a geometric mean. Recalling that the length of the altitude drawn to the hypotenuse of a right triangle is the mean proportional or geometric mean between the lengths of two segments of the hypotenuse, students may be challenged to discover a way to construct this altitude for segments of given lengths a and b. The construction is accomplished by drawing the locus of points for the vertex of the right angle of any right triangle whose hypotenuse is the segment of length $a + b$. The locus is a circle whose radius is $\frac{1}{2}(a + b)$ and whose center is the midpoint of the hypotenuse. The mean proportional is the length of the perpendicular segment from point P on the circle to point Q that separates the diameter of the circle into segments of lengths a and b.

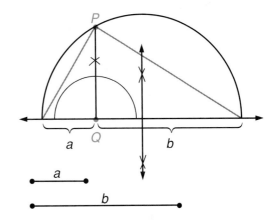

PQ is the mean proportional between the lengths of segments a and b.

4.4 Geometric Series

To further understanding of how an infinite sequence can converge to a finite sum, it is helpful to show examples of sequences that do not converge to a sum. The harmonic series

$$1 + \tfrac{1}{2} + \tfrac{1}{3} + \tfrac{1}{4} + \cdots$$

although not geometric, provides a good counterexample, since students will expect this series to converge. A geometric sequence in which $|r| \geq 1$ is not as convincing, since the absolute values of the terms increase.

4.5 Mathematical Induction

An introduction to the principle of mathematical induction is important, both for its nature and its power. Recall previous experience with an inductive proof in which all of the possibilities for a given situation are stated, examined in turn, and all but one are discarded by contradiction. Note the use of an assumption. Elicit that such an inductive proof depends upon a finite number of possibilities, all of which are known. Compare those circumstances to the situation for which a proof by the principle of mathematical induction applies: a general statement about an infinite set of numbers. The power of the principle of mathematical induction lies in reducing the number of cases that must be proved to exactly two.

In this section, the statements about natural numbers are limited to sums so that students can gain a level of comfort before they move on to work with sentences that require different algebraic techniques for showing S_{k+1}.

You might point out that Example 1 follows easily from the result of the previous text discussion. That is:

$$2 + 4 + 6 + \cdots + 2n = 2(1 + 2 + 3 + \cdots + n)$$
$$= 2 \cdot \frac{n(n+1)}{2} = n(n+1)$$

After students have become comfortable with the requirements for applying the principle of induction to given statements, they should be encouraged to make conjectures and validate them by mathematical induction (see Exercises 25 – 26). As a challenge, ask students to write a general term for pentagonal numbers. An analysis follows:

$$1, 5, 12, 22, 35, 51, 70, \ldots$$

can be written as:

$$1, (1 + 4), (1 + 4 + 7), (1 + 4 + 7 + 10),$$
$$(1 + 4 + 7 + 10 + 13), \ldots$$

then as:

$$1, (1 + 1 + 3), (1 + 1 + 3 + 1 + 2 \cdot 3),$$
$$(1 + 1 + 3 + 1 + 2 \cdot 3 + 1 + 3 \cdot 3), \ldots$$

now as:

$$(1 \cdot 1), (2 \cdot 1 + 1 \cdot 3), (3 \cdot 1 + 1 \cdot 3 + 2 \cdot 3),$$
$$(4 \cdot 1 + 1 \cdot 3 + 2 \cdot 3 + 3 \cdot 3) \cdots,$$
$$(n \cdot 1 + 1 \cdot 3 + 2 \cdot 3 + 3 \cdot 3 + \cdots + (n-1)3)$$

factoring:

$$1 + 3(0), 2 + 3(1), 3 + 3(1 + 2),$$
$$4 + 3(1 + 2 + 3), 5 + 3(1 + 2 + 3 + 4), \ldots.$$
$$n + 3[1 + 2 + 3 + 4 + \cdots + (n-1)]$$
$$\text{or } n + 3 \, \frac{[n(n-1)]}{2} = \frac{n(3n-1)}{2}$$

The technique of finite differences may also be used to demonstrate how the general term for a sequence may be found. For example, the nth term of the sequence

$$4, 11, 18, 25, \ldots$$

may be found by finding the differences between successive terms and noting that the first-order differences are constant.

$$
\begin{array}{lcccc}
n & 1 & 2 & 3 & 4 \\
 & 4, & 11, & 18, & 25 \\
 & & \vee & \vee & \vee \\
\text{differences} & & 7 & 7 & 7
\end{array}
$$

This suggests that the general term is linear and of the form $an + b$:

$$
\begin{array}{lcccc}
n & 1 & 2 & 3 & 4 \\
an + b & a + b & 2a + b & 3a + b & 4a + b \\
\text{differences} & & a & a & a
\end{array}
$$

Therefore, for the sequence being considered, $a = 7$, $a + b = 4$ or $7 + b = 4$, so $b = -3$. This suggests that the general term, $an + b$, in this case is $7n - 3$.

A similar analysis can be used for a sequence with second-order finite differences that are constant. In this case, the general term is quadratic and in the form $an^2 + bn + c$. For example, in the sequence below, the common differences between successive terms indicates that a constant is obtained for the second set of differences.

$$-1, \ 2, \ 9, \ 20, \ 35, \ 54, \ \dots$$

first-order differences $\quad 3 \quad 7 \quad 11 \quad 15 \quad 19$

second-order differences $\quad 4 \quad 4 \quad 4 \quad 4$

For the general nth term $an^2 + bn + c$:

n	$an^2 + bn + c$	first-order differences	second-order differences
1	$a + b + c$		
		$3a + b$	
2	$4a + 2b + c$		$2a$
		$5a + b$	
3	$9a + 3b + c$		$2a$
		$7a + b$	
4	$16a + 4b + c$		$2a$
		$9a + b$	
5	$25a + 5b + c$		

The solution of the system
$$2a = 4$$
$$3a + b = 3$$
$$a + b + c = -1$$
shows that a = 2, b = –3, c = 0. Therefore,
$$an^2 + bn + c = 2n^2 - 3n$$

In general, if the nth-order differences are constant for a sequence, the general term can be written as a polynomial of nth degree. More able students might appreciate an alternative type of proof by mathematical induction, which is based on the well-ordering axiom for the set of natural numbers. The axiom states that any non-empty set of natural numbers has a least element. Using this postulate, a proof by contradiction can replace a proof by mathematical induction. A sample of this type of proof follows.

Prove
$$1 + 2 + 3 + \cdots + n = \frac{n(n + 1)}{2}$$

Proof

Assume that the formula is not true for all n in N. Then there exists a natural number k for which:
$$1 + 2 + 3 + \cdots + k \neq \frac{k(k + 1)}{2}$$

Let F be the set of natural numbers for which the statement to be proved is false and T be the set of numbers for which the statement is true. By inspection, the statement is true for 1, 2, 3, so 1, 2, and 3 are in T. But k is in F. Therefore, F is non-empty and contains a least element, say L (by the well-ordering axiom). Since L is in F,
$$1 + 2 + 3 + \cdots + L \neq \frac{L(L + 1)}{2}.$$

If L is the least element in F, we know that $L - 1$ is at least 3 from the trials mentioned earlier of 1, 2, and 3. This implies that:
$$1 + 2 + 3 + \cdots + (L - 1)$$
$$= \frac{(L - 1)[(L - 1) + 1]}{2}$$

$$1 + 2 + 3 + \cdots + (L - 1) + L$$
$$= \frac{(L - 1)[(L - 1) + 1]}{2} + L$$
$$= \frac{(L - 1)L + 2L}{2}$$
$$= \frac{L(L + 1)}{2}$$

But this contradicts the assumption. Therefore, the statement holds for all natural numbers n.

4.6 Additional Applications of Mathematical Induction

In this section, proof by mathematical induction is extended to examples involving divisibility, inequalities, and sequences that are defined recursively. Since recursive definitions may be new to some students, you may wish to review simple examples of sequences that are defined by recursion formulas, such as the Fibonacci numbers.

Now that students are faced with a variety of statements to be proved by mathematical induction, the algebraic techniques needed to show S_{k+1} vary. In general, two starting methods are appropriate:

1. Begin with the inductive hypothesis S_k, and operate on it to lead up to S_{k+1}.

   This method is useful when S_n involves a sum, in which case the algebraic technique is usually adding the $(k + 1)$st term to each side of S_k.

2. Begin with S_{k+1}, operating on it until the desired conclusion is clear.

   This method is useful when S_n is a statement of divisibility.

4.7 Expansion of Binomials

For most students, it would be useful to review the standard form of a polynomial and multiplication by a binomial before discussing the Pascal Triangle.

Although combinations are not discussed here, the $\binom{n}{r}$ notation is introduced.

The proof of the binomial theorem by mathematical induction, though complex, is an excellent example of where a proof by mathematical induction is useful in the logical development of mathematics. The proof is included here, should you wish to present it.

Proof of the Binomial Theorem

1. Verify that the theorem is true for $n = 1$.

$$(x + y)^1 = \binom{1}{0}x^1 + \binom{1}{1}y^1$$
$$= 1 \cdot x + 1 \cdot y$$
$$(x + y)^1 = x + y$$

2. Assume that the theorem is true for $n = k$.

   Suppose $(x + y)^k = \binom{k}{0}x^k + \binom{k}{1}x^{k-1}y + \binom{k}{2}x^{k-2}y^2 + \cdots + \binom{k}{k}y^k$.

   It must be shown that this implies validity for $(x + y)^{k+1}$, namely that:

$$(x + y)^{k+1} = \binom{k+1}{0}x^{k+1} + \binom{k+1}{1}x^{k}y + \binom{k+1}{2}x^{k-1}y^2 + \cdots + \binom{k+1}{k+1}y^{k+1}$$

   In order to prove this statement, first the following two auxiliary theorems (called *lemmas*) about binomial coefficients must be proved.

☐ **Lemma 1:** $\binom{n}{r} = \dfrac{n!}{r!(n-r)!}$

   Proof

   Begin with the original definition for $\binom{n}{r}$:

$$\binom{n}{r} = \frac{n(n-1)(n-2)\cdots(n-r+1)}{r!}$$

Multiply the numerator and denominator by $(n - r)!$:

$$\binom{n}{r} = \frac{n(n - 1)(n - 2) \cdots (n - r + 1)(n - r)!}{r!(n - r)!}$$

Note that in the product for $n!$ the next factor after $(n - r + 1)$ is $(n - r)$. Thus $n(n - 1) \cdots (n - r + 1)(n - r)! = n!$. Therefore:

$$\binom{n}{r} = \frac{n(n - 1)(n - 2) \cdots (n - r + 1)(n - r)!}{r!(n - r)!} = \frac{n!}{r!(n - r)!}$$

☐ **_Lemma 2:_** $\binom{n}{r} + \binom{n}{r - 1} = \binom{n + 1}{r}$

Proof

Using Lemma 1:

$$\binom{n}{r} + \binom{n}{r - 1} = \frac{n!}{r!(n - r)!} + \frac{n!}{(r - 1)![n - (r - 1)]!}$$

$$= \frac{n!}{(r - 1)!(n - r)!}\left[\frac{1}{r} + \frac{1}{n - (r - 1)}\right]$$

$$= \frac{n!}{(r - 1)!(n - r)!}\left[\frac{n + 1}{r(n - r + 1)}\right]$$

$$= \frac{(n + 1)!}{r!(n - r + 1)!}$$

$$= \binom{n + 1}{r}$$

Returning to the proof of the binomial theorem, expand $(x + y)^{k + 1} = (x + y)(x + y)^k$ by multiplying the expansion of $(x + y)^k$ by x and by y and then adding the results.

$$(x + y)(x + y)^k = x\left[\binom{k}{0}x^k + \binom{k}{1}x^{k - 1}y + \cdots + \binom{k}{k}y^k\right] + y\left[\binom{k}{0}x^k + \binom{k}{1}x^{k - 1}y + \cdots + \binom{k}{k}y^k\right]$$

$$= \left[\binom{k}{0}x^{k + 1} + \binom{k}{1}x^k y + \cdots + \binom{k}{k}xy^k\right] + \left[\binom{k}{0}x^k y + \binom{k}{1}x^{k - 1}y^2 + \cdots + \binom{k}{k}y^{k + 1}\right]$$

Rearranging the terms and using the distributive property:

$$(x + y)^{k + 1} = \binom{k}{0}x^{k + 1} + \left[\binom{k}{1} + \binom{k}{0}\right]x^k y + \left[\binom{k}{2} + \binom{k}{1}\right]x^{k - 1}y^2 + \cdots$$

$$\cdots + \left[\binom{k}{k} + \binom{k}{k - 1}\right]xy^k + \binom{k}{k}y^{k + 1}$$

Using Lemma 2 in each term of the right side except the first and last:

$$(x + y)^{k + 1} = \binom{k}{0}x^{k + 1} + \binom{k + 1}{1}x^k y + \binom{k + 1}{2}x^{k - 1}y^2 + \cdots + \binom{k + 1}{k}xy^k + \binom{k}{k}y^{k + 1}$$

Except for the first and last terms, this is what was to be proved.

However, since $\binom{k}{0} = \binom{k + 1}{0} = 1$, $\binom{k + 1}{0}$ can be substituted for the first coefficient, and since $\binom{k}{k} = \binom{k + 1}{k + 1} = 1$, the last coefficient can be replaced with $\binom{k + 1}{k + 1}$, giving the needed expansion and completing the proof.

Activity 1 (4.7) _____

The answers to the following questions will lead you to an interesting sequence.

1. Generate the first 10 terms of $a_n = 8n - 7$.   (1,9,17,25, ..., 73)

2. What pattern do you notice in the sequence of terms?   ($d = 8$)

3. What kind of sequence is this?   (arithmetic)

4. Write the next ten terms.   (81, 89, 97, ..., 153)

5. Which perfect squares appear in the sequence?   (1, 9, 25, 49, 81, 121)

6. What perfect square is the 56th term of the sequence?   (21^2)

7. Write a sequence indicating the positions of the
   perfect squares in the original sequence.   (1, 2, 4, 7, 11, 16, ...)

8. Is the last sequence that you wrote arithmetic,
   geometric, or neither?   (neither; differences are successive integers)

9. Write the first 15 terms of the last sequence.   (1, 2, 4, 7, ..., 79, 92, 106)

10. If $a_i = \dfrac{i^2 - i + 2}{2}$, write each a_i for $1 \le i \le 10$.   (see #9)

11. Verify that $\dfrac{i^2 - i + 2}{2} + i$ is a_{i+1}.   (see #8)

12. Verify that for $n = \dfrac{i^2 - i + 2}{2}$, $8n - 7$ is always the square of an odd number.   (see #5)

Activity 2 (4.7) _____

Using a Programmable Calculator to Study Two Series

To study the partial sums of the series:

$$1 + \tfrac{1}{2} + \tfrac{1}{4} + \tfrac{1}{8} + \tfrac{1}{16} + \cdots$$

Key in the following sequence of programming steps on a programmable calculator. (This program is written for the TI-81 but can be adapted to other calculators.)

```
:Input T
:0-->S
:0-->N
:Lbl 1
:1/(2^N)-->R
:S+R-->S
:2-S-->D
:N+1-->N
:If N-2-T
:Goto 1
:Disp S
:Disp D
```

:Input T	Input number of terms to be added.
:0-->S	Initialize the sum at 0.
:0-->N	Initialize N at 0.
:Lbl 1	Start loop.
:1/(2^N)-->R	Compute the term of the sequence.
:S+R-->S	Add R to the cumulative total.
:2-S-->D	Compute the difference between 2 and the sum.
:N+1-->N	Increment N by 1.
:If N-2-T	When the previous value of N, $(N - 1)$, is one less than the number of terms, skip the next command.
:Goto 1	Return to loop.
:Disp S	Show value of S.
:Disp D	Show the difference.

What If ?

1. $T = 5$
2. $T = 10$
3. $T = 15$, $T = 30$, $T = 32$

Find the smallest value of T that results in a sum within one millionth of the value 2.

To what limit does the value of the series converge?

What If?

$T = 100$

Did your calculator show a sum of 2? Explain.

What If ?

Line 5 of the program is changed to `1/N-->R` and N is initialized at 1?

Observations

1. The first geometric series converges to a limit since it is a geometric series whose ratio is less than 1.

2. The second sequence, using `1/N-->R`, is the harmonic sequence, and diverges.

3. If a series converges, the sum may be taken as close as desired to a limit provided more terms are added on; but, the sum never actually reaches the limit.

Suggested Test Items

1. Write the first 4 terms of the sequence whose general term is:

   a. $a_n = \dfrac{3n}{n + 1}$

   b. $a_n = \begin{cases} 2n \text{ if } n \text{ is odd} \\ 2n + 1 \text{ if } n \text{ is even} \end{cases}$

2. Find the value of:

   a. $\displaystyle\sum_{i=0}^{3} (2 + 4i)$ **b.** $\displaystyle\sum_{n=2}^{4} \dfrac{n!}{(n + 1)!}$

3. Write the 4th term of $a_n = \dfrac{n^2 + 2}{n - 1}$.

4. Use sigma notation to represent:

   $$\frac{1}{2^3} + \frac{1}{3^3} + \frac{1}{4^3}$$

5. Use the formula $A = P\left(1 + \dfrac{r}{n}\right)^{nt}$ to write an expression for the amount of money accumulated when \$500 is invested at 8% compounded monthly for $2\frac{1}{2}$ years.

6. Find the 16th term of an arithmetic sequence whose first term is –7, and whose common difference is 2.5.

7. Express the first term of an arithmetic sequence in terms of x if the 6th term is $3x + 2$ and the common difference is $x - 1$.

8. If the terms of an arithmetic sequence are represented by a_i, $1 \le i \le n$, find a_{30} when $a_1 = -3$ and $a_{10} = 24$.

9. Use the recursion formula $a_1 = 1$ and $a_n = a_{n-1} - \frac{1}{2}$ to generate the first 5 terms of an arithmetic sequence.

10. Insert 4 arithmetic means between −6 and 14.

11. Find the sum of the first 10 terms of an arithmetic sequence whose first term is 2 and whose 10th term is 100.

12. Classify each sequence as arithmetic, geometric, or neither.

    a. $\sqrt{3}, 3, 3\sqrt{3}, 9, \ldots$

    b. $\frac{1}{2}, \frac{1}{3}, \frac{1}{4}, \frac{1}{5}, \ldots$

    c. $2 + \sqrt{2}, 4 + \sqrt{2}, 6 + \sqrt{2}$

    d. $3^{-1}, 3^{-2}, 3^{-3}, \ldots$

13. Write the first 5 terms of a geometric sequence whose first term is $\sqrt{5}$ and whose common ratio is $\sqrt{5}$.

14. If a_i represents the ith term of a geometric sequence, $1 \le i \le n$, find a_3 if $a_1 = 2$ and $a_4 = 10$.

15. A house that initially cost $250,000 has appreciated 10% per year since the owner bought it six years ago. How much is it worth now? Answer to the nearest thousand dollars.

16. Insert 3 geometric means between 5 and 405.

17. Find the sum of the first 6 terms of the geometric sequence where $a_1 = 3$ and $r = \sqrt{5}$.

18. To the nearest hundredth, evaluate:
$$\sum_{n=1}^{4} \left(\tfrac{3}{4}\right)^n$$

19. Mr. Black contributes $100 to an annuity at the beginning of each quarter. If the annuity pays 6% interest compounded quarterly, what will his balance be right before he makes his 5th payment?

20. Find the sum of the infinite geometric series:
$$\frac{4}{5} + \frac{16}{25} + \frac{64}{125} + \cdots$$

21. Express $0.531\overline{31}$ as a rational fraction.

22. Use mathematical induction to prove that $n^3 + 2n$ is divisible by 3 for all natural numbers.

23. Use mathematical induction to prove that $3^n > n$ for all natural numbers.

24. Prove by mathematical induction:
$$1 + 4 + 7 + \cdots + 3n - 2 = \frac{n}{2}(3n - 1)$$

25. Find the numerical value of:

    a. $\binom{5}{2}$ **b.** $\binom{15}{15}$

26. Expand $(x + 2y)^4$.

27. Find the 8th term of the expansion of $(2x - y)^{10}$.

Bonus

Find the sum of the digits of all integers from 1 to 999 inclusive.

Answers to Suggested Test Items

1. a. $\frac{3}{2}, 2, \frac{9}{4}, \frac{12}{5}$ **b.** 2, 5, 6, 9

2. a. 32   **b.** $\frac{47}{60}$

3. 6   **4.** $\displaystyle\sum_{n=1}^{3} \frac{1}{(n+1)^3}$

5. $500\left(1 + \frac{0.08}{12}\right)^{30}$ **6.** $30\frac{1}{2}$

7. $-2x + 7$ **8.** 84

9. $1, \frac{1}{2}, 0, -\frac{1}{2}, -1$ **10.** $-6, -2, 2, 6, 10, 14$

11. 510

12. a. geometric   **b.** neither
   c. arithmetic   **d.** geometric

13. $\sqrt{5}, 5, 5\sqrt{5}, 25, 25\sqrt{5}$

14. $2(\sqrt[3]{5})^2$ **15.** \$443,000

16. 15, 45, 135   **17.** $\frac{-372}{(1-\sqrt{5})}$

18. 2.05   **19.** \$415.23

20. 4   **21.** $\frac{263}{495}$

22. S_1: ✔

Assume S_k: $k^3 + 2k$ is divisible by 3.

Then S_{k+1}: $(k+1)^3 + 2(k+1) = (k^3 + 2k) + 3(k^2 + k + 1)$ is divisible by 3.

23. S_1: ✔

Assume S_k: $3^k > k$.

Then S_{k+1}: $3 \cdot 3^k > 3k \rightarrow 3^{k+1} > 3 \cdot k$

$3^{k+1} > k + 1$ (Since k is a natural number, $3k > k + 1$.)

24. S_1: ✔

Assume S_k: $1 + 4 + 7 + \cdots + 3k - 2 = \frac{k}{2}(3k - 1)$

Then S_{k+1}: $1 + 4 + 7 + \cdots + 3k - 2 + 3(k + 1) - 2 = \frac{k}{2}(3k - 1) + 3(k + 1) - 2$

$$= \frac{(k+1)}{2}[3(k+1) - 1]$$

25. a. 10   **b.** 1   **26.** $x^4 + 8x^3y + 24x^2y^2 + 32xy^3 + 16y^4$

27. $-960x^3y^7$

Bonus 13,500

$1-9 = 10(0) + 45$	$100-109 = 10(1) + 45$	$200-209 = 10(2) + 45$	$... 900-909 = 10(9) + 45$
$10-19 = 10(1) + 45$	$110-119 = 10(2) + 45$	$210-219 = 10(3) + 45$	$... 910-919 = 10(10) + 45$
$20-29 = 10(2) + 45$	$120-129 = 10(3) + 45$	$220-229 = 10(4) + 45$	$... 920-929 = 10(11) + 45$
$30-39 = 10(3) + 45$	$130-139 = 10(4) + 45$	$230-239 = 10(5) + 45$	$... 930-939 = 10(12) + 45$
$\vdots$	$\vdots$	$\vdots$	
$90-99 = 10(9) + 45$	$190-199 = 10(10) + 45$	$290-299 = 10(11) + 45$	$... 990-999 = 10(18) + 45$

$$
\begin{aligned}
&= \quad 10[0 + 1 + 2 \cdots + 7 + 8 + 9] + 10(45) \\
&\quad + 10[1 + 2 + 3 \cdots + 8 + 9 + 10] + 10(45) \\
&\quad + 10[2 + 3 + 4 \cdots + 9 + 10 + 11] + 10(45) \\
&\quad \vdots \\
&\quad + 10[9 + 10 + 11 \cdots + 16 + 17 + 18] + 10(45)
\end{aligned}
$$

$$
\begin{aligned}
&= \quad 10(45) + 450 \\
& \quad 10(55) + 450 \\
& \quad 10(65) + 450 \\
& \quad \vdots \\
& \quad + 10(135) + 450 \\
\hline
&= \quad 10[45 + 55 + \cdots + 135] + 10(450) \\
&= \quad 10(900) + 4,500 \\
&= \quad 9,000 + 4,500 \\
&= \quad 13,500
\end{aligned}
$$

Chapter 4

Sequences, Series, and Induction

The English economist Thomas Malthus published An Essay on the Principle of Population in 1798. In this major economics treatise, he predicted that the world was destined to suffer an ever-worsening food shortage based on his premise that populations increase according to a geometric sequence while the food supply increases according to an arithmetic sequence.

Modern farming methods and other technological improvements have enabled economists to modify the theory of Malthus and improve the outlook for humankind, but the concepts of arithmetic and geometric sequences remain as an important tool for studying the problems that arise in areas of economics and technology.

Chapter Table of Contents

4.1 Number Sequences and Series

The letters of the English alphabet written in alphabetical order are a sequence where a is first, b is second, c is third, and so on. In a similar way, sequences of numbers are used in mathematics. For example, the following lists of numbers are sequences:

$$2, 4, 6, 8, \ldots \qquad \frac{1}{3}, \frac{1}{9}, \frac{1}{27}, \ldots \qquad 1, 4, 7, 10, \ldots$$

The pattern of the first sequence suggests that 10 could be the 5th term, since each term is 2 more than the preceding term.

It is useful to think of a sequence as a function where the domain is the set of positive integers that indicate the position of a number in the sequence and the range is the set of values in the sequence. For example, using the idea of a function for the second sequence above:

$$f(1) = \frac{1}{3}, f(2) = \frac{1}{9}, f(3) = \frac{1}{27}, \ldots.$$

Subscripts are also used to designate the terms of a sequence. In general, any sequence takes the following form:

$$a_1, a_2, a_3, a_4, \ldots, a_n, \ldots$$

For example, in the third sequence above, $a_2 = 4$ means that the 2nd term of the sequence is 4 (or that 4 is the element of the range that corresponds to 2 in the domain). The three dots after a_n indicate that the sequence continues without end according to the same rule.

■ **Definition:** *An **infinite sequence** $\{a_n\}$ is a function whose domain is the set of positive integers and whose range is the set of values $\{a_n\}$. The terms of the sequence are $a_1, a_2, a_3, \ldots, a_n, \ldots$ where $f(1) = a_1, f(2) = a_2, f(3) = a_3, \ldots, f(n) = a_n, \ldots.$*

Example 1

Find the first three terms of the sequence whose nth term is:

a. $a_n = 3n + 1$ **b.** $a_n = 8(-1)^n + 2n$ **c.** $a_n = 2n! + 1$

Solution

a. $a_1 = 3(1) + 1 = 4$

$a_2 = 3(2) + 1 = 7$

$a_3 = 3(3) + 1 = 10$

b. $a_1 = 8(-1)^1 + 2(1) = 8(-1) + 2 = -6$

$a_2 = 8(-1)^2 + 2(2) = 8(1) + 4 = 12$

$a_3 = 8(-1)^3 + 2(3) = 8(-1) + 6 = -2$

Note
This sequence has both negative and positive terms.

c. $a_1 = 2(1!) + 1 = 2(1) + 1 = 3$

$a_2 = 2(2!) + 1 = 2(2 \cdot 1) + 1 = 5$

$a_3 = 2(3!) + 1 = 2(3 \cdot 2 \cdot 1) + 1 = 13$

> **Recall**
> $n! = n(n-1)(n-2) \cdot \cdots \cdot 3 \cdot 2 \cdot 1,$
> where n is a positive integer.

The Sum of a Sequence

A ball is dropped from a height of 6 feet and bounces up half this distance after hitting the ground. After hitting the ground a second time, it bounces up a distance that is half of its previous height and this pattern continues. The sum of a sequence may be used to represent the total distance traveled by the ball after it hits the ground the first time.

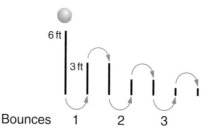

Bounces   1   2   3

The distance traveled by the ball may be represented by the sum of the sequence

↑ ↓ ↑ ↓ ↑ ↓
$3 + 3 + 1.5 + 1.5 + 0.75 + 0.75 + \dots.$

■ *Definition: The indicated sum of a sequence is called a **series**.*

It is often possible to express the sum of a sequence, even when it is infinite, in a general fashion. The Greek letter Σ (read *sigma*) is used for this purpose.

$$\sum_{i=1}^{5} a_i = a_1 + a_2 + a_3 + a_4 + a_5$$

$$\sum_{i=1}^{n} a_i = a_1 + a_2 + a_3 + \cdots + a_n$$

$$\sum_{i=3}^{6} a_i = a_3 + a_4 + a_5 + a_6$$

> **Read**
> The sum of the a_i's from $i = 1$ to infinity.

$$\sum_{i=1}^{\infty} a_i = a_1 + a_2 + a_3 + \cdots + a_n + \cdots$$

Example 2

Find the value of each sum:

a. $\displaystyle\sum_{i=1}^{3} 2i$ **b.** $\displaystyle\sum_{n=1}^{4} n!$

Solution

a. $\displaystyle\sum_{i=1}^{3} 2i = 2(1) + 2(2) + 2(3)$ **b.** $\displaystyle\sum_{n=1}^{4} n! = 1! + 2! + 3! + 4!$

$= 2 + 4 + 6$ $= 1 + (2 \cdot 1) + (3 \cdot 2 \cdot 1) + (4 \cdot 3 \cdot 2 \cdot 1)$

$= 12$ $= 1 + 2 + 6 + 24 = 33$

Example 3

Use sigma notation to write the sum of the 3rd, 4th, 5th, and 6th positive multiples of 3.

Solution

The sequence of positive multiples of 3 is 3, 6, 9, 12, ..., $3n$, ... In sigma notation, the sum of the 3rd, 4th, 5th, and 6th terms,

$9 + 12 + 15 + 18$, is: $\displaystyle\sum_{n=3}^{6} 3n$

Note: Frequently-used letters for an index are n, i, and k.

The following properties, based on the associative, commutative, and distributive properties, are helpful in dealing with algebraic expressions that use sigma notation.

1. $\displaystyle\sum_{i=1}^{n} ka_i = k \sum_{i=1}^{n} a_i$ when k is constant

2. $\displaystyle\sum_{i=1}^{n} (a_i + b_i) = \sum_{i=1}^{n} a_i + \sum_{i=1}^{n} b_i$

Example 4

Find the distance traveled by the bouncing ball, described on page 164, from the time it hits the ground the 1st time until just as it hits the ground for the 4th time.

Solution

The total distance traveled after the 1st bounce until hitting the ground the 4th time is $3 + 3 + 1.5 + 1.5 + 0.75 + 0.75$, which may be written as

$$\sum_{i=1}^{3} 2\left(\frac{3}{2^{i-1}}\right) = 2\sum_{i=1}^{3} \frac{3}{2^{i-1}}$$

$$= 2(3 + 1.5 + 0.75) = 2(5.25) = 10.5$$

Answer: The ball travels 10.5 feet between the 1st and 4th bounces.

1. 1, 2, 3, 4, 5

2. 3, 7, 11, 15, 19

3. 2, 5, 10, 17, 26

4. $1, \frac{4}{3}, \frac{3}{2}, \frac{8}{5}, \frac{5}{3}$

5. –3, 3, –3, 3, –3

6. 1, –3, 1, –3, 1

7. 3, 4, 8, 26, 122

8. 3, 6, 12, 32, 130

9. $5, 3\frac{3}{4}, 3\frac{4}{9}, 3\frac{5}{16}, 3\frac{6}{25}$

10. $2, 1\frac{1}{2}, 1\frac{1}{6}, 1\frac{1}{24}, 1\frac{1}{120}$

11. x, x^2, x^3, x^4, x^5

12. $x, x^2, x^6, x^{24}, x^{120}$

13. 1, 3, 3, 5, 5

14. 2, 4, 6, 24, 120

15. –2, 3, –4, 5, –6

16. $\frac{3}{2}, \frac{9}{4}, \frac{27}{8}, \frac{81}{16}, \frac{243}{32}$

17. 0, 1, 2, 3, 4

18. 1, 3, 7, 13, 21

19. 20

20. –65

21. –15

22. $\frac{133}{60}$

23. $\frac{47}{60}$

24. 29

25. 119

26. 54

27. 52

28. $42\frac{1}{2}$

29. $x^3 + 3x^2y + 3xy^2 + y^3$

30. $\displaystyle\sum_{n=1}^{4} (n + 3)$

31. $\displaystyle\sum_{n=1}^{4} 2n$

32. $\displaystyle\sum_{n=1}^{4} (6n + 1)$

33. $\displaystyle\sum_{n=1}^{5} \frac{1}{2^{n-1}}$

34. $\displaystyle\sum_{n=1}^{5} (-1)^{n-1} \frac{1}{2^n}$

35. $\displaystyle\sum_{n=1}^{4} (-1)^n \frac{1}{3^n}$

36. $\displaystyle\sum_{n=1}^{4} \frac{x^{2n}}{2^n}$

37. $\displaystyle\sum_{n=1}^{3} \frac{n}{n!}$

38. $\displaystyle\sum_{n=1}^{4} \frac{1}{n^{n+1}}$

39. $\displaystyle\sum_{n=1}^{3} \frac{1}{n(n+3)}$

40. $\displaystyle\sum_{n=0}^{4} \frac{4!}{n!(4-n)!} x^{4-n}y^n$

41. Average $= \dfrac{x_1 + x_2 + x_3 + \cdots + x_n}{n}$

$= \dfrac{1}{n}(x_1 + x_2 + x_3 + \ldots + x_n)$

$= \dfrac{1}{n}\displaystyle\sum_{i=1}^{n} x_i$

Exercises 4.1

In **1 – 18**, write the first five terms of the sequence whose nth term is given.

1. $a_n = n$ **2.** $a_n = 4n - 1$ **3.** $a_n = n^2 + 1$ **4.** $a_n = \dfrac{2n}{n+1}$

5. $a_n = 3(-1)^n$ **6.** $a_n = 2(-1)^{n+1} - 1$ **7.** $a_n = n! + 2$ **8.** $a_n = n! + 2n$

9. $a_n = 3 + \dfrac{1}{n} + \dfrac{1}{n^2}$ **10.** $a_n = 1 + \dfrac{1}{n!}$ **11.** $a_n = x^n$ **12.** $a_n = x^{n!}$

13. $a_n = \begin{cases} n & \text{if } n \text{ is odd} \\ n+1 & \text{if } n \text{ is even} \end{cases}$ **14.** $a_n = \begin{cases} 2n & \text{if } n < 3 \\ n! & \text{if } n \geq 3 \end{cases}$ **15.** $a_n = \begin{cases} -n - 1 & \text{if } n \text{ is odd} \\ n+1 & \text{if } n \text{ is even} \end{cases}$

16. $a_n = \dfrac{3^n}{2^n}$ **17.** $a_n = \dfrac{n^2 - 1}{n + 1}$ **18.** $a_n = \dfrac{n^3 + 1}{n + 1}$

In **19 – 29**, find the value of the sum indicated.

19. $\displaystyle\sum_{i=1}^{4} 2i$ **20.** $\displaystyle\sum_{i=1}^{5} (2 - 5i)$ **21.** $\displaystyle\sum_{i=1}^{3} (-3i + 1)$

22. $\displaystyle\sum_{i=2}^{4} \frac{i}{i+1}$ **23.** $\displaystyle\sum_{k=3}^{5} \frac{1}{k}$ **24.** $\displaystyle\sum_{n=1}^{4} (n! - 1)$

25. $\displaystyle\sum_{n=1}^{4} n(n!)$ **26.** $\displaystyle\sum_{n=2}^{10} \frac{n!}{(n-1)!}$ **27.** $\displaystyle\sum_{k=1}^{26} \frac{2^{k+1}}{2^k}$

28. $\displaystyle\sum_{n=0}^{3} 2^{2n-1}$ **29.** $\displaystyle\sum_{n=0}^{3} \frac{3!}{n!(3-n)!} x^{3-n}y^n \;\; (0! = 1)$

In **30 – 40**, use sigma notation to represent the sum. More than one correct answer may be possible.

30. $4 + 5 + 6 + 7$ **31.** $2 + 4 + 6 + 8$ **32.** $7 + 13 + 19 + 25$

33. $1 + \dfrac{1}{2} + \dfrac{1}{4} + \dfrac{1}{8} + \dfrac{1}{16}$ **34.** $\dfrac{1}{2} - \dfrac{1}{4} + \dfrac{1}{8} - \dfrac{1}{16} + \dfrac{1}{32}$ **35.** $-\dfrac{1}{3} + \dfrac{1}{9} - \dfrac{1}{27} + \dfrac{1}{81}$

36. $\dfrac{x^2}{2} + \dfrac{x^4}{4} + \dfrac{x^6}{8} + \dfrac{x^8}{16}$ **37.** $\dfrac{1}{1!} + \dfrac{2}{2!} + \dfrac{3}{3!}$ **38.** $\dfrac{1}{1^2} + \dfrac{1}{2^3} + \dfrac{1}{3^4} + \dfrac{1}{4^5}$

39. $\dfrac{1}{1 \cdot 4} + \dfrac{1}{2 \cdot 5} + \dfrac{1}{3 \cdot 6}$ **40.** $\dfrac{4!}{0!4!} x^4 + \dfrac{4!}{1!3!} x^3y + \dfrac{4!}{2!2!} x^2y^2 + \dfrac{4!}{3!1!} xy^3 + \dfrac{4!}{4!0!} y^4$

41. Prove that the average of n numbers can be written as $\dfrac{1}{n}\displaystyle\sum_{i=1}^{n} x_i$ if the set of numbers is $\{x_1, x_2, x_3, x_4, \ldots, x_n\}$.

Exercises 4.1 *(continued)*

42. The daily balance for $1,000 deposited in an account that pays 6% compounded daily can be computed using the formula $A = 1,000 \left(1 + \frac{0.06}{365}\right)^n$, where n is the number of days the money has been on deposit.

a. Write a sequence representing the daily balance for each of the first five days.

b. Find the 365th term of the sequence, which represents the balance after one year.

43. a. Use Exercise 42 to write a formula for a sequence representing the balance for $1,000 deposited in an account that pays 6% compounded quarterly.

b. Find the fourth term of the sequence (representing the balance after one year).

c. Compare your answers to Exercises **42b** and **43b**, and find how much more interest accrues on $1,000 when compounding daily.

d. If $10,000 had been deposited, how much more interest would result from compounding daily as compared to compounding quarterly?

Most calculators have a key that allows you to repeat an operation with a constant without entering the constant and the operation each time. For example, to multiply any number by 6, first enter 6 [×] [K]. Now any number can be multiplied by 6 by just entering the number and [=].

If the sum of the products is required, use the memory keys [STO], [RCL], and [SUM], or their equivalents.

For example, to evaluate $\displaystyle\sum_{i=1}^{4} i^2$, enter:

2 [yˣ] [K]

1 [=] [STO]

2 [=] [SUM]

3 [=] [SUM]

4 [=] [SUM] [RCL]

The display will give 30, the sum of $1^2 + 2^2 + 3^2 + 4^2$.

In **44 – 46**, use a calculator to find the indicated sum.

44. $\displaystyle\sum_{i=3}^{8} 5i$ **45.** $\displaystyle\sum_{n=1}^{5} \frac{n}{5}$ **46.** $\displaystyle\sum_{j=1}^{5} j^3$

42. a. $1,000.16, $1,000.33, $1,000.49, $1,000.66, $1,000.82

b. $1,061.83

43. a. $A = 1,000\left(1 + \frac{0.06}{4}\right)^n$

b. $1,061.36

c. $0.47

d. $4.67

44. 165

45. 3

46. 225

4.2 Arithmetic Sequences

The sequence 7, 10, 13, 16, ..., 34 represents the number of seats in consecutive rows of a theater from row A, the first row, to row J, the 10th row.

In this sequence, the number of seats in a given row may be obtained by adding 3 to the previous entry. This is an example of an *arithmetic sequence*.

7, 10, 13, 16, ..., 34

■ *Definition: An arithmetic sequence is a sequence in which each term after the first is obtained by adding a constant, called the common difference, to the previous term.*

In general, the n terms of an arithmetic sequence with common difference d may be represented as follows:

$$a_1 = a_1$$
$$a_2 = a_1 + d$$
$$a_3 = a_2 + d$$
$$a_4 = a_3 + d$$
$$\vdots$$
$$a_n = a_{n-1} + d$$

The representation for the nth term, which provides a rule for computing any term of the sequence (except the first) from the term preceding it, is called a *recursive formula*.

Example 1

Write a recursive formula for the nth term of each arithmetic sequence.

Solution

Sequence	Recursive Formula
a. $\frac{1}{2}$, 1, $\frac{3}{2}$, 2, $\frac{5}{2}$, 3, $\frac{7}{2}$	$a_n = a_{n-1} + \frac{1}{2}$
b. 10, 5, 0, –5, –10	$a_n = a_{n-1} - 5$

An arithmetic sequence of n terms and common difference d may also be written in terms of the first term, as follows:

$$a_1, (a_1 + d), (a_1 + 2d), (a_1 + 3d), \ldots, [a_1 + (n-1)d]$$

This formula enables you to find any term of an arithmetic sequence if you know the first term and the common difference.

Example 2

Find the number of seats in row Z of the theater described on the previous page.

Solution

The number of seats in row Z is the 26th term of the sequence 7, 10, 13, ... Since you know the number of seats in the first row, $a_1 = 7$, and the common difference, $d = 3$:

$$a_n = a_1 + (n - 1)d$$
$$a_{26} = 7 + (26 - 1)3$$
$$a_{26} = 7 + 75 = 82$$

Answer: Row Z has 82 seats.

Since there are four variables in the formula, a_1, n, d, and a_n, any one of these values may be computed if the other three are known.

Example 3

Find the first term of an arithmetic sequence if the common difference is –2 and the 80th term is –56.

Solution

$$a_n = a_1 + (n - 1)d$$
$$-56 = a_1 + (80 - 1)(-2)$$
$$-56 = a_1 + 79(-2)$$
$$-56 = a_1 - 158$$
$$a_1 = 102$$

Answer: The first term of the sequence is 102.

Example 4

The 10th term of an arithmetic sequence is 50 and the 20th term is 86. Find the 12th term of the sequence.

Solution

To find the 12th term, you must know the value of d. Since d is constant for the entire sequence, you can find its value by working with the partial sequence of eleven terms formed by the 10th through 20th terms.

The 10th term of the original sequence becomes the 1st term of a new sequence in which the 20th term of the original is the 11th term.

$$a_1, a_2, a_3, ..., a_{10}, a_{11}, ..., a_{20}$$

arithmetic sequence
of 11 terms

$$a_n = a_1 + (n - 1)d$$
$$86 = 50 + (11 - 1)d$$
$$86 = 50 + 10d$$
$$36 = 10d$$
$$d = 3.6$$

Use this value of d to find the 12th term of the original sequence.

$$a_{11} = 50 + 3.6$$
$$= 53.6$$
$$a_{12} = 53.6 + 3.6$$
$$= 57.2$$

Answer: The 12th term of the sequence is 57.2.

Arithmetic Means

For any three consecutive terms of an arithmetic sequence, the second of these three terms is the average, or *arithmetic mean*, of the other two.

To verify this statement, let the three terms be represented, based on the meaning of common difference, by $a, a + d, a + 2d$. The arithmetic mean of a and $a + 2d$ is the sum of the two terms divided by 2. Note that the result is $a + d$, the second term.

$$\frac{a + (a + 2d)}{2} = \frac{2a + 2d}{2} = a + d$$

This idea can be extended to any number of arithmetic means to be inserted between two numbers, forming an arithmetic sequence.

Example 5

Insert four arithmetic means between -8 and 12.

Solution

You must form an arithmetic sequence where -8 is the first term and 12 is the last term, with four arithmetic means between.

$$-8 \ \_\_ \ \_\_ \ \_\_ \ \_\_ \ 12$$

There are six terms in this sequence. Determine the value of d.

$$a_n = a_1 + (n - 1)d$$
$$12 = -8 + (6 - 1)d$$
$$12 = -8 + 5d$$
$$20 = 5d$$
$$d = 4$$

Use $d = 4$ to evaluate the four arithmetic means.

$$a_2 = -8 + 4 = -4$$
$$a_3 = -4 + 4 = 0$$
$$a_4 = 0 + 4 = 4$$
$$a_5 = 4 + 4 = 8$$

Answer: The four arithmetic means are $-4, 0, 4, 8$.

Example 6

The low temperatures during one week increased the same number of degrees each day beginning with –10° on Monday and ending with 2° on Friday. Write the sequence of temperatures from Monday to Friday.

Solution

Since the temperature change is constant, the sequence is arithmetic. To insert three arithmetic means between –10 and 2, find d.

$$-10 \quad \underline{\quad} \quad \underline{\quad} \quad \underline{\quad} \quad 2$$

Mon Tue Wed Thur Fri

$$a_n = a_1 + (n-1)d$$
$$2 = -10 + (5-1)d$$
$$2 = -10 + 4d$$
$$12 = 4d$$
$$d = 3$$

Answer: The sequence of low temperatures is $-10, -7, -4, -1, 2$.

The Sum of an Arithmetic Sequence

Reconsider the ten rows of the theater described on page 168. Observe a pattern that can be used to determine the total number of seats in these rows without having to find and add all 10 numbers. Notice that the sum of the 1st and 10th terms is 41, as is the sum of the 2nd and 9th terms, as is the sum of the 3rd and 8th terms, and so on.

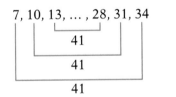

Since there are ten rows, there are five pairs whose sum is 41, for a total of 5(41) or 205 seats.

This strategy can be generalized to find the sum of the terms of any finite arithmetic sequence. Suppose the terms of an arithmetic sequence of n terms are $a_1, a_2, a_3, ..., a_n$. The sum of n terms, S_n, can be written as a series:

$$S_n = a_1 + a_2 + a_3 + \cdots + a_{n-2} + a_{n-1} + a_n$$

Using the commutative property of addition,

$$S_n = a_n + a_{n-1} + a_{n-2} + \cdots + a_3 + a_2 + a_1$$

Adding the two equations gives

$$2S_n = (a_1 + a_n) + (a_2 + a_{n-1}) + (a_3 + a_{n-2}) + \cdots + (a_n + a_1)$$

The value of each term in parentheses is the same as $(a_1 + a_n)$, as was found in the theater problem above.

Since there are n of these pairs of terms in $2S_n$:

$$2S_n = n(a_1 + a_n)$$

Therefore, the sum S_n of an arithmetic sequence of n terms is:

$$S_n = \frac{n}{2}(a_1 + a_n)$$

Since $a_n = a_1 + (n-1)d$, an alternative formula for the sum of an arithmetic sequence that can be used when a_n is not known is:

$$S_n = \frac{n}{2}[2a_1 + (n-1)d]$$

Example 7

Find the sum of an arithmetic sequence of 50 terms where the first term is -7 and the last is 104.

Solution

Since a_n is known, use the sum formula involving a_n, with $n = 50$, $a_1 = -7$, and $a_n = 104$.

$$S_n = \frac{n}{2}(a_1 + a_n)$$

$$= \frac{50}{2}(-7 + 104)$$

$$= 25(97)$$

$$= 2,425$$

Answer: The sum of the 50 terms is 2,425.

Example 8

Find the sum of the first 75 positive odd integers.

Solution

The sequence of positive odd integers, 1, 3, 5, ..., is an arithmetic sequence with the common difference 2. Since the 75th term is not known, use the sum formula that does not involve a_n, the formula $S_n = \frac{n}{2}[2a_1 + (n-1)d]$, with $a_1 = 1$, $n = 75$, and $d = 2$.

$$S_n = \frac{n}{2}[2a_1 + (n-1)d]$$

$$= \frac{75}{2}[2(1) + (75-1)2]$$

$$= \frac{75}{2}(2 + 148)$$

$$= \frac{75}{2}(150)$$

$$= 5,625$$

Answer: The sum of the first 75 positive odd numbers is 5,625.

Exercises 4.2

In **1 – 8**, determine whether the sequence is arithmetic. If the sequence is arithmetic, write a recursive formula for the nth term; otherwise write "not arithmetic."

1. 4, 8, 16, 32, ...

2. 3, 0, –3, –6, –9, ...

3. $1, \frac{5}{2}, 4, \frac{11}{2}, 7, ...$

4. 0, 3, 8, 15, 24, ...

5. –1.4, –1.0, –0.6, –0.2, ...

6. 0.2, 0.21, 0.22, ...

7. 1, 4, 9, 16, ...

8. $\sqrt{2}, \sqrt{2} + \sqrt{2}, \sqrt{2} + 2\sqrt{2}, \sqrt{2} + 3\sqrt{2}, ...$

In **9 – 17**, find a_n.

9. $a_1 = 0, \quad d = 2, \quad n = 14$

10. $a_1 = -4, \quad d = 3, \quad n = 10$

11. $a_1 = 2, \quad d = \frac{1}{3}, \quad n = 50$

12. $a_1 = 130, d = -10, n = 15$

13. $a_1 = 1.2, d = -0.3, n = 12$

14. $a_1 = \sqrt{3}, d = \sqrt{3}, n = 7$

15. $a_1 = -\frac{1}{2}, \quad d = -\frac{1}{4}, \quad n = 7$

16. $a_1 = \frac{1}{3}, \quad d = \frac{1}{5}, \quad n = 10$

17. $a_1 = 2 + \sqrt{2}, d = 2 - \sqrt{2}, n = 8$

In **18 – 19**, express in terms of z, the 5th term of the arithmetic sequence.

18. $a_1 = z, d = 3$

19. $a_1 = 2z, d = z^2$

In **20 – 21**, find the 1st term of the arithmetic sequence.

20. $a_{10} = 46, d = 4$

21. $a_{17} = 100, d = -2$

In **22 – 25**, find the number of terms in the arithmetic sequence.

22. $a_1 = 8, a_n = 140, d = 4$

23. $a_1 = -0.86, a_n = 1.24, d = 0.3$

24. 20, 16, 12, ..., –92

25. $200, 199\frac{2}{3}, 199\frac{1}{3}, ..., 100$

In **26 – 29**, find the indicated term of the arithmetic sequence.

26. If $a_1 = 3$ and $a_{30} = 177$, find a_{42}.

27. If $a_1 = -7$ and $a_{15} = 42$, find a_{32}.

28. If $a_9 = 5$ and $a_{10} = 11$, find a_{20}.

29. If $a_1 = \sqrt{5}$ and $d = 2\sqrt{5}$, find a_6.

In **30 – 32**, use the given formula to generate the first five terms of the arithmetic sequence.

30. $a_n = 2n + 1$

31. $a_n = 5n - 2$

32. $a_n = a_{n-1} - 0.7$, where $a_1 = -2$

In **33 – 36**, find three arithmetic means between the given numbers.

33. 4, 16

34. –8, 28

35. $\frac{1}{2}, \frac{13}{2}$

36. $\sqrt{2}, 7\sqrt{2}$

In **37 – 45**, find the sum of the indicated number of terms of the arithmetic sequence.

37. $a_1 = 3, a_n = 23, n = 10$

38. $a_1 = 4, a_n = -23, n = 9$

39. $a_1 = -2, a_n = 10\frac{1}{2}, n = 50$

40. $a_1 = -1.4, a_n = 1.4, n = 14$

41. $a_1 = 1, d = 2, n = 12$

42. $a_1 = 0.05, d = 0.1, n = 10$

43. $a_1 = -\sqrt{3}, d = \sqrt{3}, n = 5$

44. $a_6 = 10, a_{10} = 20, n = 18$

45. $a_{10} = 1.05, a_{20} = 3.05, n = 4$

Exercises 4.2 (continued)

In **46 – 49**, evaluate.

46. $\displaystyle\sum_{n=1}^{40} 2n$ **47.** $\displaystyle\sum_{n=1}^{150} (n+1)$ **48.** $\displaystyle\sum_{n=1}^{100} (2n+3)$ **49.** $\displaystyle\sum_{n=0}^{50} (100-2n)$

50. Given the arithmetic sequence 4.6, 4.0, 3.4, …:
   a. Find the common difference d.   **b.** Find a_{10}.
   c. Find the sum of the first 100 terms.

51. If for an arithmetic sequence, $a_{12} = 6\sqrt{2}$ and $a_{16} = 14\sqrt{2}$:
   a. Find the common difference.   **b.** Find a_{13}, a_{14}, and a_{15}.
   c. Find n if $a_n = 34\sqrt{2}$.

52. Find the sum of all the multiples of 3 between 1,000 and 2,000.

53. Find the sum of all the multiples of 13 between 100 and 200.

54. Find the sum of all the multiples of 4 that have three digits.

55. Saleem deposited $20 in the bank during the first week of the year. In each successive week, he deposited $2 more than he deposited the previous week, for a total of 52 deposits. How much did he deposit during the year?

56. A triangle of soup cans is used for a supermarket display. Twenty cans are placed on the bottom row, 19 on the second row, and so on until there is one can on the top. How many cans are there in the display?

57. Is it possible to use exactly 5,050 cans to make a triangular display similar to that in Exercise 56? If so, how many cans would be on the bottom, and how many rows would there be?

58. A reel of recording tape is 1,800 feet long. If the innermost coil is 6.28 inches long and each consecutive coil is 0.1 inch longer than the previous one, approximately how many coils are on the reel?

59. A real-estate developer placed a penalty clause in the contract for a shopping mall to be built at a cost of $15 million. The clause specified that if the mall was finished one month late, the builder would be fined $10,000, and for each additional month the project was late, the fine for that month would be $2,000 more than the fine for the previous month. If the mall was due to be completed on January 1, but was completed on April 30 of the following year, how much did the developer pay the builder after the fine was subtracted from the $15 million price?

60. A sewer-cleaning company charges an initial fee of $1.00/ft. for a 100-foot sewer line and the charge increases by 50 cents per foot for each additional 100 feet. If the company charged $1,000 for a job, what was the total length of the sewer line?

---

46. 1,640

47. 11,475

48. 10,400

49. 2,550

50. a. –0.6
   b. –0.8
   c. –2,510

51. a. $2\sqrt{2}$
   b. $8\sqrt{2}$, $10\sqrt{2}$, $12\sqrt{2}$
   c. 26

52. 499,500

53. 1,196

54. 123,300

55. $3,692

56. 210

57. Yes, 100 rows, with 100 cans in bottom row

58. 598

59. $14,600,000

60. 500 feet

4.3 Geometric Sequences

As his reward for a great service performed, legend has it that a wise man of an ancient kingdom asked only that he be given two grains of rice and that, on each succeeding day, his portion be doubled. Before long, to everyone's amazement, it became apparent that this seemingly modest request was going to empty the country's granaries! This is an example of a *geometric sequence*.

$$2, 2^2, 2^3, 2^4, ..., 2^n, ...$$

■ *Definition:* A *geometric sequence* is a sequence such that each term after the first is obtained by multiplying the previous term by a constant, called the common ratio.

The common ratio of a geometric sequence is computed by dividing any term, except the first, by the term preceding it. For example:

$$1, \tfrac{1}{2}, \tfrac{1}{4}, \tfrac{1}{8}, ...; r = \tfrac{1}{2} \qquad\qquad 5, -10, 20, -100, ...; r = -2$$

The generalized geometric sequence of n terms whose first term is a_1 and whose common ratio is r, is written $a_1, a_1r, a_1r^2, a_1r^3, ..., a_1r^{n-1}$. This suggests the following formula for finding the nth term of a geometric sequence:

$$a_n = a_1 r^{n-1}$$

Example 1

Find the fourth term of a geometric sequence whose first term is 2 and whose common ratio is $\tfrac{1}{3}$.

Solution

$$a_n = a_1 r^{n-1}$$
$$a_4 = 2\left(\tfrac{1}{3}\right)^{4-1}$$
$$= 2\left(\tfrac{1}{3}\right)^3$$
$$= 2\left(\tfrac{1}{27}\right)$$
$$= \tfrac{2}{27}$$

Note: The order of operations was followed by first raising $\tfrac{1}{3}$ to the third power and then multiplying by 2.

When the first term, a_1, and common ratio, r, are known, additional terms may be found using the following recursive formula:

$$a_n = a_{n-1} r$$

Example 2

Use the recursive formula $a_1 = -\tfrac{1}{4}$, $a_n = a_{n-1}(-3)$ to generate the first five terms of the geometric sequence.

Solution

$$a_n = a_{n-1}r$$

$$a_1 = -\frac{1}{4}$$

$$a_2 = a_1(-3) = -\frac{1}{4}(-3) = \frac{3}{4}$$

$$a_3 = a_2(-3) = \frac{3}{4}(-3) = -\frac{9}{4}$$

$$a_4 = a_3(-3) = -\frac{9}{4}(-3) = \frac{27}{4}$$

$$a_5 = a_4(-3) = \frac{27}{4}(-3) = -\frac{81}{4}$$

Answer: The first five terms of the sequence are $-\frac{1}{4}, \frac{3}{4}, -\frac{9}{4}, \frac{27}{4}, -\frac{81}{4}$.

Example 3

Identify each sequence as arithmetic or geometric, and write a recursive formula for the nth term.

Solution

Sequence	Nature	Recursive Formula
a. 1, 5, 25, 125, 625, ...	geometric	$a_n = 5a_{n-1}$
b. 15, 10, 5, 0, –5, ...	arithmetic	$a_n = a_{n-1} - 5$
c. 5, –5, 5, –5, 5, ...	geometric	$a_n = -1a_{n-1}$

Geometric sequences are used to solve problems involving population growth, compound interest, annuities, and other areas in which increase is proportional to the amount present. For example, a bank account earns interest at 6% compounded semiannually, that is, it pays 3% every half year. If an initial deposit of $1,000 is made, then the balance in the account at the end of each half-year period may be represented by a geometric sequence with a common ratio of 1.03.

Term	$\frac{1}{2}$ year	1 year	$1\frac{1}{2}$ years	2 years
Balance	$1,000(1.03)$ $1,030	$1,000(1.03)^2$ $1,060.90	$1,000(1.03)^3$ $1,092.73	$1,000(1.03)^4$ $1,125.51

Example 4

In a certain town, the population increased 5% each year over a ten-year period. If the initial population at the beginning of the ten-year period was 50,000 people, during which year did the population pass 60,000 people?

Solution

Form a geometric sequence whose common ratio is 1.05 and whose first term is 50,000:

$$50,000, \ 50,000(1.05), \ 50,000(1.05)^2, \ 50,000(1.05)^3, 50,000(1.05)^4, \ ...$$

Evaluating these expressions (to the nearest whole number) gives:

$$50,000, \ 52,500, \ 55,125, \ 57,881, \ 60,775, \ ...$$

Answer: The population passed 60,000 during the 4th year.

Geometric Means

As the arithmetic mean of two numbers is their average, the geometric mean is the mean proportional. For example, the mean proportional of 9 and 4 is 6 or –6, and it is found by setting up this proportion:

$$\frac{9}{x} = \frac{x}{4}$$

$$x^2 = 36$$

$$x = \pm 6$$

In general, the *geometric mean* of two positive numbers a and b is $\pm\sqrt{ab}$.

If one of the geometric means is placed between the two given numbers, the sequence of three numbers always forms a geometric sequence:

$$a, \sqrt{ab}, b \quad \text{or} \quad a, -\sqrt{ab}, b$$

This result can be verified by showing that the ratio of the second term to the first term is the same as the ratio of the third term to the second term.

$$\frac{a}{\sqrt{ab}} = \frac{\sqrt{ab}}{b} \quad \text{and} \quad \frac{a}{-\sqrt{ab}} = \frac{-\sqrt{ab}}{b}$$

$$ab = ab \qquad\qquad\qquad ab = ab$$

The concept of a geometric mean can be extended to any number of geometric means inserted between two numbers such that the resulting sequence is a geometric sequence.

Example 5

Insert two geometric means between 640 and 80.

Solution

The geometric sequence needed is:

$$640, a_2, a_3, 80$$

Determine the value of r.

$$a_n = a_1 r^{n-1}$$

$$80 = 640(r^{4-1})$$

$$80 = 640 r^3$$

$$\frac{1}{8} = r^3$$

$$r = \frac{1}{2}$$

Use $r = \frac{1}{2}$ to evaluate the two geometric means.

$$a_2 = a_1 r = 640\left(\tfrac{1}{2}\right) = 320$$

$$a_3 = a_2 r = 320\left(\tfrac{1}{2}\right) = 160$$

Answer: The two geometric means are 320 and 160.

1. 16, 32, 64

2. $\frac{1}{32}, \frac{1}{64}, \frac{1}{128}$

3. 1, –1, 1

4. 32, –64, 128

5. 0.123, 0.0123, 0.00123

6. $2\sqrt{6}, 4\sqrt{3}, 4\sqrt{6}$

7. $2^{-4}, 2^{-5}, 2^{-6}$

8. $8^{-1}, 8^{0}, 8^{1}$

9. 0.0625, 0.03125, 0.015625

10. 4, 12, 36, 108, 324, 972

11. –1, –1, –1, –1, –1, –1

12. –3, 6, –12, 24, –48, 96

13. 0.2, 0.8, 3.2, 12.8, 51.2, 204.8

14. $\frac{1}{4}, \frac{3}{4}, \frac{9}{4}, \frac{27}{4}, \frac{81}{4}, \frac{243}{4}$

15. $\sqrt{2}, 2\sqrt{2}, 4\sqrt{2}, 8\sqrt{2}, 16\sqrt{2}, 32\sqrt{2}$

16. $\sqrt{3}, \sqrt{6}, 2\sqrt{3}, 2\sqrt{6}$

17. $\frac{1}{2}, -\frac{1}{3}, \frac{2}{9}, -\frac{4}{27}$

18. 0.02, 0.002, 0.0002, 0.00002

19. geometric:
$a_n = 2a_{n-1}$

20. arithmetic:
$a_n = a_{n-1} + 2$

21. geometric:
$a_n = 3a_{n-1}$

22. geometric:
$a_n = -1a_{n-1}$

23. arithmetic:
$a_n = a_{n-1} + 3$

24. geometric:
$a_n = -2a_{n-1}$

Exercises 4.3

In **1 – 9**, write the next three terms of the given geometric sequence.

1. 1, 2, 4, 8, ...

2. $\frac{1}{2}, \frac{1}{4}, \frac{1}{8}, \frac{1}{16}, \cdots$

3. 1, –1, 1, –1, ...

4. 2, –4, 8, –16, ...

5. 123, 12.3, 1.23, ...

6. $\sqrt{3}, \sqrt{6}, 2\sqrt{3}, \ldots$

7. $2^{-1}, 2^{-2}, 2^{-3}, \ldots$

8. $8^{-4}, 8^{-3}, 8^{-2}, \ldots$

9. 0.5, 0.25, 0.125, ...

In **10 – 15**, write the first six terms of the geometric sequence whose first term and common ratio are given.

10. $a_1 = 4, \quad r = 3$

11. $a_1 = -1, r = 1$

12. $a_1 = -3, \quad r = -2$

13. $a_1 = 0.2, r = 4$

14. $a_1 = \frac{1}{4}, \quad r = 3$

15. $a_1 = \sqrt{2}, r = 2$

In **16 – 18**, use the recursive formula given to write the first four terms of the sequence it represents.

16. $a_1 = \sqrt{3}, a_n = a_{n-1}\sqrt{2}$

17. $a_1 = \frac{1}{2}, a_n = a_{n-1}\left(-\frac{2}{3}\right)$

18. $a_1 = 0.02, a_n = a_{n-1}(0.1)$

In **19 – 24**, identify whether the sequence is arithmetic or geometric. Write a recursive formula for the nth term.

19. 2, 4, 8, 16, ...

20. 2, 4, 6, 8, 10, ...

21. 3, 9, 27, 81, ...

22. –4, 4, –4, 4, ...

23. 5, 8, 11, 14, 17, ...

24. –3, 6, –12, 24, –48, ...

In **25 – 30**, identify whether the sequence is arithmetic, geometric or neither.

25. $\frac{1}{2}, \frac{1}{3}, \frac{4}{9}, \frac{8}{27}, \frac{16}{81}, \cdots$

26. $\frac{1}{2}, \frac{1}{3}, \frac{1}{4}, \frac{1}{5}, \frac{1}{6}, \cdots$

27. $25, -5, 1, -\frac{1}{5}, \frac{1}{25}, \cdots$

28. $\frac{1}{5}, \frac{2}{10}, \frac{3}{15}, \frac{4}{20}, \cdots$

29. $3, 1, \frac{1}{3}, 3, \ldots$

30. $2^2, 2^{-2}, 2^2, 2^{-2}, \ldots$

In **31 – 33**, write a sequence that represents the balances in an account at the end of each period for the entire first year, given the initial deposit and the terms of the compound interest payment. (For quarterly compounding show 4 terms, for monthly show 12 terms, and so on.)

31. $100 at 4% semiannually

32. $1,200 at 7% quarterly

33. $2,500 at 6% monthly

34. A car that cost $15,000 depreciated 20% each year. Find the value of the car three years after it was purchased.

35. It is estimated that the population of a town will increase by 2.4% for each of the next ten years. If the population is now 15,000, about how many people can be expected to live in the town at the end of the 8th year?

---

25. neither

26. neither

27. geometric

28. geometric

29. neither

30. neither

31. $102, $104.04

32. $1,221, $1,242.37, $1,264.11, $1,286.23

33. $2,512.50, $2,525.06, $2,537.69, $2,550.38, $2,563.13, $2,575.94, $2,588.82, $2601.77, $2,614.78, $2,627.85, $2,640.99, $2,654.19

34. $7,680

35. 18,134

Exercises 4.3 *(continued)*

In **36 – 39**, find the positive geometric mean between the given numbers.

36. 16 and 25   **37.** 9 and 4   **38.** 6 and 8   **39.** $\sqrt{2}$ and $\sqrt{3}$

In **40 – 43**, insert k geometric means between the given numbers.

40. 2 and 128; $k = 2$

41. 3 and 81; $k = 2$

42. $\frac{1}{4}$ and –8; $k = 4$

43. $-\frac{2}{3}$ and $-\frac{128}{375}$; $k = 2$

44. Three numbers form a geometric sequence whose common ratio is 3. If the first is decreased by 1, the second decreased by 3, and the third reduced to 2 less than half its value, the resulting three numbers form an arithmetic sequence. Find the numbers for the arithmetic and geometric sequences.

45. A 20-quart car radiator was mistakenly filled completely with antifreeze. It can be diluted by draining out 1 quart at a time; replacing it with pure water; then testing the mixture. After doing this 3 times, what percent of the remaining mixture would be antifreeze?

46. Three numbers form an arithmetic sequence whose common difference is 3. If the first number is increased by 1, the second increased by 6 and the third increased by 19, the resulting three numbers form a geometric sequence. Find the terms of the arithmetic and geometric sequences.

47. You are prepared to deposit $1,000 on March 1 and remove it five months (150 days) later for your tuition payment. Which would be the better option for you:

(1) a savings bank that pays 5%, compounded daily and paid from day of deposit to day of withdrawal, or

(2) a credit union that pays 6%, compounded quarterly?

Justify your answer.

48. When the Torres moved into their apartment 4 years ago, the rent was $6,000 a year. Each year, the rent has increased by the same percent, and is now, after four increases, $8,163 per year.

a. By what percent was the rent increased each year?

b. What was the rent, to the nearest dollar, for each of the years after the first?

---

36. 20

37. 6

38. $4\sqrt{3}$

39. $\sqrt[4]{6}$

40. 2, 8, 32, 128

41. 3, 9, 27, 81

42. $\frac{1}{4}, -\frac{1}{2}, 1, -2, 4, -8$

43. $-\frac{2}{3}, -\frac{8}{15}, -\frac{32}{75}, -\frac{128}{375}$

44. geometric sequence: 6, 18, 54
arithmetic sequence: 5, 15, 25

45. 86%

46. arithmetic sequence: 7, 10, 13
geometric sequence: 8, 16, 32

47. (1) $1,020.75 at end of 150 days at 5%

(2) $1,015.00 at end of 90 days at 6% (second quarter ends at end of 180 days)

Since option (2) gets only one quarter's interest, option (1) is better.

48. a. 8%

b. $6,480; $6,998; $7,558; $8,163

4.4 Geometric Series

To derive a formula for the sum of a *finite* number of terms of a geometric sequence, $a_1, a_1 r, a_1 r^2, \ldots, a_1 r^{n-1}$, write the sum, S_n:

$$S_n = a_1 + a_1 r + a_1 r^2 + a_1 r^3 + \cdots + a_1 r^{n-2} + a_1 r^{n-1}$$

Multiply both sides of this equation by r:

$$S_n r = a_1 r + a_1 r^2 + a_1 r^3 + \cdots + a_1 r^{n-1} + a_1 r^n$$

Subtract this equation from the first one:

$$S_n - S_n r = a_1 - a_1 r^n$$

Solve for S_n:

$$S_n(1 - r) = a_1 - a_1 r^n$$

$$S_n = \frac{a_1 - a_1 r^n}{1 - r} \text{ if } r \neq 1$$

Therefore, a formula for the sum of a geometric sequence of n terms, with first term a_1 and common ratio r ($r \neq 1$), is:

$$S_n = \frac{a_1 - a_1 r^n}{1 - r}$$

Since $a_n = a_1 r^{n-1}$, an alternate formula for S_n is:

$$S_n = \frac{a_1 - a_n r}{1 - r}$$

This formula can be used whenever the last term of the sequence is known.

Example 1

Find the sum of the first four terms of the geometric sequence $2, \frac{1}{2}, \frac{1}{8}, \ldots$.

Solution

So that you need not first find the 4th term(although that is easy here), use the formula that does not include a_n, with $a_1 = 2$, $n = 4$, and $r = \dfrac{\frac{1}{2}}{2} = \frac{1}{4}$.

$$S_n = \frac{a_1 - a_1 r^n}{1 - r}$$

$$S_4 = \frac{2 - 2\left(\frac{1}{4}\right)^4}{1 - \frac{1}{4}} = \frac{2 - 2\left(\frac{1}{256}\right)}{\frac{3}{4}} = \frac{2 - \frac{1}{128}}{\frac{3}{4}} = \frac{\frac{255}{128}}{\frac{3}{4}} = \frac{255}{96}$$

Answer: The sum of the first four terms is $\frac{255}{96}$.

Example 2

Find the sum of the first six terms of a geometric sequence for which $a_1 = 500$, $r = -\frac{1}{2}$, and $a_6 = -15\frac{5}{8}$.

Solution

Since the 6th term of the sequence is known, use the formula that includes a_n.

$$S_n = \frac{a_1 - a_n r}{1 - r}$$

$$S_6 = \frac{500 - \left(-15\frac{5}{8}\right)\left(-\frac{1}{2}\right)}{1 - \left(-\frac{1}{2}\right)}$$

$$= \frac{500 - \frac{125}{16}}{\frac{3}{2}} = \frac{\frac{7,875}{16}}{\frac{3}{2}} = 328\frac{1}{8}$$

Answer: The sum of the first 6 terms is $328\frac{1}{8}$.

Many situations in finance involve making periodic payments or deposits. Deposits earn interest for the time they have been on deposit, and payments include interest on the part of the loan that is outstanding. For example, a tax-sheltered annuity might involve depositing $100 each month into an account that pays 6% interest. On the other hand, a mortgage of $100,000 is paid off in monthly payments that include interest on the borrowed money as well as a part of the principal.

Example 3

 Yearly payments of $1,000 are put into an annuity that pays 6% interest per year. Find the value of the annuity after five years.

Solution

The first $1,000-payment would have earned 5 interest payments that were compounded each year. Therefore, at the end of five years, the cumulative value of the first $1,000-payment is $1,000(1.06)^5$. The second $1,000-payment earned 4 interest payments. Its value is $1,000(1.06)^4$. The third payment earned 3 interest payments and its value is $1,000(1.06)^3$. If you continue in this pattern, the cumulative value of the payments forms a geometric series in which the first term (representing the value of the last payment) is $1,000(1.06)$ and the 5th term (representing the value of the first payment) is $1,000(1.06)^5$.

$$1,000(1.06) + 1,000(1.06)^2 + 1,000(1.06)^3 + 1,000(1.06)^4 + 1,000(1.06)^5$$

For this series, $a_1 = 1,000(1.06)$, $a_n = 1,000(1.06)^5$, $n = 5$, and $r = 1.06$. The sum can be computed as follows:

$$S_n = \frac{a_1 - a_n r}{1 - r}$$

$$S_5 = \frac{1,000(1.06) - (1,000)(1.06)^5(1.06)}{1 - 1.06}$$

$$= \frac{1,000(1.06)(1 - 1.06^5)}{-0.06}$$

$$= \frac{1,060(1 - 1.06^5)}{-0.06}$$

Answer: The value of the annuity after five years is $5,975.32.

Infinite Geometric Series

Suppose you are standing two yards from a wall. You approach the wall by taking steps, each of which has a length equal to half the distance remaining between you and the wall. Theoretically, you would never reach the wall since there would always be some distance left between you and the wall that could be divided in half. Your distance from the wall would always be equal to the distance moved in the last step. As you take more and more steps, you get closer and closer to the wall. The sum of the lengths of all of your steps would approach (but not reach) two yards.

Although this problem is contrary to common sense (a paradox), it suggests the following infinite geometric series.

$$1 + \frac{1}{2} + \frac{1}{4} + \frac{1}{8} + \cdots$$

The value of this infinite sum is defined as the number that the sum is approaching as terms are added without end. Think of this as the sum of the lengths of your steps, which has a value of 2, the distance you began from the wall.

The sum of an infinite geometric sequence will have a finite value only if the absolute value of each successive term gets smaller and smaller, approaching 0. This is possible only when the common ratio is a proper fraction; that is, when $|r| < 1$. For example:

a. $3 + 1 + \frac{1}{3} + \frac{1}{9} + \cdots$ In this infinite geometric series, $r = \frac{1}{3}$. Therefore, the series has a finite sum since $|r| < 1$.

b. $\frac{1}{10} + \frac{3}{20} + \frac{9}{40} + \frac{27}{80} + \cdots$ In this infinite geometric series, $r = \frac{3}{2}$. Therefore, the series does not have a finite sum since $|r| > 1$.

Start with the formula for the sum of a finite number of terms of a geometric sequence.

$$S_n = \frac{a_1 - a_1 r^n}{1 - r}$$

If $|r| < 1$, and n approaches infinity, $|r^n|$ will become smaller and smaller tending toward zero. Therefore, the term $a_1 r^n$ approaches zero and the sum of an infinite geometric sequence is:

$$a_1 + a_1 r + a_1 r^2 + \cdots + a_1 r^{n-1} + \cdots = \frac{a_1}{1 - r}, \text{ when } |r| < 1$$

Using the symbol ∞ to represent infinity:

$$S_\infty = \frac{a_1}{1-r}$$

Example 4

Evaluate: $\displaystyle\sum_{n=1}^{\infty} 9(10)^{-n}$

Solution

The first few terms of this sum are $0.9 + 0.09 + 0.009 + \cdots$. This is a geometric series with $a_1 = 0.9$ and $r = 0.1$. Since $|r| < 1$, the sum of the sequence approaches a finite value. Using the formula for the sum of an infinite geometric sequence:

$$S_\infty = \frac{a_1}{1-r}$$

$$= \frac{0.9}{1-0.1} = \frac{0.9}{0.9} = 1$$

Answer: $\displaystyle\sum_{n=1}^{\infty} 9(10)^{-n} = 1$

This example has actually shown that $0.9 + 0.09 + 0.009 + \cdots = 0.99\overline{9} = 1$.

Recall that any repeating decimal approaches some rational number. One way to find this rational number is to consider the repeating decimal to be an infinite geometric series. For example:

$0.4343\overline{43} = 0.43 + 0.0043 + 0.000043 + \cdots = \frac{43}{99}$, with $a_1 = 0.43$ and $r = \frac{1}{100}$

$0.123123\overline{123} = 0.123 + 0.000123 + 0.000000123 + \cdots = \frac{123}{999}$, with $a_1 = 0.123$ and $r = \frac{1}{1,000}$

Exercises 4.4

In **1 – 5**, find the sum of n terms of the given geometric sequence.

1. $9, -27, 81, -243; n = 7$

2. $\frac{1}{3}, \frac{1}{9}, \frac{1}{27}; n = 5$

3. $0.1, 0.01, 0.001; n = 6$

4. $a_1 = 2, r = \sqrt{2}; n = 4$

5. $a_1 = 100, r\ 1.05, n = 5$
Round the answer to the nearest hundredth.

In **6 – 8**, find the value of each expression to four decimal places.

6. $\displaystyle\sum_{n=1}^{6} \left(\frac{4}{5}\right)^n$

7. $\displaystyle\sum_{k=1}^{12} (1.03)^k$

8. $\displaystyle\sum_{i=1}^{4} (1.015)^{1-i}$

9. Find the sum of a geometric sequence of six terms whose first term is $\frac{2}{3}$ and whose last term is $\frac{64}{729}$.

10. Find the sum of the first four terms of a geometric sequence where $a_1 = \frac{2}{5}$ and $a_4 = \frac{16}{625}$.

Answers 4.4

1. 4,923

2. $\frac{121}{243}$

3. 0.111111

4. $6 + 6\sqrt{2}$

5. 552.56

6. 2.9514

7. 14.6178

8. 3.9122

9. $\frac{1,330}{729}$

10. $\frac{406}{625}$

11. $2,082.58

12. No; $7,749.25 is the total.

13. $547,197.15

14. $5,115

15. 2

16. 2

17. $r = \frac{3}{2} > 1$, no sum

18. $-\frac{2}{9}$

19. $\frac{1}{3}$

20. $\frac{1}{9}$

21. $r = \sqrt{2} > 1$, no sum

22. $\frac{3(3 + \sqrt{3})}{2}$

23. $\frac{8}{9}$

24. $\frac{2}{3}$

25. $\frac{13}{99}$

26. $\frac{20}{33}$

27. $\frac{125}{999}$

28. $\frac{127}{198}$

29. $\frac{46}{33}$

30. $\frac{31}{11}$

31. 24

32. $(16 + 8\sqrt{2})$ in.

Exercises 4.4 (continued)

11. Mrs. Mendez is permitted to contribute $500 per quarter to a tax-sheltered annuity that pays a fixed rate of $6\frac{1}{2}\%$ per year compounded quarterly. What will the balance in her annuity be at the end of one year?

12. Mr. James decides to save $300 each month in order to buy a car in two years. If he can get 7% interest per year compounded monthly, will he have enough saved after two years to buy an $8,000 car? (This figure to include tax and particulars.) Explain.

13. Mr. Black needs a mortgage for $45,000. He is offered a 25-year mortgage at 10%, requiring monthly payments of $409. Mr. Black wonders how much money he could accumulate if he were able to bank $409 per month for 25 years, compounded monthly, at 10%. How much would his dream account contain?

14. A quiz program gives a prize of $5 for the first question answered correctly. A contestant's winnings are doubled for each correct answer after the first. How much would you win if you answered 10 questions correctly?

In **15 – 22**, find the sum of the infinite geometric series, if possible. Otherwise, find the value of r and state why the series has no finite sum.

15. $1 + \frac{1}{2} + \frac{1}{4} + \frac{1}{8} + \cdots$

16. $\frac{2}{3} + \frac{4}{9} + \frac{8}{27} + \frac{16}{81} + \cdots$

17. $7 + \frac{21}{2} + \frac{63}{4} + \cdots$

18. $-\frac{1}{3} + \frac{1}{6} - \frac{1}{12} + \frac{1}{24} - \frac{1}{72} + \cdots$

19. $\sum_{n=1}^{\infty} \left(\frac{1}{4}\right)^n$

20. $\sum_{n=1}^{\infty} (0.1)^n$

21. $\sum_{n=1}^{\infty} (\sqrt{2})^n$

22. $(3 + \sqrt{3}) + \left(1 + \frac{\sqrt{3}}{3}\right) + \cdots$

In **23 – 30**, express the repeating decimal as a fraction.

23. $0.888\overline{8}$

24. $0.666\overline{6}$

25. $0.1313\overline{13}$

26. $0.606\overline{060}$

27. $0.125125\overline{125}$

28. $0.6414\overline{141}$

29. $1.393\overline{939}$

30. $2.818\overline{181}$

31. The midpoints of the sides of a 3–4–5 right triangle are joined to form a smaller triangle, and then the midpoints of the sides of the smaller triangle are joined to form an even smaller triangle. Find the sum of the perimeters of all the triangles formed, if this process is continued without end.

32. A second square is formed by joining the midpoints of the 2-inch sides of a given square, a third by joining the midpoints of the second and so on, ad infinitum. Find the total of the perimeters of all the squares.

Answers 4.4

33. A rubber ball is dropped from a height of 2 meters. Each time it hits the ground it rebounds $\frac{4}{5}$ the distance from where it fell. How far does the ball travel before it stops?

34. A car is pushed and permitted to coast on level ground until it stops. If a front wheel, which is 20 inches in diameter, makes 1 revolution during the first second, and $\frac{2}{3}$ of the previous revolution in each succeeding second, how far does the car go before it stops? Round your answer to the nearest inch.

35. Evaluate: $\displaystyle\sum_{n=0}^{\infty} \left(\frac{1}{3^n} + \frac{1}{2^n} \right)$

36. Evaluate: $\displaystyle\sum_{n=0}^{\infty} \left(\frac{1}{4^n} - \frac{1}{5^n} \right)$

37. Find the total area of the squares shown in the diagram, if the length of a side of the largest square is 20 and the length of a side of each successive square is half the length of the side of the square to its left.

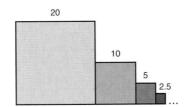

38. If the pattern below is continued indefinitely, what fractional part of the original equilateral triangle will be shaded?

39. Devise an infinite geometric sequence such that its sum is 10 and each term is equal to the sum of the terms that follow it.

40. Devise an infinite geometric sequence such that the first term is 18 and each term is equal to twice the sum of the terms that follow it.

33. 18 m

34. 188 in.

35. $3\frac{1}{2}$

36. $\frac{1}{12}$

37. $533\frac{1}{3}$

38. the entire triangle

39. $5, \frac{5}{2}, \frac{5}{4}, \frac{5}{8}, \frac{5}{16}, \cdots$

40. $18, 6, 2, \frac{2}{3}, \ldots$

4.5 *Mathematical Induction*

An important aspect of mathematics is the discovery of patterns that lead to generalizations. For example, when, at the age of ten, the 19th-century German mathematician Gauss was asked to find the sum of the counting numbers from 1 through 100, he promptly responded that the sum was 5,050. He accomplished this mentally by recognizing the pattern shown.

$$1 + 2 + 3 + 4 + \cdots + 97 + 98 + 99 + 100$$

$$101$$
$$101$$
$$101$$

$$50(101) = 5,050$$

An extension of this idea is expressed as follows:

$$1 + 2 + 3 + 4 + \cdots + n = \frac{n}{2}(n + 1)$$

In words, the sum of the first n natural numbers is the product of half the number of entries in the list and the successor of the last number.

Simply showing that a result is true for a few values of n does *not* prove that it is true for every natural number. In order to demonstrate this, consider the function $f(n) = n^2 - n + 41$. Mathematicians originally thought this function might yield a prime number for each natural number n. For example:

If $n = 1$, then $f(1) = 1^2 - 1 + 41 = 41$ Prime
If $n = 2$, then $f(2) = 2^2 - 2 + 41 = 43$ Prime
If $n = 7$, then $f(7) = 7^2 - 7 + 41 = 83$ Prime
$\vdots$
If $n = 40$, then $f(40) = 40^2 - 40 + 41 = 1,601$ Prime
but
If $n = 41$, then $f(41) = 41^2 - 41 + 41 = 41^2$ Composite

Therefore, the statement that $n^2 - n + 41$ is prime is not true for all natural numbers.

However, there are many statements in mathematics that can be proven true by an *inductive method*, in which the truth of a statement for a specific natural number relies on the truth of a previously-established case. One such method of proof is called *mathematical induction*, in which the truth of the statement for n establishes the truth of the statement for $n + 1$, the successor of n. This type of proof is in contrast to proof by the *deductive method*, in which a general statement is first proved to be true, and the truth of all specific cases follows.

The principle of *mathematical induction* is based on the following characteristics of the set of natural numbers N.

1. 1 is in the set N.
2. k is in N implies that $k + 1$ is in N.

These properties are based on Peano's postulates, which were introduced in Chapter 1.

□ **Principle of Mathematical Induction:** *For every n in N, let S_n be a statement that can be classified as true or false. If S_1 is true (the statement is true when n = 1) and whenever S_k is true it can be proven that S_{k+1} is true, then the statement is true for all n in N.*

To apply this principle to $1 + 2 + 3 + 4 + \cdots + n = \dfrac{n(n + 1)}{2}$, you must do two things:

1. Show that the statement S is true for $n = 1$.

$$S_1 = \frac{1(1 + 1)}{2} = \frac{2}{2} = 1 \qquad \text{Substitute } n = 1 \text{ in the given equation.}$$

Therefore, S_1 is true.

2. Show that if S_k is true, then S_{k+1} can be proven true.

Assume $S_k = 1 + 2 + 3 + \cdots + k = \dfrac{k(k + 1)}{2}$

and use this assumption to prove that S_{k+1} must be true.

Prove $S_{k+1} = 1 + 2 + 3 + \cdots + k + (k + 1) = \dfrac{(k + 1)[(k + 1) + 1]}{2} = \dfrac{(k + 1)(k + 2)}{2}$.

This part of a proof by mathematical induction requires careful thought. The *induction hypothesis* above states that $1 + 2 + 3 + \cdots + k = \dfrac{k(k + 1)}{2}$.

Add $k + 1$ to both sides of the equation to find the sum of $k + 1$ terms.

$$1 + 2 + 3 + \cdots + k + (k + 1) = \frac{k(k + 1)}{2} + k + 1$$

$$S_{k+1} = \frac{(k^2 + k)}{2} + k + 1$$

$$= \frac{k^2 + k + 2k + 2}{2}$$

$$= \frac{k^2 + 3k + 2}{2}$$

$$S_{k+1} = \frac{(k + 1)(k + 2)}{2} \quad \checkmark$$

For $n = k + 1$ $\qquad S_n = \dfrac{n(n + 1)}{2}$

The proof is complete and it can be concluded that the statement S is true for all natural numbers.

Example 1

S_n is the statement $2 + 4 + 6 + \cdots + 2n = n(n + 1)$.
Prove that S_n is true for all n in N.

Solution

1. Show that S_1 is true.

$$S_1 = 1(2) = 2 \quad \checkmark$$

In **1 – 6**, only the sum term is given.

1. a. $\dfrac{k(k + 1)}{2}$

   b. $\dfrac{(k + 1)(k + 2)}{2}$

2. a. $2k^2$

   b. $2(k + 1)^2$

3. a. $(k + 1)! - 1$

   b. $(k + 2)! - 1$

4. a. $\dfrac{(k)(k + 1)(2k + 1)}{6}$

   b.
   $\dfrac{(k + 1)(k + 2)(2k + 3)}{6}$

5. a. $\dfrac{(k)(3k - 1)}{2}$

   b. $\dfrac{(k + 1)(3k + 2)}{2}$

6. a. $\dfrac{k^2(k + 1)^2}{4}$

   b. $\dfrac{(k + 1)^2(k + 2)^2}{4}$

2. Show that whenever S_k is true, then $S_{k + 1}$ can be proven true. Assume that S_k is true. S_k: $2 + 4 + \cdots + 2k = k(k + 1)$.

This is the induction hypothesis. Now use this statement to prove that $S_{k + 1}$ is true, that is:

$$2 + 4 + \cdots + 2k + 2(k + 1) = (k + 1)[(k + 1) + 1] = (k + 1)(k + 2).$$

$2 + 4 + \cdots + 2k = k(k + 1)$	Assume S_k.
$2 + 4 + \cdots + 2k + 2(k + 1) = k(k + 1) + 2(k + 1)$	Add $2(k + 1)$ to each side.
$2 + 4 + \cdots + 2k + 2(k + 1) = (k + 1)(k + 2)$	Simplify the right side.

But the left side of this is $S_{k + 1}$. Therefore, the proof is complete and it can be concluded that $2 + 4 + \cdots + 2n = n(n + 1)$ for all natural numbers n.

The Greeks used *figurate numbers* to help make the transition from geometry to algebra.

Consider the diagrams to see why 1, 3, 6, 10, ... are called *triangular numbers*.

Note that the second triangular number is obtained by adding 2 to the first triangular number, that the third triangular number is obtained by adding 3 to the second, and in general, that the nth triangular number is obtained by adding n to the $(n - 1)$st number.

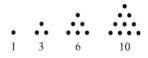

Similarly, 1, 4, 9, 16, ... are *square numbers*. The second square number is obtained by adding 3 to the first, the third square number is obtained by adding 5 (the next odd number after 3) to the second, and the fourth square number is obtained by adding 7 (the next odd number after 5) to the third. In general, the nth square number is obtained by adding $2n - 1$ to the $(n - 1)$st number.

In addition, the numbers 1, 5, 12, 22, 35, ... are *pentagonal numbers*. The $(n + 1)$st pentagonal number is obtained by adding $3n + 1$ to the nth pentagonal number. Suppose one wishes to form a pyramid of cannonballs by forming equilateral triangles with the balls and stacking one triangle on top of another (as in the diagram below). A *pyramidal number*, P_n, can be defined as the sum of the first n triangular numbers. For example, letting T_n be the nth triangular number,

$P_1 = T_1$, $P_2 = T_1 + T_2$, $P_3 = T_1 + T_2 + T_3$, and so on. Therefore:

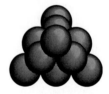

$P_1 = 1 \qquad P_2 = 1 + 3 = 4 \qquad P_3 = 1 + 3 + 6 = 10$

7. S_1: ✔

Assume S_k: $1 + 3 + 5 + \cdots + (2k - 1) = k^2$

Then $S_{k + 1}$: $1 + 3 + 5 + \cdots + (2k - 1) + (2(k + 1) - 1) = k^2 + (2(k + 1) - 1)$
$= (k + 1)^2$

8. S_1: ✔

Assume S_k: $4 + 8 + 12 + \cdots + 4k = 2k^2 + 2k$

Then $S_{k + 1}$: $4 + 8 + 12 + \cdots + 4k + 4(k + 1) = 2k^2 + 2k + 4(k + 1)$
$= 2(k + 1)^2 + 2(k + 1)$

Example 2

Prove that the sum of n triangular numbers $P_n = \frac{n}{6}(n + 1)(n + 2)$.

Solution

The nth triangular number is $\frac{n(n + 1)}{2}$.

Apply mathematical induction to show that:

$$1 + 3 + 6 + \cdots + \frac{n(n + 1)}{2} = \frac{n}{6}(n + 1)(n + 2)$$

Note: Building the pyramid from top to bottom, each pyramid is the sum of n triangular numbers. For example, P_4 is made up of $(1 + 3 + 6 + 10)$ balls, with the ten balls in the bottom layer forming an equilateral triangle having four balls on each side.

1. Show that P_1 is true.

$$P_1 = \frac{1}{6}(1 + 1)(1 + 2) = \frac{1}{6}(2)(3) = 1 \ \checkmark$$

2. Show that if P_k is true, then $P_{k + 1}$ is true.

Since P_k is true, $1 + 3 + 6 + 10 + \cdots + \frac{k(k + 1)}{2} = \frac{k}{6}(k + 1)(k + 2)$.

Add $\frac{(k + 1)(k + 2)}{2}$, the $(k + 1)$st pyramidal number, to each side of P_k:

$$1 + 3 + 6 + 10 + \cdots + \frac{k(k + 1)}{2} + \frac{(k + 1)(k + 2)}{2}$$

$$= \frac{k}{6}(k + 1)(k + 2) + \frac{(k + 1)(k + 2)}{2}$$

Simplify the right side of the equation:

$$= \frac{k(k + 1)(k + 2) + 3(k + 1)(k + 2)}{6}$$

$$= \frac{(k + 1)(k + 2)(k + 3)}{6}$$

Thus, $P_{k + 1}$ is true. By the principle of mathematical induction, it can be concluded that $P_n = \frac{n}{6}(n + 1)(n + 2)$ for all natural numbers.

Exercises 4.5

In **1 – 6**, for the given statement S_n, write: **a.** S_k **b.** $S_{k + 1}$

1. $1 + 2 + 3 + \cdots + n = \frac{n(n + 1)}{2}$

2. $2 + 6 + 10 + \cdots + (4n - 2) = 2n^2$

3. $1 \cdot 1! + 2 \cdot 2! + 3 \cdot 3! + \cdots + n \cdot n! = (n + 1)! - 1$

4. $1^2 + 2^2 + 3^2 + \cdots + n^2 = \frac{n(n + 1)(2n + 1)}{6}$

5. $1 + 4 + 7 + \cdots + (3n - 2) = \frac{n(3n - 1)}{2}$

6. $1^3 + 2^3 + 3^3 + \cdots + n^3 = \frac{n^2(n + 1)^2}{4}$

---

9. $\qquad S_1: \ \checkmark$

Assume S_k: $1 + 5 + 9 + \cdots + (4k - 3) = k(2k - 1)$

Then $S_{k + 1}$: $1 + 5 + 9 + \cdots + (4k - 3) + 4(k + 1) - 3 = k(2k - 1) + 4(k + 1) - 3$

$$= (k + 1)((2k + 1) - 1)$$

10. $\qquad S_1: \ \checkmark$

Assume S_k: $1^2 + 2^2 + 3^2 + \cdots + k^2 = \frac{k}{6}(k + 1)(2k + 1)$

Then $S_{k + 1}$: $1^2 + 2^2 + 3^2 + \cdots + k^2 + (k + 1)^2 = \frac{k}{6}(k + 1)(2k + 1) + (k + 1)^2$

$$= \frac{k + 1}{6}((k + 1) + 1)(2(k + 1) + 1)$$

Exercises 4.5 *(continued)*

In **7 – 23**, use mathematical induction to prove that the statement is true for all natural numbers, n.

7. $1 + 3 + 5 + \cdots + (2n - 1) = n^2$

8. $4 + 8 + 12 + \cdots + 4n = 2n^2 + 2n$

9. $1 + 5 + 9 + \cdots + (4n - 3) = n(2n - 1)$

10. $1^2 + 2^2 + 3^2 + \cdots + n^2 = \dfrac{n}{6}(n + 1)(2n + 1)$

11. $2^0 + 2^1 + 2^2 + 2^3 + \cdots + 2^{n-1} = 2^n - 1$

12. $1^3 + 2^3 + 3^3 + \cdots + n^3 = \dfrac{n^2(n + 1)^2}{4}$

13. $1(2) + 2(3) + 3(4) + \cdots + n(n + 1) = \dfrac{n}{3}(n + 1)(n + 2)$

14. $2 + 5 + 8 + \cdots + (3n - 1) = \dfrac{n}{2}(3n + 1)$

15. $1 + 6 + 11 + \cdots + (5n - 4) = \dfrac{n}{2}(5n - 3)$

16. $\displaystyle\sum_{i=1}^{n} (2i + 3) = n(n + 4)$

17. $\displaystyle\sum_{i=1}^{n} (4i - 3) = 2n^2 - n$

18. $\displaystyle\sum_{i=1}^{n} 4^i = \frac{4}{3}(4^n - 1)$

19. $\displaystyle\sum_{i=1}^{n} ar^{i-1} = \dfrac{a - ar^n}{1 - r}$

20. $1 \cdot 2 + 2 \cdot 2^2 + 3 \cdot 2^3 + \cdots + n \cdot 2^n = 2 + (n - 1)2^{n+1}$

21. $1 \cdot 2 \cdot 3 + 2 \cdot 3 \cdot 4 + 3 \cdot 4 \cdot 5 + \cdots + n(n + 1)(n + 2) = \dfrac{n(n + 1)(n + 2)(n + 3)}{4}$

22. $2^{-1} + 2^{-2} + 2^{-3} + \cdots + 2^{-n} = 1 - 2^{-n}$

23. In an arithmetic sequence:
$$a_1 + (a_1 + d) + (a_1 + 2d) + \cdots + (a_1 + (n - 1)d) = \frac{n}{2}[2a_1 + (n - 1)d]$$

24. The $(n + 1)$st pentagonal number, P_{n+1}, can be generated recursively from the nth number by using the following formula.
$$P_{n+1} = P_n + (n + 1) + 2n$$
For example, $P_3 = P_2 + (2 + 1) + 2(2) = 5 + 3 + 4 = 12$. Prove by mathematical induction that the nth pentagonal number, P_n, is $\dfrac{n}{2}(3n - 1)$.

1   5   12   22

25. a. For $n = 1, 2, 3, 4, 5$, evaluate:

  (1) $\left(1 + \dfrac{1}{n}\right)$ **(2)** $\left(1 + \dfrac{1}{1}\right)\left(1 + \dfrac{1}{2}\right)\left(1 + \dfrac{1}{3}\right) \cdots \left(1 + \dfrac{1}{n}\right)$

  b. Write a single term, in n, that is the equivalent of the product in part **a (2)**.

  c. Verify your answer to part **b** by mathematical induction.

11. S_1: ✔

Assume S_k: $2^0 + 2^1 + 2^2 + \cdots + 2^{k-1} = 2^k - 1$

Then S_{k+1}: $2^0 + 2^1 + 2^2 + \cdots + 2^{k-1} + 2^{(k+1)-1} = 2^k - 1 + 2^k$
$$= 2 \cdot 2^k - 1 = 2^{k+1} - 1$$

12. S_1: ✔

Assume S_k: $1^3 + 2^3 + 3^3 + \cdots + k^3 = \dfrac{k^2(k + 1)^2}{4}$

Then S_{k+1}: $1^3 + 2^3 + 3^3 + \cdots + k^3 + (k + 1)^3 = \dfrac{k^2(k + 1)^2}{4} + (k + 1)^3 = \dfrac{(k + 1)^2(k + 2)^2}{4}$

13. S_1: ✔

Assume S_k: $1(2) + 2(3) + 3(4) + \cdots + k(k + 1) = \frac{k}{3}(k + 1)(k + 2)$

Then S_{k+1}: $1(2) + 2(3) + 3(4) + \cdots + k(k + 1) + (k + 1)((k + 1) + 1) = \frac{k}{3}(k + 1)(k + 2) + (k + 1)((k + 1) + 1)$

$$= \frac{k + 1}{3}((k + 1) + 1)((k + 1) + 2)$$

14. S_1: ✔

Assume S_k: $2 + 5 + 8 + \cdots + (3k - 1) = \frac{k}{2}(3k + 1)$

Then S_{k+1}: $2 + 5 + 8 + \cdots + (3k - 1) + (3(k + 1) - 1) = \frac{k}{2}(3k + 1) + 3(k + 1) - 1$

$$= \frac{k + 1}{2}(3(k + 1) + 1)$$

15. S_1: ✔

Assume S_k: $1 + 6 + 11 + \cdots + (5k - 4) = \frac{k}{2}(5k - 3)$

Then S_{k+1}: $1 + 6 + 11 + \cdots + (5k - 4) + (5(k + 1) - 4)$

$$= \frac{k}{2}(5k - 3) + 5(k + 1) - 4$$

$$= \frac{k + 1}{2}(5(k + 1) - 3)$$

16. S_1: ✔

Assume S_k: $\displaystyle\sum_{i=1}^{k}(2i + 3) = k(k + 4)$

Then S_{k+1}: $\displaystyle\sum_{i=1}^{k+1}(2i + 3) = k(k + 4) + 2(k + 1) + 3$

$$= (k + 1)((k + 1) + 4)$$

17. S_1: ✔

Assume S_k: $\displaystyle\sum_{i=1}^{k}(4i - 3) = 2k^2 - k$

Then S_{k+1}: $\displaystyle\sum_{i=1}^{k+1}(4i - 3) = 2k^2 - k + 4(k + 1) - 3$

$$= 2(k + 1)^2 - (k + 1)$$

18. S_1: ✔

Assume S_k: $\displaystyle\sum_{i=1}^{k}4^i = \frac{4}{3}(4^k - 1)$

Then S_{k+1}: $\displaystyle\sum_{i=1}^{k+1}4^i = \frac{4}{3}(4^k - 1) + 4^{k+1}$

$$= \frac{4}{3}(4^{k+1} - 1)$$

19. S_1: ✔

Assume S_k: $\displaystyle\sum_{i=1}^{k} ar^{i-1} = \frac{a - ar^k}{1 - r}$

Then S_{k+1}: $\displaystyle\sum_{i=1}^{k+1} ar^{i-1} = \frac{a - ar^k}{1 - r} + ar^k$

$$= \frac{a - ar^{k+1}}{1 - r}$$

20. S_1: ✔

Assume S_k: $1 \cdot 2 + 2 \cdot 2^2 + 3 \cdot 2^3 + \cdots + k \cdot 2^k = 2 + (k - 1)2^{k+1}$

Then S_{k+1}: $1 \cdot 2 + 2 \cdot 2^2 + 3 \cdot 2^3 + \cdots + k \cdot 2^k + (k + 1)2^{k+1} = 2 + (k - 1) \cdot 2^{k+1} + (k + 1)2^{k+1}$

$$= 2 + ((k + 1) - 1) \cdot 2^{(k+1)+1}$$

21. S_1: ✔

Assume S_k: $1 \cdot 2 \cdot 3 + 2 \cdot 3 \cdot 4 + 3 \cdot 4 \cdot 5 + \cdots + k(k + 1)(k + 2) = \frac{k(k + 1)(k + 2)(k + 3)}{4}$

Then S_{k+1}: $1 \cdot 2 \cdot 3 + 2 \cdot 3 \cdot 4 + 3 \cdot 4 \cdot 5 + \cdots + k(k + 1)(k + 2) + (k + 1)(k + 2)(k + 3)$

$$= \frac{k(k + 1)(k + 2)(k + 3)}{4} + (k + 1)(k + 2)(k + 3)$$

$$= \frac{(k + 1)(k + 2)(k + 3)(k + 4)}{4}$$

22. S_1: ✔

Assume S_k: $2^{-1} + 2^{-2} + 2^{-3} + \cdots + 2^{-k} = 1 - 2^{-k}$

Then S_{k+1}: $2^{-1} + 2^{-2} + 2^{-3} + \cdots + 2^{-k} + 2^{-(k+1)} = 1 - 2^{-k} + 2^{-(k+1)} = 1 - 2^{-(k+1)}$

23. S_1: ✔

Assume S_k: $a_1 + (a_1 + d) + (a_1 + 2d) + \cdots + (a_1 + (k-1)d) = \frac{k}{2}(2a_1 + (k-1)d)$

Then S_{k+1}: $a_1 + (a_1 + d) + (a_1 + 2d) + \cdots + (a_1 + (k-1)d) + a_1 + kd = \frac{k}{2}(2a_1 + (k-1)d) + a_1 + kd$

$$= \frac{k+1}{2}(2a_1 + kd)$$

24. S_1: ✔

Assume S_k: $\frac{k}{2}(3k - 1)$

Then S_{k+1}: $\frac{k}{2}(3k - 1) + k + 1 + 2k = \frac{(k+1)}{2}(3(k+1) - 1)$

25. a. (1) $2, \frac{3}{2}, \frac{4}{3}, \frac{5}{4}, \frac{6}{5}$ **b.** $n + 1$

(2) $2, 3, 4, 5, 6$

c. S_1: ✔

Assume S_k: $\left(1 + \frac{1}{1}\right)\left(1 + \frac{1}{2}\right)\left(1 + \frac{1}{3}\right)\cdots\left(1 + \frac{1}{k}\right) = k + 1$

Then S_{k+1}: $\left(1 + \frac{1}{1}\right)\left(1 + \frac{1}{2}\right)\left(1 + \frac{1}{3}\right)\cdots\left(1 + \frac{1}{k}\right)\left(1 + \frac{1}{k+1}\right)$

$$= (k + 1)\left(1 + \frac{1}{k+1}\right) = (k + 1) + 1$$

26. a. (1) $\frac{1}{2}, \frac{1}{6}, \frac{1}{12}, \frac{1}{20}, \frac{1}{30}$ **b.** $\frac{n}{n+1}$

(2) $\frac{1}{2}, \frac{2}{3}, \frac{3}{4}, \frac{4}{5}, \frac{5}{6}$

c. S_1: ✔

Assume S_k: $\displaystyle\sum_{i=1}^{k} \frac{1}{i(i+1)} = \frac{k}{k+1}$

Then S_{k+1}: $\displaystyle\sum_{i=1}^{k+1} \frac{1}{i(i+1)} = \frac{k}{k+1} + \frac{1}{(k+1)(k+2)}$

$$= \frac{k+1}{k+2}$$

27. a. S_1: ✔

Assume S_k: $a_k = 3^{k-1}$

Then S_{k+1}: $a_{k+1} = 3(3^{k-1})$

$$= 3^k$$

b. S_1: ✔

Assume S_k: $b_k = 5 + 2(k - 1)$

Then S_{k+1}: $b_{k+1} = 2 + 5 + 2(k - 1)$

$$= 5 + 2((k + 1) - 1)$$

c. S_1: ✔

Assume S_k: $c_k = k \cdot 2^k$

Then S_{k+1}: $c_{k+1} = \frac{2(k+1)}{(k+1) - 1} \cdot k \cdot 2^k$

$$= (k + 1) \cdot 2^{k+1}$$

28. S_1: ✔

Assume S_k: $S_k = k^2$

Then S_{k+1}: $S_{k+1} = k^2 + 2(k + 1) - 1$

$$= (k + 1)^2$$

29. S_3: ✔

Assume S_k: $a_1 + a_2 + a_3 + \ldots + a_k = a_{k+2} - 1$

Then S_{k+1}: $= a_1 + a_2 + a_3 + \ldots + a_k + a_{k+1} = a_{k+2} - 1 + a_{k+1}$

$$= a_{k+3} - 1 \qquad (a_{k+3} = a_{k+2} + a_{k+1})$$

Exercises 4.5 *(continued)*

26. a. For $n = 1, 2, 3, 4, 5$, evaluate:

(1) $\dfrac{1}{n(n + 1)}$ **(2)** $\dfrac{1}{1(1 + 1)} + \dfrac{1}{2(2 + 1)} + \dfrac{1}{3(3 + 1)} + \cdots + \dfrac{1}{n(n + 1)}$

b. Write a single term, in n, that is the equivalent of the sum in part **a (2)**.

c. Verify your answer to part **b** by mathematical induction.

27. Prove by mathematical induction:

a. If $a_1 = 1$ and $a_n = 3a_{n-1}$, then $a_n = 3^{n-1}$.

b. If $b_1 = 5$ and $b_n = 2 + b_{n-1}$, then $b_n = 5 + 2(n - 1)$.

c. If $c_1 = 2$ and $c_n = \dfrac{2n}{n - 1} c_{n-1}$, then $c_n = n \cdot 2^n$.

28. Given $S_1 = 1$ and $S_n = S_{n-1} + 2n - 1$, prove that $S_n = n^2$.

29. 1, 1, 2, 3, 5, 8, …, is called the Fibonacci sequence, and it is formed by $a_1 = 1$, $a_2 = 1$, and $a_n = a_{n-1} + a_{n-2}$ ($n > 2$). Prove by mathematical induction that the sum of n terms of the sequence is equal to 1 less than the $(n + 2)$nd term.

30. Prove by mathematical induction that a regular convex polygon with n sides ($n \geq 3$) has $\left(\dfrac{n^2 - 3n}{2} \right)$ diagonals.

31. Let S_n represent the number of handshakes that are exchanged when each person in a group of n persons shakes hands with every other person. If one more person joins the group and shakes hands with each of the n persons already in the group, n additional handshakes are exchanged. Therefore, $S_{n+1} = S_n + n$. Prove by mathematical induction that if $S_2 = 1$, then $S_n = \dfrac{n}{2}(n - 1)$.

---

30. S_3: ✔

Assume S_k: $\dfrac{k^2 - 3k}{2}$

(for an additional vertex, add $k - 1$ diagonals)

Then S_{k+1}: $\dfrac{k^2 - 3k}{2} + k - 1 = \dfrac{(k + 1)^2 - 3(k + 1)}{2}$

31. S_2: ✔

Assume S_k: $\dfrac{k}{2}(k - 1)$

Then S_{k+1}: $\dfrac{k}{2}(k - 1) + k = \left(\dfrac{k + 1}{2} \right)k$

Exercises 4.5 *(continued)*

32. The Tower of Hanoi is a problem that has intrigued puzzle enthusiasts for centuries. The puzzle consists of a base that supports three vertical posts. On one post is a set of rings of different diameters arranged largest to smallest so that no ring is above one of smaller diameter. The task is to move the rings to a different post in the smallest number of moves under the following two conditions in each move:

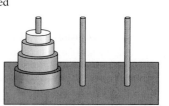

1. Only one ring may be moved at a time.   2. No larger ring may ever be placed on top of a smaller ring.

Prove that the smallest number of moves needed to move n rings from one post to another is $2^n - 1$.

Hint: Let M_n be the number of moves needed to move n rings. In order to move n rings from the first post to the second, it is necessary to move $(n - 1)$ rings from the first post to the third (M_{n-1}), then move the largest ring from the first post to the second ($M_1 = 1$), and finally move $(n - 1)$ rings from the third post to the second (M_{n-1}). Therefore, $M_n = M_{n-1} + 1 + M_{n-1} = 2M_{n-1} + 1$.

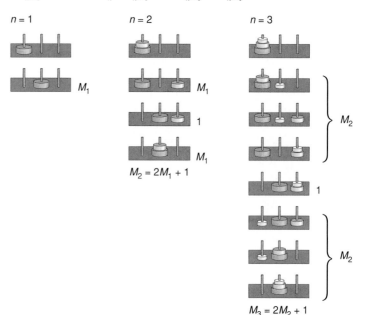

$n = 1$ $n = 2$ $n = 3$

M_1 M_1 M_2

1

M_1

$M_2 = 2M_1 + 1$ 1

M_2

$M_3 = 2M_2 + 1$

32. M_n is smallest if M_{n-1} is smallest. Since $M_1 = 1$ must be the smallest, then $M_2 = 2M_1 + 1 = 3$ must be the smallest, and so on for all n.

$S_1:$ ✔

Assume $S_k:$ $2^k - 1$ (use the recursive formula)

Then $S_{k+1}:$ $2 \cdot (2^k - 1) + 1 = 2^{k+1} - 1$

4.6 *Additional Applications of Mathematical Induction*

Mathematical induction is used to prove other types of statements about the natural numbers besides those involving sums of numbers. For example, mathematical induction can be used to prove some statements of inequality.

Example 1 _____

Prove that $2^n > n$ for all natural numbers.

Solution

1. For S_1: $2^1 > 1$ or $2 > 1$ is true.
2. The induction hypothesis S_k is $2^k > k$. This will be used to show that S_{k+1} is true; that is, to show that $2^{k+1} > k + 1$.

(1)	$2^k > k$	Assume S_k.
(2)	$2^k(2) > k(2)$	Multiply each side by 2.
(3)	$2^{k+1} > 2k$	Simplify.
(4)	$k \geq 1$	The domain is N.
(5)	$k + k \geq k + 1$	Add k to each side.
(6)	$2k \geq k + 1$	Simplify.
(7)	$2^{k+1} > 2k \geq k + 1$	Steps (3) and (6)
(8)	$2^{k+1} > k + 1$	Transitive property

By the principle of mathematical induction, it can be concluded that $2^n > n$ for all natural numbers.

Mathematical induction can also be used to prove some statements of divisibility.

Example 2 _____

Prove that, for all natural numbers, $7^n - 4^n$ is divisible by 3.

Solution

1. For S_1: $7^1 - 4^1 = 7 - 4 = 3$, which is divisible by 3.
2. The induction hypothesis S_k is that $7^k - 4^k$ is divisible by 3. Therefore, by the definition of divisibility, there exists an integer c such that $7^k - 4^k = 3c$. This will be used to show that S_{k+1} is true; that is, to show that $7^{k+1} - 4^{k+1}$ is divisible by 3.

   A convenient approach is to examine S_{k+1} and write equivalent forms until it is clearly divisible by 3.

$7^{k+1} - 4^{k+1}$	Write S_{k+1}.
$= 7 \cdot 7^k - 4 \cdot 4^k$	Rewrite the powers.
$= (3 + 4)7^k - 4 \cdot 4^k$	Rename 7.
$= 3 \cdot 7^k + 4 \cdot 7^k - 4 \cdot 4^k$	Distributive property.
$= 3 \cdot 7^k + 4(7^k - 4^k)$	Factor 4.
$= 3 \cdot 7^k + 4(3c)$	Rewrite S_k.
$= 3(7^k + 4c)$	3 is a factor of S_{k+1}.

By the principle of mathematical induction, it can be concluded that the original statement, $7^n - 4^n$ is divisible by 3, is true for all natural numbers.

Here is another example of the algebraic technique demonstrated in Example 2.

Example 3

Prove that, for all natural numbers, $n(n^2 + 2)$ is divisible by 3.

Solution

1. For S_1: $1(1^2 + 2) = 3$, which is divisible by 3.

2. The induction hypothesis S_k is:

    $k(k^2 + 2)$ is divisible by 3.

    Use this statement to show that S_{k+1} or $(k + 1)[(k + 1)^2 + 2]$ is divisible by 3.

    Rewrite S_{k+1}, trying to produce 3 as a factor.

    $(k + 1)[(k + 1)^2 + 2]$

    $$
    \begin{aligned}
    &= (k + 1)[(k^2 + 2k + 1) + 2] && \text{Expand the square.}\\
    &= (k + 1)(k^2 + 2k + 3) && \text{Combine constants.}\\
    &= k^3 + 2k^2 + 3k + k^2 + 2k + 3 && \text{Distributive property}\\
    &= k^3 + 3k^2 + 3k + 2k + 3 && \text{Combine } k^2\text{-terms.}\\
    &= k^3 + 2k + 3k^2 + 3k + 3 && \text{Commutative property}\\
    &= k(k^2 + 2) + 3(k^2 + k + 1) && \text{Factor.}
    \end{aligned}
    $$

The first part of this expression is divisible by 3 because of the induction hypothesis. The second part of the expression is divisible by 3 since 3 is one of the factors. Since each part of the sum is divisible by 3, the whole expression is divisible by 3.

By the principle of mathematical induction, it can be concluded that the original statement, $n(n^2 + 2)$ is divisible by 3, is true for all natural numbers.

Although the principle of mathematical induction is stated using 1 as the smallest number for which the relationship is to be proved true, similar reasoning can be used to prove a statement true for a subset of the natural numbers that has a least member greater than 1. The following example illustrates the proof of an inequality that is true for all natural numbers that are greater than 2.

Example 4

Prove $n! > 2^{n-1}$ for any natural number $n > 2$.

Solution

1. For $n = 3$,

$$
\begin{aligned}
3! &> 2^{3-1}\\
3 \cdot 2 \cdot 1 &> 2^2\\
6 &> 4 \ \checkmark
\end{aligned}
$$

1. S_1: $\checkmark$

    Assume S_k: $3^k \geq 3k$

    Then S_{k+1}: $3 \cdot 3^k \geq 3 \cdot 3k \qquad (3k \geq k + 1)$

    $\qquad\qquad\quad 3^{k+1} \geq 3(k + 1)$

2. S_1: $\checkmark$

    Assume S_k: $2^{k-1} \leq k!$

    Then S_{k+1}: $2 \cdot 2^{k-1} \leq 2 \cdot k!; \ 2^k \leq 2 \cdot k! \qquad (2 \leq k + 1)$

    $\qquad\qquad\quad 2^k \leq (k + 1)k! \quad ((k + 1)k! = (k + 1)!)$

    $\qquad\qquad\quad 2^k \leq (k + 1)!$

2. The induction hypothesis S_k is that $k! > 2^{k-1}$ for $k > 2$. This will be used to show that S_{k+1} is true; that is, to show that for $k > 2$, $(k+1)! > 2^{(k+1)-1}$, or $(k+1)! > 2^k$.

The object, to prove a statement that contains the expression $(k+1)$ in an inequality, suggests a start.

$k + 1 > k$	Definition of successor.
$(k+1) \cdot k! > k \cdot k!$	Multiply each side by k!.
$(k+1)! > k \cdot k!$	Rename on the left.
$k \cdot k! > 2 \cdot k!$	Given that $k > 2$.
$2 \cdot k! > 2 \cdot 2^{k-1}$	Assuming S_k.
$2 \cdot k! > 2^k$	Rename on the right.
$(k+1)! > 2^k$	Transitive property.

By the principle of mathematical induction, $n! > 2^{n-1}$ is true for all natural numbers > 2.

Exercises 4.6

In **1 – 16**, use mathematical induction to prove that the statement is true for all natural numbers n, or for the indicated subset of the natural numbers.

1. $3^n \geq 3n$

2. $2^{n-1} \leq n!$

3. $2^n < n!$ for $n \geq 4$.

4. $n^2 + n$ is divisible by 2.

5. $n^3 - n$ is divisible by 3.

6. $n^3 + 2n$ is divisible by 3.

7. $4^n - 1$ is divisible by 3.

8. $9^n - 2^n$ is divisible by 7.

9. $n^3 - n$ is divisible by 6.

10. 3 is a factor of $n(n^2 + 3n + 2)$.

11. 5 is a factor of $2^{2n-1} + 3^{2n-1}$.

12. $a^n - b^n$ is divisible by $a - b$.

13. $x + y$ is a factor of $x^{2n} - y^{2n}$.

14. $x + y$ is a factor of $x^{2n-1} + y^{2n-1}$.

15. $a^n + b^n$ is divisible by $a + b$ when n is odd.

16. $a^n - b^n$ is divisible by $a + b$ when n is even.

In **17 – 19**, disprove the given statement by showing a counterexample.

17. $n! \leq 2^n$

18. $|a + b| = a + b$

19. $n^4 - 1$ is divisible by 5.

3. $\quad S_4: \checkmark$

Assume S_k: $\quad 2^k < k! \quad k \geq 4$

Then S_{k+1}: $\quad 2 \cdot 2^k < 2 \cdot k!; \; 2^{k+1} < 2 \cdot k! \qquad$ (Since $k \geq 4, \, 2 < k + 1$)

$\qquad\qquad\qquad\qquad 2^{k+1} < (k+1)k! \quad ((k+1)k! = (k+1)!)$

$\qquad\qquad\qquad\qquad 2^{k+1} < (k+1)!$

4. S_1: ✔

Assume S_k: $k^2 + k$ is divisible by 2.

Then S_{k+1}: $(k+1)^2 + k + 1 = (k^2 + k) + 2(k+1)$ is divisible by 2.

5. S_1: ✔

Assume S_k: $k^3 - k$ is divisible by 3.

Then S_{k+1}: $(k+1)^3 - (k+1) = (k^3 - k) + 3(k^2 + k)$ is divisible by 3.

6. S_1: ✔

Assume S_k: $k^3 + 2k$ is divisible by 3.

Then S_{k+1}: $(k+1)^3 + 2(k+1) = (k^3 + 2k) + 3(k^2 + k + 1)$ is divisible by 3.

7. S_1: ✔

Assume S_k: $4^k - 1$ is divisible by 3.

Then S_{k+1}: $4^{k+1} - 1 = 4 \cdot 4^k - 1 = 4(4^k - 1) + 3$ is divisible by 3.

8. S_1: ✔

Assume S_k: $9^k - 2^k$ is divisible by 7.

Then S_{k+1}: $9^{k+1} - 2^{k+1} = 9 \cdot 9^k - 2 \cdot 2^k$

$$= (7+2)9^k - 2 \cdot 2^k$$
$$= 7 \cdot 9^k + 2 \cdot 9^k - 2 \cdot 2^k$$
$$= 7 \cdot 9^k + (9^k - 2^k)2 \text{ is divisible by 7.}$$

9. S_1: ✔

Assume S_k: $k^3 - k$ is divisible by 6.

Then S_{k+1}: $(k+1)^3 - (k+1) = (k^3 - k) + 3k(k+1)$ is divisible by 6.

(Since $k, k+1$ are consecutive numbers, one must be even. Thus, $3k(k+1)$ is divisible by 2 and 3.)

10. S_1: ✔

Assume S_k: 3 is a factor of $k(k^2 + 3k + 2)$

Then S_{k+1}: $(k+1)((k+1)^2 + 3(k+1) + 2) = k(k^2 + 3k + 2) + 3(k^2 + 3k + 2)$ is divisible by 3.

11. S_1: ✔

Assume S_k: 5 is a factor of $2^{2k-1} + 3^{2k-1}$

Then S_{k+1}: $2^{2(k+1)-1} + 3^{2(k+1)-1} = 2^2 \cdot 2^{2k-1} + 3^2 \cdot 3^{2k-1}$

$$\begin{pmatrix} 2^{2k-1} + 3^{2k-1} = 5a \\ 2^{2k-1} = 5a - 3^{2k-1} \end{pmatrix}$$

$$= 2^2(5a - 3^{2k-1}) + 3^2 \cdot 3^{2k-1}$$
$$= 2^2 \cdot 5a - 2^2 \cdot 3^{2k-1} + 3^2 \cdot 3^{2k-1}$$
$$= 5a \cdot 2^2 + (-2^2 + 3^2)3^{2k-1}$$
$$= 5(2^2 a + 3^{2k-1}) \text{ is divisible by 5.}$$

12. S_1: ✔

Assume S_k: $a^k - b^k$ is divisible by $a - b$

Then S_{k+1}: $a^{k+1} - b^{k+1} = a(a^k - b^k) + b^k(a - b)$ is divisible by $a - b$.

13. S_1: ✔

Assume S_k: $x + y$ is a factor of $x^{2k} - y^{2k}$

Then S_{k+1}: $x^{2k+2} - y^{2k+2} = x^2(x^{2k} - y^{2k}) + y^{2k}(x^2 - y^2)$

$\qquad\qquad = x^2(x^{2k} - y^{2k}) + y^{2k}(x + y)(x - y)$ is divisible by $x + y$.

14. S_1: ✔

Assume S_k: $x + y$ is a factor of $x^{2k-1} + y^{2k-1}$

Then S_{k+1}: $x^{2(k+1)-1} + y^{2(k+1)-1} = x^2 \cdot x^{2k-1} + y^2 \cdot y^{2k-1}$

$$\begin{pmatrix} x^{2k-1} + y^{2k-1} = (x+y)a \\ x^{2k-1} = (x+y)a - y^{2k-1} \end{pmatrix}$$

$\qquad\qquad = x^2((x+y)a - y^{2k-1}) + y^2 \cdot y^{2k-1}$

$\qquad\qquad = ax^2(x+y) - (x^2 - y^2)y^{2k-1}$

$\qquad\qquad = ax^2(x+y) - (x+y)(x-y)y^{2k-1}$ is divisible by $x + y$.

15. S_1: ✔

Assume S_k: $a^{2k+1} + b^{2k+1}$ is divisible by $a + b$ $\qquad$ ($2k + 1$ is an odd exponent)

Then S_{k+1}: $a^{2(k+1)+1} + b^{2(k+1)+1} = a^2 \cdot a^{2k+1} + b^2 \cdot b^{2k+1}$

$$\begin{pmatrix} a^{2k+1} + b^{2k+1} = q(a+b) \\ a^{2k+1} = q(a+b) - b^{2k+1} \end{pmatrix}$$

$\qquad\qquad = qa^2(a+b) - b^{2k+1}(a^2 - b^2)$

$\qquad\qquad = qa^2(a+b) - b^{2k+1}(a+b)(a-b)$ is divisible by $a + b$.

16. S_1: ✔

Assume S_k: $a^{2k} - b^{2k}$ is divisible by $a + b$ $\qquad$ ($2k$ is an even exponent)

Then S_{k+1}: $a^{2(k+1)} - b^{2(k+1)} = a^{2k+2} - b^{2k+2} = a^2(a^{2k} - b^{2k}) + b^{2k}(a^2 - b^2)$

$\qquad\qquad\qquad = a^2(a^{2k} - b^{2k}) + b^{2k}(a+b)(a-b)$ is divisible by $a + b$.

17. $n = 4$ $\qquad$ $n! \le 2^n$

$\qquad\qquad$ $24 \le 16$ False

18. $a = -1$ $\qquad$ $|a + b| = a + b$

$\quad b = -2$ $\quad |(-1) + (-2)| = (-1) + (-2)$

$\qquad\qquad\qquad\quad 3 = -3$ False

19. $n = 5$ $\qquad$ $n^4 - 1$

$\qquad\qquad$ $5^4 - 1 = 624$ is not divisible by 5.

4.7 Expansion of Binomials

The Pascal Triangle is formed by starting and ending each row with 1, and forming the other numbers in the row by adding each pair of adjoining numbers in the row above. For example, the 15 in the last row is the sum of the 5 and 10 above it, and the 20 is the sum of the two 10's above it. The triangle was devised by Blaise Pascal (1623–1662), who wrote a treatise on the numerous mathematical relationships that appear in this array of numbers. For example, adding across the rows yields successive powers of 2.

```
                1
            1       1
        1       2       1
    1       3       3       1
  1     4       6       4       1
 1    5    10      10      5      1
1   6    15    20     15     6     1
              ⋮
```

THE PASCAL TRIANGLE

$$1 = 1 = 2^0$$
$$1 + 1 = 2 = 2^1$$
$$1 + 2 + 1 = 4 = 2^2$$
$$1 + 3 + 3 + 1 = 8 = 2^3$$
$$1 + 4 + 6 + 4 + 1 = 16 = 2^4$$
$$\vdots$$

Important for this section is the fact that the coefficients of the expansion of $(x + y)^n$, for n a positive integer, can be obtained from the triangle. Study the expansions of $(x + y)^n$ for $n = 0$ through $n = 6$.

$$(x + y)^0 = 1$$
$$(x + y)^1 = x + y$$
$$(x + y)^2 = x^2 + 2xy + y^2$$
$$(x + y)^3 = x^3 + 3x^2y + 3xy^2 + y^3$$
$$(x + y)^4 = x^4 + 4x^3y + 6x^2y^2 + 4xy^3 + y^4$$
$$(x + y)^5 = x^5 + 5x^4y + 10x^3y^2 + 10x^2y^3 + 5xy^4 + y^5$$
$$(x + y)^6 = x^6 + 6x^5y + 15x^4y^2 + 20x^3y^3 + 15x^2y^4 + 6xy^5 + y^6$$

Expanding a binomial or finding an individual term of a binomial expansion is important for work in probability theory, calculus, genetics, and a number of other areas. In the expansions above, the following relationships should be noted.

- The expansion of $(x + y)^n$ contains $n + 1$ terms.

- The coefficients are symmetrical across each row.

- The sum of the powers of x and y in each term is n. (Consider the first term in each expansion to be $x^n y^0$ and the last term to be $x^0 y^n$.)

- In successive terms, from left to right, the exponent for x decreases by 1 and the exponent for y increases by 1.

A familiarity with these patterns makes it easy to find the literal parts of the terms of a particular binomial expansion. As you saw, the coefficients can be obtained from Pascal's Triangle. However, for higher powers, it is useful to generate the coefficients more efficiently. Thus, the following notation:

■ **Definition:** $\binom{n}{r} = \dfrac{n(n-1)(n-2)\ \ldots\ (n-r+1)}{r!}$ or $\dfrac{n!}{r!(n-r)!}$

where n and r are positive integers, and $1 \le r < n$.

■ **Definition:** $\binom{n}{0} = 1$; $\binom{n}{n} = 1$.

For example:

$$\binom{4}{3} = \frac{4 \cdot 3 \cdot 2}{3!} \qquad\qquad n - r + 1 = 2$$

$$= \frac{24}{6} = 4$$

The coefficients of the five terms in the expansion of $(x + y)^4$ are obtained by computing the values of $\binom{4}{0}$, $\binom{4}{1}$, $\binom{4}{2}$, $\binom{4}{3}$, and $\binom{4}{4}$.

$$(x + y)^4 = \binom{4}{0}x^4 + \binom{4}{1}x^3y + \binom{4}{2}x^2y^2 + \binom{4}{3}xy^3 + \binom{4}{4}y^4$$

$$= 1x^4 + 4x^3y + 6x^2y^2 + 4xy^3 + 1y^4 \qquad \text{Compare these coefficients to the 5th row of the Pascal Triangle.}$$

Example 1 _____

Expand $(x + y)^7$.

Solution

Since n is 7, this expansion contains $7 + 1$, or 8 terms.

$(x + y)^7$

$= \binom{7}{0}x^7 + \binom{7}{1}x^6y + \binom{7}{2}x^5y^2 + \binom{7}{3}x^4y^3 + \binom{7}{4}x^3y^4 + \binom{7}{5}x^2y^5 + \binom{7}{6}xy^6 + \binom{7}{7}y^7$

Remember that $\binom{7}{0}$ and $\binom{7}{7}$ are each 1. The other coefficients are evaluated by the definition of $\binom{n}{r}$. Note that the coefficient of the rth term is $\binom{n}{r-1}$.

$\binom{7}{1} = \frac{7}{1!} = 7$ $\qquad\qquad$ $\binom{7}{2} = \frac{7 \cdot 6}{2 \cdot 1} = 21$ $\qquad\qquad$ $\binom{7}{3} = \frac{7 \cdot 6 \cdot 5}{3 \cdot 2 \cdot 1} = 35$

$\binom{7}{4} = \frac{7 \cdot 6 \cdot 5 \cdot 4}{4 \cdot 3 \cdot 2 \cdot 1} = 35$ $\qquad$ $\binom{7}{5} = \frac{7 \cdot 6 \cdot 5 \cdot 4 \cdot 3}{5 \cdot 4 \cdot 3 \cdot 2 \cdot 1} = 21$ $\qquad$ $\binom{7}{6} = \frac{7 \cdot 6 \cdot 5 \cdot 4 \cdot 3 \cdot 2}{6 \cdot 5 \cdot 4 \cdot 3 \cdot 2 \cdot 1} = 7$

Answer:

$(x + y)^7 = x^7 + 7x^6y + 21x^5y^2 + 35x^4y^3 + 35x^3y^4 + 21x^2y^5 + 7xy^6 + y^7$

From Example 1, notice that $\binom{7}{1} = \binom{7}{6}$, $\binom{7}{2} = \binom{7}{5}$, and $\binom{7}{3} = \binom{7}{4}$.

In general: $\binom{n}{r} = \binom{n}{n-r}$

Example 2

Find the 4th term of the expansion of $(x + y)^{10}$.

Solution

Since $n = 10$, the expansion contains 11 coefficients:

$$\binom{10}{0}, \binom{10}{1}, \binom{10}{2}, \ldots, \binom{10}{10}$$

The coefficient of the 4th term is:

$$\binom{10}{3} = \frac{10 \cdot 9 \cdot 8}{3!}$$

$$= \frac{720}{6} = 120$$

For the 4th term:

the x-component is x^{10-3}, or x^7.

the y-component is y^3, since the sum of the exponents of the variables is 10.

Answer: The fourth term of $(x + y)^{10}$ is $120x^7y^3$.

All of the patterns and shortcuts for writing a binomial expansion are stated in the following theorem.

☐ **Binomial Theorem:**

$(x + y)^n$

$= \binom{n}{0}x^n + \binom{n}{1}x^{n-1}y + \binom{n}{2}x^{n-2}y^2 + \cdots + \binom{n}{r-1}x^{n-(r-1)}y^{r-1} + \binom{n}{r}x^{n-r}y^r + \cdots + \binom{n}{n-1}xy^{n-1} + \binom{n}{n}y^n$

Note that, since the binomial theorem can be considered a sequence of statements about $(x + y)^n$ for each positive integer n, it can be proved by mathematical induction.

The binomial theorem can be used to expand a binomial where terms are not simply x and y.

Example 3

Expand $(2a - 3b^2)^3$ using the binomial theorem.

Solution

Using $(x + y)^3 = \binom{3}{0}x^3 + \binom{3}{1}x^2y + \binom{3}{2}xy^2 + \binom{3}{3}y^3$ as a model, substitute $2a$ for x and $-3b^2$ for y.

$(2a - 3b^2)^3 = \binom{3}{0}(2a)^3 + \binom{3}{1}(2a)^2(-3b^2) + \binom{3}{2}(2a)(-3b^2)^2 + \binom{3}{3}(-3b^2)^3$

$= \binom{3}{0}8a^3 + \binom{3}{1}(4a^2)(-3b^2) + \binom{3}{2}(2a)(9b^4) + \binom{3}{3}(-27b^6)$

$= 8a^3 - 36a^2b^2 + 54ab^4 - 27b^6$

Example 4

Find the term containing a^3 in the expansion of $\left(a - \frac{2}{a}\right)^7$.

Solution

In the expansion of $(x + y)^7$, substitute a for x, and $\frac{2}{a}$ or $2a^{-1}$ for y.

For the rth term, the variable part is of the form:

$$x^{7-(r-1)}y^{(r-1)}$$

$$a^{7-(r-1)}(2a^{-1})^{(r-1)} = a^{8-r}2^{r-1}a^{-r+1} = 2^{r-1}a^{9-2r}$$

Thus, for the term containing a^3:

$$a^{9-2r} = a^3$$

Since the bases of the equal terms are the same, the exponents must be equal.

$$9 - 2r = 3$$

$$-2r = -6$$

$$r = 3$$

Since $r = 3$, it is the 3rd term of this expansion that contains a^3. The expression for the 3rd term is:

$$\binom{7}{2}(a^{7-2})(2a^{-1})^2 = 21a^5 \cdot 4a^{-2}$$
$$= 84a^3$$

Answer: $84a^3$ is the required term of the given expansion.

Example 5

Use the binomial expansion to find the value of $(1.02)^5$.

Solution

Think of $(1.02)^5$ as $(1 + .02)^5$. To use the binomial expansion for $(x + y)^5$ with $x = 1$ and $y = .02$:

$(1 + .02)^5$

$= \binom{5}{0}1^5 + \binom{5}{1}1^4(.02) + \binom{5}{2}1^3(.02)^2 + \binom{5}{3}1^2(.02)^3 + \binom{5}{4}1(.02)^4 + \binom{5}{5}(.02)^5$

$= 1 + 5(.02) + 10(.02)^2 + 10(.02)^3 + 5(.02)^4 + (.02)^5$

$= 1 + .1 + 10(.0004) + 10(.000008) + 5(.00000016) + .0000000032$

$= 1 + .1 + .004 + .00008 + .0000008 + .0000000032$

$= 1.1040808032$

1. 10

2. 21

3. 210

4. 1

5. 1

6. 34,220

7. 792

8. 792

Exercises 4.7

In **1 – 8**, find the numerical value.

1. $\binom{5}{3}$ **2.** $\binom{7}{5}$ **3.** $\binom{10}{4}$ **4.** $\binom{6}{6}$

5. $\binom{100}{0}$ **6.** $\binom{60}{57}$ **7.** $\binom{12}{7}$ **8.** $\binom{12}{5}$

In **9 – 20**, use the binomial theorem to expand the expression. Express your answer in simplest form.

9. $(x + y)^4$ **10.** $(a + b)^6$ **11.** $(s + 3)^3$ **12.** $(y - 2)^4$

13. $(a + 2b)^4$ **14.** $(2x - y)^3$ **15.** $(a + 2b^2)^4$ **16.** $(x^2 + y^3)^3$

17. $\left(y + \dfrac{1}{y}\right)^4$ **18.** $\left(x - \dfrac{2}{x}\right)^3$ **19.** $\left(y - \dfrac{1}{x}\right)^4$ **20.** $(1 + 2x^2)^4$

In **21 – 23**, use the binomial theorem to expand the expression. If $i = \sqrt{-1}$, express your answer in simplest form.

21. $(1 + i)^3$ **22.** $(1 - 2i)^4$ **23.** $(2 + i\sqrt{2})^3$

In **24 – 27**, write only the first three terms of the expansion, expressed in simplest form.

24. $(x + y)^{12}$ **25.** $(a - b)^{14}$ **26.** $\left(\dfrac{1}{x} + \dfrac{1}{y}\right)^{11}$ **27.** $(2x^2 - 3y^2)^8$

28. Find the 5th term of $(3k - 1)^6$.

29. Find the 4th term of $(x - 2y)^8$.

30. Find the term whose coefficient is $\binom{8}{6}$ in the expansion of $(a + b)^8$.

31. Find the term whose coefficient is $\dfrac{6!}{3!3!}$ in the expansion of $(x + y)^6$.

32. Write the term containing x^5 in the expansion of $(x + 2y^2)^6$.

33. Write the term containing y^4 in the expansion of $(2x + 3y)^7$.

34. Write the term containing y^8 in the expansion of $(x + 2y^2)^5$.

35. Write the term containing a^2 in the expansion of $\left(a + \dfrac{1}{a}\right)^8$.

36. Write the term containing x^6 in the expansion of $\left(x + \dfrac{2}{x}\right)^{10}$.

37. Write the term containing a^6 in the expansion of $(a^2 + 4b^3)^5$.

 In **38 – 45**, use the binomial expansion to approximate the value to four decimal places. Check with a calculator.

38. $(1.01)^3$ **39.** $(1.01)^5$ **40.** $(1.02)^4$ **41.** $(1.03)^3$

42. $(1.05)^3$ **43.** $(0.98)^5$ **44.** $(0.95)^4$ **45.** $(3.03)^3$

46. The balance for $1,000 deposited at 8% compounded quarterly for t years is $1,000(1.02)^{4t}$. Use the binomial expansion to find the balance, to the nearest dollar, after two years.

47. Use the binomial expansion to find the balance after six months for $500 deposited at 7% compounded monthly. Round your answer to the nearest penny.

9. $x^4 + 4x^3y + 6x^2y^2 + 4xy^3 + y^4$

10. $a^6 + 6a^5b + 15a^4b^2 + 20a^3b^3 + 15a^2b^4 + 6ab^5 + b^6$

11. $s^3 + 9s^2 + 27s + 27$

12. $y^4 - 8y^3 + 24y^2 - 32y + 16$

13. $a^4 + 8a^3b + 24a^2b^2 + 32ab^3 + 16b^4$

14. $8x^3 - 12x^2y + 6xy^2 - y^3$

15. $a^4 + 8a^3b^2 + 24a^2b^4 + 32ab^6 + 16b^8$

16. $x^6 + 3x^4y^3 + 3x^2y^6 + y^9$

17. $y^4 + 4y^2 + 6 + \dfrac{4}{y^2} + \dfrac{1}{y^4}$

18. $x^3 - 6x + \dfrac{12}{x} - \dfrac{8}{x^3}$

19. $y^4 - \dfrac{4y^3}{x} + \dfrac{6y^2}{x^2} - \dfrac{4y}{x^3} + \dfrac{1}{x^4}$

20. $1 + 8x^2 + 24x^4 + 32x^6 + 16x^8$

21. $-2 + 2i$

22. $-7 + 24i$

23. $-4 + 10i\sqrt{2}$

Exercises 4.7 *(continued)*

In **48 – 50**, use the following information.

If the probability of winning a game is p and the probability for losing is q where $p + q = 1$, then the probability of winning w games out of n is the term containing p^w in the expansion of $(p + q)^n$.

For example, if the Yankees have a $\frac{1}{3}$ chance of beating the Red Sox and a $\frac{2}{3}$ chance of losing, the probability of the Yankees winning 3 games out of 4 against the Red Sox is the value of the term containing $\left(\frac{1}{3}\right)^3$ in the expansion of $\left(\frac{1}{3} + \frac{2}{3}\right)^4$, which is $\binom{4}{1}\left(\frac{1}{3}\right)^3\left(\frac{2}{3}\right)^1 = 4\left(\frac{1}{27}\right)\left(\frac{2}{3}\right) = \frac{8}{81}$ (approximately 10%).

48. If the Mets have a $\frac{3}{5}$ chance of beating the Dodgers and a $\frac{2}{5}$ chance of losing, what is the probability that they will win exactly 2 games in a 3-game series?

49. Dr. J has a 90% chance of making a foul shot and a 10% chance of missing. What is the probability that he will make exactly 6 foul shots in 7 tries?

50. If in a World Series, the National League has a 30% chance of winning and the American League has a 70% chance of winning, then:

  a. If the first team to win 4 games wins the World Series, what is the probability that the World Series will be won by the American League, winning 4 out of 4 games?

  b. What is the probability of the American League winning if 5 games are played? *Hint:* They must win 3 out of 4, and then game 5.

  c. What is the probability that the American League will win:
  (1) 4 out of 6 games?
  (2) 4 out of 7 games?

  d. Assuming the American League wins, what is the most likely number of games played in the World Series?

51. There are five traffic lights on Route 31 at the town of Perinton. The lights change in a random pattern and the probability of arriving at a light when it is green is $\frac{2}{5}$.

  a. What is the probability of arriving at all five lights when they are green?

  b. What is the probability of arriving at four out of the five lights when they are green?

  c. What is the probability that a car must stop for no more than two of the five lights?

---

24. $x^{12} + 12x^{11}y + 66x^{10}y^2 \cdots$

25. $a^{14} - 14a^{13}b + 91a^{12}b^2 \cdots$

26. $\frac{1}{x^{11}} + \frac{11}{x^{10}y} + \frac{55}{x^9y^2} \cdots$

27. $256x^{16} - 3{,}072x^{14}y^2 + 16{,}128x^{12}y^4 \cdots$

28. $135k^2$　　**29.** $-448x^5y^3$

30. $28a^2b^6$　　**31.** $20x^3y^3$

32. $12x^5y^2$　　**33.** $22{,}680x^3y^4$

34. $80xy^8$

35. $56a^2$

36. $180x^6$

37. $160a^6b^6$

38. 1.0303

39. 1.0510

40. 1.0824

41. 1.0927

42. 1.1576

43. 0.9039

44. 0.8145

45. 27.8181

46. $\$1{,}172$

47. $\$517.75$

48. $\frac{54}{125}$

49. 0.3720087

50. a. 0.2401
  b. 0.28812
  c. (1) 0.21609
  　　(2) 0.129654
  d. 5

51. a. 0.01024
  b. 0.0768
  c. 0.3174

Chapter 4   Summary and Review

Sequences and Series

- An infinite sequence is a function whose domain is the set of positive integers.
- A sequence may be defined explicitly by giving an expression in terms of n, by listing every term, or recursively.
- A series is the indicated sum of a sequence.
- Sigma notation is used to indicate a sum. For example:

$$\sum_{i=1}^{n} a_i = a_1 + a_2 + a_3 + \cdots + a_n$$

Arithmetic Sequences and Series

- An arithmetic sequence is a sequence in which each term after the first is obtained by adding a constant, called the common difference, to the previous term.
- To find the nth term of an arithmetic sequence, use the formula $a_n = a_1 + (n-1)d$ or the formula $a_n = a_{n-1} + d$.
- The arithmetic mean of a and b is $\dfrac{a+b}{2}$.
- Any number of arithmetic means may be inserted between two numbers by forming an arithmetic sequence with the given numbers as the first and last terms in the sequence.
- The sum of a finite arithmetic sequence may be found by using either of the following two formulas.

  When a_n is known: When a_n is not known:

  $$S_n = \frac{n}{2}(a_1 + a_n)$$ $$S_n = \frac{n}{2}[2a_1 + (n-1)d]$$

Geometric Sequences and Series

- A geometric sequence is a sequence such that each term after the first is obtained by multiplying the previous term by a constant, called the common ratio.
- The nth term of a geometric sequence is found by using the formula $a_n = a_1 r^{n-1}$ or the formula $a_n = a_{n-1}r$.
- The geometric mean, or mean proportional, of two positive numbers a and b is $\pm\sqrt{ab}$.

- Any number of geometric means may be inserted between two numbers by forming a geometric sequence with the given numbers as the first and last terms in the sequence.

- The sum of a finite geometric sequence $\left(\text{with } r \neq 1\right)$ may be found by using either of the following two formulas.

When a_n is known:

$$S_n = \frac{a_1 - a_n r}{1 - r}$$

When a_n is not known:

$$S_n = \frac{a_1 - a_1 r^n}{1 - r}$$

- An infinite geometric sequence has a finite sum provided that $|r| < 1$. The formula $S_\infty = \frac{a_1}{1 - r}$ may be used to find the sum.

Mathematical Induction

- To prove that a statement S_n is true for all natural numbers:

    1. Show S_1 is true.
    2. Assume S_k is true, and prove that S_{k+1} is true.

- Some basic ways to handle Step 2 of the inductive proof are:

    1. Begin with the inductive hypothesis S_k, and use mathematical operations to lead up to S_{k+1}. This method is useful when S_n involves a sum.
    2. Begin with S_{k+1}, and use mathematical operations until the desired conclusion is clear. This method is useful when S_n is a statement of divisibility.
    3. The transitive property is important when S_n is a statement of inequality.

Binomial Expansion

- The expansion of $(x + y)^n$ contains $n + 1$ terms.

- $\dbinom{n}{r} = \dfrac{n(n-1)(n-2) \cdots (n-r+1)}{r!}$ or $\dfrac{n!}{r!(n-r)!}$

- $\dbinom{n}{0} = \dbinom{n}{n} = 1$

- Binomial Theorem

$$(x + y)^n = \dbinom{n}{0}x^n + \dbinom{n}{1}x^{n-1}y + \dbinom{n}{2}x^{n-2}y^2 + \cdots + \dbinom{n}{n-1}xy^{n-1} + \dbinom{n}{n}y^n$$

- The rth term of $(x + y)^n$ is: $\dbinom{n}{r-1}x^{n-(r-1)}y^{r-1}$

Review Answers

1. **a.** 3, 6, 15, 52, 245
   b. 4, 16, 34, 58, 88
2. **a.** $\frac{197}{35}$ **b.** 40

3. **a.** $\sum_{n=1}^{4}(3n+6)$

   b. $\sum_{n=2}^{5}\frac{1}{(-2)^n}$

   c. $\sum_{i=2}^{n}\frac{x^i}{i+1}$

4. **a.** 10   **b.** 17
   c. $5\sqrt{2}$ **d.** 4

5. 26

6. 21

7. $28\sqrt{3}$

8. $a_1 = 0.1$,
   $a_n = a_{n-1} + 0.02$

9. $-6, -3.4, -0.8,$
   $-1.8, 4.4, 7$

10. 330

11. 950

12. 57.5

13. 643,643

14. $5,000

15. $\frac{16}{243}$

16. 4

17. $507.54, $515.19,
    $522.96, $530.84

18. 7,800

19. 1.2, 2.4, 4.8, 9.6,
    19.2, 38.4

20. 1.2

21. $\frac{15}{8}$

22. **a.** $\frac{816}{625}$

    b. 5.15201506

23. 0.232323

24. $43,865.18

25. **a.** 2   **b.** $\sqrt{2}+1$
    c. $4 + 2\sqrt{2}$

26. **a.** $\frac{7}{9}$ **b.** $\frac{43}{99}$

    c. $6\frac{89}{99}$

27. $32 + 16\sqrt{2}$

Chapter 4   Review Exercises

1. Find the first five terms of each sequence
   a. $a_n = 2n! + n$ **b.** $a_1 = 4, a_n = a_{n-1} + 6n$

2. Evaluate: **a.** $\sum_{i=1}^{5}\frac{2i}{i+2}$ **b.** $\sum_{i=0}^{3}3^i$

3. Use sigma notation to write the series.
   a. $9 + 12 + 15 + 18$ **b.** $\frac{1}{4} - \frac{1}{8} + \frac{1}{16} - \frac{1}{32}$ **c.** $\frac{x^2}{3} + \frac{x^3}{4} + \frac{x^4}{5} + \cdots$

4. Find the value of a_n for the sequence described, where d is the common difference for an arithmetic sequence and r is the common ratio for a geometric sequence.
   a. $a_1 = 7, a_n = a_{n-1} + 1, n = 4$ **b.** $a_1 = 3, d = 2, n = 8$
   c. $a_1 = \sqrt{2}, d = \sqrt{2}, n = 5$ **d.** $a_1 = \sqrt{2}, r = \sqrt{2}, n = 4$

5. Find the first term of an arithmetic sequence if the common difference is -2 and the 8th term is 12.

6. If the first and last terms of an arithmetic sequence are 80 and 10, respectively, and the common difference is $-3\frac{1}{2}$, find the number of terms in the sequence.

7. In an arithmetic sequence, $a_3 = \sqrt{3}$ and $d = 3\sqrt{3}$. Find: a_{12}

8. Write a recursive formula that can be used to generate the arithmetic sequence 0.1, 0.12, 0.14, 0.16, ...

9. Insert four arithmetic means between -6 and 7.

10. Find the sum of the first twenty terms of the arithmetic sequence $-12, -9, -6, \ldots$.

11. For an arithmetic sequence, find the sum of a_1 through a_{25}, where $a_1 = -25$ and $a_{25} = 101$.

12. Find the sum of an arithmetic sequence where $d = \frac{1}{2}, a_2 = 4$, and $n = 10$.

13. Find the sum of the multiples of 7 between 4,000 and 5,000.

14. Find the total of the first and last month's salary of an employee whose salary increases by a given amount every month, if she has earned a total of $30,000 for the year.

15. Find the 6th term of a geometric sequence whose first term is $\frac{1}{2}$ and whose common ratio is $\frac{2}{3}$.

16. Find the 4th term of a geometric sequence, if the first term is $\sqrt{2}$ and the fifth term is $4\sqrt{2}$.

17. 🖩 Write a sequence of the quarterly balances for $500 deposited at 6% compounded monthly.

18. 🖩 The population of a town of 10,000 will decrease by 4% each year for ten years. Find, to the nearest hundred people, the population at the end of the 6th year.

28. **a.** S_1: ✓

Assume S_k: $\sum_{i=1}^{k}3^i = \frac{3}{2}(3^k - 1)$

Then S_{k+1}: $\sum_{i=1}^{k+1}3^i = \frac{3}{2}(3^k - 1) + 3^{k+1}$

$= \frac{3}{2}(3^{k+1} - 1)$

b. S_1: ✓

Assume S_k: $\sum_{i=1}^{k}(i+2) = \frac{k^2 + 5k}{2}$

Then S_{k+1}: $\sum_{i=1}^{k+1}(i+2) = \frac{k^2 + 5k}{2} + k + 1 + 2$

$= \frac{(k+1)^2 + 5(k+1)}{2}$

19. Insert four geometric means between 1.2 and 38.4.

20. Find the positive geometric mean between 3 and 0.48.

21. Find the sum of the first four terms of the geometric sequence where $a_1 = 1$ and $r = \frac{1}{2}$.

22. Evaluate:

    a. $\displaystyle\sum_{n=1}^{4} \left(\frac{3}{5}\right)^n$ **b.** $\displaystyle\sum_{i=1}^{5} (1.01)^i$

23. Find the sum of a geometric sequence whose first term is 0.23 and whose last term is 23×10^{-6}, if $r = 0.01$.

24. An annuity consists of twenty yearly deposits of $1,000, which receive 7% interest per year compounded annually. If the deposits are made at the beginning of each year, find the total value after the 20th deposit is made.

25. Find the sum:

    a. $\displaystyle\sum_{n=1}^{\infty} \left(\frac{2}{3}\right)^n$ **b.** $\displaystyle\sum_{n=1}^{\infty} \left(\sqrt{\frac{1}{2}}\right)^n$ **c.** $2 + \sqrt{2} + 1 + \frac{\sqrt{2}}{2} + \cdots$

26. Express the repeating decimal as a fraction.

    a. $0.7\overline{7}$ **b.** $0.4\overline{343}$ **c.** $6.89\overline{89}$

27. The midpoints of the sides of a square whose side is 4 cm are joined to form a smaller square. The midpoints of the new square are joined, and the process is continued without end. Find the sum of the perimeters of all the squares.

28. Prove by mathematical induction:

    a. $\displaystyle\sum_{i=1}^{n} 3^i = \frac{3}{2}(3^n - 1)$ **b.** $\displaystyle\sum_{i=1}^{n} (i + 2) = \frac{n^2 + 5n}{2}$

29. Prove by mathematical induction: $2^0 + 2^2 + 2^4 + \cdots + 2^{2n-2} = \dfrac{4^n - 1}{3}$

30. Expand $(x + 2y)^4$.

31. Write the 4th term of $(4x - 1)^7$.

32. In the expansion of $(a + 3)^5$, find the term whose coefficient is $\binom{5}{3}$.

33. Find the 3rd term in the expansion of $\left(x + \dfrac{2}{x}\right)^5$.

34. Estimate $(1.07)^6$ to the nearest hundredth by using the binomial expansion.

35. If a baseball player has a batting average of .300, what is the probability that he will get 3 hits in 4 times at bat?

---

29. S_1: ✔

    Assume S_k: $2^0 + 2^2 + 2^4 + \cdots + 2^{2k-2} = \dfrac{4^k - 1}{3}$

    Then S_{k+1}: $2^0 + 2^2 + 2^4 + \cdots + 2^{2k-2} + 2^{2(k+1)-2} = \dfrac{4^k - 1}{3} + 2^{2k}$ $(2^{2k} = 4^k)$

                            $= \dfrac{4^k - 1 + 3 \cdot 4^k}{3}$

                            $= \dfrac{4^{k+1} - 1}{3}$

30. $x^4 + 8x^3y + 24x^2y^2 + 32xy^3 + 16y^4$

31. $-8{,}960x^4$

32. $270a^2$

33. $40x$

34. 1.50

35. 0.0756

1. 5
2. 6
3. 4
4. 6,633
5. 0
6. C
7. B
8. D
9. C
10. B
11. A

☑ *Exercises for Challenge*

1. Find the integer $n \geq 3$ such that:
$$\binom{n}{1} + \binom{n}{2} + \binom{n}{3} = n^2$$

2. In a certain arithmetic sequence, the sum of the first n terms is $3n^2$ for all n. What is the common difference for the sequence?

3. Determine the largest value of n such that $2 \cdot 2^2 \cdot 2^3 \cdot \cdots \cdot 2^n$ is less than 10,000.

4. What is the sum of the integers from 1 to 200 that are divisible by 3?

5. Let S_n denote the sum of the first n terms of an arithmetic sequence that has a common difference of 4. If $S_3 = S_5$, find S_8.

6. Which of the following equations are true for all n when a_i and b_i represent real numbers?

I $\sum_{i=1}^{n} a_i + \sum_{i=1}^{n} b_i = \sum_{i=1}^{n} (a_i + b_i)$

II $\sum_{i=1}^{n} a_i \cdot \sum_{i=1}^{n} b_i = \sum_{i=1}^{n} (a_i \cdot b_i)$

III $\sum_{i=1}^{n} 3a_i = 3 \sum_{i=1}^{n} a_i$

(A) I only   (B) I and II only
(C) I and III only (D) II and III only
(E) I, II, and III

7. In the expansion of $(1 + x)^n$, the coefficient of the 4th term equals the coefficient of the 6th term. What is the value of n?

(A) 7   (B) 8   (C) 9
(D) 10   (E) 12

8. Evaluate: $\sum_{n=1}^{20} [(n+1)^3 - n^3]$

(A) 7,999   (B) 8,000
(C) 8,336   (D) 9,260
(E) 9,261

9. For which of the following sequences does $a_1 = 1$, $a_2 = 2$, and $a_3 = 3$?

I $a_n = n^3 - 6n^2 + 12n - 6$
II $a_n = -n^3 + 6n^2 - 10n + 6$
III $a_n = n^3 + 6n^2 - 8n + 2$

(A) I only   (B) III only
(C) I and II only   (D) I and III only
(E) I, II, and III

10. What is the smallest integral value of m such that
$$1 + \frac{1}{3} + \frac{1}{9} + \cdots + \frac{1}{3m} = \sum_{n=0}^{m} \frac{1}{3^n}$$
differs from $\frac{3}{2}$ by less than 10^{-7}?

(A) 14   (B) 15   (C) 16
(D) 17   (E) 18

11. The sum of the first n terms of a sequence is $3n^2 + 2n$. The 100th term of the sequence is

(A) 599   (B) 995   (C) 5,450
(D) 30,200   (E) 1,510,250

12. 🖩 $S = (1^2 - 2^2) + (3^2 - 4^2) + \cdots$
$+ [(2n-1)^2 - (2n)^2] + \cdots + (49^2 - 50^2)$

What is the value of S?

(A) $-1,250$ (B) $-1,275$
(C) $-1,299$ (D) $-5,099$
(E) $-5,000$

13. Find the coefficient of x^2 in the expansion of $\left(x\sqrt{x} + \frac{1}{\sqrt{x}}\right)^8$.

(A) 8   (B) 28   (C) 56
(D) 64   (E) 70

In **14–15**, $\{a_n\}$ is defined as $a_1 = 10$,

$a_n = \begin{cases} \dfrac{a_{n-1}}{2} & \text{if } a_{n-1} \text{ is even} \\ 3a_{n-1} + 1 & \text{if } a_{n-1} \text{ is odd} \end{cases}$

14. The 10th term of the sequence is:

(A) 1   (B) 2   (C) 4
(D) 8   (E) 16

15. Which of the following best describes the sequence?
(A) Some terms are negative.
(B) The terms of the sequence become arbitrarily small in absolute value.
(C) The terms of the sequence become arbitrarily large in absolute value.
(D) The terms become alternately arbitrarily large and small in absolute value.
(E) The terms begin to repeat.

Questions 16 – 20 each consist of two quantities, one in Column A and one in Column B. You are to compare the two quantities and choose:

A   if the quantity in Column A is greater;
B   if the quantity in Column B is greater;
C   if the two quantities are equal;
D   if the relationship cannot be determined from the information given.

1. In certain questions, information concerning one or both of the quantities to be compared is centered above the two columns.

2. In a given question, a symbol that appears in both columns represents the same thing in Column A as it does in Column B.

3. x, n, and k, etc. stand for real numbers.

Column A	Column B

16.   Coefficient of the 3rd term of the expansion of $(1 + x)^{10}$ Coefficient of the 9th term of the expansion of $(1 + x)^{10}$

Let r be any real number.

17. $\displaystyle\sum_{n=0}^{9} r^n$ $\displaystyle\sum_{n=1}^{10} r^n$

Sequences $\{a_n\}$ and $\{b_n\}$ are defined for $n > 1$ by:
$a_1 = 10$, $a_n = \frac{4}{3}a_{n-1}$
$b_1 = -2$, $b_n = b_{n-1} + 8$

18. 🖩 a_4 b_4

In an arithmetic sequence $\{a_n\}$:
$a_1 = 105$ and $a_2 - a_1 = -5$

19. $|a_{10}|$ $|a_{34}|$

For $n \geq 3$, a sequence $\{a_n\}$ is defined by:
$a_1 = 2$, $a_2 = 3$,
$a_n = 3a_{n-1} - 2a_{n-2}$

20.   32 a_6

12. B
13. C
14. A
15. E
16. C
17. D
18. A
19. C
20. B

Teacher's Chapter 5

Exponents and Logarithms

Chapter Contents

The availability of scientific calculators and computers has made computation with logarithms obsolete. But logarithms continue to be important in the development of mathematical theory, in the construction of calculators and computers, and in the solution of an important group of problems that involve change in which the increase or decrease is proportional to the present amount of the sample that is changing. Since logarithms also play an important role in differential and integral calculus, familiarity with these functions is essential to preparation for the study of calculus. Although the solutions of problems in this chapter can be found using tables provided in the book, the use of a scientific calculator is recommended.

5.1   Exponents

The material in this section is review for most students. However, students need practice with exponents, especially negative and fractional exponents.

Emphasize that when the base is positive, the value of the power is always positive, even when the exponent is negative.

The ambiguity that sometimes arises when a fractional exponent is used with a negative base makes it necessary to restrict the use of fractional exponents to powers with positive bases.

$$(-8)^{\frac{2}{3}} = (\sqrt[3]{-8})^2 = (-2)^2 = 4$$

and $\quad (-8)^{\frac{2}{3}} = \sqrt[3]{(-8)^2} = \sqrt[3]{64} = 4$

but $\quad (-4)^{\frac{3}{2}} = \sqrt{(-4)^3} = \sqrt{-64} = 8i$

and $\quad (-4)^{\frac{3}{2}} = (\sqrt{-4})^3 = (2i)^3 = 8i^3 = -8i$

5.2   The Exponential Function

This section uses an intuitive notion of a limiting value to justify the definition of the domain of an exponential function as the set of all real numbers. Point out, in the sequence of approximations of $2^{\sqrt{3}}$, the values $2^{1.7} = \sqrt[10]{2^{17}}$ and $2^{1.73} = \sqrt[100]{2^{173}}$.

The use of the fractional exponent $\frac{1}{n}$ as a

way of expressing the nth root of a number justifies the use of the y^x key of a scientific calculator to both raise a number to the nth power and to take the nth root of a number.

Activity (5.2) _____

Instructions

Use graph paper on which the edge of each large block has been divided into five or ten smaller divisions. Choose a scale so that each small division represents 0.1.

1. Draw the graph of $y = 2^x$.
2. Open a pair of compasses so that the fixed point is at the origin and the marking point is at (1, 1). Draw an arc that intersects the x-axis at a point A. Since the radius of the arc is $\sqrt{2}$, the coordinates of A are $(\sqrt{2}, 0)$.
3. Let B be the point at which the vertical line through A intersects the graph of $y = 2^x$. Read as accurately as possible the y-coordinate of B.
4. Compare the y-coordinate of B with the value of $2^{\sqrt{2}}$ obtained from a calculator.
5. Open a pair of compasses so that the fixed point is at the origin and the marking point is at the intersection of $\overleftrightarrow{AB}$ and $y = 1$. Draw an arc that intersects the x-axis at a point C. Since the radius of the arc is $\sqrt{3}$, the coordinates of C are $(\sqrt{3}, 0)$.
6. Let D be the point at which the vertical line through C intersects the graph of $y = 2^x$. Read as accurately as possible the y-coordinate of D.
7. Compare the y-coordinate of D with the value of $2^{\sqrt{3}}$ obtained from a calculator.

Discoveries

1. The graph enables us to approximate, to the nearest tenth, the values of powers of 2 with irrational exponents.
2. The accuracy of the approximation depends on the care with which the graph is drawn and the scale of the graph paper used.

What If?

You want to estimate the value of $2^{\sqrt{5}}$?

(Place the marking point of your compasses at (2, 1) or at (1, 2).)

5.3 The Number e

The number e is defined first as the value of an infinite series. The value of e can be found to be greater than $2\frac{17}{24}$ or $2.708\overline{3}$ by finding the sum of the first five terms. The value is found to be less than 3 by comparing all but the first term to the terms of a geometric series whose sum is 2. Students can accept intuitively that if we continue to add the values of the successive terms to the sum of the first five, the value of the partial sum continues to increase but since it is always less than 3, there must be some finite limit. After the 10th or 11th term, the display capability of the standard calculator is too limited to show the change in value.

The expansion of the binomial $\left(1 + \frac{1}{n}\right)^n$ approaches e as n approaches infinity. Students should accept intuitively that as n approaches infinity, $\frac{1}{n}$ approaches zero and any finite multiple of $\frac{1}{n}$ approaches zero. Students will use more rigorous proofs for these concepts in their study of calculus. However, the presentation of these ideas in an intuitive way will give them a familiarity with the concepts that will make the formal study of limits less difficult.

Computer Activity (5.3)

The following program will evaluate the series that defines e for the first 21 terms. The first column gives the value of the nth term and the second column gives the partial sum of n terms. Note that the computer prints the same value of the partial sum for the last 9 terms since the value of the term is too small to increase the value of the sum to the number of decimal places printed by the computer.

```
100  REM  THIS PROGRAM WILL APPROXIMATE
110  REM  THE VALUE OF THE BASE OF
120  REM  NATURAL LOGARITHMS TO 9 DIGITS
130  F = 1
140  T = 1
150  E = 1
160  PRINT T, E
170  FOR N = 1 TO 20
180  F = F * N
190  T = 1/F
200  E = E + T
210  PRINT T, E
220  NEXT N
230  PRINT E
```

5.4 The Logarithmic Function

This section introduces the logarithmic function as the inverse of the exponential function, emphasizing the concept in the presentation.

Since most students are more comfortable with exponents than with logarithms — more readily able to give the value of x when it is expressed as $3^x = 81$ than when it is expressed as $x = \log_3 81$ — they tend to change a logarithmic expression to an exponential expression in order to evaluate. At this level, a goal should be to increase student confidence in working with logarithmic expressions.

5.5 Logarithms With Special Bases

Common logarithms and natural logarithms are the logarithms available on calculators and computers or in books of standard tables.

In standard BASIC, the function LOG(X) will return the natural logarithm of x. Therefore, in a program that requires the use of common logarithms, the change of base formula must be used. If CL(X) is to represent the common logarithm of x in a BASIC program, then:

$$\text{DEF } CL(X) = \frac{\text{LOG}(X)}{\text{LOG}(10)}$$

can be used to define the common logarithm function.

In **1 – 8**, evaluate the given expression.

1. $(-2)^0$ **2.** $6 \cdot 3^{-2}$ **3.** $\left(\frac{1}{8}\right)^{-\frac{1}{3}}$ **4.** $\log_4 \frac{1}{2}$ **5.** $\log_3 1$

6. $\log 20 + \log 5$ **7.** $2 \log_2 6 - \log_2 9$ **8.** $e^{\ln 7}$

In **9 – 12**, write the expression without a denominator and without using radicals.

9. $\left(\frac{x^{-2}}{x^{-3}}\right)^2$ **10.** $\frac{4a^3 b}{a^{-1} b}$ **11.** $\frac{5a^3}{(5a^{-2})^2}$ **12.** $\frac{\sqrt{a^3}}{\sqrt[4]{a^3}}$

13. If $f(x) = e^{2x}$, find $f^{-1}(x)$.

14. **a.** Sketch the graph of $y = e^x$.
   b. On the same set of axes, sketch the graph of $y = e^{-x}$.
   c. Use the graphs of $y = e^x$ and $y = e^{-x}$ to sketch the graph of $y = \frac{e^x + e^{-x}}{2}$.
   (The graph of $y = \frac{e^x + e^{-x}}{2}$ is called a *catenary*. It has the same shape as that of a flexible cord suspended by its ends.)

In **15 – 22**, solve for x.

15. $4 \cdot 2^x = 16^{x-1}$ **16.** $9^{x + \frac{1}{2}} = 27^{x - \frac{1}{3}}$ **17.** $x^{-\frac{1}{2}} = \frac{1}{5}$

18. $\log_x \frac{9}{4} = -2$ **19.** $\log_5 x = 0$ **20.** $\log_4 32 = x$

21. $\log x - \log 2 = 1$ **22.** $\ln e^2 - \ln x = \ln e^3$

23. Find, to the nearest ten-thousandth, the growth rate constant for a culture of bacteria if the number of bacteria present triples every 10 minutes.

24. Find, to the nearest hundredth of a percent, the effective yield for 1 year if the interest on an investment is 7.5% compounded continuously.

25. The half-life of Plutonium-236 is 2.85 years.

   a. What is the decay constant for Plutonium-236?

   b. A sample of material originally contained 1 gram of Plutonium-236. After how many years will the sample contain 125 milligrams of Plutonium-236?

   Express your answer to the nearest tenth of a year.

Bonus

   1. Solve for x to the nearest hundredth.
   $$10^{e^x} = 52$$

   2. If $y = \ln x + c$ and $e^y = Ax$ for all x and y, express c in terms of A.

Answers to Suggested Test Items

1. 1

2. $\frac{2}{3}$

3. 2

4. $-\frac{1}{2}$

5. 0

6. 2

7. 2

8. 7

9. x^2

10. $4a^4$

11. $5^{-1}a^7$

12. $a^{\frac{3}{4}}$

13. $f^{-1}(x) = \frac{1}{2}\ln x$

14.

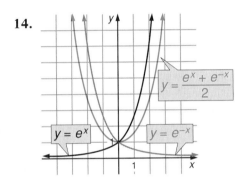

$$y = \frac{e^x + e^{-x}}{2}$$

$y = e^x$ $y = e^{-x}$

15. $x = 2$

16. $x = 2$

17. $x = 25$

18. $x = \frac{2}{3}$

19. $x = 1$

20. $x = \frac{5}{2}$

21. $x = 20$

22. $x = \frac{1}{e}$

23. $3A_0 = A_0 e^{k \cdot 10}$
$3 = e^{10k}$
$10k = \ln 3$
$k = \dfrac{\ln 3}{10}$
$k \approx 0.1099$

24. $1 + r = e^{0.075}$
$1 + r \approx 1.07788$
$r \approx 0.07788$
Approximately 7.79%

25. a. $0.5 = e^{k \cdot 2.85}$
$2.85k = \ln 0.5$
$k = \dfrac{\ln 0.5}{2.85}$
$k \approx -0.2432$

b. $125 = 1{,}000 e^{-0.2432t}$
$-0.2432t = \ln 0.125$
$t \approx 8.6$ years

Bonus

1. 0.54
$10^{e^x} = 52$
$e^x = \log 52$
$e^x = 1.716$
$x = \ln 1.716$
$x = 0.54$

2. $c = \ln A$
$e^y = Ax$
$y = \ln Ax$
$y = \ln A + \ln x$
$y = \ln x + c$
$\ln x + c = \ln A + \ln x$
$c = \ln A$

Chapter *5*

Exponents and Logarithms

While a plant is alive, it is constantly taking in carbon dioxide, a portion of which is a radio-active substance called Carbon-14. Intake and disintegration keep the proportion of Carbon-14 to the rest of the plant material relatively constant. After the plant dies, it ceases to take in Carbon-14 but that which is present continues to disintegrate. By measuring the amount of Carbon-14 present in an object made from plant material, scientists are able to approximate the time that has elapsed since the plant died, and therefore, the age of the object. The process, called carbon dating, is applied to animal life as well as to plant life. The formulas used in this scientific procedure involve powers of an irrational number, *e*, which was named for the Swiss mathematician Leonard Euler (1707-1783).

Chapter Table of Contents

5.1 Exponents

Recall that an *exponent* shows how many times a number is to be used as a factor. For example, x^3 indicates that the base x occurs 3 times as a factor: $x \cdot x \cdot x$. Recall, also, the following rules for operations involving powers with like bases.

If a, b, and c are positive integers:

$$x^a \cdot x^b = x^{a+b}$$
$$x^a \div x^b = x^{a-b} \quad \text{where } x \neq 0, a > b$$
$$(x^a)^b = x^{ab}$$

For example:

$$x^3 \cdot x^5 = x^{3+5} = x^8 \qquad\qquad 4^3 \cdot 4^5 = 4^{3+5} = 4^8$$
$$x^7 \div x^2 = x^{7-2} = x^5 \qquad\qquad 6^7 \div 6^2 = 6^{7-2} = 6^5$$
$$(x^4)^3 = x^{4(3)} = x^{12} \qquad\qquad (2^4)^3 = 2^{4(3)} = 2^{12}$$

These rules for multiplication and division of powers with like bases are stated for exponents that are positive integers. This restriction requires that the rule for division be limited to the case where the exponent of the numerator is larger than the exponent of the denominator.

Zero Exponent

If, when using the rule for division of powers with like bases, the exponent of the numerator is not required to be larger than the exponent of the denominator, then the exponents could be equal.

For $x \neq 0$, $x^a \div x^a = x^{a-a} = x^0$, using the rule for exponents

but, $x^a \div x^a = 1$, using ordinary division.

Therefore, in order for the value of x^0 to be consistent with the rules for division, the following definition is required.

■ **Definition:** $x^0 = 1$ *where* $x \neq 0$

Since the definition of x^0 is based on division by x^a, the restriction that $x \neq 0$ is necessary.

In the following examples, compare the use of x^0 and 1 when applying the rules for operations involving powers with like bases.

$$x^a \cdot x^0 = x^{a+0} = x^a \qquad\qquad x^a \cdot x^0 = x^a \cdot 1 = x^a$$
$$x^a \div x^0 = x^{a-0} = x^a \qquad\qquad x^a \div x^0 = x^a \div 1 = x^a$$
$$(x^0)^a = x^{0(a)} = x^0 = 1 \qquad\qquad (x^0)^a = 1^a = 1$$

Negative Exponents

If, when using the rule for division of powers with like bases, the exponent of the numerator is not required to be larger than the exponent of the denominator, then the exponent of the numerator could be smaller than the exponent of the denominator. For example:

$$\text{for } x \neq 0, \quad x^3 \div x^5 = x^{3-5} = x^{-2}$$

but
$$\frac{x^3}{x^5} = \frac{x^3}{x^5} \div \frac{x^3}{x^3} = \frac{x^3 \div x^3}{x^5 \div x^3} = \frac{x^0}{x^2} = \frac{1}{x^2}$$

Therefore, $x^{-2} = \frac{1}{x^2}$. Similarly, $\frac{1}{x^{-2}} = x^2$ since $\frac{1}{x^{-2}} = \frac{x^0}{x^{-2}} = x^{0-(-2)} = x^2$.

The definition of x^0 and the rule for division for powers with like bases can be used to write a general definition for a power with a negative exponent.

■ **Definition:** $\quad \dfrac{1}{x^c} = \dfrac{x^0}{x^c} = x^{0-c} = x^{-c}$

$\quad and \quad \dfrac{1}{x^{-c}} = \dfrac{x^0}{x^{-c}} = x^{0-(-c)} = x^c \quad where \ x \neq 0$

In the following examples, compare the use of x^{-c} and $\dfrac{1}{x^c}$ when applying the rules for operations involving powers with like bases.

$$x^5 \cdot x^{-3} = x^{5+(-3)} = x^2 \qquad\qquad x^5 \cdot x^{-3} = x^5 \cdot \frac{1}{x^3} = x^2$$

$$x^4 \div x^{-2} = x^{4-(-2)} = x^6 \qquad\qquad x^4 \div x^{-2} = x^4 \cdot \frac{1}{x^{-2}} = x^4 \cdot x^2 = x^6$$

$$(x^{-6})^3 = x^{-6(3)} = x^{-18} \qquad\qquad (x^{-6})^3 = \left(\frac{1}{x^6}\right)^3 = \frac{1^3}{(x^6)^3} = \frac{1}{x^{18}} = x^{-18}$$

Note that x^c and x^{-c} are multiplicative inverses, since $x^c \cdot x^{-c} = x^{c+(-c)} = x^0 = 1$.

Example 1

Simplify: $\dfrac{a^4 b^{-3} c^2}{a^5 b^{-5} c^2}$

Solution

$$\frac{a^4 b^{-3} c^2}{a^5 b^{-5} c^2} = a^{4-5} b^{-3-(-5)} c^{2-2} = a^{-1} b^2 c^0 = \frac{b^2}{a}$$

Example 2

Evaluate: $\left(\dfrac{3}{4}\right)^{-3}$

Solution

$$\left(\frac{3}{4}\right)^{-3} = \frac{3^{-3}}{4^{-3}} = 3^{-3} \cdot \frac{1}{4^{-3}} = \frac{1}{3^3} \cdot 4^3 = \left(\frac{4}{3}\right)^3 = \frac{64}{27}$$

From Example 2, note that in general, for $a \neq 0$ and $b \neq 0$:

$$\left(\frac{a}{b}\right)^{-c} = \left(\frac{b}{a}\right)^c$$

Fractional Exponents

The rules for multiplication and division of powers with like bases were stated for exponents that are integers. It is possible to use these rules to define a power with an exponent that is a fraction.

Recall that the nth root of x is defined as one of the n equal factors whose product is x. For example, for $x > 0$,

$$\sqrt{x} \cdot \sqrt{x} = x \qquad \sqrt[3]{x} \cdot \sqrt[3]{x} \cdot \sqrt[3]{x} = x \qquad \sqrt[4]{x} \cdot \sqrt[4]{x} \cdot \sqrt[4]{x} \cdot \sqrt[4]{x} = x$$

In general, for $x > 0$: $\sqrt[n]{x} \cdot \sqrt[n]{x} \cdot \sqrt[n]{x} \ldots \sqrt[n]{x} = x$, for n factors $\sqrt[n]{x}$

If there exists a number p such that $x^p = \sqrt{x}$, for $x > 0$ and $x \neq 1$, then since $\sqrt{x} \cdot \sqrt{x} = x$,

$$x^p \cdot x^p = x \quad \text{or} \quad x^{2p} = x^1$$

Thus: $\quad 2p = 1 \quad$ or $\quad p = \frac{1}{2}$

Therefore, $x^{\frac{1}{2}} = \sqrt{x}$.

In general, if n is a positive integer and $x > 0$ and $x \neq 1$,

since $\quad \underbrace{\sqrt[n]{x} \cdot \sqrt[n]{x} \cdot \sqrt[n]{x} \ldots \sqrt[n]{x}}_{n \text{ factors}} = x \quad$ and $\quad \underbrace{x^{\frac{1}{n}} \cdot x^{\frac{1}{n}} \cdot x^{\frac{1}{n}} \ldots x^{\frac{1}{n}}}_{n \text{ factors}} = x^{\frac{n}{n}} = x$

it can be concluded that $x^{\frac{1}{n}} = \sqrt[n]{x}$.

Compare the use of radicals with the use of fractional exponents.

$$\frac{5}{\sqrt{5}} = \frac{5}{\sqrt{5}} \cdot \frac{\sqrt{5}}{\sqrt{5}} = \frac{5\sqrt{5}}{5} = \sqrt{5} \qquad\qquad \frac{5}{5^{\frac{1}{2}}} = 5^{1 - \frac{1}{2}} = 5^{\frac{1}{2}} = \sqrt{5}$$

$$\sqrt{2} \cdot \sqrt{8} = \sqrt{2 \cdot 8} = \sqrt{16} = 4 \qquad 2^{\frac{1}{2}} \cdot 8^{\frac{1}{2}} = (2 \cdot 8)^{\frac{1}{2}} = 16^{\frac{1}{2}} = \sqrt{16} = 4$$

Example 3

Evaluate: $5^0 + 4^{-\frac{1}{2}} + 8^{\frac{1}{3}}$

Solution

$$5^0 + 4^{-\frac{1}{2}} + 8^{\frac{1}{3}} = 1 + \frac{1}{\sqrt{4}} + \sqrt[3]{8}$$

$$= 1 + \frac{1}{2} + 2 = 3\frac{1}{2}$$

Example 4

Evaluate: $27^{\frac{2}{3}}$

Solution

$$27^{\frac{2}{3}} = 27^{\frac{1}{3} \cdot 2} = \left(27^{\frac{1}{3}}\right)^2 = (\sqrt[3]{27})^2 = 3^2 = 9$$

$$\text{or} \quad 27^{\frac{2}{3}} = 27^{2 \cdot \frac{1}{3}} = (27^2)^{\frac{1}{3}} = (729)^{\frac{1}{3}} = \sqrt[3]{729} = 9$$

$$\text{or} \quad 27^{\frac{2}{3}} = (3^3)^{\frac{2}{3}} = 3^{3 \cdot \frac{2}{3}} = 3^2 = 9$$

In an equation that involves powers, the base may be the variable. To solve an equation such as $x^{\frac{1}{2}} = 7$, recall that

$$\left(x^{\frac{a}{b}}\right)^{\frac{b}{a}} = x^{\frac{ab}{ab}} = x^1 = x$$

Therefore, to solve for x, raise each side of the equation to the power that is the multiplicative inverse of the exponent of the variable. For example:

$$x^{\frac{1}{2}} = 7$$
$$\left(x^{\frac{1}{2}}\right)^2 = 7^2$$
$$x = 49$$

Example 5

Solve for b: $8b^{-2} = 2$

Solution

$$8b^{-2} = 2$$
$$b^{-2} = \frac{1}{4} \qquad \text{Divide by 8.}$$
$$\left(b^{-2}\right)^{-\frac{1}{2}} = \left(\frac{1}{4}\right)^{-\frac{1}{2}} \qquad \text{Raise each side to the } \left(-\frac{1}{2}\right) \text{ power.}$$
$$b = (4)^{\frac{1}{2}}$$
$$b = 2$$

The formula $A = P(1 + r)^t$ expresses the amount A accumulated on a principal P at rate r in t payments.

Example 6

After compounding interest quarterly for 8 years, each dollar of an investment had increased in value to $1.89. To the nearest percent, what was the yearly interest rate of the investment?

Solution

Let x represent the quarterly interest rate. Since interest was paid 4 times a year for 8 years, there have been 32 payments.

$$(1 + x)^{32} = 1.89$$
$$\left((1 + x)^{32}\right)^{\frac{1}{32}} = (1.89)^{\frac{1}{32}}$$
$$1 + x = (1.89)^{\frac{1}{32}}$$

Use a scientific calculator to find the value of $(1.89)^{\frac{1}{32}}$ to the nearest hundredth.

$$1 + x = 1.02$$
$$x = 0.02$$

The quarterly rate is 2% and, therefore, the yearly rate is 8%.

Exercises 5.1

In **1 – 20**, evaluate the expression.

1. 4^0 **2.** $4^{\frac{1}{2}}$ **3.** 4^{-3} **4.** $9^{-\frac{1}{4}}$ **5.** $3 \cdot 5^0$

6. $(3 \cdot 5)^0$ **7.** $3^0 \cdot 5^0$ **8.** $3^{\frac{1}{2}} \cdot 27^{\frac{1}{2}}$ **9.** $4^{\frac{3}{2}} \cdot 2^{-1}$ **10.** $8^{-\frac{5}{3}}$

11. $\left(75^{-\frac{1}{4}}\right)^0$ **12.** $5^{\frac{1}{4}} \div 5^{\frac{3}{4}}$ **13.** $3 \cdot 3^{-2}$ **14.** $64^{\frac{2}{3}}$ **15.** $(4 \cdot 8)^{\frac{2}{5}}$

16. $2 \cdot 81^{\frac{5}{2}}$ **17.** $8^{\frac{1}{2}} \cdot 64^{-\frac{1}{4}}$ **18.** $9^{\frac{1}{2}} \cdot 27^{\frac{1}{2}}$ **19.** $5 \cdot 125^{-\frac{2}{3}}$ **20.** $27^{\frac{1}{2}} \cdot 27^{\frac{1}{6}}$

In **21 – 23**, simplify the expression.

21. $\dfrac{2x^2y^{-3}}{x^{-2}y^{-3}}$ **22.** $\dfrac{6a^4b^3}{6^{-2}a^4b^{-1}}$ **23.** $\dfrac{a^{-3}b^{\frac{1}{2}}}{a^{-4}b^{\frac{3}{2}}c^6}$

24. Find the value of $2a^0 + 3a^{-\frac{1}{4}}$ if $a = 16$.   **25.** Find the value of $5b^{-1} - 3b^{-1}$ if $b = 6$.

26. Find the value of $2a^0 + (2a)^0$ when $a = 8$.   **27.** Find the value of $4b^{-\frac{1}{2}} - (4b)^{-\frac{1}{2}}$ when $b = 9$.

28. If $f(x) = x^{-\frac{1}{5}}$, find $f(32)$.   **29.** If $f(x) = 4^{-x}$, find $f(2)$.

30. If $g(x) = x^3$ and $h(x) = x^{\frac{1}{3}}$, find $g \circ h(7)$.   **31.** Use fractional exponents to evaluate $\sqrt[3]{25} \cdot \sqrt[6]{25}$.

In **32 – 37**, solve for x over the set of real numbers.

32. $x^3 = 64$ **33.** $x^{-2} = 25$ **34.** $5x^{\frac{1}{3}} = 15$

35. $9x^{-\frac{1}{2}} = 1$ **36.** $x^5 = 27x^2$ **37.** $x^{\frac{3}{2}} = 6x^2$

 In **38 – 43**, find the interest rate to the nearest percent.

38. At what rate must a sum of money be invested in order that each dollar invested will have increased to $1.79 after 10 years if interest is compounded annually?

39. At what rate must a sum of money be invested in order that each dollar invested will have increased to $1.60 after 6 years if interest is compounded semiannually?

40. An investment of $500 increased to $1,000 after 12 years. At what rate was the money invested if the interest was compounded monthly?

41. At what rate must a sum of money be invested in order that the value of the investment will have doubled after 9 years if interest is compounded quarterly?

42. Find the rate, to the nearest tenth of a percent, at which interest must be compounded daily in order that the effective yield will be 8%, that is, in order that the interest paid is equivalent to 8% paid annually.

43. Determine if the yearly rate needed to give an effective rate of 8% is greater than, equal to, or less than twice that needed to give an effective rate of 4% when compounded daily.

Answers 5.1

1. 1
2. 2
3. $\frac{1}{64}$
4. $\frac{1}{\sqrt[4]{9}}$
5. 3
6. 1
7. 1
8. 9
9. 4
10. $\frac{1}{32}$
11. 1
12. $\frac{1}{\sqrt{5}}$
13. $\frac{1}{3}$
14. 16
15. 4
16. 118,098
17. 1
18. $9\sqrt{3}$
19. $\frac{1}{5}$
20. 9
21. $2x^4$
22. 6^3b^4
23. $\frac{a}{bc^6}$
24. $3\frac{1}{2}$
25. $\frac{1}{3}$
26. 3
27. $\frac{7}{6}$
28. $\frac{1}{2}$
29. $\frac{1}{16}$
30. 7
31. 5
32. $x = 4$
33. $x = \frac{1}{5}$
34. $x = 27$
35. $x = 81$
36. $x = 3$
37. $x = \frac{1}{36}$
38. 6%
39. 8% annually
40. 5.8% ≈ 6% per year
41. 7.8% ≈ 8% per year
42. 7.7% ≈ 8% per year
43. The 8% rate yields about 7.7%. Twice the 4% rate yields about 7.8%.

5.2 The Exponential Function

Recall that an expression such as $(1.04)^x$ represents the value of each dollar invested at 4% simple interest per year for x years, where x is an integer. The sequence of terms 1.04, 1.0816, 1.124864, 1.16985856 represents the value of the investment at the end of each of the first four years, when x has the values 1, 2, 3, 4. An equation such as $y = (1.04)^x$ belongs to an important class of functions of the form $y = b^x$, where $b > 0$ and $b \neq 1$, called *exponential functions*.

As established in Section 5.1, it is not necessary to limit the value of the exponent x to positive integers. When the base b is positive, b^x represents a real number for all rational numbers x. For example, $y = 2^x$ defines a function from the rational numbers to the positive real numbers. Several ordered pairs of this function are shown in the table below. The corresponding points are shown on the graph.

x	2^x	y
-3	$2^{-3} = \left(\frac{1}{2}\right)^3$	$\frac{1}{8}$
-2	$2^{-2} = \left(\frac{1}{2}\right)^2$	$\frac{1}{4}$
-1	$2^{-1} = \left(\frac{1}{2}\right)^1$	$\frac{1}{2}$
0	2^0	1
$\frac{1}{2}$	$2^{\frac{1}{2}}$	$\sqrt{2} \approx 1.41$
1	2^1	2
$\frac{3}{2}$	$2^{\frac{3}{2}} = 2^1 \cdot 2^{\frac{1}{2}}$	$2\sqrt{2} \approx 2.83$
2	2^2	4
$\frac{5}{2}$	$2^{\frac{5}{2}} = 2^2 \cdot 2^{\frac{1}{2}}$	$4\sqrt{2} \approx 5.66$
3	2^3	8

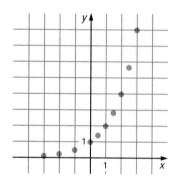

The points lie in a pattern that suggests a smooth curve. Between each pair of points, an infinite number of points could be found using other rational values of x. When a smooth curve is drawn, the resulting graph implies that the function $y = 2^x$ is defined for all real numbers x.

To establish a definition of 2^x, where x is an irrational number, consider the example $2^{\sqrt{3}}$. First examine the sequence of rational numbers x that approach $\sqrt{3}$. Note from the graph that $y = 2^x$ is an increasing function, that is, as x increases, y also increases.

$$1 < \sqrt{3} < 2 \qquad\qquad 2^1 < 2^{\sqrt{3}} < 2^2$$
$$1.7 < \sqrt{3} < 1.8 \qquad\qquad 2^{1.7} < 2^{\sqrt{3}} < 2^{1.8}$$
$$1.73 < \sqrt{3} < 1.74 \qquad\qquad 2^{1.73} < 2^{\sqrt{3}} < 2^{1.74}$$
$$1.732 < \sqrt{3} < 1.733 \qquad\qquad 2^{1.732} < 2^{\sqrt{3}} < 2^{1.733}$$
$$1.7320 < \sqrt{3} < 1.7321 \qquad\qquad 2^{1.7320} < 2^{\sqrt{3}} < 2^{1.7321}$$
$$1.73205 < \sqrt{3} < 1.73206 \qquad\qquad 2^{1.73205} < 2^{\sqrt{3}} < 2^{1.73206}$$

As the sequence of numbers 1, 1.7, 1.73, 1.732, ... approaches $\sqrt{3}$, the corresponding sequence of powers 2^1, $2^{1.7}$, $2^{1.73}$, $2^{1.732}$, ... approaches a value that is defined to be the value of $2^{\sqrt{3}}$. The graph of $y = 2^x$ helps locate $2^{\sqrt{3}}$ on the number line. That is, point A is located on the graph where $x = \sqrt{3}$, the value of y at A is the value of $2^{\sqrt{3}}$, and a horizontal line from A to the y-axis intersects the y-axis at $2^{\sqrt{3}}$. From the graph, it can be seen that $2^{\sqrt{3}}$ is approximately equal to 3.3.

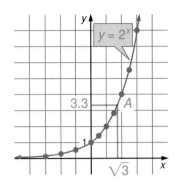

In general, for an exponential function, which is defined by the equation $y = b^x$ where $b > 0$ and $b \neq 1$, the domain is the set of real numbers and the codomain is the set of positive real numbers. The function is a 1–1 correspondence, that is, every element of the domain corresponds to one and only one element of the codomain.

Example 1

a. Sketch the graph: $y = \left(\frac{4}{3}\right)^x$

b. Find the function that is a reflection of $y = \left(\frac{4}{3}\right)^x$ over the y-axis.

Solution

a.

x	$\left(\frac{4}{3}\right)^x$	y
-3	$\left(\frac{4}{3}\right)^{-3} = \left(\frac{3}{4}\right)^3$	$\frac{27}{64} \approx 0.42$
-2	$\left(\frac{4}{3}\right)^{-2} = \left(\frac{3}{4}\right)^2$	$\frac{9}{16} \approx 0.56$
-1	$\left(\frac{4}{3}\right)^{-1} = \left(\frac{3}{4}\right)^1$	$\frac{3}{4} = 0.75$
0	$\left(\frac{4}{3}\right)^0$	1
1	$\left(\frac{4}{3}\right)^1$	$\frac{4}{3} \approx 1.33$
2	$\left(\frac{4}{3}\right)^2$	$\frac{16}{9} \approx 1.78$
3	$\left(\frac{4}{3}\right)^3$	$\frac{64}{27} \approx 2.37$

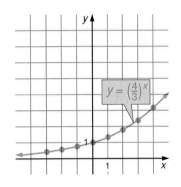

b. Under a reflection over the y-axis, $(x, y) \rightarrow (-x, y)$.

Thus, the image of $y = \left(\frac{4}{3}\right)^x$ is $y = \left(\frac{4}{3}\right)^{-x}$ or $y = \left(\frac{3}{4}\right)^x$.

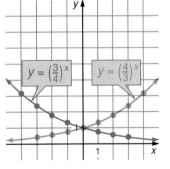

Note that the graphs in part **b** of Example 1 intersect at $(0, 1)$. The function $y = \left(\frac{4}{3}\right)^x$ is an increasing function and $y = \left(\frac{3}{4}\right)^x$ is a decreasing function. In general, if $b > 1$, then $y = b^x$ is an increasing function and if $0 < b < 1$, then $y = b^x$ is a decreasing function.

Solving an Exponential Equation

■ **Definition:** *An* **exponential equation** *is an equation that has the variable as an exponent.*

An exponential equation can be solved if each side of the equation is a power of the same base. When the bases of the two sides are the same, the exponents must be equal.

If $b \neq 0$ and $b \neq 1$, $b^p = b^q$ implies $p = q$.

For example, $5^{x+1} = 5^{3x}$ implies that

$$x + 1 = 3x$$
$$1 = 2x$$
$$\tfrac{1}{2} = x$$

Equations involving powers of different bases can sometimes be solved by expressing each power to the same base.

Example 2

Solve for a: $27^a = 9^{a-1}$

Solution

Express each side of the equation to the base 3.

$$27^a = 9^{a-1}$$
$$(3^3)^a = (3^2)^{a-1}$$
$$3^{3a} = 3^{2a-2}$$
$$3a = 2a - 2$$
$$a = -2$$

Check

$$27^a = 9^{a-1}$$
$$27^{-2} \stackrel{?}{=} 9^{-2-1}$$
$$27^{-2} \stackrel{?}{=} 9^{-3}$$
$$\frac{1}{27^2} \stackrel{?}{=} \frac{1}{9^3}$$
$$\frac{1}{729} = \frac{1}{729} \quad \vee$$

Example 3

Solve for y: $4^{y+3} = 16(8^{2y})$

Solution

Express each side of the equation to the base 2.

$$4^{y+3} = 16(8^{2y})$$
$$(2^2)^{y+3} = 2^4(2^3)^{2y}$$
$$2^{2(y+3)} = 2^4(2^{6y})$$
$$2^{2y+6} = 2^{4+6y}$$
$$2y + 6 = 4 + 6y$$
$$-4y = -2$$
$$y = \tfrac{1}{2}$$

Check

$$4^{y+3} = 16(8^{2y})$$
$$(2^2)^{\frac{1}{2}+3} \stackrel{?}{=} 2^4(2^3)^{2 \cdot \frac{1}{2}}$$
$$(2^2)^{\frac{7}{2}} \stackrel{?}{=} 2^4(2^3)$$
$$2^7 = 2^7 \quad \vee$$

Note that it is not always possible to express both sides of an equation as powers of the same base, as in $5^x = 3$. Solutions for such equations will be considered later in this chapter.

Exercises 5.2

1. a. Sketch the graph of $y = 3^x$ over the interval $-3 \le x \le 3$.
   b. From the graph, approximate the value of $3^{1.4}$ to the nearest tenth.
   c. For the domain $-3 \le x \le 3$, determine the range of $y = 3^x$.

2. a. Sketch the graph of $y = \left(\frac{2}{3}\right)^x$ over the interval $-3 \le x \le 3$.
   b. On the same set of axes, and for the same interval, sketch the graph of $y = \left(\frac{3}{2}\right)^x$.
   c. Under what transformation is the graph of $y = \left(\frac{3}{2}\right)^x$ the image of the graph of $y = \left(\frac{2}{3}\right)^x$?

3. a. Sketch the graph of $y = 4^x$ over the interval $-2 \le x \le 2$.
   b. On the same set of axes and for the same interval, sketch the image of the graph drawn in part **a** under a reflection in the y-axis.
   c. Write an equation of the graph drawn in part **b**.

In **4 – 18**, solve for x over the set of real numbers.

4. $4^x = 4^3$

5. $3^{2x} = 3^{x+2}$

6. $2^x = 4^{x-2}$

7. $25^{2x} = 125^{x+1}$

8. $9(3^x) = 81^x$

9. $8(2^x) = 4^x$

10. $49^x = 7^{x-1}$

11. $36^x = 6^{3x-1}$

12. $4^x = 8^{x-2}$

13. $9^{x-1} = 27^{2+x}$

14. $27^x = 9^{x+2}$

15. $\left(\frac{1}{2}\right)^x = 8^{2-x}$

16. $\left(\frac{1}{3}\right)^{x-3} = 9^x$

17. $\left(\frac{1}{25}\right)^{2x} = 5^{x-5}$

18. $\left(\frac{1}{4}\right)^x = \left(\frac{1}{8}\right)^{x-1}$

19. A research biologist began to grow a culture in which the number of bacteria doubled every day. Three days later, she began a second culture in a richer medium so that the number of bacteria increased fourfold every day. If each culture was begun with the same number of bacteria present, how many days after starting the second culture will the number of bacteria in the two cultures be equal?

20. A group of 5 persons met to form an organization to encourage recycling in their town. At each meeting, each person was asked to bring a friend, thus doubling the size of the group. If the plan to increase attendance is successful, at which meeting will there be 320 persons in the group?

Answers 5.2

1. a.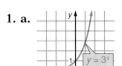

b. 4.7

c. $\frac{1}{27} \le y \le 27$

2. a. and **b.**

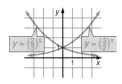

c. Reflection in the y-axis.

3. a. and **b.**

c. $y = 4^{-x}$ or $y = \left(\frac{1}{4}\right)^x$

4. $x = 3$

5. $x = 2$

6. $x = 4$

7. $x = 3$

8. $x = \frac{2}{3}$

9. $x = 3$

10. $x = -1$

11. $x = 1$

12. $x = 6$

13. $x = -8$

14. $x = 4$

15. $x = 3$

16. $x = 1$

17. $x = 1$

18. $x = 3$

19. 3rd day

20. 7th meeting

5.3 The Number e

Population growth, radioactive decay, and interest compounded continuously are examples of real-world topics in which the change in the quantity of a measurable substance is proportional to the amount present at any instant. To represent such change, an exponential function with the base e, an irrational number that is defined as the value of an infinite series, is used.

The following discussion will establish a numerical value for e. To begin, consider the definition of e.

$$e = 1 + \frac{1}{1!} + \frac{1}{2!} + \frac{1}{3!} + \cdots + \frac{1}{(n-1)!} + \cdots \quad \text{or} \quad e = \sum_{n=0}^{\infty} \frac{1}{n!} \quad (\textit{Note:} \ 0! = 1)$$

The first five terms of the series are:

$$1 + \frac{1}{1!} + \frac{1}{2!} + \frac{1}{3!} + \frac{1}{4!} = 1 + 1 + \frac{1}{2} + \frac{1}{6} + \frac{1}{24}$$

To show that this infinite series has a finite value, compare it to another series with a known value. For example, use a geometric series with $a_0 = 1$ and $r = \frac{1}{2}$.

$$\text{Geometric series: } 1 + \frac{1}{2} + \frac{1}{2^2} + \frac{1}{2^3} + \frac{1}{2^4} + \frac{1}{2^5} + \cdots$$

$$e - 1 = \frac{1}{1!} + \frac{1}{2!} + \frac{1}{3!} + \frac{1}{4!} + \frac{1}{5!} + \frac{1}{6!} + \cdots$$

Compare corresponding terms of the two series:

$$\frac{1}{1!} = 1, \quad \frac{1}{2!} = \frac{1}{2}, \quad \frac{1}{3!} < \frac{1}{2^2} \text{ since } 3 \cdot 2 \cdot 1 > 2 \cdot 2, \quad \text{and so on.}$$

In general, $(n + 1)!$ has n factors that are greater than or equal to 2 and 2^n has n factors of 2. Therefore, for $n > 2$:

$$(n + 1)! > 2^n \text{ and } \frac{1}{(n+1)!} < \frac{1}{2^n}$$

Since each term of the series that defines $e - 1$ is less than or equal to the corresponding term of the geometric series, the value of the series $e - 1$ is less than the value of the geometric series.

Now determine the value of the geometric series, where $|r| < 1$, by using the formula

$$S_\infty = \frac{a_1}{1 - r}.$$

For $a_1 = 1$ and $r = \frac{1}{2}$: $\quad \sum_{n=0}^{\infty} \frac{1}{2^n} = \frac{1}{1 - \frac{1}{2}} = \frac{1}{\frac{1}{2}} = 2$

Thus: $\quad e - 1 = \frac{1}{1!} + \frac{1}{2!} + \frac{1}{3!} + \cdots + \frac{1}{(n-1)!} + \cdots < 2$

and $\quad e = 1 + \frac{1}{1!} + \frac{1}{2!} + \frac{1}{3!} + \cdots + \frac{1}{(n-1)!} + \cdots < 3$

Therefore, the value of e is less than 3. By using a calculator to evaluate the first ten terms of the series, you can verify that, to five decimal places, $e \approx 2.71828$.

The series that is used to define e can be related to the expression $(1 + r)^t$, which is the basis of the formula used to solve problems involving compound interest, growth, and decay. The connection between e and $(1 + r)^t$ is made by expanding the binomial $\left(1 + \dfrac{1}{n}\right)^n$, as follows.

$$\left(1 + \frac{1}{n}\right)^n = \binom{n}{0} \cdot 1^n \cdot \left(\frac{1}{n}\right)^0 + \binom{n}{1} \cdot 1^{n-1} \cdot \left(\frac{1}{n}\right)^1 + \binom{n}{2} \cdot 1^{n-2} \cdot \left(\frac{1}{n}\right)^2 + \binom{n}{3} \cdot 1^{n-3} \cdot \left(\frac{1}{n}\right)^3 + \cdots$$

$$= 1 + \frac{n}{1!} \cdot \frac{1}{n} + \frac{n(n-1)}{2!} \cdot \frac{1}{n^2} + \frac{n(n-1)(n-2)}{3!} \cdot \frac{1}{n^3} + \frac{n(n-1)(n-2)(n-3)}{4!} \cdot \frac{1}{n^4} + \cdots$$

$$= 1 + \frac{1}{1!} + \frac{1}{2!} \cdot \frac{n^2 - n}{n^2} + \frac{1}{3!} \cdot \frac{n^3 - 3n^2 + 2n}{n^3} + \frac{1}{4!} \cdot \frac{n^4 - 6n^3 + 11n^2 - 6n}{n^4} + \cdots$$

$$= 1 + \frac{1}{1!} + \frac{1}{2!}\left(1 - \frac{1}{n}\right) + \frac{1}{3!}\left(1 - \frac{3}{n} + \frac{2}{n^2}\right) + \frac{1}{4!}\left(1 - \frac{6}{n} + \frac{11}{n^2} - \frac{6}{n^3}\right) + \cdots$$

If n approaches infinity, then $\frac{1}{n}$ approaches 0, and any constant times any power of $\frac{1}{n}$ approaches 0. Therefore, as $n \to \infty$:

$$\left(1 + \frac{1}{n}\right)^n = 1 + \frac{1}{1!} + \frac{1}{2!} + \frac{1}{3!} + \cdots = e$$

This relationship is used to solve problems involving investments. For example, if the interest rate per year is r, and the interest is compounded n times a year, then:

$$\text{Value per dollar} = \left(1 + \frac{r}{n}\right)^{nt}$$

Let $\dfrac{r}{n} = \dfrac{1}{k}$. Then $n = k \cdot r$ and $nt = k \cdot rt$.

$$\text{Value per dollar} = \left(1 + \frac{r}{n}\right)^{nt} = \left(1 + \frac{1}{k}\right)^{k \cdot rt} = \left[\left(1 + \frac{1}{k}\right)^k\right]^{rt}$$

If interest is said to be compounded continuously, the number of intervals into which the year is divided is considered to be approaching infinity. Thus, as n approaches infinity, $\dfrac{n}{r} = k$ approaches infinity and $\left(1 + \dfrac{1}{k}\right)^k$ approaches e. Therefore, as n approaches infinity, $\left[\left(1 + \dfrac{1}{k}\right)^k\right]^{rt}$ approaches e^{rt}.

In general, let A_0 represent an initial investment (or the size of an initial population), that is, the amount present at $t = 0$. Let A_t represent the value of the investment (or the size of the population) after t units of time, and let r represent the rate of interest (or the rate of increase or decrease) per unit of time. Then:

$$A_t = A_0 e^{rt}$$

Example 1

In the last national census, a town had a population of 5,487. It was estimated that the population would increase at a yearly rate of 2%. If this estimate is correct, what will be the population when the next national census is taken?

Solution

A national census is taken every 10 years. Since the population increase does not occur just once a year but is constantly increasing, use the formula

$$A_t = A_0 e^{rt}$$

where $t = 10$, $r = 0.02$, and $A_0 = 5,487$

$$A_{10} = 5,487 e^{0.02 \cdot 10}$$

$$A_{10} = 5,487 e^{0.2}$$

To find the value of A_{10}, you must find the value of $e^{0.2}$. Use a table such as the one in the Appendix, a part of which is shown here, or a scientific calculator.

Some calculators have a key labeled $\boxed{e^x}$ and others use the sequence of keys labeled $\boxed{\text{INV}}$ $\boxed{\text{LN}}$.

To find the value of $e^{0.2}$, enter:

.2 $\boxed{e^x}$

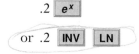
or .2 $\boxed{\text{INV}}$ $\boxed{\text{LN}}$

To four decimal places, the value of $e^{0.2}$ is 1.2214.

$$A_{10} = 5,487(1.2214)$$

$$= 6,701.8218$$

Answer: At the next census, the population is expected to be approximately 6,702.

x	e^x	e^{-x}
0.00	1.0000	1.0000
0.05	1.0513	0.9512
0.10	1.1052	0.9048
0.15	1.1618	0.8607
0.20	1.2214	0.8187
0.25	1.2840	0.7788
0.30	1.3499	0.7408
0.35	1.4191	0.7047
0.40	1.4918	0.6703
0.45	1.5683	0.6376
0.50	1.6487	0.6065
0.55	1.7333	0.5769
0.60	1.8221	0.5488
0.65	1.9155	0.5220
0.70	2.0138	0.4966
0.75	2.1170	0.4724
0.80	2.2255	0.4493
0.85	2.3396	0.4274
0.90	2.4596	0.4066
0.95	2.5857	0.3867
1.0	2.7183	0.3679

Problems involving radioactive decay are treated similarly. Since the probability of decay of every nucleus of an unstable isotope is the same, the number of nuclei that remain after a given period of time is proportional to the number of nuclei present at the beginning of the time period. As shown in the following example, the rate of decay is expressed as a decay constant that has a negative value.

Example 2

Radium has a decay constant of –0.0004 when time is expressed in years. Find, to the nearest tenth of a gram, the amount of radium that will be present after 50 years in a sample of material that contains 54 grams of radium.

Solution

Since the number of atoms and, therefore, the weight of the radium is constantly decreasing, you can find the weight of the remaining radium after 50 years by using the formula:

$$A_t = A_0 e^{rt}$$

where $t = 50$, $r = -0.0004$, and $A_0 = 54$

Note that the decrease is represented by a negative rate of change.

$$A_{50} = 54\, e^{-0.0004(50)}$$

$$= 54\, e^{-0.02}$$

The value of $e^{-0.02}$ can be found by using a calculator. To the nearest hundredth, $e^{-0.02} = 0.98$.

$$A_{50} = 54(0.98)$$

$$= 52.92$$

Answer: There will be 52.9 grams of radium remaining after 50 years.

The following example shows how the table in the Appendix can be used to find values of e^x that are not shown in the table.

Example 3

Evaluate: $e^{2.35}$

Solution

$$e^{2.35} = e^{2.0 + 0.35} = e^{2.0} \cdot e^{0.35}$$

$$= (7.3891)(1.4191)$$

$$= 10.4859$$

The value of $e^{2.35}$ can be found by using one of the following sequences on a calculator. Enter:

2.35 $\boxed{e^x}$

or   2.35 $\boxed{\text{INV}}$ $\boxed{\text{LN}}$

1.

2.

3. 7.3890561

4. 0.2231302

5. 3.3201169

6. 0.6065307

7. 0.1124882

8. 90.250135

9. $1,040.81

10. $6,549.82

11. $875.34

12. 1,062 rabbits

13. 33,641 people

14. 756 mg

15. 20.21 g

16. 4.9 g

17. 3,352 deer

18. 9,740 people

Exercises 5.3

In **1 – 2**, use values from the table in Example 1 of this section to sketch the graph of the given function.

1. $y = e^x$

2. $y = e^{-x}$

In **3 – 8**, find the value of a.

3. $a = e^2$

4. $a = e^{-1.5}$

5. $ae^5 = e^{6.2}$

6. $ea = e^{0.5}$

7. $12a = e^{0.3}$

8. $a = 10e^{2.2}$

9. An investment of $1,000 earns interest at the rate of 8% per year compounded continuously. What is the value of the investment after 6 months?

10. What is the value after 3 years of a $5,000 investment that pays interest at 9% per year if the interest is compounded continuously?

11. A credit union pays interest at the rate of 6% compounded continuously. What is the value of a deposit of $800 after $1\frac{1}{2}$ years?

12. The number of rabbits in a wildlife preserve was estimated to be increasing at the yearly rate of 12%. For each 1,000 rabbits now in the preserve, estimate the number that will be present after 6 months.

13. If a city with a population of 32,000 is growing at a rate of 1% per year, what will its population be in 5 years?

14. The hydrogen isotope tritium has a decay constant of –0.056 per year. Find, to the nearest milligram, the amount of tritium that will remain after 5 years in a sample of material that now contains 1 gram of tritium.

15. The decay constant of Uranium-227 is –0.533 when time is in minutes. Find, to the nearest hundredth of a gram, the amount remaining of a 100-gram sample after 3 minutes.

16. The decay constant of Carbon-14 is –0.000124 when time is in years. Find, to the nearest tenth of a gram, the amount of Carbon-14 present after 100 years in a piece of wood that originally contained 5 grams of Carbon-14.

17. The forest service, to balance the deer population in a national forest, is planning to reduce the present deer population of 5,000 by 4% a year. If the plan is successful, how many deer will there be after 10 years?

18. A town with a population of 12,382 is decreasing in population at the rate of 3% per year. If this decrease continues, what will be the population of the town in 8 years?

5.4 The Logarithmic Function

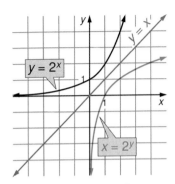

The exponential function $y = b^x$ (where $b > 0$, $b \neq 1$) is a 1 – 1 correspondence whose domain is the set of real numbers and whose codomain is the set of positive real numbers. If the graph of an exponential function, such as $y = 2^x$, is reflected over the line $y = x$, the image is the graph of a function whose domain is the set of positive real numbers and whose codomain is the set of real numbers. The equation of this image, $x = 2^y$, defines the function that is the inverse of the function $y = 2^x$.

It is customary to write an equation that defines a function so that y, the second element of the ordered pairs that satisfy that function, is expressed in terms of x, the first element. To do this for the function $x = 2^y$, consider the equation in words: "y is the exponent to which 2 must be raised to equal x."

To distinguish this inverse of the exponential equation from the exponential equation itself, the word *logarithm* is used.

y is the logarithm to the base 2 of x.

$$y = \log_2 x$$

In general, for any base b:

exponential function	inverse of the exponential function	logarithmic form
$y = b^x$	$b^y = x$ $\longrightarrow$	$y = \log_b x$

For example:

$$2^3 = 8 \quad \rightarrow \quad 3 = \log_2 8$$
$$10^{-2} = 0.01 \quad \rightarrow \quad -2 = \log_{10} 0.01$$
$$4^{\frac{1}{2}} = 2 \quad \rightarrow \quad \tfrac{1}{2} = \log_4 2$$
$$5^0 = 1 \quad \rightarrow \quad 0 = \log_5 1 \quad \text{(\textit{Note:} For any } b, \log_b 1 = 0.)$$

Example 1

Evaluate: $\log_9 3$

Solution

Let: $x = \log_9 3$

$9^x = 3$ — Definition of logarithm.

$(3^2)^x = 3^1$ — Express both sides in a common base.

$3^{2x} = 3^1$ — Apply the power rule.

$2x = 1$ — Since the bases of the two sides are the same, the exponents must be equal.

$x = \tfrac{1}{2}$

Answer: $\log_9 3 = \tfrac{1}{2}$

Example 2

If $f(x) = \left(\frac{3}{2}\right)^x$, sketch the graphs of $f(x)$ and $f^{-1}(x)$.

Solution

Make a table of values for $f(x)$.

x	$\left(\frac{3}{2}\right)^x$	y
-2	$\left(\frac{3}{2}\right)^{-2} = \left(\frac{2}{3}\right)^2$	$\frac{4}{9}$
-1	$\left(\frac{3}{2}\right)^{-1} = \left(\frac{2}{3}\right)^1$	$\frac{2}{3}$
0	$\left(\frac{3}{2}\right)^0$	1
1	$\left(\frac{3}{2}\right)^1$	$\frac{3}{2}$
2	$\left(\frac{3}{2}\right)^2$	$\frac{9}{4}$
3	$\left(\frac{3}{2}\right)^3$	$\frac{27}{8}$

To graph $f^{-1}(x)$, interchange each ordered pair, or reflect the graph of $f(x)$ over the line $y = x$. The equation of $f^{-1}(x)$ is $y = \log_{\frac{3}{2}} x$.

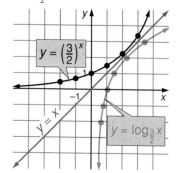

Since a logarithm is an exponent, the rules for exponents can be used to formulate the rules for logarithms.

Let: $\qquad A = x^a \qquad \rightarrow \qquad \log_x A = a$

and $\qquad B = x^b \qquad \rightarrow \qquad \log_x B = b$

Then: $\qquad AB = x^{a+b} \qquad \rightarrow \qquad \log_x AB = a + b$

$$= \log_x A + \log_x B$$

$$\frac{A}{B} = x^{a-b} \qquad \rightarrow \qquad \log_x \left(\frac{A}{B}\right) = a - b$$

$$= \log_x A - \log_x B$$

$$A^c = x^{ac} \qquad \rightarrow \qquad \log_x A^c = a \cdot c$$

$$= c \log_x A$$

These results can be summarized as follows.

$\log_x AB = \log_x A + \log_x B$	The log of a product is equal to the sum of the logs of the factors.
$\log_x \left(\frac{A}{B}\right) = \log_x A - \log_x B$	The log of a quotient is equal to the log of the numerator minus the log of the denominator.
$\log_x A^c = c \log_x A$	The log of a power is equal to the exponent times the log of the base.

For example, if $\log_5 3 = a$ and $\log_5 4 = b$, then

$$\log_5 12 = \log_5 (3 \cdot 4)$$

$$= \log_5 3 + \log_5 4$$

$$= a + b$$

The rules for operating with logarithms can be used to solve equations. Application of a rule simplifies expressions so that the definition of logarithm can be applied.

Example 3

Solve for a: $\log_3 a + \log_3 10 = 2$

Solution

$$\log_3 a + \log_3 10 = 2$$
$$\log_3 (a \cdot 10) = 2 \qquad \log_x A + \log_x B = \log_x AB$$
$$\log_3 10a = 2$$
$$10a = 3^2 \qquad \text{Definition of logarithm.}$$
$$10a = 9$$
$$a = \frac{9}{10}$$

To evaluate a multi-term expression that contains logarithms, begin by writing an equation. Again, apply the rules to simplify the log expression so that the definition of logarithm can be applied.

Example 4

Evaluate: $\log_6 8 + 2 \log_6 3 - \log_6 2$

Solution

Let: $x = \log_6 8 + 2 \log_6 3 - \log_6 2$
$$x = \log_6 8 + \log_6 3^2 - \log_6 2 \qquad c \log_x A = \log_x A^c$$

$$x = \log_6 \frac{8 \cdot 3^2}{2} \qquad \begin{aligned} \log_x A + \log_x B &= \log_x AB \\ \log_x A - \log_x B &= \log_x \left(\frac{A}{B}\right) \end{aligned}$$

$$x = \log_6 36$$

$$6^x = 36 \qquad \text{Definition of logarithm.}$$

$$6^x = 6^2 \qquad \text{Common base.}$$

$$x = 2 \qquad \text{Set exponents equal.}$$

Answer: $\log_6 8 + 2 \log_6 3 - \log_6 2 = 2$

The following relationship will frequently be used to solve a log equation.

If $\log_a x = \log_a y$, then $x = y$.

To demonstrate that this is true, replace each side of the given equation by the same variable and rewrite each equation in exponential form.

Let $\log_a x = b$ and $\log_a y = b$
$$a^b = x \qquad\qquad a^b = y$$

Therefore, by the transitive property of equality, $x = y$.

Example 5

Solve for x: $\log_2 x - \log_2 8 = \log_2 4$

Solution

$$\log_2 x - \log_2 8 = \log_2 4$$

$$\log_2 \frac{x}{8} = \log_2 4$$

Therefore, $\frac{x}{8} = 4$

$$x = 32$$

The fundamental inverse relationship between exponents and logarithms is illustrated in the following examples.

Example 6

Find the value of $2^{\log_2 7}$.

Solution

Let: $x = 2^{\log_2 7}$ and $a = \log_2 7$

$$x = 2^a$$

or $\log_2 x = a$

$$\log_2 x = \log_2 7$$

Therefore, $x = 7$

Answer: $2^{\log_2 7} = 7$

In general: $a^{\log_a b} = b$

Example 7

Evaluate: $\log_5 5^3$

Solution

Let: $x = \log_5 5^3$ and $a = 5^3$

$$x = \log_5 a$$

or $5^x = a$

$$5^x = 5^3$$

Therefore, $x = 3$

Answer: $\log_5 5^3 = 3$

In general: $\log_a a^b = b$

Exercises 5.4

1. **a.** Sketch the graph of $f(x) = \left(\frac{4}{5}\right)^x$.
   b. Sketch the graph of the inverse of $f(x)$.
   c. Write an equation for the graph drawn in part **b**.

In **2 – 9**, rewrite the equality using a logarithm.

2. $3^2 = 9$ **3.** $5^3 = 125$ **4.** $4^0 = 1$ **5.** $6^3 = 216$

6. $100^{\frac{1}{2}} = 10$ **7.** $2^{-1} = \frac{1}{2}$ **8.** $125^{-\frac{1}{3}} = \frac{1}{5}$ **9.** $25^{\frac{1}{2}} = 5$

In **10 – 13**, if $\log_a 2 = A$ and $\log_a 3 = B$, write the expression in terms of A and B.

10. $\log_a 6$ **11.** $\log_a 4$ **12.** $\log_a 12$ **13.** $\log_a \frac{2}{3}$

In **14 – 46**, evaluate the expression.

14. $\log_2 8$ **15.** $\log_8 2$ **16.** $\log_{10} 100$ **17.** $\log_3 27$

18. $\log_6 \frac{1}{6}$ **19.** $\log_4 \frac{1}{8}$ **20.** $\log_5 0.2$ **21.** $\log_{10} 0.001$

22. $\log_3 1$ **23.** $\log_2 0.25$ **24.** $\log_{27} 3$ **25.** $\log_9 1$

26. $\log_{25} \frac{1}{5}$ **27.** $\log_2 2^8$ **28.** $\log_5 5^4$ **29.** $\log_4 4$

30. $3^{\log_3 1}$ **31.** $10^{\log_{10} 5}$ **32.** $7^{2\log_7 1}$ **33.** $3^{2\log_3 4}$

34. $5^{3\log_5 2}$ **35.** $6^{-\log_6 2}$ **36.** $8^{-2\log_8 4}$ **37.** $4^{\log_2 8}$

38. $\log_{10} 5 + \log_{10} 2$ **39.** $\log_{10} 200 - \log_{10} 2$ **40.** $\log_4 10 - \log_4 5$

41. $2\log_2 6 - \log_2 9$ **42.** $10^{\log_{10} 5 + \log_{10} 3}$ **43.** $3^{\log_3 5 + \log_3 2}$

44. $10^{\log_{10} 5 - \log_{10} 3}$ **45.** $4^{-\log_2 2 \cdot 4\log_4 2}$ **46.** $9^{\log_{12} 9 + \log_{12} 16}$

In **47 – 57**, solve for x.

47. $\log_x 9 = 2$ **48.** $\log_4 x = \frac{1}{2}$ **49.** $\log_8 4 = x$

50. $\log_x 36 = -2$ **51.** $\log_2 8 + \log_2 x = 5$ **52.** $\log_2 5 + \log_2 x = \log_2 25$

53. $\log_5 x - \log_5 8 = \log_5 2$ **54.** $3\log_7 x + \log_7 2 = \log_7 8$ **55.** $-\log_2 x + 1 = \log_2 8$

56. $\log_3 20 - \log_3 x = \log_3 5 + \log_3 x$ **57.** $\log_5 x - 1 = \log_5 25$

1. **a.**
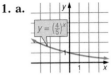
$y = \left(\frac{4}{5}\right)^x$

b.

$x = \left(\frac{4}{5}\right)^y$

c. $f^{-1}(x) = \log_{\frac{4}{5}} x$

2. $\log_3 9 = 2$
3. $\log_5 125 = 3$
4. $\log_4 1 = 0$
5. $\log_6 216 = 3$
6. $\log_{100} 10 = \frac{1}{2}$
7. $\log_2 \frac{1}{2} = -1$
8. $\log_{125} \frac{1}{5} = -\frac{1}{3}$
9. $\log_{25} 5 = \frac{1}{2}$
10. $A + B$
11. $2A$
12. $2A + B$
13. $A - B$
14. 3
15. $\frac{1}{3}$
16. 2
17. 3
18. -1
19. $-\frac{3}{2}$
20. -1
21. -3
22. 0
23. -2
24. $\frac{1}{3}$
25. 0
26. $-\frac{1}{2}$
27. 8
28. 4
29. 1
30. 1
31. 5

32. 1
33. 16
34. 8
35. $\frac{1}{2}$
36. $\frac{1}{16}$
37. 64
38. 1

39. 2
40. $\frac{1}{2}$
41. 2
42. 15
43. 10
44. $\frac{5}{3}$
45. 1

46. 81
47. $x = 3$
48. $x = 2$
49. $x = \frac{2}{3}$
50. $x = \frac{1}{6}$
51. $x = 4$
52. $x = 5$

53. $x = 16$
54. $x = \sqrt[3]{4}$
55. $x = \frac{1}{4}$
56. $x = 2$
57. $x = 125$

5.5 Logarithms With Special Bases

Logarithms were developed as an aid to computation and although they have been replaced in the performance of that function by calculators and computers, logarithms continue to be a useful concept in the solution of many problems and equations.

Common Logarithms

Since 10 is the base of our system of numeration, logarithms to the base 10, called *common logarithms*, are often used. When no base is written after the abbreviation log, the base is understood to be 10. That is,

$$\log 100 = 2 \quad \longleftrightarrow \quad \log_{10} 100 = 2$$

The table of common logarithms that appears in the Appendix gives the values of log n to four significant digits for $n = 1.00$ to $n = 9.99$. It is common practice to omit the decimal point when writing the values in the table. The use of the table, together with an understanding of scientific notation, allows you to find the common logarithm of any number that is expressed to three significant digits. A portion of the table is shown below.

Common Logarithms of Numbers

N	0	1	2	3	4	5	6	7	8	9
10	0000	0043	0086	0128	0170	0212	0253	0294	0334	0374
11	0414	0453	0492	0531	0569	0607	0645	0682	0719	0755
12	0792	0828	0864	0899	0934	0969	1004	1038	1072	1106
13	1139	1173	1206	1239	1271	1303	1335	1367	1399	1430
14	1461	1492	1523	1553	1584	1614	1644	1673	1703	1732
15	1761	1790	1818	1847	1875	1903	1931	1959	1987	2014
16	2041	2068	2095	2122	2148	2175	2201	2227	2253	2279
17	2304	2330	2355	2380	2405	2430	2455	2480	2504	2529
18	2553	2577	2601	2625	2648	2672	2695	2718	2742	2765
19	2788	2810	2833	2856	2878	2900	2923	2945	2967	2989
20	3010	3032	3054	3075	3096	3118	3139	3160	3181	3201
21	3222	3243	3263	3284	3304	3324	3345	3365	3385	3404
22	3424	3444	3464	3483	3502	3522	3541	3560	3579	3598
23	3617	3636	3655	3674	3692	3711	3729	3747	3766	3784
24	3802	3820	3838	3856	3874	3892	3909	3927	3945	3962

To use the table to find log 1.47, locate the four-digit number in the body of the table in the row labeled 14 and in the column headed 7. The entry is 1673. To understand where the decimal point belongs in these digits, note that $10^0 = 1$ or log 1 = 0, and that $10^1 = 10$ or log 10 = 1.

Since
$$1 < 1.47 < 10$$
$$\log 1 < \log 1.47 < \log 10$$
then
$$0 < \log 1.47 < 1$$

Since log 1.47 is a positive number less than 1, the four digits given in the table represent the decimal value 0.1673.

$$\log 1.47 = 0.1673 \text{ or } 10^{0.1673} = 1.47$$

Using this value, you can find the logarithms of other numbers of the form 1.47×10^n. Since $\log_x x^a = a$, $\log 10^{-3} = -3$.

$$\log 0.00147 = \log (1.47 \times 10^{-3}) = \log 1.47 + \log 10^{-3} = 0.1673 + (-3)$$
$$\log 0.0147 = \log (1.47 \times 10^{-2}) = \log 1.47 + \log 10^{-2} = 0.1673 + (-2)$$
$$\log 0.147 = \log (1.47 \times 10^{-1}) = \log 1.47 + \log 10^{-1} = 0.1673 + (-1)$$
$$\log 1.47 = \log (1.47 \times 10^{0}) = \log 1.47 + \log 10^{0} = 0.1673 + 0$$
$$\log 14.7 = \log (1.47 \times 10^{1}) = \log 1.47 + \log 10^{1} = 0.1673 + 1$$
$$\log 147 = \log (1.47 \times 10^{2}) = \log 1.47 + \log 10^{2} = 0.1673 + 2$$
$$\log 1,470 = \log (1.47 \times 10^{3}) = \log 1.47 + \log 10^{3} = 0.1673 + 3$$
$$\log 14,700 = \log (1.47 \times 10^{4}) = \log 1.47 + \log 10^{4} = 0.1673 + 4$$

The common logarithm consists of two parts: the integral part, which is called the *characteristic*, and the decimal part called the *mantissa*. The characteristic can be used to determine the general range of the number. For example, $\log 147 = 0.1673 + 2$. Since the characteristic is 2, the number is of the order 10^2, or a number in the hundreds.

Example 1 _____

If $\log 4.72 = 0.6739$, find $\log 47.2$.

Solution

$$\log 47.2 = \log (4.72 \times 10^1)$$
$$= \log 4.72 + \log 10^1$$
$$= 0.6739 + 1$$
$$= 1.6739$$

Example 2 _____

Solve for x: $10^x = 784$

Solution

$$10^x = 784 \text{ or } x = \log_{10} 784$$
$$\log 784 = \log (7.84 \times 10^2)$$
$$= \log 7.84 + \log 10^2$$
$$= 0.8943 + 2$$
$$= 2.8943$$

The value 0.8943 can be found from the entry in row 78 and column 4 of the table of common logarithms in the Appendix. The value of log 784 can also be found by using the log key of a scientific calculator.

Example 3

Find: log 0.00784

Solution

$$\log 0.00784 = \log (7.84 \times 10^{-3})$$
$$= \log 7.84 + \log 10^{-3}$$
$$= 0.8943 - 3$$

It is common practice to leave the value of the logarithm in this form when using the table. If the value of log 0.00784 is obtained from a calculator, the value will be approximately −2.1057, the sum −3 + 0.8943.

Finding the Number Whose Log Is Known

If $\log x$ is known, the value of x can be found by using the table. The mantissa of the logarithm is used to find the digits of the number x, and the characteristic is used to determine the location of the decimal point in those digits. For example, find x if $\log x = 3.7664$.

Let: $\log a = 0.7664$

Then: $\log x = 3.7664$
$$= 0.7664 + 3$$
$$= \log a + \log 10^3$$
$$= \log (a \times 10^3)$$

Therefore: $x = a \times 10^3$

Since $\log a = 0.7664$, the value of a to three significant digits can be found by locating 7664 (or the digits closest to 7664) in the body of the table of common logarithms.

Common Logarithms of Numbers

N	0	1	2	3	4	5	6	7	8	9
55	7404	7412	7419	7427	7435	7443	7451	7459	7466	7474
56	7482	7490	7497	7505	7513	7520	7528	7536	7543	7551
57	7559	7566	7574	7582	7589	7597	7604	7612	7619	7627
58	7634	7642	7649	7657	7664	7672	7679	7686	7694	7701
59	7709	7716	7723	7731	7738	7745	7752	7760	7767	7774
60	7782	7789	7796	7803	7810	7818	7825	7832	7839	7846
61	7853	7860	7868	7875	7882	7889	7896	7903	7910	7917
62	7924	7931	7938	7945	7952	7959	7966	7973	7980	7987
63	7993	8000	8007	8014	8021	8028	8035	8041	8048	8055
64	8062	8069	8075	8082	8089	8096	8102	8109	8116	8122

The first two digits of a are the label of the row in which 7664 lies and the third digit is the heading on the column in which 7664 lies. Therefore, a is a number whose significant digits are 584. Since the mantissa is the logarithm of a number between 1 and 10, $a = 5.84$. Therefore,

$$x = a \times 10^3$$
$$= 5.84 \times 10^3$$
$$= 5,840$$

The value of x if $\log x = 3.7664$ can also be found by using the INV key followed by the LOG key of a scientific calculator. The value of x can be obtained either by using the sequence

$$3.7664 \quad \boxed{\text{INV}} \quad \boxed{\text{LOG}}$$

$$\text{or} \quad \boxed{10^x} \quad 3.7664 \quad \boxed{=}$$

or by some comparable sequence depending on the particular calculator. A calculator will give the value to a larger number of significant digits. For example, a calculator may return the value 5839.827246. To understand the second sequence of keys above, note that if $\log x = 3.7664$, then $x = 10^{3.7664}$.

Example 4

Use the table of common logarithms to find the value of N if $\log N = 0.9144 - 2$.

Solution

Let: $\quad \log a = 0.9144$

Then: $\quad \log N = 0.9144 - 2$

$$= \log a + \log 10^{-2}$$

$$= \log (a \times 10^{-2})$$

Therefore: $\quad N = a \times 10^{-2}$

From the table, the closest mantissa to 0.9144 is 0.9143, which is the log of 8.21. Thus, $a = 8.21$.

$$N = 8.21 \times 10^{-2}$$

$$= 0.0821$$

A calculator would return the value 0.08211075.

Natural Logarithms

The exponential function $y = e^x$ and its inverse function $y = \log_e x$ are used in the mathematical representation of many relationships that occur in nature. Logarithms to the base e are called *natural logarithms*. A frequently-used abbreviation for the natural log is *ln*. That is,

$$\log_e a = \ln a$$

Recall the formula $A_t = A_0 e^{rt}$, which is used to model a situation in which an initial amount, A_0, is increasing at a rate that is proportional to the amount present at any time, A_t.

Example 5

A sum of money is invested at a yearly rate of 8% interest compounded continuously. Find the length of time necessary for the value of the investment to double.

Solution

If the original investment A_0 is to double, you want to find the time t at which A_t is $2A_0$.

$$A_t = A_0 e^{rt}$$

$$2A_0 = A_0 e^{0.08t} \qquad \text{Substitute } 2A_0 \text{ for } A_t \text{ and } 0.08 \text{ for } r.$$

$$2 = e^{0.08t} \qquad \text{Divide each side by } A_0.$$

$$0.08t = \ln 2 \qquad \text{Write the equation in log form.}$$

$$t = \frac{\ln 2}{0.08} \qquad \text{Solve for } t.$$

The value of t can be found by locating ln 2 in a table of natural logarithms and then dividing that value by 0.08, or the value of $\frac{\ln 2}{0.08}$ can be obtained by using a calculator.

$$t = \frac{0.6931}{0.08}$$

$$t \approx 8.66$$

Answer: It will take approximately 8 years 8 months for the investment to double.

Change of Base

Recall that some exponential equations can be solved because each power can be written in terms of the same base (for example, $3^x = 9$ is equivalent to $3^x = 3^2$, and $x = 2$). However, if the powers cannot be written to the same base, this method cannot be used. Since common logarithms can be used to express any number to base 10 and natural logarithms can be used to express any number to base e, any exponential equation can be solved with the aid of these logarithms.

For example, the solution of $10^x = 3$ is $x = \log 3$. From a calculator or a table of common logarithms, $\log 3 = 0.4471$ to four decimal places.

The solution to the following problem results in an equation in which the variable is an exponent. However, since the base is neither 10 nor e, a change of base is required.

Example 6

A sheet of cardboard is 1 mm thick. The sheet is cut in half and the pieces stacked one on the other. Then the stack is cut in half and the pieces stacked again. After how many cuts will the stack be more than 3 cm thick?

Solution

Each time the stack is cut in half and one half is stacked on the other, the thickness of the stack is doubled. Starting with the 1-mm thickness of one sheet, a geometric sequence gives the thickness after each cut.

1, 2, 4, 8, ... is equivalent to $2^0, 2^1, 2^2, 2^3, ... 2^n, ...$

where n is the number of cuts and 2^n is the thickness of the stack after n cuts. Since the thickness of the cardboard is given in millimeters, 2^n is also in millimeters. Since the final stack is to be 3 cm, or 30 mm, thick, the equation to be solved is $2^n = 30$. The solution to this exponential equation is $n = \log_2 30$. However, since no table of logarithms to base 2 is readily available, a change of base is necessary in order to approximate the value of n.

$$2^n = 30$$
$$\log 2^n = \log 30 \qquad \text{Write the common log of each side of}$$
$$n \log 2 = \log 30 \qquad \text{the equation and solve for } n.$$
$$n = \frac{\log 30}{\log 2}$$
$$n \approx \frac{1.4771}{0.3010} \approx 4.9$$

Note that $\dfrac{\log 30}{\log 2} \neq \log\left(\dfrac{30}{2}\right)$.

Answer: The stack would be more than 3 cm thick after the 5th cut.

A general formula for changing $\log_b x$ to a logarithm to the base a can be written.

Let:
$$y = \log_b x$$
$$b^y = x$$
$$\log_a b^y = \log_a x \qquad \text{Write the log to the base } a \text{ of}$$
$$y \log_a b = \log_a x \qquad \text{each side of the equation.}$$
$$y = \frac{\log_a x}{\log_a b} \qquad \text{Solve for } y.$$
$$\log_b x = \frac{\log_a x}{\log_a b} \qquad \text{Restore } \log_b x \text{ for } y.$$

For example, to find $\log_5 8$, you would want to change from base 5 to base 10 or base e, since you can find approximations for logarithms to the base 10 or the base e by using tables or a calculator. That is, in the change of base formula, replace a with 10 or e.

$$\log_b x = \frac{\log_a x}{\log_a b}$$

$$\log_5 8 = \frac{\log_{10} 8}{\log_{10} 5} \quad \text{or} \quad \log_5 8 = \frac{\ln 8}{\ln 5}$$

$$\log_5 8 \approx \frac{0.9031}{0.6990} \quad \text{or} \quad \log_5 8 \approx \frac{2.0794}{1.6094}$$

$$\log_5 8 \approx 1.292$$

To change from the common logarithm of a number to the natural logarithm, or vice versa, use these relationships:

$$\ln\ x = \frac{\log x}{\log e} \qquad \log x = \frac{\ln x}{\ln 10}$$

Example 7 _____

Use common logarithms to find the value of $\ln 57$.

Solution

$$\ln x = \frac{\log x}{\log e}$$

$$\ln 57 = \frac{\log 57}{\log e}$$

$$\approx \frac{\log 57}{\log 2.71828}$$

Answer: $\ln 57 \approx 4.04305$

Example 8 _____

Evaluate $\frac{\ln 625}{\ln 5}$ without using a table or a calculator.

Solution

$$\frac{\log_a x}{\log_a b} = \log_b x$$

$$\frac{\ln 625}{\ln 5} = \frac{\log_e 625}{\log_e 5} = \log_5 625$$

$$= \log_5 5^4 = 4$$

Answer: $\frac{\ln 625}{\ln 5} = 4$

Example 9 _____

Each time Bert's ball bounces, it rises to a height that is 0.7 of its previous height. If the ball is dropped from a height of 10 feet, after how many bounces will its height be less than 1 foot?

Solution

The sequence of heights after the first bounce is

$$10(0.7)^1,\ 10(0.7)^2,\ \ldots\ 10(0.7)^n,\ \ldots$$

where n is the number of bounces.

Let:
$$10(0.7)^n = 1$$

$$(0.7)^n = 0.1$$

$$n = \log_{0.7} 0.1$$

$$n = \frac{\log 0.1}{\log 0.7}$$

$$n \approx 6.45$$

Answer: The ball rises to a height of less than 1 foot after the 7th bounce.

Exercises 5.5

In **1 – 8**, let $\log A = x$ and $\log B = y$. Write the given logarithm in terms of x and y.

1. $\log AB$

2. $\log \left(\dfrac{A}{B} \right)$

3. $\log A^4$

4. $\log AB^2$

5. $\log (AB)^2$

6. $\log \sqrt{\dfrac{A}{B}}$

7. $\log 10A$

8. $\log \dfrac{A^3}{\sqrt{B}}$

9. If $\log 3.86 = 0.5866$, find:
  a. $\log 3860$
  b. $\log 0.386$

10. If $\log 8.22 = 0.9149$, find:
  a. x when $\log x = 1.9149$
  b. y when $\log y = 0.9149 - 3$

In **11 – 30**, evaluate the given expression.

11. $e^{\ln 1}$

12. $e^{2 \ln 1}$

13. $e^{\ln 3}$

14. $e^{\ln 3 + \ln 3}$

15. $e^{2 \ln 3}$

16. $e^{\frac{1}{2} \ln 4}$

17. $e^{-\ln 6}$

18. $e^{\frac{1}{2} \ln 9}$

19. $e^{-2 \ln 5}$

20. $e^{\log 1}$

21. $\ln e$

22. $\ln e^2$

23. $2 \ln e$

24. $\ln e^2 - \ln e^3$

25. $\ln 2e + \ln \frac{1}{2}e$

26. $\ln \sqrt{e}$

27. $\log (\ln e)$

28. $\ln (\log 10)$

29. $\ln (e \log 10)$

30. $\log (10 \ln e)$

In **31 – 34**, solve for x to the nearest thousandth.

31. $\log_3 x = \frac{1}{4}$

32. $e^x = 6$

33. $4 \log x = 1.5$

34. $8^x = 20$

35. If $f(x) = 2^{x-1}$, find $f^{-1}(x)$.

36. If $f(x) = 3^{2x}$, find $f^{-1}(x)$.

37. The decay constant for Uranium-227 is –0.533 when time is in minutes. Find, to the nearest tenth of a minute, the time needed for the Uranium-227 in a sample to decay from 2 g to 0.5 g.

38. Find, to the nearest tenth of a percent, the annual rate at which money must be invested in order to obtain an effective rate (or annual yield) of 9% when interest is compounded continuously.

39. A small town is decreasing in population because the number of deaths and the number of people who move out of the town is greater than the number of births and the number of people who move into the town. If the rate of decline is –0.7% and the current population is 4,500, in how many years will the population be less than 3,000?

40. When Patty was in kindergarten, her mother gave her 10¢ to spend each week. When Patty was in first grade, her mother tripled the amount, giving her 30¢ each week. If this pattern continued and Patty's allowance was tripled each year, in what grade would her allowance be more than $50 a week?

Chapter 5   *Summary and Review*

Operations Using Powers With Like Bases

- For all real numbers, $x, y, a,$ and b:

$$x^a \cdot x^b = x^{a+b}$$

$$x^a \div x^b = x^{a-b} \quad \text{where} \quad x \neq 0$$

$$(x^a)^b = x^{ab}$$

- If $x \neq 0$ and $y \neq 0$:

$$x^0 = 1$$

$$x^{-a} = \frac{1}{x^a} \text{ and } x^a = \frac{1}{x^{-a}}$$

$$\left(\frac{x}{y}\right)^{-a} = \left(\frac{y}{x}\right)^a$$

- If $x > 0$ and n is a positive integer: $x^{\frac{1}{n}} = \sqrt[n]{x}$
- An equation of the form $x^n = a$, where n and a are constants, can be solved for x by raising each side of the equation to the $\frac{1}{n}$ power.

Exponential Functions and Equations

- An exponential function is defined by the equation $y = b^x$, where $b > 0$ and $b \neq 1$. The domain is the set of real numbers and the codomain is the set of positive real numbers. The function is a 1–1 correspondence.
- Under a reflection over the y-axis, the image of $y = b^x$ is:

$$y = b^{-x} \text{ or } y = \left(\frac{1}{b}\right)^x$$

- An equation in which the variable is an exponent is called an exponential equation.
- An exponential equation can be solved if each side of the equation can be expressed as a power of the same base.

The Number e

- $e = 1 + \dfrac{1}{1!} + \dfrac{1}{2!} + \dfrac{1}{3!} + \cdots + \dfrac{1}{(n-1)!} + \cdots$
- $e \approx 2.71828$
- As n approaches infinity, the value of $\left(1 + \dfrac{1}{n}\right)^n$ approaches e.

- The formula $A_t = A_0 e^{rt}$ is used in problems such as those involving investment, size of population, or growth and decay. A_t represents the amount present after t units of time, A_0 the initial amount at time $t = 0$, and r the rate of change.

  Values of e^{rt} can be read from a table or obtained from a calculator.

Logarithmic Functions and Equations

- The reflection of the graph of an exponential function $f(x) = b^x$ over the line $y = x$ is the graph of the inverse function $f^{-1}(x) = \log_b x$, whose domain is the set of positive real numbers and whose codomain is the set of real numbers.

- $a^y = x \leftrightarrow y = \log_a x$

$$\log_x AB = \log_x A + \log_x B$$
$$\log_x \left(\frac{A}{B} \right) = \log_x A - \log_x B$$
$$\log_x A^c = c \log_x A$$

- Logarithms to the base 10 are called common logarithms. When no base is written for a logarithm, the base is understood to be 10.

- $a^{\log_a b} = b$ and $\log_a a^b = b$

- A common logarithm consists of two parts: the integral part called the characteristic and the decimal part called the mantissa. If $N = a \times 10^n$, then $\log N = \log a + n$. When $1 < a < 10$, then $0 < \log a < 1$ and $\log a$ is the mantissa of $\log N$. The value of $\log a$ can be found in a table of common logarithms and the value of $\log N$ can then be found by adding n to $\log a$. The value of $\log N$ can also be found by using a scientific calculator.

- If $\log N = \log a + n$ when $0 < \log a < 1$ and n is an integer, then $N = a \times 10^n$.

- Logarithms to the base e are called natural logarithms.

$$\log_e a = \ln a$$

- $\ln e^n = n$ and $e^{\ln n} = n$

- A logarithm to the base a can be used to find a logarithm to the base b by using the relationship:

$$\log_b x = \frac{\log_a x}{\log_a b}$$

- $\ln x = \dfrac{\log x}{\log e}$ and $\log x = \dfrac{\ln x}{\ln 10}$

1. 4

2. 1

3. 27

4. $\frac{1}{3}$

5. $\frac{1}{75}$

6. 1

7. 3

8. –1

9. 1

10. $\frac{1}{a^3}$

11. $\frac{y^3}{x}$

12. $\frac{b^7}{3}$

13. $1 + r^3$

14. $\frac{1}{x}$

15. $\frac{1}{3}$

16. –4

17. **a.** 9

　 b. 4

　 c. 4

18. $g(x) = \left(\frac{5}{3}\right)^{-x}$

　 or $g(x) = \left(\frac{3}{5}\right)^{x}$

19. $g(x) = \log_3 x$

20. $\frac{3}{2}$

21. 3

22. 1

23. **a.** $3a + b$

　 b. $\frac{1}{2}(a + b)$

　 c. $2(a - b)$

　 d. $a - 2$

Chapter 5　　Review Exercises

In **1 – 9**, find the value of the expression.

1. $4a^0$　　　**2.** $(4a)^0$　　　**3.** $9^{\frac{3}{2}}$　　　**4.** $2 \cdot 6^{-1}$　　　**5.** $\frac{5^{-2}}{3}$

6. $\frac{1}{2}(\log_4 80 - \log_4 5)$　　　**7.** $\log_3 9 + \log_2 8 - \log 100$

8. $\log_5 45 - 2\log_5 15$　　　**9.** $\log_2 10 + (\log_2 7 - \log_2 35)$

In **10 – 14**, write the expression in simplest form.

10. $(a \cdot a^{-2})^3$　　**11.** $\frac{x^3 y^4}{x^4 y}$　　**12.** $\frac{3ab^3}{9ab^{-4}}$　　**13.** $r^2(r^{-2} + r)$　　**14.** $\frac{x^{\frac{1}{2}}}{x^{\frac{3}{2}}}$

15. If $f(x) = 4x^{-1}$, find $f(12)$.　　　**16.** If $g(x) = \log_3 x$, find $g\left(\frac{1}{81}\right)$.

17. If $g(x) = \log_3 x$ and $h(x) = 3^x$, find:

　 a. $h \circ g(9)$　　**b.** $h \circ g(4)$　　**c.** $g \circ h(4)$

18. The graphs of $f(x) = \left(\frac{5}{3}\right)^x$ and $g(x)$, the reflection of $f(x)$ over the y-axis, are shown below. Write an equation for $g(x)$ in two different ways.

19. The graphs of $f(x) = 3^x$ and $g(x)$, the reflection of $f(x)$ over the line $y = x$, are shown below. Write an equation for $g(x)$.

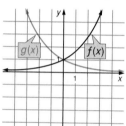

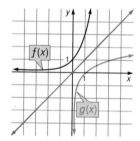

20. If, for all x, $\left(\frac{2}{3}\right)^{-x} = b^x$, what is the value of b?

21. If, for all x, $\log_b 3^x = x$, what is the value of b?

22. If, for all $x > 0$ and $y > 0$, $\log_x b = \log_y b$, what is the value of b?

23. If $\log A = a$ and $\log B = b$, write, in terms of a and b:

　 a. $\log A^3 B$　　**b.** $\log \sqrt{AB}$　　**c.** $\log \left(\frac{A}{B}\right)^2$　　**d.** $\log \left(\frac{A}{100}\right)$

In **24 – 33**, solve for x.

24. $9^x = 3^{x+3}$　　　**25.** $4^{2x} = 8^{x-1}$　　　**26.** $5 \cdot 25^{x-1} = 125^x$

27. $5 \cdot 10^x = 12$　　　**28.** $4^x = 13$　　　**29.** $3e^x - 1 = 20$

30. $\log_9 12 - \log_9 x = \log_9 3$　　　**31.** $\log_9 12 - \log_9 4 = x$

32. $2\log x - \log (x + 6) = 0$　　　**33.** $\log_3 (x + 4) - \log_3 x = \log_3 5$

24. $x = 3$　　　　　　**29.** $x = \ln 7$

25. $x = -3$　　　　　**30.** $x = 4$

26. $x = -1$　　　　　**31.** $x = \frac{1}{2}$

27. $x = \log 2.4$　　　**32.** $x = 3$

28. $x = \log_4 13$　　　**33.** $x = 1$

In **34 – 41,** evaluate the given expression.

34. $e^{2\ln 5}$ **35.** $2\ln e^5$ **36.** $10^{\log 7}$ **37.** $2\log 10^3$

38. $e^{-\ln 3}$ **39.** $\frac{1}{2}\ln e^6$ **40.** $10^{\frac{1}{3}\log 8}$ **41.** $10^{-\log 10}$

42. A sum of money is invested at 7.5% compounded continuously. At what interest rate must the money be invested in order to receive the same yearly return if interest is only compounded once a year?

43. What rate of interest compounded continuously provides a total yearly return of 8.25%?

44. Carbon-14 has a decay constant of –0.000124 when time is in years. Find, to the nearest hundred years, the age of a sample of wood that was estimated to have originally contained 12 grams of Carbon-14 and now contains 8.5 grams.

45. Research biologists estimated that when DDT was introduced into an animal's system and stored in its fatty tissue, it took 8 years for one-half of the amount to be eliminated.

   a. Find the rate at which DDT was eliminated from the animal's system.

   b. A rabbit ate some vegetables from a garden shortly after it had been sprayed with DDT. If the vegetables eaten contained 4 grams of DDT, how much of the DDT would still be present in the rabbit's body after 1 year? (Use the rate constant found in part **a** and express the answer to the nearest hundredth of a gram.)

46. The population of a town of 6,780 people is growing at the rate of 1.5% a year. How long will it take for the population to reach 8,000?

47. **a.** On the same set of axes, sketch the following functions in $R \times R$:

   $$y = e^{|x|}; \quad y = e^{-|x|}; \quad y = e^{|x|} - e^{-|x|}$$

   b. What is the domain of each function?

   c. What is the range of each function?

   d. Find, to the nearest hundredth, the value of x such that

   $$e^{-|x|} = e^{|x|} - e^{-|x|}$$

47. a.

I $y = e^{|x|}$ **II** $y = e^{-|x|}$
III $y = e^{|x|} - e^{-|x|}$

b. Set of all real numbers for each

c. Reals greater than or equal to 1 for $y = e^{|x|}$
$0 < y \le 1$ for $y = e^{-|x|}$
Nonnegative real numbers $(y \ge 0)$ for $y = e^{|x|} - e^{-|x|}$

d. ± 0.35

☑ Exercises for Challenge

1. Simplify: $\log_2\left(\log_3 \sqrt{\sqrt[4]{3}}\right)$

2. Solve for x: $\log\left(\frac{3}{2} + x\right) = \log \frac{3}{2} + \log x$

3. Determine the solution of the system:
$$8^x = 10y$$
$$2^x = 5y$$

4. Let $f(x) = 2^x - 3$. Find:
$$f^{-1}(4,194,301)$$

5. Solve for the positive integer n:
$$5^{2 + 4 + 6 + \cdots + 2n} = (0.04)^{-15}$$

6. Simplify: $\dfrac{\left(\frac{1}{2}\right)^{-3} 2^4}{2^{-6}}$

    (A) $\frac{1}{64}$ (B) 32 (C) 128

    (D) $6,144$ (E) $8,192$

7. Solve for x: $\log_x 0.125 = -2$

    (A) $\frac{2}{\sqrt{5}}$ (B) 2 (C) $2\sqrt{2}$

    (D) 5 (E) $5\sqrt{2}$

8. The graph shown could be the graph of which of the following functions?

    (A) $y = 2^x$ (B) $y = 2^{-x}$ (C) $y = 2^{|x|}$

    (D) $y = 2^{-|x|}$ (E) $y = -2^{-|x|}$

9. If m is very large, which of the following is the best approximation of $m \ln\left(1 + \frac{1}{m}\right)$?

    (A) 0 (B) $\log_{10} e$ (C) $\ln 2$

    (D) 1 (E) $\ln 10$

10. A function is of the form $f(x) = 2a^{-x}$, $a > 2$. Which of the following is equal to a?

    (A) $\frac{f(0)}{2}$ (B) $2f(1)$ (C) $\frac{2}{f(1)}$

    (D) $2f(-1)$ (E) $\frac{2}{f(-1)}$

11. $2^x \cdot 2^y + 2^x \cdot 2^y$ is equal to

    (A) 2^{2xy} (B) 4^{xy}

    (C) 8^{xy} (D) 2^{x+y+1}

    (E) $2^{2(x+y)}$

12. Let $\{a_n\}$ be a geometric progression with 6 as the first term and $\frac{1}{5}$ as the common ratio. Find the sum of the first 20 terms of the sequence $\{\log a_n\}$.

    (A) -124.23 (B) -117.24

    (C) 1.58 (D) 148.37

    (E) 255.35

Answers
Exercises for
Challenge

13. E
14. B
15. E
16. A
17. C
18. D
19. B
20. C
21. C

☑ **Exercises for Challenge** *(continued)*

13. Which statement is incorrect in the following "proof" that 3 < 2?

(A) $\frac{1}{8} < \frac{1}{4}$ → $\frac{1}{2^3} < \frac{1}{2^2}$

(B) $\frac{1}{2^3} < \frac{1}{2^2}$ → $\left(\frac{1}{2}\right)^3 < \left(\frac{1}{2}\right)^2$

(C) $\left(\frac{1}{2}\right)^3 < \left(\frac{1}{2}\right)^2$ → $\log_{10}\left(\frac{1}{2}\right)^3 < \log_{10}\left(\frac{1}{2}\right)^2$

(D) $\log_{10}\left(\frac{1}{2}\right)^3 < \log_{10}\left(\frac{1}{2}\right)^2$ → $3\log_{10}\frac{1}{2} < 2\log_{10}\frac{1}{2}$

(E) $3\log_{10}\frac{1}{2} < 2\log_{10}\frac{1}{2}$ → $3 < 2$

14. $\dfrac{\log_3 36}{\log_3 6}$ is equal to

(A) $\frac{1}{2}$ (B) 2

(C) $\log_3 2$ (D) $\log_3 6$

(E) $\log_3 30$

15. $(2^x + 2^{-x})^2$ is equal to

(A) 2
(B) $2^{x^2} + 2^{-x^2}$
(C) $2^{x^2} + 2 + 2^{-x^2}$
(D) $2^{2x} + 2^{-2x}$
(E) $2^{2x} + 2 + 2^{-2x}$

Questions 16 – 21 each consist of two quantities, one in Column A and one in Column B. You are to compare the two quantities and choose:

A if the quantity in Column A is greater;
B if the quantity in Column B is greater;
C if the two quantities are equal;
D if the relationship cannot be determined from the information given.

1. In certain questions, information concerning one or both of the quantities to be compared is centered above the two columns.

2. In a given question, a symbol that appears in both columns represents the same thing in Column A as it does in Column B.

3. x, n, and k, etc. stand for real numbers.

Column A	Column B

16. $a < b$

$\left(\frac{4}{5}\right)^a$ | $\left(\frac{4}{5}\right)^b$

17. $a = 3^{10},\ b = 6^5$

$\sqrt{a}$ | $\dfrac{b}{32}$

18. $a > 0$

$\log_a 2$ | $\log_a 3$

19. 📇

Let $\log_b 2 = 0.4419$ and $\log_b 3 = 0.7004$

The product of $\log_b 3$, $\log_b 6$, and $\log_b 9$ | The sum of $\log_b 3$, $\log_b 6$, and $\log_b 9$

20. $x^2 + y^2 = 7xy$ where $x, y > 0$

$\log\left(\dfrac{x+y}{3}\right)$ | $\dfrac{\log x + \log y}{2}$

21. $a > b$

$\dfrac{\log_a 7}{\log_a 3}$ | $\dfrac{\log_b 7}{\log_b 3}$

Teacher's Chapter 6

Vectors

Chapter Contents

Various physical phenomena, such as force, have both magnitude and direction, and can best be described by vectors. Since many real-world problems are modeled by lines and planes in space, vectors provide a method for deriving the equations of these lines and planes, and, therefore, play an important role in mathematics and in physical applications.

The purpose of this chapter is to familiarize students with the concept of vectors, the operations that apply to them, and the geometric interpretation of these operations.

A knowledge of trigonometry is not assumed in this chapter.

6.1 Operations With Vectors

After defining and naming a vector, an intuitive approach is used to find the resultant of two or more vectors geometrically. Vectors are moved to different positions in the plane, and the basic vector operations are discussed geometrically. In the exercise section, students verify that vector addition is commutative and associative, and that scalar multiplication distributes over vector addition. Exercise 20 requires knowledge of the 30°-60°-90° triangle.

Some students may have trouble with the idea that although a vector is a directed line segment, it need not be located in any particular place. Draw a number of equal vectors in different places on the chalkboard and emphasize that the direction and magnitude of a vector do not indicate a particular location.

Although force problems are not emphasized in this section, one is used to motivate thinking about the physical use of vectors. Vector applications requiring trigonometry will be introduced in Chapters 8 and 9.

6.2 Using Ordered Pairs to Represent Vectors

Ordered-pair notation is economical, a goal for all mathematical notation, and lends itself naturally to doing the various vector operations that are introduced in the remainder of the chapter.

Although other notations, such as the linear combination of unit vectors, are also used, the ordered-pair or ordered-triple notation is favored. In the beginning, some students may have difficulty distinguishing between an ordered pair representing a point in the plane and an ordered pair that represents a vector.

Point out that one way to understand why an ordered pair that represents a vector can be found by subtracting the coordinates of the endpoints is to use geometric subtraction. If two vectors are placed tail to tail, the difference vector $\vec{x}$ is drawn from the head of the second vector to the head of the first vector.

$$\vec{x} = \vec{v} - \vec{w} = (a, b) - (c, d) = (a - c, b - d)$$

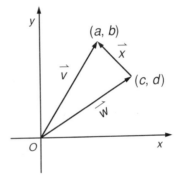

Vector addition can also be used to reinforce this concept. Since $\vec{w} + \vec{x} = \vec{v}$, then $\vec{x} = \vec{v} - \vec{w}$.

Continue drawing equal vectors in various parts of the plane to help students understand that the ordered-pair notation for a vector indicates magnitude and direction but not position. It is often convenient to draw vectors with their tails at the origin, as in the diagram above. This is sometimes referred to as a position vector, that is, if (a_1, a_2) is a vector and (a_1, a_2) are the coordinates of point A, $\overrightarrow{OA}$ is the position vector for (a_1, a_2).

Make it a point to always place vector symbols over any letters that are used to represent vectors in order to encourage students to adopt this same style. This will reduce confusion between points and vectors, particularly later in the chapter when the introduction of cross product and dot product necessitates the use of many different symbols and letters.

6.3 Vectors in 3-Space

Sufficient time should be allowed to permit students to practice drawing vectors in 3-space. Students are likely to develop more interest in this work if they can draw reasonably accurate diagrams. Encourage students to use graph paper, and to draw dashed lines parallel to the axes in order to establish the proper perspective on their 2-dimensional sheet of paper.

One of the main advantages in studying vectors is that concepts, such as perpendicularity, transfer easily from 2 dimensions to 3 dimensions. For example, the dot product is nearly as easy to use to test for perpendicularity in 3 dimensions as it is in 2 dimensions, though testing for negative reciprocal slopes is meaningless in 3-dimensional space. Students are likely to gain an appreciation for how well vector theory transfers from 2 dimensions to 3 dimensions as they complete the chapter; however, these parallels should be emphasized as they are encountered throughout the chapter. This emphasis is likely to serve as a memory aid as well as improving concept formation. In this section, the similarity of the 2- and 3-dimensional distance formulas (and their proofs), dividing a line segment proportionally, and finding a vector given its endpoints are examples of this transferability.

Activity (6.3)

1. Place a small object on a piece of graph paper so that it coincides with the origin of a set of coordinate axes that have been drawn. If the object is moved with a force of 2 ounces at an angle of 30° with the positive x-axis, draw a vector on your graph that describes the path of the object. (Let one unit on the axes represent 1 ounce of force.)

2. If the object is moved by two simultaneous forces, a horizontal force and a vertical force of 1 ounce, what would the horizontal force have to be so that the object moves in the same path as in step 1?

3. Obtain a cardboard carton whose length, width, and depth are each about the same. Remove the top and two adjacent sides of the carton so that it can serve as a model of 3-dimensional space, with the sides of the box being the coordinate planes and the bottom left corner serving as the origin. Mark a scale for the x-, y-, and z-axes using 1 inch for each unit.

   Place a small object, such as a button, at the corner of the box that represents the origin.

   Suppose the object is moved so that it follows the direction of the vector $\vec{v} = (2, 3, 4)$ over a distance equal to $|\vec{v}|$. How far from the origin will it be?

4. How can the object be moved from the origin to the position described in step 3 if you are only permitted to move it parallel to the coordinate axes?

Answers

1.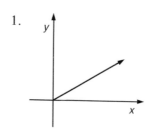

2. $\sqrt{3}$

3. $\sqrt{29}$

4. 2 units in the x direction, 3 units in the y direction, and 4 units in the z direction

6.4 The Dot Product; Planes in Space

The following connection can be made between the fact that the value of the dot product of perpendicular vectors is 0 and the fact that the value of the product of the slopes of two perpendicular lines is –1.

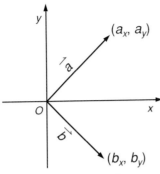

slope of $\vec{a} = \dfrac{a_y}{a_x}$; slope of $\vec{b} = \dfrac{b_y}{b_x}$

If $\vec{a}$ and $\vec{b}$ are perpendicular, then

$$\left(\frac{a_y}{a_x}\right)\left(\frac{b_y}{b_x}\right) = -1$$

$$a_y b_y = -a_x b_x$$

or $\;a_y b_y + a_x b_x = 0$

The converse can be proved by reversing the steps.

The dot product is introduced as a binary operation for vectors. In the exercises, students are asked to prove that dot product is a commutative operation and that it distributes over vector addition. Other interesting properties that can be shown include associativity and the fact that for a given vector, $\vec{v} \cdot \vec{v} = |\vec{v}|^2$.

Emphasize the fact that since the dot product of two vectors is a scalar, dot product is *not* a closed operation. This will help students distinguish between dot product and cross product, which is introduced in Section 6.5.

The derivation for the equation of a plane is closely tied to the dot product. Again, parallels between the equation $ax + by + c = 0$ in the 2-dimensional plane and the equation $Ax + By + Cz + D = 0$ in 3-dimensional space can be drawn.

Just as two points determine a unique line, a point and a normal vector determine a unique plane. To convince students, a simple model can be shown. Use a pencil point as the point , the body of the pencil as the normal vector, and a piece of cardboard as the plane. First show

that a line in space can belong to each of an infinite set of planes just as a point in the 2-dimensional plane can belong to an infinite set of lines. Then conclude that involving the normal vector to the plane establishes uniqueness.

The synonyms *orthogonal* and *normal* should be used interchangeably in class, as should *dot product* and *inner product* (also called **scalar product**), since students may meet these words in other contexts.

To reinforce the similarities between applications of vectors in 2 and 3 dimensions, point out that just as the vector (a, b) is normal to $ax + by + c = 0$ in the 2-dimensional plane, the vector (A, B, C) is normal to $Ax + By + Cz + D = 0$ in 3-dimensional space.

6.5 The Cross Product

The cross product is introduced as a vector that meets the condition of being perpendicular to the plane formed by two given vectors in space. Without dwelling on determinants, the basic symbolism is used as a convenient notation. The algorithm used for computing a cross product is a less cumbersome alternative to the expansion of a 3×3 determinant. If a computational error is made, it is easier to trace, and students need not worry about remembering to alternate signs when expanding the determinant by minors. You may wish, however, to use the 3×3 determinant, particularly in classes that have already had practice with determinants. If $\vec{a} = (a_x, a_y, a_z)$ and $\vec{b} = (b_x, b_y, b_z)$, then:

$$\vec{a} \times \vec{b} = \begin{vmatrix} \vec{i} & \vec{j} & \vec{k} \\ a_x & a_y & a_z \\ b_x & b_y & b_z \end{vmatrix}$$

Emphasize that a cross product produces a vector and, therefore, cross product is a closed binary operation. In concluding the section, students are shown that cross product is not commutative, and that if two vectors are parallel, then their cross product is $\vec{0}$. You may also choose to verify additional properties of vectors:

Cross product distributes over vector addition
$$\vec{a} \times (\vec{b} + \vec{c}) = (\vec{a} \times \vec{b}) + (\vec{a} \times \vec{c})$$

Squaring property
$$\vec{a} \times \vec{a} = \vec{0}$$

Triple cross product property
$$(\vec{a} \times \vec{b}) \times \vec{c} = (\vec{a} \cdot \vec{c})\vec{b} - (\vec{b} \cdot \vec{c})\vec{a}$$
$$\vec{a} \times (\vec{b} \times \vec{c}) = (\vec{a} \cdot \vec{c})\vec{b} - (\vec{a} \cdot \vec{b})\vec{c}$$

Also, remind students that while $(\vec{a} \times \vec{b}) \cdot \vec{c}$ has meaning, $(\vec{a} \cdot \vec{b}) \times \vec{c}$ and $\vec{a} \times (\vec{b} \cdot \vec{c})$ do not.

6.6 Lines in Space

The three parametric equations used to describe a line in space should be presented in a number of different ways so that students can make connections with topics that they have studied previously. For example, the equations

$$x = 4t + 4$$
$$y = 5t - 2$$
$$z = t + 1, \quad t \in R$$

can be interpreted as the set of vectors $(4t + 4, 5t - 2, t + 1)$, where each value of t produces a different vector of the set. Alternatively, this line may be presented as an infinite dilation of the position vector $(4, 5, 1)$ translated 4 units toward you, 2 units to the left, and 1 unit up in space. The parameter t "links" the three equations.

Students may be interested to know that planes can also be described by vector equations. If $\vec{A}$ is a vector whose tail is at the origin and whose head is point A on the plane P, and $\vec{b}$ and $\vec{c}$ are unit vectors in the plane, then

$$P = \vec{OA} + s\vec{b} + t\vec{c}$$

where s and t are parameters.

Another important connection to be made is the analogy between 2- and 3-space in the cases when a variable is missing from the equation of a line or a plane. The general equation of a line, $ax + by + c = 0$, includes the variables associated with 2-space: x and y. When the coefficient of one of these variables is 0, that equation describes a line that is parallel to an axis. For example, the line $x = 2$ is parallel to the y-axis. Note that the axis to which the line is parallel is named by the variable that is missing from the equation of the line.

Similarly, when the coefficient is 0 for one of the 3 variables in the general equation for a plane, $Ax + By + Cz + D = 0$, that equation describes a plane that is parallel to an axis. For example, the equation $x + y = 5$ describes, in 3-space, a plane that is parallel to the z-axis. The axis to which the plane is parallel is named by the variable that is missing from the equation of the plane.

Emphasize the importance of knowing the number of dimensions involved in a given situation so that an equation can be correctly interpreted. For example, the equation $x + y = 5$ represents a line in 2-space, and the same equation represents a plane in 3-space.

Still another connection between 2- and 3-space is the interpretation of an equation that declares a variable equal to 0. For example, in 2-space, the equation $x = 0$ describes the y-axis, on which the x-value of every point is 0. In 3-space, the equation $x = 0$ describes the yz-plane, in which the x-value of every point is 0.

Similar discussions will arise in Chapter 12 when surfaces are presented.

The informal proof of the formula for the distance from a point to a line in 3-space is based on the fact that $|\vec{a} \times \vec{b}|$ is the area of a parallelogram with adjacent sides a and b, a fact that is not proved in the text since the proof requires trigonometry. In classes where students have already studied trigonometry of the right triangle, you may choose to show the derivation that follows.

Note that in this chapter, the discussion of dot product is restricted to using it to test whether or not vectors are perpendicular, where the only concern is whether the dot product is 0 or not. However, it can be shown that $\vec{a} \cdot \vec{b} = |\vec{a}||\vec{b}| \cos \theta$ and $|\vec{a} \times \vec{b}| = |\vec{a}||\vec{b}| \sin \theta$, where θ is the angle between $\vec{a}$ and $\vec{b}$.

$$\vec{b} = (|\vec{b}|\cos \beta, |\vec{b}|\sin \beta)$$

$$\vec{a} = (|\vec{a}|\cos \alpha, |\vec{a}| \sin \alpha)$$

$$\vec{a} \cdot \vec{b} = |\vec{b}||\vec{a}|(\cos \alpha \cos \beta + \sin \alpha \sin \beta)$$

$$\frac{\vec{a} \cdot \vec{b}}{|\vec{a}||\vec{b}|} = \cos \alpha \cos \beta + \sin \alpha \sin \beta$$

$$\frac{\vec{a} \cdot \vec{b}}{|\vec{a}||\vec{b}|} = \cos \theta$$

$$\vec{a} \cdot \vec{b} = |\vec{a}||\vec{b}| \cos \theta$$

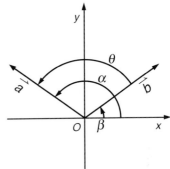

If $(\vec{a} \cdot \vec{b}) = |\vec{a}||\vec{b}| \cos \theta$, then $(\vec{a} \cdot \vec{b})^2 = |\vec{a}|^2|\vec{b}|^2 \cos^2 \theta$.

Lemma: $(\vec{a} \cdot \vec{b})^2 + |\vec{a} \times \vec{b}|^2 = |\vec{a}|^2|\vec{b}|^2$

Proof: Let $\vec{a} = (a_x, a_y, a_z)$ and $\vec{b} = (b_x, b_y, b_z)$.

$$(\vec{a} \cdot \vec{b})^2 + |\vec{a} \times \vec{b}|^2 = (a_x b_x + a_y b_y + a_z b_z)^2 + \left|\left(\begin{vmatrix} a_y & a_z \\ b_y & b_z \end{vmatrix}, \begin{vmatrix} a_z & a_x \\ b_z & b_x \end{vmatrix}, \begin{vmatrix} a_x & a_y \\ b_x & b_y \end{vmatrix}\right)\right|^2$$

$$= (a_x^2 + a_y^2 + a_z^2)(b_x^2 + b_y^2 + b_z^2)$$

Thus, $(\vec{a} \cdot \vec{b})^2 + |\vec{a} \times \vec{b}|^2 = |\vec{a}|^2|\vec{b}|^2$

$$|\vec{a}|^2|\vec{b}|^2 \cos^2 \theta + |\vec{a} \times \vec{b}|^2 = |\vec{a}|^2|\vec{b}|^2$$

$$|\vec{a} \times \vec{b}|^2 = |\vec{a}|^2|\vec{b}|^2 - |\vec{a}|^2|\vec{b}|^2 \cos^2 \theta$$

$$= |\vec{a}|^2|\vec{b}|^2(1 - \cos^2 \theta)$$

$$= |\vec{a}|^2|\vec{b}|^2 \sin^2 \theta$$

Each factor in $|\vec{a} \times \vec{b}|^2 = |\vec{a}|^2|\vec{b}|^2 \sin^2 \theta$ is positive.

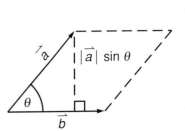

Therefore, taking the square root of each side:

$$|\vec{a} \times \vec{b}| = |\vec{a}||\vec{b}| \sin \theta$$

Thus, the area of the parallelogram with $\vec{a}$ and $\vec{b}$ as adjacent sides is:

$$|\vec{a}||\vec{b}| \sin \theta \text{ or } |\vec{a} \times \vec{b}|$$

As the final concern in this chapter, the formulas for the distance from a point to a line in 2 and 3 dimensions are presented. Students should observe that the formula for the distance from a point to a line in the 2-dimensional plane is similar to the formula in 3-dimensional space, but that the former involves the dot product and the normal vector rather than the cross product and the parallel vector. The similarity between the two formulas should be highlighted, not only as a memory aid but to underscore the usefulness of vector concepts in dealing with these calculations.

The computation of a cross product can sometimes become rather cumbersome, particularly with large numbers. There are now a number of pieces of software available, such as Derive (Soft Warehouse, Honolulu, Hawaii), which will compute a cross product given two vectors. If $\vec{a} \times \vec{b}$ is defined as

$$\vec{a} \times \vec{b} = \begin{vmatrix} \vec{i} & \vec{j} & \vec{k} \\ a_x & a_y & a_z \\ b_x & b_y & b_z \end{vmatrix}$$

then the determinant feature for a 3 × 3 matrix, which is available on some handheld calculators such as the TI-81, may also be used. Where possible, students should be exposed to this technology since it adds credibility to the topic of vectors and enables teachers to emphasize concepts and applications rather than calculations.

You may wish to introduce some vector proofs of familiar geometric theorems. Begin by asking students to study the following vector proof. Preliminary to the proof are two definitions and a theorem, which refer to the diagram shown.

■ **Definition:** *Quadrilateral ABCD is a parallelogram if $\overrightarrow{AB} = \overrightarrow{DC}$. (Ask students to explain.)*

■ **Definition:** *If M is the midpoint of $\overrightarrow{AC}$, then $\overrightarrow{AM} = \overrightarrow{MC}$.*

☐ **Theorem:** *Any vector can be expressed uniquely as the sum of scalar multiples of two given nonzero vectors, as long as the two vectors do not have the same direction.*

Prove: The diagonals of a parallelogram bisect each other.

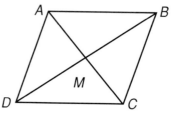

In parallelogram *ABCD*, point *M* is the intersection of the diagonals. Therefore, $\overrightarrow{AM} = s\overrightarrow{AC}$, $\overrightarrow{DM} = t\overrightarrow{DB}$, $\overrightarrow{MC} = \overrightarrow{AC} - s\overrightarrow{AC}$, and $\overrightarrow{MB} = \overrightarrow{DB} - t\overrightarrow{DB}$

By vector addition: $\overrightarrow{AD} + t\overrightarrow{DB} = \overrightarrow{AM}$ and $\overrightarrow{MB} + \overrightarrow{BC} = \overrightarrow{MC}$

By substitution: $\overrightarrow{AD} + t\overrightarrow{DB} = s\overrightarrow{AC}$ and $(\overrightarrow{DB} - t\overrightarrow{DB}) + \overrightarrow{BC} = (\overrightarrow{AC} - s\overrightarrow{AC})$

Since *ABCD* is a parallelogram, $\overrightarrow{AD} = \overrightarrow{BC}$. Therefore, $\overrightarrow{BC} + t\overrightarrow{DB} = s\overrightarrow{AC}$ and $\overrightarrow{BC} = s\overrightarrow{AC} - t\overrightarrow{DB}$

By substitution: $\overrightarrow{DB} - t\overrightarrow{DB} + s\overrightarrow{AC} - t\overrightarrow{DB} = \overrightarrow{AC} - s\overrightarrow{AC}$ or $(1 - t)\overrightarrow{DB} + s\overrightarrow{AC} = t\overrightarrow{DB} + (1 - s)\overrightarrow{AC}$

By the stated theorem, $(1 - t) = t$ and $s = (1 - s)$. Therefore, $t = \frac{1}{2}$ and $s = \frac{1}{2}$. This proves that $\overrightarrow{AM} = \frac{1}{2}\overrightarrow{AC}$ and $\overrightarrow{DM} = \frac{1}{2}\overrightarrow{DB}$ or that *M* is the midpoint of the diagonals.

Ask students to use a similar method to prove that the line joining the midpoints of two sides of a triangle is parallel to the third side of the triangle and half its length.

Since this method of proof is also useful in establishing ratios of line segments in triangles, you may wish to present the following sequence of statements as a guide through a vector proof of the theorem that the medians of a triangle intersect in a point two-thirds of the distance from any vertex.

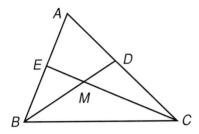

a. Write a vector equation that includes $\overrightarrow{EM}$, $\overrightarrow{EC}$, and the scalar s.

b. Write a vector equation that includes $\overrightarrow{DM}$, $\overrightarrow{DB}$, and the scalar t.

c. Write a vector equation for each side of triangle DMC.

d. Write a vector equation for each side of triangle EMB.

e. Use your equations, substitution, and other facts about the diagram to form a vector equation in terms of only $\overrightarrow{EC}$, $\overrightarrow{DB}$, and the scalars s and t.

f. Use this equation to show that $\overrightarrow{EM} = \frac{1}{3}\overrightarrow{EC}$ and $\overrightarrow{DM} = \frac{1}{3}\overrightarrow{DB}$.

Since the short segment of each median is one-third of the length of the median, the medians meet two-thirds of the distance from each vertex.

Suggested Test Items

In **1 – 5**, refer to the diagram, which shows a parallelogram with $\vec{c}$ as the diagonal.

1. Which vector is equal to $\vec{a} + \vec{b}$?

2. Which vector is equal to $\vec{a}$?

3. Which vector is equal to $\vec{c} - \vec{b}$?

4. Which vector is equal to $\vec{a} - \vec{c}$?

5. Use a pair of compasses and a marked straightedge to construct $2(\vec{b} + \vec{e})$.

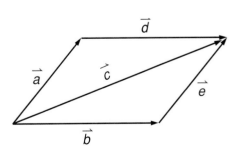

6. Two forces of 6 kg each are applied to an object. If the directions of the forces are 120° apart, describe the magnitude of the resultant and its direction with respect to each force.

7. If $\vec{v}$ = (3, 2) and $\vec{w}$ = (–2, 4), express $\vec{v} + \vec{w}$ as an ordered pair.

8. Evaluate: $|\vec{i} + \vec{j}|$

9. Sketch the vector $-3\vec{i} + 6\vec{j} + \vec{k}$ in 3-dimensional space with its tail at the origin.

In **10 – 13**, choose the best answer.
(A) dot product only
(B) cross product only
(C) dot product and cross product

10. Which operation is commutative?

11. Which operation results in 0 for two vectors if they are perpendicular?

12. Which operation results in $\vec{0}$ for two vectors if they are parallel?

13. Which operation distributes over vector addition?

14. a. Write an ordered triple that can be used to represent the vector from point P to point S in 3-space given $P(6, 1, -4)$ and $S(-1, 2, 7)$.
b. Find the magnitude of the vector found in part **a.**

15. Given points $A(0, 4, -2)$ and $C(5, -1, -2)$, find the coordinates of point B on $\overline{AC}$ such that $AB:BC = 2:3$.

16. Write as a linear combination of $\vec{i}$ and $\vec{j}$, the unit vector in the same direction as $6\vec{i} - 3\vec{j}$.

17. Write an equation for the plane that passes through the point $(1, -2, -1)$ and has $\vec{n}$ = (4, –3, 2) as its normal vector.

18. Write an equation for the plane that passes through the points $(3, 8, 2)$, $(5, 5, 8)$, and $(6, 10, 7)$.

19. Write an ordered triple that represents a normal vector to the plane whose equation is $2x - y + 4z + 7 = 0$.

20. If $\vec{a}$ = (0, 4, –3) and $\vec{b}$ = (2, –1, 0), then $\vec{a} \times \vec{b}$ equals
(A) –4
(B) 4
(C) (–3, 6, –8)
(D) (–3, –6, –8)

21. If $\vec{a}$ and $\vec{b}$ are unequal vectors, which of the following is true?
(A) $\vec{a} \cdot \vec{b}$ is perpendicular to $|\vec{a} \times \vec{b}|$.
(B) $\vec{a} \times \vec{b}$ is perpendicular to the plane determined by $\vec{a}$ and $\vec{b}$.
(C) If $\vec{a} \cdot \vec{b} = 0$, then $\vec{a} \times \vec{b} = \vec{0}$.
(D) $|\vec{a} \cdot \vec{b}|$ is the area of a parallelogram formed with $\vec{a}$ and $\vec{b}$ as adjacent sides.

22. Find the coordinates of two points on the line defined by the parametric equations $x - 4 = t$, $y + 2 = 3t$, and $z + 5 = -2t$.

23. Find the distance from the point $(3, 2, -1)$ to the line described by the parametric equations $x + 1 = t$, $y - 1 = 2t$, and $z - 2 = -t$.

24. Find the distance from the point $(2, 0, -1)$ to the plane whose equation is $6x - 5y = 7$.

Bonus

Find the equations of the lines in the 2-dimensional plane that are parallel to the line whose equation is $x + y = 6$ and 10 units from it.

Answers to Suggested Test Items

1. $\vec{c}$

2. $\vec{e}$

3. $\vec{e}$

4. $-\vec{d}$

5.

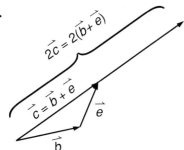

$2\vec{c} = 2(\vec{b} + \vec{e})$

$\vec{c} = \vec{b} + \vec{e}$

$\vec{e}$

$\vec{b}$

6. 6 kg; the resultant is at an angle of $60°$ with each force.

7. $(1, 6)$

8. $\sqrt{2}$

9.

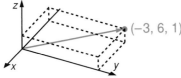

$(-3, 6, 1)$

10. A; dot product only

11. A; dot product only

12. B; cross product only

13. C; dot product and cross product

14. **a.** $\overrightarrow{PS} = (-7, 1, 11)$

　　b. $\sqrt{171}$

15. $(2, 2, -2)$

16. $\dfrac{2\sqrt{5}}{5}\vec{i} - \dfrac{\sqrt{5}}{5}\vec{j}$

17. $4x - 3y + 2z - 8 = 0$

18. $-27x + 8y + 13z - 9 = 0$

19. $(2, -1, 4)$

20. D

21. B

22. $(4, -2, -5)$ and $(5, 1, -7)$
(Other answers are possible.)

23. $\dfrac{5\sqrt{2}}{2}$

24. $\dfrac{5\sqrt{61}}{61}$

Bonus

Use the formula $d = \dfrac{|ap_x + bp_y + c|}{\sqrt{a^2 + b^2}}$.

Let $(p_x, p_y) = (x, y)$.

$$\dfrac{|x + y - 6|}{\sqrt{1^2 + 1^2}} = 10$$

$\dfrac{x + y - 6}{\sqrt{2}} = 10$　　$\dfrac{x + y - 6}{\sqrt{2}} = -10$

$x + y - 6 = 10\sqrt{2}$　　$x + y - 6 = -10\sqrt{2}$

Answer: $x + y - 6 - 10\sqrt{2} = 0$ and
$x + y - 6 + 10\sqrt{2} = 0$

Chapter 6
Vectors

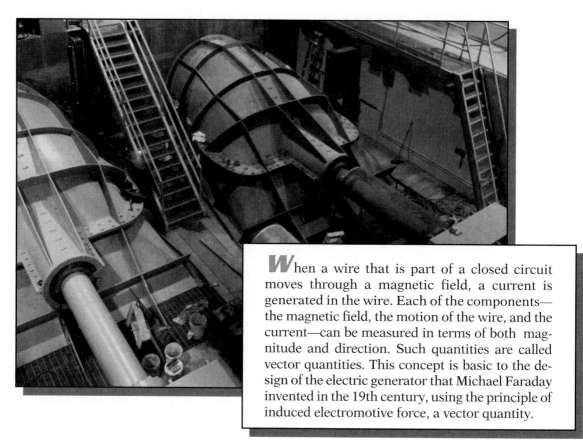

*W*hen a wire that is part of a closed circuit moves through a magnetic field, a current is generated in the wire. Each of the components—the magnetic field, the motion of the wire, and the current—can be measured in terms of both magnitude and direction. Such quantities are called vector quantities. This concept is basic to the design of the electric generator that Michael Faraday invented in the 19th century, using the principle of induced electromotive force, a vector quantity.

Chapter Table of Contents

6.1 Operations With Vectors

Suppose the puck in a hockey game is hit simultaneously by two players. One player hits the puck with a force of 100 pounds and the other player hits the puck with a force of 150 pounds, but at a 60° angle to the line of force of the first player. This situation can be modeled by the *vector* diagram at the right.

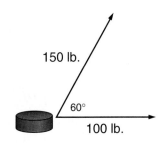

■ **Definition:** *A vector is a directed line segment.*

The length of a vector is called its *magnitude*. The direction of a vector is shown by placing an arrowhead at the terminal point of the line segment that represents the vector. The vector is named either by:

(1) using its endpoints, with the initial point listed first.

In $\overrightarrow{RS}$, R is the initial point, or tail, of the vector and S is the terminal point, or head, of the vector.

or (2) using a single letter.

The magnitude of $\vec{v}$ is also called its absolute value, written $|\vec{v}|$.

■ **Definition:** *Equal vectors are vectors with equal magnitudes and the same direction.*

Vectors have the same direction if they are parallel, or they lie on the same line, and have the same orientation. For example:

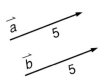

Since both vectors have a magnitude of 5, and they are parallel and are pointing in the same direction, $\vec{a} = \vec{b}$.

Since $|\overrightarrow{AB}| = 3$, and $|\overrightarrow{CD}| = 3$, and since $\overrightarrow{AB}$ and $\overrightarrow{CD}$ are collinear and pointing in the same direction, $\overrightarrow{AB} = \overrightarrow{CD}$. However, $\overrightarrow{AB} \neq \overrightarrow{EF}$ because, although they have the same magnitude and are parallel, they do not have the same direction.

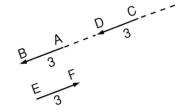

Equal vectors may be formed by using translations in either of two ways:

(1) by a translation on a line.

By moving $\overrightarrow{AB}$ along $\overleftrightarrow{AB}$ so that its initial point A is at B, a new vector, $\overrightarrow{BC}$, with $\overrightarrow{AB} = \overrightarrow{BC}$, can be formed by locating C such that $AB = BC$.

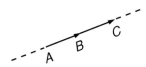

(2) by completing a parallelogram.

To locate $\overrightarrow{PQ}$ so that its initial point is at R and $\overrightarrow{PQ} = \overrightarrow{RS}$, locate S by completing parallelogram $PQSR$. Since the opposite sides of a parallelogram are equal and parallel, $\overrightarrow{PQ}$ and $\overrightarrow{RS}$ have equal magnitudes and the same direction.

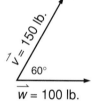

Note that $\overrightarrow{PQ} \neq \overrightarrow{SR}$, since they have opposite directions.

Addition of Vectors

Reconsidering the hockey puck mentioned at the beginning of this section, to determine which way and with how much force the puck actually moves, suggests the need to find the sum of two vectors, called the **resultant**.

For the vector diagram in which each force is measured in pounds:

Let $\overrightarrow{w}$ represent the 100-pound force and $\overrightarrow{v}$ represent the 150-pound force.

To draw the resultant of $\overrightarrow{v}$ and $\overrightarrow{w}$, first construct a vector equal to $\overrightarrow{w}$ with initial point at the terminal point of $\overrightarrow{v}$. Then draw a vector from the tail of $\overrightarrow{v}$ to the head of the newly-drawn vector. This vector is called $\overrightarrow{v} + \overrightarrow{w}$. The hockey puck would therefore move in a direction modeled by $\overrightarrow{v} + \overrightarrow{w}$ and with a magnitude equal to the absolute value of $\overrightarrow{v} + \overrightarrow{w}$ or $|\overrightarrow{v} + \overrightarrow{w}|$. (Computing the magnitude and direction of this resultant force will be discussed in a later chapter.)

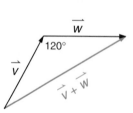

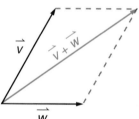

Note that the sum of two nonparallel vectors can be represented by the third side of a triangle of which two sides are the vectors being added, or by the diagonal of a parallelogram.

The method for forming the resultant of two vectors can be extended to three or more vectors by placing the tail of each successive vector that is added at the head of the previous vector. For example, the diagram below shows how this method can be used to construct $\overrightarrow{a} + \overrightarrow{b} + \overrightarrow{c} + \overrightarrow{d}$.

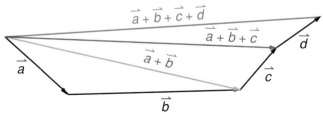

A vector with magnitude equal to that of $\vec{v}$ but in the opposite direction is commonly referred to as the *negative* or *additive inverse* of $\vec{v}$, and is written $-\vec{v}$. When a vector and its additive inverse are added, the resultant is the *zero vector*, written $\vec{0}$. The zero vector has no direction.

Example 1 _____

For each diagram, find $\vec{v} + \vec{w}$.

a. 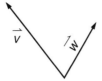　　b.

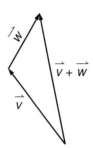

Solution

a.　In order to find $\vec{v} + \vec{w}$, the tail of $\vec{w}$ must be placed at the head of $\vec{v}$. The resultant is then formed by drawing a vector from the tail of $\vec{v}$ to the head of $\vec{w}$.

b. 　　The given vectors $\vec{v}$ and $\vec{w}$ are parallel. Therefore, when placed head to tail to find the sum, they lie on one line.

Subtraction of Vectors

In arithmetic, in order to subtract one integer from another, the additive inverse of the second is added to the first.

$$a - b = a + (-b)$$

A similar principle applies when subtracting vectors. Geometrically, the difference vector $\vec{a} - \vec{b}$ is the resultant $\vec{a} + (-\vec{b})$. As in addition of vectors, subtraction of vectors can be represented as part of a triangle or as part of a parallelogram.

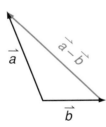　　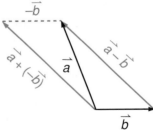

For the triangle method, place the tails of $\vec{a}$ and $\vec{b}$ together, and draw a vector from the head of $\vec{b}$ to the head of $\vec{a}$. The third side of the triangle represents the difference.

For the parallelogram method, place the tails of $\vec{a}$ and $\vec{b}$ together, and draw $-\vec{b}$ from the head of $\vec{a}$. Either of the remaining opposite sides of the completed parallelogram represents the difference.

Example 2

For each diagram, find $\vec{v} - \vec{w}$.

a.

b.

Solution

a. $\vec{v} - \vec{w}$ is the vector drawn from the head of $\vec{w}$ to the head of $\vec{v}$ when $\vec{v}$ and $\vec{w}$ are placed tail to tail as they are in the original diagram.

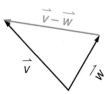

b. The given vectors $\vec{v}$ and $\vec{w}$ are parallel. Therefore, when placed tail to tail to find the difference, they lie on one line or on parallel lines.

Scalar Multiplication

Vectors can be multiplied by any real number factor, called a *scalar*. A scalar is a quantity having only magnitude. In order to increase or decrease the magnitude of a vector without changing its direction, multiply by a positive scalar. Multiplying by a negative scalar will reverse the direction of a vector.

In the diagram, $2\vec{v}$ is a vector with twice the magnitude of $\vec{v}$ and in the same direction. $-3\vec{v}$ is a vector 3 times as long as $\vec{v}$ but in the opposite direction, that is, 3 times as long as $-\vec{v}$, the opposite of $\vec{v}$.

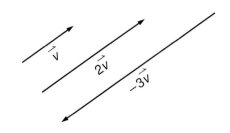

If a vector is multiplied by the scalar -1, the resulting vector is its additive inverse. For example, $-1 \cdot \vec{v} = -\vec{v}$, the additive inverse of $\vec{v}$.

Example 3

Given $\vec{m}$ and $\vec{n}$, sketch a vector equal to:

a. $2\vec{m} + 3\vec{n}$ **b.** $2\vec{m} - 3\vec{n}$

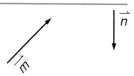

Solution

a. $2\vec{m}$ has the same direction as $\vec{m}$, but is twice as long. Similarly, $3\vec{n}$ has the same direction as $\vec{n}$, but is 3 times as long. After constructing $2\vec{m}$ and $3\vec{n}$, form the resultant $2\vec{m} + 3\vec{n}$ by placing $2\vec{m}$ and $3\vec{n}$ head to tail and drawing a vector from the tail of $2\vec{m}$ to the head of $3\vec{n}$.

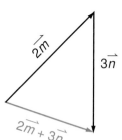

b.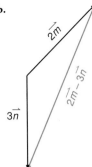

Using $2\vec{m}$ and $3\vec{n}$ as constructed in part **a**, form the difference vector $2\vec{m} - 3\vec{n}$ by placing $2\vec{m}$ and $3\vec{n}$ tail to tail and drawing the vector $2\vec{m} - 3\vec{n}$ from the head of $3\vec{n}$ to the head of $2\vec{m}$.

1.

2.

Exercises 6.1

In **1 – 15**, use the vectors shown, a marked straightedge, and a protractor to sketch the given vector.

3.

1. $\vec{a} + \vec{b}$ 2. $\vec{b} + \vec{a}$
3. $\vec{b} + \vec{c}$ 4. $\vec{a} - \vec{b}$
5. $\vec{b} - \vec{c}$ 6. $\vec{a} - \vec{c}$
7. $\vec{c} - \vec{a}$ 8. $\vec{a} + \vec{b} + \vec{c}$ 9. $2\vec{a} + \vec{b}$ 10. $2\vec{b} - \vec{c}$ 11. $\frac{1}{2}\vec{a} - 2\vec{b}$ 12. $\vec{a} - 2\vec{b} + \vec{c}$
13. $\frac{1}{3}\vec{c} - \frac{1}{2}\vec{b}$ 14. $1.5\vec{b} + 0.4\vec{a} - 0.8\vec{c}$ 15. $\vec{a} - \frac{1}{2}\vec{b} - \frac{1}{2}\vec{c}$

4.

16. a. Draw two vectors, $\vec{v}$ and $\vec{w}$, that point in different directions. Use these vectors and a diagram to verify that vector addition is commutative, that is,
$$\vec{v} + \vec{w} = \vec{w} + \vec{v}.$$

b. Add a third vector, $\vec{s}$, to $\vec{v}$ and $\vec{w}$ and verify that vector addition is associative, that is,
$$(\vec{v} + \vec{w}) + \vec{s} = \vec{v} + (\vec{w} + \vec{s}).$$

5.

17. Explain why the set of vectors is closed for the operations of vector addition and vector subtraction.

18. If m is a scalar and $\vec{v}$ and $\vec{w}$ are any two vectors, use diagrams to verify that $m(\vec{v} + \vec{w}) = m\vec{v} + m\vec{w}$ (scalar multiplication distributes over vector addition).

6.

19.

$\vec{a}$, $\vec{b}$, and $\vec{c}$ are positioned so that they form the sides of a triangle, as shown.
a. Verify that $\vec{a} + \vec{b} = \vec{c}$.
b. Why does $|\vec{a} + \vec{b}| = |\vec{c}|$?
c. Construct a line segment whose length is $|\vec{a}| + |\vec{b}|$.
d. Use facts from geometry to verify that for any two vectors, $\vec{a}$ and $\vec{b}$:
$$|\vec{a} + \vec{b}| \le |\vec{a}| + |\vec{b}|$$

7. (see figure)

20. A hockey puck is hit simultaneously by two players, each applying an 18-pound force. Find the magnitude of the resultant and describe its direction relative to the applied forces if the directions of the forces form an angle of: **a.** 120°   **b.** 60°

8.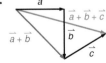

9.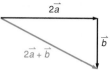

10.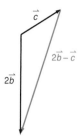

11.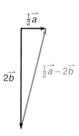

12.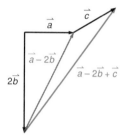

13.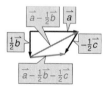

14.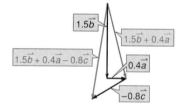

15.

16. a.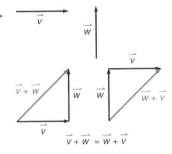

b.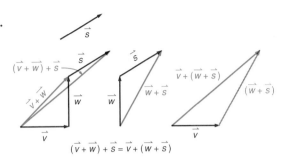

17. The sum of any 2 vectors is a vector.
The difference of any 2 vectors is a vector.

18.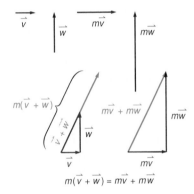

19. a. $\vec{c}$ joins the tail of $\vec{a}$ and the head of $\vec{b}$.

b. If 2 vectors are equal then they have equal magnitudes.

c. Mark off the lengths of $\vec{a}$ and $\vec{b}$ as a sum on a number line.

$$|\vec{a}| \quad + \quad |\vec{b}|$$

d. If $\vec{a}$ and $\vec{b}$ do not lie on the same line, they can be placed tail to tip so that $|\vec{a}|$, $|\vec{b}|$, and $|\vec{a} + \vec{b}|$ are the lengths of the sides of a triangle. Since the sum of the lengths of any two sides of a triangle must be greater than the length of the third side, $|\vec{a}| + |\vec{b}| > |\vec{a} + \vec{b}|$. If $\vec{a}$ and $\vec{b}$ lie on the same line or parallel lines, they can be placed tail to tip to form a line segment of length $|\vec{a} + \vec{b}|$ so that $|\vec{a}| + |\vec{b}| = |\vec{a} + \vec{b}|$. Therefore, combining both cases, $|\vec{a}| + |\vec{b}| \geq |\vec{a} + \vec{b}|$.

20. The resultant is the diagonal of the parallelogram determined by the forces.

a. The resultant forms a 60° angle with each 18-pound force, dividing the parallelogram into two equilateral triangles; therefore, its magnitude is 18.

b. The resultant forms a 30° angle with each 18-pound force. The magnitude of the resultant is $18\sqrt{3}$. (Drawing the other diagonal forms 30°-60°90° triangles.)

6.2 Using Ordered Pairs to Represent Vectors

Ordered pairs of real numbers may be used to represent vectors in the plane. The ordered pair (x, y) represents a vector whose tail may be at the origin and whose head then would be at the point (x, y). For example, the ordered pair $(3, 2)$ is used to represent $\overrightarrow{OP}$, the vector shown. A vector with its tail at the origin is called a *position vector*.

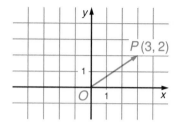

Note that since equal vectors have equal magnitudes and the same direction, the ordered pair $(3, 2)$ is also used to represent a vector anywhere in the plane with the same magnitude and direction as $\overrightarrow{OP}$.

To draw $\overrightarrow{OP}$, start at the origin and move 3 units in the positive x direction, and from there move 2 units in the positive y direction. This final position is the location for the head of the vector.

If the tail of an equal vector is located at any other point in the plane, (a, b), that is not the origin, the location of its head would be the point $(a + 3, b + 2)$. The new vector, however, would still be represented by the ordered pair $(3, 2)$ since it has the same magnitude and direction as $\overrightarrow{OP}$. For example, the coordinates of the head of a $(3, 2)$ vector that has its tail at $(-4, 1)$ would be $(-4 + 3, 1 + 2)$, or $(-1, 3)$.

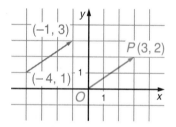

The magnitude of a vector represented by an ordered pair can be found by using the *distance formula*. For example, the magnitude of $\overrightarrow{OP}$, the $(3, 2)$ vector whose tail is at the origin, is $\sqrt{13}$. This same value, $\sqrt{13}$, is obtained when the distance formula is used to find the magnitude of the $(3, 2)$ vector from $(-4, 1)$ to $(-1, 3)$. This result is to be expected, since all vectors represented by the ordered pair $(3, 2)$ have the same magnitude.

$$
\begin{aligned}
\left|\overrightarrow{OP}\right| &= \sqrt{(x_2 - x_1)^2 + (y_2 - y_1)^2} \\
&= \sqrt{(0 - 3)^2 + (0 - 2)^2} \\
&= \sqrt{9 + 4} \\
&= \sqrt{13}
\end{aligned}
$$

Note: It is important to distinguish between an ordered pair that represents a vector and one that represents the location of a point in the plane.

Example 1

a. Find the coordinates of the head of the $(2, 5)$ vector whose tail is at the point $(-4, 3)$.

b. Find the magnitude of all $(2, 5)$ vectors.

Solution

a. If the tail of the (2, 5) vector is at
(−4, 3), then the head would be at
(−4 + 2, 3 + 5), or (−2, 8).

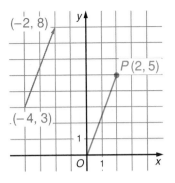

b. The magnitude of any (2, 5) vector
can be found by computing the
distance from the origin to the
point $P(2, 5)$.

$$|\overrightarrow{OP}| = \sqrt{(0 - 2)^2 + (0 - 5)^2}$$
$$= \sqrt{4 + 25}$$
$$= \sqrt{29}$$

Operations with Vectors Using Coordinates

In Section 6.1, you used geometric diagrams to perform addition, subtraction, and scalar
multiplication of vectors. These operations can also be performed using coordinates.

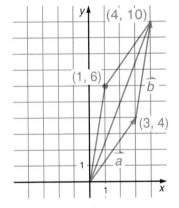

For example, if two vectors, $\vec{a}$ and $\vec{b}$, are represented
by the ordered pairs (3, 4) and (1, 6), respectively, then
the resultant, $\vec{a} + \vec{b}$, would be represented by the ordered
pair (3 + 1, 4 + 6), or (4, 10). Both the geometric and
coordinate methods of vector addition yield the same
result, as shown in the accompanying diagram.

The following are general statements for algebraic addition and subtraction of vectors.

If $\vec{a} = (a_x, a_y)$ and $\vec{b} = (b_x, b_y)$, then:
$\vec{a} + \vec{b} = (a_x + b_x, a_y + b_y)$ and $\vec{a} - \vec{b} = (a_x - b_x, a_y - b_y)$

Example 2

If $\vec{a} = (3, 7)$ and $\vec{b} = (-4, -3)$, find: **a.** $\vec{b} + \vec{a}$ **b.** $\vec{a} - \vec{b}$

Solution

a. $\vec{b} + \vec{a} = (-4, -3) + (3, 7) = (-4 + 3, -3 + 7) = (-1, 4)$

b. $\vec{a} - \vec{b} = (3, 7) - (-4, -3) = (3 - (-4), 7 - (-3)) = (7, 10)$

If a vector, such as $\vec{a} = (a_x, a_y)$, is multiplied by a scalar k, then the resulting vector is
represented by the ordered pair (ka_x, ka_y).

Example 3

If $\vec{a} = (3, 7)$, find $3\vec{a}$.

Solution

$3\vec{a} = 3(3, 7) = (3 \cdot 3, 3 \cdot 7) = (9, 21)$

1. $(1, 3)$

2. $(-3, 8)$

3. $(2, 5)$

4. $(-2, 10)$

5. $(6, 10)$

6. $(3, -1)$

7. $(9, -1)$

8. $(-1, -11)$

9. $(-2, 2)$

10. $(0, 0)$

11. $(4, 5)$

12. $(-3, 7)$

13. $(1, 1)$

14. $(5, 3)$

15. $(2, 4)$

16. $(0, 3)$

17. $(5, 16)$

18. $(2, 3)$

Any vector in a plane may be viewed as a linear combination of the unit vectors $(1, 0)$ and $(0, 1)$, where $(1, 0)$ is the horizontal unit vector and $(0, 1)$ is the vertical unit vector. For example, the vector $(3, 2)$ can be viewed as $3(1, 0) + 2(0, 1)$. The horizontal unit vector $(1, 0)$ is called $\vec{i}$ (which should not be confused with the imaginary unit i) and the vertical unit vector $(0, 1)$ is called $\vec{j}$. Therefore, the vector $(3, 2)$ may also be written $3\vec{i} + 2\vec{j}$.

In general, $\vec{a} = (a_x, a_y) = a_x\vec{i} + a_y\vec{j}$.

Example 4

Given $\vec{a} = 2\vec{i} + \vec{j}$ and $\vec{b} = -3\vec{i} + 4\vec{j}$, find:

  a. $\vec{a} + \vec{b}$ **b.** $3\vec{a} + \vec{b}$ **c.** $|\vec{a}| + |\vec{b}|$

Solution

  a. $\vec{a} + \vec{b} = (2\vec{i} + \vec{j}) + (-3\vec{i} + 4\vec{j})$

            $= (2 + (-3))\vec{i} + (1 + 4)\vec{j}$

            $= -\vec{i} + 5\vec{j}$

  b. $3\vec{a} + \vec{b} = 3(2\vec{i} + \vec{j}) + (-3\vec{i} + 4\vec{j})$

            $= (6\vec{i} + 3\vec{j}) + (-3\vec{i} + 4\vec{j})$

            $= (6 + (-3))\vec{i} + (3 + 4)\vec{j}$

            $= 3\vec{i} + 7\vec{j}$

  c. Since $\vec{a} = 2\vec{i} + \vec{j} = (2, 1)$, $|\vec{a}| = \sqrt{2^2 + 1^2} = \sqrt{5}$.

      Since $\vec{b} = -3\vec{i} + 4\vec{j} = (-3, 4)$, $|\vec{b}| = \sqrt{(-3)^2 + 4^2} = 5$.

    Answer: $|\vec{a}| + |\vec{b}| = \sqrt{5} + 5$

Exercises 6.2

In **1 – 10**, the coordinates of the tail of a vector and the vector's ordered pair representation are given. Find the coordinates of the head of the vector.

  1. tail: $(0, 0)$;   vector: $(1, 3)$ **2.** tail: $(0, 0)$;   vector: $(-3, 8)$

  3. tail: $(1, 2)$;   vector: $(1, 3)$ **4.** tail: $(1, 2)$;   vector: $(-3, 8)$

  5. tail: $(4, 7)$;   vector: $(2, 3)$ **6.** tail: $(-2, 2)$;   vector: $(5, -3)$

  7. tail: $(6, -1)$;   vector: $(3, 0)$ **8.** tail: $(-3, -5)$; vector: $(2, -6)$

  9. tail: $(-2, -4)$; vector: $(0, 6)$ **10.** tail: $(3, -1)$;   vector: $(-3, 1)$

In **11 – 18**, the coordinates of both the tail and the head of a vector are given. Give the ordered pair representation of the vector.

  11. tail: $(0, 0)$;   head: $(4, 5)$ **12.** tail: $(0, 0)$;   head: $(-3, 7)$

  13. tail: $(3, 5)$;   head: $(4, 6)$ **14.** tail: $(5, 1)$;   head: $(10, 4)$

  15. tail: $(2, -4)$;   head: $(4, 0)$ **16.** tail: $(-4, 3)$;   head: $(-4, 6)$

  17. tail: $(-5, -6)$; head: $(0, 10)$ **18.** tail: $(-2, -3)$; head: $(0, 0)$

Exercises 6.2 (continued)

In **19 – 36**, find the magnitude of the vector described.

19. the vector from $(0, 0)$ to $(10, -7)$　　**20.** the vector from $(6, 4)$ to $(1, -3)$

21. the vector from $(-3, -5)$ to $(-6, 1)$　　**22.** the vector from $(-3, -4)$ to $(0, 0)$

23. the vector $(5, 4)$　　**24.** the vector $(2, 3)$　　**25.** the vector $(-3, 9)$

26. the vector $(4, -2)$　　**27.** the vector $(-6, -3)$　　**28.** the vector $\vec{i}$

29. the vector $\vec{j}$　　**30.** the vector $\vec{i} + \vec{j}$　　**31.** the vector $\vec{i} - \vec{j}$

32. the vector $2\vec{i} + \vec{j}$　　**33.** the vector $5\vec{i} + 3\vec{j}$　　**34.** the vector $-2\vec{i} - 4\vec{j}$

35. the vector $(2, 6) + (1, -4)$　　**36.** the vector $(2\vec{i} - 3\vec{j}) - (3\vec{i} + 2\vec{j})$

In **37 – 48**, $\vec{a} = (3, -2)$ and $\vec{b} = (-1, 5)$. Find an ordered pair to represent the given vector.

37. $\vec{a} + \vec{b}$　　**38.** $\vec{a} - \vec{b}$　　**39.** $3\vec{a}$　　**40.** $4\vec{b}$

41. $-3\vec{a}$　　**42.** $-5\vec{b}$　　**43.** $2\vec{a} + \vec{b}$　　**44.** $3\vec{a} + 2\vec{b}$

45. $8\vec{a} - 2\vec{b}$　　**46.** $3\vec{b} - 4\vec{a}$　　**47.** $-\vec{b} - 2\vec{a}$　　**48.** $-(\vec{a} - \vec{b})$

In **49 – 57**, $\vec{v} = 2\vec{i} + 3\vec{j}$, and $\vec{w} = \vec{i} - 2\vec{j}$. Express the given vector as a linear combination of unit vectors in simplest form.

49. $\vec{v} + \vec{w}$　　**50.** $\vec{v} - \vec{w}$　　**51.** $4\vec{v}$

52. $2\vec{w}$　　**53.** $4\vec{v} + 2\vec{w}$　　**54.** $2\vec{v} - 3\vec{w}$

55. $5\vec{w} - 3\vec{v}$　　**56.** $5(2\vec{v} + \vec{w})$　　**57.** $-3(3\vec{w} + \vec{v})$

58. Find the ordered pair representation of the vector that has the same direction but half the magnitude of the vector represented by the ordered pair $(10, 6)$.

59. Find the ordered pair representation of the vector that has the same direction but 3 times the magnitude of the vector represented by the ordered pair $(2, 5)$.

60. Find the ordered pair representation of the vector that has twice the magnitude of the vector represented by the ordered pair $(4, -1)$, but points in the opposite direction.

61. Write the ordered pair designation for the vector that is the identity for vector addition.

19. $\sqrt{149}$

20. $\sqrt{74}$

21. $\sqrt{45} = 3\sqrt{5}$

22. 5

23. $\sqrt{41}$

24. $\sqrt{13}$

25. $\sqrt{90} = 3\sqrt{10}$

26. $\sqrt{20} = 2\sqrt{5}$

27. $\sqrt{45} = 3\sqrt{5}$

28. 1

29. 1

30. $\sqrt{2}$

31. $\sqrt{2}$

32. $\sqrt{5}$

33. $\sqrt{34}$

34. $\sqrt{20} = 2\sqrt{5}$

35. $\sqrt{13}$

36. $\sqrt{26}$

37. $(2, 3)$

38. $(4, -7)$

39. $(9, -6)$

40. $(-4, 20)$

41. $(-9, 6)$

42. $(5, -25)$

43. $(5, 1)$

44. $(7, 4)$

45. $(26, -26)$

46. $(-15, 23)$　　**52.** $2\vec{i} - 4\vec{j}$　　**58.** $(5, 3)$

47. $(-5, -1)$　　**53.** $10\vec{i} + 8\vec{j}$　　**59.** $(6, 15)$

48. $(-4, 7)$　　**54.** $\vec{i} + 12\vec{j}$　　**60.** $(-8, 2)$

49. $3\vec{i} + \vec{j}$　　**55.** $-\vec{i} - 19\vec{j}$　　**61.** $(0, 0)$

50. $\vec{i} + 5\vec{j}$　　**56.** $25\vec{i} + 20\vec{j}$

51. $8\vec{i} + 12\vec{j}$　　**57.** $-15\vec{i} + 9\vec{j}$

6.3 Vectors in 3-Space

The 2-dimensional Cartesian plane is sufficient for modeling many mathematical concepts; for others, a 3-dimensional model is required. For instance, to locate a point on the floor of a room, a 2-dimensional model can be used; to locate a point within the space of the room, a 3-dimensional model is necessary. Note that many 3-dimensional concepts are connected to work already done in 2 dimensions.

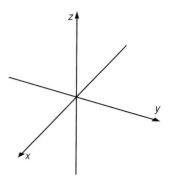

To create a 3-dimensional coordinate system, a third axis, in addition to the x- and y-axes of the Cartesian plane, is introduced. This third axis, called the z-axis, has a zero point that coincides with the origin of the xy-plane. Since it is perpendicular to the xy-plane at the origin, the z-axis represents the distance above and below the xy-plane. Visualize the xy-plane as horizontal, with the positive x-axis coming toward you.

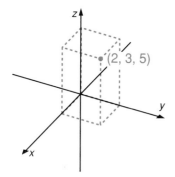

Points in 3-dimensional space are represented by *ordered triples* of the form (x, y, z). For example, the point $(2, 3, 5)$ is the point located by moving 2 units along the positive x-axis, 3 units parallel to the positive y-axis, and 5 units parallel to the positive z-axis.

Note that $(2, 3, 5)$ is the point 5 units above the point $(2, 3)$ in the xy-plane. In the diagram shown, dashed lines parallel to the axes are included to help establish the proper perspective.

Vectors in 3-dimensional space are also represented by ordered triples. For example, the vector $(2, 3, 5)$ in space may be shown with its tail at the origin and its head at the point $(2, 3, 5)$. Similar to the designations in the 2-dimensional plane, the unit vector on the x-axis, $(1, 0, 0)$, is called $\vec{i}$ and the unit vector on the y-axis, $(0, 1, 0)$, is called $\vec{j}$. The unit vector on the z-axis, $(0, 0, 1)$, is called $\vec{k}$. Therefore, the vector $(2, 3, 5)$ may be written as the linear combination of unit vectors $2\vec{i} + 3\vec{j} + 5\vec{k}$.

In 3-dimensional space, the distance between two points (x_1, y_1, z_1) and (x_2, y_2, z_2) is:

$$d = \sqrt{(x_2 - x_1)^2 + (y_2 - y_1)^2 + (z_2 - z_1)^2}$$

which is an extension of the 2-dimensional distance formula. (The derivation of this formula by using the Pythagorean Theorem twice is considered later, as an exercise.)

Similarly, the magnitude of a vector $\vec{a}$ represented by (a_x, a_y, a_z) is:

$$|\vec{a}| = \sqrt{(a_x)^2 + (a_y)^2 + (a_z)^2}$$

Example 1

a. Write the ordered triple representing the vector from $A(1, 4, 2)$ to $B(3, -1, 6)$.

b. Express $\overrightarrow{AB}$ as a linear combination of the unit vectors $\vec{i}, \vec{j}$, and $\vec{k}$.

c. Find the magnitude of $\overrightarrow{AB}$.

Solution

a. Extend the methods used in 2-space. Find the directed change in the x-, y-, and z-coordinates in going from the tail, A, to the head, B, of the vector.

$$A(1, 4, 2), B(3, -1, 6)$$
$$\text{ordered triple: } (3 - 1, \; -1 - 4, \; 6 - 2)$$

Answer: The ordered triple is $(2, -5, 4)$.

b. In general: $\quad \vec{a} = (a_x, a_y, a_z) = a_x\vec{i} + a_y\vec{j} + a_z\vec{k}$

Thus: $\quad \overrightarrow{AB} = (2, -5, 4) = 2\vec{i} - 5\vec{j} + 4\vec{k}$

Answer: As a linear combination, $\overrightarrow{AB} = 2\vec{i} - 5\vec{j} + 4\vec{k}$.

c. The magnitude can be found by using the distance formula.

$$|\vec{a}| = \sqrt{(a_x)^2 + (a_y)^2 + (a_z)^2}$$
$$|\overrightarrow{AB}| = \sqrt{2^2 + (-5)^2 + 4^2}$$
$$= \sqrt{4 + 25 + 16} = \sqrt{45} = 3\sqrt{5}$$

The same result can be obtained by finding the distance from point A to point B.

$$|\overrightarrow{AB}| = \sqrt{(3 - 1)^2 + (-1 - 4)^2 + (6 - 2)^2}$$
$$= \sqrt{2^2 + (-5)^2 + 4^2} = \sqrt{45} = 3\sqrt{5}$$

Answer: The magnitude of $\overrightarrow{AB}$ is $3\sqrt{5}$.

Example 2

Let $\vec{a} = (-2, 1, 3)$, $\vec{b} = (5, 4, 1)$, and $\vec{c} = 2\vec{a} + \vec{b}$.

a. Find the ordered triple that represents $\vec{c}$.

b. Locate $2\vec{a}, \vec{b}$, and $\vec{c}$ on a graph with their tails at the origin.

Solution

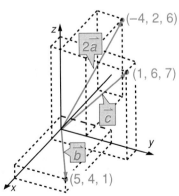

a. $\vec{c} = 2\vec{a} + \vec{b}$
$$= 2(-2, 1, 3) + (5, 4, 1)$$
$$= (-4, 2, 6) + (5, 4, 1)$$
$$= (-4 + 5, 2 + 4, 6 + 1)$$
$$= (1, 6, 7)$$

Answer:

a. The ordered triple that represents $\vec{c}$ is $(1, 6, 7)$.

b. $2\vec{a} = (-4, 2, 6)$
$\vec{b} = (5, 4, 1)$
$\vec{c} = (1, 6, 7)$

Dividing a Line Segment

It is often necessary to find the coordinates of the point that separates a given line segment or vector according to a given proportion. Both in the 2- and 3-space, the methods for adding and subtracting vectors can be used to find such points.

Example 3

Find the coordinates of point E in the diagram below, given that the distance from A to E is $\frac{2}{3}$ of the distance from A to B.

Solution

Using $A(2, 5)$ and $B(11, 1)$, $\overrightarrow{AB} = (11 - 2, 1 - 5) = (9, -4)$.

$$\overrightarrow{OA} + \overrightarrow{AE} = \overrightarrow{OE}$$

$$\overrightarrow{OA} + \tfrac{2}{3}(\overrightarrow{AB}) = \overrightarrow{OE}$$

$$(2\vec{i} + 5\vec{j}) + \tfrac{2}{3}(9\vec{i} - 4\vec{j}) = \overrightarrow{OE}$$

$$(2\vec{i} + 5\vec{j}) + \left(6\vec{i} - \tfrac{8}{3}\vec{j}\right) = \overrightarrow{OE}$$

$$(2 + 6)\vec{i} + \left(5 - \tfrac{8}{3}\right)\vec{j} = \overrightarrow{OE}$$

$$8\vec{i} + \tfrac{7}{3}\vec{j} = \overrightarrow{OE}$$

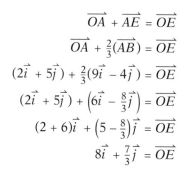

Answer: The coordinates of E are $\left(8, \tfrac{7}{3}\right)$.

The same method may be applied in 3-dimensional space.

Example 4

Given points $A(10, 6, -4)$ and $B(5, -5, 1)$, find the coordinates of point E on $\overrightarrow{AB}$ such that $AE:EB = 2:3$.

Solution

If $AE:EB = 2:3$, then the distance from A to E is $\frac{2}{5}$ of the distance from A to B.

$$\overrightarrow{AB} = (5 - 10, -5 - 6, 1 - (-4)) = (-5, -11, 5)$$

$$\overrightarrow{OA} + \tfrac{2}{5}\overrightarrow{AB} = \overrightarrow{OE}$$

$$(10\vec{i} + 6\vec{j} - 4\vec{k}) + \tfrac{2}{5}(-5\vec{i} - 11\vec{j} + 5\vec{k}) = \overrightarrow{OE}$$

$$(10\vec{i} + 6\vec{j} - 4\vec{k}) + \left(-2\vec{i} - \tfrac{22}{5}\vec{j} + 2\vec{k}\right) = \overrightarrow{OE}$$

$$(10 - 2)\vec{i} + \left(6 - \tfrac{22}{5}\right)\vec{j} + (-4 + 2)\vec{k} = \overrightarrow{OE}$$

$$8\vec{i} + \tfrac{8}{5}\vec{j} - 2\vec{k} = \overrightarrow{OE}$$

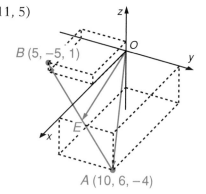

Answer: The coordinates of E are $\left(8, \tfrac{8}{5}, -2\right)$.

Exercises 6.3

In **1 – 9**, draw a graph of the vector represented by the given ordered triple, and find the magnitude of the vector.

1. $(1, 3, 5)$ **2.** $(3, 7, 2)$ **3.** $(10, 1, 6)$

4. $(2, 9, -1)$ **5.** $(3, -4, 2)$ **6.** $(-3, 2, 5)$

7. $(6, 4, -3)$ **8.** $(-1, -2, -3)$ **9.** $(-5, 0, -12)$

In **10 – 15**, write an ordered triple that can be used to represent the vector from R to S.

10. $R(2, 2, 4)$; $S(1, 2, 3)$ **11.** $R(6, 1, 1)$; $S(2, 1, 0)$

12. $R(-4, 2, 3)$; $S(-1, 0, 5)$ **13.** $R(-7, 5, 3)$; $S(3, -4, 2)$

14. $R(-1, -2, 5)$; $S(6, -1, -3)$ **15.** $R(0, 0, 3)$; $S(5, 4, 0)$

In **16 – 24**, $\vec{u} = 2\vec{i} + 3\vec{j} - \vec{k}$, and $\vec{v} = 5\vec{i} - 6\vec{j} + 2\vec{k}$. Find $\vec{w}$.

16. $\vec{w} = \vec{u} + \vec{v}$ **17.** $\vec{w} = \vec{u} - \vec{v}$ **18.** $\vec{w} = \vec{v} - \vec{u}$

19. $\vec{w} = 2\vec{u} + \vec{v}$ **20.** $\vec{w} = 3\vec{u} - 4\vec{v}$ **21.** $\vec{w} = 3(\vec{u} + \vec{v})$

22. $\vec{w} = 3\vec{u} + 3\vec{v}$ **23.** $\vec{w} = 5(\vec{u} - \vec{v})$ **24.** $\vec{w} = 5\vec{u} - 5\vec{v}$

In **25 – 30**, find the coordinates of E on $\overrightarrow{AB}$ for the given coordinates of A and B.

25. $A(18, 6, 5)$; $B(12, 0, -1)$; $\overrightarrow{AE} = \frac{5}{6}\overrightarrow{AB}$

26. $A(4, -2, 7)$; $B(1, 7, 1)$; $\overrightarrow{AE} = \frac{2}{3}\overrightarrow{AB}$

27. $A(0, -8, 7)$; $B(4, 12, 1)$; $\overrightarrow{AE} = \frac{3}{4}\overrightarrow{AB}$

28. $A(3, 7, -11)$; $B(8, 1, -16)$; $\overrightarrow{AE} = \frac{1}{5}\overrightarrow{AB}$

29. $A(30, 15, 9)$; $B(-2, -1, 5)$; $\frac{AE}{EB} = \frac{3}{1}$

30. $A(6, 3, 8)$; $B(-1, 10, -6)$; $AE:EB = 5:2$

31. Show that the magnitude of $\overrightarrow{AB}$ is the same as the magnitude of $\overrightarrow{BA}$, where A and B are each a point in space represented by an ordered triple.

 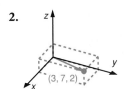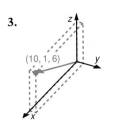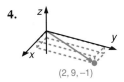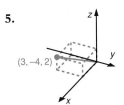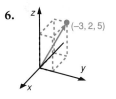
 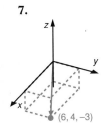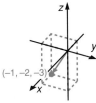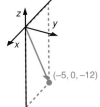

10. $(-1, 0, -1)$ **11.** $(-4, 0, -1)$ **12.** $(3, -2, 2)$

13. $(10, -9, -1)$ **14.** $(7, 1, -8)$ **15.** $(5, 4, -3)$

16. $7\vec{i} - 3\vec{j} + \vec{k}$ **17.** $-3\vec{i} + 9\vec{j} - 3\vec{k}$ **18.** $3\vec{i} - 9\vec{j} + 3\vec{k}$

19. $9\vec{i} + 0\vec{j} + 0\vec{k}$ **20.** $-14\vec{i} + 33\vec{j} - 11\vec{k}$ **21.** $21\vec{i} - 9\vec{j} + 3\vec{k}$

22. $21\vec{i} - 9\vec{j} + 3\vec{k}$ **23.** $-15\vec{i} + 45\vec{j} - 15\vec{k}$ **24.** $-15\vec{i} + 45\vec{j} - 15\vec{k}$

25. $(13, 1, 10)$ **26.** $(2, 4, 3)$ **27.** $\left(3, 7, \frac{5}{2}\right)$

28. $(4, \frac{29}{5}, -12)$ **29.** $(6, 3, 6)$ **30.** $(1, 8, -2)$

31. $A(a_x, a_y, a_z)\ B(b_x, b_y, b_z)$

$$|\overrightarrow{AB}| = \sqrt{(b_x - a_x)^2 + (b_y - a_y)^2 + (b_z - a_z)^2}$$

$$= \sqrt{b_x^2 - 2a_xb_x + a_x^2 + b_y^2 - 2a_yb_y + a_y^2 + b_z^2 - 2a_zb_z + a_z^2}$$

$$|\overrightarrow{BA}| = \sqrt{(a_x - b_x)^2 + (a_y - b_y)^2 + (a_z - b_z)^2}$$

$$= \sqrt{a_x^2 - 2a_xb_x + b_x^2 + a_y^2 - 2a_yb_y + b_y^2 + a_z^2 - 2a_zb_z + b_z^2}$$

$$|\overrightarrow{BA}| = \sqrt{b_x^2 - 2a_xb_x + a_x^2 + b_y^2 - 2a_yb_y + a_y^2 + b_z^2 - 2a_zb_z + a_z^2}$$

$$|\overrightarrow{AB}| = |\overrightarrow{BA}|$$

32. a. $\dfrac{2\vec{i}}{\sqrt{14}} + \dfrac{3\vec{j}}{\sqrt{14}} + \dfrac{\vec{k}}{\sqrt{14}}$

b. $\dfrac{-4\vec{i}}{\sqrt{26}} + \dfrac{\vec{j}}{\sqrt{26}} + \dfrac{\vec{k}}{\sqrt{26}}$

c. $\dfrac{5\vec{i}}{\sqrt{29}} + \dfrac{0\vec{j}}{\sqrt{29}} - \dfrac{2\vec{k}}{\sqrt{29}}$

d. $\dfrac{-4\vec{i}}{\sqrt{84}} - \dfrac{2\vec{j}}{\sqrt{84}} + \dfrac{8\vec{k}}{\sqrt{84}}$

$\dfrac{-2\vec{i}}{\sqrt{21}} - \dfrac{\vec{j}}{\sqrt{21}} + \dfrac{4\vec{k}}{\sqrt{21}}$

33. $(7, -1, 5)$

34.

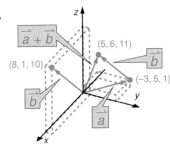

35. a. $\sqrt{\ell^2 + w^2}$

b. $\sqrt{\ell^2 + w^2 + h^2}$

36.

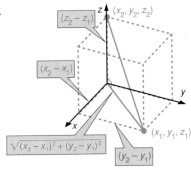

To simplify the diagram, one point is in the *xy*-plane and the other is on the *z*-axis. However, the proof is general.

$$(\text{distance})^2 = \left(\sqrt{(x_2 - x_1)^2 + (y_2 - y_1)^2}\right)^2 + (z_2 - z_1)^2$$

$$(\text{distance})^2 = (x_2 - x_1)^2 + (y_2 - y_1)^2 + (z_2 - z_1)^2$$

$$\text{distance} = \sqrt{(x_2 - x_1)^2 + (y_2 - y_1)^2 + (z_2 - z_1)^2}$$

32. If $\vec{a}$ with magnitude $|\vec{a}|$ is known, then a unit vector in the same direction as $\vec{a}$ may be found by dividing $\vec{a}$ by its magnitude. For example, the vector $\vec{a} = \vec{i} + 3\vec{j} + 5\vec{k}$ has magnitude $|\vec{a}| = \sqrt{1^2 + 3^2 + 5^2} = \sqrt{35}$. Therefore, the unit vector in the same direction as $\vec{a}$ can be represented as:

$$\frac{\vec{i}}{\sqrt{35}} + \frac{3\vec{j}}{\sqrt{35}} + \frac{5\vec{k}}{\sqrt{35}}$$

Represent the unit vector in the same direction as the given vector.

a. $2\vec{i} + 3\vec{j} + \vec{k}$ **b.** $-4\vec{i} + \vec{j} + 3\vec{k}$

c. $5\vec{i} + 0\vec{j} - 2\vec{k}$ **d.** $(-4, -2, 8)$

33. Find the ordered triple that represents the vector from the origin to the midpoint of $\overline{AB}$ for $A(8, 1, 4)$ and $B(6, -3, 6)$.

34. Construct the vector $\vec{a} + \vec{b}$ if $\vec{a} = (-3, 5, 1)$ and $\vec{b} = (8, 1, 10)$.

35. A rectangular solid has congruent bases $ABCD$ and $EFGH$, as shown in the diagram.
   a. In plane $ABCD$, find AC in terms of ℓ and w.
   b. In plane $ACGE$, use the result of part **a** to find AG in terms of ℓ, w, and h.

36. Prove that the distance from (x_1, y_1, z_1) to (x_2, y_2, z_2) is given by $d = \sqrt{(x_2 - x_1)^2 + (y_2 - y_1)^2 + (z_2 - z_1)^2}$. (You may assume the distance formula for two dimensions and the Pythagorean Theorem.)

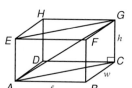

37. Let $A(a_x, a_y, a_z)$ and $B(b_x, b_y, b_z)$ be any two points in space. Prove that the midpoint of $\overline{AB}$ is $\left(\frac{a_x + b_x}{2}, \frac{a_y + b_y}{2}, \frac{a_z + b_z}{2}\right)$.

37. Let M be the midpoint of $\overline{AB}$.

$\overrightarrow{AB} = (b_x - a_x, b_y - a_y, b_z - a_z)$
$\overrightarrow{OM} = \overrightarrow{OA} + \frac{1}{2}\overrightarrow{AB}$
$= (a_x, a_y, a_z) + \frac{1}{2}(b_x - a_x, b_y - a_y, b_z - a_z)$
$= \left(a_x + \frac{b_x - a_x}{2}, a_y + \frac{b_y - a_y}{2}, a_z + \frac{b_z - a_z}{2}\right)$
$= \left(\frac{2a_x + b_x - a_x}{2}, \frac{2a_y + b_y - a_y}{2}, \frac{2a_z + b_z - a_z}{2}\right) = \left(\frac{a_x + b_x}{2}, \frac{a_y + b_y}{2}, \frac{a_z + b_z}{2}\right)$

Therefore, the midpoint of $\overline{AB} = \left(\frac{a_x + b_x}{2}, \frac{a_y + b_y}{2}, \frac{a_z + b_z}{2}\right)$.

6.4 The Dot Product; Planes in Space

It would be useful to have a method for determining whether or not two vectors are perpendicular. Consider first two perpendicular vectors in the 2-dimensional plane.

Two perpendicular vectors, $\vec{a}$ and $\vec{b}$, having position coordinates (a_x, a_y) and (b_x, b_y), respectively, are placed with their tails at the origin. If a third vector, $\vec{c}$, is drawn from the head of $\vec{a}$ to the head of $\vec{b}$, then the distance formula gives

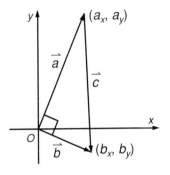

$$|\vec{c}| = \sqrt{(a_x - b_x)^2 + (a_y - b_y)^2}$$

$$|\vec{c}|^2 = (a_x - b_x)^2 + (a_y - b_y)^2$$

$$= (a_x)^2 - 2a_x b_x + (b_x)^2 + (a_y)^2 - 2a_y b_y + (b_y)^2$$

$$= (a_x)^2 + (a_y)^2 + (b_x)^2 + (b_y)^2 - 2(a_x b_x + a_y b_y)$$

But $[(a_x)^2 + (a_y)^2]$ and $[(b_x)^2 + (b_y)^2]$ are the squares of the magnitudes of $\vec{a}$ and $\vec{b}$, respectively. Therefore, the expression can be simplified further.

$$|\vec{c}|^2 = |\vec{a}|^2 + |\vec{b}|^2 - 2(a_x b_x + a_y b_y) \qquad (1)$$

Since $|\vec{a}|$ and $|\vec{b}|$ are the lengths of the legs of a right triangle and $|\vec{c}|$ is the length of the hypotenuse, applying the Pythagorean Theorem yields:

$$|\vec{c}|^2 = |\vec{a}|^2 + |\vec{b}|^2 \qquad (2)$$

Equations (1) and (2) can both be true only if $2(a_x b_x + a_y b_y) = 0$, which means that $a_x b_x + a_y b_y = 0$. The quantity $(a_x b_x + a_y b_y)$ is called the ***inner*** or ***dot product*** of the vectors $\vec{a}$ and $\vec{b}$. The inner or dot product is a *binary operation* applied to two vectors and may be written as follows:

$$\vec{a} \cdot \vec{b} = a_x b_x + a_y b_y$$

The discussion above proves the following theorem.

☐ ***Theorem:*** *Two vectors $\vec{a}$ and $\vec{b}$ in the plane are perpendicular if $\vec{a} \cdot \vec{b} = 0$.*

The dot product in 3-dimensional space is defined in a similar way. If $\vec{a} = (a_x, a_y, a_z)$ and $\vec{b} = (b_x, b_y, b_z)$, then

$$\vec{a} \cdot \vec{b} = a_x b_x + a_y b_y + a_z b_z.$$

☐ ***Theorem:*** *Two vectors $\vec{a}$ and $\vec{b}$ in space are perpendicular if $\vec{a} \cdot \vec{b} = 0$.*

The dot product of two vectors is a commutative operation, and also distributes over vector addition. You will prove these facts in the exercises for this section. Note that the dot product of two vectors is a scalar and, therefore, the dot product is not a closed operation on vectors. This should not be confused with multiplying a vector by a scalar, which results in another vector.

Example 1

Let $\vec{a} = (2, -3)$ and $\vec{b} = (5, 4)$.

a. Find the value of $\vec{a} \cdot \vec{b}$.

b. Determine whether $\vec{a}$ is perpendicular to $\vec{b}$.

Solution

a. $\vec{a} \cdot \vec{b} = a_x b_x + a_y b_y$

$= (2)(5) + (-3)(4)$

$= 10 - 12$

$= -2$

Answer: The value of $\vec{a} \cdot \vec{b}$ is -2.

b. $\vec{a}$ is *not* perpendicular to $\vec{b}$ because $\vec{a} \cdot \vec{b} \neq 0$.

Example 2

Given $\vec{v} = 4\vec{i} - 6\vec{j} + \vec{k}$ and $\vec{w} = \vec{i} - \vec{j} - 10\vec{k}$.

a. Find the value of $\vec{v} \cdot \vec{w}$.

b. Determine whether $\vec{v}$ is perpendicular to $\vec{w}$.

c. Sketch the vectors.

Solution

a. $\vec{v}$ and $\vec{w}$ are represented as the sum of unit vectors. Their coordinate representations are $(4, -6, 1)$ and $(1, -1, -10)$, respectively.

$\vec{v} \cdot \vec{w} = a_x b_x + a_y b_y + a_z b_z$

$= (4)(1) + (-6)(-1) + (1)(-10)$

$= 4 + 6 + (-10)$

$= 0$

Answer: The value of $\vec{v} \cdot \vec{w}$ is 0.

b. Since $\vec{v} \cdot \vec{w} = 0$, $\vec{v}$ is perpendicular to $\vec{w}$.

c.

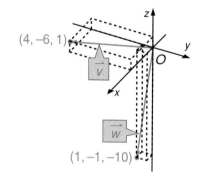

An Equation of a Plane

Just as an equation can be assigned to name a line or a curve, so can an equation be written to name a plane. We can use vectors to write this equation.

Consider a plane P that contains the point $A(1, 3, 5)$. Let $\vec{v} = (4, 2, 2)$ be a vector that is perpendicular to plane P.

Since $\vec{v}$ is perpendicular to plane P, it is also perpendicular to any line, line segment, or vector in this plane. Thus, if $B(x, y, z)$ is any point in plane P, $\vec{v}$ is perpendicular to $\overrightarrow{AB}$. Therefore, $\vec{v} \cdot \overrightarrow{AB} = 0$.

By vector addition: $\overrightarrow{OA} + \overrightarrow{AB} = \overrightarrow{OB}$.

$$\overrightarrow{AB} = \overrightarrow{OB} - \overrightarrow{OA}$$
$$= (x\vec{i} + y\vec{j} + z\vec{k}) - (\vec{i} + 3\vec{j} + 5\vec{k})$$
$$= (x - 1)\vec{i} + (y - 3)\vec{j} + (z - 5)\vec{k}$$

Since $\vec{v} \cdot \overrightarrow{AB} = 0$

$$(4, 2, 2) \cdot (x - 1, y - 3, z - 5) = 0. \qquad \text{This is the dot product of } \vec{v} \text{ and } \overrightarrow{AB}.$$

To compute the dot product of of $\vec{v}$ and $\overrightarrow{AB}$, use the general rule.

$$\vec{a} \cdot \vec{b} = a_x b_x + a_y b_y + a_z b_z$$
$$\vec{v} \cdot \overrightarrow{AB} = 4(x - 1) + 2(y - 3) + 2(z - 5) = 0$$
$$= 4x + 2y + 2z - 20 = 0$$

This, then, is an equation of plane P.

Note that the coefficients of x, y, and z are the coordinates of $\vec{v}$.

In general, if $\vec{n} = (A, B, C)$ is perpendicular to a plane, $R(r_x, r_y, r_z)$ is a given point in the plane, and $P(x, y, z)$ is any point in the plane, then:

$$\vec{n} \cdot \overrightarrow{RP} = 0$$
$$(A, B, C) \cdot (x - r_x, y - r_y, z - r_z) = 0$$
$$A(x - r_x) + B(y - r_y) + C(z - r_z) = 0$$
$$Ax + By + Cz + (-Ar_x - Br_y - Cr_z) = 0$$

when $(-Ar_x - Br_y - Cr_z)$ is a constant, D.

Therefore, an equation of a plane in 3 dimensions is:

$$Ax + By + Cz + D = 0$$

where (A, B, C) is a vector that is perpendicular to the plane. Note that such a vector is said to be *normal*, or *orthogonal*, to the plane.

Thus, when a normal vector to a plane is known, use the coordinates of that vector and the coordinates of a point in the plane to find an equation of the plane.

Example 3

$\vec{v} = (5, -2, 3)$ is orthogonal to a plane P containing point $(4, 7, 1)$. Find an equation of plane P.

Solution

Since $\vec{v}$ is perpendicular to P, $\vec{v}$ is perpendicular to some vector in P whose endpoints are (x, y, z) and $(4, 7, 1)$. This vector may be represented as $(x - 4, y - 7, z - 1)$. Therefore,

$$\vec{v} \cdot (x - 4, y - 7, z - 1) = 0$$
$$(5, -2, 3) \cdot (x - 4, y - 7, z - 1) = 0$$
$$5(x - 4) - 2(y - 7) + 3(z - 1) = 0$$
$$5x - 20 - 2y + 14 + 3z - 3 = 0$$
$$5x - 2y + 3z - 9 = 0$$

Answer: $5x - 2y + 3z - 9 = 0$ is an equation of plane P.

Exercises 6.4

In **1 – 15**, each ordered pair represents a vector. Find the value of the indicated dot product, and state whether the vectors are perpendicular.

1. $(6, 4) \cdot (1, 7)$
2. $(1, 3) \cdot (4, 0)$
3. $(2, -1) \cdot (7, 14)$
4. $(1, -2) \cdot (-5, 14)$
5. $(15, -9) \cdot (-3, -5)$
6. $(-3, -7) \cdot (-1, -4)$
7. $(0, 5) \cdot (5, 0)$
8. $(0.2, 0.4) \cdot (0.8, -0.4)$
9. $(-1.2, 0.9) \cdot (6, 8)$
10. $\left(-\frac{1}{4}, -\frac{1}{12}\right) \cdot \left(\frac{2}{3}, 2\right)$
11. $\left(\frac{2}{5}, 2\right) \cdot \left(\frac{1}{3}, -15\right)$
12. $(\sqrt{2}, -1) \cdot (2, \sqrt{8})$
13. $(\sqrt{3}, 6) \cdot (\sqrt{2}, 1)$
14. $(1 + \sqrt{2}, 0) \cdot (1 - \sqrt{2}, 0)$
15. $(2 + \sqrt{5}, 1) \cdot (2 - \sqrt{5}, 1)$

In **16 – 23**, find the value of the indicated dot product and determine whether the vectors are perpendicular.

16. $(1, 2, 3) \cdot (1, 0, 2)$
17. $(0, 0, 1) \cdot (1, 0, 0)$
18. $(0, 1, 0) \cdot (1, 0, 0)$
19. $(2, 4, 6) \cdot (1, 3, -5)$
20. $(4, 2, 1) \cdot (-4, -2, -1)$
21. $(3, -2, 3) \cdot (5, 3, -3)$
22. $(2\vec{i} + 3\vec{j} + \vec{k}) \cdot (\vec{i} - \vec{j} + \vec{k})$
23. $(6\vec{i} - 2\vec{j} - 2\vec{k}) \cdot (2\vec{i} + 4\vec{j} + 2\vec{k})$

24. Show that the dot product is a commutative operation using $\vec{a} = (a_x, a_y, a_z)$ and $\vec{b} = (b_x, b_y, b_z)$.

25. Show that the dot product distributes over vector addition.

26. Prove that if $\vec{i}, \vec{j}$, and $\vec{k}$ are unit vectors, then $\vec{i} \cdot \vec{i} = \vec{j} \cdot \vec{j} = \vec{k} \cdot \vec{k} = 1$.

27. Explain why $\vec{a} \cdot \vec{b} \cdot \vec{c}$ is meaningless.

1. 34; not perpendicular
2. 4; not perpendicular
3. 0; perpendicular
4. –33; not perpendicular
5. 0; perpendicular

6. 31; not perpendicular
7. 0; perpendicular
8. 0; perpendicular
9. 0; perpendicular
10. $-\frac{1}{3}$; not perpendicular

11. $-\frac{448}{15}$; not perpendicular
12. 0; perpendicular
13. $\sqrt{6} + 6$; not perpendicular
14. –1; not perpendicular
15. 0; perpendicular

16. 7; not perpendicular   **17.** 0; perpendicular   **18.** 0; perpendicular

19. –16; not perpendicular   **20.** –21; not perpendicular   **21.** 0; perpendicular

22. 0; perpendicular   **23.** 0; perpendicular

24. $\vec{a} = (a_x, a_y, a_z)$ $\vec{b} = (b_x, b_y, b_z)$

$\vec{a} \cdot \vec{b} = a_x b_x + a_y b_y + a_z b_z$

$\vec{b} \cdot \vec{a} = b_x a_x + b_y a_y + b_z a_z$

Since multiplication of real numbers is commutative, $\vec{a} \cdot \vec{b} = \vec{b} \cdot \vec{a}$.

25. $\vec{a} = (a_x, a_y, a_z)$ $\vec{b} = (b_x, b_y, b_z)$ $\vec{c} = (c_x, c_y, c_z)$

$\vec{a} \cdot (\vec{b} + \vec{c}) = (a_x, a_y, a_z) \cdot (b_x + c_x, b_y + c_y, b_z + c_z)$

$\qquad\qquad = a_x(b_x + c_x) + a_y(b_y + c_y) + a_z(b_z + c_z)$

$\qquad\qquad = a_x b_x + a_x c_x + a_y b_y + a_y c_y + a_z b_z + a_z c_z$

$\vec{a} \cdot \vec{b} + \vec{a} \cdot \vec{c} = (a_x, a_y, a_z) \cdot (b_x, b_y, b_z) + (a_x, a_y, a_z) \cdot (c_x, c_y, c_z)$

$\qquad\qquad = (a_x b_x + a_y b_y + a_z b_z) + (a_x c_x + a_y c_y + a_z c_z)$

$\qquad\qquad = a_x b_x + a_x c_x + a_y b_y + a_y c_y + a_z b_z + a_z c_z$

$\vec{a} \cdot (\vec{b} + \vec{c}) = \vec{a} \cdot \vec{b} + \vec{a} \cdot \vec{c}$

26. $\vec{i} = (1, 0, 0)$ $\vec{j} = (0, 1, 0)$ $\vec{k} = (0, 0, 1)$

$\vec{i} \cdot \vec{i} = 1(1) + 0(0) + 0(0) = 1$

$\vec{j} \cdot \vec{j} = 0(0) + 1(1) + 0(0) = 1$

$\vec{k} \cdot \vec{k} = 0(0) + 0(0) + 1(1) = 1$

27. The dot product of two vectors is a scalar, which cannot form a dot product with a third vector.

28. $x + y + z - 6 = 0$

29. $2x + y + 2z - 10 = 0$

30. $3x - y + z - 9 = 0$

31. $-5x + 3y + 2z + 7 = 0$

32. $-3x - 2y + 8z - 72 = 0$

33. $1.3x + 0.6y + 4z - 13.9 = 0$

34. $4x - 2y + 3z - 5.1 = 0$

35. $0.3x + 2y + 1.4z - 1.65 = 0$

36. $x = 0$ This is the yz-plane.

37. $z = 0$ This is the xy-plane.

38. $(3, 4, 2)$

39. $(1, -2, 8)$

40. $(5, -12, -7)$

41. $(6, 3, -1)$

42. $(1, 2, 1)$

43. $(0, 3, 1)$

44. $(1, -2, 0)$

45. $(1, 0, -1)$

46. a. $(2, 3, 1)$

  b. $\sqrt{14}$

  c. $\left(\dfrac{2}{\sqrt{14}}, \dfrac{3}{\sqrt{14}}, \dfrac{1}{\sqrt{14}} \right)$

47. $\left(0, \dfrac{4}{\sqrt{41}}, \dfrac{-5}{\sqrt{41}} \right)$

48. $\left(\dfrac{6}{\sqrt{41}}, \dfrac{1}{\sqrt{41}}, \dfrac{-2}{\sqrt{41}} \right)$

49. a. $x - y - z + 6 = 0$

  b. $x + y - 3z + 5 = 0$

Exercises 6.4 *(continued)*

In **28 – 37**, find an equation for the plane that contains the given point P and to which the given vector $\vec{v}$ is perpendicular.

28. $\vec{v} = (1, 1, 1);\quad P(2, 3, 1)$

29. $\vec{v} = (2, 1, 2);\qquad P(1, 2, 3)$

30. $\vec{v} = (3, -1, 1);\quad P(3, 2, 2)$

31. $\vec{v} = (-5, 3, 2);\qquad P(2, -1, 3)$

32. $\vec{v} = (-3, -2, 8);\ P(-2, -5, 7)$

33. $\vec{v} = (1.3, 0.6, 4);\qquad P(1, 1, 3)$

34. $\vec{v} = (4, -2, 3);\quad P(0.4, -0.1, 1.1)$

35. $\vec{v} = 0.3\vec{i} + 2\vec{j} + 1.4\vec{k};\ P\left(1, \frac{1}{2}, \frac{1}{4}\right)$

36. $\vec{v} = (1, 0, 0);\qquad P(0, 4, 2)$
(Describe the plane.)

37. $\vec{v} = (0, 0, 1);\qquad P(2, 3, 0)$
(Describe the plane.)

In **38 – 45**, find a normal vector for the plane whose equation is given.

38. $3x + 4y + 2z + 1 = 0$

39. $x - 2y + 8z + 7 = 0$

40. $5x - 12y - 7z + 2 = 0$

41. $6x + 3y - z + 6 = 0$

42. $x + 2y = 8 - z$

43. $3y + z = 5$

44. $x - 2y = 5$

45. $x - 9 = z$

46. The equation of a plane is $2x + 3y + z + 14 = 0$.

   a. Write an ordered triple to represent a vector that is normal to this plane.

   b. Find the magnitude of this normal vector.

   c. Write the ordered triple that represents the unit normal vector for the plane.

47. Find a unit normal vector for the plane whose equation is $4y - 5z = 0$.

48. Find a unit vector that is orthogonal to the plane $6x + y - 2z - 85 = 0$.

49. A plane is perpendicular to and passes through the midpoint of the line segment whose endpoints are given. Find an equation of the plane.

   a. endpoints: $(2, 4, 7)$ and $(4, 2, 5)$

   b. endpoints: $(3, -1, 6)$ and $(5, 1, 0)$

50. **a.** Graph position vectors $\overrightarrow{OA}$ and $\overrightarrow{OB}$ for the points $A(10, 12)$ and $B(6, -5)$.

   b. Find the slope of $\overrightarrow{OA}$ and the slope of $\overrightarrow{OB}$, and draw a conclusion about $\angle AOB$.

   c. Compute $\overrightarrow{OA} \cdot \overrightarrow{OB}$, and determine if this result is consistent with your conclusion about $\angle AOB$.

   d. Given $A(a_x, a_y)$ and $B(b_x, b_y)$ with $\overrightarrow{OA} \cdot \overrightarrow{OB} = 0$, show that the product of the slopes of $\overrightarrow{OA}$ and $\overrightarrow{OB}$ is -1.

50.

a.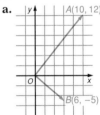

b. slope of
$$\overrightarrow{OA} = \frac{12 - 0}{10 - 0} = \frac{6}{5}$$

slope of
$$\overrightarrow{OB} = \frac{-5 - 0}{6 - 0} = \frac{-5}{6}$$

$$m\angle AOB = 90°$$

c. $\overrightarrow{OA} \cdot \overrightarrow{OB} = (10, 12) \cdot (6, -5)$

$$= [10(6) + 12(-5)]$$

$$= 60 - 60 = 0$$

Consistent; since the dot product is 0, the vectors are perpendicular.

d. $(a_x, a_y) \cdot (b_x, b_y)$

Dot product $= 0$

$$a_x b_x + a_y b_y = 0$$

$$a_y b_y = -a_x b_x$$

Divide both sides by $a_x b_x$.

$$\frac{a_y b_y}{a_x b_x} = -1 \qquad \text{This is the product of the slopes.}$$

slope of $\overrightarrow{OA} = \dfrac{a_y}{a_x}$

slope of $\overrightarrow{OB} = \dfrac{b_y}{b_x}$

product of slopes $= \dfrac{a_y b_y}{a_x b_x}$

6.5 The Cross Product

In Section 6.4, you saw how the dot product of two vectors leads to a method for finding an equation of a plane when a normal vector to the plane and a point in the plane are known. Now, you will see how another relationship between vectors leads to a second method for writing an equation of a plane.

In the diagram, nonparallel vectors $\vec{a}$ and $\vec{b}$ determine plane P. Let $\vec{c}$ be a vector perpendicular to plane P. Then $\vec{c}$ must also be perpendicular to any vector in the plane. Therefore, $\vec{c}$ is perpendicular to $\vec{a}$ and to $\vec{b}$. This relationship can be used to find $\vec{c}$ when $\vec{a}$ and $\vec{b}$ are known. Since $\vec{c}$ is perpendicular to $\vec{a}$ and to $\vec{b}$:

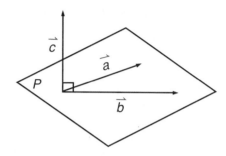

$$\vec{a} \cdot \vec{c} = 0 \text{ and } \vec{b} \cdot \vec{c} = 0$$

These relationships will be used to derive an equation of plane P.

Let $\vec{a} = (a_x, a_y, a_z)$, $\vec{b} = (b_x, b_y, b_z)$, and $\vec{c} = (c_x, c_y, c_z)$.

Then:
$$a_x c_x + a_y c_y + a_z c_z = 0 \qquad (1)$$
$$b_x c_x + b_y c_y + b_z c_z = 0 \qquad (2)$$

Since the coordinates designating $\vec{a}$ and $\vec{b}$ are known, solve equations (1) and (2) for c_x and c_y in terms of c_z.

$$a_x c_x + a_y c_y = -a_z c_z \qquad (3)$$
$$b_x c_x + b_y c_y = -b_z c_z \qquad (4)$$

Eliminate the term c_y from this system of equations by first multiplying equation (3) by b_y and equation (4) by $(-a_y)$,

$$a_x b_y c_x + a_y b_y c_y = -a_z b_y c_z$$
$$-a_y b_x c_x - a_y b_y c_y = a_y b_z c_z$$

and then adding the resulting equations.

$$(a_x b_y - a_y b_x)c_x = (a_y b_z - a_z b_y)c_z$$
$$c_x = \frac{(a_y b_z - a_z b_y)c_z}{a_x b_y - a_y b_x} \qquad (a_x b_y - a_y b_x) \neq 0$$

By a similar procedure:

$$c_y = \frac{(a_z b_x - a_x b_z)c_z}{a_x b_y - a_y b_x} \qquad (a_x b_y - a_y b_x) \neq 0$$

Since $\vec{c}$ can be any vector perpendicular to the plane, any value for c_z can be chosen. For convenience, select $a_x b_y - a_y b_x$ for c_z, simplifying the expressions for c_x and c_y.

$$c_x = a_y b_z - a_z b_y \qquad\qquad c_y = a_z b_x - a_x b_z \qquad\qquad c_z = a_x b_y - a_y b_x$$

The three equations above can be used to determine the coordinates of $\vec{c}$, a vector perpendicular to $\vec{a}$ and $\vec{b}$, if the designating coordinates of $\vec{a}$ and $\vec{b}$ are known. The expressions for c_x, c_y, and c_z can also be written in the form that follows.

For $a, b, c, d \in R$, let $\begin{vmatrix} a & b \\ c & d \end{vmatrix}$ be the symbol for the value $ad - bc$. This symbol is called a *determinant*.

In determinant form, the values for c_x, c_y, and c_z found earlier are:

$$c_x = \begin{vmatrix} a_y & a_z \\ b_y & b_z \end{vmatrix} \qquad\qquad c_y = \begin{vmatrix} a_z & a_x \\ b_z & b_x \end{vmatrix} \qquad\qquad c_z = \begin{vmatrix} a_x & a_y \\ b_x & b_y \end{vmatrix}$$

Notice that the expressions in the determinant have the remaining subscripts of the coordinates in cyclic order. For example, the determinant for c_x contains subscripts of y and z. The previous discussion has proved the following theorem.

☐ **Theorem:** *The vector*

$$\vec{c} = \left(\begin{vmatrix} a_y & a_z \\ b_y & b_z \end{vmatrix}, \begin{vmatrix} a_z & a_x \\ b_z & b_x \end{vmatrix}, \begin{vmatrix} a_x & a_y \\ b_x & b_y \end{vmatrix} \right)$$

is perpendicular to the plane determined by $\vec{a} = (a_x, a_y, a_z)$ *and* $\vec{b} = (b_x, b_y, b_z)$. *In this form,* $\vec{c}$ *is called the **cross product** of* $\vec{a}$ *and* $\vec{b}$, *written* $\vec{a} \times \vec{b}$. *Note that the cross product of two vectors is also a vector.*

Example 1 ————————————————————————

Verify that the vector $(0, 0, 1)$ is perpendicular to the plane determined by the vectors $(1, 0, 0)$ and $(0, 1, 0)$.

Solution

It must be shown that $(0, 0, 1) = (1, 0, 0) \times (0, 1, 0)$.

$$(1, 0, 0) \times (0, 1, 0) = \left(\begin{vmatrix} 0 & 0 \\ 1 & 0 \end{vmatrix}, \begin{vmatrix} 0 & 1 \\ 0 & 0 \end{vmatrix}, \begin{vmatrix} 1 & 0 \\ 0 & 1 \end{vmatrix} \right)$$
$$= (0 - 0, 0 - 0, 1 - 0) = (0, 0, 1)$$

Note: Example 1 has simply verified that $\vec{k} = \vec{i} \times \vec{j}$, meaning that the z-axis unit vector is perpendicular to the *xy*-plane.

Example 2 ————————————————————————

Let $\vec{a} = 3\vec{i} + 5\vec{j} + 6\vec{k}$ and $\vec{b} = -4\vec{i} + 7\vec{j} + 3\vec{k}$.

a. Find $\vec{a} \times \vec{b}$.

b. Write an equation of the plane determined by $\vec{a}$ and $\vec{b}$.

Solution

a. $\vec{a} = (a_x, a_y, a_z) = (3, 5, 6), \vec{b} = (b_x, b_y, b_z) = (-4, 7, 3)$.

Let $\vec{c} = \vec{a} \times \vec{b}$. Then:

$$c_x = \begin{vmatrix} a_y & a_z \\ b_y & b_z \end{vmatrix} = \begin{vmatrix} 5 & 6 \\ 7 & 3 \end{vmatrix} = 15 - 42 = -27$$

$$c_y = \begin{vmatrix} a_z & a_x \\ b_z & b_x \end{vmatrix} = \begin{vmatrix} 6 & 3 \\ 3 & -4 \end{vmatrix} = -24 - 9 = -33$$

$$c_z = \begin{vmatrix} a_x & a_y \\ b_x & b_y \end{vmatrix} = \begin{vmatrix} 3 & 5 \\ -4 & 7 \end{vmatrix} = 21 - (-20) = 41$$

Answer: $\vec{a} \times \vec{b} = (-27, -33, 41)$

b. Only one of the vectors, $\vec{a}$ or $\vec{b}$, needs to be used to find a vector in the plane. Recall that a vector from a known point (a_x, a_y, a_z) to any other point of the plane (x, y, z) can be represented by $(x - a_x, y - a_y, z - a_z)$. Since $\vec{a} \times \vec{b}$ is orthogonal to the plane and therefore to every vector in the plane, an equation can be written by letting the dot product equal 0.

$$(\vec{a} \times \vec{b}) \cdot (x - 3, y - 5, z - 6) = 0$$
$$(-27, -33, 41) \cdot (x - 3, y - 5, z - 6) = 0$$
$$-27(x - 3) - 33(y - 5) + 41(z - 6) = 0$$
$$-27x + 81 - 33y + 165 + 41z - 246 = 0$$
$$-27x - 33y + 41z = 0$$
$$27x + 33y - 41z = 0 \qquad \text{\textit{Multiply each side by} } -1.$$

Answer: An equation of the plane is $27x + 33y - 41z = 0$.

To find an equation of a plane when the coordinates of three noncollinear points are known, involve the three points two at a time to determine two distinct vectors. (Recall that two points determine a line, and that three noncollinear points determine a plane.)

Example 3

Write an equation for the plane containing the points $A(1, 4, 3)$, $B(2, 1, 5)$, and $C(3, 2, 1)$.

Solution

$$\overrightarrow{AB} = (2 - 1)\vec{i} + (1 - 4)\vec{j} + (5 - 3)\vec{k} = \vec{i} - 3\vec{j} + 2\vec{k}$$

$$\overrightarrow{BC} = (3 - 2)\vec{i} + (2 - 1)\vec{j} + (1 - 5)\vec{k} = \vec{i} + \vec{j} - 4\vec{k}$$

$\overrightarrow{AB} \times \overrightarrow{BC}$ will determine a vector that is a normal to the plane determined by $\overrightarrow{AB}$ and $\overrightarrow{BC}$. This normal vector and either $\overrightarrow{AB}$ or $\overrightarrow{BC}$ can then be used to determine an equation of the plane.

$$\overrightarrow{AB} \times \overrightarrow{BC} = \left(\begin{vmatrix} -3 & 2 \\ 1 & -4 \end{vmatrix}, \begin{vmatrix} 2 & 1 \\ -4 & 1 \end{vmatrix}, \begin{vmatrix} 1 & -3 \\ 1 & 1 \end{vmatrix} \right)$$
$$= (10, 6, 4)$$

(10, 6, 4) is the normal vector to the plane, and therefore the dot product of this normal vector with any vector in the plane containing point B is 0.

$$(10, 6, 4) \cdot (x - 2, y - 1, z - 5) = 0$$
$$10(x - 2) + 6(y - 1) + 4(z - 5) = 0$$
$$10x - 20 + 6y - 6 + 4z - 20 = 0$$
$$10x + 6y + 4z - 46 = 0$$
$$5x + 3y + 2z - 23 = 0$$

Answer: $5x + 3y + 2z - 23 = 0$ is an equation of the plane.

The xy-plane contains the vectors $\vec{i}$ and $\vec{j}$ and is perpendicular to the vector $\vec{k}$ at $(0, 0, 0)$.

If (x, y, z) is any point of the plane, then $(x - 0, y - 0, z - 0)$ is a vector in the plane and $\vec{k} = (0, 0, 1)$ is perpendicular to the plane. Therefore, an equation of the plane is:

$$(0, 0, 1) \cdot (x - 0, y - 0, z - 0) = 0$$
$$0 + 0 + z = 0$$
$$z = 0$$

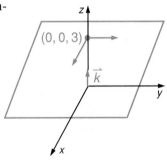

If a plane is parallel to one of the coordinate planes, the coefficients of two of the variables are 0.

A plane parallel to the xy-plane and 3 units above it contains the point $(0, 0, 3)$ and is perpendicular to the vector $\vec{k}$. If (x, y, z) is any point of the plane, then $(x - 0, y - 0, z - 3)$ is a vector in the plane, and $(0, 0, 1)$ is perpendicular to the plane. Therefore, an equation of the plane is:

$$(0, 0, 1) \cdot (x - 0, y - 0, z - 3) = 0$$
$$0 + 0 + z - 3 = 0$$
$$z = 3$$

Since $\vec{a} \times \vec{b}$ is a vector, it is important to specify its direction, which can be determined by using three fingers of your right hand. Arrange your thumb, index finger, and middle finger as in the diagram. If $\vec{a}$ is your index finger and $\vec{b}$ your middle finger, then your thumb points in the direction of $\vec{a} \times \vec{b}$. The vector $\vec{b} \times \vec{a}$ is in the opposite direction. This model can be verified by selecting two vectors, $\vec{a}$ and $\vec{b}$, in the xy-plane and showing that $\vec{a} \times \vec{b} = -(\vec{b} \times \vec{a})$, that is, that $\vec{a} \times \vec{b}$ results in a vector that is opposite in direction to $\vec{b} \times \vec{a}$.

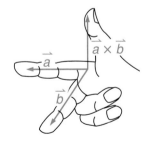

1. $(4, -4, -4)$

2. $(-6, -2, 13)$

3. $(47, -4, 33)$

4. $(0, 0, -7)$

5. $(-30, 0, 0)$

6. $-8\vec{i} + 4\vec{j} + 16\vec{k}$

7. $15\vec{i} + 76\vec{j} - 22\vec{k}$

8. $-11\vec{k}$

9. $-2\vec{i} - 7\vec{j} + 14\vec{k}$

10. $6\vec{j}$

Example 4

Let $\vec{a} = (1, 3, 2)$ and $\vec{b} = (-1, 1, 5)$. Find $\vec{a} \times \vec{b}$ and $\vec{b} \times \vec{a}$.

Solution

$$\vec{a} \times \vec{b} = \left(\begin{vmatrix} 3 & 2 \\ 1 & 5 \end{vmatrix}, \begin{vmatrix} 2 & 1 \\ 5 & -1 \end{vmatrix}, \begin{vmatrix} 1 & 3 \\ -1 & 1 \end{vmatrix} \right) = (15 - 2, -2 - 5, 1 - (-3))$$

$$= (13, -7, 4)$$

$$\vec{b} \times \vec{a} = \left(\begin{vmatrix} 1 & 5 \\ 3 & 2 \end{vmatrix}, \begin{vmatrix} 5 & -1 \\ 2 & 1 \end{vmatrix}, \begin{vmatrix} -1 & 1 \\ 1 & 3 \end{vmatrix} \right) = (2 - 15, 5 - (-2), -3 - 1)$$

$$= (-13, 7, -4)$$

Therefore, $\vec{a} \times \vec{b} = -(\vec{b} \times \vec{a})$. Cross product is not a commutative operation.

However, cross product does distribute over vector addition.

$$\vec{a} \times (\vec{b} + \vec{c}) = (\vec{a} \times \vec{b}) + (\vec{a} \times \vec{c})$$

If $\vec{a}$ and $\vec{b}$ have the same direction, then one vector is a scalar multiple of the other. In that case, $\vec{a} \times \vec{b} = \vec{0}$. (Recall that $\vec{0}$, represented by $(0, 0, 0)$ is the zero vector, and it has no length and no direction.)

Exercises 6.5

In **1 – 10**, find $\vec{a} \times \vec{b}$.

1. $\vec{a} = (1, 3, -2)$; $\vec{b} = (1, -1, 2)$

2. $\vec{a} = (4, 1, 2)$; $\vec{b} = (-1, 3, 0)$

3. $\vec{a} = (-3, 6, 5)$; $\vec{b} = (-2, -7, 2)$

4. $\vec{a} = (2, 3, 0)$; $\vec{b} = (1, -2, 0)$

5. $\vec{a} = (0, 6, 0)$; $\vec{b} = (0, 0, -5)$

6. $\vec{a} = 3\vec{i} + 2\vec{j} + \vec{k}; \vec{b} = \vec{i} + 6\vec{j} - \vec{k}$

7. $\vec{a} = -2\vec{i} + 3\vec{j} + 9\vec{k}; \vec{b} = 8\vec{i} - \vec{j} + 2\vec{k}$

8. $\vec{a} = 3\vec{i} + 5\vec{j}$; $\vec{b} = \vec{i} - 2\vec{j}$

9. $\vec{a} = 7\vec{i} - 2\vec{j}$; $\vec{b} = 2\vec{j} + \vec{k}$

10. $\vec{a} = 3\vec{i}$; $\vec{b} = -2\vec{k}$

11. Show that the vector found as the answer in Exercise 1 is perpendicular to the given $\vec{a}$ and $\vec{b}$.

12. Show that the vector found as the answer in Exercise 2 is perpendicular to the given $\vec{a}$ and $\vec{b}$.

13. Find $\vec{c}$ such that it is perpendicular to $\vec{a} = 4\vec{i} + 2\vec{j} - 3\vec{k}$ and $\vec{b} = 9\vec{i} - \vec{j} + 2\vec{k}$.

14. Find $\vec{r}$ such that it is perpendicular to plane P containing $\vec{a} = 6\vec{i} - 3\vec{j} + \vec{k}$ and $\vec{b} = \vec{i} - \vec{j} + 2\vec{k}$.

---

11. $\vec{a} = (1, 3, -2)$ $\vec{b} = (1, -1, 2)$ $\vec{n} = (4, -4, -4)$

$\vec{a} \cdot \vec{n} = (1, 3, -2) \cdot (4, -4, -4) = 1(4) + 3(-4) + (-2)(-4)$
$= 4 - 12 + 8$
$= 0$

$\vec{b} \cdot \vec{n} = (1, -1, 2) \cdot (4, -4, -4) = 1(4) + (-1)(-4) + 2(-4)$
$= 4 + 4 - 8$
$= 0$

$\vec{n} \perp \vec{a}$ and $\vec{n} \perp \vec{b}$ since the respective dot products are 0.

12. $\vec{a} = (4, 1, 2)$ $\vec{b} = (-1, 3, 0)$ $\vec{n} = (-6, -2, 13)$

$\vec{a} \cdot \vec{n} = (4, 1, 2) \cdot (-6, -2, 13) = 4(-6) + 1(-2) + 2(13)$
$= -24 - 2 + 26$
$= 0$

$\vec{b} \cdot \vec{n} = (-1, 3, 0) \cdot (-6, -2, 13) = -1(-6) + 3(-2) + 0(13)$
$= 6 - 6 + 0$
$= 0$

$\vec{n} \perp \vec{a}$ and $\vec{n} \perp \vec{b}$ since the respective dot products are equal to 0.

Exercises 6.5 (continued)

In **15 – 18**, write an equation for the plane containing the given points.

15. (2, 4, 5), (1, 4, 1), (2, 1, 3)

16. (3, –1, 4), (2, 1, –7), (4, 8, –3)

17. (3, 0, 2), (4, –1, 0), (–4, 3, 1)

18. (5, –2, –1), (6, 4, –3), (2, 1, 0)

19. Write an equation of the plane parallel to the *xy*-plane and 5 units above it.

20. Write an equation of a plane parallel to the *yz*-plane and passing through (–2, 0, 0).

21. Write an equation of the plane parallel to the *z*-axis and passing through (0, 2, 0) and (5, 0, 0).

22. Write an equation of the plane parallel to the *y*-axis and passing through (0, 0, 3) and (12, 0, 0).

23. The vertices of triangle *ABC* are *A*(6, 0, 1), *B*(–2, 3, 4), and *C*(1, 7, –2). Find an equation of the plane containing triangle *ABC*.

24. Show that if $\vec{a} = (a_x, a_y, a_z)$ and $\vec{b} = k\vec{a}$, where *k* is a scalar constant, then $\vec{a} \times \vec{b} = \vec{0}$.

25. It can be shown that $|\vec{a} \times \vec{b}|$ is equal to the area of the parallelogram formed with $\vec{a}$ and $\vec{b}$ as adjacent sides. Given the points *A*(0, 0, 0), *B*(1, 1, 2), and *C*(1, 1, –2), find the area of parallelogram *ABCD* where *D* is the fourth point needed to complete the parallelogram.

26. Find the area of the parallelogram whose adjacent sides are $\vec{a} = (1, -6, 4)$ and $\vec{b} = (5, -10, 7)$. (Express your answer to the nearest integer.)

27. Find the area of parallelogram *ABCD*, given the points *A*(0, 0, 0), *B*(4, 2, –3), and *C*(1, 4, –1).

28. Find the area of the triangle described in Exercise 23.

29. Find an equation of the *yz*-plane.

30. Find an equation of the plane through (–4, 0, 0) parallel to the *yz*-plane.

31. A plane through (2, 0, 0) and (0, 2, 0) is parallel to the *z*-axis. Find its equation. *Hint*: Any point (2, 0, *a*) is also a point of the plane.

32. A plane through (0, –1, 0) and (0, 0, 3) is parallel to the *x*-axis. Find its equation.

Answers 6.5

13. $\vec{i} - 35\vec{j} - 22\vec{k}$

14. $-5\vec{i} - 11\vec{j} - 3\vec{k}$

15. $-12x - 2y + 3z + 17 = 0$

16. $85x - 18y - 11z - 229 = 0$

17. $7x + 15y - 4z - 13 = 0$

18. $12x + 5y + 21z - 29 = 0$

19. $z = 5$

20. $x = -2$

21. $2x + 5y - 10 = 0$

22. $x + 4z - 9 = 0$

23. $30x + 39y + 41z - 221 = 0$

24.

$$\vec{a} = (a_x, a_y, a_z) \qquad \vec{b} = k\vec{a} = (ka_x, ka_y, ka_z)$$

$$\vec{a} \times \vec{b} = (a_x, a_y, a_z) \times (ka_x, ka_y, ka_z)$$

$$= \left(\begin{vmatrix} a_y & a_z \\ ka_y & ka_z \end{vmatrix}, \begin{vmatrix} a_z & a_x \\ ka_z & ka_x \end{vmatrix}, \begin{vmatrix} a_x & a_y \\ ka_x & ka_y \end{vmatrix} \right)$$

$$= (a_y ka_z - a_z ka_y, a_z ka_x - ka_z a_x, a_x ka_y - ka_x a_y)$$

$$= (k(a_y a_z - a_z a_y), k(a_z a_x - a_z a_x), k(a_x a_y - a_x a_y))$$

$$= (k(0), k(0), k(0)) = (0, 0, 0)$$

25. $4\sqrt{2}$

26. 24

27. $\sqrt{297}$

28. $\frac{1}{2}\sqrt{4,102}$

29. $x = 0$

30. $x = -4$

31. $2x + 2y - 4 = 0$

32. $3y - z + 3 = 0$

6.6 Lines in Space

Just as in the 2-dimensional plane, two points in 3-dimensional space determine a line. Therefore, the equation of a line can be written in terms of the coordinates of two points, $A(a_x, a_y, a_z)$ and $B(b_x, b_y, b_z)$. The line determined by $\overrightarrow{AB}$ is the line that contains $\overrightarrow{AB}$.

Let $P(x, y, z)$ be any point other than A or B on $\overleftrightarrow{AB}$ such that $\overrightarrow{AP}$ is a scalar multiple of $\overrightarrow{AB}$. Then the line can be described in terms of $\overrightarrow{AP}$.

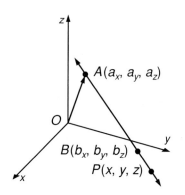

$$\overrightarrow{AB} = (b_x - a_x)\vec{i} + (b_y - a_y)\vec{j} + (b_z - a_z)\vec{k}$$

$$\overrightarrow{AP} = (x - a_x)\vec{i} + (y - a_y)\vec{j} + (z - a_z)\vec{k}$$

Since $\overrightarrow{AP}$ is a scalar multiple of $\overrightarrow{AB}$:

$$\overrightarrow{AP} = t(\overrightarrow{AB}), \ t \in R$$

$$\overrightarrow{AP} = t(b_x - a_x)\vec{i} + t(b_y - a_y)\vec{j} + t(b_z - a_z)\vec{k}$$

Therefore: $x - a_x = t(b_x - a_x)$ or $x = t(b_x - a_x) + a_x$

$y - a_y = t(b_y - a_y)$ or $y = t(b_y - a_y) + a_y$

$z - a_z = t(b_z - a_z)$ or $z = t(b_z - a_z) + a_z$

Together, these three equations are the *vector equations* of a line in space. That is, all three equations are needed to describe one line in space. The variable t, to which arbitrary values may be assigned, is called a *parameter*, and the equations are called *parametric equations*. In these vector or parametric equations, note that the coefficients of t are the coordinates of $\overrightarrow{AB}$, and the constant terms are the coordinates of $\overrightarrow{OA}$.

Another form of the equations for a line in space can be found by solving each equation above for $-t$ and replacing $-t$ by t since t can be any number.

$$t = \frac{x - a_x}{a_x - b_x} \qquad\qquad t = \frac{y - a_y}{a_y - b_y} \qquad\qquad t = \frac{z - a_z}{a_z - b_z}$$

Therefore: $\dfrac{x - a_x}{a_x - b_x} = \dfrac{y - a_y}{a_y - b_y} = \dfrac{z - a_z}{a_z - b_z}$

This is called the *two-point form* of the equation of a line in space.

Example 1

a. Write the parametric equations of the line passing through $A(2, -1, 4)$ and $B(-3, 2, 0)$.

b. Write the two-point form of the equation for $\overleftrightarrow{AB}$.

Solution

a. $x - a_x = t(b_x - a_x)$ $\qquad$ $y - a_y = t(b_y - a_y)$ $\qquad$ $z - a_z = t(b_z - a_z)$

$\quad$ $x - 2 = t(-3 - 2)$ $\qquad$ $y - (-1) = t(2 - (-1))$ $\qquad$ $z - 4 = t(0 - 4)$

$\quad$ $x - 2 = -5t$ $\qquad\qquad$ $y + 1 = 3t$ $\qquad\qquad\qquad$ $z - 4 = -4t$

Answer: Parametric equations of $\overleftrightarrow{AB}$ are:

$$x = -5t + 2, \quad y = 3t - 1, \quad z = -4t + 4$$

b. $\dfrac{x - 2}{2 - (-3)} = \dfrac{y - (-1)}{-1 - 2} = \dfrac{z - 4}{4 - 0}$

Answer: The two-point form is: $\dfrac{x - 2}{5} = \dfrac{y + 1}{-3} = \dfrac{z - 4}{4}$

The equations of the line are not unique. An equivalent set of equations could be obtained by using $\overrightarrow{BA}$ and $\overrightarrow{BP}$.

Replacing t by any constant in the parametric equations of a line yields the coordinates of a point on that line. For example, reconsider $\overleftrightarrow{AB}$ of Example 1. If t is replaced by 3 in each of the parametric equations, the result is the point $(-13, 8, -8)$, which is on $\overleftrightarrow{AB}$. Note that the coordinates of A can be found by letting $t = 0$ and the coordinates of B by letting $t = 1$.

Furthermore, the parametric equations of a line can be used to find the coordinates of any point on the line when one coordinate is known. For example, to find the coordinates of C on $\overleftrightarrow{AB}$ of Example 1, given that the x-coordinate of C is 7:

$\quad$ $7 = -5(t) + 2$ $\qquad\qquad$ Substitute in the equation for x to

$\quad$ $5 = -5t$ $\qquad\qquad\qquad$ find the value of t.

$\quad$ $t = -1$

$\quad$ $y = 3(-1) - 1 = -4$ $\qquad$ Use the value of t to evaluate the

$\quad$ $z = -4(-1) + 4 = 8$ $\qquad$ other coordinates.

$\quad$ $C(7, -4, 8)$.

Example 2

For the line passing through $R(1, -2, 5)$ and $S(2, -2, 3)$:
a. Find the parametric equations.
b. Find the two-point form.
c. Describe the position of the line in space.

Solution

a. $x - 1 = (2 - 1)t$ $\qquad$ or $\qquad$ $x = t + 1$

$\quad$ $y + 2 = (-2 - [-2])t$ $\quad$ or $\qquad$ $y = -2$

$\quad$ $z - 5 = (3 - 5)t$ $\qquad$ or $\qquad$ $z = -2t + 5$

Answer: The parametric equations of $\overleftrightarrow{RS}$ are:

$$x = t + 1, \quad y = -2, \quad z = -2t + 5$$

b. $\dfrac{x - 1}{1 - 2} = \dfrac{y - (-2)}{-2 - (-2)} = \dfrac{z - 5}{5 - 3}$

Note that since 0 cannot be the value of a denominator, the equation must be written in two parts.

Answer: The two-point form is: $\dfrac{x - 1}{-1} = \dfrac{z - 5}{2}, \quad y = -2$

c. $\overleftrightarrow{RS}$ is in the plane $y = -2$, which is a plane parallel to the xz-plane.

Example 3 _____

For the line passing through $C(1, 2, 3)$ and $D(5, 2, 3)$:

a. Write the parametric equations.

b. Describe the position of the line in space.

Solution

a. $x - 1 = 4t$ $y - 2 = 0t$ $z - 3 = 0t$

Answer: The parametric equations of $\overleftrightarrow{CD}$ are:

$$x = 4t + 1, \quad y = 2, \quad z = 3$$

b. Note that for all values of t in the equations, $y = 2$ and $z = 3$. Thus, this line is the intersection of the plane $y = 2$, which is parallel to the xz-plane, and the plane $z = 3$, which is parallel to the xy-plane. Such a line is parallel to the x-axis.

Answer: $\overleftrightarrow{CD}$ is a line that is parallel to the x-axis.

Using Cross Products to Find an Equation of a Line

If two planes, $A_x + B_y + C_z + D = 0$ and $A'_x + B'_y + C'_z + D' = 0$, are neither parallel nor identical, then they intersect in a line. If planes P and P' have normal vectors $\vec{n}$ and $\vec{n'}$, respectively, and the planes intersect, then $\vec{n}$ and $\vec{n'}$ are each perpendicular to the line of intersection formed by the planes. Therefore, $\vec{n} \times \vec{n'}$ is a vector perpendicular to $\vec{n}$ and $\vec{n'}$ and parallel to the line of intersection of planes P and P'. Let $C(c_x, c_y, c_z)$ be a point on line ℓ, the intersection of planes P and P', and $D(x, y, z)$ be any other point on line ℓ. Then:

$$\overrightarrow{CD} = (x - c_x)\vec{i} + (y - c_y)\vec{j} + (z - c_z)\vec{k}.$$

But $\overrightarrow{CD} \parallel (\vec{n} \times \vec{n'})$, and when two vectors are parallel, one is a scalar multiple of the other. Thus, $\overrightarrow{CD} = t(\vec{n} \times \vec{n'})$, where t is a scalar. Therefore:

$$\overrightarrow{CD} = (x - c_x)\vec{i} + (y - c_y)\vec{j} + (z - c_z)\vec{k} = t(\vec{n} \times \vec{n'})$$

This last equation can be used to define line ℓ. By simplifying this equation and recalling that $\vec{n} = (A, B, C)$ and $\vec{n'} = (A', B', C')$ then:

$$(x - c_x)\vec{i} + (y - c_y)\vec{j} + (z - c_z)\vec{k} = t\left(\begin{vmatrix} B & C \\ B' & C' \end{vmatrix} \vec{i} + \begin{vmatrix} C & A \\ C' & A' \end{vmatrix} \vec{j} + \begin{vmatrix} A & B \\ A' & B' \end{vmatrix} \vec{k} \right)$$

Therefore, the parametric equations of the line can be written in terms of three equations describing the individual vector components.

$$x - c_x = t \begin{vmatrix} B & C \\ B' & C' \end{vmatrix} \qquad y - c_y = t \begin{vmatrix} C & A \\ C' & A' \end{vmatrix} \qquad z - c_z = t \begin{vmatrix} A & B \\ A' & B' \end{vmatrix}$$

Example 4

  a. Find the parametric equations that describe the line formed by the intersection of planes P and P' whose equations are $2x + y + z - 5 = 0$ and $x + 3y + 2z + 1 = 0$, respectively.

  b. Find another point on the line different from the one found in part **a** and show that it lies on both planes.

Solution

  a. The desired line ℓ will be described in terms of the vector $\vec{\ell}$, which is the cross product of the normal vectors $\vec{n}$ and $\vec{n}'$ to the two planes P and P', respectively.

$$\vec{\ell} = \vec{n} \times \vec{n}'$$

From the coefficients of x, y, and z in the equations of the planes,

$$\vec{n} = (2, 1, 1) \text{ and } \vec{n}' = (1, 3, 2)$$

Therefore:

$$\vec{\ell} = \vec{n} \times \vec{n}' = \left(\begin{vmatrix} 1 & 1 \\ 3 & 2 \end{vmatrix}, \begin{vmatrix} 1 & 2 \\ 2 & 1 \end{vmatrix}, \begin{vmatrix} 2 & 1 \\ 1 & 3 \end{vmatrix} \right)$$
$$= (2 - 3, 1 - 4, 6 - 1)$$
$$= (-1, -3, 5)$$

Thus, $\vec{\ell} = -\vec{i} - 3\vec{j} + 5\vec{k}$ is a vector contained in or parallel to line ℓ, that is, having the same direction as line ℓ.

One point on the line is now needed in order to form the three parametric equations. By letting $x = 0$ in the equations representing planes P and P', the point where the line intersects the yz-plane can be found.

$$2(0) + y + z - 5 = 0$$
$$y + z = 5 \qquad \text{From the equation of plane } P$$
$$0 + 3y + 2z + 1 = 0$$
$$3y + 2z = -1 \qquad \text{From the equation of plane } P'$$

Solving the system above, $z = 16$

$$y = -11$$

Therefore, $(0, -11, 16)$ is a point on the line and

$$x - 0 = t(-1)$$
$$y - (-11) = t(-3)$$
$$z - 16 = t(5)$$

Answer: The three parametric equations, for $t \in R$, are:

$$x = -t$$
$$y + 11 = -3t$$
$$z - 16 = 5t$$

b. By selecting a value for t, additional points on ℓ can be found. Let $t = 2$.

$$x = -(2) \quad \text{or} \quad x = -2$$
$$y + 11 = -3(2) \quad \text{or} \quad y = -17$$
$$z - 16 = 5(2) \quad \text{or} \quad z = 26$$

Check: The point $(-2, -17, 26)$ can be checked in the equation representing each plane to verify that it is a point lying in each plane.

$2x + y + z - 5 = 0$	$x + 3y + 2z + 1 = 0$
$2(-2) + (-17) + 26 - 5 = 0$	$-2 + 3(-17) + 2(26) + 1 = 0$
$-4 - 17 + 26 - 5 = 0$	$-2 - 51 + 52 + 1 = 0$
$0 = 0$ ✔	$0 = 0$ ✔

Answer: The point $(-2, -17, 26)$ is another point on ℓ.

Suppose the equations of two planes are $x + 2y + 3z + 4 = 0$ and $2x + 4y + 6z + 5 = 0$. Then their normal vectors are $\vec{n} = (1, 2, 3)$ and $\vec{n'} = (2, 4, 6)$. Since $\vec{\ell} = \vec{n} \times \vec{n'}$:

$$\vec{\ell} = \left(\begin{vmatrix} 2 & 3 \\ 4 & 6 \end{vmatrix}, \begin{vmatrix} 3 & 1 \\ 6 & 2 \end{vmatrix}, \begin{vmatrix} 1 & 2 \\ 2 & 4 \end{vmatrix} \right) = (0, 0, 0)$$

But $(0, 0, 0)$ is the zero vector, $\vec{0}$, suggesting that there is no line ℓ in which the planes intersect and, therefore, that the planes are parallel. This was predictable from the fact that the coefficients, but not the constants, in the equation of one of the planes are twice the coefficients in the other equation.

Exercises 6.6

Note: When the solution to an exercise is an equation of a line or of a plane, there are many possible answers.

In **1 – 6**, find parametric equations of $\overrightarrow{AB}$ for the given coordinates of points A and B.

1. $A(0, 2, 3); B(1, 0, 4)$ **2.** $A(1, -1, 2); B(-3, 2, -2)$ **3.** $A(4, -2, 5); B(3, 0, -4)$

4. $A(4, 3, 2); B(1, 3, 4)$ **5.** $A(2, -3, 1); B(2, -3, 3)$ **6.** $A(5, 0, 3); \quad B(-1, 0, 3)$

7. Write the two-point form of an equation of the line passing through the points $(2, -3, 4)$ and $(5, 2, -1)$.

8. a. Write the two-point form of the equation of the line passing through the points $(4, -5, 3)$ and $(2, -1, 1)$.

     b. Given that the point $(6, y, z)$ is on the line, find the values of y and z.

9. Find an equation of the plane parallel to the plane whose equation is $5x + 2y - z = 0$ and passing through the point $(6, 1, -3)$.

10. Find an equation of the plane parallel to the plane whose equation is $5x + y - 7 = 0$ and passing through the point $(3, 1, 4)$.

---

1. $x = t + 1, y = -2t, z = t + 4$

2. $x = -4t - 3, y = 3t + 2, z = -4t - 2$

3. $x = -t + 3, y = 2t, z = -9t - 4$

4. $x = -3t + 1, y = 3, z = 2t + 4$

5. $x = 2, y = -3, z = 2t + 3$

6. $x = -6t - 1, y = 0, z = 3$

7. $\frac{x-5}{3} = \frac{y-2}{5} = \frac{z+1}{-5}$ or $\frac{x-2}{-3} = \frac{y+3}{-5} = \frac{z-4}{5}$

8. a. $\frac{x-2}{-2} = \frac{y+1}{4} = \frac{z-1}{-2}$ or $\frac{x-4}{2} = \frac{y+5}{-4} = \frac{z-3}{2}$

     b. If $x = 6$

$$\frac{6-2}{-2} = \frac{y+1}{4} \qquad y = -9$$

$$\frac{6-2}{-2} = \frac{z-1}{-2} \qquad z = 5$$

9. $5x + 2y - z - 35 = 0$

10. $5x + y - 16 = 0$

Exercises 6.6 *(continued)*

11. If two planes are perpendicular, then their normal vectors are perpendicular. Explain how you would find one of the normal vectors perpendicular to the normal vector for the plane whose equation is $2x + y + 3z + 1 = 0$.

12. Find an equation of a plane perpendicular to $2x + y + 3z + 1 = 0$ and passing through the point $(1, 4, 3)$.

13. Find an equation of a plane perpendicular to $6x - 4y + 3z + 10 = 0$ and passing through the point $(-4, 3, 2)$.

14. Find an equation of a plane perpendicular to $6x - 4y + 3z + 10 = 0$, passing through the point $(-2, 1, 0)$, and not parallel to the plane whose equation was found in Exercise 13.

In **15 – 22**, find a set of equations that can be used to describe the line formed by the intersection of the planes whose equations or descriptions are given.

15. $x + y + z + 4 = 0, 2x + 4y - 3z + 1 = 0$

16. $5x - 6y + z - 2 = 0, 2x - y + z = 0$

17. $6x - y + 3z = 0, x - 2y + 6z - 7 = 0$

18. $8x - 5y + 4 = 0, 3x + 6z - 7 = 0$

19. $x - 9y - 4z - 3 = 0, 2x - 5y + z + 9 = 0$

20. $x + y + 2z + 3 = 0, 2x - y - 8 = 0$

21. the plane parallel to the z-axis and passing through the points $(1, 1, 0)$ and $(2, 4, 0)$; the plane parallel to the x-axis and passing through the points $(0, 6, 1)$ and $(0, 3, -1)$

22. the plane parallel to the y-axis and passing through the points $(0, 0, 5)$ and $(6, 0, 0)$; the plane parallel to the x-axis and passing through the points $(0, 0, 5)$ and $(0, 4, 0)$

11. If 2 vectors are perpendicular, then their dot product is 0.
$$\vec{N} \cdot (2, 1, 3) = 0$$
$$(N_x, N_y, N_z) \cdot (2, 1, 3) = 0$$
$$2N_x + N_y + 3N_z = 0$$
Choose numbers for N_x, N_y, and N_z to make the dot product zero.

12. $x - 2y + 7 = 0$

13. $3y + 4z - 17 = 0$

14. $x - 2z + 2 = 0$

15. $x = -7t, y = 5t - \frac{13}{7}$, $z = 2t - \frac{15}{7}$

16. $x = -5t, y = -3t - \frac{2}{5}$, $z = 7t - \frac{2}{5}$

17. $x = -\frac{7}{11}, y = -33t, z = -11t + \frac{14}{11}$

18. $x = -30t - \frac{1}{2}, y = -48t, z = 15t + \frac{17}{2}$

19. $x = -29t - \frac{11}{3}, y = -9t, z = 13t - \frac{5}{3}$

20. $x = 2t + 4, y = 4t, z = -3t - \frac{7}{2}$

21. $x = 3t, y = 9t - 2, z = 6t - \frac{13}{3}$

22. $x = -30t, y = -20t, z = 25t + 5$

6.7 Distance From a Point

Suppose you wish to find the distance from a point P to a line ℓ in space as shown below. Let A and B be points on ℓ and let $\vec{v}$ be a vector parallel to ℓ such that $\overrightarrow{AB} = \vec{v}$.

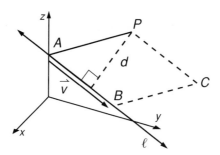

Since A is a point on ℓ and $\vec{v}$ is a vector parallel to ℓ, then it can be shown that $|\overrightarrow{AP} \times \vec{v}|$ is the area of parallelogram $ABCP$. However, the area of parallelogram $ABCP$ can also be found by multiplying the length of the base times the height, or $|\vec{v}|d$. ($|\vec{v}|$ is the magnitude of $\vec{v}$, and thus equal to the length of $\overline{AB}$). Therefore:

$$|\overrightarrow{AP} \times \vec{v}| = |\vec{v}|d, \text{ or } d = \frac{|\overrightarrow{AP} \times \vec{v}|}{|\vec{v}|}$$

The distance d from a point P to a line ℓ containing a point A is:

$$\frac{|\overrightarrow{AP} \times \vec{v}|}{|\vec{v}|}, \text{ where } \vec{v} \text{ is parallel to } \ell.$$

Example 1

Find the distance from the point $P\,(1, 2, 3)$ to the line described by the following parametric equations, where $t \in R$:

$$x + 2 = 4t$$
$$y - 1 = 2t$$
$$z + 1 = t$$

Solution

If $t = 1$, then $x = 2$, $y = 3$, and $z = 0$. The point $A\,(2, 3, 0)$ is on the line. Therefore, the vector from $A(2, 3, 0)$ on the line to $P(1, 2, 3)$ is $\overrightarrow{AP} = (1 - 2, 2 - 3, 3 - 0) = (-1, -1, 3)$. A vector parallel to the line is $\vec{v} = (4, 2, 1)$, and

$$|\vec{v}| = \sqrt{4^2 + 2^2 + 1} = \sqrt{21}$$

$$\overrightarrow{AP} \times \vec{v} = (-1, -1, 3) \times (4, 2, 1)$$

$$= \left(\begin{vmatrix} -1 & 3 \\ 2 & 1 \end{vmatrix}, \begin{vmatrix} 3 & -1 \\ 1 & 4 \end{vmatrix}, \begin{vmatrix} -1 & -1 \\ 4 & 2 \end{vmatrix} \right)$$

$$= (-7, 13, 2)$$

$$|\overrightarrow{AP} \times \vec{v}| = \sqrt{(-7)^2 + 13^2 + 2^2} = \sqrt{222}$$

$$d = \frac{|\overrightarrow{AP} \times \vec{v}|}{|\vec{v}|}$$

$$d = \frac{\sqrt{222}}{\sqrt{21}} = \frac{\sqrt{3} \cdot \sqrt{74}}{\sqrt{3} \cdot \sqrt{7}}$$

$$d = \frac{\sqrt{74}}{\sqrt{7}} \cdot \frac{\sqrt{7}}{\sqrt{7}} = \frac{\sqrt{518}}{7}$$

Answer: The distance from P to the line described is $\frac{\sqrt{518}}{7}$.

The distance from a point P to a line in the 2-dimensional plane can be found by using the dot product and the normal vector. Suppose the equation of the line is $ax + by + c = 0$, where (x, y) is any point on the line. A normal vector to the line $ax + by + c = 0$ can be found by selecting points on the line in order to establish a vector on or parallel to the line, and then using the dot product. The points $\left(0, -\frac{c}{b}\right)$ and $\left(-\frac{c}{a}, 0\right)$ are on the line $ax + by + c = 0$. Therefore, $ax + by + c = 0$ contains the vector $\left(-\frac{c}{a}, \frac{c}{b}\right)$. If $\vec{n} = (n_x, n_y)$ is normal to the line, then

$$\left(-\frac{c}{a}, \frac{c}{b}\right) \cdot (n_x, n_y) = 0$$

$$-\frac{c}{a} n_x + \frac{c}{b} n_y = 0$$

$$\frac{c}{b} n_y = \frac{c}{a} n_x$$

$$\frac{n_y}{b} = \frac{n_x}{a}$$

Therefore, $n_y = b$ and $n_x = a$ will always satisfy the equation, and thus

(a, b) is always a normal vector for the line $ax + by + c = 0$.

In the 2-dimensional plane, the distance d from a point P to a line ℓ containing point A is:

$$d = \frac{|\overrightarrow{AP} \cdot \vec{n}|}{|\vec{n}|}, \text{ where } \vec{n} \text{ is the vector normal to the line}$$

The expression $|\overrightarrow{AP} \cdot \vec{n}|$ is the absolute value of the real number that is the dot product of $\overrightarrow{AP}$ and the normal $\vec{n}$.

Example 2

Find the distance from the point $P(2, 1)$ to the line whose equation is $3x = 4y + 10$.

Solution

The vector normal to the line $3x = 4y + 10$ is (a, b) when the equation is in the form $ax + by + c = 0$. Therefore, the vector normal to $3x - 4y - 10 = 0$ is $(3, -4)$. The point $A(2, -1)$ is also on the line. Therefore:

$$\overrightarrow{AP} = (2 - 2, 1 - (-1))$$

$$d = \frac{|\overrightarrow{AP} \cdot \vec{n}|}{|\vec{n}|}$$

$$= \frac{|(0, 2) \cdot (3, -4)|}{\sqrt{3^2 + (-4)^2}}$$

$$= \frac{|0 + (-8)|}{\sqrt{25}} = \frac{8}{5}$$

Answer: The distance from P to the line is $\frac{8}{5}$.

The dot product formula for d, the distance from a point to a line, can be transformed into a formula involving only the coefficients of the equation of the line $ax + by + c = 0$ and the coordinates of the point from which the distance d is measured. Using point $P(p_x, p_y)$ and the point A on the line as before:

$$d = \frac{|\overrightarrow{AP} \cdot \vec{n}|}{|\vec{n}|}$$

$$= \frac{|(\overrightarrow{OP} - \overrightarrow{OA}) \cdot \vec{n}|}{|\vec{n}|}$$

$$= \frac{|(\overrightarrow{OP} \cdot \vec{n}) - (\overrightarrow{OA} \cdot \vec{n})|}{|\vec{n}|} \qquad \text{Dot product distributes over subtraction.}$$

$$= \frac{|(p_x, p_y) \cdot (a, b) - \overrightarrow{OA} \cdot (a, b)|}{|\vec{n}|}$$

Since $(x, y) \cdot (a, b) = ax + by$, and, for any point on the line $ax + by + c = 0$, $ax + by = -c$ then $(x, y) \cdot (a, b) = -c$. Therefore, since A is a point on the line, $\overrightarrow{OA} \cdot \vec{n} = \overrightarrow{OA} \cdot (a, b) = -c$. Therefore:

$$d = \frac{|(p_x, p_y) \cdot (a, b) - (-c)|}{|\vec{n}|}$$

$$= \frac{|ap_x + bp_y + c|}{\sqrt{a^2 + b^2}}$$

The formula for the distance d from the point whose coordinates are (p_x, p_y) to the line whose equation is $ax + by + c = 0$ is:

$$d = \frac{|ap_x + bp_y + c|}{\sqrt{a^2 + b^2}}$$

Example 3

Find the distance from the point $(-3, 5)$ to the line whose equation is $x + y + 8 = 0$.

Solution

If $x + y + 8 = 0$, then $a = 1$, $b = 1$, and $c = 8$.

$$d = \frac{|ap_x + bp_y + c|}{\sqrt{a^2 + b^2}}$$

$$= \frac{|(1)(-3) + (1)(5) + 8|}{\sqrt{1^2 + 1^2}}$$

$$= \frac{10}{\sqrt{2}} = \frac{10}{\sqrt{2}} \cdot \frac{\sqrt{2}}{\sqrt{2}} = 5\sqrt{2}$$

Answer: The distance from the point to the line is $5\sqrt{2}$.

A similar analysis can be used to derive a formula in 3-dimensional space. The distance from a point $P(p_x, p_y, p_z)$ to a plane $Ax + By + Cz + D = 0$ is:

$$d = \frac{|Ap_x + Bp_y + Cp_z + D|}{\sqrt{A^2 + B^2 + C^2}}, \text{ where } (A, B, C) \text{ is a vector normal to the plane.}$$

Example 4

Find the distance from the point $P(6, 1, -2)$ to the plane whose equation is $8x - 12y - 9z - 28 = 0$.

Solution

$$d = \frac{|Ap_x + Bp_y + Cp_z + D|}{\sqrt{A^2 + B^2 + C^2}}$$

$$= \frac{|8(6) + (-12)(1) + (-9)(-2) - 28|}{\sqrt{8^2 + (-12)^2 + (-9)^2}}$$

$$= \frac{|48 - 12 + 18 - 28|}{\sqrt{64 + 144 + 81}} = \frac{26}{\sqrt{289}} = \frac{26}{17}$$

Answer: The distance from the point to the plane is $\frac{26}{17}$.

Exercises 6.7

In **1 – 8**, find the distance from the given point to the line described by the given parametric equations.

1. $P(6, 2, 0)$; $x - 4 = t$, $y - 2 = -t$, $z = 0$
2. $P(0, -1, 4)$; $x - 2 = t$, $y + 3 = t$, $z = 4$
3. $P(0, 0, 0)$; $x - 2 = -6t$, $y - 4 = 3t$, $z - 1 = t$
4. $P(1, 1, 1)$; $x = 7t$, $y + 3 = 5t$, $z - 8 = t$
5. $P(1, 1, 2)$; $x - 1 = 2t$, $y + 4 = t$, $z = 3t$
6. $P(-3, 5, 0)$; $x + 4 = -6t$, $y + 4 = -6t$, $z = 0$
7. $P(4, -3, 2)$; $x - 1 = 2t$, $y + 3 = t$, $z - 2 = 3t$
8. $P(1, 3, 2)$; $x - 5 = t$, $y - 7 = -2t$, $z - 3 = 5t$

Answers 6.7

1. $d = \sqrt{2}$

2. $d = 2\sqrt{2}$

3. $d = \sqrt{\dfrac{965}{46}}$

$= \dfrac{\sqrt{44{,}390}}{46}$

4. $d = \sqrt{\dfrac{182}{3}}$

$= \dfrac{\sqrt{546}}{3}$

5. $d = \sqrt{\dfrac{285}{14}}$

$= \dfrac{\sqrt{3{,}990}}{14}$

6. $d = 4\sqrt{2}$

7. $d = 3\sqrt{\dfrac{5}{7}}$

$= \dfrac{3\sqrt{35}}{7}$

8. $d = \sqrt{\dfrac{989}{30}}$

$= \dfrac{\sqrt{29{,}670}}{30}$

9. $d = \dfrac{4\sqrt{5}}{5}$

10. $d = \dfrac{7}{13}$

11. $d = 0$

12. $d = \dfrac{7\sqrt{13}}{13}$

13. $d = \dfrac{5\sqrt{2}}{2}$

14. $d = \dfrac{8\sqrt{29}}{29}$

15. $d = 4\sqrt{2}$

16. $d = \dfrac{4\sqrt{5}}{5}$

17. $d = \dfrac{15\sqrt{13}}{13}$

18. $d = \dfrac{21\sqrt{26}}{26}$

19. $d = \dfrac{5\sqrt{74}}{37}$

20. $d = \dfrac{12\sqrt{17}}{17}$

21. $d = \dfrac{26}{17}$

22. $d = \dfrac{3\sqrt{14}}{14}$

23. $d = \dfrac{3\sqrt{6}}{4}$

24. $d = 4$

25. $d = \dfrac{3\sqrt{10}}{5}$

26. $d = \dfrac{7\sqrt{2}}{10}$

27. $d = \dfrac{3\sqrt{14}}{14}$

28. $d = \dfrac{3\sqrt{33}}{11}$

29. $d = \dfrac{8\sqrt{6}}{9}$

30. $d = \dfrac{8}{\sqrt{234}}$

$\qquad = \dfrac{4\sqrt{26}}{39}$

In **9 – 14**, use the formula $d = \dfrac{|\overrightarrow{AP} \cdot \vec{n}|}{|\vec{n}|}$ to find the distance from the given point to the line whose equation is given.

9. $P(0, 0);\ 2x + y + 4 = 0$ **10.** $P(3, 1);\quad 5x - 12y - 10 = 0$

11. $P(-3, 5); 2x - y + 11 = 0$ **12.** $P(0, 0);\quad 2x = 3y - 7$

13. $P(3, -2);\ x - y = 0$ **14.** $P(-1, -3); 5x - 2y + 7 = 0$

In **15 – 20**, use the formula $d = \dfrac{|ap_x + bp_y + c|}{\sqrt{a^2 + b^2}}$ to find the distance from the given point to the line whose equation is given.

15. $P(6, 1);\quad x + y + 1 = 0$ **16.** $P(2, 3);\quad 2x - y + 3 = 0$

17. $P(5, -2); 3x - 2y - 4 = 0$ **18.** $P(0, -4);\quad x - 5y + 1 = 0$

19. $P(-2, 0); 5x - 7y = 0$ **20.** $P(-4, -1); 4x - y + 3 = 0$

In **21 – 26**, find the distance from the given point to the plane whose equation is given.

21. $P(6, 1, -2); 8x - 12y - 9z - 28 = 0$ **22.** $P(1, 2, -1); 3x - 2y + z - 1 = 0$

23. $P(-1, 2, 1); 4x - 2y + 2z - 3 = 0$ **24.** $P(5, 2, 4);\ 2x + y + \dfrac{z}{4} - 4 = 0$

25. $P(3, -1, -2); x + 3y + 6 = 0$ **26.** $P(-3, 0, 2); 5x - 3y + 4z = 0$

In **27 – 30**, find the distance between the two parallel planes whose equations are given.

27. $3x + 2y - z + 2 = 0,\ 3x + 2y - z + 5 = 0$ **28.** $2x - 5y + 2z - 6 = 0,\ 2x - 5y + 2z + 3 = 0$

29. $x + y - 2z - 3 = 0,\ 3x + 3y - 6z + 7 = 0$ **30.** $5x - y + 4 = 0,\ 15x - 3y + 4 = 0$

31. Derive a general rule for finding the distance from $(0, 0, 0)$ to the plane $Ax + By + Cz + D = 0$.

32. The distance between two lines, ℓ_1 and ℓ_2, in 3-dimensional space can be determined by finding the length of a perpendicular line segment drawn between the two lines. This length can be found by using the formula

$$d = \dfrac{|(\vec{v}_1 \times \vec{v}_2) \cdot \overrightarrow{P_1 P_2}|}{|\vec{v}_1 \times \vec{v}_2|}$$

where P_1 is the endpoint of the segment on ℓ_1, P_2 is the endpoint on ℓ_2, $\vec{v}_1$ is a vector parallel to ℓ_1 and $\vec{v}_2$ is a vector parallel to ℓ_2. Use this formula to find the distance, to the nearest tenth, between the lines whose parametric equations are:

$x - 4 = 20t, y - 3 = 12t, z - 5 = 9t$ and $x - 1 = 2t, y - 6 = 5t, z - 2 = 14t$

33. If line ℓ_1 passes through the points $(2, 1, -2)$ and $(1, 2, 3)$, and line ℓ_2 passes through the points $(2, -2, 2)$ and $(3, -1, -2)$, find the shortest distance between ℓ_1 and ℓ_2, to the nearest hundredth. *Hint:* Use the formula in Exercise 32.

31. $P(0, 0, 0)\qquad Ax + By + Cz + D = 0$ **32.** 4.6

$d = \dfrac{|A(0) + B(0) + C(0) + D|}{\sqrt{A^2 + B^2 + C^2}}$ **33.** 1.19

$d = \dfrac{|D|}{\sqrt{A^2 + B^2 + C^2}}$

Chapter 6   Summary and Review

Equality, Magnitude, and Direction of Vectors

- A vector is a directed line segment.
- The length of a vector is called its magnitude.
- $\overrightarrow{AB}$ has initial point A and terminal point B.
- Equal vectors are vectors with equal magnitudes that point in the same direction. Equal vectors are either parallel or lie on the same line.
- Vectors can be added by using the triangle rule or the parallelogram rule.

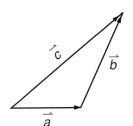

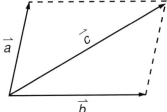

- The sum of two or more vectors is called their resultant.
- $-\vec{v}$ has the same magnitude but the opposite direction of $\vec{v}$.
- The magnitude of a vector is increased or decreased when the vector is multiplied by a scalar $k \neq \pm 1$.
- The ordered pair (x, y) represents a vector whose terminal point is (x, y) when its initial point is the origin.
- The magnitude of $\vec{v}$, written $|\vec{v}|$, can be found using the distance formula to compute the distance between its initial and terminal points. The magnitude of a vector represented by the ordered pair (x, y) is $\sqrt{x^2 + y^2}$.

Addition, Subtraction, and Scalar Multiplication

- Two vectors, $\vec{a} = (a_x, a_y)$ and $\vec{b} = (b_x, b_y)$, may be added, subtracted, or multiplied by a scalar k:
$$\vec{a} + \vec{b} = (a_x + b_x, a_y + b_y)$$
$$\vec{a} - \vec{b} = (a_x - b_x, a_y - b_y)$$
$$k\vec{a} = (ka_x, ka_y)$$
- Any vector in the 2-dimensional plane can be expressed as a linear combination of the unit vectors $\vec{i} = (1, 0)$ and $\vec{j} = (0, 1)$.

- Vectors in 3-dimensional space can be represented by an ordered triple, or as a linear combination of the unit vectors $\vec{i} = (1, 0, 0), \vec{j} = (0, 1, 0)$, and $\vec{k} = (0, 0, 1)$.

- In 3-dimensional space, the distance, d, between two points (x_1, y_1, z_1) and (x_2, y_2, z_2) is given by the formula $d = \sqrt{(x_2 - x_1)^2 + (y_2 - y_1)^2 + (z_2 - z_1)^2}$.

- The formula for the magnitude of a vector, $\vec{a} = (a_x, a_y, a_z)$, is:
$$|\vec{a}| = \sqrt{(a_x)^2 + (a_y)^2 + (a_z)^2}$$

- The coordinates of point E, which divides AB so that $AE:EB = m:n$, may be found from the vector equation:
$$\overrightarrow{OA} + \frac{m}{m + n}(\overrightarrow{AB}) = \overrightarrow{OE}$$

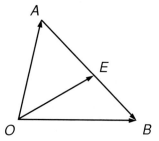

Dot Product of Vectors

- The dot product $\vec{a} \cdot \vec{b}$ of $\vec{a} = (a_x, a_y)$ and $\vec{b} = (b_x, b_y)$ is defined as:
$$\vec{a} \cdot \vec{b} = a_x b_x + a_y b_y$$

- The two vectors, $\vec{a}$ and $\vec{b}$, are perpendicular if and only if $\vec{a} \cdot \vec{b} = 0$.

- In 3-dimensional space, if $\vec{a} = (a_x, a_y, a_z)$ and $\vec{b} = (b_x, b_y, b_z)$, then:
$$\vec{a} \cdot \vec{b} = a_x b_x + a_y b_y + a_z b_z$$

- $\vec{a} \cdot \vec{b} = \vec{b} \cdot \vec{a}$

- $\vec{a} \cdot (\vec{b} + \vec{c}) = \vec{a} \cdot \vec{b} + \vec{a} \cdot \vec{c}$

- The equation of a plane in 3-dimensional space is of the form $Ax + By + Cz + D = 0$, a linear equation in x, y, and z.

- The vector (A, B, C) is normal to the plane whose equation is $Ax + By + Cz + D = 0$.

Cross Product of Vectors

- If $\vec{a} = (a_x, a_y, a_z)$ and $\vec{b} = (b_x, b_y, b_z)$, then the vector that is the cross product of $\vec{a}$ and $\vec{b}$ is defined as:

$$\vec{a} \times \vec{b} = \left(\begin{vmatrix} a_y & a_z \\ b_y & b_z \end{vmatrix}, \begin{vmatrix} a_z & a_x \\ b_z & b_x \end{vmatrix}, \begin{vmatrix} a_x & a_y \\ b_x & b_y \end{vmatrix} \right)$$

- The vector $\vec{a} \times \vec{b}$ is perpendicular to the plane determined by $\vec{a}$ and $\vec{b}$.
- $\vec{a} \times \vec{b} = -(\vec{b} \times \vec{a})$

Equation of a Plane

- Three noncollinear points determine a plane.

- If $A(a_x, a_y, a_z)$, $B(b_x, b_y, b_z)$, and $P(x, y, z)$ are 3 points on a line in 3-space, the line can be described by the parametric equations

$$x - a_x = t(b_x - a_x) \quad y - a_y = t(b_y - a_y) \quad z - a_z = t(b_z - a_z) \qquad (t \in R)$$

- If $\vec{n} = (A, B, C)$ and $\vec{n'} = (A', B', C')$ are vectors normal to intersecting planes P and P', respectively, then the parametric equations for the line of intersection are

$$x - c_x = t \begin{vmatrix} B & C \\ B' & C' \end{vmatrix} \qquad y - c_y = t \begin{vmatrix} C & A \\ C' & A' \end{vmatrix} \qquad z - c_z = t \begin{vmatrix} A & B \\ A' & B' \end{vmatrix}$$

where (c_x, c_y, c_z) is a point on the line and $t \in R$.

Distance from a Point to a Line or a Plane

- In 3-space, the distance d from a point P to a line ℓ containing point A can be determined by the formula

$$d = \frac{|\vec{AP} \times \vec{v}|}{|\vec{v}|}$$

where $\vec{v}$ is a vector parallel to ℓ.

- In 2-space, the distance d from a point P to a line ℓ containing point A can be determined by the formula

$$d = \frac{|\vec{AP} \cdot \vec{n}|}{|\vec{n}|}$$

where $\vec{n}$ is a vector normal to ℓ.

- In the 2-dimensional plane, the distance d from point $P(p_x, p_y)$ to the line whose equation is $ax + by + c = 0$ can be determined by the formula

$$d = \frac{|ap_x + bp_y + c|}{\sqrt{a^2 + b^2}}$$

- In 3-dimensional space, the distance d from point $P(p_x, p_y, p_z)$ to a plane whose equation is $Ax + By + Cz + D = 0$ can be determined by the formula

$$d = \frac{|Ap_x + Bp_y + Cp_z + D|}{\sqrt{A^2 + B^2 + C^2}}$$

Chapter 6   Review Exercises

In **1 – 4**, use the vectors shown, a marked straightedge, and a protractor to sketch the given vector.

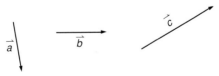

1. $\vec{a} + \vec{c}$ **2.** $\vec{c} - 2\vec{b}$

3. $\vec{a} + 2\vec{b} + 3\vec{c}$ **4.** $2\vec{c} - \vec{a}$

1.

2.

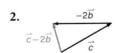

3.

4.

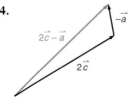

5. $2\sqrt{34}$

6. $(8, -4)$

7. $(-3, -11)$

8. $\sqrt{17}$

9. $3\sqrt{5}$

10. $\vec{i} + 7\vec{j}$

11. a. always

    b. sometimes, when $\vec{v} = \vec{w}$

    c. sometimes, when $\vec{v} = \vec{0}$

    d. always

    e. always

    f. sometimes, when $a = b$

    g. always

    h. never

12a.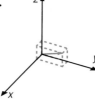

b.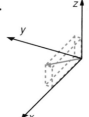

c.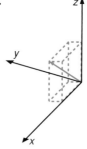

5. Forces of 10 pounds and 6 pounds are applied at a 90° angle to each other. Find the magnitude of the resultant.

6. Find the coordinates of the head of the vector $(2, -3)$ if its tail is at the point $(6, -1)$.

7. Write the ordered pair notation for a vector whose head and tail are at the points $(-1, -8)$ and $(2, 3)$, respectively.

8. Find the magnitude of the vector represented by the ordered pair $(-1, 4)$.

9. Find the magnitude of the vector whose head and tail are at the points $(4, 2)$ and $(10, -1)$, respectively.

10. Express the difference vector $(3\vec{i} + \vec{j}) - (2\vec{i} - 6\vec{j})$ as a single vector.

11. Determine whether the given statement is *always*, *sometimes*, or *never* true. If you answer *sometimes* true, give an example of when the statement is true.

    a. $\vec{v} + \vec{w} = \vec{w} + \vec{v}$

    b. $\vec{v} - \vec{w} = \vec{w} - \vec{v}$

    c. $\vec{v} = -\vec{v}$

    d. $k(\vec{v} + \vec{w}) = k\vec{v} + k\vec{w}, \; k \in R$

    e. $\vec{v} + (-\vec{v}) = \vec{0}$

    f. $a\vec{i} + b\vec{j} = b\vec{i} + a\vec{j}, \; a, b \in R$

    g. $|\overrightarrow{ST}| = |\overrightarrow{TS}|$

    h. $\vec{i} = \vec{j}$

12. Sketch the given vector in standard position in 3-space.

    a. $(1, 4, 2)$ **b.** $(6, -1, 3)$ **c.** $4\vec{i} - 2\vec{j} + 6\vec{k}$

13. Find the magnitude of the given vector.

    a. $\vec{v} = (-1, 4, 3)$

    b. $-6\vec{i} + 3\vec{j} - \vec{k}$

14. Write the ordered triple that represents the vector whose tail is at the point $(-1, 0, 4)$ and whose head is at the point $(-3, 2, -5)$.

15. If $\vec{u} = (1, 4, -2)$ and $\vec{v} = (6, -2, -1)$, find $\vec{w}$.

    a. $\vec{w} = 2\vec{u} + \vec{v}$

    b. $\vec{w} = 2(\vec{u} - \vec{v})$

    c. $\vec{w} = \frac{1}{2}(3\vec{u} - 2\vec{v})$

16. Given the points $A(6, 3, -12)$ and $B(4, 10, -2)$, find the coordinates of point E such that the distance from A to E is $\frac{3}{4}$ of the distance from A to B.

17. Given the points $A(3, 6, -9)$ and $B(2, 4, -7)$, find the coordinates of point E such that $AE:EB = 1:2$.

18. Write an ordered triple that represents the unit vector in the same direction as $3\vec{i} + 4\vec{j} + \sqrt{2}\vec{k}$.

13. a. $\sqrt{26}$

    b. $\sqrt{46}$

14. $(-2, 2, -9)$

15. a. $(8, 6, -5)$

    b. $(-10, 12, -2)$

    c. $\left(\frac{-9}{2}, 8, -2\right)$

16. $\left(\frac{9}{2}, \frac{33}{4}, \frac{-9}{2}\right)$

17. $\left(\frac{8}{3}, \frac{16}{3}, \frac{-25}{3}\right)$

18. $\left(\frac{\sqrt{3}}{3}, \frac{4\sqrt{3}}{9}, \frac{\sqrt{6}}{9}\right)$

Review Answers

19. Find the indicated dot product and state whether the given vectors are perpendicular.

a. $(2, -3) \cdot (6, -2)$ **b.** $(2, -2) \cdot (3, 3)$

c. $(4, 3, -5) \cdot (-1, 3, 7)$ **d.** $(4, -2, 1) \cdot (2, 3, -2)$

20. Write an equation for a plane that contains the point $(2, -1, 3)$ and for which $2\vec{i} + 7\vec{j} + \vec{k}$ is a normal vector.

21. Write, as a linear combination of $\vec{i}$, $\vec{j}$, and $\vec{k}$, a vector normal to the plane whose equation is $x + 2y - 3z + 7 = 0$.

22. Find a unit normal vector for the plane whose equation is $6x - y + 12 = 0$.

23. If $\vec{a} = (-1, 4, 2)$ and $\vec{b} = (1, -3, 5)$, find $\vec{a} \times \vec{b}$.

24. Write an ordered triple that represents the vector equal to:
$$(\vec{i} + \vec{j} + 2\vec{k}) \times (-3\vec{i} - \vec{j} + 2\vec{k})$$

25. Find a vector that is perpendicular to the vectors $(4, 3, -1)$ and $(1, 5, -2)$.

26. Write an equation for the plane that is determined by the points $(1, 6, 3)$, $(2, -4, 7)$, and $(1, 1, 2)$.

27. Find an equation for the plane with normal vector $4\vec{i} - 2\vec{j} + 3\vec{k}$ and passing through the point $(4, 0, -2)$.

28. Find an equation for a plane parallel to the plane whose equation is $3x - y + 2z + 1 = 0$ and that passes through the point $(2, 1, -1)$.

29. Write the equation of the yz-plane in 3-space.

30. Write parametric equations that describe the line of intersection of the two planes whose equations are $x + 2y + z = 0$ and $3x - y + 2z - 3 = 0$.

31. Write parametric equations that describe the line passing through the points $(-3, 5, 7)$ and $(0, -1, 4)$.

32. Write parametric equations that describe the x-axis in 3-space.

33. Find the distance from the point $(1, 1)$ to the line whose equation is $y = 3x + 2$.

34. Find the distance from the point $(4, 3, -1)$ to the line described by the parametric equations $x - 1 = t$, $y + 2 = 2t$, and $z + 3 = t$.

35. The equation for a line in 2-space is of the form $ax + by + c = 0$. If $a = 2$, $b = -3$, and $c = 1$, find the distance from the origin to the line.

36. Find the distance from the plane whose equation is $x + y + 2z + 1 = 0$ to the point $(1, 0, 4)$.

37. Find the distance between two parallel planes whose equations are $2x + 3y - z + 5 = 0$ and $2x + 3y - z - 1 = 0$.

Review Answers

19. a. 18, no
b. 0, yes
c. −30, no
d. 0, yes

20. $2x + 7y + z = 0$

21. $\vec{i} + 2\vec{j} - 3\vec{k}$

22. $\left(\dfrac{6}{\sqrt{37}}, \dfrac{-1}{\sqrt{37}}, 0 \right)$

23. $(26, 7, -1)$

24. $(4, -8, 2)$

25. $(-1, 7, 17)$

26. $30x + y - 5z - 21 = 0$

27. $4x - 2y + 3z - 10 = 0$

28. $3x - y + 2z - 3 = 0$

29. $x = 0$

30. $x = 5t$, $y + \dfrac{3}{5} = t$, $z - \dfrac{6}{5} = -7t$

31. $x = 3t - 3$, $y = -6t + 5$, $z = -3t + 7$

32. $x = t$, $y = 0$, $z = 0$

33. $\dfrac{2\sqrt{10}}{5}$

34. $\dfrac{\sqrt{2}}{2}$

35. $\dfrac{\sqrt{13}}{13}$

36. $\dfrac{5\sqrt{6}}{3}$

37. $\dfrac{3\sqrt{14}}{7}$

☑ **Exercises for Challenge**

1. Find an equation of the plane parallel to the xz-plane that contains the point $(2, -5, 3)$.

2. The points $A(3, 2, 5)$, $B(5, 1, 2)$, and $C(1, 0, 2)$ are the vertices of a triangle. If $\overrightarrow{CD}$ is perpendicular to $\overrightarrow{AB}$, find the coordinates of D on $\overrightarrow{AB}$ and the length of the vector $\overrightarrow{CD}$.

3. A vector, $\vec{c}$, is perpendicular to the plane determined by $\vec{a}$ and $\vec{b}$. If $\vec{a} = (1, 2, 3)$, $\vec{b} = (4, 5, 6)$, and $|\vec{c}| = 5$, express the coordinates of $\vec{c}$ to the nearest ten-thousandth.

4. The midpoint of $\overline{AB}$ is $(.32, .26)$. Point $C(4, -1)$ is on $\overline{AB}$. If $|\overrightarrow{AC}| = \frac{1}{3}|\overrightarrow{AB}|$, find the coordinates of A and B.

5. The distance from the origin to the line $y = mx + 6$ is 3. What are the possible values of m?

Questions 6 – 11 each consist of two quantities, one in Column A and one in Column B. You are to compare the two quantities and choose:

A if the quantity in Column A is greater;
B if the quantity in Column B is greater;
C if the two quantities are equal;
D if the relationship cannot be determined from the information given.

1. In certain questions, information concerning one or both of the quantities to be compared is centered above the two columns.

2. In a given question, a symbol that appears in both columns represents the same thing in Column A as it does in Column B.

3. x, n, and k, etc. stand for real numbers.

	Column A	Column B

In **6-7**, $\vec{v}$ is a nonzero vector, and k is a real number.

$$|k\vec{v}| < |\vec{v}|$$

	Column A	Column B
6.	k	-1

$$|k\vec{v}| = |\vec{v}|$$

| **7.** | k | 1 |

$$\vec{a} = (2, 3, 0), \vec{b} = (4, 2, 0), \vec{c} = (1, 2, 0)$$

| **8.** | $|\vec{a} \times \vec{c}|$ | $|\vec{b} \times \vec{c}|$ |

$\vec{a}$ and $\vec{b}$ are vectors in a plane.

| **9.** | $\vec{a} \cdot \vec{b}$ | $(2\vec{a}) \cdot \vec{b}$ |

$\vec{a}$ and $\vec{b}$ are nonzero vectors in 3-space such that
$$\vec{a} \cdot \vec{b} = 0$$

| **10.** | $(\vec{a} + \vec{b}) \cdot (\vec{a} + \vec{b})$ | $|\vec{a}|^2 + |\vec{b}|^2$ |

ℓ_1 is the line $x + 4y + z - 3 = 0$.
ℓ_2 is the line $20x - 21y + 102 = 0$.
P is the point $(-2, 1, 7)$.

| **11.** | The number of points on ℓ_1 that lie 1.414 units from P | The number of points on ℓ_2 that lie 1.414 units from P |

1. $y + 5 = 0$ or $y = -5$

2. $D = \left(4, \frac{3}{2}, \frac{7}{2}\right)$, $|\overrightarrow{CD}| = \frac{3}{2}\sqrt{6}$

3. $(2.041, -4.082, 2.041)$ or $(-2.041, 4.082, -2.041)$

4. $A(11.36, -3.52)$, $B(-10.72, 4.04)$

5. $\pm\sqrt{3}$

6. A

7. D

8. B

9. D

10. C

11. B

☑ **Exercises for Challenge** *(continued)*

12.

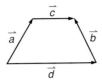

If $\vec{a}, \vec{b}, \vec{c}, \vec{d}$ are related as shown in the figure, which of the following statements is true?

(A) $\vec{a} + \vec{b} + \vec{c} + \vec{d} = 0$

(B) $\vec{a} + \vec{b} + \vec{c} = \vec{d}$

(C) $\vec{a} + \vec{b} = \vec{c} + \vec{d}$

(D) $\vec{a} + \vec{c} = \vec{b} + \vec{d}$

(E) $\vec{a} + \vec{d} = \vec{b} + \vec{c}$

13. A point (a, b) is equidistant from the lines $3y = 4x$ and $4y = 3x$. Which of the following could be (a, b)?

   I $(0, 0)$

   II $(3, 4)$

   III $(1, -1)$

(A) I only

(B) II only

(C) III only

(D) I and II only

(E) I and III only

14. Choose the statement that must be true if $\vec{a} \cdot \vec{b} = \vec{a} \cdot \vec{c}$.

(A) $\vec{b} = \vec{c}$

(B) $\vec{a} \neq \vec{0}$

(C) $\vec{a}$ is perpendicular to $\vec{b}$ or to $\vec{c}$.

(D) $\vec{a}$ is perpendicular to $\vec{b} - \vec{c}$.

(E) $\vec{b}$ is parallel to $\vec{c}$.

15. The coordinates of $\vec{v} = (x_1, y_1)$ and $\vec{w} = (x_2, y_2)$ are all nonnegative. If $\vec{v}$ is perpendicular to $\vec{w}$, which of the following must be true?

(A) Either $\vec{v}$ or $\vec{w}$ is $\vec{0}$.

(B) Either x_1 or x_2 is 0.

(C) Either x_1 or y_1 is 0.

(D) Either x_1 or y_2 is 0.

(E) Either x_2 or y_1 is 0.

16. Which of the following planes does not intersect the x-axis?

(A) $3x - 5z + 1 = 0$

(B) $4y - 4 = 0$

(C) $x + 2y - 7 = 0$

(D) $x + 2y - 7z = 0$

(E) $9y + 4z = 0$

17. A function f is defined by $f(\vec{v}, \vec{w}) = \vec{v} \cdot (\vec{w} + \vec{v})$. Which of the following is not a property of f?

(A) $f(\vec{v}, -\vec{v}) = 0$

(B) $f(\vec{v}, \vec{w})$ is a scalar.

(C) $f(\vec{v}, \vec{w}) = f(\vec{w}, \vec{v})$

(D) $f(\vec{v}, \vec{v}) = 2f(\vec{v}, \vec{0})$

(E) $f(\vec{v}, \vec{w}) \geq 0$ if v is perpendicular to w.

12. D

13. E

14. D

15. B

16. B

17. C

Teacher's Chapter 7

Matrices

Chapter Contents

7.1 Introduction to Matrices

A matrix is presented as a convenient notation for displaying data. The procedures for the basic matrix operations are introduced by verbal situations for which the arithmetic procedures are familiar to students. After each basic matrix operation is detailed, students are shown how to perform that operation on a calculator that works with matrices.

Matrices are then applied to a variety of algebraic, geometric, and real-life situations.

The method of expansion by minors along a row or column is presented for finding the determinant of an $n \times n$ matrix where $n \geq 3$. You may wish to demonstrate the following second method for finding the determinant of a 3×3 matrix.

Called the diagonal method, this procedure, applicable only to 3×3 matrices, is accomplished by copying columns 1 and 2 as the 4th and 5th columns of the matrix. From the sum of the products of the elements on each southeast diagonal, subtract the sum of the products of the elements on each southwest diagonal.

$$A = \begin{bmatrix} a & b & c \\ d & e & f \\ g & h & i \end{bmatrix} = \begin{bmatrix} a & b & c & a & b \\ d & e & f & d & e \\ g & h & i & g & h \end{bmatrix} \begin{matrix} \text{Column} \\ 1 \quad 2 \end{matrix}$$

$$| A | = aei + bfg + cdh - (ceg + afh + bdi)$$

An alternate approach used for determining the sign of each product in an expansion is to think of any determinant as an alternating array of + and – signs, with the sign in the first row, first column always being positive. For example:

$$\begin{bmatrix} + & - & + & - \\ - & + & - & + \\ + & - & + & - \\ - & + & - & + \end{bmatrix}$$

The appropriate sign can then be determined according to the sign of the element whose minor is being expanded on.

Classes with access to modern technology can appreciate the effectiveness of the tools now available. If tools that can compute a determinant are not available, note that, with the exception of Exercises 32 and 33, all of the exercises can be accomplished by hand using the procedures presented in the text. A scientific calculator, although not able to automatically evaluate a determinant, can be used to expedite calculations.

The formula presented in Exercise 44 applies a determinant to find the area of a triangle in 2-space given the coordinates of its vertices. Students should note that this formula is an alternative to the method of constructing a rectangle around an obliquely-placed triangle to find its area.

7.2 Operations on Matrices

Although the text compares the procedure for matrix multiplication with the dot product for vectors, it is possible to study matrices before vectors. If the chapter on vectors is not covered in the course, or if vectors are studied after matrices, the traditional method for matrix multiplication can be taught, that is, to have the student move his left index finger across the ith row of the matrix on the left and his right index finger down the jth column of the matrix on the right in order to obtain the ijth element of the product.

To reinforce the properties of operations within the set of matrices, make comparisons with the real number system and other mathematical systems that students may have studied. For example, in the real number system, if $ab = 0$, then either a or b must be 0; but in the set of integers modulo 6, if $ab = 0$, neither a nor b must be 0; and for matrices, if $AB = 0$, neither A nor B must be the zero matrix. The set of real functions under the operation of composition serves as an additional example of a noncommutative system.

7.3 Using Matrices to Solve Systems of Equations

To show how the solution of a system of linear equations is accomplished by dealing only with the coefficients, parallel the operations on the augmented matrix with the algebraic solution.

Example: $2x + y = 5$
$\qquad\quad 3x - 2y = 4$

$$R_1 \begin{bmatrix} 2 & 1 & 5 \\ 3 & -2 & 4 \end{bmatrix}$$
$$R_2$$

(I) $2x + y = 5$
(II) $3x - 2y = 4$

$$2R_1 + R_2 \begin{bmatrix} 7 & 0 & 14 \\ 3 & -2 & 4 \end{bmatrix}$$

2(I) $4x + 2y = 10$
(II) $\underline{3x - 2y = 4}$

2(I) + (II) $7x = 14$ Multiply by $\frac{1}{7}$.
(III) $x = 2$

$$\tfrac{1}{7}R_1 \begin{bmatrix} 1 & 0 & 2 \\ 3 & -2 & 4 \end{bmatrix}$$

$$-3R_1 + R_2 \begin{bmatrix} 1 & 0 & 2 \\ 0 & -2 & -2 \end{bmatrix}$$

(II) $3x - 2y = 4$
−3(III) $\underline{-3x \qquad = -6}$

$$-\tfrac{1}{2}R_2 \begin{bmatrix} 1 & 0 & 2 \\ 0 & 1 & 1 \end{bmatrix}$$

−3(III) + (II) $-2y = -2$ Multiply by $\frac{1}{2}$.
                  $y = 1$

In addition to a calculator that can perform matrix operations and the available commercial computer software, the following program will reduce a matrix.

Computer Activity (7.3) _____

```
  1 PRINT "SOLVING SYSTEMS USING
          MATRICES"
  5 PRINT "INPUT THE NUMBER OF
          VARIABLES."
 10 INPUT N
 15 PRINT "ENTER A MATRIX BY ROWS.
          PRESS RETURN AFTER EACH ENTRY."
 20 DIM A(N + 1, N + 1)
 30 FOR R = 1 TO N
 40 FOR C = 1 TO N + 1
 60 INPUT A(R, C)
 70 NEXT C
 80 NEXT R
100 FOR R = 1 TO N
110 IF A(R, R) <> 0 THEN 260
120 FOR S = R + 1 TO N
130 IF A(S, S) <> 0 THEN 205
140 NEXT S
150 PRINT "MORE THAN ONE SOLUTION"
200 GOTO 600
205 FOR I = 1 TO N + 1
210 LET T = A(R, I)
230 LET A(R, I) = A(S, I)
240 LET A(S, I) = T
```

```
250 NEXT I
260 IF A(R, R) = 1 THEN 330
270 LET B = A(R, R)
300 FOR C = 1 TO N + 1
310 LET A(R, C) = A(R, C)/B
320 NEXT C
330 FOR W = 1 TO N
340 IF W = R THEN 440
350 IF A(W, R) = 0 THEN 440
400 LET W1 = A(W, R)
410 FOR C = R TO N + 1
420 LET A(W, C) = A(W, C) - W1*A(R, C)
430 NEXT C
440 NEXT W
450 NEXT R
460 PRINT "THIS IS THE REDUCED
          MATRIX." : PRINT
500 FOR W = 1 TO N
510 FOR C = 1 TO N + 1
520 PRINT A(W, C) ; " " ;
530 NEXT C : PRINT
540 NEXT W
600 END
```

7.4 An Additional Method for Solving Systems

The matrix solution for a system of equations presented in this section depends on determining the inverse of the coefficient matrix. For higher-order matrices, using a calculator or software makes the method efficient. Students should be aware that the technical tools will sometimes offer an approximation when the actual matrix entry should be 0. Such an example was discussed in Section 7.3 when the calculator was used to compute an augmented matrix. In general, as the entries of a matrix get closer to those of a singular matrix, the round-off error increases.

You may wish to mention transposing matrices, and note that the matrix in Example 4 can be seen to be singular because one column is a multiple of another.

For complex real-world problems that are modeled algebraically, such as launching a rocket, it may be necessary to solve systems of dozens of equations. Major breakthroughs have been made by mathematicians on this topic in the 20th century. Students might be asked to do a short research paper related to this area of study.

7.5 Matrices and Transformations

Reviewing the definitions of the common transformations previously presented in Chapter 3 affords those students who have not studied transformations before this course a level of comfort. Scale changes and shears will be new to most students.

A good way to demonstrate a scale change is to stretch a printed plastic bag vertically and horizontally, allowing the print to stretch unequally. An airplane take-off is a good model of a shear. As the plane travels forward in a line defined by its angle with the ground, the plane is subject to the downward vertical force of gravity or the force of a prevailing wind (wind shear).

An important point in this section is that the operation of composition for reflection, rotation, and dilation follows an arithmetic that can be modeled using matrix multiplication. As is the case for matrix multiplication, composition of transformations is not commutative, but is associative.

Exercise 39 is an interesting example of a transformation described by a 2×2 matrix. Encourage students to create 2×2 transformation matrices of their own.

7.6 Additional Applications of Matrices

From the uses of matrices presented in this section, students should sense, in addition to the utility of mathematics as a tool, that mathematics is a dynamic subject with room for new applications and discoveries.

Challenge students to create examples within their own experience where matrices can be helpful for dealing with the data. For example, if freshmen, sophomores, juniors, and seniors pay yearly dues for various school activities as represented by the matrix at the right, find the total collected for each fee if there are 300 freshmen, 250 sophomores, 200 juniors, and 175 seniors.

	F	S	J	SN
Student organization	10	12	15	15
Lab fees	2	6	10	15
Dances	5	5	5	5
Yearbook	0	0	5	20

Answer: $11,625 in fees for student organization, $6,725 for lab fees, $4,625 for dances, and $4,500 for the yearbook.

$$\begin{bmatrix} 10 & 12 & 15 & 15 \\ 2 & 6 & 10 & 15 \\ 5 & 5 & 5 & 5 \\ 0 & 0 & 5 & 20 \end{bmatrix} \cdot \begin{bmatrix} 300 \\ 250 \\ 200 \\ 175 \end{bmatrix} = \begin{bmatrix} 11,625 \\ 6,725 \\ 4,625 \\ 4,500 \end{bmatrix}$$

Students should be encouraged to do some library research on the Leontief model, which is described in the April 1965 issue of Scientific American, as well as on the use of mathematics in creating codes. The use of modular arithmetic for creating security codes for semiconductor chips is an area of current interest, as is the use of bar codes for identification of packaged items such as those sold in supermarkets. Encourage students to invent their own coding matrices and create messages for classmates to decode.

An additional application of matrices involving probabilities may be presented. When the outcome of a trial of an experiment depends on the immediately preceding trial, a mathematical system called a Markov Chain is created. Two examples of Markov Chains follow.

Example 1:

Three boys, Alan, Bob, and Carl, are having a 3-way catch. Alan only throws to Bob and Bob always throws to Carl. Carl, however, is as likely to throw to Alan as he is to Bob. As the ball is thrown again and again, the probability that it will be thrown to any one of the boys depends on whom it was thrown to last.

P, the matrix of probabilities below, summarizes the probability of any of the three boys having the ball thrown to him.

$$P = \begin{array}{c} \\ A \\ B \\ C \end{array} \begin{array}{ccc} A & B & C \\ \left[\begin{array}{ccc} 0 & 1 & 0 \\ 0 & 0 & 1 \\ 0.5 & 0.5 & 0 \end{array}\right] \end{array}$$

Examine this matrix to understand its entries. For example, since Alan always throws to Bob, Alan's row contains 1 in Bob's column and 0 in the other columns. Since Carl is as likely to throw to Alan as to Bob, both probabilities are 0.5. The matrix represents the probabilities for the first throw.

Question:

What is the likelihood of Carl receiving the ball the second time it is thrown?

Solution

The product of $P \cdot P = P^2$ will show the probability of any of the three boys having the ball thrown to him on the second throw.

$$P^2 = \begin{bmatrix} 0 & 1 & 0 \\ 0 & 0 & 1 \\ 0.5 & 0.5 & 0 \end{bmatrix}^2 = \begin{bmatrix} 0 & 0 & 1 \\ 0.5 & 0.5 & 0 \\ 0 & 0.5 & 0.5 \end{bmatrix}$$

Thus, if Alan throws the ball first, the probability of the ball being thrown to Carl on the second throw, $P(AC_2)$, is 1, and if Carl throws the ball first, the probability of the ball being thrown to Bob on the second throw, $P(CB_2)$, is 0.5. A tree diagram verifies the entries in the table.

Answer: The likelihood of Carl receiving the ball on the second throw depends on who throws first. If Alan, the probability is 1; if Bob, 0; if Carl, 0.5.

Note that on the third throw, the matrix is

$$P \cdot P^2 = \begin{bmatrix} 0 & 1 & 0 \\ 0 & 0 & 1 \\ 0.5 & 0.5 & 0 \end{bmatrix} \cdot \begin{bmatrix} 0 & 0 & 1 \\ 0.5 & 0.5 & 0 \\ 0 & 0.5 & 0.5 \end{bmatrix} = \begin{bmatrix} 0.5 & 0.5 & 0 \\ 0 & 0.5 & 0.5 \\ 0.25 & 0.25 & 0.5 \end{bmatrix}$$

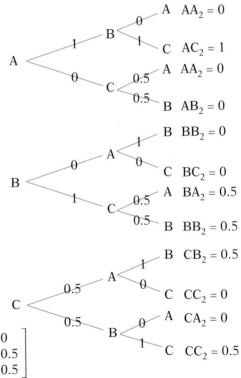

Thus, the probability that if Carl threw the ball the first time, it was thrown to Alan on the third throw, (CA_3), is 0.25.

In general, the probability for each person to receive the ball on the nth throw can be displayed in the matrix P^n.

Other questions that can be asked are:

1. Use the square of the original probability matrix to evaluate

   a. $P(AC_2)$ **b.** $P(AA_2)$ **c.** $P(AB_2)$ **d.** $P(BA_2)$ **e.** $P(CA_2)$

   Answers: **a.** 1   **b.** 0   **c.** 0   **d.** 0.5   **e.** 0

2. What matrix could be used to find P_3, the probability of a boy catching the third throw if a particular boy starts the catch?

   Answer: $\begin{bmatrix} 0 & 1 & 0 \\ 0 & 0 & 1 \\ 0.5 & 0.5 & 0 \end{bmatrix}^3$

Example 2:

A trucking company serves three regions, New York City, Philadelphia, and Washington, D.C. To operate efficiently, the number of deliveries made to each region, the number of trucks and drivers available at a given time, and other factors are studied.

Of the items picked up by trucks in New York City, 50% must be delivered to other parts of N.Y.C., 20% go to Philadelphia, and 30% go to Washington. Similarly, for the items picked up in Philadelphia, 40% remain in Philadelphia, 10% go to New York, and 50% go to Washington. For the packages picked up in Washington, 40% go to other parts of Washington, 30% go to New York, and 30% go to Philadelphia.

The matrix T summarizes the delivery probability of a truck on its first trip.

$$T = \begin{array}{c c} & \begin{array}{c c c} \text{NY} & \text{P} & \text{W} \end{array} \\ \begin{array}{c} \text{NY} \\ \text{P} \\ \text{W} \end{array} & \begin{bmatrix} 0.5 & 0.2 & 0.3 \\ 0.1 & 0.4 & 0.5 \\ 0.3 & 0.3 & 0.4 \end{bmatrix} \end{array}$$

The tree diagram shows the probabilities for a truck starting in New York.

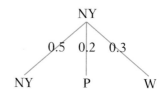

An extended diagram tracks the second round.

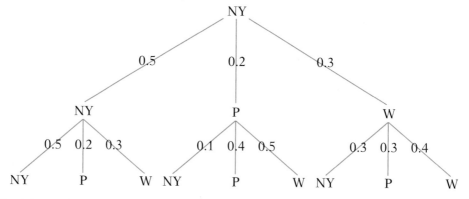

The probability that a truck that starts in New York will be in New York after delivering a second pickup is: $0.5(0.5) + 0.2(0.1) + 0.3(0.3) = 0.36$

The probability of that truck being in Philadelphia after completing a second delivery is: $0.5(0.2) + 0.2(0.4) + 0.3(0.3) = 0.27$

Finally, if a truck starts in New York, the probability of being in Washington after delivering the second pickup is: $0.5(0.3) + 0.2(0.5) + 0.3(0.4) = 0.37$

Note: The sum of the three probabilities, 0.36, 0.27, and 0.37, is 1 since they represent all possible locations of the truck.

The calculations to compute the probabilities lend themselves to the use of matrices. Note that in each line of calculations, the factors 0.5, 0.2, and 0.3 appear on the left of each product.

$$\begin{bmatrix} 0.5 & 0.2 & 0.3 \end{bmatrix} \begin{bmatrix} 0.5 & 0.2 & 0.3 \\ 0.1 & 0.4 & 0.5 \\ 0.3 & 0.3 & 0.4 \end{bmatrix} = \begin{matrix} NY & P & W \\ \begin{bmatrix} 0.36 & 0.27 & 0.37 \end{bmatrix} \end{matrix}$$

The probability of being in New York, Washington, or Philadelphia after delivering a third package would be

$$\begin{bmatrix} 0.36 & 0.27 & 0.37 \end{bmatrix} T = \begin{bmatrix} 0.318 & 0.291 & 0.391 \end{bmatrix}$$

The matrix $\begin{bmatrix} 0.5 & 0.2 & 0.3 \end{bmatrix}$ is the first row of T and represents the probability of a truck that starts in New York making its first delivery to one of the three cities. The matrix

$$T^2 = \begin{matrix} & \begin{matrix} NY & P & W \end{matrix} \\ \begin{matrix} NY \\ P \\ W \end{matrix} & \begin{bmatrix} 0.5 & 0.2 & 0.3 \\ 0.1 & 0.4 & 0.5 \\ 0.3 & 0.3 & 0.4 \end{bmatrix}^2 \end{matrix} = \begin{bmatrix} 0.36 & 0.27 & 0.37 \\ 0.24 & 0.33 & 0.43 \\ 0.3 & 0.3 & 0.4 \end{bmatrix}$$

gives all possible probabilities for the whereabouts of the trucks after completing the delivery of a second pickup. For example, 0.33, the entry in row 2, column 2, is the probability that a truck starting in Philadelphia will be in Philadelphia after completing delivery of the second package.

In general, t_{ij}, an entry in T^n, is the probability that a truck that started in city i is in city j after n pickups and deliveries.

Example 3: Each of a fleet of 100 trucks is expected to make 2 deliveries per day. If 10 trucks start in New York, 50 start in Philadelphia, and 40 start in Washington, how many trucks are likely to be in each of the cities at the end of their routes for that day?

Solution

$$T^2 = \begin{bmatrix} 0.36 & 0.27 & 0.37 \\ 0.24 & 0.33 & 0.43 \\ 0.3 & 0.3 & 0.4 \end{bmatrix}$$

$$\begin{bmatrix} 10 & 50 & 40 \end{bmatrix} T^2 = \begin{bmatrix} 27.6 & 31.2 & 41.2 \end{bmatrix}$$

Answer: Approximately 28 trucks would be in New York, 31 in Philadelphia, and 41 in Washington.

1. If $S = \begin{bmatrix} 2 & 7 \\ 3 & -8 \\ 4 & 1 \end{bmatrix}$ and $T = \begin{bmatrix} 6 & 4 \\ -3 & 2 \\ 4 & 5 \end{bmatrix}$, find $2S - 3T$.

2. If $S = \begin{bmatrix} 6 & 4 \\ 2 & -3 \end{bmatrix}$ and $T = \begin{bmatrix} 3 & -2 \\ 1 & 0 \end{bmatrix}$, find ST.

3. If $A = \begin{bmatrix} 1 & 7 \\ -3 & -25 \end{bmatrix}$, find A^{-1}.

4. Evaluate: $\begin{vmatrix} 7 & 15 \\ 2 & 5 \end{vmatrix}$

5. Evaluate: $\begin{vmatrix} 1 & 2 & 0 & 0 \\ 4 & 0 & -2 & 0 \\ 2 & 1 & 0 & 3 \\ 1 & -1 & 0 & 0 \end{vmatrix}$

6. Write a matrix equation that can be used to solve the following system.
$$x + y + z = 3$$
$$2x + z = 3$$
$$3y - 2z = 1$$

7. Which of the following matrices is singular?

   (A) $\begin{bmatrix} 2 & 4 \\ 6 & 8 \end{bmatrix}$ (B) $\begin{bmatrix} 1 & 3 \\ 5 & 9 \end{bmatrix}$

   (C) $\begin{bmatrix} 12 & 8 \\ 9 & 6 \end{bmatrix}$ (D) none of these

8. Solve the following system using Cramer's Rule.
$$3x - 2y + z = 11$$
$$2x + 4y - 3z = -7$$
$$4x + y = 6$$

9. If $\begin{bmatrix} 6 & 7 & 1 & 4 \\ 3 & -1 & 0 & 8 \end{bmatrix} = \begin{bmatrix} 2x & 7 & 1 & 2y \\ (z-1) & -1 & 0 & 8 \end{bmatrix}$, find $x, y,$ and z.

10. Find the product: $\begin{bmatrix} 2 & 3 & 4 \\ -1 & 0 & 2 \end{bmatrix} \begin{bmatrix} -1 & 4 \\ 0 & 1 \\ 5 & 2 \end{bmatrix}$

11. If the system
$$a + 2b = 1$$
$$3a + 8b + 5d = 0$$
$$a + 4b + 3c + 10d = -1$$
$$-3c = -1$$
has a coefficient matrix C

and $C^{-1} = \begin{bmatrix} 6 & -2 & 1 & 1 \\ -\frac{5}{2} & 1 & -\frac{1}{2} & -\frac{1}{2} \\ 0 & 0 & 0 & -\frac{1}{3} \\ \frac{2}{5} & -\frac{1}{5} & \frac{1}{5} & \frac{1}{5} \end{bmatrix}$

find $a, b, c,$ and d.

12. A code involving 6 letters assigns the following numbers:
0 = space, 1 = D, 3 = S, 5 = T, 7 = B, 9 = R, 11 = L

If the matrix $\begin{bmatrix} 6 & 2 \\ 4 & 1 \end{bmatrix}$ was used to code the message, decode the message.

Message: $\begin{bmatrix} 18 & 5 \end{bmatrix} \begin{bmatrix} 74 & 23 \end{bmatrix} \begin{bmatrix} 12 & 3 \end{bmatrix}$
$\begin{bmatrix} 86 & 25 \end{bmatrix} \begin{bmatrix} 6 & 2 \end{bmatrix}$

13. The inventory of a company that sells 3 different types of sweatshirts, A, B, and C, in 3 sizes, S, M, and L, is represented by the following matrix.

$$\begin{array}{c} \\ S \\ M \\ L \end{array} \begin{array}{ccc} A & B & C \\ \left[\begin{array}{ccc} 10 & 8 & 0 \\ 4 & 1 & 2 \\ 3 & 7 & 5 \end{array}\right] \end{array}$$

If the costs of the sweatshirts are $5 for A, $7 for B, and $10 for C, use matrices to find the amount of money that will be taken in if:

a. all size S are sold

b. all size M are sold

c. all size L are sold

14. In a 2-sector economy consisting of oil and steel, production of 1 unit of steel requires 0.2 unit of steel and 0.1 unit of oil. One unit of oil requires 0.03 unit of steel and 0.2 unit of oil.

a. Write a technology matrix to represent the given information.

b. If the production matrix is $\begin{array}{c} \text{oil} \\ \text{steel} \end{array} \left[\begin{array}{c} 10 \\ 5 \end{array}\right]$, how much is left that can be sold after internal consumption?

c. If the demand matrix is $\left[\begin{array}{c} 1 \\ 2 \end{array}\right]$, find the intensity matrix with the values estimated to the nearest tenth.

Bonus

Find a 3×3 matrix A, other than $\left[\begin{array}{ccc} 1 & 0 & 0 \\ 0 & 1 & 0 \\ 0 & 0 & 1 \end{array}\right]$, which is a left identity for the matrix $\left[\begin{array}{cc} 1 & 2 \\ 1 & 3 \\ 2 & 1 \end{array}\right]$.

Answers to Suggested Test Items

1. $\left[\begin{array}{cc} -14 & 2 \\ 15 & -22 \\ -4 & -13 \end{array}\right]$
 2. $\left[\begin{array}{cc} 22 & -12 \\ 3 & -4 \end{array}\right]$

3. $\left[\begin{array}{cc} \frac{25}{4} & \frac{7}{4} \\ -\frac{3}{4} & -\frac{1}{4} \end{array}\right]$
 4. 5
 5. 18

6. $\left[\begin{array}{c} x \\ y \\ z \end{array}\right] = \left[\begin{array}{ccc} 1 & 1 & 1 \\ 2 & 0 & 1 \\ 0 & 3 & -2 \end{array}\right]^{-1} \cdot \left[\begin{array}{c} 3 \\ 3 \\ 1 \end{array}\right]$

7. C
 8. $x = 2$
 $y = -2$
 $z = 1$
 9. $x = 3$
 $y = 2$
 $z = 4$

10. $\left[\begin{array}{cc} 18 & 19 \\ 11 & 0 \end{array}\right]$
 11. $a = 2$
 $b = -\frac{1}{2}$
 $c = 1$
 $d = -\frac{2}{5}$

12. DSRT_SHLD
 13. a. $106
 b. $47
 c. $114

14. a. $\left[\begin{array}{cc} 0.2 & 0.1 \\ 0.03 & 0.2 \end{array}\right]$
 b. $\left[\begin{array}{c} 7.5 \\ 3.7 \end{array}\right]$
 c. $\left[\begin{array}{c} 1.6 \\ 2.6 \end{array}\right]$

Bonus: $\left[\begin{array}{ccc} -4 & 3 & 1 \\ 0 & 1 & 0 \\ 0 & 0 & 1 \end{array}\right]$

There are many possible answers. In this case, only the first row of the identity matrix was changed.

$$\left[\begin{array}{ccc} a & b & c \\ 0 & 1 & 0 \\ 0 & 0 & 1 \end{array}\right] \left[\begin{array}{cc} 1 & 2 \\ 1 & 3 \\ 2 & 1 \end{array}\right] = \left[\begin{array}{cc} 1 & 2 \\ 1 & 3 \\ 2 & 1 \end{array}\right]$$

(I) $a + b + 2c = 1$
(II) $2a + 3b + c = 2$

$-2\text{(I)} + \text{(II)}$ $b - 3c = 0$
                  $b = 3c$

Since there are fewer equations than unknowns, there are infinitely many solutions.

Let $c = 1$.
Then $b = 3$ and $a = -4$.

Chapter 7

Matrices

*H*ow should telephone lines be located to minimize construction and maintenance? What route should a salesperson follow to visit each of a number of customers with a minimum of time and travel expense? What is the most efficient route for a fleet of trucks to follow to make deliveries? Problems such as these, requiring the analysis of networks, can be modeled mathematically, with special attention to the organization and display of the large amount of data that needs to be processed.

Chapter Table of Contents

7.1 Introduction to Matrices

The cost of three different types of computers from mail-order stores in three different cities could be shown in a rectangular array. For example, if Commo, JCN, and Orange are sold by mail-order stores in New York, Los Angeles, and Honolulu, then the array

	Commo	JCN	Orange
New York	600	2,500	1,600
Los Angeles	650	2,450	1,690
Honolulu	620	2,560	1,580

might be used to display the prices of the computers. This method of exhibiting data in an array called a *matrix*, combined with various mathematical principles, will, in many cases, provide a convenient way to discover information about the data and solve various types of problems.

■ **Definition:** *A **matrix** is a rectangular array of numbers displayed in rows and columns. A matrix that has m rows and n columns is said to have **order** or **dimension** m × n, read m by n.*

Any particular element of a matrix may be identified by its row and column. For example in the 3 × 4 matrix shown, the element 6 is located in the first row, third column.

$$\begin{bmatrix} 1 & 5 & 6 & 4 \\ 3 & 2 & 1 & 0 \\ 8 & 4 & 7 & 2 \end{bmatrix}$$

The elements of an $m \times n$ matrix are written with double-subscripted variables to designate their positions. For example, a_{45} is the element in the fourth row, fifth column, assuming that $m \geq 4$ and $n \geq 5$. In general a_{ij} represents the element in the ith row and jth column.

$$\begin{bmatrix} a_{11} & a_{12} & a_{13} & \cdots & a_{1n} \\ a_{21} & a_{22} & a_{23} & \cdots & a_{2n} \\ a_{31} & a_{32} & a_{33} & \cdots & a_{3n} \\ \vdots & \vdots & \vdots & \vdots & \vdots \\ a_{m1} & a_{m2} & a_{m3} & \cdots & a_{mn} \end{bmatrix}$$

A matrix that has an equal number of rows and columns, $m = n$, is called a ***square matrix***. For example, S is a square matrix of dimension 3 × 3. A square $n \times n$ matrix has a main diagonal of n elements in positions $a_{11}, a_{22}, a_{33}, \ldots a_{nn}$. The elements of the main diagonal of S are 1, 4 and 7.

$$S = \begin{bmatrix} 1 & 8 & 9 \\ 6 & 4 & 2 \\ 3 & 5 & 7 \end{bmatrix}$$

Example 1

Refer to matrices A, B, and C.

$$A = \begin{bmatrix} 1 & 0 & 4 & 3 \end{bmatrix} \qquad B = \begin{bmatrix} 0 & -1 & 2 \\ -7 & 6 & 4 \\ -8 & \frac{1}{2} & 1 \end{bmatrix} \qquad C = \begin{bmatrix} \pi \\ \sqrt{2} \\ \sqrt{3} \end{bmatrix}$$

a. Find the order of each matrix.

b. Identify the elements of the main diagonal of the square matrix.

c. If the elements of A, B, and C are a_{ij}, b_{ij}, and c_{ij}, respectively, find

    (1) a_{14} **(2)** b_{31} **(3)** c_{11}

Solution

a. Matrix A, with 1 row and 4 columns, has order 1 × 4.
Matrix B, with 3 rows and 3 columns, has order 3 × 3.
Matrix C, with 3 rows and 1 column, has order 3 × 1.

b. *B* is a square matrix since it has an equal number of rows and columns. The elements of the main diagonal, corresponding to $b_{11}, b_{22},$ and b_{33} are 0, 6, and 1.

c. (**1**) a_{14}, the element of A in the first row, fourth column, is 3.
(**2**) $b_{31} = -8$
(**3**) $c_{11} = \pi$

A matrix, all of whose elements are 0, is called a *zero matrix*. Zero matrices may have any dimension. For example,

$$\begin{bmatrix} 0 \end{bmatrix} \qquad \begin{bmatrix} 0 & 0 & 0 \end{bmatrix} \qquad \begin{bmatrix} 0 & 0 & 0 \\ 0 & 0 & 0 \end{bmatrix} \qquad \begin{bmatrix} 0 & 0 \\ 0 & 0 \end{bmatrix}$$

are all zero matrices.

The Determinant of a Square Matrix

Associated with each square matrix is a number called its *determinant*. The determinant of a square matrix is designated by placing vertical bars around the array of numbers of the matrix. For example:

$$\text{if } A = \begin{bmatrix} 6 & 8 \\ 3 & 5 \end{bmatrix}, \text{ then } \begin{vmatrix} 6 & 8 \\ 3 & 5 \end{vmatrix} = |A| = \det(A)$$

Thus, det(A) is a function that maps all $n \times n$ matrices for a given n to the set of real numbers.

Determinants are calculated according to specific procedures. For example, the determinant of a 1 × 1 matrix is its single element, as in:

$$\det[-5] = |-5| = -5$$

The determinant of a 2 × 2 matrix is calculated as follows:

$$\det\begin{bmatrix} a_{11} & a_{12} \\ a_{21} & a_{22} \end{bmatrix} = \begin{vmatrix} a_{11} & a_{12} \\ a_{21} & a_{22} \end{vmatrix} = a_{11}a_{22} - a_{12}a_{21}$$

Example 2

Calculate the determinant of matrix *T*.

$$T = \begin{bmatrix} \sqrt{2} & 3\sqrt{3} \\ 0 & 4 \end{bmatrix}$$

Solution

$$\det(T) = \begin{vmatrix} \sqrt{2} & 3\sqrt{3} \\ 0 & 4 \end{vmatrix} = (\sqrt{2})(4) - (3\sqrt{3})(0) = 4\sqrt{2}$$

The determinant of a 3 × 3 or higher order square matrix is computed by a method called *expansion by minors*. In order to compute the determinant of the 3 × 3 matrix A,

$$A = \begin{bmatrix} 1 & 2 & 8 \\ 3 & -1 & 6 \\ 5 & 4 & 7 \end{bmatrix}, \text{ follow the steps outlined below.}$$

Step 1
Select any row or column of the matrix. Use row 1 in this example. Write the first element of the row or column chosen and then cross out the entire row and column to which this element belongs. Multiply the determinant of the remaining 2 × 2 array, which is called the minor of the element, by this element.

$$\begin{vmatrix} 1 & 2 & 8 \\ 3 & -1 & 6 \\ 5 & 4 & 7 \end{vmatrix} = 1 \begin{vmatrix} -1 & 6 \\ 4 & 7 \end{vmatrix} \cdots$$

Step 2
Write the next element of row 1, which is 2. Cross out its row and column and multiply by the remaining 2 × 2 determinant, subtracting this result from the earlier one.

$$\begin{vmatrix} 1 & 2 & 8 \\ 3 & -1 & 6 \\ 5 & 4 & 7 \end{vmatrix} = 1 \begin{vmatrix} -1 & 6 \\ 4 & 7 \end{vmatrix} - 2 \begin{vmatrix} 3 & 6 \\ 5 & 7 \end{vmatrix} \cdots$$

Step 3
Finally, write the last element of the chosen row, 8. Multiply by its minor and add the result to what has been done so far.

$$\begin{vmatrix} 1 & 2 & 8 \\ 3 & -1 & 6 \\ 5 & 4 & 7 \end{vmatrix} = 1 \begin{vmatrix} -1 & 6 \\ 4 & 7 \end{vmatrix} - 2 \begin{vmatrix} 3 & 6 \\ 5 & 7 \end{vmatrix} + 8 \begin{vmatrix} 3 & -1 \\ 5 & 4 \end{vmatrix}$$

Step 4
Evaluate each 2 × 2 determinant.

$$\begin{aligned} |A| &= 1[(-1)(7) - (6)(4)] - 2[(3)(7) - (6)(5)] + 8[(3)(4) - (-1)(5)] \\ &= 1(-31) - 2(-9) + 8(17) \\ &= -31 + 18 + 136 \\ &= 123 \end{aligned}$$

Observe that as each minor is written into the expansion, the sign pattern alternates. In the preceding example:

the sign of the 1st product was +

the sign of the 2nd product was −

the sign of the 3rd product was +

Note that for a 3 × 3 determinant, the sign pattern may be + − + or − + −.

The first sign is dependent upon the position *ij* of the first element on which you are expanding, as follows:

(1) Calculate: $(-1)^{i+j}$

(2) If $(-1)^{i+j}$ is positive, begin with +.

If $(-1)^{i+j}$ is negative, begin with −.

Thus, in the preceding example, the expansion began with + because the first element chosen was in row 1, column 1; that is, $i + j = 1 + 1 = 2$ and $(-1)^2$ is positive.

To verify that but a single result is obtained as the value of a given determinant, recalculate the previous example by expanding on column 2.

$$\begin{vmatrix} 1 & 2 & 8 \\ 3 & -1 & 6 \\ 5 & 4 & 7 \end{vmatrix} = -2\begin{vmatrix} 3 & 6 \\ 5 & 7 \end{vmatrix} + (-1)\begin{vmatrix} 1 & 8 \\ 5 & 7 \end{vmatrix} - 4\begin{vmatrix} 1 & 8 \\ 3 & 6 \end{vmatrix}$$

$$= -2(-9) - 1(-33) - 4(-18)$$
$$= 18 + 33 + 72$$
$$= 123$$

Note that, in this case, the alternating sign pattern begins with a negative sign. This is true because the first element of the expansion is in the position a_{12}, and $(-1)^{1+2}$ is negative.

When evaluating a determinant, any row or column may be chosen to expand on; however, it is convenient to choose the row or column with the greatest number of zeros in order to minimize the arithmetic.

Example 3

Evaluate the determinant of:

$$B = \begin{bmatrix} -3 & 1 & 2 & 0 \\ 4 & 1 & 0 & 0 \\ 1 & -1 & 2 & 0 \\ 0 & -2 & 1 & 4 \end{bmatrix}$$

Solution

Since column 4 has the greatest number of zeros of any of the rows or columns, expand by minors along column 4. Since the first element of column 4 in matrix B corresponds to b_{14} and $(-1)^{1+4}$ is negative, begin the expansion with a negative sign.

$$|B| = \begin{vmatrix} -3 & 1 & 2 & 0 \\ 4 & 1 & 0 & 0 \\ 1 & -1 & 2 & 0 \\ 0 & -2 & 1 & 4 \end{vmatrix}$$

$$= -0\begin{vmatrix} 4 & 1 & 0 \\ 1 & -1 & 2 \\ 0 & -2 & 1 \end{vmatrix} + 0\begin{vmatrix} -3 & 1 & 2 \\ 1 & -1 & 2 \\ 0 & -2 & 1 \end{vmatrix} - 0\begin{vmatrix} -3 & 1 & 2 \\ 4 & 1 & 0 \\ 0 & -2 & 1 \end{vmatrix} + 4\begin{vmatrix} -3 & 1 & 2 \\ 4 & 1 & 0 \\ 1 & -1 & 2 \end{vmatrix}$$

To continue the calculation, only the last term needs to be expanded since the others are each zero. Do this expansion on column 3 of the 3 × 3 determinant. Begin this sign pattern with + because the first element of this expansion is in row 1, column 3, of the minor, and $(-1)^{1+3}$ is positive.

$$|B| = 4\begin{vmatrix} -3 & 1 & 2 \\ 4 & 1 & 0 \\ 1 & -1 & 2 \end{vmatrix} = 4\left(2\begin{vmatrix} 4 & 1 \\ 1 & -1 \end{vmatrix} - 0\begin{vmatrix} -3 & 1 \\ 1 & -1 \end{vmatrix} + 2\begin{vmatrix} -3 & 1 \\ 4 & 1 \end{vmatrix}\right)$$

$$= 4(2(-5) - 0 + 2(-7))$$
$$= 4(-10 - 14) = -96$$

Answer: $|B| = -96$

Matrices and the Calculator or Computer

Calculators or computer programs that do matrix operations require that the matrix be entered by rows after the dimension of the matrix is keyed in. The following discussion is based on the TI-81 calculator and can easily be adapted for software or other calculators.

To enter the matrix $\begin{bmatrix} 1 & 4 & 8 & 1 \\ 6 & 8 & -1 & 2 \\ 0 & 1 & 3 & 4 \end{bmatrix}$

Step 1
Place the calculator in matrix mode by pressing MATRIX. Press ▶ to highlight EDIT.

Step 2
Press 1, 2, or 3 to declare the matrix as A, B, or C, respectively. In this case, declare the matrix A.

Step 3
Enter the dimension of the matrix by keying in the number of rows followed by ENTER and the number of columns followed by ENTER.

3 ENTER 4 ENTER

Step 4
Enter the elements by starting with a_{11} and keying in each element followed by ENTER. Continue across each row.

1 ENTER 4 ENTER 8 ENTER 1 ENTER 6 ENTER ... 3 ENTER 4 ENTER

Step 5
To display the matrix to determine that it was entered correctly, key in

2nd QUIT 2nd [A] ENTER

Enter each of the following matrices and display them to check that they were entered correctly. Remember that (-) is used for a negative entry (not − , which is for the operation of subtraction).

$$A = \begin{bmatrix} -6 & 2 \\ 1 & 7 \end{bmatrix} \qquad B = \begin{bmatrix} -3 & 1 & 2 & 0 \\ 4 & 1 & 0 & 0 \\ 1 & -1 & 2 & 0 \\ 0 & -2 & 1 & 4 \end{bmatrix} \qquad C = \begin{bmatrix} 1 & 4 & 9 & -6 \\ 2 & 8.4 & 7 & 0 \\ 1 & 7 & -6 & 3.2 \end{bmatrix}$$

Evaluating a determinant by the process of expanding by minors can be an exacting procedure for higher-order determinants, especially when there is no row or column containing several zeros. Techniques for producing zeros in a determinant or matrix will be discussed later in this chapter. Accessing today's technology to perform extensive calculations enhances the study of mathematics.

Reconsider Example 3 to see how a calculator or computer program that works with matrices can be used. Enter matrix B, and use the following key sequence to evaluate $\det(B)$.

MATRX 5 2nd [B] ENTER

Triangular Matrices

A square matrix that has only zeros above or below its main diagonal is called a *triangular matrix*. For example, S and T are triangular matrices.

$$S = \begin{bmatrix} 6 & 0 & 0 \\ 4 & 3 & 0 \\ -1 & 2 & 5 \end{bmatrix} \qquad T = \begin{bmatrix} 1 & -2 & 3 & 4 \\ 0 & 6 & -1 & 3 \\ 0 & 0 & 8 & 2 \\ 0 & 0 & 0 & 7 \end{bmatrix}$$

The determinant of a triangular matrix is equal to the product of the elements of the main diagonal. For example, the determinant of S is $6 \cdot 3 \cdot 5$ or 90. Use your calculator to verify that $\det(S) = 90$ and $\det(T) = 336$.

A general verification for a 3×3 matrix can be done by expansion. Beginning with the triangular matrix $A = \begin{bmatrix} a_{11} & a_{12} & a_{13} \\ 0 & a_{22} & a_{23} \\ 0 & 0 & a_{33} \end{bmatrix}$:

Expand by the first column. $\quad |A| = a_{11} \begin{vmatrix} a_{22} & a_{23} \\ 0 & a_{33} \end{vmatrix} - 0 \begin{vmatrix} a_{12} & a_{13} \\ 0 & a_{33} \end{vmatrix} + 0 \begin{vmatrix} a_{12} & a_{13} \\ a_{22} & a_{23} \end{vmatrix}$

$$= a_{11}(a_{22}a_{33})$$
$$= a_{11}a_{22}a_{33} \qquad \text{This is the main diagonal.}$$

If the zeros were above the main diagonal, the same result would be obtained. Mathematical induction can be used to extend the proof for any $n \times n$ matrix.

Example 4

Evaluate $|A|$ if $A = \begin{bmatrix} 6 & 4 & 1 & 8 & 3 \\ 0 & 3 & 10 & 4 & 2 \\ 0 & 0 & 2 & 1 & 5 \\ 0 & 0 & 0 & 7 & 3 \\ 0 & 0 & 0 & 0 & 8 \end{bmatrix}$

Solution

A is a triangular matrix since all of the elements below the main diagonal are 0. Therefore, the determinant of A is the product of the elements of the main diagonal.

$$|A| = 6(3)(2)(7)(8) = 2{,}016$$

Exercises 7.1

In **1 – 5**: **a.** Indicate the order of the matrix.
b. For the square matrices, list the elements of the main diagonal.

1. $\begin{bmatrix} -1 & 2 \\ 1 & 4 \\ 5 & 8 \end{bmatrix}$ **2.** $\begin{bmatrix} 7 & 4 & 2 & 6 \\ 0 & -1 & 4 & 2 \end{bmatrix}$ **3.** $\begin{bmatrix} 2 & 4 \\ 6 & 8 \end{bmatrix}$ **4.** $\begin{bmatrix} 2 \\ 3 \\ 8 \\ 1 \end{bmatrix}$ **5.** $\begin{bmatrix} 0 & 1 & 0 & -1 \\ 2 & 4 & 3 & 0 \\ \sqrt{3} & 1 & \sqrt{6} & 4 \\ 7 & 6 & -3 & 1 \end{bmatrix}$

1. 3×2 **2.** 2×4 **3. a.** 2×2 **4.** 4×1 **5. a.** 4×4

b. $2, 8$ **b.** $0, 4, \sqrt{6}, 1$

Exercises 7.1 *(continued)*

In **6 – 11**, use the matrices A, B, and C with elements a_{ij}, b_{ij}, and c_{ij}, respectively, to find the value that corresponds to the element.

$$A = \begin{bmatrix} -1 & 2 \\ 3 & 4 \\ -2 & 7 \end{bmatrix} \quad B = \begin{bmatrix} 2 & 1 & 3 \\ 5 & 6 & 4 \\ 9 & 7 & 8 \end{bmatrix} \quad C = \begin{bmatrix} 3 & -2 & 1 & 7 & 12 \\ 5 & 4 & -3 & 6 & 2 \\ 1 & 8 & 16 & 11 & 9 \\ 3 & -4 & 13 & -6 & 10 \end{bmatrix}$$

6. a_{22} **7.** a_{32} **8.** b_{13} **9.** b_{31} **10.** c_{45} **11.** c_{34}

12. Write a matrix A of order 2×3 whose elements are $a_{11} = 2$, $a_{12} = 3$, $a_{13} = 4$, $a_{21} = -3$, $a_{22} = -4$ and $a_{23} = 5$.

13. Write a 3×4 zero matrix.

In **14 – 21**, evaluate the determinant of the matrix.

14. $\begin{bmatrix} 7 \end{bmatrix}$ **15.** $\begin{bmatrix} -1 \end{bmatrix}$ **16.** $\begin{bmatrix} 1 & 2 \\ 4 & 7 \end{bmatrix}$ **17.** $\begin{bmatrix} 10 & 3 \\ 7 & 4 \end{bmatrix}$

18. $\begin{bmatrix} -5 & 2 \\ 1 & -3 \end{bmatrix}$ **19.** $\begin{bmatrix} 3 & -1 \\ -2 & 2 \end{bmatrix}$ **20.** $\begin{bmatrix} -5 & 1 \\ -4 & 7 \end{bmatrix}$ **21.** $\begin{bmatrix} 0 & -3 \\ 8 & 6 \end{bmatrix}$

In **22 – 29**, expand by minors to evaluate the determinant of the matrix.

22. $\begin{bmatrix} 3 & 4 & 8 \\ 7 & 1 & 6 \\ 2 & 9 & 5 \end{bmatrix}$ **23.** $\begin{bmatrix} 1 & -2 & 5 \\ 1 & 4 & -1 \\ 2 & 6 & 3 \end{bmatrix}$ **24.** $\begin{bmatrix} 0 & 4 & 1 \\ 2 & 3 & -2 \\ 4 & -1 & -3 \end{bmatrix}$ **25.** $\begin{bmatrix} -2 & 1 & 9 \\ 4 & 0 & 7 \\ 0 & 0 & 12 \end{bmatrix}$

26. $\begin{bmatrix} 1 & 0 & 0 \\ 40 & 26 & 14 \\ 18 & -6 & 9 \end{bmatrix}$ **27.** $\begin{bmatrix} -3 & 5 & 7 \\ 0 & 12 & 3 \\ 0 & 0 & 1 \end{bmatrix}$ **28.** $\begin{bmatrix} 17 & 0 & 0 \\ 36 & 2 & 0 \\ -4 & 16 & 1 \end{bmatrix}$ **29.** $\begin{bmatrix} 1 & 4 & 3 & -1 \\ 0 & 0 & 4 & 3 \\ 0 & 0 & 7 & 6 \\ 8 & 0 & 2 & -3 \end{bmatrix}$

 In **30 – 33**, use a calculator to evaluate the determinant of the matrix.

30. $\begin{bmatrix} 4 & -2 & 1 & 5 \\ 3 & 8 & 0 & 0 \\ 0 & 0 & 7 & 0 \\ -1 & -3 & 6 & 2 \end{bmatrix}$ **31.** $\begin{bmatrix} 1 & 3 & 4 & 6 & 2 \\ 8 & 0 & 0 & 0 & 0 \\ -1 & 2 & 9 & 7 & 3 \\ 2 & 1 & 0 & 0 & 0 \\ 5 & 0 & 6 & 1 & 0 \end{bmatrix}$

32. $\begin{bmatrix} -1 & 2 & 5 & 3 \\ 4 & 6 & -1 & 2 \\ 8 & 7 & 0 & -6 \\ 1 & 2 & -2 & -1 \end{bmatrix}$ **33.** $\begin{bmatrix} 3 & -2 & 4 & 8 & -7 \\ 6 & 3 & 1 & -2 & -9 \\ 0 & 4 & -1 & -6 & 6 \\ 5 & 3 & 0 & 4 & -1 \\ 8 & 2 & -1 & 0 & 2 \end{bmatrix}$

34. $6x^2 + 4$ **35.** $-5xy$ **36.** $13x^3$ **37.** $x^2 - 2x - 1$

38. 30   **39.** $-4, 5$ **40.** $1 \pm \sqrt{2}$ **41.** $\pm \sqrt{5}$

42. Whatever row or column is used in expanding the determinant, the result is the same. Assume that one row i contains all zeros. Then $a_{i1} = 0$, $a_{i2} = 0$, $a_{i3} = 0$, ..., $a_{in} = 0$. In using this row to expand the determinant, each minor would be multiplied by 0, giving a final sum of 0.

43. The determinant of a triangular matrix is equal to the product of the elements in the main diagonal. If one of the elements of the main diagonal is 0, then the product will be 0.

44. a. $21\frac{1}{2}$ **b.** $71\frac{1}{2}$

45.

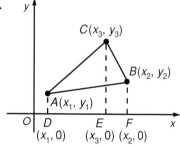

The area can be found by using trapezoids.

Area of $\triangle ABC$ = Area of trapezoid $ACED$ + Area of trapezoid $ECBF$ – Area of trapezoid $ABFD$

Area of trapezoid $ACED = \frac{1}{2}h(b_1 + b_2)$

$= \frac{1}{2}DE(EC + DA)$

$= \frac{1}{2}(x_3 - x_1)(y_3 + y_1)$

Area of trapezoid $ECBF = \frac{1}{2}EF(EC + FB)$

$= \frac{1}{2}(x_2 - x_3)(y_3 + y_2)$

Area of trapezoid $ABFD = \frac{1}{2}DF(AD + BF)$

$= \frac{1}{2}(x_2 - x_1)(y_1 + y_2)$

$$\text{Area of } \triangle ABC = \frac{1}{2}\left[(x_3 - x_1)(y_3 + y_1) + (x_2 - x_3)(y_3 + y_2) - (x_2 - x_1)(y_1 + y_2)\right]$$

$$= \frac{1}{2}\left[x_3y_3 - x_1y_3 + x_3y_1 - x_1y_1 + x_2y_3 - x_3y_3 + x_2y_2 - x_3y_2 - x_2y_1 + x_1y_1 - x_2y_2 + x_1y_2\right]$$

$$= \frac{1}{2}\left[-x_1y_3 + x_1y_2 + x_2y_3 - x_2y_1 + x_3y_1 - x_3y_2\right]$$

$$= \frac{1}{2}\left[x_1(y_2 - y_3) - x_2(y_1 - y_3) + x_3(y_1 - y_2)\right]$$

The sign of the determinant depends on the orientation of points A, B, and C. Choose the sign of $\pm\frac{1}{2}$ so that the area is positive.

$$\pm\frac{1}{2}\begin{vmatrix} x_1 & y_1 & 1 \\ x_2 & y_2 & 1 \\ x_3 & y_3 & 1 \end{vmatrix} = \pm\frac{1}{2}\left[x_1(y_2 - y_3) - x_2(y_1 - y_3) + x_3(y_1 - y_2)\right]$$

46. a. The general equation for a line containing the points (x_1, y_1) and (x_2, y_2) is $y - y_1 = m(x - x_1)$, where m is the slope of the line, defined as $\dfrac{y_2 - y_1}{x_2 - x_1}$.

$$\begin{vmatrix} x & y & 1 \\ x_1 & y_1 & 1 \\ x_2 & y_2 & 1 \end{vmatrix} = 0$$

$$x(y_1 - y_2) - x_1(y - y_2) + x_2(y - y_1) = 0$$

$$xy_1 - xy_2 - x_1y + x_1y_2 + x_2y - x_2y_1 = 0$$

$$x_2y - x_1y - x_2y_1 = xy_2 - x_1y_2 - xy_1 \qquad \text{Add } x_1y_1 \text{ to both sides.}$$

$$x_2y - x_1y - x_2y_1 + x_1y_1 = xy_2 - x_1y_2 - xy_1 + x_1y_1$$

$$(x_2 - x_1)y - (x_2 - x_1)y_1 = (x - x_1)y_2 - (x - x_1)y_1$$

$$(x_2 - x_1)(y - y_1) = (x - x_1)(y_2 - y_1)$$

$$y - y_1 = \frac{y_2 - y_1}{x_2 - x_1}(x - x_1)$$

$$y - y_1 = m(x - x_1)$$

b. (1) $2x + 5y = 22$

    (2) $6x - 3y = -24$

Exercises 7.1 (continued)

In **34 – 37**, simplify the expression.

34. $\begin{vmatrix} 2x & -1 \\ 4 & 3x \end{vmatrix}$ **35.** $\begin{vmatrix} 2x & 3x \\ y & -y \end{vmatrix}$ **36.** $\begin{vmatrix} x^2 & -2x^3 \\ 6 & x \end{vmatrix}$ **37.** $\begin{vmatrix} (x+1) & x \\ 2 & (x-1) \end{vmatrix}$

In **38 – 41**, solve for all real values of x.

38. $\begin{vmatrix} 2 & 10 \\ 5 & (x-4) \end{vmatrix} = 2$ **39.** $\begin{vmatrix} x & 4 \\ 5 & (x-1) \end{vmatrix} = 0$ **40.** $\begin{vmatrix} x & 2 \\ x+\frac{1}{2} & x \end{vmatrix} = 0$ **41.** $\begin{vmatrix} x+1 & 1 \\ x+5 & x \end{vmatrix} = 0$

42. Prove that if one row or column of a square matrix has zeros for all entries, then the determinant of the matrix is 0.

43. Prove that if one or more entries of the diagonal of a triangular matrix is zero, then the determinant of the matrix is 0.

44. If the coordinates of a triangle are $A(x_1, y_1)$, $B(x_2, y_2)$, and $C(x_3, y_3)$, then the area of the triangle is

$$\pm\frac{1}{2}\begin{vmatrix} x_1 & y_1 & 1 \\ x_2 & y_2 & 1 \\ x_3 & y_3 & 1 \end{vmatrix}$$

where the sign is chosen in order to obtain a positive area. Using the formula, find the area of each of the following triangles ABC given the coordinates of their vertices.

a. $A(1, 4)$, $B(6, 0)$, $C(-1, -3)$

b. $A(0, 10)$, $B(10, 1)$, $C(3, -7)$

45. Use analytic geometry to prove the formula for the area of a triangle given in Exercise 44.

Hint: Place the triangle in quadrant I.

46. a. If two points on a line have coordinates (x_1, y_1) and (x_2, y_2), prove that

$$\begin{vmatrix} x & y & 1 \\ x_1 & y_1 & 1 \\ x_2 & y_2 & 1 \end{vmatrix} = 0$$

b. Use the determinant in part **a** to find an equation of the line passing through the given points.

(1) $(6, 2)$, $(1, 4)$ **(2)** $(-3, 2)$, $(0, 8)$

47. If three points (x_1, y_1), (x_2, y_2), (x_3, y_3) are collinear, then

$$\begin{vmatrix} x_1 & y_1 & 1 \\ x_2 & y_2 & 1 \\ x_3 & y_3 & 1 \end{vmatrix} = 0$$

Use this relationship to determine if the 3 points given are collinear.

a. $(0, 3)$, $(4, 11)$, $(-1, 1)$

b. $(1, 5)$, $(7, 11)$, $(-6, -10)$

7.2 Operations on Matrices

Reconsider the application of a matrix, mentioned in Section 6.1, in which the costs of three different types of computers from mail-order stores in three different cities were recorded (see matrix C below). The charges that these stores might add on for shipping and handling are recorded in a second matrix, S.

$$C = \begin{bmatrix} 600 & 2{,}500 & 1{,}600 \\ 650 & 2{,}450 & 1{,}690 \\ 620 & 2{,}560 & 1{,}580 \end{bmatrix} \qquad S = \begin{bmatrix} 20 & 50 & 35 \\ 10 & 25 & 20 \\ 45 & 45 & 45 \end{bmatrix}$$

Then matrix T, the sum of matrices C and S, can be written to conveniently compare the total costs of purchasing a computer from the stores listed.

$$T = C + S = \begin{bmatrix} 600 & 2{,}500 & 1{,}600 \\ 650 & 2{,}450 & 1{,}690 \\ 620 & 2{,}560 & 1{,}580 \end{bmatrix} + \begin{bmatrix} 20 & 50 & 35 \\ 10 & 25 & 20 \\ 45 & 45 & 45 \end{bmatrix} = \begin{bmatrix} 620 & 2{,}550 & 1{,}635 \\ 660 & 2{,}475 & 1{,}710 \\ 665 & 2{,}605 & 1{,}625 \end{bmatrix}$$

This example suggests a definition for matrix addition.

■ **Definition:** *If matrix A and matrix B each have dimension m × n, then A + B is an m × n matrix each of whose entries is the sum of the corresponding entries in A and B.*

Note that addition is defined only for matrices that have the same dimension.

An $m \times n$ zero matrix is an identity matrix for addition for the set of $m \times n$ matrices.

For example: $\begin{bmatrix} 4 & 7 & 1 \\ 2 & 3 & 4 \end{bmatrix} + \begin{bmatrix} 0 & 0 & 0 \\ 0 & 0 & 0 \end{bmatrix} = \begin{bmatrix} 4 & 7 & 1 \\ 2 & 3 & 4 \end{bmatrix}$

In general, if O is an $m \times n$ zero matrix and A is an $m \times n$ matrix, then:

$$\underset{A}{\begin{bmatrix} a_{11} & a_{12} & \cdots & a_{1n} \\ a_{21} & a_{22} & \cdots & a_{2n} \\ \vdots & \vdots & \vdots & \vdots \\ a_{m1} & a_{m2} & \cdots & a_{mn} \end{bmatrix}} + \underset{O}{\begin{bmatrix} 0 & 0 & \cdots & 0 \\ 0 & 0 & \cdots & 0 \\ \vdots & \vdots & \vdots & \vdots \\ 0 & 0 & \cdots & 0 \end{bmatrix}} = \underset{A}{\begin{bmatrix} a_{11} & a_{12} & \cdots & a_{1n} \\ a_{21} & a_{22} & \cdots & a_{2n} \\ \vdots & \vdots & \vdots & \vdots \\ a_{m1} & a_{m2} & \cdots & a_{mn} \end{bmatrix}}$$

The additive inverse of an $m \times n$ matrix A is an $m \times n$ matrix $-A$ such that for each a_{ij} in A, the corresponding entry in $-A$ is $-a_{ij}$.

$$\underset{A}{\begin{bmatrix} a_{11} & a_{12} & \cdots & a_{1n} \\ a_{21} & a_{22} & \cdots & a_{2n} \\ \vdots & \vdots & \vdots & \vdots \\ a_{m1} & a_{m2} & \cdots & a_{mn} \end{bmatrix}} + \underset{(-A)}{\begin{bmatrix} -a_{11} & -a_{12} & \cdots & -a_{1n} \\ -a_{21} & -a_{22} & \cdots & -a_{2n} \\ \vdots & \vdots & \vdots & \vdots \\ -a_{m1} & -a_{m2} & \cdots & -a_{mn} \end{bmatrix}} = \underset{O}{\begin{bmatrix} 0 & 0 & \cdots & 0 \\ 0 & 0 & \cdots & 0 \\ \vdots & \vdots & \vdots & \vdots \\ 0 & 0 & \cdots & 0 \end{bmatrix}}$$

The additive inverse is used to find the difference of two matrices A and B.

$$A - B = A + (-B)$$

$A - B$ is only defined when A and B have the same dimension.

Example 1 _____

Given: $A = \begin{bmatrix} 5 & 7 & 1 & 2 \\ 3 & 6 & 4 & 8 \end{bmatrix}$ and $B = \begin{bmatrix} 0 & 6 & -2 & 1 \\ 3 & 7 & 4 & 9 \end{bmatrix}$

Find: **a.** $A + B$ **b.** $B - A$

Solution

a. Since A and B are both 2×4 matrices, add the corresponding elements of A and B.

$$A + B = \begin{bmatrix} 5+0 & 7+6 & 1+(-2) & 2+1 \\ 3+3 & 6+7 & 4+4 & 8+9 \end{bmatrix}$$

$$= \begin{bmatrix} 5 & 13 & -1 & 3 \\ 6 & 13 & 8 & 17 \end{bmatrix}$$

b. $B - A$ is found by adding $-A$ to B.

$$B - A = \begin{bmatrix} 0 & 6 & -2 & 1 \\ 3 & 7 & 4 & 9 \end{bmatrix} - \begin{bmatrix} 5 & 7 & 1 & 2 \\ 3 & 6 & 4 & 8 \end{bmatrix}$$

$$B + (-A) = \begin{bmatrix} 0 & 6 & -2 & 1 \\ 3 & 7 & 4 & 9 \end{bmatrix} + \begin{bmatrix} -5 & -7 & -1 & -2 \\ -3 & -6 & -4 & -8 \end{bmatrix}$$

$$B - A = \begin{bmatrix} -5 & -1 & -3 & -1 \\ 0 & 1 & 0 & 1 \end{bmatrix}$$

The sum of two matrices may be found on a calculator by using the following key sequence after matrices A and B have been entered:

| 2nd | [A] | + | 2nd | [B] | ENTER |

Scalar Multiplication

Suppose the business manager for a company is planning to purchase 6 computers from one of the mail-order companies previously mentioned. Assuming no discounts for purchases in quantities of 6 or less, multiplying each entry in the matrix that displays the total costs will show the comparative costs of buying 6 computers of the same kind from the different companies.

$$6 \begin{bmatrix} 620 & 2{,}550 & 1{,}635 \\ 660 & 2{,}475 & 1{,}710 \\ 665 & 2{,}605 & 1{,}625 \end{bmatrix} = \begin{bmatrix} 6 \cdot 620 & 6 \cdot 2{,}550 & 6 \cdot 1{,}635 \\ 6 \cdot 660 & 6 \cdot 2{,}475 & 6 \cdot 1{,}710 \\ 6 \cdot 665 & 6 \cdot 2{,}605 & 6 \cdot 1{,}625 \end{bmatrix} = \begin{bmatrix} 3{,}720 & 15{,}300 & 9{,}810 \\ 3{,}960 & 14{,}850 & 10{,}260 \\ 3{,}990 & 15{,}630 & 9{,}750 \end{bmatrix}$$

The constant multiplier 6 is called a *scalar*

Scalar multiplication of matrices is similar to scalar multiplication of a vector. If a matrix A is multiplied by a scalar, k, then the result is a matrix kA. Each entry of kA is k times the corresponding element of A.

34.

$$\begin{bmatrix} 6 & 6 \\ 2 & 2 \end{bmatrix} \cdot \begin{bmatrix} a_{11} & a_{12} \\ a_{21} & a_{22} \end{bmatrix} = \begin{bmatrix} 0 & 0 \\ 0 & 0 \end{bmatrix}$$

$$\begin{bmatrix} 6(a_{11}) + 6(a_{21}) & 6(a_{12}) + 6(a_{22}) \\ 2(a_{11}) + 2(a_{21}) & 2(a_{12}) + 2(a_{22}) \end{bmatrix} = \begin{bmatrix} 0 & 0 \\ 0 & 0 \end{bmatrix}$$

$6a_{11} + 6a_{21} = 0$ $a_{11} + a_{21} = 0$
$6a_{12} + 6a_{22} = 0$ $a_{12} + a_{22} = 0$
$2a_{11} + 2a_{21} = 0$
$2a_{12} + 2a_{22} = 0$

Any matrix in which the 2 elements of each column are additive inverses will produce the zero matrix.

One example would be $\begin{bmatrix} 1 & 1 \\ -1 & -1 \end{bmatrix}$.

35. a. $A(BC) = \begin{bmatrix} a_1 & a_2 \\ a_3 & a_4 \end{bmatrix} \cdot \left(\begin{bmatrix} b_1 & b_2 \\ b_3 & b_4 \end{bmatrix} \cdot \begin{bmatrix} c_1 & c_2 \\ c_3 & c_4 \end{bmatrix} \right)$

$$= \begin{bmatrix} a_1 & a_2 \\ a_3 & a_4 \end{bmatrix} \cdot \left(\begin{bmatrix} b_1 c_1 + b_2 c_3 & b_1 c_2 + b_2 c_4 \\ b_3 c_1 + b_4 c_3 & b_3 c_2 + b_4 c_4 \end{bmatrix} \right)$$

$$= \begin{bmatrix} a_1 b_1 c_1 + a_1 b_2 c_3 + a_2 b_3 c_1 + a_2 b_4 c_3 & a_1 b_1 c_2 + a_1 b_2 c_4 + a_2 b_3 c_2 + a_2 b_4 c_4 \\ a_3 b_1 c_1 + a_3 b_2 c_3 + a_4 b_3 c_1 + a_4 b_4 c_3 & a_3 b_1 c_2 + a_3 b_2 c_4 + a_4 b_3 c_2 + a_4 b_4 c_4 \end{bmatrix}$$

$(AB)C = \left(\begin{bmatrix} a_1 & a_2 \\ a_3 & a_4 \end{bmatrix} \cdot \begin{bmatrix} b_1 & b_2 \\ b_3 & b_4 \end{bmatrix} \right) \cdot \begin{bmatrix} c_1 & c_2 \\ c_3 & c_4 \end{bmatrix}$

$$= \left(\begin{bmatrix} a_1 b_1 + a_2 b_3 & a_1 b_2 + a_2 b_4 \\ a_3 b_1 + a_4 b_3 & a_3 b_2 + a_4 b_4 \end{bmatrix} \right) \cdot \begin{bmatrix} c_1 & c_2 \\ c_3 & c_4 \end{bmatrix}$$

$$= \begin{bmatrix} a_1 b_1 c_1 + a_2 b_3 c_1 + a_1 b_2 c_3 + a_2 b_4 c_3 & a_1 b_1 c_2 + a_2 b_3 c_2 + a_1 b_2 c_4 + a_2 b_4 c_4 \\ a_3 b_1 c_1 + a_4 b_3 c_1 + a_3 b_2 c_3 + a_4 b_4 c_3 & a_3 b_1 c_2 + a_4 b_3 c_2 + a_3 b_2 c_4 + a_4 b_4 c_4 \end{bmatrix}$$

b. $A \cdot (B + C) = A \cdot B + A \cdot C$

$A \cdot (B + C) = \begin{bmatrix} a_1 & a_2 \\ a_3 & a_4 \end{bmatrix} \cdot \left(\begin{bmatrix} b_1 & b_2 \\ b_3 & b_4 \end{bmatrix} + \begin{bmatrix} c_1 & c_2 \\ c_3 & c_4 \end{bmatrix} \right)$

$$= \begin{bmatrix} a_1 & a_2 \\ a_3 & a_4 \end{bmatrix} \cdot \left(\begin{bmatrix} (b_1 + c_1) & (b_2 + c_2) \\ (b_3 + c_3) & (b_4 + c_4) \end{bmatrix} \right)$$

$$= \begin{bmatrix} a_1(b_1 + c_1) + a_2(b_3 + c_3) & a_1(b_2 + c_2) + a_2(b_4 + c_4) \\ a_3(b_1 + c_1) + a_4(b_3 + c_3) & a_3(b_2 + c_2) + a_4(b_4 + c_4) \end{bmatrix}$$

$A \cdot B + A \cdot C = \begin{bmatrix} a_1 & a_2 \\ a_3 & a_4 \end{bmatrix} \cdot \begin{bmatrix} b_1 & b_2 \\ b_3 & b_4 \end{bmatrix} + \begin{bmatrix} a_1 & a_2 \\ a_3 & a_4 \end{bmatrix} \cdot \begin{bmatrix} c_1 & c_2 \\ c_3 & c_4 \end{bmatrix}$

$$= \begin{bmatrix} a_1 b_1 + a_2 b_3 & a_1 b_2 + a_2 b_4 \\ a_3 b_1 + a_4 b_3 & a_3 b_2 + a_4 b_4 \end{bmatrix} + \begin{bmatrix} a_1 c_1 + a_2 c_3 & a_1 c_2 + a_2 c_4 \\ a_3 c_1 + a_4 c_3 & a_3 c_2 + a_4 c_4 \end{bmatrix}$$

$$= \begin{bmatrix} a_1 b_1 + a_1 c_1 + a_2 b_3 + a_2 c_3 & a_1 b_2 + a_1 c_2 + a_2 b_4 + a_2 c_4 \\ a_3 b_1 + a_3 c_1 + a_4 b_3 + a_4 c_3 & a_3 b_2 + a_3 c_2 + a_4 b_4 + a_4 c_4 \end{bmatrix}$$

$$= \begin{bmatrix} a_1(b_1 + c_1) + a_2(b_3 + c_3) & a_1(b_2 + c_2) + a_2(b_4 + c_4) \\ a_3(b_1 + c_1) + a_4(b_3 + c_3) & a_3(b_2 + c_2) + a_4(b_4 + c_4) \end{bmatrix}$$

36. Matrix multiplication is not commutative. $BA \neq AB$

37. $(A + B)(A - B) \neq A^2 - B^2$

$A^2 + BA - AB - B^2 \neq A^2 - B^2$ 　　　　Since $BA \neq AB$, $BA - AB \neq 0$.

38. $A = \begin{bmatrix} a_1 & a_2 & a_3 \\ 0 & a_4 & a_5 \\ 0 & 0 & a_6 \end{bmatrix}$ 　　　 $B = \begin{bmatrix} b_1 & b_2 & b_3 \\ 0 & b_4 & b_5 \\ 0 & 0 & b_6 \end{bmatrix}$

$A + B = \begin{bmatrix} a_1 + b_1 & a_2 + b_2 & a_3 + b_3 \\ 0 & a_4 + b_4 & a_5 + b_5 \\ 0 & 0 & a_6 + b_6 \end{bmatrix}$

$AB = \begin{bmatrix} a_1 b_1 & a_1 b_2 + a_2 b_4 & a_1 b_3 + a_2 b_5 + a_3 b_6 \\ 0 & a_4 b_4 & a_4 b_5 + a_5 b_6 \\ 0 & 0 & a_6 b_6 \end{bmatrix}$

The sum and product are triangular matrices.

39. 　　　 $A = \begin{bmatrix} a_1 & a_2 \\ a_3 & a_4 \end{bmatrix}$ 　　　 $B = \begin{bmatrix} b_1 & b_2 \\ b_3 & b_4 \end{bmatrix}$

$A^{-1} = \begin{bmatrix} a_4 & -a_2 \\ -a_3 & a_1 \end{bmatrix} \dfrac{1}{a_1 a_4 - a_2 a_3}$

$B^{-1} = \begin{bmatrix} b_4 & -b_2 \\ -b_3 & b_1 \end{bmatrix} \dfrac{1}{b_1 b_4 - b_2 b_3}$

$B^{-1} \cdot A^{-1} = \dfrac{\begin{bmatrix} a_4 b_4 + b_2 a_3 & -a_2 b_4 - a_1 b_2 \\ -a_4 b_3 - b_1 a_3 & a_2 b_3 + a_1 b_1 \end{bmatrix}}{(a_1 a_4 - a_2 a_3)(b_1 b_4 - b_2 b_3)}$

$A \cdot B = \begin{bmatrix} a_1 b_1 + a_2 b_3 & a_1 b_2 + a_2 b_4 \\ a_3 b_1 + a_4 b_3 & a_3 b_2 + a_4 b_4 \end{bmatrix}$

$(A \cdot B)^{-1} = \dfrac{\begin{bmatrix} a_3 b_2 + a_4 b_4 & -a_1 b_2 - a_2 b_4 \\ -a_3 b_1 - a_4 b_3 & a_1 b_1 + a_2 b_3 \end{bmatrix}}{(a_1 b_1 + a_2 b_3)(a_3 b_2 + a_4 b_4) - (a_3 b_1 + a_4 b_3)(a_1 b_2 + a_2 b_4)}$

$(A \cdot B)^{-1}$ and $B^{-1} \cdot A^{-1}$ have equal numerators;
　　　show denominators are equal.

$(a_1 b_1 + a_2 b_3)(a_3 b_2 + a_4 b_4) - (a_3 b_1 + a_4 b_3)(a_1 b_2 + a_2 b_4)$
$= (a_1 a_3 b_1 b_2 + a_1 a_4 b_1 b_4 + a_2 a_3 b_3 b_2 + a_2 a_4 b_3 b_4)$
$\quad - (a_1 a_3 b_1 b_2 + a_2 a_3 b_1 b_4 + a_1 a_4 b_2 b_3 + a_2 a_4 b_3 b_4)$
$= a_1 a_4 (b_1 b_4 - b_2 b_3) + a_2 a_3 (b_2 b_3 - b_1 b_4)$
$= (a_1 a_4 - a_2 a_3)(b_1 b_4 - b_2 b_3)$

7.3 Using Matrices to Solve Systems of Equations

Recall that a system of linear equations consists of two or more linear equations that may have a common solution. In earlier courses, you learned to solve such a system by working with pairs of equations to eliminate a variable. The Swiss mathematician Gabriel Cramer demonstrated, in 1750, an efficient method, using determinants, for the solution of a square system, that is, a system that has the same number of equations as variables.

A linear system can be rewritten in matrix form.

$$
\begin{matrix} x + y + z = 1 \\ 2x - 3y + 2z = 7 \\ x + 6y = -4 \end{matrix} \quad \text{can be written as} \quad \begin{bmatrix} 1 & 1 & 1 \\ 2 & -3 & 2 \\ 1 & 6 & 0 \end{bmatrix} \begin{bmatrix} x \\ y \\ z \end{bmatrix} = \begin{bmatrix} 1 \\ 7 \\ -4 \end{bmatrix}
$$

A general system of two linear equations with two unknowns, x and y, may be written

$$
\begin{matrix} ax + by = e \\ cx + dy = f \end{matrix} \quad \text{or} \quad \begin{bmatrix} a & b \\ c & d \end{bmatrix} \begin{bmatrix} x \\ y \end{bmatrix} = \begin{bmatrix} e \\ f \end{bmatrix}
$$

A solution for the system in terms of $a, b, c, d, e,$ and f may be determined, if it exists, by using matrix algebra as follows.

Let: $A = \begin{bmatrix} a & b \\ c & d \end{bmatrix}$ $\qquad A \begin{bmatrix} x \\ y \end{bmatrix} = \begin{bmatrix} e \\ f \end{bmatrix}$

Multiply both sides by A^{-1}. $\qquad A^{-1}A \begin{bmatrix} x \\ y \end{bmatrix} = A^{-1} \begin{bmatrix} e \\ f \end{bmatrix}$

Substitute for $A^{-1}A$ and A^{-1}. $\qquad I \begin{bmatrix} x \\ y \end{bmatrix} = \frac{1}{|A|} \begin{bmatrix} d & -b \\ -c & a \end{bmatrix} \begin{bmatrix} e \\ f \end{bmatrix}$

Definition of I.
Matrix multiplication. $\qquad \begin{bmatrix} x \\ y \end{bmatrix} = \frac{1}{|A|} \begin{bmatrix} de - bf \\ -ce + af \end{bmatrix}$

Meaning of equal matrices. $\qquad x = \frac{1}{|A|}(de - bf) \quad y = \frac{1}{|A|}(af - ce)$

The expressions for x and y can each be written as the ratio of two determinants.

$$
x = \frac{\begin{vmatrix} e & b \\ f & d \end{vmatrix}}{\begin{vmatrix} a & b \\ c & d \end{vmatrix}} \qquad y = \frac{\begin{vmatrix} a & e \\ c & f \end{vmatrix}}{\begin{vmatrix} a & b \\ c & d \end{vmatrix}} \quad \text{where } ad - bc \neq 0
$$

Note that the determinant in each denominator is the same for x and y, with the entries following the same pattern as the coefficients for x and y in the system of linear equations. The determinant used in the numerator of the expression for x is the coefficient matrix with the constant terms e and f replacing the coefficients of x. Similarly, the determinant used in the numerator of the expression for y is the coefficient matrix with the constants e and f replacing the coefficients of y.

□ **Cramer's Rule:** *Given a system of n linear equations with n unknowns, which has a unique solution, and*

$$\text{coefficient matrix} \qquad \text{variable matrix} \qquad \text{constant matrix}$$

$$A = \begin{bmatrix} a_{11} & a_{12} & a_{13} & \cdots & a_{1n} \\ a_{21} & a_{22} & a_{23} & \cdots & a_{2n} \\ \vdots & \vdots & \vdots & \ddots & \vdots \\ a_{n1} & a_{n2} & a_{n3} & \cdots & a_{nn} \end{bmatrix} \qquad X = \begin{bmatrix} x_1 \\ x_2 \\ \vdots \\ x_n \end{bmatrix} \qquad D = \begin{bmatrix} d_1 \\ d_2 \\ \vdots \\ d_n \end{bmatrix}$$

then the n solutions of the system are the elements of X:

$$x_1 = \frac{\begin{vmatrix} d_{11} & a_{12} & a_{13} & \cdots & a_{1n} \\ d_{21} & a_{22} & a_{23} & \cdots & a_{2n} \\ \vdots & \vdots & \vdots & \ddots & \vdots \\ d_{n1} & a_{n2} & a_{n3} & \cdots & a_{nn} \end{vmatrix}}{|A|}$$

$$x_2 = \frac{\begin{vmatrix} a_{11} & d_{12} & a_{13} & \cdots & a_{1n} \\ a_{21} & d_{22} & a_{23} & \cdots & a_{2n} \\ \vdots & \vdots & \vdots & \ddots & \vdots \\ a_{n1} & d_{n2} & a_{n3} & \cdots & a_{nn} \end{vmatrix}}{|A|}$$

$$\vdots$$

$$x_n = \frac{\begin{vmatrix} a_{11} & a_{12} & a_{13} & \cdots & d_{1n} \\ a_{21} & a_{22} & a_{23} & \cdots & d_{2n} \\ \vdots & \vdots & \vdots & \ddots & \vdots \\ a_{n1} & a_{n2} & a_{n3} & \cdots & d_{nn} \end{vmatrix}}{|A|}$$

Example 1 _____

Use Cramer's Rule to solve the system:
$$2x - 5y = 2$$
$$3x - 7y = 1$$

Solution

The determinant of the coefficient matrix for the system is $\begin{vmatrix} 2 & -5 \\ 3 & -7 \end{vmatrix}$.
Therefore:

$$x = \frac{\begin{vmatrix} 2 & -5 \\ 1 & -7 \end{vmatrix}}{\begin{vmatrix} 2 & -5 \\ 3 & -7 \end{vmatrix}} = \frac{-14 - (-5)}{1} = -9 \qquad y = \frac{\begin{vmatrix} 2 & 2 \\ 3 & 1 \end{vmatrix}}{\begin{vmatrix} 2 & -5 \\ 3 & -7 \end{vmatrix}} = \frac{2 - 6}{1} = -4$$

Observe that in the numerators of the expressions for x and y the constants in the system, 2 and 1, are in the same positions as the coefficients of the corresponding variable.

Note that in order to apply Cramer's Rule to solve a system of equations, each equation must be written with the constant term alone on one side.

Example 2

Use Cramer's Rule to solve the system:
$$6x - 2y + z = 16$$
$$2x - y + 5z = -2$$
$$2x - 3z = 8$$

Solution

The coefficient matrix for the system is:
$$\begin{bmatrix} 6 & -2 & 1 \\ 2 & -1 & 5 \\ 2 & 0 & -3 \end{bmatrix}$$

Replace the coefficients of each variable with the constants to form the determinants for the numerators, and use the determinant of the coefficient matrix for each denominator.

$$x = \frac{\begin{vmatrix} 16 & -2 & 1 \\ -2 & -1 & 5 \\ 8 & 0 & -3 \end{vmatrix}}{\begin{vmatrix} 6 & -2 & 1 \\ 2 & -1 & 5 \\ 2 & 0 & -3 \end{vmatrix}} \qquad y = \frac{\begin{vmatrix} 6 & 16 & 1 \\ 2 & -2 & 5 \\ 2 & 8 & -3 \end{vmatrix}}{\begin{vmatrix} 6 & -2 & 1 \\ 2 & -1 & 5 \\ 2 & 0 & -3 \end{vmatrix}} \qquad z = \frac{\begin{vmatrix} 6 & -2 & 16 \\ 2 & -1 & -2 \\ 2 & 0 & 8 \end{vmatrix}}{\begin{vmatrix} 6 & -2 & 1 \\ 2 & -1 & 5 \\ 2 & 0 & -3 \end{vmatrix}}$$

$$x = \frac{-12}{-12} \qquad\qquad y = \frac{72}{-12} \qquad\qquad z = \frac{24}{-12}$$

$$x = 1 \qquad\qquad y = -6 \qquad\qquad z = -2$$

Review Section 7.1 if you have forgotten how to find the determinant of a 3×3 matrix. The solution $(1, -6, -2)$ checks in each of the three equations.

To solve this system using a calculator that performs matrix operations, enter the coefficient matrix as [A]. The determinant can be found by using the following key sequence.

$\boxed{\text{MATRX}}$ 5 $\boxed{\text{2nd}}$ $\boxed{\text{[A]}}$ $\boxed{\text{ENTER}}$ The calculator will return the value -12.

Now replace the x coefficients in the first column of A by the constants, that is, $a_{11} = 16$, $a_{21} = -2$ and $a_{31} = 8$. Replace the coefficients of y by the constants and enter the resulting matrix as [B]. Then replace the coefficients of z by the constants and enter the matrix as [C]. Find the values of x, y, and z.

$\boxed{\text{MATRX}}$ 5 $\boxed{\text{2nd}}$ $\boxed{\text{[A]}}$ $\boxed{\div}$ $\boxed{(-)}$ 12 $\boxed{\text{ENTER}}$

$\boxed{\text{MATRX}}$ 5 $\boxed{\text{2nd}}$ $\boxed{\text{[B]}}$ $\boxed{\div}$ $\boxed{(-)}$ 12 $\boxed{\text{ENTER}}$

$\boxed{\text{MATRX}}$ 5 $\boxed{\text{2nd}}$ $\boxed{\text{[C]}}$ $\boxed{\div}$ $\boxed{(-)}$ 12 $\boxed{\text{ENTER}}$

Types of Systems

Consider the following three systems of equations.

$$\begin{aligned} x + y &= 5 \\ x - y &= 3 \end{aligned} \qquad\qquad \begin{aligned} x + y &= 5 \\ 2x + 2y &= 10 \end{aligned} \qquad\qquad \begin{aligned} x + y &= 5 \\ x + y &= 3 \end{aligned}$$

The first system has one solution $(x = 4, y = 1)$ and it is called a *consistent system* Since the second system contains two equations that are equivalent, an infinite number of ordered pairs will satisfy both equations; this is a *dependent system*. The third system has no solution because if $x + y = 5$ then $x + y$ cannot also equal 3; this is an *inconsistent system*

Observe the behavior of the graphs in each of the three possible types of systems.

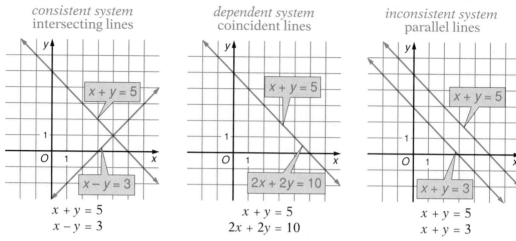

consistent system intersecting lines	*dependent system* coincident lines	*inconsistent system* parallel lines
$x + y = 5$ $x - y = 3$	$x + y = 5$ $2x + 2y = 10$	$x + y = 5$ $x + y = 3$

Note, for each system:
the coefficient matrix

$$\begin{bmatrix} 1 & 1 \\ 1 & -1 \end{bmatrix} \qquad \begin{bmatrix} 1 & 1 \\ 2 & 2 \end{bmatrix} \qquad \begin{bmatrix} 1 & 1 \\ 1 & 1 \end{bmatrix}$$

the value of the corresponding determinant

$$\begin{vmatrix} 1 & 1 \\ 1 & -1 \end{vmatrix} = -2 \qquad \begin{vmatrix} 1 & 1 \\ 2 & 2 \end{vmatrix} = 0 \qquad \begin{vmatrix} 1 & 1 \\ 1 & 1 \end{vmatrix} = 0$$

and the number of solutions

one   infinite   none

Only a system with a nonzero determinant will have a unique solution.

The Augmented Matrix

Recall that the solution set of a system of linear equations remains the same even when any equation of the system is multiplied by a nonzero constant or when a multiple of one equation of the system is added to another. These principles are applied to eliminate a variable between pairs of equations of the system, eventually leading to a solution for the system.

These principles can also be applied to a matrix that is used to represent a system of equations. Since each row of the matrix represents the coefficients of one of the equations in the system, the term *row operations* is used. Using row operations, a matrix is simplified so that the solution for the system can more easily be determined. The procedure is an alternate to the method of solving by Cramer's Rule.

Following are the row operations that can be performed:

1. Multiply a row by a nonzero constant.
2. Add a multiple of one row to another.
3. Interchange two rows.

The procedure for solving a system of equations using matrices begins with the coefficient matrix for the system. For example, the coefficient matrix for the system

$$\begin{aligned} x - 3z &= -2 \\ 3x + y - 2z &= 5 \\ 2x + 2y + z &= 4 \end{aligned} \quad \text{is} \quad \begin{bmatrix} 1 & 0 & -3 \\ 3 & 1 & -2 \\ 2 & 2 & 1 \end{bmatrix}$$

If a column that includes the constants of the system is added to the coefficient matrix, an *augmented matrix* is formed.

$$\begin{bmatrix} 1 & 0 & -3 & -2 \\ 3 & 1 & -2 & 5 \\ 2 & 2 & 1 & 4 \end{bmatrix}$$

The objective is to transform the augmented matrix, using row operations, so that it is of the form $\begin{bmatrix} 1 & 0 & 0 & a \\ 0 & 1 & 0 & b \\ 0 & 0 & 1 & c \end{bmatrix}$, which is equivalent to the solution $\begin{aligned} x &= a \\ y &= b \\ z &= c \end{aligned}$

The first three rows and columns of the final matrix form the identity matrix.

Since there is a considerable amount of arithmetic involved, it is important to make it a practice to make notations of what is being done.

Augmented Matrix

$$\begin{matrix} R_1 \\ R_2 \\ R_3 \end{matrix} \begin{bmatrix} 1 & 0 & -3 & -2 \\ 3 & 1 & -2 & 5 \\ 2 & 2 & 1 & 4 \end{bmatrix}$$

The row operations are equivalent to operations on equations, as shown below.

To obtain a zero as the first entry of the second row, add –3 times Row 1 above to Row 2.

$$-3R_1 + R_2 \begin{bmatrix} 1 & 0 & -3 & -2 \\ 0 & 1 & 7 & 11 \\ 2 & 2 & 1 & 4 \end{bmatrix}$$

$$\begin{aligned} 1x + 0y - 3z &= -2 \\ -3(1x + 0y - 3z) + 3x + 1y - 2z &= -3(-2) + 5 \\ 2x + 2y + 1z &= 4 \end{aligned}$$

To obtain a zero in the third row, second column, add –2 times Row 2 to Row 3.

$$-2R_2 + R_3 \begin{bmatrix} 1 & 0 & -3 & -2 \\ 0 & 1 & 7 & 11 \\ 2 & 0 & -13 & -18 \end{bmatrix}$$

$$\begin{aligned} 1x + 0y - 3z &= -2 \\ 0x + 1y + 7z &= 11 \\ -2(0x + 1y + 7z) + 2x + 2y + 1z &= -2(11) + 4 \end{aligned}$$

To obtain a zero in the third row, first column, add –2 times Row 1 to Row 3.

$$-2R_1 + R_3 \begin{bmatrix} 1 & 0 & -3 & -2 \\ 0 & 1 & 7 & 11 \\ 0 & 0 & -7 & -14 \end{bmatrix}$$

$$\begin{aligned} 1x + 0y - 3z &= -2 \\ 0x + 1y + 7z &= 11 \\ -2(1x + 0y - 3z) + 2x + 0y - 13z &= -2(-2) - 18 \end{aligned}$$

To obtain a zero in the second row, third column, add Row 3 to Row 2.

$$R_3 + R_2 \begin{bmatrix} 1 & 0 & -3 & -2 \\ 0 & 1 & 0 & -3 \\ 0 & 0 & -7 & -14 \end{bmatrix}$$

$$\begin{aligned} 1x + 0y - 3z &= -2 \\ 0x + 0y - 7z + 0x + 1y + 7z &= -14 + 11 \\ 0x + 0y - 7z &= -14 \end{aligned}$$

Multiply Row 3 by $-\frac{1}{7}$.

$$-\frac{1}{7}R_3 \begin{bmatrix} 1 & 0 & -3 & -2 \\ 0 & 1 & 0 & -3 \\ 0 & 0 & 1 & 2 \end{bmatrix} \qquad \begin{array}{l} 1x + 0y - 3z = -2 \\ 0x + 1y + 0z = -3 \\ -\frac{1}{7}(0x + 0y - 7z) = -\frac{1}{7}(-14) \end{array}$$

Add 3 times Row 3 to Row 1.

$$3R_3 + R_1 \begin{bmatrix} 1 & 0 & 0 & 4 \\ 0 & 1 & 0 & -3 \\ 0 & 0 & 1 & 2 \end{bmatrix} \qquad \begin{array}{r} 3(0x + 0y + 1z) + 1x + 0y - 3z = 3(2) - 2 \\ 0x + 1y + 0z = -3 \\ 0x + 0y + 1z = 2 \end{array}$$

The last matrix represents the set of equations

$$x = 4 \qquad\qquad y = -3 \qquad\qquad z = 2$$

which is equivalent to the original system and makes the solution apparent. As noted earlier, the first three rows and columns of the final matrix form the identity matrix.

The sequence of steps outlined above is one of a number of different ways of using row operations to obtain an equivalent matrix from which solutions can be obtained. A strategy that often minimizes the number of steps is to try to obtain the necessary zeros in one column at a time. This method of solving a system is sometimes called *Gaussian elimination*. It is essentially an algebraic solution that is done without writing the variables.

A calculator that performs matrix operations will do row operations as well, and can be used to solve the previous example. The result of each of the following 7 steps on the calculator can be checked against the previous text.

Step 1
Enter the augmented matrix as [A].

Step 2

Objective : $-3R_1 + R_2$ to replace element a_{21} by 0

Calculator function : ***Row+**, which will multiply a row by a scalar value, add the result to a second row, and store the result in the second row.

Key sequence : **MATRX** 4 **(−)** 3 **ALPHA** **,** **2nd** **[A]** **ALPHA** **,** 1 **ALPHA** **,** 2 **)** **STO►** **2nd** **[A]** **ENTER**

Step 3

Objective : $-2R_2 + R_3$ to replace element a_{32} by 0

Key sequence : **MATRX** 4 **(−)** 2 **ALPHA** **,** **2nd** **[A]** **ALPHA** **,** 2 **ALPHA** **,** 3 **)** **STO►** **2nd** **[A]** **ENTER**

Step 4

Objective : $-2R_1 + R_3$ to replace element a_{31} by 0

Key sequence : **MATRX** 4 **(−)** 2 **ALPHA** **,** **2nd** **[A]** **ALPHA** **,** 1 **ALPHA** **,** 3 **)** **STO►** **2nd** **[A]** **ENTER**

Example 2

Given: $A = \begin{bmatrix} 5 & 7 & 1 & 2 \\ 3 & 6 & 4 & 8 \end{bmatrix}$ and $B = \begin{bmatrix} 0 & 6 & -2 & 1 \\ 3 & 7 & 4 & 9 \end{bmatrix}$

Find: $3B - 2A$

Solution

$$3B - 2A = 3\begin{bmatrix} 0 & 6 & -2 & 1 \\ 3 & 7 & 4 & 9 \end{bmatrix} - 2\begin{bmatrix} 5 & 7 & 1 & 2 \\ 3 & 6 & 4 & 8 \end{bmatrix}$$

$$= \begin{bmatrix} 0 & 18 & -6 & 3 \\ 9 & 21 & 12 & 27 \end{bmatrix} - \begin{bmatrix} 10 & 14 & 2 & 4 \\ 6 & 12 & 8 & 16 \end{bmatrix}$$

$$= \begin{bmatrix} 0 & 18 & -6 & 3 \\ 9 & 21 & 12 & 27 \end{bmatrix} + \begin{bmatrix} -10 & -14 & -2 & -4 \\ -6 & -12 & -8 & -16 \end{bmatrix}$$

$$= \begin{bmatrix} -10 & 4 & -8 & -1 \\ 3 & 9 & 4 & 11 \end{bmatrix}$$

Example 2 can be evaluated on a calculator using the matrices A and B entered for Example 1. Since the calculator cannot evaluate $3B - 2A$ directly, an intermediate step must be used as shown in the following key sequence.

3 **2nd** **[B]** **ENTER**

2nd **ANS** **−** 2 **2nd** **[A]** **ENTER**

Matrix Multiplication

A theater owner wants to compare the revenues for one weekend using the present ticket prices and proposed ticket prices. The data for prices and numbers of admissions is given in the following two matrices.

$$\begin{array}{c} \\ \textbf{old} \\ \textbf{new} \end{array} \begin{array}{cc} \textbf{adult} & \textbf{child} \\ \begin{bmatrix} 6.00 & 3.50 \\ 5.00 & 4.50 \end{bmatrix} \end{array} \qquad \begin{array}{c} \\ \textbf{adult} \\ \textbf{child} \end{array} \begin{array}{ccc} \textbf{Fri.} & \textbf{Sat.} & \textbf{Sun.} \\ \begin{bmatrix} 30 & 70 & 100 \\ 45 & 95 & 160 \end{bmatrix} \end{array}$$

The total revenue for one day, under either price scale, is the sum of the monies taken in from the adult admissions and from the children admissions. Each of these values is calculated from the product of the price per ticket and the number of tickets sold. The daily revenues based on old prices are:

Fri.: $(6.00)(30)$ + $(3.50)(45)$ = 337.50
Sat.: $(6.00)(70)$ + $(3.50)(95)$ = 752.50
Sun.: $(6.00)(100)$ + $(3.50)(160)$ = $1,160.00$

The daily revenues based on new prices are:

Fri.: $(5.00)(30)$ + $(4.50)(45)$ = 352.50
Sat.: $(5.00)(70)$ + $(4.50)(95)$ = 777.50
Sun.: $(5.00)(100)$ + $(4.50)(160)$ = $1,220.00$

This data about the total daily revenues can be arranged in a matrix.

$$
\begin{array}{c}
 \\
\textbf{old} \\
\textbf{new}
\end{array}
\begin{array}{ccc}
\textbf{Fri.} & \textbf{Sat.} & \textbf{Sun.}
\end{array}
\left[
\begin{array}{ccc}
337.50 & 752.50 & 1{,}160.00 \\
352.50 & 777.50 & 1{,}220.00
\end{array}
\right]
$$

Each entry in the final matrix is the sum of the products obtained by multiplying a row of the first matrix by a column of the second matrix. The process suggests the procedure for matrix multiplication.

Matrix multiplication is similar to the dot product of vectors, because a dot product is a sum of products. Recall the dot product of two vectors (a, b, c) and (d, e, f):

$$(a, b, c) \cdot (d, e, f) = ad + be + cf$$

In matrix multiplication, think of a row of the first matrix as one vector and the corresponding column of the second matrix as the second vector. For example, a 1×3 matrix may be multiplied by a 3×1 matrix as follows:

$$
\begin{bmatrix} 1 & 6 & 4 \end{bmatrix}
\begin{bmatrix} 2 \\ 3 \\ 7 \end{bmatrix}
= \begin{bmatrix} 1(2) + 6(3) + 4(7) \end{bmatrix}
= \begin{bmatrix} 48 \end{bmatrix}
$$

Note that the product matrix is a 1×1 matrix.

This procedure can be extended to matrices with higher dimension provided that the number of columns in the first matrix is the same as the number of rows of the second. In general, an $m \times n$ matrix may be multiplied by an $n \times p$ matrix and the product will be an $m \times p$ matrix. To illustrate, multiply S, a 2×3 matrix, by T, a 3×2 matrix.

$$
S = \begin{bmatrix} 1 & 4 & 7 \\ 2 & 1 & 3 \end{bmatrix}
\qquad
T = \begin{bmatrix} 6 & 1 \\ 3 & -2 \\ 0 & 4 \end{bmatrix}
$$

The first row of S and the first column of T may be treated as vectors. Their dot product gives a_{11}, the entry for the first row, first column in the product matrix.

$$
\begin{array}{ccc}
2 \times 3 & 3 \times 2 & 2 \times 2
\end{array}
$$

$$
ST = \begin{bmatrix} 1 & 4 & 7 \\ 2 & 1 & 3 \end{bmatrix} \cdot \begin{bmatrix} 6 & 1 \\ 3 & -2 \\ 0 & 4 \end{bmatrix} = \begin{bmatrix} a_{11} & a_{12} \\ a_{21} & a_{22} \end{bmatrix}
$$

$$
a_{11} = s_{11}t_{11} + s_{12}t_{21} + s_{13}t_{31}
$$

$$
= 1(6) + 4(3) + 7(0) = 18
$$

$$
ST = \begin{bmatrix} 18 & a_{12} \\ a_{21} & a_{22} \end{bmatrix}
$$

To obtain a_{12}, the entry for the first row, second column of the product matrix, form the dot product of the first row of S and the second column of T.

$$a_{12} = s_{11}t_{12} + s_{12}t_{22} + s_{13}t_{32}$$
$$= 1(1) + 4(-2) + 7(4) = 21$$

$$ST = \begin{bmatrix} 18 & 21 \\ a_{21} & a_{22} \end{bmatrix}$$

Follow a similar procedure for the remaining entries: $a_{21} = s_{21}t_{11} + s_{22}t_{21} + s_{23}t_{31}$

$$= 2(6) + 1(3) + 3(0) = 15$$
$$a_{22} = s_{21}t_{12} + s_{22}t_{22} + s_{23}t_{32}$$
$$= 2(1) + 1(-2) + 3(4) = 12$$
$$ST = \begin{bmatrix} 18 & 21 \\ 15 & 12 \end{bmatrix}$$

The procedure for multiplying two matrices may be summarized as follows:

Procedure for Multiplying Two Matrices

The product of an $m \times n$ matrix A and an $n \times p$ matrix B is the $m \times p$ matrix AB whose element in the ith row and jth column is computed by taking the dot product of the ith row of A and the jth column of B.

Example 3

Given: $A = \begin{bmatrix} 0 & 1 & 3 & 2 \\ 8 & 6 & 2 & 7 \end{bmatrix}$ and $B = \begin{bmatrix} -1 & 6 \\ 2 & 1 \\ 5 & 8 \\ 3 & 7 \end{bmatrix}$

a. Find AB. **b.** Find BA. **c.** Is matrix multiplication commutative?

Solution

a. A, a 2×4 matrix, and B, a 4×2 matrix, may be multiplied to form the product matrix AB, of dimension 2×2.

$$
\underset{2 \times 4}{AB = \begin{bmatrix} 0 & 1 & 3 & 2 \\ 8 & 6 & 2 & 7 \end{bmatrix}} \underset{4 \times 2}{\begin{bmatrix} -1 & 6 \\ 2 & 1 \\ 5 & 8 \\ 3 & 7 \end{bmatrix}} = \underset{2 \times 2}{\begin{bmatrix} c_{11} & c_{12} \\ c_{21} & c_{22} \end{bmatrix}}
$$

$$c_{ij} = a_{i1}b_{1j} + a_{i2}b_{2j} + \cdots + a_{in}b_{nj}$$
$$c_{11} = 0(-1) + 1(2) + 3(5) + 2(3) = 23$$
$$c_{12} = 0(6) + 1(1) + 3(8) + 2(7) = 39$$
$$c_{21} = 8(-1) + 6(2) + 2(5) + 7(3) = 35$$
$$c_{22} = 8(6) + 6(1) + 2(8) + 7(7) = 119$$
$$AB = \begin{bmatrix} 23 & 39 \\ 35 & 119 \end{bmatrix}$$

b. B, a 4×2 matrix, and A, a 2×4 matrix, may be multiplied to form BA, of dimension 4×4.

$$BA = \begin{bmatrix} -1 & 6 \\ 2 & 1 \\ 5 & 8 \\ 3 & 7 \end{bmatrix} \begin{bmatrix} 0 & 1 & 3 & 2 \\ 8 & 6 & 2 & 7 \end{bmatrix}$$

$$= \begin{bmatrix} -1(0) + 6(8) & -1(1) + 6(6) & -1(3) + 6(2) & -1(2) + 6(7) \\ 2(0) + 1(8) & 2(1) + 1(6) & 2(3) + 1(2) & 2(2) + 1(7) \\ 5(0) + 8(8) & 5(1) + 8(6) & 5(3) + 8(2) & 5(2) + 8(7) \\ 3(0) + 7(8) & 3(1) + 7(6) & 3(3) + 7(2) & 3(2) + 7(7) \end{bmatrix}$$

$$= \begin{bmatrix} 48 & 35 & 9 & 40 \\ 8 & 8 & 8 & 11 \\ 64 & 53 & 31 & 66 \\ 56 & 45 & 23 & 55 \end{bmatrix}$$

c. Clearly, $AB \neq BA$.

Thus, matrix multiplication is not commutative.

Example 3 demonstrates that matrix multiplication is *not* a commutative operation because the product matrix AB is different from the product matrix BA. It is also possible that, because of particular dimensions, only one product matrix may exist. For example, suppose A is a 1×4 matrix and B is a 4×2 matrix. Then the product matrix AB, of dimension 1×2, could be found; but, BA would not be defined because it is impossible to multiply a 4×2 matrix by a 1×4 matrix.

Matrix multiplication can be done on a calculator by using the following key sequence after matrices A and B have been entered:

| 2nd | [A] | 2nd | [B] | ENTER |

Reconsider matrices A and B of Example 3, and use a calculator to evaluate AB and BA.

An *identity matrix* I for multiplication is a square matrix each of whose main diagonal elements is 1 and whose remaining elements are 0. For example, the second- and third-order identity matrices are:

$$\begin{bmatrix} 1 & 0 \\ 0 & 1 \end{bmatrix} \quad \text{and} \quad \begin{bmatrix} 1 & 0 & 0 \\ 0 & 1 & 0 \\ 0 & 0 & 1 \end{bmatrix}$$

Thus, for any second-order matrix A

$$A \times \begin{bmatrix} 1 & 0 \\ 0 & 1 \end{bmatrix} = A \quad \text{and} \quad \begin{bmatrix} 1 & 0 \\ 0 & 1 \end{bmatrix} \times A = A$$

and for any third-order matrix B

$$B \times \begin{bmatrix} 1 & 0 & 0 \\ 0 & 1 & 0 \\ 0 & 0 & 1 \end{bmatrix} = B \quad \text{and} \quad \begin{bmatrix} 1 & 0 & 0 \\ 0 & 1 & 0 \\ 0 & 0 & 1 \end{bmatrix} \times B = B$$

If A is a square matrix and A^{-1} exists such that $A \cdot A^{-1} = A^{-1} \cdot A = I$, then A^{-1} is called the *inverse* of A. For example:

$$\begin{bmatrix} 4 & 7 \\ 1 & 5 \end{bmatrix} \text{ and } \begin{bmatrix} \frac{5}{13} & -\frac{7}{13} \\ -\frac{1}{13} & \frac{4}{13} \end{bmatrix} \text{ are inverses of each other}$$

because $\begin{bmatrix} 4 & 7 \\ 1 & 5 \end{bmatrix} \begin{bmatrix} \frac{5}{13} & -\frac{7}{13} \\ -\frac{1}{13} & \frac{4}{13} \end{bmatrix} = \begin{bmatrix} \frac{5}{13} & -\frac{7}{13} \\ -\frac{1}{13} & \frac{4}{13} \end{bmatrix} \begin{bmatrix} 4 & 7 \\ 1 & 5 \end{bmatrix} = \begin{bmatrix} 1 & 0 \\ 0 & 1 \end{bmatrix}$

$$A \cdot A^{-1} \qquad = \qquad A^{-1} \cdot A \qquad = I.$$

If the arrangements of the numbers of A and A^{-1} are inspected carefully, a method for finding the inverse of a 2 × 2 matrix is suggested. Note that the fractions in A^{-1} each have denominators of 13, which is the determinant of the first matrix. The numerators of the fractions of A^{-1} consist of the elements of A but their positions or signs are different than in A.

Method for Finding the Inverse of a 2 × 2 Matrix

To find the inverse of a 2 × 2 matrix, interchange the elements of the main diagonal and change the signs of the elements of the other diagonal. Then multiply the resulting matrix by the multiplicative inverse of the determinant of the original matrix.

Note that if the determinant of a 2 × 2 matrix A is 0, then A does not have an inverse since division by 0 is undefined.

In general, the multiplicative inverse of A, a 2 × 2 matrix $\begin{bmatrix} a & b \\ c & d \end{bmatrix}$, is the matrix

$$A^{-1} = \frac{1}{|A|} \begin{bmatrix} d & -b \\ -c & a \end{bmatrix}$$

This statement may be verified as follows. Recall that $|A| = ad - bc \neq 0$.

$$A \cdot A^{-1} = \begin{bmatrix} a & b \\ c & d \end{bmatrix} \begin{bmatrix} \dfrac{d}{|A|} & \dfrac{-b}{|A|} \\ \dfrac{-c}{|A|} & \dfrac{a}{|A|} \end{bmatrix} = \begin{bmatrix} \dfrac{ad-bc}{ad-bc} & \dfrac{-ab+ab}{ad-bc} \\ \dfrac{cd-dc}{ad-bc} & \dfrac{-bc+ad}{ad-bc} \end{bmatrix}$$

$$= \begin{bmatrix} 1 & 0 \\ 0 & 1 \end{bmatrix}$$

Example 4 _____

Find A^{-1}, the inverse of $A = \begin{bmatrix} 5 & -4 \\ 3 & 1 \end{bmatrix}$, and verify by matrix multiplication that $A \cdot A^{-1} = I$ and $A^{-1} \cdot A = I$.

A^{-1} is determined by interchanging the elements on the main diagonal, changing the signs of the other elements and then dividing by $|A|$, where $|A| = (5)(1) - (-4)(3) = 17$.

$$A^{-1} = \frac{\begin{bmatrix} 1 & 4 \\ -3 & 5 \end{bmatrix}}{17}$$

$$A \cdot A^{-1} = \begin{bmatrix} 5 & -4 \\ 3 & 1 \end{bmatrix} \begin{bmatrix} \frac{1}{17} & \frac{4}{17} \\ -\frac{3}{17} & \frac{5}{17} \end{bmatrix}$$

$$= \begin{bmatrix} \frac{5(1) + (-4)(-3)}{17} & \frac{5(4) + (-4)(5)}{17} \\ \frac{3(1) + 1(-3)}{17} & \frac{3(4) + 1(5)}{17} \end{bmatrix}$$

$$= \begin{bmatrix} 1 & 0 \\ 0 & 1 \end{bmatrix}$$

$$A^{-1} \cdot A = \begin{bmatrix} \frac{1}{17} & \frac{4}{17} \\ -\frac{3}{17} & \frac{5}{17} \end{bmatrix} \begin{bmatrix} 5 & -4 \\ 3 & 1 \end{bmatrix}$$

$$= \begin{bmatrix} \frac{1(5) + (4)(3)}{17} & \frac{1(-4) + 4(1)}{17} \\ \frac{-3(5) + 5(3)}{17} & \frac{-3(-4) + 5(1)}{17} \end{bmatrix}$$

$$= \begin{bmatrix} 1 & 0 \\ 0 & 1 \end{bmatrix}$$

Therefore, $A \cdot A^{-1} = A^{-1} \cdot A = I$.

The procedure for computing inverses of square matrices of order higher than 2 will be described later in this chapter. On a calculator, the inverse of matrix A may be obtained from the following key sequence: [2nd] [[A]] [X^{-1}] [ENTER]

Evaluate $A \cdot A^{-1}$.

[2nd] [[A]] [2nd] [[A]] [X^{-1}] [ENTER] **Display:** $\begin{bmatrix} 1 & 0 \\ 0 & 1 \end{bmatrix}$

Note that for many matrices, the entries for the inverse matrix may be infinite decimals that are approximated to ten decimal places. To view such an inverse, it is necessary to use [▶] to scroll to the left. The product of a matrix and its inverse may display approximate values of 0 and 1. For example, when $A = \begin{bmatrix} 1 & 4 \\ -3 & 5 \end{bmatrix}$ is entered, $A \cdot A^{-1}$ may be displayed as $\begin{bmatrix} 1 & 0 \\ -2E - 13 & 1 \end{bmatrix}$.

The element in the second row, first column is $-2 \times 10^{-13} \approx 0$.

Exercises 7.2

In **1 – 12**, use the matrices below to find the matrix that is equal to the given expression.

$$S = \begin{bmatrix} 4 & 6 \\ 1 & 2 \end{bmatrix} \quad T = \begin{bmatrix} 0 & -1 \\ 1 & 2 \end{bmatrix} \quad U = \begin{bmatrix} 1 & 4 \\ 6 & -3 \\ 2 & -1 \\ 5 & 0 \end{bmatrix} \quad V = \begin{bmatrix} 9 & 6 & 4 & -1 \end{bmatrix}$$

1. $S + T$ **2.** $T + S$ **3.** $S - T$ **4.** $T - S$

5. $2U$ **6.** $3V$ **7.** $-S$ **8.** $5T$

9. $5T - S$ **10.** $4S + 3T$ **11.** $7T - 4S$ **12.** $\frac{1}{2}S + 0.2T$

In **13 – 24**, use the matrices below to find the matrix equal to the given expression. If the expression is undefined, explain why.

$$A = \begin{bmatrix} 1 \\ 4 \\ 5 \end{bmatrix} \quad B = \begin{bmatrix} -2 & 3 & 6 \end{bmatrix} \quad C = \begin{bmatrix} 2 & 4 & 0 & 6 \\ 1 & 5 & -2 & 3 \end{bmatrix}$$

$$D = \begin{bmatrix} -2 & 3 \\ 7 & 5 \\ -3 & 2 \\ 6 & 8 \end{bmatrix} \quad E = \begin{bmatrix} 6 & 5 \\ 1 & 0 \\ -4 & 7 \end{bmatrix} \quad F = \begin{bmatrix} 0 & 2 & -2 \\ 1 & 3 & -3 \\ 4 & 7 & -7 \end{bmatrix}$$

13. AB **14.** BA **15.** CD

16. DC **17.** BE **18.** EB

19. FA **20.** EC **21.** $AB + F$

22. $2A \cdot B$ **23.** F^2 **24.** $FE + E$

25. The costs of an alternator and a battery for two different cars are shown by the matrix:

	Alt.	Batt.
Car 1	100	50
Car 2	50	60

The labor costs for installation of these parts in these cars are shown by:

	Alt.	Batt.
Car 1	25	5
Car 2	32	7

Compute a matrix that shows the total cost for parts and labor to install an alternator or battery in the cars.

Answers 7.2

1. $\begin{bmatrix} 4 & 5 \\ 2 & 4 \end{bmatrix}$

2. $\begin{bmatrix} 4 & 5 \\ 2 & 4 \end{bmatrix}$

3. $\begin{bmatrix} 4 & 7 \\ 0 & 0 \end{bmatrix}$

4. $\begin{bmatrix} -4 & -7 \\ 0 & 0 \end{bmatrix}$

5. $\begin{bmatrix} 2 & 8 \\ 12 & -6 \\ 4 & -2 \\ 10 & 0 \end{bmatrix}$

6. $\begin{bmatrix} 27 & 18 & 12 & -3 \end{bmatrix}$

7. $\begin{bmatrix} -4 & -6 \\ -1 & -2 \end{bmatrix}$

8. $\begin{bmatrix} 0 & -5 \\ 5 & 10 \end{bmatrix}$

9. $\begin{bmatrix} -4 & -11 \\ 4 & 8 \end{bmatrix}$

10. $\begin{bmatrix} 16 & 21 \\ 7 & 14 \end{bmatrix}$

11. $\begin{bmatrix} -16 & -31 \\ 3 & 6 \end{bmatrix}$

12. $\begin{bmatrix} 2 & 2.8 \\ 0.7 & 1.4 \end{bmatrix}$

13. $\begin{bmatrix} -2 & 3 & 6 \\ -8 & 12 & 24 \\ -10 & 15 & 30 \end{bmatrix}$

14. $\begin{bmatrix} 40 \end{bmatrix}$

15. $\begin{bmatrix} 60 & 74 \\ 57 & 48 \end{bmatrix}$

16. $\begin{bmatrix} -1 & 7 & -6 & -3 \\ 19 & 53 & -10 & 57 \\ -4 & -2 & -4 & -12 \\ 20 & 64 & -16 & 60 \end{bmatrix}$

17. $\begin{bmatrix} -33 & 32 \end{bmatrix}$

18. undefined; number of columns of $E \neq$ number of rows of B.

19. $\begin{bmatrix} -2 \\ -2 \\ -3 \end{bmatrix}$

20. $\begin{bmatrix} 17 & 49 & -10 & 51 \\ 2 & 4 & 0 & 6 \\ -1 & 19 & -14 & -3 \end{bmatrix}$

21. $\begin{bmatrix} -2 & 5 & 4 \\ -7 & 15 & 21 \\ -6 & 22 & 23 \end{bmatrix}$

22. $\begin{bmatrix} -4 & 6 & 12 \\ -16 & 24 & 48 \\ -20 & 30 & 60 \end{bmatrix}$

23. $\begin{bmatrix} -6 & -8 & 8 \\ -9 & -10 & 10 \\ -21 & -20 & 20 \end{bmatrix}$

24. $\begin{bmatrix} 16 & -9 \\ 22 & -16 \\ 55 & -22 \end{bmatrix}$

25. $\begin{bmatrix} 125 & 55 \\ 82 & 67 \end{bmatrix}$

26.
$\begin{bmatrix} 14{,}000{,}000{,}000 \\ 9{,}600{,}000{,}000 \\ 5{,}500{,}000{,}000 \end{bmatrix}$

27. AB, BC, CA can be computed.

28. a. $A(BC) = (AB)C$

$B(AC) = (BA)C$

b. Matrix multiplication appears to be associative.

29. $\begin{bmatrix} -\frac{1}{5} & \frac{2}{5} \\ \frac{4}{5} & -\frac{3}{5} \end{bmatrix}$

30. $\begin{bmatrix} 2 & -\frac{1}{4} \\ 1 & 0 \end{bmatrix}$

31. $\begin{bmatrix} 12 & -20 \\ -15 & 30 \end{bmatrix}$

32. $\begin{bmatrix} 10 & 12.5 \\ 0 & 2.5 \end{bmatrix}$

33. No inverse; the determinant = 0.

---

26. In the first of three consecutive years, an automobile manufacturer sells $\frac{1}{2}$ million Zooms, $\frac{3}{4}$ million Speedos, and 1 million Treks. In the second year, the manufacturer sells 600,000 Zooms, $\frac{1}{2}$ million Speedos, and 400,000 Treks. In the third year, the manufacturer sells 200,000 Zooms, 100,000 Speedos, and 700,000 Treks. If the prices of Zooms, Speedos, and Treks are $6,000, $8,000, and $5,000, respectively, use matrices to find the total revenue produced by each car in each of the three years.

27. Assume the following dimensions for 3 matrices:

A is 3×2, B is 2×4, C is 4×3

Tell which of the following products can be computed:

AB, BA, AC, CA, BC, CB

28. For any three 3×3 matrices A, B, and C:

a. Compute and compare $A(BC)$ and $(AB)C$

$B(AC)$ and $(BA)C$

b. What appears to be true about matrix multiplication?

In **29 – 33**, find the inverse of each matrix or explain why there is no inverse.

29. $\begin{bmatrix} 3 & 2 \\ 4 & 1 \end{bmatrix}$ **30.** $\begin{bmatrix} 0 & 1 \\ -4 & 8 \end{bmatrix}$ **31.** $\begin{bmatrix} \frac{1}{2} & \frac{1}{3} \\ \frac{1}{4} & \frac{1}{5} \end{bmatrix}$ **32.** $\begin{bmatrix} 0.1 & -0.5 \\ 0 & 0.4 \end{bmatrix}$ **33.** $\begin{bmatrix} 2 & 8 \\ 1 & 4 \end{bmatrix}$

34. If a and b are real numbers and $ab = 0$, then either $a = 0$ or $b = 0$, or $a = 0$ and $b = 0$. Show that this property is not generally true for matrices by finding a 2×2 nonzero matrix that, multiplied by the matrix $\begin{bmatrix} 6 & 6 \\ 2 & 2 \end{bmatrix}$, results in the zero matrix.

35. Let:

$A = \begin{bmatrix} a_1 & a_2 \\ a_3 & a_4 \end{bmatrix}$ $B = \begin{bmatrix} b_1 & b_2 \\ b_3 & b_4 \end{bmatrix}$ $C = \begin{bmatrix} c_1 & c_2 \\ c_3 & c_4 \end{bmatrix}$

a. Show that: $A(BC) = (AB)C$

b. Using A, B, and C, show that matrix multiplication distributes over matrix addition for square 2×2 matrices.

36. If A and B are square matrices, find the error in the following "proof."

Prove: $(A + B)^2 = A^2 + 2AB + B^2$

$(A + B)^2 = (A + B)(A + B)$

$= A^2 + BA + AB + B^2$

$= A^2 + 2AB + B^2$

37. Use a similar analysis to that in Exercise 36 to show that, in general, the product of the sum and difference of two $n \times n$ matrices is *not* equal to the difference of their squares.

38. Verify that the set of 3×3 triangular matrices with zeros below the diagonal is closed for matrix addition and multiplication.

39. Prove that for any 2×2 matrices A and B with inverses A^{-1} and B^{-1}:

$(A \cdot B)^{-1} = B^{-1} \cdot A^{-1}$

Step 5

Objective : $R_3 + R_2$ to replace element a_{23} by 0

Calculator function : **Row+**, which will add two rows and store the result in the second row.

Completed function : Row+(matrix,row 1,row 2) Store as new [A].

Key sequence : MATRX 2 2nd [A] ALPHA , 3 ALPHA , 2) STO► 2nd [A] ENTER

Step 6

Objective : $-\frac{1}{7}R_3$

Calculator function : ***Row**, which will multiply a row by a scalar value and store the result in that row.

Completed function : *Row(scalar,matrix,row) Store as new [A].

Key sequence : MATRX 3 (−) 1 ÷ 7 ALPHA , 2nd [A] ALPHA , 3) STO► 2nd [A] ENTER

Step 7

Objective : $3R_3 + R_1$ to replace element a_{13} by 0

Key sequence : MATRX 4 3 ALPHA , 2nd [A] ALPHA , 3 ALPHA , 1) ENTER

Note that the calculator will not produce exactly 0 as the value for element a_{13}. The results depend on the entries obtained in step 6, where the calculator multiplied by the decimal equivalent of $-\frac{1}{7}$. The entries a_{33} and a_{34} were displayed as 1 and 2 respectively, but were stored as the approximate results of the computation.

Example 3

$100,000 is invested in three bank accounts and the total income after 1 year is $6,900. If the accounts pay 6%, 7%, and 9% interest each year, and twice as much is invested at 7% as at 6%, find the amount of money that was invested at each rate of interest.

Solution

If x is used to represent the number of dollars invested at 6%, y the amount at 7%, and z the amount at 9%, the following equations can be used to solve the problem.

$$x + y + z = 100,000$$
$$y = 2x \text{ or } 2x - y = 0$$
$$0.06x + 0.07y + 0.09z = 6,900$$

The augmented matrix for this system is

$$\begin{bmatrix} 1 & 1 & 1 & 100{,}000 \\ 2 & -1 & 0 & 0 \\ 0.06 & 0.07 & 0.09 & 6{,}900 \end{bmatrix}$$

$$\begin{matrix} \\ R_1 + R_2 \\ 100R_3 \end{matrix} \begin{bmatrix} 1 & 1 & 1 & 100{,}000 \\ 3 & 0 & 1 & 100{,}000 \\ 6 & 7 & 9 & 690{,}000 \end{bmatrix}$$

$$\begin{matrix} \\ \\ -2R_2 + R_3 \end{matrix} \begin{bmatrix} 1 & 1 & 1 & 100{,}000 \\ 3 & 0 & 1 & 100{,}000 \\ 0 & 7 & 7 & 490{,}000 \end{bmatrix}$$

$$-\tfrac{1}{7}R_3 + R_1 \begin{bmatrix} 1 & 0 & 0 & 30{,}000 \\ 3 & 0 & 1 & 100{,}000 \\ 0 & 7 & 7 & 490{,}000 \end{bmatrix}$$

$$-3R_1 + R_2 \begin{bmatrix} 1 & 0 & 0 & 30{,}000 \\ 0 & 0 & 1 & 10{,}000 \\ 0 & 7 & 7 & 490{,}000 \end{bmatrix}$$

$$-7R_2 + R_3 \begin{bmatrix} 1 & 0 & 0 & 30{,}000 \\ 0 & 0 & 1 & 10{,}000 \\ 0 & 7 & 0 & 420{,}000 \end{bmatrix}$$

$$\tfrac{1}{7}R_3 \begin{bmatrix} 1 & 0 & 0 & 30{,}000 \\ 0 & 0 & 1 & 10{,}000 \\ 0 & 1 & 0 & 60{,}000 \end{bmatrix}$$

interchanging rows 2 and 3

$$\begin{bmatrix} 1 & 0 & 0 & 30{,}000 \\ 0 & 1 & 0 & 60{,}000 \\ 0 & 0 & 1 & 10{,}000 \end{bmatrix}$$

This represents the equation

$$\begin{bmatrix} 1 & 0 & 0 \\ 0 & 1 & 0 \\ 0 & 0 & 1 \end{bmatrix} \cdot \begin{bmatrix} x \\ y \\ z \end{bmatrix} = \begin{bmatrix} 30{,}000 \\ 60{,}000 \\ 10{,}000 \end{bmatrix}$$

Answer: $x = \$30{,}000$ @ 6%, $y = \$60{,}000$ @ 7%, and $z = \$10{,}000$ @ 9%.

Check: $\$30{,}000$ @ 6% for 1 year = $\$1{,}800$
$\$60{,}000$ @ 7% for 1 year = $\$4{,}200$
$\$10{,}000$ @ 9% for 1 year = $\$900$
Total income = $\$6{,}900$

Row operations can also be used to solve a system in which the number of independent equations is less than the number of variables. The solution of such a system is not unique and is usually expressed in terms of a parameter or in terms of one or more of the variables.

Example 4

Solve the following system: $\quad 2x - 3y - 8z = 4$
$\qquad\qquad\qquad\qquad\qquad\quad x - 2y + z = 1$

Solution

Write an augmented matrix for the system and use row operations to transform the first two rows and columns into an identity matrix.

$$\begin{bmatrix} 2 & -3 & -8 & 4 \\ 1 & -2 & 1 & 1 \end{bmatrix}$$

$-R_2 + R_1 \quad \begin{bmatrix} 1 & -1 & -9 & 3 \\ 1 & -2 & 1 & 1 \end{bmatrix}$

$-R_1 + R_2 \quad \begin{bmatrix} 1 & -1 & -9 & 3 \\ 0 & -1 & 10 & -2 \end{bmatrix}$

$\begin{matrix} -R_2 + R_1 \\ -R_2 \end{matrix} \quad \begin{bmatrix} 1 & 0 & -19 & 5 \\ 0 & 1 & -10 & 2 \end{bmatrix}$

The augmented matrix represents the equations:

$$x - 19z = 5 \quad \text{or} \quad x = 5 + 19z$$
$$y - 10z = 2 \quad \text{or} \quad y = 2 + 10z$$

Therefore, when $z = 0$, the solution is $(5, 2, 0)$, and when $z = 1$, another solution is $(24, 12, 1)$. For each value of z there is a solution.

Exercises 7.3

In **1 – 8**, use Cramer's Rule to solve the system. If the determinant of the coefficient matrix is zero, tell how this affects the system.

1. $2a + b = 4$
$a - b = 2$

2. $3a + 2b = 2$
$a + 2b = -1$

3. $6x + 2y = 7$
$x - 2y = 5$

4. $x_1 = 2x_2 + 1$
$x_1 + 2x_2 = 4$

5. $-5a + 16b = -7$
$3a - 9b = 4$

6. $x + y = 5$
$3x + 3y = 15$

7. $5x - 3y = -46$
$4x + 3z = -14$
$-8y + 4z = -48$

8. $a - b + 5c = 9$
$a - 2b + 4c = 8$
$6a - b - 3c = 2$

 In **9 – 14**, apply Cramer's Rule to solve.

9. $x - z = -2$
$3x + y - 2z = 5$
$2x + 2y + z = 4$

10. $a + b + c = 3$
$2a - b + 2c = 6$
$5a + 2b + 2c = 0$

11. $5x + 4y + z = 11$
$3x - y + 2z = -1$
$4x - y - 3z = -2$

12. $x + y + z = 0$
$x + y - z = -3$
$3x - 2y + 3z = 10$

13. $5a - 3b + c = 4$
$4a + b - 5c = -2$
$a - 4b - 2c = 0$

14. $2a + b - c + 2d = -6$
$3a + 4b + d = 1$
$a + 5b + 2c + 6d = -3$
$5a + 2b - c - d = 3$

In **15 – 23**, use row operations on the augmented matrix to solve.

15. $a + b = 4$
$5a + 2b = 11$

16. $x - y = 11$
$4x + 7y = -22$

17. $x_1 - 8x_2 = 8\frac{1}{2}$
$4x_1 - 5x_2 = 7$

18. $8x = 41 - 3y$
$6x + 5y = 39$

19. $2x + 4y + 3z = 6$
$-5x + 3y + z = -7$
$-3x + 7y + 5z = 3$

20. $-2x_1 + 3x_2 + 5x_3 = -17$
$4x_1 + 2x_2 - 3x_3 = 29$
$-6x_1 + 5x_2 - 7x_3 = 7$

21. $x + 5y + 2z = 5$
$y + z = 1.4$
$4x - 6y - 3z = -4.4$

22. $a - 2c = -1$
$b + c = -1$
$b + 3d = -1$
$3c - d = 8$

23. $a + b - 2c - 5d = 4$
$6a + 2b + 5c + d = 9$
$a + 4c + 3d = 5$
$2a - b + c + 2d = -4$

Answers 7.3

1. $a = 2$
$b = 0$

2. $a = \frac{3}{2}$
$b = -\frac{5}{4}$

3. $x = \frac{12}{7}$
$y = -\frac{23}{14}$

4. $x_1 = \frac{5}{2}$
$x_2 = \frac{3}{4}$

5. $a = \frac{1}{3}$
$b = -\frac{1}{3}$

6. infinite number of solutions

7. $x = -5$
$y = 7$
$z = 2$

8. $a = 1$
$b = -\frac{1}{2}$
$c = \frac{3}{2}$

9. $x = -16$
$y = 25$
$z = -14$

10. $a = -2$
$b = 0$
$c = 5$

11. $x = \frac{13}{47}$
$y = \frac{110}{47}$
$z = \frac{12}{47}$

12. $x = 0.5$
$y = -2$
$z = 1.5$

13. $a = 0.5$
$b = -0.25$
$c = 0.75$

14. $a = 1$
$b = 0$
$c = 4$
$d = -2$

15. $a = 1$
$b = 3$

16. $x = 5$
$y = -6$

17. $x_1 = \frac{1}{2}$
$x_2 = -1$

18. $x = 4$
$y = 3$

19. $x = 1$
$y = -2$
$z = 4$

20. $x_1 = 4$
$x_2 = 2$
$x_3 = -3$

Section 7.3 Using Matrices to Solve Systems of Equations   **313**

Exercises 7.3 *(continued)*

24. Find a quadratic function $f(x)$ such that $f(1) = 4$, $f(-1) = 0$, and $f(6) = 49$. *Hint:* Represent the function as $f(x) = ax^2 + bx + c$.

25. Find a third-degree function $f(x)$ such that $f(2) = 10$, $f(-2) = -14$, $f(0) = -2$, and $f(1) = 1$.

26. Pennies, nickels, dimes, and quarters are combined to make a total of $4.44. If the number of nickels is twice the number of pennies, there are 2 more quarters than nickels, and there are 37 coins in all, how many of each type of coin are used?

27. Three different companies, x, y, and z, buy various quantities of three different computers, A, B, and C. If company x buys one of each and pays a total of $2,700, company y buys 5 of A and 3 of B and pays $5,500, and company z buys 1 of B and 2 of C and pays $3,400, what is the unit price of each computer?

28. The inverse of a square matrix may be found by augmenting the matrix with an identity matrix of the same dimension and then using row operations until the identity matrix appears in the left half of the matrix.

  a. Use this method to find the inverse of $A = \begin{bmatrix} 4 & 2 \\ 1 & 5 \end{bmatrix}$

  b. To check, use the method discussed in this chapter for finding the inverse of a 2 × 2 matrix.

29. Use the method mentioned in Exercise 28 to find the inverse of

$$\begin{bmatrix} 1 & -1 & 0 \\ 1 & 0 & -1 \\ 6 & -2 & -3 \end{bmatrix}$$

Check your answer by using matrix multiplication.

30. The graph at the right shows the cost, in millions of dollars, of goods sold in each of three departments of a supermarket. The total markup (the difference between selling price and cost) was $3.315 million in January, $3.295 million in February, and $2.955 million in March.

  a. Use the data from the graph to write a 3 × 3 matrix A that displays the cost of merchandise sold. Represent months as rows and departments as columns.

  b. Write a 3 × 1 matrix M that displays the markup for each month.

  c. Let x_1 be the rate of markup on produce, x_2 the rate of markup on meat, and x_3 the rate of markup on nonperishables. Determine the rate of markup for each department by solving the equation $A \cdot \begin{bmatrix} x_1 \\ x_2 \\ x_3 \end{bmatrix} = M$.

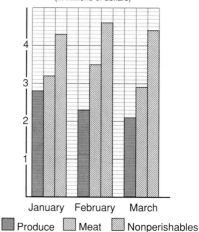

Cost of Merchandise Sold
(in millions of dollars)

January   February   March

■ Produce ▨ Meat ▨ Nonperishables

---

21. $x = 0.4$

    $y = 0.6$

    $z = 0.8$

22. $a = 5$

    $b = -4$

    $c = 3$

    $d = 1$

23. $a = -1$

    $b = 1$

    $c = 3$

    $d = -2$

24. $f(x) = x^2 + 2x + 1$

25. $f(x) = x^3 + 2x - 2$

26. 4 pennies

    8 nickels

    15 dimes

    10 quarters

27. Computer A – $500

    Computer B – $1,000

    Computer C – $1,200

28. $\begin{bmatrix} \frac{5}{18} & -\frac{1}{9} \\ -\frac{1}{18} & \frac{2}{9} \end{bmatrix}$

29. $\begin{bmatrix} -2 & -3 & 1 \\ -3 & -3 & 1 \\ -2 & -4 & 1 \end{bmatrix}$

30. a.
$$\begin{array}{c} \\ J \\ F \\ M \end{array} \begin{array}{ccc} P & M & N \\ \begin{bmatrix} 2.8 & 3.2 & 4.3 \\ 2.3 & 3.5 & 4.6 \\ 2.1 & 2.9 & 4.4 \end{bmatrix} \end{array}$$

b.
$$\begin{array}{c} J \\ F \\ M \end{array} \begin{bmatrix} 3.315 \\ 3.295 \\ 2.955 \end{bmatrix}$$

c. $x_1 = 0.4$

    $x_2 = 0.35$

    $x_3 = 0.25$

7.4 An Additional Method for Solving Systems

In Section 7.3, you learned to solve a square system by using Cramer's Rule or by using row operations on the augmented matrix. Another method for solving a square system uses the inverse of the coefficient matrix if it exists. Consider the form of this matrix solution as shown for a square system of order 2.

The system $\begin{aligned} ax + by &= e \\ cx + dy &= f \end{aligned}$ can be written in the form $\begin{bmatrix} a & b \\ c & d \end{bmatrix}\begin{bmatrix} x \\ y \end{bmatrix} = \begin{bmatrix} e \\ f \end{bmatrix}$, where $\begin{bmatrix} a & b \\ c & d \end{bmatrix}$ is the coefficient matrix, $\begin{bmatrix} x \\ y \end{bmatrix}$ is the variable matrix, and $\begin{bmatrix} e \\ f \end{bmatrix}$ is the constant matrix.

Left-multiply each side of the matrix equation by the inverse of the coefficient matrix if it exists. (Note that, since matrix multiplication is not commutative, the order of multiplication on each side is important.)

$$\begin{bmatrix} a & b \\ c & d \end{bmatrix}^{-1}\begin{bmatrix} a & b \\ c & d \end{bmatrix}\begin{bmatrix} x \\ y \end{bmatrix} = \begin{bmatrix} a & b \\ c & d \end{bmatrix}^{-1}\begin{bmatrix} e \\ f \end{bmatrix}$$

$$I\begin{bmatrix} x \\ y \end{bmatrix} = \begin{bmatrix} a & b \\ c & d \end{bmatrix}^{-1}\begin{bmatrix} e \\ f \end{bmatrix}$$

In general, when X = the variable matrix, A = the coefficient matrix with inverse A^{-1}, and B = the constant matrix, then $X = A^{-1}B$.

$$\begin{bmatrix} x \\ y \end{bmatrix} = \begin{bmatrix} a & b \\ c & d \end{bmatrix}^{-1}\begin{bmatrix} e \\ f \end{bmatrix}$$

$$X = A^{-1}B$$

Thus, matrix $A^{-1}B$ represents a unique solution for a square system if A^{-1} exists.

Example 1

Use the inverse matrix to solve the following system.

$$2x - y = -2$$
$$4x + y = 5$$

Solution

Rewrite the system in matrix form. $\qquad AX = B$

$$\begin{bmatrix} 2 & -1 \\ 4 & 1 \end{bmatrix}\begin{bmatrix} x \\ y \end{bmatrix} = \begin{bmatrix} -2 \\ 5 \end{bmatrix}$$

Find A^{-1}. Recall that

if $\qquad A = \begin{bmatrix} a & b \\ c & d \end{bmatrix}$

then $A^{-1} = \dfrac{1}{|A|}\begin{bmatrix} d & -b \\ -c & a \end{bmatrix}$

$A^{-1} = \dfrac{1}{2(1) - 4(-1)}\begin{bmatrix} 1 & 1 \\ -4 & 2 \end{bmatrix}$

$A^{-1} = \dfrac{1}{6}\begin{bmatrix} 1 & 1 \\ -4 & 2 \end{bmatrix}$

Left-multiply each side of the matrix equation by A^{-1}.

$$\frac{1}{6}\begin{bmatrix} 1 & 1 \\ -4 & 2 \end{bmatrix}\begin{bmatrix} 2 & -1 \\ 4 & 1 \end{bmatrix}\begin{bmatrix} x \\ y \end{bmatrix} = \frac{1}{6}\begin{bmatrix} 1 & 1 \\ -4 & 2 \end{bmatrix}\begin{bmatrix} -2 \\ 5 \end{bmatrix}$$

$$I\begin{bmatrix} x \\ y \end{bmatrix} = \frac{1}{6}\begin{bmatrix} 1(-2) + 1(5) \\ (-4)(-2) + 2(5) \end{bmatrix}$$

$$\begin{bmatrix} x \\ y \end{bmatrix} = \frac{1}{6}\begin{bmatrix} 3 \\ 18 \end{bmatrix} = \begin{bmatrix} \frac{1}{2} \\ 3 \end{bmatrix}$$

Answer: The solution is $x = \frac{1}{2}, y = 3$.

To solve square systems of order 3 or higher, a general method for finding the inverse of a matrix is needed. Note that the previously-discussed method for finding the inverse of a 2 × 2 matrix is unique to that order and cannot be used for any other order square matrix.

A 2 × 2 matrix will be used to demonstrate the more general method of finding an inverse of a square matrix and this method will then be extended to higher-order square matrices.

Suppose $A = \begin{bmatrix} 4 & 2 \\ 3 & 1 \end{bmatrix}$. If $AX = I$, where I is the identity matrix, then $X = A^{-1}$.

Assume that A^{-1} exists and is equal to $\begin{bmatrix} a_{11} & a_{12} \\ a_{21} & a_{22} \end{bmatrix}$. Then:

$$\begin{bmatrix} 4 & 2 \\ 3 & 1 \end{bmatrix}\begin{bmatrix} a_{11} & a_{12} \\ a_{21} & a_{22} \end{bmatrix} = \begin{bmatrix} 1 & 0 \\ 0 & 1 \end{bmatrix}$$

Using matrix multiplication:

$$\begin{bmatrix} 4a_{11} + 2a_{21} & 4a_{12} + 2a_{22} \\ 3a_{11} + 1a_{21} & 3a_{12} + 1a_{22} \end{bmatrix} = \begin{bmatrix} 1 & 0 \\ 0 & 1 \end{bmatrix}$$

If two matrices are equal, their corresponding entries are equal. Therefore:

$$4a_{11} + 2a_{21} = 1 \qquad\qquad 4a_{12} + 2a_{22} = 0$$
$$3a_{11} + 1a_{21} = 0 \qquad\qquad 3a_{12} + 1a_{22} = 1$$

The set of equations used to find a_{11} and a_{21} has the same coefficient matrix as the set of equations used to find a_{12} and a_{22}. Therefore, they may be represented by the following augmented matrices.

$$\begin{bmatrix} 4 & 2 & \vdots & 1 \\ 3 & 1 & \vdots & 0 \end{bmatrix} \text{ and } \begin{bmatrix} 4 & 2 & \vdots & 0 \\ 3 & 1 & \vdots & 1 \end{bmatrix}$$

Since both systems have the same coefficient matrix, they can be solved together by using Gaussian elimination on the matrix that consists of the original 2 × 2 matrix *adjoined* to the 2 × 2 identity matrix.

$$\begin{bmatrix} 4 & 2 & | & 1 & 0 \\ 3 & 1 & | & 0 & 1 \end{bmatrix}$$

$$\begin{matrix} 3R_1 \\ -4R_2 \end{matrix} \begin{bmatrix} 12 & 6 & | & 3 & 0 \\ -12 & -4 & | & 0 & -4 \end{bmatrix}$$

$$R_2 + R_1 \begin{bmatrix} 12 & 6 & | & 3 & 0 \\ 0 & 2 & | & 3 & -4 \end{bmatrix}$$

$$-3R_2 \begin{bmatrix} 12 & 6 & | & 3 & 0 \\ 0 & -6 & | & -9 & 12 \end{bmatrix}$$

$$R_1 + R_2 \begin{bmatrix} 12 & 0 & | & -6 & 12 \\ 0 & -6 & | & -9 & 12 \end{bmatrix}$$

$$\begin{matrix} \frac{1}{12}R_1 \\ -\frac{1}{6}R_2 \end{matrix} \begin{bmatrix} 1 & 0 & | & -\frac{1}{2} & 1 \\ 0 & 1 & | & \frac{3}{2} & -2 \end{bmatrix}$$

This result represents

$$\begin{bmatrix} 1 & 0 \\ 0 & 1 \end{bmatrix} \begin{bmatrix} a_{11} \\ a_{21} \end{bmatrix} = \begin{bmatrix} -\frac{1}{2} \\ \frac{3}{2} \end{bmatrix} \text{ and } \begin{bmatrix} 1 & 0 \\ 0 & 1 \end{bmatrix} \begin{bmatrix} a_{12} \\ a_{22} \end{bmatrix} = \begin{bmatrix} 1 \\ -2 \end{bmatrix}$$

$$\begin{matrix} a_{11} = -\frac{1}{2} & \qquad a_{12} = 1 \\ a_{21} = \frac{3}{2} & \qquad a_{22} = -2 \end{matrix}$$

Therefore, the inverse matrix A^{-1} has been determined:

$$\begin{bmatrix} a_{11} & a_{12} \\ a_{21} & a_{22} \end{bmatrix} = \begin{bmatrix} -\frac{1}{2} & 1 \\ \frac{3}{2} & -2 \end{bmatrix}$$

This method is useful for finding the inverse, A^{-1}, of any square matrix A, if A^{-1} exists. Although the use of the inverse matrix for square systems with three or more variables is more time-consuming than other methods, it is particularly useful when finding the solutions of each of a group of systems where the coefficient matrix for each system is the same.

Example 2 _____

Find the solution for each of the following systems.

a. $\quad\begin{aligned} x + y + z &= 6 \\ 2x - 3y + 2z &= 10 \\ x - y + 3z &= 8 \end{aligned}$ **b.** $\quad\begin{aligned} x + y + z &= 1 \\ 2x - 3y + 2z &= 7 \\ x - y + 3z &= 7 \end{aligned}$ **c.** $\quad\begin{aligned} x + y + z &= 4 \\ 2x - 3y + 2z &= 6 \\ x - y + 3z &= 1 \end{aligned}$

Solution

Since all three sets of equations have the same coefficient matrix A, use the inverse matrix A^{-1} to find the solution for each system. A^{-1} may be found by adjoining the 3×3 identity matrix to A and using row operations to transform the resulting matrix so that the identity matrix appears on the left.

$$
\begin{array}{cc}
A & I
\end{array}
$$

$$
\left[\begin{array}{ccc|ccc}
1 & 1 & 1 & 1 & 0 & 0 \\
2 & -3 & 2 & 0 & 1 & 0 \\
1 & -1 & 3 & 0 & 0 & 1
\end{array}\right]
$$

Step 1

$$
\begin{array}{c}
 \\
-2R_1 + R_2 \\
R_3 - R_1
\end{array}
\left[\begin{array}{ccc|ccc}
1 & 1 & 1 & 1 & 0 & 0 \\
0 & -5 & 0 & -2 & 1 & 0 \\
0 & -2 & 2 & -1 & 0 & 1
\end{array}\right]
$$

Step 2

$$
\begin{array}{c}
 \\
2R_2 \\
-5R_3
\end{array}
\left[\begin{array}{ccc|ccc}
1 & 1 & 1 & 1 & 0 & 0 \\
0 & -10 & 0 & -4 & 2 & 0 \\
0 & 10 & -10 & 5 & 0 & -5
\end{array}\right]
$$

Step 3

$$
\begin{array}{c}
 \\
 \\
R_2 + R_3
\end{array}
\left[\begin{array}{ccc|ccc}
1 & 1 & 1 & 1 & 0 & 0 \\
0 & -10 & 0 & -4 & 2 & 0 \\
0 & 0 & -10 & 1 & 2 & -5
\end{array}\right]
$$

Step 4

$$
\begin{array}{c}
10R_1 \\
 \\
 \\
\end{array}
\left[\begin{array}{ccc|ccc}
10 & 10 & 10 & 10 & 0 & 0 \\
0 & -10 & 0 & -4 & 2 & 0 \\
0 & 0 & -10 & 1 & 2 & -5
\end{array}\right]
$$

Step 5

$$
R_1 + R_2 + R_3
\left[\begin{array}{ccc|ccc}
10 & 0 & 0 & 7 & 4 & -5 \\
0 & -10 & 0 & -4 & 2 & 0 \\
0 & 0 & -10 & 1 & 2 & -5
\end{array}\right]
$$

Step 6

$$
\begin{array}{c}
\frac{1}{10}R_1 \\
-\frac{1}{10}R_2 \\
-\frac{1}{10}R_3
\end{array}
\left[\begin{array}{ccc|ccc}
1 & 0 & 0 & 0.7 & 0.4 & -0.5 \\
0 & 1 & 0 & 0.4 & -0.2 & 0 \\
0 & 0 & 1 & -0.1 & -0.2 & 0.5
\end{array}\right]
$$

Thus: $A^{-1} = \begin{bmatrix} 0.7 & 0.4 & -0.5 \\ 0.4 & -0.2 & 0 \\ -0.1 & -0.2 & 0.5 \end{bmatrix}$

a. The solution of the first set of equations is

$$
\begin{bmatrix} x \\ y \\ z \end{bmatrix} = \begin{bmatrix} 0.7 & 0.4 & -0.5 \\ 0.4 & -0.2 & 0 \\ -0.1 & -0.2 & 0.5 \end{bmatrix} \cdot \begin{bmatrix} 6 \\ 10 \\ 8 \end{bmatrix} = \begin{bmatrix} 4.2 \\ 0.4 \\ 1.4 \end{bmatrix}
$$

$$x = 4.2, y = 0.4, z = 1.4$$

b. The solution of the second set of equations is

$$
\begin{bmatrix} x \\ y \\ z \end{bmatrix} = \begin{bmatrix} 0.7 & 0.4 & -0.5 \\ 0.4 & -0.2 & 0 \\ -0.1 & -0.2 & 0.5 \end{bmatrix} \cdot \begin{bmatrix} 1 \\ 7 \\ 7 \end{bmatrix} = \begin{bmatrix} 0 \\ -1 \\ 2 \end{bmatrix}
$$

$$x = 0, y = -1, z = 2$$

c. The solution of the third set of equations is

$$
\begin{bmatrix} x \\ y \\ z \end{bmatrix} = \begin{bmatrix} 0.7 & 0.4 & -0.5 \\ 0.4 & -0.2 & 0 \\ -0.1 & -0.2 & 0.5 \end{bmatrix} \cdot \begin{bmatrix} 4 \\ 6 \\ 1 \end{bmatrix} = \begin{bmatrix} 4.7 \\ 0.4 \\ -1.1 \end{bmatrix}
$$

$$x = 4.7, y = 0.4, z = -1.1$$

In Section 7.2, you saw how a calculator that does matrix operations can be used to display A^{-1} after matrix A has been entered:

Enter: ENTER

Now, to find the solution for the first set of equations in Example 2, enter $\begin{bmatrix} 6 \\ 10 \\ 8 \end{bmatrix}$ as [B], and use the following key sequence.

Enter: 2nd [A] x^{-1} × 2nd [B] ENTER

The solutions of the other systems may be obtained in a similar way.

Example 3

The prices of 3 penny stocks are shown for a 3-month period. Ms. Smith bought the same number of shares of each stock in December and in January as she bought in November, and Mr. Jones did the same. Given their total monthly costs, find how many shares these two investors bought each month.

Cost in Cents	Stock A	B	C
Nov.	2	2	3
Dec.	3	3	2
Jan.	3	2	1

Cost in Dollars	Smith	Jones
Nov.	65	170
Dec.	85	230
Jan.	65	200

Solution

Let a, b, and c be the respective numbers of shares of stocks A, B, and C bought by Ms. Smith and by Mr. Jones in each of the 3 months. Thus, in cents, the cost equations for each investor for each month are:

	Smith	Jones
Nov.	$2a + 2b + 3c = 6{,}500$	$2a + 2b + 3c = 17{,}000$
Dec.	$3a + 3b + 2c = 8{,}500$	$3a + 3b + 2c = 23{,}000$
Jan.	$3a + 2b + c = 6{,}500$	$3a + 2b + c = 20{,}000$

Note that the values of a, b, and c for Smith are different from those for Jones. But, the coefficient matrix for the two systems of equations is the same.

$$\begin{bmatrix} 2 & 2 & 3 \\ 3 & 3 & 2 \\ 3 & 2 & 1 \end{bmatrix}$$

The matrix solution for each system is:

$$\begin{bmatrix} a \\ b \\ c \end{bmatrix} = \begin{bmatrix} 2 & 2 & 3 \\ 3 & 3 & 2 \\ 3 & 2 & 1 \end{bmatrix}^{-1} \cdot \begin{bmatrix} 6{,}500 \\ 8{,}500 \\ 6{,}500 \end{bmatrix} \qquad \begin{bmatrix} a \\ b \\ c \end{bmatrix} = \begin{bmatrix} 2 & 2 & 3 \\ 3 & 3 & 2 \\ 3 & 2 & 1 \end{bmatrix}^{-1} \cdot \begin{bmatrix} 17{,}000 \\ 23{,}000 \\ 20{,}000 \end{bmatrix}$$

Use row operations on the adjoined matrix to rewrite it so that the identity matrix appears on the left.

$$\begin{bmatrix} 2 & 2 & 3 & | & 1 & 0 & 0 \\ 3 & 3 & 2 & | & 0 & 1 & 0 \\ 3 & 2 & 1 & | & 0 & 0 & 1 \end{bmatrix} \begin{matrix} \text{can be} \\ \text{transformed} \\ \text{into} \end{matrix} \begin{bmatrix} 1 & 0 & 0 & | & 0.2 & -0.8 & 1 \\ 0 & 1 & 0 & | & -0.6 & 1.4 & -1 \\ 0 & 0 & 1 & | & 0.6 & -0.4 & 0 \end{bmatrix}$$

Thus, the value of the inverse of the coefficient matrix is:

$$\begin{bmatrix} 2 & 2 & 3 \\ 3 & 3 & 2 \\ 3 & 2 & 1 \end{bmatrix}^{-1} = \begin{bmatrix} 0.2 & -0.8 & 1 \\ -0.6 & 1.4 & -1 \\ 0.6 & -0.4 & 0 \end{bmatrix}$$

Substitute the value of the inverse into the matrix solution and complete the matrix multiplication.

$$\text{Smith} \quad \begin{bmatrix} a \\ b \\ c \end{bmatrix} = \begin{bmatrix} 0.2 & -0.8 & 1 \\ -0.6 & 1.4 & -1 \\ 0.6 & -0.4 & 0 \end{bmatrix} \cdot \begin{bmatrix} 6,500 \\ 8,500 \\ 6,500 \end{bmatrix} = \begin{bmatrix} 1,000 \\ 1,500 \\ 500 \end{bmatrix}$$

$$\text{Jones} \quad \begin{bmatrix} a \\ b \\ c \end{bmatrix} = \begin{bmatrix} 0.2 & -0.8 & 1 \\ -0.6 & 1.4 & -1 \\ 0.6 & -0.4 & 0 \end{bmatrix} \cdot \begin{bmatrix} 17,000 \\ 23,000 \\ 20,000 \end{bmatrix} = \begin{bmatrix} 5,000 \\ 2,000 \\ 1,000 \end{bmatrix}$$

Answer: Stock purchases for each of the 3 months:

	A	B	C
Smith	1,000	1,500	500
Jones	5,000	2,000	1,000

Invertible vs. Singular Matrices

A square matrix that has an inverse is called an *invertible matrix*; a square matrix that has no inverse is called a *singular matrix*.

Note that a square matrix may be invertible or singular. However, a nonsquare matrix is always singular. The following indirect argument demonstrates that a nonsquare matrix does not have an inverse.

Suppose A is an $m \times n$ matrix and $m \neq n$. If A^{-1} did exist, it would have to be an $n \times m$ matrix. Thus, AA^{-1} would be an $m \times m$ matrix, and $A^{-1}A$ would be an $n \times n$ matrix. But, this is a contradiction since AA^{-1} must equal $A^{-1}A$. Therefore, A^{-1} does not exist.

Using the method of an adjoined matrix to find an inverse of a square matrix when one does not exist will result in the appearance of a row of zeros in the left half of the adjoined matrix.

Example 4

Use the adjoined matrix to determine if the given matrix is invertible or singular.

$$\begin{bmatrix} 1 & 2 & 4 \\ 2 & 4 & 4 \\ 1 & 2 & 6 \end{bmatrix}$$

Solution

Using row operations on the adjoined matrix (for example, $-R_1 + R_3$, $-\frac{1}{2}R_2 + R_1$, $-R_1 + R_3$) results in a row of zeros in the left half.

$$\begin{bmatrix} 1 & 2 & 4 & | & 1 & 0 & 0 \\ 2 & 4 & 4 & | & 0 & 1 & 0 \\ 1 & 2 & 6 & | & 0 & 0 & 1 \end{bmatrix} \begin{matrix} \text{can be} \\ \text{transformed} \\ \text{into} \end{matrix} \begin{bmatrix} 0 & 0 & 2 & | & 1 & -0.5 & 0 \\ 2 & 4 & 4 & | & 0 & 1 & 0 \\ 0 & 0 & 0 & | & -2 & 0.5 & 1 \end{bmatrix}$$

Thus, the given matrix has no inverse. Note also that an inverse cannot exist because the value of the determinant is 0.

Answer: The given matrix is singular.

Exercises 7.4

In **1 – 4**, indicate for the given system if:
- **a.** the determinant of the coefficient matrix = 0 or $\neq 0$
- **b.** the graphs of the lines intersect, coincide, or are parallel
- **c.** the system has 0, 1, or infinite solutions

1. $x + y = 5$
$x - y = 7$

2. $4x + y = 3$
$12x + 3y = 9$

3. $a + b = 4$
$2a + 2b = 10$

4. $5x + 4y = 10$
$6x - y = 8$

In **5 – 14**, use the inverse matrix to solve the system.

5. $x + 5y = 13$
$-2x + y = -4$

6. $3x + y = 2$
$2x - y = 3$

7. $4a + 3b = 24$
$2a - b = 2$

8. $7a + b = 12$
$-5a + 3b = 10$

9. $4x + 5y = 5$
$8x + 10y = 10$

10. $6x + 4y = 3$
$3x + 2y = 2$

11. $a - 3b = -2$
$3a - 2b + c = 5$
$2a + b + 2c = 4$

12. $x + y + z = -2$
$3x + 3y + z = -18$
$4x + 2y + z = -20$

13. $x_1 + x_2 + x_3 + x_4 = 0$
$2x_1 - x_2 + 3x_3 - x_4 = 11$
$3x_1 + 2x_2 - x_3 + x_4 = -3$
$x_2 - 5x_3 - x_4 = -9$

14. $2x_1 + x_2 - x_3 = 1$ *Hint:* The inverse of the coefficient
$2x_2 + x_3 - x_4 = 7$ matrix is $\begin{bmatrix} \frac{26}{9} & -\frac{16}{9} & -\frac{7}{9} & \frac{1}{9} \\ -\frac{37}{9} & \frac{29}{9} & \frac{11}{9} & -\frac{2}{3} \\ \frac{2}{3} & -\frac{1}{3} & -\frac{1}{3} & 0 \\ -\frac{68}{9} & \frac{46}{9} & \frac{19}{9} & -\frac{4}{3} \end{bmatrix}$
$4x_1 - 6x_3 + x_4 = -11$
$-5x_1 + 2x_2 - 3x_4 = 3$

In **15 – 16**, find the solution for each set of equations by using one coefficient matrix.

15. a. $3x - 4y + 5z = -9$
$2x - y + 2z = -2$
$3x + 2y - z = 3$

b. $3x - 4y + 5z = 16$
$2x - y + 2z = 13$
$3x + 2y - z = 22$

c. $3x - 4y + 5z = 20$
$2x - y + 2z = 10$
$3x + 2y - z = 8$

16. a. $6x_1 + x_2 - x_3 = 22$
$4x_1 - 2x_2 = 0$
$-3x_1 + 2x_3 = -18$

b. $6x_1 + x_2 - x_3 = 19$
$4x_1 - 2x_2 = 8$
$-3x_1 + 2x_3 = -7$

c. $6x_1 + x_2 - x_3 = 1$
$4x_1 - 2x_2 = 2$
$-3x_1 + 2x_3 = -4$

---

13. $x_1 = 1$
$x_2 = -1$
$x_3 = 2$
$x_4 = -2$

14. $x_1 = 0$
$x_2 = 3$
$x_3 = 2$
$x_4 = 1$

15. a. $x = -1, y = 4, z = 2$
b. $x = 6, y = 3, z = 2$
c. $x = 4, y = -2, z = 0$

16. a. $x_1 = 2, x_2 = 4, x_3 = -6$
b. $x_1 = 3, x_2 = 2, x_3 = 1$
c. $x_1 = 0, x_2 = -1, x_3 = -2$

1. a. determinant $\neq 0$
   b. lines intersect
   c. one solution

2. a. determinant $= 0$
   b. lines coincide
   c. infinite number of solutions

3. a. determinant $= 0$
   b. lines are parallel
   c. no solutions

4. a. determinant $\neq 0$
   b. lines intersect
   c. one solution

5. $x = 3$
   $y = 2$

6. $x = 1$
   $y = -1$

7. $a = 3$
   $b = 4$

8. $a = 1$
   $b = 5$

9. infinite number of solutions

10. no solution

11. $a = 4$
   $b = 2$
   $c = -3$

12. $x = -5$
   $y = -3$
   $z = 6$

Exercises 7.4 (continued)

In **17 – 22**, determine if the matrix is invertible or singular.

17. $\begin{bmatrix} 1 & 1 \\ 2 & 2 \end{bmatrix}$

18. $\begin{bmatrix} 3 & 4 & 1 \\ 6 & 8 & 2 \\ 1 & 3 & 7 \end{bmatrix}$

19. $\begin{bmatrix} 10 & 16 & 14 \\ 12 & 24 & 21 \\ 1 & 1 & 1 \end{bmatrix}$

20. $\begin{bmatrix} 1 & 2 & 3 \\ -4 & 5 & -6 \\ 7 & 8 & 9 \end{bmatrix}$

21. $\begin{bmatrix} -4 & -3 & -2 \\ 0 & 1 & 2 \\ 3 & 4 & 5 \end{bmatrix}$

22. $\begin{bmatrix} 3 & 1 & -2 \\ 0 & 4 & 0 \\ 6 & 1 & -4 \end{bmatrix}$

23. By showing that the value of its determinant is 0, prove that a 3 × 3 matrix has no inverse if its elements are consecutive integers beginning with the element in the first row first column.

24. Postage for a first-class letter used to cost $.25 for the first ounce and $.20 for each additional ounce. A proposed increase would change the rate to $.45 for the first ounce and $.15 for each additional ounce. If a letter costs the same under the old and new rates, how much does it weigh?

25. The Advanced Devices Company produces 3 different microchips x, y, and z. The unit prices of the chips fluctuated according to supply and demand during the year according to the table shown.

	Jan.	May	Sept.
x	1¢	2¢	1¢
y	3	4	4
z	4	4	4

Two computer manufacturers, ABC and MBI, place orders for chips in January, May, and September. ABC buys the same number of chip x each time, also the same numbers of chips y and z, and MBI does the same. If the total cost for their orders is as shown, find how many of each chip ABC and MBI order each time.

	Jan.	May	Sept.
ABC	$2,200	$3,000	$2,500
MBI	1,500	2,100	1,800

17. singular

18. singular

19. invertible

20. invertible

21. singular

22. singular

26. For the function
$$f(x) = ax^3 + bx^2 + cx + d$$
let $f(1) = 1$, $f(2) = 4$, $f(3) = 10$, and $f(4) = 20$. Find the values of $a, b, c,$ and d.

27. The sequence $-2, 0, 12, 40, 90, \ldots$ has the general term $an^3 + bn^2 + cn + d$. Find the values of $a, b, c,$ and d.

28. The grades that each of 3 students obtained on each of 3 math tests and the final grade that they received are shown. Assuming that the final grade represents a weighted average of the 3 tests, determine how much each test was weighted as a percentage of the final grade.

	Test 1	Test 2	Test 3	Final Grade
Ann	100	80	100	96
Howard	100	80	80	86
Phyllis	100	75	90	90

23.

$$\begin{vmatrix} x & x+1 & x+2 \\ x+3 & x+4 & x+5 \\ x+6 & x+7 & x+8 \end{vmatrix}$$

$$= x\begin{vmatrix} x+4 & x+5 \\ x+7 & x+8 \end{vmatrix} -(x+1)\begin{vmatrix} x+3 & x+5 \\ x+6 & x+8 \end{vmatrix} +(x+2)\begin{vmatrix} x+3 & x+4 \\ x+6 & x+7 \end{vmatrix}$$

$$= x((x+4)(x+8) - (x+5)(x+7)) - (x+1)((x+3)(x+8) - (x+5)(x+6))$$
$$+ (x+2)((x+3)(x+7) - (x+4)(x+6))$$

$$= x(x^2 + 12x + 32 - x^2 - 12x - 35) - (x+1)(x^2 + 11x + 24 - x^2 - 11x - 30)$$
$$+ (x+2)(x^2 + 10x + 21 - x^2 - 10x - 24)$$

$$= -3x + 6x + 6 - 3x - 6$$

$$= 0$$

24. 5 oz.

25. ABC: $x = 500, y = 300,$
$z = 200$
MBI: $x = 300, y = 300,$
$z = 75$

26. $a = \frac{1}{6}, b = \frac{1}{2}, c = \frac{1}{3}, d = 0$

27. $a = 1, b = -1, c = -2, d = 0$

28. 30%, 20%, 50%

7.5 Matrices and Transformations

Matrices provide a convenient way to represent transformations of the plane, based on the representation of the coordinates (x, y) by a matrix, $\begin{bmatrix} x \\ y \end{bmatrix}$.

Under a transformation of the plane, each point is associated with an image point. Before discussing matrix representations of the common transformations, reconsider the familiar consequence of multiplying by the 2×2 identity matrix.

$\begin{bmatrix} 1 & 0 \\ 0 & 1 \end{bmatrix} \begin{bmatrix} x \\ y \end{bmatrix} = \begin{bmatrix} x \\ y \end{bmatrix}$ Thus, the 2×2 identity matrix can be considered the *identity transformation* for the coordinates (x, y).

Reflection

Each of the common reflections r of the point $P(x, y)$ can be represented by a matrix.

reflection in the x-axis	reflection in the y-axis	reflection in the line $y = x$
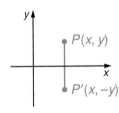	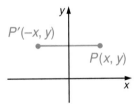	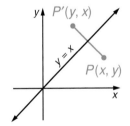
under $r_{x\text{-}axis}$:	under $r_{y\text{-}axis}$:	under $r_{y=x}$:
$(x, y) \to (x, -y)$	$(x, y) \to (-x, y)$	$(x, y) \to (y, x)$
$\begin{bmatrix} 1 & 0 \\ 0 & -1 \end{bmatrix}$	$\begin{bmatrix} -1 & 0 \\ 0 & 1 \end{bmatrix}$	$\begin{bmatrix} 0 & 1 \\ 1 & 0 \end{bmatrix}$
represents $r_{x\text{-}axis}$ since	represents $r_{y\text{-}axis}$ since	represents $r_{y=x}$ since
$\begin{bmatrix} 1 & 0 \\ 0 & -1 \end{bmatrix} \begin{bmatrix} x \\ y \end{bmatrix} = \begin{bmatrix} x \\ -y \end{bmatrix}$	$\begin{bmatrix} -1 & 0 \\ 0 & 1 \end{bmatrix} \begin{bmatrix} x \\ y \end{bmatrix} = \begin{bmatrix} -x \\ y \end{bmatrix}$	$\begin{bmatrix} 0 & 1 \\ 1 & 0 \end{bmatrix} \begin{bmatrix} x \\ y \end{bmatrix} = \begin{bmatrix} y \\ x \end{bmatrix}$

Example 1

Use matrices to find the image of △RST, with vertices R(1, 4), S(3, 7), and T(6, –1), when it is reflected in the y-axis.

Solution

△RST may be represented by the matrix that displays the coordinates of each vertex as a column of the matrix. $\begin{bmatrix} 1 & 3 & 6 \\ 4 & 7 & -1 \end{bmatrix}$

To produce a reflection of △RST in the y-axis, left-multiply the matrix representing the triangle by the matrix that represents the $r_{y\text{-}axis}$.

$$\begin{bmatrix} -1 & 0 \\ 0 & 1 \end{bmatrix} \begin{bmatrix} 1 & 3 & 6 \\ 4 & 7 & -1 \end{bmatrix} = \begin{bmatrix} -1(1) + 0(4) & -1(3) + 0(7) & -1(6) + 0(-1) \\ 0(1) + 1(4) & 0(3) + 1(7) & 0(6) + 1(-1) \end{bmatrix}$$

$$= \begin{bmatrix} -1 & -3 & -6 \\ 4 & 7 & -1 \end{bmatrix}$$
$$R'S'T'$$

Therefore, the image triangle, △R'S'T', has vertices R'(–1, 4), S'(–3, 7), and T'(–6, –1).

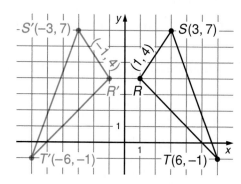

Recall that a composite transformation of the plane maps a set of points P into a set of points P' under one transformation and then maps P' into a set of points P'' under a second transformation. Matrices may be used to accomplish the composite transformation

$$r_{x\text{-}axis} \circ r_{y\text{-}axis} (x, y) = r_{x\text{-}axis} (-x, y) = (-x, -y)$$

by left-multiplying $\begin{bmatrix} x \\ y \end{bmatrix}$ by the $r_{y\text{-}axis}$ matrix and then left-multiplying the result by the $r_{x\text{-}axis}$ matrix.

$$\begin{bmatrix} 1 & 0 \\ 0 & -1 \end{bmatrix} \left(\begin{bmatrix} -1 & 0 \\ 0 & 1 \end{bmatrix} \begin{bmatrix} x \\ y \end{bmatrix} \right) = \begin{bmatrix} 1 & 0 \\ 0 & -1 \end{bmatrix} \begin{bmatrix} -x \\ y \end{bmatrix} = \begin{bmatrix} -x \\ -y \end{bmatrix}$$

Since matrix multiplication is associative, the same result can be obtained as follows.

$$\left(\begin{bmatrix} 1 & 0 \\ 0 & -1 \end{bmatrix} \begin{bmatrix} -1 & 0 \\ 0 & 1 \end{bmatrix} \right) \begin{bmatrix} x \\ y \end{bmatrix} = \begin{bmatrix} -1 & 0 \\ 0 & -1 \end{bmatrix} \begin{bmatrix} x \\ y \end{bmatrix} = \begin{bmatrix} -x \\ -y \end{bmatrix}$$

Example 2

Find the image of quadrilateral $ABCD$, $A (0, 0)$, $B (3, -3)$, $C (2, 5)$, and $D (-6, 1)$ under the transformation $r_{x\text{-axis}} \circ r_{x = y}$.

Solution

To apply the required composite transformation, left-multiply the matrix that represents $ABCD$ first by the matrix that represents $r_{x = y}$ and then by the matrix that represents $r_{x\text{-axis}}$.

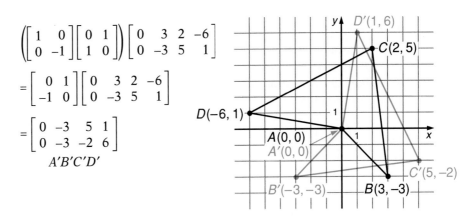

$$\begin{array}{ccc} r_{x\text{-axis}} \circ r_{x = y} & ABCD \\ \begin{bmatrix} 1 & 0 \\ 0 & -1 \end{bmatrix} \begin{bmatrix} 0 & 1 \\ 1 & 0 \end{bmatrix} \begin{bmatrix} 0 & 3 & 2 & -6 \\ 0 & -3 & 5 & 1 \end{bmatrix} \end{array}$$

$$\left(\begin{bmatrix} 1 & 0 \\ 0 & -1 \end{bmatrix} \begin{bmatrix} 0 & 1 \\ 1 & 0 \end{bmatrix} \right) \begin{bmatrix} 0 & 3 & 2 & -6 \\ 0 & -3 & 5 & 1 \end{bmatrix}$$

$$= \begin{bmatrix} 0 & 1 \\ -1 & 0 \end{bmatrix} \begin{bmatrix} 0 & 3 & 2 & -6 \\ 0 & -3 & 5 & 1 \end{bmatrix}$$

$$= \begin{bmatrix} 0 & -3 & 5 & 1 \\ 0 & -3 & -2 & 6 \end{bmatrix}$$
$$A'B'C'D'$$

Answer: $A'B'C'D'$ with $A'(0, 0)$, $B'(-3, -3)$, $C'(5, -2)$, $D'(1, 6)$.

Dilation and Scale Change

The dilation D_k, with k a real number, maps (x, y) onto (kx, ky). This transformation may be accomplished using matrices:

$$\begin{bmatrix} k & 0 \\ 0 & k \end{bmatrix} \begin{bmatrix} x \\ y \end{bmatrix} = \begin{bmatrix} kx \\ ky \end{bmatrix}$$

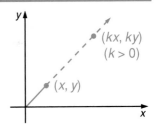

Example 3

Find the image of $\triangle ABC$, with vertices $A(6, 0)$, $B(-6, 0)$, and $C(0, 6)$, under D_2.

Solution

Left-multiply the matrix that represents $\triangle ABC$ by the matrix that represents D_2.

$$\begin{array}{ccc} D_2 & ABC & A'B'C' \\ \begin{bmatrix} 2 & 0 \\ 0 & 2 \end{bmatrix} \begin{bmatrix} 6 & -6 & 0 \\ 0 & 0 & 6 \end{bmatrix} = \begin{bmatrix} 12 & -12 & 0 \\ 0 & 0 & 12 \end{bmatrix} \end{array}$$

Answer: $\triangle A'B'C'$

$A'(12, 0)$,

$B'(-12, 0)$,

$C'(0, 12)$

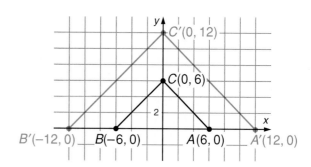

Note that in a dilation D_k, both the horizontal and vertical dimensions of a plane figure are changed by the same factor k, as represented by the matrix $\begin{bmatrix} k & 0 \\ 0 & k \end{bmatrix}$. Consider the matrix $\begin{bmatrix} h & 0 \\ 0 & v \end{bmatrix}$, which would effect a transformation that changes the horizontal and vertical dimensions of a figure by different factors. This type of transformation is called a *scale change*. For example, observe how the figure shown undergoes a scale change, by a horizontal factor of 2 and a vertical factor of 1.5, by matrix operation and diagrammatically.

$$\begin{bmatrix} 2 & 0 \\ 0 & 1.5 \end{bmatrix} \begin{bmatrix} \overset{A}{1} & \overset{B}{1} & \overset{C}{4} & \overset{D}{3} & \overset{E}{1} & \overset{F}{1} & \overset{G}{-1} & \overset{H}{-1} & \overset{I}{-4} & \overset{J}{-3} & \overset{K}{-1} & \overset{L}{-1} \\ 2 & 1 & 1 & -1 & -1 & -2 & -2 & -1 & -1 & 1 & 1 & 2 \end{bmatrix}$$

$$= \begin{bmatrix} \overset{A'}{2} & \overset{B'}{2} & \overset{C'}{8} & \overset{D'}{6} & \overset{E'}{2} & \overset{F'}{2} & \overset{G'}{-2} & \overset{H'}{-2} & \overset{I'}{-8} & \overset{J'}{-6} & \overset{K'}{-2} & \overset{L'}{-2} \\ 3 & 1.5 & 1.5 & -1.5 & -1.5 & -3 & -3 & -1.5 & -1.5 & 1.5 & 1.5 & 3 \end{bmatrix}$$

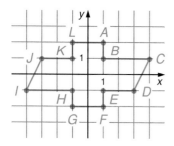 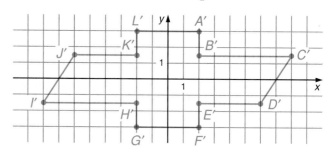

Rotation

A counterclockwise rotation of 90° about the origin maps (x, y) onto $(-y, x)$. Observe that this single transformation is equivalent to two line reflections, and that the transformation can be accomplished by using matrices:

$$r_{y = x} \circ r_{x\text{-}axis} = R_{90°}$$

$$\begin{bmatrix} 0 & 1 \\ 1 & 0 \end{bmatrix} \begin{bmatrix} 1 & 0 \\ 0 & -1 \end{bmatrix} = \begin{bmatrix} 0 & -1 \\ 1 & 0 \end{bmatrix}$$

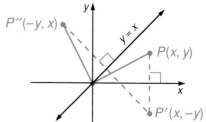

Note that, by convention, a counterclockwise rotation is indicated by a positive angle, and a clockwise rotation by a negative angle. Although a rotation may have any point in the plane as its center, all of the rotations discussed in this section will assume that the origin is the center.

Some rotations can be thought of as the result of two successive rotations. For example: $R_{180°} = R_{90°} \circ R_{90°}$ Thus, the matrix for $R_{180°}$ can be found from the product of two $R_{90°}$ matrices.

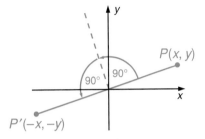

$$R_{90°} \circ R_{90°} = R_{180°}$$

$$\begin{bmatrix} 0 & -1 \\ 1 & 0 \end{bmatrix}\begin{bmatrix} 0 & -1 \\ 1 & 0 \end{bmatrix} = \begin{bmatrix} -1 & 0 \\ 0 & -1 \end{bmatrix}$$

The result of applying the $R_{180°}$ matrix to the matrix that represents the point (x, y) corresponds to the diagrammatic result shown.

$$\begin{bmatrix} -1 & 0 \\ 0 & -1 \end{bmatrix}\begin{bmatrix} x \\ y \end{bmatrix} = \begin{bmatrix} -x \\ -y \end{bmatrix}$$

Note that $R_{180°}$ is equivalent to a point reflection in the origin.

Another example of a rotation that is a composite of other rotations is:

$$R_{270°} = R_{90°} \circ R_{90°} \circ R_{90°} \quad \text{or} \quad R_{270°} = R_{180°} \circ R_{90°}$$

Example 4

Use matrices to find the image of $\triangle ABC$, with vertices $A(5, 0)$, $B(-2, -3)$, and $C(7, -1)$ if it is rotated $90°$ in a clockwise direction.

Solution

A clockwise rotation of $90°$ is equivalent to a counter-clockwise rotation of $270°$.

$$R_{-90°} = R_{270°} = R_{180°} \circ R_{90°}$$

$$= \begin{bmatrix} -1 & 0 \\ 0 & -1 \end{bmatrix}\begin{bmatrix} 0 & -1 \\ 1 & 0 \end{bmatrix} = \begin{bmatrix} 0 & 1 \\ -1 & 0 \end{bmatrix}$$

Left-multiply the matrix that represents $\triangle ABC$ by the matrix that represents $R_{270°}$.

$$\begin{array}{ccc} R_{270°} & ABC & = & A'B'C' \end{array}$$

$$\begin{bmatrix} 0 & 1 \\ -1 & 0 \end{bmatrix}\begin{bmatrix} 5 & -2 & 7 \\ 0 & -3 & -1 \end{bmatrix} = \begin{bmatrix} 0 & -3 & -1 \\ -5 & 2 & -7 \end{bmatrix}$$

Answer: $\triangle A'B'C'$

$A'(0, -5)$,

$B'(-3, 2)$,

$C'(-1, -7)$.

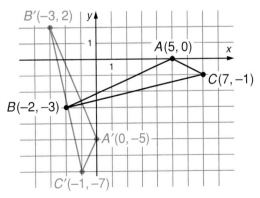

In a later chapter, you will see how a rotation of any angle may be accomplished using trigonometry.

Translation

Unlike the other common transformations, a translation cannot be uniquely represented by a matrix.

For example, consider the result of the translation $T_{2,3}$ on the quadrilateral $ABCD$ with vertices $A(4, 0)$, $B(-4, 0)$, $C(-4, 4)$, $D(4, 4)$.

The following matrix operation would produce the vertices of the image.

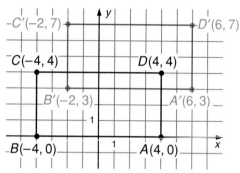

$$\underset{T_{2,3}}{\begin{bmatrix} 2 & 2 & 2 & 2 \\ 3 & 3 & 3 & 3 \end{bmatrix}} + \underset{ABCD}{\begin{bmatrix} 4 & -4 & -4 & 4 \\ 0 & 0 & 4 & 4 \end{bmatrix}}$$

$$= \underset{A'B'C'D'}{\begin{bmatrix} 6 & -2 & -2 & 6 \\ 3 & 3 & 7 & 7 \end{bmatrix}}$$

However, the 2 × 4 matrix used to apply $T_{2,3}$ to a quadrilateral would have to be rewritten with different dimensions in order to apply to other geometric figures. Thus, it is impractical to use matrices for a translation.

Simple Shear

Another transformation to which matrices can be applied is called a *simple shear*. To describe such a transformation, consider the following model.

Quadrilateral $ABCD$ represents a rubber band that is held in place by 4 pins, A, B, C, and D. If vertical forces of different magnitudes and opposite directions are applied to relocate 2 of the vertices, B and D, keeping the other 2 vertices stationary, the quadrilateral is transformed. $AB'CD'$ represents the image of $ABCD$ under a simple shear.

Note that under this particular shear, the positions of all the points on the y-axis that are interior to $ABCD$, as well as A and C, are preserved. All other points, in or on $ABCD$, are transformed so that

(1) the x-coordinates remain fixed, and

(2) the y-coordinate of the image is such that the vertical distance from the point to its image is 5 times the distance from the y-axis to the point.

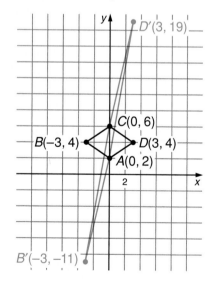

Using $\begin{bmatrix} 1 & 0 \\ 5 & 1 \end{bmatrix}$ to represent the shear described, left-multiply the matrix that represents ABCD to obtain the coordinates of the vertices of the image.

$$\text{Shear} \quad\quad ABCD \quad\quad\quad\quad AB'CD'$$
$$\begin{bmatrix} 1 & 0 \\ 5 & 1 \end{bmatrix}\begin{bmatrix} 0 & -3 & 0 & 3 \\ 2 & 4 & 6 & 4 \end{bmatrix} = \begin{bmatrix} 0 & -3 & 0 & 3 \\ 2 & -11 & 6 & 19 \end{bmatrix}$$

In general, if (x', y') is the image of (x, y) under a shear transformation, then for

a vertical shear   a horizontal shear
$x' = x$ $y' = y$ where ℓ is a constant.
$y' = \ell x + y$ $x' = \ell y + x$

Exercises 7.5

In **1 – 7**, indicate the dimension of the matrix that would be used to represent the vertices of the given figure.

1. a point   **2.** a line segment   **3.** a triangle

4. a trapezoid   **5.** a pentagon   **6.** a hexagon   **7.**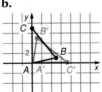

In **8 – 19**, given $\triangle ABC$ represented by $\begin{bmatrix} 0 & 5 & 0 \\ 0 & 1 & 7 \end{bmatrix}$.

a. Write the matrix that describes $\triangle A'B'C'$, the image of $\triangle ABC$ under the transformation indicated by the given matrix.

b. Graph $\triangle ABC$ and $\triangle A'B'C'$.

c. Identify the type of transformation: reflection, dilation, scale change, rotation, shear, or the identity transformation.

8. $\begin{bmatrix} 1 & 0 \\ 0 & -1 \end{bmatrix}$ **9.** $\begin{bmatrix} -1 & 0 \\ 0 & 1 \end{bmatrix}$ **10.** $\begin{bmatrix} 0 & 1 \\ 1 & 0 \end{bmatrix}$ **11.** $\begin{bmatrix} -1 & 0 \\ 0 & -1 \end{bmatrix}$

12. $\begin{bmatrix} 0 & -1 \\ 1 & 0 \end{bmatrix}$ **13.** $\begin{bmatrix} 1 & 0 \\ 0 & 1 \end{bmatrix}$ **14.** $\begin{bmatrix} 0 & 1 \\ -1 & 0 \end{bmatrix}$ **15.** $\begin{bmatrix} 5 & 0 \\ 0 & 5 \end{bmatrix}$

16. $\begin{bmatrix} 2 & 0 \\ 0 & 3 \end{bmatrix}$ **17.** $\begin{bmatrix} 1 & 2 \\ 0 & 1 \end{bmatrix}$ **18.** $\begin{bmatrix} 2 & 0 \\ 3 & 2 \end{bmatrix}$ **19.** $\begin{bmatrix} 1 & 0 \\ 3 & 1 \end{bmatrix}$

In **20 – 31**, given $ABCD$ with vertices $A(0, 0)$, $B(5, 0)$, $C(5, 3)$, and $D(0, 3)$. Use a matrix to find $A'B'C'D'$, the image of $ABCD$ under the given transformation.

20. $r_{x\text{-}axis}$ **21.** $r_{y\text{-}axis}$ **22.** $r_{y = x}$ **23.** D_3 **24.** D_{-2}

25. a scale change of a factor of 2 horizontally and 3 vertically   **26.** a point reflection in the origin

27. $R_{90°}$ **28.** $R_{-180°}$ **29.** $R_{360°}$ **30.** $R_{-450°}$ **31.** $R_{540°}$

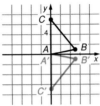
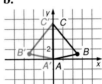
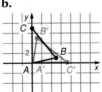
 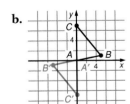

12. a. $\begin{bmatrix} 0 & -1 & -7 \\ 0 & 5 & 0 \end{bmatrix}$

b.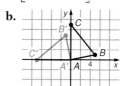

c. rotation of 90°

13. a. $\begin{bmatrix} 0 & 5 & 0 \\ 0 & 1 & 7 \end{bmatrix}$

b.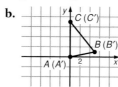

c. identity

14. a. $\begin{bmatrix} 0 & 1 & 7 \\ 0 & -5 & 0 \end{bmatrix}$

b.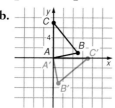

c. rotation of –90°

15. a. $\begin{bmatrix} 0 & 25 & 0 \\ 0 & 5 & 35 \end{bmatrix}$

b.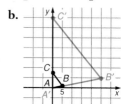

c. dilation of 5

16. a. $\begin{bmatrix} 0 & 10 & 0 \\ 0 & 3 & 21 \end{bmatrix}$

b.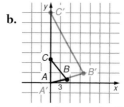

c. scale change of a factor of 2 horizontally and 3 vertically

17. a. $\begin{bmatrix} 0 & 7 & 14 \\ 0 & 1 & 7 \end{bmatrix}$

b.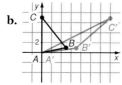

c. shear

18. a. $\begin{bmatrix} 0 & 10 & 0 \\ 0 & 17 & 14 \end{bmatrix}$

b.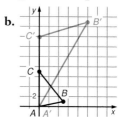

c. shear

19. a. $\begin{bmatrix} 0 & 5 & 0 \\ 0 & 16 & 7 \end{bmatrix}$

b.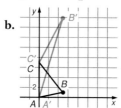

c. shear

20. $A'(0, 0)$, $B'(5, 0)$, $C'(5, -3)$, $D'(0, -3)$

21. $A'(0, 0)$, $B'(-5, 0)$, $C'(-5, 3)$, $D'(0, 3)$

22. $A'(0, 0)$, $B'(0, 5)$, $C'(3, 5)$, $D'(3, 0)$

23. $A'(0, 0)$, $B'(15, 0)$, $C'(15, 9)$, $D'(0, 9)$

24. $A'(0, 0)$, $B'(-10, 0)$, $C'(-10, -6)$, $D'(0, -6)$

25. $A'(0, 0)$, $B'(10, 0)$, $C'(10, 9)$, $D'(0, 9)$

26. $A'(0, 0)$, $B'(-5, 0)$, $C'(-5, -3)$, $D'(0, -3)$

27. $A'(0, 0)$, $B'(0, 5)$, $C'(-3, 5)$, $D'(-3, 0)$

28. $A'(0, 0)$, $B'(-5, 0)$, $C'(-5, -3)$, $D'(0, -3)$

29. $A'(0, 0)$, $B'(5, 0)$, $C'(5, 3)$, $D'(0, 3)$

30. $A'(0, 0)$, $B'(0, -5)$, $C'(3, -5)$, $D'(3, 0)$

31. $A'(0, 0)$, $B'(-5, 0)$, $C'(-5, -3)$, $D'(0, -3)$

32. $\begin{bmatrix} -\frac{\sqrt{2}}{2} & -\frac{\sqrt{2}}{2} \\ \frac{\sqrt{2}}{2} & -\frac{\sqrt{2}}{2} \end{bmatrix}$

33. a. $\begin{bmatrix} -\frac{1}{2} & -\frac{\sqrt{3}}{2} \\ \frac{\sqrt{3}}{2} & -\frac{1}{2} \end{bmatrix}$

b. $\begin{bmatrix} \frac{-\sqrt{2}+\sqrt{6}}{4} & \frac{-\sqrt{2}-\sqrt{6}}{4} \\ \frac{\sqrt{2}+\sqrt{6}}{4} & \frac{-\sqrt{2}+\sqrt{6}}{4} \end{bmatrix}$

34. a. unchanged
b. multiplied by 9
c. unchanged
d. unchanged
e. multiplied by 6

35. $\begin{bmatrix} 2 & 0 \\ 0 & -2 \end{bmatrix}$

36. $\begin{bmatrix} 0 & 3 \\ 2 & 0 \end{bmatrix}$

37. $d = \dfrac{|ap_x + bp_y + c|}{\sqrt{a^2 + b^2}}$

$y = x$

$-x + y = 0 \qquad a = -1, b = 1, c = 0$

$P(x_1, y_1)$

$d = \dfrac{|-1x_1 + 1y_1 + 0|}{\sqrt{(-1)^2 + 1^2}} = \dfrac{|-x_1 + y_1|}{\sqrt{2}}$

$P(y_1, x_1)$

$d = \dfrac{|-1y_1 + 1x_1 + 0|}{\sqrt{(-1)^2 + 1^2}} = \dfrac{|-y_1 + x_1|}{\sqrt{2}}$

Since $|-x_1 + y_1| = |-y_1 + x_1|$, the distances are equal.

Exercises 7.5 (continued)

32. If $R_{45°} = \begin{bmatrix} \frac{\sqrt{2}}{2} & -\frac{\sqrt{2}}{2} \\ \frac{\sqrt{2}}{2} & \frac{\sqrt{2}}{2} \end{bmatrix}$, use $R_{90°}$ and matrix multiplication to find $R_{135°}$.

33. If $R_{30°} = \begin{bmatrix} \frac{\sqrt{3}}{2} & -\frac{1}{2} \\ \frac{1}{2} & \frac{\sqrt{3}}{2} \end{bmatrix}$, find: **a.** $R_{120°}$ **b.** $R_{75°}$

34. If the transformations represented by the given matrices are applied to a triangle, how would its area change?

a. $\begin{bmatrix} 1 & 0 \\ 0 & 1 \end{bmatrix}$ **b.** $\begin{bmatrix} 3 & 0 \\ 0 & 3 \end{bmatrix}$ **c.** $\begin{bmatrix} 1 & 0 \\ 0 & -1 \end{bmatrix}$ **d.** $\begin{bmatrix} 0 & -1 \\ 1 & 0 \end{bmatrix}$ **e.** $\begin{bmatrix} 2 & 0 \\ 0 & 3 \end{bmatrix}$

In **35 – 36**, find a single transformation matrix for $T \circ T'$.

35. $T = \begin{bmatrix} 2 & 0 \\ 0 & 2 \end{bmatrix}$ and $T' = \begin{bmatrix} 1 & 0 \\ 0 & -1 \end{bmatrix}$

36. $T = \begin{bmatrix} 3 & 0 \\ 0 & 2 \end{bmatrix}$ and $T' = \begin{bmatrix} 0 & 1 \\ 1 & 0 \end{bmatrix}$

37. Use the vector formula for the distance from a point to a line to prove that the points (x_1, y_1) and (y_1, x_1) are equidistant from the line $y = x$.

38. Given: $D_k = \begin{bmatrix} k & 0 \\ 0 & k \end{bmatrix}$, a dilation of factor k, and any transformation of the plane, $T = \begin{bmatrix} a & b \\ c & d \end{bmatrix}$.

Prove: $D_k \circ T = T \circ D_k$.

39. Describe what the transformation $\begin{bmatrix} 1 & 1 \\ 1 & 1 \end{bmatrix}$ does to any point in the plane.

In **40 – 41**, use the point (10, 10), a carefully drawn diagram, and a protractor to estimate, to the nearest 5°, the angle in the rotation described by the given matrix.

40. $R_x = \begin{bmatrix} 0.65 & -0.75 \\ 0.75 & 0.65 \end{bmatrix}$

41. $R_x = \begin{bmatrix} 0.96 & -0.25 \\ 0.25 & 0.96 \end{bmatrix}$

In **42 – 43**, find a matrix to represent the shear described.

42. a shear that moves a point horizontally a directed distance equal to twice its directed distance from the x-axis

43. a vertical shear that transforms points to the right of the y-axis up by an amount 1.5 times their original distance from the y-axis

In **44 – 45**, devise a 3 × 3 matrix that will produce, for any point (x, y, z) in space, the transformation described.

44. D_3

45. $r_{x\text{-}axis}$ *Hint*: Analyze (1, 0, 0), (0, 1, 0), (0, 0, 1).

---

38. $D_k \circ T \overset{?}{=} T \circ D_k$

Multiply the matrices.

$\begin{bmatrix} k & 0 \\ 0 & k \end{bmatrix} \cdot \begin{bmatrix} a & b \\ c & d \end{bmatrix} \overset{?}{=} \begin{bmatrix} a & b \\ c & d \end{bmatrix} \cdot \begin{bmatrix} k & 0 \\ 0 & k \end{bmatrix}$

$\begin{bmatrix} ka & kb \\ kc & kd \end{bmatrix} = \begin{bmatrix} ka & kb \\ kc & kd \end{bmatrix}$

39. All images lie on the line $y = x$.

40. 50°

41. 15°

42. $\begin{bmatrix} 1 & 2 \\ 0 & 1 \end{bmatrix}$

43. $\begin{bmatrix} 1 & 0 \\ 1.5 & 1 \end{bmatrix}$

44. $\begin{bmatrix} 3 & 0 & 0 \\ 0 & 3 & 0 \\ 0 & 0 & 3 \end{bmatrix}$

45. $\begin{bmatrix} 1 & 0 & 0 \\ 0 & -1 & 0 \\ 0 & 0 & -1 \end{bmatrix}$

7.6 *Additional Applications of Matrices*

Decision making in business and industry relies on such elements as supply and demand, and optimal productivity. The simplified examples discussed in this text will serve to familiarize you with the use of matrices to solve such problems.

Example 1

To produce a particular model of automobile, A, it is known that 1 engine, 5 tires, 2 units of steel, and 3 units of miscellaneous other items are necessary. A smaller subcompact model, B, uses the same engine and tires but requires only 1 unit of steel and 2 units of miscellaneous items.

a. Write a matrix that summarizes this information, using the columns to represent each type of material.

b. If monthly output is to be 1,000 of model A and 2,000 of B, determine the total number of units of each of the 4 components necessary for monthly production.

c. If unit costs to the company are $2,000 per engine, $15 per tire, $500 for steel, and $1,000 for miscellaneous items, determine the total cost of monthly production for models A and B.

Solution

a.

	Engine	Tires	Steel	Misc.
Model A	1	5	2	3
Model B	1	5	1	2

Answer: The matrix at the right can be used to represent this data.

$$\begin{bmatrix} 1 & 5 & 2 & 3 \\ 1 & 5 & 1 & 2 \end{bmatrix}$$

b. To compute the total number of units for production of 1,000 of model A and 2,000 of model B, multiply matrices.

$$\begin{bmatrix} 1,000 & 2,000 \end{bmatrix} \begin{bmatrix} 1 & 5 & 2 & 3 \\ 1 & 5 & 1 & 2 \end{bmatrix} = \begin{bmatrix} 3,000 & 15,000 & 4,000 & 7,000 \end{bmatrix}$$

Answer: 3,000 engines, 15,000 tires, 4,000 units of steel, and 7,000 units of miscellaneous items are needed.

c. To compute the total cost, multiply the production, materials, and cost matrices together, in that order. (Use the result of part **b** for the first product.)

$$\begin{bmatrix} 1,000 & 2,000 \end{bmatrix} \begin{bmatrix} 1 & 5 & 2 & 3 \\ 1 & 5 & 1 & 2 \end{bmatrix} \begin{bmatrix} 2,000 \\ 15 \\ 500 \\ 1,000 \end{bmatrix} = \begin{bmatrix} 15,225,000 \end{bmatrix}$$

Answer: The total cost is $15,225,000.

Economists use matrices to study trends and create models of the economy of a particular locale. An example of this type of model is the Leontief input-output model, named for Wassily Leontief, a Russian mathematician who received a Nobel Prize in economics in 1973 for its creation.

The Leontief Input-Output Model

The economy of a locality or country may be thought of as consisting of items that are produced and the processes necessary for production. The processes used to produce items also make use of these same items. For example, it is necessary to use computers to design and manufacture other computers and it is necessary to use machinery to build other machinery.

The items of production that are referred to in this model usually include major items such as steel, power, labor, or food. A *technology input-output matrix* is used to represent the interrelationships among the items produced and the processes necessary to produce them. Leontief's original model included 81 industrial sectors of the economy, but for this discussion, in order to understand the theory, we will use an example with 4 sectors. The sectors will be basic metal production (BM), basic nonmetal production (BN), energy (E), and services (S).

Study the input-output matrix shown. The entries are in millions of dollars. For example, the entry 0.01 in row 1 column 3 indicates that 0.01 × $1,000,000 or $10,000 worth of energy input is necessary for $1,000,000 of basic metal output. Similarly, 0.05 or $50,000 worth of services are necessary to produce 1 unit of $1,000,000 worth of basic nonmetal.

$$
\begin{array}{c}
\text{O} \\ \text{U} \\ \text{T} \\ \text{P} \\ \text{U} \\ \text{T}
\end{array}
\begin{array}{c}
\\ \text{BM} \\ \text{BN} \\ \text{E} \\ \text{S}
\end{array}
\overset{\text{INPUT}}{
\begin{array}{cccc}
\text{BM} & \text{BN} & \text{E} & \text{S}
\end{array}}
\\
\begin{bmatrix}
0.5 & 0.01 & 0.01 & 0.02 \\
0.02 & 0.4 & 0.01 & 0.05 \\
0.04 & 0.03 & 0.4 & 0.03 \\
0.1 & 0.1 & 0.2 & 0.2
\end{bmatrix}
$$

In order to produce x units of E, we would have to input $0.04x$ of BM, $0.03x$ of BN, $0.4x$ of E, and $0.03x$ of S. If we represent the production needs of each of the 4 sectors by $x_{BM}, x_{BN}, x_E,$ and x_S, where x is the number of units and each unit represents $1,000,000, then it is possible to define what is called an *intensity matrix X*,

$$
X = \begin{bmatrix} x_{BM} \\ x_{BN} \\ x_E \\ x_S \end{bmatrix}, \text{ to summarize needs of production.}
$$

Suppose $X = \begin{bmatrix} 10 \\ 5 \\ 2 \\ 4 \end{bmatrix}$. The amount of each of the sectors being used up to produce each of

the others can be determined by multiplying the input-output matrix by the intensity matrix.

$$
\begin{bmatrix}
0.5 & 0.01 & 0.01 & 0.02 \\
0.02 & 0.4 & 0.01 & 0.05 \\
0.04 & 0.03 & 0.4 & 0.03 \\
0.1 & 0.1 & 0.2 & 0.2
\end{bmatrix}
\begin{bmatrix} 10 \\ 5 \\ 2 \\ 4 \end{bmatrix}
=
\begin{bmatrix} 5.15 \\ 2.42 \\ 1.47 \\ 2.7 \end{bmatrix}
$$

The column matrix obtained indicates the total number of units of each sector being used to meet the needs of production indicated by the intensity matrix. This column matrix, C, representing the number of units being consumed, will be referred to as a *consumption matrix*. Therefore $C = T \cdot X$, where T is the technology input-output matrix and X is the intensity matrix.

The total value of goods being consumed in production, represented by C, can be subtracted from X to obtain the *demand matrix D*. In our example,

$$D = X - C = \begin{bmatrix} 10 \\ 5 \\ 2 \\ 4 \end{bmatrix} - \begin{bmatrix} 5.15 \\ 2.42 \\ 1.47 \\ 2.7 \end{bmatrix} = \begin{bmatrix} 4.85 \\ 2.58 \\ 0.53 \\ 1.3 \end{bmatrix}$$

Using matrix algebra we can derive a useful relationship between the technology input-output matrix and the demand matrix D.

$D = X - C$	The demand matrix is the difference of the intensity and consumption matrices.
$= X - TX$	Substitute for C.
$= IX - TX$	Rewrite X using I, the identity matrix.
$= (I - T)X$	Factor.
$(I - T)^{-1}D = (I - T)^{-1}(I - T)X$	Left-multiply by the inverse of the coefficient of X.
$(I - T)^{-1}D = IX$	Definition of inverse.
$(I - T)^{-1}D = X$	Definition of identity.

Thus, $X = (I - T)^{-1}D$ when $(I - T)^{-1}$ exists.

In other words, if the technology matrix, T, and the demand are known, the intensity matrix X, which indicates the number of units of each sector necessary, can be computed.

Example 2

Use the accompanying technology input-output matrix, T, for the economy of three sectors M, N, and P.

$$T = \begin{array}{c} \\ \text{O} \\ \text{U} \\ \text{T} \\ \text{P} \\ \text{U} \\ \text{T} \end{array} \begin{array}{c} \\ \\ M \\ N \\ P \\ \\ \end{array} \begin{array}{ccc} \text{INPUT} & & \\ M & N & P \\ \end{array} \begin{bmatrix} 0.2 & 0.5 & 0.1 \\ 0.0 & 0.2 & 0.6 \\ 0.2 & 0.3 & 0.3 \end{bmatrix}$$

a. Find how many units of M are necessary to produce 1 unit of N.

b. How many units of P are necessary to produce 1 unit of M?

c. Determine how many units of N are necessary to produce 4, 6, and 5 units of M, N, and P, respectively.

d. Use the formula $X = (I - T)^{-1}D$ to determine the intensity matrix X, given a demand matrix D of $\begin{bmatrix} 5 \\ 6 \\ 1 \end{bmatrix}$. Estimate each value to the nearest tenth.

e. Determine the amount of each of the sectors being consumed to produce each of the others.

Solution

a. The first column of entries indicates how many units of M produce 1 unit of $M, N,$ and P. The second row first column indicates that 0 units of M are necessary to produce 1 unit of N.

b. The third column shows the number of units of P necessary to produce 1 unit of each sector. Thus, 0.1 units of P produce 1 unit of M.

c. Reading down column N and multiplying by 4, 6, and 5, respectively:

$$4(0.5) + 6(0.2) + 5(0.3) = 4.7 \text{ units of } N$$

The same result is obtained using matrix multiplication.

$$\begin{bmatrix} 4 & 6 & 5 \end{bmatrix} \begin{bmatrix} 0.5 \\ 0.2 \\ 0.3 \end{bmatrix} = \begin{bmatrix} 4.7 \end{bmatrix}$$

d. $(I - T)^{-1}D = X$

$$\left(\begin{bmatrix} 1 & 0 & 0 \\ 0 & 1 & 0 \\ 0 & 0 & 1 \end{bmatrix} - \begin{bmatrix} 0.2 & 0.5 & 0.1 \\ 0.0 & 0.2 & 0.6 \\ 0.2 & 0.3 & 0.3 \end{bmatrix} \right)^{-1} \cdot \begin{bmatrix} 5 \\ 6 \\ 1 \end{bmatrix}$$

$$\begin{bmatrix} 0.8 & -0.5 & -0.1 \\ 0 & 0.8 & -0.6 \\ -0.2 & -0.3 & 0.7 \end{bmatrix}^{-1} \cdot \begin{bmatrix} 5 \\ 6 \\ 1 \end{bmatrix}$$

The inverse exists and may be obtained by using a calculator. Values are shown to the nearest tenth.

$$\begin{bmatrix} 1.7 & 1.7 & 1.7 \\ 0.5 & 2.4 & 2.1 \\ 0.7 & 1.5 & 2.8 \end{bmatrix} \cdot \begin{bmatrix} 5 \\ 6 \\ 1 \end{bmatrix} = \begin{bmatrix} 20.0 \\ 19.0 \\ 15.3 \end{bmatrix} = X$$

e. The column matrix C that represents the number of units consumed in production may be obtained from the product of the input-output matrix and the intensity matrix

$$T \qquad \cdot \quad X \quad = \quad C$$

$$\begin{bmatrix} 0.2 & 0.5 & 0.1 \\ 0.0 & 0.2 & 0.6 \\ 0.2 & 0.3 & 0.3 \end{bmatrix} \begin{bmatrix} 20.0 \\ 19.0 \\ 15.3 \end{bmatrix} = \begin{bmatrix} 15.0 \\ 13.0 \\ 14.3 \end{bmatrix}$$

or from the difference between the intensity matrix and the demand matrix.

$$X \quad - \quad D \quad = \quad C$$

$$\begin{bmatrix} 20.0 \\ 19.0 \\ 15.3 \end{bmatrix} - \begin{bmatrix} 5 \\ 6 \\ 1 \end{bmatrix} = \begin{bmatrix} 15.0 \\ 13.0 \\ 14.3 \end{bmatrix}$$

Therefore, approximately 15.0 units of M, 13.0 units of N, and 14.3 units of P are consumed.

Secret Codes

Codes are used in many areas. For example, codes are used to secure the design of semiconductor chips and to allow depositors access to bank-teller machines. Virtually all codes involve some use of mathematics. Matrices are a mathematical tool often used for coding and decoding.

Think of a code as having two components, the set of instructions for encoding a message and the instructions for translating the message.

For example, suppose numbers are assigned to each letter of the alphabet as follows, with 0 used for a blank space.

A = 1	E = 5	I = 13	M = 21	Q = 30	U = 34	Y = 43
B = 2	F = 10	J = 14	N = 22	R = 31	V = 40	Z = 44
C = 3	G = 11	K = 15	O = 23	S = 32	W = 41	space = 0
D = 4	H = 12	L = 20	P = 24	T = 33	X = 42	

This set of instructions could then be used to translate the following message:

$$M \quad A \quad T \quad H \; \_ \; I \quad S \; \_ \; B \quad E \quad A \quad U \quad T \quad I \quad F \quad U \quad L \; \_$$
$$21/1/33/12/0/13/32/0/2/5/1/34/33/13/10/34/20/0$$

This row of numbers can then be divided into row matrices of a particular dimension, say 2, with an extra space after the message to complete the last row matrix.

$$\begin{bmatrix} 21 & 1 \end{bmatrix}\begin{bmatrix} 33 & 12 \end{bmatrix}\begin{bmatrix} 0 & 13 \end{bmatrix}\begin{bmatrix} 32 & 0 \end{bmatrix}\begin{bmatrix} 2 & 5 \end{bmatrix}\begin{bmatrix} 1 & 34 \end{bmatrix}\begin{bmatrix} 33 & 13 \end{bmatrix}\begin{bmatrix} 10 & 34 \end{bmatrix}\begin{bmatrix} 20 & 0 \end{bmatrix}$$

A 2 × 2 invertible matrix can be used to code the message. For example, given the encoding matrix D. In order to code the message, multiply each of the 1 × 2 matrices by the 2 × 2 coding matrix D.

$$D = \begin{bmatrix} 13 & 4 \\ 6 & 2 \end{bmatrix}$$

$$\begin{bmatrix} 21 & 1 \end{bmatrix}D\begin{bmatrix} 33 & 12 \end{bmatrix}D\begin{bmatrix} 0 & 13 \end{bmatrix}D\begin{bmatrix} 32 & 0 \end{bmatrix}D\begin{bmatrix} 2 & 5 \end{bmatrix}D\begin{bmatrix} 1 & 34 \end{bmatrix}D\begin{bmatrix} 33 & 13 \end{bmatrix}D$$
$$\begin{bmatrix} 10 & 34 \end{bmatrix}D\begin{bmatrix} 20 & 0 \end{bmatrix}D$$

$$=\begin{bmatrix} 279 & 86 \end{bmatrix}\begin{bmatrix} 501 & 156 \end{bmatrix}\begin{bmatrix} 78 & 26 \end{bmatrix}\begin{bmatrix} 416 & 128 \end{bmatrix}\begin{bmatrix} 56 & 18 \end{bmatrix}\begin{bmatrix} 217 & 72 \end{bmatrix}\begin{bmatrix} 507 & 158 \end{bmatrix}$$
$$\begin{bmatrix} 334 & 108 \end{bmatrix}\begin{bmatrix} 260 & 80 \end{bmatrix}$$

The coded message would be received in the form of these last nine 1 × 2 matrices.

The person or machine receiving the message could then decode it by using D^{-1}. For example, to decode the first two coded matrices:

$$D^{-1} = \frac{1}{26-24}\begin{bmatrix} 2 & -4 \\ -6 & 13 \end{bmatrix}$$

$$= \begin{bmatrix} 1 & -2 \\ -3 & 6.5 \end{bmatrix}$$

$$\begin{bmatrix} 279 & 86 \end{bmatrix}\begin{bmatrix} 1 & -2 \\ -3 & 6.5 \end{bmatrix} = \begin{bmatrix} 21 & 1 \end{bmatrix}$$
$$\qquad\qquad\qquad\qquad\qquad\;\; M \quad A$$

$$\begin{bmatrix} 501 & 156 \end{bmatrix}\begin{bmatrix} 1 & -2 \\ -3 & 6.5 \end{bmatrix} = \begin{bmatrix} 33 & 12 \end{bmatrix}$$
$$\qquad\qquad\qquad\qquad\qquad\;\; T \quad H$$

Exercises 7.6

In **1 – 2**, develop a matrix that can be used to summarize the materials needed to produce each of the items described. Each row of the matrix should be used to represent the components for a single item.

For example, the matrix shown states that 20 grams of gold and 10 grams of silver are needed for a necklace, while 20 grams of gold and 40 grams of silver are needed for a bracelet.

$$\begin{array}{c} \\ \text{necklace} \\ \text{bracelet} \end{array} \begin{array}{cc} \textbf{Gold} & \textbf{Silver} \\ \begin{bmatrix} 20 & 10 \\ 20 & 40 \end{bmatrix} \end{array}$$

1. One unit of a therapeutic multi-vitamin requires 6 units of vitamin A, 3 of vitamin C, and 12 of stabilizers. One unit of a lesser strength vitamin requires 3 units of A, 2 of C, and 10 of stabilizers.

2. An instrument manufacturer produces three kinds of flutes. The least expensive requires 10 units of nickel and 1 unit of silver. The middle-priced flute uses 5 units each of nickel and silver. The costliest uses 7 units of silver, 2 of nickel, and 1 of gold.

3. A manufacturer produces a 40 megabyte computer and an 80 megabyte computer. The 40MB computer uses 1 unit of steel, 2 units of electronic components, and 3 units of miscellaneous items. The 80MB computer requires 1.1 units of steel, 3 of electronic components, and 5 of miscellaneous items.

If the manufacturer expects to produce 50 of the 40MB computers and 20 of the 80MB computers each week, use matrices to determine how many units of steel, electronic components, and miscellaneous components are needed for the week's production.

4. An independent automobile manufacturer wishes to analyze the labor cost in designing and manufacturing two types of automobiles, the Knight and the Zephyr. He estimates that the Knight will require 2,000 hours of engineering design time and the Zephyr, a faster model, will require 2,200 hours of engineering design time. Manufacturing the Knight will require 1,500 person-hours of assembly-line work and the Zephyr will require 2,000 person-hours of assembly-line work. The Knight will require 5,000 hours of robot time and the Zephyr, 7,000 hours of robot time.

a. If the projected monthly production is to be 50 Knights and 30 Zephyrs, use matrices to compute the total number of design hours, person-hours for assembly, and robot hours for assembly.

b. If the manufacturer estimates design costs at $60/hour, the assembly costs at $25/hour, and the robot costs at $5/hour, compute the costs for these three items at the end of the first month.

1. $\begin{bmatrix} 6 & 3 & 12 \\ 3 & 2 & 10 \end{bmatrix}$

2. $\begin{bmatrix} 10 & 1 & 0 \\ 5 & 5 & 0 \\ 7 & 2 & 1 \end{bmatrix}$

3. 72 units of steel
160 units of electronic components
250 units of miscellaneous components

4. a. 166,000 hours of engineering time
135,000 hours of assembly-line time
460,000 hours of robot time

b. $9,960,000 for engineering time
$3,375,000 for assembly-line time
$2,300,000 for robot time

Exercises 7.6 (continued)

5. Given a technology input-output matrix for two sectors of the economy, goods (G) and services (S).

$$T = \begin{array}{c} \\ G \\ S \end{array} \begin{array}{cc} G & S \\ \left[\begin{array}{cc} 0.4 & 0.2 \\ 0.5 & 0.1 \end{array} \right] \end{array}$$

a. How many units of goods are needed to produce 1 unit of services?

b. How many units of services are needed to produce 1 unit of goods?

c. How many units of services must be input to produce 4 units of goods and 3 units of services?

d. Compute $(I - T)^{-1}$ and use it to find the intensity matrix of goods and services that must be input to meet a demand for 5 units of goods and 8 units of services. Answer to the nearest tenth.

6. Given a technology matrix involving energy (E) and transportation (P), with each unit representing $1,000,000.

$$T = \begin{array}{c} \\ E \\ P \end{array} \begin{array}{cc} E & P \\ \left[\begin{array}{cc} 0.5 & 0.1 \\ 0.2 & 0.1 \end{array} \right] \end{array}$$

a. How many dollars worth of transportation must be used to produce $1,000,000 worth of energy?

b. How many dollars worth of energy must be input to the system to produce $3,000,000 worth of energy?

c. How much energy input, in dollars, is necessary to produce $6 million worth of energy and $5 million worth of transportation?

d. Compute each entry of $(I - T)^{-1}$ to the nearest tenth and use it to find the intensity matrix X, which reflects a demand of $10 million each of energy and transportation.

7. Given a technology matrix T for 3 sectors of the post-World-War II U.S. economy, agriculture (A), manufacturing (M), and domestic goods (D).

$$T = \begin{array}{c} \\ A \\ M \\ D \end{array} \begin{array}{ccc} A & M & D \\ \left[\begin{array}{ccc} 0.2 & 0.1 & 0.05 \\ 0.1 & 0.3 & 0.3 \\ 0.4 & 0.4 & 0.01 \end{array} \right] \end{array}$$

a. How many units of agriculture need to be input to the system to produce 3 units of agriculture and 5 units each of manufacturing and domestic goods?

b. For the intensity matrix $\begin{bmatrix} 4 \\ 6 \\ 2 \end{bmatrix}$, find the column matrix C that represents the number of units of each sector consumed.

c. If $(I - T)^{-1} = \begin{bmatrix} 1.3 & 0.3 & 0.2 \\ 0.5 & 1.8 & 0.6 \\ 0.8 & 0.9 & 1.3 \end{bmatrix}$, compute an intensity matrix for a demand of 6 units of agriculture, 2 of manufacturing, and 1 of domestic goods.

5. a. 0.5

b. 0.2

c. 1.1

d. $\begin{bmatrix} 13.9 \\ 16.6 \end{bmatrix}$

6. a. $100,000

b. $1,500,000

c. $4,000,000

d. $(I - T)^{-1} = \begin{bmatrix} 2.1 & 0.2 \\ 0.5 & 1.2 \end{bmatrix}$

$X = \begin{bmatrix} 23 \\ 17 \end{bmatrix}$

7. a. 3.1

b. $\begin{bmatrix} 1.5 \\ 2.8 \\ 4.02 \end{bmatrix}$

c. $\begin{bmatrix} 8.6 \\ 7.2 \\ 7.9 \end{bmatrix}$

Exercises 7.6 *(continued)*

8. For 1958, Leontief used numbers approximately equal to those in the technology matrix shown to describe apparel, food, drugs, and furniture, which he called final nonmetal (FN); motor vehicles, aircraft, and communications, which he called final metal (FM) and transportation, insurance, real estate, and finance, which he called services (S). Each unit equals $1 million.

$$T = FM \begin{array}{c} \\ FN \\ FM \\ S \end{array} \begin{bmatrix} \overset{FN}{0.170} & \overset{FM}{0.004} & \overset{S}{0.008} \\ 0.003 & 0.295 & 0.016 \\ 0.120 & 0.074 & 0.234 \end{bmatrix}$$

a. According to this model, what was the value of services (S) that had to be input in 1958 to produce $3 million worth of final nonmetal and $2 million worth of final metal?

b. If the matrix $X = \begin{bmatrix} 100 \\ 50 \\ 10 \end{bmatrix}$ is an intensity matrix for some time during 1958, find the column matrix C that represents the number of units of each sector consumed.

c. If $(I - T)^{-1} = \begin{bmatrix} 1.200 & 0.008 & 0.01 \\ 0.009 & 1.400 & 0.03 \\ 0.190 & 0.140 & 1.30 \end{bmatrix}$, find the value of each sector from the intensity matrix that would occur if the demand matrix was $\begin{bmatrix} 200 \\ 100 \\ 10 \end{bmatrix}$.

Answer to the nearest million dollars.

In **9 – 11**, use the number assignments listed below.

A = 10	E = 20	I = 30	M = 41	Q = 51	U = 61	Y = 71
B = 11	F = 21	J = 31	N = 42	R = 52	V = 62	Z = 72
C = 12	G = 22	K = 32	O = 43	S = 53	W = 63	space = 0
D = 13	H = 23	L = 40	P = 50	T = 60	X = 70	

9. a. Assign numbers to the message:
"MATRIX MULT IS NOT COMM"

b. Use the coding matrix $\begin{bmatrix} 10 & -5 \\ 10 & 5 \end{bmatrix}$ to code the message.

c. Find the matrix necessary to decode the message.

d. Use this matrix to decode the coded message:
$$\begin{bmatrix} 720 & -160 \end{bmatrix} \begin{bmatrix} 130 & -65 \end{bmatrix}$$

10. The coding matrix $\begin{bmatrix} 4 & 3 \\ 5 & 4 \end{bmatrix}$ was used to code the message:

$$\begin{bmatrix} 290 & 220 \end{bmatrix} \begin{bmatrix} 228 & 176 \end{bmatrix} \begin{bmatrix} 60 & 48 \end{bmatrix} \begin{bmatrix} 482 & 377 \end{bmatrix} \begin{bmatrix} 340 & 268 \end{bmatrix}$$

a. Find the decoding matrix and use it to decipher the message.

b. Use the same decoding matrix to decipher the following message:
499 385 244 183 300 238 80 60
417 323 300 238 240 180

11. The coding matrix $\begin{bmatrix} 1 & 2 & 3 \\ 1 & -1 & 1 \\ 1 & 1 & -1 \end{bmatrix}$ was used to code the following message:

111 132 73 53 76 39 75 128 137 82 94 166 60 120 180
Decode the message.

8. a. $56,000

b. $\begin{bmatrix} 17.28 \\ 15.21 \\ 18.04 \end{bmatrix}$

c. $241,000,000 of final nonmetal
$142,000,000 of final metal
$65,000,000 of services

9. a. 41|10|60|52|30|70|0|41|61|40|60|0|
30|53|0|42|43|60|0|12|43|41|41|0

b. 510|–155|1,120|–40|1,000|200|410|205|
1,010|–105|600|–300|830|115|420|210|
1,030|85|120|60|840|–10|410|–205

9. c. $\begin{bmatrix} 0.05 & 0.05 \\ -0.1 & 0.1 \end{bmatrix}$

d. RED_

10. a. $\begin{bmatrix} 4 & -3 \\ -5 & 4 \end{bmatrix}$ TAKE COVER

b. YOU ARE SMART

11. MATH IS GREAT

Chapter 7   *Summary and Review*

Matrix Notation

- An $m \times n$ matrix is a rectangular array of numbers with m rows and n columns.
- A double-subscripted variable such as a_{ij} is used to represent the entry in the ith and jth column of the matrix.
- A square matrix is a matrix that has an equal number of rows and columns.
- The main diagonal of a square matrix with dimension $n \times n$ consists of the elements of the matrix $a_{11}, a_{22}, \dots a_{nn}$.
- A zero matrix is a matrix each of whose entries is zero.

Determinants

- The determinant of a square matrix is a number associated with the matrix and designated by placing vertical bars around the array of numbers of the matrix.
$$| A | = \det (A)$$
- $\det \begin{bmatrix} a_{11} \end{bmatrix} = a_{11}$

$$\det \begin{bmatrix} a_{11} & a_{12} \\ a_{21} & a_{22} \end{bmatrix} = a_{11} \cdot a_{22} - a_{12} \cdot a_{21}$$
- Determinants of square matrices with order greater than 2 are computed by using expansion by minors. In order to minimize computation, the expansion should be done using the row or column in the matrix with the greatest number of zeros.
- A triangular matrix is a matrix that has only zeros for all the entries above or below its main diagonal.
- The determinant of a triangular matrix is equal to the product of the elements of the main diagonal of the matrix.

Matrix Operations

- If S and T are each $m \times n$ matrices, then $S + T$ is an $m \times n$ matrix, where each entry is the sum of the corresponding entries in S and T.
- If a matrix S is multiplied by a scalar k, then the result is kS, where each entry is k times the corresponding entry of S.
- The additive inverse of a matrix T is $-T$, where each entry is the additive inverse of the corresponding element of T.

- Matrix subtraction: $S - T = S + (-T)$

- Matrix multiplication: The product of an $m \times n$ matrix S and an $n \times p$ matrix T is the $m \times p$ matrix ST, where the entry in the ith row and jth column is computed as the dot product of the ith row of S and the jth column of T.

- Matrix multiplication is *not* commutative.

- An identity matrix is a square matrix with elements of 1 on the main diagonal and 0 as all other entries.

- The $n \times n$ identity matrix is the multiplicative identity for the set of $n \times n$ matrices. If A is an $n \times n$ matrix and I is the $n \times n$ identity matrix, then
    $$AI = IA = A$$

- If S and T are $n \times n$ matrices and $ST = I$, then S and T are inverses of each other.

- The inverse of the 2×2 matrix $A = \begin{bmatrix} a & b \\ c & d \end{bmatrix}$ is
    $$A^{-1} = \frac{1}{|A|} \begin{bmatrix} d & -b \\ -c & a \end{bmatrix}$$
    where $|A| = ad - bc$ is the determinant of A.

Solving Systems Using Matrices

- Cramer's Rule: If a set of n linear equations in n variables has a coefficient matrix C with a nonzero determinant $|C|$, then the solution of the system is
    $$x_1 = \frac{C_1}{|C|}, \, x_2 = \frac{C_2}{|C|}, \, \dots, \, x_n = \frac{C_n}{|C|}$$
    where C_i is the matrix formed by replacing the ith column of C by the constants in the system.

- The augmented matrix for a system of equations consists of the coefficient matrix with the constants of the system affixed as an additional column.

- To obtain a matrix of an equivalent system, the following row operations may be performed on a matrix that represents a system of linear equations:
    1. Multiplying a row by a nonzero constant.
    2. Adding a multiple of one row to another.
    3. Interchanging two rows.

- If $AX = C$ is a matrix equation of a system of linear equations with A the coefficient matrix, X the variable matrix, and C the constant matrix, and if A^{-1} exists, then the solution for the system is
    $$X = A^{-1}C$$

- The inverse of a matrix A may be found by using row operations on the matrix $\left[A \vdots I \right]$, where I is the identity matrix whose order is the same as A, until it is transformed into $\left[I \vdots A^{-1} \right]$.

- An invertible matrix is a matrix that has an inverse. A singular matrix is a matrix that has no inverse.

Transformations Using Matrices

- A point P and its image P' can be represented by matrices.

$$P = \begin{bmatrix} x \\ y \end{bmatrix} \qquad P' = \begin{bmatrix} x' \\ y' \end{bmatrix}$$

- To obtain the image of a point under a given transformation, left-multiply by the required transformation matrix.

Reflection over the x-axis: $\begin{bmatrix} 1 & 0 \\ 0 & -1 \end{bmatrix} \begin{bmatrix} x \\ y \end{bmatrix} = \begin{bmatrix} x \\ -y \end{bmatrix}$

Reflection over the y-axis: $\begin{bmatrix} -1 & 0 \\ 0 & 1 \end{bmatrix} \begin{bmatrix} x \\ y \end{bmatrix} = \begin{bmatrix} -x \\ y \end{bmatrix}$

Reflection over the line $y = x$: $\begin{bmatrix} 0 & 1 \\ 1 & 0 \end{bmatrix} \begin{bmatrix} x \\ y \end{bmatrix} = \begin{bmatrix} y \\ x \end{bmatrix}$

Rotation of $90°$ with center of rotation at the origin: $\begin{bmatrix} 0 & -1 \\ 1 & 0 \end{bmatrix} \begin{bmatrix} x \\ y \end{bmatrix} = \begin{bmatrix} -y \\ x \end{bmatrix}$

Rotation of $180°$ with center of rotation at the origin: $\begin{bmatrix} -1 & 0 \\ 0 & -1 \end{bmatrix} \begin{bmatrix} x \\ y \end{bmatrix} = \begin{bmatrix} -x \\ -y \end{bmatrix}$

Dilation of k: $\begin{bmatrix} k & 0 \\ 0 & k \end{bmatrix} \begin{bmatrix} x \\ y \end{bmatrix} = \begin{bmatrix} kx \\ ky \end{bmatrix}$

Scale change: $\begin{bmatrix} a & 0 \\ 0 & b \end{bmatrix} \begin{bmatrix} x \\ y \end{bmatrix} = \begin{bmatrix} ax \\ by \end{bmatrix}$

Simple shear: $\begin{bmatrix} 1 & a \\ 0 & 1 \end{bmatrix} \begin{bmatrix} x \\ y \end{bmatrix} = \begin{bmatrix} x + ay \\ y \end{bmatrix}$ or $\begin{bmatrix} 1 & 0 \\ b & 1 \end{bmatrix} \begin{bmatrix} x \\ y \end{bmatrix} = \begin{bmatrix} x \\ bx + y \end{bmatrix}$

Other Applications of Matrices

- The Leontief input-output model is used to study an economy. The important matrix elements of this model are:

  T, a technology matrix that shows the number of units of each component needed to produce one unit of another

  X, the intensity matrix that shows the production need for each component

  C, the consumption matrix that shows the numbers of units used up during production

  D, the demand matrix that shows the number of units produced to meet anticipated demand

- If an invertible matrix A and a numbering scheme that is a one-to-one correspondence between the letters of the alphabet and a subset of the integers is used to code messages, then A^{-1} can be used to decode them.

1. a. 6
   b. 1
2. 62
3. 7
4. 30
5. 9
6. −192
7. 6

8. $\begin{bmatrix} 1 & 3 \\ 8 & 4 \end{bmatrix}$

9. $\begin{bmatrix} 3 & -7 \\ 0 & 8 \end{bmatrix}$

10. $\begin{bmatrix} -6 & 3 \\ 18 & 11 \end{bmatrix}$

11. $\begin{bmatrix} 8 & 12 \\ 8 & -3 \end{bmatrix}$

12. $\begin{bmatrix} 32 & 28 \\ 2 & -17 \end{bmatrix}$

13. $\begin{bmatrix} 12 & 18 \\ 43 & 16 \end{bmatrix}$

14. undefined;
different dimensions

15. $\begin{bmatrix} 18 & 19 & 23 \\ 56 & 20 & 8 \\ -18 & -11 & -10 \end{bmatrix}$

16. $\begin{bmatrix} -4 & 5 \\ 5 & -6 \end{bmatrix}$

17. $\begin{bmatrix} 3 & 1 \\ -\frac{1}{2} & 0 \end{bmatrix}$

18. no inverse;
determinant = 0

19. $x = \frac{15}{4}, y = -\frac{19}{4}$

20. $x = 1, y = \frac{1}{2}$

21. $x = 3, y = 2$

22. $a = \frac{1}{4}, b = \frac{9}{4}, c = -1$

23. $x = 4, y = -5$

24. $x = 1, y = -\frac{1}{2}, z = \frac{3}{2}$

25. $a = -\frac{8}{5}, b = -\frac{3}{2}, c = \frac{4}{5}$

26. $x = 2, y = \frac{1}{5}, z = 0, w = \frac{4}{5}$

27. $f(x) = x^3 + 9x^2 - 11x + 5$

28. 1 small crate
3 medium crates
5 large crates

29. consistent

30. dependent

31. inconsistent

32. $x = 6, y = 7$

33. $a = \frac{3}{2}, b = \frac{1}{4}$

34. $x = 1, y = 3, z = -4$

35. $x = 6, y = -1, z = 2$

36. singular

37. a. $\begin{bmatrix} 0 & 12 & 15 & 9 \\ 0 & 0 & 18 & 6 \end{bmatrix}$

37. b.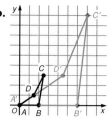

c. dilation of scale factor 3

Chapter 7   Review Exercises

1. If $A = \begin{bmatrix} 1 & 4 & 6 & 5 \\ -1 & 3 & 7 & 8 \\ 4 & 1 & 6 & -2 \end{bmatrix}$ find: **a.** a_{13} **b.** a_{32}

In **2 – 7**, find the value of the determinant of the matrix.

2. $\begin{bmatrix} 10 & 8 \\ 1 & 7 \end{bmatrix}$
    3. $\begin{bmatrix} -1 & 6 \\ -2 & 5 \end{bmatrix}$
    4. $\begin{bmatrix} 5 & 0 & 7 \\ 1 & 6 & 2 \\ -1 & 4 & 0 \end{bmatrix}$

5. $\begin{bmatrix} 3 & 0 & 0 \\ 6 & 1 & 8 \\ -4 & 0 & 3 \end{bmatrix}$
  6. $\begin{bmatrix} 1 & 6 & 0 & 0 \\ 3 & -1 & 0 & 4 \\ 2 & 0 & 0 & 0 \\ 9 & 1 & 4 & 3 \end{bmatrix}$
  7. $\begin{bmatrix} 1 & 18 & 20 \\ 0 & 3 & 44 \\ 0 & 0 & 2 \end{bmatrix}$

In **8 – 15**, use the matrices below to find the matrix in the expression. If the expression is undefined, explain why.

$S = \begin{bmatrix} 1 & -1 \\ 2 & 3 \end{bmatrix}$ $T = \begin{bmatrix} 0 & 4 \\ 6 & 1 \end{bmatrix}$ $U = \begin{bmatrix} 2 & 3 & 4 \\ 6 & 1 & -1 \end{bmatrix}$ $V = \begin{bmatrix} 6 & 1 \\ 4 & 8 \\ -3 & -2 \end{bmatrix}$

8. $S + T$ **9.** $3S - T$ **10.** ST **11.** TS
12. TS^2 **13.** UV **14.** $S + V$ **15.** VU

In **16 – 18**, find the inverse of the matrix or explain why there is no inverse.

16. $\begin{bmatrix} 6 & 5 \\ 5 & 4 \end{bmatrix}$
    17. $\begin{bmatrix} 0 & -2 \\ 1 & 6 \end{bmatrix}$
    18. $\begin{bmatrix} 6 & 3 \\ 4 & 2 \end{bmatrix}$

In **19 – 22**, solve the system by using Cramer's Rule.

19. $\begin{aligned} x + y &= -1 \\ 3x - 5y &= 35 \end{aligned}$
 20. $\begin{aligned} 5x + 6y &= 8 \\ -13x + 12y &= -7 \end{aligned}$
 21. $\begin{aligned} x &= y + 1 \\ 3x + 5y &= 19 \end{aligned}$
 22. $\begin{aligned} a - b &= -2 \\ 3a + b - 2c &= 5 \\ 2a + 2b + c &= 4 \end{aligned}$

In **23 – 26**, use row operations on the augmented matrix to solve the system.

23. $\begin{aligned} 3x + 2y &= 2 \\ x + y &= -1 \end{aligned}$
    24. $\begin{aligned} 6x - y - 3z &= 2 \\ 2x - 2y + 10z &= 18 \\ x - 2y + 4z &= 8 \end{aligned}$

25. $\begin{aligned} -3a + 2b - c &= 1 \\ a + 2c &= 0 \\ 4a - 4b + 3c &= 2 \end{aligned}$
  26. $\begin{aligned} 2x + 2y + 2z - 3w &= 2 \\ 3x + 6y - 2z + w &= 8 \\ x + y - 3z - 4w &= -1 \\ 2x + y + 5z + w &= 5 \end{aligned}$

27. If $f(x)$ is a third-degree function and $f(0) = 5$, $f(1) = 4$, $f(2) = 27$, $f(3) = 80$, find $f(x)$.

28. A company manufactures Model 286, Model 386, and Model 486 computers. Depending on the size of the order, the computers are transported in small, medium, or large crates. A small crate can hold 15 Model 286, 10 Model 386, and 8 Model 486. A medium-size crate can hold 70 Model 286, 50 Model 386, and 35 Model 486. A large crate can hold 130 Model 286, 80 Model 386, and 85 Model 486. If an order of 875 Model 286, 560 Model 386, and 538 Model 486 computers must be shipped to one place, how many of each type of crate should be used to fill the order?

In **29 – 31**, determine whether the system is consistent, inconsistent or dependent.

29. $x + y = 14$
    $x - y = 10$

30. $3x + 4y = 1$
    $6x + 8y = 2$

31. $4x - y = 3$
    $4x - y = 6$

In **32 – 35**, use the inverse matrix to solve the system.

32. $4x + y = 31$
    $3x - 2y = 4$

33. $4a - 2b = 5.5$
    $-3a + 6b = -3$

34. $x + y + z = 0$
    $x - y + 2z = -10$
    $2x - y - z = 3$

35. $2x - y = 13$
    $x - 2y + z = 10$
    $3x + z = 20$

36. Determine if the matrix $\begin{bmatrix} 10 & 12 \\ 2 & 2.4 \end{bmatrix}$ is invertible or singular.

In **37 – 39**:

a. Determine the image $A'B'C'D'$ of quadrilateral $ABCD$, $A(0, 0)$, $B(4, 0)$, $C(5, 6)$, $D(3, 2)$ under the given transformation.

b. Draw $ABCD$ and $A'B'C'D'$ on a graph.

c. Indicate the type of transformation.

37. $\begin{bmatrix} 3 & 0 \\ 0 & 3 \end{bmatrix}$

38. $\begin{bmatrix} 0 & 1 \\ -1 & 0 \end{bmatrix}$

39. $\begin{bmatrix} 1 & 0 \\ 3 & 1 \end{bmatrix}$

40. Write a single matrix that can be used to rotate a point $-270°$.

41. Write a product of three matrices equivalent to $D_3 \circ r_{x\text{-}axis}(x, y)$.

42. Given: $R_{-135°} = \begin{bmatrix} -\frac{\sqrt{2}}{2} & -\frac{\sqrt{2}}{2} \\ \frac{\sqrt{2}}{2} & -\frac{\sqrt{2}}{2} \end{bmatrix}$

If $R_x = \begin{bmatrix} -\frac{\sqrt{2}}{2} & -\frac{\sqrt{2}}{2} \\ \frac{\sqrt{2}}{2} & -\frac{\sqrt{2}}{2} \end{bmatrix}^3$, find x.

43. If $T = \begin{bmatrix} -1 & 0 \\ 0 & 1 \end{bmatrix}$ and $T' = \begin{bmatrix} 3 & 0 \\ 0 & 3 \end{bmatrix}$, write $, T' \cdot T$ as a single matrix.

Review Answers

38. a.
$$\begin{bmatrix} 0 & 0 & 6 & 2 \\ 0 & -4 & -5 & -3 \end{bmatrix}$$

b.
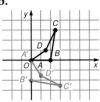

c. rotation of $-90°$ or $270°$

39. a.
$$\begin{bmatrix} 0 & 4 & 5 & 3 \\ 0 & 12 & 21 & 11 \end{bmatrix}$$

b.
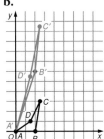

c. shear

40. $\begin{bmatrix} 0 & -1 \\ 1 & 0 \end{bmatrix}$

41.
$$\begin{bmatrix} 3 & 0 \\ 0 & 3 \end{bmatrix} \cdot \begin{bmatrix} 1 & 0 \\ 0 & -1 \end{bmatrix} \cdot \begin{bmatrix} x \\ y \end{bmatrix}$$

42. $R_{-405°}$

43. $\begin{bmatrix} -3 & 0 \\ 0 & 3 \end{bmatrix}$

---

44. a. $\begin{array}{c} \quad\ B \quad\ S \\ \begin{array}{c} L \\ ST \end{array} \begin{bmatrix} 3 & 5 \\ 1 & 2.5 \end{bmatrix} \end{array}$

b. Let x = number of lots of baseballs, and y = number of lots of softballs.

$\begin{array}{c} L \\ ST \end{array} \begin{bmatrix} 3 & 5 \\ 1 & 2.5 \end{bmatrix} \begin{bmatrix} x \\ y \end{bmatrix} = \begin{bmatrix} 58 \\ 26 \end{bmatrix}$

c. $x = \dfrac{\begin{vmatrix} 58 & 5 \\ 26 & 2.5 \end{vmatrix}}{\begin{vmatrix} 3 & 5 \\ 1 & 2.5 \end{vmatrix}}$ $y = \dfrac{\begin{vmatrix} 3 & 58 \\ 1 & 26 \end{vmatrix}}{\begin{vmatrix} 3 & 5 \\ 1 & 2.5 \end{vmatrix}}$

$x = 6$ $y = 8$

Chapter 7   Review Exercises *(continued)*

44. To produce 100 baseballs requires 3 units of leather and 1 unit of string and to produce 100 softballs requires 5 units of leather and 2.5 units of string.

   a. Develop a matrix that can be used to summarize the materials needed to produce baseballs and softballs in lots of 100.

   b. Write a matrix equation that can be used to find the number of lots of 100 balls of each kind that can be made from 58 units of leather and 26 units of string.

   c. Solve the equation written in part **b.**

45. A photographer takes pictures of the area where he lives and sells them to tourists who visit during the summer months. During June, he sold 40 small, 27 medium, and 8 large prints for $654. During July, he sold 32 small, 38 medium, and 12 large prints for $788. During August, he sold 42 small, 23 medium, and 6 large prints for $590.

   a. Write a matrix that can be used to show the number of each size print that he sold during each of the three months.

   b. Write a matrix equation that can be used to find the price of each size print.

   c. Solve the equation written in part **b.**

In **46 – 47**, use the numbering assignment 0 = space, 1 = A, 2 = B, 3 = C, 4 = D, 5 = E, …

46. Use $\begin{bmatrix} 3 & 11 \\ 1 & 4 \end{bmatrix}$ to decode the message $\begin{bmatrix} 7 & -19 \end{bmatrix}\begin{bmatrix} 16 & -44 \end{bmatrix}$.

47. If matrix $\begin{bmatrix} 4 & -5 \\ -2 & 3 \end{bmatrix}$ was used to code the message $\begin{bmatrix} 10 & -12 \end{bmatrix}\begin{bmatrix} 16 & -20 \end{bmatrix}$, decode the message.

48. Two metals, *A* and *B*, are used in the filament of a light bulb. A 100-watt bulb requires 1 unit of *A* and 2 units of *B*, and a 50-watt bulb requires 0.5 units of *A* and 1.5 units of *B*.

   a. Write a matrix to represent this data using one row for each wattage.

   b. If the company wants to produce 1 million 100-watt bulbs and 3 million 50-watt bulbs, use matrices to determine how many units of each metal will be needed.

49. Given the technology matrix
$$T = \begin{array}{c} G \\ S \end{array}\begin{bmatrix} \overset{G}{0.6} & \overset{S}{0.2} \\ 0.1 & 0.3 \end{bmatrix}$$

   a. Determine how many units of goods must be input to produce 4 units of goods and 6 units of services.

   b. Compute the intensity matrix necessary to meet anticipated demand for 100 units of goods and 200 units of services. (Round entries to the nearest tenth of a unit.)

45. a. $\begin{bmatrix} 40 & 27 & 8 \\ 32 & 38 & 12 \\ 42 & 23 & 6 \end{bmatrix}$

b. Let x = price of small print, y = price of medium print, and z = price of large print.

$$\begin{bmatrix} 40 & 27 & 8 \\ 32 & 38 & 12 \\ 42 & 23 & 6 \end{bmatrix} \cdot \begin{bmatrix} x \\ y \\ z \end{bmatrix} = \begin{bmatrix} 654 \\ 788 \\ 590 \end{bmatrix}$$

c. $\begin{bmatrix} 40 & 27 & 8 & 654 \\ 32 & 38 & 12 & 788 \\ 42 & 23 & 6 & 590 \end{bmatrix} = \begin{bmatrix} 1 & 0 & 0 & 6 \\ 0 & 1 & 0 & 10 \\ 0 & 0 & 1 & 18 \end{bmatrix}$

$x = 6, y = 10, z = 18$

46. BAD_
47. CAD_
48. a. $\begin{bmatrix} 1 & 2 \\ 0.5 & 1.5 \end{bmatrix}$

   b. 2,500,000 of *A*
     6,500,000 of *B*

49. a. 3   **b.** $\begin{bmatrix} 430 \\ 340 \end{bmatrix}$

☑ Exercises for Challenge

1. Find the transformation matrix for which the image of any point (x, y) is $(0, y)$.

2. By what percent does the area of a square with vertices at $(1, 0)$, $(0, 1)$, $(-1, 0)$ and $(0, -1)$ increase under the transformation $\begin{bmatrix} 3 & 0 \\ 0 & 3 \end{bmatrix}$?

3. Find the transformation matrix under which the image of $(1, 2)$ is $(3, 14)$ and the image of $(7, 4)$ is $(31, 58)$.

4. For what values of a and b does the following system have infinitely many solutions?
$$x + ay = 4$$
$$3x - a^2y = b$$

5. A and B are 4×4 matrices such that $A = kB$ and $|A| = 4|B|$. Find the value of k.

6. An equation of the set of points that are the images of the points on the line $y = x$ under the transformation matrix $\begin{bmatrix} 2 & 1 \\ 0 & 1 \end{bmatrix}$ is
   (A) $y = \frac{1}{3}x$ (B) $y = 3x$ (C) $y = 2x$
   (D) $y = \frac{1}{2}x$ (E) $x + y = 2$

7. Which point of the graph is the image of $(2, 3)$ under the transformation $\begin{bmatrix} 1 & 0 \\ 0 & -2 \end{bmatrix}$?

   (A) A (B) B (C) C (D) D (E) E

8. For what value of a does $\begin{bmatrix} a & 9 \\ -3 & 80 \end{bmatrix}$ have no multiplicative inverse?
   (A) 0.1500 (B) −0.1500
   (C) 0.2250 (D) 0.3375
   (E) −0.3375

9. Determine the general solution of the given system. $\quad 8x + 7y = a$ $\quad 7x + 6y = b$
   (A) $x = 6a - 7b$, $y = -7a + 8b$
   (B) $x = -6a + 7b$, $y = -7a + 8b$
   (C) $x = -6a + 7b$, $y = 7a - 8b$
   (D) $x = -\frac{6}{97}a + \frac{7}{97}b$, $y = \frac{7}{97}a - \frac{8}{97}b$
   (E) $x = \frac{6}{97}a + \frac{7}{97}b$, $y = \frac{7}{97}a + \frac{8}{97}b$

10. The set of all 3×3 matrices does not form a group under multiplication because
    (A) multiplication is not commutative
    (B) the set is not closed under multiplication
    (C) there is no multiplicative identity
    (D) not all nonzero elements have multiplicative inverses
    (E) multiplication is not associative

11. Which transformation matrix maps every point to a point on the x-axis?
    (A) $\begin{bmatrix} 0 & 0 \\ 0 & 1 \end{bmatrix}$ (B) $\begin{bmatrix} 0 & 0 \\ 1 & 1 \end{bmatrix}$
    (C) $\begin{bmatrix} 1 & 1 \\ 0 & 0 \end{bmatrix}$ (D) $\begin{bmatrix} -1 & 0 \\ 0 & -1 \end{bmatrix}$
    (E) $\begin{bmatrix} 0 & -1 \\ -1 & 0 \end{bmatrix}$

Answers
Exercises for
Challenge

1. $\begin{bmatrix} 0 & 0 \\ 0 & 1 \end{bmatrix}$

2. 800

3. $\begin{bmatrix} 5 & -1 \\ 6 & 4 \end{bmatrix}$

4. $a = 0$ or $a = -3, b = 12$

5. $k = \pm\sqrt{2}$

6. A

7. E

8. E

9. C

10. D

11. C

12. B

13. A

14. B

15. A

16. C

17. D

18. C

19. B

☑ Exercises for Challenge (continued)

12. Under the transformation $\begin{bmatrix} 1 & 0 \\ 0 & -2 \end{bmatrix}$,
which points remain fixed?

(A) There are no fixed points.
(B) all points on the x-axis
(C) all points on the y-axis
(D) all points on the line $y = x$
(E) all points on the line $y = -2$

13. A sequence $\{a_n\}$ is defined for $n \geq 2$ as shown by the matrices. If $a_1 = 1$, find a_4.

$$\begin{bmatrix} a_n \\ a_{n-1} \end{bmatrix} = \begin{bmatrix} 6 & 4 \\ 1 & 0 \end{bmatrix}^n \begin{bmatrix} 1 \\ 1 \end{bmatrix}$$

(A) 2,800 (B) 2,008 (C) 1,216
(D) 424 (E) 160

14. Let a function from the complex numbers to the set of 2 × 2 matrices be defined by $f(a + bi) = \begin{bmatrix} a & -b \\ b & a \end{bmatrix}$, when a and b are real numbers. Which statement is false?

(A) $f(a + 0i) = aI$

(B) $f(0 + bi) = -bI$

(C) $f((a + bi) + (c + di))$
$= f(a + bi) + f(c + di)$

(D) $f((a + bi)(c + di)) = f(a + bi) \cdot f(c + di)$

(E) f is one-to-one

Questions 15 – 19 each consist of two quantities, one in Column A and one in Column B. You are to compare the two quantities and choose:

A if the quantity in Column A is greater;
B if the quantity in Column B is greater;
C if the two quantities are equal;
D if the relationship cannot be determined from the information given.

1. In certain questions, information concerning one or both of the quantities to be compared is centered above the two columns.

2. In a given question, a symbol that appears in both columns represents the same thing in Column A as it does in Column B.

3. x, n, and k, etc. stand for real numbers.

Column A	Column B
$A = \begin{bmatrix} 3 & 2 \\ -4 & 6 \end{bmatrix}$	

15. det (A) det (A^{-1})

$A = \begin{bmatrix} 4 & 0 \\ -2 & 3 \end{bmatrix}$ and $B = \begin{bmatrix} 7 & -1 \\ 9 & 4 \end{bmatrix}$

16. det (AB) det $(A) \cdot$ det (B)

A is an $n \times n$ matrix.

17. det (A) det (A^{-1})

Let A denote the transformation matrix that rotates the plane 30° clockwise about the origin.

18. The distance from (2, 4) to the origin.   The distance from the image of (2, 4) to the origin.

$A = \begin{bmatrix} 2 & 0 & 0 \\ 0 & 4 & 0 \\ 0 & 0 & 5 \end{bmatrix}$ and $A^{-1} = \begin{bmatrix} a_{11} & a_{12} & a_{13} \\ a_{21} & a_{22} & a_{23} \\ a_{31} & a_{32} & a_{33} \end{bmatrix}$

19. a_{22} 1

Teacher's Chapter 8

The Trigonometric Functions

Chapter Contents

The trigonometric functions are defined for a domain that includes values that are larger and smaller than the measures of the acute angles of right triangles. The right-triangle definitions for these functions are consistent with the broader definitions.

The six trigonometric functions are initially defined in terms of the coordinates of points on a unit circle, and are then redefined in terms of angles. At first, work with angles beyond quadrant I is accomplished through a right triangle whose hypotenuse is the terminal ray of the angle. Later, reference angles are introduced.

Applications of indirect measure in right triangles are presented. Finally, the graphs of the trigonometric functions are discussed, with transformations of graphs used to make connections among the functions.

8.1   Circular Functions

To further emphasize the concept of a periodic function, elicit examples from the experience of the student, such as the recording of time on a clock or calendar.

To complement the text presentation that the wrapping function has its argument arc length measured in linear units around a circle, you may wish to demonstrate bending a number line around a circle. Use a string that is first made to coincide with a given length on a number line. Wrapping the string around circles of different radii reinforces the idea that the circumference formula is applicable to a circle of any radius and that π is simply an irrational constant.

This is also a good time to discuss the necessity for using sensible rational approximations for irrational numbers such as π and, later, for the values of the trigonometric functions.

The following activity can serve as a springboard for such a discussion.

Activity (8.1) _____

Directions:

1. For several objects of different sizes, such a flower pot or the plastic lid of a coffee can, measure the outer rim and the diameter. Record the measurements.

2. Calculate the ratio of the circumference to the diameter for each object measured.

3. Compare the values of the ratio to a rational approximation of π.

4. Find the average of the ratios and compare that value to an approximation for π.

Discoveries

1. The ratios of the circumferences to the diameters of the circular objects are approximately equal.

2. The value of each ratio is approximately equal to π.

3. The average of the values of the ratios is even closer to the approximate value of π.

What If?

All of the data for the class were entered on a computer spreadsheet and the average of each ratio determined? (The value would be a good approximation for π. The accuracy of the results would depend upon the care with which measurements were taken.)

_____

Since some students initially tend to see *s* as having values only in the interval [0, 2π], it is important to emphasize that *s* can be any real number. It is essential that students connect the development of the wrapping function to the function concepts in Chapter 3. Eliciting examples in addition to those of the text will reinforce the concepts that the wrapping function is not one-to-one and that the inverse of the wrapping function is not a function.

8.2 Using Angles to Define the Trigonometric Functions

To convert between radians and degrees, the text uses a proportion rather than a conversion factor because the proportion is less likely to be confusing in that the setup is always the same. If you choose to introduce the conversion factor approach, be sure to do some dimensional analysis to show how the units work out, thus giving guidance for the selection of the appropriate factor. For example, to change 80° to radians, choose the conversion factor that will cancel the degree unit:

$$\overset{4}{\cancel{80}}{}^{\circ} \times \frac{\pi \text{ radians}}{\underset{9}{\cancel{180}^{\circ}}} = \frac{4\pi}{9} \text{ radians}$$

Similarly, to change $\frac{5\pi}{2}$ radians to degrees, choose the conversion factor that will cancel the radian unit:

$$\frac{5\pi}{\cancel{2}} \cancel{\text{ radians}} \times \frac{\overset{90}{\cancel{180}^{\circ}}}{\cancel{\pi \text{ radians}}} = 450°$$

Students are often slow in acclimating themselves to the use of radian measure. When defining a radian, emphasize that it has a more natural relationship to the length of the intercepted arc than degree measure. Avoid always placing the equivalent number of degrees on the board when radians are used and be sure to do most applications in radians. Encourage students to think in terms of radians and reinforce the natural use of π for expressing radian measure.

Using dilations, the definitions of the trigonometric functions are extended from the unit circle to a circle with radius r.

To determine the values of various trigonometric functions given the value of one such function, the text, in Example 9, uses the equation of a unit circle with center at the origin. An alternate approach is to use an equation of a circle with radius r. For example, reconsider the conditions of Example 9: given that θ is in quadrant III and that $\cos \theta = -\frac{1}{4}$, find $\sin \theta$ and $\tan \theta$.

Solution

The equation of a circle with center at the origin and radius r is $x^2 + y^2 = r^2$, where $r > 0$. In quadrant III, both x and y are negative.

$$\cos \theta = -\frac{1}{4}$$
$$= \frac{-1}{4} = \frac{x}{r}$$
$$x^2 + y^2 = r^2$$
$$(-1)^2 + y^2 = 4^2$$
$$1 + y^2 = 16$$
$$y^2 = 15$$
$$y = -\sqrt{15}$$
$$\sin \theta = \frac{y}{r} = \frac{-\sqrt{15}}{4}$$
$$\tan \theta = \frac{y}{x} = \frac{-\sqrt{15}}{-1} = \sqrt{15}$$

Knowing $r > 0$, use $x = -1$ and $r = 4$. (See note below.)

Note: Students should be aware that, in general, because two quantities are in the ratio $\frac{-1}{4}$, it does not mean that one quantity is –1 and the other is 4. The quantities could be –2 and 8, or –3 and 12, etc.; that is, $-a$ and $4a$. However, because the trigonometric functions are defined in terms of ratios, we can allow the numerator and denominator of the ratio in simplest form to be taken as values for placement in a reference triangle. The advantages to using this approach are: students get perspective from the given ratio, of the placement and values of two of the three sides of the reference triangle; they can visualize what to expect for the sign of the 3rd side; they can continue to think in terms of one set of formulas for the trigonometric functions; and, they can bypass some work with fractions.

The values of the trigonometric functions of 30°, 45° and 60° are discussed here as an outgrowth of the applications of the trigonometric functions to right triangles and the students' knowledge of the ratios of the sides of 30°-60°-90° and 45°-45°-90° triangles. Point out that we know from geometry, not trigonometry, that the hypotenuse of a 30°-60°-90° triangle is twice the shorter leg and this enables us to find the trigonometric functions of 30° and 60°. The fact that the hypotenuse is twice the shorter leg may be demonstrated by using an equilateral triangle with an altitude drawn to one side.

Assuming ABC is the given 30°-60°-90° triangle,

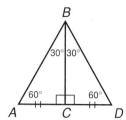

1. $\triangle ABD$ is equilateral.
2. $AD = AB$
3. $AC = CD$
4. $AD = 2(AC)$
5. $AB = 2(AC)$

The Pythagorean Theorem can then be used to find the relationship between the longer leg and the shorter leg.

In the 45°-45°-90° triangle, the Pythagorean Theorem is used to establish a relationship between the length of the leg and the hypotenuse.

Emphasize that, by the properties of similar triangles, these ratios and the trigonometric functions of 30°, 45°, and 60° have fixed values. Once these ratios are established, they are used to find functions of related angles in other quadrants.

Activity (8.2)

Directions

1. On a sheet of graph paper on which each large division is separated into 5 or 10 smaller divisions, let 50 smaller divisions equal 1 so that each small division equals 0.02. Draw x- and y-axes that intersect at the lower left. With the origin as center, draw a quarter of a unit circle.

2. With the origin as vertex and the x-axis as the initial ray, use a protractor to draw an angle of 30°. Approximate, to the nearest hundredth, the coordinates of the point at which the terminal ray of the angle intersects the arc of the unit circle. Compare these coordinates with the values of cos 30° and sin 30° found by using a calculator.

3. Repeat step 2 for several acute angles.

Discoveries

The x-coordinate of the point at which the terminal ray intersects the unit circle is approximately equal to the cosine of the angle and the y-coordinate is approximately equal to the sine of the angle.

What If?

The activity were repeated using a full circle and the measures of angles greater than 90°. (The same relationship between the coordinates and the sines and cosines would be true.)

8.3 Approximating Values of the Trigonometric Functions

The use of a scientific calculator is an integral part of this chapter. Students should be encouraged to have a calculator available at all times and to use calculators on classroom tests, even though it is possible to study all of the major ideas in the chapter without a calculator.

Because calculators vary in the ways in which keys are labeled, in the order in which keys must be pressed, and in the number of digits in the display, the calculator directions in the text may need to be modified. Students should consult the manuals for their calculators or experiment with different key strokes if they do not obtain the correct results when they follow the text.

In addition to degrees and radians, some calculators have a third way of measuring angles, called *gradients*. Gradients are usually used for measuring the ascent or descent of roads and are expressed in terms of *grads*, where 1 grad $= \frac{1}{100}$ of a right angle. A road that is said to have a 10% grade rises at an angle of 9°.

8.4 Applying Trigonometry in Right Triangles

The main purpose of this section is to show how right-triangle trigonometry can be used to do problems involving indirect measurement. Even if students have previously studied right-triangle trigonometry, they will be challenged by some of the exercises that require more analysis than those generally done when right-triangle trigonometry is introduced in earlier courses.

Note with the class that if one side and one acute angle of a right triangle are known, congruence postulates insure that the remaining two sides and two angles are determined. Thus, their measures can be found by using trigonometry.

In this section, since the trigonometric functions are defined in terms of the sides of a right triangle without mention of a circle or coordinate system, the domain is restricted to the interval (0°, 90°). Thus, degrees, rather than radians, are used.

Students can be assigned, individually or in small groups, the task of finding the height of a local building by using indirect measurement. A large wooden protractor is excellent for estimating angles of elevation. Students might also be asked to estimate, and then verify, the angle of elevation of a building of known height from a known distance from the foot of the building. Also, have students propose problems.

To reinforce understanding of the reciprocal functions, encourage the use of the secant, cosecant, and cotangent functions where they naturally occur. For example, $\cot 40° = \dfrac{x}{5}$ is "better" than $\tan 40° = \dfrac{5}{x}$ since x can be found without using long division. Knowing the difference between function inverse and reciprocal (multiplicative inverse) is particularly important when using a calculator. Encourage students to experiment with their calculators to verify that $\tan^{-1} x \neq \dfrac{1}{\tan x}$.

Unless otherwise indicated, the rounding of an answer should be done only as the very last step rather than at any intermediate result. Students should be encouraged to solve for the desired variable in terms of all the others and then perform all of the calculations at one time, utilizing the full level of accuracy available. This is demonstrated in Example 2 where BC is expressed as $(\tan 40°)12$, leading to

$$AC = (\cot 25°)(BC)$$
$$AC = (\cot 25°)(\tan 40°)(12)$$

Example 3 provides an illustration of a practical problem when the traditional trigonometry table would not provide sufficient accuracy for a meaningful answer. Since calculators often give more place values than are appropriate, remind students that an answer to a problem should not have a greater number of significant digits than the data used in the problem.

8.5 Extending Trigonometry

The concept of a reference angle between 0° and 90° allows trigonometry to be extended to an angle of any size. Emphasize that the reference angle is the angle that the terminal side of θ makes with the positive or negative _x-axis_. It is a common error for students to use the angle that the terminal side of θ makes with the _y_-axis as the reference angle. To dispel this notion, point out that the definitions of the trigonometric functions used angles in standard position, that is, with the _x_-axis as the initial side. All reference angles are determined with respect to that axis.

Transformations are presented to verify that the trigonometric function values of the reference angle θ_r are the absolute values of the function values of θ. These relationships give enough information to find function values for any measure in a table that lists function values only for 0° through 90°. The use of the calculator diminishes the importance of these relationships since the calculator will return function values for any measure directly.

Use Exercise 72 to call attention to the fact that the periodic nature of the trigonometric functions makes them useful for devising formulas used in situations involving data that repeat periodically.

8.6   *The Graphs of the Sine and Cosine Functions*

A separate section is devoted to the graphs of sine and cosine since most of the important concepts related to trigonometric graphs can be taught within this framework. Also, most of the applications to areas in science involve sine and cosine graphs.

The sine graph is developed here as the graph of any other unknown function might be—by taking a representative sample of points, plotting them, and then connecting the points with a smooth curve. The periodic nature of the function is visually apparent. An alternate development of the graph makes use of the definition of sin x as the ordinate of a point on the unit circle. These ordinates can be projected onto the y-axis for each value of x as a single point rotates around the circle and its (x, y) values change.

Establishing a consistent relationship of units between the x- and y-axes (one radian on the x-axis about the same as one unit on the y-axis) insures that different students drawing the same graphs correctly will have graphs that are approximately the same shape.

The even or odd properties of function were discussed in detail in Chapter 3 and are referred to here regarding the trigonometric functions. This is important in determining the symmetries for the graphs of each of the functions. Also discussed in Chapter 3 were the general effects on a graph when its function has constant multipliers in different positions, that is, $y = a\ f(x)$ and $y = f(ax)$. Now students see these transformations applied to the basic sine and cosine functions.

The exercises include a variety of physical situations featuring the sine curve. In Exercise 58, mention that, in order to improve the fit of the theoretical sine or cosine curve to the scatter plot, the equation can be changed to adjust the amplitude and period of the curve.

8.7   *The Graphs of the Other Trigonometric Functions*

Each of the remaining four trigonometric graphs has asymptotes, which distinguishes them as a group from the two covered in the previous section. Thus, the important first step in drawing these graphs is to determine and place the asymptotes. Emphasize that vertical asymptotes occur if a denominator were to be 0 when the numerator is not 0. For tangent and cotangent, this means thinking in terms of the quotient relationships:

$$\tan \theta = \frac{\sin \theta}{\cos \theta} \qquad \cot \theta = \frac{\cos \theta}{\sin \theta}$$

For secant and cosecant, this means thinking in terms of the reciprocal relationships:

$$\csc \theta = \frac{1}{\sin \theta} \qquad \sec \theta = \frac{1}{\cos \theta}$$

Point out that the same mutual relationship that exists between the sine and cosine graphs exists between the graphs of secant and cosecant, namely, each is the translation of the other by $\pm \frac{\pi}{2}$ units horizontally.

The general technique for obtaining composite graphs is to identify the component functions to get some idea about the graph's shape without making a detailed table of values. Addition of ordinates is shown as a method for determining key values.

1. Find the value of the wrapping function for each of the following values of arc lengths.

   a. $-\frac{3\pi}{2}$ **b.** $\frac{17\pi}{6}$

2. Find $\sin\theta$, $\cos\theta$, and $\tan\theta$ if the terminal side of θ has a point with the given coordinates.

   a. $(7, 24)$ **b.** $(-11, 60)$ **c.** $(1, -\sqrt{3})$

3. Find the numerical value of each.

   a. $\tan\frac{\pi}{3}$ **b.** $\csc\left(-\frac{\pi}{6}\right)$ **c.** $\tan\frac{9\pi}{4}$

4. Use a calculator to find x to the nearest thousandth if $(x, 0.641)$ are the coordinates of a point on a unit circle.

5. Which one of the following correctly pairs degree measure with the equivalent number of radians?

   (A) $(1°, 1)$ (B) $\left(\frac{1°}{5}, \frac{\pi}{5}\right)$

   (C) $\left(135°, \frac{3\pi}{4}\right)$ (D) $\left(45°, \frac{1}{4}\right)$

6. Find the length of an arc on a circle that is intercepted by a central angle of 1.1 radians if the radius of the circle is 7 inches.

7. **a.** Express in terms of π the number of minutes that pass when the minute hand of a clock rotates through an angle of $\frac{1}{2}$ radian.

   b. Is this more or less than 5 minutes?

8. Find the exact value of each of the six trigonometric functions of an angle θ if the terminal side of θ passes through the point $(-6, 7)$.

9. Find the length of $\overline{AB}$ and $\overline{BC}$ in each of the triangles.

   a.   **b.**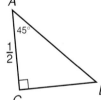

10. Use a scientific calculator to find the value of each function to 6 significant digits.

    a. $\cos 73°12'$

    b. $\csc 0°13'$

    c. $\cot 1.06$

11. Use the trigonometric table to find the value of each function to the nearest hundredth.

    a. $\sin 45°20'$

    b. $\csc 0.0524$

    c. $\tan 1.277$

12. Use a scientific calculator to find θ to the nearest tenth of a radian in the interval $\left[0, \frac{\pi}{2}\right)$.

    a. $\cos\theta = 0.124$

    b. $\cot\theta = 6.6$

    c. $\tan\theta = \frac{\pi}{7}$

13. Which trigonometric function is paired with another that has the same range?

    (A) (tan, cos)   (B) (cot, sin)

    (C) (sec, csc)   (D) (tan, sec)

14. Find the value of the secant of the smaller acute angle in a right triangle whose legs are 4 cm and 7 cm.

15. A safe angle for a ladder is 75° with the ground. To the nearest tenth of a foot, how high up the side of a building will the top of a 16′ ladder reach if it makes a 75° angle with the ground?

16. Find x to the nearest tenth.

a. **b.**

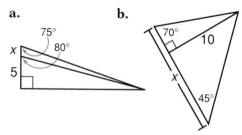

17. To estimate the height of a building, the angle of elevation of the top of the building is determined to be 53° at a point on the ground that is 100′ from the building. Based on this sighting, find the height of the building to the nearest foot.

18. Find the value of each function.

a. sec 135°

b. cot 480°

c. csc (−150)°

19. Find two positive values of θ, $\theta \in [0, 2\pi]$, that satisfy each equation.

a. $\tan \theta = -1$ **b.** $\csc \theta = \dfrac{2}{\sqrt{3}}$

20. Sketch the graph of each composite function, showing at least two periods.

a. $y = 3 + \cos \frac{1}{3} x$

b. $y = 1 - 2 \sin \left(x - \frac{\pi}{4} \right)$

21. Write an equation of each of the given graphs.

a. **b.**

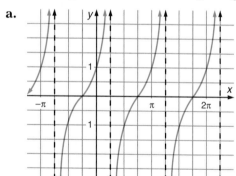

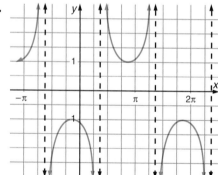

c.

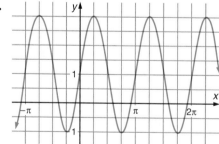

22. Sketch $y = \frac{x}{2} - \sin x$ from -2π to 2π.

23. Find the amplitude and period.

   a. $y = 0.7 \cos \left(\frac{1}{2}x - 3 \right)$

   b. $y = -5 \sin \frac{3\pi}{2}x$

Bonus I

A rotating light on top of an offshore lighthouse casts a ray of light along a straight shoreline so that the center of light is s units from a tree at a fixed point on the shore. When the light is directed toward the shore, the distance s is determined by the equation

$$s = 20 \tan 3\pi t$$

where t is the time in minutes, with $t = 0$ when the light shines directly on the tree.

   a. How soon after the light is centered on one spot will it again shine on that spot?

   b. For what values of t in the interval $\left[0, \frac{2}{3} \right]$ is s defined?

Bonus II

Find the shorter base of a trapezoid if the following four facts are known about the trapezoid.

1. One of the angles formed by the longer base and one of the non-parallel sides is 30°.

2. The median of the trapezoid is 7 units larger than the shorter base.

3. The nonparallel side that makes a 30° angle with the base of the trapezoid is 8 units less than twice the median.

4. The area of the trapezoid is 102.

Answers to Suggested Test Items

1. a. $\left(-\frac{3\pi}{2}, (0, 1) \right)$ **b.** $\left(\frac{17\pi}{6}, \left(-\frac{\sqrt{3}}{2}, \frac{1}{2} \right) \right)$

2. a. $\sin \theta = \frac{24}{25}$ $\cos \theta = \frac{7}{25}$ $\tan \theta = \frac{24}{7}$

   b. $\sin \theta = \frac{60}{61}$ $\cos \theta = \frac{-11}{61}$ $\tan \theta = \frac{-60}{11}$

   c. $\sin \theta = -\frac{\sqrt{3}}{2}$ $\cos \theta = \frac{1}{2}$ $\tan \theta = -\sqrt{3}$

3. a. $\sqrt{3}$ **b.** -2 **c.** 1

4. 0.768   **5.** C   **6.** 7.7 in.

7. a. $\frac{15}{\pi}$ min.   **b.** less

8. $\sin \theta = \frac{7\sqrt{85}}{85}$ $\cos \theta = \frac{-6\sqrt{85}}{85}$ $\tan \theta = -\frac{7}{6}$

   $\cot \theta = -\frac{6}{7}$ $\sec \theta = \frac{-\sqrt{85}}{6}$ $\csc \theta = \frac{\sqrt{85}}{7}$

9. a. $AB = 14\sqrt{3}$ **b.** $AB = \frac{\sqrt{2}}{2}$

   $BC = 21$ $BC = \frac{1}{2}$

10. a. 0.289032   **b.** 264.443

   c. 0.560405

11. a. 0.71   **b.** 19.09   **c.** 3.31

12. a. 1.4   **b.** 0.2   **c.** 0.4

13. C   **14.** $\frac{\sqrt{65}}{7}$ **15.** 15.5 ft.

16. a. $x \approx 2.6$ **b.** 13.6

17. 133 ft.

18. a. $-\sqrt{2}$ **b.** $-\frac{\sqrt{3}}{3}$ **c.** -2

19. a. $\frac{3\pi}{4}, \frac{7\pi}{4}$ **b.** $\frac{\pi}{3}, \frac{2\pi}{3}$

Answers to Suggested Test Items (continued)

20. a.

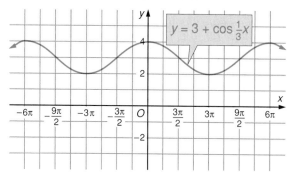

b.

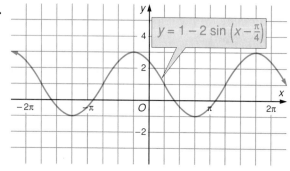

21. a. $y = \tan\left(x + \frac{\pi}{4}\right)$

   b. $y = \csc\left(x - \frac{\pi}{3}\right)$

   c. $y = 1 + 2\sin 2x$

22.

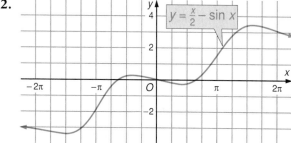

23. a. amplitude = 0.7   period = 4π

   b. amplitude = 5   period = $\frac{4}{3}$

Answers to Suggested Test Items (*continued*)

Bonus I

In the picture, s is positive when the light shines on a spot above the tree and negative below the tree.

The symmetry of the figure requires that, assuming a constant speed of rotation, the line from the lighthouse to the tree is perpendicular to the shore.

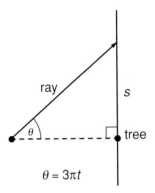

$$\theta = 3\pi t$$

a. The light only shines on the shore during half of its rotation and out to sea for the other half. It must make one complete rotation before it again shines on the tree.

$$3\pi t = 2\pi$$
$$t = \frac{2}{3}$$

b. s is undefined when the ray of light is parallel to or away from the shore.

$$\frac{\pi}{2} \le 3\pi t \le \frac{3\pi}{2}$$
$$\frac{1}{6} \le t \le \frac{1}{2}$$

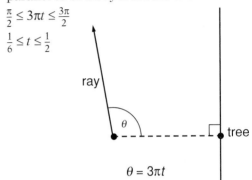

$$\theta = 3\pi t$$

s is defined when $0 \le t < \frac{1}{6}$ or $\frac{1}{2} < t \le \frac{2}{3}$.

Bonus II

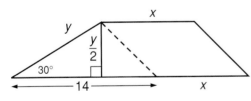

$$\text{median} = x + 7$$

$$y = 2(x + 7) - 8$$

$$y = 2x + 6$$

$$\frac{y}{2} = x + 3$$

$$\text{Area} = \text{median} \times \text{height}$$

$$102 = (x + 7)(x + 3)$$

$$102 = x^2 + 10x + 21$$

$$x^2 + 10x - 81 = 0$$

$$x = \frac{-10 \pm \sqrt{(10)^2 - 4(-81)}}{2}$$

$$x = \frac{-10 \pm \sqrt{100 + 324}}{2}$$

$$x = \frac{-10 \pm \sqrt{424}}{2}$$

$$x = \frac{-10 \pm 2\sqrt{106}}{2}$$

$$x = \sqrt{106} - 5$$

Chapter 8

The Trigonometric Functions

Sound waves bring the energy of a vibrating string or column of air to your ear. The number of waves per second, that is, the *frequency* of the wave, determines the pitch of a sound. Musical instruments have different sounds even when playing the same note, because most sounds are made up of different frequencies—the fundamental frequencies and higher frequencies called *overtones*. The quality of the sound depends on the relative intensities of the frequencies and overtones.

Sound is just one of many types of waves. Light, radio, television, x-rays, and microwaves are other examples. The trigonometric functions are used to classify and compare the properties of these waves.

Chapter Table of Contents

8.1 Circular Functions

Physical phenomena such as sound, heat, light, and electricity can be classified as forms of electromagnetic radiation, and thought of as traveling in wave-like patterns. For example, if a sound of constant pitch is generated at a steady volume, a wave pattern would be seen to continually repeat itself on the screen of an oscilloscope.

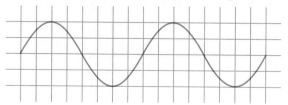

Or imagine yourself on a Ferris Wheel. You enter when the seat is at the lowest point of its travel. As the wheel rotates, you rise higher and higher until your seat reaches the highest point of its travel. You then descend, repeating the previous heights in reverse. As you pass the starting position, the cycle begins to repeat itself. Your distance from the ground, which repeats a given set of values, can be described by the following graph.

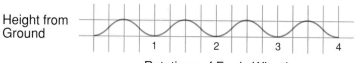

Height from Ground

Rotations of Ferris Wheel

Like the graphs of the sound wave and the path of a seat on a Ferris Wheel, graphs that continually repeat themselves are called *periodic*. Some periodic functions can be described in terms of a circle.

Consider a *unit circle*, where the center is at the origin and the measure of the radius is 1 unit. With the initial point at $(1, 0)$ and measuring in a counterclockwise direction to the terminal point (x, y), describe an arc of length s.

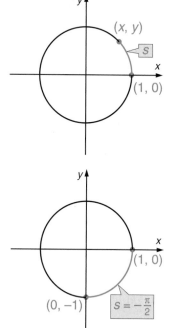

Since the circumference of the circle is $2\pi r$ and $r = 1$, s may have any value from 0 to 2π depending on where (x, y) is chosen. For example, if (x, y) is $(1, 0)$, then $s = 0$; and if (x, y) is $(-1, 0)$, then $s = \pi$. The domain of s is extended to include all positive real numbers by considering arcs that wrap around the circle more than once. For example, if s were the length of an arc of $2\frac{1}{2}$ revolutions, then the value of s would be $2\frac{1}{2} \times 2\pi$ or 5π. The negative real numbers are added to the domain of s to represent arc lengths measured in a clockwise direction. For example, the arc length from $(1, 0)$ to $(0, -1)$ on the unit circle, measured in a clockwise direction, is $-\frac{\pi}{2}$. Thus, the value of s can be any real number, leading to the following definition.

■ **Definition:** *The* wrapping function *is a function that maps every real number s that is the length of an arc with its initial point at* (1, 0) *on a unit circle to an ordered pair* (x, y) *that is the terminal point of the arc of length s.*

The discussion above suggests that the following ordered pairs belong to the wrapping function.

$$(0, (1, 0)) \quad (\pi, (-1, 0)) \quad (5\pi, (-1, 0))$$

Observe from the second and third ordered pairs that different values of *s* may correspond to the same second element, with the result that the wrapping function is not a one-to-one function.

Example 1

Find the value (ordered pair) in the range of the wrapping function that corresponds to the real number $\frac{3\pi}{2}$ in the domain.

Solution

The circumference of a unit circle is 2π. Find what part is represented by the given value.

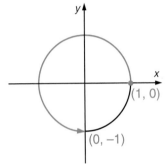

$$\frac{\frac{3\pi}{2}}{2\pi} = \frac{3}{4}$$

Thus, *s* is the length of an arc of $\frac{3}{4}$ revolution.

Answer: $\frac{3\pi}{2}$ corresponds to the point $(0, -1)$ in the range.

Example 2

Find the value of the wrapping function when $s = -\pi$.

Solution

Use a diagram to locate the terminal point of the arc whose measure is represented by the value $-\pi$, that is, an arc in the clockwise direction from (1, 0) with length that is $\frac{1}{2}$ the circumference.

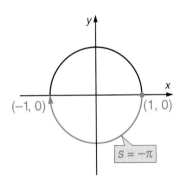

Answer: The value of the wrapping function at $s = -\pi$ is $(-1, 0)$.

Example 3

Evaluate the wrapping function at $s = \frac{\pi}{4}$.

Solution

Since $\frac{\pi}{2}$ is a quarter of the unit circle, $\frac{\pi}{4}$ is midway between $(1, 0)$ and $(0, 1)$. The central angle corresponding to $\frac{\pi}{4}$ is 45°, thus forming an isosceles right triangle whose hypotenuse is 1.

Use the Pythagorean Theorem to determine the lengths of the legs of this triangle.

$$x^2 + y^2 = 1^2$$
$$x^2 + x^2 = 1 \qquad y = x$$
$$2x^2 = 1$$
$$x^2 = \frac{1}{2}$$
$$x = \sqrt{\frac{1}{2}} = \frac{\sqrt{2}}{2}$$

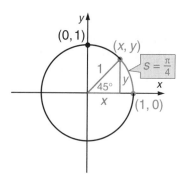

Answer: $s = \frac{\pi}{4}$ corresponds to $\left(\frac{\sqrt{2}}{2}, \frac{\sqrt{2}}{2}\right)$ in the wrapping function.

The Trigonometric Functions

For any arc of length s, you know that its terminal point is an ordered pair (x, y). Suppose a new function is formed that maps s onto the y-value of the terminal point of an arc of length s. For example, the terminal point of an arc of length $\frac{\pi}{2}$ is $(0, 1)$. Since $y = 1$ in this ordered pair, the new function, known as the *sine function*, maps $\frac{\pi}{2}$ onto 1. This is written as:

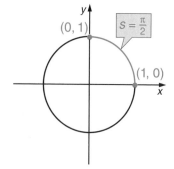

$$\sin \frac{\pi}{2} = 1$$

In general:

$$\sin s = y$$

The function that maps arc length s onto the x-value of the terminal point of the arc is called the *cosine function*. This is written:

$$\cos s = x$$

For example, if $s = \frac{\pi}{4}$, then the coordinates of the endpoint of s on the unit circle are $\left(\cos \frac{\pi}{4}, \sin \frac{\pi}{4}\right)$.

In Example 3, it was shown that when $s = \frac{\pi}{4}$, then $(x, y) = \left(\frac{\sqrt{2}}{2}, \frac{\sqrt{2}}{2}\right)$. Therefore:

$$\cos \frac{\pi}{4} = \frac{\sqrt{2}}{2} \text{ and } \sin \frac{\pi}{4} = \frac{\sqrt{2}}{2}$$

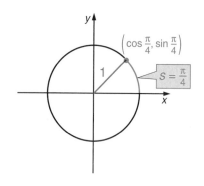

Four additional functions are defined as follows:

The *tangent function* maps s to $\frac{y}{x}$ if $x \neq 0$.

$\tan s = \frac{y}{x}$, where $x \neq 0$

The *cosecant function* maps s to $\frac{1}{y}$ if $y \neq 0$.

$\csc s = \frac{1}{y}$, where $y \neq 0$

The *secant function* maps s to $\frac{1}{x}$ if $x \neq 0$.

$\sec s = \frac{1}{x}$, where $x \neq 0$

The *cotangent function* maps s to $\frac{x}{y}$ if $y \neq 0$.

$\cot s = \frac{x}{y}$, where $y \neq 0$

These six functions, which are studied in the branch of mathematics called *trigonometry*, are referred to as *trigonometric functions*. They are also called *circular functions* because of their dependence on the unit circle and their periodic nature.

Example 4

Find the value of each trigonometric function at the given point.

a. $\sin \frac{\pi}{2}$ **b.** $\tan \pi$ **c.** $\cos 0$

Solution

a. If $s = \frac{\pi}{2}$, then $(0, 1)$ is the endpoint of this arc of length $\frac{\pi}{2}$. By definition, $\sin s = y$. Therefore, $\sin \frac{\pi}{2} = 1$.

b. If $s = \pi$, then $(-1, 0)$ is the endpoint of the arc. Since $\tan s = \frac{y}{x}$ (where $x \neq 0$), then $\tan \pi = \frac{0}{-1} = 0$.

c. $(1, 0)$ is the terminal point as well as the initial point of an arc of length 0. Since $\cos s = x$, then $\cos 0 = 1$.

In this chapter, arc lengths of $\frac{\pi}{6}$, $\frac{\pi}{4}$, and $\frac{\pi}{3}$ and their multiples are used frequently. They represent $\frac{1}{12}$, $\frac{1}{8}$, and $\frac{1}{6}$ of a unit circle and correspond to central angles of 30°, 45°, and 60°. The coordinates of the terminal points of these arcs can be obtained by examining the 30°-60°-90° and the 45°-45°-90° right triangles.

Since all 30°-60°-90° triangles are similar, the ratio of their sides is the same. Triangle *ABC* at the right is a 30°-60°-90° triangle whose hypotenuse is 1 unit long. The lengths of the legs of the triangle can be found by extending side *AC* to a point *D*, where *AC* = *CD*, and drawing segment *BD*. Since triangles *BCD* and *BCA* are congruent, the measure of angle *D* is 60° and the measure of angle *CBD* is 30°. Triangle *ABD*, therefore, is an equilateral triangle each of whose sides has length 1. Since *AC* = *CD*, *AC* = $\frac{1}{2}$. The length *BC* can be found by using the Pythagorean Theorem, as follows:

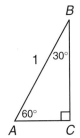 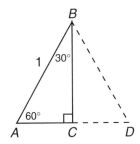

$$(AC)^2 + (BC)^2 = (AB)^2$$
$$\left(\tfrac{1}{2}\right)^2 + (BC)^2 = 1^2$$
$$\tfrac{1}{4} + (BC)^2 = 1$$
$$(BC)^2 = \tfrac{3}{4}$$
$$BC = \frac{\sqrt{3}}{2}$$

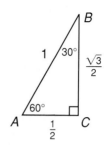

Thus:

The lengths of the legs of the 30°-60°-90° triangle whose hypotenuse is 1 are $\frac{1}{2}$ and $\frac{\sqrt{3}}{2}$.

To find the ratio of the sides of a 45°-45°-90° right triangle, draw an isosceles right triangle whose hypotenuse is 1 unit long. Represent the lengths of the equal legs by ℓ. By the Pythagorean Theorem:

$$\ell^2 + \ell^2 = 1^2$$
$$2\ell^2 = 1$$
$$\ell^2 = \tfrac{1}{2}$$
$$\ell = \frac{1}{\sqrt{2}} = \frac{\sqrt{2}}{2}$$

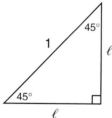

Thus:

The lengths of the legs of the 45°-45°-90° triangle whose hypotenuse is 1 are each $\frac{\sqrt{2}}{2}$.

Example 5

Find the value of each of the six trigonometric functions when $s = \frac{\pi}{3}$.

Solution

$\frac{\pi}{3}$ is $\frac{1}{6}$ of 2π, or $\frac{1}{6}$ of the arc length of a unit circle. Therefore, the radius drawn from the terminal point of an arc of length $\frac{\pi}{3}$ forms a 60° angle with the x-axis and is the hypotenuse of a 30°-60°-90° triangle.

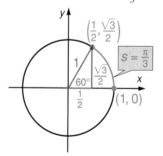

$$\sin \frac{\pi}{3} = y = \frac{\sqrt{3}}{2}$$

$$\csc \frac{\pi}{3} = \frac{1}{y} = \frac{2}{\sqrt{3}} = \frac{2\sqrt{3}}{3}$$

$$\cos \frac{\pi}{3} = x = \frac{1}{2}$$

$$\sec \frac{\pi}{3} = \frac{1}{x} = 2$$

$$\tan \frac{\pi}{3} = \frac{y}{x} = \frac{\frac{\sqrt{3}}{2}}{\frac{1}{2}} = \sqrt{3}$$

$$\cot \frac{\pi}{3} = \frac{x}{y} = \frac{1}{\sqrt{3}} = \frac{\sqrt{3}}{3}$$

If 2π, or any integral multiple of 2π, is added to s, the corresponding ordered pair on the circumference of the circle is the same as it is for s. In general, if (a, b) is the terminal point of an arc with length s, then (a, b) is the terminal point of the arc with length $s + 2\pi n$ for all integral values of n. Therefore:

$$\text{trig } (s) = \text{trig } (s + 2\pi n), \text{ for } n \text{ an integer}$$

where "trig" represents any one of the six trigonometric functions.

If f is a function and there exists a positive number p such that

$$f(x) = f(x + p)$$

for every real number x in the domain, then $f(x)$ is a periodic function, with the smallest positive value for p as the period. For example, since the coordinates of the terminal point of $s = \frac{\pi}{3}$ are $\left(\cos \frac{\pi}{3}, \sin \frac{\pi}{3}\right)$,

$$s = \frac{\pi}{3} + 2\pi = \frac{7\pi}{3}$$

$$s = \frac{\pi}{3} + 4\pi = \frac{13\pi}{3}$$

$$s = \frac{\pi}{3} + 6\pi = \frac{19\pi}{3}$$

$$\vdots$$

$$s = \frac{\pi}{3} + 2\pi n$$

represent arc lengths whose endpoints have the coordinates $\left(\cos \frac{\pi}{3}, \sin \frac{\pi}{3}\right)$.

Example 6

Find the period of

a. the wrapping function   **b.** the tan function

Solution

a. The range of the wrapping function is the set of all ordered pairs on the circumference of the unit circle, and s must vary from 0 to 2π before the values repeat.

Answer: The period of the wrapping function is 2π.

b. Use a table to track changes in s and the corresponding changes in x, y, and $\frac{y}{x}$, which is the value of tan s.

s	x	y	$\frac{y}{x} = \tan s$
$0 \rightarrow \frac{\pi}{2}$	$1 \rightarrow 0$	$0 \rightarrow 1$	increasing: $\quad 0 \rightarrow \infty$
$\frac{\pi}{2} \rightarrow \pi$	$0 \rightarrow -1$	$1 \rightarrow 0$	increasing: $-\infty \rightarrow 0$
$\pi \rightarrow \frac{3\pi}{2}$	$-1 \rightarrow 0$	$0 \rightarrow -1$	increasing: $\quad 0 \rightarrow \infty$
$\frac{3\pi}{2} \rightarrow 2\pi$	$0 \rightarrow 1$	$-1 \rightarrow 0$	increasing: $-\infty \rightarrow 0$

Observe that as s increases from 0 to 2π, the value $\frac{y}{x}$ is always increasing. However, as s increases beyond π, the values of $\frac{y}{x}$ begin to repeat. For example, $\tan \frac{\pi}{3} = \tan\left(\frac{\pi}{3} + \pi\right)$.

$$\tan \frac{\pi}{3} = \frac{\frac{\sqrt{3}}{2}}{\frac{1}{2}}$$

$$= \sqrt{3}$$

$$\tan\left(\frac{\pi}{3} + \pi\right) = \frac{-\frac{\sqrt{3}}{2}}{-\frac{1}{2}}$$

$$= \sqrt{3}$$

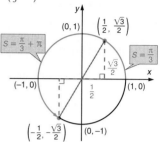

Answer: The period of the tangent function is π.

Exercises 8.1

In **1 – 8**, find the value of the wrapping function for the given arc length s.

1. 0 **2.** π **3.** $\frac{3\pi}{2}$ **4.** $-\frac{\pi}{2}$

5. -2π **6.** $-\frac{5\pi}{2}$ **7.** $-\frac{\pi}{6}$ **8.** $\frac{3\pi}{4}$

9. If $\left(\frac{3}{5}, \frac{4}{5}\right)$ is the point on the unit circle that is the terminal point of an arc of length $s = a$, use symmetry to find the coordinates of the wrapping function value of $s = -a$.

10. If $\left(\frac{12}{13}, \frac{5}{13}\right)$ is the value of the wrapping function when $s = b$, find the value of the wrapping function when $s = b + \pi$.

11. If $\left(\frac{\sqrt{3}}{2}, -\frac{1}{2}\right)$ is the value of the wrapping function when $s = c$, find the value of the wrapping function when $s = c - 2\pi$.

12. If $(0.28, 0.96)$ is the value of the wrapping function when $s = d$, find the value of the wrapping function when $s = d - \pi$.

In **13 – 20**, find $\sin s$ and $\cos s$ for the given value of s.

13. π **14.** $\frac{\pi}{3}$ **15.** $\frac{2\pi}{3}$ **16.** $\frac{3\pi}{4}$

17. $-\frac{5\pi}{6}$ **18.** 66π **19.** 37π **20.** -41π

In **21 – 28**, find the value of the function.

21. $\sec \frac{\pi}{3}$ **22.** $\csc\left(-\frac{\pi}{6}\right)$ **23.** $\tan \frac{\pi}{4}$ **24.** $\cot\left(-\frac{\pi}{4}\right)$

25. $\csc \frac{\pi}{2}$ **26.** $\cot\left(-\frac{7\pi}{4}\right)$ **27.** $\cos \frac{5\pi}{6}$ **28.** $\cot \frac{3\pi}{2}$

1. $(0, (1, 0))$

2. $(\pi, (-1, 0))$

3. $\left(\frac{3\pi}{2}, (0, -1)\right)$

4. $\left(-\frac{\pi}{2}, (0, -1)\right)$

5. $(-2\pi, (1, 0))$

6. $\left(-\frac{5\pi}{2}, (0, -1)\right)$

7. $\left(-\frac{\pi}{6}, \left(\frac{\sqrt{3}}{2}, -\frac{1}{2}\right)\right)$

8. $\left(\frac{3\pi}{4}, \left(-\frac{\sqrt{2}}{2}, \frac{\sqrt{2}}{2}\right)\right)$

9. $\left(-a, \left(\frac{3}{5}, -\frac{4}{5}\right)\right)$

10. $\left(b + \pi, \left(-\frac{12}{13}, -\frac{5}{13}\right)\right)$

11. $\left(c - 2\pi, \left(\frac{\sqrt{3}}{2}, -\frac{1}{2}\right)\right)$

12. $(d - \pi, (-0.28, -0.96))$

13. $\sin \pi = 0, \cos \pi = -1$

14. $\sin \frac{\pi}{3} = \frac{\sqrt{3}}{2}, \cos \frac{\pi}{3} = \frac{1}{2}$

15. $\sin \frac{2\pi}{3} = \frac{\sqrt{3}}{2}, \cos \frac{2\pi}{3} = -\frac{1}{2}$

16. $\sin \frac{3\pi}{4} = \frac{\sqrt{2}}{2}$, $\cos \frac{3\pi}{4} = -\frac{\sqrt{2}}{2}$

17. $\sin\left(-\frac{5\pi}{6}\right) = -\frac{1}{2}$, $\cos\left(-\frac{5\pi}{6}\right) = -\frac{\sqrt{3}}{2}$

18. $\sin 66\pi = \sin 2\pi = 0$, $\cos 66\pi = \cos 2\pi = 1$

19. $\sin 37\pi = \sin \pi = 0$, $\cos 37\pi = \cos \pi = -1$

20. $\sin(-41\pi) = \sin(-\pi) = 0$, $\cos(-41\pi) = \cos(-\pi) = -1$

21. $\sec \frac{\pi}{3} = 2$

22. $\csc\left(-\frac{\pi}{6}\right) = -2$

23. $\tan \frac{\pi}{4} = 1$

24. $\cot\left(-\frac{\pi}{4}\right) = -1$

25. $\csc \frac{\pi}{2} = 1$

26. $\cot\left(-\frac{7\pi}{4}\right) = 1$

27. $\cos \frac{5\pi}{6} = -\frac{\sqrt{3}}{2}$

28. $\cot\left(\frac{3\pi}{2}\right) = 0$

29. $\sin s = a^2$

30. $\tan s = \frac{b^3}{b^2} = b$

31. a. $\frac{7}{25}$

    b. $\sin s = \frac{24}{25}$, $\cos s = \frac{7}{25}$, $\tan s = \frac{24}{7}$

32. $\tan s = \frac{3}{4}$, $\cot s = \frac{4}{3}$, $\sec s = \frac{5}{4}$, $\csc s = \frac{5}{3}$

33. $\frac{12}{5}$

34. a. ± 0.791
    b. ± 0.954
    c. ± 0.910
    d. ± 0.975

35. a. Quadrant III
    b. Quadrants I and III

36. $\sin s = 0.71$, $\tan s = 1.02$

37. $\sin s = -0.70$, $\tan s = -0.99$

38. $\sin s = 0.4$, $\tan s = 0.44$

39. $\sin s = -0.98$, $\tan s = 4.66$

40. a. $\sin s = y$

    $\csc s = \frac{1}{y}$

    $y = \frac{1}{\frac{1}{y}}$

    $\sin s = \frac{1}{\csc s}$

    b. $\cos s = x$ $\sec s = \frac{1}{x}$

       $x = \frac{1}{\frac{1}{x}}$

       $\cos s = \frac{1}{\sec s}$

41. a. $\sin s = y \quad \cos s = x \quad \tan s = \dfrac{y}{x}$

$$\dfrac{y}{x} = \dfrac{y}{x}$$

$$\tan s = \dfrac{\sin s}{\cos s}$$

b. $\sin s = y \quad \cos s = x \quad \cot s = \dfrac{x}{y}$

$$\dfrac{x}{y} = \dfrac{x}{y}$$

$$\cot s = \dfrac{\cos s}{\sin s}$$

42. a. $\sin s = y \quad \cos s = x$

$$x^2 + y^2 = 1$$

$$\cos^2 s + \sin^2 s = 1$$

$$\sin^2 s + \cos^2 s = 1$$

b. $\sec s = \dfrac{1}{x} \quad \tan s = \dfrac{y}{x}$

$$x^2 + y^2 = 1$$

$$x^2 = 1 - y^2$$

$$\left(\dfrac{1}{x}\right)^2 - \left(\dfrac{y}{x}\right)^2 = \dfrac{1 - y^2}{x^2} = \dfrac{x^2}{x^2} = 1$$

$$\sec^2 s - \tan^2 s = 1$$

c. $\csc s = \dfrac{1}{y} \quad \cot s = \dfrac{x}{y}$

$$x^2 + y^2 = 1$$

$$y^2 = 1 - x^2$$

$$\left(\dfrac{1}{y}\right)^2 - \left(\dfrac{x}{y}\right)^2 = \dfrac{1 - x^2}{y^2} = \dfrac{y^2}{y^2} = 1$$

$$\csc^2 s - \cot^2 s = 1$$

43. If n is even, then $\sin\left(n\pi + \dfrac{\pi}{2}\right)$ can be written as $\sin\left(2k\pi + \dfrac{\pi}{2}\right)$, where k is a nonnegative integer. When n is even, $(-1)^n = 1$. Since $2k\pi$ represents k complete revolutions, $\left(2k\pi + \dfrac{\pi}{2}\right)$ corresponds to $(0, 1)$ in the wrapping function. Therefore, if $s = n\pi + \dfrac{\pi}{2}$ and n is even, then $\sin s = (-1)^n = 1$.

If n is odd, then $\sin\left(n\pi + \dfrac{\pi}{2}\right)$ can be written as $\sin\left((2k + 1)\pi + \dfrac{\pi}{2}\right)$, where k is a nonnegative integer.

$$\sin\left((2k + 1)\pi + \dfrac{\pi}{2}\right)$$
$$= \sin\left(2k\pi + \pi + \dfrac{\pi}{2}\right)$$
$$= \sin\left(2k\pi + \dfrac{3\pi}{2}\right)$$

$2k\pi$ is k complete revolutions and $\dfrac{3\pi}{2}$ corresponds to $(0, -1)$ in the wrapping function.

$$\sin\left(2k\pi + \dfrac{3\pi}{2}\right) = -1$$

When n is odd, $(-1)^n = -1$. Therefore, when n is odd,

$$\sin\left(n\pi + \dfrac{\pi}{2}\right) = (-1)^n.$$

29. Find sin s in terms of a if s corresponds to the point (a, a^2) on a unit circle.

30. Find tan s in terms of b if s corresponds to the point (b^2, b^3) on a unit circle.

31. a. If the point $\left(x, \frac{24}{25}\right)$, where $x > 0$, is on a unit circle, find x.

    b. If s corresponds to the point in part **a**, determine the values of sin s, cos s, and tan s.

32. If sin $s = \frac{3}{5}$ and cos $s = \frac{4}{5}$, find the four other trigonometric functions of s.

33. If sec $s = \frac{13}{5}$ and csc $s = \frac{13}{12}$, find tan s.

34. Find all possible values of x or y, to the nearest thousandth, if the coordinates represent a point on a unit circle.

    a. $(x, 0.612045)$ **b.** $(0.30124, y)$
    c. $(x, -0.415)$ **d.** $(-0.22354, y)$

35. In which quadrants are there points on a unit circle that are terminal points of an arc of length s such that, for the values of s in that quadrant:

    a. sin s and cos s are both negative   **b.** tan s is positive

In **36 – 39**, for the given conditions, find the values of sin s and tan s to the nearest hundredth.

36. cos $s = 0.7$ and s terminates in quadrant I

37. sec $s = 1.41$ and s terminates in quadrant IV

38. csc $s = 2.5$ and s terminates in quadrant I

39. cos $s = -0.21$ and s terminates in quadrant III

In **40 – 42**, use the definitions of the trigonometric functions to prove the given statement.

40. a. $\sin s = \dfrac{1}{\csc s}$

    b. $\cos s = \dfrac{1}{\sec s}$

41. a. $\tan s = \dfrac{\sin s}{\cos s}$, where cos $s \neq 0$

    b. $\cot s = \dfrac{\cos s}{\sin s}$, where sin $s \neq 0$

42. a. $\sin^2 s + \cos^2 s = 1$

    b. $\sec^2 s - \tan^2 s = 1$

    c. $\csc^2 s - \cot^2 s = 1$

43. Prove that for any nonnegative integer n: $\sin\left(n\pi + \frac{\pi}{2}\right) = (-1)^n$

8.2 Using Angles to Define the Trigonometric Functions

In addition to defining the trigonometric functions in terms of arc length on a unit circle, real numbers that represent the measures of angles can be used. To understand what is meant by an angle in this context, consider an angle formed by rotating a *ray* around its endpoint. This endpoint is called the *vertex* of the angle. The position of the ray before the rotation begins is the *initial side* of the angle and the position of the ray when the rotation ends is the *terminal side*. If the rotation is in a counterclockwise direction, the measure of the angle is positive; if the rotation is clockwise, the measure is negative.

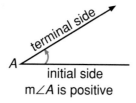

m∠A is positive

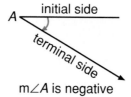

m∠A is negative

Measuring Angles Using Degrees

If a point (other than the endpoint) on a ray traces a complete rotation, a circle is formed. If the circumference of the circle is divided into 360 equal arcs, an angle with vertex at the center of the circle that intercepts one of these arcs has a measure of one *degree*, written 1°. The arc also has a measure of 1°.

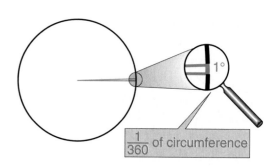

$\frac{1}{360}$ of circumference

The idea of using $\frac{1}{360}$ of the circle was devised by the Babylonians because 360 has many divisors, thus allowing many common fractional parts of a rotation to correspond to a whole number of degrees. For example, $\frac{1}{2}, \frac{1}{3}, \frac{1}{4}, \frac{1}{6}, \frac{1}{8}$, and $\frac{1}{12}$ of a rotation can be expressed as 180, 120, 90, 60, 45, and 30 degrees, respectively. Each degree can be further divided into 60 minutes and each minute into 60 seconds. To put these measures into perspective, consider that a central angle of one second would intercept an arc about 100 feet long on the Earth's circumference, which is approximately 25,000 miles. An arc of one minute would be a little more than 1 mile, and 1 degree, about 70 miles. An angle A with measure 40 degrees 20 minutes 15 seconds is written $m\angle A = 40°20'15''$.

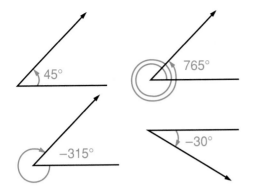

The increased use of calculators has made it practical to express parts of a degree using decimals. For example, 36.5° is often used rather than 36°30′. Similarly, 40.9° is the same as 40°54′ (0.9 × 60 minutes = 54 minutes). Many scientific calculators have keys that convert angle measures expressed as decimals into degrees, minutes, and seconds, and vice versa.

Examples of various angles found by rotation and their measures in degrees are shown at the left.

Measuring Angles Using Radians

An angle has a measure of 1 radian when its vertex is at the center of a circle and it intercepts an arc equal in length to the length of the radius of the circle. The measure of a central angle in radians is equal to the ratio of the length of its arc to the length of the radius of the circle. A central angle of b radians will intercept an arc of length b times the radius of the circle.

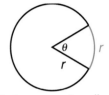

θ measures 1 radian

Since the circumference of a circle has a length of $2\pi r$, and $2\pi r$ is 2π times the length of the radius of the circle, a complete rotation of 360° is equivalent to 2π radians. Therefore, a straight angle or one-half of a complete rotation has a measure of 180° or π radians.

$$180° = \pi \text{ radians}$$

Using this relationship, it is easy to convert angle measures given in degrees to equivalent angle measures in radians. Some frequently used equivalents are shown in the table.

Measure in Degrees	360	270	180	90	60	45	30	1
Measure in Radians	2π	$\dfrac{3\pi}{2}$	π	$\dfrac{\pi}{2}$	$\dfrac{\pi}{3}$	$\dfrac{\pi}{4}$	$\dfrac{\pi}{6}$	$\dfrac{\pi}{180}$

To convert between degrees and radians the following proportion may be used:

$$\frac{\text{measure in degrees}}{\text{measure in radians}} = \frac{180}{\pi}$$

Example 1

Convert 15 degrees to radian measure.

Solution

Use the proportion $\dfrac{\text{degrees}}{\text{radians}} = \dfrac{180}{\pi}$

$$\frac{15}{\text{radians}} = \frac{180}{\pi}$$

$$\text{radians} = \frac{15\pi}{180} = \frac{\pi}{12}$$

Answer: $15° = \frac{\pi}{12}$ radians

Example 2 _____

Convert $\dfrac{3\pi}{4}$ radians to degree measure.

Solution

$$\frac{\text{degrees}}{\text{radians}} = \frac{180}{\pi}$$

$$\frac{\text{degrees}}{\frac{3\pi}{4}} = \frac{180}{\pi}$$

$$\text{degrees} = \frac{180\left(\frac{3\pi}{4}\right)}{\pi} = 135$$

Answer: $\dfrac{3\pi}{4}$ radians = 135°

Example 3 _____

Convert 3.2 radians to degree measure.

Solution

$$\frac{\text{degrees}}{\text{radians}} = \frac{180}{\pi}$$

$$\frac{\text{degrees}}{3.2} = \frac{180}{\pi}$$

$$\text{degrees} = \frac{180(3.2)}{\pi} = \frac{576}{\pi}$$

Using $\pi \approx 3.14$,

$$\frac{576}{\pi} \approx \frac{576}{3.14} \approx 183.44$$

Answer: 3.2 radians = $\left(\dfrac{576}{\pi}\right)^{\circ} \approx 183.44°$

Example 4 _____

 Convert 1 radian to degree measure, rounding the answer to the nearest thousandth of a degree.

Solution

$$\frac{\text{degrees}}{\text{radians}} = \frac{180}{\pi}$$

$$\frac{\text{degrees}}{1} = \frac{180}{\pi}$$

$$\text{degrees} = \frac{180}{\pi} \approx 57.296°$$

Answer: 1 radian ≈ 57.296°

It is convenient to remember that 1 radian is about 57.3°. Scientific calculators have a key for π, which uses a value accurate to at least eight decimal places.

There are a number of useful formulas involving radian measure, such as the relationship between a central angle measured in radians and the length of the arc it intercepts on the circumference of a circle. Note that the formula indicates that the measure of the angle in radians is equal to the number of segments of length r into which the intercepted arc can be divided.

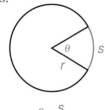

$$\theta = \frac{s}{r}$$

Example 5

Find the length of the radius of a circle in which an angle of 2.5 radians intercepts an arc 10 cm long.

Solution

$$\theta = \frac{s}{r}$$

$$2.5 = \frac{10}{r}$$

$$r = \frac{10}{2.5} = 4$$

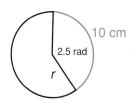

10 cm

2.5 rad

r

Answer: The length of the radius is 4 cm.

Using Angles as the Domain of Trigonometric Functions

The figure shows an angle θ in **standard position** with its vertex at the origin and its initial side coinciding with the positive x-axis. A unit circle is drawn with its center at the origin. Since the measure of θ is s radians, the domain of the trigonometric functions may be either the set of real numbers representing the arc length s or the set of real numbers representing the measures of angle θ.

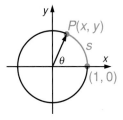

Previously, the coordinates of point P on the unit circle were used to generate the values of the trigonometric functions. The values of these functions can also be expressed in terms of the coordinates of a point on a circle with radius r.

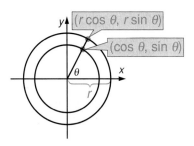

In the second figure, θ is the measure of an angle in standard position. The smaller of the two concentric circles is a unit circle and the larger is the image of the unit circle under the dilation D_r.

As defined earlier, on the unit circle, $(x, y) = (\cos \theta, \sin \theta)$. Under this dilation D_r:

$$D_r(x, y) = (x', y')$$

$$D_r(\cos \theta, \sin \theta) = (r \cos \theta, r \sin \theta)$$

Therefore, $r \cos \theta = x'$

$$\cos \theta = \frac{x'}{r}$$

and $r \sin \theta = y'$

$$\sin \theta = \frac{y'}{r}$$

A similar method can be used to generalize each of the other trigonometric functions in terms of the coordinates of a point on a circle with radius r. This can be summarized as follows:

If (x, y) (x and y not both 0) are the coordinates of a point on the terminal side of an angle θ in standard position and r is the radius of a circle passing through (x, y) with center at the origin, then:

$$\sin \theta = \frac{y}{r} \qquad\qquad \csc \theta = \frac{r}{y}\ (y \neq 0)$$

$$\cos \theta = \frac{x}{r} \qquad\qquad \sec \theta = \frac{r}{x}\ (x \neq 0)$$

$$\tan \theta = \frac{\frac{y}{r}}{\frac{x}{r}} = \frac{y}{x}\ (x \neq 0) \qquad \cot \theta = \frac{\frac{x}{r}}{\frac{y}{r}} = \frac{x}{y}\ (y \neq 0)$$

Example 6

The point $(-7, -24)$ is on the terminal side of angle θ.
Find $\sin \theta$, $\cos \theta$, and $\tan \theta$.

Solution

In order to find the trigonometric functions, find r. The point $(-7, -24)$ is on a circle with center at the origin. The equation of the circle is $x^2 + y^2 = r^2$.

$$(-7)^2 + (-24)^2 = r^2$$

$$49 + 576 = r^2$$

$$625 = r^2$$

$$25 = r$$

$$\sin \theta = \frac{y}{r} = \frac{-24}{25} = -\frac{24}{25}$$

$$\cos \theta = \frac{x}{r} = \frac{-7}{25} = -\frac{7}{25}$$

$$\tan \theta = \frac{y}{x} = \frac{-24}{-7} = \frac{24}{7}$$

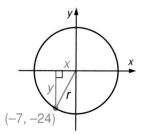

Defining the Trigonometric Functions in Terms of the Sides of a Right Triangle

In right triangle OPA with right angle at A, y is the length of the leg opposite angle θ, x is the length of the leg adjacent to angle θ, and r is the hypotenuse of triangle OPA. Previously, $\sin \theta$ has been defined as $\frac{y}{r}$. Therefore, in terms of the right triangle with acute angle θ, $\sin \theta$ is the ratio of the length of the side opposite θ and the length of the hypotenuse:

$$\sin \theta = \frac{\text{length of opposite side}}{\text{length of hypotenuse}}$$

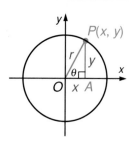

The six trigonometric function values of the acute angle θ are defined below using abbreviations "opp" for length of the opposite leg, "adj" for length of the adjacent leg, and "hyp" for length of the hypotenuse.

$$\sin \theta = \frac{y}{r} = \frac{\text{opp}}{\text{hyp}} \qquad \cos \theta = \frac{x}{r} = \frac{\text{adj}}{\text{hyp}} \qquad \tan \theta = \frac{y}{x} = \frac{\text{opp}}{\text{adj}}$$

$$\csc \theta = \frac{r}{y} = \frac{\text{hyp}}{\text{opp}} \qquad \sec \theta = \frac{r}{x} = \frac{\text{hyp}}{\text{adj}} \qquad \cot \theta = \frac{x}{y} = \frac{\text{adj}}{\text{opp}}$$

Example 7

Find the values of sin 30°, cos 30°, and tan 30° by using a right triangle with a 30° angle.

Solution

Since all right triangles with a 30° acute angle are similar, choose such a triangle with 1 as the length of the shorter leg. Then, according to the ratios for the sides of a 30°-60°-90° triangle, the length of the hypotenuse is 2 and the length of the longer leg is $\sqrt{3}$.

Therefore, $\sin 30° = \dfrac{\text{opp}}{\text{hyp}} = \dfrac{1}{2}$

$\cos 30° = \dfrac{\text{adj}}{\text{hyp}} = \dfrac{\sqrt{3}}{2}$

$\tan 30° = \dfrac{\text{opp}}{\text{adj}} = \dfrac{1}{\sqrt{3}} = \dfrac{\sqrt{3}}{3}$

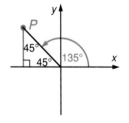

Example 8

Find sin 135°, cos 135°, and tan 135°.

Solution

Since a right triangle with a 135° angle cannot be drawn, place the angle in standard position and compute the coordinates of a point on the terminal ray.

A right triangle is formed in the second quadrant. Since one acute angle is 45°, the other must also be 45°, and the triangle is isosceles. Since the legs are equal in length, the x- and y-coordinates of P have the same absolute value. In quadrant II, the x-coordinate of P is negative and the y-coordinate is positive. Thus, if you choose a value of x, the value of y is determined. For example, if $x = -2$, y must be 2 and P is the point $(-2, 2)$.

Use $x^2 + y^2 = r^2$, the equation of the circle passing through P with center at the origin, to determine the value of r.

$$(-2)^2 + (2)^2 = r^2$$

$$4 + 4 = r^2$$

$$r = 2\sqrt{2}$$

$$\sin 135° = \frac{y}{r} = \frac{2}{2\sqrt{2}} = \frac{\sqrt{2}}{2}$$

$$\cos 135° = \frac{x}{r} = \frac{-2}{2\sqrt{2}} = \frac{-\sqrt{2}}{2}$$

$$\tan 135° = \frac{y}{x} = \frac{2}{-2} = -1$$

The previous example suggests that the definitions of the trigonometric functions of an angle in terms of the coordinates of a point on the terminal ray of the angle can be used to determine the sign of each of the six functions for angles in each of the four quadrants. For example, if an arc of length s on the unit circle terminates in quadrant III, x and y are both negative, thus: $\sin s = y$ is negative, $\cos s = x$ is negative and $\tan s = \dfrac{y}{x}$ is positive. The signs of the trigonometric functions in the four quadrants are summarized in the table below. Note that each pair of functions is positive in quadrant I and one other quadrant.

Quadrant	I	II	III	IV
sin and csc	+	+	−	−
cos and sec	+	−	−	+
tan and cot	+	−	+	−

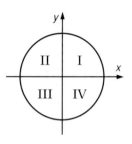

Example 9

An angle in standard position in quadrant III has a cosine of $-\frac{1}{4}$. Find the sine and tangent of the angle.

Solution

Let the angle be θ. If $\cos \theta = -\frac{1}{4}$, then let the point $\left(-\frac{1}{4}, y\right)$ be on the terminal ray of the angle and on the unit circle. The value of y can be found by using the equation of the unit circle with center at the origin, $x^2 + y^2 = 1$.

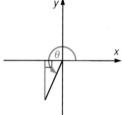

$$\left(-\tfrac{1}{4}\right)^2 + y^2 = 1$$

$$\tfrac{1}{16} + y^2 = 1$$

$$y^2 = \tfrac{15}{16}$$

$$y = -\frac{\sqrt{15}}{4} \quad \text{In quadrant III, } y \text{ is negative.}$$

$$\sin \theta = y = -\frac{\sqrt{15}}{4}$$

$$\tan \theta = \frac{y}{x} = \frac{-\dfrac{\sqrt{15}}{4}}{-\dfrac{1}{4}} = \sqrt{15}$$

Exercises 8.2

In **1 – 12**, convert the degree measure to radians. Express answers in terms of π where necessary.

1. 135°   **2.** 120°   **3.** 330°   **4.** 150°

5. 405°   **6.** 540°   **7.** –60°   **8.** –390°

9. –600°   **10.** 140°   **11.** $\frac{100°}{\pi}$ **12.** $-\frac{110°}{\pi}$

In **13 – 28**, the given value represents the radian measure of an angle. Write an equivalent degree measure (express your answer to the nearest tenth of a degree), using a calculator as needed.

13. $\frac{2\pi}{3}$ **14.** $\frac{\pi}{6}$ **15.** $-\frac{4\pi}{3}$ **16.** $-\frac{5\pi}{3}$

17. 5π **18.** 5   **19.** 6.3   **20.** 3.15

21. $\frac{2}{3}$ **22.** $-\frac{1}{4}$ **23.** $\frac{10}{7}$ **24.** $-\frac{5}{6}$

25. $\frac{3}{\pi}$ **26.** $\frac{7}{5\pi}$ **27.** $-\frac{8}{\pi}$ **28.** $-\frac{7}{2\pi}$

29. Find the measure in radians of a central angle in a circle whose radius is 8 cm if the angle intercepts an arc 4 cm long.

30. Find the measure in radians of a central angle in a circle whose radius is 1 inch if the angle intercepts an arc 2.45 inches long.

31. A central angle in a circle has a measure of 1.5 radians. If the radius of the circle is 6 inches, find the length of the arc on the circle intercepted by the angle.

32. A central angle in a circle has a measure of $\frac{2\pi}{3}$ radians. If the radius of the circle is 12 inches, find the length of the arc on the circle intercepted by the angle.

33. A central angle of 1.4 radians intercepts an arc 5.6 inches long on the circumference of a circle. Find the length of the radius of the circle.

34. A central angle of 5 radians intercepts an arc 32 inches long on the circumference of a circle. Find the length of the radius of the circle.

35. How much time passes if the minute hand of a clock moves through an angle of:
a. $-\frac{\pi}{2}$ radians   **b.** $-\frac{2\pi}{3}$ radians

36. How many radians does a clock's hour hand pass through in:
a. 2 hours   **b.** 30 minutes

In **37 – 41**, find the exact value of sin θ, cos θ, and tan θ if the terminal ray of θ passes through the given point.

37. (–3, 4)   **38.** (–7, 24)   **39.** (15, 8)   **40.** (4, 5)   **41.** (–5, 6)

41. $\sin \theta = \frac{6\sqrt{61}}{61}$, $\cos \theta = -\frac{5\sqrt{61}}{61}$, $\tan \theta = -\frac{6}{5}$

42. $\sin \theta = -\frac{3\sqrt{13}}{13}$, $\cos \theta = -\frac{2\sqrt{13}}{13}$, $\tan \theta = \frac{3}{2}$, $\csc \theta = -\frac{\sqrt{13}}{3}$, $\sec \theta = -\frac{\sqrt{13}}{2}$, $\cot \theta = \frac{2}{3}$

43. $\sin 45° = \frac{\sqrt{2}}{2}$, $\cos 45° = \frac{\sqrt{2}}{2}$, $\tan 45° = 1$, $\csc 45° = \sqrt{2}$, $\sec 45° = \sqrt{2}$, $\cot 45° = 1$

44. $\sin 120° = \frac{\sqrt{3}}{2}$, $\cos 120° = -\frac{1}{2}$, $\tan 120° = -\sqrt{3}$, $\csc 120° = \frac{2\sqrt{3}}{3}$, $\sec 120° = -2$, $\cot 120° = -\frac{\sqrt{3}}{3}$

Answers 8.2

1. $\frac{3\pi}{4}$ **2.** $\frac{2\pi}{3}$

3. $\frac{11\pi}{6}$ **4.** $\frac{5\pi}{6}$

5. $\frac{9\pi}{4}$ **6.** 3π

7. $-\frac{\pi}{3}$ **8.** $-\frac{13\pi}{6}$

9. $-\frac{10\pi}{3}$ **10.** $\frac{7\pi}{9}$

11. $\frac{5}{9}$ **12.** $-\frac{11}{18}$

13. 120°   **14.** 30°

15. –240°   **16.** –300°

17. 900°   **18.** 286.5°

19. 361.0°   **20.** 180.5°

21. 38.2°   **22.** –14.3°

23. 81.9°   **24.** –47.7°

25. 54.7°   **26.** 25.5°

27. –145.9°   **28.** –63.8°

29. $\frac{1}{2}$ radian

30. 2.45 radians

31. 9 in.   **32.** 8π in.

33. 4 in.   **34.** 6.4 in.

35. a. 15 minutes
    b. 20 minutes

36. a. $-\frac{\pi}{3}$
    b. $-\frac{\pi}{12}$

37. $\sin \theta = \frac{4}{5}$,
    $\cos \theta = -\frac{3}{5}$,
    $\tan \theta = -\frac{4}{3}$

38. $\sin \theta = \frac{24}{25}$,
    $\cos \theta = -\frac{7}{25}$,
    $\tan \theta = -\frac{24}{7}$

39. $\sin \theta = \frac{8}{17}$,
    $\cos \theta = \frac{15}{17}$,
    $\tan \theta = \frac{8}{15}$

40. $\sin \theta = \frac{5\sqrt{41}}{41}$,
    $\cos \theta = \frac{4\sqrt{41}}{41}$,
    $\tan \theta = \frac{5}{4}$

45. $\sin(-30°) = -\frac{1}{2}$,

$\cos(-30°) = \frac{\sqrt{3}}{2}$,

$\tan(-30°) = -\frac{\sqrt{3}}{3}$,

$\csc(-30°) = -2$,

$\sec(-30°) = \frac{2\sqrt{3}}{3}$,

$\cot(-30°) = -\sqrt{3}$

46. $\sin\frac{5\pi}{6} = \frac{1}{2}$,

$\cos\frac{5\pi}{6} = -\frac{\sqrt{3}}{2}$,

$\tan\frac{5\pi}{6} = -\frac{\sqrt{3}}{3}$,

$\csc\frac{5\pi}{6} = 2$,

$\sec\frac{5\pi}{6} = -\frac{2\sqrt{3}}{3}$,

$\cot\frac{5\pi}{6} = -\sqrt{3}$

47. $\sin\pi = 0$,

$\cos\pi = -1$,

$\tan\pi = 0$

$\csc\pi$ is undefined,

$\sec\pi = -1$,

$\cot\pi$ is undefined

48. $\sin\left(-\frac{\pi}{2}\right) = -1$,

$\cos\left(-\frac{\pi}{2}\right) = 0$,

$\tan\left(-\frac{\pi}{2}\right)$ is undefined,

$\csc\left(-\frac{\pi}{2}\right) = -1$,

$\sec\left(-\frac{\pi}{2}\right)$ is undefined,

$\cot\left(-\frac{\pi}{2}\right) = 0$

49. $\sin 420° = \frac{\sqrt{3}}{2}$,

$\cos 420° = \frac{1}{2}$,

$\tan 420° = \sqrt{3}$,

$\csc 420° = \frac{2\sqrt{3}}{3}$,

$\sec 420° = 2$,

$\cot 420° = \frac{\sqrt{3}}{3}$

50. $\sin\theta = \frac{5}{13}$, $\cos\theta = -\frac{12}{13}$,

$\tan\theta = -\frac{5}{12}$, $\csc\theta = \frac{13}{5}$,

$\sec\theta = -\frac{13}{12}$, $\cot\theta = -\frac{12}{5}$

Exercises 8.2 *(continued)*

In **42 – 49**, find the exact values of the six trigonometric functions of the given angle.

42. The terminal side of angle θ passes through the point $(-2, -3)$.

43. 45°

44. 120° **45.** –30° **46.** $\frac{5\pi}{6}$ radians

47. π radians **48.** $-\frac{\pi}{2}$ radians **49.** 420°

In **50 – 53**, find the exact value of the remaining five trigonometric functions if θ has its terminal side in the given quadrant and θ has the given function value.

50. $\sin\theta = \frac{5}{13}$; θ in quadrant II

51. $\tan\theta = -\frac{3}{4}$; θ in quadrant IV

52. $\cos\theta = -\frac{2}{3}$; θ in quadrant III

53. $\sec\theta = \frac{3}{2}$; θ in quadrant I

The prefix "co" in the functions cosine, cosecant, and cotangent stands for *complementary*. For example, if $\sin\theta = y$, then the cosine of the complement of θ is also y. This can be written in symbols using the fact that the complement of θ can be expressed as $(90° - \theta)$. Thus, $\sin\theta = \cos(90° - \theta)$, $\sec\theta = \csc(90° - \theta)$, and $\tan\theta = \cot(90° - \theta)$.

In **54 – 57**, verify the given relationship by computing the value of each side of the equation independently.

54. $\sin 30° = \cos 60°$ **55.** $\sin 45° = \cos 45°$

56. $\tan\frac{\pi}{3} = \cot\frac{\pi}{6}$ **57.** $\sec\frac{\pi}{2} = \csc 0$

In **58 – 64**, find the lengths of the other sides of the figure.

58.

59.

60.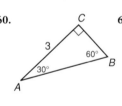

61.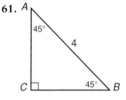

62.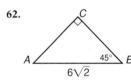

63.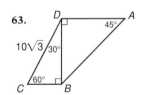

64.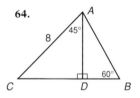

---

51. $\sin\theta = -\frac{3}{5}$, $\cos\theta = \frac{4}{5}$, $\tan\theta = -\frac{3}{4}$,

$\csc\theta = -\frac{5}{3}$, $\sec\theta = \frac{5}{4}$, $\cot\theta = -\frac{4}{3}$

52. $\sin\theta = -\frac{\sqrt{5}}{3}$, $\cos\theta = -\frac{2}{3}$,

$\tan\theta = \frac{\sqrt{5}}{2}$, $\csc\theta = -\frac{3\sqrt{5}}{5}$,

$\sec\theta = -\frac{3}{2}$, $\cot\theta = \frac{2\sqrt{5}}{5}$

53. $\sin\theta = \frac{\sqrt{5}}{3}$, $\cos\theta = \frac{2}{3}$,

$\tan\theta = \frac{\sqrt{5}}{2}$, $\csc\theta = \frac{3\sqrt{5}}{5}$,

$\sec\theta = \frac{3}{2}$, $\cot\theta = \frac{2\sqrt{5}}{5}$

Exercises 8.2 *(continued)*

In mechanics, it is often necessary to measure the speed of a point that is rotating in a circle. This measure is called ***angular velocity*** and is expressed in radians (or degrees or revolutions) per unit of time. That is, since $\theta = s \div r$, then θ/unit of time = $(s \div r)$/unit of time. For example, on a circular track whose radius is $\frac{1}{10}$ mile, a car traveling at 60 miles/hour (or 1 mile/minute) would have an angular velocity of $60 \div \frac{1}{10} = 600$ radians/hour (or 10 radians/minute).

65. John is traveling on a bicycle at a speed of 880 feet/minute. If the outer radius of the front wheel is 1.2 feet, find its angular velocity in radians/minute.

66. The wheel on a unicycle has an outer diameter of 1 foot. How many revolutions per minute does the wheel turn if someone riding the unicycle is moving 10 miles/hour? Express the answer to the nearest hundredth.

67. The fan belt on a car connects two pulleys that are 4 inches and 6 inches in diameter. When the engine is idling, the larger pulley rotates at 400 revolutions per minute.

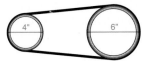

    a. To the nearest tenth, find the speed, in inches per minute, of a single crack that has begun to form in the belt.

    b. Find the speed of the smaller pulley.

68. Assuming Earth is a sphere with a radius of 3,960 miles, find the speed in miles/hour of any stationary building along the equator as Earth rotates. *Hint:* Earth makes 1 revolution in 24 hours.

69. Rochester, NY and Richmond, VA are located on approximately the same line of longitude. If Rochester is at 43.5° north latitude and Richmond is at 37.5° north latitude, find the distance between the two cities. Assume Earth is a sphere whose radius is 3,960 miles. Answer to the nearest mile.

70. New York City and Montreal are located at approximately 73.5° longitude. If their latitudes are 40.5° north and 45.5° north, find the distance between them, to the nearest tenth of a kilometer. Assume that Earth is a sphere whose diameter is 12,670 km.

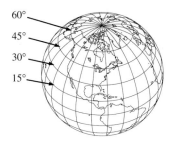

57.

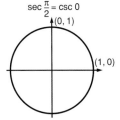

$\sec \frac{\pi}{2} = \csc 0$

$\sec \frac{\pi}{2} = \frac{1}{0}$ undefined $\csc 0 = \frac{1}{0}$ undefined

58. $BC = 4$ $AC = 4\sqrt{3}$

59. $AB = 6$ $AC = 3\sqrt{3}$

60. $BC = \sqrt{3}$ $AB = 2\sqrt{3}$

61. $AC = BC = 2\sqrt{2}$

62. $AC = BC = 6$

63. $BC = 5\sqrt{3}$ $DB = 15$

    $AD = 15$

    $AB = 15\sqrt{2}$

64. $AD = DC = 4\sqrt{2}$

    $AB = \frac{8\sqrt{6}}{3}$ $DB = \frac{4\sqrt{6}}{3}$

65. 733.3 rad/min.

66. 280.11 rev./min.

67. a. 7,539.8 in./min.

    b. 600 rev./min.

68. 1,036.7 mph

69. 415 mi.

70. 552.8 km

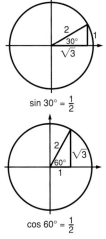
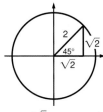
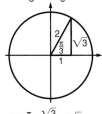
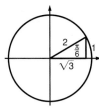

8.3 Approximating Values of the Trigonometric Functions

The exact values of the trigonometric functions for some angles such as 30°, 45°, and 60° can be found by using geometry. However, for most values of the trigonometric functions, only a rational estimate of their irrational values can be made. This is usually done by using a calculator or a table of values.

Most scientific calculators accept radians or degrees as the argument, or the input value, of the function, and it is important that the user select the correct mode in which the calculator is to operate before entering data. It is also important to distinguish between the decimal-degree mode and the degree-minute-second mode if your calculator can express results in either mode.

Example 1

 Find the sine and cosine of 46.25° to the nearest thousandth.

Solution

The trigonometric function keys on most scientific calculators are `SIN`, `COS`, and `TAN`. The argument of the function is entered first on most, but not all, calculators. Check that your calculator is in decimal-degree mode.

Enter:	Display:
46.25 `SIN` or `SIN` 46.25	0.7223640
46.25 `COS` or `COS` 46.25	0.69151306

The number of digits displayed by your calculator may differ from what is shown here.

Answer: sin 46.25° = 0.722 and cos 46.25° = 0.692

Example 2

 Find tan 26°12′ correct to five significant digits.

Solution

For most calculators, it is necessary to express the 12′ of 26°12′ as a fractional part of a degree. Since there are 60 minutes in a degree, divide 12 by 60.

Enter: 12 `÷` 60 `+` 26 `=` `TAN`

Display: 0.492061

Answer: tan 26°12′ = 0.49206

Note: Some calculators have a key to convert from minutes and seconds to a fractional part of a degree, expressed as a decimal value. This key may be labeled DMS/DD or ° ′ ″. If your calculator has such a key, it may be used in place of the division by 60.

You have seen that a trigonometric function can have the same value for many different angles. For example, sin 30° = sin 390° = 0.5. In this section, however, you need only be concerned with acute angles, all of which have positive trigonometric values. In many problems, it is necessary only to find the acute angle that corresponds to a given function value. For example, if θ is acute and sin θ = 0.5, then θ = 30°.

Example 3 _____

 If cos θ = 0.75, find the acute angle θ to the nearest degree.

Solution

Place the calculator in degree mode.

Enter: .75 INV COS

or .75 COS⁻¹

or 2nd COS⁻¹ .75

Display: | 41.409622 |

Answer: To the nearest degree, θ = 41°.

Example 4 _____

 If csc θ = 1.41, find acute angle θ to the nearest tenth of a degree.

Solution

Place the calculator in degree mode. Since there is no calculator key that will return cosecant directly, use the relationship sin $\theta = \dfrac{1}{\csc\theta}$. Therefore, sin $\theta = \dfrac{1}{1.41}$.

Enter: 1.41 ¹/ₓ INV SIN

or 1.41 X⁻¹ SIN⁻¹

or 2nd SIN⁻¹ 1.41 X⁻¹

Display: | 45.171476 |

Answer: To the nearest tenth of a degree, θ = 45.2°.

If a calculator is not available, a table such as Table 5 in the Appendix can be used to find the approximate value of a trigonometric function or the angle that corresponds to a given value of a trigonometric function. Note that, due to space limitations, most tables give values accurate to fewer decimal places than the values computed by a calculator.

The values below represent a portion of Table 5. The table makes use of the relationships between functions and cofunctions (sine and cosine, tangent and cotangent, secant and cosecant). Recall, for example, that $\sin \theta = \cos (90° - \theta)$. The left side of the table lists degree measures from 0° to 45°, and the right side lists measures from 45° to 90°. To find the values of the trigonometric functions of θ in the interval (45°, 90°], the values in the right-hand columns must be read from bottom to top, using the headings at the bottom of the table.

VALUES OF TRIGONOMETRIC FUNCTIONS

θ Degrees	θ Radians	$\sin \theta$	$\csc \theta$	$\tan \theta$	$\cot \theta$	$\sec \theta$	$\cos \theta$		
0°00′	.0000	.0000	Undefined	.0000	Undefined	1.000	1.0000	1.5708	90°00′
10′	.0029	.0029	343.8	.0029	343.8	1.000	1.0000	1.5679	50′
20′	.0058	.0058	171.9	.0058	171.9	1.000	1.0000	1.5650	40′
30′	.0087	.0087	114.6	.0087	114.6	1.000	1.0000	1.5621	30′
40′	.0116	.0116	85.95	.0116	85.94	1.000	.9999	1.5592	20′
50′	.0145	.0145	68.76	.0145	68.75	1.000	.9999	1.5563	10′
1°00′	.0175	.0175	57.30	.0175	57.29	1.000	.9998	1.5533	89°00′
10′	.0204	.0204	49.11	.0204	49.10	1.000	.9998	1.5504	50′
20′	.0233	.0233	42.98	.0233	42.96	1.000	.9997	1.5475	40′
30′	.0262	.0262	38.20	.0262	38.19	1.000	.9997	1.5446	30′
40′	.0291	.0291	34.38	.0291	34.37	1.000	.9996	1.5417	20′
50′	.0320	.0320	31.26	.0320	31.24	1.001	.9995	1.5388	10′
2°00′	.0349	.0349	28.65	.0349	28.64	1.001	.9994	1.5359	88°00′
10′	.0378	.0378	26.45	.0378	26.43	1.001	.9993	1.5330	50′
20′	.0407	.0407	24.56	.0407	24.54	1.001	.9992	1.5301	40′
30′	.0436	.0436	22.93	.0437	22.90	1.001	.9990	1.5272	30′
40′	.0465	.0465	21.49	.0466	21.47	1.001	.9989	1.5243	20′
50′	.0495	.0494	20.23	.0495	20.21	1.001	.9988	1.5213	10′
3°00′	.0524	.0523	19.11	.0524	19.08	1.001	.9986	1.5184	87°00′
10′	.0553	.0552	18.10	.0553	18.07	1.002	.9985	1.5155	50′
20′	.0582	.0581	17.20	.0582	17.17	1.002	.9983	1.5126	40′
30′	.0611	.0610	16.38	.0612	16.35	1.002	.9981	1.5097	30′
40′	.0640	.0640	15.64	.0641	15.60	1.002	.9980	1.5068	20′
50′	.0669	.0669	14.96	.0670	14.92	1.002	.9978	1.5039	10′
4°00′	.0698	.0698	14.34	.0699	14.30	1.002	.9976	1.5010	86°00′
10′	.0727	.0727	13.76	.0729	13.73	1.003	.9974	1.4981	50′
20′	.0756	.0756	13.23	.0758	13.20	1.003	.9971	1.4952	40′
30′	.0785	.0785	12.75	.0787	12.71	1.003	.9969	1.4923	30′
40′	.0814	.0814	12.29	.0816	12.25	1.003	.9967	1.4893	20′
50′	.0844	.0843	11.87	.0846	11.93	1.004	.9964	1.4864	10′
5°00′	.0873	.0872	11.47	.0875	11.43	1.004	.9962	1.4835	85°00′
10′	.0902	.0901	11.10	.0904	11.06	1.004	.9959	1.4806	50′
20′	.0931	.0929	10.76	.0934	10.71	1.004	.9957	1.4777	40′
30′	.0960	.0958	10.43	.0963	10.39	1.005	.9954	1.4748	30′
		$\cos \theta$	$\sec \theta$	$\cot \theta$	$\tan \theta$	$\csc \theta$	$\sin \theta$	θ Radians	θ Degrees

Example 5

Use the table in the Appendix to find sec 23°40′ to four significant digits.

Solution

Locate 23°40′ on the left side of the table between 23° and 24°. Read across to the column labeled **sec** at the *top* of the table.

sec 23°40′ ≈ 1.092

Answer: To four significant digits, sec 23°40′ = 1.092.

Example 6

Use the table to find a value of θ to the nearest ten minutes (0° < θ < 90°) such that tan θ = 2.282.

Solution

Use the columns in the table having the heading **tan** at the top or bottom. Since 2.282 is in the column headed **tan** at the *bottom*, read the angle from the column at the right.

Answer: θ = 66°20′

Note: Since the angles on the right increase from bottom to top, 66°20′ is located above 66°00′.

Exercises 8.3

 In **1 – 12**, find the value of the function to the nearest thousandth.

1. sin 82°	**2.** cos 46.9°	**3.** tan 1.71°	**4.** cos 0.6°
5. sin 60.32°	**6.** tan 12.3°	**7.** sin 15°30′	**8.** cos 42°10′
9. sec 42°35′	**10.** sec 42.7°	**11.** csc 16.37°	**12.** cot 18°04′

 In **13 – 21**, find the measure of the acute angle θ to the nearest tenth of a degree.

13. sin θ = 0.5	**14.** cos θ = 0.66	**15.** sin θ = 0.471
16. tan θ = 1.612	**17.** sin θ = 0.8124	**18.** tan θ = 16.15
19. sec θ = 3.14	**20.** csc θ = 1.243	**21.** cot θ = 2.6234

22. Find the value of θ to the nearest minute if 0° < θ < 90°.

    a. sin θ = 0.6500   **b.** cos θ = 0.6000

In **23 – 34**, use the trigonometric table to find the value of the function to the nearest hundredth. Note that when the measure of the angle is in degrees, the degree symbol is used. When no unit of measure is written, the measure is understood to be radians.

23. cos 41°	**24.** sin 36°10′	**25.** sec 40°20′	**26.** csc 32°50′
27. cot 66°	**28.** tan 68°10′	**29.** sec 84°30′	**30.** sin 45°10′
31. cos 46°40′	**32.** csc 0.1018	**33.** cos 0.2996	**34.** tan 1.2741

Answers 8.3

1. 0.990
2. 0.683
3. 0.030
4. 1.000
5. 0.869
6. 0.218
7. 0.267
8. 0.741
9. 1.358
10. 1.361
11. 3.548
12. 3.066
13. 30.0°
14. 48.7°
15. 28.1°
16. 58.2°
17. 54.3°
18. 86.5°
19. 71.4°
20. 53.6°
21. 20.9°
22. a. 40°32′
   b. 53°08′
23. 0.75
24. 0.59
25. 1.31
26. 1.84
27. 0.45
28. 2.50
29. 10.43
30. 0.71
31. 0.69
32. 9.84
33. 0.96
34. 3.27

Answers 8.3

35. 8°40′

36. 82°00′

37. 72°00′

38. 57°20′

39. 76°10′

40. 31°30′

41. 0.28

42. 0.46

43. 0.32

44. 1.42

45. 1.03

46. 1.10

47. sin, tan, sec

48. sin, cos

49. tan, cot

50. sec, csc

Exercises 8.3 *(continued)*

In **35 – 40**, use the trigonometric table to find θ to the nearest ten minutes in the interval $[0°, 90°]$.

35. $\sin \theta = 0.1507$ **36.** $\cos \theta = 0.1392$ **37.** $\tan \theta = 3.078$

38. $\sec \theta = 1.852$ **39.** $\csc \theta = 1.03$ **40.** $\cot \theta = 1.634$

In **41 – 46**, use the trigonometric table to find s in the interval $\left[0, \frac{\pi}{2}\right]$. Express your answer to the nearest hundredth of a radian.

41. $\tan s = 0.2899$ **42.** $\cos s = 0.8963$ **43.** $\sin s = 0.3118$

44. $\sec s = 6.659$ **45.** $\csc s = 1.167$ **46.** $\cot s = 0.5095$

In **47 – 50**, use the trigonometric table to answer each of the following questions. Assume that the angle measure lies in the interval $[0°, 90°]$.

47. Which trigonometric functions increase as the acute angle increases in measure?

48. Which trigonometric functions have values only in the interval $[0, \pm1]$?

49. Which trigonometric functions have a range that is the real numbers?

50. Which trigonometric functions have no values in the interval $(0, 1)$?

In calculus, you will learn that infinite series are used to express the values of various functions in mathematics. The sine and cosine functions are among these. If s is the measure of an angle in radians, then

$$\sin s = s - \frac{s^3}{3!} + \frac{s^5}{5!} - \frac{s^7}{7!} + \cdots + (-1)^{n-1}\frac{s^{2n-1}}{(2n-1)!} + \cdots$$

$$\cos s = 1 - \frac{s^2}{2!} + \frac{s^4}{4!} - \frac{s^6}{6!} + \cdots + (-1)^{n-1}\frac{s^{2n-2}}{(2n-2)!} + \cdots$$

51. Evaluate the first four terms of the series for each of the following; then compare your result with the value in the trigonometric table and record the difference between the value you found with your calculator and the one in the table. What can you conclude regarding the accuracy of using the first four terms of the series for these values?

     a. $\cos 0.4$ **b.** $\sin 0.21$

52. Find the value of the trigonometric function to the nearest thousandth. (Measures are in radians.)

     a. $\cos (\tan 1)$ **b.** $\tan (\cos 1)$

     c. $\cos (\tan 0.3)$ **d.** $\tan (\tan 0.41)$

51. $\cos 0.4 \approx 0.9211$ from series, $\cos 0.4 \approx 0.9211$ from table; $\sin 0.21 \approx 0.2085$ from series, $\sin 0.21 \approx 0.2085$ from table. The series is accurate to 4 decimal places for these values.

52. a. 0.013

     b. 0.600

     c. 0.953

     d. 0.464

Exercises 8.3 *(continued)*

One of the reasons for studying mathematics is to be able to understand technical explanations that occur in other areas. For example, a physics book gives the following explanation for Snell's law of refraction of light.

> ... when a light ray strikes a surface making an angle i with its normal (a line perpendicular to the surface), the refracted ray, which lies in the same plane with the incident ray and the normal, makes an angle with the normal in accordance with the relation,
>
> $$n_1 \sin i = n_2 \sin r$$
>
> where i and r refer to the angles (measured with the normal) of incidence and refraction, respectively. n_1 is the refractive index of the medium in which the incident ray originates and n_2 is the refractive index of the medium into which the ray passes.

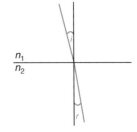

In **53 – 55**, refer to Snell's law.

53. Assume $n_1 = 1.5$ (the index of refraction for glass), and $n_2 = 1.33$ (the index of refraction for water). If i is 38°, find r to the nearest degree.

54. Light entering a diamond from the air, which has an index of 1, enters the diamond at an angle of 40° with the normal and has an angle of refraction of 15.4°. To the nearest tenth, determine the refractive index for the diamond.

55. Refer to the diagram below. Find the index of refraction, to the nearest hundredth, for light going from air into substance x.

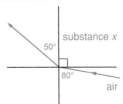

56. In the diagram at the right, if AB is experimentally observed to be 15 when CD is 10, verify that $\dfrac{n_2}{n_1} = 1.5$.

57. Refer to the diagram at the right.
Verify: $\dfrac{\sin i}{\sin r} = \dfrac{AB}{CD}$

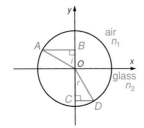

56. $n_1 \sin i = n_2 \sin r$

$\dfrac{n_2}{n_1} = \dfrac{\sin i}{\sin r}$

$= \dfrac{\dfrac{15}{OA}}{\dfrac{10}{OD}}$ 　 $OA = OD$
Radii are equal in length.

$= \dfrac{15}{10} = 1.5$

57. Draw $\overline{AE}$ and $\overline{DF}$ perpendicular to the horizontal. $AB = OE, CD = OF$.

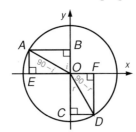

$\sin i = \cos (90 - i)$ 　 $\sin r = \cos(90 - r)$

$= \dfrac{OE}{OA}$ 　 $= \dfrac{OF}{OD}$

$= \dfrac{AB}{OA}$ 　 $= \dfrac{CD}{OD}$

$\dfrac{\sin i}{\sin r} = \dfrac{\dfrac{AB}{OA}}{\dfrac{CD}{OD}} = \dfrac{AB}{CD}$

8.4 *Applying Trigonometry in Right Triangles*

You have seen that the trigonometric functions can be defined in terms of the sides of a right triangle provided that the domain is restricted to angles in the interval $(0°, 90°)$ or $\left(0, \frac{\pi}{2}\right)$. For these angles, the values of the trigonometric functions are always positive.

For example, in the right triangle shown, the six trigonometric functions of acute angle θ are:

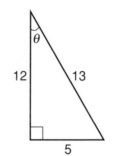

$$\sin \theta = \frac{\text{opp}}{\text{hyp}} = \frac{5}{13} \qquad \csc \theta = \frac{\text{hyp}}{\text{opp}} = \frac{13}{5}$$

$$\cos \theta = \frac{\text{adj}}{\text{hyp}} = \frac{12}{13} \qquad \sec \theta = \frac{\text{hyp}}{\text{adj}} = \frac{13}{12}$$

$$\tan \theta = \frac{\text{opp}}{\text{adj}} = \frac{5}{12} \qquad \cot \theta = \frac{\text{adj}}{\text{opp}} = \frac{12}{5}$$

Note that the trigonometric function values of the other acute angle of the right triangle are different. For example, the sine of the other acute angle is $\frac{12}{13}$.

The measure of angle θ in the right triangle above can be found from the value of any one of the six trigonometric functions and a calculator or table. For example, to find the measure of θ, using $\sin \theta = \frac{5}{13}$, on a calculator:

Enter: 5 ÷ 13 = INV SIN

Display: `22.619865`

The measure of θ is about 22.62°.

Recall that for some acute angles, such as 30°, 45°, and 60°, the values of the trigonometric functions can be stated exactly. By the use of special formulas, which will be discussed later, you will be able to include additional angles whose trigonometric values can be expressed exactly. However, in most cases, a rational approximation that is appropriate for the work being done is used in solving problems.

Indirect Measurement

The trigonometric functions are often used to measure distances or heights that are not directly accessible. For example, to measure the height of a tall building, it is impractical to try to use a ruler or tape measure. An estimate of the height, however, could be made using indirect measurement, as follows:

Suppose an observer 1,000 ft. from the building determines that the angle of elevation to the top of the building (measured close to the ground) is 51°. The tangent or cotangent ratio can be used to find the height of the building. If h represents the height, then in the right triangle formed:

$$\tan 51° = \frac{h}{1,000} \text{ or } \cot 51° = \frac{1,000}{h}$$

Use the tangent equation since it is generally easier to solve an equation when the variable is in the numerator and because tangent is simpler to work with on a scientific calculator.

$$\tan 51° = \frac{h}{1,000}$$
$$h = (\tan 51°)1,000$$
$$h \approx (1.235)1,000$$
$$h \approx 1,235$$

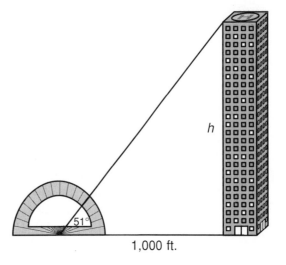

1,000 ft.

Use a table or calculator to find the value of tan 51°.

The actual height of the building is about 1,235 feet.

Example 1

An ice cream stand is located diagonally across a rectangular field from where a child is standing. Instead of walking along the walkways around the field, the child walks across the field to get to the stand. If the measurements are as given in the diagram, how much distance does the child save by walking across the field? Answer to the nearest foot.

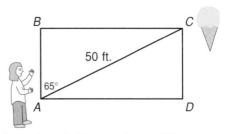

Solution

In the diagram, AB and BC represent walkway distances. To find AB, use a cosine function.

$$\cos 65° = \frac{\text{leg adjacent to } 65° \text{ angle}}{\text{hypotenuse}}$$
$$\cos 65° = \frac{AB}{50}$$
$$(\cos 65°)50 = AB$$
$$(0.4226)50 \approx AB$$
$$AB \approx 21.1 \text{ feet}$$

To find BC, use the Pythagorean Theorem (since the lengths of two sides of $\triangle ABC$ are now known), or the sine function.

$$\sin 65° = \frac{\text{leg opposite 65° angle}}{\text{hypotenuse}} = \frac{BC}{50}$$

$$\sin 65°(50) = BC$$

$$(0.9063)(50) \approx BC$$

$$BC \approx 45.3 \text{ feet}$$

Total distance $AB + BC \approx 21.1 + 45.3 = 66.4$ feet.

The difference is $66.4 - 50 = 16.4$ feet.

Answer: To the nearest foot, the child saves 16 feet by walking across the field.

Example 2

To the nearest hundredth, find AD.

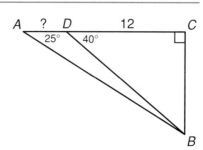

Solution

To find AD, which is not the side of a right triangle, you must first find BC in right $\triangle BCD$ and AC in right $\triangle ABC$.

In right $\triangle BCD$,

$$\tan 40° = \frac{BC}{12}$$

$$BC = (\tan 40°)12$$

In right $\triangle ABC$,

$$\cot 25° = \frac{AC}{BC}$$

$$\cot 25° = \frac{AC}{(\tan 40°)12}$$

$$AC = (\cot 25°)(\tan 40°)12$$

$$AC \approx 21.593$$

$$AC = AD + 12$$

$$AD \approx 21.593 - 12$$

$$AD \approx 9.593$$

To find cot 25° on a calculator, in the degree mode, enter:

25 [TAN] [1/x]

Answer: To the nearest hundredth, $AD \approx 9.59$.

Example 3

For every mile traveled along its length, a road rises 41.40 feet from the horizontal. Find the angle of elevation of the road to the nearest minute. (*Note:* The trigonometric table does not provide sufficient accuracy for this problem.)

1 mi.

θ

41.40 ft.

Solution

In the diagram, θ represents the angle of elevation of the road. To have consistent units, convert 1 mile to 5,280 feet. Since the lengths of the hypotenuse and the leg opposite angle θ are known, use the sine ratio.

$$\sin \theta = \frac{41.4}{5,280}$$

$$\sin \theta \approx 0.00784091$$

$$\theta \approx 0.4493°$$

$$0.4493(60) \approx 27 \text{ minutes} \qquad 1° = 60 \text{ minutes}$$

Answer: To the nearest minute, the angle of elevation of the road is $0°27'$.

Exercises 8.4

In **1 – 3**, find the value of each of the six trigonometric functions of θ.

1.

2.
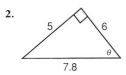

3.
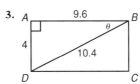

4. If θ is an acute angle and $\cos \theta = \frac{4}{5}$, find the values of the other five trigonometric functions of θ.

5. If θ is an acute angle and $\tan \theta = \frac{\sqrt{2}}{3}$, find the values of the other five trigonometric functions of θ.

6. Find the sine, cosine, and tangent of the smaller acute angle in a right triangle whose sides are 7, 24, and 25.

7. Find the sine, cosine, and tangent of the larger acute angle in a right triangle whose legs measure 8 and 17.

 In **8 – 18**, find x to the nearest whole number.

8.
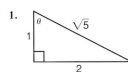

9.
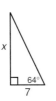

10.

11.

12.
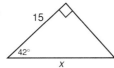

13.
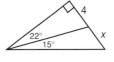

Answers 8.4

1. $\sin \theta = \frac{2\sqrt{5}}{5}$,

$\cos \theta = \frac{\sqrt{5}}{5}$,

$\tan \theta = 2$,

$\csc \theta = \frac{\sqrt{5}}{2}$,

$\sec \theta = \sqrt{5}$,

$\cot \theta = \frac{1}{2}$

2. $\sin \theta = \frac{25}{39}$,

$\cos \theta = \frac{10}{13}$,

$\tan \theta = \frac{5}{6}$,

$\csc \theta = \frac{39}{25}$,

$\sec \theta = \frac{13}{10}$,

$\cot \theta = \frac{6}{5}$

3. $\sin \theta = \frac{5}{13}$,

$\cos \theta = \frac{12}{13}$,

$\tan \theta = \frac{5}{12}$,

$\csc \theta = \frac{13}{5}$,

$\sec \theta = \frac{13}{12}$,

$\cot \theta = \frac{12}{5}$

4. $\sin \theta = \frac{3}{5}$,

$\tan \theta = \frac{3}{4}$,

$\csc \theta = \frac{5}{3}$,

$\sec \theta = \frac{5}{4}$,

$\cot \theta = \frac{4}{3}$

5. $\sin \theta = \frac{\sqrt{22}}{11}$, $\cos \theta = \frac{3\sqrt{11}}{11}$,

$\csc \theta = \frac{\sqrt{22}}{2}$, $\sec \theta = \frac{\sqrt{11}}{3}$,

$\cot \theta = \frac{3\sqrt{2}}{2}$

6. $\sin \theta = \frac{7}{25}$, $\cos \theta = \frac{24}{25}$, $\tan \theta = \frac{7}{24}$

7. $\sin \theta = \frac{17}{\sqrt{353}}$, $\cos \theta = \frac{8}{\sqrt{353}}$, $\tan \theta = \frac{17}{8}$

8. 14

9. 14

10. 7

11. 13

12. 20

13. 3

Exercises 8.4 (continued)

14.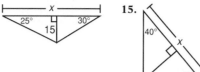

15.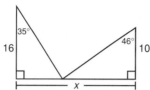

16.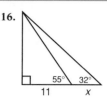

17.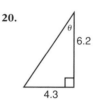

18.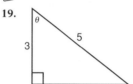

In **19 – 21**, find the measure of θ to the nearest degree.

19.

20.

21.

22. Find the measure, to the nearest degree, of the smaller acute angle of a right triangle whose legs are 7 cm and 10 cm.

23. Find the measure, to the nearest ten minutes, of the larger acute angle of a right triangle whose hypotenuse is 6 m and whose shorter leg is $5\frac{1}{3}$ m.

In **24 – 27**, find the measures of the remaining sides and angles of right $\triangle ABC$ with legs a and b and hypotenuse c. Answer to the nearest whole number or degree.

24. $a = 8, b = 12$

25. $c = 30, b = 10$

26. m∠A = 40°, $c = 16$

27. m∠B = 15°, $b = 7.2$

In **28 – 30**, two vectors, $\vec{a}$ and $\vec{b}$, are perpendicular to each other. To the nearest degree, find the measure of the angle that the resultant, $\vec{a} + \vec{b}$, makes with $\vec{a}$.

28. $|\vec{a}| = 5, |\vec{b}| = 5$

29. $|\vec{a}| = 12, |\vec{b}| = 9$

30. $|\vec{a}| = 72, |\vec{b}| = 3$

In **31 – 33**, to the nearest degree, find the angle that $\vec{a}$ makes with the nonnegative ray of the x-axis.

31. $\vec{a} = 3\vec{i} + 2\vec{j}$

32. $\vec{a} = \vec{i} - 3\vec{j}$

33. $\vec{a} = -5\vec{i} - 2\vec{j}$

14. 58

15. 19

16. 14

17. 22

18. 5

19. 53°

20. 35°

21. 41°

22. 35°

23. 27°20′

24. $c = 4\sqrt{13}$, m∠A = 34°, m∠B = 56°

25. $a = 20\sqrt{2}$, m∠A = 71°, m∠B = 19°

26. m∠B = 50°, BC = 10, AC = 12

27. m∠A = 75°, BC = 27, AB = 28

28. 45°

29. 37°

30. 2°

31. 34°

32. –72°

33. 202°

Exercises 8.4 (continued)

34. From a point 100 feet from the base of a building, the angle of elevation of the top of the building (the upward angle from the ground) is 36°. Find the height of the building to the nearest foot.

35. A ladder 16 feet long leaning against the side of a building makes an angle of 75° with the ground. How far up along the building, to the nearest foot, is the top of the ladder?

36. To the nearest degree, find the angle of elevation of the sun when a 20-foot pole casts a 25-foot shadow.

37. An airplane that is 30 meters above the ground begins to descend at a constant angle just as it passes the perimeter of an airport. What is the angle of descent with the horizontal if the plane touches the ground 200 meters into the airport? Answer to the nearest degree.

38. 🖩 An observer at the top of a tall building wishes to estimate the distance from his position to the farthest point on the surface of Earth that he can see on a clear day. Estimating the radius of Earth at 4,000 miles and the height of the building at 1,250 feet, he sketches a diagram.

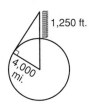
1,250 ft.

4,000 mi.

  a. Find the distance, to the nearest mile, from the observer's location to the farthest object on the ground that he can see.

  b. To the nearest degree, find the angle that his line of sight makes with the building.

39. A 25-foot ladder can be swung between two buildings in a narrow alleyway without moving the foot of the ladder. If the ladder makes an angle of 60° with one building and 40° with the other, how far apart, to the nearest foot, are the two buildings?

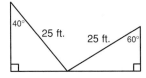

40° · 25 ft. · 25 ft. · 60°

40. A television antenna on a 20-foot pole is supported by two guy wires as shown. If the wires make angles of 60° and 40° with the roof on which the antenna rests, how far apart, to the nearest foot, are the points where the wires are attached to the roof?

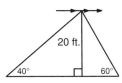

20 ft.

40° · 60°

34. 73 feet

35. 15 feet

36. 39°

37. 9°

38. a. 44 miles

  b. 89°

39. 38 feet

40. 35 feet

Exercises 8.4 *(continued)*

41. Using an analysis similar to that in Exercise 38:

    a. Find, to the nearest tenth of a mile, how high up in an airplane you must be in order to see a city that is 100 miles from where you are flying.

    b. Find, to the nearest ten minutes, the measure of the angle that your line of sight makes with the line of flight of the airplane.

42. An observer makes two sightings of the top of a 500-foot building from the ground. At the first sighting, the angle of elevation is 46°. Pacing back a certain distance along a line perpendicular to the building, she determines the angle of elevation to be 32°. Find the distance, to the nearest foot, between the two sightings.

43. At an angle of 35°, an airplane takes off from a point 1 mile from the edge of an airport. If the plane just passes the edge of the airport after 30 seconds, what is its average speed as it passes over the airport? Answer to the nearest whole number of miles per hour.

44. In order to estimate the height of a building standing on a hill that is 100 m high, an observer at the base of the hill determines that the angle of elevation of the top of the building is 24°10′ and the angle of the elevation of the bottom of the building is 21°40′. Find the height of the building to the nearest meter.

45. A closed rectangular box has a square base 8 inches on each side and a height of 5 inches.

    a. Find the length of the longest diagonal between two corners of the box.

    b. To the nearest degree, find the measure of the angle that this diagonal makes with the bottom of the box.

In **46 – 47**, find a general expression for length *DB* in terms of the trigonometric functions of angles α and β and side *AC* of the triangles in the diagram.

46.

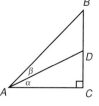

47.

41. a. 1.2 miles

    b. 0°40′

42. 317 feet

43. 146 mph

44. 13 m

45. a. $\sqrt{153}$

    b. 24°

46. In $\triangle ADC$, $\tan \alpha = \dfrac{DC}{AC}$

$$DC = AC\,(\tan \alpha)$$

In $\triangle ABC$, $\tan(\alpha + \beta) = \dfrac{BC}{AC}$

$$BC = AC\,(\tan(\alpha + \beta))$$

$$DB = BC - DC$$
$$DB = AC\,(\tan(\alpha + \beta)) - AC\,(\tan \alpha)$$
$$DB = AC\,[\tan(\alpha + \beta) - \tan \alpha]$$

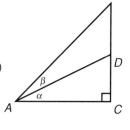

Exercises 8.4 (continued)

Force is a vector quantity. It is often convenient to represent a force as the sum of two vectors, called components, that are perpendicular to each other. When these two components lie in the horizontal and vertical directions, they are called the horizontal and vertical components of the vector.

In **48 – 50**, find, to the nearest hundredth, the magnitude of the horizontal and vertical components of vector $\vec{a}$ when θ is the measure of the angle that $\vec{a}$ makes with the vertical component.

48. $|\vec{a}| = 10$, $\theta = 30°$ **49.** $|\vec{a}| = 12$, $\theta = 45°$ **50.** $|\vec{a}| = 62$, $\theta = 75°$

51. A child pulls a wagon by exerting a force of 8.0 pounds in the direction that makes an angle of 35° with the horizontal. To the nearest tenth, find the magnitude of the horizontal and vertical components of the applied force.

52. A truck driver is unloading 40-pound boxes of canned goods by sliding them down a ramp from the back of the truck to the ground. The weight of the box is a force that acts in a direction perpendicular to the ground.

  a. Find, to the nearest tenth, the magnitude of the component of the 40-pound force that acts parallel to the ramp if the ramp makes an angle of 25° with the ground.

  b. The boxes move down the ramp at the safest speed when the component of the 40-pound force that is parallel to the ramp has a magnitude of 10 pounds. Find, to the nearest tenth of a degree, the angle that the ramp should make with the ground to achieve this.

53. In the diagram, *ABO* and *BCD* are circular segments. If $\overleftrightarrow{BD} \parallel \overleftrightarrow{OA}$, $BD = 4$, $OA = 5$, m∠*AOB* = 72°, and m∠*BDC* = 36°, find, to the nearest hundredth, the distance from *C* to $\overline{OA}$.

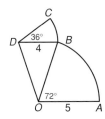

54. In the diagram, *P* lies on a unit circle and on $\overline{OC}$, m∠*BOC* = 55°, and $\overline{BC}$ is tangent to the circle at *B*. To the nearest thousandth, find *PC*.

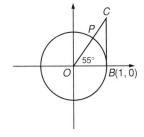

47. $\cot \beta = \dfrac{CD}{AC}$

$CD = AC\,(\cot \beta)$

$\cot \alpha = \dfrac{CB}{AC}$

$CB = AC\,(\cot \alpha)$

$DB = CB - CD$

$DB = AC\,(\cot \alpha) - AC\,(\cot \beta)$

$DB = AC\,(\cot \alpha - \cot \beta)$

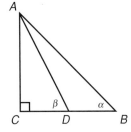

48. 5.00, 8.66

49. 8.49, 8.49

50. 59.89, 16.05

51. 6.6, 4.6

52. a. 16.9

   b. 14.5°

53. 7.11

54. 0.743

8.5 Extending Trigonometry

Earlier in the chapter, it was explained how a point on a unit circle can be used to define a trigonometric function of an angle of any size. Now you will see another method for finding the trigonometric functions of an angle of any size—a method that references any angle to a particular angle in quadrant I.

Consider an angle that terminates in quadrant II, such as an angle of 125°. Let P be a point on the terminal ray of that angle such that $OP = 1$. Therefore, the coordinates of P are (cos 125°, sin 125°). Let A be the foot of the perpendicular from P to the x-axis. Triangle PAO is a right triangle with a 55° angle at O.

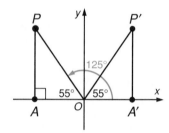

When $\triangle PAO$ is reflected in the y-axis, its image is $\triangle P'A'O$, a right triangle with a 55° angle at O. Since P' is the image of P under a reflection in the y-axis, the coordinates of P' are (−cos 125°, sin 125°). Since P' is on the terminal side of a 55° angle and $OP' = 1$, the coordinates of P' are (cos 55°, sin 55°). This suggests that −cos 125° = cos 55° or cos 125° = −cos 55° and sin 125° = sin 55°.

Furthermore: $\tan 125° = \dfrac{\sin 125°}{\cos 125°} = \dfrac{\sin 55°}{-\cos 55°} = -\tan 55°$

Thus, all of the trigonometric functions of 125° can be expressed in terms of the corresponding functions of 55°, which is called the *reference angle* for the 125° angle.

In general, the reference angle θ_r for any angle θ that terminates in quadrant II can be found by subtracting the degree measure of θ from 180° or the radian measure of θ from π radians.

For quadrant II: $\theta_r = 180° - \theta$

$\sin \theta = \sin \theta_r$ $\qquad$ $\cos \theta = -\cos \theta_r$ $\qquad$ $\tan \theta = -\tan \theta_r$

A similar analysis in quadrant III can be done using 220° as an example. If $\triangle OPA$ is reflected over the origin, the image is $\triangle OP'A'$, which is congruent to $\triangle OPA$ and contains a 40° angle. Since $OA = OA'$, cos 220° = −cos 40° and since $AP = A'P'$, sin 220° = −sin 40°. Also,

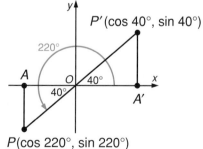

$\tan 220° = \dfrac{\sin 220°}{\cos 220°} = \dfrac{-\sin 40°}{-\cos 40°} = \tan 40°$. Thus, if θ is an angle in quadrant III and θ_r is the reference angle for θ, then sin $\theta = -\sin \theta_r$, cos $\theta = -\cos \theta_r$, and tan $\theta = \tan \theta_r$. The measure of θ_r can be obtained by subtracting 180° from the degree measure of θ or π radians from the radian measure of θ; that is:

For quadrant III: $\theta_r = \theta - 180°$

For an arbitrary angle θ in quadrant IV, $\triangle OAP$ can be reflected over the x-axis. Since $AP = AP'$, $\sin \theta = -\sin \theta_r$, and since $OA = OA$, $\cos \theta = \cos \theta_r$. Since $\tan \theta = \dfrac{\sin \theta}{\cos \theta}$, $\tan \theta = \dfrac{-\sin \theta_r}{\cos \theta_r} = -\tan \theta_r$. The measure of θ_r can be obtained by subtracting the degree measure of θ from $360°$ or the radian measure of θ from 2π radians.

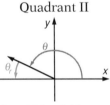
$P'(\cos \theta_r, \sin \theta_r)$
$P(\cos \theta, \sin \theta)$

Therefore, for any angle θ, there is an acute angle θ_r such that:

$$|\sin \theta| = \sin \theta_r \qquad |\cos \theta| = \cos \theta_r \qquad |\tan \theta| = \tan \theta_r$$

The quadrant in which the terminal side of θ lies determines the sign of its function value and the relationship between θ and θ_r. For easy reference, the signs of the trigonometric function values are shown again in the following table.

Quadrant	I	II	III	IV
sin and csc	+	+	−	−
cos and sec	+	−	−	+
tan and cot	+	−	+	−

The following diagrams show the corresponding reference angles for angles in each of the four quadrants. Note that an acute angle has the same measure as its reference angle. For non-acute angles, the reference angle is the smallest positive angle between the terminal side of the angle and the x-axis. If θ is a positive angle in the interval $(90°, 360°)$ or $\left(\frac{\pi}{2}, 2\pi\right)$, the formulas below the appropriate diagram can be used to find θ_r.

Quadrant I

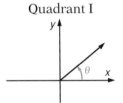

$\theta_r = \theta$ (degrees)
$\theta_r = \theta$ (radians)

Quadrant II

$\theta_r = 180 - \theta$ (degrees)
$\theta_r = \pi - \theta$ (radians)

Quadrant III

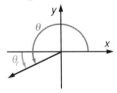

$\theta_r = \theta - 180$ (degrees)
$\theta_r = \theta - \pi$ (radians)

Quadrant IV

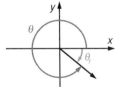

$\theta_r = 360 - \theta$ (degrees)
$\theta_r = 2\pi - \theta$ (radians)

Example 1

Find the reference angle, θ_r, for $\theta = 115°$.

Solution

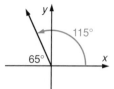

$115°$ terminates in quadrant II.
$\theta_r = 180° - \theta = 180° - 115° = 65°$

Example 2

Find θ_r for $\theta = 307°$.

Solution

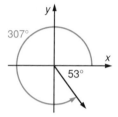

$307°$ terminates in quadrant IV.
$\theta_r = 360° - \theta = 360 - 307 = 53°$

Example 3

Find θ_r for $\theta = -\dfrac{5\pi}{6}$.

Solution

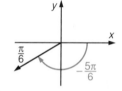

$-\dfrac{5\pi}{6}$ terminates in quadrant III.
Convert to the positive coterminal angle and use $\theta_r = \theta - \pi$.

$$2\pi - \frac{5\pi}{6} = \frac{7\pi}{6}$$

$$\theta_r = \theta - \pi = \frac{7\pi}{6} - \pi = \frac{\pi}{6}$$

Example 4

Find θ_r for $\theta = 402°$.

Solution

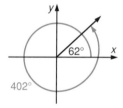

$402°$ terminates in quadrant I and is coterminal with $62°$. Since $62°$ is in quadrant I, $\theta_r = 62°$.

Procedure for Using a Trigonometric Function Table

1. If the value of θ is greater than $360°$ (or 2π) or less than 0, add or subtract multiples of $360°$ (or 2π) to find the measure of a coterminal angle between 0 and $360°$ (or 2π).

2. Determine the quadrant in which θ lies. Then:

   a. Find the sign of the function value for angles in that quadrant.

   b. Determine the reference angle for θ or for its coterminal angle.

3. Find the function value of the reference angle in the table.

4. Write the function value for θ by using the function value of the reference angle, including the required sign.

Example 5

Find the value of cos 136° by using the trigonometric table.

Solution

Since 136° is in quadrant II, $\theta_r = 180° - \theta = 180° - 136° = 44°$. In quadrant II, since cosine is negative, cos 136° = −cos 44° = −0.7193.

Example 6

Find the value of tan 6.8032 by using the trigonometric table.

Solution

Since 6.8032 is greater than 2π, find a coterminal angle in $(0, 2\pi)$.

$$6.8032 - 2\pi = 6.8032 - 2(3.1416) = 0.52$$

Since $\frac{\pi}{2} = \frac{3.1416}{2} = 1.5708$, the value 0.52 is less than $\frac{\pi}{2}$ and is in quadrant I. Thus, tan θ = tan θ_r, or tan 6.8032 = tan 0.52 = 0.5726.

The value of a trigonometric function of any angle can also be found by using a scientific calculator, which will usually provide a result that reflects the sign of the function for an angle in any one of the four quadrants. Recall that a scientific calculator can accept an angle in radians or degrees.

Example 7

 Find the value of sin 277° to six significant digits.

Solution

Place the calculator in degree mode.

Enter: 277 [SIN]

Display: $\boxed{-0.9925462}$

Answer: sin 277° ≈ −0.992546

Example 8

 Find the value of cos 3.62 to six significant digits.

Solution

Place the calculator in radian mode.

Enter: 3.62 [COS]

Display: $\boxed{-0.8877293}$

Answer: cos 3.62 ≈ −0.887729

Example 9

 Find the value of cot 5 to six significant digits.

Solution

Since most scientific calculators do not have a cotangent key, use the relationship $\cot \theta = \dfrac{1}{\tan \theta}$. Place the calculator in radian mode.

Enter: 5 TAN ¹/x

Display: $\boxed{-0.2958129}$

Answer: $\cot 5 \approx -0.295813$.

Example 10

Find two values of θ in the interval $(0°, 360°)$ that satisfy the equation $\cos \theta = \frac{1}{2}$.

Solution

The cosine function has positive values in quadrants I and IV. Since $\cos \theta = \frac{1}{2}$, θ is the measure of the angle whose terminal ray intersects a unit circle at a point with x-coordinate $\frac{1}{2}$. There are two such points, A in quadrant I and A' in quadrant IV. Since A' and A are on the unit circle:

$$x^2 + y^2 = 1$$
$$\left(\tfrac{1}{2}\right)^2 + y^2 = 1^2$$
$$\tfrac{1}{4} + y^2 = 1$$
$$y^2 = \tfrac{3}{4}$$
$$y = \pm \tfrac{\sqrt{3}}{2}$$

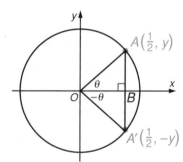

The sides of $\triangle OAB$ and $\triangle OA'B$ are in the ratio $\frac{1}{2} : \frac{\sqrt{3}}{2} : 1$. Thus, these are $30°$-$60°$-$90°$ triangles and $\theta = 60°$ in quadrant I or $\theta = 360° - 60° = 300°$ in quadrant IV.

Answer: The values of θ in the interval $(0°, 360°)$ that satisfy the equation $\cos \theta = \frac{1}{2}$ are $60°$ and $300°$.

Exercises 8.5

In **1 – 16**, for the given value of θ, find θ_r and sketch the angle in standard position. Where necessary, use π = 3.14 as an approximation. Also, 180° = 179°60′ or 179°59′60″.

1. 135° **2.** 207° **3.** 49° **4.** 307°

5. −14° **6.** 97°20′ **7.** −161°27′ **8.** 456°

9. 184°50′20″ **10.** 2.6 **11.** 6 **12.** −0.64

13. 5.71 **14.** −1.64 **15.** 6.50 **16.** 8.85

In **17 – 32**, use the trigonometric table to find the best possible estimate of the given trigonometric function. Use the approximation π ≈ 3.1416 where necessary.

17. sin 47°10′ **18.** tan 107°50′ **19.** cos 192° **20.** sin (−16°40′)

21. sec 306°10′ **22.** cot (−100°50′) **23.** csc 475°20′ **24.** sin 0.4363

25. cos 1.3701 **26.** tan 1.4137 **27.** sec 3.6128 **28.** csc 6.6264

29. tan (−2.4551) **30.** cot (−5.7741) **31.** cos $\left(\frac{\pi}{8}\right)$ **32.** sin $\left(\frac{2\pi}{5}\right)$

 In **33 – 52**, find the value of the trigonometric function to six significant digits.

33. cos 86° **34.** sin 137° **35.** tan 96° **36.** sin 400°

37. tan 586° **38.** sin 105.3° **39.** cos 614.61° **40.** tan 0.27

41. cos 6.451 **42.** tan (−214°) **43.** sin (−116.8°) **44.** tan $\left(-\frac{13\pi}{5}\right)$

45. sin (−4.32) **46.** sin 65°30′ **47.** cos 100°20′ **48.** tan 406°15′

49. sec 40° **50.** csc 105°54′ **51.** cot 6.65 **52.** sec 114°58′16″

In **53 – 58**, find two values of θ in the interval (0°, 360°) that satisfy the equation.

53. sin θ = $\frac{1}{2}$ **54.** cos θ = $\frac{\sqrt{3}}{2}$ **55.** tan θ = −$\sqrt{3}$

56. sec θ = −$\frac{2}{\sqrt{3}}$ **57.** tan θ = 1 **58.** csc θ = −2

In **59 – 61**, find two values of θ in the interval (0, 2π) that satisfy the equation.

59. cos θ = $\frac{\sqrt{2}}{2}$ **60.** cot θ = −1 **61.** csc θ = $\sqrt{2}$

---

1. 45°

2. 27°

3. 49°

4. 53°

5. 14°

6. 82°40′

7. 18°33′

8. 84°

9. 4°50′20″

10. 0.54

11. 0.28

12. 0.64

13. 0.57

14. 1.50

15. 0.22

16. 0.57

17. 0.7333

18. −3.1084

19. −0.9781

20. −0.2868

21. 1.6945

22. 0.1914

23. 1.1064

24. 0.4226

25. 0.1994

26. 6.3131

27. −1.1223

28. 2.9716

29. 0.8195

30. 1.7916

31. 0.9239

32. 0.9511

33. 0.069756

34. 0.681998

35. −9.51436

36. 0.642788

37. 1.03553

38. 0.964557

39. −0.265388

40. 0.276758

41. 0.985952

42. −0.674509

43. −0.892586

44. 3.07768

45. 0.923998   **46.** 0.909961   **47.** –0.179375   **48.** 1.04461

49. 1.30541   **50.** 1.03978   **51.** 2.60279   **52.** –2.36876

53. 30°, 150°   **54.** 30°, 330°   **55.** 120°, 300°   **56.** 150°, 210°

57. 45°, 225°   **58.** 210°, 330°   **59.** $\dfrac{\pi}{4}, \dfrac{7\pi}{4}$ **60.** $\dfrac{3\pi}{4}, \dfrac{7\pi}{4}$

61. $\dfrac{\pi}{4}, \dfrac{3\pi}{4}$ **62.** 38°, 142°   **63.** 60°, 240°   **64.** 99°, 261°

65. 68°, 292°

66. Choose points P and P'
so that $OP = OP'$ and
$\text{m}\angle BOP' = \text{m}\angle AOP = \theta_r$

     $\triangle OPA \cong \triangle OP'B$ by AAS

     $x' = -x, y' = -y$

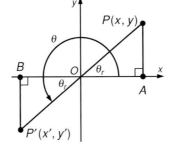

By definition,

$$\sin \theta_r = \frac{y}{OP} \qquad \csc \theta_r = \frac{OP}{y}$$

$$\cos \theta_r = \frac{x}{OP} \qquad \sec \theta_r = \frac{OP}{x}$$

$$\tan \theta_r = \frac{y}{x} \qquad \cot \theta_r = \frac{x}{y}$$

Also, by definition,

$$\sin \theta = \frac{y'}{OP'} \qquad \csc \theta = \frac{OP'}{y'}$$

$$\cos \theta = \frac{x'}{OP'} \qquad \sec \theta = \frac{OP'}{x'}$$

$$\tan \theta = \frac{y'}{x'} \qquad \cot \theta = \frac{x'}{y'}$$

Therefore,

$$\sin \theta = \frac{y'}{OP'} = \frac{-y}{OP} = -\sin \theta_r \qquad\qquad \csc \theta = \frac{OP'}{y'} = \frac{OP}{-y} = -\csc \theta_r$$

$$\cos \theta = \frac{x'}{OP'} = \frac{-x}{OP} = -\cos \theta_r \qquad\qquad \sec \theta = \frac{OP'}{x'} = \frac{OP}{-x} = -\sec \theta_r$$

$$\tan \theta = \frac{y'}{x'} = \frac{-y}{-x} = \frac{y}{x} = \tan \theta_r \qquad\qquad \cot \theta = \frac{x'}{y'} = \frac{-x}{-y} = \frac{x}{y} = \cot \theta_r$$

The only difference between the functions of θ and θ_r is the sign,
therefore $|T(\theta)| = |T(\theta_r)|$.

Answers 8.5

67. 0.85, 3.99

68. 3.95, 5.47

69. 0.62, 3.76

70. a. −0.000882164
  b. −0.004525335

71. 0.84

72. June, 25;
  December, 21

Exercises 8.5 *(continued)*

 In **62 – 65**, approximate two values of θ in the interval $(0°, 360°)$ that satisfy the equation. Answer to the nearest degree.

62. $\sin \theta = 0.6142$ **63.** $\tan \theta = 1.7149$ **64.** $\cos \theta = -0.1642$ **65.** $\sec \theta = 2.68$

66. Prove that if θ is in quadrant III and T is any trigonometric function, then: $|T(\theta)| = |T(\theta_r)|$

 In **67 – 69**, find two values of θ in the interval $(0, 2\pi)$ that satisfy the equation. Answer to the nearest hundredth of a radian.

67. $\tan \theta = 1.131$ **68.** $\sin \theta = -0.724$ **69.** $\cot \theta = 1.4$

70. Evaluate the first five terms of the series below for cos 2.25. Then compare your result with the value for cos 2.25 that is obtained by using the **cos** key.

$$\cos x = 1 - \frac{x^2}{2!} + \frac{x^4}{4!} - \frac{x^6}{6!} + \frac{x^8}{8!} + \cdots + (-1)^{n-1} \frac{x^{2n-2}}{(2n-2)!} \cdots$$

   a. What is the difference between the two values?

   b. Repeat the procedure for cos 100°. (Use radians.)

71. Use the first three terms of the infinite sum below for $\sin x$ to estimate sin 1 to the nearest hundredth. Do not use a calculator.

$$\sin x = x - \frac{x^3}{3!} + \frac{x^5}{5!} - \frac{x^7}{7!} + \cdots + (-1)^{n-1} \frac{x^{2n-1}}{(2n-1)!} \cdots$$

72. For a certain industry, monthly demand is described by the following formula. D is the number of units required and m is the month of the year, where January = 1, February = 2, and so on.

$$D = 6.2 \sin \frac{\pi m}{12} + 0.36m + 16.7$$

Use the formula to estimate, to the nearest whole unit, the number of units in demand in June and in December.

8.6 The Graphs of the Sine and Cosine Functions

The general procedure for drawing the graph of a function is to compile and plot a representative list of ordered pairs (x, y) that are elements of the function and then to draw a smooth curve through these points. This procedure can be used to obtain the graphs of the trigonometric functions. Note that a complete understanding of the graph of $y = \sin x$ serves as a foundation for understanding the other trigonometric graphs.

The Graph of $y = \sin x$

If x represents an angle in radians, the following table of values of ordered pairs is a representative sample of the elements of the function $y = \sin x$.

x	0	$\frac{\pi}{6}$	$\frac{\pi}{4}$	$\frac{\pi}{3}$	$\frac{\pi}{2}$	$\frac{3\pi}{4}$	π	$\frac{5\pi}{4}$	$\frac{3\pi}{2}$	$\frac{7\pi}{4}$	2π
y	0	$\frac{1}{2}$	$\frac{\sqrt{2}}{2}$	$\frac{\sqrt{3}}{2}$	1	$\frac{\sqrt{2}}{2}$	0	$-\frac{\sqrt{2}}{2}$	-1	$-\frac{\sqrt{2}}{2}$	0

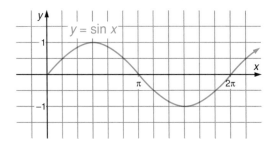

For convenience, the table contains values of the function that can easily be computed without the use of a calculator or table. The values of x between 0 and 2π will produce one period of the graph. Recall that for any function $f(x)$, if $f(x) = f(x + p)$, then the function is periodic. Since the sine function is of that form, with $\sin x = \sin (x + 2\pi n)$, where n is an integer, the ordered pairs repeat in each successive interval of 2π.

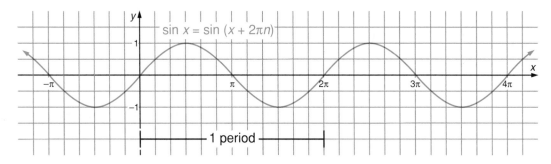

Notice that since π is about equal to 3, the scales on the x- and y-axes of this graph are approximately the same. The range of $y = \sin x$ is the set of all real numbers in the interval $[-1, 1]$. The graph of $y = \sin x$ is sometimes called a *sine wave* or *sinusoid*.

Recall that *odd functions*, for which $f(-x) = -f(x)$, are symmetric about the origin. Thus, since *sin* $(-x) = -sin\ x$, $y = \sin x$ is an odd function, and its graph has symmetry with respect to the origin.

The Graph of $y = \cos x$

Since $\cos x = \sin\left(\frac{\pi}{2} - x\right)$, this suggests that the graph of $y = \cos x$ is related to the graph of $y = \sin x$. The graph of $y = \cos x$ has the same wave-like shape as $y = \sin x$ and can be obtained by translating every point of the graph of $y = \sin x$ a distance of $\frac{\pi}{2}$ units to the left. The table below verifies this relationship for given values of x.

x	0	$\frac{\pi}{4}$	$\frac{\pi}{2}$	$\frac{3\pi}{4}$	π	$\frac{5\pi}{4}$	$\frac{3\pi}{2}$	$\frac{7\pi}{4}$	2π
$\sin x$	0	$\frac{\sqrt{2}}{2}$	1	$\frac{\sqrt{2}}{2}$	0	$-\frac{\sqrt{2}}{2}$	-1	$-\frac{\sqrt{2}}{2}$	0
$\cos x$	1	$\frac{\sqrt{2}}{2}$	0	$-\frac{\sqrt{2}}{2}$	-1	$-\frac{\sqrt{2}}{2}$	0	$\frac{\sqrt{2}}{2}$	1

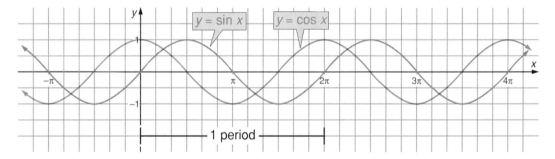

The graph of $y = \cos x$ also has a period of 2π, repeating its values over an interval of 2π. The range of $y = \cos x$ is the set of all real numbers in the interval $[-1, 1]$.

Recall that *even functions*, for which $f(x) = f(-x)$, are symmetric about the y-axis. Thus, since *cos x = cos (–x)*, $y = \cos x$ is an even function, and its graph has symmetry with respect to the y-axis.

It is important for you to feel familiar with the basic sine and cosine curves before you go on to study their variations. Note that the basic curves can be sketched from key points, that is, points at which the curves have their greatest and least values, and the y-intercepts. For one cycle of each curve, the following are the key points.

$$y = \sin x: \ (0, 0), \left(\tfrac{\pi}{2}, 1\right), (\pi, 0), \left(\tfrac{3\pi}{2}, -1\right), (2\pi, 0)$$

$$y = \cos x: \ (0, 1), \left(\tfrac{\pi}{2}, 0\right), (\pi, -1), \left(\tfrac{3\pi}{2}, 0\right), (2\pi, 1)$$

Summary of Properties of $y = \sin x$ and $y = \cos x$					
	Domain	Range	Period	Function	Symmetry
$y = \sin x$	R	$[-1, 1]$	2π	odd	origin
$y = \cos x$	R	$[-1, 1]$	2π	even	y-axis

From the previous graph, observe the intersection points of the curves $y = \sin x$ and $y = \cos x$. Thus, in the interval $[0, 2\pi]$, the equation $\sin x = \cos x$ has two solutions: $x = \frac{\pi}{4}$ or $x = \frac{5\pi}{4}$.

Graphs of $y = a \sin x$ and $y = a \cos x$

A graphing calculator or a comparable computer program will help you to visualize the effect of multiplying a function value by a constant.

Enter each set of related functions and describe how the graph of the first function of each set is changed to obtain the other functions of that set. Let the range of values of x be from –0.2 to 6.3 and the range of values of y from –4 to 4.

Set 1: $y_1 = \sin x$ $y_2 = 2y_1$ $y_3 = 3y_1$ $y_4 = 3.5y_1$

Set 2: $y_1 = \cos x$ $y_2 = 2y_1$ $y_3 = 3y_1$ $y_4 = 3.5y_1$

Set 3: $y_1 = \sin x$ $y_2 = 0.8y_1$ $y_3 = 0.5y_1$ $y_4 = 0.3y_1$

Set 4: $y_1 = \cos x$ $y_2 = 0.8y_1$ $y_3 = 0.5y_1$ $y_4 = 0.3y_1$

Set 5: $y_1 = \sin x$ $y_2 = -y_1$ $y_3 = -2y_1$ $y_4 = -0.5y_1$

Set 6: $y_1 = \cos x$ $y_2 = -y_1$ $y_3 = -2y_1$ $y_4 = -0.5y_1$

You should have observed the following relationships:

1. Multiplying the function value by a positive constant greater than 1 stretches the graph in the vertical direction.

2. Multiplying the function value by a positive constant less than 1 shrinks the graph in the vertical direction.

3. The graph of $y = -\sin x$ is the image of $y = \sin x$ under a reflection in the x-axis and the graph of $y = -\cos x$ is the reflection of $y = \cos x$ under a reflection in the x-axis.

4. Multiplying the function value by a negative constant stretches or shrinks the graph in the vertical direction after a reflection in the x-axis.

The equation $y = 3 \sin x$ implies that for a given value of x, the value of y is 3 times what it was for $y = \sin x$. Thus, to obtain the graph of $y = 3 \sin x$, consider the key points for the basic curve $y = \sin x$. Allow the x-values to remain the same, and multiply the y-values by 3.

$y = \sin x$	$(0, 0)$	$\left(\frac{\pi}{2}, 1\right)$	$(\pi, 0)$	$\left(\frac{3\pi}{2}, -1\right)$	$(2\pi, 0)$
$y = 3 \sin x$	$(0, 0)$	$\left(\frac{\pi}{2}, 3\right)$	$(\pi, 0)$	$\left(\frac{3\pi}{2}, -3\right)$	$(2\pi, 0)$

The key points for $y = 3 \sin x$ will produce one cycle of the curve. These values repeat in intervals of 2π, thus keeping 2π as the period of the transformed curve, the same as that of the basic curve.

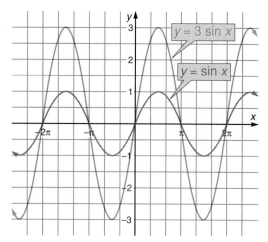

Observe that the constant 3 has transformed the basic sine graph by stretching it vertically, extending the range from $[-1, 1]$ to $[-3, 3]$. In general, the graph of $y = a \sin x$ or $y = a \cos x$, with $a \neq 0$, has a range of $[-|a|, |a|]$, where $|a|$ is called the *amplitude* of the graph. Thus, for example, $y = 4 \sin x$ has an amplitude of $|4|$ or 4, and $y = -\frac{1}{2} \sin x$ has an amplitude of $\left|-\frac{1}{2}\right|$ or $\frac{1}{2}$.

Similarly, the graph of $y = 2 \cos x$ transforms the graph of $y = \cos x$ by stretching the graph vertically by a factor of 2, thus extending the range from $[-1, 1]$ to $[-2, 2]$. The amplitude of the transformed curve is $|2|$ or 2.

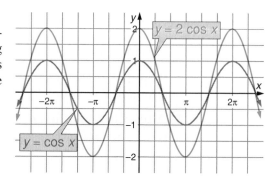

Graphs of $y = \sin ax$ and $y = \cos ax$

Before finding trigonometric function values, use a graphing calculator or a computer program to visualize graphs of those functions in which the x-coordinate is multiplied by a constant.

Enter each set of functions and describe how the graph of each function is related to the graph of the first function of that set. Let the range of values of x be from -0.2 to 12.6 and the range of values of y from -2 to 2.

Set 1: $y_1 = \sin x$ $y_2 = \sin 2x$ $y_3 = \sin 3x$ $y_4 = \sin 4x$

Set 2: $y_1 = \cos x$ $y_2 = \cos 2x$ $y_3 = \cos 3x$ $y_4 = \cos 4x$

Set 3: $y_1 = \sin x$ $y_2 = \sin \frac{1}{2}x$ $y_3 = \sin \frac{1}{3}x$ $y_4 = \sin \frac{1}{4}x$

Set 4: $y_1 = \cos x$ $y_2 = \cos \frac{1}{2}x$ $y_3 = \cos \frac{1}{3}x$ $y_4 = \cos \frac{1}{4}x$

You should have observed the following relationships:

1. The graphs of $y = \sin ax$ and of $y = \cos ax$ are the graphs of $y = \sin x$ and of $y = \cos x$ compressed horizontally when $a > 1$.

2. The graphs of $y = \sin ax$ and of $y = \cos ax$ are the graphs of $y = \sin x$ and of $y = \cos x$ stretched horizontally when $0 < a < 1$.

To obtain the graph of $y = \cos 2x$, consider the key points for the basic curve $y = \cos x$. Allow the y-values to remain the same, and divide the x-values by 2.

$y = \cos x$	$(0, 1)$	$\left(\frac{\pi}{2}, 0\right)$	$(\pi, -1)$	$\left(\frac{3\pi}{2}, 0\right)$	$(2\pi, 1)$
$y = \cos 2x$	$(0, 1)$	$\left(\frac{\pi}{4}, 0\right)$	$\left(\frac{\pi}{2}, -1\right)$	$\left(\frac{3\pi}{4}, 0\right)$	$(\pi, 1)$

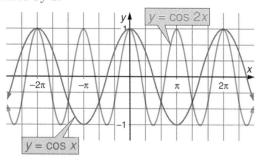

Observe that the key points for $y = \cos 2x$ produced one cycle of the curve in the interval $[0, \pi]$, or two cycles of the curve in the interval $[0, 2\pi]$. In general, the nonzero constant a in the equation $y = \sin ax$ or $y = \cos ax$ affects the period so that the period of the transformed curve is $\left|\frac{2\pi}{a}\right|$. Thus, for example, $y = \sin 3x$ has a period of $\left|\frac{2\pi}{3}\right|$ or $\frac{2\pi}{3}$, and $y = \cos\left(-\frac{1}{4}x\right)$ has a period of $\left|\frac{2\pi}{-\frac{1}{4}}\right|$ or 8π.

Example 1

Sketch the graph of $y = -2 \sin \frac{1}{3}x$ from 0 to 6π.

Solution

Here both the amplitude and the period of the graph of $y = \sin x$ are transformed. The coefficient -2 indicates that the amplitude is 2 and the coefficient $\frac{1}{3}$ indicates that the period is $\left|\frac{2\pi}{\frac{1}{3}}\right|$ or 6π.

Modify the values of $y = \sin x$ by multiplying each y-value by -2 and multiplying each x-value by 3.

$y = \sin x$	$(0, 0)$	$\left(\frac{\pi}{2}, 1\right)$	$(\pi, 0)$	$\left(\frac{3\pi}{2}, -1\right)$	$(2\pi, 0)$
$y = -2 \sin \frac{1}{3}x$	$(0, 0)$	$\left(\frac{3\pi}{2}, -2\right)$	$(3\pi, 0)$	$\left(\frac{9\pi}{2}, 2\right)$	$(6\pi, 0)$

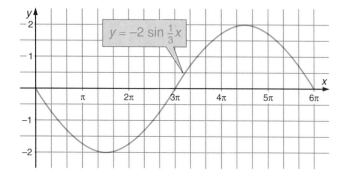

Note that the graph of $y = -a \sin bx$ (or $y = -a \cos bx$) is a reflection of $y = a \sin bx$ (or $y = a \cos bx$) over the x-axis, as shown.

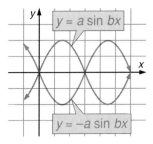

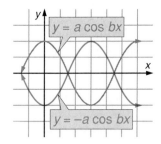

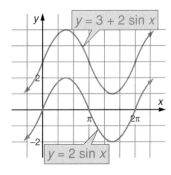

You have seen how the graph of $y = \sin x$ or $y = \cos x$ can be stretched or shrunk vertically or horizontally. Other transformations include translations, which are a vertical or horizontal shift of the graph. A vertical shift (up or down) can be accomplished by adding a constant to the y values of the graph. For example, the graph of $y = 3 + 2 \sin x$ is a translation of the graph of $y = 2 \sin x$ by 3 units in the vertical direction.

In general, $y = k + a \sin bx$ (or $y = k + a \cos bx$) shifts $y = a \sin bx$ (or $y = a \cos bx$) vertically k units. If k is positive, the shift is upward, and if k is negative, the shift is downward.

The graph of $y = \sin x$ or $y = \cos x$ is shifted to the left or right, if the equation has a constant added to the x value. For example, the graph of $y = \sin\left(x - \frac{\pi}{4}\right)$ is a translation by $\frac{\pi}{4}$ units to the right of the graph of $y = \sin x$, and the graph of $y = 2 \cos\left(x + \frac{\pi}{6}\right)$ is a translation by $\frac{\pi}{6}$ units to the left of the graph of $y = 2 \cos x$.

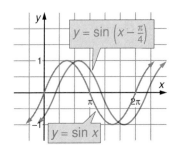

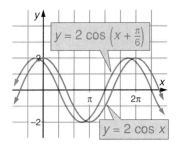

In the context of trigonometric graphs, a horizontal translation is called a *phase shift*. In general, the graph of $y = a \sin b(x + h)$ [or $y = a \cos b(x + h)$] is the same as $y = a \sin bx$ (or $y = a \cos bx$) except that it is translated h units to the left if $h > 0$ and h units to the right if $h < 0$. Note that the graphs of $y = \cos x$ and $y = \sin\left(x + \frac{\pi}{2}\right)$ are identical.

Example 2

Sketch the graph of $y = 1 + \frac{1}{2} \cos 2\left(x - \frac{\pi}{3}\right)$ over the interval $[0, 2\pi]$.

Solution

This graph is a transformation of the graph of $y = \cos x$ in four ways. Using the general equation $y = k + a \cos b(x + h)$ as a reference:

1. Since $a = \frac{1}{2}$, the amplitude $|a|$ is $\frac{1}{2}$.

2. Since $b = 2$, the period $\left|\frac{2\pi}{b}\right|$ is $\left|\frac{2\pi}{2}\right|$ or π.

3. Since $k = 1$, the graph is raised 1 unit.

4. Since $h = -\frac{\pi}{3}$, the graph is translated $\frac{\pi}{3}$ units to the right.

Since each of the four transformations is independent of the others, start with the graph of $y = \cos x$ and develop the final graph in a series of steps.

Step 1

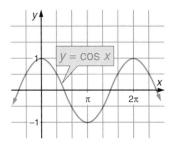

Step 2

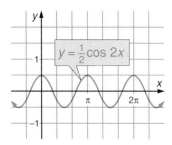

Step 3

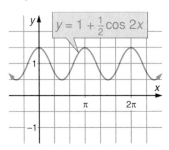

Step 4

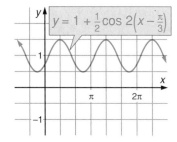

Sine and cosine curves with appropriate values of a, b, h, and k can be used to visualize observed data such as temperatures that change in a periodic pattern. The table below shows the average temperatures for Mason City.

Jan.	Feb.	Mar.	Apr.	May	June	July	Aug.	Sept.	Oct.	Nov.	Dec.
−5	0	24	41	62	79	85	78	69	48	31	9

Using a graphing calculator, enter the data. To do this, use the **STAT** function of the calculator, highlighting **DATA**.

$$\boxed{\textbf{2nd}} \quad \boxed{\text{STAT}} \quad \boxed{\blacktriangleright} \quad \boxed{\blacktriangleright} \quad \boxed{\text{ENTER}}$$

The calculator is now ready to accept data. The x-values will refer to the months by number and the y-values will refer to the average temperatures. For January, let $x_1 = 1$ and $y_1 = -5$; for February, $x_2 = 2$ and $y_2 = 0$; and so on.

After all of the data has been entered, use the calculator to display the data in a graph called a *scatter plot*.

First enter a suitable range; use 0 to 13 for x and -8 to 88 for y.

$$\boxed{\text{RANGE}} \;\; 0 \;\; \boxed{\text{ENTER}} \;\; 13 \;\; \boxed{\text{ENTER}} \;\; 1 \;\; \boxed{\text{ENTER}} \;\; -8$$

$$\boxed{\text{ENTER}} \;\; 88 \;\; \boxed{\text{ENTER}} \;\; 1 \;\; \boxed{\text{ENTER}} \;\; 1$$

$$\boxed{\textbf{2nd}} \;\; \boxed{\text{STAT}} \;\; \boxed{\blacktriangleright} \;\; 2 \;\; \boxed{\text{ENTER}}$$

Note the pattern into which the data points fall. Knowing that the temperature values will repeat each year, the pattern of points suggests a sine or cosine graph.

To determine the equation that best approximates the values, consider the following:

1. The curve starts with its lowest value at $x = 1$. Thus, use a function of the form $-a \cos bx$ translated 1 unit to the right and an appropriate number of units up.

2. The amplitude a is one-half the difference between the largest and smallest values. Thus, $a = \left| \frac{85 - (-5)}{2} \right| = 45$.

3. The curve completes one cycle in 12 months. Thus, $\frac{2\pi}{b} = 12$ or $b = \frac{\pi}{6}$.

4. Since the lowest point is $y = -5$ and the highest point is $y = 85$, the curve must be translated $\frac{-5 + 85}{2} = 40$ units up.

Thus, an appropriate equation is:

$$y = -45 \cos \left(\tfrac{\pi}{6}(x - 1) \right) + 40$$

To use the calculator to compare the graph of this function with the scatter plot of the given data, first enter the equation of the function as y_1:

$$y_1 = -45 \cos (\pi(x - 1)/6) + 40$$

and press $\boxed{\text{GRAPH}}$. Now recall the scatter plot onto the same screen as the cosine curve:

$$\boxed{\textbf{2nd}} \;\; \boxed{\text{STAT}} \;\; \boxed{\blacktriangleright} \;\; 2 \;\; \boxed{\text{ENTER}}$$

To further distinguish the cosine graph from the individual points that were plotted using **2:scatter**, redisplay the given data points using **3:xyline** rather than **2:scatter**. Observe how closely the cosine curve fits the curve that the calculator draws to connect the data points.

Note that another appropriate equation can be written in terms of sine. Since the temperature closest to the middle is 41 for April, the 4th month, a sine graph that is translated 4 units to the right and 40 units up can be used.

$$y = 45 \sin \left(\tfrac{\pi}{6}(x - 4) \right) + 40$$

Exercises 8.6

In **1 – 16**: **a.** State the amplitude.

  b. State the period.

  c. Sketch the graph over the interval $[0, 2\pi]$.

1. $y = 3 \sin x$ **2.** $y = \frac{1}{2} \cos x$ **3.** $y = -\sin x$ **4.** $y = -2 \cos x$

5. $y = \frac{1}{3} \sin x$ **6.** $y = -\frac{1}{4} \cos x$ **7.** $y = \frac{4}{5} \sin x$ **8.** $y = -\frac{2}{3} \cos x$

9. $y = \cos \frac{1}{2}x$ **10.** $y = \sin 2x$ **11.** $y = 2 \cos 3x$ **12.** $y = 2 \sin \frac{1}{3}x$

13. $y = -\sin 2x$ **14.** $y = -2 \cos \frac{1}{4}x$ **15.** $y = -4 \sin 3x$ **16.** $y = \frac{2}{3} \cos \frac{1}{3}x$

In **17 – 40**: **a.** State the number of units and the direction of any vertical translation, or write "none" if there is no vertical translation.

  b. State the number of units and direction of any phase shift, or write "none" if there is no phase shift.

  c. Sketch the graph over an interval one period in length.

17. $y = 3 + \sin x$ **18.** $y = 2 + 2 \cos x$ **19.** $y = 1 - \cos x$

20. $y = -2 + \frac{1}{2} \sin x$ **21.** $y = -1 + 2 \cos \frac{1}{3}x$ **22.** $y = \frac{1}{2} + 3 \sin 2x$

23. $y = \cos \left(x - \frac{\pi}{4}\right)$ **24.** $y = \sin (x + \pi)$ **25.** $y = 2 \sin \left(x + \frac{\pi}{3}\right)$

26. $y = 3 \cos \left(x + \frac{2\pi}{3}\right)$ **27.** $y = \frac{1}{2} \sin \left(x - \frac{\pi}{6}\right)$ **28.** $y = 1 + \sin \left(x + \frac{\pi}{4}\right)$

29. $y = 2 + \cos (x + \pi)$ **30.** $y = -1 + \cos \left(x - \frac{\pi}{2}\right)$ **31.** $y = \frac{1}{2} + \sin 2\left(x + \frac{\pi}{3}\right)$

32. $y = 3 + \cos \frac{1}{2}\left(x - \frac{3\pi}{2}\right)$ **33.** $y = -2 + 2 \sin 2\left(x + \frac{\pi}{2}\right)$ **34.** $y = \sin (2x + \pi)$

35. $y = \cos (3x - \pi)$ **36.** $y = \frac{1}{3} \sin \left(2x + \frac{\pi}{2}\right)$ **37.** $y = 3 - 4 \cos (x + \pi)$

38. $y = -\frac{3}{2} + \sin \left(3x - \frac{\pi}{2}\right)$ **39.** $y = 3 - 2 \sin \left(x + \frac{\pi}{2}\right)$ **40.** $y = 3 - \cos 2\left(x + \frac{\pi}{2}\right)$

In **41 – 49**, write an equation for the graph. (Answers may be given in terms of sine or cosine.)

41. 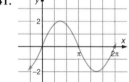   **42.** 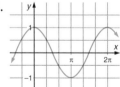   **43.**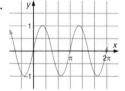

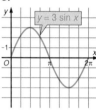

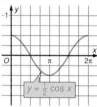

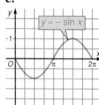

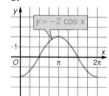

5. a. amplitude = $\frac{1}{3}$ **6. a.** amplitude = $\frac{1}{4}$ **7. a.** amplitude = $\frac{4}{5}$

  b. period = 2π **b.** period = 2π **b.** period = 2π

  c. 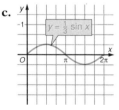   **c.** 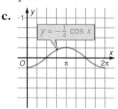   **c.**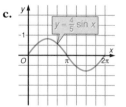

8. a. amplitude $= \frac{2}{3}$
  b. period $= 2\pi$
  c.
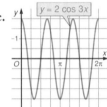

9. a. amplitude $= 1$
  b. period $= \dfrac{2\pi}{\frac{1}{2}} = 4\pi$
  c.
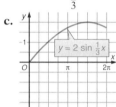

10. a. amplitude $= 1$
  b. period $= \dfrac{2\pi}{2} = \pi$
  c.
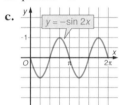

11. a. amplitude $= 2$
  b. period $= \dfrac{2\pi}{3}$
  c.
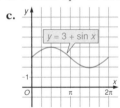

12. a. amplitude $= 2$
  b. period $= \dfrac{2\pi}{\frac{1}{3}} = 6\pi$
  c.
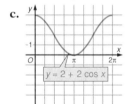

13. a. amplitude $= 1$
  b. period $= \pi$
  c.
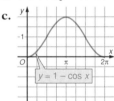

14. a. amplitude $= 2$
  b. period $= 8\pi$
  c.

15. a. amplitude $= 4$
  b. period $= \dfrac{2\pi}{3}$
  c.
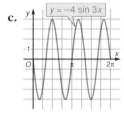

16. a. amplitude $= \frac{2}{3}$
  b. period $= 6\pi$
  c.

17. a. 3 units up   **b.** none
  c.

18. a. 2 units up   **b.** none
  c.

19. a. 1 unit up   **b.** none
  c.
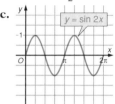

20. a. 2 units down   **b.** none
  c.
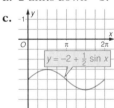

21. a. 1 unit down   **b.** none
  c.
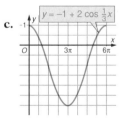

22. a. $\frac{1}{2}$ unit up   **b.** none
  c.
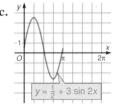

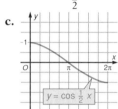

23. a. none **b.** $\frac{\pi}{4}$ to the right

c.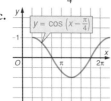

24. a. none **b.** π to the left

c.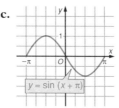

25. a. none **b.** $\frac{\pi}{3}$ to the left

c.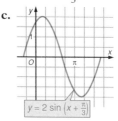

26. a. none **b.** $\frac{2\pi}{3}$ to the left

c.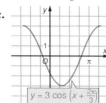

27. a. none **b.** $\frac{\pi}{6}$ to the right

c.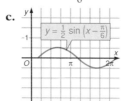

28. a. 1 unit up **b.** $\frac{\pi}{4}$ to the left

c.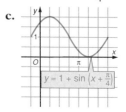

29. a. 2 units up **b.** π to the left

c.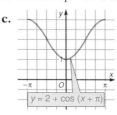

30. a. 1 unit down
b. $\frac{\pi}{2}$ to the right

c.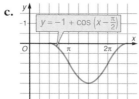

31. a. $\frac{1}{2}$ unit up **b.** $\frac{\pi}{3}$ to the left

c.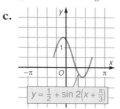

32. a. 3 units up **b.** $\frac{3\pi}{2}$ to the right

c.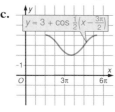

33. a. 2 units down
b. $\frac{\pi}{2}$ to the left

c.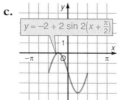

34. a. none **b.** $\frac{\pi}{2}$ to the left

c.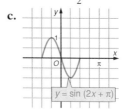

35. a. none **b.** $\frac{\pi}{3}$ to the right

c.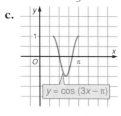

36. a. none **b.** $\frac{\pi}{4}$ to the left

c.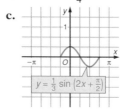

37. a. 3 units up **b.** π to the left

c.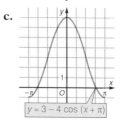

38. a. $\frac{3}{2}$ units down

b. $\frac{\pi}{6}$ to the right

c.

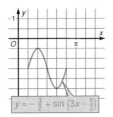

$y = -\frac{3}{2} + \sin\left(3x - \frac{\pi}{2}\right)$

39. a. 3 units up

b. $\frac{\pi}{2}$ to the left

c.

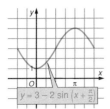

$y = 3 - 2\sin\left(x + \frac{\pi}{2}\right)$

40. a. 3 units up

b. $\frac{\pi}{2}$ to the left

c.

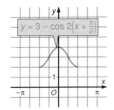

$y = 3 - \cos 2\left(x + \frac{\pi}{2}\right)$

41. $y = 2\sin x$

42. $y = \cos x$

43. $y = \sin 2x$

44. $y = \cos 6x$

45. $y = 2 + \sin x$

46. $y = -3 + \cos x$

47. $y = -\cos x$

48. $y = 100 \sin \dfrac{x}{60}$

49. $y = 4 + 2\sin x$

50. $k = 1, a = 1, b = 2, h = \frac{\pi}{4}$

51. $\frac{\pi}{2}$

52. $\frac{\pi}{2}$

Exercises 8.6 *(continued)*

44.

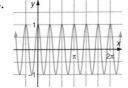

45.

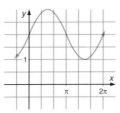

46.

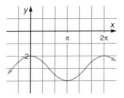

47.

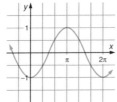

48.

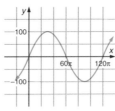

49.

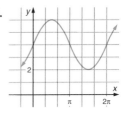

50. Find the values of a, b, k and h so that the graph of $y = k + a \sin b(x + h)$ is the graph shown.

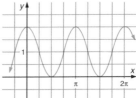

51. In the interval $[0, 2\pi]$, if $y = \sin(x + h)$ has the same graph as $y = \cos x$, what is the value of h?

52. In the interval $[0, 2\pi]$, if $y = \cos(x - h)$ has the same graph as $y = \sin x$, what is the value of h?

53. The sound wave generated by a tone on the piano can be described by the equation

$$y = 0.001 \sin 440\pi t$$

where t is the time in seconds.

a. The *frequency* of a sound wave is defined as the number of cycles or periods per second. Find the frequency of the tone.

b. Sketch the wave for this tone, with t as the horizontal axis.

54. The oscillation of a spring causes the distance from one end of the spring to a fixed point to vary sinusoidally with time. Let the distance be maximized at 8 in. when $t = 0.2$ second and minimized at 2 in. when $t = 0.6$ second.

a. Sketch the graph of the function.

b. Write an equation describing the distance d in inches in terms of the time t.

c. Find the value of d when $t = 1.0$.

---

53. a. 220 cycles/sec.

b.

$y = 0.001 \sin 440\pi t$

54. a.

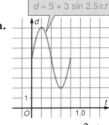

$d = 5 + 3 \sin 2.5\pi t$

b. $d = 5 + 3 \sin \dfrac{2\pi t}{0.8}$ **c.** 8

Exercises 8.6 (continued)

55. Biorhythm theory states that a person is likely to be at his/her best when three independent sinusoidal functions are simultaneously at a relatively high value. These functions are: the physical, which has a period of 23 days; the emotional, which has a period of 28 days; and the intellectual, which has a period of 33 days.

a. *P*, the physical function, *E*, the emotional function, and *I*, the intellectual function, each have an amplitude of 10. Write an equation for each function in terms of *d*, the number of days, if, when *d* = 0, each function is at 0 and increasing.

b. Draw the graphs of the three functions on the same set of axes, using three different colors.

c. Predict from your graph what the value of your intellectual index will be the next time that your physical index peaks.

d. Predict from your graph, to the nearest day, the next time that your physical and emotional indexes will have the same value.

e. Verify your predictions in parts **c** and **d** using your equations of the functions.

56. One theory states that the effectiveness of certain drugs on a person follows a pattern of biorhythm for that person. Assume that the effectiveness rating of a drug ranges from 1 to 10, with 10 indicating maximum effectiveness. The graph shown describes this situation for a particular person over a 90-day period.

a. How many days will pass after the drug is most effective before it is most effective again?

b. Write an equation of the relationship, using *E* as the effectiveness index (1 – 10) and *d* for the number of days.

c. What is the effectiveness index, to the nearest whole number, of the drug on the 105th day?

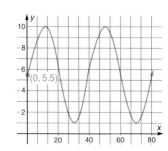

---

55. a. $P = 10 \sin \frac{2\pi}{23} d$

$E = 10 \sin \frac{2\pi}{28} d$

$I = 10 \sin \frac{2\pi}{33} d$

b.

c. $I \approx 9$ **d.** at day 6

e. In part **c**, *P* peaks at about day $5\frac{1}{2}$.

$I(5.5) = 10 \sin \frac{2\pi}{33}(5.5) \approx 8.7$, close to 9.

In part **d**, when *d* = 6:

$P = 10 \sin \frac{2\pi}{23}(6)$
≈ 9.98

$E = 10 \sin \frac{2\pi}{28}(6)$
≈ 9.75

The values of *P* and *E* are very close.

Exercises 8.6 *(continued)*

57. In calculus, the value of $\frac{\sin x}{x}$ is studied.

   a. Draw the graph of $y = \sin x$ for values of x from 0 to 0.1 in increments of 0.01.

   b. On the same set of axes, draw the graph of $y = x$.

   c. Predict the value that $\frac{\sin x}{x}$ approaches as $|x|$ gets smaller.

Note: Engineers often use the fact that, in radians, $\sin x$ is approximately equal to x to estimate values of $\sin x$ when $|x|$ is small.

58. Refer to the calculator discussion in this section, in which graphed temperature data was compared to a cosine curve.

   a. Reenter the given temperature data, and repeat the data for a second year (from $x_{13} = 13$ to $x_{24} = 24$).

   b. Adjust the range for x, and compare two cycles of the scatter plot to two cycles of the suggested *sine* curve.

59. Given a list of the number of hours between sunrise and sunset at 40° north latitude at weekly intervals for one year:

   a. Using a graphing calculator, enter the data and draw a scatter plot to show the variation in time between sunrise and sunset.

   b. Write an equation that can be used to describe the data. Enter the equation and compare the scatter plot of the data with the graph of the equation that you wrote.

Winter	Spring	Summer	Fall
9.3	12.4	15.0	12.1
9.4	12.7	14.95	11.8
9.5	13.0	14.9	11.5
9.65	13.3	14.7	11.2
9.8	13.6	14.55	10.9
10.0	13.9	14.3	10.6
10.3	14.15	14.15	10.35
10.6	14.35	13.9	10.1
10.9	14.55	13.6	9.9
11.15	14.7	13.3	9.7
11.5	14.9	13.0	9.5
11.8	14.95	12.7	9.4
12.0	15.0	12.4	9.3

56. a. 40 days

   b. $E = 5.5 + 4.5 \sin \frac{2\pi}{40} d$

   or

   $E = 5.5 + 4.5 \cos \frac{2\pi}{40}(d - 10)$

   c. 2

57. a. and b.

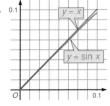

   c. 1

58. a. $y_{13} = y_1, y_{14} = y_2$, etc.

58. b. Use 0 to 24 for x-values, keep the same y-values. Scatter plot and graph of

$$y = 45 \sin\left(\pi \frac{(x-4)}{6}\right) + 40$$

should almost coincide.

59. a. Use 0 to 53 for x-values and 0 to 18 for y-values

   b. $y = -2.85 \cos \frac{\pi}{26} x + 12.15$

8.7 The Graphs of the Other Trigonometric Functions

The Graph of y = tan x

The relationship $\tan x = \dfrac{\sin x}{\cos x}$ can be used to determine a number of properties about the graph of $y = \tan x$. First, since $\tan x$ is expressed as a quotient with denominator $\cos x$, values of $\tan x$ do not exist when $\cos x = 0$. Thus, $\tan x$ is undefined when $x = \frac{\pi}{2} + k\pi$, where k is an integer, and the graph has vertical asymptotes at these values. Since $\tan(-x) = -\tan x$, then $y = \tan x$ is an odd function, which has symmetry with respect to the origin. Using these properties and the values shown, the graph of $y = \tan x$ from $x = -2\pi$ to $x = 2\pi$ can be drawn.

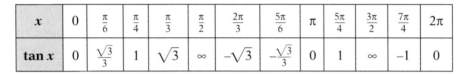

x	0	$\frac{\pi}{6}$	$\frac{\pi}{4}$	$\frac{\pi}{3}$	$\frac{\pi}{2}$	$\frac{2\pi}{3}$	$\frac{5\pi}{6}$	π	$\frac{5\pi}{4}$	$\frac{3\pi}{2}$	$\frac{7\pi}{4}$	2π
$\tan x$	0	$\frac{\sqrt{3}}{3}$	1	$\sqrt{3}$	∞	$-\sqrt{3}$	$-\frac{\sqrt{3}}{3}$	0	1	∞	-1	0

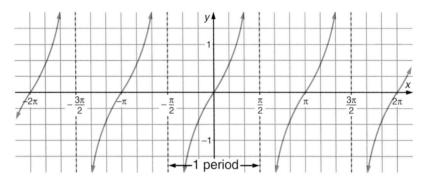

Since the table suggests that π is the smallest value of p for which $\tan(x + p) = \tan x$, the period of $y = \tan x$ is π. Notice that the value of $\tan x$ increases very quickly as x approaches $\frac{\pi}{2}$. The trigonometric table or a calculator will confirm this; for example,

$\tan 1.5 \approx 14.1 \qquad \tan 1.55 \approx 48.1 \qquad \tan 1.56 \approx 92.6 \qquad \tan 1.57 \approx 1{,}255.8$

Since the range of $y = \tan x$ is the set of real numbers, stretching or shrinking the curve in a vertical direction to obtain the graph of $y = a \tan x$ will affect each y-value but the range remains unchanged. Observe how the y-values are affected by comparing the graphs of $y = \frac{1}{4} \tan x$ and $y = 2 \tan x$ to the graph of $y = \tan x$.

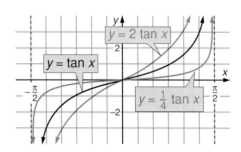

The period of the basic tangent curve is changed when x is multiplied by a constant so that the period of $y = \tan bx$ is $\left|\dfrac{\pi}{b}\right|$. Compare the graphs of $y = \tan \dfrac{1}{2}x$ and $y = \tan 3x$.

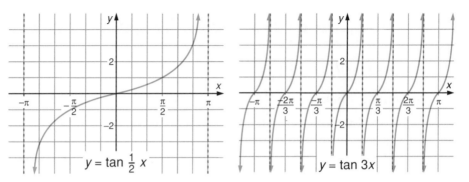

Note that to complete one cycle, it takes $\left|\dfrac{\pi}{\frac{1}{2}}\right|$ or 2π radians for $y = \tan \dfrac{1}{2}x$ and $\left|\dfrac{\pi}{3}\right|$ or $\dfrac{\pi}{3}$ radians for $y = \tan 3x$.

Notice that on both graphs, the scale along the x- and y-axes are not the same. The length that represents 1 on the y-axis is about one-half as long as the length that represents 1 on the x-axis. Different scales on the x- and y-axes are frequently used for "fit," for example, when the graph is drawn in a predetermined space such as on a calculator.

The Graph of $y = \cot x$

Since $\cot x = \dfrac{1}{\tan x}$, the graph of $y = \cot x$ has asymptotes at the values of x for which $\tan x = 0$, namely, when $x = k\pi$, for k an integer. Similarly, $\cot x = 0$ at the values for which $y = \tan x$ is undefined. Since $\cot (-x) = \dfrac{\cos (-x)}{\sin (-x)} = \dfrac{\cos x}{-\sin x} = -\cot x$, $y = \cot x$ is an odd function, and is symmetric with respect to the origin. Notice that the period of the graph of $y = \cot x$ is also π, the same as for the graph of $y = \tan x$. Key points include $\left(-\dfrac{\pi}{4}, -1\right)$, $\left(\dfrac{\pi}{2}, 0\right)$, $\left(\dfrac{\pi}{4}, 1\right)$.

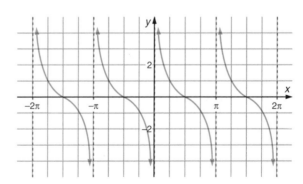

The Graph of $y = \sec x$

Since $\sec x = \dfrac{1}{\cos x}$, the graph of $y = \sec x$ has asymptotes at the values of x for which $\cos x = 0$, namely, when $x = \dfrac{\pi}{2} + k\pi$, for k an integer. Since

$$\sec(-x) = \frac{1}{\cos(-x)} = \frac{1}{\cos x} = \sec x,$$

$y = \sec x$ is an even function and is symmetric with respect to the y-axis. The period of $y = \sec x$ is the same as that of the cosine function, 2π. Key values for this graph include $(-\pi, -1)$, $(0, 1)$, $(\pi, -1)$. The graph of $y = \cos x$ is helpful in determining the shape of $y = \sec x$. The graphs coincide whenever $\cos x = \pm 1$, and the absolute value of one function decreases as that of the other increases. As $\cos x$ is positive and approaches 0, $\sec x$ approaches infinity. As $\cos x$ is negative and approaches 0, $\sec x$ approaches negative infinity.

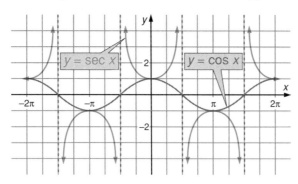

The Graph of $y = \csc x$

Since $\csc x = \dfrac{1}{\sin x}$, the relationship between $\csc x$ and $\sin x$ is analogous to the relationship between $\sec x$ and $\cos x$. Since $y = \sin x$ is an odd function, $y = \csc x$ is also an odd function, with symmetry to the origin. The period of $y = \csc x$ is 2π, and its asymptotes occur at $x = k\pi$, for k an integer.

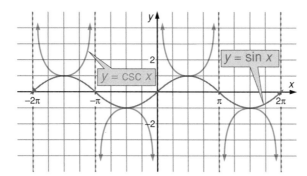

Notice that the graph of $y = \csc x$ is actually the graph of $y = \sec x$ translated $\dfrac{\pi}{2}$ units to the right, just as the graph of $y = \sin x$ is the graph of $y = \cos x$ translated $\dfrac{\pi}{2}$ units to the right.

Example 1 _____

Sketch the graph of $y = 2 \sec \left(x - \dfrac{\pi}{4} \right)$ on the interval $\left(-\dfrac{5\pi}{4}, \dfrac{7\pi}{4} \right)$.

Solution

The expression $x - \frac{\pi}{4}$ translates the graph of $y = \sec x$ $\frac{\pi}{4}$ to the right, and the coefficient 2 doubles each of the y values of

$$y = \sec \left(x - \frac{\pi}{4}\right).$$

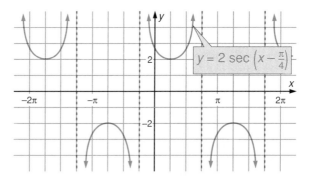

Additional Graphs Related to the Trigonometric Functions

Some trigonometric graphs consist of a sum, difference, product, or quotient of two functions, one or both of which are trigonometric functions. An understanding of the graphs of the individual functions may be helpful in graphing the composite.

Example 2

Sketch the graph of $y = x + 2 - \sin x$ on the interval $(-\pi, \pi)$.

Solution

Since the ordinates of this graph are the sum of the ordinates for $y = x + 2$ and $y = -\sin x$, sample ordinates of the graph of $y = x + 2 - \sin x$ may be obtained by adding the ordinates of the two graphs. For example, when $x = 0$, $x + 2 = 2$, and $-\sin x = 0$; thus, $(0, 2)$ is on the graph of $y = x + 2 - \sin x$. When $x = \frac{\pi}{2}$, $x + 2 \approx 3.57$, and $-\sin x = -1$; thus, $\left(\frac{\pi}{2}, 2.57\right)$ is on the graph.

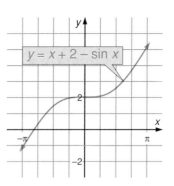

x	$-\pi$	$-\frac{3\pi}{4}$	$-\frac{\pi}{2}$	$-\frac{\pi}{4}$	0	$\frac{\pi}{4}$	$\frac{\pi}{2}$	$\frac{3\pi}{4}$	π
$x + 2$	-1.14	-0.36	0.43	1.21	2	2.79	3.57	4.36	5.14
$-\sin x$	0	0.71	1	0.71	0	-0.71	-1	-0.71	0
y	-1.14	0.35	1.43	1.92	2	2.08	2.57	3.65	5.14

Example 3 _____

Sketch the graph of $y = \sin x + \tan x$ on the interval $[-\pi, \pi]$.

Solution

Since the ordinates for the graph of $y = \sin x + \tan x$ are the sum of the ordinates of $y = \sin x$ and $y = \tan x$, first examine these basic graphs simultaneously.

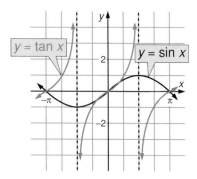

At $x = -\pi$ and at $x = \pi$, the ordinates are both 0; thus, $(-\pi, 0)$ and $(\pi, 0)$ are on the graph of $y = \sin x + \tan x$. For $x = \frac{\pi}{2} + 2k\pi$, where k is an integer, $\tan x$ is undefined, and the graph of $y = \sin x + \tan x$ has vertical asymptotes at these values of x. At values of x where $\sin x = \tan x$, the graph of $y = \sin x + \tan x$ has an ordinate that is twice the ordinate of either graph.

From $-\pi$ to $-\frac{\pi}{2}$, $\tan x$ is positive and $\sin x$ is negative; thus, the graph of $y = \sin x + \tan x$ has ordinates that are less than those for $\tan x$. From $-\frac{\pi}{2}$ to 0, $\tan x$ and $\sin x$ are both negative; thus, $\sin x + \tan x$ is less than either. From 0 to $\frac{\pi}{2}$, both functions are positive; thus, $\sin x + \tan x$ is greater than either function. From $\frac{\pi}{2}$ to π, $\sin x$ is positive and $\tan x$ is negative; thus, $\sin x + \tan x$ is between them.

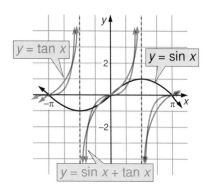

1. Copy and complete the table.

	Domain	Range	Period	Asymptotes (if any)	For the given intervals, Increasing (I) or Decreasing (D)			
					$\left(0,\frac{\pi}{2}\right)$	$\left(\frac{\pi}{2},\pi\right)$	$\left(\pi,\frac{3\pi}{2}\right)$	$\left(\frac{3\pi}{2},2\pi\right)$
$y = \sin x$								
$y = \cos x$								
$y = \tan x$								
$y = \cot x$								
$y = \sec x$								
$y = \csc x$								

2. Classify each of the six trigonometric functions as even or odd.

3. Classify each of the six trigonometric functions as having symmetry with respect to the origin or the y-axis.

In **4 – 15**, graph the function over the interval $(-2\pi, 2\pi)$ and state the period of the graph.

4. $y = 3 \tan x$ 5. $y = \frac{1}{2} \tan x$ 6. $y = 4 \cot x$ 7. $y = -2 \tan x$

8. $y = 1 + \cot x$ 9. $y = -2 + \tan x$ 10. $y = -\csc x$ 11. $y = 3 \sec x$

12. $y = 1 + \sec x$ 13. $y = 3 - \tan x$ 14. $y = 1 + \frac{1}{2} \csc x$ 15. $y = -1 + \cot x$

In **16 – 21**, graph the function over an interval that includes one period on either side of the origin.

16. $y = \sec 2x$ 17. $y = 3 \tan \frac{1}{2}x$ 18. $y = -\sec \frac{1}{2}x$

19. $y = -\csc 3x$ 20. $y = 2 \cot \frac{1}{3}x$ 21. $y = 2 \tan 2x$

In **22 – 30**, draw one period of the graph and state, in a sentence, how the graph has been transformed from the graph of the basic function.

22. $y = \tan\left(x + \frac{\pi}{2}\right)$ 23. $y = \cot\left(x - \frac{\pi}{3}\right)$ 24. $y = 2 \sec (x - \pi)$

25. $y = \frac{1}{2} \csc\left(x - \frac{\pi}{4}\right)$ 26. $y = 2 + \tan\left(x - \frac{\pi}{6}\right)$ 27. $y = 1 + \cot\left(x - \frac{\pi}{2}\right)$

28. $y = \csc (x + \pi) - 1$ 29. $y = \sec\left(\frac{1}{2}x + \pi\right)$ 30. $y = 1 + \frac{1}{3} \tan\left(\frac{1}{4}x + \pi\right)$

1.

	Domain*	Range	Period	Asymptotes*	Increasing or Decreasing $(0, \frac{\pi}{2})$	$(\frac{\pi}{2}, \pi)$	$(\pi, \frac{3\pi}{2})$	$(\frac{3\pi}{2}, 2\pi)$
$y = \sin x$	R	$[-1, 1]$	2π	none	I	D	D	I
$y = \cos x$	R	$[-1, 1]$	2π	none	D	D	I	I
$y = \tan x$	$R/\{\frac{\pi}{2} + k\pi\}$	R	π	$x = \frac{\pi}{2} + k\pi$	I	I	I	I
$y = \cot x$	$R/\{k\pi\}$	R	π	$x = \pi + k\pi$	D	D	D	D
$y = \sec x$	$R/\{\frac{\pi}{2} + k\pi\}$	$R/(-1, 1)$	2π	$x = \frac{\pi}{2} + k\pi$	I	I	D	D
$y = \csc x$	$R/\{k\pi\}$ †	$R/(-1, 1)$	2π	$x = \pi + k\pi$	D	I	I	D

† $R/\{k\pi\}$ = Reals except $k\pi$ *k is an integer

2.

Function	odd or even
sine	odd
cosine	even
tangent	odd
cotangent	odd
secant	even
cosecant	odd

3.

Function	Symmetry Origin	y-axis
sine	yes	no
cosine	no	yes
tangent	yes	no
cotangent	yes	no
secant	no	yes
cosecant	yes	no

4.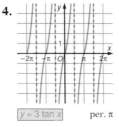
$y = 3 \tan x$ per. π

5.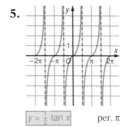
$y = \frac{1}{2} \tan x$ per. π

6.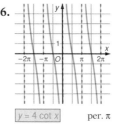
$y = 4 \cot x$ per. π

7.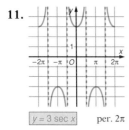
$y = -2 \tan x$ per. π

8.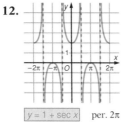
$y = 1 + \cot x$ per. π

9.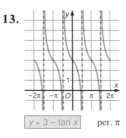
$y = -2 + \tan x$ per. π

10.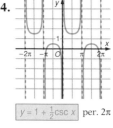
$y = -\csc x$ per. 2π

11.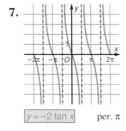
$y = 3 \sec x$ per. 2π

12.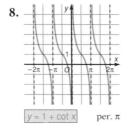
$y = 1 + \sec x$ per. 2π

13.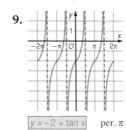
$y = 3 - \tan x$ per. π

14.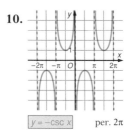
$y = 1 + \frac{1}{2}\csc x$ per. 2π

15.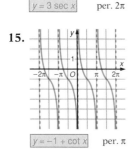
$y = -1 + \cot x$ per. π

16.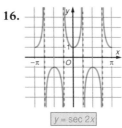
$y = \sec 2x$

17.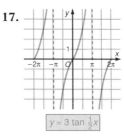
$y = 3 \tan \frac{1}{2}x$

18.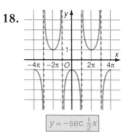
$y = -\sec \frac{1}{2}x$

19.

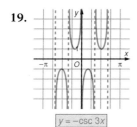

$$y = -\csc 3x$$

20.

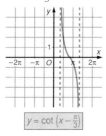

$$y = 2 \cot \tfrac{1}{3} x$$

21.

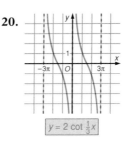

$$y = 2 \tan 2x$$

22. The graph has been shifted $\frac{\pi}{2}$ to the left.

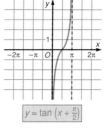

$$y = \tan\left(x + \tfrac{\pi}{2}\right)$$

23. The graph has been shifted $\frac{\pi}{3}$ to the right.

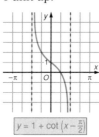

$$y = \cot\left(x - \tfrac{\pi}{3}\right)$$

24. The graph has been shifted π to the right and the amplitude is doubled.

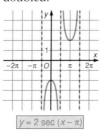

$$y = 2 \sec (x - \pi)$$

25. The graph is shifted $\frac{\pi}{4}$ to the right and the amplitude is multiplied by $\frac{1}{2}$.

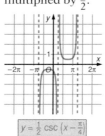

$$y = \tfrac{1}{2} \csc \left(x - \tfrac{\pi}{4}\right)$$

26. The graph is shifted $\frac{\pi}{6}$ to the right and 2 units up.

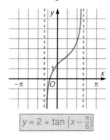

$$y = 2 + \tan\left(x - \tfrac{\pi}{6}\right)$$

27. The graph is shifted $\frac{\pi}{2}$ to the right and 1 unit up.

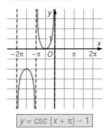

$$y = 1 + \cot\left(x - \tfrac{\pi}{2}\right)$$

28. The graph is shifted π to the left and 1 unit down.

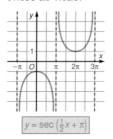

$$y = \csc (x + \pi) - 1$$

29. The graph is shifted 2π to the left and is twice as wide.

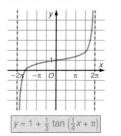

$$y = \sec \left(\tfrac{1}{2} x + \pi\right)$$

30. The graph is moved 4π to the left and is 4 times as wide and $\frac{1}{3}$ as high. It has been raised 1 unit.

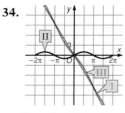

$$y = 1 + \tfrac{1}{3} \tan \left(\tfrac{1}{4} x + \pi\right)$$

31.

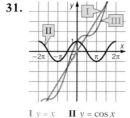

I $y = x$ **II** $y = \cos x$
III $y = x + \cos x$

32.

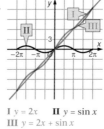

I $y = 2x$ **II** $y = \sin x$
III $y = 2x + \sin x$

33.

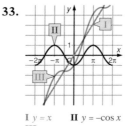

I $y = x$ **II** $y = -\cos x$
III $y = x - \cos x$

34.

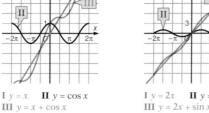

I $y = -3x$ **II** $y = \sin x$
III $y = -3x + \sin x$

Exercises 8.7 *(continued)*

In **31 – 39**, consider the given function as two separate functions. Use the graphs of the individual functions to sketch the graph of the function required. Show enough of the graph so that its complete shape is established.

31. $y = x + \cos x$ **32.** $y = 2x + \sin x$ **33.** $y = x - \cos x$

34. $y = -3x + \sin x$ **35.** $y = \sin x + \cos x$ **36.** $y = \tan x - \sin x$

37. $y = 2 \sin x + \cos \frac{1}{2}x$ **38.** $y = 2x + 1 - \sin x$ **39.** $y = \frac{1}{2}x + \cos x + 1$

40. A rotating TV camera in the ceiling of a super-market is 16 feet directly above a shopper. In its continuous clockwise rotation, the camera focuses half the time into the store, and half the time above the ceiling. Let s be the distance, in feet, between the shopper and a point on the floor on which the camera is focused, and assume that the width of the store is large enough to be considered infinite without significantly changing the accuracy of the problem (when the angle of focus is 85° with the vertical, the camera is focused about 180 feet in front of the shopper). The following equation describes the value of s for those values of t, the time in seconds, for which s is defined.

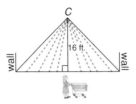

$$s = 16 \tan \frac{\pi}{2}t$$

Assume that $t = 0$ when the camera is focused on the shopper and that the values of s are positive as the camera focuses in front of the shopper and negative when it focuses behind.

a. To the nearest foot, find s, if it exists (the camera is focused on the floor), when:

    (1) $t = 0.1.$ **(2)** $t = 0.2.$ **(3)** $t = 3.5.$

b. For what values of t between 0 and 4 are there no values for s that are consistent with the physical con-straints of the problem?

41. A man 6 feet tall observes an oak tree 25 feet tall. θ, the angle that the man's line of sight to the top of the tree makes with a line parallel to the ground, can be expressed in terms of s, the distance from the tree, as: $s = 19 \cot \theta$.

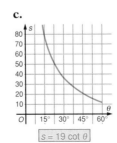

a. Derive this equation from the diagram.

b. Find s when $\theta = 60°$.

c. Graph the function on the interval (0°, 60°).

40. a. (1) $s \approx 3$ feet
    (2) $s \approx 5$ feet
    (3) $s \approx -16$ feet

b. $1 \le t \le 3$
The camera's focus is at the ceiling or above when the angle is between $\frac{\pi}{2}$ and $\frac{3\pi}{2}$; that is, $\frac{\pi}{2} \le \frac{\pi}{2}t \le \frac{3\pi}{2}$, or $1 \le t \le 3$.

41. a. $\cot \theta = \frac{s}{19}$, or
$$s = 19 \cot \theta$$

b. $\frac{19\sqrt{3}}{3}$

c.

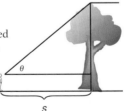

$s = 19 \cot \theta$

35.

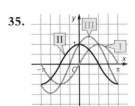

I $y = \sin x$ II $y = \cos x$
III $y = \sin x + \cos x$

36.

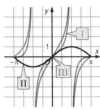

I $y = \tan x$ II $y = \sin x$
III $y = \tan x - \sin x$

37.

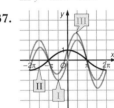

I $y = 2 \sin x$
II $y = \cos \frac{1}{2}x$
III $y = 2 \sin x + \cos \frac{1}{2}x$

38.

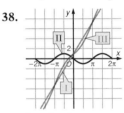

I $y = 2x + 1$
II $y = -\sin x$
III $y = 2x + 1 - \sin x$

39.

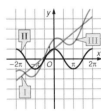

I $y = \frac{1}{2}x + 1$ II $y = \cos x$
III $y = \frac{1}{2}x + \cos x + 1$

The Wrapping Function

- A unit circle is a circle whose center is at the origin and whose radius is 1.

- The circumference of a circle with radius r is $2\pi r$.

- The wrapping function is a function whose domain is the set of real numbers, representing the length of an arc of the unit circle, and whose range is the set of ordered pairs (x, y), the coordinates of the endpoints of the arc.

- The wrapping function is not a one-to-one function.

- If $(s, (x, y))$ is a member of the wrapping function, then the six trigonometric functions are defined in terms of s as follows:

$$\sin s = y \qquad\qquad \csc s = \frac{1}{y} \;\; (y \neq 0)$$

$$\cos s = x \qquad\qquad \sec s = \frac{1}{x} \;\; (x \neq 0)$$

$$\tan s = \frac{y}{x} \;\; (x \neq 0) \qquad\qquad \cot s = \frac{x}{y} \;\; (y \neq 0)$$

- A function is periodic if for every real number x in the domain there exists a positive number p such that $f(x) = f(x + p)$. The smallest positive value of p is called the period of the function.

- Each of the six trigonometric functions is periodic.

- The period of the wrapping function is 2π.

Angles and Angle Measure

- An angle can be formed by rotating a ray around its endpoint.

- Angles formed by a *counterclockwise* rotation have *positive* measures, and angles formed by a *clockwise* rotation have *negative* measures.

- Angles can be measured in degrees ($1° = \frac{1}{360}$ of a complete revolution) or radians (when the vertex of the angle is at the center of the circle, 1 radian = a rotation that intercepts an arc equal in length to 1 radius on the circle).

- π radians = $180°$

- To convert between radians and degrees, use the proportion: $\dfrac{\text{degrees}}{\text{radians}} = \dfrac{180}{\pi}$

- If a central angle whose measure is θ cuts off an arc of length s on a circle whose radius is r, then, when s and r are expressed in terms of the same units of measure: $\theta = \dfrac{s}{r}$

Right Triangle Trigonometry

- The trigonometric functions may be defined in terms of a right triangle with an acute angle θ as follows:

$$\sin \theta = \frac{\text{opp}}{\text{hyp}} \qquad \cos \theta = \frac{\text{adj}}{\text{hyp}} \qquad \tan \theta = \frac{\text{opp}}{\text{adj}}$$

$$\csc \theta = \frac{\text{hyp}}{\text{opp}} \qquad \sec \theta = \frac{\text{hyp}}{\text{adj}} \qquad \cot \theta = \frac{\text{adj}}{\text{opp}}$$

where opp, adj, and hyp represent the length of the leg opposite angle θ, the length of the leg adjacent to angle θ, and the length of the hypotenuse of the right triangle, respectively.

- The values of the trigonometric functions for $30°$, $45°$, and $60°$ $\left(\text{or } \frac{\pi}{6}, \frac{\pi}{4}, \text{ and } \frac{\pi}{3}\right)$ can be found exactly by using geometry.

	sin	cos	tan	csc	sec	cot
30°	$\frac{1}{2}$	$\frac{\sqrt{3}}{2}$	$\frac{\sqrt{3}}{3}$	2	$\frac{2\sqrt{3}}{3}$	$\sqrt{3}$
45°	$\frac{\sqrt{2}}{2}$	$\frac{\sqrt{2}}{2}$	1	$\sqrt{2}$	$\sqrt{2}$	1
60°	$\frac{\sqrt{3}}{2}$	$\frac{1}{2}$	$\sqrt{3}$	$\frac{2\sqrt{3}}{3}$	2	$\frac{\sqrt{3}}{3}$

Relationships Among Trigonometric Functions

- There are three pairs of *cofunctions*:

    sine, cosine   tangent, cotangent   secant, cosecant

- Cofunctions of complementary angles are equal; for example, $\sin x = \cos \left(\frac{\pi}{2} - x\right)$

- There are three pairs of *reciprocal functions*. Assuming no denominator is 0:

$$\csc x = \frac{1}{\sin x} \qquad \sec x = \frac{1}{\cos x} \qquad \cot x = \frac{1}{\tan x}$$

- Tangent and cotangent can be expressed in terms of sine and cosine. Assuming no denominator is 0:

$$\tan x = \frac{\sin x}{\cos x} \qquad \cot x = \frac{\cos x}{\sin x}$$

Function Values of Any Angle Measure

- The signs of the trigonometric function values vary by quadrant.

Quadrant	I	II	III	IV
sin and csc	+	+	−	−
cos and sec	+	−	−	+
tan and cot	+	−	+	−

- When the measure of an acute angle is given in radians or degrees, the approximate values of the trigonometric functions can be found by using a table of trigonometric function values or a scientific calculator.

- Trigonometry can be used to find measurements, such as distances or heights of structures perpendicular to the ground, by indirect measurement.

- Reference angles can be used to find the values of trigonometric functions for angles of any measure.

- The reference angle, θ_r, for any angle θ in a particular quadrant, is computed as follows:

	In Degrees	In Radians
Quadrant I	$\theta_r = \theta$	$\theta_r = \theta$
Quadrant II	$\theta_r = 180 - \theta$	$\theta_r = \pi - \theta$
Quadrant III	$\theta_r = \theta - 180$	$\theta_r = \theta - \pi$
Quadrant IV	$\theta_r = 360 - \theta$	$\theta_r = 2\pi - \theta$

Graphs of Trigonometric Functions

- The graphs of the trigonometric functions are:

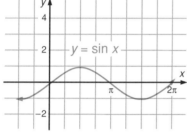

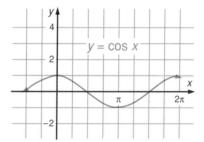

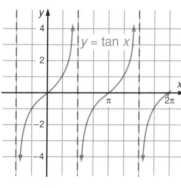

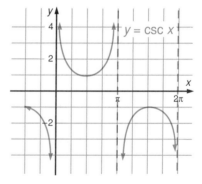

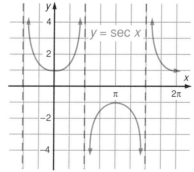

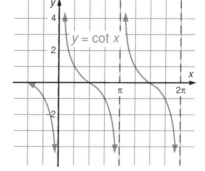

- The sine, cosecant, tangent, and cotangent functions are odd functions and have symmetry with respect to the origin.
- The cosine and secant functions are even functions and have symmetry with respect to the y-axis.
- The period of the graphs of $y = \sin x$, $y = \cos x$, $y = \csc x$, and $y = \sec x$ is 2π. The period of the graphs of $y = \tan x$ and $y = \cot x$ is π.
- The graph of $y = a \sin bx$ (or $y = a \cos bx$), where a is a real number and b is a positive real number, has an amplitude of $|a|$ and a period of $\frac{2\pi}{b}$.
- The graph of $y = k + a \sin bx$ (or $y = k + a \cos bx$), where k is a real number, is the graph of $y = a \sin bx$ (or $y = a \cos bx$) translated k units upward if k is positive and k units downward if k is negative.
- The graph of $y = a \sin b(x + h)$ [or $y = a \cos b(x + h)$], where h is a real number, is the graph of $y = a \sin bx$ (or $y = a \cos bx$) translated h units to the left if h is positive and h units to the right if h is negative.
- The following is a summary of the locations of asymptotes for the graphs of those trigonometric functions that have asymptotes. In each expression, k is an integer.

$$y = \tan x \qquad x = \frac{\pi}{2} + k\pi$$

$$y = \cot x \qquad x = k\pi$$

$$y = \sec x \qquad x = \frac{\pi}{2} + k\pi$$

$$y = \csc x \qquad x = k\pi$$

- The graph of a function formed by the sum or difference of two functions may be drawn by adding or subtracting the ordinates of the individual functions to find ordinates for the function being drawn.

Chapter 8   Review Exercises

In **1 – 6**, find the value of the wrapping function for the given arc length s.

1. $\frac{\pi}{2}$ **2.** 0 **3.** $\frac{\pi}{4}$ **4.** $-\frac{\pi}{4}$ **5.** 2π **6.** 5π

In **7 – 18**, find the exact value of the given function.

7. $\sin 45°$ **8.** $\sin 120°$ **9.** $\cos 225°$ **10.** $\tan 135°$

11. $\sec \frac{\pi}{6}$ **12.** $\csc \left(-\frac{\pi}{3}\right)$ **13.** $\cot \frac{3\pi}{4}$ **14.** $\tan \pi$

15. $\sin \left(-\frac{5\pi}{3}\right)$ **16.** $\cos \frac{13\pi}{6}$ **17.** $\tan \left(-\frac{10\pi}{3}\right)$ **18.** $\cot \frac{7\pi}{4}$

20. $\sin s = y = \frac{\sqrt{5}}{5}$, $\cos s = x = -\frac{2\sqrt{5}}{5}$,

$\tan s = \frac{y}{x} = -\frac{1}{2}$, $\csc s = \frac{1}{y} = \sqrt{5}$,

$\sec s = \frac{1}{x} = -\frac{\sqrt{5}}{2}$, $\cot s = \frac{x}{y} = -2$

21. $\tan s = \frac{\sqrt{14}}{7}$

22. a. $32.6833°$ **b.** $114.4042°$

23. 0.7604

24. 0.0029	**32.** 1.002	**40.** 1.489575	**48.** 72°00′
25. 2.112	**33.** −1.022	**41.** 0.774	**49.** 46°10′
26. 85.95	**34.** 1.530	**42.** 0.025	**50.** 68°20′
27. 1.410	**35.** 0.906308	**43.** 0.540	**51.** 33°40′
28. 1.0000	**36.** −0.393407	**44.** 0.598	**52.** 58°30′
29. −0.2896	**37.** 0.327731	**45.** 0.074	**53.** $\frac{1}{3}$ radian
30. 0.9063	**38.** 3.280144	**46.** 0.253	**54.** 15 cm
31. 3.732	**39.** 0.642093	**47.** 6°00′	

55. $\frac{40}{3\pi}$ cm

56. 8π cm

57. 4.07, 5.36

58. 0.39, 3.54

59. 1.81, 4.96

60. 6.14

61. 37 in.

62. 60°

63. 54°

64.

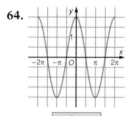

$y = 2 \cos x$

65.

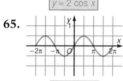

$y = \frac{1}{2} \sin x$

66.

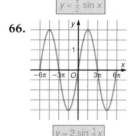

$y = 2 \sin \frac{1}{3}x$

67.

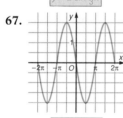

$y = -2 \sin x$

68.

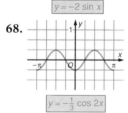

$y = -\frac{1}{3} \cos 2x$

Chapter 8   Review Exercises *(continued)*

19. If $\cos s = \frac{5}{13}$ and s corresponds to (x, y) in quadrant I, find the five remaining trigonometric functions of s.

20. If $\tan s = -\frac{1}{2}$ and s corresponds to (x, y) in quadrant II, find the five remaining trigonometric functions of s.

21. If $\sin s = -\frac{\sqrt{2}}{3}$ and $\cos s < 0$, find $\tan s$.

22. Express each angle measure to the nearest ten-thousandth of a degree.

    a. 32°41′   **b.** 114°24′15″

In **23 – 34**, use the trigonometric table to find the value.

23. cos 40°30′   **24.** sin 0°10′   **25.** tan 64°40′   **26.** sec (–89°20′)

27. csc 45°10′   **28.** sin 89°50′   **29.** sin (–0.2938)   **30.** cos 0.4363

31. tan 1.3090 ·   **32.** sec 0.0698   **33.** csc (–1.3614)   **34.** cot 0.5789

In **35 – 40**, find the value of the given function, rounding to six decimal places.

35. sin 115°   **36.** cos 113°10′   **37.** tan 0.3167

38. sec (–1.261)   **39.** cot 1   **40.** csc 42°10′12″

In **41 – 46**, for the interval $\left[0, \frac{\pi}{2}\right)$, find θ to the nearest thousandth of a radian.

41. $\cos \theta = 0.7150$ **42.** $\sin \theta = 0.0247$ **43.** $\tan \theta = 0.6$

44. $\sec \theta = 1.21$ **45.** $\cot \theta = 13.4$ **46.** $\csc \theta = 4$

In **47 – 52**, use the trigonometric table to find θ to the nearest ten minutes in the interval [0°, 90°).

47. $\sin \theta = 0.1046$ **48.** $\cos \theta = 0.3101$ **49.** $\tan \theta = 1.04$

50. $\sec \theta = 2.7$ **51.** $\csc \theta = 1.8041$ **52.** $\cot \theta = 0.613$

53. Find the measure of a central angle that intercepts an arc of 1.6 inches in a circle whose radius is 4.8 inches.

54. Find the length of an arc intercepted by an angle of 0.3 radians in a circle whose radius is 50 cm.

55. Find the radius of a circle in which a central angle of 135° cuts off an arc 10 cm long.

56. Find the length of the arc intercepted by an angle of 120° in a circle whose radius is 12 cm.

69.

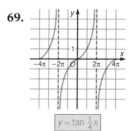

$y = \tan \frac{1}{4}x$

70.

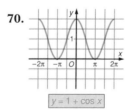

$y = 1 + \cos x$

71.

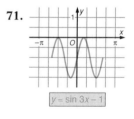

$y = \sin 3x - 1$

In **57 – 59**, for the interval $(0, 2\pi)$, find two values of θ, to the nearest hundredth, that satisfy the equation.

57. $\sin \theta = -0.8$ **58.** $\cot \theta = 2.4$ **59.** $\sec \theta = -4.16$

60. Using the diagram below, find x to the nearest hundredth if $m\angle A = 15°$, $m\angle B = 41°$, and $AB = 30$.

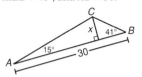

61. From a point 20 feet from the base of a building, the angle of elevation to the bottom of a window on an upper floor is 22°10′ and the angle of elevation to the top of the window is 29°20′. Find the height of the window to the nearest inch.

62. To the nearest degree, find the angle of elevation of the sun if a goalpost 20 feet high casts a shadow 11.6 feet long.

63. To the nearest degree, find the measure of the acute angle that $\overrightarrow{OP}$ makes with the x-axis if O is the origin and P is the point $(5, 7)$.

In **64 – 75**, graph the function over a two-period interval that contains the origin.

64. $y = 2 \cos x$ **65.** $y = \frac{1}{2} \sin x$ **66.** $y = 2 \sin \frac{1}{3}x$

67. $y = -2 \sin x$ **68.** $y = -\frac{1}{3} \cos 2x$ **69.** $y = \tan \frac{1}{4}x$

70. $y = 1 + \cos x$ **71.** $y = \sin 3x - 1$ **72.** $y = \cos \left(x - \frac{\pi}{3}\right)$

73. $y = \sec \frac{1}{2}\left(x + \frac{\pi}{4}\right)$ **74.** $y = 3 + 2 \sin \left(2x + \frac{\pi}{6}\right)$ **75.** $y = 3 + 2 \csc \left(2x + \frac{\pi}{6}\right)$

In **76 – 79**, graph the function over a one-period interval that contains the origin.

76. $y = 3x + \sin x$ **77.** $y = \tan x - x$ **78.** $y = \cos x + 2 \sin x$ **79.** $y = \sec x + \csc x$

In **80 – 83**, find, to the nearest degree, the measure of the angle that the resultant, $\vec{a} + \vec{b}$, makes with $\vec{a}$.

80. $\vec{a} = 2\vec{i}, \vec{b} = 2\vec{j}$ **81.** $\vec{a} = 2\sqrt{3}\,\vec{i}, \vec{b} = 2\vec{j}$

82. $\vec{a} = 4\vec{i}, \vec{b} = \vec{i} + 3\vec{j}$ **83.** $\vec{a} = 3\vec{i} + 2\vec{j}, \vec{b} = \vec{i} + \vec{j}$ *Hint:* Find the difference between the measures of the angles that $\vec{a}$ and $\vec{a} + \vec{b}$ make with the x-axis.

Review Answers

72.

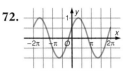

$y = \cos \left(x - \frac{\pi}{3}\right)$

73.

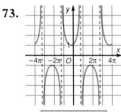

$y = \sec \frac{1}{2}\left(x + \frac{\pi}{4}\right)$

74.

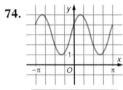

$y = 3 + 2 \sin \left(2x + \frac{\pi}{6}\right)$

75.

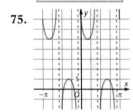

$y = 3 + 2 \csc \left(2x + \frac{\pi}{6}\right)$

76.

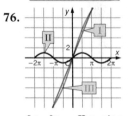

I $y = 3x$ II $y = \sin x$
III $y = 3x + \sin x$

77.

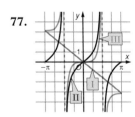

I $y = -x$ II $y = \tan x$
III $y = \tan x - x$

78.

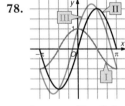

I $y = \cos x$ II $y = 2 \sin x$
III $y = \cos x + 2 \sin x$

79.

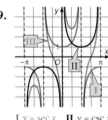

I $y = \sec x$ II $y = \csc x$
III $y = \sec x + \csc x$

80. 45°
81. 30°
82. 31°
83. 3°

☑ Exercises for Challenge

1. How many solutions are there to the equation $\cos x = x$?

2. Find the degree measures of the acute angles α and β if $\sin (\alpha - \beta) = \frac{1}{2}$ and $\cos (\alpha + \beta) = \frac{1}{2}$.

3. 🖩 To the nearest thousandth, find the cosine of the smaller of two acute angles of a right triangle if the measure of one of the legs is 30% of the measure of the other leg.

4. Which of the following is equal to $\cos 520°$?
(A) $\cos (-20°)$ (B) $\sin 70°$
(C) $-\sin 20°$ (D) $-\cos 70°$
(E) $-\sin 70°$

5. In the diagram, $P(a, b)$, $R(c, 1)$, and $T(1, t)$ are on the terminal ray of angle θ. P is on a unit circle, R is on a line tangent to the circle at the y-axis, and T is on a line tangent to the circle at the x-axis. Which are equal to c?

I $\dfrac{1}{t}$ II $\dfrac{a}{b}$ III $1 - \sqrt{a^2 + b^2}$

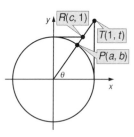

(A) I only   (B) II only   (C) III only
(D) I and II only   (E) I, II, and III

6. The functions $y = 2 \sin (3x + 4)$ and $y = 4 \cos (3x + 2)$ have the same
(A) amplitude   (B) period
(C) range   (D) roots
(E) none of the above

7. Which equation has no solution?
(A) $\sin x = -1$ (B) $\cos x = -2$
(C) $\tan x = \pi$ (D) $\cot x = 5$
(E) $\sec x = 2$

8. The graph of which function has no vertical asymptotes?

(A) $y = \dfrac{\sin x}{\cos x}$ (B) $y = \dfrac{1}{1 + \sin x}$

(C) $y = \dfrac{1}{2 + \sin x}$ (D) $y = \dfrac{\cos x}{2 \sin x}$

(E) $y = \dfrac{1}{\cos 2x}$

9. Which of the following is *false* for the graph of the function $f(x) = \cos x + \sec x$?

(A) The graph intersects the x-axis.

(B) The graph is symmetric with respect to the y-axis.

(C) The graph has vertical asymptotes.

(D) In the interval $-\frac{\pi}{2} < x < \frac{\pi}{2}$, the graph lies above the graph of $y = \cos x$.

(E) In the interval $-\frac{\pi}{2} < x < \frac{\pi}{2}$, the graph lies above the graph of $y = \sec x$.

10. Which functions are *not* the same as $y = \sin x$?

   I $y = \cos (x + \pi)$

   II $y = \frac{1}{2} \sin 2x$

   III $y = \sqrt{1 - \cos^2 x}$

  (A) I only   (B) II only   (C) III only

  (D) I and II only   (E) I, II, and III

11. 🖩 The graph of $y = \sin x$ is shifted 1 unit up and 2 units to the left. The resulting graph is that of a function whose value to the nearest thousandth at $x = 3$ is

  (A) 0.0174   (B) 0.0411   (C) 0.0872

  (D) 1.8415   (E) 1.9900

12. If $-\frac{\pi}{4} < x < \frac{\pi}{4}$, $\tan x$ *cannot* be equal to

  (A) 0   (B) $\frac{1}{2}$ (C) $\frac{\sqrt{2}}{2}$

  (D) $\frac{\sqrt{3}}{2}$ (E) 1

13. 🖩 Which of the following statements best describes the behavior of $y = \frac{\tan x}{x}$ when x is very close to 0 but not equal to 0?

  (A) y is very close to 1.

  (B) y is very close to 0.

  (C) $y = 1$.

  (D) y is very large.

  (E) y is undefined.

Questions 14 – 18 each consist of two quantities, one in Column A and one in Column B. You are to compare the two quantities and choose:

  A if the quantity in Column A is greater;
  B if the quantity in Column B is greater;
  C if the two quantities are equal;
  D if the relationship cannot be determined from the information given.

1. In certain questions, information concerning one or both of the quantities to be compared is centered above the two columns.

2. In a given question, a symbol that appears in both columns represents the same thing in Column A as it does in Column B.

3. x, n, and k, etc. stand for real numbers.

Column A	Column B

14. The period of $y = \sin x$ The period of $y = \tan \frac{1}{2} x$

$$\tan x = -\frac{1}{3}$$

15. $\sec^2 x$ $\csc^2 x$

The measure of an angle is x degrees or y radians, where $0 < x < 360$ and $0 < y < 2\pi$.

16. x y

17. $\sin x + \cos x$ $\sqrt{2}$

The measure of the three angles of a triangle are α, β and γ.

18. $\sin (\alpha + \beta)$ $\sin \gamma$

10. E

11. B

12. E

13. A

14. C

15. B

16. A

17. D

18. C

Teacher's Chapter 9

Applications of Trigonometry

Chapter Contents

In the previous chapter, the trigonometric functions were defined for any real number by using the unit circle and the result was applied to the right triangle. In this chapter, the study of the interrelationships among function values is continued through the identities and formulas. The use of the trigonometric functions to solve problems involving the right triangle is extended to include applications to oblique triangles and to vectors.

9.1 Trigonometric Identities

To emphasize the difference between an identity and an equation, elicit additional examples of algebraic identities, such as $x \cdot x = x^2$, or $(x + 1)^2 = x^2 + 2x + 1$.

When students prove identities, encourage them to work on one side of the identity only. This, of course, is always possible since the steps in verifying an identity are always reversible, which is a concept that should be demonstrated. Provide the opportunity for students to present different ways of solving the same identity.

As examples and exercises are considered, point out strategies. For example, in Exercise 37, the right side of the equation is in terms of tangent; look for ways to express the left side in terms of tangent. Students should understand that no single strategy works all the time, but, with practice, they can develop a sense of how to proceed. Remind them that the side of the equation they are trying to match acts as a guide for what needs to be done. The exercises provide a variety of types of identities to improve students' algebraic skills.

9.2 The Sum and Difference Formulas

Emphasize that in these formulas, it is the argument of the function that is a sum or difference. Shortly, there will be formulas for converting the sum of two function values such as $\sin A + \sin B$ to a product.

Since the derivations in this section require students to use the fact that $A + B = A - (-B)$ and that $\sin A = \cos (90° - A)$, as well as the distance formula, you may wish to begin the lesson with a set of short exercises such as:

1. Express $A + B$ as a difference.
2. For what values of A does $\sin A = \cos (90° - A)$?
3. Express the length of the line segment between the points whose coordinates are $(3, 2)$ and (a, b).

Obvious applications of the sum and difference formulas are that they provide ways to find exact values of the trigonometric functions for angles such as 15° and 75°.

Exercises 1 – 6 demonstrate that the sum and difference formulas are valid for given angle measures. These exercises offer intuitive examples, not rigorous proofs. Exercise 5 is included for completeness, to show an instance where both sides of the equation are undefined.

Since Exercises 19 – 26 require exact answers, remind students not to use the table or a calculator to answer these questions. Emphasize that the exact value of $\sqrt{3}$ is written only as $\sqrt{3}$ and that any decimal approximation, no matter how long, is not exact.

9.3 The Double-Angle, Half-Angle, and Sum-to-Product Formulas

The graphs of $y = \sin 2x$ and $y = 2 \sin x$ drawn on the same set of axes can be used to demonstrate that $\sin 2x \neq 2 \sin x$.

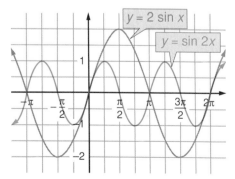

This is a good opportunity to emphasize the usefulness of a graph for studying the behavior of a function. While the graph is displayed, ask students to determine those values of x in $[0, 2\pi]$ for which sin $2x$ does equal 2 sin x.

Now that students have formulas for sin $(A + B)$ and sin $2A$, they can be challenged to derive new formulas for sin $3A$ or sin $4A$, and so on. Does sin $2x = 2$ sin x cos x imply that sin $3x = 3$ sin x cos x?

The half-angle formulas are each derived from a different form of the identity for cos $2A$. Each form is useful but can be easily derived when needed. The derivations for sin $\frac{1}{2}\theta$ and tan $\frac{1}{2}\theta$ are done in the text but cos $\frac{1}{2}\theta$ is left as an exercise. It is important for students to have a clear understanding of the need to choose the sign of the function value when using these identities. By studying a number of examples, demonstrate that θ and $\frac{\theta}{2}$ may not be in the same quadrant. A calculator may be used to compare the exact value of sin $\frac{\pi}{8}$, $\frac{\sqrt{2 - \sqrt{2}}}{2}$, with the approximate value given by the calculator.

The sum-to-product formulas are derived by using the identities for the function of $(A \pm B)$ and the product-to-sum formulas by using an algebraic substitution in the sum-to-product formulas. Students need not memorize all of the formulas introduced in this section and the previous one. You may ask students to memorize the most frequently used identities, and, when they need them, use a table of formulas for the others. Students should know that such formulas do exist and when they are useful.

In Exercises 62 – 64, students may at first attempt to derive a formula for sin $3x$ or cos $3y$ to solve the identity. However, a careful look at the whole problem suggests that the sum-to-product formulas are more appropriate. Encourage students to present and compare different methods of solution.

9.4 Trigonometric Equations

Emphasize the connection between the methods used to solve algebraic equations and those used to solve trigonometric equations. To avoid incomplete solutions, be sure students understand that they are solving for an angle and cannot stop when they have isolated only the function of the angle. For example, in Example 5, the solution set is $\left\{\frac{\pi}{3}, \pi, \frac{5\pi}{3}\right\}$ and not cos $x = \frac{1}{2}$, cos $x = -1$.

As they apply the trigonometric identities and formulas to make substitutions toward solution of trigonometric equations, considerable decision making is necessary. Additional judgment and care are necessary in matching the solution to the given domain.

9.5 The Inverses of the Trigonometric Functions

It is a good idea to review the concept of inverse function with some introductory exercises, such as:

Determine which of the following functions have inverses and, if they do, describe the inverse function.

1. $y = 3x + 2$
2. $y = x^2$
3. $\{(2, 3), (4, 3), (-1, 0)\}$

Answers
1. $y = \dfrac{x - 2}{3}$
2. no inverse if the domain and codomain are the set of real numbers; $y = \sqrt{x}$ if the domain and codomain are the nonnegative reals
3. no inverse

In this section, $\sin^{-1} x$ and arc sin x are used for the inverse function of $y = \sin x$ and similar notation is used for the inverses of the other trigonometric functions. This notation is used only for the principal values stated for each trigonometric function. It is unnecessary to distinguish between, for example, "Arc sin x" and "arc sin x" or "Sin^{-1} x" and "sin^{-1} x".

However, it may be useful to discuss this distinction since students may encounter it in other texts. That is, in some texts $y = \arcsin x$ is not a function and y can be any value such that $\sin y = x$, while $y = \text{Arc} \sin x$ is a function and $y \in \left[-\frac{\pi}{2}, \frac{\pi}{2}\right]$.

Students should understand that if the domains of the trigonometric functions are not restricted, each function has no inverse, and interchanging x and y or reflecting over the line $y = x$ forms a relation with each member of the domain of the relation having multiple images. This can be demonstrated by drawing a horizontal line through the graph of any of the trigonometric functions or a vertical line through the graph of the reflection over $y = x$ and noting the multiple intersections of the graph and the line.

A scientific calculator is helpful in demonstrating how values of an inverse function can be found within the range that is defined for it. Note that when asked to compute arc sin (-0.5), the calculator correctly displays -30, not 210 or 330. Using a calculator to find values of the inverses of the reciprocal functions may, however, cause some problems. Have students examine their own calculators for uniformity with the definitions given in the text.

In Section 8.3, the text showed how to deal with the reciprocal functions of sine, cosine, and tangent on a calculator. That is, since most scientific calculators do not have keys for cosecant, secant, or cotangent, the key $\boxed{1/x}$ must be used. In like manner, to work with the reciprocals of the inverse functions, again the key $\boxed{1/x}$ must be used. For example, the value of arc sec (-2) may be found by using the following key sequence, or a similar one according to the calculator being used.

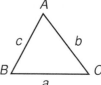

Display: $\boxed{120.}$

(in degree mode)

However, arc cot presents a problem since the range of arc cot x is not the same as the range of arc tan $\left(\frac{1}{x}\right)$. Students may be challenged to resolve this inconsistency. One method is to use the identity arc cot $x = 90° - \text{arc tan } x$. For example, to find arc cot $(-\sqrt{3})$, the following key sequence can be used when the calculator is in degree mode.

Display: $\boxed{150.}$

9.6 Solving Oblique Triangles Using the Law of Sines

An alternate proof for the Law of Sines uses the formula for the area of a triangle, $\frac{1}{2}ab \sin C$.

In $\triangle ABC$, with sides a, b, c:
$$\frac{a}{\sin A} = \frac{b}{\sin B} = \frac{c}{\sin C}$$

Proof:

$$\text{Area of } ABC = \tfrac{1}{2}bc \sin A$$
$$= \tfrac{1}{2}ac \sin B$$
$$= \tfrac{1}{2}ab \sin C$$

Dividing by $\frac{1}{2}abc$ gives the Law of Sines in the equivalent form:

$$\frac{\sin A}{a} = \frac{\sin B}{b} = \frac{\sin C}{c}$$

Emphasize the verbal form of the theorem to reinforce the importance of establishing the ratio between the measure of each side and the sine of the angle *opposite* that side.

Congruence postulates and their implications should be reviewed before the solution of oblique triangles is considered. A clear understanding of congruent triangles will help students to remember whether to use the Law of Sines or the Law of Cosines, and will also aid them in understanding the ambiguous case.

Since it is appropriate to have students solve for all of the missing sides or angles of a triangle, most of the exercises require a good deal of computation. Point out that it may not be necessary to evaluate a solution at every step. Carrying a trigonometric expression through several steps may eliminate intermediate calculations.

To avoid some difficulties with the ambiguous case, which students find confusing, suggest that students sketch the triangle uniformly; for example, with the given angle at the lower left of the sketch, they can analyze each example by asking a similar series of questions. Given $a = 4$, $b = 7$, $m\angle A = 30°$, and asked to determine the possible number of triangles, students might do the following.

1. Draw a sketch with the side opposite 30° dangling, suggesting that first they must determine, by computing the altitude, whether or not a length of 4 will reach the opposite side.

$$\sin 30° = \frac{\text{altitude}}{7}$$

$$\text{altitude} = 3\tfrac{1}{2}$$

Since $4 > 3\tfrac{1}{2}$, a triangle exists.

2. Consider the obtuse and acute triangles with the given dimensions and determine if the obtuse triangle causes a contradiction. (Is the larger of the two sides opposite the obtuse angle?) Since there is no contradiction, there are two possible triangles.

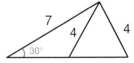

The most important reason for solving triangles is to apply indirect measurement to situations in which constraints make the use of right-triangle trigonometry impossible. For example, the following activity shows how to

find the height of a tall building whose top can only be sighted from a considerable and unknown distance from the foot of the building.

Activity (9.6) _____

Directions

1. Identify the tallest building in your city.

2. Approximate the height of the building by using indirect measurement as follows:

   a. Stand far enough away from the building so that you can comfortably see the top.

   b. Use a protractor to determine the angle of elevation to the top of the building.

   c. Move to a point further away from the building and note the distance from your last observation point to this one. With the protractor, take a reading of the angle of elevation to the top of the building from the second point.

3. Use the two angles of elevation and the distance between the two sighting points to determine the approximate height of the building. Remember to consider your own height in your analysis. Use your calculator to do the computation.

4. Research the actual height of the building and compare it to the height that you calculated using indirect measurement.

9.7 Solving Triangles Using the Law of Cosines

The proof in the text derives the relationship in a triangle known as the Law of Cosines by assuming that two sides and the included angle of the triangle are known. Once the formula is established, students should be able to see that given three sides (SSS), any angle can be found. Elicit that since SAS or SSS establishes congruence, the answers found for the other parts of the triangle are unique.

Note that the Pythagorean Theorem may be presented as a special case of the Law of Cosines where one angle of the triangle is a right angle. In addition, if $a^2 + b^2$ and c^2 are known, then the relative sizes of $a^2 + b^2$ and c^2 may be used to determine whether the triangle is obtuse or acute. Be sure students understand that the Law of Cosines is not a proof of the Pythagorean Theorem, since the proof of the Law of Cosines depends on the distance formula that, in turn, was derived using the Pythagorean Theorem.

So that one series of keystrokes can be used on a calculator rather than a series of intermediate calculations, encourage students to use alternate forms of the Law of Cosines such as:

$$\cos C = \frac{a^2 + b^2 - c^2}{2ab}$$

A number of different letter patterns should be used so that students understand the cosine law in words rather than in terms of the letters a, b and c.

Although the ambiguous case is usually associated with the Law of Sines, the Law of Cosines can also be used to find the possible measure of the third side of a triangle when SSA is given. As seen in Example 3, the Law of Cosines can be used to determine the number of triangles possible as well as the possible lengths of the third side. The calculations needed in applying the Law of Cosines to the ambiguous case make the calculator a must for these questions.

9.8 Simple Harmonic Motion

Since students see different physical applications of behaviors that can be approximated by a sine or cosine function, here is a good opportunity to assign an independent writing project. Ask students to summarize an article from a technical magazine such as *Scientific American* or a chapter from a physics book that describes an example of simple harmonic motion. One excellent source on waves is *Almost All About Waves* by J. R. Pierce, published by the Massachusetts Institute of Technology, 1974.

The examples and exercises of this section also provide opportunity for dimensional analysis. Since position is a function of time, students should be aware of appropriate units of measure.

Suggested Test Items

1. If $\csc x = \frac{5}{2}$ and $\tan x < 0$, find $\cos x$.

2. Simplify: $\dfrac{1}{1 + \sin x} + \dfrac{1}{1 - \sin x}$

3. Simplify as much as possible.

a. $\dfrac{\sec y}{\csc y}$

b. $\dfrac{1}{1 + \cot^2 x}$

4. Evaluate: $(5 \sin x + 5i \cos x)(5 \sin x - 5i \cos x)$, where $i = \sqrt{-1}$

In **5 – 8**, prove that the equation is an identity.

5. $\dfrac{\sin x}{1 - \cos x} = \cot x + \dfrac{1}{\sin x}$

6. $\csc y - \csc y \cos^2 y = \sin y$

7. $\cos^2 x - \sin^2 x = 2 \cos^2 x - 1$

8. $\csc \left(\frac{\pi}{2} - x \right) \cos x = 1$

9. Express $\cos\left(\frac{\pi}{4} + x\right)$ in terms of $\sin x$ and $\cos x$.

10. By using the formula for $\tan (A + B)$, determine the value of $\tan 75°$ in radical form.

11. By using the formula for $\sin (A - B)$, express $\sin \frac{\pi}{12}$ in radical form.

12. If $\cos x = \frac{2}{3}$ with $x \in$ IV and $\cos y = -\frac{1}{3}$ with $y \in$ II, find the value of $\sin (x + y)$.

13. Write $\sin (\theta + 60°) + \cos (\theta + 30°)$ in simplest form.

14. Prove: $\tan (u + v) = \dfrac{\cot u + \cot v}{\cot u \cot v - 1}$

15. Write $4 \sin 5\theta \cos 5\theta$ in terms of sine only.

16. Express $\dfrac{2 \tan \frac{\pi}{8}}{1 - \tan^2 \frac{\pi}{8}}$ as a tangent of a single angle.

In **17 – 18**, find the exact numerical value.

17. $\cos^2 \frac{7\pi}{12} - \sin^2 \frac{7\pi}{12}$

18. $2 \sin 67.5° \cos 67.5°$

19. Use a half-angle formula to find the value of $\cot \frac{\pi}{8}$.

20. If $x \in$ III and $\tan x = \frac{7}{24}$, find $\cos 2x$.

21. Transform $10 \sin 4x \cos 5x$ into an equivalent sum or difference.

22. Transform $\cos \frac{\pi}{3} + \cos \frac{5\pi}{6}$ into an equivalent product.

23. Prove that $\cot 2y = \dfrac{\sin 3y - \sin y}{\cos y - \cos 3y}$ is an identity.

24. Prove that $\tan \theta (1 + \cos 2\theta) = \sin 2\theta$ is an identity.

25. Solve $2 \sin (x - 30°) = 1$ for $x \in [0°, 360°)$.

26. Solve $\cos^2 x = 3 \sin^2 x$ for $x \in [0, 2\pi)$.

27. Write the general solution for:
$\cos^2 y - \sin^2 y = \dfrac{\sqrt{3}}{2}$

28. Write the general solution for:
$\cot^3 \theta = 3 \cot \theta$

In **29 – 30**, use a calculator or the table to solve.

29. Find the general solution, to the nearest tenth of a radian, for:
$2 \tan^2 x - 7 \tan x - 4 = 0$

30. Estimate the value, to the nearest hundredth of a radian, for:
arc tan (7.1) + arc sin (-0.1)

In **31 – 34**, find the value if possible, or write "no value."

31. $\arctan\sin\left(-\frac{\sqrt{3}}{2}\right)$

32. $\operatorname{arc sec}(-2)$

33. $\arcsin\left(\sin\frac{\pi}{4}\right)$

34. $\sec(\operatorname{arc sec}(-0.2))$

35. Find the exact value of:
$\sec\left(\arctan\frac{3}{2}\right)$

36. Solve for y: $x = \arccos(y - 6)$

In **37 – 39**, refer to the diagram to write the expression in terms of a.

37. $\tan x$

38. $\tan\theta$

39. $\tan y$ (Use the formula for $\tan(\theta - x)$.)

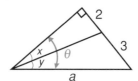

40. Find the remaining sides and angles in $\triangle ABC$ if $A = 24°$, $B = 72°$ and $a = 6$. (Answer to the nearest degree or integer.)

41. Find each of the angles of $\triangle RST$ to the nearest degree if $r = 8$, $s = 9$, and $t = 15$.

42. If $A = 30°$, $b = 10$ and $a = 6$, find the missing parts of all possible triangles. Answer to the nearest degree or integer.

43. If $A(t)$, the amperage of an electric current, t in seconds, is $A(t) = 20\sin 240\pi t$, find:

    a. The strongest value of the current.

    b. The current's period and frequency.

    c. The strength of the current at $t = 0$, $t = 0.01$ seconds (answer to the nearest ampere), and $t = 10$ seconds.

44. To find the distance from point A on the ground to an inaccessible point on the top of a lighthouse at point B, angles of elevation are computed from two points on the ground 150 ft. apart.

Find the distance to the nearest foot.

45. A 15-foot ladder reaches to the top of a 20-foot pool slide. The top of the slide is 12 feet above the ground. By moving the foot of the ladder closer to the foot of the slide, the height of the top of the slide is increased by 1 foot. To the nearest degree, by how much is the angle between the ladder and the ground increased?

46. A weight attached to a spring is raised 3 cm, then released, causing the spring to oscillate up and down. If it takes the spring $\frac{1}{10}$ of a second to travel from its lowest to its highest point, write an equation of the function, $f(t)$, that describes the directed vertical distance of the spring from its point of equilibrium. $f(t)$ is in cm and t is in seconds. Assume the distance is positive when the weight is above the point of equilibrium.

47. Three towns, Alton, Balton, and Calton, are situated at the intersections of three roads that intersect in pairs to form a triangle. The Balton-Calton road is 60 km long and the Alton-Calton road is 50 km long. The two roads from Balton leading to Alton and Calton intersect at an angle of 20°. A radio station in Calton broadcasts over a 50-km radius. If a car is going from Alton to Balton, for what distance along the road can the broadcast be received? Answer to the nearest kilometer. *Hint:* Find the acute angle at Alton. Draw the triangle and the broadcast range.

48. Use Heron's formula to find the area of a triangle whose sides are 26, 28 and 30.

Bonus

a. Express arc sin x and arc cos x in terms of arc tan x.

b. Use the results of part **a** and only the arc tan or tan $^{-1}$ function on your calculator to find arc sin 0.2 and arc cos 0.7 to the nearest hundredth of a radian.

Answers to Suggested Test Items

1. $-\frac{\sqrt{21}}{5}$ **2.** $\frac{2}{\cos^2 x}$

3. a. $\tan y$ **4.** 25
   b. $\sin^2 x$

5. $\frac{\cos x}{\sin x} + \frac{1}{\sin x} = \frac{\cos x + 1}{\sin x} \cdot \frac{1 - \cos x}{1 - \cos x}$

$= \frac{1 - \cos^2 x}{\sin x \, (1 - \cos x)} = \frac{\sin^2 x}{\sin x \, (1 - \cos x)}$

$= \frac{\sin x}{1 - \cos x}$

6. $\csc y(1 - \cos^2 y) = \frac{1}{\sin y} \cdot \sin^2 y = \sin y$

7. $\cos^2 x - (1 - \cos^2 x) = 2 \cos^2 x - 1$

8. $\frac{1}{\sin\left(\frac{\pi}{2} - x\right)} \cdot \cos x = \frac{1}{\cos x} \cdot \cos x = 1$

9. $\frac{\sqrt{2}}{2} (\cos x - \sin x)$

10. $2 + \sqrt{3}$ **11.** $\frac{\sqrt{6} - \sqrt{2}}{4}$

12. $\frac{4\sqrt{2} + \sqrt{5}}{9}$ **13.** $\sqrt{3} \cos \theta$

14. $\frac{\sin u \cos v + \cos u \sin v}{\cos u \cos v - \sin u \sin v}$

$= \frac{\dfrac{\sin u \cos v}{\sin u \sin v} + \dfrac{\cos u \sin v}{\sin u \sin v}}{\dfrac{\cos u \cos v}{\sin u \sin v} - \dfrac{\sin u \sin v}{\sin u \sin v}}$

$= \frac{\cot v + \cot u}{\cot u \cot v - 1}$

15. $2 \sin 10\theta$ **16.** $\tan \frac{\pi}{4}$

17. $-\frac{\sqrt{3}}{2}$ **18.** $\frac{\sqrt{2}}{4}$

19. $\sqrt{3 + 2\sqrt{2}}$

20. $\frac{527}{625}$ **21.** $5 \sin 9x - 5 \sin x$

22. $2 \cos \frac{7\pi}{12} \cos \frac{\pi}{4}$

Answers to Suggested Test Items (continued)

23. $\dfrac{2 \cos 2y \sin y}{-2 \sin 2y \sin (-y)} = \dfrac{2 \cos 2y \sin y}{2 \sin 2y \sin y} = \cot 2y$

24. $\dfrac{\sin \theta}{\cos \theta} (1 + 2 \cos^2 \theta - 1) = \dfrac{\sin \theta}{\cos \theta} (2 \cos^2 \theta)$

$= 2 \sin \theta \cos \theta = \sin 2\theta$

25. $60°, 180°$

26. $\dfrac{\pi}{6}, \dfrac{5\pi}{6}, \dfrac{7\pi}{6}, \dfrac{11\pi}{6}$

27. $y = \dfrac{\pi}{12} + \pi k, y = \dfrac{13}{12} \pi + \pi k$, k an integer

28. $\theta = \dfrac{\pi}{2} + \pi k, \theta = \dfrac{\pi}{6} + \pi k, \theta = \dfrac{5\pi}{6} + \pi k$,

k an integer

29. $x = 2.7 + \pi k, x = 1.3 + \pi k$, k an integer

30. 1.33

31. $-\dfrac{\pi}{3}$

32. $\dfrac{2\pi}{3}$

33. $\dfrac{\pi}{4}$

34. arc sec (-0.2) is undefined. No solution.

35. $\dfrac{\sqrt{13}}{2}$

36. $y = \cos x + 6$

37. $\tan x = \dfrac{2}{\sqrt{a^2 - 25}}$

38. $\tan \theta = \dfrac{5}{\sqrt{a^2 - 25}}$

39. $\tan y = \dfrac{3\sqrt{a^2 - 25}}{a^2 - 15}$

40. $C = 84°, b = 14, c = 15$

41. $T = 124°, S = 30°, R = 26°$

42. 2 triangles: 1. $B = 56°, C = 94°, c = 12$

                 2. $B = 124°, C = 26°, c = 5$

43. **a.** 20

     b. period $= \dfrac{1}{120}$ second, frequency $= 120$

     c.   at $t = 0$:   0

        at $t = 0.01$ seconds: 19

         at $t = 10$ seconds: 0

44. 430 feet   **45.** $7°$

46. $f(t) = 3 \sin 10\pi t$

47. 91 km   **48.** 336

Bonus

a.

$y = $ arc $\sin x$

$x = \sin y$

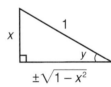

$\tan y = \dfrac{x}{\pm \sqrt{1 - x^2}}$

$y = $ arc $\tan \left(\dfrac{x}{\pm \sqrt{1 - x^2}} \right)$

arc $\sin x = $ arc $\tan \left(\dfrac{x}{\pm \sqrt{1 - x^2}} \right)$

$y = $ arc $\cos x$

$x = \cos y$

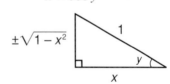

$\tan y = \dfrac{\pm \sqrt{1 - x^2}}{x}$

$y = $ arc $\tan \left(\dfrac{\pm \sqrt{1 - x^2}}{x} \right)$

arc $\cos x = $ arc $\tan \left(\dfrac{\pm \sqrt{1 - x^2}}{x} \right)$

b. arc $\sin 0.2 = $ arc $\tan \left(\dfrac{0.2}{0.9798} \right) = $ arc $\tan 0.2041 = 0.20$

     arc $\cos 0.7 = $ arc $\tan \left(\dfrac{0.7141}{0.7} \right) = $ arc $\tan 1.0202 = 0.80$

Chapter 9

Applications of Trigonometry

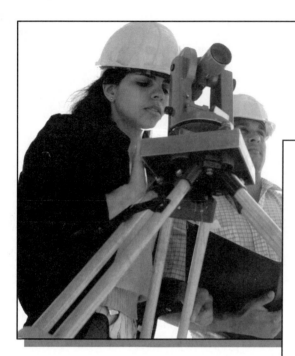

*T*he triangle is perhaps one of the most fundamental shapes in the universe. Numerous theorems relate the measures of the sides and angles, determine the rigidity of the figure, and establish the conditions for congruence and similarity. Each of these aspects of the figures plays an important role in the use of triangles in measurement.

Measurements obtained by using the precisely calibrated instruments of a surveyor, combined with the use of trigonometry as it relates to the right triangle, enable engineers to build the roads, bridges, and office spaces essential to our technological society.

Chapter Table of Contents

9.1 Trigonometric Identities

Frequently, mathematical applications involve the use of complex expressions that are simplified to make them more convenient to use.

To simplify algebraic expressions, identities such as $x + x = 2x$ or $\frac{x}{x} = 1$, $(x \neq 0)$ are utilized. Recall that an *identity* is an expression that is true when the variable is replaced by any element from the domain of that variable. For example, $\frac{x}{x} = 1$ is true for all nonzero real numbers but $\frac{x}{5} = 1$ is only true when $x = 5$. Therefore, $\frac{x}{x} = 1$ is an identity and $\frac{x}{5} = 1$ is a conditional equation.

In trigonometry, identities are also used to simplify expressions. A number of these fundamental trigonometric identities have been verified in the previous chapter. They and others are listed below. Each identity is valid for values of θ for which each function is defined and no denominator is equal to zero.

Trigonometric Identities

Reciprocal Relationships

$$\sin \theta = \frac{1}{\csc \theta} \qquad \cos \theta = \frac{1}{\sec \theta} \qquad \tan \theta = \frac{1}{\cot \theta}$$

Quotient Relationships

$$\tan \theta = \frac{\sin \theta}{\cos \theta} \qquad\qquad \cot \theta = \frac{\cos \theta}{\sin \theta}$$

Pythagorean Relationships

$$\sin^2 \theta + \cos^2 \theta = 1$$
$$\sec^2 \theta - \tan^2 \theta = 1$$
$$\csc^2 \theta - \cot^2 \theta = 1$$

Complementary Relationships

$$\sin \theta = \cos \left(\frac{\pi}{2} - \theta \right)$$
$$\sec \theta = \csc \left(\frac{\pi}{2} - \theta \right)$$
$$\tan \theta = \cot \left(\frac{\pi}{2} - \theta \right)$$

Symmetry Relationships

$$\cos (-x) = \cos x \qquad \sin (-x) = -\sin x \qquad \tan (-x) = -\tan x$$
$$\sec (-x) = \sec x \qquad \csc (-x) = -\csc x \qquad \cot (-x) = -\cot x$$

These fundamental identities are used to simplify trigonometric expressions, to find the value of trigonometric functions, to factor expressions containing trigonometric terms, to write expressions in terms of other trigonometric functions, and to prove other identities.

Example 1 _____

If $\cot x = \frac{1}{3}$ and $\cos x$ is negative, find the value of $\sec x$.

Solution

Observe that there is no fundamental identity that relates cotangent, the given value, to secant, the value to be found. Since, however, there is a fundamental identity that relates tangent to secant, first use a reciprocal relation to rewrite cotangent in terms of tangent.

$$\tan x = \frac{1}{\cot x}$$

$$\tan x = \frac{1}{\frac{1}{3}}$$

$$\tan x = 3$$

Now use the Pythagorean identity that relates tangent to secant.

$$\sec^2 x - \tan^2 x = 1$$
$$\sec^2 x - 3^2 = 1$$
$$\sec^2 x = 10$$
$$\sec x = \pm\sqrt{10}$$

To determine the sign of secant, you must know the quadrant in which angle x lies. Since $\cot x > 0$, and $\cos x < 0$, x must lie in quadrant III. Sec x is negative in quadrant III.

Answer: $\sec x = -\sqrt{10}$

When simplifying a trigonometric expression, it is often helpful to first rewrite the expression in terms of sine and cosine only.

Example 2 _____

Write each expression in terms of a single function.

a. $\cos \theta \csc \theta$ **b.** $\cos \theta (\cot \theta + \tan \theta)$

Solution

a. Begin by changing $\csc \theta$ to $\dfrac{1}{\sin \theta}$.

$$\cos \theta \csc \theta = \cos \theta \; \frac{1}{\sin \theta}$$

$$= \frac{\cos \theta}{\sin \theta}$$

$$= \cot \theta$$

Answer: $\cos \theta \csc \theta = \cot \theta$

b. Use the quotient relationships to convert the expression to one in terms of sine and cosine only.

$$\cos\theta(\cot\theta + \tan\theta) = \cos\theta\left(\frac{\cos\theta}{\sin\theta} + \frac{\sin\theta}{\cos\theta}\right)$$

$$= \frac{\cos^2\theta}{\sin\theta} + \sin\theta$$

$$= \frac{\cos^2\theta + \sin^2\theta}{\sin\theta}$$

$$= \frac{1}{\sin\theta} \qquad \text{Use the Pythagorean relation } \cos^2\theta + \sin^2\theta = 1.$$

$$= \csc\theta \qquad \text{Use the reciprocal relation.}$$

Answer: $\cos\theta(\cot\theta + \tan\theta) = \csc\theta$

Proving Identities

Proving an identity is different from solving an equation. Solving an equation means finding the conditions under which the equation is true. Since an identity is true for all values of the variable for which the equation is defined, proving that an equation is an identity means showing that the two sides of the equation are equivalent.

While no single strategy will work for every proof, some general principles can be followed. Generally, it is more efficient to begin working with the more complex side of the equality. Use the fundamental identities to make substitutions, usually rewriting expressions in terms of sine and cosine only. Use algebraic techniques to transform expressions until they are identical to the other side.

In the large part, your success at proving identities will depend on your algebraic skills, especially factoring, combining fractions, and simplifying complex fractions. It may be beneficial to review these skills at this time. Keep in mind that there are usually many ways to prove a trigonometric identity. It is not necessary that you be able to visualize the entire proof before you begin writing. Use the fundamental identities and algebraic techniques on one side of the equality until you discover a clear series of steps that will transform it into an equivalent expression that is identical to the other side.

Example 3 _____

Prove that $\csc x \cos x = \cot x$ is an identity.

Solution

Begin by working with the left side.

$$\csc x \cos x = \cot x$$

$$\left(\frac{1}{\sin x}\right)\cos x = \qquad \text{Rewrite the left side in terms of sine and cosine.}$$

$$\frac{\cos x}{\sin x} =$$

$$\cot x = \cot x$$

Thus: $\csc x \cos x = \cot x$

Since a proof proceeds from what is known to what is to be proved, the steps of the solution in Example 3 should be written in reverse order to constitute a proof. However, the given form is generally accepted as the proof since it indicates that you could start at $\cot x = \cot x$ and arrive at the given identity. Notice that $\csc x \cos x = \cot x$ is true for all values of x except $k\pi$, where k is an integer, since $\csc x$ and $\cot x$ are undefined for those values.

Example 4 _____

Prove that $\dfrac{\cos t}{\sec t} + \dfrac{\sin t}{\csc t} = 1$ is an identity.

Solution

Begin with the left side.

$$\frac{\cos t}{\sec t} + \frac{\sin t}{\csc t} = 1$$

$$\frac{\cos t \csc t + \sin t \sec t}{\sec t \csc t} =$$ Combine fractions using the common denominator $\sec t \csc t$.

$$\frac{\cos t\left(\dfrac{1}{\sin t}\right) + \sin t\left(\dfrac{1}{\cos t}\right)}{\left(\dfrac{1}{\cos t}\right)\left(\dfrac{1}{\sin t}\right)} =$$ Change to an equivalent expression in terms of sine and cosine using the reciprocal relationships.

$$\frac{\dfrac{\cos t}{\sin t} + \dfrac{\sin t}{\cos t}}{\dfrac{1}{\cos t \sin t}} =$$

$$\frac{\cos^2 t + \sin^2 t}{1} =$$ Simplify the complex fraction by multiplying numerator and denominator by $\cos t \sin t$.

$$\frac{1}{1} = 1 \qquad \sin^2 t + \cos^2 t = 1$$

Thus: $$\frac{\cos t}{\sec t} + \frac{\sin t}{\csc t} = 1$$

Example 5 _____

Prove that $\dfrac{\sin\left(\frac{\pi}{2} - y\right)}{\cos(-y) + 1} = \dfrac{\sin y}{\sin y + \tan y}$ is an identity.

Solution

Begin with the right side.

$$\frac{\sin\left(\frac{\pi}{2}-y\right)}{\cos\left(-y\right)+1}=\frac{\sin y}{\sin y+\tan y}$$

$$=\frac{\sin y}{\sin y+\dfrac{\sin y}{\cos y}} \qquad \text{Rewrite in terms of sine and cosine.}$$

$$=\frac{\sin y\cos y}{\sin y\cos y+\sin y} \qquad \text{Multiply numerator and denominator by } \cos y.$$

$$=\frac{\sin y\cos y}{\sin y\left(\cos y+1\right)} \qquad \text{Factor the denominator.}$$

$$=\frac{\cos y}{\cos y+1} \qquad \text{Cancel the common factor.}$$

$$\frac{\sin\left(\frac{\pi}{2}-y\right)}{\cos\left(-y\right)+1}=\frac{\sin\left(\frac{\pi}{2}-y\right)}{\cos\left(-y\right)+1} \qquad \begin{array}{l}\cos y=\sin\left(\frac{\pi}{2}-y\right)\\[1ex]\cos y=\cos\left(-y\right)\end{array}$$

Thus: $\dfrac{\sin\left(\frac{\pi}{2}-y\right)}{\cos\left(-y\right)+1}=\dfrac{\sin y}{\sin y+\tan y}$

Example 6

Prove that $\dfrac{1-\cos x}{1+\cos x}=\dfrac{(1-\cos x)^2}{\sin^2 x}$ is an identity.

Solution

Since the right side is in terms of two functions and the left is only in terms of $\cos x$, work with the right side and try to change it to an equivalent expression in terms of $\cos x$.

$$\frac{1-\cos x}{1+\cos x}=\frac{(1-\cos x)^2}{\sin^2 x}$$

$$=\frac{(1-\cos x)^2}{1-\cos^2 x} \qquad \begin{array}{l}\text{Since } \sin^2 x+\cos^2 x=1,\\ \sin^2 x=1-\cos^2 x.\end{array}$$

$$=\frac{(1-\cos x)^2}{(1+\cos x)(1-\cos x)} \qquad \text{Factor the denominator.}$$

$$\frac{1-\cos x}{1+\cos x}=\frac{1-\cos x}{1+\cos x} \qquad \text{Cancel the common factor.}$$

Thus: $\dfrac{1-\cos x}{1+\cos x}=\dfrac{(1-\cos x)^2}{\sin^2 x}$

1. $\dfrac{\sqrt{15}}{15}$

2. $-\dfrac{\sqrt{10}}{10}$

3. $-2\sqrt{6}$

4. $\dfrac{5}{6}$

5. $-\dfrac{5}{3}$

6. 0

7. $\dfrac{3}{2}$

8. $\dfrac{2\sqrt{3}}{3}$

9. $\sec x$

10. $\cot y$

11. $\cos x$

12. $1 + \cos s$

13. $\tan x + 1$

14. $\sin y$

15. 1

16. $\sin^2 y$

17. $\csc y$

18. $2(\cos t + 1)$

19. $2(\sin x + 1)$

20. $\cot x$

21. $\sin y$

22. $\csc^2 z$

23. $\tan y$

24. $\sec^2 x$

25. 1

26. $6 \sin y$

27. $-2 \cos x$

28. $\tan y$

29. $-\sec \theta$

30. $\sec^2 x$

31. $\csc^2 y$

32. $\sec^2 \theta$

In **33 – 67**, one side is transformed, to be identical to the other side.

33. $\dfrac{1}{\cos \theta} \cos \theta =$

$\qquad 1 =$

34. $\cos y \dfrac{1}{\sin y} =$

$\qquad \dfrac{\cos y}{\sin y} =$

$\qquad \cot y =$

In **1 – 8**, use the given information and the fundamental identities to find the function value indicated.

1. If $\sin x = \frac{1}{4}$ and $\sec x > 0$, find $\tan x$.

2. If $\tan \theta = 3$ and $\csc \theta < 0$, find $\cos \theta$.

3. If $\cos \theta = \frac{1}{5}$ and $\sin \theta < 0$, find $\tan \theta$.

4. If $\sec y = 1.2$ and $\tan y < 0$, find $\cos y$.

5. If $\tan x = -\frac{3}{4}$ and $\cos x = \frac{4}{5}$, find $\csc x$.

6. If $\sec x = 1$ and $\sin x = 0$, find $\tan x$.

7. If $\cos(-x) = \frac{\sqrt{5}}{3}$ and $\cot x = \frac{\sqrt{5}}{2}$, find $\csc x$.

8. If $\sin(-x) = -\frac{1}{2}$ and $\cot x = \sqrt{3}$, find $\sec x$.

In **9 – 25**, simplify the expression.

9. $\dfrac{\tan x}{\sin x}$

10. $\dfrac{\csc y}{\sec y}$

11. $\cot x \sin x$

12. $\cot s \tan s + \cot s \sin s$

13. $\sec x(\sin x + \cos x)$

14. $\cos y \tan y$

15. $\dfrac{\sin x}{\csc x} + \dfrac{\cos x}{\sec x}$

16. $\dfrac{\tan^2 y}{1 + \tan^2 y}$

17. $\cot y + \dfrac{\sin y}{1 + \cos y}$

18. $(\cos t + 1)^2 + \sin^2 t$

19. $(1 + \sin x)^2 + \dfrac{1}{\sec^2 x}$

20. $\sin\left(\frac{\pi}{2} - x\right) \csc x$

21. $\cot\left(\frac{\pi}{2} - y\right) \cos y$

22. $(\cot z + 1)^2 - 2 \cot z$

23. $\dfrac{1}{\sin y \cos y} - \dfrac{\cos y}{\sin y}$

24. $\dfrac{(\sec^2 x + \tan^2 x)^2}{\sec^4 x - \tan^4 x} - \tan^2 x$

25. $(\csc y + \cot y)^4 (\csc y - \cot y)^4$

In **26 – 32**, transform the given expression to an equivalent expression in terms of the stated function.

26. $6 \cos y \tan y$, in terms of $\sin y$

27. $-2 \sin x \cot x$, in terms of $\cos x$

28. $\csc y \sec y \sin^2 y$, in terms of $\tan y$

29. $\dfrac{1 - \sec \theta}{1 - \cos \theta}$, in terms of $\sec \theta$

30. $\tan^2 x (\sin x + \cot x \cos x)^2$, in terms of $\sec x$

31. $\cos y (\sec y + \cos y \csc^2 y)$, in terms of $\csc y$

32. $(1 - \tan \theta)^2 + 2 \tan \theta$, in terms of $\sec \theta$

In **33 – 67**, prove that the equation is an identity.

33. $\sec \theta \cos \theta = 1$

34. $\cos y \csc y = \cot y$

35. $\tan x + \cot x = \csc x \sec x$

36. $\cos x \cot x = \csc x - \sin x$

37. $\sec y(\sec y - \cos y) = \tan^2 y$

38. $\cos x(\sec x - \cos x) = \sin^2 x$

39. $(1 + \sin y)(1 - \sin y) = \cos^2 y$

40. $(1 + \cos y)(1 - \cos y) = \sin^2 y$

41. $\sin^2 B - \cos^2 B = 1 - 2 \cos^2 B$

42. $\cos^2 A - \sin^2 A = 1 - 2 \sin^2 A$

43. $\sin^2 y + 3 = 4 - \cos^2 y$

44. $\sec^2 x + 5 = \tan^2 x + 6$

---

35. $\dfrac{\sin x}{\cos x} + \dfrac{\cos x}{\sin x} =$

$\dfrac{\sin^2 x + \cos^2 x}{\sin x \cos x} =$

$\dfrac{1}{\sin x \cos x} =$

$\dfrac{1}{\sin x} \dfrac{1}{\cos x} =$

$\csc x \sec x =$

36. $= \dfrac{1}{\sin x} - \sin x$

$= \dfrac{1 - \sin^2 x}{\sin x}$

$= \dfrac{\cos^2 x}{\sin x}$

$= \cos x \dfrac{\cos x}{\sin x}$

$= \cos x \cot x$

37. $\dfrac{1}{\cos y}\left(\dfrac{1}{\cos y} - \cos y\right) =$

$\dfrac{1}{\cos y}\left(\dfrac{1 - \cos^2 y}{\cos y}\right) =$

$\dfrac{1}{\cos y}\left(\dfrac{\sin^2 y}{\cos y}\right) =$

$\dfrac{\sin^2 y}{\cos^2 y} =$

$\tan^2 y =$

38. $\cos x \left(\dfrac{1}{\cos x} - \cos x \right) =$

$\dfrac{\cos x}{\cos x} - \cos^2 x =$

$1 - \cos^2 x =$

$\sin^2 x =$

39. $1 - \sin^2 y =$

$\cos^2 y =$

40. $1 - \cos^2 y =$

$\sin^2 y =$

41. $= 1 - \cos^2 B - \cos^2 B$

$= \sin^2 B - \cos^2 B$

42. $= 1 - \sin^2 A - \sin^2 A$

$= \cos^2 A - \sin^2 A$

43. $= 3 + 1 - \cos^2 y$

$= 3 + \sin^2 y$

$= \sin^2 y + 3$

44. $= \tan^2 x + 1 + 5$

$= \sec^2 x + 5$

45. $\cos^2 A \, (\sec^2 A) =$

$\cos^2 A \left(\dfrac{1}{\cos^2 A} \right) =$

$1 =$

46. $(\csc^2 B)(\sin^2 B) =$

$\left(\dfrac{1}{\sin^2 B} \right) \sin^2 B =$

$1 =$

47. $\dfrac{\frac{\cos x}{\sin x}}{\cos^2 x} =$

$\dfrac{\cos x}{\sin x \cos^2 x} =$

$\dfrac{1}{\sin x \cos x} =$

$\dfrac{1}{\sin x} \cdot \dfrac{1}{\cos x} = \csc x \sec x$

48. $= \dfrac{\frac{1}{\cos y}}{\cot^2 y}$

$= \dfrac{\frac{1}{\cos y}}{\frac{\cos^2 y}{\sin^2 y}}$

$= \dfrac{\sin^2 y}{\cos^3 y}$

$= \dfrac{\sin^2 y}{\cos^3 y} \cdot \dfrac{\sin y}{\sin y}$

$= \dfrac{\sin^3 y}{\cos^3 y} \cdot \dfrac{1}{\sin y}$

$= \tan^3 y \cdot \csc y$

$= \dfrac{1}{\cot^3 y} \cdot \csc y$

$= \dfrac{\csc y}{\cot^3 y}$

49. $\dfrac{1 - \cos^2 x}{\sin x \cos x} =$

$\dfrac{\sin^2 x}{\sin x \cos x} =$

$\dfrac{\sin x}{\cos x} =$

$\tan x =$

50. $\dfrac{1}{\csc x} - \csc x =$

$\dfrac{1 - \csc^2 x}{\csc x} =$

51. $\dfrac{\cos^2 \theta + (1 + \sin \theta)^2}{\cos \} \theta \, (1 + \sin \theta)} =$

$\dfrac{\cos^2 \theta + 1 + 2 \sin \theta + \sin^2 \theta}{\cos \theta \, (1 + \sin \theta)} =$

$\dfrac{1 + 1 + 2 \sin \theta}{\cos \theta \, (1 + \sin \theta)} =$

$\dfrac{2 + 2 \sin \theta}{\cos \theta \, (1 + \sin \theta)} =$

$\dfrac{2(1 + \sin \theta)}{\cos \theta \, (1 + \sin \theta)} =$

$\dfrac{2}{\cos \theta} =$

52. $\dfrac{\tan^2 x + \sec x + 1}{\sec x + 1} =$

$\dfrac{\sec^2 x + \sec x}{\sec x + 1} =$

$\dfrac{\sec x(\sec x + 1)}{\sec x + 1} =$

$\sec x =$

53. $\dfrac{\sin s + \cos s}{1 - \frac{\sin^2 s}{\cos^2 s}} =$

$\dfrac{\cos^2 s(\sin s + \cos s)}{\cos^2 s - \sin^2 s} =$

$\dfrac{\cos^2 s \, (\sin s + \cos s)}{(\cos s + \sin s)(\cos s - \sin s)} =$

$\dfrac{\cos^2 s}{\cos s - \sin s} =$

54. $\dfrac{\cos x - \frac{1}{\cos x}}{1 + \cos x} =$

$\dfrac{\cos^2 x - 1}{\cos x \, (1 + \cos x)} =$

$\dfrac{(\cos x + 1)(\cos x - 1)}{\cos x \, (\cos x + 1)} =$

$\dfrac{\cos x - 1}{\cos x} =$

$\dfrac{\cos x}{\cos x} - \dfrac{1}{\cos x} =$

$1 - \sec x =$

55.

$$2\,\frac{1}{\cos^2 y} - \sin^2 y - 2\,\frac{1}{\cos^2 y}\,\sin^2 y =$$

$$\frac{2 - \sin^2 y\,\cos^2 y - 2\,\sin^2 y}{\cos^2 y} =$$

$$\frac{2(1 - \sin^2 y) - \sin^2 y\,\cos^2 y}{\cos^2 y} =$$

$$\frac{2\,\cos^2 y - \sin^2 y\,\cos^2 y}{\cos^2 y} =$$

$$\frac{\cos^2 y\,(2 - \sin^2 y)}{\cos^2 y} =$$

$$2 - \sin^2 y =$$

$$1 + 1 - \sin^2 y =$$

$$1 + \cos^2 y =$$

56.

$$(\cos x + \sin x)(\sec^2 x) =$$

$$(\cos x + \sin x)\!\left(\frac{1}{\cos^2 x}\right) =$$

$$\frac{\cos x}{\cos^2 x} + \frac{\sin x}{\cos^2 x} =$$

$$\frac{1}{\cos x} + \frac{\sin x}{\cos x}\cdot\frac{1}{\cos x} =$$

$$\sec x + \tan x \cdot \sec x =$$

$$\sec x\,(1 + \tan x) =$$

$$\sec x\,(\tan x + 1) =$$

57.

$$\frac{(\tan^2 x + \sec^2 x)(\tan^2 x + \sec^2 x)}{(\sec^2 x + \tan^2 x)(\sec^2 x - \tan^2 x)} =$$

$$\frac{\tan^2 x + \sec^2 x}{\sec^2 x - \tan^2 x} =$$

$$\frac{\tan^2 x + \sec^2 x}{1} =$$

$$\tan^2 x + \sec^2 x =$$

58.

$$(\csc^2 x - 1)^2 =$$
$$(\cot^2 x)^2 =$$
$$\cot^4 x =$$

59.

$$[(\csc\theta + \cot\theta)(\csc\theta - \cot\theta)]^4 =$$
$$(\csc^2\theta - \cot^2\theta)^4 =$$
$$1^4 =$$
$$1 =$$

60.

$$\frac{(\tan^2 x + 1) - 4\tan x + 2}{(\tan^2 x + 1) - 10} =$$

$$\frac{\tan^2 x - 4\tan x + 3}{\tan^2 x - 9} =$$

$$\frac{(\tan x - 3)(\tan x - 1)}{(\tan x + 3)(\tan x - 3)} =$$

$$\frac{\tan x - 1}{\tan x + 3} =$$

61.

$$\frac{6\sin x + 8 + 1 - \cos^2 x}{-3 - \cos^2 x} =$$

$$\frac{\sin^2 x + 6\sin x + 8}{-4 + 1 - \cos^2 x} =$$

$$\frac{(\sin x + 4)(\sin x + 2)}{\sin^2 x - 4} =$$

$$\frac{(\sin x + 4)(\sin x + 2)}{(\sin x + 2)(\sin x - 2)} =$$

$$\frac{\sin x + 4}{\sin x - 2} =$$

62.

$$\cos x\,\frac{1}{\cos x} =$$
$$1 =$$

63.

$$\frac{1}{\sin x}\,\sin x =$$
$$1 =$$

64.

$$\frac{\sec y}{-\csc y} =$$

$$\frac{\dfrac{1}{\cos y}}{-\dfrac{1}{\sin y}} =$$

$$-\frac{\sin y}{\cos y} =$$

$$-\tan y =$$

65.

$$\frac{1}{\sec x + \tan x} =$$

$$\frac{1}{\dfrac{1}{\cos x} + \dfrac{\sin x}{\cos x}} =$$

$$\frac{\cos x}{1 + \sin x} =$$

$$\frac{\cos(-x)}{1 + \sin x} =$$

66.

$$\frac{(\sin y - \cos y)(\sin^2 y + \sin y\cos y + \cos^2 y)}{\sin y - \cos y} =$$

$$\sin^2 y + \sin y\cos y + \cos^2 y =$$

$$\sin y\cos y + \sin^2 y + \cos^2 y =$$

$$\sin y\cos y + 1 =$$

67.

$$(\csc^2 x)^3 - \cot^6 x - 1 =$$
$$(\cot^2 x + 1)^3 - \cot^6 x - 1 =$$
$$\cot^6 x + 3\cot^4 x + 3\cot^2 x + 1 - \cot^6 x - 1 =$$
$$3(\cot^4 x + \cot^2 x) =$$
$$3\cot^2 x\,(\cot^2 x + 1) =$$
$$3\cot^2 x\,\csc^2 x =$$

Exercises 9.1 *(continued)*

45. $(1 - \sin^2 A)(1 + \tan^2 A) = 1$

46. $(1 + \cot^2 B)(1 - \cos^2 B) = 1$

47. $\dfrac{\cot x}{\cos^2 x} = \csc x \sec x$

48. $\dfrac{\csc y}{\cot^3 y} = \dfrac{\sec y}{\csc^2 y - 1}$

49. $\dfrac{1}{\sin x \cos x} - \dfrac{\cos x}{\sin x} = \tan x$

50. $\dfrac{1}{\csc x} - \dfrac{1}{\sin x} = \dfrac{1 - \csc^2 x}{\csc x}$

51. $\dfrac{\cos \theta}{1 + \sin \theta} + \dfrac{1 + \sin \theta}{\cos \theta} = \dfrac{2}{\cos \theta}$

52. $\dfrac{\tan^2 x}{\sec x + 1} + 1 = \sec x$

53. $\dfrac{\sin s + \cos s}{1 - \tan^2 s} = \dfrac{\cos^2 s}{\cos s - \sin s}$

54. $\dfrac{\cos x - \sec x}{1 + \cos x} = 1 - \sec x$

55. $2 \sec^2 y - \sin^2 y - 2 \sec^2 y \sin^2 y = 1 + \cos^2 y$

56. $(\cos x + \sin x)(\tan^2 x + 1) = \sec x \, (\tan x + 1)$

57. $\dfrac{(\tan^2 x + \sec^2 x)^2}{\sec^4 x - \tan^4 x} = \tan^2 x + \sec^2 x$

58. $1 + \csc^4 x - 2 \csc^2 x = \cot^4 x$

59. $(\csc \theta + \cot \theta)^4 (\csc \theta - \cot \theta)^4 = 1$

60. $\dfrac{\sec^2 x - 4 \tan x + 2}{\sec^2 x - 10} = \dfrac{\tan x - 1}{\tan x + 3}$

61. $\dfrac{6 \sin x - \cos^2 x + 9}{-(3 + \cos^2 x)} = \dfrac{\sin x + 4}{\sin x - 2}$

62. $\sin \left(\dfrac{\pi}{2} - x \right) \sec x = 1$

63. $\csc x \cos \left(\dfrac{\pi}{2} - x \right) = 1$

64. $\dfrac{\sec (-y)}{\csc (-y)} = -\tan y$

65. $\dfrac{1}{\sec (-x) + \tan x} = \dfrac{\cos (-x)}{1 + \sin x}$

66. $\dfrac{\sin^3 y - \cos^3 y}{\sin y - \cos y} = \sin y \cos y + 1$

67. $\csc^6 x - \cot^6 x - 1 = 3(\cot^2 x \csc^2 x)$

In **68 – 74**, verify that the expression is *not* an identity by finding at least one real-number value of the variable that makes the equation false.

68. $\cos t = \sqrt{1 - \sin^2 t}$

69. $\csc t = \sqrt{1 + \cot^2 t}$

70. $\cos t \sec t = 0$

71. $\sqrt{\sin^2 s} = \sin s$

72. $\log |\tan \theta| + \log |\cos \theta| = \log |\sin \theta|$

73. $\log |\csc y + \cot y| + \log |\csc y - \cot y| = 0$

74. $\log (1 + \cos \theta) + \log (1 - \cos \theta) = \log \sin^2 \theta$

In **75 – 76**, find a real-number value of the variable for which the equation is true.

75. $\cos \theta = \sqrt{1 - \sin^2 \theta}$

76. $\sqrt{\sec^2 x - \tan^2 x} = \sec x + \tan x$

68. If $t = \pi$

$\cos \pi \overset{?}{=} \sqrt{1 - \sin^2 \pi}$

$-1 \overset{?}{=} \sqrt{1 - 0}$

$-1 \overset{?}{=} \sqrt{1}$

$-1 \neq 1$

69. If $t = \dfrac{7\pi}{4}$

$\csc \dfrac{7\pi}{4} \overset{?}{=} \sqrt{1 - \cot^2 \dfrac{7\pi}{4}}$

$-\sqrt{2} \overset{?}{=} \sqrt{1 + (-1)^2}$

$-\sqrt{2} \neq \sqrt{2}$

70. If $t = \dfrac{\pi}{3}$

$\cos \dfrac{\pi}{3} \left(\sec \dfrac{\pi}{3} \right) \overset{?}{=} 0$

$\dfrac{1}{2} (2) \overset{?}{=} 0$

$1 \neq 0$

71. If $s = \dfrac{7\pi}{6}$

$\sqrt{\sin^2 \dfrac{7\pi}{6}} \overset{?}{=} \sin \dfrac{7\pi}{6}$

$\sqrt{\dfrac{1}{4}} \overset{?}{=} -\dfrac{1}{2}$

$\dfrac{1}{2} \neq -\dfrac{1}{2}$

72. undefined at $\theta = 270°$

73. undefined at $y = 0°$

74. undefined at $\theta = 0°$

75. Since $\sqrt{1 - \sin^2 \theta} \geq 0$,

$\cos \theta \geq 0$

$-\dfrac{\pi}{2} \geq \theta \geq \dfrac{\pi}{2}$

76. $\sqrt{\sec^2 x - \tan^2 x} = \sec x + \tan x$

$\sqrt{1} = \sec x + \tan x$

$1 = \dfrac{1}{\cos x} + \dfrac{\sin x}{\cos x}$

$\cos x = \dfrac{\cos x}{\cos x} + \dfrac{\sin x \cos x}{\cos x}$

$\cos x = 1 + \sin x$

If $x = 0$ $\cos 0 = 1 + \sin 0$

$1 = 1 + 0$

$1 = 1$

9.2 The Sum and Difference Formulas

It is sometimes useful to express the sine or cosine of the sum or difference of two angles in terms of the sines and cosines of the individual angles. First, note that sin $(A + B)$ is not simply sin A + sin B. For example, if $A = 30°$ and $B = 60°$:

$$\sin(30° + 60°) = \sin 90° = 1 \text{ and } \sin 30° + \sin 60° = \tfrac{1}{2} + \tfrac{\sqrt{3}}{2} \neq 1$$

In this section, the formulas for sin $(A \pm B)$ and cos $(A \pm B)$, where A and B are the measures of angles, are derived and applied, beginning with the formula for cos $(A - B)$.

Both of the diagrams below show a unit circle and a central angle whose measure is $A - B$. Under a rotation of B about O, T' and S', the images of T and S, are on the terminal rays of angles whose respective measures are B and $(A - B) + B$ or A. Since distance is preserved under rotation, $ST = S'T'$.

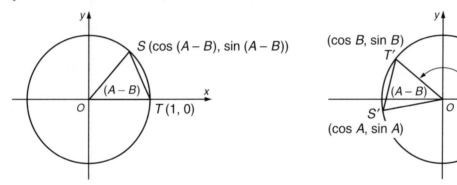

Using the distance formula for ST and for $S'T'$:

$$ST = \sqrt{[\cos(A - B) - 1]^2 + \sin^2(A - B)}$$

$$(ST)^2 = [\cos(A - B) - 1]^2 + \sin^2(A - B)$$

$$= \cos^2(A - B) - 2\cos(A - B) + 1 + \sin^2(A - B)$$

$$= [\cos^2(A - B) + \sin^2(A - B)] + 1 - 2\cos(A - B)$$

$$= 1 + 1 - 2\cos(A - B)$$

$$(ST)^2 = 2 - 2\cos(A - B)$$

$$S'T' = \sqrt{(\cos A - \cos B)^2 + (\sin A - \sin B)^2}$$

$$(S'T')^2 = (\cos A - \cos B)^2 + (\sin A - \sin B)^2$$

$$= \cos^2 A - 2\cos A \cos B + \cos^2 B + \sin^2 A - 2\sin A \sin B + \sin^2 B$$

$$= (\cos^2 A + \sin^2 A) + (\sin^2 B + \cos^2 B) - 2\cos A \cos B - 2\sin A \sin B$$

$$(S'T')^2 = 2 - 2(\cos A \cos B + \sin A \sin B)$$

Since $ST = S'T'$, $(ST)^2 = (S'T')^2$.

$$2 - 2 \cos (A - B) = 2 - 2(\cos A \cos B + \sin A \sin B)$$

Solving for $\cos (A - B)$:

$$\cos (A - B) = \cos A \cos B + \sin A \sin B$$

The previous diagram showed A and B as positive rotations with $A > B$; however, a similar proof can be done for any values of A and B. Therefore, it can be concluded that the formula is true for all real numbers.

A formula for $\cos (A + B)$ can be derived by writing $\cos (A + B)$ as $\cos [A - (-B)]$. Recall that cosine is an even function and sine is an odd function, that is, $\cos (-x) = \cos x$ and $\sin (-x) = -\sin x$.

Use the formula for the cosine of the difference of two angle measures.

$$\cos [(A - (-B)] = \cos A \cos (-B) + \sin A \sin (-B)$$
$$= \cos A \cos B + \sin A (-\sin B)$$
$$\cos (A + B) = \cos A \cos B - \sin A \sin B$$

To derive a formula for $\sin (A - B)$, use the cofunction relationship $\sin \theta = \cos \left(\frac{\pi}{2} - \theta \right)$.

$$\sin (A - B) = \cos \left[\frac{\pi}{2} - (A - B) \right]$$
$$= \cos \left[\left(\frac{\pi}{2} - A \right) + B \right]$$

Apply the formula for the cosine of the sum of two angle measures.

$$\cos \left[\left(\frac{\pi}{2} - A \right) + B \right] = \cos \left(\frac{\pi}{2} - A \right) \cos B - \sin \left(\frac{\pi}{2} - A \right) \sin B$$
$$= \sin A \cos B - \cos A \sin B$$
$$\sin (A - B) = \sin A \cos B - \cos A \sin B$$

A method similar to the one used to derive the identity for $\cos (A + B)$ produces the following formula:

$$\sin (A + B) = \sin A \cos B + \cos A \sin B$$

Example 1

Find the exact value of $\sin \frac{13\pi}{12}$.

Solution

Express $\sin \frac{13\pi}{12}$ as $\sin \left(\frac{5\pi}{6} + \frac{\pi}{4} \right)$.

Note that since $\pi - \frac{5\pi}{6} = \frac{\pi}{6}$, then $\sin \frac{5\pi}{6} = \sin \frac{\pi}{6}$ and $\cos \frac{5\pi}{6} = -\cos \frac{\pi}{6}$.

$$\sin \left(\frac{5\pi}{6} + \frac{\pi}{4} \right) = \sin \frac{5\pi}{6} \cos \frac{\pi}{4} + \cos \frac{5\pi}{6} \sin \frac{\pi}{4}$$
$$= \left(\frac{1}{2} \right)\left(\frac{\sqrt{2}}{2} \right) + \left(-\frac{\sqrt{3}}{2} \right)\left(\frac{\sqrt{2}}{2} \right)$$
$$= \frac{\sqrt{2}}{4} - \frac{\sqrt{6}}{4}$$

Answer: $\sin \frac{13\pi}{12} = \frac{\sqrt{2} - \sqrt{6}}{4}$

Use a calculator to find the approximate values of $\sin \frac{13\pi}{12}$ and $\frac{\sqrt{2}-\sqrt{6}}{4}$, and compare the results.

A formula for tan $(A - B)$ can be derived using the identity $\tan \theta = \frac{\sin \theta}{\cos \theta}$.

$$\tan (A - B) = \frac{\sin (A - B)}{\cos (A - B)}$$

$$= \frac{\sin A \cos B - \cos A \sin B}{\cos A \cos B + \sin A \sin B}$$

This expression can be converted to one in terms of tan A and tan B by dividing the numerator and denominator by cos A cos B. Thus,

$$\tan (A - B) = \frac{\dfrac{\sin A \cos B}{\cos A \cos B} - \dfrac{\cos A \sin B}{\cos A \cos B}}{\dfrac{\cos A \cos B}{\cos A \cos B} + \dfrac{\sin A \sin B}{\cos A \cos B}}$$

$$= \frac{\tan A - \tan B}{1 + \tan A \tan B}$$

$$\tan (A - B) = \frac{\tan A - \tan B}{1 + \tan A \tan B}$$

The formula for tan $(A + B)$ is:

$$\tan (A + B) = \frac{\tan A + \tan B}{1 - \tan A \tan B}$$

Example 2

If $\tan x = \frac{3}{4}$ and $\tan y = -\frac{5}{13}$, find tan $(x + y)$.

Solution

$$\tan (x + y) = \frac{\tan x + \tan y}{1 - \tan x \tan y}$$

$$= \frac{\frac{3}{4} + \left(-\frac{5}{13}\right)}{1 - \frac{3}{4}\left(-\frac{5}{13}\right)}$$

$$= \frac{\frac{19}{52}}{1 + \frac{15}{52}}$$

$$= \frac{19}{52 + 15}$$

Answer: $\tan (x + y) = \frac{19}{67}$

Example 3

Let x be the measure of an angle in quadrant II, y the measure of an angle in quadrant III, $\sin x = \frac{1}{3}$, and $\sin y = -\frac{3}{4}$. Find cos $(x + y)$.

Solution

To use the formula for cos $(x + y)$, the values of cos x and cos y must be known. These values can be found by the Pythagorean identity $\sin^2 \theta + \cos^2 \theta = 1$.

For cos x: $\sin^2 x + \cos^2 x = 1$

$$\left(\tfrac{1}{3}\right)^2 + \cos^2 x = 1$$

$$\cos^2 x = \tfrac{8}{9}$$

$$\cos x = \pm\sqrt{\tfrac{8}{9}} = \pm\tfrac{2\sqrt{2}}{3}$$

Since cos x is negative in quadrant II:

$$\cos x = -\tfrac{2\sqrt{2}}{3}$$

For cos y: $\sin^2 y + \cos^2 y = 1$

$$\left(-\tfrac{3}{4}\right)^2 + \cos^2 y = 1$$

$$\cos^2 y = \tfrac{7}{16}$$

$$\cos y = \pm\tfrac{\sqrt{7}}{4}$$

Since cos y is negative in quadrant III:

$$\cos y = -\tfrac{\sqrt{7}}{4}$$

Therefore,

$$\cos (x + y) = \cos x \cos y - \sin x \sin y$$

$$= \left(-\tfrac{2\sqrt{2}}{3}\right)\left(-\tfrac{\sqrt{7}}{4}\right) - \left(\tfrac{1}{3}\right)\left(-\tfrac{3}{4}\right)$$

$$= \tfrac{2\sqrt{14}}{12} + \tfrac{3}{12}$$

Answer: $\cos (x + y) = \tfrac{2\sqrt{14} + 3}{12}$

Exercises 9.2

In **1 – 6**, demonstrate that the formula is true for the values of A and B given.

1. $\cos (A - B) = \cos A \cos B + \sin A \sin B$
   for $A = 45°$ and $B = 45°$

2. $\sin (A - B) = \sin A \cos B - \cos A \sin B$
   for $A = 60°$ and $B = 30°$

3. $\cos (A + B) = \cos A \cos B - \sin A \sin B$
   for $A = \tfrac{\pi}{4}$ and $B = \tfrac{3\pi}{4}$

4. $\sin (A + B) = \sin A \cos B + \cos A \sin B$
   for $A = \tfrac{\pi}{2}$ and $B = 0$

5. $\tan (A - B) = \dfrac{\tan A - \tan B}{1 + \tan A \tan B}$
   for $A = \tfrac{\pi}{3}$ and $B = \tfrac{5\pi}{6}$

6. $\tan (A + B) = \dfrac{\tan A + \tan B}{1 - \tan A \tan B}$
   for $A = 0$ and $B = \tfrac{\pi}{4}$

1. $\cos (45° - 45°) = \cos 45° \cos 45° + \sin 45° \sin 45°$

$$\cos 0° = \tfrac{\sqrt{2}}{2} \cdot \tfrac{\sqrt{2}}{2} + \tfrac{\sqrt{2}}{2} \cdot \tfrac{\sqrt{2}}{2}$$

$$1 = \tfrac{2}{4} + \tfrac{2}{4}$$

$$1 = 1$$

2. $\sin (60° - 30°) = \sin 60° \cos 30° - \cos 60° \sin 30°$

$$\sin 30° = \tfrac{\sqrt{3}}{2} \cdot \tfrac{\sqrt{3}}{2} - \tfrac{1}{2} \cdot \tfrac{1}{2}$$

$$\tfrac{1}{2} = \tfrac{3}{4} - \tfrac{1}{4}$$

$$\tfrac{1}{2} = \tfrac{2}{4}$$

$$\tfrac{1}{2} = \tfrac{1}{2}$$

3. $\cos\left(\frac{\pi}{4} + \frac{3\pi}{4}\right) = \cos\frac{\pi}{4}\cos\frac{3\pi}{4} - \sin\frac{\pi}{4}\sin\frac{3\pi}{4}$

$\cos\pi = \left(\frac{\sqrt{2}}{2}\right)\left(-\frac{\sqrt{2}}{2}\right) - \left(\frac{\sqrt{2}}{2}\right)\left(\frac{\sqrt{2}}{2}\right)$

$-1 = -\frac{2}{4} - \frac{2}{4}$

$-1 = -\frac{4}{4}$

$-1 = -1$

4. $\sin\left(\frac{\pi}{2} + 0\right) = \sin\frac{\pi}{2}\cos 0 + \cos\frac{\pi}{2}\sin 0$

$\sin\frac{\pi}{2} = (1)(1) + (0)(0)$

$1 = 1 + 0$

$1 = 1$

5. $\tan\left(\frac{\pi}{3} - \frac{5\pi}{6}\right) = \dfrac{\tan\frac{\pi}{3} - \tan\frac{5\pi}{6}}{1 + \tan\frac{\pi}{3}\tan\frac{5\pi}{6}}$

$\tan\left(-\frac{\pi}{2}\right) = \dfrac{\sqrt{3} - \left(-\frac{\sqrt{3}}{3}\right)}{1 + (\sqrt{3})\left(-\frac{\sqrt{3}}{3}\right)}$

$= \dfrac{3\sqrt{3} + \sqrt{3}}{3 - 3}$ both sides undefined

6. $\tan\left(0 + \frac{\pi}{4}\right) = \dfrac{\tan 0 + \tan\frac{\pi}{4}}{1 - \tan 0\tan\frac{\pi}{4}}$

$\tan\frac{\pi}{4} = \dfrac{0 + 1}{1 - (0)(1)}$

$1 = \dfrac{1}{1 - 0}$

$1 = \dfrac{1}{1}$

$1 = 1$

7. $\cos\left(\frac{\pi}{3} + \frac{\pi}{6}\right) \neq \cos\frac{\pi}{3} + \cos\frac{\pi}{6}$

$\cos\frac{\pi}{2} \neq \frac{1}{2} + \frac{\sqrt{3}}{2}$

$0 \neq \frac{1 + \sqrt{3}}{2}$

8. $\sin\left(\frac{5\pi}{4} + \frac{\pi}{2}\right) \neq \sin\frac{5\pi}{4} + \sin\frac{\pi}{2}$

$\sin\frac{7\pi}{4} \neq -\frac{\sqrt{2}}{2} + 1$

$-\frac{\sqrt{2}}{2} \neq \frac{2 - \sqrt{2}}{2}$

9. $\tan\left(\frac{\pi}{6} + \frac{5\pi}{3}\right) \neq \tan\frac{\pi}{6} + \tan\frac{5\pi}{3}$

$\tan\frac{11\pi}{6} \neq \frac{\sqrt{3}}{3} + (-\sqrt{3})$

$-\frac{\sqrt{3}}{3} \neq \frac{\sqrt{3} - 3\sqrt{3}}{3}$

$-\frac{\sqrt{3}}{3} \neq -\frac{2}{3}\sqrt{3}$

10. $\sin\left(\frac{\pi}{3} + \left(-\frac{\pi}{6}\right)\right) \neq \sin\frac{\pi}{3} + \sin\left(-\frac{\pi}{6}\right)$

$\sin\frac{\pi}{6} \neq \frac{\sqrt{3}}{2} + \left(-\frac{1}{2}\right)$

$\frac{1}{2} \neq \frac{\sqrt{3} - 1}{2}$

11. $\cos\left(\frac{7\pi}{3} + \left(-\frac{4\pi}{3}\right)\right) \neq \cos\frac{7\pi}{3} + \cos\left(-\frac{4\pi}{3}\right)$

$\cos\pi \neq \frac{1}{2} + \left(-\frac{1}{2}\right)$

$-1 \neq 0$

12. $\tan\left[\left(-\frac{\pi}{6}\right) + \left(-\frac{\pi}{3}\right)\right] \neq \tan\left(-\frac{\pi}{6}\right) + \tan\left(-\frac{\pi}{3}\right)$

$\tan\left(-\frac{\pi}{2}\right) \neq -\frac{\sqrt{3}}{3} + (-\sqrt{3})$

$\text{undefined} \neq \frac{-\sqrt{3} - 3\sqrt{3}}{3}$

$\text{undefined} \neq -\frac{4\sqrt{3}}{3}$

13. $\sin(A - B) = \sin A\cos B - \cos A\sin B$

$\sin(A - (-B)) = \sin A\cos(-B) - \cos A\sin(-B)$

$= \sin A\cos B - \cos A(-\sin B)$

$\sin(A + B) = \sin A\cos B + \cos A\sin B$

14. $\tan(A-B) = \dfrac{\tan A - \tan B}{1 + \tan A \tan B}$

$\tan(A-(-B)) = \dfrac{\tan A - \tan(-B)}{1 + \tan A \tan(-B)}$

$= \dfrac{\tan A - (-\tan B)}{1 + (\tan A)(-\tan B)}$

$\tan(A+B) = \dfrac{\tan A + \tan B}{1 - \tan A \tan B}$

15. $\dfrac{\sqrt{2}}{2}(\cos\theta + \sin\theta)$

16. $\dfrac{\sqrt{3}}{2}\cos\theta + \dfrac{1}{2}\sin\theta$

17. $\dfrac{\tan\theta + \sqrt{3}}{1 - \sqrt{3}\tan\theta}$

18. $-\sin\theta$

19. $2-\sqrt{3}$

20. $\dfrac{\sqrt{6}+\sqrt{2}}{4}$

21. $\dfrac{\sqrt{2}-\sqrt{6}}{4}$

22. $-2-\sqrt{3}$

23. $\dfrac{\sqrt{6}+\sqrt{2}}{4}$

24. $\dfrac{\sqrt{6}+\sqrt{2}}{4}$

25. $2-\sqrt{3}$

26. $\dfrac{\sqrt{2}-\sqrt{6}}{4}$

27. $\dfrac{1}{8}$

28. 8

29. $\dfrac{4-3\sqrt{3}}{10}$

30. $\dfrac{33}{65}$

31. $\sin 50°$

32. $\tan 0.6$

33. $\cos 1$

34. $\tan(\pi + \theta) = \dfrac{0 + \tan\theta}{1 - (0)\tan\theta}$

$= \tan\theta$

In **35 – 40**, one side of the identity is transformed, to be made identical to the other side.

35. $\cos\dfrac{\pi}{2}\cos x - \sin\dfrac{\pi}{2}\sin x =$

$(0)\cos x - (1)\sin x =$

$-\sin x =$

36. $\dfrac{\tan 3\pi - \tan x}{1 + \tan 3\pi \tan x} =$

$\dfrac{0 - \tan x}{1 + (0)\tan x} =$

$\dfrac{-\tan x}{1} =$

$-\tan x =$

37. $\cos 135° \cos x + \sin 135° \sin x =$

$\left(-\dfrac{\sqrt{2}}{2}\right)\cos x + \left(\dfrac{\sqrt{2}}{2}\right)\sin x =$

$\dfrac{\sqrt{2}}{2}(\sin x - \cos x) =$

38.

$= (\sin x \cos y + \cos x \sin y)(\sin x \cos y - \cos x \sin y)$

$= \sin^2 x \cos^2 y - \cos^2 x \sin^2 y$

$= \sin^2 x (1 - \sin^2 y) - \cos^2 x \sin^2 y$

$= \sin^2 x - \sin^2 x \sin^2 y - \cos^2 x \sin^2 y$

$= \sin^2 x - \sin^2 y (\sin^2 x + \cos^2 x)$

$= \sin^2 x - \sin^2 y (1)$

$= \sin^2 x - \sin^2 y$

$= (\sin x + \sin y)(\sin x - \sin y)$

39.

$= (\cos x \cos y - \sin x \sin y)(\cos x \cos y + \sin x \sin y)$

$= \cos^2 x \cos^2 y - \sin^2 x \sin^2 y$

$= \cos^2 x (1 - \sin^2 y) - \sin^2 x \sin^2 y$

$= \cos^2 x - \cos^2 x \sin^2 y - \sin^2 x \sin^2 y$

$= \cos^2 x - \sin^2 y (\cos^2 x + \sin^2 x)$

$= \cos^2 x - \sin^2 y$

40. $= \dfrac{\dfrac{1}{\tan u}\dfrac{1}{\tan v} - 1}{\dfrac{1}{\tan u} + \dfrac{1}{\tan v}}$

$= \dfrac{1 - \tan u \tan v}{\tan v + \tan u}$

$= \dfrac{1}{\dfrac{\tan u + \tan v}{1 - \tan u \tan v}}$

$= \dfrac{1}{\tan(u+v)}$

$= \cot(u+v)$

41. $\sin\theta = \dfrac{3\sqrt{3}-4}{10}$,

$\cos\theta = \dfrac{4\sqrt{3}+3}{10}$

42. $\sin\theta = 0.6$,

$\cos\theta = 0.8$

43.

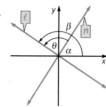

$$\theta = \beta - \alpha$$
$$m_n = \tan \alpha$$
$$m_\ell = \tan \beta$$
$$\tan \theta = \tan (\beta - \alpha)$$
$$= \frac{\tan \beta - \tan \alpha}{1 + \tan \beta \tan \alpha}$$
$$= \frac{m_\ell - m_n}{1 + m_\ell\, m_n}$$

44. 135°

45. 25 feet

46. a.
$$\sin (x + y) = \frac{\sqrt{7} + 1}{4}$$

$$\cos (x + y) = \frac{\sqrt{7} - 1}{4}$$

$$\tan (x + y) = \frac{\sqrt{7} + 1}{\sqrt{7} - 1}$$

b.
$$\frac{\sqrt{7} + 1}{\sqrt{7} - 1} \overset{?}{=} \frac{\dfrac{\sqrt{7} + 1}{4}}{\dfrac{\sqrt{7} - 1}{4}}$$

$$\frac{\sqrt{7} + 1}{\sqrt{7} - 1} = \frac{\sqrt{7} + 1}{\sqrt{7} - 1}$$

47.

$$\tan \angle ABC = \frac{x}{3x} = \frac{1}{3}$$

$$\tan \angle AEC = \frac{x}{2x} = \frac{1}{2}$$

$$\tan \angle ADC = \frac{x}{x} = 1$$

$$\tan (\angle ABC + \angle AEC) = \frac{\tan \angle ABC + \tan \angle AEC}{1 - \tan \angle ABC \tan \angle AEC}$$

$$= \frac{\dfrac{1}{3} + \dfrac{1}{2}}{1 - \left(\dfrac{1}{3}\right)\left(\dfrac{1}{2}\right)}$$

$$= \frac{\dfrac{5}{6}}{1 - \dfrac{1}{6}} = \frac{\dfrac{5}{6}}{\dfrac{5}{6}} = 1$$

$$\tan \angle ADC = \tan (\angle ABC + \angle AEC) = 1$$
$$m\angle ADC = m\angle ABC + m\angle AEC$$

Exercises 9.2 (continued)

In **7 – 12**, verify that $T(A + B) \neq T(A) + T(B)$ for the given trigonometric function T.

7. $A = \frac{\pi}{3}$, $B = \frac{\pi}{6}$; T: cosine

8. $A = \frac{5\pi}{4}$, $B = \frac{\pi}{2}$; T: sine

9. $A = \frac{\pi}{6}$, $B = \frac{5\pi}{3}$; T: tangent

10. $A = \frac{\pi}{3}$, $B = -\frac{\pi}{6}$; T: sine

11. $A = \frac{7\pi}{3}$, $B = -\frac{4\pi}{3}$; T: cosine

12. $A = -\frac{\pi}{6}$, $B = -\frac{\pi}{3}$; T: tangent

13. Derive the formula
$$\sin (A + B) = \sin A \cos B + \cos A \sin B$$
from the formula for $\sin (A - B)$.

14. Derive the formula
$$\tan (A + B) = \frac{\tan A + \tan B}{1 - \tan A \tan B}$$
from the formula for $\tan (A - B)$.

15. Express $\sin \left(\frac{\pi}{4} + \theta\right)$ in terms of $\sin \theta$ and $\cos \theta$.

16. Express $\cos \left(\frac{\pi}{6} - \theta\right)$ in terms of $\sin \theta$ and $\cos \theta$.

17. Express $\tan \left(\theta + \frac{7\pi}{3}\right)$ in terms of $\tan \theta$.

18. Express $\sin (\theta + \pi)$ in terms of $\sin \theta$.

In **19 – 26**, find the exact value of the trigonometric function.

19. $\tan 15°$

20. $\sin 75°$

21. $\sin 195°$

22. $\tan (-75°)$

23. $\cos \frac{\pi}{12}$

24. $\sin \frac{7\pi}{12}$

25. $\tan \left(-\frac{11\pi}{12}\right)$

26. $\cos \frac{17\pi}{12}$

27. If $\tan x = \frac{2}{3}$ and $\tan y = \frac{1}{2}$, find the value of $\tan (x - y)$.

28. If $\tan x = \frac{3}{2}$ and $\tan y = \frac{1}{2}$, find the value of $\tan (x + y)$.

29. Find the value of $\sin (x - y)$ if $\sin x = \frac{1}{2}$ with x in quadrant II and $\cos y = \frac{4}{5}$ with y in quadrant IV.

30. Find the value of $\cos (x - y)$ if $\sin x = -\frac{5}{13}$ with x in quadrant III and $\sin y = \frac{3}{5}$ with y in quadrant II.

31. Express in terms of one function:
$$\sin 10° \cos 40° + \sin 40° \cos 10°$$

32. Express $\dfrac{\tan 1.2 - \tan 0.6}{1 + \tan 1.2 \tan 0.6}$ in terms of one function.

33. Write, in terms of one function:
$$\cos (x - 1) \cos x + \sin (x - 1) \sin x$$
Express the result in simplest form.

34. By using the formula for the tangent of the sum of two angles, verify:
$$\tan (\pi + \theta) = \tan \theta$$

In **35 – 40**, verify each identity.

35. $\cos \left(\frac{\pi}{2} + x\right) = -\sin x$

36. $\tan (3\pi - x) = -\tan x$

37. $\cos (135° - x) = \frac{\sqrt{2}}{2} (\sin x - \cos x)$

38. $(\sin x + \sin y)(\sin x - \sin y)$
$= \sin (x + y) \sin (x - y)$

39. $\cos^2 x - \sin^2 y = \cos (x + y) \cos (x - y)$

40. $\cot (u + v) = \dfrac{\cot u \cot v - 1}{\cot u + \cot v}$

In **41 – 42**, θ is the interval $\left[0, \frac{\pi}{2}\right]$. Find sin θ and cos θ, given the value of a trigonometric function of a transformation of θ.

41. $\sin\left(\theta + \frac{\pi}{6}\right) = 0.6$

Hint: Use a Pythagorean relation to find $\cos\left(\theta + \frac{\pi}{6}\right)$.

42. $\cos(\theta + \pi) = -0.8$

43. Two nonperpendicular intersecting lines n and ℓ have an angle θ between them measured counterclockwise from n to ℓ. If the slopes of the lines are m_ℓ and m_n, show that:

$$\tan \theta = \frac{m_\ell - m_n}{1 + m_\ell m_n}$$

44. If the line $y = 4x + 4$ is rotated through an angle θ about its point of intersection with the line $y = \frac{3}{5}x - 7$, the two lines will coincide. Use the formula in Exercise 43 to find the degree measure of θ.

45. A sign is mounted on a base that is 15 feet high. When viewed from a point 30 feet from the foot of the base, the angle between the line of sight to the bottom of the sign and the ground is equal in measure to the angle between the line of sight to the bottom of the sign and the line of sight to the top of the sign. Find the height of the sign.

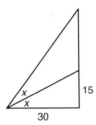

46. In right $\triangle ABC$ with right $\angle C$, $AC = BC = 1$, $\overline{BE} \perp \overline{AB}$, $BE = \frac{1}{2}$, and $\overrightarrow{AE}$ intersects $\overrightarrow{CB}$ at D. Let $m\angle CAB = x$ and $m\angle BAD = y$.

a. Use formulas to find $\sin(x + y)$, $\cos(x + y)$, and $\tan(x + y)$.

b. Show that: $\tan(x + y) = \dfrac{\sin(x + y)}{\cos(x + y)}$

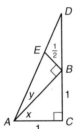

47. In right $\triangle ABC$ with right $\angle C$, $\overline{BC}$ is 3 times as long as $\overline{AC}$. Trisection points D and E are chosen on $\overline{BC}$ with D closer to C. Prove that:

$$m\angle ABC + m\angle AEC = m\angle ADC$$

48. Use the formulas for $\cos\left(A + \frac{\pi}{2}\right)$ and $\sin\left(A + \frac{\pi}{2}\right)$ to show that the image of (x, y) under a rotation of $\frac{\pi}{2}$ about the origin is $(-y, x)$.

48.

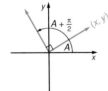

$\sin A = y \qquad \cos A = x$

$\sin\left(A + \frac{\pi}{2}\right) = \sin A \cos \frac{\pi}{2} + \cos A \sin \frac{\pi}{2}$

$\qquad = \sin A \, (0) + \cos A \, (1)$

$\qquad = \cos A$

$\qquad = x$

$\cos\left(A + \frac{\pi}{2}\right) = \cos A \cos \frac{\pi}{2} - \sin A \sin \frac{\pi}{2}$

$\qquad = \cos A \, (0) - \sin A \, (1)$

$\qquad = -\sin A$

$\qquad = -y$

The coordinates of the rotation are $(-y, x)$.

9.3 The Double-Angle, Half-Angle, and Sum-to-Product Formulas

It is useful to have formulas that express $\sin 2\theta$ and $\cos 2\theta$ in terms of $\sin \theta$ and $\cos \theta$. First note that $\sin 60° \neq 2 \sin 30°$ since:

$$\sin 60° = \frac{\sqrt{3}}{2} \text{ and } 2 \sin 30° = 2\left(\tfrac{1}{2}\right) = 1$$

In general then, for any angle θ, $\sin 2\theta \neq 2 \sin \theta$. It can also be shown that $\cos 2\theta \neq 2 \cos \theta$, and $\tan 2\theta \neq 2 \tan \theta$.

Since the formula for $\sin (A + B)$ is true for all replacements of the variables, it is true when $A = B$, and a formula for $\sin 2A$ can be derived.

$$\sin 2A = \sin (A + A)$$
$$= \sin A \cos A + \cos A \sin A$$
$$\sin 2A = 2 \sin A \cos A$$

Similarly, a formula for $\cos 2A$ can be derived.

$$\cos 2A = \cos (A + A)$$
$$= \cos A \cos A - \sin A \sin A$$
$$\cos 2A = \cos^2 A - \sin^2 A$$

The formula for $\cos 2A$ can be written in two other forms by using the Pythagorean identity $\sin^2 x + \cos^2 x = 1$.

Substituting $1 - \sin^2 A$ for $\cos^2 A$ in the formula above:

$$\cos 2A = (1 - \sin^2 A) - \sin^2 A$$
$$\cos 2A = 1 - 2 \sin^2 A$$

Substituting $1 - \cos^2 A$ for $\sin^2 A$ in the first formula for $\cos 2A$:

$$\cos 2A = \cos^2 A - (1 - \cos^2 A)$$
$$\cos 2A = 2 \cos^2 A - 1$$

The formula for $\tan 2A$ can be obtained by using the relationship

$$\tan 2A = \frac{\sin 2A}{\cos 2A} \text{ for } A \neq \frac{\pi}{4} + \frac{\pi}{2} k, \text{ where } k \text{ is an integer.}$$

The resulting formula, which you will prove later, is:

$$\tan 2A = \frac{2 \tan A}{1 - \tan^2 A}$$

Example 1

If θ is an angle in quadrant III and $\cos\theta = -\frac{4}{5}$, find the value of the six trigonometric functions of 2θ.

Solution

From the diagram, $\sin\theta = -\frac{3}{5}$ and $\tan\theta = \frac{3}{4}$.

Therefore,

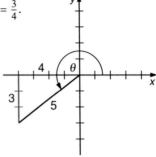

$\sin 2\theta = 2\sin\theta\cos\theta$

$\qquad = 2\left(-\frac{3}{5}\right)\left(-\frac{4}{5}\right) = \frac{24}{25}$

$\cos 2\theta = 2\cos^2\theta - 1$

$\qquad = 2\left(-\frac{4}{5}\right)^2 - 1 = \frac{7}{25}$

$\tan 2\theta = \dfrac{2\tan\theta}{1-\tan^2\theta}$

Note: $\dfrac{\sin 2\theta}{\cos 2\theta}$ may also be used to find $\tan 2\theta$.

$\qquad = \dfrac{2\left(\frac{3}{4}\right)}{1-\left(\frac{3}{4}\right)^2} = \dfrac{\frac{6}{4}}{\frac{7}{16}} = \frac{24}{7}$

$\csc 2\theta = \dfrac{1}{\sin 2\theta} = \dfrac{25}{24}$ $\qquad$ $\sec 2\theta = \dfrac{1}{\cos 2\theta} = \dfrac{25}{7}$ $\qquad$ $\cot 2\theta = \dfrac{1}{\tan 2\theta} = \dfrac{7}{24}$

Example 2

Find the value of $\sin\frac{\pi}{8}\,\cos\frac{\pi}{8}$.

Solution

The expression contains the product of the sine and cosine of the same angle and, therefore, suggests the formula for sine 2θ.

$$\sin 2\theta = 2\sin\theta\cos\theta$$

$$\tfrac{1}{2}\sin 2\theta = \sin\theta\cos\theta$$

If $\theta = \frac{\pi}{8}$, then $2\theta = \frac{\pi}{4}$.

$$\tfrac{1}{2}\sin\tfrac{\pi}{4} = \sin\tfrac{\pi}{8}\cos\tfrac{\pi}{8}$$

$$\frac{\frac{\sqrt{2}}{2}}{2} = \sin\tfrac{\pi}{8}\cos\tfrac{\pi}{8}$$

Answer: $\sin\frac{\pi}{8}\,\cos\frac{\pi}{8} = \frac{\sqrt{2}}{4}$

Half-Angle Formulas

Formulas for half angles can be derived from the formulas for cos 2A, since if 2A = θ, then $A = \dfrac{\theta}{2}$.

$$\cos 2A = 1 - 2 \sin^2 A$$

$$2 \sin^2 A = 1 - \cos 2A$$

$$\sin^2 A = \frac{1 - \cos 2A}{2}$$

$$\sin A = \pm\sqrt{\frac{1 - \cos 2A}{2}}$$

Letting 2A = θ, then $A = \dfrac{\theta}{2}$ and: $\sin \dfrac{\theta}{2} = \pm\sqrt{\dfrac{1 - \cos \theta}{2}}$

Since the formula indicates that the value of $\sin \dfrac{\theta}{2}$ can be either positive or negative, the sign is chosen according to the quadrant in which $\dfrac{\theta}{2}$ terminates. For example, if θ is 450°, then $\dfrac{\theta}{2}$ is 225°, which terminates in quadrant III where $\sin \dfrac{\theta}{2}$ is negative.

The formula for $\cos \dfrac{\theta}{2}$ can be derived in a similar manner, using $\cos 2A = 2 \cos^2 A - 1$ and solving for cos A in terms of cos 2A. The resulting formula is:

$$\cos \frac{\theta}{2} = \pm\sqrt{\frac{1 + \cos \theta}{2}}$$

The formula for $\tan \dfrac{\theta}{2}$ can be found using the formulas for $\sin \dfrac{\theta}{2}$ and $\cos \dfrac{\theta}{2}$.

$$\tan \frac{\theta}{2} = \frac{\sin \dfrac{\theta}{2}}{\cos \dfrac{\theta}{2}}$$

$$= \frac{\pm\sqrt{\dfrac{1 - \cos \theta}{2}}}{\pm\sqrt{\dfrac{1 + \cos \theta}{2}}}$$

$$= \pm\sqrt{\frac{1 - \cos \theta}{1 + \cos \theta}}$$

$$\tan \frac{\theta}{2} = \pm\sqrt{\frac{1 - \cos \theta}{1 + \cos \theta}} \ , \ \theta \neq \pi + 2\pi k, \text{ where } k \text{ is an integer.}$$

Example 3 _____

Find the exact value of $\sin \frac{\pi}{8}$.

Solution

Since $\frac{\pi}{8} = \frac{1}{2} \cdot \frac{\pi}{4}$, use the formula:

$$\sin \frac{\theta}{2} = \pm \sqrt{\frac{1 - \cos \theta}{2}}$$

$$\sin \frac{1}{2}\left(\frac{\pi}{4}\right) = \pm \sqrt{\frac{1 - \cos \frac{\pi}{4}}{2}}$$

$$= \pm \sqrt{\frac{1 - \frac{\sqrt{2}}{2}}{2}}$$

$$= \pm \sqrt{\frac{2 - \sqrt{2}}{4}}$$

$$= \frac{\pm \sqrt{2 - \sqrt{2}}}{2}$$

Since $\frac{\pi}{8}$ is an angle in quadrant I, use the positive value of the square root.

Answer: $\sin \dfrac{\pi}{8} = \dfrac{\sqrt{2 - \sqrt{2}}}{2}$

Example 4 _____

Find the exact value of $\tan 165°$.

Solution

$$\tan \frac{\theta}{2} = \pm \sqrt{\frac{1 - \cos \theta}{1 + \cos \theta}}$$

$$\tan 165° = \pm \sqrt{\frac{1 - \cos 330°}{1 + \cos 330°}}$$

$$= \pm \sqrt{\frac{1 - \frac{\sqrt{3}}{2}}{1 + \frac{\sqrt{3}}{2}}}$$

$$= \pm \sqrt{\frac{2 - \sqrt{3}}{2 + \sqrt{3}}}$$

$$= \pm \sqrt{\frac{2 - \sqrt{3}}{2 + \sqrt{3}} \cdot \frac{2 - \sqrt{3}}{2 - \sqrt{3}}} = \pm \sqrt{\frac{4 - 4\sqrt{3} + 3}{4 - 3}}$$

$$= \pm \sqrt{7 - 4\sqrt{3}}$$

The tangent function is negative in quadrant II.

Answer: $\tan 165° = -\sqrt{7 - 4\sqrt{3}}$

Sum-to-Product Formulas

It is sometimes useful to write a sum or difference of two sines or two cosines as the product of two sines or cosines. This can be done by using the formulas for the sine or cosine of the sum and difference of two angles.

$$(1) \ \sin (A + B) = \sin A \cos B + \cos A \sin B$$

$$(2) \ \sin (A - B) = \sin A \cos B - \cos A \sin B$$

Subtracting equation (2) from equation (1):

$$\sin (A + B) - \sin (A - B) = 2 \cos A \sin B$$

Adding equations (1) and (2):

$$\sin (A + B) + \sin (A - B) = 2 \sin A \cos B$$

Similarly, using the formulas for cos $(A + B)$ and cos $(A - B)$:

$$\cos (A + B) - \cos (A - B) = -2 \sin A \sin B$$

$$\cos (A + B) + \cos (A - B) = 2 \cos A \cos B$$

To express the terms on the left more simply, let $A + B = x$ and $A - B = y$, and solve for A and B.

$A + B = x$	$A + B = x$
$A - B = y$	$A - B = y$
$2A = x + y$	$2B = x - y$
$A = \frac{1}{2}(x + y)$	$B = \frac{1}{2}(x - y)$

Replace A by $\frac{1}{2}(x + y)$ and B by $\frac{1}{2}(x - y)$ in sin $(A + B) - $ sin $(A - B) = 2 \cos A \sin B$.

$$\sin x - \sin y = 2 \cos \tfrac{1}{2}(x + y) \sin \tfrac{1}{2}(x - y)$$

Similarly, it can be shown that:

$$\sin x + \sin y = 2 \sin \tfrac{1}{2}(x + y) \cos \tfrac{1}{2}(x - y)$$

$$\cos x - \cos y = -2 \sin \tfrac{1}{2}(x + y) \sin \tfrac{1}{2}(x - y)$$

$$\cos x + \cos y = 2 \cos \tfrac{1}{2}(x + y) \cos \tfrac{1}{2}(x - y)$$

It is difficult to remember each of these formulas accurately since they are so similar. Therefore, it is best to remember how they are derived from the formulas for sin $(A \pm B)$ and cos $(A \pm B)$ and to derive them when you need them.

Example 5

Express $\cos 3° - \cos 74°$ as a product of trigonometric functions.

Solution

$$\cos x - \cos y = -2 \sin \tfrac{1}{2}(x + y) \sin \tfrac{1}{2}(x - y)$$

$$\cos 3° - \cos 74° = -2 \sin \tfrac{1}{2}(3° + 74°) \sin \tfrac{1}{2}(3° - 74°)$$

$$= -2 \sin 38.5° \sin (-35.5°)$$

$$= -2 \sin 38.5°(-\sin 35.5°) \quad \text{Sine is an odd function.}$$

Answer: $\cos 3° - \cos 74° = 2 \sin 38.5° \sin 35.5°$

Example 6

Express $2 \sin 12° \cos 55°$ as a sum of trigonometric functions.

Solution

Use the relationship $\sin (A + B) + \sin (A - B) = 2 \sin A \cos B$, with A equal to $12°$ and B equal to $55°$.

$$\sin (12° + 55°) + \sin (12° - 55°) = 2 \sin 12° \cos 55°$$

$$\sin 67° + \sin (-43°) = 2 \sin 12° \cos 55°$$

Answer: $2 \sin 12° \cos 55° = \sin 67° + (-\sin 43°)$

or

$$2 \sin 12° \cos 55° = \sin 67° - \sin 43°$$

Example 7

Simplify: $\dfrac{\cos 20 + \cos 54}{\sin 20 - \sin 54}$

Solution

Use the formulas for the sum of two cosines and the difference of two sines.

$$\frac{\cos 20 + \cos 54}{\sin 20 - \sin 54} = \frac{2 \cos \tfrac{1}{2}(20 + 54) \cos \tfrac{1}{2}(20 - 54)}{2 \cos \tfrac{1}{2}(20 + 54) \sin \tfrac{1}{2}(20 - 54)}$$

$$= \frac{2 \cos 37 \cos (-17)}{2 \cos 37 \sin (-17)}$$

$$= \frac{\cos 17}{-\sin 17}$$

$$= -\cot 17$$

Answer: $\dfrac{\cos 20 + \cos 54}{\sin 20 - \sin 54} = -\cot 17$

1. sin 30°

2. sin 46°

3. sin 4θ

4. cos 32°

5. cos 6

6. cos 6θ

7. −cos $\frac{\pi}{8}$

8. tan 30°

9. cot $\frac{\pi}{4}$

10. sin 5°

11. tan$\frac{1}{2}$

12. cot $\frac{1}{2}$

13. cos² $\frac{1}{2}A$

14. sin² 2θ

15. tan² $\frac{3θ}{2}$

16. $\frac{\sqrt{2}}{2}$

17. $-\frac{\sqrt{2}}{4}$

18. $-\frac{\sqrt{3}}{2}$

19. $-\frac{\sqrt{3}}{2}$

20. $\frac{\sqrt{3}}{2}$

21. −1

22. 1

23. $\frac{2\sqrt{3}}{3}$

24. −2

25. $\frac{\sqrt{2+\sqrt{3}}}{2}$

26. $-\frac{\sqrt{2+\sqrt{3}}}{2}$

27. $\sqrt{3-2\sqrt{2}}$

28. $\frac{\sqrt{2-\sqrt{3}}}{2}$

29. $\sqrt{3+2\sqrt{2}}$

30. $-\frac{\sqrt{2-\sqrt{2}}}{2}$

31. $-2\sqrt{2+\sqrt{3}}$

32. $-2\sqrt{2+\sqrt{3}}$

33. $\sqrt{3+2\sqrt{2}}$

Exercises 9.3

In **1 – 15**, write the expression as a single function of an angle.

1. $2\sin 15° \cos 15°$

2. $2\cos 23° \sin 23°$

3. $2\sin 2θ \cos 2θ$

4. $\cos^2 16° - \sin^2 16°$

5. $2\cos^2 3 - 1$

6. $1 - 2\sin^2 3θ$

7. $\sin^2 \frac{\pi}{16} - \cos^2 \frac{\pi}{16}$

8. $\frac{2\tan 15°}{1 - \tan^2 15°}$

9. $\frac{1 - \tan^2 \frac{\pi}{8}}{2\tan \frac{\pi}{8}}$

10. $\pm\sqrt{\frac{1-\cos 10°}{2}}$

11. $\pm\sqrt{\frac{1-\cos 1}{1+\cos 1}}$

12. $\pm\sqrt{\frac{1+\cos 1}{1-\cos 1}}$

13. $\frac{1+\cos A}{2}$

14. $\frac{1-\cos 4θ}{2}$

15. $\frac{1-\cos 3θ}{1+\cos 3θ}$

In **16 – 24**, find the exact value.

16. $2\sin \frac{\pi}{8} \cos \frac{\pi}{8}$

17. $\sin \frac{5\pi}{8} \cos \frac{5\pi}{8}$

18. $\cos^2 \frac{5\pi}{12} - \sin^2 \frac{5\pi}{12}$

19. $2\cos^2 105° - 1$

20. $1 - 2\sin^2 165°$

21. $\frac{2\tan 67° 30'}{1 - \tan^2 67° 30'}$

22. $\frac{1 - \tan^2 \frac{\pi}{8}}{2\tan \frac{\pi}{8}}$

23. $\frac{1}{2\cos^2 \frac{\pi}{12} - 1}$

24. $\frac{1}{2\sin 105° \cos 105°}$

In **25 – 33**, use a half-angle formula to evaluate the given expression.

25. $\sin 105°$

26. $\cos 165°$

27. $\tan \frac{\pi}{8}$

28. $\sin \frac{\pi}{12}$

29. $\tan 67° 30'$

30. $\cos 112° 30'$

31. $\sec 105°$

32. $\csc 195°$

33. $\cot 202.5°$

In **34 – 41**, x is in the quadrant given, with a known function value. Find the value of $\sin 2x$, $\cos 2x$, and $\tan 2x$.

34. x is in quadrant I, $\sin x = \frac{4}{5}$

35. x is in quadrant II, $\cos x = -\frac{8}{17}$

36. x is in quadrant III, $\cot x = 2$

37. x is in quadrant IV, $\tan x = -\frac{2}{3}$

38. x is in quadrant II, $\sec x = -\frac{3}{2}$

39. x is in quadrant II, $\sin x = \frac{1}{4}$

40. x is in quadrant III, $\csc x = -4$

41. x is in quadrant IV, $\cos x = \frac{5}{13}$

In **42 – 47**, express the given sum or difference as a product of trigonometric functions.

42. $\sin 45° + \sin 30°$

43. $\cos 150° + \cos 60°$

44. $\sin 2θ - \sin θ$

45. $\cos 5y + \cos 2y$

46. $\cos (x - y) - \cos (x + y)$

47. $\sin \frac{3\pi}{4} - \sin \frac{\pi}{4}$

34. $\sin 2x = \frac{24}{25}$, $\cos 2x = -\frac{7}{25}$,

$\tan 2x = -\frac{24}{7}$

35. $\sin 2x = -\frac{240}{289}$, $\cos 2x = -\frac{161}{289}$,

$\tan 2x = \frac{240}{161}$

36. $\sin 2x = \frac{4}{5}$, $\cos 2x = \frac{3}{5}$, $\tan 2x = \frac{4}{3}$

37. $\sin 2x = -\frac{12}{13}$, $\cos 2x = \frac{5}{13}$, $\tan 2x = -\frac{12}{5}$

38. $\sin 2x = -\frac{4\sqrt{5}}{9}$, $\cos 2x = -\frac{1}{9}$,

$\tan 2x = 4\sqrt{5}$

39. $\sin 2x = -\frac{\sqrt{15}}{8}$, $\cos 2x = \frac{7}{8}$, $\tan 2x = -\frac{\sqrt{15}}{7}$

40. $\sin 2x = \frac{\sqrt{15}}{8}$, $\cos 2x = \frac{7}{8}$, $\tan 2x = \frac{\sqrt{15}}{7}$

41. $\sin 2x = -\frac{120}{169}$, $\cos 2x = -\frac{119}{169}$, $\tan 2x = \frac{120}{119}$ **42.** $2 \sin 37.5° \cos 7.5°$ **43.** $2 \cos 105° \cos 45°$

44. $2 \cos \frac{3}{2}\theta \sin \frac{1}{2}\theta$ **45.** $2 \cos \frac{7}{2} y \cos \frac{3}{2} y$ **46.** $2 \sin x \sin y$ **47.** $2 \cos \frac{\pi}{2} \sin \frac{\pi}{4}$

48. $\sin 60° + \sin 0°$ **49.** $2 \sin \frac{\pi}{2} + 2 \sin 0$ **50.** $4 \sin 210° - 4 \sin 90°$ **51.** $\frac{1}{2} \sin 10 + \frac{1}{2} \sin 4$

52. $-\frac{5}{2} \cos 9a + \frac{5}{2} \cos a$ **53.** $-\frac{1}{2} \cos 8a + \frac{1}{2} \cos 4a$ **54.** $-\frac{1}{2} \cos 2a + \frac{1}{2} \cos 2b$ **55.** $\frac{1}{4} \cos 80° + \frac{1}{4} \cos 60°$

In **56 – 68**, one side of the identity is transformed,
to be made identical to the other side.

56.
$$= \frac{2}{\sec^2 x} - 1$$
$$= 2 \cos^2 x - 1$$
$$= \cos 2x$$

57.
$$= \frac{1 + \frac{\sin^2 \theta}{\cos^2 \theta}}{1 - \frac{\sin^2 \theta}{\cos^2 \theta}}$$
$$= \frac{\cos^2 \theta + \sin^2 \theta}{\cos^2 \theta - \sin^2 \theta}$$
$$= \frac{1}{\cos 2\theta}$$
$$= \sec 2\theta$$

58.
$$\frac{2 \sin^2 \theta}{\frac{\sin \theta}{\cos \theta}} =$$
$$\frac{2 \sin^2 \theta \cos \theta}{\sin \theta} =$$
$$2 \sin \theta \cos \theta =$$
$$\sin 2\theta =$$

59.
$$(1 - \tan^2 x)\, \frac{2 \tan x}{1 - \tan^2 x} =$$
$$2 \tan x =$$

60.
$$= \frac{2 \frac{\sin x}{\cos x}}{1 + \frac{\sin^2 x}{\cos^2 x}}$$
$$= \frac{2 \sin x \cos x}{\cos^2 x + \sin^2 x}$$
$$= \frac{\sin 2x}{1}$$
$$= \sin 2x$$

61.
$$\frac{\sin \theta}{\cos \theta} + \frac{\sin \theta}{\cos \theta} (2 \cos^2 \theta - 1) =$$
$$\frac{\sin \theta}{\cos \theta} + \frac{2\sin \theta \cos^2 \theta - \sin \theta}{\cos \theta} =$$
$$\frac{\sin \theta}{\cos \theta} + \frac{2\sin \theta \cos^2 \theta}{\cos \theta} - \frac{\sin \theta}{\cos \theta} =$$
$$2 \sin \theta \cos \theta =$$
$$\sin 2\theta =$$
$$\frac{1}{\csc 2\theta} =$$

62.
$$\frac{2 \sin \frac{1}{2}(8x) \cos \frac{1}{2}(-2x)}{-2 \sin \frac{1}{2}(8x) \sin \frac{1}{2}(-2x)} =$$
$$\frac{2 \sin 4x \cos (-x)}{-2 \sin 4x \sin (-x)} =$$
$$\frac{\cos x}{-(-\sin x)} =$$
$$\frac{\cos x}{\sin x} =$$
$$\cot x =$$

63.
$$= \frac{-2 \sin \frac{1}{2}(4y) \sin \frac{1}{2}(-2y)}{2 \cos \frac{1}{2}(4y) \sin \frac{1}{2}(2y)}$$
$$= \frac{-2 \sin 2y \sin (-y)}{2 \cos 2y \sin y}$$
$$= \frac{-2 \sin 2y (-\sin y)}{2 \cos 2y \sin y}$$
$$= \frac{2 \sin 2y \sin y}{2 \cos 2y \sin y}$$
$$= \frac{\sin 2y}{\cos 2y}$$
$$= \tan 2y$$

64.
$$\frac{\left[2 \cos \frac{1}{2}(8\theta) \sin \frac{1}{2}(2\theta)\right]^2}{\left[2 \cos \frac{1}{2}(8\theta) \sin \frac{1}{2}(2\theta)\right]\left[2 \sin \frac{1}{2}(8\theta) \cos \frac{1}{2}(2\theta)\right]} =$$
$$\frac{\left[2 \cos 4\theta \sin \theta\right]^2}{\left[2 \cos 4\theta \sin \theta\right]\left[2 \sin 4\theta \cos \theta\right]} =$$
$$\frac{4 \cos^2 4\theta \sin^2 \theta}{4 \sin 4\theta \cos 4\theta \sin \theta \cos \theta} =$$
$$\frac{\cos 4\theta \sin \theta}{\sin 4\theta \cos \theta} =$$
$$\cot 4\theta \tan \theta =$$
$$\frac{1}{\tan 4\theta} \tan \theta =$$
$$\frac{\tan \theta}{\tan 4\theta} =$$

65. $\left[2 \cos \frac{1}{2}(x + y) \sin \frac{1}{2}(x - y)\right]\left[2 \sin \frac{1}{2}(x + y) \cos \frac{1}{2}(x - y)\right] =$

$\left[2 \sin \frac{1}{2}(x + y) \cos \frac{1}{2}(x + y)\right]\left[2 \sin \frac{1}{2}(x - y) \cos \frac{1}{2}(x - y)\right] =$

$\sin(x + y) \sin(x - y) =$

66. $\sin^2 x + 2 \sin x \cos x + \cos^2 x =$

$\sin 2x + \sin^2 x + \cos^2 x =$

$\sin 2x + 1 =$

67. $(\sin^2 \theta + \cos^2 \theta)(\sin^2 \theta - \cos^2 \theta) =$

$1[-(\cos^2 \theta - \sin^2 \theta)] =$

$- \cos 2\theta =$

68. $= \dfrac{\sin 2A}{\cos 2A}$

$= \dfrac{2 \sin A \cos A}{\cos^2 A - \sin^2 A}$

$= \dfrac{\dfrac{2 \sin A}{\cos A}}{1 - \dfrac{\sin^2 A}{\cos^2 A}}$

$= \dfrac{2 \tan A}{1 - \tan^2 A}$

69. $\cos 2A = 2 \cos^2 A - 1$

$\cos^2 A = \dfrac{1 + \cos 2A}{2}$

$\cos A = \pm \sqrt{\dfrac{1 + \cos 2A}{2}}$

$A = \dfrac{\theta}{2}$

$\cos \dfrac{\theta}{2} = \pm \sqrt{\dfrac{1 + \cos \theta}{2}}$

70.

(1) $\qquad \cos(A + B) = \cos A \cos B - \sin A \sin B$

(2) $\qquad \cos(A - B) = \cos A \cos B + \sin A \sin B$

$\rule{6cm}{0.4pt}$

$\cos(A + B) - \cos(A - B) = -2 \sin A \sin B$

Subtract (1) and (2).

71.

(1) $\qquad \cos(A + B) = \cos A \cos B - \sin A \sin B$

(2) $\qquad \cos(A - B) = \cos A \cos B + \sin A \sin B$

$\rule{6cm}{0.4pt}$

$\cos(A + B) + \cos(A - B) = 2 \cos A \cos B$

Add (1) and (2).

72. $\sin(A + B) + \sin(A - B) = 2 \sin A \cos B$

$A + B = x \qquad A - B = y$

$x + y = 2A \qquad x - y = 2B$

$A = \frac{1}{2}(x + y) \quad B = \frac{1}{2}(x - y)$

Substituting,

$\sin x + \sin y = 2 \sin \frac{1}{2}(x + y) \cos \frac{1}{2}(x - y)$

73. $\cos(A + B) - \cos(A - B) = -2 \sin A \sin B$

$A + B = x \qquad A - B = y$

$x + y = 2A \qquad x - y = 2B$

$A = \frac{1}{2}(x + y) \quad B = \frac{1}{2}(x - y)$

Substituting,

$\cos x - \cos y = -2 \sin \frac{1}{2}(x + y) \sin \frac{1}{2}(x - y)$

74. a. $a \sin \theta + b \cos \theta = \sqrt{a^2 + b^2} \sin(\theta + \alpha)$

$a = \sqrt{3}$

$b = 1$

$\sqrt{a^2 + b^2} = \sqrt{(\sqrt{3})^2 + (1)^2}$

$= \sqrt{3 + 1}$

$= \sqrt{4}$

$= 2$

$\sin \alpha = \frac{1}{2}$

$\cos \alpha = \dfrac{\sqrt{3}}{2}$

$\alpha = 30°$

$\sqrt{3} \sin \theta + \cos \theta = 2 \sin(\theta + 30°)$

Exercises 9.3 (continued)

In **48 – 55**, express the given product as a sum (or difference) of trigonometric functions.

48. $2 \sin 30° \cos 30°$

49. $4 \sin \frac{\pi}{4} \cos \frac{\pi}{4}$

50. $8 \sin 60° \cos 150°$

51. $\sin 7 \cos 3$

52. $5 \sin 4a \sin 5a$

53. $\sin 6a \sin 2a$

54. $\sin (a + b) \sin (a - b)$

55. $\frac{1}{2} \cos 70° \cos 10°$

In **56 – 73**, prove that the given equation is an identity.

56. $\cos 2x = \dfrac{2 - \sec^2 x}{\sec^2 x}$

57. $\sec 2\theta = \dfrac{1 + \tan^2 \theta}{1 - \tan^2 \theta}$

58. $\dfrac{2 \sin^2 \theta}{\tan \theta} = \sin 2\theta$

59. $(1 + \tan x)(1 - \tan x) \tan 2x = 2 \tan x$

60. $\sin 2x = \dfrac{2 \tan x}{1 + \tan^2 x}$

61. $\tan \theta + \tan \theta \cos 2\theta = \dfrac{1}{\csc 2\theta}$

62. $\dfrac{\sin 3x + \sin 5x}{\cos 3x - \cos 5x} = \cot x$

63. $\tan 2y = \dfrac{\cos y - \cos 3y}{\sin 3y - \sin y}$

64. $\dfrac{(\sin 5\theta - \sin 3\theta)^2}{(\sin 5\theta - \sin 3\theta)(\sin 5\theta + \sin 3\theta)} = \dfrac{\tan \theta}{\tan 4\theta}$

65. $(\sin x - \sin y)(\sin x + \sin y) = \sin (x + y) \sin (x - y)$

66. $(\sin x + \cos x)^2 = \sin 2x + 1$

67. $\sin^4 \theta - \cos^4 \theta = -\cos 2\theta$

68. $\tan 2A = \dfrac{2 \tan A}{1 - \tan^2 A}$

69. $\cos \dfrac{\theta}{2} = \pm \sqrt{\dfrac{1 + \cos \theta}{2}}$

70. $\cos (A + B) - \cos (A - B) = -2 \sin A \sin B$ **71.** $\cos (A + B) + \cos (A - B) = 2 \cos A \cos B$

72. $\sin x + \sin y = 2 \sin \frac{1}{2}(x + y) \cos \frac{1}{2}(x - y)$ **73.** $\cos x - \cos y = -2 \sin \frac{1}{2}(x + y) \sin \frac{1}{2}(x - y)$

As seen earlier, a function that is the sum of two functions, such as $y = a \sin \theta + b \cos \theta$, can be graphed by adding the ordinates of the individual functions. An alternate method is to rewrite the equation using the formula

$a \sin \theta + b \cos \theta = \sqrt{a^2 + b^2} \sin (\theta + \alpha)$, where

α is the smallest positive angle satisfying $\sin \alpha = \dfrac{b}{\sqrt{a^2 + b^2}}$

and $\cos \alpha = \dfrac{a}{\sqrt{a^2 + b^2}}$.

74. a. Use the formula to verify: $\sqrt{3} \sin \theta + \cos \theta = 2 \sin (\theta + 30°)$

    b. Use the formula to sketch the graphs of $y = \sin x - \cos x$ and $y = -3 \sin x + 4 \cos x$.

---

74. b. $\sin x - \cos x = \sqrt{2} \sin (x + 315°)$ $y = -3 \sin x + 4 \cos x = 5 \sin (x + 127°)$

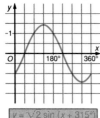

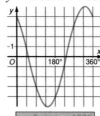

9.4 Trigonometric Equations

The procedures for solving equations that involve trigonometric functions are similar to those used for solving algebraic equations. Basically, you must isolate the term that contains the variable. In a trigonometric equation, the variable is an angle measure.

For example, to solve the equation $\sqrt{2} \cos x = 1$ for $x \in [0, 2\pi)$, isolate the term $\cos x$ by dividing each side of the equation by $\sqrt{2}$.

$$\sqrt{2} \cos x = 1$$

$$\cos x = \frac{1}{\sqrt{2}} = \frac{\sqrt{2}}{2}$$

$$\cos x = \frac{\sqrt{2}}{2}$$

Now solve for x by finding all angle measures in the interval $[0, 2\pi)$ that have the cosine function value of $\frac{\sqrt{2}}{2}$. There are two such angles; thus, $x = \frac{\pi}{4}$ and $x = \frac{7\pi}{4}$. Observe that the intersections of the graphs of $y = \sqrt{2} \cos x$ and $y = 1$ confirm these solutions of $\sqrt{2} \cos x = 1$.

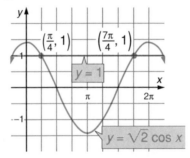

For all solutions, the domain of the function must be considered when determining the solution set. If the domain of the previous equation, $\sqrt{2} \cos x = 1$, were restricted to acute angles, that is, $x \in \left[0, \frac{\pi}{2}\right)$, then $x = \frac{\pi}{4}$ would be the only solution. A general solution for all real numbers x would be $x = \pm \frac{\pi}{4} + 2\pi n$, where n is an integer.

Trigonometric equations occur in forms that are similar to the algebraic equations you studied in Chapter 2. Trigonometric equations that are in quadratic form will require taking the square root of both sides, factoring, or the use of the quadratic formula in order to be solved. Other equations will require substitution, using the identities and formulas established in the previous sections.

Example 1 _____

Solve $\tan x + \sqrt{3} = 0$ for $x \in [0, 2\pi)$.

Solution

$$\tan x + \sqrt{3} = 0 \qquad \text{Subtract } \sqrt{3} \text{ from each side.}$$
$$\tan x = -\sqrt{3}$$

Determine the values of x in the interval $[0, 2\pi)$ for which the tangent function has a value of $-\sqrt{3}$. Since the value of $\tan x$ is negative, x must be the measure of an angle in quadrants II or IV. In quadrant II, when $\tan x = -\sqrt{3}$, $x = \frac{2\pi}{3}$ and in quadrant IV, $x = \frac{5\pi}{3}$. The solution can be checked by substituting in the original equation.

Answer: The solution set is $\left\{ \frac{2\pi}{3}, \frac{5\pi}{3} \right\}$.

Example 2

Find the general solution for $4 \cos^2 \theta = 3$.

Solution

Isolate $\cos^2 \theta$ on one side of the equation and then take the square root of each side.

$$4 \cos^2 \theta = 3$$
$$\cos^2 \theta = \frac{3}{4}$$
$$\cos \theta = \pm \frac{\sqrt{3}}{2}$$

Since the period of $\cos \theta$ is 2π, first find all the solutions in the interval $[0, 2\pi)$ and then add $2\pi n$, where n is an integer, to each of these solutions. If $\cos x = \pm \frac{\sqrt{3}}{2}$, then $x = \frac{\pi}{6}, \frac{5\pi}{6}, \frac{7\pi}{6}, \frac{11\pi}{6}$ in $[0, 2\pi)$.

Answer: The general solution for the equation is

$$x = \left\{ \frac{\pi}{6} + 2\pi n, \frac{5\pi}{6} + 2\pi n, \frac{7\pi}{6} + 2\pi n, \frac{11\pi}{6} + 2\pi n \right\} \text{ or } \left\{ \pm \frac{\pi}{6} + \pi n \right\},$$
where n is an integer.

Example 3

Solve $\sin^2 x + \sin x = 0$ for $x \in [0°, 360°)$.

Solution

This is a quadratic equation in $\sin x$. Since the right side of the equation is 0, factor the left side and set each factor equal to 0.

$$\sin^2 x + \sin x = 0$$
$$\sin x \,(\sin x + 1) = 0$$

$\sin x = 0$	$\sin x + 1 = 0$
	$\sin x = -1$
$x = 0°, 180°$	$x = 270°$

Answer: The solution set is $\{0°, 180°, 270°\}$.

Example 4

Find the general solution for $\tan^2 x - 2 \tan x + 1 = 0$.

Solution

Factor the trinomial in $\tan x$.

$$\tan^2 x - 2 \tan x + 1 = 0$$

$$(\tan x - 1)(\tan x - 1) = 0$$

$$\tan x - 1 = 0$$

$$\tan x = 1$$

Since $\tan x$ has a period of π, find the solution of $\tan x = 1$ in the interval $[0, \pi)$ and add πn for the general solution. The only solution in the interval $[0, \pi)$ is $\frac{\pi}{4}$.

Answer: The general solution for the equation is $\left\{ \frac{\pi}{4} + \pi n \right\}$, where n is an integer.

Example 5

Solve $2 \sin^2 x - 1 = \cos x$ for $x \in [0, 2\pi)$.

Solution

Use a trigonometric identity or formula to transform the equation into one containing a single function.

Using the identity $\sin^2 x + \cos^2 x = 1$, substitute $1 - \cos^2 x$ for $\sin^2 x$.

$$2 \sin^2 x - 1 = \cos x$$

$$2(1 - \cos^2 x) - 1 = \cos x$$

$$2 - 2 \cos^2 x - 1 = \cos x$$

$$2 \cos^2 x + \cos x - 1 = 0$$

$$(2 \cos x - 1)(\cos x + 1) = 0 \qquad \text{Factor the left side.}$$

$2 \cos x - 1 = 0$	$\cos x + 1 = 0$
$\cos x = \frac{1}{2}$	$\cos x = -1$
$x = \frac{\pi}{3}, \frac{5\pi}{3}$	$x = \pi$

Answer: The solution set is $\left\{ \frac{\pi}{3}, \pi, \frac{5\pi}{3} \right\}$.

Example 6

Use a calculator or trigonometric table to estimate the general solution of $3 \csc^2 x + 4 \csc x - 5 = 0$ to the nearest hundredth.

Solution

Since the expression on the left side cannot be factored, use the quadratic formula.

$$\csc x = \frac{-4 \pm \sqrt{4^2 - 4(3)(-5)}}{2(3)}$$

$$= \frac{-4 \pm \sqrt{76}}{6}$$

$\csc x = 0.786$ or $\csc x = -2.119...$

There are no solutions.

From the trigonometric table or a calculator, the reference angle is approximately 0.49. Since the cosecant is negative in quadrant III and IV,
$x \approx 0.49 + \pi \approx 3.63$, or
$x \approx 2\pi - 0.49 \approx 5.79$.

Since the period of the cosecant function is 2π, the general solution of the equation to the nearest hundredth is:

Answer: $\{3.63 + 2\pi n, 5.79 + 2\pi n\}$ where n is an integer.

Example 7

Find the general solution of $\sin 3x = 1$.

Solution

If $\sin 3x = 1$, then $3x = \frac{\pi}{2} + 2\pi n$, where n is an integer. Now divide both sides of this last equation by 3.

Answer: The general solution for the equation is $\left\{ \frac{\pi}{6} + \frac{2\pi n}{3} \right\}$, where n is an integer.

Observe that the intersections of the graphs of $y = \sin 3x$ and $y = 1$ confirm these solutions.

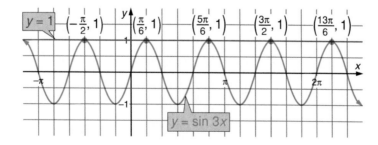

1. $\frac{7\pi}{6}, \frac{11\pi}{6}$

2. $\frac{\pi}{3}, \frac{5\pi}{3}$

3. $\frac{\pi}{6}, \frac{11\pi}{6}$

4. $\frac{\pi}{3}, \frac{4\pi}{3}$

5. no solution

6. $\frac{\pi}{3}, \frac{2\pi}{3}$

7. $\frac{11\pi}{24}, \frac{43\pi}{24}$

8. $\frac{\pi}{12}, \frac{3\pi}{4}$

9. $\frac{\pi}{6}, \frac{5\pi}{6}, \frac{7\pi}{6}, \frac{11\pi}{6}$

10. $\frac{\pi}{6}, \frac{5\pi}{6}, \frac{7\pi}{6}, \frac{11\pi}{6}$

11. $\frac{\pi}{3}, \frac{2\pi}{3}, \frac{4\pi}{3}, \frac{5\pi}{3}$

12. $\frac{\pi}{6}, \frac{5\pi}{6}, \frac{7\pi}{6}, \frac{11\pi}{6}$

13. $\frac{\pi}{4}, \frac{3\pi}{4}, \frac{5\pi}{4}, \frac{7\pi}{4}$

14. $\frac{\pi}{4}, \frac{3\pi}{4}, \frac{5\pi}{4}, \frac{7\pi}{4}$

15. $\frac{\pi}{3}, \frac{2\pi}{3}, \frac{4\pi}{3}, \frac{5\pi}{3}$

16. $\frac{\pi}{6}, \frac{\pi}{4}, \frac{5\pi}{6}, \frac{7\pi}{4}$

17. $\frac{\pi}{3}, \pi, \frac{5\pi}{3}$

18. $0, \frac{\pi}{4}, \pi, \frac{5\pi}{4}$

19. $\frac{\pi}{6}, \frac{5\pi}{6}, \frac{7\pi}{6}, \frac{11\pi}{6}, \frac{\pi}{3},$
$\frac{2\pi}{3}, \frac{4\pi}{3}, \frac{5\pi}{3}$

20. $\frac{\pi}{6}, \frac{5\pi}{6}, \frac{7\pi}{6}, \frac{11\pi}{6}$

21. $\frac{\pi}{6}, \frac{5\pi}{6}, \frac{7\pi}{6}, \frac{11\pi}{6}, \frac{\pi}{3},$
$\frac{2\pi}{3}, \frac{4\pi}{3}, \frac{5\pi}{3}$

22. $\theta = \frac{\pi}{4} + \pi n$,
n an integer

23. $\theta = \frac{\pi}{2} + 2\pi n$,
$\theta = \frac{7\pi}{6} + 2\pi n, \theta = \frac{3\pi}{2} + 2\pi n$,
$\theta = \frac{11\pi}{6} + 2\pi n$, n an integer

24. $\theta = \frac{2\pi}{3} + 2\pi n, \theta = \frac{4\pi}{3} + 2\pi n$,
$\theta = \pi + 2\pi n$, n an integer

25. $\theta = \frac{\pi}{6} + 2\pi n, \theta = \frac{5\pi}{6} + 2\pi n$,
n an integer

26. $\theta = \frac{\pi}{6} + 2\pi n, \theta = \frac{5\pi}{6} + 2\pi n$,
$\theta = \frac{3\pi}{2} + 2\pi n$, n an integer

27. $\theta = \pi n, \theta = \frac{\pi}{3} + \pi n$,
$\theta = \frac{2\pi}{3} + \pi n$, n an integer

28. $\theta = \frac{\pi}{2} + \pi n, \theta = \frac{\pi}{6} + 2\pi n$,
$\theta = \frac{5\pi}{6} + 2\pi n$, n an integer

29. $\theta = \frac{\pi}{2} + \pi n, \theta = \frac{\pi}{4} + \frac{\pi}{2}n$, n an integer

30. $x = 1.2 + \pi n, x = 1.1 + \pi n$, n an integer

31. $x = -0.9 + 2\pi n, x = 4.0 + 2\pi n$,
n an integer

32. $x = 1.3 + \pi n, x = 2.0 + \pi n$, n an integer

33. $x = 1.3 + 2\pi n, x = 5.0 + 2\pi n$,
n an integer

Exercises 9.4

In **1 – 21**, find all the solutions in $[0, 2\pi)$.

1. $2 \sin x + 1 = 0$
2. $2 \cos x = 1$
3. $\sqrt{3} \sec x = 2$

4. $\tan x - \sqrt{3} = 0$
5. $2 \sec x = 1$
6. $2 \sin x - \sqrt{3} = 0$

7. $2 \cos \left(x - \frac{\pi}{8}\right) = 1$
8. $\csc \left(x + \frac{\pi}{12}\right) = 2$
9. $4 \sin^2 \theta = 1$

10. $4 \cos^2 \theta = 3$
11. $\tan^2 x - 3 = 0$
12. $3 \sec^2 x = 4$

13. $\sec^2 x - 2 = 0$
14. $\tan^2 x - 1 = 0$
15. $\sin^2 x - 3 \cos^2 x = 0$

16. $(2 \sin x - 1)(2 \cos x - \sqrt{2}) = 0$
17. $\sin^2 x = \frac{1 + \cos x}{2}$

18. $\sec^2 x - 1 = \tan x$
19. $(\cot^2 x - 3)(3 \cot^2 x - 1) = 0$

20. $\sin^4 x = \frac{1}{16}$
21. $3 \tan^4 x - 10 \tan^2 x + 3 = 0$

In **22 – 29**, find the general solution for θ, where θ is in radians.

22. $2 \tan \theta - \sec^2 \theta = 0$
23. $2 \cos \theta \sin \theta + \cos \theta = 0$

24. $2 \cos^2 \theta + 3 \cos \theta + 1 = 0$
25. $2 \sin^2 \theta - 5 \sin \theta + 2 = 0$

26. $\cot^2 \theta - \csc \theta - 1 = 0$
27. $\tan^3 \theta - 3 \tan \theta = 0$

28. $2 \cot^2 \theta \sin \theta = \cot^2 \theta$
29. $\csc^2 \theta \cot \theta = 2 \cot \theta$

In **30 – 33**, find the general solution for x to the nearest tenth of a radian.

30. $\tan x + 6 \cot x - 5 = 0$
31. $2(\sin^2 x - 1) = \sin x$

32. $\tan^2 x - 2 \tan x - 10 = 0$
33. $\cos^2 x + 3 \cos x - 1 = 0$

In **34 – 41**, solve for x to the nearest degree, $x \in [0°, 360°)$.

34. $\cos 2x = \frac{\sqrt{3}}{2}$
35. $\cot 3x = 1$
36. $\csc 4x = 2$

37. $2 \sin 2x = \sqrt{3}$
38. $\sin \frac{x}{2} = 1$
39. $\cos \frac{x}{2} = 0.4$

40. $\cos^2 2x - \sin^2 2x = 2 \sin^2 2x$
41. $2 \sin^2 3x = 3 \cos 3x$

42. Solve $\csc x + \cot x = 1$ for $x \in [0, 2\pi)$.
Hint: Square both sides and check for extraneous solutions.

43. Solve $\sin x + 1 = \cos x$ for $x \in [0, 2\pi)$.

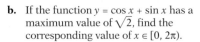

44. a. Use the graph to determine the value of x for which $\cos x + \sin x$ has a maximum value.

b. If the function $y = \cos x + \sin x$ has a maximum value of $\sqrt{2}$, find the corresponding value of $x \in [0, 2\pi)$.

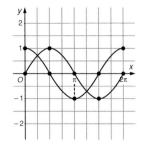

A ball is hit with an initial velocity v, and its path makes an angle θ with the horizontal. An equation of motion that describes the horizontal distance d that the ball travels before returning to its initial height is:

$$d = \frac{1}{32} v^2 \sin 2\theta$$

In **45 – 46**, use this equation of motion and find θ to the nearest degree.

45. A golf ball is hit at ground level at an angle of θ with the horizontal and has an initial velocity of 120 feet per second. The ball travels 100 feet horizontally before it hits the ground.

46. For a home run in a certain stadium, a baseball must travel a horizontal distance of 400 feet. The ball is hit at an initial velocity of 150 feet per second.

47. An inclined ramp is used to transport debris from the third floor of a building that is being remodeled. The bottom of the ramp empties into a truck that is 15 feet from the base of the building. The third floor is 30 feet above the truck bed. By how much will the angle of a similar ramp have to increase to deliver debris from the fourth floor window, which is 10 feet higher than the third floor window? Answer to the nearest degree.

48. The legs of a right triangle are in the ratio 3 to 1. Verify that if the trisection points of the shorter leg are joined to the opposite vertex, the resulting segments *do not* trisect the angle.

34. $15°, 165°, 195°,$ $345°$

35. $15°, 75°, 135°,$ $195°, 255°, 315°$

36. $7.5°, 37.5°, 97.5°,$ $127.5°, 187.5°,$ $217.5°, 277.5°,$ $307.5°$

37. $30°, 60°, 210°, 240°$

38. $180°$

39. $132°$

40. $15°, 75°, 105°,$ $165°, 195°, 255°,$ $285°, 345°$

41. $20°, 100°, 140°,$ $220°, 260°, 340°$

42. $\dfrac{\pi}{2}$

43. $0, \dfrac{3\pi}{2}$

44. a. $\dfrac{\pi}{4}$

b. $\dfrac{\pi}{4}$

45. $6°$

46. $17°$

47. $6°$

48.

$\tan a = \dfrac{x}{9x} = \dfrac{1}{9}$

$\tan (a + b) = \dfrac{2x}{9x} = \dfrac{2}{9}$

$\tan (a + b + c) = \dfrac{3x}{9x} = \dfrac{1}{3}$

$a \approx 6.3°$

$(a + b) \approx 12.5°$

$(a + b + c) \approx 18.4°$

$a + b \approx 12.5°$

$6.3° + b \approx 12.5°$

$b \approx 6.2°$

$a + b + c \approx 18.4°$

$12.5° + c \approx 18.4°$

$c \approx 5.9°$

$a, b,$ and c are not congruent. Therefore, the angle is *not* trisected.

9.5 *The Inverses of the Trigonometric Functions*

In Chapter 3, you learned that a function has an inverse only if it is a $1-1$ correspondence. For the domain of definition of the six trigonometric functions, none are $1-1$. For a codomain of the set of real numbers, only the tangent and cotangent functions are onto. This can be seen from the graphs of the functions.

For example, using your graphing calculator, a computer program, or ordinary graph paper, display the graphs of $y = \sin x$ and $y = 0.5$ for values of x from -6.5 to 6.5 and values of y from -1.5 to 1.5. Is $y = \sin x$ a $1-1$ correspondence? You can see that it is not $1-1$ since there are many values of x for which y is 0.5, that is, points of intersection of the two graphs. You can also see that it is not onto since there are no values of x that correspond to values of y that are greater than 1 or less than -1.

For the same sets of values for x and y given above, display the graphs of $y = \cos x$ and $y = 0.5$. What conclusions can you draw from these graphs? You can see from the graphs that, like $y = \sin x$, $y = \cos x$ is neither $1-1$ nor onto.

Finally, using the same sets of values for x and y, display the graphs of $y = \tan x$ and $y = 0.5$. Are your conclusions from these graphs the same as for $y = \sin x$ and $y = \cos x$? Note that, like $y = \sin x$ and $y = \cos x$, $y = \tan x$ is not $1-1$. However, for every element of the codomain, there is at least one element of the domain. Although you have used only a limited codomain, this is true for all real numbers. Therefore, the function $y = \tan x$ is onto, but is not a $1-1$ correspondence since it is not $1-1$.

Although the trigonometric functions with the domain and range for which they are defined have no inverses, a subset of each function may be chosen that is a $1-1$ correspondence and that, therefore, has an inverse. For a given function, *any* subset that includes each element of the range exactly once would be a $1-1$ function. It is convenient, however, to restrict each function to a fixed set of values for that function.

Recall that to form the equation of the inverse of a function, x and y are interchanged. The equation of the inverse of $y = \sin x$ is, therefore, $x = \sin y$. If this equation is solved for y, the notation **$y = \arc \sin x$** is used. The function $y = \arc \sin x$ means that y is the length of an arc of the unit circle whose sine is x. An alternate notation, often used on calculators, is $\sin^{-1} x$ where the -1 exponent means the function inverse, not the multiplicative inverse $\dfrac{1}{\sin x}$. For the sine function, all values of the range occur once in the interval $\left[-\frac{\pi}{2}, \frac{\pi}{2}\right]$. Therefore, this is a convenient interval to use to define the inverse of the sine function.

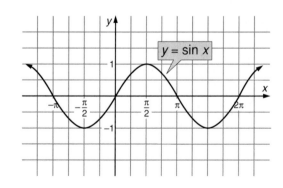

On your graphing calculator, computer, or graph paper, display the sine graph and its inverse. Use values of x from –1.6 to 1.6 and values of y from –1.6 to 1.6.

Compare the graphs. Observe that the graph of $y = \text{arc sin } x$ (or $y = \sin^{-1} x$) is the reflection over the line $y = x$ of the graph of the chosen subset of $y = \sin x$.

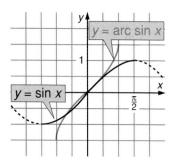

For the cosine function, all values of the range occur once in the interval $[0, \pi]$. Therefore, this is a convenient interval to use to define the inverse of the cosine function.

On your graphing calculator, computer, or graph paper, display the cosine graph and its inverse. Use values of x from –1.6 to 3.2 and values of y from –1.6 to 3.2.

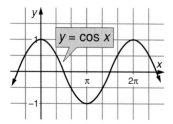

On the graphing calculator, in order to include all of the graph of $y = \text{arc cos } x$ (or $y = \cos^{-1} x$), it is necessary to include more of the cosine graph than that subset used to form the inverse function. Compare the two graphs. Observe that the graph of $y = \text{arc cos } x$ is the reflection over the line $y = x$ of the graph of the chosen subset of $y = \cos x$.

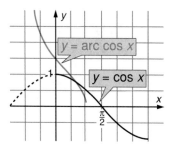

For the tangent function, all values of the range occur once in the interval $\left(-\frac{\pi}{2}, \frac{\pi}{2}\right)$. Therefore, this is a convenient interval to use to define the inverse of the tangent function.

On your graphing calculator, computer, or graph paper, display the tangent graph and its inverse. Use values of x from –1.6 to 1.6 and values of y from –1.6 to 1.6.

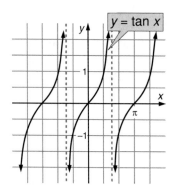

Note that since the range of the tangent function is the set of all real numbers, only a portion of the graphs can be displayed. Compare the two graphs. Observe that the graph of $y = \text{arc tan } x$ (or $y = \tan^{-1} x$) is the reflection over the line $y = x$ of the chosen subset of $y = \tan x$.

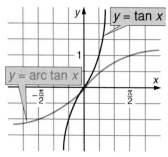

To form an inverse function for $y = \csc x$, use the same subset as is used for the reciprocal function, $y = \sin x$, except that 0 is omitted, that is:

$$-\frac{\pi}{2} \le x \le \frac{\pi}{2}, \text{ where } x \ne 0$$

Since $\sin 0 = 0$, the reciprocal is undefined.

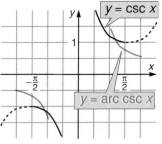
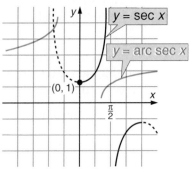

To form the inverse function for $y = \sec x$, use the same subset as is used for the reciprocal function, $y = \cos x$, except that $\frac{\pi}{2}$ is omitted, that is:

$$0 \le x \le \pi, \text{ where } x \ne \frac{\pi}{2}$$

Since $\cos \frac{\pi}{2} = 0$, the reciprocal is undefined.

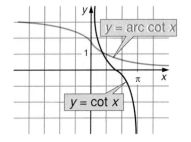

To form the inverse function for $y = \cot x$, the same interval as the reciprocal function, $y = \tan x$, could be used. However, by using the interval $0 < x < \pi$, the function $y = \text{arc cot } x$ is a continuous function.

Remember that the choice of specific intervals is arbitrary. In some books, the range of arc csc x, arc sec x, and arc cot x differ from what is given here.

Example 1

If possible, evaluate the expression. If the expression has no value, explain why not.

a. arc sin $\frac{\sqrt{3}}{2}$ **b.** arc sec (-2) **c.** arc csc $\frac{1}{2}$

Solution

a. If $y = $ arc sin $\frac{\sqrt{3}}{2}$, then $\sin y = \frac{\sqrt{3}}{2}$ where $y \in \left[-\frac{\pi}{2}, \frac{\pi}{2}\right]$.

Answer: arc sin $\frac{\sqrt{3}}{2} = \frac{\pi}{3}$

b. If $y = $ arc sec (-2), then $\sec y = -2$ where $y \in [0, \pi]$, and

$\cos y = -\frac{1}{2}$, where $y \in [0, \pi]$.

Answer: arc sec $(-2) = \frac{2\pi}{3}$

c. If $y = $ arc csc $\frac{1}{2}$, then $\csc y = \frac{1}{2}$, or $\sin y = 2$ where $y \in \left[-\frac{\pi}{2}, \frac{\pi}{2}\right]$.

There is no angle whose sine is 2.

Answer: arc csc $\frac{1}{2}$ has no value.

Earlier, you learned to find the measure of an acute angle of a right triangle by using trigonometry. For example, for the equation sin y = 0.75, you can find y by using a table of trigonometric function values or a calculator. The sequence of calculator keys is:

.75 **INV** **SIN** or **SIN⁻¹** .75

Note that the inverse of the sine function is used since you are looking for an angle whose sine is 0.75. Similarly, if you look for the closest value to 0.75 in the sine column of a table of trigonometric function values and determine that it corresponds to approximately 49°, you are also using the inverse of the sine function.

Example 2

 Use a calculator or the trigonometric table to find each value to the nearest hundredth of a radian.

a. arc sin 0.61   **b.** arc tan (–2.7)

Solution

a. Let y = arc sin 0.61. Then sin y = 0.61 for $y \in \left[-\frac{\pi}{2}, \frac{\pi}{2}\right]$. In the sine column of the trigonometric table, 0.6111 is the value closest to 0.61. The corresponding angle is 0.6574 radians, or, to the nearest hundredth, 0.66 radians.

If a calculator is used, be sure that the calculator is in radian mode and use the following key sequence, or the equivalent on your calculator.

.61 **INV** **SIN**

Answer: To the nearest hundredth, arc sin 0.61 = 0.66 radians.

b. Let y = arc tan (–2.7). Then tan y = –2.7 where $y \in \left(-\frac{\pi}{2}, \frac{\pi}{2}\right)$. In the tangent column of the trigonometric table, the closest value to 2.7 is 2.699, which corresponds to 1.2159 or approximately 1.22 radians. However, since the argument of the function is –2.7 and tan (–x) = –tan x, the value of y is –1.22, which is in the range of the arc tan function.

If a calculator is used, be sure that it is in radian mode, then use the following key sequence.

–2.7 **INV** **TAN**

Answer: To the nearest hundredth, arc tan (–2.7) = –1.22 radians.

Note: A trigonometric table displays only positive values, while a calculator allows both positive and negative input and displays both positive and negative results.

In Chapter 3, it was established that for a function f that has an inverse, $f(f^{-1}(x)) = x$ and $f^{-1}(f(x)) = x$. This principle can be applied to trigonometric functions and their inverses. For example, if $x \in [-1, 1]$ and $y \in \left[-\frac{\pi}{2}, \frac{\pi}{2}\right]$:

$$\cos (\text{arc cos } x) = x \quad \text{or} \quad \text{arc cos } (\cos y) = y$$

This inverse-function property can be used to solve certain types of problems.

Example 3

If possible, find the value of each of the following.

a. sin (arc sin 0.3)   **b.** cos (arc cos 1.2)

c. arc sec $\left(\sec \frac{5\pi}{4}\right)$ **d.** tan $\left(\text{arc sin}\left(-\frac{\sqrt{2}}{2}\right)\right)$

Solution

a. Since 0.3 is in the domain of y = arc sin x, and the value of arc sin 0.3 is in the domain of the sine function, the inverse property may be used.

Answer: sin (arc sin 0.3) = 0.3

b. The domain of y = arc cos x does not contain 1.2.

Answer: cos (arc cos 1.2) has no value.

c. Although $\frac{5\pi}{4}$ is not in the restricted domain of the secant function, it does have a value that is in the domain of the arc secant function. Therefore, although the inverse property cannot be used, a value can be found.

$$\sec \frac{5\pi}{4} = -\sqrt{2}$$

$$\text{arc sec}\left(-\sqrt{2}\right) = \frac{3\pi}{4}$$

Answer: arc sec $\left(\sec \frac{5\pi}{4}\right) = \frac{3\pi}{4}$

d. Let y = arc sin $\left(-\frac{\sqrt{2}}{2}\right)$. Then sin $y = -\frac{\sqrt{2}}{2}$ where $y \in \left[-\frac{\pi}{2}, \frac{\pi}{2}\right]$.

Therefore, $y = -\frac{\pi}{4}$ and tan $\left(-\frac{\pi}{4}\right) = -1$.

Answer: tan $\left(\text{arc sin}\left(-\frac{\sqrt{2}}{2}\right)\right) = -1$

Example 4

Find the value of cos $\left(\text{arc tan}\left(-\frac{1}{2}\right)\right)$.

Solution

Let θ = arc tan $\left(-\frac{1}{2}\right)$. Then tan $\theta = -\frac{1}{2}$, where $\theta \in \left(-\frac{\pi}{2}, \frac{\pi}{2}\right)$. Since tan θ is negative, θ is the measure of an angle in quadrant IV. By the Pythagorean Theorem, the length of the hypotenuse of the reference triangle is $\sqrt{5}$. Therefore, cos $\theta = \frac{2}{\sqrt{5}}$ or $\frac{2\sqrt{5}}{5}$.

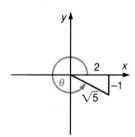

Answer: cos $\left(\text{arc tan}\left(-\frac{1}{2}\right)\right) = \frac{2\sqrt{5}}{5}$

Exercises 9.5

In **1 – 6**, find the domain and range of the function.

1. $y = \text{arc sin } x$

2. $y = \text{arc cos } x$

3. $y = \text{arc tan } x$

4. $y = \text{arc cot } x$

5. $y = \text{arc sec } x$

6. $y = \text{arc csc } x$

In **7 – 21**, find the value of the expression.

7. $\text{arc cos } \frac{1}{2}$

8. $\text{arc sin } \frac{\sqrt{3}}{2}$

9. $\text{arc tan } 1$

10. $\text{arc sec } \sqrt{2}$

11. $\text{arc csc } \frac{2\sqrt{3}}{3}$

12. $\text{arc sin } \left(-\frac{1}{2}\right)$

13. $\text{arc cos } \left(-\frac{\sqrt{3}}{2}\right)$

14. $\text{arc tan } (-1)$

15. $\text{arc cot } (-1)$

16. $\text{arc tan } \sqrt{3}$

17. $\text{arc sin } 0$

18. $\text{arc cos } 0$

19. $\text{arc tan } 0$

20. $\text{arc cot } (-\sqrt{3})$

21. $\text{arc sin } \left(-\frac{\sqrt{2}}{2}\right)$

In **22 – 33**, use a calculator or the trigonometric table to find the value of the expression to the nearest hundredth of a radian. If no such value exists, state why not.

22. $\text{arc sin } 0.6$

23. $\text{arc cos } 0.41$

24. $\text{arc tan } 1.7$

25. $\text{arc sec } 3.25$

26. $\text{arc csc } 0.6$

27. $\text{arc cos } (-0.14)$

28. $\text{arc cot } 6.8$

29. $\text{arc sin } (-0.1)$

30. $\text{arc sin } 2$

31. $\text{arc cos } (-1.7)$

32. $\text{arc sin } \pi$

33. $\text{arc sec } 0$

In **34 – 42**, if possible, find the value of the expression. Otherwise, write "no value".

34. $\cos \left(\text{arc cos } \frac{1}{2}\right)$

35. $\sin (\text{arc sin } 0.12)$

36. $\tan (\text{arc tan } (-1.2))$

37. $\sec (\text{arc sec } 2.714)$

38. $\text{arc sin } \left(\sin \left(-\frac{\pi}{2}\right)\right)$

39. $\text{arc sin } \left(\sin \frac{3\pi}{4}\right)$

40. $\text{arc tan } \left(\tan \frac{\pi}{4}\right)$

41. $\text{arc cot } \left(\cot \frac{\pi}{6}\right)$

42. $\text{arc cos } \left(\cos \left(-\frac{\pi}{4}\right)\right)$

In **43 – 51**, find the value of the expression without using a calculator or trigonometric table.

43. $\tan \left(\text{arc sin } \frac{2}{3}\right)$

44. $\sin \left(\text{arc tan } \frac{1}{7}\right)$

45. $\sin (\text{arc tan } (-3))$

46. $\csc \left(\text{arc sin } \left(-\frac{1}{4}\right)\right)$

47. $\cot \left(\text{arc cos } \frac{1}{10}\right)$

48. $\cos \left(\text{arc cot } \left(-\frac{2}{5}\right)\right)$

49. $\text{arc cos } \left(\tan \frac{\pi}{4}\right)$

50. $\text{arc tan } \left(\sin \frac{\pi}{2}\right)$

51. $\text{arc cos } \left(\sin \left(-\frac{\pi}{6}\right)\right)$

Function	Domain	Range
1. $y = \text{arc sin } x$	$[-1, 1]$	$\left[-\frac{\pi}{2}, \frac{\pi}{2}\right]$
2. $y = \text{arc cos } x$	$[-1, 1]$	$[0, \pi]$
3. $y = \text{arc tan } x$	R	$\left(-\frac{\pi}{2}, \frac{\pi}{2}\right)$
4. $y = \text{arc cot } x$	R	$(0, \pi)$
5. $y = \text{arc sec } x$	$R \neq (-1, 1)$	$\left[0, \frac{\pi}{2}\right) \cup \left(\frac{\pi}{2}, \pi\right]$
6. $y = \text{arc csc } x$	$R \neq (-1, 1)$	$\left[-\frac{\pi}{2}, 0\right) \cup \left(0, \frac{\pi}{2}\right]$

7. $\frac{\pi}{3}$

8. $\frac{\pi}{3}$

9. $\frac{\pi}{4}$

10. $\frac{\pi}{4}$

11. $\frac{\pi}{3}$

12. $-\frac{\pi}{6}$

13. $\frac{5\pi}{6}$

14. $-\frac{\pi}{4}$

15. $\frac{3\pi}{4}$

16. $\frac{\pi}{3}$

17. 0

18. $\frac{\pi}{2}$

19. 0

20. $\frac{5\pi}{6}$

21. $-\frac{\pi}{4}$

22. 0.64

23. 1.15

24. 1.04

25. 1.26

26. no value; arc sec is not defined between -1 and 1

27. 1.71

28. 0.15

29. -0.10

30. no value; domain of arc sin is $[-1, 1]$

Answers 9.5

31. no value; domain of arc cos is [–1, 1]

32. no value; domain of arc sin is [–1, 1] and $\pi \approx 3.14$

33. no value; arc sec is not defined in (–1, 1)

34. $\frac{1}{2}$ **35.** 0.12

36. –1.2 **37.** 2.714

38. $-\frac{\pi}{2}$ **39.** $\frac{\pi}{4}$

40. $\frac{\pi}{4}$ **41.** $\frac{\pi}{6}$

42. $\frac{\pi}{4}$ **43.** $\frac{2\sqrt{5}}{5}$

44. $\frac{\sqrt{2}}{10}$ **45.** $-\frac{3\sqrt{10}}{10}$

46. –4 **47.** $\frac{\sqrt{11}}{33}$

48. $-\frac{2\sqrt{29}}{29}$ **49.** 0

50. $\frac{\pi}{4}$ **51.** $\frac{2\pi}{3}$

52. 0.10 **53.** 0.99

54. –1.33 **55.** –1.00

56. 7.47 **57.** 1.06

58. $x = \cos y$

59. $x = \cos 4y$

60. $x = 2 \tan 3y$

61. $x = \sin y - 4$

62. $x = \frac{1}{2}(\tan 2y + 6)$

63. $x = \cos 2y + 3$

64. a. odd **b.** neither **c.** odd

65.

sec (arc tan 2x) = y

$$y = \sqrt{1 + 4x^2}$$

66. $\frac{4\sqrt{2}}{9}$

67. 0

68. $\frac{32\sqrt{2} + 9\sqrt{15}}{-7}$

 In **52 – 57**, use a calculator or trigonometric table to find the value of the expression to two decimal places.

52. sin (arc tan 0.1) **53.** cos (arc sin 0.14) **54.** cot (arc cos (–0.8))

55. tan (arc sin (–0.7071)) **56.** sec (arc tan (–7.4)) **57.** csc (arc sec 3)

In **58 – 63**, solve the equation for x in terms of y.

58. $y = \text{arc cos } x$ **59.** $4y = \text{arc cos } x$ **60.** $3y = \text{arc tan } \frac{1}{2}x$

61. $y = \text{arc sin } (x + 4)$ **62.** $2y = \text{arc tan } (2x - 6)$ **63.** $y = \frac{1}{2} \text{arc cos } (x - 3)$

64. Classify each function as even, odd, or neither.

a. $y = \text{arc sin } x$ **b.** $y = \text{arc cos } x$ **c.** $y = \text{arc tan } x$

65. Write an algebraic expression in terms of x that is equal to sec (arc tan 2x). *Hint:* Form a right triangle such that one acute angle has a tangent of 2x.

66. Find the value of $\sin \left(2 \text{ arc sin } \frac{1}{3}\right)$. *Hint:* Use the formula $\sin 2x = 2 \sin x \cos x$.

67. Evaluate:
$$\cos \left(\text{arc tan } \sqrt{3} + \text{arc cos } \frac{\sqrt{3}}{2}\right)$$

68. Use the formula for tan (A + B) to evaluate:
$$\tan \left(\text{arc sin } \frac{1}{3} + \text{arc cos } \frac{1}{4}\right)$$

69. Refer to the diagram.

a. Express each of the following functions in terms of x.

(1) tan α

(2) tan θ

(3) tan β

Hint: $\beta = \theta - \alpha$

b. Using arc tan, express β in terms of x.

c. To the nearest degree, find β:

(1) if $x = 3$

(2) if $x = 10$

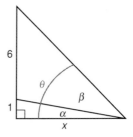

---

69. a. (1) $\tan \alpha = \dfrac{1}{x}$

(2) $\tan \theta = \dfrac{7}{x}$

(3) $\tan \beta = \dfrac{6x}{x^2 + 7}$

b. $\beta = \text{arc tan} \left(\dfrac{6x}{x^2 + 7}\right)$

c. (1) $\beta = 48°$

(2) $\beta = 29°$

9.6 Solving Oblique Triangles Using the Law of Sines

You have seen how trigonometry can be used to find unknown measures in right triangles. In this section, you will see how trigonometry can be used to find unknown measures in triangles that do not contain a right angle.

In the figures below, an acute and an obtuse triangle each have an altitude $\overline{BD}$ of length h drawn to side $\overline{AC}$, intersecting $\overleftrightarrow{AC}$ at D, to form triangles ADB and BCD. The capital letter at each vertex is matched by a lowercase letter representing the length of the opposite side.

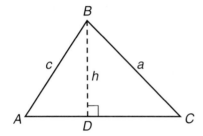   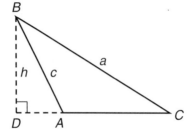

Both figures show triangles ADB and BCD.

In $\triangle ADB$: $\sin A = \dfrac{h}{c}$

In $\triangle BCD$: $\sin C = \dfrac{h}{a}$

Note: $\angle BAD$ and $\angle BAC$ are supplementary. Since $\sin x = \sin (\pi - x)$, it is true that $\sin A = \dfrac{h}{c}$ for both angles with vertex at A.

Therefore, $h = c \sin A$ and $h = a \sin C$, and $c \sin A = a \sin C$.

This last equation is equivalent to the proportion:

(1) $\qquad \dfrac{a}{\sin A} = \dfrac{c}{\sin C}$

By constructing the altitude of length h to side $\overline{AB}$ of each triangle, a similar process will show that:

(2) $\qquad \dfrac{a}{\sin A} = \dfrac{b}{\sin B}$

Combining (1) and (2) yields:

$$\dfrac{a}{\sin A} = \dfrac{b}{\sin B} = \dfrac{c}{\sin C} \qquad \textbf{\textit{The Law of Sines}}$$

☐ **Law of Sines:** *For any triangle, the ratio of the measure of a side to the sine of the angle opposite that side is a constant.*

Other equivalent forms of the Law of Sines include:

$$\frac{\sin A}{a} = \frac{\sin B}{b} = \frac{\sin C}{c} \quad \text{and} \quad \frac{a}{b} = \frac{\sin A}{\sin B}$$

Note that the second of these equations states that in a triangle, the ratio of the lengths of two sides is equal to the ratio of the sines of the angles opposite these sides.

The Law of Sines is used to solve a triangle when the measures of two angles and the side opposite one of them (AAS) are known.

Example 1 _____

Find the unknown measures of the sides and angles in the triangle shown.

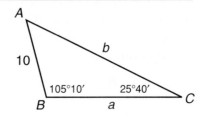

Solution

To find b, use the form of the Law of Sines that contains b and the three known measures of the triangle.

$$\frac{b}{\sin B} = \frac{c}{\sin C}$$

$$b = \frac{c \sin B}{\sin C}$$

$$= \frac{10 \sin 105° \ 10'}{\sin 25° \ 40'}$$

If a calculator is used to evaluate, be sure to save all intermediate results and to round the final result. To work with the trigonometric table, use the reference angle 74° 50′.

$$b \approx 22.3, \text{ to three significant digits}$$

Since the sum of the measures of the angles of a triangle is 180°:

$$m\angle A = 180° - m\angle B - m\angle C$$

$$= 180° - 105° \ 10' - 25° \ 40' = 49° \ 10'$$

To find the length of side a, again apply the Law of Sines.

$$\frac{a}{\sin A} = \frac{c}{\sin C}$$

$$a = \frac{c \sin A}{\sin C}$$

$$= \frac{10 \sin 49° \ 10'}{\sin 25° \ 40'} \approx 17.5$$

Answer: $a \approx 17.5, b \approx 22.3$ and $m\angle A = 49° \ 10'$

The Law of Sines can also be used to solve a triangle when the measures of two angles and the included side (ASA) are given.

Example 2 _____

Two observers standing 50 feet apart along the shoreline of a lake observe a boat in the water. If their lines of sight make angles of 27° and 55° with the shoreline, how far, to the nearest foot, is each observer from the boat?

Solution

Since the measures of two angles and the included side of the triangle (ASA) are known, the Law of Sines can be used. To establish a ratio between the length of a side and the sine of the angle opposite that side, first find the third angle of the triangle.

$$m\angle C = 180 - m\angle A - m\angle B$$
$$= 180 - 27 - 55 = 98°$$

Find b, the distance from observer A to the boat.

$$\frac{c}{\sin C} = \frac{b}{\sin B}$$

$$b = \frac{c \sin B}{\sin C}$$

$$= \frac{50 \sin 55°}{\sin 98°} \approx 41$$

Find a. Since b is an approximate value, it is not used in the calculation.

$$\frac{c}{\sin C} = \frac{a}{\sin A}$$

$$a = \frac{c \sin A}{\sin C}$$

$$= \frac{50 \sin 27°}{\sin 98°} \approx 23$$

Answer: Observer A is about 41 feet and observer B is about 23 feet from the boat.

The Ambiguous Case

Recall that the congruence of two sides and the angle opposite one of them to the corresponding parts of another triangle (SSA) is not sufficient to prove congruence for two triangles. Similarly, if the lengths of two sides of a triangle and the measure of the angle opposite one of them are known, the triangle may not be uniquely determined. That is, given the measures of two sides and an angle opposite one of them as parts of a proposed triangle, there are three possible outcomes.

1. It may not be possible to form such a triangle.
2. There may be two possible triangles.
3. There may be a uniquely-determined triangle.

This situation is often referred to as the *ambiguous case*.

Suppose a, b, and m∠A are known. To construct any triangle at all, the side of length a must be long enough to intersect the side of length c.

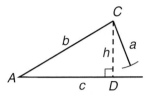

The shortest distance from C to the side of length c is CD (or h), the length of the perpendicular from C to that side. Therefore, a must be greater than or equal to CD.

$$\sin A = \frac{h}{b}$$

$$h = b \sin A$$

If $a < b \sin A$, no triangle can be formed. For example, if $a = 3$, $b = 7$, and m∠$A = 30°$, the length h of the perpendicular segment from C to the opposite side of the proposed triangle is determined as follows.

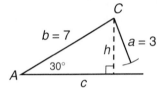

$$h = 7 \sin 30°$$

$$h = 7\left(\tfrac{1}{2}\right)$$

$$h = 3\tfrac{1}{2}$$

Since a is less than h, no triangle is possible.

If $a = 3\tfrac{1}{2}$, then $a = h$ and a right triangle is formed.

Now suppose that A and b remain the same, and $3\tfrac{1}{2} < a < 7$. Then, the side of length a could intersect the side of length c in two places, thus forming either an acute triangle or an obtuse triangle.

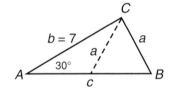

The two remaining cases are $a > 7$ and $a = 7$.

If $a > 7$, then only one triangle is possible.

If $a = 7$, then an isosceles triangle is formed.

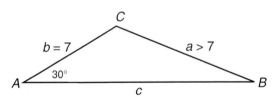

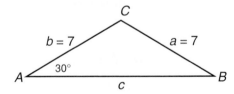

The previous discussion can be generalized for ∠A as any acute angle, and is summarized as follows.

Given a, b, and m∠A, where ∠A is acute:

    1. If $a < b \sin A$, then no triangle is possible.

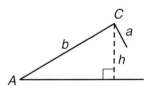

2. If $a = b \sin A$, then only one triangle,
a right triangle, is possible.

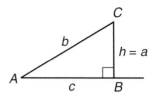

3. If $b \sin A < a < b$, then two triangles are possible.

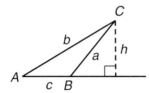

4. If $a \geq b$, then only one triangle is possible.

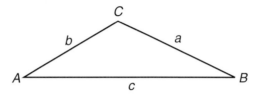

Given a, b, and m$\angle A$, where $\angle A$ is obtuse, then a must be the longest side.

If $a \leq b$, then no
triangle is possible.

If $a > b$, then one
triangle is possible.

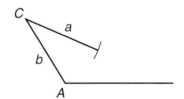

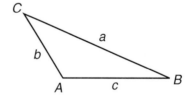

Example 3

If $a = 10$, $b = 4$, and m$\angle A = 40°$, find the number of distinct triangles
that can be drawn.

Solution

Since $\angle A$ is acute and $a > b$, only
one triangle is possible.

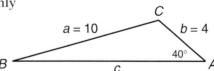

1. $C = 71°, b = 20,$
   $c = 18$

2. $C = 50°, a = 16,$
   $b = 7$

3. $B = 46°, C = 24°,$
   $c = 35$

4. $A = 19°, C = 23°,$
   $c = 15$

5. $a = 10, C = 110°,$
   $c = 16$

6. $a = 2, C = 163°30',$
   $c = 5$

7. $T = 4°, S = 81°,$
   $s = 119$

8. $Z = 172.8°, x = 2,$
   $y = 4$

9. $7\sqrt{2}$

10. $\dfrac{17\sqrt{6}}{2}$

11. $6\sqrt{6}$

12. $8\sqrt{2}$

13. no triangle exists

14. **(1)** $B = 72.7°,$
    $C = 46.3°,$
    $c = 10$

    (2) $B = 107.3°,$
    $C = 11.7°,$
    $c = 3$

15. **(1)** $B = 76.9°,$
    $C = 42.1°,$
    $c = 3$

    (2) $B = 103.1°,$
    $C = 15.9°,$
    $c = 1$

16. no triangle
    possible

17. $S = 42.3°,$
    $T = 22.7°,$
    $t = 26$

18. $Y = 90°,$
    $Z = 30°,$
    $z = 15$

19. 82

20. **a.** $b > \dfrac{20\sqrt{3}}{3}$

    b. $10 < b < \dfrac{20\sqrt{3}}{3}$

    c. $b = \dfrac{20\sqrt{3}}{3}$
    or $b < 10$

21. **a.** $a > 7\sqrt{2}$

    b. $7 < a < 7\sqrt{2}$

    c. $a = 7\sqrt{2}$
    or $a < 7$

22. 1.1 miles

23. 9'8''

24. 52.8'

25. 3.5'

Example 4

If $a = 40$, $b = 65$, and $m\angle A = 26°$, find the measures of the remaining sides and angles of all possible triangles, to the nearest degree or whole number.

Solution

Since $\angle A$ is acute and $b > a$, first determine whether a is greater than h to be sure that a triangle is possible.

$$\sin 26° = \frac{h}{65}$$
$$h = 65 \sin 26°$$
$$h \approx 28.49$$

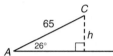

Therefore, $h < a < b$ and two triangles are possible. Thus, it will be necessary to solve for both measures of $\angle B$.

$$\frac{a}{\sin A} = \frac{b}{\sin B}$$
$$\sin B = \frac{b \sin A}{a}$$
$$= \frac{65 \sin 26°}{40}$$
$$= 0.7124$$
$$m\angle B \approx 45° \text{ or } 135°$$

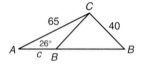

If $m\angle B \approx 45°$:

$$m\angle C = 180 - m\angle A - m\angle B$$
$$\approx 180 - 26 - 45$$
$$\approx 109°$$

$$\frac{a}{\sin A} = \frac{c}{\sin C}$$
$$c = \frac{a \sin C}{\sin A}$$
$$= \frac{40 \sin 109°}{\sin 26°}$$
$$\approx 86$$

If $m\angle B \approx 135°$:

$$m\angle C = 180 - m\angle A - m\angle B$$
$$\approx 180 - 26 - 135$$
$$\approx 19°$$

$$\frac{a}{\sin A} = \frac{c}{\sin C}$$
$$c = \frac{a \sin C}{\sin A}$$
$$= \frac{40 \sin 19°}{\sin 26°}$$
$$\approx 30$$

Answer: There are two possible triangles, both obtuse.

$$m\angle B \approx 135°, m\angle C \approx 19°, c \approx 30$$
$$\text{or } m\angle B \approx 45°, m\angle C \approx 109°, c \approx 86$$

Example 5

In $\triangle RST$, $m\angle R = 110.0°$, $r = 15.0$, and $s = 10.0$. Find the number of distinct triangles that can be drawn and all possible values of $m\angle S$.

Solution

Since $\angle R$ is obtuse and $r > s$, only one triangle is possible.

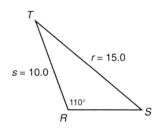

$$\frac{r}{\sin R} = \frac{s}{\sin S}$$

$$\sin S = \frac{s \sin R}{r}$$

$$= \frac{10 \sin 110°}{15}$$

$$\sin S \approx 0.6265$$

$$m\angle S \approx 38.8°$$

Answer: Only one triangle is possible, in which $m\angle S \approx 38.8°$.

Exercises 9.6

In **1 – 8**, find the measures of the remaining sides and angles of the triangle. Where an exact answer cannot be found, give your answers to the nearest degree or whole number.

1. $m\angle A = 21°$, $m\angle B = 88°$, $a = 7$

2. $m\angle A = 105°$, $m\angle B = 25°$, $c = 13$

3. $m\angle A = 110°$, $a = 82$, $b = 63$

4. $m\angle B = 138°$, $a = 12$, $b = 25$

5. $m\angle A = 35°$, $m\angle B = 35°$, $b = 10$

6. $m\angle A = 6° \ 20'$, $m\angle B = 10° \ 10'$, $b = 3.2$

7. $m\angle R = 95°$, $r = 120$, $t = 9$

8. $m\angle X = 2.6°$, $m\angle Y = 4.6°$, $z = 6.1$

In **9 – 12**, without using a calculator or trigonometric table, find the exact measure of the indicated side.

9. If $m\angle A = 30°$, $m\angle B = 45°$, and $a = 7$, then find b.

10. If $m\angle B = 45°$, $m\angle C = 120°$, and $b = 17$, then find c.

11. If $m\angle C = 120°$, $m\angle B = 45°$, and $c = 18$, then find b.

12. If $m\angle A = 135°$, $m\angle C = 30°$, and $c = 8$, then find a.

Answers 9.6

26. 45′

27. 59 mm

28. 33 pounds, 72 pounds

29. 70

30. 10.9

31. 685 feet

32.

$$\sin A = \frac{h}{c}$$

$$\frac{b}{\sin B} = \frac{a}{\sin A}$$

$$b = \frac{a \sin B}{\sin A}$$

Area of $\triangle ABC$

$$= \tfrac{1}{2}bh$$

$$= \tfrac{1}{2} \cdot \frac{a \sin B}{\sin A} \cdot h$$

$$= \tfrac{1}{2} \frac{a \sin B}{h} \cdot h$$
$$\qquad\quad \frac{}{c}$$

$$= \tfrac{1}{2}ac \sin B$$

33.

$$\frac{a}{\sin A} = \frac{b}{\sin B}$$

$$a \cdot \frac{1}{\sin A} = \frac{b}{\sin B}$$

$$a \cdot \csc A = \frac{b}{\sin B}$$

$$b = a \csc A \sin B$$

$$\sin C = \frac{h}{a}$$

$$h = a \sin C$$

Area of $\triangle CBA = \tfrac{1}{2}bh$

$$= \tfrac{1}{2}a \csc A \sin B \cdot a \sin C$$

$$= 0.5 \, a^2 \csc A \sin B \sin C$$

Exercises 9.6 *(continued)*

In **13 – 18**, find the measures of the remaining sides and angles of all possible triangles, if a triangle exists. If there are two possible triangles, solve both. Give lengths to the nearest tenth and angle measures to the nearest tenth of a degree.

13. $m\angle A = 61°$, $a = 3.2$, $b = 13.1$

14. $m\angle A = 61°$, $a = 12$, $b = 13.1$

15. $m\angle A = 61°$, $a = 4.4$, $b = 4.9$

16. $m\angle R = 115°$, $r = 62$, $s = 95$

17. $m\angle R = 115°$, $r = 62$, $s = 46$

18. $m\angle X = 60°$, $y = 30$, $x = 15\sqrt{3}$

19. Find $|\vec{b}|$ to the nearest integer if $|\vec{a}| = 37$, the measure of the angle between $\vec{a}$ and $\vec{b}$ is 50°, and the measure of the angle between $\vec{a}$ and $\vec{a} + \vec{b}$ is 35°.

20. If $m\angle A = 60°$ and $a = 10$,
   a. for what values of b are no triangles possible?
   b. for what values of b are two triangles possible?
   c. for what values of b is only one triangle possible?

21. If $m\angle B = 45°$ and $b = 7$,
   a. for what values of a are no triangles possible?
   b. for what values of a are two triangles possible?
   c. for what values of a is only one triangle possible?

22. Point A on the shoreline of a large lake is $\frac{1}{2}$ mile from a lighthouse at the end of a pier. Point B is 0.6 mile from point A along the shore. A revolving beacon at the top of the lighthouse, after shining on point A, must rotate 18° to shine on point B. How far, to the nearest tenth of a mile, is point B from the lighthouse?

23. An attic room has a sloped ceiling that meets the floor on two sides of the room at angles of 58° and 35°. The two sides where the floor and ceiling meet are 20 feet apart. To the nearest inch, find the height of the room at the highest point.

24. To estimate an inaccessible distance RS, Jack paces off a distance of 210 feet from S at an angle of 108° with $\overline{RS}$ and stations himself at T. He then determines that $\overline{TR}$ and $\overline{ST}$ form an angle of 12.5°. To the nearest tenth of a foot, how long is the distance from R to S?

25. A horizontal door canopy extends 2.0 feet out from a building. It is supported by a brace that extends from the front of the canopy to form an angle of 35° with the building and is fastened at a point on the building 3.0 feet below where the canopy is fastened to the building. To the nearest tenth of a foot, how long is the brace?

34.

$$\frac{a}{\sin A} = \frac{b}{\sin B}$$

$$a \sin B = b \sin A$$

$$a \sin B - b \sin A = b \sin A - a \sin B$$

$$a \sin B - b \sin A + a \sin A - b \sin B = b \sin A - a \sin B + a \sin A - b \sin B$$

$$a (\sin A + \sin B) - b (\sin A + \sin B) = (a + b) \sin A - (a + b) \sin B$$

$$(a - b)(\sin A + \sin B) = (a + b)(\sin A - \sin B)$$

$$\frac{a - b}{a + b} = \frac{\sin A - \sin B}{\sin A + \sin B}$$

Exercises 9.6 *(continued)*

26. A tree grows vertically on a slope that makes an angle of 8° with the horizontal. From a point 75 feet down the slope from the tree, the angle between a line of sight to the top of the tree and the slope measures 28.8°. Find the height of the tree to the nearest foot.

27. A chord in a circle that is subtended by a 50° central angle is 5 cm long. To the nearest millimeter, find the radius of the circle.

28. Two forces act at a point to produce a resultant of 100 pounds, which makes an angle of 27° with the smaller force and an angle of 12° with the larger. Find the magnitude of the applied forces.

29. Two vectors of equal magnitude have an angle of 110° between them. Find, to the nearest integer, the magnitude of each vector if the magnitude of the sum of the vectors is 80.

30. To the nearest tenth, find the shortest distance from D to $\overline{AC}$.

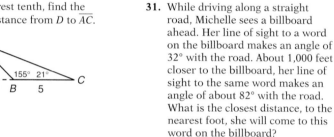

31. While driving along a straight road, Michelle sees a billboard ahead. Her line of sight to a word on the billboard makes an angle of 32° with the road. About 1,000 feet closer to the billboard, her line of sight to the same word makes an angle of about 82° with the road. What is the closest distance, to the nearest foot, she will come to this word on the billboard?

32. The area of a triangle is equal to one-half the product of the lengths of two sides and the sine of the included angle. Use the Law of Sines to prove this. (Recall the area formula $A = \frac{1}{2} \cdot$ base $\cdot$ height.)

33. Verify that given $a, b, c, \text{m}\angle A,$ $\text{m}\angle B,$ and $\text{m}\angle C$ for a triangle, the area can be found by computing $0.5a^2 \csc A \sin B \sin C$.

34. Prove that for any $\triangle ABC$:
$$\frac{a - b}{a + b} = \frac{\sin A - \sin B}{\sin A + \sin B}$$

35. Prove that for any $\triangle ABC$, $\dfrac{a}{\sin A}$ is the length of the diameter of the circle that can be circumscribed about the triangle. *Hint:* Draw diameter BOA' to use right $\triangle A'BC$.

---

35. Draw diameter $\overline{BOA'}$.
Draw $\overline{A'C}$.

Inscribed angles A and A' both intercept $\overset{\frown}{BC}$, therefore $\angle A \cong \angle A'$. $\angle A'CB$ is a right angle.

$a = BC \qquad d = \text{diameter}$

$$\sin A' = \frac{BC}{A'B} = \frac{a}{d}$$

$$\sin A = \sin A'$$

$$\sin A = \frac{a}{d}$$

$$d = \frac{a}{\sin A}$$

9.7 Solving Triangles Using the Law of Cosines

The Law of Sines is effective for solving a triangle when the measures of two angles and the side opposite one of them (AAS) are known. For other circumstances, another formula is needed. Assume that you know the lengths of two sides a and b, and the measure of the included $\angle C$. Place $\triangle ABC$ in the xy-plane with $\overrightarrow{BC}$ coinciding with the x-axis and C at the origin; $\angle C$ may be obtuse or acute.

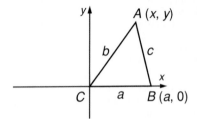
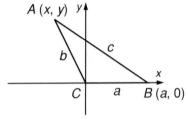

Vertex A has arbitrary coordinates (x, y). Use the distance formula to express c in terms of a and the coordinates of point A, (x, y).

$$c = \sqrt{(a - x)^2 + (-y)^2}$$
$$c^2 = (a - x)^2 + y^2$$

Express x and y in terms of b and $m\angle C$: $x = b \cos C$, $y = b \sin C$

Now substitute the values of x and y in the distance formula and simplify the result.

$$c^2 = (a - b \cos C)^2 + (b \sin C)^2$$
$$= a^2 - 2ab \cos C + b^2 \cos^2 C + b^2 \sin^2 C$$
$$= b^2(\sin^2 C + \cos^2 C) + a^2 - 2ab \cos C$$
$$= b^2(1) + a^2 - 2ab \cos C$$
$$c^2 = a^2 + b^2 - 2ab \cos C \qquad \text{The Law of Cosines}$$

☐ **Law of Cosines:** *In any triangle, the square of the length of any side is equal to the sum of the squares of the lengths of the remaining two sides minus twice the product of those lengths and the cosine of the included angle.*

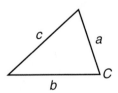
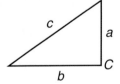
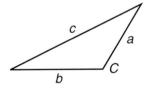

If $\angle C$ is acute, cos C is positive and $-2ab \cos C$ is negative.

$$c^2 < a^2 + b^2$$

If $\angle C$ is a right angle, cos C is 0 and $-2ab \cos C$ is 0.

$$c^2 = a^2 + b^2$$

If $\angle C$ is obtuse, cos C is negative and $-2ab \cos C$ is positive.

$$c^2 > a^2 + b^2$$

The Law of Cosines is used to find the third side of a triangle when the measures of two sides and the included angle (SAS) are known.

Example 1

If in $\triangle ABC$, $a = 6$, $b = 13$, and $m\angle C = 60°$, find c to the nearest integer.

Solution

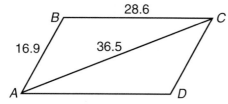

$$c^2 = a^2 + b^2 - 2ab \cos C$$

$$c = \sqrt{a^2 + b^2 - 2ab \cos C}$$

$$= \sqrt{6^2 + 13^2 - 2(6)(13) \cos 60°}$$

$$= \sqrt{36 + 169 - 156\left(\tfrac{1}{2}\right)}$$

$$= \sqrt{205 - 78}$$

$$= \sqrt{127}$$

$$c \approx 11$$

Note: To find the remaining two angles, use the Law of Sines.

The Law of Cosines can also be used to solve a triangle when the lengths of three sides (SSS) are known.

Example 2

The lengths of the sides of a parallelogram are 16.9 cm and 28.6 cm. If the longer diagonal of the parallelogram is 36.5 cm, find, to the nearest ten minutes, the degree measure of the obtuse angle of the parallelogram.

Solution

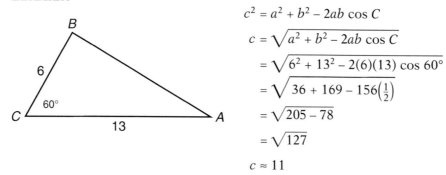

In parallelogram $ABCD$, the longer diagonal, $\overline{AC}$, forms a triangle with the sides of the parallelogram. Since the lengths of the three sides of $\triangle ABC$ are known, use the Law of Cosines to find $m\angle B$.

Write the Law of Cosines in a form that includes $\angle B$.

$$b^2 = a^2 + c^2 - 2ac \cos B$$

$$(36.5)^2 = (16.9)^2 + (28.6)^2 - 2(16.9)(28.6) \cos B$$

$$\cos B = \frac{(36.5)^2 - (16.9)^2 - (28.6)^2}{-2(16.9)(28.6)}$$

$$= \frac{228.68}{-966.68}$$

$$\cos B \approx -0.2366$$

Since $\cos B$ is negative, $\angle B$ is obtuse.

Note: A calculator will return the degree measure of $\angle B$ in decimal form as $103.68596°$. To change the decimal part of this degree measure to minutes, subtract 103 from the entry on the screen and then multiply the resulting decimal by 60.

Answer: $m\angle B \approx 103°40'$

Recall that in the ambiguous case, SSA is given and it must be determined whether one, two, or no triangles are possible for the given dimensions. If SSA is given, the Law of Cosines can be used to determine the possible lengths for the third side of the triangle.

Example 3

In $\triangle ABC$, $a = 10$, $b = 7$, and $m\angle B = 20°$. Determine the possible lengths of c, to the nearest tenth.

Solution

Use the Law of Cosines beginning with b.

$$b^2 = a^2 + c^2 - 2ac \cos B$$

$$7^2 = 10^2 + c^2 - 2(10)c \cos 20°$$

$$c^2 = 7^2 - 10^2 + (20 \cos 20°)c$$

or $\quad c^2 - (20 \cos 20°)c + 51 = 0$ 　　This is a quadratic equation in c that can be solved by the quadratic formula.

$$c = \frac{20 \cos 20° \pm \sqrt{(20 \cos 20°)^2 - 4(1)(51)}}{2(1)}$$

$$\approx \frac{18.79 \pm \sqrt{149.2088}}{2}$$

$$\approx \frac{18.79 \pm 12.21}{2}$$

$$\approx 15.5 \text{ or } 3.3$$

Answer: $c \approx 15.5$ or $c \approx 3.3$

Note: Two possible values for c were found, indicating that two triangles are possible for the given dimensions. If one of the results had been negative, then only one triangle would have been possible. If the roots of the quadratic had been imaginary, then no triangle would have been possible.

Exercises 9.7

In **1 – 8**, find the measures of the remaining sides and angles of the triangle to the nearest degree or whole number.

1. m∠A = 30°, b = 20, c = 14

2. m∠C = 150°, a = 4, b = 5

3. m∠B = 60°, c = 10, a = 18

4. m∠A = 120°, b = 3, c = 3

5. r = 7, s = 8, t = 9

6. x = 12, y = 15, z = 10

7. r = 16.4, s = 4.2, t = 19.6

8. a = 5.2, b = 7.3, c = 9.4

In **9 – 14**, find the measure of the part of the triangle indicated by x, to the nearest tenth or to the nearest tenth of a degree.

9.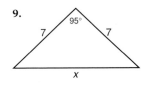

10.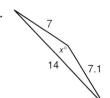

11.

12.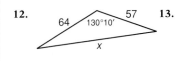

13.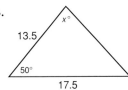

14.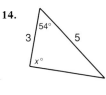

In **15 – 20**, find the possible lengths for the third side of the triangle. Where necessary, give your answer to the nearest tenth.

15. a = 6, b = 8, m∠B = 29°

16. r = 3, s = 4, m∠S = 5°

17. x = 10, y = 8, m∠Y = 20°

18. a = 7, b = 3, m∠A = 60°

19. $r = \frac{1}{10}$, $s = \frac{1}{4}$, m∠S = 10°

20. a = 10.6, b = 6.7, m∠B = 30°

In **21 – 23**, find, to the nearest degree, the measure of the angle between $\vec{a}$ and $\vec{b}$.

21. $|\vec{a}| = 3.7$, $|\vec{b}| = 4.2$, $|\vec{a} + \vec{b}| = 7.1$

22. $|\vec{a}| = 7$, $|\vec{b}| = 9$, $|\vec{a} + \vec{b}| = 4$

23. $|\vec{a}| = 37$, $|\vec{b}| = 92$, $|\vec{a} + \vec{b}| = 67$

Answers

1. a = 11, B = 108°, C = 42°

2. c = 9, A = 13°, B = 17°

3. b = 16, C = 34°, A = 86°

4. a = 5, B = 30°, C = 30°

5. R = 48°, S = 58°, T = 74°

6. X = 53°, Y = 85°, Z = 42°

7. R = 36°, S = 9°, T = 135°

8. A = 33°, B = 51°, C = 96°

9. 10.3

10. 166.3°

11. 157.4°

12. 109.8

13. 80.5°

14. 89.1°

15. 12.7

16. 7.0

17. 16.6 or 2.2

18. 8

19. 0.3

20. 13.3 or 5.1

21. 52°

22. 155°

23. 142°

24. acute, $A = 83°$

25. obtuse, $C = 110°$

26. obtuse, $R = 136°$

27. obtuse, $Y = 110°$

28. a. 353 miles
   b. 312 miles

29. 43 miles

30. 143 pounds

31. 1,044 meters

Exercises 9.7 *(continued)*

In **24 – 27**, indicate whether the triangle is acute or obtuse, and then find the measure of the largest angle to the nearest degree.

24. $a = 6, b = 5, c = 4$

25. $a = 4.5, b = 6.4, c = 9$

26. $r = 70, s = 28.2, t = 46.8$

27. $x = \frac{1}{10}, y = \frac{1}{5}, z = \frac{1}{7}$

28.

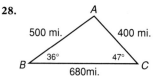

Three cities, A, B, and C, are located at the vertices of a triangle with the distances between them as shown in the figure. If m$\angle B = 36°$ and m$\angle C = 47°$, how far from A, to the nearest mile, will a plane be one hour after passing B if it is flying directly from B to C at a constant speed of

   a. 600 m.p.h.   **b.** 300 m.p.h

29. Two ships leave the same port at noon and sail away from each other at an angle of 105°. If one travels at 8 m.p.h. and the other at 10 m.p.h., how far apart will they be at 3 P.M.? Answer to the nearest mile.

30. Find the magnitude of the resultant force if two forces of 82 pounds and 76 pounds are applied at a point so that the measure of the angle between them is 50°.

31. In order to measure the distance across a swamp, a surveyor paces off 500 meters in a straight line from one end of the swamp. He then paces off 700 meters in a straight line at an angle of 120° to the first line. How far is it across the swamp, to the nearest meter?

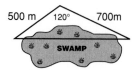

32. Use the Law of Cosines to show that:

   a. $1 + \cos A = \dfrac{(a + b + c)(-a + b + c)}{2bc}$

   b. $1 - \cos A = \dfrac{(a + b - c)(a - b + c)}{2bc}$

The semiperimeter s of triangle ABC is given by the formula $s = \dfrac{a + b + c}{2}$. This quantity appears in a number of mathematical formulas.

33. Use the half-angle formulas to show that:

   a. $\cos \dfrac{A}{2} = \pm\sqrt{\dfrac{s(s - a)}{bc}}$

   b. $\sin \dfrac{A}{2} = \pm\sqrt{\dfrac{(s - b)(s - c)}{bc}}$

   c. $\tan \dfrac{A}{2} = \pm\sqrt{\dfrac{(s - b)(s - c)}{s(s - a)}}$

32. a.
$$a^2 = b^2 + c^2 - 2bc \cos A$$
$$-2bc \cos A = a^2 - b^2 - c^2$$
$$2bc \cos A = b^2 + c^2 - a^2$$
$$\cos A = \frac{b^2 + c^2 - a^2}{2bc}$$
$$\cos A = \frac{b^2 + 2bc + c^2 - a^2 - 2bc}{2bc}$$
$$\cos A = \frac{(b + c)^2 - a^2}{2bc} - \frac{2bc}{2bc}$$
$$\cos A = \frac{[(b + c) + a][(b + c) - a]}{2bc} - 1$$
$$\cos A = \frac{(b + c + a)(b + c - a)}{2bc} - 1$$
$$1 + \cos A = \frac{(a + b + c)(-a + b + c)}{2bc}$$

b.
$$a^2 = b^2 + c^2 - 2bc \cos A$$
$$-2bc \cos A = a^2 - b^2 - c^2$$
$$-\cos A = \frac{a^2 - (b^2 + c^2)}{2bc}$$
$$-\cos A = \frac{a^2 - (b^2 - 2bc + c^2) - 2bc}{2bc}$$
$$-\cos A = \frac{a^2 - (b - c)^2}{2bc} - \frac{2bc}{2bc}$$
$$-\cos A = \frac{[a + (b - c)][a - (b - c)]}{2bc} - 1$$
$$1 - \cos A = \frac{(a + b - c)(a - b + c)}{2bc}$$

33. a. $\cos \dfrac{A}{2} = \pm \sqrt{\dfrac{1 + \cos A}{2}}$ $\qquad 1 + \cos A = \dfrac{(a + b + c)(-a + b + c)}{2bc}$, from Exercise **32a.**

$$= \pm \sqrt{\frac{\dfrac{(a + b + c)(-a + b + c)}{2bc}}{2}}$$

$$= \pm \sqrt{\frac{\dfrac{(a + b + c)}{2} \cdot \dfrac{(a + b + c - 2a)}{2}}{bc}}$$

$$= \pm \sqrt{\frac{\dfrac{(a + b + c)}{2} \cdot \left[\dfrac{(a + b + c)}{2} - \dfrac{2a}{2} \right]}{bc}}$$

$$= \pm \sqrt{\frac{s(s - a)}{bc}}$$

33. b. $\sin \dfrac{A}{2} = \pm \sqrt{\dfrac{1 - \cos A}{2}}$

$1 - \cos A = \dfrac{(a + b - c)(a - b + c)}{2bc}$,

from Exercise **32b.**

c. $\tan \dfrac{A}{2} = \dfrac{\sin \frac{A}{2}}{\cos \frac{A}{2}}$

$= \pm \sqrt{\dfrac{\frac{(a + b - c)(a - b + c)}{2bc}}{2}}$

$= \pm \sqrt{\dfrac{\frac{(a + b - c)}{2} \cdot \frac{(a - b + c)}{2}}{bc}}$

$= \pm \sqrt{\dfrac{\left[\frac{(a + b + c) - 2c}{2}\right] \cdot \left[\frac{(a + b + c) - 2b}{2}\right]}{bc}}$

$= \pm \sqrt{\dfrac{(s - c)(s - b)}{bc}}$

$= \pm \sqrt{\dfrac{(s - b)(s - c)}{bc}}$

$= \dfrac{\pm \sqrt{\dfrac{(s - b)(s - c)}{bc}}}{\pm \sqrt{\dfrac{s(s - a)}{bc}}}$

$= \pm \sqrt{\dfrac{\frac{(s - b)(s - c)}{bc}}{\frac{s(s - a)}{bc}}}$

$= \pm \sqrt{\dfrac{(s - b)(s - c)}{s(s - a)}}$

34. a. 63.7 feet

  b. 87°

35.

$s = \dfrac{a + b + c}{2}$

$2s = a + b + c$

$a + b = 2s - c$

$b + c = 2s - a$

$a + c = 2s - b$

$A = \frac{1}{2}bc \sin A$

$A^2 = \frac{1}{4}b^2c^2 \sin^2 A$

$A^2 = \frac{1}{4}b^2c^2(1 - \cos^2 A)$

$A^2 = \frac{1}{4}b^2c^2(1 + \cos A)(1 - \cos A)$

Note: On the left side of the equations, A = Area.

See Exercise **32**.

$A^2 = \frac{1}{4}b^2c^2 \dfrac{(a + b + c)(-a + b + c)}{2bc} \cdot \dfrac{(a + b - c)(a - b + c)}{2bc}$

$A^2 = \dfrac{b^2c^2}{4} \cdot \dfrac{(a + b + c)(-a + b + c)(a + b - c)(a - b + c)}{4b^2c^2}$

$A^2 = \dfrac{(a + b + c)(-a + b + c)(a + b - c)(a - b + c)}{16}$

$A^2 = \dfrac{a + b + c}{2} \cdot \dfrac{(-a + 2s - a)(2s - c - c)(2s - b - b)}{8}$

$A^2 = s \cdot \dfrac{(2s - 2a)(2s - 2c)(2s - 2b)}{8}$

$A^2 = s \cdot \dfrac{2(s - a)2(s - c)2(s - b)}{8}$

$A^2 = s(s - a)(s - b)(s - c)$

$A = \sqrt{s(s - a)(s - b)(s - c)}$

Exercises 9.7 *(continued)*

34. The infield of a baseball field is in the shape of a square with sides 90 feet long. Home plate and the three bases are at the vertices of the square. The pitcher's mound is 60 feet from home plate on the diagonal from home plate to second base.

a. Find, to the nearest tenth of a foot, how far the pitcher must throw the ball to first base.

b. Use your answer from part **a** to determine the degree measure of the angle, to the nearest degree, through which the pitcher must turn if she is facing home plate and wants to throw to first.

Note: The pitcher's mound is *not* at the center of the square.

The following formula, known as Heron's formula, can be used to find the area of a triangle when only the lengths of the sides are known.

$$A = \sqrt{s(s-a)(s-b)(s-c)} \quad \text{where } s = \frac{a+b+c}{2}$$

35. Heron's formula can be proved by using the area formula, $A = \frac{1}{2}bc \sin A$ for $\triangle ABC$, together with the Law of Cosines. Complete the proof below.

$$A = \tfrac{1}{2}bc \sin A$$

$$A^2 = \tfrac{1}{4}b^2c^2 \sin^2 A$$

$$A^2 = \tfrac{1}{4}b^2c^2(1 - \cos^2 A)$$

Hint: Factor, and use the formulas in Exercise 32.

36. Use Heron's formula to find the area, to the nearest tenth, of each triangle whose dimensions are given.

a. $a = 13, b = 14, c = 15$

b. $x = 16.2, y = 18.6, z = 21$

c. $a = 0.8, b = 0.64, c = 1.2$

d. $a = \sqrt{8}, b = \sqrt{2}, c = \frac{3\sqrt{2}}{2}$

37. Use Heron's formula to find the length of the altitude to the longest side of a triangle with sides of lengths 13, 14, and 15 cm.

36. a. 84
   b. 144.7
   c. 0.2
   d. 1.5

37. 11.2

9.8 Simple Harmonic Motion

The oscillation of a spring caused by a weight applied to its end; the vertical movement of a piston due to the rotation of a crankshaft in an internal combustion engine; the rise and fall of a buoy tied to a string that bobs up and down because of the tide—all of these are examples of *simple harmonic motion*. Functions with the form $f(t) = a \sin \theta t$ or $f(t) = a \cos \theta t$, where $a, \theta, t \in R$, provide mathematical models to describe such motion.

When a weight w is suspended from a spring, the point at which the weight hangs without any motion is the *resting position* of the spring. After the weight is pulled downward and released, causing the spring to alternately stretch and contract, the vertical position of the weight at time t is $f(t)$, where $f(t)$ is positive when the weight is above the resting position and negative when the weight is below the resting position. The value of $f(t)$ over a range of values of t takes the form $f(t) = a \sin bt$ or $f(t) = a \cos bt$, where a is the amplitude and $\dfrac{2\pi}{b}$ is the period. The values a and b are dependent on the weight w, on the elasticity of the spring, and on the force used to set the spring in motion. A strong spring vibrates more rapidly, that is, $f(t)$ has a smaller period. Similarly, a greater force placed on the weight causes the spring to oscillate over a greater range of values for $f(t)$, that is, $f(t)$ has a greater amplitude.

Example 1

A weight hanging from the bottom of a spring is pulled downward and released, causing the spring to oscillate 5 cm above and below its resting position. It takes 2 seconds for the spring to return to its resting position. Assuming that there is no air resistance to slow down the spring, write an equation that describes the position, $f(t)$, of the weight with respect to time.

Solution

Since, at the resting position, $t = 0$ and $f(t) = 0$, use the general equation $f(t) = a \sin bt$. Given that the spring stretches to a maximum value of 5 cm means that the amplitude is 5. Since the time the spring takes to complete one oscillation is 2 seconds, the period $\dfrac{2\pi}{b}$ is 2, or $b = \pi$.

Answer: $f(t) = 5 \sin \pi t$

Example 2

A spring with a weight attached to its end is vibrating up and down according to the equation

$$f(t) = 10 \cos \frac{\pi}{4} t$$

where, at t seconds, $f(t)$ is the directed distance, in centimeters, of the weight from the resting position of the spring.

a. Find the maximum displacement of the weight above its resting position.

b. Find the period of oscillation.

c. Find the frequency of oscillation.

d. Find the position of the weight at $t = 0$ seconds, $t = 4$ seconds, $t = 6$ seconds, and $t = 8$ seconds.

Solution

a. The maximum displacement is found by determining the amplitude, which is 10. Therefore, the weight will move 10 cm above and below its resting position.

b. Since the period is $\frac{2\pi}{b}$, where $b = \frac{\pi}{4}$, the period of oscillation is $\frac{2\pi}{\frac{\pi}{4}} = 8$. Thus, it takes 8 seconds for one complete oscillation.

c. The frequency of oscillation is the reciprocal of the period. Thus, the frequency is $\frac{1}{8}$; that is, $\frac{1}{8}$ of an oscillation occurs every second.

d. At $t = 0$: $f(t) = 10 \cos \frac{\pi}{4}(0) = 10 \cos 0 = 10(1) = 10$ cm

At $t = 4$: $f(t) = 10 \cos \frac{\pi}{4}(4) = 10 \cos \pi = 10(-1) = -10$ cm

At $t = 6$: $f(t) = 10 \cos \frac{\pi}{4}(6) = 10 \cos \frac{3\pi}{2} = 10(0) = 0$ cm

At $t = 8$: $f(t) = 10 \cos \frac{\pi}{4}(8) = 10 \cos 2\pi = 10(1) = 10$ cm

If the position is some value other than the minimum, the maximum, or 0 when $t = 0$, then the function describing the position of the weight is modified by adding a constant to t and becomes

$$f(t) = a \sin b(t + c) \text{ or } f(t) = a \cos b(t + c)$$

where c may be positive or negative. Recall that if c is given in radians and it is positive, then the graph of the function $f(t) = a \sin b(t + c)$ is a translation of c radians to the left of the graph of $f(t) = a \sin bt$.

Example 3

A wheel whose center is at the origin and whose radius is 4 feet is rotating at an angular velocity of 2 radians per second. At $t = 0$, point P on the wheel has rotated through an angle of $\frac{\pi}{3}$ radians.

a. Write an equation of the form $f(t) = a \sin b(t + c)$ to describe the y-coordinate of P with respect to time t.

b. Sketch a graph of the function in part **a.**

Solution

a. At time $t = 0$, the coordinates of P are $\left(4 \cos \frac{\pi}{3}, 4 \sin \frac{\pi}{3}\right)$. After t seconds, P has rotated through an angle of $2t$ radians. Therefore, the coordinates of P after t seconds are:

$$\left(4 \cos \left(2t + \tfrac{\pi}{3}\right), 4 \sin \left(2t + \tfrac{\pi}{3}\right)\right)$$

Since the maximum y-coordinate of P is 4, the amplitude of the function describing the y-coordinate of P is 4.

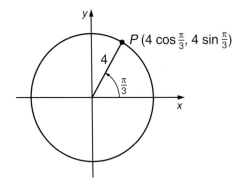

Answer: An equation of the y-coordinate is:

$$f(t) = 4 \sin \left(2t + \tfrac{\pi}{3}\right) \text{ or } f(t) = 4 \sin 2\left(t + \tfrac{\pi}{6}\right)$$

b.

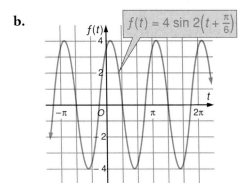

$$f(t) = 4 \sin 2\left(t + \tfrac{\pi}{6}\right)$$

Exercises 9.8

In **1 – 12**, state whether the given function can or cannot be used to represent a simple harmonic motion.

1. $f(t) = \cos t$

2. $f(t) = 2 \sin 3t$

3. $f(t) = -5 \cos 2t$

4. $f(t) = 6 \sin \frac{1}{2}t$

5. $f(t) = 3 \tan t$

6. $f(t) = 4 \sin \frac{\pi}{3}t$

7. $f(t) = 2 \cot \pi t$

8. $f(t) = 2 \sin (2t - 3)$

9. $f(t) = 10 \cos 3\left(t - \frac{\pi}{6}\right)$

10. $f(t) = 6 \sin \pi\left(2t + \frac{1}{3}\right)$

11. $f(t) = 3 + \sin \pi t$

12. $f(t) = 3 \cos \left(\frac{\pi}{3} - 2t\right)$

13. The amperage of an alternating electric current is a simple harmonic motion with respect to time. If $A(t)$ is the amperage at t seconds and $A(t) = 10 \sin 120\pi t$, find the following.

  a. The number of amperes when the current is strongest.

  b. The period of the current.

  c. The frequency of the current.

  d. To the nearest tenth, the number of amperes when:
    (1) $t = 0$ **(2)** $t = 0.005$ **(3)** $t = 0.04$

14. A weight attached to a spring is raised 2 inches and then released, causing the spring to stretch and contract 2 inches above and below its resting position.

  a. If it takes the spring $\frac{1}{3}$ of a second for one complete oscillation, write an equation of a function $f(t)$ to describe the position of the weight relative to its resting position.

  b. What is the position of the weight at:
    (1) $t = 1$ **(2)** $t = \frac{4}{3}$ **(3)** $t = \frac{1}{4}$

  c. Write an equation to describe the position of the weight if at $t = 0$, it is at a point 2 inches below the resting point of the spring.

15. As the ocean waves pass, a buoy bobs up and down, following a simple harmonic motion. At $t = 2$ seconds, the buoy passes a point midway between its lowest and highest vertical position, and then arrives at its lowest position after traveling down 18 inches. As it rises, it again passes its middle position, when $t = 3.5$ seconds.

Let the equation describing the position of the buoy be of the form $f(t) = a \sin b(t + c)$, where $f(t)$ is the directed distance from the middle position.

  a. Find the values of a, b, and c.

  b. Write an equation for the function $f(t)$.

  c. Sketch the function.

  d. If the equation for the buoy is $f(t) = a \cos b(t + c)$, find a, b, and c.

Answers 9.8

1. Yes

2. Yes

3. Yes

4. Yes

5. No

6. Yes

7. No

8. Yes

9. Yes

10. Yes

11. No

12. Yes

13. a. 10

  b. $\frac{1}{60}$ sec.

  c. 60

  d. (1) 0

    (2) 9.5

    (3) 5.9

14. a. $f(t) = 2 \sin 6\pi t$

  b. (1) at equilibrium

    (2) at equilibrium

    (3) 2″ below equilibrium

  c. $f(t) = -2 \sin 6\pi t$

15. a. $a = 18, b = \frac{2\pi}{3}, c = -0.5$

  b. $f(t) = 18 \sin \frac{2\pi}{3}(t - 0.5)$

  c.

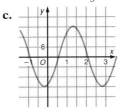

  d. $a = 18, b = \frac{2\pi}{3}, c = -1.25$

Exercises 9.8 (continued)

The sounds heard from a musical instrument are created by changes in air pressure caused by the vibration of parts of the instrument, such as strings, a reed, or the skin of a drum. These sound waves follow a simple harmonic motion.

16. In the equation $D(t) = 0.01 \cos 220\pi t$, $D(t)$ represents the change in the ambient air pressure in dynes at time t due to the vibration. If standard atmospheric pressure is 10^{10} dynes, how many dynes of pressure reach the ear of the listener at $t = 1.2$ seconds?

17. The region enclosed by a circle whose radius is 3 is turning in a counterclockwise direction with the origin as the center of rotation. A point P on the circumference has coordinates $(0, -3)$ at $t = 0$.

   a. If the velocity of P is 4 radians per second, write an expression for $f(t)$, the y-coordinate of P at time t in seconds.

   b. Suppose a 7-inch-long string is suspended vertically from P and a weight is attached to the other end to keep it taut. Modify the equation of part **a** so that $f(t)$ represents the y-coordinate of the point where the weight is attached to the string.

18. A Ferris wheel rotates at 3 revolutions per minute in a counterclockwise direction. Let the wheel have a diameter of 20 feet, and a person sitting at the 3-o'clock position be 10 feet above the ground at $t = 0$.

   a. Write an expression for the person's height above the ground (in feet) at time t (in minutes).

   b. How high off the ground is the person at:

     (1) $t = 2\frac{1}{2}$ seconds

     (2) $t = 5$ seconds

     (3) $t = 4$ minutes

19. A piston in an internal combustion engine travels up and down, approximating simple harmonic motion, and rotates a crankshaft below. At $t = 0$ seconds, the piston is at its lowest point with the crankshaft rotating at 1 revolution per second. The radius of the crankshaft is 5 inches and the piston travels through a distance of 10 inches from its highest to its lowest point.

   a. Write an equation to approximate $f(t)$, the height of the piston above its lowest point, in terms of the time t.

   b. Find the least value of t when the piston is 5 inches above its lowest point.

   c. How many seconds does it take for the piston to move from its highest to its lowest point?

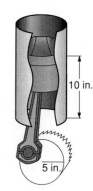

10 in.

5 in.

---

16. $10^{10} + 0.01$ dynes

17. a. $f(t) = 3 \sin 4\left(t + \frac{3\pi}{8}\right)$

   b. $f(t) = 3 \sin 4\left(t + \frac{3\pi}{8}\right) - 7$

18. a. $h = 10 \sin 6\pi t + 10$

   b. (1) $(5\sqrt{2} + 10)$ feet

     (2) 20 feet

     (3) 10 feet

19. a. $f(t) = 5 \sin 2\pi\left(t - \frac{1}{4}\right) + 5$

   b. $\frac{1}{4}$ sec. **c.** $\frac{1}{2}$ sec.

Note: Identities and formulas are true for all values for which the particular functions are defined.

Basic Identities

- Reciprocal Relationships

$$\sin \theta = \frac{1}{\csc \theta} \qquad \cos \theta = \frac{1}{\sec \theta} \qquad \tan \theta = \frac{1}{\cot \theta}$$

- Quotient Relationships

$$\tan \theta = \frac{\sin \theta}{\cos \theta} \qquad \cot \theta = \frac{\cos \theta}{\sin \theta}$$

- Pythagorean Relationships

$$\sin^2 \theta + \cos^2 \theta = 1 \qquad \sec^2 \theta - \tan^2 \theta = 1 \qquad \csc^2 \theta - \cot^2 \theta = 1$$

- Complementary Relationships

$$\sin \theta = \cos \left(\frac{\pi}{2} - \theta\right) \qquad \sec \theta = \csc \left(\frac{\pi}{2} - \theta\right) \qquad \tan \theta = \cot \left(\frac{\pi}{2} - \theta\right)$$

- Symmetry Relationships

$$\cos (-x) = \cos x \qquad \sin (-x) = -\sin x \qquad \tan (-x) = -\tan x$$
$$\sec (-x) = \sec x \qquad \csc (-x) = -\csc x \qquad \cot (-x) = -\cot x$$

- To prove a trigonometric identity, work on one side of the identity and make substitutions by using the fundamental identities.

The Sum and Difference Formulas

$$\cos (A - B) = \cos A \cos B + \sin A \sin B \qquad \cos (A + B) = \cos A \cos B - \sin A \sin B$$
$$\sin (A - B) = \sin A \cos B - \cos A \sin B \qquad \sin (A + B) = \sin A \cos B + \cos A \sin B$$
$$\tan (A - B) = \frac{\tan A - \tan B}{1 + \tan A \tan B} \qquad \tan (A + B) = \frac{\tan A + \tan B}{1 - \tan A \tan B}$$

Double-Angle Formulas

$$\sin 2A = 2 \sin A \cos A$$
$$\cos 2A = \cos^2 A - \sin^2 A \quad \text{or} \quad 1 - 2 \sin^2 A \quad \text{or} \quad 2 \cos^2 A - 1$$
$$\tan 2A = \frac{2 \tan A}{1 - \tan^2 A}$$

Half-Angle Formulas

$$\sin \frac{\theta}{2} = \pm \sqrt{\frac{1 - \cos \theta}{2}} \qquad \cos \frac{\theta}{2} = \pm \sqrt{\frac{1 + \cos \theta}{2}} \qquad \tan \frac{\theta}{2} = \pm \sqrt{\frac{1 - \cos \theta}{1 + \cos \theta}}$$

Sum-to-Product Formulas

$$\sin (A + B) - \sin (A - B) = 2 \cos A \sin B \qquad \sin (A + B) + \sin (A - B) = 2 \sin A \cos B$$
$$\cos (A + B) - \cos (A - B) = -2 \sin A \sin B \qquad \cos (A + B) + \cos (A - B) = 2 \cos A \cos B$$
$$\sin x - \sin y = 2 \cos \tfrac{1}{2}(x + y) \sin \tfrac{1}{2}(x - y) \qquad \sin x + \sin y = 2 \sin \tfrac{1}{2}(x + y) \cos \tfrac{1}{2}(x - y)$$
$$\cos x - \cos y = -2 \sin \tfrac{1}{2}(x + y) \sin \tfrac{1}{2}(x - y) \qquad \cos x + \cos y = 2 \cos \tfrac{1}{2}(x + y) \cos \tfrac{1}{2}(x - y)$$

Trigonometric Equations

- A trigonometric equation can be solved within a given domain or for the general angle.
- The trigonometric identities can be used to solve trigonometric equations containing more than one trigonometric function.

Inverse Trigonometric Functions

- Subsets of the six trigonometric functions that are 1 – 1 functions are formed by restricting the domain appropriately for each of the six functions. These 1 – 1 functions have inverse functions. The function restrictions are:

$y = \sin x:\ -\frac{\pi}{2} \le x \le \frac{\pi}{2}$ $\qquad$ $y = \csc x:\ -\frac{\pi}{2} \le x \le \frac{\pi}{2}, x \ne 0$

$y = \cos x:\quad 0 \le x \le \pi$ $\qquad$ $y = \sec x:\quad 0 \le x \le \pi, x \ne \frac{\pi}{2}$

$y = \tan x:\ -\frac{\pi}{2} < x < \frac{\pi}{2}$ $\qquad$ $y = \cot x:\quad 0 < x < \pi$

- The inverse functions are:

$y = \text{arc sin } x$ or $\sin^{-1} x$, where $x \in [-1, 1]$

$y = \text{arc csc } x$ or $\csc^{-1} x$, where $x \in (-\infty, -1]$ or $x \in [1, \infty)$

$y = \text{arc cos } x$ or $\cos^{-1} x$, where $x \in [-1, 1]$

$y = \text{arc sec } x$ or $\sec^{-1} x$, where $x \in (-\infty, -1]$ or $x \in [1, \infty)$

$y = \text{arc tan } x$ or $\tan^{-1} x$, where $x \in R$

$y = \text{arc cot } x$ or $\cot^{-1} x$, where $x \in R$

For any $\triangle ABC$ with sides of lengths a, b, and c opposite angles A, B, and C, respectively:

- Law of Sines: In any triangle, the ratio of the measure of a side to the sine of the angle opposite that side is a constant.

$$\frac{a}{\sin A} = \frac{b}{\sin B} = \frac{c}{\sin C}$$

- The Law of Sines can be used to solve a triangle if the degree measures of two angles and the length of one side of a triangle are known (AAS or ASA).

- If the lengths of two sides and the degree measure of an angle opposite one of these sides (SSA) are known, there may not be a triangle with these dimensions or there may be one or two triangles with these dimensions. This is called the ambiguous case.

Given a, b, and m$\angle A$ with $\angle A$ acute:

1. $a < b \sin A$, no triangle is possible.
2. $a = b \sin A$, one right triangle.
3. $b \sin A < a < b$, two triangles are possible.
4. $a \geq b$, only one triangle is possible.

Given a, b, and m$\angle A$ with $\angle A$ obtuse:

1. $a \leq b$, no triangle is possible.
2. $a > b$, one triangle is possible.

- Law of Cosines: In any triangle, the square of the length of any side is equal to the sum of the squares of the lengths of the remaining two sides minus twice the product of those lengths and the cosine of the included angle.

$$c^2 = a^2 + b^2 - 2ab \cos C$$

- If $\angle C$ is the largest angle of $\triangle ABC$ and $\angle C$ is obtuse, then $c^2 > a^2 + b^2$; if $\angle C$ is a right angle, then $c^2 = a^2 + b^2$; and if $\angle C$ is acute, then $c^2 < a^2 + b^2$.

- The Law of Cosines can be used to solve a triangle if the lengths of three sides of the triangle (SSS) or the lengths of two sides and the degree measure of the angle between them (SAS) are known.

- The Law of Cosines can be used to determine the possible lengths of the third side of a triangle when the lengths of two sides and the degree measure of the angle opposite one of them are known (the ambiguous case).

In **1 – 12**, one side of the identity is transformed, to be made identical to the other side.

1. $\dfrac{1}{\sin \theta} \cdot \cos \theta =$

$\qquad \dfrac{\cos \theta}{\sin \theta} =$

$\qquad \cot \theta =$

2. $\cos x \, \dfrac{\sin x}{\cos x} =$

$\qquad \sin x =$

3. $\dfrac{\cos^2 y}{\sin^2 y} \cdot \dfrac{1}{\cos^2 y} =$

$\qquad \dfrac{1}{\sin^2 y} =$

$\qquad \csc^2 y =$

$\qquad \cot^2 y + 1 =$

4. $\dfrac{1}{\sin x} - \sin x =$

$\qquad \dfrac{1 - \sin^2 x}{\sin x} =$

$\qquad \dfrac{\cos^2 x}{\sin x} =$

5. $\dfrac{1}{\dfrac{1}{\cos \theta} - \cos \theta} =$

$\qquad \dfrac{\cos \theta}{1 - \cos^2 \theta} =$

$\qquad \dfrac{\cos \theta}{\sin^2 \theta} =$

6. $\sec^2 x - 1 + \cot^2 x + 1 =$

$\qquad \sec^2 x + \cot^2 x =$

Simple Harmonic Motion

- Functions of the form $f(t) = a \sin bt$ or $a \cos bt$, where a, b, and $t \in R$, are models for simple harmonic motion.
- Examples of simple harmonic motion include the oscillation of a spring, the motion of a pendulum, or the vertical change in position of a point on a rotating circle, such as the height of a particular seat on a Ferris wheel.

Chapter 9 *Review Exercises*

In **1 – 12**, prove the given identity.

1. $\csc \theta \cos \theta = \cot \theta$

2. $\cos x \tan x = \sin x$

3. $\cot^2 y \sec^2 y = \cot^2 y + 1$

4. $\csc x - \sin x = \dfrac{\cos^2 x}{\sin x}$

5. $\dfrac{1}{\sec \theta - \cos \theta} = \dfrac{\cos \theta}{\sin^2 \theta}$

6. $\tan^2 x + \csc^2 x = \sec^2 x + \cot^2 x$

7. $\dfrac{\cos \theta}{1 - \tan \theta} + \dfrac{\sin \theta}{1 - \cot \theta} = \cos \theta + \sin \theta$

8. $\cos^2 y \sin^2 y = \sin^2 y - \sin^4 y$

9. $\tan \frac{1}{2}x = \csc x - \cot x$

10. $\sin x \sin 3x = \frac{1}{2}(\cos 2x - \cos 4x)$

11. $\dfrac{1 + \cos 2y}{1 - \cos 2y} = \cot^2 y$

12. $\dfrac{\csc \theta - \sec \theta}{\cot \theta - \tan \theta} = \dfrac{\cos \theta - \sin \theta}{\cos 2\theta}$

13. Express $\sin \left(\frac{\pi}{3} - \theta \right)$ in terms of $\sin \theta$ and $\cos \theta$.

14. Express $\tan \left(\theta + \frac{\pi}{4} \right)$ in terms of $\tan \theta$.

In **15 – 24**, find the exact value of the expression without using a trigonometric table or a calculator.

15. $\sin 15°$

16. $\cot 195°$

17. $\tan 285°$

18. $\csc (-105°)$

19. $\sec \frac{7\pi}{12}$

20. $\csc \left(-\frac{11\pi}{12} \right)$

21. $\tan \frac{7\pi}{12}$

22. $\sin 157°30'$

23. $2 \sin 37.5° \cos 37.5°$

24. $\dfrac{1 - \tan^2 22.5°}{2 \tan 22.5°}$

25. If $\sin x = \frac{1}{3}$ and $\cos y = \frac{1}{5}$ where x is in quadrant II and y is in quadrant IV, find the value of $\sin (x - y)$.

26. If $\sin x = -\frac{1}{\sqrt{5}}$ and $\sin y = \frac{2}{3}$ where x is in quadrant IV and y is in quadrant II, find the value of $\cos (x - y)$.

27. Express $\sin 1 \cos 0.1 + \cos 1 \sin 0.1$ using a single trigonometric function.

28. Simplify: $2 \sin \frac{\pi}{8} \cos \frac{\pi}{8}$

29. Simplify $\cos^2 18° - \sin^2 18°$ without using a trigonometric table or a calculator.

30. Write an equivalent expression for $1 - 2 \sin^2 12°$ as the cosine of an angle.

7.
$$\frac{\cos\theta}{1 - \frac{\sin\theta}{\cos\theta}} + \frac{\sin\theta}{1 - \frac{\cos\theta}{\sin\theta}} =$$

$$\frac{\cos^2\theta}{\cos\theta - \sin\theta} + \frac{\sin^2\theta}{\sin\theta - \cos\theta} =$$

$$\frac{\cos^2\theta}{\cos\theta - \sin\theta} + \frac{-\sin^2\theta}{\cos\theta - \sin\theta} =$$

$$\frac{\cos^2\theta - \sin^2\theta}{\cos\theta - \sin\theta} =$$

$$\frac{(\cos\theta + \sin\theta)(\cos\theta - \sin\theta)}{\cos\theta - \sin\theta} =$$

$$\cos\theta + \sin\theta =$$

8. $(1 - \sin^2 y)\sin^2 y =$

$$\sin^2 y - \sin^4 y =$$

9. $= \dfrac{1}{\sin x} - \dfrac{\cos x}{\sin x}$

$$= \frac{1 - \cos x}{\sin x}$$

$$= \frac{1 - \cos x}{\sqrt{1 - \cos^2 x}}$$

$$= \frac{(1 - \cos x)\sqrt{1 - \cos^2 x}}{1 - \cos^2 x}$$

$$= \frac{(1 - \cos x)\sqrt{1 - \cos^2 x}}{(1 + \cos x)(1 - \cos x)}$$

$$= \frac{\sqrt{1 - \cos^2 x}}{1 + \cos x}$$

$$= \frac{\sqrt{(1 + \cos x)(1 - \cos x)}}{\sqrt{(1 + \cos x)^2}}$$

$$= \frac{\sqrt{(1 + \cos x)(1 - \cos x)}}{\sqrt{(1 + \cos x)(1 + \cos x)}}$$

$$= \sqrt{\frac{(1 + \cos x)(1 - \cos x)}{(1 + \cos x)(1 + \cos x)}}$$

$$= \sqrt{\frac{1 - \cos x}{1 + \cos x}}$$

$$= \tan\tfrac{1}{2}x$$

10. $= \frac{1}{2}\left[-2\sin\frac{1}{2}(6x)\sin\frac{1}{2}(-2x)\right]$

$$= \frac{1}{2}\left[-2\sin 3x\sin(-x)\right]$$

$$= -\sin 3x\,(-\sin x)$$

$$= \sin 3x\sin x$$

$$= \sin x\sin 3x$$

11. $\dfrac{1 + (\cos^2 y - \sin^2 y)}{1 - (\cos^2 y - \sin^2 y)} =$

$$\frac{1 + \cos^2 y - \sin^2 y}{1 - \cos^2 y + \sin^2 y} =$$

$$\frac{\cos^2 y + \cos^2 y}{\sin^2 y + \sin^2 y} =$$

$$\frac{2\cos^2 y}{2\sin^2 y} =$$

$$\frac{\cos^2 y}{\sin^2 y} =$$

$$\cot^2 y =$$

12.
$$\frac{\dfrac{1}{\sin\theta} - \dfrac{1}{\cos\theta}}{\dfrac{\cos\theta}{\sin\theta} - \dfrac{\sin\theta}{\cos\theta}} =$$

$$\frac{\cos\theta - \sin\theta}{\cos^2\theta - \sin^2\theta} =$$

$$\frac{\cos\theta - \sin\theta}{\cos 2\theta} =$$

13. $\dfrac{\sqrt{3}\cos\theta - \sin\theta}{2}$

14. $\dfrac{1 + \tan\theta}{1 - \tan\theta}$

15. $\dfrac{\sqrt{6}-\sqrt{2}}{4}$ **16.** $2+\sqrt{3}$ **17.** $-2-\sqrt{3}$ **18.** $\sqrt{2}-\sqrt{6}$

19. $-(\sqrt{2}+\sqrt{6})$ **20.** $-(\sqrt{2}+\sqrt{6})$ **21.** $-2-\sqrt{3}$ **22.** $\dfrac{\sqrt{2-\sqrt{2}}}{2}$

23. $\dfrac{\sqrt{6}+\sqrt{2}}{4}$ **24.** 1

25. $\dfrac{1-8\sqrt{3}}{15}$ **26.** $\dfrac{-10-2\sqrt{5}}{15}$ **27.** $\sin 1.1$

28. $\dfrac{\sqrt{2}}{2}$ **29.** $\cos 36°$ **30.** $\cos 24°$

31. $\tan^2 \dfrac{5}{2}\theta$ **32.** $\tan \dfrac{3\pi}{4}$ or -1 **33.** $\dfrac{2\sqrt{46}}{25}$

34. $\dfrac{4}{5}$ **35.** $-\dfrac{24}{7}$ **36.** $2 \cos \dfrac{3x}{2} \cos \dfrac{x}{2}$

37. $2 \sin \dfrac{9y}{2} \cos \dfrac{3y}{2}$ **38.** $\dfrac{1}{2}\cos 6a + \dfrac{1}{2}\cos 2a$ **39.** $\dfrac{1}{2}\sin \dfrac{5}{6}x + \dfrac{1}{2}\sin \dfrac{1}{6}x$

40. $-\tan 2x$ **41.**
$$2 \cos 3x \cos x + \cos 6x =$$
$$2 \cos 3x \cos x + 2 \cos^2 3x - 1 =$$
$$2 \cos 3x \,(\cos x + \cos 3x) - 1 =$$
$$2 \cos 3x \,(2 \cos 2x \cos x) - 1 =$$
$$4 \cos x \cos 2x \cos 3x - 1 =$$

42. $\dfrac{2\pi}{3}, \dfrac{4\pi}{3}$ **43.** $\dfrac{4\pi}{3}, \dfrac{5\pi}{3}$ **44.** $\dfrac{\pi}{3}, \dfrac{2\pi}{3}, \dfrac{4\pi}{3}, \dfrac{5\pi}{3}$

45. $0, \dfrac{\pi}{3}, \dfrac{5\pi}{3}$ **46.** $\dfrac{\pi}{8}, \dfrac{3\pi}{8}, \dfrac{5\pi}{8}, \dfrac{7\pi}{8}, \dfrac{9\pi}{8}, \dfrac{11\pi}{8}, \dfrac{13\pi}{8}, \dfrac{15\pi}{8}$ **47.** $\dfrac{2\pi}{3}, \dfrac{4\pi}{3}$

48. $\dfrac{\pi}{6}, \dfrac{5\pi}{6}, \dfrac{\pi}{2}$ **49.** $\theta = \pi k, \theta = \dfrac{\pi}{2} + 2\pi k, k$ an integer **50.** $-\dfrac{\pi}{6}$

51. $-\dfrac{\pi}{4}$ **52.** $\dfrac{\pi}{6}$ **53.** $\dfrac{2\sqrt{5}}{5}$

54. 0 **55.** $\dfrac{1}{2}$ **56.** 0.995

57. $x = \sin 3y$ **58.** $y = \cos x - 2$ **59.** $\dfrac{7}{8}$

31. Simplify: $\dfrac{1 - \cos 5\theta}{1 + \cos 5\theta}$

32. Simplify: $\dfrac{\sin \frac{3\pi}{8} \cos \frac{3\pi}{8}}{\frac{1}{2} - \sin^2 \frac{3\pi}{8}}$

33. If x is in quadrant I and $\cos x = \frac{\sqrt{2}}{5}$, find $\sin 2x$.

34. If x is in quadrant III and $\tan x = \frac{1}{3}$, find $\cos 2x$.

35. If x is in quadrant IV and $\csc x = -\frac{5}{3}$, find $\tan 2x$.

36. Write $\cos 2x + \cos x$ as a product.

37. Write $\sin 6y + \sin 3y$ as a product.

38. Write $\cos 4a \cos 2a$ as a sum.

39. Write $\sin \frac{1}{2}x \cos \frac{1}{3}x$ as a sum.

40. Write $\dfrac{\cos 3x - \cos x}{\sin 3x - \sin x}$ as a single function.

41. Prove the identity:
$\cos 2x + \cos 4x + \cos 6x = 4 \cos x \cos 2x \cos 3x - 1$

In **42 – 47**, find all solutions for x if $x \in [0, 2\pi)$.

42. $2 \cos x + 1 = 0$

43. $\sqrt{3} \csc x = -2$

44. $\sin^2 x = \frac{3}{4}$

45. $\tan^2 x = 3 \sec x - 3$

46. $\csc^2 2x = 2$

47. $\tan^2 \frac{1}{2}x = 3$

48. Solve $2 \sin \theta + \csc \theta = 3$ for θ, where $\theta \in [0, 2\pi)$.

49. Find a general solution for θ if $\cos^2 \theta + \sin \theta - 1 = 0$.

In **50 – 55**, find the value of the expression.

50. arc csc (-2)

51. arc tan (-1)

52. arc sec $\frac{2\sqrt{3}}{3}$

53. $\cot \left(\text{arc cos } \frac{2}{3} \right)$

54. arc tan $\left(\cos \frac{\pi}{2} \right)$

55. $\cos \left(\text{arc sin} \left(-\frac{\sqrt{3}}{2} \right) \right)$

56. Use a calculator or trigonometric table to find sin (arc cos 0.1) to the nearest thousandth.

57. Solve $3y = $ arc sin x for x in terms of y.

58. Solve $x = $ arc cos $(y + 2)$ for y in terms of x.

59. Evaluate: $\cos \left(2 \text{ arc sin } \frac{1}{4} \right)$

In **60 – 67**, find the measures of the remaining sides and angles of all possible triangles, if a triangle exists. Where an exact answer cannot be found, answer to the nearest degree or whole number.

60. $m\angle A = 23°$, $m\angle B = 60°$, $a = 4$

61. $m\angle A = 100°$, $m\angle B = 18°$, $c = 5$

62. $a = 7$, $b = 8$, $m\angle C = 40°$

63. $m\angle B = 30°$, $m\angle C = 120°$, $b = 10$

64. $a = 13$, $b = 14$, $c = 15$

65. $a = 18$, $b = 15$, $m\angle C = 40°$

66. $m\angle B = 120°$, $a = 10$, $b = 12$

67. $a = 37$, $b = 37$, $c = 60$

Chapter 9   Review Exercises *(continued)*

68. If the lengths of the sides of a triangle are 60, 64, and 89, is the triangle acute or obtuse? Explain.

69. If m∠A = 50°, a = 10, and b = 7, find c, m∠B, and m∠C for all possible triangles.

70. Find the possible lengths of the third side of a triangle if the lengths of two sides are 10 and 8, and the degree measure of the angle opposite the side of length 8 is 35°.

71. The lengths of the diagonals of a parallelogram are 5 cm and 8 cm. If the diagonals intersect at an angle of 152°, find the lengths of the sides of the parallelogram to the nearest centimeter.

72. Find the height, to the nearest foot, of a tree growing at an angle of 85° with the ground if an observer standing 60 feet from the base of the tree determines that the measure of the angle of elevation from the ground to the top of the tree is 35°.

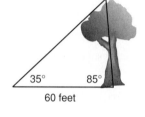

35°   85°

60 feet

73. A weight attached to a spring is pulled down 3 cm from its resting position and then released, causing the spring to oscillate. If each oscillation takes $\frac{1}{4}$ of a second, write an equation for the function $f(t)$, where $f(t)$ is in centimeters and t is in seconds, that describes the position of the weight relative to the resting position. Assume the distance is positive above the resting position and that $f(t) = 0$ at $t = 0$.

74. A Ferris wheel rotates at 2 revolutions per minute in a counterclockwise direction. Let the wheel have a radius of 15 feet and a person sitting at the 9-o'clock position be 15 feet above the ground at $t = 0$.

  a. Write an expression for that person's height above the ground (in feet) at time t (in minutes).

  b. How high above the ground is this person at $t = 7.5$ minutes?

68. obtuse; $89^2 > 60^2 + 64^2$

69. $B = 32°$, $C = 98°$, $c = 13$

70. 3 or 14

71. 2 cm, 6 cm

72. 40 feet

73. $f(t) = 3 \sin 8\pi t$

74. a. $f(t) = 15 \sin 4\pi \left(t + \frac{1}{4}\right) + 15$

  b. 15 feet

☑ *Exercises for Challenge*

1. 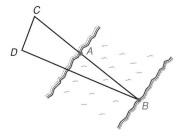 To find AB, the width of a river, a surveyor chooses points C and D so that BAC is a line segment. If m$\angle BCD = 74°29'$, m$\angle BDC = 86°56'$, $CD = 36$ m, and $AC = 50$ m, find the width of the river, to the nearest tenth of a meter.

2. Find the exact value of:

$$\frac{\sin 5° \cos 25° + \cos 5° \sin 25°}{\cos 5° \cos 35° + \sin 5° \sin 35°}$$

3. If $\sin x = \frac{1}{6}$, find the the exact value of $\sin 3x$.

4. The bisectors of $\angle A$ and $\angle B$ of $\triangle ABC$ meet at D. If $AD = 7$, $DB = 12$, and m$\angle C = 100°$, find AB to the nearest integer.

5. If $\cos x = -\frac{1}{10}$, find the exact value of $\cos 4x$.

6. When $\sin (\pi - \theta) \neq 0$, $\dfrac{\cos \left(\frac{\pi}{2} - \theta\right)}{\sin (\pi - \theta)}$ equals

(A) 1 (B) $\sin \theta - 1$
(C) $\csc \theta$ (D) $\cot \theta$
(E) $-\cot \theta$

7. For $\frac{\pi}{2} \leq x < \frac{3\pi}{2}$, solve:

$$3 \sin x = 2 \cos^2 x$$

(A) $\frac{\pi}{6}$ (B) $\frac{2\pi}{3}$ (C) $\frac{3\pi}{4}$

(D) $\frac{5\pi}{6}$ (E) $\frac{7\pi}{6}$

8. How many solutions are there to the equation $4 \sin x - \sin^4 x = 0$ when $-\pi \leq x \leq 0$?

(A) 0 (B) 1 (C) 2
(D) 3 (E) 4

9. $\overline{AC}$ and $\overline{BD}$ are tangent to the unit circle at A and B respectively.

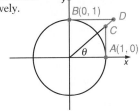

The length of $\overline{CD}$ is equal to
(A) $1 - \tan \theta$ (B) $\cot \theta$
(C) $\csc \theta$ (D) $\csc \theta - \sec \theta$
(E) $1 + \csc \theta$

10. Which of the following is a false statement?

(A) $\sin \left(2 \arccos \left(-\frac{3}{5}\right)\right) = -\frac{24}{25}$

(B)
$\cos \left(2\left(\arcsin \frac{1}{\sqrt{2}} + \arctan \sqrt{3}\right)\right) = \frac{\sqrt{3}}{2}$

(C) $\tan \left(2 \arcsin \frac{\sqrt{3}}{2}\right) = -\sqrt{3}$

(D) $\tan \left(\arctan \frac{1}{2} + \arctan \frac{1}{3}\right) = 1$

(E) $\arccos\left(-\frac{1}{2}\right) = 2 \arccos \frac{1}{2}$

**Answers
Exercises for
Challenge**

1. 62.8 m

2. $\frac{\sqrt{3}}{3}$

3. $\frac{13}{27}$

4. 18

5. 0.9208

6. A

7. D

8. C

9. D

10. B

11. C

12. A

13. D

14. B

15. A

16. A

17. C

18. B

19. C

20. D

☑ **Exercises for Challenge** *(continued)*

11. Which of the following is not true in $\triangle ABC$?

   (A) $AB = \dfrac{AC \sin (A + B)}{\sin B}$

   (B) $AB = \dfrac{AC \sin (A + B)}{\sin (A + C)}$

   (C) $AB = \dfrac{AC \sin (B + C)}{\sin (A + B)}$

   (D) $AB = \dfrac{AC \cos \left(\frac{\pi}{2} - C\right)}{\cos \left(\frac{\pi}{2} - B\right)}$

   (E) $AB = AC (\sin A \cot B + \cos A)$

12. If a, b, and c are the measures of the sides of $\triangle ABC$ and p is the measure of the perimeter, which of the following is true?

   (A) $\cos C = \dfrac{p}{2ab}(a + b - c) - 1$

   (B) $\cos C = 1 - \dfrac{p}{2ab}(a + b - c)$

   (C) $\cos C = 1 - \dfrac{p}{2ab}(c - a - b)$

   (D) $\cos C = \dfrac{p}{2ab}(a + b - c)$

   (E) $\cos C = \dfrac{p}{2ab}(c - a - b)$

13. 🖩 The measure of one side of an isosceles triangle is 10 and the measure of the vertex angle is 98°. To the nearest tenth, what is the area of the smallest triangle that can be drawn?

   (A) 99.0   (B) 49.5   (C) 43.5
   (D) 21.7   (E) 7.0

14. The area of an equilateral triangle inscribed in a circle is what percent of the area of the circle?

   (A) 27.6%   (B) 41.3%   (C) 47.7%
   (D) 82.7%   (E) 87.9%

Questions 15 – 20 each consist of two quantities, one in Column A and one in Column B. You are to compare the two quantities and choose:

   A if the quantity in Column A is greater;
   B if the quantity in Column B is greater;
   C if the two quantities are equal;
   D if the relationship cannot be determined from the information given.

1. In certain questions, information concerning one or both of the quantities to be compared is centered above the two columns.

2. In a given question, a symbol that appears in both columns represents the same thing in Column A as it does in Column B.

3. x, n, and k, etc. stand for real numbers.

Column A	Column B
15. $\sin 1 + \cos 1$	$\sin \left(\frac{\pi}{4} + 1\right)$

$$0 < \alpha < \frac{\pi}{4}$$

Column A	Column B
16. $\cos^4 \frac{\alpha}{2} - \sin^4 \frac{\alpha}{2}$	$\cos 2\alpha$

$$\alpha = \arctan \tfrac{3}{4} \text{ and}$$
$$\beta = \arctan \tfrac{1}{7}$$

Column A	Column B
17. $\alpha + \beta$	$\frac{\pi}{4}$

$$\tan^2 \alpha = 1 + 2 \tan^2 \beta,$$
$$0 < \alpha, \beta < \frac{\pi}{2}$$

Column A	Column B
18. $\cos^2 \alpha$	$\cos^2 \beta$

Column A	Column B
19. $\tan 10° + \tan 20°$	$\dfrac{1}{2 \cos 10° \cos 20°}$

The measures of the angles of $\triangle ABC$ are 80°, 60° and 40°.

Column A	Column B
20. The perimeter of $\triangle ABC$	The area of $\triangle ABC$

Teacher's Chapter *10*

The Conic Sections

Chapter Contents

This topic clearly demonstrates the beauty, simplicity, and unity of mathematics. The conics are developed in this chapter using eccentricity as the constant ratio of the distance of a point from a fixed point to the distance of that point from a fixed line because this definition highlights the common bond among all conics. From this definition and the symmetry of the ellipse and the hyperbola, the two-focus definition of these two conics can be formulated.

10.1 The Parabola

The parabola is familiar to most students. In addition to graphing experience with the parabola $y = ax^2 + bx + c$, some students may be familiar with the locus definition and its use in the derivation of the equation of the parabola.

When studying functions and their graphs in Chapter 3, students graphed $y = x^2$ and its reflection in the x-axis, $y = -x^2$. Ask students to determine the coordinates of the focus and the equation of the directrix of these curves. (For $y = x^2$, the focus is at $\left(0, \frac{1}{4}\right)$ and the directrix is the line $y = -\frac{1}{4}$, since $4d = 1$. For $y = -x^2$, the focus is at $\left(0, -\frac{1}{4}\right)$ and the directrix is the line $y = \frac{1}{4}$.)

A parabola can be constructed from string that forms the tangents to the curve. On a bulletin board or heavy piece of cardboard, place an equal number of pins along each of two intersecting lines. Number the pins along one line with the odd numbers so that the smaller numbers are farther from the intersection. Number the pins along the other line with the even numbers so that the smaller numbers are closer to the intersection. Wrap string from one pin to the next following the sequence of the numbers.

1, 2, 1; 3, 4, 3; 5, 6, 5; 7 ...

Vary the size of the angle between the lines to form different parabolas.

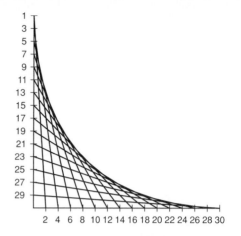

Activity (10.1)

Directions

1. On a large sheet of tracing paper, place a point F about 2 inches from the bottom edge.

2. Choose some point A_1 on the bottom edge of the paper. Fold the paper so that A_1 lies on F and crease the paper.

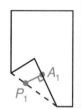

3. With the paper folded, draw a line from F, perpendicular to the edge, and mark P_1, the point at which this perpendicular meets the crease.

4. Choose a new point A_2 and repeat the process to find P_2.

5. Repeat for A_3 through A_{12} to locate points P_1 through P_{12}. Draw a smooth curve through the points P_n.

Discoveries:

1. The points P_1 through P_{12} are points on a parabola.

2. The edge of the paper is the directrix of the parabola.

3. The point F is the focus of the parabola.

4. For $n = 1$ to 12, $P_n F = P_n A_n$.

What If ?

The point F is farther from the edge of the paper. (The sides of the parabola will slant upwards more gradually.)

When working with equations of conics under translation, it may be necessary to review the procedure for completing the square of a binomial.

10.2 The Ellipse

This section presents a general equation for an ellipse with center at the origin by deriving the equations of specific curves from the coordinates of the focus, the equation of the directrix, and the value of the eccentricity. In Section 10.4, the general equation will be derived in terms of eccentricity and length of the major axis. The focus-directrix definition is used because it is simpler and because it unifies the conics. The two-focus definition is introduced in Section 10.4 as a consequence of the symmetry of the curve.

10.3 The Hyperbola

Elicit from students where asymptotes have been seen in other contexts (e.g., in graphs of rational, logarithmic, and trigonometric functions). Earlier, asymptotes were generally vertical or horizontal lines.

Another way of writing the equations of the asymptotes is to replace 1 by 0 in the intercept form of the equation of the hyperbola.

$$\frac{x^2}{a^2} - \frac{y^2}{b^2} = 0$$

$$\left(\frac{x}{a} - \frac{y}{b}\right)\left(\frac{x}{a} + \frac{y}{b}\right) = 0$$

$$\frac{x}{a} - \frac{y}{b} = 0 \qquad \frac{x}{a} + \frac{y}{b} = 0$$

$$\frac{y}{b} = \frac{x}{a} \qquad \frac{y}{b} = \frac{-x}{a}$$

$$y = \frac{b}{a}x \qquad y = -\frac{b}{a}x$$

While this process obtains the correct equations for the asymptotes, it does not demonstrate why these lines are asymptotes. Therefore, this procedure, if used, is only an aid for obtaining the equations of the asymptotes, not a derivation.

After students understand the concept of asymptotes, a mechanical method of graphing the asymptotes of $\frac{x^2}{a^2} - \frac{y^2}{b^2} = \pm 1$ can be used. Construct a rectangle with one pair of sides parallel to the y-axis through $(-a, 0)$ and $(a, 0)$ and the other pair of sides parallel to the x-axis through $(0, b)$ and $(0, -b)$. The lines through opposite vertices of the rectangle are the asymptotes of the hyperbola.

10.4 Relating the Ellipse and Hyperbola

In the previous sections, the coordinates of the focus and the equation of the directrix for the conics that were used in the derivations were chosen so that the conic would belong to the set of conics with axes of symmetry on the coordinate axes. Now, this section begins by establishing the relationship between the coordinates of the focus and the equation of the directrix that will make this possible. In the derivation, the value of a is defined to be positive. This does not limit the derivation since the symmetry of the curve shows that $(ae, 0)$ can be the focus and $x = \frac{a}{e}$ the directrix or $(-ae, 0)$ can be the focus and $x = -\frac{a}{e}$ the directrix. These are the same two points whether a is positive or any nonzero number.

The inequality $\frac{x^2}{9} + \frac{y^2}{4} < 1$ represents the region enclosed by the ellipse $\frac{x^2}{9} + \frac{y^2}{4} = 1$. The area of this region can be approximated by using probability, first approximating the area in the first quadrant. If a point in the region enclosed by the rectangle $ABCD$ is chosen at random, the probability that that point lies in the region enclosed by the ellipse is:

$$\frac{\text{Area of the ellipse}}{\text{Area of } ABCD}$$

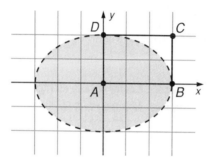

This value can be found empirically by generating the coordinates of T random points in the region enclosed by the rectangle $ABCD$ and counting the number, S, that lie in the interior of the ellipse. The probability that a point lies in the region enclosed by the ellipse is $\frac{S}{T}$. Therefore:

$$\frac{\text{Area of the ellipse}}{\text{Area of } ABCD} = \frac{S}{T}$$

$$\text{Area of the ellipse} = \frac{S}{T} \times \text{Area of } ABCD$$

This method of finding an area is called the *Monte Carlo Method*.

The following computer program can be used to generate the coordinates of 100 points, test the coordinates in the inequality that defines the area enclosed by the ellipse, and count the number of points that satisfy that inequality. The pairs are chosen for that portion of the ellipse in the first quadrant by generating x-coordinates from 0 to 3 and y-coordinates from 0 to 2. Since the area of

$ABCD$ is 6, the area of one-quarter of the ellipse is $6\left(\frac{S}{100}\right)$, and the area of the entire region enclosed by the ellipse is $4(6)\left(\frac{S}{100}\right)$ or $24\left(\frac{S}{T}\right)$.

```
100  REM  THIS PROGRAM WILL
          GENERATE RANDOM NUMBERS,
110  REM  TEST THE PAIR AS A SOLUTION
          OF AN INEQUALITY,
120  REM  AND USE A PROPORTION TO
          FIND THE AREA UNDER A
          CURVE.
130  S = 0
140  FOR I = 1 TO 100
150  X = RND(1) * 3
160  Y = RND(1) * 2
170  IF X^2 / 9 + Y^2 / 4 < 1 THEN
          S = S + 1
180  NEXT I
190  A = 24 * S / 100
200  PRINT "A = ";A
210  END
```

Run the program several times and find the average of the areas generated. Compare this average with the theoretical area given by the formula $A = \pi a b$.

The method described above can be used to approximate the area bounded above by any curve or curves whose equations are known and below by the x-axis. Change statements 150 and 160, entering the horizontal and vertical dimensions of the first-quadrant rectangle in place of 3 and 2 respectively. Change statement 170, entering the equation of the inequality that defines the region, and statement 190, entering the area of the full rectangle in place of 24. For example, to find the area between the x-axis and the curve $y = -x^2 + 16$, change the following statements in the program.

```
150  X = RND(1) * 4
160  Y = RND(1) * 16
170  IF Y < - X^2 + 16 THEN
          S = S + 1
190  A = 128 * S / 100
```

Directions:

Place a sheet of paper on a bulletin board or piece of heavy cardboard. Draw a set of coordinate axes and place tacks at *F* and *F'*, two points on the *x*-axis equidistant from the origin.

Attach at *F* and *F'* the ends of a piece of string that is longer than the distance between the two tacks. Using a pencil, pull the string taut. Draw a curve by moving the pencil while keeping the string taut. Include all possible points, above and below the *x*-axis.

Move the tacks closer together and repeat the procedure two or three times, using a different color pencil each time.

Let *F* and *F'* coincide at the origin and repeat the procedure.

Discoveries:

1. When the two points are distinct, the curve is an ellipse.

2. The length of the major axis remains unchanged when the length of the string remains unchanged.

3. The length of the major axis is the length of the string.

4. As the points *F* and *F'* are moved closer together, the length of the minor axis increases.

5. When the two fixed points coincide, the curve is a circle.

What If ?

The points *F* and *F'* are placed on the *y*-axis. (The major axis is a segment of the *y*-axis.)

10.5 *The Circle; Special Cases*

The circle can be defined as a limiting case of the ellipse as the value of *e* approaches 0. If $e \to 0$, $ae \to 0$ and $\frac{a}{e} \to \infty$. Since the distances of the foci from the center of the curve approach 0 and the distance of the directrix from the center of the curve approaches infinity, the circle is more simply defined as the limiting case when the foci of the ellipse coincide.

Most students are familiar with the equation of the circle in terms of the center and radius of the circle. In this section, *a* is used as the radius in order to associate the curve with that of an ellipse with major axis of length 2*a*.

Note that every conic has an equation of the form $Ax^2 + Bxy + Cy^2 + Dx + Ey + F = 0$ but not every equation of that form is a conic.

For example: $x^2 + y^2 + 2x + 4y + 5 = 0$ can be written as: $(x + 1)^2 + (y + 2)^2 = 0$ This is a degenerate conic, a single point:
$$(-1, -2)$$
$x^2 + y^2 + 2x + 4y + 6 = 0$ can be written as:
$$(x + 1)^2 + (y + 2)^2 = -1$$

This equation has no solution in the real coordinate plane.

Activity (*10.5*)_____

Directions:

Draw a large circle on paper or light cardboard and cut it out or use a lightweight paper plate that has been flattened.

1. Choose, in the interior of the circle, any point that is not the center and label it *F*.

2. Choose some point on the circle. Place the point you have chosen on *F* and crease the fold.

3. Repeat step 2 for many more points around the circle.

4. Observe the shape outlined by the creases.

Discoveries:

1. The shape outlined by the creases is an ellipse.

2. The other focus of the ellipse is the center of the circle.

What If ?

You wanted to prove that one point on each crease is a point on an ellipse with one focus at *F* and one at the center of the circle.

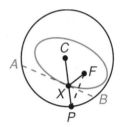

Proof: Let *C* be the center of the circle. Let *P* be a point on the circle and $\overline{AB}$ the crease that was made when *P* was placed on *F*. Therefore, the crease is the perpendicular bisector of $\overline{PF}$. Draw $\overline{PC}$ intersecting $\overline{AB}$ at *X*. $PX = FX$ and $PC = PX + XC$. Therefore, $PC = FX + XC$. But for all *P*, *PC* is the length of the radius of the circle and therefore a constant. The figure is an ellipse since for all *X*, the sum of the distances from a point on the curve to the two foci is a constant.

You may wish to show students how to use a rotation to transform an equation of a conic with an oblique axis of symmetry into one with an axis of symmetry parallel to the *x*- or *y*-axis.

To write the equations of a rotation through an angle of measure α, draw a new set of axes so that the *x'*-axis makes an angle of α with the *x*-axis. Let *P* be any point at distance *r* from the origin. Therefore,

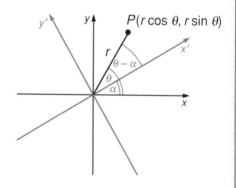

$$P(x, y) = (r \cos \theta, r \sin \theta)$$

That is,
$$x = r \cos \theta$$
$$y = r \sin \theta$$

Then,
$$P(x', y') = (r \cos (\theta - \alpha), r \sin (\theta - \alpha))$$
$$x' = r \cos (\theta - \alpha)$$
$$= r(\cos \theta \cos \alpha + \sin \theta \sin \alpha)$$
$$= r \cos \theta \cos \alpha + r \sin \theta \sin \alpha$$
$$= x \cos \alpha + y \sin \alpha$$
$$y' = r \sin (\theta - \alpha)$$
$$= r(\sin \theta \cos \alpha - \cos \theta \sin \alpha)$$
$$= r \sin \theta \cos \alpha - r \cos \theta \sin \alpha$$
$$= y \cos \alpha - x \sin \alpha$$

These equations for *x'* and *y'* can be written in matrix form:

$$\begin{bmatrix} x' \\ y' \end{bmatrix} = \begin{bmatrix} \cos \alpha & \sin \alpha \\ -\sin \alpha & \cos \alpha \end{bmatrix} \cdot \begin{bmatrix} x \\ y \end{bmatrix}$$

Since the inverse of $\begin{bmatrix} \cos\alpha & \sin\alpha \\ -\sin\alpha & \cos\alpha \end{bmatrix}$ is $\begin{bmatrix} \cos\alpha & -\sin\alpha \\ \sin\alpha & \cos\alpha \end{bmatrix}$, this matrix equation can be rewritten by multiplying both sides by the inverse.

$$\begin{bmatrix} \cos\alpha & -\sin\alpha \\ \sin\alpha & \cos\alpha \end{bmatrix} \cdot \begin{bmatrix} x' \\ y' \end{bmatrix} = \begin{bmatrix} \cos\alpha & -\sin\alpha \\ \sin\alpha & \cos\alpha \end{bmatrix} \cdot \begin{bmatrix} \cos\alpha & \sin\alpha \\ -\sin\alpha & \cos\alpha \end{bmatrix} \cdot \begin{bmatrix} x \\ y \end{bmatrix}$$

$$\begin{bmatrix} \cos\alpha & -\sin\alpha \\ \sin\alpha & \cos\alpha \end{bmatrix} \cdot \begin{bmatrix} x' \\ y' \end{bmatrix} = \begin{bmatrix} x \\ y \end{bmatrix}$$

Thus: $x = x'\cos\alpha - y'\sin\alpha$

$y = x'\sin\alpha + y'\cos\alpha$

These expressions for x and y can be substituted in the general form of a conic to determine the relationship between the given values of the coefficients and the values after the rotation.

$$\begin{aligned}
Ax^2 & + Bxy + Cy^2 + Dx + Ey + F \\
&= A(x'\cos\alpha - y'\sin\alpha)^2 \\
&\quad + B(x'\cos\alpha - y'\sin\alpha)(x'\sin\alpha + y'\cos\alpha) \\
&\quad + C(x'\sin\alpha + y'\cos\alpha)^2 \\
&\quad + D(x'\cos\alpha - y'\sin\alpha) \\
&\quad + E(x'\sin\alpha + y'\cos\alpha) \\
&\quad + F \\
&= A(x'^2\cos^2\alpha - 2x'y'\cos\alpha\sin\alpha + y'^2\sin^2\alpha) \\
&\quad + B(x'^2\cos\alpha\sin\alpha + x'y'\cos^2\alpha - x'y'\sin^2\alpha - y'^2\sin\alpha\cos\alpha) \\
&\quad + C(x'^2\sin^2\alpha + 2x'y'\sin\alpha\cos\alpha + y'^2\cos^2\alpha) \\
&\quad + D(x'\cos\alpha - y'\sin\alpha) \\
&\quad + E(x'\sin\alpha + y'\cos\alpha) \\
&\quad + F
\end{aligned}$$

Thus, when the conic $Ax^2 + Bxy + Cy^2 + Dx + Ey + F = 0$ is rotated through an angle of measure α, the equation of the rotated conic can be written as:

$A'x^2 + B'xy + C'y^2 + D'x + E'y + F = 0$, where,

A' is the sum of the coefficients of the terms in x'^2, that is,

$A' = A\cos^2\alpha + B\sin\alpha\cos\alpha + C\sin^2\alpha$.

B' is the sum of the coefficients of the terms in $x'y'$, that is,

$B' = -2A\sin\alpha\cos\alpha + B\cos^2\alpha - B\sin^2\alpha + 2C\sin\alpha\cos\alpha$

$\quad = B(\cos^2\alpha - \sin^2\alpha) - 2(A - C)\sin\alpha\cos\alpha$.

C' is the sum of the coefficients of the terms in y'^2, that is,

$C' = A\sin^2\alpha - B\sin\alpha\cos\alpha + C\cos^2\alpha$.

D' is the sum of the coefficients of the terms in x', that is,

$D' = D\cos\alpha + E\sin\alpha$.

E' is the sum of the coefficients of the terms in y', that is,

$E' = E\cos\alpha - D\sin\alpha$.

F' is the constant term, that is,

$F' = F$.

Usually α is chosen so that $B' = 0$, and the directrix of the conic is perpendicular to a coordinate axis.

$$B' = B(\cos^2 \alpha - \sin^2 \alpha) - 2(A - C) \sin \alpha \cos \alpha$$

$$0 = B \cos 2\alpha - (A - C) \sin 2\alpha$$

$$(A - C) \sin 2\alpha = B \cos 2\alpha$$

$$\frac{\sin 2\alpha}{\cos 2\alpha} = \frac{B}{A - C}$$

$$\tan 2\alpha = \frac{B}{A - C}$$

$$\alpha = \frac{1}{2} \text{arc tan} \left(\frac{B}{A - C} \right)$$

$\cos^2 x - \sin^2 x = \cos 2x;$
$2 \sin x \cos x = \sin 2x$

For example, consider the equation $3x^2 + 2xy - y^2 + 12 = 0$. Since $B^2 - 4AC = 2^2 - 4(3)(1) = -8$ is negative, the conic is an ellipse. Under a rotation of $\alpha = \frac{1}{2} \text{arc tan} \frac{2}{3-1} = \frac{1}{2} \cdot \frac{\pi}{4} = \frac{\pi}{8}$, the axes of the ellipse will be parallel to the coordinate axes.

If $A = C$, then the fraction $\dfrac{B}{A - C}$ is undefined and $\alpha = \frac{1}{2}\left(\pm \frac{\pi}{2} \right)$ or $\pm \frac{\pi}{4}$.

Computer Activity (10.5)

The following program computes the coefficients of a conic after a rotation of axes that makes the x-axis coincide with an axis of symmetry of the conic.

```
100  REM  THIS PROGRAM WILL FIND THE
110  REM  COEFFICIENTS OF A CONIC AFTER
120  REM  A ROTATION THAT MAKES THE
130  REM  X-AXIS COINCIDE WITH AN AXIS
140  REM  OF SYMMETRY OF THE CONIC.
150  PRINT "ENTER THE COEFFICIENTS
     OF THE STANDARD"
160  PRINT "FORM OF THE EQUATION OF
     THE GIVEN CONIC."
170  PRINT "ENTER A."
180  INPUT A
190  PRINT "ENTER B."
200  INPUT B
210  PRINT "ENTER C."
220  INPUT C
230  PRINT "ENTER D."
240  INPUT D
250  PRINT "ENTER E."
260  INPUT E
270  PRINT "ENTER F."
280  INPUT F
290  IF A = C THEN G = 3.14159/4
300  IF A > C OR A < C THEN
     G = (1/2) * ATN(B/(A - C))
310  AT = A * (COS(G)) ^ 2 + B *
     SIN(G) * COS(G) + C *
     (SIN(G)) ^ 2
320  CT = A * (SIN(G)) ^ 2 - B *
     SIN(G) * COS(G) + C *
     (COS(G)) ^ 2
330  DT = D * COS(G) + E * SIN(G)
340  ET = E * COS(G) - D * SIN(G)
350  FT = F
360  PRINT "A' = "; AT
370  PRINT "C' = "; CT
380  PRINT "D' = "; DT
390  PRINT "E' = "; ET
400  PRINT "F' = "; FT
410  END
```

Suggested Test Items

In **1 – 6**, use the value of $B^2 - 4AC$ to identify the conic as an ellipse, a hyperbola, a parabola, a circle, or a degenerate conic.

1. $x^2 - y^2 = 4$

2. $x^2 + y^2 + 8x = 0$

3. $x^2 - 4x - 8y + 1 = 0$

4. $9x^2 + y^2 + 18x + 9 = 0$

5. $xy - 2 = 0$

6. $4x^2 + 12y^2 - 12x - 12y + 3 = 0$

7. a. Write an equation of the ellipse with foci at $(\pm 6, 0)$ and eccentricity $\frac{3}{4}$.

  b. Find the coordinates of the points at which the conic intersects its axes of symmetry.

  c. Find an equation of a directrix.

  d. Sketch the curve.

8. a. Write an equation of the hyperbola with center at $(1, 1)$, focus at $(6, 1)$ and directrix at $x = 2$.

  b. Find the length of the transverse axis.

  c. Find equations of the asymptotes.

  d. Sketch the hyperbola.

9. a. Write an equation of the parabola with focus at $(3, 1)$ and directrix $y = 3$.

  b. Find the coordinates of the turning point.

  c. Sketch the curve.

10. An equation of a hyperbola is $x^2 - 4y^2 - 2x + 40y - 199 = 0$.

  a. Find the coordinates of the center.

  b. Find the eccentricity.

  c. Find the coordinates of the foci.

  d. Find equations of the directrices.

  e. Find the length of the transverse axis.

  f. Find the equations of the asymptotes.

  g. Sketch the curve.

In **11 – 12**, for the given ellipse:

  a. Find the coordinates of the center.

  b. Find the eccentricity.

  c. Find the coordinates of the foci.

  d. Find equations of the directrices.

  e. Find the length of the major axis.

  f. Find the length of the minor axis.

  g. Sketch the curve.

11. $x^2 + 5y^2 - 20 = 0$

12. $x^2 + 12y^2 - 8x - 20 = 0$

13. A golf ball is hit so that it rises to a maximum height of 30 feet and touches the ground 300 feet from the point at which it was hit. Write an equation for the parabolic path of the ball if x represents the horizontal distance and y the vertical distance of the ball from its starting point.

14. A carpenter wants to draw a pattern for an elliptical table top by using a string with its ends attached at the foci of the ellipse. The ellipse is to have a major axis that is 4 feet long and a minor axis that is 3 feet long. How far apart are the foci and what is the length of the string?

Bonus

The light ray that strikes a mirrored surface is called the ray of incidence and the light ray that is reflected is called the ray of reflection. The angle between the ray of incidence and the normal (perpendicular) to the surface is called the angle of incidence and the angle between the normal and the ray of reflection is called the angle of reflection. When light is reflected, the angle of incidence is congruent to the angle of reflection.

Let $y = x^2$ be the cross section of a mirrored surface. A source of light placed at $\left(0, \frac{1}{4}\right)$, the focus of the parabola, will be reflected parallel to the axis of symmetry of the parabola.

a. Find an equation of the line tangent to the parabola at $(1, 1)$ by writing the equation in terms of m, the slope, and finding the value of m such that the line intersects the parabola in exactly one point.

b. Find an equation of the normal to the parabola at $(1, 1)$, that is, the equation of the line that is perpendicular to the tangent line at that point.

c. Show that the angle of incidence of a ray from the focus is equal to the angle of reflection of that ray parallel to the axis of symmetry.

d. Repeat parts **a** through **c** for the point $(3, 9)$ and for the point $(-5, 25)$.

Answers to Suggested Test Items

1. hyperbola

2. circle

3. parabola

4. degenerate ellipse

5. hyperbola

6. ellipse

7. a. $\dfrac{x^2}{64} + \dfrac{y^2}{28} = 1$ **b.** $(\pm 8, 0), (0, \pm 2\sqrt{7})$

   c. $x = \dfrac{32}{3}$ **d.**

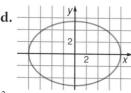

8. a. $\dfrac{(x-1)^2}{5} - \dfrac{(y-1)^2}{20} = 1$

   b. $2\sqrt{5}$ **c.** $y = \pm 2(x-1) + 1$

   d.

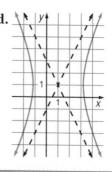

9. a. $(y - 2) = -\frac{1}{4}(x - 3)^2$

   b. $(3, 2)$ **c.**

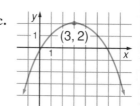

10. a. $(1, 5)$ **b.** $\dfrac{\sqrt{5}}{2}$

   c. $(1 \pm 5\sqrt{5}, 5)$ **d.** $x = 1 \pm 4\sqrt{5}$

   e. 20 **f.** $y = \pm \frac{1}{2}(x - 1) + 5$

   g.

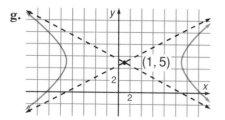

11. a. $(0, 0)$ **b.** $\sqrt{\frac{4}{5}} = \frac{2}{5}\sqrt{5}$ **c.** $(\pm 4, 0)$ **d.** $x = \pm 5$

    e. $4\sqrt{5}$ **f.** 4 **g.**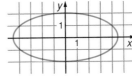

12. a. $(4, 0)$ **b.** $\frac{1}{6}\sqrt{33}$ **c.** $(4 \pm \sqrt{33}, 0)$ **d.** $x = 4 \pm \frac{12}{11}\sqrt{33}$

    e. 12 **f.** $2\sqrt{3}$ **g.**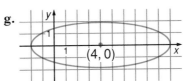

13. $y - 30 = -\frac{1}{750}(x - 150)^2$ **14.** $\sqrt{7}, 4$

Bonus

a. tangent at $(1, 1)$ $\dfrac{y - 1}{x - 1} = m$

$$y = m(x - 1) + 1$$

At point of intersection with $y = x^2$,

$$m(x - 1) + 1 = x^2$$
$$x^2 - mx + (m - 1) = 0$$
$$x = \frac{m \pm \sqrt{m^2 - 4(m - 1)}}{2}$$

If there is only one point of intersection,

$$m^2 - 4(m - 1) = 0$$
$$(m - 2)^2 = 0$$
$$m = 2$$

Equation of the tangent: $y = 2x - 1$

b. Equation of the normal: $\dfrac{y - 1}{x - 1} = -\dfrac{1}{2}$

$$y = -\frac{1}{2}x + \frac{3}{2}$$

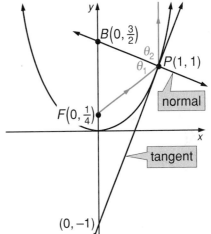

c. For $y = x^2$, focus is at $\left(0, \frac{1}{4}\right)$

  θ_1 = angle of incidence
  θ_2 = angle of reflection
  $BF = PF$ By distance formula
  $\theta_1 = \mathrm{m}\angle PBF$ $\triangle PBF$ is isosceles
  $\theta_2 = \mathrm{m}\angle PBF$ parallel lines;
               alternate interior angles
  $\theta_1 = \theta_2$ Transitive property.

d. $y = 6x - 9$ $y = -10x - 25$

   $y = -\frac{1}{6}x + \frac{19}{2}$ $y = -\frac{1}{10}x + \frac{51}{2}$

   $BF = PF = \frac{37}{4}$ $BF = PF = \frac{101}{4}$

Chapter 10

The Conic Sections

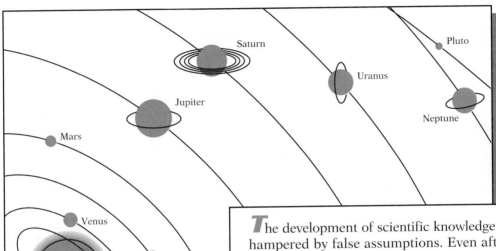

*T*he development of scientific knowledge is often hampered by false assumptions. Even after early observers of the heavens had proposed that Earth and the other planets revolve about the sun, progress in determining their orbits was hindered by the assumption that the orbits were circular. In the sixteenth century, the Danish astronomer Tycho Brahe spent many years carefully observing and recording the positions of the planets and stars. Using the results of Brahe's observations, his student Johannes Kepler proposed three laws of planetary motion. The first of these laws identified the orbits of the planets as ellipses.

Chapter Table of Contents

Introduction

When a cone is intersected by a plane, four types of cross sections can occur.

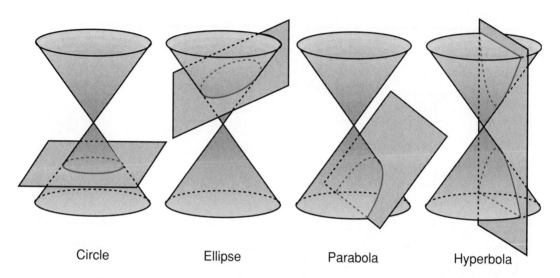

Circle   Ellipse   Parabola   Hyperbola

Known as the *conic sections*, these curves have many real-world applications. For example, they are used in furniture design, in architectural models for buildings and bridges, and to describe natural phenomena like the path of a projectile or planet.

The conic sections can also be defined as a locus of points.

■ **Definition:** *A conic is the locus of points such that the ratio of the distances of any point on the locus to a fixed point and a fixed line is a constant.*

Let F be the fixed point (the *focus*) and ℓ the fixed line (the *directrix*). Then, by definition, from any point P on the locus:

$$\frac{\text{distance from } P \text{ to the focus}}{\text{distance from } P \text{ to the directrix}} = \text{a constant}$$

$$\frac{PF}{PM} = e$$

where M is the projection of P on ℓ.

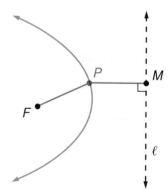

The value of the constant ratio e (the *eccentricity*) determines the shape of the curve.

Consider the meaning of e to determine its possible values. Since e represents the ratio of two distances, e must be positive.

The general locus definition of a conic can be rewritten for specific conics when:

$$e = 1, \quad e < 1, \quad e > 1$$

10.1 The Parabola

When the value of the eccentricity is 1, meaning that $PF : PM = 1$ or $PF = PM$, the locus is a *parabola*.

■ **Definition:** *A **parabola** is the locus of points equidistant from a fixed point and a fixed line.*

The simplest equation for a parabola can be derived by placing the focus on the y-axis and the directrix perpendicular to the y-axis so that the origin is equidistant from the focus and the directrix.

Let the focus be the point $F(0, d)$, the directrix the line $y = -d$, and $P(x, y)$ any point on the locus. The distance from P to the focus is PF. The distance from P to the directrix is the length of the perpendicular segment from P to the directrix. If M is the projection of P on the directrix, the distance from P to the directrix is PM.

Therefore:

$$PF = PM$$

$$\sqrt{(x - 0)^2 + (y - d)^2} = |y - (-d)|$$

$$(x - 0)^2 + (y - d)^2 = (y + d)^2$$

$$x^2 + y^2 - 2dy + d^2 = y^2 + 2dy + d^2$$

$$x^2 = 4dy$$

$$\frac{1}{4d}x^2 = y$$

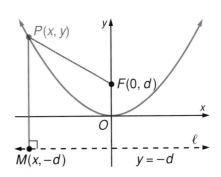

Note that this function is an even function and is, therefore, symmetric with respect to the y-axis. Its **axis of symmetry** is a line through the focus perpendicular to the directrix. The turning point, the point on the axis of symmetry that is equidistant from the focus and directrix, is the origin. Compare the equation of this parabola with the general quadratic function $y = ax^2 + bx + c$: $a = \frac{1}{4d}$, $b = 0$, and $c = 0$. The value of a is the reciprocal of $2(2d)$, where $|2d|$ is the distance between the focus and directrix. The values of b and c depend on the coordinates of the focus and the equation of the directrix.

The following example shows how to derive the equation of a parabola whose focus is not on the y-axis.

Example 1 _____

Write an equation of the parabola whose focus is (4, 1) and whose directrix is $y = -3$.

Solution

Let $P(x, y)$ be any point of the locus. The coordinates of F are $(4, 1)$ and the coordinates of M, the projection of P on $y = -3$, are $(x, -3)$.

$$PF = PM$$
$$\sqrt{(x-4)^2 + (y-1)^2} = |y - (-3)|$$
$$(x-4)^2 + (y-1)^2 = (y+3)^2$$
$$x^2 - 8x + 16 + y^2 - 2y + 1 = y^2 + 6y + 9$$
$$x^2 - 8x + 8 = 8y$$
$$\frac{x^2}{8} - \frac{8x}{8} + \frac{8}{8} = y$$
$$\frac{1}{8}x^2 - x + 1 = y$$

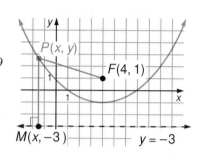

The coordinates of the turning point, or *vertex*, can be read from an alternate form of this equation. Begin at the third line of the derivation above.

$$(x-4)^2 + (y-1)^2 = (y+3)^2$$
$$(x-4)^2 + y^2 - 2y + 1 = y^2 + 6y + 9$$
$$(x-4)^2 = 8y + 8$$
$$(x-4)^2 = 8(y+1)$$
$$y + 1 = \tfrac{1}{8}(x-4)^2$$

The expressions $x - 4$ and $y + 1$ or $y - (-1)$ give the coordinates of the vertex as $(4, -1)$.

Answer: The equation is $y = \tfrac{1}{8}x^2 - x + 1$, or $y + 1 = \tfrac{1}{8}(x-4)^2$.

Note that the distance d from the vertex to the focus is 2, and the distance from the directrix to the focus is $2d$, or 4. Twice that value, namely 8, appears as the denominator of the coefficient of the squared expression. Note also that the parabola $y + 1 = \tfrac{1}{8}(x-4)^2$ is the image of the parabola $y = \tfrac{1}{8}x^2$ under the translation $T_{4,-1}$, which replaces x by $x - 4$ and y by $y + 1$.

In general, the equation of a parabola with vertex at (h, k), focus at $(h, k + d)$, and horizontal directrix $y = k - d$ is:

$$y - k = \frac{1}{4d}(x - h)^2$$

This parabola is the image of $y = \dfrac{1}{4d}x^2$ under the translation $T_{h, k}$.

Example 2

For the parabola with focus at $(3, -2)$ and directrix $y = 6$:

a. Write an equation of the axis of symmetry.
b. Find the coordinates of the turning point and determine its nature.
c. Write an equation of the parabola.
d. Determine the coordinates of two points on the parabola other than the turning point.

Solution

a. The axis of symmetry is perpendicular to the directrix. Therefore, the axis of symmetry is a vertical line whose equation is $x = c$, where c is a constant.

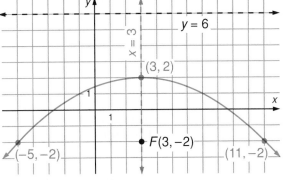

Since the focus is on the axis of symmetry and the x-coordinate of the focus is 3, the equation of the axis of symmetry is $x = 3$.

b. Since the turning point is also on the axis of symmetry, its x-coordinate is 3. Since the turning point is equidistant from the focus and the directrix, its y-coordinate is the average of the y-coordinates of the focus and the point at which the axis of symmetry intersects the directrix, $\frac{-2 + 6}{2}$ or 2.

Since the directrix is above the focus, the turning point (3, 2) is a *maximum* point, and the curve opens downward.

c. Use the locus definition of a parabola to write the equation.

$$PF = PM$$
$$\sqrt{(x - 3)^2 + (y - (-2))^2} = |y - 6|$$
$$(x - 3)^2 + (y + 2)^2 = (y - 6)^2$$
$$(x - 3)^2 + y^2 + 4y + 4 = y^2 - 12y + 36$$
$$(x - 3)^2 = -16y + 32$$
$$(x - 3)^2 = -16(y - 2)$$
$$-\frac{1}{16}(x - 3)^2 = y - 2$$

d. Consider the line segment through the focus, parallel to the directrix, with endpoints on the parabola. This line segment is called the *latus rectum* of the parabola.

Since, in this case, the latus rectum is horizontal, the ordinates of its endpoints must be the same as the ordinate of the focus, –2. To determine the abscissas of these endpoints, note that each of them is 8 units from the directrix and must, therefore, be 8 units from the focus.

Thus, the points (–5, –2) and (11, –2) are on the parabola.

Compare the parabolas of Examples 1 and 2 to note an additional characteristic of a parabola whose directrix is parallel to the x-axis.

- If the focus is above the directrix, d is positive, the curve opens upward, and the y-coordinate of the turning point is the minimum value of the range of the function.

- If the focus is below the directrix, d is negative, the curve opens downward, and the y-coordinate of the turning point is the maximum value of the range of the function.

The following example discusses a parabola whose directrix is parallel to the y-axis.

Example 3 _____

Find an equation of the parabola whose focus is (1, 4) and whose directrix is $x = -3$.

Solution

Let $P(x, y)$ be any point of the locus. The coordinates of F are (1, 4) and the coordinates of M, the projection of P on $x = -3$, are (-3, y). Therefore:

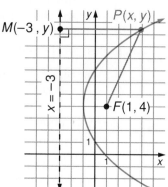

$$PF = PM$$
$$\sqrt{(x-1)^2 + (y-4)^2} = |x - (-3)|$$
$$(x-1)^2 + (y-4)^2 = (x+3)^2$$
$$x^2 - 2x + 1 + (y-4)^2 = x^2 + 6x + 9$$
$$(y-4)^2 = 8x + 8$$
$$(y-4)^2 = 8(x+1)$$
$$\tfrac{1}{8}(y-4)^2 = x + 1$$

Answer: The equation is $x + 1 = \tfrac{1}{8}(y-4)^2$.

Notice that the equation above is not that of a function, since for all values of x except –1 in the domain, there are two values of y in the range. For example, (1, 0) and (1, 8) are both ordered pairs of the relation.

In general, the equation of a parabola with vertex at (h, k), focus at $(h + d, k)$, and vertical directrix $x = h - d$ is:

$$x - h = \frac{1}{4d}(y-k)^2$$

If d is positive, the curve opens to the right. If d is negative, the curve opens to the left.

Transformations of Parabolas

Transformations in a plane can be used to show that all parabolas are similar, that is, lengths of corresponding chords are proportional and angles formed by corresponding chords are congruent. You have seen the image of a parabola under the translation $T_{h,k}$. To consider the image of a parabola under a dilation, recall that under a dilation with center at the origin and constant of dilation k, the image of (x, y) is (kx, ky).

Since any point on $y = x^2$ is of the form (x, x^2), any point on the image of $y = x^2$ under the dilation $D_{\frac{1}{a}}$ is of the form $\left(\tfrac{1}{a}x, \tfrac{1}{a}x^2\right)$, shown at the right to be a point of the equation $y = ax^2$.

Let:
$$y = ax^2$$
$$x = \frac{1}{a}x$$
$$y = a\left(\frac{1}{a}x\right)^2$$
$$y = \frac{1}{a}x^2$$

Therefore, $\left(\tfrac{1}{a}x, \tfrac{1}{a}x^2\right)$ is a point on $y = ax^2$.

Thus, any parabola whose equation is of the form $y = ax^2$ is the image of the parabola $y = x^2$ under the dilation $D_{\frac{1}{a}}$ and is, therefore, similar to $y = x^2$.

For example, the image of $y = x^2$ under the dilation D_2 is $y = \frac{1}{2}x^2$. The image of $(-1, 1)$ is $(-2, 2)$, the image of $(2, 4)$ is $(4, 8)$, and the image of (a, b) is $(2a, 2b)$.

Note that the chords joining the vertex of $y = \frac{1}{2}x^2$ to $(4, 8)$ and $(-2, 2)$ are twice as long as the chords joining the vertex of $y = x^2$ to the corresponding points $(2, 4)$ and $(-1, 1)$. The same angle is formed by these chords in both parabolas.

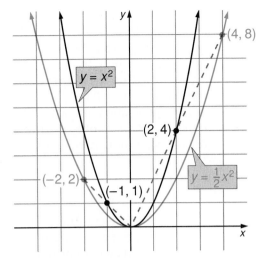

Any parabola whose equation is of the form $y = a(x - h)^2 + k$ is similar to the parabola whose equation is $y = x^2$ under a dilation $D_{\frac{1}{a}}$ followed by a translation $T_{h, k}$.

Example 4

Under what transformations is the parabola $y = 2x^2 - 4x - 3$ the image of the parabola $y = x^2$?

Solution

$$y = 2x^2 - 4x - 3$$
$$y = 2(x^2 - 2x) - 3 \qquad \text{Complete the square.}$$
$$y = 2(x^2 - 2x + 1) - 3 - 2$$
$$y = 2(x - 1)^2 - 5$$

Working backward from the image equation, first apply the translation $T_{1, -5}$.

$$y = 2x^2 - 4x - 3$$
$$y - 5 = 2(x + 1)^2 - 4(x + 1) - 3$$
$$y - 5 = 2(x^2 + 2x + 1) - 4x - 4 - 3$$
$$y = 2x^2 + 4x + 2 - 4x - 4 - 3 + 5$$
$$y = 2x^2$$

Then apply the dilation $D_{\frac{1}{2}}$.

$$y = 2x^2$$
$$\tfrac{1}{2}y = 2\left(\tfrac{1}{2}x\right)^2$$
$$\tfrac{1}{2}y = 2\left(\tfrac{1}{4}x^2\right)$$
$$\tfrac{1}{2}y = \tfrac{1}{2}x^2$$
$$y = x^2$$

Thus $y = x^2 \xrightarrow{\;D_{\frac{1}{2}}\;} y = 2x^2 \xrightarrow{\;T_{1, -5}\;} y - (-5) = 2(x - 1)^2$

Answer: The parabola $y = 2x^2 - 4x - 3$ is the image of the parabola $y = x^2$ under the translation $T_{1, -5}$ following the dilation $D_{\frac{1}{2}}$.

A mirror in the shape of a *paraboloid*, a surface formed by rotating a parabola about its axis of symmetry, has the property that light rays from a source of light at the focus of the parabola are reflected by the mirrored surface parallel to the axis of symmetry of the parabola. Flashlights, searchlights, and high beam of automobile headlights use a parabolic, mirrored surface to produce a beam of parallel light rays.

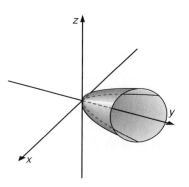

Example 5

The parabolic cross section of a searchlight is 30 cm wide and 20 cm deep. Locate the position at which the source of light should be placed to produce parallel rays of light.

Solution:

If the parabolic cross section is positioned in the coordinate plane with the turning point of the parabola at the origin and the axis of symmetry along the y-axis as shown in the diagram, the equation of the parabola is $y = \frac{1}{4d}x^2$ where d is the distance

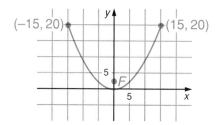

from the turning point to the focus. Since the distance across the parabola at its widest point is 30, the x-coordinates of the endpoints of the parabola are –15 and 15. Since the depth of the parabola is 20, the y-coordinates of each of these points is 20.

To find the value of d, substitute the coordinates of either endpoint of the parabola into the equation.

$$y = \frac{1}{4d}\, x^2$$

$$20 = \frac{1}{4d}\, (15)^2$$

$$80d = 225$$

$$d = \frac{225}{80} = \frac{45}{16} \approx 2.8$$

Answer: The source of light should be placed on the axis of symmetry 2.8 cm from the turning point of the parabola.

A paraboloid can also be used to direct incoming sound waves, radio waves, or light rays to the focus of the parabola. For example, a parabolic reflector can be used by a solar stove to concentrate the rays of the sun in order to change solar energy into heat energy, and a dish antenna can be used to reflect radio waves to a receiver placed at the focus.

Inequalities

Every point on a curve satisfies the equation of the curve. All other points in the plane satisfy an inequality related to the curve.

For example, for every real number x, there is a point on the graph of $y = x^2 - 4x + 3$ of the form $(x, x^2 - 4x + 3)$. Any point above the curve has a y-value that is greater than that of the corresponding point on the curve, that is, $y > x^2 - 4x + 3$. Therefore, the region above the curve is the graph of this inequality. Similarly, any point below the curve has a y-value that is less than that of the point on the curve, that is, $y < x^2 - 4x + 3$, and the region below the curve is the graph of the less-than inequality.

Example 6

Sketch the graph of: $y < -x^2 + 2x$

Solution

Work with the related equality. First sketch the graph of $y = -x^2 + 2x$.

$$y = -x^2 + 2x$$
$$y = -(x^2 - 2x + 1) + 1$$
$$y = -(x-1)^2 + 1$$

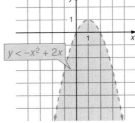

The graph of $y < -x^2 + 2x$ is the shaded region below the parabola. Note that $(1, -1)$ is a point in the shaded region and its coordinates satisfy the inequality: $-1 < -(1)^2 + 2(1)$.

The values of x in the solution set of $y < -x^2 + 2x$ for which y equals 0 lie on the x-axis. Setting y equal to 0 gives the related inequality $-x^2 + 2x > 0$, whose solution set can be read from the graph as $0 < x < 2$.

Exercises 10.1

In **1 – 6**, derive an equation of the parabola having the given point as focus and the given line as directrix.

1. $(0, -3), y = 3$ **2.** $(0, 2), y = -2$ **3.** $(1, 1), y = 5$

4. $(1, 1), y = -3$ **5.** $(2, 0), x = -2$ **6.** $(1, 1), x = 5$

7. Write an equation of the axis of symmetry of a parabola whose focus is $(3, 4)$ and whose directrix is:
    a. $y = -1$ **b.** $x = -1$

8. Find the coordinates of the turning point of a parabola whose focus is $(-2, 1)$ and whose directrix is:
    a. $y = -3$ **b.** $x = 4$

9. Find the coordinates of the focus of a parabola if the turning point is $(2, -1)$ and the directrix is:
    a. $y = 5$ **b.** $x = 1$

10. Write an equation of a parabola whose focus is $(1, 1)$ and whose turning point is:
    a. $(1, -5)$ **b.** $(5, 1)$

11. Find the coordinates of the endpoints of the latus rectum of the parabola determined by: **a.** focus $(0, 5)$, directrix $y = 3$ **b.** focus $(2, -1)$, directrix $x = 5$

1. $y = -\frac{1}{12}x^2$

2. $y = \frac{1}{8}x^2$

3. $y - 3 = -\frac{1}{8}(x - 1)^2$

4. $y + 1 = \frac{1}{8}(x - 1)^2$

5. $x = \frac{1}{8}y^2$

6. $x - 3 = -\frac{1}{8}(y - 1)^2$

7. a. $x = 3$ **b.** $y = 4$

8. a. $(-2, -1)$
   b. $(1, 1)$

9. a. $(2, -7)$
   b. $(3, -1)$

10. a.
   $y + 5 = \frac{1}{24}(x - 1)^2$
   b.
   $x - 5 = -\frac{1}{16}(y - 1)^2$

11. a. $(-2, 5), (2, 5)$
   b. $(2, 2), (2, -4)$

12. a. 16 m
   b. 8 m
   c. 16 m

13. a. $\frac{5}{2}$ sec.
   b. 25 feet
   c. $\frac{5}{4}$ sec.

14. a. 3 sec.
   b. 11.025 m
   c. 1.5 sec.

15. 125 feet

16. D_3

17. $D_{\frac{1}{4}}$ and $T_{1, 0}$

18. $T_{-5, 1}$

19. D_2 and $r_{x\text{-axis}}$

20. $D_{\frac{1}{3}}$ and $T_{-1,-5}$

21. $D_{\frac{1}{3}}, r_{x\text{-axis}}, T_{\frac{3}{2}, \frac{27}{4}}$

22.

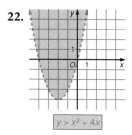

$y > x^2 + 4x$

23.

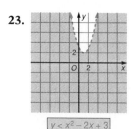

$y < x^2 - 2x + 3$

24.

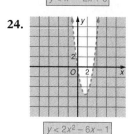

$y < 2x^2 - 6x - 1$

25.

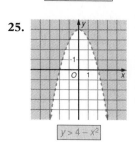

$y > 4 - x^2$

Exercises 10.1 *(continued)*

12. A ball is thrown so that its position is given by the equation $y = -\frac{1}{4}x^2 + 4x$, where x represents the horizontal and y the vertical distance, in meters, of the ball from the point at which it was thrown.

   a. Find the maximum height to which the ball rises.

   b. Find the horizontal distance the ball has traveled when it reaches its maximum height.

   c. What horizontal distance has the ball traveled when it returns to the vertical height at which it was thrown?

13. When a stone is thrown upward, its height y, in feet, is given by the equation $y = -16t^2 + 40t$ where t represents the number of seconds since the stone was thrown. Assume the stone was thrown from ground level ($y = 0$) and falls back to the ground.

   a. After how many seconds does the stone hit the ground?

   b. What is the maximum height to which the stone rises?

   c. How many seconds does it take for the stone to reach its maximum height?

14. A golf ball is hit so that its vertical height y, in meters, at any time t, in seconds, is given by $y = -4.9t^2 + 14.7t$.

   a. After how many seconds does the ball hit the ground?

   b. What is the maximum height to which the ball rises?

   c. After how many seconds does the ball reach its maximum height?

A flexible chain suspended by its ends forms a curve called a *catenary*. The curve looks very much like a parabola. When weights are attached to the chain at appropriate intervals, the shape of the chain does become a parabola. The cables supporting the roadway of a suspension bridge such as the Golden Gate Bridge in San Francisco are parabolas, formed when vertical supports are suspended from the catenary cable.

15. A suspension bridge over a small stream has supports that are 500 ft. apart. The supports rise 130 ft. above the roadway and the supporting cable is 5 ft. above the roadway at its lowest point. Find the distance from the lowest point of the cable to the focus of the parabola formed by the cable.

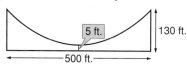

5 ft.   130 ft.

500 ft.

In **16 – 21**, determine the transformations under which the parabola having the given equation is the image of the parabola $y = x^2$.

   16. $y = \frac{1}{3}x^2$ **17.** $y = 4(x - 1)^2$ **18.** $y = (x + 5)^2 + 1$

   19. $y = -\frac{1}{2}x^2$ **20.** $y = 3x^2 + 6x - 2$ **21.** $y = -3x^2 + 9x$

In **22 – 27**, graph the solution set of the inequality.

   22. $y > x^2 + 4x$ **23.** $y < x^2 - 2x + 3$ **24.** $y < 2x^2 - 6x - 1$

   25. $y > 4 - x^2$ **26.** $y < 3x - x^2$ **27.** $y > \frac{1}{4}x^2 + x$

28. Use the graph of $y > \frac{1}{4}x^2 + x$ to determine the solution set of $0 > \frac{1}{4}x^2 + x$.

29. a. Sketch the graph of $y > x^2 - 2x + 2$.

   b. What is the solution set of $0 > x^2 - 2x + 2$?

26.

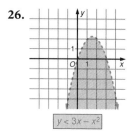

$y < 3x - x^2$

27.

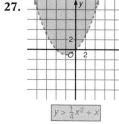

$y > \frac{1}{4}x^2 + x$

29. a.

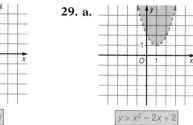

$y > x^2 - 2x + 2$

b. empty set

28. $-4 < x < 0$

10.2 The Ellipse

Recall that for a locus that is a conic, the eccentricity is the ratio of two distances. These distances relate point P on the locus to a fixed point F (the focus) and a fixed line ℓ (the directrix). If M is the projection of P on ℓ, then the eccentricity is the constant ratio $PF : PM$. The value of this ratio determines the shape of the curve.

■ **Definition:** *An* ellipse *is a conic whose eccentricity is a positive constant that is less than* 1.

If $PF : PM < 1$, the conic is an ellipse.

In the following examples, the focus and the directrix have been chosen so that the graph of the ellipse is symmetric with respect to the coordinate axes.

Example 1

Write an equation of the ellipse with focus $F(4, 0)$, directrix $x = 9$, and eccentricity $\frac{2}{3}$.

Solution

Let $P(x, y)$ be any point of the ellipse.

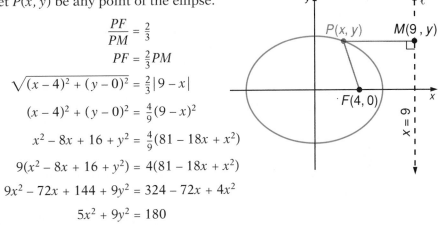

$$\frac{PF}{PM} = \frac{2}{3}$$

$$PF = \frac{2}{3}PM$$

$$\sqrt{(x - 4)^2 + (y - 0)^2} = \frac{2}{3}|9 - x|$$

$$(x - 4)^2 + (y - 0)^2 = \frac{4}{9}(9 - x)^2$$

$$x^2 - 8x + 16 + y^2 = \frac{4}{9}(81 - 18x + x^2)$$

$$9(x^2 - 8x + 16 + y^2) = 4(81 - 18x + x^2)$$

$$9x^2 - 72x + 144 + 9y^2 = 324 - 72x + 4x^2$$

$$5x^2 + 9y^2 = 180$$

Answer: An equation of the ellipse is $5x^2 + 9y^2 = 180$.

If both sides of the equation found in Example 1 are divided by the constant term, the equation is in *intercept form* because the x- and y-intercepts of the graph can easily be determined.

$$\frac{5x^2}{180} + \frac{9y^2}{180} = \frac{180}{180}$$

$$\frac{x^2}{36} + \frac{y^2}{20} = 1 \qquad \text{The intercept form of the equation.}$$

The y-intercepts are found by letting $x = 0$ and solving for y.

$$\frac{0^2}{36} + \frac{y^2}{20} = 1$$

$$\frac{y^2}{20} = 1$$

$$y^2 = 20$$

$$y = \pm\sqrt{20} = \pm 2\sqrt{5}$$

The x-intercepts are found by letting $y = 0$ and solving for x.

$$\frac{x^2}{36} + \frac{0^2}{20} = 1$$

$$\frac{x^2}{36} = 1$$

$$x^2 = 36$$

$$x = \pm\sqrt{36} = \pm 6$$

To draw the graph of the ellipse, first solve the equation for y in terms of x. Then choose values of x between the x-intercepts, +6 and –6, and find the corresponding values of y. Finally, plot the ordered pairs and connect the points with a smooth curve.

$$5x^2 + 9y^2 = 180$$

$$9y^2 = 180 - 5x^2$$

$$y^2 = \frac{180 - 5x^2}{9}$$

$$y = \frac{\pm\sqrt{180 - 5x^2}}{3}$$

x	$\dfrac{\pm\sqrt{180 - 5x^2}}{3}$	y
0	$\dfrac{\pm\sqrt{180 - 5(0)^2}}{3} = \dfrac{\pm\sqrt{180}}{3}$	$\pm 2\sqrt{5} \approx \pm 4.5$
± 2	$\dfrac{\pm\sqrt{180 - 5(2)^2}}{3} = \dfrac{\pm\sqrt{160}}{3}$	$\pm\dfrac{4}{3}\sqrt{10} \approx \pm 4.2$
± 4	$\dfrac{\pm\sqrt{180 - 5(4)^2}}{3} = \dfrac{\pm\sqrt{100}}{3}$	$\pm\dfrac{10}{3} \approx \pm 3.3$
± 6	$\dfrac{\pm\sqrt{180 - 5(6)^2}}{3} = \dfrac{\pm\sqrt{0}}{3}$	0

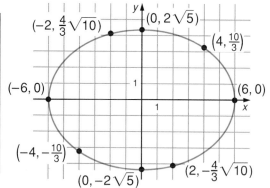

Note that if x is replaced by a number with absolute value greater than 6, the value of y is imaginary. Since, for each value of x in the interval $-6 < x < 6$, there are two values of y, the equation $5x^2 + 9y^2 = 180$ is a relation that is not a function. The domain is $\{x \mid -6 \le x \le 6\}$ and the range is $\{y \mid -2\sqrt{5} \le y \le 2\sqrt{5}\}$.

To display the graph of an ellipse on a calculator, it is necessary to enter the equation in two parts. That is, the ellipse of Example 1 is the union of the two functions:

$$y = \frac{+\sqrt{180 - 5x^2}}{3} \qquad \text{and} \qquad y = \frac{-\sqrt{180 - 5x^2}}{3}$$

Step 1 Set the range. For x, go from –7.5 to 7.5, and for y, from –5 to 5.

Step 2 Key in the two parts of the ellipse. For Y_1, enter:

| 2nd | √ | (| 180 | − | 5 | X|T | X² |) | ÷ | 3 |

For Y_2, enter:

| (−) | 2nd | √ | (| 180 | − | 5 | X|T | X² |) | ÷ | 3 |

Step 3 Graph.

There are two lines of symmetry for an ellipse, and one point of symmetry. The segments of the lines of symmetry whose endpoints are on the ellipse are called the *major axis* and *minor axis* of the ellipse, with the longer segment as the major axis and shorter segment as the minor axis. The point of intersection of the major and minor axes is the *center* of the ellipse. The endpoints of the major and minor axes are the *vertices* of the ellipse.

For the ellipse in Example 1, the lines of symmetry are the *x*-axis and the *y*-axis, and the point of symmetry is the origin. The major axis is the segment whose endpoints are $(-6, 0)$ and $(6, 0)$, with length $|6 - (-6)|$ or 12. The minor axis is the segment whose endpoints are $(0, -2\sqrt{5})$ and $(0, 2\sqrt{5})$, with length $|2\sqrt{5} - (-2\sqrt{5})|$ or $4\sqrt{5}$.

In general, the intercept form of an equation of an ellipse with center at the origin, focus on the *x*-axis, and vertical directrix is:

$$\frac{x^2}{a^2} + \frac{y^2}{b^2} = 1 \quad \text{where } |a| > |b|$$

The *x*-intercepts of the ellipse are $(a, 0)$ and $(-a, 0)$, with the length of the major axis equal to $2|a|$. The *y*-intercepts are $(0, b)$ and $(0, -b)$, with the minor axis of length $2|b|$.

Example 2

Sketch the graph of the ellipse whose equation is $2x^2 + 9y^2 = 18$.

Solution

To sketch the graph, obtain the intercepts by rewriting the equation.

$$\frac{2x^2}{18} + \frac{9y^2}{18} = \frac{18}{18}$$

$$\frac{x^2}{9} + \frac{y^2}{2} = 1$$

$$\begin{array}{c|c} a^2 = 9 & b^2 = 2 \\ a = \pm 3 & b = \pm\sqrt{2} \end{array}$$

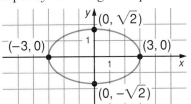

Thus, the *x*-intercepts are ± 3 and the *y*-intercepts are $\pm\sqrt{2}$.

In the next example, the focus is on the *y*-axis and the directrix is a horizontal line.

Example 3

Write an equation of the ellipse with focus $F(0, 4)$, directrix $y = 9$, and eccentricity $\frac{2}{3}$.

Solution

Let $P(x, y)$ be any point of the ellipse.

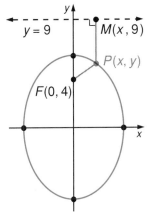

$$\frac{PF}{PM} = \frac{2}{3}$$

$$PF = \frac{2}{3}PM$$

$$\sqrt{(x - 0)^2 + (y - 4)^2} = \frac{2}{3}|9 - y|$$

$$(x - 0)^2 + (y - 4)^2 = \frac{4}{9}(9 - y)^2$$

$$x^2 + y^2 - 8y + 16 = \frac{4}{9}(81 - 18y + y^2)$$

$$9(x^2 + y^2 - 8y + 16) = 4(81 - 18y + y^2)$$

$$9x^2 + 9y^2 - 72y + 144 = 324 - 72y + 4y^2$$

$$9x^2 + 5y^2 = 180$$

Answer: An equation of the ellipse is $9x^2 + 5y^2 = 180$.

1.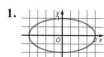

2.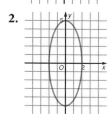

3.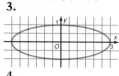

4.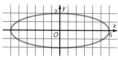

5.

6.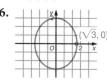

7. $\frac{x^2}{25} + \frac{y^2}{4} = 1$

8. $\frac{x^2}{81} + \frac{y^2}{49} = 1$

9. $\frac{x^2}{1} + \frac{y^2}{9} = 1$

10. $\frac{y^2}{16} + \frac{4x^2}{9} = 1$

11. $\frac{x^2}{5} + \frac{y^2}{81} = 1$

12. $\frac{x^2}{3} + \frac{y^2}{8} = 1$

13. $\frac{x^2}{36} + \frac{y^2}{16} = 1$

14. $\frac{2}{5}$

15. a. $3x^2 + 4y^2 = 108$

b. $3x^2 + 4y^2 = 108$

Divide both sides of the equation of the ellipse found in Example 3 by the constant term to write the equation in intercept form.

$$\frac{9x^2}{180} + \frac{5y^2}{180} = \frac{180}{180}$$

$$\frac{x^2}{20} + \frac{y^2}{36} = 1$$

The x-intercepts are $\pm\sqrt{20}$ or $\pm2\sqrt{5}$ and the y-intercepts are $\pm\sqrt{36}$ or ±6. The major axis is a segment of the y-axis from $y = -6$ to $y = 6$ and the minor axis is a segment of the x-axis from $x = -2\sqrt{5}$ to $x = 2\sqrt{5}$.

Note that if the graph of $5x^2 + 9y^2 = 180$ is reflected over the line $y = x$, the image is the graph of $9x^2 + 5y^2 = 180$.

In general, the intercept form of an equation of an ellipse with center at the origin, focus on the y-axis, and horizontal directrix is:

$$\frac{x^2}{b^2} + \frac{y^2}{a^2} = 1 \qquad \text{where } |a| > |b|$$

The y-intercepts of the ellipse are $(0, a)$ and $(0, -a)$, with the length of the major axis equal to $2|a|$. The x-intercepts are $(b, 0)$ and $(-b, 0)$, with the minor axis of length $2|b|$.

An ellipse whose center is not at the origin will be considered later in this chapter.

Exercises 10.2

In **1 – 6**, sketch the graph of the ellipse with the given equation.

1. $x^2 + 4y^2 = 4$ **2.** $25x^2 + 4y^2 = 100$ **3.** $x^2 + 9y^2 = 9$

4. $x^2 + 9y^2 = 36$ **5.** $16x^2 + 4y^2 = 64$ **6.** $4x^2 + 3y^2 = 12$

In **7 – 12**, write an equation of the ellipse with the given points as vertices.

7. $(\pm5, 0), (0, \pm2)$ **8.** $(\pm9, 0), (0, \pm7)$ **9.** $(\pm1, 0), (0, \pm3)$

10. $(\pm1.5, 0), (0, \pm4)$ **11.** $(\pm\sqrt{5}, 0), (0, \pm9)$ **12.** $(\pm\sqrt{3}, 0), (0, \pm2\sqrt{2})$

13. Write an equation of the ellipse whose major axis is a segment of length 12 on the x-axis and whose minor axis is a segment of length 8 on the y-axis.

14. The focus of an ellipse is at $(4, 0)$ and the directrix is the line whose equation is $x = 25$. Find the eccentricity if $(10, 0)$ are the coordinates of a point on the ellipse.

15. a. Write an equation of the ellipse whose focus is at $(3, 0)$, whose directrix has the equation $x = 12$, and whose eccentricity is $\frac{1}{2}$.

b. Write an equation of the ellipse with focus at $(-3, 0)$, directrix $x = -12$, and eccentricity equal to $\frac{1}{2}$.

c. What do you observe about the equations derived in parts **a** and **b**?

16. a. Write an equation of the ellipse with focus at $(9, 0)$, directrix $x = 16$, and eccentricity equal to $\frac{3}{4}$.

b. Sketch the graph of the equation derived in part **a**.

c. Find the equation of the image of the ellipse drawn in part **b** under a reflection over the line $y = x$.

d. What are the coordinates of the focus and the equation of the directrix of the ellipse whose equation was found in part **c**?

15. c. The equations are the same. There are two possible foci and two possible directrices for the same ellipse.

16. a. $7x^2 + 16y^2 = 1{,}008$

b. $\frac{x^2}{144} + \frac{y^2}{63} = 1$
The endpoints of the major axis: $(\pm12, 0)$
The endpoints of the minor axis: $(\pm\sqrt{63}, 0)$

16. b.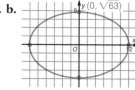

c. $\frac{x^2}{63} + \frac{y^2}{144} = 1$

d. $F(0, 9)$ directrix $y = 16$

10.3 The Hyperbola

Recall that for a conic, the nature of the locus of points P is determined by the value of its eccentricity, which is $PF : PM$, where F is a fixed point (the focus) and M is the projection of point P on a fixed line (the directrix).

■ **Definition:** *A hyperbola is a conic whose eccentricity is greater than* 1.

If $PF : PM > 1$, the conic is a hyperbola.

In the following examples, the focus and the directrix have been chosen so that the graph of the hyperbola is symmetric with respect to the coordinate axes.

Example 1

Write an equation of the hyperbola with the focus $F(4, 0)$, directrix $x = 1$, and eccentricity 2.

Solution

Let $P(x, y)$ be any point of the hyperbola.

$$\frac{PF}{PM} = 2$$

$$PF = 2PM$$

$$\sqrt{(x - 4)^2 + (y - 0)^2} = 2\,|x - 1|$$

$$(x - 4)^2 + (y - 0)^2 = 4(x - 1)^2$$

$$x^2 - 8x + 16 + y^2 = 4(x^2 - 2x + 1)$$

$$x^2 - 8x + 16 + y^2 = 4x^2 - 8x + 4$$

$$-3x^2 + y^2 = -12$$

$$3x^2 - y^2 = 12$$

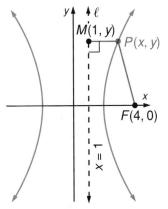

Answer: An equation of the hyperbola is $3x^2 - y^2 = 12$.

As with the ellipse, the equation of a hyperbola can be written in intercept form, shown for Example 1.

$$\frac{3x^2}{12} - \frac{y^2}{12} = \frac{12}{12}$$

$$\frac{x^2}{4} - \frac{y^2}{12} = 1$$

To find any y-intercepts, let $x = 0$ and solve the resulting equation for y.

$$\frac{0^2}{4} - \frac{y^2}{12} = 1$$

$$-\frac{y^2}{12} = 1$$

$$y^2 = -12$$

Since there is no real number whose square is negative, the graph has no y-intercepts.

To find any x-intercepts, let $y = 0$, and solve the resulting equation for x.

$$\frac{x^2}{4} - \frac{0^2}{12} = 1$$

$$\frac{x^2}{4} = 1$$

$$x^2 = 4$$

$$x = \pm\sqrt{4} = \pm 2$$

The x-intercepts are $+2$ and -2.

To draw the graph of the hyperbola, first solve the equation for y in terms of x. Then choose values of x that are greater than 2 or less than –2, and find the corresponding values of y. Finally, plot the ordered pairs and connect all those points with positive values of x, while separately connecting all those points with negative values of x.

$$3x^2 - y^2 = 12$$
$$-y^2 = -3x^2 + 12$$
$$y^2 = 3x^2 - 12$$
$$y = \pm\sqrt{3x^2 - 12}$$

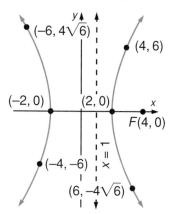

x	$\pm\sqrt{3x^2 - 12}$	y
± 2	$\pm\sqrt{3(2)^2 - 12} = \pm\sqrt{0}$	0
± 4	$\pm\sqrt{3(4)^2 - 12} = \pm\sqrt{36}$	± 6
± 6	$\pm\sqrt{3(6)^2 - 12} = \pm\sqrt{96}$	$\pm 4\sqrt{6} \approx \pm 9.8$

Note that for values of x in the interval $-2 < x < 2$, the value of y is imaginary. Since for each value of x with absolute value greater than 2, there are two values of y, the equation $3x^2 - y^2 = 12$ is a relation that is not a function. The domain is $\{x \mid |x| \geq 2\}$ and the range is the set of real numbers.

As with the ellipse, a hyperbola has two lines of symmetry and one point of symmetry. In the ellipse, both lines of symmetry intersect the ellipse, creating the major axis and the minor axis. In the hyperbola, only one of the lines of symmetry intersects the curve. The segment of the line of symmetry that intersects the curve is called the *transverse axis*. The line of symmetry that does not intersect the hyperbola is called the *conjugate axis*. The conjugate axis of the hyperbola is perpendicular to the transverse axis at its midpoint. This point is the *center* of the hyperbola.

For the hyperbola in Example 1, the lines of symmetry are the x-axis and the y-axis, and the point of symmetry is the origin. The endpoints of the transverse axis are $(-2, 0)$ and $(2, 0)$, and the length of the transverse axis is $|2 - (-2)|$ or 4. Note that the focus of this curve is a point on the x-axis, which is the line of symmetry that contains the transverse axis. The y-axis is the conjugate axis and the origin is the center.

In general, the intercept form of an equation of a hyperbola with center at the origin, focus on the x-axis, and vertical directrix is:

$$\frac{x^2}{a^2} - \frac{y^2}{b^2} = 1$$

The graph intersects the x-axis at $(a, 0)$ and $(-a, 0)$. The length of the transverse axis is $|2a|$.

If the focus is on the y-axis and the directrix is a horizontal line, the transverse axis will be a segment of the y-axis, as shown in the following example.

Example 2

Write an equation of the hyperbola with focus $F(0, 4)$, directrix $y = 1$, and eccentricity 2.

Solution

Let $P(x, y)$ be any point of the hyperbola.

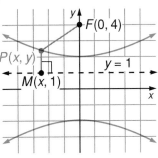

$$\frac{PF}{PM} = 2$$

$$PF = 2PM$$

$$\sqrt{(x - 0)^2 + (y - 4)^2} = 2|y - 1|$$

$$(x - 0)^2 + (y - 4)^2 = 4(y - 1)^2$$

$$x^2 + y^2 - 8y + 16 = 4(y^2 - 2y + 1)$$

$$x^2 + y^2 - 8y + 16 = 4y^2 - 8y + 4$$

$$x^2 - 3y^2 = -12$$

Answer: An equation of the hyperbola is $3y^2 - x^2 = 12$.

Divide both sides of the equation of the hyperbola in Example 2 by 12 to put the equation into intercept form.

$$\frac{3y^2}{12} - \frac{x^2}{12} = \frac{12}{12}$$

$$\frac{y^2}{4} - \frac{x^2}{12} = 1$$

Since the coefficient of x^2 is negative, the x-axis is the conjugate axis and there are no x-intercepts. The y-intercepts are ±2.

Note that the graph of $3y^2 - x^2 = 12$ is the image of the graph of $3x^2 - y^2 = 12$ under a reflection over the line $y = x$.

Example 3

Find the length of the transverse axis of the hyperbola $8y^2 - 2x^2 = 16$.

Solution

Write the equation in intercept form.

$$8y^2 - 2x^2 = 16$$

$$\frac{8y^2}{16} - \frac{2x^2}{16} = \frac{16}{16}$$

$$\frac{y^2}{2} - \frac{x^2}{8} = 1$$

Since the coefficient of x^2 is negative, there are no x-intercepts. The y-intercepts are $\pm\sqrt{2}$. Therefore, the transverse axis is a segment of the y-axis of length $|\sqrt{2} - (-\sqrt{2})|$ or $2\sqrt{2}$.

In general, the intercept form of an equation of a hyperbola with center at the origin, focus on the y-axis, and horizontal directrix is:

$$\frac{y^2}{a^2} - \frac{x^2}{b^2} = 1$$

The graph intersects the y-axis at $(0, a)$ and $(0, -a)$. The length of the transverse axis is $|2a|$.

Asymptotes of a Hyperbola

The four points that are the endpoints of the major and minor axes of an ellipse enabled you to use the intercept form of the equation to sketch the ellipse. Since the hyperbola has only two points of intersection with an axis, more information about the curve is needed in order to sketch it.

A hyperbola has two important boundary lines called *asymptotes* that can be used to sketch the curve. To find equations of these asymptotes, write the equation of the hyperbola in intercept form and then solve for y in terms of x.

$$\frac{x^2}{a^2} - \frac{y^2}{b^2} = 1$$

$$-\frac{y^2}{b^2} = -\frac{x^2}{a^2} + 1$$

$$y^2 = \frac{b^2}{a^2}x^2 - b^2$$

$$y^2 = \frac{b^2}{a^2}x^2\left(1 - \frac{a^2}{x^2}\right)$$

$$y = \pm\frac{b}{a}x\sqrt{1 - \frac{a^2}{x^2}}$$

Examine the expression to note that as the absolute value of x increases, the value of $\frac{a^2}{x^2}$ approaches 0 and, thus, the expression under the radical sign approaches 1. Therefore, the values of y approach $\pm\frac{b}{a}x$, meaning that, for the hyperbola with center at the origin, focus on the x-axis, and vertical directrix, the equations of the two asymptotes are:

$$y = \frac{b}{a}x \quad \text{and} \quad y = -\frac{b}{a}x$$

Example 4

Sketch the graph of the hyperbola whose equation is $4x^2 - 9y^2 = 36$.

Solution

Step 1   From the equation in intercept form, find the intercepts.

$$4x^2 - 9y^2 = 36$$

$$\frac{x^2}{9} - \frac{y^2}{4} = 1$$

The x-intercepts are ±3. There are no y-intercepts.

Step 2   Use $y = \pm\frac{b}{a}x$ to determine equations of the asymptotes.

From the intercept form, get the values of a and b.

$$a^2 = 9 \qquad\qquad b^2 = 4$$
$$a = \pm3 \qquad\qquad b = \pm2$$

Therefore, the equations of the asymptotes are:

$$y = \frac{2}{3}x \qquad \text{and} \qquad y = -\frac{2}{3}x$$

Step 3   Plot the *x*-intercepts and sketch the asymptotes.

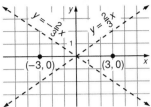

Step 4   Draw the hyperbola.

The vertices are at the *x*-intercepts and the branches approach the asymptotes as the absolute values of *x* increase.

Note that the point of intersection of the asymptotes is the center of the hyperbola.

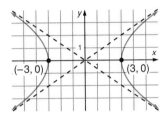

As with the ellipse, the graph of a hyperbola must be entered on a graphing calculator in two parts.

Step 1   Solve the equation for *y* in terms of *x*.

$$4x^2 - 9y^2 = 36$$

$$y = \pm \frac{\sqrt{4x^2 - 36}}{3}$$

Step 2   Use values for *x* from –7.5 to 7.5, and for *y*, from –5 to 5.

Step 3   Key in the two parts of the hyperbola and the two asymptotes.

For Y_1, enter:

For Y_2, enter:

| (-) | 2nd | √ | (| 4 | X|T | X² | − | 36 |) | ÷ | 3 |

For Y_3, enter: 2 X|T ÷ 3

For Y_4, enter: (-) 2 X|T ÷ 3

Step 4   Graph.

A hyperbola whose center is not at the origin will be considered in the next section.

Answers 10.3

1. a. *x*-axis   **b.** 4
   c. Domain: $|x| \geq 2$
      Range: Real
      numbers
   d. $y = \pm 2x$
   e.

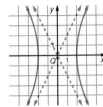

2. a. *x*-axis   **b.** $4\sqrt{2}$
   c. Domain:
      $|x| \geq 2\sqrt{2}$
      Range: Real
      numbers
   d. $y = \pm \frac{1}{2}x$
   e.

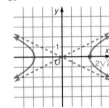

3. a. *x*-axis   **b.** 6
   c. Domain: $|x| \geq 3$
      Range: Real
      numbers
   d. $y = \pm x$
   e.

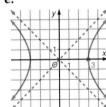

4. a. *y*-axis   **b.** $2\sqrt{3}$
   c. Domain: Real numbers
      Range: $|y| \geq \sqrt{3}$
   d. $y = \pm \frac{\sqrt{3}}{3}x$ **e.**

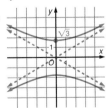

5. a. *x*-axis   **b.** 2
   c. Domain: $|x| \geq 1$
      Range: Real numbers
   d. $y = \pm x$ **e.**

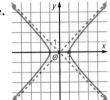

6. a. y-axis   **b.** 16
   c. Domain:
      Real numbers
      Range: $|y| \geq 8$
   d. $y = \pm 2\sqrt{2}x$
   e.

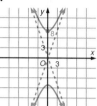

7. a.

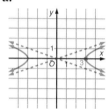

b.

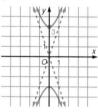

   c. $y^2 - 9x^2 = 9$

8. $\dfrac{x^2}{9} - \dfrac{y^2}{72} = 1$

9. a. $\dfrac{x^2}{36} - \dfrac{y^2}{45} = 1$

   b.

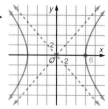

   c. $\dfrac{y^2}{36} - \dfrac{x^2}{45} = 1$

   d. (0, 9) $y = 4$

In **1 – 6**, answer each question for the given hyperbola:
   a. Of which coordinate axis is the transverse axis a segment?
   b. What is the length of the transverse axis?
   c. What are the domain and range of the equation?
   d. What are the equations of the asymptotes?
   e. Sketch the graph of the hyperbola.

1. $4x^2 - y^2 = 16$ **2.** $3x^2 - 12y^2 = 24$ **3.** $x^2 - y^2 = 9$

4. $x^2 - 3y^2 = -9$ **5.** $x^2 - y^2 = 1$ **6.** $y^2 - 8x^2 = 64$

7. a. Sketch the graph of the hyperbola $x^2 - 9y^2 = 9$.

   b. Sketch the reflection of this graph over the line $y = x$.

   c. Write an equation of the reflection.

8. Write an equation of the hyperbola whose focus is at (9, 0), whose directrix has the equation $x = 1$, and whose eccentricity is 3.

9. a. Write an equation of the hyperbola whose focus is at (9, 0), whose directrix has the equation $x = 4$, and whose eccentricity is $\frac{3}{2}$.

   b. Sketch the graph of the equation derived in part **a**.

   c. Find an equation of the image of the hyperbola drawn in part **b** under a reflection over the line $y = x$.

   d. What are the coordinates of the focus and an equation of the directrix of the hyperbola whose equation was found in part **c**?

10. The focus of a hyperbola is at (25, 0) and the directrix is a line whose equation is $x = 4$. Find the eccentricity of the hyperbola if (10, 0) are the coordinates of a point of the hyperbola.

11. The focus of a hyperbola is at (4, 0) and the directrix is a line whose equation is $x = 1$. Find the eccentricity of the hyperbola if (4, 6) are the coordinates of a point of the hyperbola.

In **12 – 13**, for the hyperbola described, find an equation of the directrix.

12. focus (8, 0), eccentricity = 4, (2, 0) are the coordinates of an endpoint of the transverse axis

13. focus (0, 10), eccentricity = $\frac{10}{3}$, (0, 3) are the coordinates of an endpoint of the transverse axis

In **14 – 15**, for the hyperbola described, find the coordinates of the focus.

14. eccentricity = $\frac{5}{2}$, directrix is $y = 4$, (0, 10) are the coordinates of an endpoint of the transverse axis

15. eccentricity = 6, directrix is $x = \frac{1}{2}$, (3, 0) are the coordinates of an endpoint of the transverse axis

---

10. $\frac{5}{2}$

11. 2

12. $x = \frac{1}{2}$

13. $y = \frac{9}{10}$

14. (0, 25)

15. (18, 0)

10.4 Relating the Ellipse and Hyperbola

If values for the focus, directrix, and eccentricity are known, the equation of a conic can be derived from the definition of eccentricity. Since, as seen in the preceding sections, the algebra of such derivations for the equations of an ellipse and a hyperbola are identical, it is possible to derive one general equation for both. Although this equation will refer to the ellipse and hyperbola that are centered at the origin and that have a vertical directrix, the equations of all other ellipses and hyperbolas can be obtained from this general equation by translations, reflections, and rotations.

If the directrix is a vertical line, the horizontal line through the focus is an axis of symmetry for the ellipse or hyperbola. By choosing a point $(c, 0)$ on the x-axis as the focus, the curve will be symmetric with respect to the x-axis. Let $x = d$ be the equation of the directrix, $P(a, 0)$ be a point on the curve such that $a > 0$, and M be the projection of P on the directrix so that $PF : PM = e$, the eccentricity.

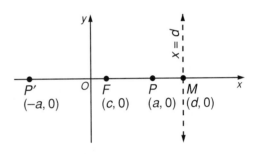

The curve will be symmetric with respect to the y-axis if the values of c and d are chosen correctly. Because of the positioning and the symmetry, c and d can be expressed in terms of a and e.

Since $P(a, 0)$ is on the curve:

$$PF = ePM$$
$$a - c = e(d - a)$$
$$a - c = de - ae \quad (1)$$

Add equations (1) and (2), and solve for d.

$$a - c = de - ae$$
$$a + c = de + ae$$
$$\overline{}$$
$$2a = 2de$$
$$\frac{a}{e} = d$$

Thus, the equation of the directrix is $x = \dfrac{a}{e}$.

By symmetry, $P'(-a, 0)$ is also on the curve:

$$P'F = eP'M$$
$$c - (-a) = e(d - (-a))$$
$$c + a = de + ae \quad (2)$$

Subtract equation (2) from equation (1), and solve for c.

$$a - c = de - ae$$
$$a + c = de + ae$$
$$\overline{}$$
$$-2c = -2ae$$
$$c = ae$$

Thus, the coordinates of the focus are $(ae, 0)$.

Although the coordinates of the focus and directrix are expressed in the same way for both the ellipse and the hyperbola, the relative positions of the focus and directrix depend on the value of e. Consider $P(a, 0)$ to be the vertex to the right of the origin. That is, $a > 0$.

Ellipse

$$0 < e < 1$$
$$0 < ae < a$$
$$0 < a < \frac{a}{e}$$

Thus, the focus is to the left and the directrix to the right of $(a, 0)$.

Hyperbola

$$e > 1$$
$$ae > a$$
$$a > \frac{a}{e}$$

Thus, the directrix is to the left and the focus to the right of $(a, 0)$.

Now, let the point $P(x, y)$ be any point on the curve. Note that $(a, 0)$ and $(-a, 0)$ are the vertices of the curves.

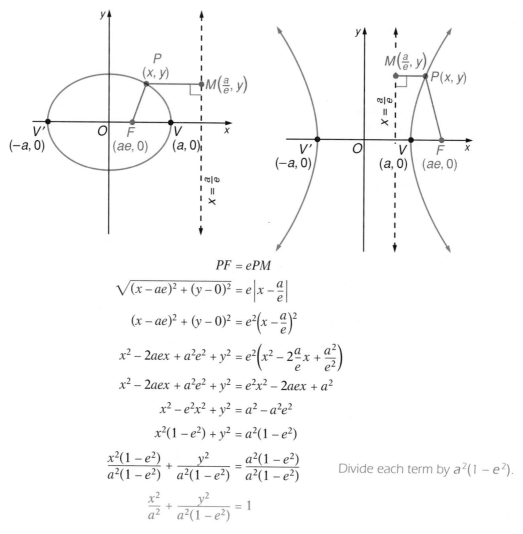

$$PF = ePM$$
$$\sqrt{(x - ae)^2 + (y - 0)^2} = e\left|x - \frac{a}{e}\right|$$
$$(x - ae)^2 + (y - 0)^2 = e^2\left(x - \frac{a}{e}\right)^2$$
$$x^2 - 2aex + a^2e^2 + y^2 = e^2\left(x^2 - 2\frac{a}{e}x + \frac{a^2}{e^2}\right)$$
$$x^2 - 2aex + a^2e^2 + y^2 = e^2x^2 - 2aex + a^2$$
$$x^2 - e^2x^2 + y^2 = a^2 - a^2e^2$$
$$x^2(1 - e^2) + y^2 = a^2(1 - e^2)$$
$$\frac{x^2(1 - e^2)}{a^2(1 - e^2)} + \frac{y^2}{a^2(1 - e^2)} = \frac{a^2(1 - e^2)}{a^2(1 - e^2)} \qquad \text{Divide each term by } a^2(1 - e^2).$$
$$\frac{x^2}{a^2} + \frac{y^2}{a^2(1 - e^2)} = 1$$

This is the general equation, in intercept form, for both the ellipse and the hyperbola.

The equations of the ellipse and the hyperbola differ by the sign of $1 - e^2$. As shown below, $1 - e^2$ is positive for the ellipse and negative for the hyperbola.

Ellipse	**Hyperbola**
Since e is a positive number less than 1, e^2 is also a positive number less than 1, and $1 - e^2$ is positive.	Since e is greater than 1, e^2 is also greater than 1, and $1 - e^2$ is a negative number.

Note how the familiar intercept forms showing the x- and y-intercepts are derived.

Since $a^2(1 - e^2)$ is positive, let: $$a^2(1 - e^2) = b^2$$	Since $a^2(1 - e^2)$ is negative, let: $$a^2(1 - e^2) = -b^2$$
$$\frac{x^2}{a^2} + \frac{y^2}{b^2} = 1$$	$$\frac{x^2}{a^2} - \frac{y^2}{b^2} = 1$$
The x-intercepts are a and $-a$.	The x-intercepts are a and $-a$.
The y-intercepts are b and $-b$.	There are no y-intercepts.
Length of the major axis: $\lvert 2a \rvert$	Length of the transverse axis: $\lvert 2a \rvert$
Length of the minor axis: $\lvert 2b \rvert$	Asymptotes: $y = \pm \dfrac{b}{a}x$

Example 1

Find the coordinates of the focus, the equation of the directrix, and the eccentricity of the ellipse whose equation is $9x^2 + 25y^2 = 225$.

Solution

Write the equation in intercept form.

$$9x^2 + 25y^2 = 225$$

$$\frac{9x^2}{225} + \frac{25y^2}{225} = \frac{225}{225}$$

$$\frac{x^2}{25} + \frac{y^2}{9} = 1$$

Thus, the x-intercepts are ± 5, the y-intercepts are ± 3, and the major axis is a portion of the x-axis. Now compare the equation above with the general equation of the ellipse having focus on the x-axis.

$$\frac{x^2}{a^2} + \frac{y^2}{b^2} = 1$$

$$\frac{x^2}{25} + \frac{y^2}{9} = 1 \qquad\qquad \frac{x^2}{a^2} + \frac{y^2}{a^2(1 - e^2)} = 1$$

$$a^2 = 25 \qquad\qquad a^2(1 - e^2) = 9$$

$$a = \pm 5 \qquad\qquad 25(1 - e^2) = 9$$

$$1 - e^2 = \frac{9}{25}$$

$$e^2 = \frac{16}{25}$$

$$e = \frac{4}{5} \quad (e \text{ is positive.})$$

Since, in general, the focus is at $(ae, 0)$ and the directrix is the line $x = \frac{a}{e}$, and since $a = \pm 5$ and $e = \frac{4}{5}$, either the focus is the point $(4, 0)$ and the equation of the directrix is $x = \frac{25}{4}$, or the focus is the point $(-4, 0)$ and the equation of the directrix is $x = -\frac{25}{4}$.

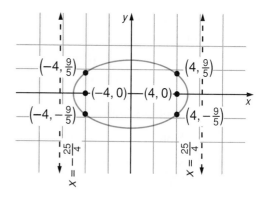

Answer: There are two possibilities for the focus, either $(4, 0)$ or $(-4, 0)$. If the focus is $(4, 0)$, the equation of the directrix is $x = \frac{25}{4}$, and if the focus is $(-4, 0)$, the equation of the directrix is $x = -\frac{25}{4}$. In either case, the eccentricity is $\frac{4}{5}$.

Since a can be positive or negative, every ellipse or hyperbola has two foci and two directrices. This is evident from the symmetry of the curves. If the ellipse or the hyperbola (along with its focus and directrix) are reflected over the y-axis, the image of $F(ae, 0)$ is $F'(-ae, 0)$ and the image of the directrix $x = \frac{a}{e}$ is the line $x = -\frac{a}{e}$.

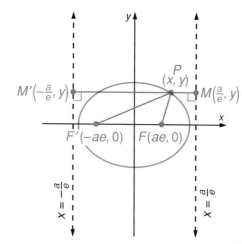

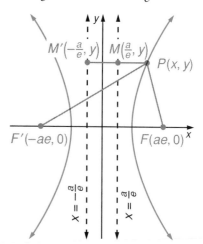

Ellipse

$$PF = ePM$$
$$PF' = ePM'$$

Adding these equations:

$$PF + PF' = ePM + ePM'$$
$$PF + PF' = e(PM + PM')$$

$(PM + PM')$ is the distance between the directrices.

Hyperbola

$$PF = ePM$$
$$PF' = ePM'$$

Subtracting:

$$|PF' - PF| = |ePM' - ePM|$$
$$|PF' - PF| = e|PM' - PM|$$

$|PM' - PM|$ is the distance between the directrices. Use absolute value since the difference could be positive or negative.

$$PM + PM' = \left|\left[\left(\frac{a}{e} - x\right) + \left(x - \left(-\frac{a}{e}\right)\right)\right]\right|$$

$$= \left|\frac{a}{e} + \frac{a}{e}\right| = \left|2\frac{a}{e}\right|$$

$$\left|PM' - PM\right| = \left|\left[\left(x - \left(-\frac{a}{e}\right)\right) - \left(x - \frac{a}{e}\right)\right]\right|$$

$$= \left|\frac{a}{e} + \frac{a}{e}\right| = \left|2\frac{a}{e}\right|$$

Since $PF + PF' = e(PM + PM')$:

$$PF + PF' = e\left|2\frac{a}{e}\right| = |2a|$$

Since $|PF' - PF| = e|PM' - PM|$:

$$|PF' - PF| = e\left|2\frac{a}{e}\right| = |2a|$$

These results can be used to state new definitions of the ellipse and the hyperbola.

An ellipse is the locus of points such that the sum of the distances from any point on the locus to two fixed points is a constant. The constant sum is equal to the length of the major axis.

A hyperbola is the locus of points such that the absolute value of the difference of the distances from any point on the locus to two fixed points is a constant. The constant difference is equal to the length of the transverse axis.

Example 2

The point $P\left(3, \frac{16}{5}\right)$ lies on the ellipse $16x^2 + 25y^2 = 400$.

Show that the sum of the distances from P to the foci is equal to the length of the major axis.

Solution

Write the equation in intercept form.

$$\frac{16x^2}{400} + \frac{25y^2}{400} = \frac{400}{400}$$

$$\frac{x^2}{25} + \frac{y^2}{16} = 1$$

$$a^2 = 25 \qquad a^2(1 - e^2) = 16$$

$$a = \pm 5 \qquad 25(1 - e^2) = 16$$

$$1 - e^2 = \frac{16}{25}$$

$$e^2 = \frac{9}{25}$$

$$e = \frac{3}{5} \text{ (e is positive.)}$$

The foci, $(\pm ae, 0)$, are at $F(3, 0)$ and $F'(-3, 0)$. Now find the sum of the distances from P to the foci.

$$PF + PF' = \sqrt{(3 - 3)^2 + \left(\frac{16}{5} - 0\right)^2} + \sqrt{(3 - (-3))^2 + \left(\frac{16}{5} - 0\right)^2}$$

$$= \sqrt{0 + \frac{256}{25}} + \sqrt{36 + \frac{256}{25}}$$

$$= \frac{16}{5} + \frac{34}{5} = 10$$

Since the length of the major axis is $|2a| = |2(5)|$ or 10, the sum of the distances from P to the foci is equal to the length of the major axis.

Example 3

Write an equation of the hyperbola whose foci are $F(6, 0)$ and $F'(-6, 0)$, and whose transverse axis has length 4.

Solution

The length of the transverse axis is $|2a| = 4$. Therefore, $a = \pm 2$.

Since the coordinates of one focus are $(6, 0)$, $ae = 6$.
Using $a = 2$: $ae = 6$, $2e = 6$, $e = 3$

Therefore,
$$\frac{x^2}{a^2} + \frac{y^2}{a^2(1 - e^2)} = 1$$

$$\frac{x^2}{2^2} + \frac{y^2}{2^2(1 - 3^2)} = 1$$

$$\frac{x^2}{4} + \frac{y^2}{4(1 - 9)} = 1$$

$$\frac{x^2}{4} + \frac{y^2}{(-32)} = 1$$

$$\frac{x^2}{4} - \frac{y^2}{32} = 1$$

Answer: An equation of the hyperbola is $\dfrac{x^2}{4} - \dfrac{y^2}{32} = 1$.

Conics With a Horizontal Directrix

The general equation of an ellipse or a hyperbola with focus at $(0, ae)$ and directrix $y = \dfrac{a}{e}$, can be found using a reflection over the line $y = x$.

Under a reflection over the line $y = x$, $(x, y) \rightarrow (y, x)$.

Ellipse	**Hyperbola**
The image of $\dfrac{x^2}{a^2} + \dfrac{y^2}{b^2} = 1$	The image of $\dfrac{x^2}{a^2} - \dfrac{y^2}{b^2} = 1$
is $\dfrac{y^2}{a^2} + \dfrac{x^2}{b^2} = 1$.	is $\dfrac{y^2}{a^2} - \dfrac{x^2}{b^2} = 1$.
The x-intercepts are b and $-b$.	There are no x-intercepts.
The y-intercepts are a and $-a$.	The y-intercepts are a and $-a$.
Length of the major axis: $|2a|$	Length of the transverse axis: $|2a|$
Length of the minor axis: $|2b|$	Asymptotes: $y = \pm \dfrac{a}{b} x$

Example 4

Sketch the graph of the hyperbola whose equation is $y^2 - 4x^2 = 1$.

Solution

Write the equation in intercept form.

$$y^2 - 4x^2 = 1$$

$$\frac{y^2}{1} - \frac{x^2}{\frac{1}{4}} = 1$$

$$a^2 = 1 \qquad b^2 = \frac{1}{4}$$

$$a = \pm 1 \qquad b = \pm \frac{1}{2}$$

Thus, the intercepts of the graph are (0, 1) and (0, –1), and the asymptotes of the graph are $y = \pm \frac{1}{\frac{1}{2}} x$, or $y = \pm 2x$.

Now sketch the two branches of the curve passing through the intercepts and approaching the asymptotes as $|x|$ increases.

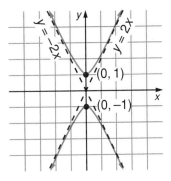

Translation of Axes

Translations can be used to derive the equation of an ellipse or a hyperbola with center at (h, k). If, through the point (h, k), an x'-axis is drawn parallel to the x-axis and a y'-axis is drawn parallel to the y-axis, (h, k) is the origin of this new coordinate system.

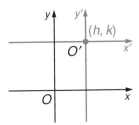

In terms of x' and y':

the equation of the ellipse with center at (h, k) is:

$$\frac{x'^2}{a^2} + \frac{y'^2}{b^2} = 1$$

But $x' = x - h$ and $y' = y - k$. Therefore:

the equation of the ellipse is:

$$\frac{(x - h)^2}{a^2} + \frac{(y - k)^2}{b^2} = 1$$

the equation of the hyperbola with center at (h, k) is:

$$\frac{x'^2}{a^2} - \frac{y'^2}{b^2} = 1$$

the equation of the hyperbola is:

$$\frac{(x - h)^2}{a^2} - \frac{(y - k)^2}{b^2} = 1$$

Example 5

Find the coordinates of the center, any turning points, and the foci, and the equations of the directrices for the conic whose equation is $x^2 + 4y^2 - 2x + 24y + 33 = 0$. Sketch the graph of the conic.

Solution

To write the equation in intercept form, complete the squares in x and y.

$$x^2 + 4y^2 - 2x + 24y + 33 = 0$$

$$x^2 - 2x + 4(y^2 + 6y) = -33$$

$$x^2 - 2x + 1 + 4(y^2 + 6y + 9) = -33 + 1 + 36$$

$$(x - 1)^2 + 4(y + 3)^2 = 4$$

$$\frac{(x-1)^2}{4} + \frac{4(y+3)^2}{4} = \frac{4}{4}$$

$$\frac{(x-1)^2}{4} + \frac{(y+3)^2}{1} = 1$$

$$a^2 = 4 \qquad\qquad a^2(1 - e^2) = 1$$

$$a = \pm 2 \qquad\qquad 4(1 - e^2) = 1$$

$$1 - e^2 = \frac{1}{4}$$

$$e^2 = \frac{3}{4}$$

$$e = \frac{\sqrt{3}}{2}$$

$$ae = \pm 2\left(\frac{\sqrt{3}}{2}\right) = \pm\sqrt{3}$$

Since $e = \frac{\sqrt{3}}{2} < 1$, the conic is an ellipse. Since $h = 1$ and $k = -3$, the center is at $C(1, -3)$.

The foci are on a horizontal line through the center at a distance of $ae = \pm\sqrt{3}$ from the center. Therefore, the foci are at $F(1 + \sqrt{3}, -3)$ and $F'(1 - \sqrt{3}, -3)$.

The endpoints of the major axis are on a horizontal line through the center at a distance of $a = \pm 2$ from the center. Thus, the endpoints of the major axis are at $A(3, -3)$ and $A'(-1, -3)$. The endpoints of the minor axis are on a vertical line through the center at a distance of $b = \pm 1$ from the center. Thus, the endpoints of the minor axis are at $B(1, -2)$ and $B'(1, -4)$. The equations of the directrices are:

$$x = h \pm \frac{a}{e} \quad \text{or} \quad x = 1 \pm \frac{4\sqrt{3}}{3}$$

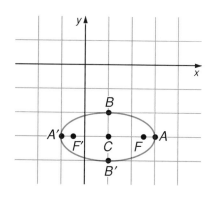

Graphs of Inequalities

To graph an inequality that is bounded by an ellipse or hyperbola, work with the related equality.

For example, every point on the ellipse $\frac{x^2}{9} + \frac{y^2}{4} = 1$ satisfies the equation of the ellipse. The remaining points in the plane are either inside the ellipse, satisfying the inequality $\frac{x^2}{9} + \frac{y^2}{4} < 1$, or outside the ellipse, satisfying the inequality $\frac{x^2}{9} + \frac{y^2}{4} > 1$.

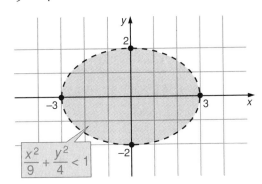
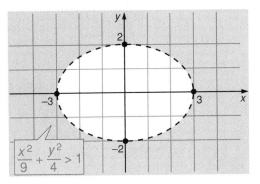

Recall that to display the graph of an ellipse on a calculator, it is necessary to enter the equation in two parts. To graph the ellipse $\frac{x^2}{9} + \frac{y^2}{4} = 1$, it is first necessary to solve for y.

$$\frac{x^2}{9} + \frac{y^2}{4} = 1$$

$$\frac{y^2}{4} = 1 - \frac{x^2}{9}$$

$$y^2 = 4\left(1 - \frac{x^2}{9}\right)$$

$$y = \pm 2\sqrt{1 - \frac{x^2}{9}}$$

On the calculator, enter the two parts of the equation of the ellipse.

Under Y_1, enter:

2 [2nd] [√] [(] 1 [−] [X|T] [X²] [÷] 9 [)]

Under Y_2, enter:

[(−)] 2 [2nd] [√] [(] 1 [−] [X|T] [X²] [÷] 9 [)]

Set the range for x from -3.5 to 3.5 and for y from -3 to 3.

The draw function on the calculator can be used to display the solution sets of the inequalities related to the ellipse $\frac{x^2}{9} + \frac{y^2}{4} = 1$. Press [2nd] [DRAW] to display the menu. Then highlight **7:Shade(** and press [ENTER]. The instruction that appears on the screen must be completed by expressions for the lower boundary and upper boundary of the curve.

1. a. ellipse
   b. vertices: $(\pm6, 0)$
   $(0, \pm2)$
   c.

2. a. hyperbola
   b. vertices:
   $(\pm\sqrt{2}, 0)$
   c.

3. a. ellipse
   b. vertices: $(\pm3, 0)$
   $(0, \pm\sqrt{3})$
   c.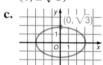

4. a. hyperbola
   b. vertices: $(\pm2, 0)$
   c.

5. a. ellipse
   b. vertices: $(\pm5, 0)$
   $(0, \pm10)$
   c.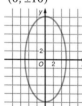

6. a. hyperbola
   b. vertices: $(0, \pm1)$
   c.

$$\frac{x^2}{9} + \frac{y^2}{4} < 1$$

$$\frac{y^2}{4} < 1 - \frac{x^2}{9}$$

$$y^2 < 4\left(1 - \frac{x^2}{9}\right)$$

$$y < 2\sqrt{1 - \frac{x^2}{9}} \quad \text{or} \quad y > -2\sqrt{1 - \frac{x^2}{9}}$$

Note that for the inequality, taking the positive square root leaves the inequality unchanged and taking the negative square root, like dividing a negative number, reverses the inequality.

Thus, the lower boundary for this inequality is the lower portion of the ellipse and the upper boundary is the upper portion of the ellipse. Since these parts of the ellipse have already been entered in the calculator as Y_2 and Y_1, respectively, they can be recalled by the key [Y-VARS].

The calculator will shade the region inside the ellipse.

To display the solution set for the inequality $\frac{x^2}{9} + \frac{y^2}{4} > 1$, it is necessary to shade the regions above the upper portion of the ellipse, Y_1, and below the lower portion of the ellipse, Y_2. The shaded region must be considered in two parts. The upper shaded region is bounded below by the upper half of the ellipse and by the x-axis. The calculator requires that the upper region be also bounded above. An appropriate value for the upper boundary is greater than the maximum y-value in the range of the ellipse, say $y = 4$. Enter this constant on the calculator as Y_3. Similarly, an appropriate constant for the lower boundary is smaller than the minimum y-value in the range of the ellipse, say $y = -4$. Enter this constant on the calculator as Y_4.

Thus, to display the solution set for the inequality $\frac{x^2}{9} + \frac{y^2}{4} > 1$, enter the following key sequence:

The calculator will redraw the ellipse and shade the upper region of the inequality. Continue with the following key sequence.

The calculator will shade the lower region of the inequality.

7. a. ellipse
   b. vertices:
   $(6 \pm 6, 0)$
   $(6, \pm3)$
   c.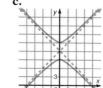

8. a. hyperbola
   b. vertices:
   $(0, 12 \pm 2\sqrt{2})$
   c.

9. a. ellipse
   b. vertices:
   $(-1 \pm 6, 1)$
   $(-1, 1 \pm 2)$
   c.

10. $\dfrac{x^2}{25} + \dfrac{y^2}{9} = 1$

11. $\dfrac{x^2}{36} + \dfrac{y^2}{4} = 1$

Exercises 10.4

In **1 – 9**: **a.** State whether the graph of the equation is an ellipse or a hyperbola.
  b. Find the coordinates of the vertices.
  c. Sketch the curve.

1. $x^2 + 9y^2 = 36$ **2.** $3x^2 - 2y^2 = 6$ **3.** $x^2 = 9 - 3y^2$
4. $x^2 = y^2 + 4$ **5.** $4x^2 + y^2 = 100$ **6.** $5y^2 - x^2 = 5$
7. $x^2 + 4y^2 - 12x = 0$ **8.** $x^2 - y^2 + 24y = 136$ **9.** $x^2 + 9y^2 + 2x - 18y = 26$

10. Write an equation of the ellipse that intersects the x-axis at $(\pm 5, 0)$ and the y-axis at $(0, \pm 3)$.

11. An ellipse is symmetric with respect to the x and y axes. The major axis is a segment of the x-axis with length 12 and the length of the minor axis is 4. Write an equation of the ellipse.

12. The equation of an ellipse is $9x^2 + 25y^2 = 225$.

  a. Find the x-intercepts of the ellipse.
  b. Find the y-intercepts of the ellipse.
  c. What is the length of the major axis?
  d. What is the length of the minor axis?
  e. What are the coordinates of the foci?
  f. What are the equations of the directrices?
  g. Sketch the ellipse.

In **13 – 14**, for the ellipse described:

  a. Write an equation.
  b. Find the x-intercepts.
  c. Find the y-intercepts.
  d. Find the length of the major axis.
  e. Find the length of the minor axis.
  f. Sketch the curve.

13. focus $(4, 0)$, directrix $x = 9$, and eccentricity $\frac{2}{3}$

14. focus $(0, 4)$, directrix $y = 9$, and eccentricity $\frac{2}{3}$

15. The ellipse whose equation is $5x^2 + 9y^2 = 180$ is reflected over the line $y = x$. Find the equation of the ellipse that is the image under this reflection.

16. Under a reflection in $y = -x$, the image of (x, y) is $(-y, -x)$.
  a. Write an equation of the ellipse $5x^2 + 9y^2 = 180$ under a reflection in the line $y = -x$.
  b. Compare your answer with the answer to Exercise 15.

17. The equation of a hyperbola is $3x^2 - y^2 = 12$.

  a. Find the x-intercepts of the hyperbola if they exist.
  b. Find the y-intercepts of the hyperbola if they exist.
  c. What is the length of the transverse axis?
  d. Find the coordinates of the foci.
  e. Write the equations of the directrices.
  f. Write the equations of the asymptotes.
  g. Sketch the hyperbola.

Answers 10.4

12. a. x-intercepts ± 5
  b. y-intercepts ± 3
  c. 10   **d.** 6
  e. $(\pm 4, 0)$
  f. $x = \pm\frac{25}{4}$
  g.

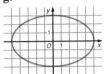

13. a. $\dfrac{x^2}{36} + \dfrac{y^2}{20} = 1$
  b. x-intercepts ± 6
  c. y-intercepts $\pm 2\sqrt{5}$
  d. 12
  e. $4\sqrt{5}$
  f.

14. a. $\dfrac{x^2}{20} + \dfrac{y^2}{36} = 1$
  b. x-intercepts $\pm 2\sqrt{5}$
  c. y-intercepts ± 6
  d. 12
  e. $4\sqrt{5}$
  f.

15. $5y^2 + 9x^2 = 180$
16. a. $5y^2 + 9x^2 = 180$
  b. Answers are the same.

17. a. x-intercepts ± 2
  b. none   **c.** 4
  d. $(\pm 4, 0)$
  e. $x = \pm 1$
  f. $y = \pm\sqrt{3}x$

17. g.

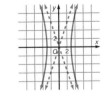

18. a. $\dfrac{x^2}{9} - \dfrac{y^2}{72} = 1$
  b. x-intercepts ± 3
  c. none   **d.** 6
  e. $y = \pm 2\sqrt{2}x$

18. f.

Answers 10.4

19. a. $\dfrac{y^2}{9} - \dfrac{x^2}{72} = 1$

b. none

c. y-intercepts ± 3

d. 6

e. $y = \pm\dfrac{\sqrt{2}}{4}x$

f.
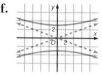

20. a. $\dfrac{3}{5}$ **b.** ellipse

c. $\dfrac{x^2}{25} + \dfrac{y^2}{16} = 1$

21. a. 3

b. hyperbola

c. $\dfrac{x^2}{9} - \dfrac{y^2}{72} = 1$

22. a. $\dfrac{4}{5}$ **b.** ellipse

c. $\dfrac{x^2}{36} + \dfrac{y^2}{100} = 1$

23. a. $\dfrac{3}{2}$

b. hyperbola

c. $-\dfrac{x^2}{5} + \dfrac{y^2}{4} = 1$

24. a. $\sqrt{5}$

b. hyperbola

c. $\dfrac{x^2}{20} - \dfrac{y^2}{80} = 1$

25. a. $\dfrac{\sqrt{3}}{3}$ **b.** ellipse

c. $\dfrac{x^2}{27} + \dfrac{y^2}{18} = 1$

26. Tacks should be 24 inches apart and the string should be 26 inches long.

Exercises 10.4 (continued)

In **18 – 19**, for the hyperbola described:

a. Write an equation.

b. Find the x-intercepts if they exist.

c. Find the y-intercepts if they exist.

d. Find the length of the transverse axis.

e. Write the equations of the two asymptotes.

f. Sketch the hyperbola.

18. focus (9, 0), directrix $x = 1$, and eccentricity 3

19. focus (0, 9), directrix $y = 1$, and eccentricity 3

In **20 – 25**, the given points are the foci of a conic and the given value of $2a$ is the length of the major or transverse axis of that conic.

a. Find the eccentricity of the conic.

b. Identify the conic as an ellipse or a hyperbola.

c. Write the equation of the conic in intercept form.

20. $(\pm 3, 0)$, $2a = 10$

21. $(\pm 9, 0)$, $2a = 6$

22. $(0, \pm 8)$, $2a = 20$

23. $(0, \pm 3)$, $2a = 4$

24. $(\pm 10, 0)$, $2a = 4\sqrt{5}$

25. $(\pm 3, 0)$, $2a = 6\sqrt{3}$

26. An ellipse can be drawn by attaching the ends of a string to tacks placed at the two foci. The string should be equal in length to the major axis of the ellipse. Using a pencil, pull the string taut and draw the ellipse. A carpenter wants to draw a pattern for a table top that is to be in the shape of an ellipse. If the length of the major axis is to be 26 inches and the length of the minor axis is to be 10 inches, how far apart should the tacks be placed and how long should the string be?

27. Write an equation of the ellipse with center at (2, –3), focus at (5, –3) and whose directrix has the equation $x = \dfrac{31}{3}$.

28. Write an equation of the hyperbola whose transverse axis has endpoints at (1, 1) and (1, 5), and which has eccentricity 3.

29. Write an equation of the parabola with focus at (0, 0) and turning point at (0, –4).

30. Write an equation of the hyperbola with focus at the origin and whose transverse axis has endpoints at (0, 2) and (0, 12).

In **31 – 34**, graph the inequality.

31. $x^2 + 4y^2 < 4$

32. $x^2 + 20y^2 > 100$

33. $x^2 - y^2 < 1$

34. $x^2 - 9y^2 > 36$

27.
$$\dfrac{(x-2)^2}{25} + \dfrac{(y+3)^2}{16} = 1$$

28.
$$-\dfrac{(x-1)^2}{32} + \dfrac{(y-3)^2}{4} = 1$$

29. $y + 4 = \dfrac{1}{16}x^2$

30.
$$-\dfrac{x^2}{24} + \dfrac{(y-7)^2}{25} = 1$$

31.

$x^2 + 4y^2 < 4$

32.

$x^2 + 20y^2 > 100$

33.
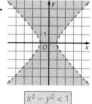
$x^2 - y^2 < 1$

34.

$x^2 - 9y^2 > 36$

10.5 The Circle; Special Cases

Recall that the eccentricity of an ellipse is a positive constant less than 1. As the value of the eccentricity approaches 0, the length of the minor axis of the ellipse approaches the length of the major axis and the shape of the ellipse approaches the shape of a circle. In terms of the two-focus definition of an ellipse, if $e = 0$ then $ae = 0$ and $-ae = 0$, and the two foci coincide. This result leads to the following definition.

■ **Definition:** *A circle is the locus of points equidistant from a fixed point.*

Let $P(x, y)$ be any point of a circle with center at (h, k) and let a be the distance from the center to any point of the circle. An equation of the circle can be obtained by using the distance formula:

$$\sqrt{(x-h)^2 + (y-k)^2} = a$$

$$(x-h)^2 + (y-k)^2 = a^2$$

This equation can also be derived from the general equation of an ellipse with $e = 0$ and center at (h, k).

$$\frac{(x-h)^2}{a^2} + \frac{(y-k)^2}{a^2(1-0^2)} = 1$$

$$\frac{(x-h)^2}{a^2} + \frac{(y-k)^2}{a^2} = 1 \quad \text{or} \quad (x-h)^2 + (y-k)^2 = a^2$$

Example 1

Write an equation of the locus of points 3 units from $(2, -1)$.

Solution

Use the formula: $(x-h)^2 + (y-k)^2 = a^2$

Let: $h = 2$, $k = -1$, and $a = 3$

$$(x-2)^2 + (y-(-1))^2 = 3^2$$
$$(x-2)^2 + (y+1)^2 = 9$$

Answer: An equation of the circle is $(x-2)^2 + (y+1)^2 = 9$.

Example 2

Find the coordinates of the center and the length of the radius of the circle whose equation is $x^2 + y^2 - 4x + 6y - 12 = 0$.

Solution

Complete the squares in x and y in order to write the equation in the form $(x - h)^2 + (y - k)^2 = a^2$.

$$x^2 + y^2 - 4x + 6y - 12 = 0$$
$$x^2 - 4x + y^2 + 6y = 12$$
$$x^2 - 4x + 4 + y^2 + 6y + 9 = 12 + 4 + 9$$
$$(x - 2)^2 + (y + 3)^2 = 25$$
$$(x - 2)^2 + (y + 3)^2 = 5^2$$

Therefore, $h = 2$, $k = -3$, and $a = 5$.

Answer: The center is at $(2, -3)$ and the length of the radius is 5.

General Equation for a Conic

In the preceding sections, we have derived the equations of conics with vertical or horizontal directrices by using the eccentricity. Each equation contained the squares of either x or y or of both x and y. In general, the equation of a conic can be expressed as follows:

$$Ax^2 + Bxy + Cy^2 + Dx + Ey + F = 0 \quad \text{where } A, B, \text{ and } C \text{ are not all 0.}$$

The coefficients of this equation can be used to differentiate among the conics.

If $B \neq 0$, an axis of symmetry is an oblique line. If $B = 0$, an axis of symmetry of the conic is parallel to one of the coordinates axes.

If $B^2 - 4AC < 0$, the conic is an ellipse. For example:

$$x^2 + 4y^2 - 2x + 24y + 33 = 0$$
$$\frac{(x - 1)^2}{4} + \frac{(y + 3)^2}{1} = 1$$

$B^2 - 4AC = 0^2 - 4(1)(4) = -16 < 0$
Completing the squares leads to the equation of an ellipse.

If $B^2 - 4AC = 0$, the conic is a parabola. For example:

$$\tfrac{1}{8}x^2 - x - y + 1 = 0$$
$$\tfrac{1}{8}x^2 - x + 1 = y$$

$B^2 - 4AC = 0^2 - 4\left(\tfrac{1}{8}\right)(0) = 0$
For a parabola, when $B = 0$ either A or C must be 0, but not both.

If $B^2 - 4AC > 0$, the conic is a hyperbola. For example:

$$4x^2 - y^2 - 8x - 4y - 4 = 0$$
$$\frac{(x - 1)^2}{1} - \frac{(y + 2)^2}{4} = 1$$

$B^2 - 4AC = 0^2 - 4(4)(-1) = 16 > 0$
Completing the squares leads to the equation of a hyperbola.

Moreover, when $B = 0$, the coefficients A and C play an important role.

If $A = C$, the conic is a circle. For example:

$$x^2 + y^2 - 4x + 6y - 12 = 0 \qquad B = 0, A = C$$
$$(x - 2)^2 + (y + 3)^2 = 5^2 \qquad \text{Completing the squares leads to the equation of a circle.}$$

If $A \neq C$ and they have the same sign, the conic is an ellipse. For example:

$$9x^2 + 4y^2 - 54x - 8y + 49 = 0 \qquad B = 0, A \neq C, \text{ both positive}$$
$$\frac{(x - 3)^2}{4} + \frac{(y - 1)^2}{9} = 1 \qquad \text{Completing the squares leads to the equation of an ellipse.}$$

If A and C have opposite signs, the conic is a hyperbola. For example:

$$9x^2 - 4y^2 - 54x + 8y + 41 = 0 \qquad B = 0, A \neq C, \text{ opposite signs}$$
$$\frac{(x - 3)^2}{4} - \frac{(y - 1)^2}{9} = 1 \qquad \text{Completing the squares leads to the equation of a hyperbola.}$$

Degenerate Conics

To write the equation of a conic in intercept form, you must complete the squares in x and y and then divide both sides of the equation by the resulting constant. However, if the constant is 0, this last step is not possible and the equation represents what is called a *degenerate conic*.

Example 3

Graph the equation $x^2 - y^2 + 2x - 8y - 15 = 0$.

Solution

$$x^2 - y^2 + 2x - 8y - 15 = 0$$
$$x^2 + 2x - (y^2 + 8y) = 15 \qquad \text{Complete the square.}$$
$$x^2 + 2x + 1 - (y^2 + 8y + 16) = 15 + 1 - 16$$
$$(x + 1)^2 - (y + 4)^2 = 0$$

Factor the difference of two squares on the left side of the equation.

$$[(x + 1) + (y + 4)][(x + 1) - (y + 4)] = 0$$
$$[x + y + 5][x - y - 3] = 0$$

Now, since the constant on the right side is 0, set each factor equal to 0.

$$x + y + 5 = 0 \quad \text{or} \quad x - y - 3 = 0$$
$$x + y = -5 \qquad\qquad x - y = 3$$

But, since $x + y = -5$ and $x - y = 3$ are each a linear equation, the graph is a degenerate hyperbola. The graph consists of two lines that intersect at $(-1, -4)$.

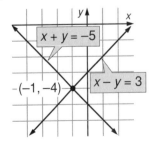

In general, the equation of a degenerate hyperbola is:

$$b^2(x - h)^2 - a^2(y - k)^2 = 0$$

Since this equation is the equivalent of two linear equations, a degenerate hyperbola is two intersecting lines, which are the asymptotes of a corresponding actual hyperbola.

The equation of a degenerate ellipse has the form:

$$b^2(x - h)^2 + a^2(y - k)^2 = 0$$

Since the only solution of this equation in the set of real numbers is (h, k), a degenerate ellipse is a point, which is the center of a corresponding actual ellipse.

The equation of a degenerate parabola has the form:

$$0(y - k) = (x - h)^2$$

Since this is the equation of the line $x = h$, a degenerate parabola is a line, which is the axis of symmetry of a corresponding actual parabola.

Exercises 10.5

In **1 – 6,** write an equation of the locus of points at the given distance, a, from the given point.

1. $a = 2$, $(0, 0)$ **2.** $a = 5$, $(1, 3)$ **3.** $a = 1$, $(-1, 5)$

4. $a = 3$, $(0, -3)$ **5.** $a = 12$, $(-4, 0)$ **6.** $a = 6$, $(-2, -2)$

In **7 – 14,** find the coordinates of the center and the length of the radius of the circle whose equation is given.

7. $x^2 + y^2 = 16$ **8.** $(x - 3)^2 + (y - 1)^2 = 49$

9. $(x - 2)^2 + (y + 1)^2 = 4$ **10.** $(x + 6)^2 + (y + 5)^2 = 8$

11. $x^2 + y^2 - 2y = 0$ **12.** $x^2 + y^2 - 2x - 8y = -1$

13. $x^2 + y^2 + 12y + 20 = 0$ **14.** $x^2 + y^2 + 5x - 10y = 11$

15. Write an equation of the circle with center at $(2, -3)$ that is tangent to the *x*-axis.

16. A circle whose center is in the second quadrant is tangent to both axes. Write an equation of the circle if the length of the radius is 6.

17. Write an equation of the circle with center at $(-2, 5)$ that passes through $(0, 8)$.

18. Write an equation of a circle that has a chord with endpoints at $(1, -3)$ and $(1, 7)$ which is 12 units from the center of the circle. (Two circles are possible.)

19. Find an equation of the circle that passes through the points $(1, -2)$, $(1, 4)$, and $(5, 6)$.

1. $x^2 + y^2 = 4$ **2.** $(x - 1)^2 + (y - 3)^2 = 25$ **3.** $(x + 1)^2 + (y - 5)^2 = 1$ **4.** $x^2 + (y + 3)^2 = 9$

5. $(x + 4)^2 + y^2 = 144$ **6.** $(x + 2)^2 + (y + 2)^2 = 36$ **7.** $(0, 0)$; 4   **8.** $(3, 1)$; 7

9. $(2, -1)$; 2   **10.** $(-6, -5)$; $2\sqrt{2}$ **11.** $(0, 1)$; 1   **12.** $(1, 4)$; 4

13. $(0, -6)$; 4   **14.** $\left(-\frac{5}{2}, 5\right)$; $\frac{13}{2}$ **15.** $(x - 2)^2 + (y + 3)^2 = 9$ **16.** $(x + 6)^2 + (y - 6)^2 = 36$

17. $(x + 2)^2 + (y - 5)^2 = 13$

18. $(x - 13)^2 + (y - 2)^2 = 169$ and
$(x + 11)^2 + (y - 2)^2 = 169$

19. $(x - 5)^2 + (y - 1)^2 = 25$

20. a. circle

b. center: $(0, 3)$
intercepts on horizontal and vertical axes
of symmetry: $(0 \pm 2, 3)$ and $(0, 3 \pm 2)$

21. a. ellipse

b. center: $(1, 0)$
vertices: $(1 \pm \sqrt{2}, 0), (1, \pm \sqrt{6})$

22. a. hyperbola

b. center: $(6, 0)$
vertices: $(6 \pm 6, 0)$

23. a. parabola

b. vertex: $(1, -2)$

24. a. degenerate conic

25. a. hyperbola

b. center: $(0, 0)$
vertices: $(0, \pm 1)$

26. a. hyperbola

b. center: $\left(0, \frac{1}{2}\right)$
vertices: $\left(0, \frac{1}{2} \pm \frac{1}{2}\right)$

27. a. circle

b. center: $(-4, 3)$
intercepts on horizontal and vertical axes of
symmetry: $(-4 \pm 5, 3)$ and $(-4, 3 \pm 5)$

28. a. parabola

b. vertex: $(1, 2)$

29. a. degenerate conic

30. a. $\dfrac{(x + 2)^2}{8} + \dfrac{(y - 2)^2}{4} = 1$

b. center: $(-2, 2)$
endpoints of major axis: $(-2 \pm 2\sqrt{2}, 2)$
endpoints of minor axis: $(-2, 0)$ and $(-2, 4)$
foci: $(-2 \pm 2, 2)$
directrices: $x = -2 \pm 4$

31. **a.** $\dfrac{(x-6)^2}{36} - \dfrac{y^2}{36} = 1$

b. center: $(6, 0)$
endpoints of transverse axis: $(6 \pm 6, 0)$
foci: $(6 \pm 6\sqrt{2}, 0)$
directrices: $x = 6 \pm 3\sqrt{2}$

32. **a.** $2(y + 4) = (x + 4)^2$

b. turning point: $(-4, -4)$
focus: $\left(-4, -\dfrac{7}{2}\right)$
directrix: $y = -\dfrac{9}{2}$

33. **a.** $\dfrac{(x+1)^2}{4} + \dfrac{(y+1)^2}{6} = 1$

b. center: $(-1, -1)$
endpoints of major axis: $(-1, -1 \pm \sqrt{6})$
endpoints of minor axis: $(1, -1), (-3, -1)$
foci: $(-1, -1 \pm \sqrt{2})$
directrices: $y = -1 \pm 3\sqrt{2}$

34. **a.** $2(x + 7) = (y + 3)^2$

b. turning point: $(-7, -3)$
focus: $\left(-\dfrac{13}{2}, -3\right)$
directrix: $x = -\dfrac{15}{2}$

35. **a.** $-\dfrac{(x+1)^2}{256} + \dfrac{(y+16)^2}{256} = 1$

b. center: $(-1, -16)$
endpoints of transverse axis: $(-1, -16 \pm 16)$
foci: $(-1, -16 \pm 16\sqrt{2})$
directrices: $y = -16 \pm 8\sqrt{2}$

36. **a.** $x^2 - y^2 = 2$
$$\left(x \cos \tfrac{\pi}{4} + y \sin \tfrac{\pi}{4}\right)^2 - \left(-x \sin \tfrac{\pi}{4} + y \cos \tfrac{\pi}{4}\right)^2 = 2$$
$$\left(x^2 \cdot \tfrac{1}{2} + 2xy \cdot \tfrac{1}{2} + y^2 \cdot \tfrac{1}{2}\right) - \left(x^2 \cdot \tfrac{1}{2} - 2xy \cdot \tfrac{1}{2} + y^2 \cdot \tfrac{1}{2}\right) = 2$$
$$2xy = 2$$
$$xy = 1$$

b. $xy = 1$
$$\left(x \cos \tfrac{\pi}{4} + y \sin \tfrac{\pi}{4}\right)\left(-x \sin \tfrac{\pi}{4} + y \cos \tfrac{\pi}{4}\right) = 1$$
$$-x^2 \cdot \tfrac{1}{2} + xy \cdot \tfrac{1}{2} - xy \cdot \tfrac{1}{2} + y^2 \cdot \tfrac{1}{2} = 1$$
$$\tfrac{1}{2}y^2 - \tfrac{1}{2}x^2 = 1$$
$$y^2 - x^2 = 2$$

c. $x^2 - y^2 = 2$
$$\left(x \cos \tfrac{\pi}{2} + y \sin \tfrac{\pi}{2}\right)^2 - \left(-x \sin \tfrac{\pi}{2} + y \cos \tfrac{\pi}{2}\right)^2 = 2$$
$$(0 + y)^2 - (-x + 0)^2 = 2$$
$$y^2 - x^2 = 2$$

d. $x = y^2$
$$\left(x \cos \tfrac{\pi}{4} + y \sin \tfrac{\pi}{4}\right) = \left(-x \sin \tfrac{\pi}{4} + y \cos \tfrac{\pi}{4}\right)^2$$
$$x \cdot \tfrac{\sqrt{2}}{2} + y \cdot \tfrac{\sqrt{2}}{2} = x^2 \cdot \tfrac{1}{2} - xy + y^2 \cdot \tfrac{1}{2}$$
$$x^2 - 2xy + y^2 - \sqrt{2}x - \sqrt{2}y = 0$$

e. $Ax^2 + Bxy + Cy^2 + Dx + Ey + F = 0$
Since $B^2 - 4AC = (-2)^2 - 4(1)(1) = 0$, the curve is a parabola.

In **20 – 29**: **a.** Identify the equation as that of a circle, ellipse, hyperbola, parabola, or a degenerate conic.

b. If the conic is not degenerate, find the center and vertex (or vertices) on the axis (or axes) of symmetry.

20. $x^2 + y^2 - 6y + 5 = 0$

21. $3x^2 + y^2 - 6x - 3 = 0$

22. $x^2 - y^2 - 12x = 0$

23. $y^2 + 4x + 4y = 0$

24. $x^2 + 5y^2 + 2x + 1 = 0$

25. $x^2 - 4y^2 + 4 = 0$

26. $x^2 - 4y^2 + 4y = 0$

27. $x^2 + y^2 + 8x - 6y = 0$

28. $x^2 - 2x - y + 3 = 0$

29. $5x^2 + 3y^2 + 10x - 12y + 17 = 0$

In **30 – 35**: **a.** Write the equation in intercept form.

b. If the conic is an ellipse or hyperbola, find the coordinates of the center, the coordinates of the endpoints of the major and minor axes or of the transverse axis, the coordinates of the foci, and the equations of the directrices. If the conic is a parabola, find the coordinates of the turning point, the coordinates of the focus, and the equation of the directrix.

30. $x^2 + 2y^2 + 4x - 8y + 4 = 0$

31. $x^2 - y^2 - 12x = 0$

32. $x^2 + 8x - 2y + 8 = 0$

33. $3x^2 + 2y^2 + 6x + 4y - 7 = 0$

34. $-y^2 + 2x - 6y + 5 = 0$

35. $x^2 - y^2 + 2x - 32y + 1 = 0$

36. When a set of points with a given equation is rotated about the origin through an angle whose measure is α, the equation of the image can be found by replacing x by $x \cos \alpha + y \sin \alpha$ and y by $-x \sin \alpha + y \cos \alpha$ in the original equation.

a. Show that the image of $x^2 - y^2 = 2$ under a rotation of $\frac{\pi}{4}$ is $xy = 1$.

b. Show that the image of $xy = 1$ under a rotation of $\frac{\pi}{4}$ is $y^2 - x^2 = 2$.

c. Show that the image of $x^2 - y^2 = 2$ under a rotation of $\frac{\pi}{2}$ is $y^2 - x^2 = 2$.

d. What is the image of $x = y^2$ under a rotation of $\frac{\pi}{4}$?

e. Use the value of $B^2 - 4AC$ to show that the result of part **d** is a parabola.

Chapter 10   *Summary and Review*

Defining a Conic

- A conic is the locus of points such that the ratio of the distances of any point on the locus to a fixed point (the focus) and a fixed line (the directrix) is a constant.

- In the locus definition of a conic, the constant ratio of distances is the eccentricity.

  Let F be the fixed point and let M be the point where the perpendicular from any point P on the locus meets the fixed line. Then the eccentricity is:

  $$\frac{PF}{PM} = e$$

  e is always positive since it represents the ratio of two distances.

- The value of e determines the nature of the curve:

  If $0 < e < 1$, the curve is an ellipse.

  If $e = 1$, the curve is a parabola.

  If $e > 1$, the curve is a hyperbola.

- The definition of the eccentricity, $PF : PM$, and its known value are used to derive the equations of three conics.

The Parabola

- A parabola is a conic whose eccentricity is 1, or the locus of points equidistant from a fixed point and a fixed line.

- The simplest equation for a parabola is derived by placing the focus on the y-axis and the directrix perpendicular to the y-axis so that the origin is d units from both the focus and the directrix. The equation is:

  $$y = \frac{1}{4d} x^2$$

- Different positions for the focus and directrix affect the general equation of a parabola with vertex, or turning point, at (h, k).

  focus at $(h, k + d)$
  directrix $y = k - d$ parabola: $y - k = \dfrac{1}{4d}(x - h)^2$

  focus at $(h + d, k)$
  directrix $x = h - d$ parabola: $x - h = \dfrac{1}{4d}(y - k)^2$

- A parabola with a vertical axis of symmetry:

    opens upward when d is positive,
    opens downward when d is negative.

    A parabola with horizontal axis of symmetry:

    opens to the right when d is positive,
    opens to the left when d is negative.

- The latus rectum of a parabola is the line segment through the focus, parallel to the directrix, with endpoints on the curve.

The Ellipse

- An ellipse is

    a conic whose eccentricity is a positive constant that is less than 1, or

    the locus of points such that the sum of the distances from any point on the locus to two fixed points is a constant. The constant sum is equal to the length of the major axis.

- In intercept form, the equation of an ellipse is:

    center at $(0, 0)$ $\quad \dfrac{x^2}{a^2} + \dfrac{y^2}{b^2} = 1$ $\qquad$ center at (h, k) $\quad \dfrac{(x - h)^2}{a^2} + \dfrac{(y - k)^2}{b^2} = 1$

The Hyperbola

- A hyperbola is

    a conic whose eccentricity is greater than 1, or

    the locus of points such that the absolute value of the difference of the distances from any point on the locus to two fixed points is a constant. The constant difference is equal to the length of the transverse axis.

- In intercept form, the equation of a hyperbola is:

    center at $(0, 0)$ $\quad \dfrac{x^2}{a^2} - \dfrac{y^2}{b^2} = 1$ $\qquad$ center at (h, k) $\quad \dfrac{(x - h)^2}{a^2} - \dfrac{(y - k)^2}{b^2} = 1$

The Circle

- A circle is the locus of points equidistant from a fixed point. The equation of a circle with radius a is:

    center at $(0, 0)$ $\quad x^2 + y^2 = a^2$ $\qquad$ center at (h, k) $\quad (x - h)^2 + (y - k)^2 = a^2$

- A single equation is used to describe the conics:

$$Ax^2 + Bxy + Cy^2 + Dx + Ey + F = 0$$

If $B = 0$, the axis of symmetry is parallel to one of the coordinate axes. The following tests can be used to determine the type of conic.

If $B^2 - 4AC < 0$, the conic is an ellipse.

If $B^2 - 4AC = 0$, the conic is a parabola.

If $B^2 - 4AC > 0$, the conic is a hyperbola.

When $B = 0$:

If $A = C$, the conic is a circle.

If $A \neq C$, and they have the same sign, the conic is an ellipse.

If $A \neq C$, and they have different signs, the conic is a hyperbola.

- Summary of Conics

Parabola

Equation:	$y - k = \dfrac{1}{4d}(x - h)^2$	$x - h = \dfrac{1}{4d}(y - k)^2$
Focus:	$(h, k + d)$	$(h + d, k)$
Directrix:	$y = k - d$	$x = h - d$
Axis of symmetry:	$x = h$	$y = k$
Turning point:	(h, k)	(h, k)

Ellipse

Equation:	$\dfrac{(x - h)^2}{a^2} + \dfrac{(y - k)^2}{b^2} = 1$	$\dfrac{(x - h)^2}{b^2} + \dfrac{(y - k)^2}{a^2} = 1$

$$\text{where } b^2 = a^2(1 - e^2)$$

Foci:	$(h \pm ae, 0)$	$(0, k \pm ae)$
Directrices:	$x = h \pm \dfrac{a}{e}$	$y = k \pm \dfrac{a}{e}$
Axes of symmetry:	$x = h$ and $y = k$	$x = h$ and $y = k$
Center:	(h, k)	(h, k)

Circle

Equation:	$(x - h)^2 + (y - k)^2 = a^2$
Axes of symmetry:	Any line through (h, k)
Center:	(h, k)

Hyperbola

Equation:	$\dfrac{(x - h)^2}{a^2} - \dfrac{(y - k)^2}{b^2} = 1$	$\dfrac{(y - k)^2}{a^2} - \dfrac{(x - h)^2}{b^2} = 1$

$$\text{where } b^2 = a^2(1 - e^2)$$

Foci:	$(h \pm ae, 0)$	$(0, k \pm ae)$
Directrices:	$x = h \pm \dfrac{a}{e}$	$y = k \pm \dfrac{a}{e}$
Axes of symmetry:	$x = h$ and $y = k$	$x = h$ and $y = k$
Center:	(h, k)	(h, k)
Asymptotes:	$y = \pm \dfrac{b}{a}(x - h) + k$	$y = \pm \dfrac{a}{b}(x - h) + k$

- A degenerate hyperbola consists of two lines that intersect. A degenerate ellipse is a point. A degenerate parabola is a line.

Chapter 10   Review Exercises

In **1–6**:   **a.** Write an equation of the conic with the given points as foci, the given lines as directrices and the given eccentricity.
         b. Identify the conic.

1. $\left(0, \frac{1}{2}\right), y = -\frac{1}{2}, e = 1$ **2.** $(\pm 3, 0), x = \pm 12, e = \frac{1}{2}$ **3.** $(\pm 9, 0), x = \pm 1, e = 3$

4. $(0, 0), x = 4, e = 1$ **5.** $(0, \pm 4), y = \pm 25, e = \frac{2}{5}$ **6.** $(0, \pm 3), y = \pm 3, e = 2$

In **7–10**:   **a.** Identify the conic.   **b.** Find the eccentricity.
         c. Find the coordinates of the focus.   **d.** Find an equation of the directrix.

7. $\frac{x^2}{4} + y^2 = 1$ **8.** $\frac{(x-3)^2}{2} - \frac{(y+1)^2}{2} = 1$

9. $x^2 + 8y^2 + 6x + 1 = 0$ **10.** $3x^2 - y^2 + 12x + 2y + 2 = 0$

11. a. Write an equation of the ellipse with foci at $(\pm 4, 0)$ and major axis of length 10.
      b. Sketch the graph of the ellipse.

12. a. Write an equation of the hyperbola with foci at $(\pm 2, 0)$ and eccentricity 2.
      b. Sketch the graph of the hyperbola.

13. a. Write an equation of the parabola with turning point $(1, 3)$ and focus $(1, 5)$.
      b. Sketch the graph of the parabola.

14. Find the coordinates of the foci of an ellipse that intersects the x-axis at $(\pm 10, 0)$ and the y-axis at $(0, \pm 4)$.

15. Write an equation of the circle, with center at $(5, 6)$, that passes through the point $(0, -6)$.

16. Find the center and radius of the circle whose equation is $x^2 + y^2 - 4x - 3y = 0$.

17. a. Show that the image of the graph of the equation
$$\frac{x^2}{1^2} + \frac{y^2}{1^2(1 - e^2)} = 1 \text{ under the dilation } D_a \text{ is the graph}$$
      whose equation is $\dfrac{x^2}{a^2} + \dfrac{y^2}{a^2(1 - e^2)} = 1$.
      b. What conclusion can be drawn from part **a**?

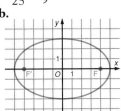

Chapter 10 Review Exercises (continued)

A *whisper chamber* or *whispering gallery* is a domed space in which sound that originates at one focus is reflected to the other focus. The dome may be half of the surface formed by rotating an ellipse about its major axis or by rotating two parabolas that face in opposite directions about a common axis of symmetry. Sound that originates at one focus is reflected from the surface and can be clearly heard at the other focus.

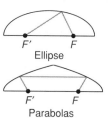

One famous whispering gallery is Statuary Hall of the United States Capitol in Washington, D.C. John Quincy Adams discovered this property of the hall when he was a member of the House of Representatives, which met in Statuary Hall until 1857. Other famous whispering galleries are in the Mormon Tabernacle in Salt Lake City, Utah, and St. Paul's Cathedral in London, England.

18. A whispering gallery has a dome that has a cross-sectional shape of an ellipse. The length of the major axis of the ellipse is 100 m and the length of the minor axis is 80 m. Locate the two points at which sound from one point can be heard at the other.

Each planet in the solar system moves in an elliptical orbit about the sun, with the sun at a focus of the ellipse. The point in a planet's orbit when it is closest to the sun is called *perihelion* and the point when it is farthest from the sun is called *aphelion*. The perihelion and aphelion are the endpoints of the major axis of the ellipse.

19. At perihelion, Mercury is 46,000,000 km from the sun and at aphelion it is 69,800,000 km from the sun.

   a. Find the length of the major axis of Mercury's orbit.

   b. Find the distance from the center of the orbit to the sun.

   c. Find the eccentricity of Mercury's orbit.

20. The eccentricity of Earth's orbit is 0.0167 and the sun is 2,500,000 km from the center of the orbit. Find, to the nearest hundred-thousand kilometers, the distance from Earth to the sun at perihelion and at aphelion.

21. The Gateway Arch in St. Louis, Missouri, was built to be a symbol of the city as the "gateway to the west." The arch is approximately the shape of a parabola that opens downward, with the turning point of the parabola 590 ft. above the ground. The parabola has a focus that is 37.5 ft. below the turning point. Find, to the nearest foot, the span of the arch at its base.

---

17. a. Under the dilation D_a the image of (x, y) is (ax, ay). When (ax, ay) is a point on $\dfrac{(ax)^2}{a^2} + \dfrac{(ay)^2}{a^2(1 - e^2)} = 1$, the equation becomes $\dfrac{x^2}{1^2} + \dfrac{y^2}{1^2(1 - e^2)} = 1$ and (x, y) is a point on $\dfrac{x^2}{1^2} + \dfrac{y^2}{1^2(1 - e^2)} = 1$.

  b. If two conics have the same eccentricity, each is the image of the other under a dilation, as well as under a translation or a rotation. All conics with the same eccentricity are similar.

18. The two points are 30 m from the center, on the major axis.

19. a. 115,800,000 km
    b. 11,900,000 km   **c.** 0.21

20. at perihelion: 147,200,000 km
   at aphelion: 152,200,000 km

21. 595 feet

☑ Exercises for Challenge

1. The endpoints of a diameter of the circle whose equation is
$x^2 + 4x + y^2 + ky = 0$
are $(0, 0)$ and $(-4, 6)$. Find k.

2. 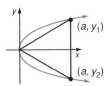 Mars moves in an ellipse with the sun as one focus. The smallest distance from Mars to the sun is 206.8 million km and the greatest distance is 249.2 million km. To the nearest thousandth, what is the eccentricity of the orbit of Mars?

3. Find the positive value of a such that the three points, $(0, 0)$, (a, y_1), and (a, y_2) on the parabola $y^2 = 3x$ form the vertices of an equilateral triangle.

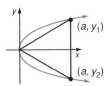

4. Determine the distance from a focus of the hyperbola $\dfrac{x^2}{25} - \dfrac{y^2}{16} = 1$ to one of its asymptotes.

5. If the hyperbola $x^2 - y^2 = 1$ is rotated $45°$ counterclockwise, the resulting curve has the equation $xy = k$. Find k.

6. Describe the set of points that are twice as far from the line $x = -4$ as from the point $(-1, 0)$.

(A) two lines that intersect at $(-2.5, 0)$
(B) an ellipse with center at $(-2.5, 0)$
(C) an ellipse with center at $(0, 0)$
(D) a hyperbola with center at $(-2.5, 0)$
(E) a hyperbola with center at $(0, 0)$

7. Which of the following are the equations of circles that are tangent to both coordinate axes and pass through $(1, 2)$?

I $(x - 1)^2 + (y - 1)^2 = 1$
II $(x - 2)^2 + (y - 2)^2 = 1$
III $(x - 5)^2 + (y - 5)^2 = 25$

(A) I only   (B) II only
(C) III only   (D) I and II only
(E) I and III only

8. A new curve is formed by tripling the y-coordinates of all points on the circle $x^2 + y^2 = 16$. An equation of the new curve is

(A) $\dfrac{x^2}{16} + \dfrac{9y^2}{16} = 1$ (B) $\dfrac{9x^2}{16} + \dfrac{y^2}{16} = 1$

(C) $\dfrac{x^2}{16} + \dfrac{y^2}{144} = 1$ (D) $\dfrac{x^2}{144} + \dfrac{y^2}{16} = 1$

(E) $\dfrac{x^2}{144} + \dfrac{y^2}{144} = 1$

9. Find, to the nearest thousandth, the eccentricity of an ellipse if the distance between the foci equals the distance between an endpoint of the major axis and an endpoint of the minor axis.

(A) 0.816   (B) 0.775   (C) 0.707
(D) 0.632   (E) 0.577

10. Let $A(0, 0)$ and $B(4, 0)$ be two fixed points, and P a point that moves so that $\angle APB$ is always a right angle. Which equation describes all possible coordinates of P?

(A) $x^2 + y^2 = 16$
(B) $(x - 2)^2 + y^2 = 4$
(C) $(x - 2)^2 + (y - 1)^2 = 5$
(D) $(x - 3)^2 + y^2 = 9$
(E) $(x - 3)^2 + (y - 1)^2 = 10$

11. E

12. D

13. A

14. D

15. C

16. D

17. A

18. B

---

11. Which two curves are similar but not congruent?

   I $y = x^2$ and $y = 2x^2$
   II $y = x^2$ and $y = (x - 1)^2 + 3$
   III $y = x^2$ and $3y = (x - 1)^2 + 6$

   (A) I only   (B) II only
   (C) III only   (D) I and II only
   (E) I and III only

12. Which type of curve could be the graph of a function from a subset of the reals to the reals?

   I   ellipse
   II   parabola
   III   hyperbola

   (A) I only   (B) II only
   (C) III only   (D) II and III only
   (E) I, II, and III

13. The vertices of a rectangle are points of the ellipse $x^2 + 4y^2 = 36$. Two sides are parallel to the y-axis through the foci. What is the area of the rectangle?

   (A) $18\sqrt{3}$ (B) $9\sqrt{3}$ (C) $12\sqrt{3}$
   (D) 36   (E) 27

Questions 14 – 18 each consist of two quantities, one in Column A and one in Column B. You are to compare the two quantities and choose:

A   if the quantity in Column A is greater;
B   if the quantity in Column B is greater;
C   if the two quantities are equal;
D   if the relationship cannot be determined from the information given.

1.   In certain questions, information concerning one or both of the quantities to be compared is centered above the two columns.

2.   In a given question, a symbol that appears in both columns represents the same thing in Column A as it does in Column B.

3. x, n, and k, etc. stand for real numbers.

	Column A	Column B

The distance from the focus to the directrix for the parabola $y = ax^2$ is greater than the corresponding distance for the parabola $y = bx^2$.

14. a b

$H_1 = \left\{ (x, y) \,\middle|\, \dfrac{x^2}{3} - \dfrac{y^2}{4} = 1 \right\}$

$H_2 = \left\{ (x, y) \,\middle|\, \dfrac{4x^2}{3} - y^2 = 1 \right\}$

15.   The eccentricity of H_1 The eccentricity of H_2

---

16.   The area of an ellipse with foci at $(-1, 0)$ and $(1, 0)$ The area of an ellipse with foci at $(-2, 0)$ and $(2, 0)$

An ellipse has eccentricity $\frac{1}{3}$ and minor axis of length 6.

17. 🖩   The length of the major axis $6 + \frac{1}{3}$

Let m_1 be the slope of an asymptote of $x^2 - 4y^2 = 16$ and m_2 be the slope of an asymptote of $4x^2 - y^2 = 16$.

18. $|m_1|$ $|m_2|$

Teacher's Chapter *11*

Polar Coordinates

Chapter Contents

This chapter is important in itself, because of the new approach to graphing that it presents, demonstrating to students that there is more than one coordinate system and that ease of presentation will dictate which should be used in a particular context.

This chapter is also important because of its relationship to other topics, bringing together many of the concepts that students encountered in earlier chapters, such as trigonometric identities, rotations, and mathematical induction.

11.1 The Polar Coordinate System

A coordinate system is an arbitrary method of picturing a function or a number. Students, familiar with the rectangular coordinate system, should recognize that the polar coordinate system is useful in navigation and map making.

The plotting of points in the polar coordinate system makes use of many of the identities involving trigonometric functions that have been developed in the previous sections. Emphasize that although a point can be represented by many pairs of polar coordinates, a given pair of polar coordinates uniquely determines a point.

11.2 Connecting the Coordinate Systems

Some procedures, such as finding the coordinates of an image under translation, are more easily performed using rectangular coordinates. Other procedures, such as finding an image under a rotation, are more easily performed using polar coordinates. Therefore, it is useful to be able to change from one system to the other.

Most of the examples and exercises use angle measures that are multiples of $\frac{\pi}{6}$ or $\frac{\pi}{4}$ so that exact values, which do not require a calculator or table, may be used. Exercises 21–28 do require the use of a calculator to find approximate values of r and θ. These exercises also offer practice using the inverse tangent function on a scientific calculator.

Computer Activities (11.2)

The following programs will change rectangular coordinates to polar coordinates and polar coordinates to rectangular coordinates.

```
100 REM  THIS PROGRAM CHANGES
            RECTANGULAR
110 REM  COORDINATES TO POLAR
            COORDINATES
120 PRINT "ENTER RECTANGULAR
            COORDINATES AND THE"
130 PRINT "COMPUTER WILL RETURN
            POLAR COORDINATES."
140 PRINT "ENTER THE X COORDINATE."
150 INPUT X
160 PRINT "ENTER THE Y COORDINATE."
170 INPUT Y
180 PI = 3.14159265
190 R = SQR(X^2 + Y^2)
200 IF X = 0 AND Y >= 0 THEN
        A = PI/2
210 IF X = 0 AND Y < 0 THEN
        A = 3 * PI/2
220 IF X < > 0 THEN A = ATN(Y/X)
230 IF X < 0 THEN A = A + PI
240 IF A < 0 AND X > 0 THEN
        A = A + 2 * PI
250 PRINT "THE POLAR COORDINATES
            OF ("; X; ","; Y; ")"
260 PRINT "ARE ("; R; ","; A; ")."
270 PRINT "THE ANGLE MEASURE IS
            GIVEN IN RADIANS."
280 END
```

```
100  REM  THIS PROGRAM CHANGES POLAR
          COORDINATES
110  REM  TO RECTANGULAR COORDINATES.
120  PRINT "ENTER THE POLAR
          COORDINATES AND THE"
130  PRINT "COMPUTER WILL RETURN
          RECTANGULAR"
140  PRINT "COORDINATES."
150  PRINT "ENTER THE R COORDINATE."
160  INPUT R
170  PRINT "ENTER THE ANGLE MEASURE
          IN DEGREES."
180  INPUT T
190  PI = 3.14159265
200  X = INT(R * COS(T * PI/180) *
     100 + .5)/100
210  Y = INT(R * SIN(T * PI/180) *
     100 + .5)/100
220  IF T = 90 OR T = 270 THEN X = 0
230  IF T = 0 OR T = 180 THEN Y = 0
240  PRINT "RECTANGULAR COORDINATES
          OF ("; R; ","; T; ")"
250  PRINT "ARE ("; X; ","; Y; "),"
260  PRINT "ROUNDED TO THE NEAREST
          HUNDREDTH."
270  END
```

Note that the following activity can be used as reinforcement later in the unit or as motivation at the outset of the topic.

Activity (11.2) _____

Directions

Use graph paper on which 1 inch is divided into 5 or 10 smaller divisions so that each division equals .2 or .1 inch. Draw the x and y axes. Let the polar ray coincide with the nonnegative ray of the x-axis.

1. Draw an angle of 50° with a protractor, using the polar ray as one ray.

2. Locate P, a point 2 inches from the origin on that ray. The polar coordinates of P are $(2, 50°)$.

3. Use a calculator to find $x = 2 \cos 50°$ and $y = 2 \sin 50°$.

4. Compare the values of x and y found in step 3 with the rectangular coordinates P read from the graph.

5. Repeat steps 1 – 4 using a point that is 3 inches from the pole along the terminal ray of a 110° angle.

6. Draw a ray from the origin through Q whose rectangular coordinates are $(4, 2)$.

7. Use a ruler to measure $\overline{OQ}$ and a protractor to measure the angle that $\overline{OQ}$ makes with the polar axis. Write the polar coordinates of Q.

8. Use $r^2 = x^2 + y^2$ to find OQ and $\theta = \arctan \dfrac{y}{x}$ to find θ.
   Compare these values of r and θ with the values obtained by measurement.

9. Repeat steps 6 – 8 using $Q(-3, 1)$.

Discoveries

If (r, θ) are the polar coordinates and (x, y) are the rectangular coordinates, $x = r \cos \theta$, $y = r \sin \theta$, $r = \pm\sqrt{x^2 + y^2}$, and $\theta = \arctan \dfrac{y}{x}$.

What If?

The point is in the third or the fourth quadrant. (Measure the angle using the nonpositive ray of the x-axis as the initial side and add 180° to the value.)

11.3 Derivations of Polar Equations

This section applies the student's knowledge of trigonometric identities and triangle formulas to derive the equations of circles and lines in the polar plane.

The derivation of the equation of a line uses the cosine value of an acute-angle measure as the ratio of sides of a right triangle. Students should understand that right triangle trigonometry uses positive lengths of sides. When r or a represent negative numbers, $-r$ and $-a$ are the positive values of the lengths of the sides of the right triangle.

A vertical line in quadrants II and III can be seen as a reflection in the pole of a vertical line in quadrants I and IV. For example, the image of a point $P(r, \theta)$ in quadrant IV is $P'(r, \theta + \pi)$ in quadrant II, and the image of $A(a, 0)$ on the polar axis is $A'(a, \pi)$ on the ray opposite the polar axis. In $\triangle P'OA'$ in quadrant II, $m\angle P'OA' = \pi - (\theta + \pi) = -\theta$. Since $-\frac{\pi}{2} < \theta < 0$ in quadrant IV, $-\theta$ is positive.

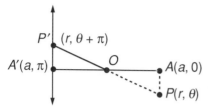

Thus, in quadrant II:

$$\cos(-\theta) = \frac{a}{r}$$

$$\cos\theta = \frac{a}{r}$$

$$r\cos\theta = a$$

A similar derivation applies in reflecting quadrant I points in quadrant III.

The equation $r\cos\theta = a$ can be solved for r and written as $r = \dfrac{a}{\cos\theta}$ or $r = a\sec\theta$. These two forms of the equation emphasize that the set of allowable values for θ excludes values of the form $\frac{\pi}{2} + k\pi$, for k an integer.

Exercises 21 and 22 ask for equations of lines under rotation. You may wish to ask students to write the equation of a circle through the pole with center at $A(a, \alpha)$. If the polar axis is rotated through an angle α, the new coordinates of any point P are $(r, \theta - \alpha)$. Therefore, the equation of the circle is $r = a\cos(\theta - \alpha)$.

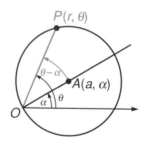

Compare the use of the translation $T_{a,b}$ in rectangular coordinates and the rotation R_α in polar coordinates. For example, the graph of $y - b = (x - a)^2$ is the image of the graph of $y = x^2$ under $T_{a,b}$ and $r\cos(\theta - \alpha)$ is the image of $r = \cos\theta$ under R_α.

The relationship between polar and rectangular coordinates can be used to express the new coordinates of the point $P(x, y)$, or $P(r, \theta)$, after the axes have been rotated through an angle α. The new coordinates of P with respect to the x'- and y'-axes are (x', y') or $(r, \theta - \alpha)$.

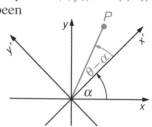

Thus, since $x = r\cos\theta$ and $y = r\sin\theta$:

$$x' = r\cos(\theta - \alpha)$$
$$= r(\cos\theta\cos\alpha + \sin\theta\sin\alpha)$$
$$= x\cos\alpha + y\sin\alpha$$

$$y' = r\sin(\theta - \alpha)$$
$$= r(\sin\theta\cos\alpha - \cos\theta\sin\alpha)$$
$$= y\cos\alpha - x\sin\alpha$$

This can be written as a matrix equation:

$$\begin{bmatrix} x' \\ y' \end{bmatrix} = \begin{bmatrix} \cos\alpha & \sin\alpha \\ -\sin\alpha & \cos\alpha \end{bmatrix} \cdot \begin{bmatrix} x \\ y \end{bmatrix}$$

11.4 Graphs of Polar Equations

This section introduces simple equations that define r as a function of θ and investigates their graphs. The text plots, in detail, some basic equations. It is important that students understand the relationship between the quadrant in which the angle of a given measure θ lies and the quadrant in which the point (r, θ) lies when r is positive and when r is negative.

A graphing calculator or computer generates graphics that display the order in which the points are plotted as θ increases from 0 to 2π. This can be effective in enabling students to experience a variety of related graphs without the detailed work of pointwise plotting.

In the text, $r = \sin 2\theta$ is plotted by finding critical points in each quadrant. However, after the first loop of the curve is graphed using values of θ in the interval $\left[0, \frac{\pi}{2}\right]$, the symmetry of the graph is sufficient to complete the drawing of the curve. The graphs required in the exercises should be plotted using symmetries and a knowledge of the expected shape of the curve rather than pointwise location of many points.

Ask students the kind of symmetry that can be expected in an equation involving $\cos \theta$. (Symmetry with respect to the polar axis since $\cos(-\theta) = \cos \theta$.) What kind of symmetry can be expected in an equation involving $\sin \theta$? (Symmetry with respect to a vertical line through the pole since $\sin(\pi - \theta) = \sin \theta$.)

11.5 Polar Equations of the Conics

When writing the equation of a conic in polar form, the standard conic is positioned with the focus at the pole to allow a simple representation for the distance from a point on the conic to the focus.

Polar coordinates are often used to write the equations of the orbits of satellites.

The systems of polar equations in Exercise 17 provide an interesting topic for further discussion. In rectangular coordinates, the algebraic solutions to a system of equations correspond to the points of intersection of their graphs. In polar coordinates, however, some points of intersection may not represent a solution of the system of equations. Since the polar coordinates of a point are not unique, it may happen that the polar equations are satisfied by different coordinates for the same point. Therefore, the coordinates of points of intersection must be checked in the given equations and only pairs that satisfy both equations accepted as solutions.

Example

Solve the system: $r = 2 \sin \theta$ and $\theta = \frac{\pi}{6}$

Substituting $\theta = \frac{\pi}{6}$ into the sine equation gives $r = 1$. Thus, the only point resulting from the algebraic solution is $\left(1, \frac{\pi}{6}\right)$. The graphs also intersect at the pole, where the coordi-

nates are $(0, 0)$ for the circle and $\left(0, \frac{\pi}{6}\right)$ for the line. This point of intersection is not a solution of the system.

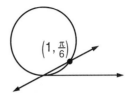

11.6 Complex Numbers in Polar Form

Addition and subtraction of complex numbers are more easily performed using $a + bi$ form but products, quotients, powers, and roots are more easily found using polar form. Since these latter operations are frequently encountered, polar form of a complex number is an important concept.

Be sure that students understand the distinction between the graphing of a function that relates two real numbers, r as a function of θ, and the complex number plane in which each point represents a distinct complex number.

This section also provides the students with an opportunity to use the principle of mathematical induction.

The Fundamental Theorem of Algebra states that every polynomial function of degree n has n roots in the set of complex numbers. Since $p(x) = x^n - c$ is one such polynomial function, there are n complex roots of the equation $x^n - c = 0$ or $x^n = c$. Therefore, in the set of complex numbers, there are n nth roots of any constant c.

In the set of real numbers, we distinguish between the positive and negative square roots of a positive number. Thus, $\sqrt{a}$ is the symbol for the positive, or principal, square root of a and $-\sqrt{a}$ is the symbol for the negative square root of a. Since the complex field is not ordered, in the set of complex numbers, $\sqrt{a + bi}$ represents either of the square roots.

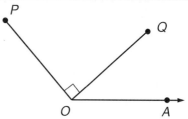

1. In the diagram, $m\angle AOP = 140°$, $OP = OQ = 3$ and $\overline{OP} \perp \overline{OQ}$.

   a. Write the polar coordinates of P.

   b. Write the polar coordinates of Q.

2. If $\left(5, \frac{\pi}{4}\right)$ and $(-5, \theta)$ are polar coordinates of the same point, what is the value of θ in the interval $(-\pi, \pi]$?

3. Write the polar equation of a circle through the pole with center at $\left(2, \frac{\pi}{2}\right)$.

In **4 – 15**, sketch the graph on polar graph paper.

4. $r = 2$

5. $r = 8 \sin \theta$

6. $r = -4 \cos \theta$

7. $r \cos \theta = -1$

8. $r \sin \theta = 6$

9. $\theta = -\frac{\pi}{6}$

10. $r = 4 + 4 \sin \theta$

11. $r = 5 - 5 \cos \theta$

12. $r = 3 + 2 \sin \theta$

13. $r = 2 - 3 \cos \theta$

14. $r = 2 \sin 4\theta$

15. $r = \sin 3\theta$

In **16 – 19**, write the polar coordinates of the point whose rectangular coordinates are given.

16. $(-5, 0)$

17. $(0, 3)$

18. $(-1, \sqrt{3})$

19. $(-4, -4)$

In **20 – 23**, write the rectangular coordinates of the point whose polar coordinates are given.

20. $(1, 0)$

21. $(-3, \pi)$

22. $\left(4, \frac{3\pi}{4}\right)$

23. $\left(-2, \frac{3\pi}{2}\right)$

24. Write, in polar coordinates, an equation of a parabola with focus at the pole and directrix a vertical line through $(3, \pi)$.

25. Write, in polar coordinates, an equation of an ellipse with focus at the pole, eccentricity $\frac{1}{4}$, and directrix 15 units to the left of the pole.

26. An equation of a conic with focus at the pole of the polar coordinate system is: $r = \frac{6}{3 - \sin \theta}$

   a. Identify the curve as an ellipse, a parabola, or a hyperbola.

   b. Find the endpoints of an axis of symmetry of the curve.

   c. Sketch the curve.

27. Write, in terms of x and y, an equation of the ellipse whose equation in polar coordinates is: $r = \frac{9}{2 + \cos \theta}$

28. Write, in terms of r and θ, the equation of the circle whose equation in rectangular coordinates is $(x - 3)^2 + y^2 = 9$.

Bonus

   When $a + bi = r(\cos \theta + i \sin \theta)$, $\sqrt{a + bi} = \sqrt{r}\left(\cos \frac{\theta}{2} + i \sin \frac{\theta}{2}\right)$.

   If $\sqrt{a + bi} = x + iy$, express x and y in terms of a and b.

   (Express $\cos \frac{\theta}{2}$ and $\sin \frac{\theta}{2}$ in terms of $\cos \theta$.)

Answers to Suggested Test Items

1. a. $(3, 140°)$

   b. $(3, 50°)$

2. $\theta = -\frac{3\pi}{4}$

3. $r = 4 \sin \theta$

4.

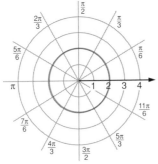

5.

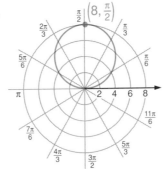

6.

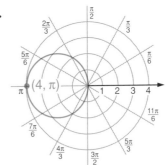

7.

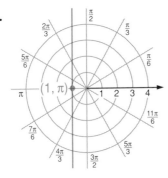

8.

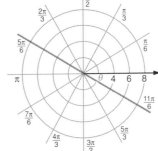

9.

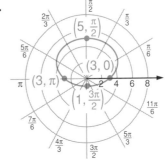

10.

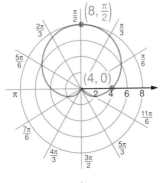

11.

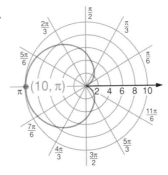

12.

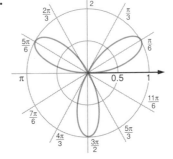

13.

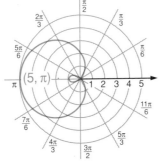

14.

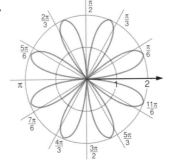

15.

16. $(5, \pi)$ **17.** $\left(3, \frac{\pi}{2}\right)$ **18.** $\left(2, \frac{2\pi}{3}\right)$ **19.** $\left(4\sqrt{2}, \frac{5\pi}{4}\right)$ **20.** $(1, 0)$

21. $(3, 0)$ **22.** $(-2\sqrt{2}, 2\sqrt{2})$ **23.** $(0, 2)$ **24.** $r = \frac{3}{1 - \cos\theta}$ **25.** $r = \frac{15}{4 - \cos\theta}$

26. a. ellipse   **b.** $\left(3, \frac{\pi}{2}\right), \left(\frac{3}{2}, \frac{3\pi}{2}\right)$ **27.** $\frac{(x + 3)^2}{36} + \frac{y^2}{27} = 1$

c.

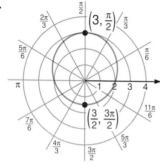

$$3(x + 3)^2 + 4y^2 = 108$$

28. $r = 6 \cos\theta$

Bonus

$$x + yi = \sqrt{a + bi}$$

$$x + yi = \sqrt{r}\left(\cos\frac{\theta}{2} + i \sin\frac{\theta}{2}\right)$$

$$x + yi = \sqrt{r}\sqrt{\frac{1 + \cos\theta}{2}} + i\sqrt{r}\sqrt{\frac{1 - \cos\theta}{2}}$$

$$x = \sqrt{\frac{r + r\cos\theta}{2}} \qquad y = \sqrt{\frac{r - r\cos\theta}{2}}$$

Substitute $r = \sqrt{a^2 + b^2}$.

Also, since $a + bi = r(\cos\theta + i \sin\theta)$, $a = r \cos\theta$.

$$x = \sqrt{\frac{\sqrt{a^2 + b^2} + a}{2}} \qquad y = \sqrt{\frac{\sqrt{a^2 + b^2} - a}{2}}$$

Chapter *11*

Polar Coordinates

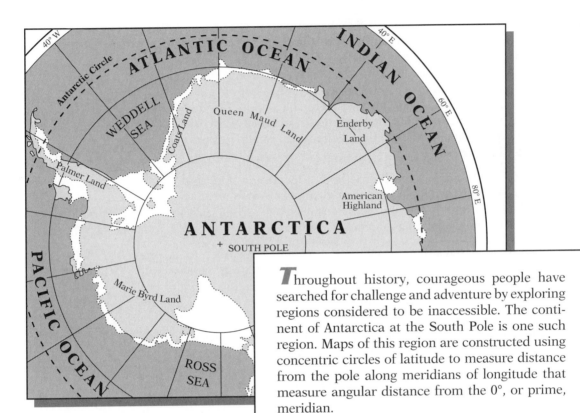

*T*hroughout history, courageous people have searched for challenge and adventure by exploring regions considered to be inaccessible. The continent of Antarctica at the South Pole is one such region. Maps of this region are constructed using concentric circles of latitude to measure distance from the pole along meridians of longitude that measure angular distance from the 0°, or prime, meridian.

Chapter Table of Contents

11.1 The Polar Coordinate System

After years of working as a fishing guide on Seneca Lake, Marsha has learned the best places to fish at various times of the day and under various weather conditions. On a cool morning before the sun was up, she set off in her boat to the northwest at an angle of 60° with the shoreline, stopping at a point one-half mile from her dock. Marsha's technique for locating her fishing spots demonstrates a method of identifying every point in a plane by specifying an angle measure and a distance.

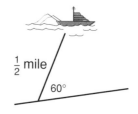

You are familiar with the method of locating a point by means of horizontal and vertical distances. An alternate method uses Marsha's technique of locating her fishing spot, that is, by the use of an angle and a distance.

A system that uses an angle measure and a distance to locate a point begins with a fixed ray and its endpoint as a reference. Consider a horizontal ray, $\overrightarrow{OA}$, where the endpoint O is called the *pole* and $\overrightarrow{OA}$ is the *polar axis*. Any point P in the plane can be located by specifying r, the directed distance from O to P, and θ, the directed measure of $\angle AOP$. The values of r and θ for a specific point P are the *polar coordinates* of P, and are written as the ordered pair (r, θ). The polar coordinates of a point uniquely determine that point.

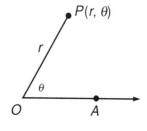

To locate the point P with polar coordinates (1, 30°), first find the ray that is the image of $\overrightarrow{OA}$ under a rotation of 30° in the counterclockwise direction. P is the point on this ray that is 1 unit from O.

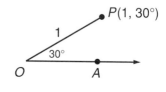

If point R lies on the polar axis, the degree measure of $\angle AOR$ is 0. If R is 2 units from O, the polar coordinates of R are (2, 0°).

Consider a point S located on a line perpendicular to the polar axis at the pole and 5 units above the pole. Since $\overrightarrow{OS}$ is the image of $\overrightarrow{OA}$ under a counterclockwise rotation of 90°, or $\frac{\pi}{2}$ radians, the polar coordinates of S are (5, 90°) or $\left(5, \frac{\pi}{2}\right)$. But point S is also the image of $\overrightarrow{OA}$ under a clockwise rotation of –270° or $-\frac{3\pi}{2}$ radians. Therefore the polar coordinates of S can also be written as (5, –270°) or $\left(5, -\frac{3\pi}{2}\right)$.

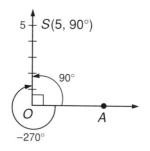

While a given pair of polar coordinates uniquely determines a point in a plane, the polar coordinates of a given point are not unique. If the absolute values of two rotations differ by 360°, or 2π radians, then the two rotations determine the same ray. For example, the rotations of 90° and –270° determine the same ray since $|90-(-270)|=360$. When measured in radians, the rotations of $\frac{\pi}{2}$ and $-\frac{3\pi}{2}$ determine the same ray since $\left|\frac{\pi}{2}-\left(-\frac{3\pi}{2}\right)\right|=2\pi$. Therefore, the polar coordinates of the point S discussed above can be any ordered pair of the form $(5, (90 + 360k)°)$ or $\left(5, \frac{\pi}{2}+2\pi k\right)$ where k is an integer. The most commonly used ordered pair is the one for which $k = 0$.

Example 1

Write three different polar coordinate representations of a point P that is 0.5 units from O on the ray that is the image of $\overrightarrow{OA}$ under a rotation of 120°.

Solution

The general form of the polar coordinates is $(0.5, (120 + 360k)°)$.

If $k = 0$: $(0.5, (120 + 360k)°) = (0.5, (120 + 0)°) = (0.5, 120°)$

If $k = 1$: $(0.5, (120 + 360k)°) = (0.5, (120 + 360)°) = (0.5, 480°)$

If $k = -1$: $(0.5, (120 + 360k)°) = (0.5, (120 + (-360))°) = (0.5, -240°)$

Note: When the angle measure is expressed in radians, the general form of the polar coordinates of P is $\left(0.5, \frac{2\pi}{3}+2\pi k\right)$, and the polar coordinates found above are $\left(0.5, \frac{2\pi}{3}\right)$, $\left(0.5, \frac{8\pi}{3}\right)$, and $\left(0.5, -\frac{4\pi}{3}\right)$.

With this system, to help locate the point for a given ordered pair of polar coordinates, polar graph paper is used. To help measure distances from the pole, polar graph paper shows concentric circles with radii that are integral multiples of the radius of a unit circle. To help show angle measures, rays that represent common rotations of the polar axis are shown. In the diagram, the rays form angles with the polar axis having measures that are multiples of $\frac{\pi}{6}$ radians, or 30°. The point whose polar coordinates are $\left(3, \frac{5\pi}{6}\right)$ can be located by finding the point where the ray that makes an angle of $\frac{5\pi}{6}$ radians with the polar axis intersects the circle of radius 3 units.

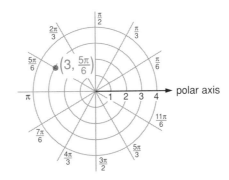

Example 2

Plot the following points on a polar graph.

a. $\left(2, \frac{7\pi}{6}\right)$ **b.** $\left(4, -\frac{\pi}{3}\right)$ **c.** $\left(-2, \frac{2\pi}{3}\right)$ **d.** $(-1, \pi)$

Solution

a. The point $\left(2, \frac{7\pi}{6}\right)$ lies at the intersection of the ray labeled $\frac{7\pi}{6}$ and a circle of radius 2 units.

b. The point $\left(4, -\frac{\pi}{3}\right)$ lies at the intersection of the ray that is the image of the polar axis under a rotation of $\frac{\pi}{3}$ in the clockwise direction and a circle of radius 4 units. Note that $\left(4, -\frac{\pi}{3}\right)$ could also be written as $\left(4, \frac{5\pi}{3}\right)$.

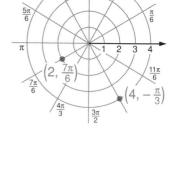

c. Just as on a number line, a negative distance is measured in the direction opposite that of a positive distance. Therefore, to locate the point $\left(-2, \frac{2\pi}{3}\right)$, first find the ray that is the image of the polar axis under a rotation of $\frac{2\pi}{3}$. Then locate $\left(-2, \frac{2\pi}{3}\right)$ on the opposite ray, 2 units from the pole. Notice that this point could also be described by the polar coordinates $\left(2, -\frac{\pi}{3}\right)$ or $\left(2, \frac{5\pi}{3}\right)$.

d. Similarly, to locate the point $(-1, \pi)$, find the ray that is the image of the polar axis under a rotation of π. Locate $(-1, \pi)$ on the opposite ray, 1 unit from the pole. Note that the point $(-1, \pi)$ also has the polar coordinates $(1, 0)$.

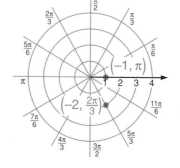

In general, where k is an integer:

$$(r, \theta) = (-r, \theta + (2k + 1)\pi)$$

Example 3

Rewrite the polar coordinates $\left(-3, \frac{5\pi}{6}\right)$ using a positive value of r.

Solution

Since the opposite ray of a rotation of $\frac{5\pi}{6}$ is a rotation of $\frac{5\pi}{6} + \pi = \frac{11\pi}{6}$, the polar coordinates $\left(-3, \frac{5\pi}{6}\right)$ may be rewritten as $\left(3, \frac{11\pi}{6}\right)$.

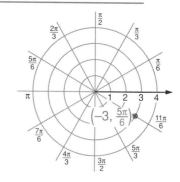

Exercises 11.1

1. Write the polar coordinates of each point A through H, using a positive value of r and the least positive value of θ.

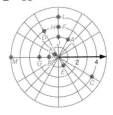

In **2 – 13**, plot the point whose polar coordinates are given.

2. $A(2, 60°)$ **3.** $B(1, 180°)$ **4.** $C(4, -30°)$ **5.** $D(3, 120°)$

6. $E(-1, 120°)$ **7.** $F(-3, 270°)$ **8.** $G(2, \pi)$ **9.** $H\left(3, \frac{\pi}{2}\right)$

10. $J\left(-2, -\frac{\pi}{2}\right)$ **11.** $K\left(\frac{1}{2}, \frac{5\pi}{6}\right)$ **12.** $L\left(-4, \frac{3\pi}{2}\right)$ **13.** $M(-5, 0)$

In **14 – 25**, to describe the given point, rewrite the polar coordinates in two other ways. Use degree measure if degrees are given and radian measure if radians are given.

14. $(1, 20°)$ **15.** $(-1, 200°)$ **16.** $(-6, -90°)$ **17.** $(4, -100°)$

18. $(6, 3\pi)$ **19.** $\left(2, \frac{\pi}{4}\right)$ **20.** $\left(5, \frac{5\pi}{6}\right)$ **21.** $\left(-3, \frac{7\pi}{6}\right)$

22. $(-1, 0)$ **23.** $(7, -\pi)$ **24.** $\left(4, \frac{13\pi}{6}\right)$ **25.** $(3, 2\pi)$

In **26 – 29**, both pairs of coordinates represent the same point. Find the least positive value of θ.

26. $\left(2, \frac{\pi}{3}\right)$ and $(-2, \theta)$

27. $\left(-3, \frac{\pi}{6}\right)$ and $(3, \theta)$

28. $\left(4, -\frac{\pi}{4}\right)$ and $(-4, \theta)$

29. $\left(-1, -\frac{\pi}{2}\right)$ and $(1, \theta)$

In **30 – 33**, both pairs of coordinates represent the same point. Find the value of r.

30. $\left(1, \frac{3\pi}{4}\right)$ and $\left(r, \frac{7\pi}{4}\right)$

31. $\left(2, \frac{2\pi}{3}\right)$ and $\left(r, \frac{8\pi}{3}\right)$

32. $\left(-3, \frac{5\pi}{6}\right)$ and $\left(r, -\frac{\pi}{6}\right)$

33. $\left(-2, -\frac{\pi}{2}\right)$ and $\left(r, -\frac{3\pi}{2}\right)$

In **34 – 36**, describe the locus of points that satisfy the given equation.

34. Each point that satisfies the equation $r = 2$ has polar coordinates of the form $(2, \theta)$ where θ is any number.

35. Each point that satisfies the equation $\theta = \frac{\pi}{2}$ has polar coordinates of the form $\left(r, \frac{\pi}{2}\right)$ where r is any number.

36. Each point that satisfies the equation $\theta = \frac{\pi}{6}$ has polar coordinates of the form $\left(r, \frac{\pi}{6}\right)$ where r is any number.

Answers 11.1

1. $A(4, 0)$ $B\left(3, \frac{\pi}{6}\right)$

     $C\left(1, \frac{\pi}{2}\right)$ $D\left(2, \frac{2\pi}{3}\right)$

     $E(3, \pi)$ $F\left(2, \frac{4\pi}{3}\right)$

     $G\left(3, \frac{3\pi}{2}\right)$ $H\left(4, \frac{11\pi}{6}\right)$

2 – 13

14. $(1, -340°)$, $(-1, 200°)$

15. $(1, 20°)$, $(-1, -160°)$

16. $(6, 90°)$, $(-6, 270°)$

17. $(-4, 80°)$, $(4, 260°)$

18. $(6, \pi)$, $(6, -\pi)$

19. $\left(2, -\frac{7\pi}{4}\right)$, $\left(-2, \frac{5\pi}{4}\right)$

20. $\left(5, -\frac{7\pi}{6}\right)$, $\left(-5, \frac{11\pi}{6}\right)$

21. $\left(3, \frac{13\pi}{6}\right)$, $\left(3, \frac{\pi}{6}\right)$

22. $(1, \pi)$, $(1, -\pi)$

23. $(7, \pi)$, $(-7, 0)$

24. $\left(4, \frac{\pi}{6}\right)$, $\left(4, -\frac{11\pi}{6}\right)$

25. $(3, 0)$, $(-3, \pi)$

26. $\frac{4\pi}{3}$

27. $\frac{7\pi}{6}$

28. $\frac{3\pi}{4}$

29. $\frac{\pi}{2}$

30. -1

31. 2

32. 3

33. 2

34. A circle with center at the pole and radius 2.

35. A straight line passing through the pole and perpendicular to the polar axis.

36. A straight line passing through the pole and forming an angle of $\frac{\pi}{6}$ with the polar axis.

11.2 Connecting the Coordinate Systems

To compare polar coordinates to rectangular coordinates, note that the reference points and axes of the two systems can be made to coincide by letting the nonnegative ray of the *x*-axis correspond to the polar axis. Then, the origin of the rectangular coordinate system will coincide with the pole of the polar coordinate system, and the *y*-axis will be perpendicular to the polar axis at the pole.

One way to establish a relationship between the polar coordinates (r, θ) and the rectangular coordinates (x, y) of a point P is through a reference right triangle. This method is used to convert from rectangular coordinates to polar coordinates.

Example 1

Find polar coordinates of the point P whose rectangular coordinates are $(2\sqrt{3}, 2)$.

Solution

To write polar coordinates, you must find values for r and θ.

r is the hypotenuse of a right triangle whose legs are $x = 2\sqrt{3}$ and $y = 2$.

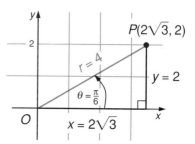

By the Pythagorean Theorem:

$r^2 = x^2 + y^2$

$r^2 = (2\sqrt{3})^2 + 2^2$

$r^2 = 12 + 4 = 16$

$r = 4$

Using Trigonometry:

$\tan \theta = \dfrac{y}{x}$

$\tan \theta = \dfrac{2}{2\sqrt{3}} = \dfrac{\sqrt{3}}{3}$

$\theta = \dfrac{\pi}{6}$

(Since P is in quadrant I)

Thus, polar coordinates of P are $\left(4, \frac{\pi}{6}\right)$. Note that since polar coordinates are not unique, other pairs, such as $\left(4, -\frac{11\pi}{6}\right)$ or $\left(-4, \frac{7\pi}{6}\right)$, are also polar coordinates of P. Commonly, the positive value of r is used.

Answer: For P, whose rectangular coordinates are $(2\sqrt{3}, 2)$, polar coordinates are $\left(4, \frac{\pi}{6}\right)$.

In general, when changing from rectangular coordinates to polar coordinates, begin by finding the value of r.

$$r^2 = x^2 + y^2$$

or $\quad r = \pm\sqrt{x^2 + y^2}$

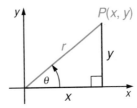

Use the tangent ratio to determine θ:

$$\tan \theta = \frac{y}{x}$$

There are infinitely many values of θ for a given point (x, y). By convention, a value of θ between -2π and 2π is chosen. The sign of r is determined by the value chosen for θ.

Example 2

Find two pairs of polar coordinates for the point P whose rectangular coordinates are $(-2, 2)$.

Solution

$$r = \pm\sqrt{x^2 + y^2}$$
$$= \pm\sqrt{(-2)^2 + 2^2}$$
$$= \pm\sqrt{8} = \pm 2\sqrt{2}$$
$$\tan \theta = \frac{2}{-2} = -1$$

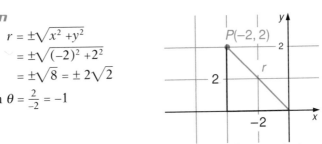

Since $\tan \theta$ is negative, θ is the measure of an angle in quadrant II or in quadrant IV whose reference angle is $\frac{\pi}{4}$.

$$\theta = \pi - \frac{\pi}{4} \qquad \text{or} \qquad \theta = 2\pi - \frac{\pi}{4}$$
$$= \frac{3\pi}{4} \qquad\qquad\qquad = \frac{7\pi}{4} \text{ or } -\frac{\pi}{4}$$

Since the point $(-2, 2)$ is in quadrant II, either the positive value of r is used with the value of θ that is the measure of an angle in quadrant II or the negative value of r is used with the value of θ that is the measure of an angle in quadrant IV. Therefore, the polar coordinates of the point $(-2, 2)$ are $\left(2\sqrt{2}, \frac{3\pi}{4}\right)$ or $\left(-2\sqrt{2}, \frac{7\pi}{4}\right)$. The polar coordinates can also be written using a value of θ that differs from one of the given values by a multiple of 2π.

Answer: For the point P, whose rectangular coordinates are $(-2, 2)$, polar coordinates are $\left(2\sqrt{2}, \frac{3\pi}{4}\right)$ or $\left(-2\sqrt{2}, \frac{7\pi}{4}\right)$.

If a calculator is used to perform the computations, approximate values of r and θ will be displayed. To find the value of r, enter:

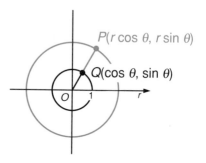

Display: 2.8284271

To find the value of θ, set the calculator in radian mode and enter:

| (| 2 | ÷ | 2 | +/− |) | INV | TAN | = |

Display: -0.7853982

This is the value of $-\frac{\pi}{4}$ to seven decimal places.

Another way to establish a relationship between the polar coordinates (r, θ) and the rectangular coordinates (x, y) of a point P is through a unit circle.

Let Q be a point of the unit circle on the terminal ray of an angle in standard position whose measure is θ. Then the rectangular coordinates of Q are $(\cos \theta, \sin \theta)$.

If P is a point on $\overrightarrow{OQ}$ at a distance r from O, then P is the image of Q under a dilation D_r and the rectangular coordinates (x, y) of P are $(r \cos \theta, r \sin \theta)$. Therefore, to change from polar coordinates to rectangular coordinates, use the equations:

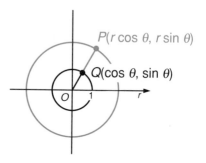

$$x = r \cos \theta \quad \text{and} \quad y = r \sin \theta$$

Example 3

Find the rectangular coordinates of the point P whose polar coordinates are $\left(3, -\frac{\pi}{6}\right)$.

Solution

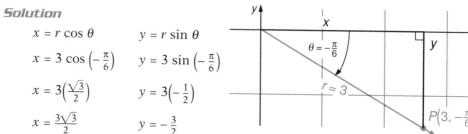

$x = r \cos \theta \qquad y = r \sin \theta$

$x = 3 \cos \left(-\frac{\pi}{6}\right) \qquad y = 3 \sin \left(-\frac{\pi}{6}\right)$

$x = 3\left(\frac{\sqrt{3}}{2}\right) \qquad y = 3\left(-\frac{1}{2}\right)$

$x = \frac{3\sqrt{3}}{2} \qquad y = -\frac{3}{2}$

Answer: The rectangular coordinates are $\left(\frac{3\sqrt{3}}{2}, -\frac{3}{2}\right)$.

Note that the rectangular coordinates of a point are unique.

If a calculator is used to perform these computations, approximate values of x and y will be displayed. To find x, put the calculator in radian mode and enter:

| (| π | ÷ | 6 |) | +/− | COS | × | 3 | = |

Display: 2.5980762

To find y, put the calculator in radian mode and enter:

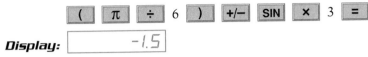

Display: -1.5

Some calculators have a key that will change from polar to rectangular and from rectangular to polar coordinates.

The equations $x = r \cos \theta$ and $y = r \sin \theta$ can also be used to change an equation in x and y to a polar equation in r and θ.

Example 4

Write an equation of the line $y = 2x - 1$ in terms of r and θ.

Solution

Using $x = r \cos \theta$ and $y = r \sin \theta$, the equation $y = 2x - 1$ is equivalent to

$$r \sin \theta = 2(r \cos \theta) - 1$$

$$r \sin \theta - 2r \cos \theta = -1$$

$$r(\sin \theta - 2 \cos \theta) = -1$$

$$r = \frac{-1}{\sin \theta - 2 \cos \theta}$$

Answer: The equation $y = 2x - 1$ is equivalent to $r = \dfrac{-1}{\sin \theta - 2 \cos \theta}$.

Note that the points $\left(\frac{1}{2}, 0\right)$ and $\left(-\frac{1}{2}, \pi\right)$ are the same point, the x-intercept of the graph of the equation. The point $\left(-1, \frac{\pi}{2}\right)$ is the y-intercept of the graph.

At $\theta = 0$:   At $\theta = \pi$:   At $\theta = \frac{\pi}{2}$:

$r = \frac{-1}{-2}$ $r = \frac{-1}{2}$ $r = \frac{-1}{1}$

$= \frac{1}{2}$ $= -\frac{1}{2}$ $= -1$

Example 5

The rectangular coordinates of A are $(1, \sqrt{3})$. Find the rectangular coordinates of A', the image of A under a rotation of $30°$.

Solution

Step 1: Find the polar coordinates of A.
Since A is in quadrant I, find an acute angle θ and a positive value of r.

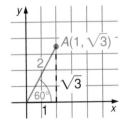

$r = \sqrt{1^2 + (\sqrt{3})^2}$ $\tan \theta = \frac{\sqrt{3}}{1}$

$= \sqrt{1 + 3}$ $\tan \theta = \sqrt{3}$

$= \sqrt{4} = 2$ $\theta = 60°$

Thus, the polar coordinates of A are $(2, 60°)$.

1. $(1, 0)$

2. $(0, 1)$

3. $(0, -4)$

4. $(-3, 0)$

5. $(1, 0)$

6. $\left(\frac{3}{2}, -\frac{3\sqrt{3}}{2}\right)$

7. $\left(-\frac{\sqrt{2}}{2}, -\frac{\sqrt{2}}{2}\right)$

8. $(3\sqrt{3}, -3)$

9. $(3, 0)$ or $(-3, \pi)$

10. $\left(5, \frac{\pi}{2}\right)$ or $\left(-5, -\frac{\pi}{2}\right)$

11. $(-2, 0)$ or $(2, \pi)$

12. $\left(3, -\frac{\pi}{2}\right)$ or $\left(-3, \frac{\pi}{2}\right)$

13. $\left(5\sqrt{2}, \frac{\pi}{4}\right)$ or $\left(-5\sqrt{2}, \frac{5\pi}{4}\right)$

14. $\left(-5\sqrt{2}, \frac{\pi}{4}\right)$ or $\left(5\sqrt{2}, \frac{5\pi}{4}\right)$

15. $\left(6, \frac{\pi}{6}\right)$ or $\left(-6, \frac{7\pi}{6}\right)$

16. $\left(8, \frac{\pi}{3}\right)$ or $\left(-8, \frac{4\pi}{3}\right)$

17. $\left(-3\sqrt{2}, -\frac{\pi}{4}\right)$ or $\left(3\sqrt{2}, \frac{3\pi}{4}\right)$

18. $\left(12, \frac{5\pi}{6}\right)$ or $\left(-12, -\frac{\pi}{6}\right)$

19. $\left(12, -\frac{\pi}{3}\right)$ or $\left(-12, \frac{2\pi}{3}\right)$

20. $\left(2\sqrt{2}, -\frac{\pi}{3}\right)$ or $\left(-2\sqrt{2}, \frac{2\pi}{3}\right)$

21. $(5, 127°)$ or $(-5, -53°)$

22. $(13, -67°)$ or $(-13, 113°)$

23. $(25, 254°)$ or $(-25, 74°)$

24. $(5.39, 112°)$ or $(-5.39, -68°)$

25. $(4, 41°)$, or $(-4, 221°)$

26. $(10, -37°)$ or $(-10, 143°)$

Step 2: Find the polar coordinates of A'. Under a rotation of 30°, the image of $A(2, 60°)$ is $A'(2, (60 + 30)°)$ or $A'(2, 90°)$.

Step 3: Find the rectangular coordinates of A'.

$x = r \cos \theta \qquad y = r \sin \theta$

$= 2 \cos 90° \qquad = 2 \sin 90°$

$= 2(0) = 0 \qquad = 2(1) = 2$

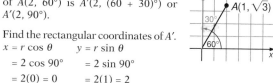

Answer: The rectangular coordinates of A' are $(0, 2)$.

Exercises 11.2

In **1 – 8**, write the rectangular coordinates of the points whose polar coordinates are given.

1. $(1, 0)$

2. $\left(1, \frac{\pi}{2}\right)$

3. $\left(4, -\frac{\pi}{2}\right)$

4. $(3, \pi)$

5. $(-1, 3\pi)$

6. $\left(-3, -\frac{4\pi}{3}\right)$

7. $\left(-1, \frac{\pi}{4}\right)$

8. $\left(6, -\frac{13\pi}{6}\right)$

In **9 – 20**, write the polar coordinates of the points whose rectangular coordinates are given.

9. $(3, 0)$

10. $(0, 5)$

11. $(-2, 0)$

12. $(0, -3)$

13. $(5, 5)$

14. $(-5, -5)$

15. $(3\sqrt{3}, 3)$

16. $(4, 4\sqrt{3})$

17. $(-3, 3)$

18. $(-6\sqrt{3}, 6)$

19. $(6, -6\sqrt{3})$

20. $(\sqrt{2}, -\sqrt{6})$

In **21 – 28**, write the polar coordinates of the point whose rectangular coordinates are given. Use a calculator or a table of values of trigonometric functions to find the approximate value of θ to the nearest degree.

21. $(-3, 4)$

22. $(5, -12)$

23. $(-7, -24)$

24. $(-2, 5)$

25. $(3, \sqrt{7})$

26. $(8, -6)$

27. $(-8, -4)$

28. $(6, -10)$

29. Find the rectangular coordinates of A', the image of point A under a rotation of 45°, if the polar coordinates of A are $(2, 135°)$.

30. Find the polar coordinates of B', the image of point B under a rotation of –30°, if the rectangular coordinates of B are $(0, 4)$.

31. Find the rectangular coordinates of C', the image of point C under a rotation of 120°, if the rectangular coordinates of C are $(-3, 0)$.

32. Express the rectangular coordinates of P', the image of $P(x, y)$ under a rotation of α, in terms of $x, y,$ and α.

33. Rewrite the equation of the line $x = 2$ in terms of r and θ.

34. Rewrite the equation of the line $y = -1$ in terms of r and θ.

35. Rewrite the equation of the circle $x^2 - 2x + y^2 = 0$ in terms of r and θ.

36. Rewrite the equation of the circle $x^2 + y^2 - 2y = 0$ in terms of r and θ.

---

27. $(8.94, 207°)$ or $(-8.94, 27°)$

28. $(11.66, -59°)$, or $(-11.66, 121°)$

29. $A'(-2, 0)$

30. $B'(4, 60°)$

31. $\left(\frac{3}{2}, -\frac{3\sqrt{3}}{2}\right)$

32. $x' = x \cos \alpha - y \sin \alpha$
$y' = y \cos \alpha + x \sin \alpha$

33. $r \cos \theta = 2$

34. $r \sin \theta = -1$

35. $r = 2 \cos \theta$

36. $r = 2 \sin \theta$

11.3 Derivations of Polar Equations

The set of points plotted in the polar plane satisfying a given equation in r and θ is the *polar graph* of that equation.

Circles

The simplest polar graph is a circle whose center is the pole. Since this is the set of points whose distance r from the pole is constant, the equation of a circle whose center is the pole is $r = c$, where the constant c is the length of the radius of the circle.

Example 1

Sketch the graph of the polar equation $r = 3$.

Solution

The solution of $r = 3$ is the set of points that are 3 units from the pole for all values of θ.

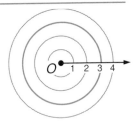

When the center of the circle is not at the pole, the equation can be found by considering the polar coordinates of any point on the circle. To derive the polar equation of a circle that passes through the pole O and has its center on the line of the polar axis, let $A(a, 0)$ be the center. Then the length of the radius $\overline{AO}$ is $|a|$. First consider the case where a is positive.

Let $P(r, \theta)$ where $r > 0$ and $0 < \theta < \frac{\pi}{2}$, be a point on the circle. In the diagram, $\triangle OAP$ is isosceles with $OA = AP = a$ and $m\angle AOP = m\angle APO = \theta$.

$$m\angle AOP + m\angle APO + m\angle OAP = \pi$$
$$\theta \quad + \quad \theta \quad + m\angle OAP = \pi$$
$$m\angle OAP = \pi - 2\theta$$

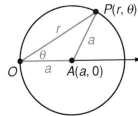

Using the Law of Sines,

$$\frac{OP}{\sin \angle OAP} = \frac{OA}{\sin \angle OPA}$$

$$\frac{r}{\sin (\pi - 2\theta)} = \frac{a}{\sin \theta}$$

$$r = \frac{a \sin (\pi - 2\theta)}{\sin \theta} \qquad \text{Use the identity } \sin (\pi - A) = \sin A.$$

$$r = \frac{a \sin 2\theta}{\sin \theta} \qquad \text{Use the identity } \sin 2A = 2 \sin A \cos A.$$

$$r = \frac{a (2 \sin \theta \cos \theta)}{\sin \theta}$$

$$r = 2a \cos \theta$$

If $-\frac{\pi}{2} < \theta < 0$ in the previous derivation, then $-\theta$ would be the positive measure of $\angle POA$ and $\angle OPA$ in $\triangle OAP$.

$$\frac{OP}{\sin \angle OAP} = \frac{OA}{\sin \angle OPA}$$

$$\frac{r}{\sin (\pi - 2(-\theta))} = \frac{a}{\sin (-\theta)}$$

$$r = \frac{a \sin (\pi + 2\theta)}{\sin (-\theta)}$$

Use the identities $\sin (\pi + A) = -\sin A$ and $\sin (-A) = -\sin A$.

$$r = \frac{a (-\sin 2\theta)}{-\sin \theta}$$

$$r = \frac{a \sin 2\theta}{\sin \theta}$$

$$r = \frac{a (2 \sin \theta \cos \theta)}{\sin \theta}$$

Use the identity $\sin 2A = 2 \sin A \cos A$.

$$r = 2a \cos \theta$$

Therefore, any point of the circle in quadrant IV can be described by the equation $r = 2a \cos \theta$.

Since all points of the circle are either in quadrant I or quadrant IV, $r = 2a \cos \theta$ is the polar equation of a circle through the pole with center on the polar axis at $(a, 0)$ when $a > 0$. The diameter of the circle is $2a$.

To sketch the graph, values of θ in the interval $\left(-\frac{\pi}{2}, \frac{\pi}{2}\right]$ are sufficient. Other values of θ produce alternate coordinates of the same points. For example:

If $\theta = \frac{\pi}{3}$, $r = 2a \cos \frac{\pi}{3} = 2a\left(\frac{1}{2}\right) = a$ and $\left(a, \frac{\pi}{3}\right)$ are the coordinates of a point of the circle.

If $\theta = \frac{4\pi}{3}$, $r = 2a \cos \frac{4\pi}{3} = 2a\left(-\frac{1}{2}\right) = -a$ and $\left(-a, \frac{4\pi}{3}\right)$ are coordinates of the same point.

Example 2

Sketch the graph of $r = 2 \cos \theta$ for $-\frac{\pi}{2} < \theta \le \frac{\pi}{2}$.

Solution

$r = 2 \cos \theta$ is the polar equation of a circle with diameter 2 and center on the polar axis at $(1, 0)$. The polar coordinates of four points on the circle are derived below.

If $\theta = \frac{\pi}{6}$, then $r = 2 \cos \frac{\pi}{6}$

$$= 2\left(\frac{\sqrt{3}}{2}\right) = \sqrt{3} \rightarrow \left(\sqrt{3}, \frac{\pi}{6}\right)$$

If $\theta = -\frac{\pi}{3}$, then $r = 2 \cos \left(-\frac{\pi}{3}\right)$

$$= 2\left(\frac{1}{2}\right) = 1 \rightarrow \left(1, -\frac{\pi}{3}\right)$$

If $\theta = \frac{\pi}{2}$, then $r = 2 \cos \frac{\pi}{2}$

$$= 2(0) = 0 \rightarrow \left(0, \frac{\pi}{2}\right)$$

If $\theta = -\frac{\pi}{4}$, then $r = 2 \cos \left(-\frac{\pi}{4}\right)$

$$= 2\left(\frac{\sqrt{2}}{2}\right) = \sqrt{2} \rightarrow \left(\sqrt{2}, -\frac{\pi}{4}\right)$$

To display polar graphs on a graphing calculator, the equations must be entered in *parametric form*, that is, each of the variables X and Y is expressed in terms of a third variable, T, which the calculator uses in place of θ. Use radian mode for T.

Step 1: Prepare the calculator to accept parametric equations.

Press `MODE`; highlight **Param**, `ENTER`

Choose either rectangular or polar coordinates, depending on the type of coordinates desired when `TRACE` is pressed, for example, when finding the points of intersection of two or more curves.

Step 2: Set the range.

For T, the variable that is used in place of θ from $-\frac{\pi}{2}$ to $\frac{\pi}{2}$, use -1.8 to 1.8.

For X: 0 to 4   For Y: -3 to 3

Step 3: Write the function in parametric form.

To obtain the graph of $r = 2 \cos \theta$, replace r by $2 \cos \theta$ in the relationships $x = r \cos \theta$ and $y = r \sin \theta$.

$x = r \cos \theta$ $y = r \sin \theta$
$x = (2 \cos \theta)(\cos \theta)$ $y = 2 \cos \theta \sin \theta$
$x = 2 \cos^2 \theta$

Step 4: Enter the function in parametric form.

Step 5: Graph.

Note that the graph displayed does not look like a circle because the scale in the horizontal direction is not equal to the scale in the vertical direction.

To adjust the scale:

press `ZOOM`; highlight **5:Square**, `ENTER`; `GRAPH`

The graph that is now displayed is a circle.

To demonstrate that the values of T, that is, θ, can be any interval of length π, change the range values of T. Use 0 to 3.2 or 3.2 to 6.4.

The equation of the circle through the pole with center at $(a, 0)$ was derived for $a > 0$. If the center of the circle passing through the pole is the point $A(a, 0)$ and $a < 0$, then $-a$, the length of a radius, is a positive number.

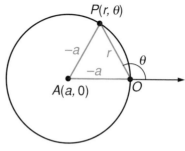

$$OA = AP = -a$$
$$m\angle AOP = m\angle APO = \pi - \theta$$
$$m\angle AOP + m\angle APO + m\angle OAP = \pi$$
$$(\pi - \theta) + (\pi - \theta) + m\angle OAP = \pi$$
$$m\angle OAP = 2\theta - \pi$$

Using the Law of Sines,

$$\frac{r}{\sin(2\theta - \pi)} = \frac{-a}{\sin(\pi - \theta)}$$

$$r = \frac{-a\sin(2\theta - \pi)}{\sin(\pi - \theta)}$$
Use the identities $\sin(A - \pi) = -\sin A$ and $\sin(\pi - A) = \sin A$.

$$r = \frac{-a(-\sin 2\theta)}{\sin \theta}$$

$$r = \frac{2a\sin\theta\cos\theta}{\sin\theta}$$
Use the identity $\sin 2A = \sin A \cos A$.

$$r = 2a\cos\theta$$

A point $P(r, \theta)$, where $r > 0$ and $-\pi < \theta < -\frac{\pi}{2}$, can also be used to verify that the polar equation of a circle with center at $A(a, 0)$ where $a < 0$, and radius $|a|$ is $r = 2a \cos \theta$. This case is left as an exercise.

Thus, in general, the polar equation of a circle that passes through the pole and has center at $A(a, 0)$, where a is any nonzero real number, is:

$$r = 2a\cos\theta$$

Example 3

Describe the graph of the polar equation $r = -4\cos\theta$.

Solution

The graph of $r = -4\cos\theta$ is a circle with diameter 4 and center at $(-2, 0)$ (or $(2, \pi)$).

Lines

Since every point on a vertical line through the pole has the polar coordinates $\left(r, \frac{\pi}{2}\right)$, the polar equation of this line is $\theta = \frac{\pi}{2}$.

To derive the polar equation of any other vertical line, consider a line through $A(a, 0)$. When $a > 0$, A is on the polar axis and the line lies in quadrants I and IV.

Let $P(r, \theta)$ be any point in quadrant I that is on the line, that is, $0 < \theta < \frac{\pi}{2}$. Using right $\triangle OAP$ in the diagram,

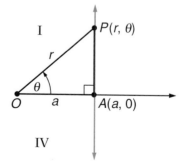

$$\cos\theta = \frac{a}{r}$$
$$r\cos\theta = a$$

Let $P(r, \theta)$ be any point in quadrant IV that is on the line, that is, $-\frac{\pi}{2} < \theta < 0$. Thus, θ is negative and $-\theta$ is the positive measure of $\angle POA$ in $\triangle OAP$.

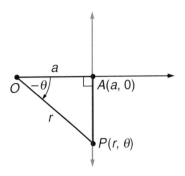

$$\cos(-\theta) = \frac{a}{r}$$

$$r\cos(-\theta) = a$$

$$r\cos\theta = a \qquad \text{Use the identity} \\ \cos(-A) = \cos A.$$

When a is negative, $A(a, 0)$ is on the ray opposite the polar axis and the line lies in quadrants II and III. Since a is negative, $-a$ is the length of side OA in right $\triangle OAP$. Let $P(r, \theta)$ be any point in quadrant II that is on the line.

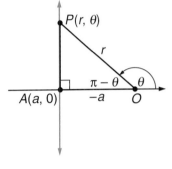

$$\cos(\pi - \theta) = \frac{-a}{r} \qquad \text{Use the identities } \cos(-A) = \cos A \\ \text{and } \cos(\pi - A) = -\cos A.$$

$$-\cos\theta = \frac{-a}{r}$$

$$\cos\theta = \frac{a}{r}$$

$$r\cos\theta = a$$

If $P(r, \theta)$ is a point in quadrant III, then $\theta - \pi$ is the measure of $\angle POA$ in right $\triangle OAP$. The derivation of the equation $r\cos\theta = a$ for this case is left as an exercise.

Thus, in general, the polar equation of a vertical line that passes through the point $A(a, 0)$, where a is any real number, is:

$$r\cos\theta = a$$

Example 4

Sketch the graph of $r\cos\theta = -3$ for $-\frac{\pi}{2} < \theta \leq \frac{\pi}{2}$.

Solution

$r\cos\theta = -3$ is the polar equation of a vertical line through $(-3, 0)$.
To obtain polar coordinates of two points on the line:

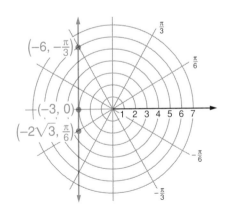

Let $\theta = \frac{\pi}{6}$

$$r\cos\frac{\pi}{6} = -3$$

$$r\left(\frac{\sqrt{3}}{2}\right) = -3$$

$$r = -3 \cdot \frac{2}{\sqrt{3}}$$

$$= -2\sqrt{3}$$

$\left(-2\sqrt{3}, \frac{\pi}{6}\right)$ is a point on the line.

Let $\theta = -\frac{\pi}{3}$

$$r\cos\left(-\frac{\pi}{3}\right) = -3$$

$$r\left(\frac{1}{2}\right) = -3$$

$$r = -3(2)$$

$$= -6$$

$\left(-6, -\frac{\pi}{3}\right)$ is a point on the line.

1.

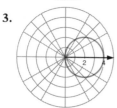

2.

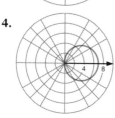

3.

4.

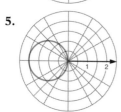

5.

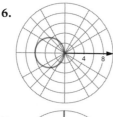

6.

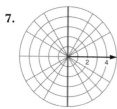

7.

8.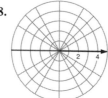

9.

Exercises 11.3

In **1 – 12**, sketch the graph of the given polar equation:
  a. on polar graph paper   **b.** on a calculator

1. $r = 5$ **2.** $r = 2$ **3.** $r = 4 \cos \theta$

4. $r = 7 \cos \theta$ **5.** $r = -2 \cos \theta$ **6.** $r = -6 \cos \theta$

7. $\theta = \frac{\pi}{2}$ **8.** $\theta = 0$ **9.** $r \cos \theta = 4$

10. $r \cos \theta = 12$ **11.** $r \cos \theta = -2$ **12.** $r \cos \theta = -5$

13. A circle with center at $C\left(a, \frac{\pi}{2}\right)$ passes through the pole.
  a. Let $P(r, \theta)$ be a point in quadrant I that is on the circle. Show that the polar equation of the half-circle in quadrant I is $r = 2a \sin \theta$.
  b. Let $P(r, \theta)$ be a point in quadrant II that is on the circle. In $\triangle OCP$, $m\angle POC = \theta - \frac{\pi}{2}$. Show that the polar equation of the half-circle in quadrant II is $r = 2a \sin \theta$.
  c. What is the polar equation of the circle through the pole with center at $\left(2, \frac{\pi}{2}\right)$?

In **14 – 16**, sketch the graph.

14. $r = \sin \theta$ **15.** $r = 2 \sin \theta$ **16.** $r = -2 \sin \theta$

17. Let k be a vertical line through $A(a, 0)$. Let k' be the image of k under a rotation through an angle α about O. Let $P(r, \theta)$ be any point on k'.
  a. Show the equation of k' is $r \cos (\theta - \alpha) = a$.
  b. Express in terms of a and α the coordinates of Q, the point at which k' intersects the polar axis.
  c. Use the result in part **a** to show that the equation of a horizontal line through $\left(a, \frac{\pi}{2}\right)$ is $r \sin \theta = a$.

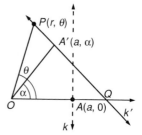

In **18 – 20**, sketch the graph.

18. $r \sin \theta = 3$ **19.** $r \sin \theta = 5$ **20.** $r \sin \theta = -4$

21. A vertical line through $A(4, 0)$ has been rotated 30° about O. Write the polar equation of the image line and find its point of intersection with the polar axis.

22. A vertical line through $(2, \pi)$ has been rotated 60° about O. Write a polar equation of the image line.

23. Use a point in quadrant III to derive the equation of a circle through the pole if the center of the circle is at $(a, 0)$ and $a < 0$.

24. Derive the equation of a vertical line through $(a, 0)$ if a is negative and $P(r, \theta)$ is a point on the line in quadrant III.

10.

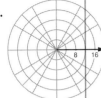

11.

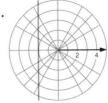

12.

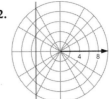

13. a. In isosceles triangle OCP, $OC = PC = a$ and $OP = r$.

$$m\angle POC = m\angle OPC = \frac{\pi}{2} - \theta$$

$$m\angle POC + m\angle OPC + m\angle OCP = \pi$$

$$\left(\frac{\pi}{2} - \theta\right) + \left(\frac{\pi}{2} - \theta\right) + m\angle OCP = \pi$$

$$m\angle OCP = 2\theta$$

Use the Law of Sines.

$$\frac{OP}{\sin \angle OCP} = \frac{OC}{\sin \angle OPC}$$

$$\frac{r}{\sin 2\theta} = \frac{a}{\sin \left(\frac{\pi}{2} - \theta\right)}$$

$$r = \frac{a \sin 2\theta}{\sin \left(\frac{\pi}{2} - \theta\right)}$$

$$= \frac{a(2 \sin \theta \cos \theta)}{\cos \theta}$$

$$= 2a \sin \theta$$

b. In isosceles triangle OCP, $OC = PC = a$ and $OP = r$.

$$m\angle POC = m\angle OPC = \theta - \frac{\pi}{2}$$

$$m\angle POC + m\angle OPC + m\angle OCP = \pi$$

$$\left(\theta - \frac{\pi}{2}\right) + \left(\theta - \frac{\pi}{2}\right) + m\angle OCP = \pi$$

$$m\angle OCP = 2\pi - 2\theta$$

Use the Law of Sines.

$$\frac{OP}{\sin \angle OCP} = \frac{OC}{\sin \angle OPC}$$

$$\frac{r}{\sin (2\pi - 2\theta)} = \frac{a}{\sin \left(\theta - \frac{\pi}{2}\right)}$$

$$r = \frac{a \sin (2\pi - 2\theta)}{\sin \left(\theta - \frac{\pi}{2}\right)}$$

$$= \frac{a \sin (-2\theta)}{-\cos \theta}$$

$$= \frac{a (-2 \sin \theta \cos \theta)}{-\cos \theta}$$

$$= 2a \sin \theta$$

c. $r = 4 \sin \theta$

14.

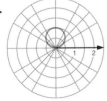

15.

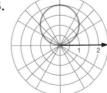

16.

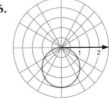

17. a. $\angle OA'P$ is a right angle, $OP = r$,
$OA' = OA = a$, and $m\angle A'OP = (\theta - \alpha)$

$$\cos \angle A'OP = \frac{OA'}{OP}$$

$$\cos (\theta - \alpha) = \frac{a}{r}$$

$$r \cos (\theta - \alpha) = a$$

b. In right $\triangle OQA'$, $OA' = a$. Therefore,

$$\cos \alpha = \frac{a}{OQ}$$

$$OQ = \frac{a}{\cos \alpha}$$

$$OQ = a \sec \alpha$$

Q is the point $(a \sec \alpha, 0)$.

c.
$$r \cos (\theta - \alpha) = a$$
$$r(\cos \theta \cos \alpha + \sin \theta \sin \alpha) = a$$

If $\alpha = \frac{\pi}{2}$, $r\left(\cos \theta \cos \frac{\pi}{2} + \sin \theta \sin \frac{\pi}{2}\right) = a$
$$r(\cos \theta(0) + \sin \theta(1)) = a$$
$$r \sin \theta = a$$

18.

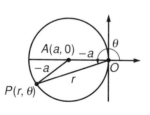

19.

20.

21. $r \cos (\theta - 30°) = 4$; $\left(\frac{8\sqrt{3}}{3}, 0\right)$

22. $r \cos (\theta - 60°) = 2$

23. Since a is negative, $OA = -a$

$$PA = OA = -a$$
$$OP = r$$
$$m\angle P = m\angle AOP = \theta - \pi$$
$$m\angle A + m\angle P + m\angle AOP = \pi$$
$$m\angle A + (\theta - \pi) + (\theta - \pi) = \pi$$
$$m\angle A = 3\pi - 2\theta$$

$$\frac{OP}{\sin A} = \frac{OA}{\sin P}$$

$$\frac{r}{\sin (3\pi - 2\theta)} = \frac{-a}{\sin (\theta - \pi)}$$

$$\frac{r}{\sin (2\theta)} = \frac{-a}{-\sin \theta}$$

$$r = \frac{-a(2 \sin \theta \cos \theta)}{-\sin \theta}$$

$$r = 2a \cos \theta$$

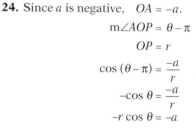

24. Since a is negative, $OA = -a$.

$$m\angle AOP = \theta - \pi$$
$$OP = r$$
$$\cos (\theta - \pi) = \frac{-a}{r}$$
$$-\cos \theta = \frac{-a}{r}$$
$$-r \cos \theta = -a$$
$$r \cos \theta = a$$

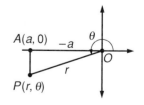

11.4 Graphs of Polar Equations

Equations that express r in terms of multiples of the sine or cosine of an angle θ have interesting graphs in the polar plane. To consider the graphs of some equations of the form $r = a \pm b \cos \theta$ and $r = a \pm b \sin \theta$, some rules concerning symmetry in the polar plane will be considered first.

Symmetry With Respect to the Line Containing the Polar Axis

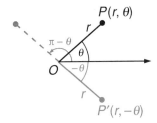

When point P is reflected over the line containing the polar axis, the image of $P(r, \theta)$ is $P'(r, -\theta)$ or $P'(-r, \pi - \theta)$. Therefore, the graph of an equation is symmetric with respect to the line containing the polar axis if for each point (r, θ) on the graph, the point $(r, -\theta)$ or $(-r, \pi - \theta)$ is also a point on the graph.

For example, recall the graph of the equation $r = a \cos \theta$, a circle through the pole with its center on the polar axis. Note that this graph is symmetric with respect to the line containing the polar axis since the equation is unchanged when $(r, -\theta)$ is substituted for (r, θ). That is:

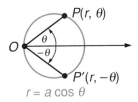

$r = a \cos \theta$

$r = a \cos (-\theta)$ Substitute $(r, -\theta)$.

$r = a \cos \theta$ $\cos (-\theta) = \cos \theta$

$r = a \cos \theta$

Symmetry With Respect to the Line Perpendicular to the Polar Axis at the Pole

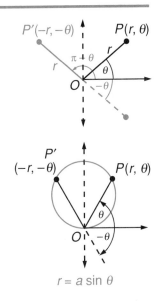

When point P is reflected over the line that is perpendicular to the polar axis at the pole, the image of $P(r, \theta)$ is $P'(-r, -\theta)$ or $P'(r, \pi - \theta)$. Therefore, the graph of an equation is symmetric with respect to the line perpendicular to the polar axis at the pole if for each point (r, θ) on the graph, the point $(-r, -\theta)$ or $(r, \pi - \theta)$ is also a point on the graph.

For example, the graph of $r = a \sin \theta$ is a circle passing through the pole with its center on the line perpendicular to the polar axis at the pole. This graph is symmetric with respect to the line perpendicular to the polar axis at the pole since the equation is unchanged when $(-r, -\theta)$ is substituted for (r, θ). That is:

$r = a \sin \theta$

$-r = a \sin (-\theta)$ Substitute $(-r, -\theta)$.

$-r = a (-\sin \theta)$ $\sin (-\theta) = -\sin \theta$

$r = a \sin \theta$

$r = a \sin \theta$

Symmetry With Respect to the Pole

When point P is reflected through the pole, the image of $P(r, \theta)$ is $P'(-r, \theta)$ or $P'(r, \pi + \theta)$. Therefore, the graph of an equation is symmetric with respect to the pole if for each point (r, θ) on the graph, the point $(-r, \theta)$ or $(r, \pi + \theta)$ is also a point on the graph.

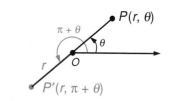

For example, the graph of $r^2 = 4 \cos^2 \theta$ is a pair of circles through the pole with centers at $(2, 0)$ and $(-2, 0)$ since $r^2 = 4 \cos^2 \theta$ is equivalent to $r = 2 \cos \theta$ or $r = -2 \cos \theta$. This graph is symmetric with respect to the pole since its equation is unchanged when $(-r, \theta)$ is substituted for (r, θ). That is:

$$r^2 = 4 \cos^2 \theta$$
$$(-r)^2 = 4 \cos^2 \theta \qquad \text{Substitute } (-r, \theta).$$
$$r^2 = 4 \cos^2 \theta$$

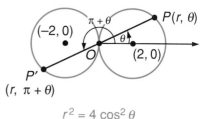

$$r^2 = 4 \cos^2 \theta$$

Limaçons

How do the graphs of $r = a \cos \theta$ and $r = a \sin \theta$ change if a constant is added to the right side of the equation? For example, what is the graph of the equation $r = 2 + 2 \cos \theta$? To sketch this curve, first look for symmetry and then plot several points of the curve. Notice that since $\cos (-\theta) = \cos \theta$, replacing (r, θ) by $(r, -\theta)$ yields the original equation. Therefore, the graph is symmetric with respect to the line containing the polar axis.

Make a table including values of θ in quadrants I and II. Use the symmetry of the graph to complete the curve for values of θ in quadrants III and IV.

θ	$\cos \theta$	$2 + 2 \cos \theta$
0	1	$2 + 2(1) = 4$
$\frac{\pi}{6}$	$\frac{\sqrt{3}}{2}$	$2 + 2\left(\frac{\sqrt{3}}{2}\right) \approx 3.73$
$\frac{\pi}{3}$	$\frac{1}{2}$	$2 + 2\left(\frac{1}{2}\right) = 3$
$\frac{\pi}{2}$	0	$2 + 2(0) = 2$
$\frac{2\pi}{3}$	$-\frac{1}{2}$	$2 + 2\left(-\frac{1}{2}\right) = 1$
$\frac{5\pi}{6}$	$-\frac{\sqrt{3}}{2}$	$2 + 2\left(-\frac{\sqrt{3}}{2}\right) \approx 0.27$
π	-1	$2 + 2(-1) = 0$

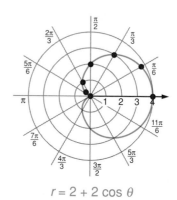

$r = 2 + 2 \cos \theta$

Because of the heart-shaped form of the curve, the graph is called a *cardioid*.

The graph of any equation of the form $r = a \pm a \cos \theta$ or $r = a \pm a \sin \theta$ is a cardioid. Note that for a given value of a, the graph of $r = a \pm a \sin \theta$ is the image of $r = a \pm a \cos \theta$ under a rotation of $\frac{\pi}{2}$.

In the equation of the cardioid, the constant term and the coefficient of cos θ or sin θ have the same absolute value. The following example shows how the shape of the curve is changed when these absolute values are not the same.

Example 1

Sketch the graph of $r = 1 + 2 \sin \theta$.

Solution

Check for symmetry. Replace (r, θ) in the equation by $(r, \pi - \theta)$:

$$r = 1 + 2 \sin \theta$$
$$r = 1 + 2 \sin (\pi - \theta) \qquad \sin (\pi - \theta) = \sin \theta$$
$$r = 1 + 2 \sin \theta$$

Thus, the graph is symmetric with respect to the line perpendicular to the polar axis at the pole.

Choose values of θ in quadrants IV and I. Note that for some values of θ in quadrant IV, r is negative and the points are located in quadrant II. Use the symmetry of the graph to complete the curve.

θ	$\sin \theta$	$1 + 2 \sin \theta$
$-\frac{\pi}{2}$	-1	$1 + 2(-1) = -1$
$-\frac{\pi}{3}$	$-\frac{\sqrt{3}}{2}$	$1 + 2\left(-\frac{\sqrt{3}}{2}\right) \approx -0.73$
$-\frac{\pi}{6}$	$-\frac{1}{2}$	$1 + 2\left(-\frac{1}{2}\right) = 0$
0	0	$1 + 2(0) = 1$
$\frac{\pi}{6}$	$\frac{1}{2}$	$1 + 2\left(\frac{1}{2}\right) = 2$
$\frac{\pi}{3}$	$\frac{\sqrt{3}}{2}$	$1 + 2\left(\frac{\sqrt{3}}{2}\right) \approx 2.73$
$\frac{\pi}{2}$	1	$1 + 2(1) = 3$

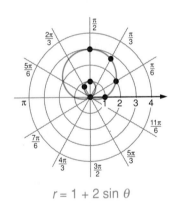

$r = 1 + 2 \sin \theta$

Example 2

Sketch the graph of $r = 2 + \sin \theta$.

Solution

Check for symmetry by replacing (r, θ) in the equation by $(r, \pi - \theta)$. As in Example 1, the graph of $r = 2 + \sin \theta$ is symmetric with respect to the line perpendicular to the polar axis at the pole.

Choose values of θ in quadrants IV and I. Use the symmetry of the graph to complete the curve.

θ	$\sin\theta$	$2 + \sin\theta$
$-\dfrac{\pi}{2}$	-1	$2 + (-1) = 1$
$-\dfrac{\pi}{3}$	$-\dfrac{\sqrt{3}}{2}$	$2 + \left(-\dfrac{\sqrt{3}}{2}\right) \approx 1.13$
$-\dfrac{\pi}{6}$	$-\dfrac{1}{2}$	$2 + \left(-\dfrac{1}{2}\right) = \dfrac{3}{2}$
0	0	$2 + 0 = 2$
$\dfrac{\pi}{6}$	$\dfrac{1}{2}$	$2 + \dfrac{1}{2} = \dfrac{5}{2}$
$\dfrac{\pi}{3}$	$\dfrac{\sqrt{3}}{2}$	$2 + \dfrac{\sqrt{3}}{2} \approx 2.87$
$\dfrac{\pi}{2}$	1	$2 + 1 = 3$

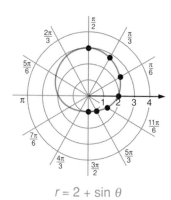

$r = 2 + \sin\theta$

The graph of any equation of the form $r = a \pm b\cos\theta$ or $r = a \pm b\sin\theta$ is called a $\texttt{limaçon}$. If $|a| < |b|$, the limaçon has two loops (for example, $r = 1 + 2\sin\theta$ in Example 1). If $|a| = |b|$, the limaçon is a cardioid (for example, $r = 2 + 2\cos\theta$ on page 538). If $|a| > |b|$, the limaçon is one flattened loop (for example, $r = 2 + \sin\theta$ in Example 2).

Rose Curves

How do the graphs of $r = a\cos\theta$ and $r = a\sin\theta$ change if θ is multiplied by an integer greater than 1? For example, what is the shape of the graph of $r = \sin 2\theta$, where $a = 1$? To sketch this curve, first look for symmetry and then plot several points.

When (r, θ) in the equation $r = \sin 2\theta$ is replaced by $(-r, \pi - \theta)$,

$$-r = \sin 2(\pi - \theta)$$
$$-r = \sin (2\pi - 2\theta) \qquad \sin (2\pi - 2\theta) = \sin (-2\theta)$$
$$-r = \sin (-2\theta) \qquad\qquad \sin (-2\theta) = -\sin 2\theta$$
$$-r = -\sin 2\theta$$
$$r = \sin 2\theta$$

Thus, the curve is symmetric with respect to the line containing the polar axis.

When (r, θ) is replaced by $(-r, -\theta)$,

$$r = \sin 2\theta$$
$$-r = \sin 2(-\theta)$$
$$-r = \sin (-2\theta) \qquad \sin (-2\theta) = -\sin 2\theta$$
$$-r = -\sin 2\theta$$
$$r = \sin 2\theta$$

Thus, the curve is symmetric with respect to the line perpendicular to the polar axis at the pole.

The curve is also symmetric with respect to the pole. The verification of this symmetry is left as an exercise.

Choose values of θ in quadrant I.

θ	2θ	$\sin 2\theta$
0	0	0
$\dfrac{\pi}{6}$	$\dfrac{\pi}{3}$	$\dfrac{\sqrt{3}}{2} \approx 0.87$
$\dfrac{\pi}{4}$	$\dfrac{\pi}{2}$	1
$\dfrac{\pi}{3}$	$\dfrac{2\pi}{3}$	$\dfrac{\sqrt{3}}{2} \approx 0.87$
$\dfrac{\pi}{2}$	π	0

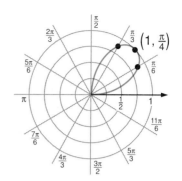

Since the graph is symmetric with respect to the line containing the polar axis, the portion of the curve located in quadrant IV can be drawn.

Since the graph is symmetric with respect to the line perpendicular to the polar axis at the pole, the remaining portions of the curve located in quadrants II and III can be drawn.

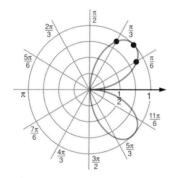

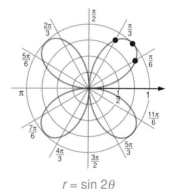

$r = \sin 2\theta$

The graph of any equation of the form $r = a \cos n\theta$ or $r = a \sin n\theta$ is called a *rose*.

A graphing calculator can be helpful in sketching other graphs of the form $r = a \cos n\theta$ and $r = a \sin n\theta$. Put the calculator in parametric mode and in radian mode. Let the range of T be from 0 to 6.283185307, that is from 0 to 2π, and the values of X and Y from -4 to 4. Graph the equation $r = 2 \cos 4\theta$ by using $x = r \cos \theta$ and $y = r \sin \theta$ and replacing r by $2 \cos 4\theta$.

Y= 2 **COS** 4 **X|T** **COS** **X|T** **ENTER**

2 **COS** 4 **X|T** **SIN** **X|T** **GRAPH**

Note that the curve is a rose with 8 petals.

In general, if $r = a \cos n\theta$ or $r = a \sin n\theta$ and n is even, the number of petals is $2n$.

Use a graphing calculator to sketch the graph of $r = 4 \sin 3\theta$. Use the same values for T, X, and Y as were used for $r = 2 \cos 4\theta$.

Y= 4 **COS** 3 **X|T** **COS** **X|T** **ENTER**

4 **COS** 3 **X|T** **SIN** **X|T** **GRAPH**

Note that the rose has 3 petals.

Repeat for $r = 2 \sin 5\theta$. Note that the rose has 5 petals. In general, if $r = a \cos n\theta$ or $r = a \sin n\theta$ and n is odd, the number of petals is n.

Exercises 11.4

In **1 – 12**, answer these questions: **a.** Is the graph of the equation a circle, a limaçon or a rose? **b.** What symmetries does the graph have? **c.** Sketch the graph on polar graph paper or on a graphing calculator.

1. $r = 3 \cos \theta$ **2.** $r = 3 + 3 \cos \theta$ **3.** $r = 3 - \cos \theta$ **4.** $r = 3 \cos 3\theta$

5. $r = 1 - \sin \theta$ **6.** $r = 2 + 4 \sin \theta$ **7.** $r = 1 + 2 \cos \theta$ **8.** $r = 4 \sin \theta$

9. $r = 2 + \cos \theta$ **10.** $r = 2 \sin 2\theta$ **11.** $r = 5 \cos 2\theta$ **12.** $r = 2 - 5 \sin \theta$

In **13 – 24**, sketch the graph on polar graph paper or on a graphing calculator.

13. $r = 2 + 2 \sin \theta$ **14.** $r = 1 - \cos \theta$ **15.** $r = 1 - 2 \sin \theta$ **16.** $r = 1 - 4 \cos \theta$

17. $r = 4 + 2 \sin \theta$ **18.** $r = 2 \cos 2\theta$ **19.** $r = 4 \sin 3\theta$ **20.** $r = 5 \cos 4\theta$

21. $r^2 = 9 \sin^2 \theta$ **22.** $r^2 = 4 \cos^2 \theta$ **23.** $r^2 = \cos^2 \theta$ **24.** $r^2 = \sin^2 \theta$

25. A ladder that is 20 feet long is placed against a vertical wall so that the ladder makes an angle of θ degrees with the wall.

   a. Express the distance r from the base of the wall to the ladder as a function of θ.
   b. Sketch the graph of the function found in part **a** in the polar plane.

26. A Ferris wheel of radius 20 feet is mounted with its center 25 feet above the ground. Let O be the center of the wheel, $\overrightarrow{OA}$ be a horizontal ray such that A is to the right of O and P be any point on the wheel. Express h, the height of P above the ground, as a function of θ, where $\theta = \mathrm{m}\angle AOP$.

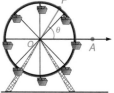

27. A clock standing on a table has a 6-inch pendulum. At the highest point of its swing, the free end of the pendulum is 7 inches from the table. The fixed end of the pendulum is 10 inches from the table.

   a. Let θ be the angle between a vertical center line through the fixed point of the pendulum and the pendulum. If the pendulum is to the right of the center line, θ is positive. If the pendulum is to the left of the center line, θ is negative. What are the possible values of θ?
   b. Express r, the distance from the free end of the pendulum to the table, as a function of θ.
   c. Sketch the graph of the function found in part **b** in the polar plane.

28. Show that the graph of $r = \sin 2\theta$ is symmetric with respect to the pole.

1. a. circle   **b.** polar axis

  c.
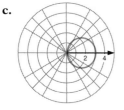

2. a. limaçon   **b.** polar axis

  c.
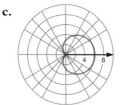

3. a. limaçon   **b.** polar axis

  c.
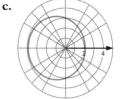

4. a. rose

   b. polar axis

   c.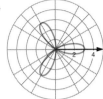

5. a. limaçon

   b. vertical axis through pole

   c.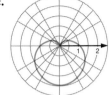

6. a. limaçon

   b. vertical axis through pole

   c.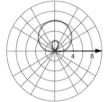

7. a. limaçon

   b. polar axis

   c.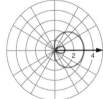

8. a. circle

   b. vertical axis through pole

   c.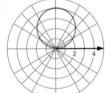

9. a. limaçon

   b. polar axis

   c.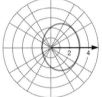

10. a. rose

   b. polar axis
   vertical axis through pole
   pole

   c.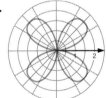

11. a. rose

   b. polar axis
   vertical axis through pole
   pole

   c.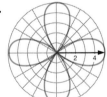

12. a. limaçon

   b. vertical axis through pole

   c.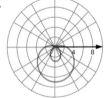

13.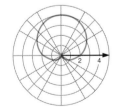

14.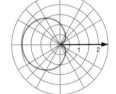

15.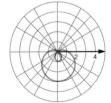

16.

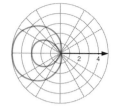

17.

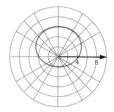

18.

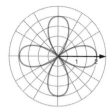

19.

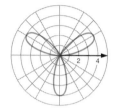

20.

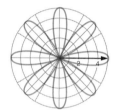

21.

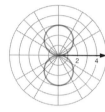

22.

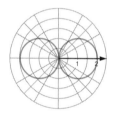

23.

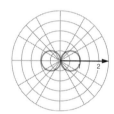

24.

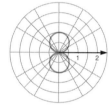

25. a. $r = 10 \sin 2\theta$

b.

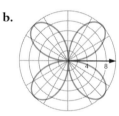

26. $h = 25 + 20 \sin \theta$

27. a. maximum value is $\frac{\pi}{3}$;

minimum value is $-\frac{\pi}{3}$

b. $r = 10 - 6 \cos \theta$

c.

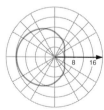

28. Replace (r, θ) by $(r, \pi + \theta)$.

$r = \sin 2\theta$

$r = 2 \sin \theta \cos \theta$

$r = 2 \sin (\pi + \theta) \cos (\pi + \theta)$

$r = 2(-\sin \theta)(-\cos \theta)$

$r = \sin 2\theta$

Since the equation remains unchanged when (r, θ) is replaced by $(r, \pi + \theta)$, the graph is symmetric with respect to the pole.

11.5 Polar Equations of the Conics

The path of Halley's Comet is an ellipse with the sun at one focus. To derive the equation of such an ellipse, distances from the sun can be more easily expressed if the focus of the ellipse is placed at the pole of a polar coordinate system and the equation of the ellipse is written in polar coordinates.

Let $P(r, \theta)$ be any point on a conic with a vertical directrix and its focus at the pole. Let p be the distance from the pole to the directrix. Let P' be the projection of the point P onto the polar axis. To find the distance PM, where M is the projection of the point P onto the directrix, two cases must be considered:

The directrix is to the left of the pole.

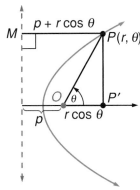

$$OP' = r \cos \theta$$
$$PM = p + OP'$$
$$PM = p + r \cos \theta$$

The directrix is to the right of the pole.

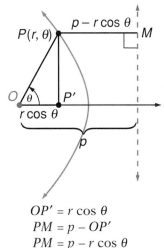

$$OP' = r \cos \theta$$
$$PM = p - OP'$$
$$PM = p - r \cos \theta$$

In the diagrams, θ is an acute angle and $r \cos \theta$ is positive. Each relationship also holds when θ is an obtuse angle and $r \cos \theta$ is negative.

For a conic with eccentricity e: $\dfrac{OP}{PM} = e$

When the directrix is to the left of the pole:

$$OP = ePM$$
$$r = e(p + r \cos \theta)$$
$$r = ep + er \cos \theta$$
$$r - er \cos \theta = ep$$
$$r(1 - e \cos \theta) = ep$$
$$r = \frac{ep}{1 - e \cos \theta}$$

When the directrix is to the right of the pole:

$$OP = ePM$$
$$r = e(p - r \cos \theta)$$
$$r = ep - er \cos \theta$$
$$r + er \cos \theta = ep$$
$$r(1 + e \cos \theta) = ep$$
$$r = \frac{ep}{1 + e \cos \theta}$$

Thus, in general, the standard form of the equation of a conic with vertical directrix and focus at the pole is:

$$r = \frac{ep}{1 \pm e \cos \theta}$$

The various conics are then distinguished by the value of e.

Example 1

Describe and sketch the conic whose equation is:

$$r = \frac{4}{3 - 3 \cos \theta}$$

Solution

Put the equation in the standard form. Since the constant in the denominator must be 1, divide numerator and denominator of the right side of the equation by the constant term in the denominator.

$$r = \frac{4}{3 - 3 \cos \theta}$$

$$r = \frac{\frac{4}{3}}{\frac{3}{3} - \frac{3}{3} \cos \theta} \qquad \textit{Divide by the constant term.}$$

$$r = \frac{\frac{4}{3}}{1 - 1 \cos \theta} \qquad \textit{Standard form.}$$

When the equation of a conic is in standard form, the eccentricity is the absolute value of the coefficient of $\cos \theta$. In this conic, $e = |-1|$, or 1, indicating that this conic is a parabola.

Use the numerator of the equation in standard form to find the value of p, the distance from the focus at the pole to the directrix.

$$ep = \tfrac{4}{3}$$

$$1 \cdot p = \tfrac{4}{3}$$

$$p = \tfrac{4}{3}$$

Since the coefficient of $\cos \theta$ is negative, the directrix is to the left of the pole, and passes through the point $(-p, 0)$, or $(-\tfrac{4}{3}, 0)$. Thus, the equation of the directrix is $r \cos \theta = -\tfrac{4}{3}$.

The turning point of a parabola is located on a line through the focus that is perpendicular to the directrix. Therefore, the turning point for this parabola is on the line of the polar axis. The value of θ at the turning point may be 0 or π.

If $\theta = 0$, $r = \dfrac{\frac{4}{3}}{1 - \cos 0}$

$$r = \frac{\frac{4}{3}}{1 - 1} \quad \text{or} \quad \frac{\frac{4}{3}}{0}$$

At $\theta = 0$, r is undefined.

If $\theta = \pi$, $r = \dfrac{\frac{4}{3}}{1 - \cos \pi}$

$$r = \frac{\frac{4}{3}}{1 - (-1)} = \frac{\frac{4}{3}}{2} = \frac{2}{3}$$

Therefore, the turning point is $\left(\frac{2}{3}, \pi\right)$. This is the expected result, since the turning point is equidistant from the focus (the pole) and the directrix, which are $\frac{4}{3}$ units apart.

To sketch the graph, other points on the parabola are found by choosing values for θ and computing the corresponding values of r.

Let $\theta = \frac{\pi}{2}$: $\qquad r = \dfrac{\frac{4}{3}}{1 - \cos \frac{\pi}{2}}$

$$= \dfrac{\frac{4}{3}}{1 - 0} = \frac{4}{3}$$

Thus, $\left(\frac{4}{3}, \frac{\pi}{2}\right)$ is a point of the parabola.

Let $\theta = \frac{3\pi}{2}$: $\qquad r = \dfrac{\frac{4}{3}}{1 - \cos \frac{3}{2}\pi}$

$$= \dfrac{\frac{4}{3}}{1 - 0} = \frac{4}{3}$$

Thus, $\left(\frac{4}{3}, \frac{3\pi}{2}\right)$ is a point of the parabola.

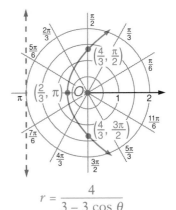

$$r = \frac{4}{3 - 3 \cos \theta}$$

To display this parabola on a graphing calculator:

Step 1: Put the calculator in the following modes: parametric and radian.

Step 2: Set the range.

For the variable T that is used in place of θ, use 0 to 6.283185307 to approximate 0 to 2π.

For X: -2 to 6 $\qquad$ For Y: -5 to 5

Step 3: Write the function in parametric form.

Use the relationships $x = r \cos \theta$ and $y = r \sin \theta$, replacing r by $\dfrac{4}{3 - 3 \cos \theta}$ for this particular parabola.

$x = r \cos \theta$ $\qquad\qquad\qquad$ $y = r \sin \theta$

$x = \left(\dfrac{4}{3 - 3 \cos \theta}\right) \cos \theta$ $\qquad$ $y = \left(\dfrac{4}{3 - 3 \cos \theta}\right) \sin \theta$

$x = \dfrac{4 \cos \theta}{3 - 3 \cos \theta}$ $\qquad\qquad\quad$ $y = \dfrac{4 \sin \theta}{3 - 3 \cos \theta}$

Step 4: Enter the function in parametric form and graph.

| Y= | 4 | COS | X\|T | ÷ | (| 3 | − | 3 | COS | X\|T |) | ENTER |

| | 4 | SIN | X\|T | ÷ | (| 3 | − | 3 | COS | X\|T |) | GRAPH |

Example 2

Write the equation of the hyperbola with focus at the pole and for which the endpoints of the transverse axis are (5, 0) and (–15, π).

Solution

To write the equation in standard form, you must find values for e and p.

The endpoint (–15, π) of the transverse axis can be written (15, 0). The center of the hyperbola is at (10, 0), the midpoint of the transverse axis. Use a as the distance from the center to an endpoint of the transverse axis. Thus, $a = 5$.

ae is the distance from the center to the focus. Thus:

$$ae = 10$$
$$5e = 10$$
$$e = 2$$

p is the distance from the focus to the directrix and $\dfrac{a}{e}$ is the distance from the center to the directrix. Thus:

$$p = ae - \frac{a}{e}$$
$$= 10 - \frac{5}{2} = \frac{15}{2}$$

Thus far, the equation is: $r = \dfrac{2\left(\frac{15}{2}\right)}{1 \pm 2 \cos \theta} = \dfrac{15}{1 \pm 2 \cos \theta}$

Use (5, 0) and (–15, π), the endpoints of the transverse axis, to determine the sign of the coefficient of cos θ.

If $(r, \theta) = (5, 0)$, then:

$$r = \frac{15}{1 - 2 \cos \theta} \quad \text{or} \quad r = \frac{15}{1 + 2 \cos \theta}$$

$$5 = \frac{15}{1 - 2 \cos 0} \qquad 5 = \frac{15}{1 + 2 \cos 0}$$

$$5 = \frac{15}{1 - 2(1)} \qquad 5 = \frac{15}{1 + 2(1)}$$

$$5 = \frac{15}{-1} \text{ false} \qquad 5 = \frac{15}{3} \text{ true}$$

If $(r, \theta) = (-15, \pi)$, then:

$$r = \frac{15}{1 - 2 \cos \theta} \quad \text{or} \quad r = \frac{15}{1 + 2 \cos \theta}$$

$$-15 = \frac{15}{1 - 2 \cos \pi} \qquad -15 = \frac{15}{1 + 2 \cos \pi}$$

$$-15 = \frac{15}{1 - 2(-1)} \qquad -15 = \frac{15}{1 + 2(-1)}$$

$$-15 = \frac{15}{3} \text{ false} \qquad -15 = \frac{15}{-1} \text{ true}$$

Thus, the coefficient of cos θ is positive.

Answer: The equation of the hyperbola is: $r = \dfrac{15}{1 + 2 \cos \theta}$

Example 3

Write in rectangular coordinates the equation $r = \dfrac{15}{1 + 2 \cos \theta}$.

Solution

$$r = \frac{15}{1 + 2 \cos \theta}$$

$$r(1 + 2 \cos \theta) = 15$$

$$r + 2r \cos \theta = 15$$

$$\sqrt{x^2 + y^2} + 2x = 15 \qquad\qquad r = \sqrt{x^2 + y^2};\ r \cos \theta = x$$

$$\sqrt{x^2 + y^2} = 15 - 2x$$

$$x^2 + y^2 = (15 - 2x)^2$$

$$x^2 + y^2 = 225 - 60x + 4x^2$$

$$-3x^2 + 60x + y^2 = 225$$

$$-3(x^2 - 20x) + y^2 = 225$$

$$-3(x^2 - 20x + 100) + y^2 = 225 - 300 \qquad \text{Complete the square in } x.$$

$$-3(x - 10)^2 + y^2 = -75$$

$$\frac{-3(x - 10)^2}{-75} + \frac{y^2}{-75} = \frac{-75}{-75}$$

Answer: $\dfrac{(x - 10)^2}{25} - \dfrac{y^2}{75} = 1$ $\qquad$ This is the equation of a hyperbola.

Conics With a Horizontal Directrix

A conic with a horizontal directrix is obtained by rotating, $\frac{\pi}{2}$ about the pole, a conic with a vertical directrix.

$$r = \frac{ep}{1 \pm e \cos \theta}$$

Thus, to derive the equation for a conic with a horizontal directrix, begin with the standard form for a conic with a vertical directrix and replace θ by $\theta - \frac{\pi}{2}$, and use the identity $\cos\left(\theta - \frac{\pi}{2}\right) = \sin \theta$.

$$r = \frac{ep}{1 \pm e \cos\left(\theta - \frac{\pi}{2}\right)}$$

$$r = \frac{ep}{1 \pm e \sin \theta}$$

The sign of the term $e \sin \theta$ is negative if the directrix is below the focus and positive if the directrix is above the focus.

Example 4

Write an equation of the ellipse with focus at the pole, eccentricity $\frac{2}{3}$, and a horizontal directrix that passes through $\left(3, \frac{\pi}{2}\right)$.

Solution

Since the directrix is above the focus, the general form of the equation is:

$$r = \frac{ep}{1 + e \sin \theta}$$

Since the distance, p, from the pole to the directrix is 3 and $e = \frac{2}{3}$,

$$r = \frac{\frac{2}{3}(3)}{1 + \frac{2}{3} \sin \theta}$$

$$r = \frac{2}{1 + \frac{2}{3} \sin \theta} \cdot \frac{3}{3}$$

Answer: $r = \dfrac{6}{3 + 2 \sin \theta}$

1. a. $e = 1$ b. parabola
   c. vertical

2. a. $e = 2$ b. hyperbola
   c. vertical

3. a. $e = 3$ b. hyperbola
   c. horizontal

4. a. $e = \frac{1}{2}$ b. ellipse
   c. vertical

5. a. $e = \frac{2}{3}$ b. ellipse
   c. horizontal

6. a. $e = \frac{5}{3}$ b. hyperbola
   c. horizontal

7.

8.

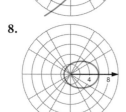

9.

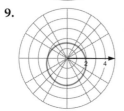

10.

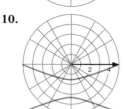

11.

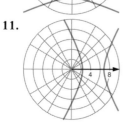

For the ellipse in Example 4, since the major axis is perpendicular to the directrix, it is a segment of a vertical line through the pole. Thus, the values of θ are $\frac{\pi}{2}$ and $\frac{3\pi}{2}$ for the endpoints of the major axis.

If $\theta = \frac{\pi}{2}$: $r = \dfrac{6}{3 + 2 \sin \frac{\pi}{2}}$

$\qquad = \dfrac{6}{3 + 2(1)} = \dfrac{6}{5}$

If $\theta = \frac{3\pi}{2}$: $r = \dfrac{6}{3 + 2 \sin \frac{3\pi}{2}}$

$\qquad = \dfrac{6}{3 + 2(-1)} = 6$

The endpoints of the major axis are $\left(\frac{6}{5}, \frac{\pi}{2}\right)$ and $\left(6, \frac{3\pi}{2}\right)$. The endpoints of the minor axis, which lies along the line of the polar axis, can be found in a similar fashion. These four endpoints are used to sketch the ellipse.

Example 5

Describe and sketch the graph of the conic whose equation is:

$$r = \frac{12}{3 + 4 \sin \theta}$$

Solution

Put the equation into standard form.

$$r = \dfrac{\frac{12}{3}}{\frac{3}{3} + \frac{4}{3} \sin \theta}$$

$$r = \dfrac{4}{1 + \frac{4}{3} \sin \theta}$$

Since $e = \frac{4}{3}$, the conic is a hyperbola.

$ep = 4$
$\frac{4}{3} p = 4$
$p = 3$

The directrix is 3 units above the focus at the pole.

Determine several points on the curve by choosing values of θ and computing the corresponding values of r.

If $\theta = 0$, $r = \dfrac{12}{3 + 4(0)} = 4$

If $\theta = \frac{\pi}{2}$, $r = \dfrac{12}{3 + 4(1)} = \dfrac{12}{7}$

If $\theta = \pi$, $r = \dfrac{12}{3 + 4(0)} = 4$

If $\theta = \frac{3\pi}{2}$, $r = \dfrac{12}{3 + 4(-1)} = -12$

If $\theta = \frac{4\pi}{3}$, $r \approx \dfrac{12}{3 + 4(-0.866)} = \dfrac{12}{-0.464} \approx -25.8$

If $\theta = \frac{5\pi}{3}$, $r \approx \dfrac{12}{3 + 4(-0.866)} = \dfrac{12}{-0.464} \approx -25.8$

Note that the center of the hyperbola is at $\left(\dfrac{12 + \frac{12}{7}}{2}, \dfrac{\pi}{2}\right)$, or $\left(\dfrac{48}{7}, \dfrac{\pi}{2}\right)$.

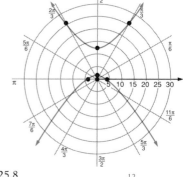

$$r = \frac{12}{3 + 4 \sin \theta}$$

12.

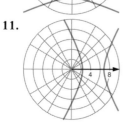

13. a. $r = \dfrac{12}{3 + 2 \cos \theta}$ b. $r = \dfrac{18}{2 + 3 \cos \theta}$

14. a. $r \cos \theta = 3$ b. $r \sin \theta = 3$

15. a. endpoints: $\left(4, \frac{\pi}{2}\right), \left(-\frac{4}{3}, \frac{\pi}{2}\right)$ b. endpoints: $(4, 0), \left(-\frac{4}{3}, 0\right)$

    center: $\left(\frac{4}{3}, \frac{\pi}{2}\right)$ center: $\left(\frac{4}{3}, 0\right)$

16. a. (1) 51 million miles   b. 4,950 million miles
    (2) 5,000 million miles

Exercises 11.5

In **1 – 6**:   **a.** Find the value of e.
         b. Identify the conic as a parabola, an ellipse, or a hyperbola.
         c. Is the directrix a vertical line or a horizontal line?

1. $r = \dfrac{4}{1 - \cos \theta}$ **2.** $r = \dfrac{4}{1 + 2 \cos \theta}$ **3.** $r = \dfrac{5}{1 + 3 \sin \theta}$

4. $r = \dfrac{4}{2 - \cos \theta}$ **5.** $r = \dfrac{5}{3 - 2 \sin \theta}$ **6.** $r = \dfrac{6}{3 - 5 \sin \theta}$

In **7 – 12**, sketch the graph of the given equation.

7. $r = \dfrac{3}{1 + \cos \theta}$ **8.** $r = \dfrac{6}{3 - 2 \cos \theta}$ **9.** $r = \dfrac{8}{4 + \sin \theta}$

10. $r = \dfrac{10}{2 - 5 \sin \theta}$ **11.** $r = \dfrac{7}{1 + 2 \cos \theta}$ **12.** $r = \dfrac{20}{4 - 4 \sin \theta}$

13. Write in polar coordinates an equation of a conic for which the focus is at the pole, the equation of the directrix is $r \cos \theta = 6$, and the eccentricity is:

    a. $\frac{2}{3}$ **b.** $\frac{3}{2}$

14. Find the equation of the directrix of the parabola whose equation is:

    a. $r = \dfrac{3}{1 + \cos \theta}$ **b.** $r = \dfrac{3}{1 + \sin \theta}$

15. Find the coordinates of the endpoints of the major axis and the center of the ellipse whose equation is:

    a. $r = \dfrac{4}{2 - \sin \theta}$ **b.** $r = \dfrac{4}{2 - \cos \theta}$

16. The equation of the elliptical orbit of Halley's Comet can be written as

    $r = \dfrac{100}{1 + 0.98 \cos \theta}$ when r is given in millions of miles.

    a. Find the distance of the comet from the Sun located at a focus when **(1)** $\theta = 0$ **(2)** $\theta = \pi$

    b. Find the length of the major axis of the path of Halley's Comet.

17. **a.** Solve the system of equations algebraically.
    b. Graph both equations on the same polar graph.
    c. Find the coordinates of the points of intersection of the graphs.
    d. Are the results of **a** and **c** the same? If not, explain.

    (1) $r = 4 \cos \theta,\ \theta = \dfrac{\pi}{3}$ **(2)** $r = 3,\ r = \dfrac{3}{1 - 2 \cos \theta}$

    (3) $r = 4 \sin \theta,\ r = \dfrac{4}{1 + 2 \sin \theta}$ **(4)** $r = 1 + \sin \theta,\ r = -\sin \theta$

Answers

17.
(1) a. $\left(2, \dfrac{\pi}{3}\right)$

b.

c. $(0, 0),\ \left(2, \dfrac{\pi}{3}\right)$

d. The point $(0, 0)$ or $\left(0, \dfrac{\pi}{3}\right)$ is a point of intersection but not a common solution since different forms of the polar coordinates satisfy the equations.

(2) a. $\left(3, \dfrac{\pi}{2}\right),\ \left(3, -\dfrac{\pi}{2}\right)$

b.
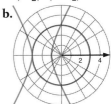

c. $\left(3, \dfrac{\pi}{2}\right),\ (3, \pi),$ $\left(3, -\dfrac{\pi}{2}\right)$

d. The point $(3, \pi)$ or $(-3, 0)$ is a point of intersection but not a common solution.

(3) a. $\left(-4, -\dfrac{\pi}{2}\right),\ \left(2, \dfrac{\pi}{6}\right),\ \left(2, \dfrac{5\pi}{6}\right)$ **b.**
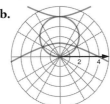

c. $\left(-4, -\dfrac{\pi}{2}\right),\ \left(2, \dfrac{\pi}{6}\right),\ \left(2, \dfrac{5\pi}{6}\right)$ **d.** results are the same

(4) a. $\left(\dfrac{1}{2}, \dfrac{7\pi}{6}\right),\ \left(\dfrac{1}{2}, \dfrac{11\pi}{6}\right)$ **b.**
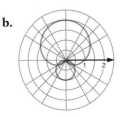

c. $\left(\dfrac{1}{2}, \dfrac{7\pi}{6}\right),\ \left(\dfrac{1}{2}, \dfrac{11\pi}{6}\right),\ (0, 0)$ **d.** The point $(0, 0)$ or $\left(0, -\dfrac{\pi}{2}\right)$ is a point of intersection but not a common solution.

11.6 *Complex Numbers in Polar Form*

Since the coordinates of every point in the real plane can be expressed in either rectangular or polar form, it seems reasonable to expect that the coordinates of points in the complex plane can also be expressed in polar form.

In the complex plane, a is the first coordinate and b is the second coordinate of the point $P(a, b)$ that is the graph of the complex number $a + bi$. With r and θ defined as in the polar plane, that is, $r = \sqrt{a^2 + b^2}$ and $\theta = $ arc tan $\dfrac{b}{a}$:

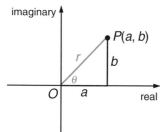

$$a = r \cos \theta \qquad b = r \sin \theta$$

$$a + bi = r \cos \theta + ri \sin \theta$$

$$a + bi = r(\cos \theta + i \sin \theta)$$

When the complex number $a + bi$ is written in polar form, $r(\cos \theta + i \sin \theta)$, θ is called the *argument* and r is called the *absolute value*, or *modulus*.

Like the absolute value of a real number, the modulus of a complex number is the distance between the graph of 0 and the graph of the complex number, without respect to direction, and is always positive.

Example 1

Express the complex number $2 + 2i$ in polar form.

Solution

Use the values of a and b to find the values of r and θ.

In $2 + 2i$, $a = 2$ and $b = 2$. $\qquad \tan \theta = \dfrac{b}{a}$

$r = \sqrt{a^2 + b^2}$ $\qquad\qquad\qquad \tan \theta = \dfrac{2}{2} = 1$

$ = \sqrt{2^2 + 2^2} = \sqrt{8} = 2\sqrt{2} \qquad \theta = \dfrac{\pi}{4} + k\pi$, where k is an integer

Thus, the modulus of $2 + 2i$ is $2\sqrt{2}$ and, since $(2, 2)$ is in quadrant I, the argument is $\dfrac{\pi}{4}$.

Answer: $2 + 2i = 2\sqrt{2}\left(\cos \dfrac{\pi}{4} + i \sin \dfrac{\pi}{4}\right)$

Note that, in an interval of 2π, there are always two values of θ that satisfy the equation $\tan \theta = \frac{b}{a}$. The appropriate value depends upon the quadrant in which the complex number lies, or upon the signs of a and b. One way of finding the value of θ is to use the following rules:

If $a > 0$, $\theta = \text{arc tan } \frac{b}{a}$. 　　　　If $a < 0$, $\theta = \text{arc tan } \frac{b}{a} + \pi$.

If $a = 0$ and $b > 0$, $\theta = \frac{\pi}{2}$. 　　　　If $a = 0$ and $b < 0$, $\theta = \frac{3\pi}{2}$.

Or, θ may be defined in terms of the following two equations:

$$\sin \theta = \frac{b}{r} \quad \text{and} \quad \cos \theta = \frac{a}{r}$$

Example 2

Express $3\left(\cos \frac{5\pi}{6} + i \sin \frac{5\pi}{6}\right)$ in $a + bi$ form.

Solution

Evaluate the trigonometric functions.

$$\cos \frac{5\pi}{6} = \cos \left(\pi - \frac{\pi}{6}\right) = -\cos \frac{\pi}{6} = -\frac{\sqrt{3}}{2}$$

$$\sin \frac{5\pi}{6} = \sin \left(\pi - \frac{\pi}{6}\right) = \sin \frac{\pi}{6} = \frac{1}{2}$$

Thus, $3\left(\cos \frac{5\pi}{6} + i \sin \frac{5\pi}{6}\right) = 3\left(-\frac{\sqrt{3}}{2} + \frac{1}{2}i\right) = -\frac{3\sqrt{3}}{2} + \frac{3}{2}i$

Answer:　$3\left(\cos \frac{5\pi}{6} + i \sin \frac{5\pi}{6}\right) = -\frac{3\sqrt{3}}{2} + \frac{3}{2}i$

A major advantage of writing complex numbers in polar form is that some of the operations with complex numbers are simplified, particularly multiplication, division, raising to a power, and extracting a root.

Multiplication of Complex Numbers

Consider the product of two complex numbers, $3\sqrt{3} + 3i$ and $\frac{1}{4} + \frac{\sqrt{3}}{4}i$, first in $a + bi$ form. Using the distributive property:

$$(3\sqrt{3} + 3i)\left(\frac{1}{4} + \frac{\sqrt{3}}{4}i\right) = \frac{3\sqrt{3}}{4} + \frac{9}{4}i + \frac{3}{4}i + \frac{3\sqrt{3}i^2}{4}$$

$$= \frac{3\sqrt{3}}{4} + \frac{3\sqrt{3}}{4}(-1) + \frac{9}{4}i + \frac{3}{4}i \qquad i^2 = -1$$

$$= \frac{3\sqrt{3}}{4} - \frac{3\sqrt{3}}{4} + \left(\frac{9}{4} + \frac{3}{4}\right)i$$

$$= 0 + \frac{12}{4}i$$

$$= 0 + 3i$$

Multiplying the polar form of the same complex numbers involves a lengthier computation at the outset, but leads to a simplified generalization.

Following the method of Example 1, you can verify that $3\sqrt{3} + 3i = 6\left(\cos \frac{\pi}{6} + i \sin \frac{\pi}{6}\right)$ and $\frac{1}{4} + \frac{\sqrt{3}}{4}i = \frac{1}{2}\left(\cos \frac{\pi}{3} + i \sin \frac{\pi}{3}\right)$.

Now write the product, and evaluate.

$$6\left(\cos\tfrac{\pi}{6} + i\sin\tfrac{\pi}{6}\right)\left[\tfrac{1}{2}\left(\cos\tfrac{\pi}{3} + i\sin\tfrac{\pi}{3}\right)\right]$$

$$= 6 \cdot \tfrac{1}{2}\left(\cos\tfrac{\pi}{6}\cos\tfrac{\pi}{3} + i\cos\tfrac{\pi}{6}\sin\tfrac{\pi}{3} + i\sin\tfrac{\pi}{6}\cos\tfrac{\pi}{3} + i^2\sin\tfrac{\pi}{6}\sin\tfrac{\pi}{3}\right)$$

$$= 3\left(\underbrace{\cos\tfrac{\pi}{6}\cos\tfrac{\pi}{3} - \sin\tfrac{\pi}{6}\sin\tfrac{\pi}{3}}_{\cos\left(\tfrac{\pi}{6} + \tfrac{\pi}{3}\right)} + i\left(\underbrace{\cos\tfrac{\pi}{6}\sin\tfrac{\pi}{3} + \sin\tfrac{\pi}{6}\cos\tfrac{\pi}{3}}_{\sin\left(\tfrac{\pi}{6} + \tfrac{\pi}{3}\right)}\right)\right)$$

$$= 3\left(\cos\tfrac{\pi}{2} + i\sin\tfrac{\pi}{2}\right)$$

Note that this result can be verified by converting it to $a + bi$ form, thus obtaining $0 + 3i$, which is the product found previously.

Examine the product in polar form to note that the modulus of the product is the product of the moduli of the factors and the argument of the product is the sum of the arguments of the factors. Is this example a special case, or will this always be true? To answer this question, multiply two general complex numbers in polar form.

$$r_1(\cos\theta + i\sin\theta)[r_2(\cos\phi + i\sin\phi)]$$
$$= r_1r_2(\cos\theta\cos\phi + i\cos\theta\sin\phi + i\sin\theta\cos\phi + i^2\sin\theta\sin\phi)$$
$$= r_1r_2(\cos\theta\cos\phi - \sin\theta\sin\phi + i(\cos\theta\sin\phi + \sin\theta\cos\phi))$$
$$= r_1r_2[\cos(\theta + \phi) + i\sin(\theta + \phi)]$$

Therefore, the product of any two complex numbers in polar form has:

1. a modulus that is the product of the moduli of the factors, and
2. an argument that is the sum of the arguments of the factors.

The polar form of a complex number can be abbreviated.

$$r(\cos\theta + i\sin\theta) \longrightarrow r \operatorname{cis} \theta.$$

Example 3

Multiply $2\left(\cos\tfrac{\pi}{4} + i\sin\tfrac{\pi}{4}\right)(\cos\pi + i\sin\pi)$.

Solution

$$\left(2\operatorname{cis}\tfrac{\pi}{4}\right)\left(1\operatorname{cis}\pi\right) = 2 \cdot 1 \operatorname{cis}\left(\tfrac{\pi}{4} + \pi\right)$$

$$= 2\operatorname{cis}\tfrac{5\pi}{4}$$

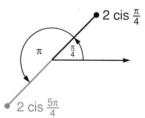

Notice that the multiplication of $2\operatorname{cis}\tfrac{\pi}{4}$ by $1\operatorname{cis}\pi$ has the effect of rotating the graph of $2\operatorname{cis}\tfrac{\pi}{4}$ through π radians, the argument of $1\operatorname{cis}\pi$.

The square of a complex number is obtained by applying the rule for multiplication.

$$(r\operatorname{cis}\theta)^2 = (r\operatorname{cis}\theta)(r\operatorname{cis}\theta) = r^2\operatorname{cis}2\theta$$

For example:

$$\left(2\operatorname{cis}\tfrac{\pi}{3}\right)^2 = 4\operatorname{cis}\tfrac{2\pi}{3}$$

The rules for multiplication can be used twice to find the cube of a complex number.

$$(r \text{ cis } \theta)^3 = (r \text{ cis } \theta)^2(r \text{ cis } \theta)$$
$$= (r^2 \text{ cis } 2\theta)(r \text{ cis } \theta)$$
$$= r^3 \text{ cis } 3\theta$$

The pattern suggests that it might always be true that:

$$(r \text{ cis } \theta)^n = r^n \text{ cis } n\theta$$

Mathematical induction can be used to prove that the statement above is true for all natural numbers n.

Proof

The statement is true for $n = 1$.

$$(r \text{ cis } \theta)^1 = r \text{ cis } \theta$$

Assume that the statement is true for $n = k$. Prove that it is true for $n = k + 1$.

$$(r \text{ cis } \theta)^{k+1} = (r \text{ cis } \theta)^k(r \text{ cis } \theta)$$
$$= (r^k \text{ cis } k\theta)(r \text{ cis } \theta)$$
$$= r^{k+1} \text{ cis } (k\theta + \theta)$$
$$= r^{k+1} \text{ cis } ((k+1)\theta)$$

Thus, for $n = k + 1$:

$$(r \text{ cis } \theta)^n = r^n \text{ cis } n\theta$$

Therefore, the statement is true for all natural numbers n.

The statement $(r \text{ cis } \theta)^n = r^n \text{ cis } n\theta$ is called **DeMoivre's Theorem**. Although the statement was only proved for positive integers above, it is also true for all negative integers n, which you are asked to prove later, as an exercise.

Example 4

Find $(1 + i)^5$.

Solution

Step 1: Write $1 + i$ in polar form.

$$r = \sqrt{1^2 + 1^2} = \sqrt{1 + 1} = \sqrt{2} \qquad \tan \theta = \frac{1}{1} = 1$$
Thus, $1 + i = \sqrt{2} \text{ cis } \frac{\pi}{4}$ $\qquad \qquad \qquad \theta = \frac{\pi}{4}$

Step 2: Find $\left(\sqrt{2} \text{ cis } \frac{\pi}{4}\right)^5$.

$$\left(\sqrt{2} \text{ cis } \frac{\pi}{4}\right)^5 = (\sqrt{2})^5 \text{ cis } 5\left(\frac{\pi}{4}\right)$$
$$= \sqrt{32} \text{ cis } \frac{5\pi}{4} = 4\sqrt{2} \text{ cis } \frac{5\pi}{4}$$

Step 3: Write $4\sqrt{2} \text{ cis } \frac{5\pi}{4}$ in $a + bi$ form.

$$a = r \cos \theta \qquad \qquad \qquad b = r \sin \theta$$
$$= 4\sqrt{2} \cos \frac{5\pi}{4} \qquad \qquad = 4\sqrt{2} \sin \frac{5\pi}{4}$$
$$= 4\sqrt{2}\left(-\frac{\sqrt{2}}{2}\right) \qquad \qquad = 4\sqrt{2}\left(-\frac{\sqrt{2}}{2}\right)$$
$$a = -4 \qquad \qquad \qquad \qquad b = -4$$

Thus, $4\sqrt{2} \text{ cis } \frac{5\pi}{4} = a + bi = -4 - 4i$.

Answer: $(1 + i)^5 = -4 - 4i$

Division of Complex Numbers

Recall that the procedure for dividing complex numbers in $a + bi$ form involves multiplication by the conjugate of the denominator.

For example, to divide -2 by $1 + i$:

$$\frac{-2}{1+i} = \frac{-2}{1+i} \cdot \frac{1-i}{1-i}$$

$$= \frac{-2+2i}{1-i^2}$$

$$= \frac{-2+2i}{1-(-1)} = \frac{-2+2i}{2}$$

$$= -1 + i$$

Write the dividend, divisor, and quotient in polar form, and note that the modulus of the quotient is the quotient of the moduli of the dividend and divisor, and that the argument of the quotient is the difference of the arguments of the dividend and divisor.

Dividend: $\qquad -2 = 2 \text{ cis } \pi$

Divisor: $\qquad 1 + i = \sqrt{2} \text{ cis } \frac{\pi}{4}$

Quotient: $\qquad -1 + i = \sqrt{2} \text{ cis } \frac{3\pi}{4}$ $\qquad 2 \div \sqrt{2} = \sqrt{2}$

$$\pi - \frac{\pi}{4} = \frac{3\pi}{4}$$

A general proof follows.

$$\frac{r_1(\cos\theta + i\sin\theta)}{r_2(\cos\phi + i\sin\phi)} = \frac{r_1(\cos\theta + i\sin\theta)}{r_2(\cos\phi + i\sin\phi)} \cdot \frac{\cos\phi - i\sin\phi}{\cos\phi - i\sin\phi} \qquad \cos\phi = \cos(-\phi);$$
$$-\sin\phi = \sin(-\phi)$$

$$= \frac{r_1(\cos\theta + i\sin\theta)}{r_2(\cos\phi + i\sin\phi)} \cdot \frac{\cos(-\phi) + i\sin(-\phi)}{\cos(-\phi) + i\sin(-\phi)}$$

$$= \frac{r_1}{r_2} \cdot \frac{\cos(\theta-\phi) + i\sin(\theta-\phi)}{\cos(\phi-\phi) + i\sin(\phi-\phi)}$$

$$= \frac{r_1}{r_2} \cdot \frac{\cos(\theta-\phi) + i\sin(\theta-\phi)}{\cos 0 + i\sin 0}$$

$$= \frac{r_1}{r_2} \cdot \frac{\cos(\theta-\phi) + i\sin(\theta-\phi)}{1 + 0i}$$

$$\frac{r_1(\cos\theta + i\sin\theta)}{r_2(\cos\phi + i\sin\phi)} = \frac{r_1}{r_2}[\cos(\theta-\phi) + i\sin(\theta-\phi)]$$

Therefore, the quotient of two complex numbers in polar form has:

1. a modulus that is the quotient of the moduli of the dividend and divisor, and
2. an argument that is the difference of the arguments of the dividend and divisor.

Example 5

Find: $(6 \text{ cis } 200°) \div (2 \text{ cis } 20°)$

Solution

The modulus of the quotient is $(6 \div 2)$ or 3. The argument is $(200 - 20)°$ or $180°$.

Therefore, the quotient is: $\quad 3 \text{ cis } 180° = 3(\cos 180° + i \sin 180°)$
$$= 3(-1 + 0i) = -3$$

Answer: $(6 \text{ cis } 200°) \div (2 \text{ cis } 20°) = -3$

Roots of Complex Numbers

Every complex number $r \operatorname{cis} \theta$ can be expressed as the product of two equal factors since, using the rule for multiplying two complex numbers in polar form:

$$\left(\sqrt{r} \operatorname{cis} \tfrac{\theta}{2}\right)\left(\sqrt{r} \operatorname{cis} \tfrac{\theta}{2}\right) = \left(\sqrt{r}\right)^2 \operatorname{cis}\left(\tfrac{\theta}{2} + \tfrac{\theta}{2}\right) = r \operatorname{cis} \theta$$

Thus, $\sqrt{r} \operatorname{cis} \tfrac{\theta}{2}$ is a square root of $r \operatorname{cis} \theta$.

In general:
$$\left(\sqrt[n]{r} \operatorname{cis} \tfrac{\theta}{n}\right)^n = \left(\sqrt[n]{r}\right)^n \operatorname{cis} n\left(\tfrac{\theta}{n}\right) = r \operatorname{cis} \theta$$
$$\sqrt[n]{r \operatorname{cis} \theta} = \sqrt[n]{r} \operatorname{cis} \tfrac{\theta}{n}$$

Therefore, each of the n nth roots of a complex number in polar form has:

1. a modulus that is the nth root of the modulus of the original number, and

2. an argument that is $\frac{1}{n}$ times the argument of the original number (or times the measure of an angle with the same terminal as the angle whose measure is the argument).

$$\sqrt[n]{r \operatorname{cis} \theta} = \sqrt[n]{r} \operatorname{cis}\left(\tfrac{\theta + 2\pi k}{n}\right)$$

Example 6

Find the three complex cube roots of 8.

Solution

Step 1: Change 8 to polar form.
$$8 = 8 + 0i$$
$$= 8(\cos 0 + i \sin 0) \quad \text{or} \quad 8 \operatorname{cis} 0$$

Step 2: Apply the identities.
$$\cos(\theta + 2\pi k) = \cos \theta \quad \text{and} \quad \sin(\theta + 2\pi k) = \sin \theta$$
$$8 \operatorname{cis} 0 = 8 \operatorname{cis}(0 + 2\pi k)$$

Step 3: Apply the rule for finding nth roots in polar form.
$$\sqrt[n]{r \operatorname{cis} \theta} = \sqrt[n]{r} \operatorname{cis} \tfrac{\theta}{n}$$
$$\sqrt[3]{8 \operatorname{cis} 0} = \sqrt[3]{8} \operatorname{cis} \tfrac{0 + 2\pi k}{3}$$

Step 4: Evaluate $\sqrt[3]{8} \operatorname{cis} \tfrac{0 + 2\pi k}{3}$.

If $k = 0$: $\sqrt[3]{8} \operatorname{cis} 0 \quad$ or $\quad 2 \operatorname{cis} 0$

If $k = 1$: $\sqrt[3]{8} \operatorname{cis} \tfrac{2\pi}{3} \quad$ or $\quad 2 \operatorname{cis} \tfrac{2\pi}{3}$

If $k = 2$: $\sqrt[3]{8} \operatorname{cis} \tfrac{4\pi}{3} \quad$ or $\quad 2 \operatorname{cis} \tfrac{4\pi}{3}$

Note that the values repeat when $k \geq 3$.

Step 5: Rewrite the roots in $a + bi$ form.
$$2 \operatorname{cis} 0 = 2(\cos 0 + i \sin 0) = 2(1 + i(0)) = 2$$
$$2 \operatorname{cis} \tfrac{2\pi}{3} = 2\left(\cos \tfrac{2\pi}{3} + i \sin \tfrac{2\pi}{3}\right) = 2\left(-\tfrac{1}{2} + i\left(\tfrac{\sqrt{3}}{2}\right)\right) = -1 + i\sqrt{3}$$
$$2 \operatorname{cis} \tfrac{4\pi}{3} = 2\left(\cos \tfrac{4\pi}{3} + i \sin \tfrac{4\pi}{3}\right) = 2\left(-\tfrac{1}{2} + i\left(-\tfrac{\sqrt{3}}{2}\right)\right) = -1 - i\sqrt{3}$$

Note that there is one real cube root of 8, namely 2, and that the complex roots occur as a conjugate pair.

Answer: The three cube roots of 8 are 2, $-1 \pm i\sqrt{3}$.

The graphs of the three cube roots of 8 are shown in the figure. Notice the symmetric arrangement of the graphs in the complex plane. The roots are the coordinates of the vertices of an equilateral triangle. Each root is the image of one of the other roots under a rotation of $\frac{2\pi}{3}$.

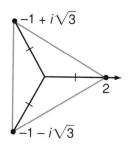

It will sometimes be necessary to use a calculator or a table of trigonometric values to evaluate the roots of a number. For example, to find the four complex fourth roots of i, which are the solution to the equation $x^4 = i$, the procedure detailed in Example 6 yields the following solutions rounded to the nearest hundredth:

$$1 \text{ cis } \tfrac{\pi}{8} = 1\left(\cos \tfrac{\pi}{8} + i \sin \tfrac{\pi}{8}\right) \approx 0.92 + 0.38i$$

$$1 \text{ cis } \tfrac{5\pi}{8} = 1\left(\cos \tfrac{5\pi}{8} + i \sin \tfrac{5\pi}{8}\right) \approx -0.38 + 0.92i$$

$$1 \text{ cis } \tfrac{9\pi}{8} = 1\left(\cos \tfrac{9\pi}{8} + i \sin \tfrac{9\pi}{8}\right) \approx -0.92 - 0.38i$$

$$1 \text{ cis } \tfrac{13\pi}{8} = 1\left(\cos \tfrac{13\pi}{8} + i \sin \tfrac{13\pi}{8}\right) \approx 0.38 - 0.92i$$

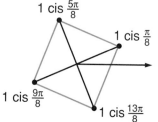

Note that the four roots are the coordinates of the vertices of a square. Each root is the image of one of the roots under a rotation of $\frac{\pi}{2}$.

You have seen that the graph of the 3 cube roots of a number are the vertices of an equilateral triangle and that the 4 fourth roots are the vertices of a square. A graphing calculator can be used to show that the graph of the n nth roots of a complex number are the vertices of a regular polygon of n sides.

Consider the graph of a circle of radius 3 and center at the origin. To display this graph on the calculator:

Step 1: Put the calculator in the following modes:

parametric and degree (easier for this discussion)

Step 2: Set the range.

For T, the variable that is used in place of θ, use 0 to 360 degrees. Let **Tstep** = 15, meaning that the calculator will graph in intervals of 15 degrees.

For X: –7.5 to 7.5   For Y: –5 to 5

Note that this range will compensate for the unequal size of the screen in the vertical and horizontal directions, thereby producing a circle that looks like a circle. Recall that you can also compensate for the difference by using item 5 of the zoom feature.

Step 3: Enter the parametric equations of the circle, $x = 3 \cos \theta$ and $y = 3 \sin \theta$, and graph the complete circle.

| Y= | 3 | COS | X|T | ENTER |
| 3 | SIN | X|T | GRAPH |

Step 4: Experiment with changes in the value for `Tstep`.

Change `Tstep` to 120 and press GRAPH.

Note that the calculator displays an equilateral triangle whose vertices are the graphs of the 3 complex roots of the number $3 + 0i$.

Change:

`Tstep` to 90	square,	m central $\angle = 90°$
`Tstep` to 72	regular pentagon,	m central $\angle = 72°$
`Tstep` to 60	regular hexagon,	m central $\angle = 60°$
`Tstep` to 51.4	regular heptagon,	m central $\angle \approx 51.4°$
`Tstep` to 45	regular octagon,	m central $\angle = 45°$

Note that the calculator displays for figures of 9 or more sides resemble the complete circle.

Exercises 11.6

In **1 – 18**, express the indicated product or quotient in r cis θ form.

1. $\left[3\left(\cos \frac{\pi}{2} + i \sin \frac{\pi}{2}\right)\right]\left[4\left(\cos \frac{\pi}{2} + i \sin \frac{\pi}{2}\right)\right]$ **2.** $\left[5\left(\cos \frac{\pi}{3} + i \sin \frac{\pi}{3}\right)\right]\left[0.2\left(\cos \frac{\pi}{6} + i \sin \frac{\pi}{6}\right)\right]$

3. $\left[8\left(\cos\left(-\frac{\pi}{3}\right) + i \sin\left(-\frac{\pi}{3}\right)\right)\right]\left[\frac{1}{2}\left(\cos \frac{\pi}{3} + i \sin \frac{\pi}{3}\right)\right]$

4. $(1 \text{ cis } \pi)\left(3 \text{ cis }\left(-\frac{\pi}{6}\right)\right)$ **5.** $\left(2 \text{ cis } \frac{5\pi}{6}\right)\left(2 \text{ cis }\left(-\frac{\pi}{6}\right)\right)$ **6.** $\left(6 \text{ cis } \frac{3\pi}{2}\right)\left(\frac{1}{3} \text{ cis } \frac{7\pi}{4}\right)$

7. $(8 \text{ cis } (-\pi))\left(3 \text{ cis }\left(-\frac{\pi}{2}\right)\right)$ **8.** $(4 \text{ cis } \pi)\left(\frac{1}{4} \text{ cis } \frac{3\pi}{2}\right)$ **9.** $\left(9 \text{ cis } \frac{\pi}{12}\right)\left(2 \text{ cis } \frac{11\pi}{12}\right)$

10. $(4 \text{ cis } 20°) \div (2 \text{ cis } 65°)$ **11.** $(9 \text{ cis } \pi) \div \left(3 \text{ cis }\left(-\frac{\pi}{2}\right)\right)$ **12.** $(5 \text{ cis } 10°) \div \left(\frac{1}{5} \text{ cis } 100°\right)$

13. $(6 \text{ cis } 240°) \div (4 \text{ cis } 30°)$ **14.** $(12 \text{ cis } 80°) \div (8 \text{ cis } (-10°))$ **15.** $(1 \text{ cis } 300°) \div (2 \text{ cis } 60°)$

16. $\dfrac{8 \text{ cis } \frac{\pi}{2}}{8 \text{ cis } \frac{\pi}{3}}$ **17.** $\dfrac{15 \text{ cis } 300°}{3 \text{ cis } 90°}$ **18.** $\dfrac{3 \text{ cis } 200°}{1 \text{ cis } (-10°)}$

In **19 – 26**, express the given power in $a + bi$ form.

19. $\left(1 \text{ cis } \frac{3\pi}{2}\right)^3$ **20.** $\left(2 \text{ cis } \frac{\pi}{6}\right)^6$ **21.** $\left(2 \text{ cis } \frac{5\pi}{4}\right)^4$ **22.** $\left(1 \text{ cis } \frac{\pi}{3}\right)^9$

23. $\left(1 \text{ cis } \frac{\pi}{4}\right)^8$ **24.** $(-2)^5$ **25.** $(2 + 2i)^5$ **26.** $(1 + \sqrt{3})^6$

27. If $\frac{1}{x^n} = x^{-n}$, show that:
$(r \text{ cis } \theta)^{-n} = r^{-n} \text{ cis } (-n\theta)$

28. If $1 + i = \sqrt{2} \text{ cis } \frac{\pi}{4}$, what is the additive inverse of $1 + i$ in polar form?

29. What is the additive inverse of $4 \text{ cis } \frac{\pi}{4}$ in polar form?

30. What is the additive inverse of $r \text{ cis } \theta$ in polar form?

Answers 11.6

1. $12 \text{ cis } \pi$
2. $1 \text{ cis } \frac{\pi}{2}$
3. $4 \text{ cis } 0$
4. $3 \text{ cis } \frac{5\pi}{6}$
5. $4 \text{ cis } \frac{2\pi}{3}$
6. $2 \text{ cis } \frac{5\pi}{4}$
7. $24 \text{ cis } \left(-\frac{3}{2}\pi\right)$
8. $1 \text{ cis } \frac{\pi}{2}$
9. $18 \text{ cis } \pi$
10. $2 \text{ cis } (-45°)$
11. $3 \text{ cis } \frac{3\pi}{2}$
12. $25 \text{ cis } (-90°)$
13. $\frac{3}{2} \text{ cis } 210°$
14. $\frac{3}{2} \text{ cis } 90°$
15. $\frac{1}{2} \text{ cis } 240°$
16. $1 \text{ cis } \frac{\pi}{6}$
17. $5 \text{ cis } 210°$
18. $3 \text{ cis } 210°$
19. $0 + i$
20. $-64 + 0i$
21. $-16 + 0i$
22. $-1 + 0i$
23. $1 + 0i$
24. $-32 + 0i$
25. $-128 - 128i$
26. $64 + 0i$

27. $(r \operatorname{cis} \theta)^{-n} = \dfrac{1}{(r \operatorname{cis} \theta)^n}$

$\qquad = \dfrac{1}{r^n \operatorname{cis} n\theta}$

$\qquad = \dfrac{1}{r^n (\cos n\theta + i \sin n\theta)} \cdot \dfrac{\cos n\theta - i \sin n\theta}{\cos n\theta - i \sin n\theta}$

$\qquad = \dfrac{(\cos (-n\theta) + i \sin (-n\theta))}{r^n(\cos^2 n\theta - i^2 \sin^2 n\theta)}$

$\qquad = \dfrac{(\cos (-n\theta) + i \sin (-n\theta))}{r^n(\cos^2 n\theta + \sin^2 n\theta)}$

$\qquad = r^{-n} \operatorname{cis} (-n\theta)$

28. $\sqrt{2} \operatorname{cis} \dfrac{5\pi}{4}$ **29.** $4 \operatorname{cis} \dfrac{5\pi}{4}$ **30.** $r \operatorname{cis} (\theta + \pi)$

31. a. $2\sqrt{2} \operatorname{cis} \dfrac{\pi}{4}$
    b. $2\sqrt{2} \operatorname{cis} \left(-\dfrac{\pi}{4}\right)$
    c. $r \operatorname{cis} (-\theta)$

32. $r \operatorname{cis} \theta \cdot r \operatorname{cis} (-\theta) = r^2 \operatorname{cis} 0$
$r^2 \operatorname{cis} 0 = r^2(1 + 0i) = r^2$

33. a. $(-1 + i\sqrt{3})^3 = (-1 + i\sqrt{3})^2(-1 + i\sqrt{3})$
$\qquad\qquad\qquad = (1 - 2i\sqrt{3} + 3i^2)(-1 + i\sqrt{3})$
$\qquad\qquad\qquad = (-2 - 2i\sqrt{3})(-1 + i\sqrt{3})$
$\qquad\qquad\qquad = 2 - 2i\sqrt{3} + 2i\sqrt{3} - 6i^2$
$\qquad\qquad\qquad = 8$

    b. $(-1 - i\sqrt{3})^3 = (-1 - i\sqrt{3})^2(-1 - i\sqrt{3})$
$\qquad\qquad\qquad = (1 + 2i\sqrt{3} + 3i^2)(-1 - i\sqrt{3})$
$\qquad\qquad\qquad = (-2 + 2i\sqrt{3})(-1 - i\sqrt{3})$
$\qquad\qquad\qquad = 2 + 2i\sqrt{3} - 2i\sqrt{3} - 6i^2$
$\qquad\qquad\qquad = 8$

In **34 – 37**, $(r \operatorname{cis} \theta)^{\frac{1}{2}} = \sqrt{r} \operatorname{cis} \dfrac{\theta}{2}$ and $\sqrt{r} \operatorname{cis} \dfrac{\theta + 2\pi}{2} = \sqrt{r} \operatorname{cis} \left(\dfrac{\theta}{2} + \pi\right)$

34. $1 \operatorname{cis} \dfrac{\pi}{6}$ and $1 \operatorname{cis} \dfrac{7\pi}{6}$

35. $3 \operatorname{cis} \dfrac{3\pi}{4}$ and $3 \operatorname{cis} \dfrac{7\pi}{4}$

36. $2 \operatorname{cis} 150°$ and $2 \operatorname{cis} 330°$

37. $1 \operatorname{cis} 120°$ and $1 \operatorname{cis} 300°$

38. a. $1 \operatorname{cis} 0$; $1 \operatorname{cis} \dfrac{\pi}{2}$; $1 \operatorname{cis} \pi$; $1 \operatorname{cis} \dfrac{3\pi}{2}$

    b.

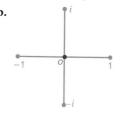

    c. $1 + 0i$; $0 + i$; $-1 + 0i$; $0 - i$

39. a. $1 \operatorname{cis} 0$; $1 \operatorname{cis} \dfrac{\pi}{3}$; $1 \operatorname{cis} \dfrac{2\pi}{3}$; $1 \operatorname{cis} \pi$;
$\qquad 1 \operatorname{cis} \dfrac{4}{3}\pi$; $1 \operatorname{cis} \dfrac{5}{3}\pi$

    b.

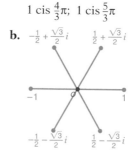

    c. $1 + 0i$; $\dfrac{1}{2} + \dfrac{\sqrt{3}}{2}i$; $-\dfrac{1}{2} + \dfrac{\sqrt{3}}{2}i$; $-1 + 0i$;
$\qquad -\dfrac{1}{2} - \dfrac{\sqrt{3}}{2}i$; $\dfrac{1}{2} - \dfrac{\sqrt{3}}{2}i$

40. a. $1 \text{ cis } 0$; $1 \text{ cis } \frac{\pi}{4}$; $1 \text{ cis } \frac{\pi}{2}$; $1 \text{ cis } \frac{3\pi}{4}$; $1 \text{ cis } \pi$;

$1 \text{ cis } \frac{5\pi}{4}$; $1 \text{ cis } \frac{3\pi}{2}$; $1 \text{ cis } \frac{7\pi}{4}$

b.
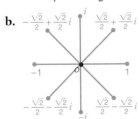

c. $1 + 0i$; $\frac{\sqrt{2}}{2} + \frac{\sqrt{2}}{2}i$; $0 + i$; $-\frac{\sqrt{2}}{2} + \frac{\sqrt{2}}{2}i$;

$-1 + 0i$; $-\frac{\sqrt{2}}{2} - \frac{\sqrt{2}}{2}i$; $0 - i$; $\frac{\sqrt{2}}{2} - \frac{\sqrt{2}}{2}i$

41. a. $3 \text{ cis } 0$; $3 \text{ cis } \frac{2\pi}{3}$; $3 \text{ cis } \frac{4\pi}{3}$

b. $-\frac{3}{2} + \frac{3\sqrt{3}}{2}i$

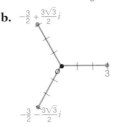

c. $3 + 0i$; $-\frac{3}{2} + \frac{3\sqrt{3}}{2}i$; $-\frac{3}{2} - \frac{3\sqrt{3}}{2}i$

42. a. $2 \text{ cis } \frac{\pi}{3}$; $2 \text{ cis } \pi$; $2 \text{ cis } \frac{5\pi}{3}$

b.
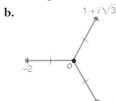

c. $1 + i\sqrt{3}$; $-2 + 0i$; $1 - i\sqrt{3}$

43. a. $4\sqrt{2} \text{ cis } \frac{5\pi}{6}$; $4\sqrt{2} \text{ cis } \frac{11\pi}{6}$

b. $-2\sqrt{6} + 2i\sqrt{2}$

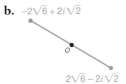

c. $-2\sqrt{6} + 2i\sqrt{2}$; $2\sqrt{6} - 2i\sqrt{2}$

44. a. $1 \text{ cis } \frac{\pi}{2}$; $1 \text{ cis } \frac{7\pi}{6}$; $1 \text{ cis } \frac{11\pi}{6}$

b.
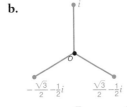

c. $0 + i$; $-\frac{\sqrt{3}}{2} - \frac{1}{2}i$; $\frac{\sqrt{3}}{2} - \frac{1}{2}i$

45. a. $\sqrt{2\sqrt{2}} \text{ cis } \frac{3\pi}{8}$; $\sqrt{2\sqrt{2}} \text{ cis } \frac{11\pi}{8}$

b. approximate values

$0.64 + 1.55i$

$-0.64 - 1.55i$

c. $\frac{\sqrt{2\sqrt{2}}}{2}\sqrt{2 - \sqrt{2}} + \frac{\sqrt{2\sqrt{2}}}{2}i\sqrt{2 + \sqrt{2}}$;

$-\frac{\sqrt{2\sqrt{2}}}{2}\sqrt{2 - \sqrt{2}} - \frac{\sqrt{2\sqrt{2}}}{2}i\sqrt{2 + \sqrt{2}}$

Exercises 11.6 *(continued)*

31. The complex conjugate of $a + bi$ is $a - bi$.

    a. Express the complex number $2 + 2i$ in polar form.

    b. Express the complex conjugate of $2 + 2i$ in polar form.

    c. Express the complex conjugate of $r \operatorname{cis} \theta$ in polar form.

32. Use polar coordinates to show that the product of a complex number and its conjugate is always a real number.

33. Use multiplication to verify that:

    a. $(-1 + i\sqrt{3})^3 = 8$ b. $(-1 - i\sqrt{3})^3 = 8$

In **34 – 37**, find the square roots of the given expression.

34. $1 \operatorname{cis} \frac{\pi}{3}$ 35. $9 \operatorname{cis} \frac{3\pi}{2}$ 36. $4 \operatorname{cis} 300°$ 37. $1 \operatorname{cis} 240°$

In **38 – 45**: a. Express the roots in polar form.

    b. Graph the roots in the polar plane and draw a line segment from the pole to each graph.

    c. Rewrite the roots in $a + bi$ form.

38. the fourth roots of unity (i.e. of 1)

39. the sixth roots of unity

40. the eighth roots of unity

41. the cube roots of 27

42. the cube roots of -8

43. the square roots of $16 - 16i\sqrt{3}$

44. the cube roots of $-i$

45. the square roots of $(1 + i)^3$

In **46 – 51**, solve the equation for all values of x in the set of complex numbers.

46. $x^3 = 125$ 47. $x^4 - 16 = 0$ 48. $x^6 + 1 = 0$

49. $x^3 + 1 = 0$ 50. $9x^4 - 1 = 0$ 51. $x^2 + 4 = -4i\sqrt{3}$

52. Find: $\sqrt{(-i)^5}$

53. Find the approximate values of the cube roots of $1 \operatorname{cis} 72°$. Answer in $a + bi$ form, with decimal values rounded to the nearest hundredth.

54. If $x^{\frac{p}{q}} = \sqrt[q]{x^p}$, show that:

    $(r \operatorname{cis} \theta)^{\frac{p}{q}} = r^{\frac{p}{q}} \operatorname{cis}\left(\frac{p}{q}\theta\right)$

46. $5; \ -\dfrac{5}{2} + \dfrac{5i\sqrt{3}}{2}; \ -\dfrac{5}{2} - \dfrac{5i\sqrt{3}}{2}$

47. $2; \ 0 + 2i; \ -2 + 0i; \ 0 - 2i$

48. $\dfrac{\sqrt{3}}{2} + \dfrac{1}{2}i; \ 0 + i; \ -\dfrac{\sqrt{3}}{2} + \dfrac{1}{2}i; \ -\dfrac{\sqrt{3}}{2} - \dfrac{1}{2}i; \ 0 - i; \ \dfrac{\sqrt{3}}{2} - \dfrac{1}{2}i$

49. $\dfrac{1}{2} + \dfrac{\sqrt{3}}{2}i; \ -1 + 0i; \ \dfrac{1}{2} - \dfrac{\sqrt{3}}{2}i$

50. $\dfrac{\sqrt{3}}{3}; \ 0 + \dfrac{\sqrt{3}}{3}i; \ -\dfrac{\sqrt{3}}{3} + 0i; \ 0 - \dfrac{\sqrt{3}}{3}i$

51. $-\sqrt{2} + i\sqrt{6}; \ \sqrt{2} - i\sqrt{6}$

52. $-\dfrac{\sqrt{2}}{2} + \dfrac{\sqrt{2}}{2}i; \ \dfrac{\sqrt{2}}{2} - \dfrac{\sqrt{2}}{2}i$

53. $0.91 + 0.41i; \ -0.81 + 0.59i; \ -0.10 - 0.99i$

54. $(r \operatorname{cis} \theta)^{\frac{p}{q}} = \sqrt[q]{(r \operatorname{cis} \theta)^p}$

    $(r \operatorname{cis} \theta)^p = r^p \operatorname{cis} p\theta$

    $\sqrt[q]{r^p \operatorname{cis} p\theta} = (r^p \operatorname{cis} (p\theta + 2\pi k))^{\frac{1}{q}}$

    $= r^{\frac{p}{q}} \operatorname{cis}\left(\dfrac{p}{q}\theta + \dfrac{2\pi k}{q}\right)$ for $k = 0, 1, ..., q - 1$

Definitions

- A horizontal ray, $\overrightarrow{OA}$, where O is the endpoint and A is any other point on the ray to the right of O, is the polar axis and the endpoint O is the pole.

- Any point P in the plane can be located by specifying r, the directed distance from O to P, and θ, the directed measure of $\angle AOP$. The values of r and θ for a specific point P are the polar coordinates of P, written as the ordered pair (r, θ).

- If point R lies on the polar axis, the degree measure of $\angle AOR$ is 0. If point S is on a line perpendicular to the polar axis at the pole, the coordinates of S are $\left(r, \frac{\pi}{2}\right)$.

- The polar coordinates of a given point are not unique. The polar coordinates of a given point P can be any pair of the form $(r, (\theta + 360k)^\circ)$, or $(r, \theta + 2\pi k)$, where k is an integer. The most commonly used ordered pair is the one given for $k = 0$. The coordinates of (r, θ) can also be written as $(-r, (\theta \pm 180)^\circ)$ if θ is in degrees or as $(-r, \theta \pm \pi)$ if θ is in radians.

- To compare polar coordinates to rectangular coordinates, the nonnegative ray of the x-axis corresponds to the polar axis, the origin of the rectangular coordinate system coincides with the pole of the polar coordinate system, and the y-axis is perpendicular to the polar axis at the pole.

- If P has rectangular coordinates (x, y) and polar coordinates (r, θ), then $x = r \cos \theta$, $y = r \sin \theta$, $r = \pm \sqrt{x^2 + y^2}$, and $\theta = \text{arc} \tan \frac{y}{x} + \pi k$.

- Polar Equations for circles

equation	conditions
$r = c$	center at the pole radius = constant c
$r = 2a \cos \theta$	center at $(a, 0)$ diameter = $2a$ passes through the pole
$r = 2a \sin \theta$	center on the line perpendicular, at the pole, to the polar axis diameter = $2a$

- Polar equations for lines

equation	conditions
$\theta = 0$	horizontal line through the pole
$\theta = \frac{\pi}{2}$	vertical line through the pole
$r \cos \theta = a$	vertical line through $(a, 0)$
$r \sin \theta = a$	horizontal line through $\left(a, \frac{\pi}{2}\right)$

1. **a.** $P\left(2, \frac{5\pi}{4}\right)$

   b. $P'\left(2, \frac{\pi}{4}\right)$

2. $300°$

3. $r = 6 \cos \theta$

4.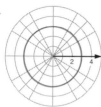

5.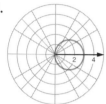

6.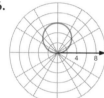

7.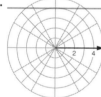

8.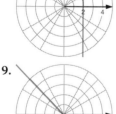

9.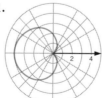

- Polar Symmetry

if the following point satisfies the given equation	then the following type of symmetry exists
$(-r, \theta)$ or $(r, \theta + \pi)$	to the pole
$(r, -\theta)$ or $(-r, \pi - \theta)$	to the line containing the polar axis
$(-r, -\theta)$ or $(r, \pi - \theta)$	to the line perpendicular, at the pole, to the polar axis

Polar Graphs

- The graph of any equation of the form $r = a \pm b \cos \theta$ or $r = a \pm b \sin \theta$ is called a limaçon. If $|a| < |b|$, the limaçon has two loops. If $|a| = |b|$, the limaçon is a cardioid. If $|a| > |b|$, the limaçon is one flattened loop.

- The graph of any equation of the form $r = a \cos n\theta$ or $r = a \sin n\theta$ is called a rose. If n is odd, the graph of $r = a \cos n\theta$ is symmetric with respect to the line containing the polar axis and has n petals, and the graph of $r = a \sin n\theta$ is symmetric with respect to the line perpendicular to the polar axis at the pole and has n petals. If n is even, the graphs of both $r = a \cos n\theta$ and $r = a \sin n\theta$ are symmetric with respect to the line containing the polar axis, to the line perpendicular to the polar axis at the pole, and to the pole; these graphs have $2n$ petals.

- The equation in polar coordinates of a conic with focus at the pole and a vertical directrix at a distance p from the pole is: $r = \dfrac{ep}{1 \pm e \cos \theta}$

- The equation in polar coordinates of a conic with focus at the pole and a horizontal directrix at a distance p from the pole is: $r = \dfrac{ep}{1 \pm e \sin \theta}$

Complex Numbers in Polar Form

- The polar form of the complex number $a + bi$ is $r(\cos \theta + i \sin \theta)$, abbreviated r cis θ, where θ is the argument and r is the modulus, or absolute value.

- The product of two complex numbers can be obtained by multiplying moduli and adding arguments.
$$(r_1 \text{ cis } \theta) \cdot (r_2 \text{ cis } \phi) = r_1 r_2 \text{ cis } (\theta + \phi)$$

- The quotient of two complex numbers can be obtained by dividing moduli and subtracting arguments.
$$(r_1 \text{ cis } \theta) \div (r_2 \text{ cis } \phi) = \frac{r_1}{r_2} \text{ cis } (\theta - \phi)$$

- DeMoivre's Theorem states that a complex number in polar form can be raised to a power n by raising the modulus to the power n and multiplying the argument by the power n.
$$(r \text{ cis } \theta)^n = r^n \text{ cis } n\theta$$

- DeMoivre's Theorem is extended to include nth roots.
$$\sqrt[n]{r \text{ cis } \theta} = \sqrt[n]{r} \text{ cis } \left(\frac{\theta + 2\pi k}{n}\right)$$
where k is an integer; evaluate from $k = 0$ to $k = n - 1$.

10.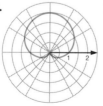

11.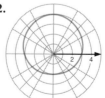

12.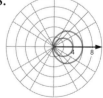

13.

1. Let $m\angle AOP = \frac{5\pi}{4}$ and $OP = 2$.
   a. Write the polar coordinates of P.
   b. Write the polar coordinates of P'.

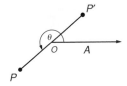

2. If $(3, 120°)$ and $(-3, \theta)$ are the polar coordinates of the same point, what is the least positive value of θ?

3. Derive the polar equation of a circle passing through the pole with center at $(-3, \pi)$.

In **4 – 15**, sketch the graph of the given polar equation.

4. $r = 3$
5. $r = 3 \cos \theta$
6. $r = 6 \sin \theta$
7. $r \sin \theta = 4$
8. $r \cos \theta = 2$
9. $\theta = \frac{3\pi}{4}$
10. $r = 1 + \sin \theta$
11. $r = 2 - 2 \cos \theta$
12. $r = 3 + \sin \theta$
13. $r = 1 + 5 \cos \theta$
14. $r = 1 - 3 \sin \theta$
15. $r = 3 \sin 4\theta$

16. Write in polar coordinates the equation of a parabola with focus at the pole, directrix a vertical line through $(4, 0)$.

17. Write in polar coordinates the equation of an ellipse with focus at the pole, directrix a horizontal line through $\left(-4, \frac{\pi}{2}\right)$, and eccentricity $\frac{2}{3}$.

18. a. Sketch the graph of the conic whose equation is $r = \frac{4}{2 + \cos \theta}$.
    b. Is the conic an ellipse, a hyperbola, or a parabola?

19. Find the rectangular coordinates of P', the image of point P under a rotation of $60°$ if the rectangular coordinates of P are $(0, 3)$.

In **20 – 23**, write the polar coordinates of the point whose rectangular coordinates are given.

20. $(3, 0)$
21. $(0, -2)$
22. $(-4, 4)$
23. $(3\sqrt{3}, -3)$

In **24 – 27**, write the rectangular coordinates of the point whose polar coordinates are given.

24. $(-2, 0)$
25. $(2, \pi)$
26. $\left(5, \frac{\pi}{2}\right)$
27. $\left(6, -\frac{\pi}{6}\right)$

28. Express $3 \text{ cis } 270°$ in $a + bi$ form.

29. Express the complex number $-4 - 4i$ in polar form.

30. Find the product $(3 \text{ cis } 24°)(2 \text{ cis } 96°)$. Express the answer in $a + bi$ form.

31. Find the quotient
    $(12 \text{ cis } 230°) \div (2 \text{ cis } 80°)$.
    Express the answer in $a + bi$ form.

32. Express $(2 - 2i)^8$ in $a + bi$ form.

33. a. Express the fifth roots of $32 \text{ cis } 100°$ in polar form.
    b. Rewrite the five roots found in part **a** in $a + bi$ form. Use a calculator or a table to approximate the decimal values of a and b to the nearest hundredth.

34. If P and Q are the graphs of two of the fifth roots of $r \text{ cis } \theta$, what is the least positive value of $m\angle POQ$?

Review Answers

14.
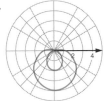

15.
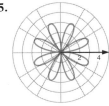

16. $r = \dfrac{4}{1 + \cos \theta}$

17. $r = \dfrac{8}{3 - 2 \sin \theta}$

18. $r = \dfrac{2}{1 + \frac{1}{2} \cos \theta}$

a.
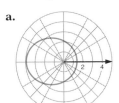

b. an ellipse

19. $\left(-\dfrac{3\sqrt{3}}{2}, \dfrac{3}{2}\right)$

20. $(3, 0)$

21. $\left(2, \dfrac{3\pi}{2}\right)$

22. $\left(4\sqrt{2}, \dfrac{3\pi}{4}\right)$

23. $\left(6, \dfrac{11\pi}{6}\right)$

24. $(-2, 0)$

25. $(-2, 0)$

26. $(0, 5)$

27. $(3\sqrt{3}, -3)$

28. $0 - 3i$

29. $4\sqrt{2} \text{ cis } \dfrac{5\pi}{4}$

30. $-3 + 3\sqrt{3}i$

31. $-3\sqrt{3} + 3i$

32. $4{,}096 + 0i$

33. a. 2 cis 20°   b. $1.88 + 0.68i$
        2 cis 92° $-0.07 + 2.00i$
        2 cis 164° $-1.92 + 0.55i$
        2 cis 236° $-1.12 - 1.66i$
        2 cis 308° $1.23 - 1.58i$

34. $\dfrac{2\pi}{5}$

1. 60°
2. 8.02
3. 4
4. 1
5. 8
6. B
7. B
8. D
9. C
10. E
11. B
12. A
13. B

☑ Exercises for Challenge

1. If $\dfrac{1}{1 - i\sqrt{3}}$ is written in the form $r\ \text{cis}\ \theta$, find the least possible positive value of θ, in degrees.

2. 🖩 A circle with center on the polar axis passes through both the pole and the point whose polar coordinates are $\left(5, \frac{2\pi}{7}\right)$. To the nearest hundredth, what is the diameter of the circle?

3. Find the smallest positive integer n such that:
$$(1 + i)^n = (1 - i)^n$$

4. If $z^2 - \sqrt{3}z + 1 = 0$, find $z^{10} + \dfrac{1}{z^{10}}$.

5. How many different points on the curve $r = \cos\theta\sin\theta$ lie 0.25 units from the origin?

6. Which pair of polar coordinates represent the same point?
   (A) $(1, \pi)$ and $(-1, -\pi)$
   (B) $\left(2, \frac{\pi}{2}\right)$ and $\left(-2, \frac{3\pi}{2}\right)$
   (C) $\left(3, \frac{\pi}{4}\right)$ and $\left(4, \frac{\pi}{3}\right)$
   (D) $(5, 0)$ and $(-5, 0)$
   (E) $\left(6, \frac{\pi}{4}\right)$ and $\left(6, -\frac{\pi}{4}\right)$

7. $\left[\sqrt{5}(\cos 75° + i\sin 75°)\right]^6 =$
   (A) 125   (B) $125i$ (C) $-125i$
   (D) $\frac{125}{2} + \frac{125\sqrt{3}}{2}i$ (E) $\frac{125\sqrt{3}}{2} + \frac{125}{2}i$

8. Which complex number is not a 6th root of 1?
   (A) $\text{cis}\ \frac{\pi}{3}$ (B) $\text{cis}\ \frac{2\pi}{3}$ (C) $-\text{cis}\ \frac{2\pi}{3}$
   (D) $\text{cis}\ \frac{5\pi}{6}$ (E) $\text{cis}\ \pi$

9. Which of the following graphs does not pass through the origin?
   (A) $r = 2\sin\theta$ (B) $r = \frac{1}{2} + \cos\theta$
   (C) $r = 3 - 2\sin\theta$ (D) $\theta = \frac{\pi}{2}$
   (E) $r^2 = 8\cos\theta$

10. Which of the following is a parametric representation of the ellipse $9x^2 + y^2 = 9$?

    I $\{(\cos\theta, 3\sin\theta); 0 \le \theta < 2\pi\}$
    II $\{(\sin\theta, 3\cos\theta); 0 \le \theta < 2\pi\}$
    III $\{(\cos 2\theta, 3\sin 2\theta); 0 \le \theta < \pi\}$

    (A) I only   (B) II only
    (C) III only   (D) I and II only
    (E) I, II, and III

11. The complex number z lies twice as far from the origin as the complex number w. Their product zw is 5 units from the origin. Determine $|z|$.
    (A) $2\sqrt{5}$ (B) $\sqrt{10}$ (C) $\sqrt{5}$
    (D) $\frac{1}{2}\sqrt{10}$ (E) $\frac{1}{2}\sqrt{5}$

12. 🖩 The graphs of $r = 2\cos\theta$ and $r = 3\sin\theta$ intersect at $(0, 0)$ and at a point whose polar coordinates are closest to
    (A) $(1.7, 34°)$ (B) $(3.6, 34°)$
    (C) $(3.6, 56°)$ (D) $(1.7, 56°)$
    (E) $(1.7, 214°)$

13. How many 7th roots of 1 are of the form $a + bi$, where a and b are real numbers and $b > 0$?
    (A) 2   (B) 3   (C) 4
    (D) 5   (E) 6

14. The distance between the points with polar coordinates (r_1, θ_1) and (r_2, θ_2) can be represented by

(A) $\sqrt{r_1{}^2 + r_2{}^2 + 2r_1r_2\cos(\theta_1 - \theta_2)}$

(B) $\sqrt{r_1{}^2 + r_2{}^2 - 2r_1r_2\sin(\theta_1 - \theta_2)}$

(C) $\sqrt{r_1{}^2 + r_2{}^2 - 2r_1r_2\cos(\theta_1 + \theta_2)}$

(D) $\sqrt{r_1{}^2 + r_2{}^2 + 2r_1r_2\sin(\theta_1 - \theta_2)}$

(E) $\sqrt{r_1{}^2 + r_2{}^2 - 2r_1r_2\cos(\theta_1 - \theta_2)}$

15. If $1 - \cos\alpha + i\sin\alpha = r\operatorname{cis}\theta$ and $0 \le \alpha \le \pi$, then r equals

(A) $2\sin\frac{\alpha}{2}$ (B) $2\sin\alpha$

(C) $\sqrt{2}\sin\alpha$ (D) $\sqrt{2}\sin\frac{\alpha}{2}$

(E) $2\sqrt{2}\sin\frac{\alpha}{4}$

16. The center of the conic $r = \dfrac{2}{2 - \cos\theta}$ is the point whose polar coordinates are:

(A) $(0, 0)$ (B) $\left(\frac{2}{3}, 0\right)$ (C) $\left(\frac{4}{3}, 0\right)$

(D) $\left(\frac{2}{3}, \pi\right)$ (E) $\left(\frac{1}{3}, \pi\right)$

Questions 17 – 22 each consist of two quantities, one in Column A and one in Column B. You are to compare the two quantities and choose:

A if the quantity in Column A is greater;
B if the quantity in Column B is greater;
C if the two quantities are equal;
D if the relationship cannot be determined from the information given.

1. In certain questions, information concerning one or both of the quantities to be compared is centered above the two columns.

2. In a given question, a symbol that appears in both columns represents the same thing in Column A as it does in Column B.

3. *x*, *n*, and *k*, etc. stand for real numbers.

Column A	Column B
17. $\left|(10 + 2i)^5\right|$	$\left|(2 - 3i)^9\right|$

Let α_1, α_2, α_3 be the three cube roots of 4.

18. $\left|\alpha_1 + \alpha_2 + \alpha_3\right|$	4

19.

$\left|3\operatorname{cis}40° + 6\operatorname{cis}50°\right|$	$\left|4\operatorname{cis}45° + 5\operatorname{cis}45°\right|$

The number of petals on the rose:

20. $r = 3\sin 2\theta$	$r = 2\sin 3\theta$

The function $r = f(\theta)$ is symmetric with respect to the pole and with respect to the polar axis.

21. $f\left(\frac{\pi}{3}\right)$	$f\left(\frac{2\pi}{3}\right)$

Eccentricity of the conic whose polar equation is:

22. $r = \dfrac{3}{6 - 5\cos\theta}$	$r = \dfrac{1}{2 + 4\sin\theta}$

14. E
15. A
16. B
17. A
18. B
19. B
20. A
21. C
22. B

Teacher's Chapter 12

Curved Surfaces

Chapter Contents

This chapter builds on the work with the 3-dimensional coordinate system and the conics studied in earlier chapters.

A computer program that sketches 3-dimensional surfaces is useful in studying this topic since students have difficulty in visualizing and sketching the graphs of these equations. Well-drawn figures are an important factor in developing familiarity with the curves described here.

12.1 Lines and Planes in Space

This section reexamines the equations of lines and planes in space, giving special attention to the intersections with the coordinate planes and axes and the planes parallel to a coordinate plane or axis. The use of the traces in the coordinate planes helps students visualize the position of a plane whose equation is given.

This section also provides the opportunity to review the vector equation of a line and of a plane and the use of determinants in the solution of equations.

12.2 The Sphere

The sphere is perhaps the most familiar of the 3-dimensional surfaces. The distance formula for 3-dimensional space is recalled and then applied to the derivation of the equation of a sphere with radius r and center at (h, k, l). Students should recognize that $x^2 + y^2 + z^2 = r^2$ is a special case in which the center of the sphere is at the origin.

The text extends the idea of a translation in a plane to the 3-dimensional translation in space by considering the composition of the translation $T_{h,k,0}$ in the xy-plane, followed by the translation $T_{0,0,l}$ in the plane $x = h$ parallel to the yz-plane or in the plane $y = k$ parallel to the xz-plane.

The formulas for the volume and surface area of a sphere are stated and used without derivation.

12.3 The Cylinder

Most students think of a cylinder as a right circular cylinder, such as a soup can. This section will help to broaden their perspective.

When a cylindrical surface is defined in terms of a generating line that remains parallel to its original position as it moves along a fixed curve, the cylinder is usually named for its defining curve. That is, for example, if the defining curve is a parabola, the surface is a parabolic cylinder. The text considers only cylinders with a defining line that is parallel to a coordinate axis.

Emphasize that the graph in 3-dimensional space of an equation in two variables is a cylindrical surface generated by a line parallel to the axis of the third variable. An equation such as $2x^2 + 5y^2 = 50$ is the equation of an elliptical cylinder generated by the line defined by $x = 5$, $y = 0$, $z = t$ moving along the ellipse whose equation is $2x^2 + 5y^2 = 50$, $z = 0$. Stress that the equation of the generating line is not unique. Any line parallel to the z-axis that intersects the curve could be a generating line. Also, the defining curve could be any curve that is the intersection of the surface and a plane, but it is convenient to use a curve in a coordinate plane or in a plane parallel to a coordinate plane.

A cylinder is defined as that portion of a closed cylindrical surface between parallel planes and the portions of the parallel planes enclosed by the cylindrical surface. Although a cylinder is defined as a portion of a cylindrical surface, the term cylinder is often used when referring to the entire surface.

The formula for the volume of a cylinder is stated and used without derivation. Note that the general formula $V = Ah$ applies also to a prism, where the base is a polygon.

12.4 Surfaces of Revolution

Surfaces generated by revolving a curve about an axis are commonly used in calculus. If the curve is a closed curve, the axis of rotation is usually an axis of symmetry. All of the examples used in this section are of this type. You may wish to show students examples of surfaces, such as a torus or ring, which are generated

when a closed curve is rotated about a line in the plane of the curve when this line does not intersect the curve.

The circular cylindrical surface and the circular conical surface are surfaces of revolution for which the generating curve is a line. For the cylindrical surface, the line is parallel to the axis of rotation, and for the conical surface, the line intersects the axis of rotation. Although it is possible to have a cone whose cross section is any closed curve, the term usually applies to the circular conical surface. If a circular conical surface is intersected by a plane that is not perpendicular to the axis of rotation, the intersection is one of the plane conics studied in Chapter 10.

Quadric Surfaces

This section classifies the common quadric surfaces, most of which have been discussed in previous sections. This work also extends the scope of study of the conic sections.

Quadric surfaces are commonly used in architectural design, sometimes producing interesting structures, such as the cooling towers on Three Mile Island in Pennsylvania. Tunnels, bridges, overpasses, and underpasses on highways are often in the shape of parabolic or elliptical cylinders. Encourage students to find pictures of other examples of quadric surfaces.

Suggested Test Items

1. For the plane whose equation is $x - 2y + 3z - 9 = 0$, find:
   a. the equations of the traces in the coordinate planes
   b. the coordinates of the points of intersection with the coordinate axes

2. Find an equation of the plane that intersects the coordinate axes in the points $(-5, 0, 0)$, $(0, 4, 0)$, and $(0, 0, -1)$.

3. Write an equation of the plane for which the equation of its trace in the xy-plane is $x + y = 4$ and the equation of its trace in the yz-plane is $3y + 2z = 12$.

4. Find AB, the distance from $A(0, -3, 5)$ to $B(-5, 4, 1)$.

5. Write an equation of the sphere with center at $(10, -1, 0)$ and radius $2\sqrt{5}$.

6. Write an equation of the sphere that has its center at $(2, 3, 4)$ and that is tangent to the x-axis.

7. A cylindrical surface is generated by a line parallel to the y-axis and a circle in the xz-plane with center at the origin and radius 3.
   a. Write an equation of the cylindrical surface.
   b. Find the volume of the cylinder formed by the intersection of the cylindrical surface with the planes whose equations are $y = -1$ and $y = 4$.

In **8 – 11**, **a.** Identify the equation as that of a cylindrical surface, a surface of revolution, both a cylindrical surface and a surface of revolution, or neither a cylindrical surface nor a surface of revolution.
   b. Describe the traces of the surface in each of the coordinate planes or planes parallel to the coordinate planes.

8. $x^2 + y^2 + z^2 = 1$ 9. $3x^2 - 9y^2 - 9z^2 = 0$ 10. $3x^2 - 9y^2 - 9z^2 = 1$ 11. $y = 3z^2$

12. Write an equation of the surface of revolution formed when an ellipse with vertices at $(\pm 4, 0, 0)$ and $(0, 0, \pm 2)$ is rotated about the z-axis.

13. Write an equation of the cylinder whose defining curve is the parabola $y = z^2$, $x = 0$ and whose generating line is a line parallel to the x-axis.

Bonus

1. The defining curve of a cylinder is $x^2 + y^2 = 100$, $z = 0$ and the generating line is $y + z = 10$. Write an equation for the surface. *Hint:* Find the center of the circular cross section in the plane whose equation is $z = t$.

2. Find the equation of the line that is the intersection of the planes whose equations are $x + y - z = 5$ and $2x - y - z = 7$.

Answers to Suggested Test Items

1. **a.** $x - 2y - 9 = 0$ **b.** $(9, 0, 0)$
   $-2y + 3z - 9 = 0$ $(0, -\frac{9}{2}, 0)$
   $x + 3z - 9 = 0$ $(0, 0, 3)$

2. $\frac{x}{-5} + \frac{y}{4} + \frac{z}{-1} = 1$ or $4x - 5y + 20z + 20 = 0$

3. $3x + 3y + 2z = 12$ **4.** $3\sqrt{10}$

5. $(x - 10)^2 + (y + 1)^2 + z^2 = 20$

6. $(x - 2)^2 + (y - 3)^2 + (z - 4)^2 = 25$

7. **a.** $x^2 + z^2 = 9$ **b.** 45π

8. **a.** surface of revolution
   b. circles of radius 1 with center at the origin

9. **a.** surface of revolution
   b. parallel to the xy-plane, a hyperbola
   parallel to the xz-plane, a hyperbola
   parallel to the yz-plane, circle

10. **a.** surface of revolution
    b. xy-plane, a hyperbola
    xz-plane, a hyperbola
    parallel to the yz-plane, circles

11. **a.** cylindrical surface
    b. parallel to the xy-plane, 1 line
    parallel to the xz-plane, 2 lines
    yz-plane, a parabola

12. $\frac{x^2}{16} + \frac{y^2}{16} + \frac{z^2}{4} = 1$

13. $y = z^2$

Bonus

1. $x^2 + (y + z)^2 = 100$

$$x^2 + y^2 = 100, z = 0$$
$$y + z = 10$$

Axis of rotation: $z = t$, $y = -t$, $x = 0$
Distance between general point (x, y, z) and center:

$$\sqrt{(x - 0)^2 + (y - (-t))^2 + (z - t)^2} = 10$$

At $z = t$: $\sqrt{x^2 + (y + z)^2 + 0^2} = 10$
$$x^2 + (y + z)^2 = 100$$

2. $\frac{x}{2} = \frac{y + 1}{1} = \frac{z + 6}{3}$

$$x + y - z = 5$$
$$2x - y - z = 7$$

$$3x - 2z = 12$$

$$x = \frac{2z + 12}{3}$$

$$\frac{x}{2} = \frac{z + 6}{3}$$

$$\frac{2z + 12}{3} + y - z = 5$$

$$\frac{2z}{3} + 4 - z - 5 = -y$$

$$\frac{2z}{3} - \frac{3z}{3} = -y + 1$$

$$-\frac{z}{3} = -y + 1$$

$$\frac{z}{3} + 2 = y - 1 + 2$$

$$\frac{z + 6}{3} = \frac{y + 1}{1}$$

$$\frac{x}{2} = \frac{z + 6}{3} = \frac{y + 1}{1}$$

Chapter *12*
Curved Surfaces

A dish antenna contains a curved parabolic surface that is used to receive and transmit electromagnetic waves such as television or radio waves. When receiving a signal, parallel waves are reflected by the parabolic surface to the receiver, which is located at the focus of the parabolic cross section. When sending a signal, waves from a transmitter located at the focus are reflected outward in parallel rays.

We live in a 3-dimensional world. The shapes of solid figures play an important role in the technology and art of that world.

Chapter Table of Contents

12.1 Lines and Planes in Space

Recall that the equation of a plane in 3-dimensional space is written $Ax + By + Cz + D = 0$. One way of visualizing a plane is to investigate the intersection of the plane with the coordinate planes. The lines of intersection are called *traces*.

To find the equation of the linear trace in, for example, the xy-plane, let $z = 0$. Therefore, $Ax + By + C(0) + D = 0$, or $Ax + By + D = 0$, is the equation of the trace in the xy-plane. The equations of the traces in the other two coordinate planes are found in a similar way.

Example 1

Find the equations of the traces of the plane whose equation is $2x + 5y - z + 10 = 0$.

Solution

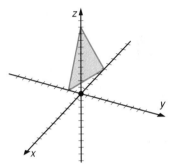

On the xy-plane, $z = 0$. Therefore, the equation of the trace in the xy-plane is $2x + 5y + 10 = 0$.

On the yz-plane, $x = 0$. Therefore, the equation of the trace in the yz-plane is $5y - z + 10 = 0$.

On the xz-plane, $y = 0$. Therefore, the equation of the trace in the xz-plane is $2x - z + 10 = 0$.

The intersection of a plane (or of its traces in the coordinate planes) with the axes are the points where the values of two of the variables are zero. That is, a plane intersects the x-axis at the point $(a, 0, 0)$, the y-axis at $(0, b, 0)$, and the z-axis at $(0, 0, c)$.

To determine the intersection points for a particular plane, substitute the general coordinates into the equation of that plane. For example, reconsider the plane of Example 1. Substitute:

$(a, 0, 0)$	$(0, b, 0)$	$(0, 0, c)$
into	into	into
$2x + 5y - z + 10 = 0$	$2x + 5y - z + 10 = 0$	$2x + 5y - z + 10 = 0$
$2a + 0 - 0 + 10 = 0$	$0 + 5b - 0 + 10 = 0$	$0 + 0 - c + 10 = 0$
$a = -5$	$b = -2$	$c = 10$

Thus, the plane $2x + 5y - z + 10 = 0$ intersects the x-axis at $(-5, 0, 0)$, the y-axis at $(0, -2, 0)$, and the z-axis at $(0, 0, 10)$, which can be seen on the graph shown above.

Recall that, in 3-dimensional space, for a line through (a_x, a_y, a_z) and (b_x, b_y, b_z), the equation may be written:

in general form

$$\frac{x - a_x}{a_x - b_x} = \frac{y - a_y}{a_y - b_y} = \frac{z - a_z}{a_z - b_z}$$

in parametric form

$$x = t(a_x - b_x) + a_x$$
$$y = t(a_y - b_y) + a_y$$
$$z = t(a_z - b_z) + a_z$$

Reconsider the equations of the traces found for the plane of Example 1. Substituting its points of intersection with the axes into the parametric form of a line confirms the original results.

The trace of $2x + 5y - z + 10 = 0$ in the xy-plane is the line passing through $(-5, 0, 0)$ and $(0, -2, 0)$. Therefore, the parametric equations of the trace are

$$\begin{array}{ccc}
x = t(-5 - 0) - 5 & & x = -5t - 5 \\
y = t(0 - (-2)) + 0 & \text{or} & y = 2t \\
z = t(0 - 0) + 0 & & z = 0
\end{array}$$

The equation $y = 2t$ can now be solved for t and substituted into the first equation.

$$x = -5\left(\frac{y}{2}\right) - 5$$
$$2x = -5y - 10$$
$$2x + 5y + 10 = 0$$

Similarly, you can verify that the parametric equations for the line through $(-5, 0, 0)$ on the x-axis and $(0, 0, 10)$ on the z-axis are equivalent to the equation of the trace $2x - z + 10 = 0$ in the xz-plane, and that the parametric equations of the line through $(0, -2, 0)$ on the y-axis and $(0, 0, 10)$ on the z-axis are equivalent to the equation of the trace $5y - z + 10 = 0$ in the yz-plane.

Example 2

For the plane whose equation is $3x - 6y + 0z - 12 = 0$, find:
a. the equations of the traces in the coordinate planes
b. the coordinates of the points of intersection with the coordinate axes

Solution

a. On the xy-plane, $z = 0$. Therefore, the equation of the trace in the xy-plane is $3x - 6y - 12 = 0$, or $x - 2y - 4 = 0$.

On the yz-plane, $x = 0$. Therefore, the equation of the trace in the yz-plane is $-6y - 12 = 0$, or $y = -2$.

On the xz-plane, $y = 0$. Therefore, the equation of the trace in the xz-plane is $3x - 12 = 0$, or $x = 4$.

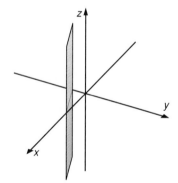

b. The intersection of the plane with the x-axis is the point $(4, 0, 0)$. The intersection with the y-axis is $(0, -2, 0)$. There is no point of intersection with the z-axis, since its coordinates must be of the form $(0, 0, c)$, which when substituted in the equation of the plane gives $3(0) - 6(0) + 0(c) - 12 = 0$ or $-12 = 0$, a contradiction. Therefore, the plane is parallel to the z-axis. Note, also, that the traces in the yz-plane and the xz-plane are each parallel to the z-axis.

The equation of the plane in Example 2 is usually written $3x - 6y - 12 = 0$, or $x - 2y - 4 = 0$. Thus, the equation of this plane, which is parallel to the z-axis, is the same as the equation of the trace in the xy-plane, which is perpendicular to the z-axis. This example suggests that if the coefficient of one of the variables in the equation of a plane in 3-dimensional space is zero, then the plane is parallel to the axis of that variable.

Example 3

For the plane whose equation is $0x + 0y + z - 3 = 0$, find:
a. the equations of the traces in the coordinate planes
b. the coordinates of the points of intersection with the coordinate axes

Solution

a. On the *xy*-plane, $z = 0$. Substituting into the equation yields the contradiction $-3 = 0$. Therefore, there is no trace; that is, the plane is parallel to the *xy*-plane.

Both in the *yz*-plane where $x = 0$ and in the *xz*-plane where $y = 0$, the equation of the trace is $z - 3 = 0$ or $z = 3$.

b. The coordinates of the point at which the plane intersects the *z*-axis is $(0, 0, 3)$. The plane does not intersect the *x*-axis or the *y*-axis.

The equation of the plane in Example 3 is usually written $z - 3 = 0$, or $z = 3$. Thus, the equation of this plane, which is parallel to the *xy*-plane, is the same as the equation of the trace in either of the other two planes. This example suggests that if the coefficients of two of the variables in the equation of a plane in 3-dimensional space are zero, then the plane is parallel to the plane of those variables.

Recall that vectors can be used to find an equation of a plane when the coordinates of three noncollinear points in the plane are given. If the coordinates of any three noncollinear points of a plane are known, then an equation of the plane may also be found by solving a set of simultaneous equations or by using determinants.

Example 4

Find an equation of the plane that passes through the points $(1, 7, 1)$, $(3, 4, 1)$ and $(-1, -2, 5)$.

Solution

Method 1   The equation of a plane may be written in the form

$$Ax + By + Cz + D = 0.$$

Each of the points of the plane must satisfy the equation of the plane.

$$A(1) + B(7) + C(1) + D = 0$$

$$A(3) + B(4) + C(1) + D = 0$$

$$A(-1) + B(-2) + C(5) + D = 0$$

This leads to a set of three equations in four variables.

$$A + 7B + C + D = 0$$

$$3A + 4B + C + D = 0$$

$$-A - 2B + 5C + D = 0$$

Solve for B, C, and D in terms of A.

$$B = \frac{2A}{3} \qquad C = 2A \qquad D = -\frac{23A}{3}$$

1. a. xy-plane:
$x + y - 5 = 0$
xz-plane:
$x + z - 5 = 0$
yz-plane:
$y + z - 5 = 0$

b. $(5, 0, 0), (0, 5, 0),$
$(0, 0, 5)$

2. a. xy-plane:
$3x - 2y + 18 = 0$
xz-plane:
$3x + z + 18 = 0$
yz-plane:
$-2y + z + 18 = 0$

b. $(-6, 0, 0),$
$(0, 9, 0),$
$(0, 0, -18)$

3. a. xy-plane:
$2x - 2y - 12 = 0$
xz-plane:
$2x - z - 12 = 0$
yz-plane:
$-2y - z - 12 = 0$

b. $(6, 0, 0),$
$(0, -6, 0),$
$(0, 0, -12)$

4. a. xy-plane:
$6x - y = 10$
xz-plane:
$6x + 8z = 10$
yz-plane:
$-y + 8z = 10$

b. $\left(\frac{5}{3}, 0, 0\right),$
$(0, -10, 0),$
$\left(0, 0, \frac{5}{4}\right)$

5. a. xy-plane:
$5x = 10 - y$
xz-plane:
$x = 2$
yz-plane:
$y = 10$

b. $(2, 0, 0),$
$(0, 10, 0),$
no z intercept

Substitute these values in $Ax + By + Cz + D = 0$ and divide by A.

$$x + \tfrac{2}{3}y + 2z - \tfrac{23}{3} = 0$$

$$3x + 2y + 6z - 23 = 0$$

Answer: $3x + 2y + 6z - 23 = 0$

Method 2 The equation of the plane passing through three noncollinear points (x_1, y_1, z_1), (x_2, y_2, z_2), and (x_3, y_3, z_3) is

$$\begin{vmatrix} x & y & z & 1 \\ x_1 & y_1 & z_1 & 1 \\ x_2 & y_2 & z_2 & 1 \\ x_3 & y_3 & z_3 & 1 \end{vmatrix} = 0$$

For the given coordinates, the equation is

$$\begin{vmatrix} x & y & z & 1 \\ 1 & 7 & 1 & 1 \\ 3 & 4 & 1 & 1 \\ -1 & -2 & 5 & 1 \end{vmatrix} = 0$$

Expand the determinant by minors using row 1.

$$x\begin{vmatrix} 7 & 1 & 1 \\ 4 & 1 & 1 \\ -2 & 5 & 1 \end{vmatrix} - y\begin{vmatrix} 1 & 1 & 1 \\ 3 & 1 & 1 \\ -1 & 5 & 1 \end{vmatrix} + z\begin{vmatrix} 1 & 7 & 1 \\ 3 & 4 & 1 \\ -1 & -2 & 1 \end{vmatrix} - 1\begin{vmatrix} 1 & 7 & 1 \\ 3 & 4 & 1 \\ -1 & -2 & 5 \end{vmatrix} = 0$$

A calculator that evaluates determinates can be used to simplify the equation.

$$-12x \quad\quad -8y \quad\quad -24z \quad\quad +92 \quad = 0$$

Answer: $3x + 2y + 6z - 23 = 0$

Exercises 12.1

In **1 – 6**, for the plane whose equation is given, find, if they exist:

a. the equations of the traces in the coordinate planes
b. the coordinates of the points of intersection with the coordinates axes

1. $x + y + z - 5 = 0$ **2.** $3x - 2y + z + 18 = 0$ **3.** $2x - 2y - z - 12 = 0$

4. $6x - y + 8z = 10$ **5.** $5x = 10 - y$ **6.** $0x + 3y + 0z = 8$

In **7 – 10**, write an equation of the plane that intersects the coordinate axes in the given points.

7. $(-1, 0, 0), (0, -1, 0), (0, 0, 1)$ **8.** $(4, 0, 0), (0, 4, 0), (0, 0, -1)$

9. $(3, 0, 0). \left(0, -\tfrac{1}{2}, 0\right), \left(0, 0, \tfrac{3}{2}\right)$ **10.** $(-2, 0, 0), (0, 3, 0),$
no intersection with the z-axis

6. a. xy-plane: $3y = 8$
xz-plane: no trace
yz-plane: $3y = 8$

b. no x intercept, $\left(0, \tfrac{8}{3}, 0\right),$
no z intercept

7. $x + y - z = -1$

8. $x + y - 4z = 4$

9. $x - 6y + 2z = 3$

10. $3x - 2y = -6$

Exercises 12.1 (continued)

In **11 – 14**, find an equation of the plane through the given points.

11. $(1, 0, 0)$, $(2, 3, 0)$, $(-1, 3, 1)$

12. $(2, 0, 6)$, $(3, 1, 4)$, $(3, 2, 4)$

13. $(1, -2, 3)$, $(3, 2, 1)$, $(1, 3, 2)$

14. $(-1, 0, 2)$, $(3, -4, -1)$, $(5, 4, 0)$

15. The points $(3, -2, 1)$ and $(5, 1, 6)$ lie on the plane $x + y - z = 0$ and the plane $2x - 3y + z = 13$.

   a. Write an equation of the line of intersection of the planes.

   b. Write the coordinates of another point on both planes, and show that it lies on the line of intersection.

   c. Show that the point $(1, 1, 2)$ is on $x + y - z = 0$ but not on $2x - 3y + z = 13$ and, therefore, not on the line of intersection.

In **16 – 17**, find the coordinates of the points of intersection with the coordinate axes of the plane whose matrix equation is given.

16.
$$\begin{vmatrix} x & y & z & 1 \\ 1 & -1 & 2 & 1 \\ 1 & 1 & -1 & 1 \\ 2 & -2 & 2 & 1 \end{vmatrix} = 0$$

17.
$$\begin{vmatrix} x & y & z & 1 \\ 1 & 0 & 2 & 1 \\ 3 & 5 & 2 & 1 \\ 4 & 10 & 4 & 1 \end{vmatrix} = 0$$

In **18 – 23**, write an equation of the plane in the form $Ax + By + Cz + D = 0$.

18. A plane parallel to the z-axis that contains the line whose equation is:
$$\frac{x - 3}{2} = \frac{y - 1}{2} = \frac{z + 1}{1}$$

19. A plane parallel to the y-axis that contains the line whose equation is:
$$x = 3t + 1$$
$$y = t - 1$$
$$z = 2t + 3$$

20. A plane that intersects the z-axis at $(0, 0, 2)$, and contains the line whose equation is:
$$\frac{x + 1}{3} = \frac{y - 3}{1} = \frac{z + 8}{5}$$

21. A plane that intersects the y-axis at $(0, 3, 0)$, and contains the line whose equation is:
$$x = t - 2$$
$$y = t + 3$$
$$z = -t - 1$$

22. A plane that is parallel to the yz-plane and contains the point $(4, 1, -5)$.

23. A plane that is parallel to the xz-plane and contains the point $(2, -7, 6)$.

15. a. $\dfrac{x - 3}{2} = \dfrac{y + 2}{3} = \dfrac{z - 1}{5}$

   b. $(1, -5, -4)$

   $\dfrac{x - 3}{2} = \dfrac{y + 2}{3} = \dfrac{z - 1}{5}$

   $\dfrac{1 - 3}{2} = \dfrac{-5 + 2}{3} = \dfrac{-4 - 1}{5}$

15. c. $(1, 1, 2)$

   $1 + 1 - 2 = 0$

   $2(1) - 3(1) + 1(2) \neq 13$

   $\dfrac{1 - 3}{2} \neq \dfrac{1 + 2}{3} \neq \dfrac{2 - 1}{5}$

16. $\left(\dfrac{4}{3}, 0, 0\right)$, $\left(0, \dfrac{4}{3}, 0\right)$, $(0, 0, 2)$

17. $(2, 0, 0)$, $(0, -5, 0)$, $(0, 0, 4)$

18. $x - y = 2$

19. $2x - 3z + 7 = 0$

20. $5x - 5y - 2z + 4 = 0$

21. $x - 3y - 2z + 9 = 0$

22. $x = 4$

23. $y = -7$

12.2 The Sphere

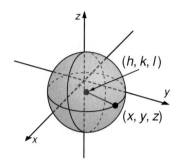

■ **Definition:** *A sphere is the locus of points in space that are equidistant from a fixed point called the center.*

The equation of a sphere can be derived from the formula for the distance between two points in space. Recall that this formula, which is derived from the Pythagorean Theorem, establishes the distance d from $A(x_1, y_1, z_1)$ to $B(x_2, y_2, z_2)$ as:

$$d = \sqrt{(x_1 - x_2)^2 + (y_1 - y_2)^2 + (z_1 - z_2)^2}$$

Let (h, k, l) be the center of a sphere and r be the distance from the center to any point (x, y, z) of the locus. Then the equation of the sphere is

$$r = \sqrt{(x - h)^2 + (y - k)^2 + (z - l)^2}$$
$$(x - h)^2 + (y - k)^2 + (z - l)^2 = r^2$$

For example, the equation of the sphere with radius 2 and center at (1, –1, 3) is

$$(x - 1)^2 + (y - (-1))^2 + (z - 3)^2 = 2^2$$
$$(x - 1)^2 + (y + 1)^2 + (z - 3)^2 = 4$$

Example 1 _____

Find the center and the length of the radius of the sphere whose equation is $x^2 + y^2 + z^2 + 4y - 2z = 11$.

Solution

Write the equation in standard form by completing the squares in y and z.

$$x^2 + y^2 + z^2 + 4y - 2z = 11$$
$$x^2 + (y^2 + 4y + \quad) + (z^2 - 2z + \quad) = 11$$
$$x^2 + (y^2 + 4y + 4) + (z^2 - 2z + 1) = 11 + 4 + 1$$
$$x^2 + (y + 2)^2 + (z - 1)^2 = 16$$
$$(x - 0)^2 + (y - (-2))^2 + (z - 1)^2 = 4^2$$

Compare this equation with the general equation of the sphere, to note that $h = 0$, $k = -2$, $l = 1$, and $r = 4$.

Answer: The center of the sphere is at (0, –2, 1) and the length of the radius is 4.

Translation

Recall that $T_{h,k}$ is a translation in the plane that maps (x, y) to $(x + h, y + k)$. A translation in 3-space can be thought of as the translation $T_{h,k,0}$ in the xy-plane followed by the translation $T_{0,0,l}$ in a plane parallel to the xz-plane or to the yz-plane. The composition of these two translations, $T_{h,k,l}$, is a translation in 3-space of h units parallel to the x-axis, k units parallel to the y-axis, and l units parallel to the z-axis.

Example 2

Under the translation $T_{1,-1,3}$, write the equation of the image of the sphere of radius 2 and center at the origin.

Solution

An equation of the sphere of radius 2 with center at the origin is

$$x^2 + y^2 + z^2 = 2^2$$

Under the translation $T_{1,-1,3}$, the equation is

$$(x - 1)^2 + (y - (-1))^2 + (z - 3)^2 = 2^2$$

$$(x - 1)^2 + (y + 1)^2 + (z - 3)^2 = 2^2$$

Answer: $(x - 1)^2 + (y + 1)^2 + (z - 3)^2 = 4$

Note that this is the sphere with radius 2 and center at $(1, -1, 3)$, whose equation was derived earlier in this section.

Volume and Surface Area

The *volume* of a sphere is the number of cubic units enclosed by the sphere. If the length of the radius of the sphere is r, then the volume V is given by the formula:

$$V = \tfrac{4}{3}\pi r^3$$

For example, the volume of a sphere with a radius of 12 cm is exactly $2{,}304\pi$ cm³, or approximately 7,238 cm³.

$$V = \tfrac{4}{3}\pi r^3$$
$$= \tfrac{4}{3}\pi(12)^3$$
$$= 2{,}304\pi$$

The number of square units in the *surface area* S of a sphere with radius r is given by the formula:

$$S = 4\pi r^2$$

For example, the surface area of a sphere with a radius of 12 cm is exactly 576π cm², or approximately 1,810 cm².

$$S = 4\pi r^2$$
$$= 4\pi(12)^2$$
$$= 576\pi$$

Example 3

The diameter of Earth is about 8,000 miles. If the planet were a perfect sphere, what would the surface area be to the nearest million square miles?

1. $2\sqrt{2}$

2. $\sqrt{26}$

3. $\sqrt{54}$

4. $x^2 + y^2 + z^2 = 9$

Solution

If the diameter is 8,000, then the radius is 4,000.

$$S = 4\pi r^2$$
$$= 4\pi(4,000)^2$$
$$= 64,000,000\pi$$
$$\approx 64,000,000(3.14)$$

Answer: The surface area of the Earth is approximately 201,000,000 square miles.

Exercises 12.2

In **1 – 3**, find the distance between the given points.

1. $(0, 0, 2)$ and $(0, -2, 0)$ **2.** $(1, 3, 2)$ and $(1, 2, 7)$ **3.** $(6, 2, 1)$ and $(-1, 1, -1)$

In **4 – 9**, write the equation of the sphere with the given point as center and the given radius, r.

4. $(0, 0, 0)$, $r = 3$ **5.** $(2, 5, 3)$, $r = 1$ **6.** $(1, 0, -2)$, $r = 5$

7. $(-2, -2, 0)$, $r = 9$ **8.** $(-3, 0, 0)$, $r = 2$ **9.** $(0, 5, -2)$, $r = 10$

In **10 – 13**, find the center and radius of the sphere with the given equation.

10. $x^2 + y^2 + z^2 = 25$

11. $(x - 2)^2 + (y - 1)^2 + z^2 = 9$

12. $x^2 + y^2 + z^2 + 6x - 8z = 0$

13. $x^2 + y^2 + z^2 - x + 4z = 2$

In **14 – 17**, find the equation of the image of the sphere under the indicated translation.

14. $x^2 + y^2 + z^2 = 16$, $T_{3,2,0}$

15. $x^2 + y^2 + z^2 = 9$, $T_{1,3,-4}$

16. $(x - 1)^2 + (y - 3)^2 + (z + 4)^2 = 9$, $T_{-1,-3,4}$

17. $x^2 - 4x + y^2 - 2y + z^2 = 0$, $T_{2,2,-3}$

18. Write an equation of the sphere with center at the origin that contains the point $(0, 0, 5)$.

19. Write an equation of the sphere with center at $(1, -2, 5)$ that contains the point $(5, 2, 3)$.

20. Find the range of possible values for c such that $x^2 + y^2 + z^2 - 2y + 8z = c$ is the equation of a sphere.

21. Find, to the nearest tenth, the volume of a sphere with center at $(0, -2, 1)$ that contains the point $(3, -2, -3)$.

22. The length of the radius of a sphere is 4 m.

   a. Find the volume of the sphere to the nearest hundredth.

   b. Find the surface area of the sphere to the nearest hundredth.

23. Write an equation of the sphere with center at $(1, -2, 6)$ and volume 36π cubic units.

24. A plane is tangent to a sphere if it is perpendicular to a radius at a point on the sphere. Find the equation of the tangent, at the point $(6, 3, 2)$, to a sphere with center at the origin.

5. $(x - 2)^2 + (y - 5)^2 + (z - 3)^2 = 1$

6. $(x - 1)^2 + y^2 + (z + 2)^2 = 25$

7. $(x + 2)^2 + (y + 2)^2 + z^2 = 81$

8. $(x + 3)^2 + y^2 + z^2 = 4$

9. $x^2 + (y - 5)^2 + (z + 2)^2 = 100$

10. $(0, 0, 0)$, $r = 5$

11. $(2, 1, 0)$, $r = 3$

12. $(-3, 0, 4)$, $r = 5$

13. $\left(\frac{1}{2}, 0, -2\right)$, $r = \frac{5}{2}$

14. $(x - 3)^2 + (y - 2)^2 + z^2 = 16$

15. $(x - 1)^2 + (y - 3)^2 + (z + 4)^2 = 9$

16. $x^2 + y^2 + z^2 = 9$

17. $(x - 4)^2 + (y - 3)^2 + (z + 3)^2 = 5$

18. $x^2 + y^2 + z^2 = 25$

19. $(x - 1)^2 + (y + 2)^2 + (z - 5)^2 = 36$

20. $c > -17$

21. 523.6

22. **a.** 268.08 m^3

   b. 201.06 m^2

23. $(x - 1)^2 + (y + 2)^2 + (z - 6)^2 = 9$

24. $2x + y = 15$

12.3 The Cylinder

The container commonly used to can products is a familiar shape called a *cylinder*. Since the base of this cylinder is a circle, an investigation of the equation that describes this cylinder begins by recalling the equation of a circle.

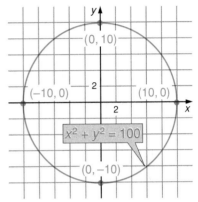

In the 2-dimensional plane, the equation $x^2 + y^2 = 100$ is the equation of a circle with center at the origin and radius of 10.

In 3-dimensional space, the equations $x^2 + y^2 = 100$ and $z = 0$ define a circle in the xy-plane, and the equations $x^2 + y^2 = 100$ and $z = 3$ define a circle in a plane parallel to and 3 units above the xy-plane.

In 3-dimensional space, the equation $x^2 + y^2 = 100$ represents the pair of equations $x^2 + y^2 = 100$ and $z = c$, where c can be any real number.

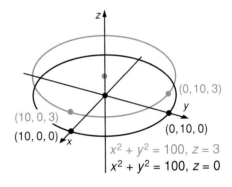

In 3-dimensional space, the equation $x^2 + y^2 = 100$ is the equation of a *cylindrical surface*. Note that z can take on any value but the values of x and y must satisfy the equation $x^2 + y^2 = 100$. For example, the points whose coordinates are shown in the diagram are all points of the surface defined by the equation $x^2 + y^2 = 100$.

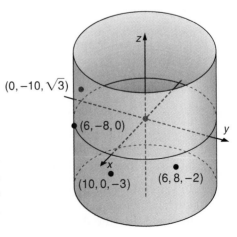

■ **Definition:** *A* cylindrical surface *is any surface in space generated by a line moving so that it always intersects a fixed plane curve in one point and so that the line is always parallel to its original position.*

Note that the generating line is called the *generatrix* and the fixed curve is called the *defining curve*, or *directrix*. If the directrix is a closed curve, the resulting surface is a closed cylindrical surface. However, the directrix may also be an open curve, as simple as a straight line.

For a given cylindrical surface, the defining curve is not unique. Any plane parallel to the plane of the defining curve intersects the cylindrical surface in a curve congruent to the defining curve, and this congruent curve could also be considered a defining curve. In addition, any curve that is the intersection of the surface with a plane that also intersects the generating line in one point could be a defining curve.

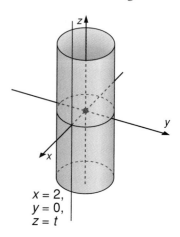

It is convenient, however, to choose a defining curve in a plane that is perpendicular to the generating line. For example, the graph of the equation $x^2 + y^2 = 4$ is a circle of radius 2 in the xy-plane. The surface generated by a line parallel to the z-axis and intersecting $x^2 + y^2 = 4$, such as the line $x = 2$, $y = 0$, $z = t$ (where t is any real number), is a circular cylindrical surface whose equation is $x^2 + y^2 = 4$.

$$x = 2,$$
$$y = 0,$$
$$z = t$$

Example 1

Describe the graph in 3-dimensional space whose equation is:

$$\frac{x^2}{9} + \frac{z^2}{4} = 1$$

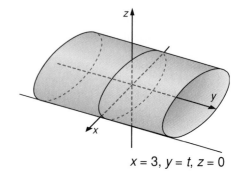

Solution

The equation defines a cylindrical surface. It is convenient to choose the plane of the defining curve to be one of the coordinate planes.

In the xz-plane, the defining curve, $\frac{x^2}{9} + \frac{z^2}{4} = 1$, $y = 0$, is an ellipse with center at the origin.

$$x = 3, y = t, z = 0$$

When the equation is compared to the general equation of an ellipse, $\frac{x^2}{a^2} + \frac{z^2}{b^2} = 1$, the major axis is a segment of the x-axis, with length $2a$, or 6, and the minor axis is a segment of the z-axis, with length $2b$, or 4.

Thus, the surface is an elliptical cylinder. The generating line is any line parallel to the y-axis that intersects the ellipse in a point, for example, $x = 3$, $y = t$, $z = 0$.

In general, in 3-dimensional space, any equation in two variables defines a cylindrical surface whose generating line is a line parallel to the axis of the third variable. It is understood that the third variable assumes every real number as its values.

Example 2

Write an equation of the cylindrical surface generated by a line parallel to the x-axis if the defining curve is a hyperbola in the yz-plane with vertices at $(0, \pm 1, 0)$ and eccentricity 2.

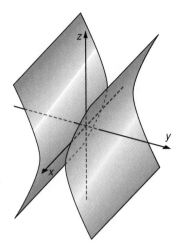

Solution

Since the vertices of the defining curve are on the y-axis, the equation of the hyperbola is of the form $\dfrac{y^2}{a^2} - \dfrac{z^2}{b^2} = 1$.

Since the vertices are at $(0, \pm 1, 0)$, $a^2 = 1$.

To determine b^2, substitute the value found for a^2 and the value given for e into the relation:

$$-b^2 = a^2(1 - e^2)$$
$$-b^2 = 1(1 - 2^2)$$
$$-b^2 = -3$$
$$b^2 = 3$$

Thus, an equation of a defining curve is $\dfrac{y^2}{1} - \dfrac{z^2}{3} = 1$, $x = 0$, and, therefore, the equation of the cylindrical surface is $\dfrac{y^2}{1} - \dfrac{z^2}{3} = 1$.

■ **Definition:** *A* cylinder *is a solid figure that consists of the portion of a closed cylindrical surface between two parallel planes that intersect the cylindrical surface in a closed curve and the portions of the two parallel planes enclosed by the cylindrical surface.*

The parts of the planes enclosed by the cylindrical surface are called the *bases* of the cylinder. If the bases are perpendicular to the generating line, the cylinder is a *right cylinder*.

In general, the volume, V, of a cylinder is equal to the area of a base, A, times the distance between the bases (or the altitude), h:

$$V = Ah$$

Thus, for example, the volume of a right cylinder whose base is a circle is:

$$V = \pi r^2 h$$

Example 3

Find the volume of the cylinder bounded by the cylindrical surface whose equation is $x^2 + y^2 = 25$ and the planes whose equations are $z = 0$ and $z = 8$.

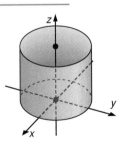

Solution

The base of the cylinder is a circle of radius 5, and the distance between the bases is 8. Therefore, the volume of the cylinder is:

$$V = Ah$$
$$= \pi r^2 h$$
$$= \pi(5)^2(8) = 200\pi$$

Answer: The volume of the cylinder is 200π, or about 628, cubic units.

Exercises 12.3

In **1 – 6**, identify each cylindrical surface by describing a defining curve and a generating line.

1. $x^2 + y^2 = 1$ **2.** $4x^2 + 12z^2 = 48$ **3.** $(x - 1)^2 + (z + 1)^2 = 4$

4. $y = 2z^2$ **5.** $y^2 + 9z^2 = 9$ **6.** $x = y^2 - 2y$

7. Write an equation of the cylindrical surface generated by a line parallel to the z-axis if the defining curve is a circle in the xy-plane with center at the origin and radius of length 3.

8. Write an equation of the cylindrical surface generated by a line parallel to the x-axis if the defining curve is an ellipse with endpoints of the major axis at (0, -5, 0) and (0, 5, 0) and endpoints of the minor axis at (0, 0, –3) and (0, 0, 3).

9. Write an equation of the cylindrical surface generated by a line parallel to the x-axis if the defining curve is a parabola with focus at (0, 0, 1) and directrix $x = 0$, $z = -1$.

10. Write an equation of the cylindrical surface generated by a line parallel to the y-axis if the defining curve is a hyperbola in the xz-plane with vertices at (±2, 0, 0) and eccentricity $\frac{3}{2}$.

11. Find the volume of the cylinder formed by the cylindrical surface whose equation is $x^2 + z^2 = 4$ and the planes whose equations are $y = -2$ and $y = 3$.

12. The diameter of a soft-drink can is determined by the size that is most comfortable for the user to hold. If 6 cm is a convenient diameter, how tall must the can be so that it will hold 12 fl. oz.? (1 fl. oz. = 30 cm³)

13. A sardine can is in the shape of a cylinder whose base is a rectangle 14 cm by 7 cm with rounded corners of radius 1 cm. Find the volume of the can if the height of the cylinder is 3 cm.

14. A chocolate bar is in the shape of a cylinder with a trapezoidal base. Find the volume of the bar if the parallel sides of the trapezoid measure 5 cm and 3 cm, the height of the trapezoid measures 2 cm, and the bar is 12 cm long.

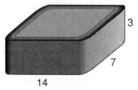

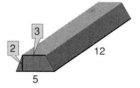

1. circular cylinder
$x^2 + y^2 = 1$, $z = 0$
$x = 1$, $y = 0$, $z = t$

2. elliptical cylinder
$4x^2 + 12z^2 = 48$, $y = 0$
$x = 0$, $z = 2$, $y = t$

3. circular cylinder
$(x - 1)^2 + (z + 1)^2 = 4$, $y = 0$
$x = 3$, $z = -1$, $y = t$

4. parabolic cylinder
$y = 2z^2$, $x = 0$
$y = 0$, $z = 0$, $x = t$

5. elliptical cylinder
$y^2 + 9z^2 = 9$, $x = 0$
$y = 3$, $z = 0$, $x = t$

6. parabolic cylinder
$x = y^2 - 2y$, $z = 0$
$x = -1$, $y = 1$, $z = t$

7. $x^2 + y^2 = 9$, $z = 0$

8. $\dfrac{y^2}{25} + \dfrac{z^2}{9} = 1$, $x = 0$

9. $z = \frac{1}{4} y^2$, $x = 0$

10. $\dfrac{x^2}{4} - \dfrac{z^2}{5} = 1$, $y = 0$

11. 20π

12. $\dfrac{40}{\pi} \approx 12.7$ cm

13. 291.4 cm³

14. 96 cm³

12.4 Surfaces of Revolution

Many common objects, such as lamp shades, mixing bowls, or the handles of tools, have circular cross sections. These shapes are an important class of 3-dimensional surfaces.

If a plane curve is rotated about a line in the plane of the curve, the resulting surface is a *surface of revolution*. The line is called the *axis of rotation* and the curve is called the *generating curve*. If a plane perpendicular to the axis of rotation intersects the surface of revolution, the intersection is either a circle or a point.

Consider the surface defined by

$$\frac{x^2}{9} + \frac{y^2}{4} + \frac{z^2}{4} = 1$$

The equations of the intersections of the surface with the coordinate planes give an indication of its shape, and help to locate the axis of rotation.

On the xy-plane, $z = 0$. The trace in the xy-plane has the equation $\frac{x^2}{9} + \frac{y^2}{4} = 1$, which is an ellipse with center at the origin and vertices at $(\pm 3, 0, 0)$ and $(0, \pm 2, 0)$.

On the xz-plane, $y = 0$. The equation of the trace is $\frac{x^2}{9} + \frac{z^2}{4} = 1$, which is an ellipse with center at the origin and vertices at $(\pm 3, 0, 0)$ and $(0, 0, \pm 2)$.

On the yz-plane, $x = 0$. The equation of the trace is $\frac{y^2}{4} + \frac{z^2}{4} = 1$, which is a circle with center at the origin and radius 2. Therefore, this identifies a circular cross section. The axis of rotation is a line through the center of the circle, perpendicular to the plane of the circle. Any plane parallel to the yz-plane has the equation $x = c$. The equation of the intersection of a plane $x = c$ and the surface is

$$\frac{c^2}{9} + \frac{y^2}{4} + \frac{z^2}{4} = 1$$

$$\frac{y^2}{4} + \frac{z^2}{4} = 1 - \frac{c^2}{9}$$

When $|c| < 3$, the intersection is a circle with center on the x-axis. For example, if the plane is $x = 2$, the equation of the intersection is:

$$\frac{y^2}{4} + \frac{z^2}{4} = 1 - \frac{2^2}{9}$$

$$\frac{y^2}{4} + \frac{z^2}{4} = \frac{5}{9}$$

$$y^2 + z^2 = \frac{20}{9}$$

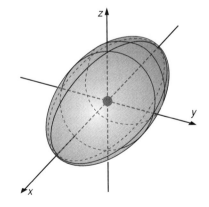

Thus, the intersection is a circle with center on the x-axis and radius $r = \sqrt{\frac{20}{9}} = \frac{2\sqrt{5}}{3}$.

When $|c| = 3$, the intersection is a point. When $|c| > 3$, there are no points of intersection. Therefore, since any plane perpendicular to the x-axis that intersects the surface, intersects it in a circle or a point, the surface is a surface of rotation with the x-axis as the axis of rotation. The generating curve can be the ellipse in either the xy-plane or the xz-plane.

To show that an equation, for example $4x^2 - y^2 + 4z^2 = 0$, is the equation of a surface of revolution, you must be able to find a line such that every plane perpendicular to that line that intersects the surface intersects it in a circle or a point. Since the coefficients of the terms in x and z are equal, consider planes parallel to the xz-plane. If $y = c$, where c is any constant, the equation of the intersection is:

$$4x^2 - c^2 + 4z^2 = 0$$
$$4x^2 + 4z^2 = c^2$$

If $c = 0$, this is the equation of a point. If $c \neq 0$, this is the equation of a circle. Therefore, the curve is a surface of revolution whose axis of rotation is the y-axis.

To find the equation of the curve that generated the surface of revolution just discussed, find the intersection of the equation of the surface of revolution with a plane that contains the axis of rotation. Since the axis of rotation is the y-axis, you can use the yz-plane, whose equation is $x = 0$. The intersection of $x = 0$ and $4x^2 - y^2 + 4z^2 = 0$ is:

$$4(0)^2 - y^2 + 4z^2 = 0$$
$$y = \pm 2z$$

Thus, the surface of revolution is generated by either the line $y = 2z$, with parametric equations $x = 0$, $y = 2t$, $z = t$ or the line $y = -2z$, with parametric equations $x = 0$, $y = -2t$, $z = t$. The axis of rotation is the y-axis. If you use the xy-plane instead of the yz-plane, the resulting equations $y = \pm 2x$ produce the same surface of revolution.

In general, a surface of revolution generated by a line that intersects the axis of rotation is a circular conical surface.

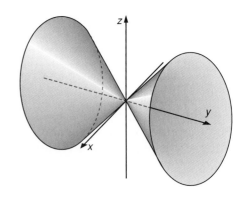

Example 1

For the surface of revolution defined by $8z = x^2 + y^2$:
a. Identify the axis of rotation.
b. Identify a generating curve.

Solution

a. For every positive value of z, $8z = x^2 + y^2$ is the equation of a circle with center at $(0, 0, z)$ on the z-axis. Therefore, the z-axis is the axis of rotation.

b. The generating curve can be the trace in any plane that contains the z-axis. It is convenient to use a curve in a coordinate plane.

In the yz-plane, $x = 0$ and $8z = 0 + y^2$. Therefore, the generating curve can be the parabola $z = \frac{1}{8}y^2$, $x = 0$.

In the xz-plane, $y = 0$ and $8z = x^2 + 0$. Therefore, the generating curve can be the parabola $z = \frac{1}{8}x^2$, $y = 0$.

■ *Definition:* *A* *right circular cone* *is a solid figure that consists of that portion of a circular conical surface between a plane perpendicular to the axis of rotation and the point at which the generating line intersects the axis of rotation and the portion of the plane enclosed by the conical surface.*

Note that the part of the plane enclosed by the conical surface is the base of the cone, and the intersection of the generating line and the axis of rotation is the vertex.

The number of cubic units in the volume of a right circular cone is given by:

$$V = \frac{1}{3}\pi r^2 h$$

where r is the radius of the base and h is the distance from the base to the vertex.

Example 2

Find the volume of the cone formed by the conical surface whose equation is $x^2 + y^2 = 2z^2$ and the plane whose equation is $z = 3$.

Solution

The conical surface has its vertex at $(0, 0, 0)$. The plane intersects the conical surface in a circle whose equation is $x^2 + y^2 = 2(3)^2$ or $x^2 + y^2 = 18$. The radius of the base is $\sqrt{18}$ or $3\sqrt{2}$. The distance from the base to the vertex is 3. Thus, the volume of the cone is:

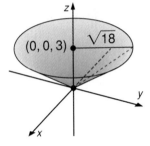

$$V = \frac{1}{3}\pi r^2 h$$
$$= \frac{1}{3}\pi (3\sqrt{2})^2(3) = 18\pi$$

Answer: The volume of the cone is 18π, or approximately 56.5, cubic units.

Example 3

Write an equation of the surface of revolution formed when the parabola $y = x^2 - 4$, $z = 0$ is rotated about the y-axis.

Solution

When the parabola is rotated about the y-axis, the surface is a set of points such that the trace on any plane perpendicular to the y-axis, that is, any plane $y = t$ parallel to the xz-plane, is a circle. The equation of that circle is $x^2 + z^2 = r^2$, where r is the distance from the y-axis to the surface. Since r is the same for every point on the circle in the plane $y = t$, a typical point, $(x, t, 0)$, can be considered. The point is also on the rotated parabola, so $y = x^2 - 4$ or $|x| = \sqrt{y + 4}$. Therefore, the distance, r, from the point $(\sqrt{y + 4}, y, 0)$ to the center of the circle on the y-axis is $\sqrt{y + 4}$. The equation of the surface of revolution is:

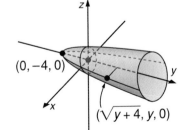

$$x^2 + z^2 = (\sqrt{y + 4})^2$$

Answer: $x^2 + z^2 = y + 4$

Note that the same surface of revolution is generated when the parabola $y = z^2 - 4$ in the yz-plane is rotated about the y-axis.

1.

a.   **b.**

x-axis $x^2 + y^2 = 1$,
$z = 0$, or
$x^2 + z^2 = 1$,
$y = 0$

y-axis $x^2 + y^2 = 1$,
$z = 0$, or
$y^2 + z^2 = 1$,
$x = 0$

z-axis $x^2 + z^2 = 1$,
$y = 0$, or
$y^2 + z^2 = 1$,
$x = 0$

2. a. x-axis

b. $x = z^2$, $y = 0$, or
$x = y^2$, $z = 0$

3. a. z-axis

b. $\dfrac{y^2}{9} - \dfrac{z^2}{4} = 1$,
$x = 0$, or
$\dfrac{x^2}{9} - \dfrac{z^2}{4} = 1$,
$y = 0$

4. a. y-axis

b. $x^2 + 4y^2 = 16$,
$z = 0$, or
$4y^2 + z^2 = 16$,
$x = 0$

5. a. z-axis

b. $x = z$, $y = 0$, or
$y = z$, $x = 0$
$x = -z$, $y = 0$, or
$y = -z$, $x = 0$

6. a. z-axis

b. $x^2 - z^2 = 1$,
$y = 0$, or
$y^2 - z^2 = 1$,
$x = 0$

Exercises 12.4

In **1 – 6**, for the surface of revolution defined by the given equation:

    a. Identify the axis of rotation.

    b. Identify the generating curve.

1. $x^2 + y^2 + z^2 = 1$ **2.** $x = y^2 + z^2$ **3.** $4x^2 + 4y^2 - 9z^2 = 36$

4. $x^2 + 4y^2 + z^2 = 16$ **5.** $x^2 + y^2 = z^2$ **6.** $x^2 + y^2 - z^2 = 1$

In **7 – 14**, write an equation of the surface of revolution with the given curve as the generating curve and the given axis as the axis of rotation.

7. $y = x^2$, $z = 0$; y-axis   **8.** $x^2 + 4y^2 = 16$, $z = 0$; y-axis

9. $x^2 + y^2 = 16$, $z = 0$; y-axis   **10.** $x^2 - z^2 = 1$, $y = 0$; x-axis

11. $x^2 - z^2 = 1$, $y = 0$; z-axis   **12.** $x = 2z$, $y = 0$; x-axis

13. $x = 2z$, $y = 0$; z-axis   **14.** $x = 4$, $y = 0$; z-axis

15. A hyperbola is the intersection of a conical surface and a plane parallel to the axis of the cone.

    a. Find the equation of the hyperbola that is the intersection of the plane whose equation is $x = 2$ and the conical surface whose equation is:
$$4x^2 - y^2 + 4z^2 = 0$$

    b. Determine the eccentricity of the hyperbola.

16. A parabolic reflector in a car headlight is in the shape of a surface generated by rotating a parabola about its axis of symmetry.

    a. Write an equation of the parabolic reflector if the equation of the generating parabola is $y = \frac{1}{4}x^2$.

    b. Find the coordinates of the focus, the point at which the source of light is located.

17. A dish antenna is in the form of a surface generated by rotating a parabola about its axis of symmetry.

    a. If the diameter of the outer edge of the dish is 200 cm and the depth of the dish is 100 cm, write an equation of the surface. *Hint*: Place the vertex of the parabola at the origin and let the z-axis be the axis of rotation.

    b. Find the coordinates of the focus, the point at which the receiver is located.

18. Find the volume of a right circular cone whose base has a diameter of 12 m and whose height is 5 m.

19. A right circular cone has a base with a radius of 5 cm. The slant height of the cone (the distance from any point on the circumference of the base to the vertex of the cone) is 13 cm.

    a. Find the height of the cone.

    b. Find the volume of the cone.

---

7. $y = x^2 + z^2$

8. $x^2 + 4y^2 + z^2 = 16$

9. $x^2 + y^2 + z^2 = 16$

10. $x^2 - y^2 + z^2 = 1$

11. $x^2 - y^2 - z^2 = 1$

12. $x^2 - 4y^2 - 4z^2 = 0$

13. $x^2 + y^2 - 4z^2 = 0$

14. $x^2 + y^2 = 16$

15. a. $y^2 - 4z^2 = 16$ **b.** $\dfrac{\sqrt{5}}{2}$

16. a. $x^2 + z^2 = 4y$ **b.** $(0, 1, 0)$

17. a. $x^2 + y^2 = 100z$ **b.** $(0, 0, 25)$

18. 60π m^3

19. a. 12 cm   **b.** 100π cm^3

12.5 Quadric Surfaces

The graph of any second-degree equation in 3-dimensional space is called a *quadric surface*. You have already seen some quadric surfaces, such as the sphere and surfaces of revolution. In this section, you will learn the names and equations of the most common quadric surfaces. In studying the graphs, consider their traces in the coordinate planes.

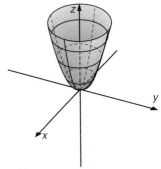

For example, the graph of $2x^2 + 2y^2 - z = 0$ is a quadric surface. On the xy-plane, $z = 0$ and the trace is a point at the origin. In any plane parallel to the xy-plane, $z = c$, where c is a constant. If $c > 0$, the plane intersects the surface in a circle. Therefore, the surface is a surface of revolution whose axis of rotation is the z-axis. In the xz-plane, $y = 0$ and the trace is the parabola $z = 2x^2$. In the yz-plane, $x = 0$ and the trace is the parabola $z = 2y^2$. This quadric surface is called a *paraboloid*.

The Gateway Arch in St. Louis is an example of
a quadric surface used in architectural design.

Following are the equations and names of some common quadric surfaces with center or turning point at the origin.

1. $x^2 + y^2 + z^2 = r^2$

The surface is a *sphere*.

Each trace is a circle of radius $|r|$.

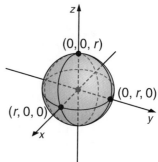

2. $\dfrac{x^2}{a^2} + \dfrac{y^2}{b^2} + \dfrac{z^2}{c^2} = 1$

The surface is an *ellipsoid*.

Each trace is an ellipse when $|a| \neq |b| \neq |c|$.

Note that the sphere is a special case of the ellipsoid when $|a| = |b| = |c|$. If any two of the constants, a, b, or c, are equal, the ellipsoid is a surface of revolution.

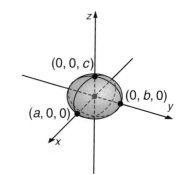

3. $\dfrac{x^2}{a^2} + \dfrac{y^2}{b^2} - \dfrac{z^2}{c^2} = 1$

The surface is a *hyperboloid of one sheet*.

The trace in the xy-plane is an ellipse if $|a| \neq |b|$ or a circle if $|a| = |b|$.

The traces in the xz-plane and the yz-plane are hyperbolas.

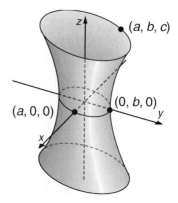

4. $\dfrac{x^2}{a^2} - \dfrac{y^2}{b^2} - \dfrac{z^2}{c^2} = 1$

The surface is a *hyperboloid of two sheets*.

The traces in the xy-plane and the xz-plane are hyperbolas.

There is no trace in the yz-plane.

The trace in a plane parallel to the yz-plane whose equation is $x = d$, where d is a constant and $|d| > |a|$, is an ellipse if $|b| \neq |c|$ or a circle if $|b| = |c|$.

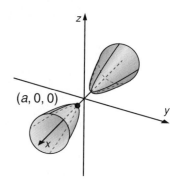

5. $\dfrac{x^2}{a^2} + \dfrac{y^2}{b^2} = cz$

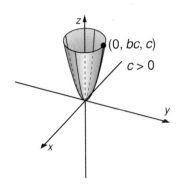

The surface is an *elliptical paraboloid*.

The traces in the xz-plane and the yz-plane are parabolas.

The trace in the xy-plane is a point.

The trace in a plane parallel to the xy-plane whose equation is $z = d$, where d is a constant that has the same sign as c, is an ellipse if $|a| \neq |b|$ or a circle if $|a| = |b|$.

6. $\dfrac{x^2}{a^2} - \dfrac{y^2}{b^2} = cz$

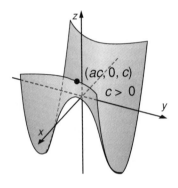

The surface is a *hyperbolic paraboloid*.

The traces in the xz-plane and the yz-plane are parabolas.

The trace in the xy-plane is a pair of lines that intersect at the origin.

The trace in a plane parallel to the xy-plane whose equation is $z = d$, where d is a constant, is a hyperbola.

7. $\dfrac{x^2}{a^2} + \dfrac{y^2}{b^2} - \dfrac{z^2}{c^2} = 0$

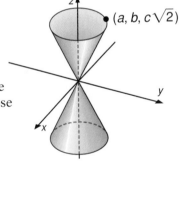

The surface is an *elliptical* or *circular cone*.

The trace in the xy-plane is a point.

The trace in a plane parallel to the xy-plane whose equation is $z = d$, where d is a constant, is an ellipse if $|a| \neq |b|$ or a circle if $|a| = |b|$.

The traces in the xz-plane and the yz-plane are pairs of intersecting lines.

8. $\dfrac{x^2}{a^2} \pm \dfrac{y^2}{b^2} = 1$

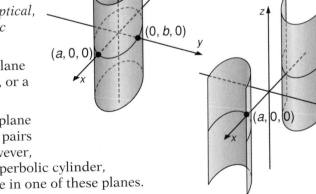

The surface is an *elliptical,
circular,* or *hyperbolic
cylinder*.

The trace in the xy-plane
is an ellipse, a circle, or a
hyperbola.

The traces in the xz-plane
and the yz-plane are pairs
of parallel lines; however,
if the surface is a hyperbolic cylinder,
there will be no trace in one of these planes.

9. $y = ax^2$

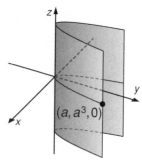

The surface is a *parabolic cylinder*.

The trace on the xy-plane is a parabola.

The traces in a plane parallel to the xz-plane whose equation is $y = c$ are a pair of lines if $a > 0$ and $c > 0$ or if $a < 0$ and $c < 0$.

The trace in a plane parallel to the yz-plane is a line.

In this list of quadric surfaces, a distinction is made between an ellipse and a circle. Even though a circle can be considered a special case of the ellipse, it is often useful to note when the ellipse is a circle in order to identify surfaces of revolution.

The cooling towers of the nuclear power plant at Three Mile Island in Pennsylvania are examples of elliptical hyperboloids of one sheet.

Example 1

a. Identify the quadric surface whose equation is $\dfrac{x^2}{25} + \dfrac{y^2}{9} - \dfrac{z^2}{4} = 1$.

b. Describe the traces in the coordinate planes or in planes parallel to the coordinate planes.

Solution

a. The quadric surface is a hyperboloid of one sheet.

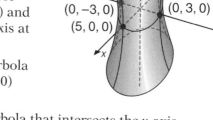

b. The trace in the xy-plane is an ellipse that intersects the x-axis at $(5, 0, 0)$ and $(-5, 0, 0)$, or $(\pm 5, 0, 0)$, and the y-axis at $(0, \pm 3, 0)$.

The trace in the xz-plane is a hyperbola that intersects the x-axis at $(\pm 5, 0, 0)$ and does not intersect the z-axis.

The trace in the yz-plane is a hyperbola that intersects the y-axis at $(0, \pm 3, 0)$ and does not intersect the z-axis.

Each of the quadric surfaces listed has its center (if the traces are circles, ellipses, or hyperbolas), or its turning point (if the traces are parabolas) at the origin. To write equations with the center or turning point at (h, k, l), replace x by $(x - h)$, y by $(y - k)$, and z by $(z - l)$ in the equations previously listed.

Example 2

Write an equation of the ellipsoid with center at $(2, -3, 5)$, if the trace in a plane through the center parallel to the xy-plane is a circle of radius 5, and the trace in a plane through the center parallel to the xz-plane is an ellipse with a major axis of length 10 and minor axis of length 8.

Solution

First, write the equation of the surface with the given characteristics and center at the origin. The equation has the form:

$$\frac{x^2}{a^2} + \frac{y^2}{b^2} + \frac{z^2}{c^2} = 1$$

The trace in the xy-plane is $x^2 + y^2 = 5^2$, or $\frac{x^2}{5^2} + \frac{y^2}{5^2} = 1$.

Therefore, $a^2 = 5^2$ and $b^2 = 5^2$.

The equation of the trace in the xz-plane is $\frac{x^2}{a^2} + \frac{z^2}{c^2} = 1$. Since $|a| = 5$, then $2|a| = 10$ and the major axis lies on the x-axis. Thus, the minor axis lies on the z-axis and $2|c| = 8$, or $|c| = 4$. Therefore, the equation of the surface with center at the origin and the given characteristics is:

$$\frac{x^2}{5^2} + \frac{y^2}{5^2} + \frac{z^2}{4^2} = 1$$

To translate the surface so that its center is at $(2, -3, 5)$, replace x by $x - 2$, y by $y + 3$, and z by $z - 5$.

Answer: $\dfrac{(x - 2)^2}{5^2} + \dfrac{(y + 3)^2}{5^2} + \dfrac{(z - 5)^2}{4^2} = 1$

Exercises 12.5

In **1 – 12**, identify the quadric surface whose equation is given, and describe the traces in the coordinate planes or in planes parallel to the coordinate planes.

1. $x^2 + y^2 + z^2 = 9$

2. $\dfrac{x^2}{36} + \dfrac{y^2}{9} + \dfrac{z^2}{4} = 1$

3. $\dfrac{x^2}{36} + \dfrac{y^2}{9} - \dfrac{z^2}{4} = 1$

4. $\dfrac{x^2}{36} - \dfrac{y^2}{9} - \dfrac{z^2}{4} = 1$

5. $\dfrac{x^2}{36} + \dfrac{y^2}{9} - \dfrac{z^2}{4} = 0$

6. $\dfrac{x^2}{4} + y^2 = z$

7. $\dfrac{x^2}{4} - y^2 = z$

8. $x^2 + y^2 = 12$

9. $\dfrac{x^2}{4} + \dfrac{y^2}{12} = 1$

10. $\dfrac{x^2}{4} - \dfrac{y^2}{12} = 1$

11. $4y = z^2$

12. $\dfrac{(x - 1)^2}{9} + \dfrac{(y + 3)^2}{9} + \dfrac{(z - 2)^2}{4} = 1$

Answers 12.5

1. sphere of radius 3; circles of radius 3 in each coordinate plane

2. ellipsoid; ellipse in each coordinate plane

3. hyperboloid of one sheet; hyperbola in the xz- and yz-planes, ellipse in the xy-plane

4. hyperboloid of two sheets; no trace in the yz-plane, hyperbolas in the xy- and xz-planes

5. elliptical cone; ellipses in planes parallel to the xy-plane, lines in the xz- and yz-planes

6. elliptical paraboloid; parabolas in the xz- and yz-planes, ellipses in planes parallel to the xy-plane

7. hyperbolic paraboloid; parabolas in the xz- and yz-planes, hyperbolas in planes parallel to the xy-plane

8. circular cylinder; circle in the xy-plane, two parallel lines in the xz- and in the yz-planes

9. elliptical cylinder; ellipse in the xy-plane, two parallel lines in the xz- and in the yz-planes

10. hyperbolic cylinder; hyperbola in the xy-plane, two parallel lines in the xz- and in the yz-planes

11. parabolic cylinder; parabola in the yz-plane, two parallel lines in planes parallel to the xz-plane, one line in planes parallel to the xy-plane

12. ellipsoid with center at $(1, -3, 2)$; ellipse in planes parallel to the coordinate planes that intersect the figure in more than one point

13. a. $\dfrac{x^2}{16} + \dfrac{y^2}{4} = \dfrac{z^2}{9}$

b. $(0, 4, 6)$,
$(8, 0, 6)$

14.

$$\dfrac{x^2}{144} + \dfrac{y^2}{576} - \dfrac{z^2}{900} = 1$$

15. a. 100 m

b. 85.4 m

c. 80 m

Exercises 12.5 *(continued)*

13. The vertex of an elliptical cone is at the origin. Its trace on the plane whose equation is $z = 3$ is an ellipse with vertices at $(\pm 4, 0, 3)$ and $(0, \pm 2, 3)$.

   a. Write an equation of the cone.

   b. Find the coordinates of the vertices of the trace on the plane $z = 6$.

14. A decorative end table is in the shape of a hyperboloid of one sheet, generated by rotating a hyperbola about the axis of symmetry that does not intersect the curve. The radius of the table is 12 cm at its narrowest part and 24 cm at the top and the bottom. The height of the table is 60 cm. Find the equation of the hyperboloid.
Hint: Let the center of the surface of revolution be the origin of the coordinate system.

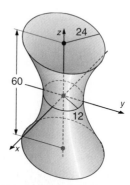

15. An architect designed a circular building whose outer shell, which is in the shape of a hyperboloid of one sheet, has the equation:

$$\dfrac{x^2}{80^2} + \dfrac{y^2}{80^2} - \dfrac{(z - 18)^2}{24^2} = 1$$

If z represents the height above the ground of a point on the shell of the building in meters, find the radius of the floor space at

   a. $z = 0$

   b. $z = 9$

   c. $z = 18$

St. Louis Science Center
Planetarium Building

Planes in 3-Space

- Equations of the traces in the coordinate planes can be found by replacing each variable in turn by 0. The intercepts can be found by finding the value of each variable when the other two variables equal 0.

- The planes parallel to the yz-plane, the xz-plane, and the xy-plane have the equations $x = c$, $y = c$, and $z = c$ respectively, where c is a constant. If $c = 0$, the plane is a coordinate plane.

- If in the equation of a plane, $Ax + By + Cz + D = 0$, the coefficient of x, or of y, or of z is zero, the plane is parallel to the axis of that variable.

Distance in 3-Space

- Let $A(x_1, y_1, z_1)$ and $B(x_2, y_2, z_2)$ be any two points in 3-dimensional space. The distance, d, from A to B is $d = \sqrt{(x_1 - x_2)^2 + (y_1 - y_2)^2 + (z_1 - z_2)^2}$.

The Sphere

- A sphere is the locus of points in space that are equidistant from a fixed point called the center. Let (h, k, l) be the center of the sphere and r be the distance from the center to any point (x, y, z) of the locus. Then the equation of the sphere is $(x - h)^2 + (y - k)^2 + (z - l)^2 = r^2$.

- The number of cubic units in the volume of a sphere with radius r is given by the formula $V = \frac{4}{3}\pi r^3$.

- The number of square units in the surface area of a sphere with radius r is given by the formula $S = 4\pi r^2$.

The Cylinder

- A cylindrical surface is any surface in space generated by a line, the generatrix, moving so that it always intersects a fixed plane curve, the directrix, in one point and so that the line is always parallel to its original position.

  If the directrix is a closed curve, the resulting surface is a closed cylindrical surface. Any plane parallel to the plane of the directrix intersects the cylindrical surface in a curve congruent to the directrix.

- In 3-dimensional space, any equation in two variables defines a cylindrical surface whose generating line is a line parallel to the axis of the third variable.

- A cylinder is a solid figure that consists of the portion of a closed cylindrical surface between two parallel planes that intersect the cylindrical surface in a closed curve and the portions of the two parallel planes enclosed by the cylindrical surface. The parts of the planes enclosed by the cylindrical surface are called the bases of the cylinder. The volume, V, of the cylinder is equal to the area of a base, A, times the distance between the bases, h, that is, $V = Ah$.

Surfaces of Revolution

- If a plane curve is rotated about a line in the plane of the curve, the resulting surface is a surface of revolution. The line is called the axis of rotation and the curve is called the generating curve. If a plane perpendicular to the axis of rotation intersects the surface of revolution, the intersection is either a circle or a point.

Quadric Surfaces

- The graph of any second-degree equation in 3-dimensional space is called a quadric surface.

- The equation of a quadric surface with the center or turning point at (h, k, l) can be written by replacing x by $(x - h)$, y by $(y - k)$, and z by $(z - l)$ in the equation of the congruent quadric surface with center or turning point at the origin.

1. 6

2. 12

3. 5

4. $\dfrac{x}{3} - \dfrac{y}{5} + \dfrac{z}{7} = 1$ or $35x - 21y + 15z = 105$

5. $23x - 9y + 8z + 8 = 0$

6. $x = 1$

7. $x - 2y + 7z - 3 = 0$

8. $(x - 3)^2 + y^2 + (z + 1)^2 = 81$

9. $(x + 2)^2 + (y - 4)^2 + (z - 1)^2 = 20$

10. a. $\dfrac{256}{3}\pi$ **b.** 64π

11. a. $\dfrac{343}{6}\pi$ **b.** 49π

12. $x^2 + 9y^2 = 144$, an ellipse in the xy-plane; $x^2 - 16z^2 = 144$, a hyperbola in the xz-plane; $9y^2 - 16z^2 = 144$, a hyperbola in the yz-plane

Chapter 12   Review Exercises

In **1 – 3**, find the distance between the given points.

1. $(0, 0, 0)$, $(4, -4, 2)$ **2.** $(1, 3, -1)$, $(9, -1, 7)$ **3.** $(1, 0, -3)$, $(0, 2\sqrt{5}, -1)$

In **4 – 7**, write an equation of the plane determined by the points.

4. $(3, 0, 0)$, $(0, -5, 0)$, $(0, 0, 7)$ **5.** $(1, 7, 4)$, $(0, 0, -1)$, $(2, 6, 0)$

6. $(1, 2, -1)$, $(1, -2, 3)$, $(1, 4, 2)$ **7.** $(-4, 7, 3)$, $(1, -1, 0)$, $(7, 2, 0)$

8. Write an equation of the sphere with center at $(3, 0, -1)$ and radius 9.

9. Write an equation of the sphere with center at $(-2, 4, 1)$ that contains the point $(2, 6, 1)$.

10. An equation of a sphere is $x^2 + y^2 + z^2 - 4x + 6z = 3$.
   a. Find the volume of the sphere.
   b. Find the surface area of the sphere.

11. A tennis ball has a diameter of 7 cm.
   a. Find the volume of the ball.
   b. Find the surface area of the ball.

In **12 – 17**, find the trace in each of the coordinate planes.

12. $x^2 + 9y^2 - 16z^2 = 144$ **13.** $3x - 7y - 4z - 42 = 0$

14. $3x^2 + y^2 = z$ **15.** $x^2 - 5y^2 - z^2 = 20$

16. $4y - z = 12$ **17.** $x^2 + 4y^2 + 12z^2 = 48$

In **18 – 23**, identify each of the curves as a cylindrical surface, a surface of revolution, both a cylindrical surface and a surface of revolution, or neither a cylindrical surface nor a surface of revolution.

18. $x^2 + y^2 + z^2 = 9$ **19.** $x^2 + y^2 - z^2 = 9$ **20.** $x^2 + 2z^2 = 2y$

21. $x^2 + 2y^2 = 18$ **22.** $x^2 + z^2 = 1$ **23.** $x^2 - 9y^2 = z$

24. The equation of an ellipsoid is $x^2 + 12y^2 + 4z^2 = 36$.
   a. Find the equation of the image of the ellipsoid under the translation $T_{2, -5, 3}$.
   b. Find the coordinates of the intersections of the image curve with its axes of symmetry.

25. The volume of 1.5 pounds of raisins is about 1,200 cm³. The raisins are to be packaged in a cylindrical box with a diameter of 10 cm. Find, to the nearest tenth, the height of the box.

Review Answers

13. $3x - 7y - 42 = 0$, a line in the xy-plane; $3x - 4z - 42 = 0$, a line in the xz-plane; $-7y - 4z - 42 = 0$, a line in the yz-plane

14. $3x^2 + y^2 = 0$, the point $(0, 0, 0)$ in the xy-plane; $3x^2 = z$, a parabola in the xz-plane; $y^2 = z$, a parabola in the yz-plane

15. $x^2 - 5y^2 = 20$, a hyperbola in the xy-plane; $x^2 - z^2 = 20$, a hyperbola in the xz-plane; $-5y^2 - z^2 = 20$, no trace in the yz-plane

16. $y = 3$, a line parallel to the x-axis in the xy-plane; $z = -12$, a line parallel to the x-axis in the xz-plane; $4y - z = 12$, a line in the yz-plane

17. $x^2 + 4y^2 = 48$; $x^2 + 12z^2 = 48$; $4y^2 + 12z^2 = 48$ an ellipse in each coordinate plane

---

18. surface of revolution
19. surface of revolution
20. neither
21. cylindrical surface
22. both
23. neither

24. a. $(x - 2)^2 + 12(y + 5)^2 + 4(z - 3)^2 = 36$
   b. $(-4, -5, 3)$ and $(8, -5, 3)$;
   $(2, -5 - \sqrt{3}, 3)$ and $(2, -5 + \sqrt{3}, 3)$;
   $(2, -5, 0)$ and $(2, -5, 6)$

25. 15.3 cm

Answers
Exercises for
Challenge

1. $x^2 + y^2 = z$

2. $z = 2$

3. $\sqrt{34}$

4. D

5. C

6. E

7. C

8. D

9. A

10. E

☑ Exercises for Challenge

1. What is the equation of the quadric surface obtained by rotating the parabola defined by $z = y^2$, $x = 0$ about the z-axis?

2. What is an equation of the plane that contains the intersection of the quadric surfaces whose equations are $x^2 + y^2 + (z-1)^2 = 3$ and $x^2 + y^2 = z$?

3. A sphere with center at $(3, -2, 5)$ intersects the xy-plane in a circle of radius 3. What is the length of the radius of the sphere?

4. A spherical balloon with center at $(5, -2, 3)$ has a radius of length 3. If the volume of the balloon increases by a factor of 9, by what factor does the surface area increase?
   (A) $\sqrt[3]{3}$ (B) $\sqrt[3]{9}$ (C) 3
   (D) $3\sqrt[3]{3}$ (E) 6

5. The area A of an ellipse with major and minor axes of length $2a$ and $2b$ is $A = \pi ab$. Find the volume of the region bounded by the cylindrical surface $x^2 + 4z^2 = 1$ and the planes $y = 2$ and $y = -3$.
   (A) $\frac{\pi}{2}$ (B) 4π (C) $\frac{5}{2}\pi$
   (D) 5π (E) 20π

6. Which of the following is an equation of the quadric surface containing all points such that $\sqrt{x} + \sqrt{y} = \sqrt{z}$ and $x, y, z \neq 0$?
   (A) $x + y = z$
   (B) $4xy = z^2 - x^2 + y^2$
   (C) $4xy = z^2 - x^2 - y^2$
   (D) $x^2 + y^2 + z^2 + 2xz + 2xy + 2yz = 0$
   (E) $x^2 + y^2 + z^2 - 2xy - 2xz - 2yz = 0$

7. Which of the following is *not* an axis of rotation for the sphere whose equation is $(x-1)^2 + (y+1)^2 + z^2 = 4$?
   (A) $x = t, y = -1, z = 0$
   (B) $x = 1, y = t, z = 0$
   (C) $x = 1, y = -1, z = 0$
   (D) $x = 1, y = -1, z = t$
   (E) All are axes of rotation.

8. The surface area of Sphere I is 10 times as large as the surface area of Sphere II. To the nearest tenth, the volume of Sphere I is how many times as large as the volume of Sphere II?
   (A) 3.1 (B) 3.2 (C) 10.0
   (D) 31.6 (E) 32.0

9. To the nearest hundredth, what is the least positive value of x for a point on the quadric surface whose equation is $\frac{x^2}{12} - \frac{y^2}{14} - \frac{z^2}{16} = 1$?
   (A) 3.46 (B) 3.74 (C) 4.00
   (D) 4.32 (E) 51.85

10. The intersection of the surface whose equation is $\frac{x^2}{a^2} - \frac{y^2}{b^2} - \frac{z^2}{c^2} = 0$ with the plane whose equation is $y = b$ is
    (A) two intersecting lines
    (B) a circle (C) an ellipse
    (D) a parabola (E) a hyperbola

Answers
Exercises for
Challenge

11. B

12. A

13. C

14. A

15. A

16. C

17. B

18. B

☑ *Exercises for Challenge* (continued)

Questions **11** and **12** refer to:

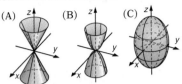

(A)   (B)   (C)

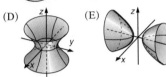

(D)   (E)

11. Which of the choices above could be the quadric surface defined by $x^2 + 9y^2 = |4z|$?

12. Which of the choices above could be the quadric surface defined by $x^2 + 9y^2 = 4z^2$?

13. An ant travels from $A(0, 0, 5)$ to $B(3, 4, 0)$ by walking down the yz-plane to the y–axis and across the xy-plane. What is the length of the shortest possible path?

(A) $\sqrt{50}$ (B) $\sqrt{74}$ (C) $\sqrt{80}$
(D) 10   (E) 12

Questions 14 – 18 each consist of two quantities, one in Column A and one in Column B. You are to compare the two quantities and choose:

A if the quantity in Column A is greater;
B if the quantity in Column B is greater;
C if the two quantities are equal;
D if the relationship cannot be determined from the information given.

1. In certain questions, information concerning one or both of the quantities to be compared is centered above the two columns.

2. In a given question, a symbol that appears in both columns represents the same thing in Column A as it does in Column B.

3. x, n, and k, etc. stand for real numbers.

Column A	Column B

S_1 is the hyperboloid
$$x^2 + \frac{y^2}{9} - \frac{z^2}{4} = -1.$$
S_2 is the hyperboloid
$$x^2 - \frac{y^2}{9} - \frac{z^2}{4} = -1.$$

14. Number of sheets in S_1 | Number of sheets in S_2

---

d is the distance from the origin to the center of the sphere
$x^2 + y^2 + z^2 - 2x + 8y - 4z + 18 = 0.$

15. d^2 | 20

A sphere of radius 10 has surface area S and volume V. Let R be the ratio $V : S$, rounded to the nearest hundredth.

16. R | 3.33

Let t be a parameter so that, for all t, the point $(t, 2\cos t, 3\sin t)$ lies on the elliptical cylinder:

$$\frac{y^2}{a^2} + \frac{z^2}{b^2} = 1$$

17. $|a|$ | $|b|$

Let V be the volume of the region bounded by the circular cylinder $4y^2 + 4z^2 - 8y + 16z = 16$ and the planes $x = 1$ and $x = 4$.

18. V | 86.4

Teacher's Chapter 13

Statistics

Chapter Contents

As technology makes possible the accumulation of more and more data and as the world becomes more and more complex, the need for processes and procedures that enable us to make intelligent use of the available data becomes more critical. The science of statistics provides those processes and procedures and sets the constraints and cautions that, if used correctly, prevent the misinterpretation of data.

This chapter presents the fundamental concepts of statistics, which will be a review for many students, and utilizes those concepts to provide an understanding of statistical methods and applications.

13.1 Organizing Data

Organization is an aid to understanding. Large numbers of unorganized data are meaningless. This section presents methods of listing and grouping data to make its scope easier to grasp. Calculators and computers are important tools for this task. Computer spread sheets enable students to enter, organize, and revise data, and make it possible to store sets of data for more detailed investigation as new concepts are presented.

When data represents the number of people or things, the data is represented by discrete variables, that is, integers. For example, if a distribution lists the number of persons residing in single-family dwellings, the variables are natural numbers. On the other hand, if the data represents measurements, such as heights, for which a more precise measuring instrument would give more exact data, the variables are considered to be continuous. However, for calculation purposes, a height rounded to 140 cm is taken to be exactly 1 cm less than a height rounded to 141 cm.

Computer Activity (13.1) _____

The program will organize data entered by the user into a table that lists the entries in order, with the frequency of each.

```
100 REM  THIS PROGRAM WILL MAKE A
110 REM  FREQUENCY TABLE FOR A SET OF
120 REM  NUMBERS WITH A RANGE LESS
         THAN
130 REM  ONE HUNDRED.
140 DIM C(100)
150 PRINT "THIS IS A PROGRAM THAT
         WILL HELP YOU TO"
160 PRINT "MAKE A FREQUENCY TABLE
         FOR A SET OF"
170 PRINT "NUMBERS IF THE
         DIFFERENCE BETWEEN THE"
180 PRINT "LARGEST AND SMALLEST
         NUMBERS IS"
190 PRINT "NOT MORE THAN 100."
200 PRINT
210 PRINT "TO EACH QUESTION,
         RESPOND BY TYPING"
220 PRINT "THE ANSWER AND PRESSING
         THE RETURN KEY."
230 PRINT
240 FOR K = 0 TO 100
250     C(K) = 0
260 NEXT K
270 PRINT "HOW MANY NUMBERS ARE
         IN THE SET"
280 PRINT "OF DATA?"
290 INPUT N
300 PRINT
310 PRINT "WHAT IS THE LARGEST
         NUMBER?"
320 INPUT L
330 PRINT
340 PRINT "WHAT IS THE SMALLEST
         NUMBER?"
350 INPUT S
360 PRINT
370 IF L < = S THEN 310
380 IF L - S > 100 THEN 630
390 PRINT "ENTER THE NUMBERS FROM
         THE SET OF DATA,"
400 PRINT "PRESSING THE RETURN KEY
         AFTER EACH."
410 PRINT
420 PRINT "YOU MUST ENTER "; N; "
         NUMBERS BEFORE"
430 PRINT "THE FREQUENCY TABLE WILL
         BE PRINTED."
440 PRINT
```

```
450 PRINT "NUMBERS THAT ARE
            LARGER OR SMALLER
            THAN"
460 PRINT "THE LIMITS SET IN THE
            ABOVE STATEMENTS"
470 PRINT "WILL BE REJECTED."
480 PRINT
490 FOR K = 1 TO N
500     INPUT X
510     IF X > L THEN PRINT "TOO
        BIG!": GOTO 500
520     IF X < S THEN PRINT "TOO
        SMALL!": GOTO 500
530     C(X - S) = C(X - S) + 1
540 NEXT K
550 PRINT
560 PRINT TAB(10); "NUMBER";
            TAB(20); "FREQUENCY"
570 PRINT
580 FOR K = L - S TO 0 STEP -1
590     IF C(K) = 0 THEN 610
600     PRINT TAB(12); K + S;
        TAB(23); C(K)
610 NEXT K
620 GOTO 650
630 PRINT "THE RANGE OF THIS SET
            OF NUMBERS IS"
640 PRINT "TOO BIG FOR THIS
            PROGRAM TO ORGANIZE."
650 PRINT
660 PRINT "DO YOU WANT TO DO
            ANOTHER? (Y OR N)"
670 INPUT R$
680 IF R$ = "Y" OR R$ = "YES"
            THEN 230
690 END
```

13.2 Measures of Central Tendency

Most students are familiar with mean, median and mode. To these familiar three, a fourth, midrange, is added in this section.

Quartiles and interquartile range, together with the whisker-box plot, are used to achieve some sense of the pattern in which the data of a distribution occurs. The method of finding quartiles given in the text is an intuitive approach. A more formula-oriented approach is frequently used by statistical computer programs in common use.

One method is to divide the data into quarters by using the entry at the $\frac{n + 1}{4}$ position for the first-quartile value and the entry at the $\frac{3(n + 1)}{4}$ position for the third quartile. If $\frac{n + 1}{4}$ and $\frac{3(n + 1)}{4}$ are not integers, interpolation is used. For example, if a set of data has 11 data points, then the first-quartile value is the $\frac{11 + 1}{4}$, or 3rd, data point and the third-quartile value is the $\frac{3(11 + 1)}{4}$, or 9th, data point. If a set of data has 12 data points, then the first-quartile value is at the $\frac{12 + 1}{4}$, or 3.25, position. That is, interpolate to find the number that is 0.25, or $\frac{1}{4}$, of the way between the 3rd and 4th data points. The third quartile value is at the $\frac{3(12 + 1)}{4}$, or 9.75, position. That is, interpolate to find the number that is 0.75, or $\frac{3}{4}$, of the way between the 9th and 10th data points.

At this level, students may be permitted to think of the 1st quartile as equivalent to the 25th percentile, and the 3rd quartile as the 75th percentile. In practice, since statistical methods used to find quartiles differ from those for percentiles, they are not exact equivalents.

Computer Activity (13.2) _____

The program will calculate the mean of a set of data.

```
100 REM  THIS PROGRAM WILL
            PERFORM THE
110 REM  COMPUTATION TO FIND
            THE MEAN
120 PRINT
130 PRINT "THIS IS A PROGRAM THAT
            WILL FIND THE"
140 PRINT "MEAN OF A SET OF DATA."
150 PRINT
160 S = 0
170 PRINT "HOW MANY NUMBERS ARE
            THERE IN THE"
180 PRINT "SET OF DATA?"
```

```
190 INPUT A
200 PRINT
210 PRINT "ENTER THE NUMBERS FROM
            THE SET OF DATA,"
220 PRINT "PRESSING THE RETURN KEY
            AFTER EACH."
230 PRINT
240 PRINT "THE MEAN WILL BE
            CALCULATED AFTER ALL"
250 PRINT A; " NUMBERS HAVE BEEN
            ENTERED."
260 PRINT
270 FOR K = 1 TO A
280    INPUT D
290    S = S + D
300 NEXT K
310 PRINT
320 PRINT "THE SUM OF THE "; A;
            " NUMBERS IS "; S; "."
330 PRINT
340 PRINT "THE MEAN = "; S; "/";
            A; " = "; S / A
350 PRINT
360 PRINT "HAVE YOU ANOTHER SET OF
            DATA FOR WHICH"
370 PRINT "YOU WANT TO FIND THE
            MEAN? (Y OR N)"
380 INPUT R$
390 IF R$ = "Y" OR R$ = "YES"
            THEN 120
400 END
```

13.3 Measures of Dispersion

The standard deviation is the most frequently used measure of dispersion in statistical studies today. Standard deviation is particularly important because of its relationship to the normal curve, as explained in the next section.

When students receive information about their scores on standardized tests, these scores are usually the result of converting raw scores (for example, the number of correct answers, or the number of correct answers adjusted for correct answers that result from guessing) to a standard range of values using z-scores. For example, one method, which results in a top score of 800, converts to standard scores by multiplying z-scores by 100 and adding 500. Thus, for the test scores, the mean is 500 and the standard deviation is 100. Another method converts test scores to standard scores by multiplying the z-scores by 5 and adding 15. Thus, these test scores have a mean of 15 and a standard deviation of 5.

Computer Activity (13.3) _____

The program will compute the mean and standard deviation for a set of data entered by the user.

```
100 REM  THIS PROGRAM WILL MAKE A
110 REM  FREQUENCY TABLE FOR
            A SET OF
120 REM  NUMBERS WITH A RANGE
            LESS THAN
130 REM  ONE HUNDRED.
140 REM  IT WILL DETERMINE THE
            MEAN AND
150 REM  THE STANDARD DEVIATION
160 REM  FOR THE SET OF DATA.
170 DIM C(100)
180 PRINT "THIS IS A PROGRAM THAT
            WILL MAKE A"
190 PRINT "FREQUENCY TABLE FOR A
            SET OF NUMBERS"
200 PRINT "IF THE DIFFERENCE
            BETWEEN THE LARGEST"
210 PRINT "AND SMALLEST NUMBERS IS
            LESS THAN 100."
220 PRINT
230 PRINT "IT WILL CALCULATE THE
            MEAN AND THE"
240 PRINT "STANDARD DEVIATION."
250 PRINT
260 PRINT "TO EACH QUESTION,
            RESPOND BY TYPING"
270 PRINT "THE ANSWER AND PRESSING
            THE RETURN KEY."
280 PRINT
290 FOR K = 0 TO 100
300    C(K) = 0
310 NEXT K
320 T = 0
330 R = 0
340 PRINT "HOW MANY NUMBERS ARE
            IN THE SET"
350 PRINT "OF DATA?"
360 INPUT N
370 PRINT
380 PRINT "WHAT IS THE LARGEST
            NUMBER?"
390 INPUT L
400 PRINT
```

```
410   PRINT "WHAT IS THE SMALLEST
          NUMBER?"
420   INPUT S
430   PRINT
440   IF L < = S THEN 380
450   IF L - S > 100 THEN 890
460   PRINT "ENTER THE NUMBERS FROM
          THE SET OF DATA,"
470   PRINT "PRESSING THE RETURN KEY
          AFTER EACH."
480   PRINT "YOU MUST ENTER "; N; "
          NUMBERS BEFORE"
490   PRINT "THE FREQUENCY TABLE
          WILL BE PRINTED."
500   PRINT "NUMBERS THAT ARE LARGER
          OR SMALLER THAN"
510   PRINT "LIMITS SET IN THE
          ABOVE STATEMENTS WILL"
520   PRINT "BE REJECTED."
530   PRINT
540   FOR K = 1 TO N
550      INPUT X
560      IF X > L THEN PRINT "TOO
         BIG!": GOTO 550
570      IF X < S THEN PRINT "TOO
         SMALL!": GOTO 550
580      T = T + X
590      C(X - S) = C(X - S) + 1
600   NEXT K
610   PRINT
620   M = INT(T / N * 10 + .5) / 10
630   PRINT "TO THE NEAREST TENTH,"
640   PRINT "THE MEAN IS "; M
650   PRINT
660   PRINT "X(I) = A DATA ENTRY"
670   PRINT "F(I) = FREQUENCY OF X(I)."
680   PRINT "D(I) = DEVIATION FROM
          THE MEAN"
690   PRINT "S(I) = SQUARE OF D(I)."
700   PRINT
710   PRINT "X(I)"; TAB(7); "F(I)";
          TAB(14); "D(I)";
720   PRINT TAB(21); "S(I)";
          TAB(28); "F(I) * S(I)"
730   FOR K = L - S TO 0 STEP -1
740      IF C(K) = 0 THEN 820
750      Q = K + S - M
760      P = Q * Q
770      R = R + P * C(K)
780      Q = INT(Q * 10 + .5) / 10
790      P = INT(P * 100 + .5) / 100
800      PRINT K + S; TAB(8); C(K);
         TAB(15); Q;
810      PRINT TAB(22); P; TAB(32);
         P * C(K)
820   NEXT K
830   PRINT
840   R = (R / N) ^ .5
850   D = INT(R * 10 + .5) / 10
860   PRINT "THE STANDARD DEVIATION,"
870   PRINT "TO THE NEAREST TENTH,
          IS "; D
880   GOTO 910
890   PRINT "THE RANGE OF THIS SET OF
          NUMBERS IS"
900   PRINT "TOO BIG FOR THIS PROGRAM
          TO ORGANIZE."
910   PRINT
920   PRINT "DO YOU WANT TO DO
          ANOTHER? (Y OR N)"
930   INPUT R$
940   IF R$ = "Y" OR R$ = "YES" THEN 280
950   END
```

13.4 The Normal Distribution

The standard normal curve is defined by the equation $y = \dfrac{1}{\sqrt{2\pi}} e^{-\frac{z^2}{2}}$. Since the derivation of this exponential function requires a degree of mathematical knowledge and sophistication well beyond the student who uses this book, the curve is introduced through the more familiar concept of probability. Note that the probability model is a discrete distribution, used to model distributions of continuous variables.

Students should be aware that not all sets of data are normal distributions. For example, sets of test grades for a single class are usually not normal distributions. Remind students that the assumption of a normal distribution is made for the examples and exercises of this section.

13.5 Sampling and Estimation

When the group under discussion includes a finite number of subjects, it may be possible to assemble information about every member of the group. It is more frequently the case that the group consists of a very large number of subjects, even a potentially infinite number (all existing or possible future members of a group), or consists of inaccessible members. Such groups are studied by considering information obtained from samples.

The description of what constitutes a random sample is one that is frequently used in statistics texts and is an intuitive way of introducing the concept. A more rigorous definition considers the set of all possible samples of n data points taken from the population and the random selection of one of those samples so that each sample is equally likely to be selected. This more rigorous definition assures not only that each data point is equally likely to be in the chosen sample but that each pair of data points is equally likely to be in the chosen sample.

The explanation in the text of the method of obtaining the standard deviation based on sample data is an intuitive justification of the formula. Another common approach is to use the fact that the number of degrees of freedom in a set of n data points for a given mean is $n - 1$, to justify dividing by $n - 1$.

In the previous section, only sets of data that were approximately normal distributions were used. In this section, the Central Limit Theorem allows application of the probabilities of the standard normal curve to samples of 30 or more taken from any distribution. Remind students that the mean of a sample is one of the data points in a normally distributed set of means.

When finding the endpoints of a confidence interval, some students may prefer to start with the formula for z each time and solve that formula for μ since the formula for z is already familiar. Other students may prefer to learn the formula for μ in terms of z and $\sigma_{\bar{x}}$.

13.6 Chi-Square Distribution

The text uses two different approaches to finding the value of χ^2. Some students may wish to use the expanded table that lists observed value and expected value in separate columns. Call attention to the fact that the sum of the deviations of observed value from expected value is always 0.

Discuss conclusions that are reached and note that no specific percentage can be given as a cutoff point separating likely and unlikely occurrences; in fact, we can never say with absolute certainty that given results have not occurred by chance. Thus, in Example 3, while we conclude that the coin is probably not a fair coin since the 32 heads in 100 tosses will occur less than 1% of the time, it is still possible that the coin is fair.

Suggested Test Items

1. Following are the ages of the 25 employees of Mico:
   23, 33, 54, 25, 27, 63, 28, 31, 34, 51, 37, 29, 30,
   27, 31, 40, 26, 31, 47, 22, 34, 37, 45, 61, 59.

   a. Use a stem-and-leaf diagram to organize the data.

   b. Find the mean of the ungrouped data.

   c. Find the median of the ungrouped data.

   d. Find the first and third quartiles and draw a whisker-box plot.

   e. What is the mode (or modes) of the data?

   f. Find the midrange of the data.

   g. What is the percentile rank of 37? of 28?

   h. Organize the data in intervals of 5, beginning with 20-24, and draw a histogram.

   i. Find the mean of the grouped data and compare it with the mean found for the ungrouped data.

   j. Find the median of the grouped data and compare it with the median found for the ungrouped data.

2. A new energy-efficient light bulb was developed and a sample of 500 bulbs were tested for the length of time before burning out. The results are shown in the chart.

   a. Find the mean number of hours that the bulbs burned.

   b. Find the median number of hours that the bulbs burned.

   c. Find the standard deviation of the bulb-life.

Hours	Frequency
1,100 – 1,199	13
1,000 – 1,099	38
900 – 999	76
800 – 899	98
700 – 799	142
600 – 699	87
500 – 599	27
400 – 499	12
300 – 399	7

3. A college determined that the mean height for incoming freshmen was 65 inches with a standard deviation of 4 inches. If the heights are approximately a normal distribution, what percent of the incoming freshmen can be expected to be taller than 72 inches? Answer to the nearest whole percent.

4. The mean wage for a retail store was $5.25 per hour with a standard deviation of $2.00. Is it appropriate to use z-scores to determine the percent of the employees that receive an hourly wage of more than $8.25? Explain.

5. The receiving department of a chain of grocery stores received a shipment of butternut squash. A sample of 50 squash had the following weights:

   3.0, 2.7, 4.2, 3.7, 3.5, 2.6, 4.8, 4.0, 3.1, 3.3, 3.1, 3.2, 3.5, 3.2, 3.8, 3.5, 3.1, 1.9, 2.6, 4.2, 3.7, 3.3, 3.0, 3.7, 3.3, 3.1, 3.2, 1.9, 3.6, 3.3, 3.1, 3.8, 3.6, 3.9, 3.4, 2.7, 3.9, 4.4, 2.1, 4.5, 3.3, 3.6, 3.8, 3.2, 4.5, 4.1, 3.3, 3.6, 2.6, 3.5.

   a. Find the mean weight of the sample.

   b. Find, to the nearest thousandth, the standard deviation of the sample.

   c. Find, to the nearest thousandth, the standard deviation of the population.

   d. Find the 95% confidence interval for the mean weight of the squash in the shipment.

6. A sample of 30 pieces of sewing thread were tested for breaking strength. The sample mean was 12.5 ounces and the standard deviation was estimated at 2.8 ounces. Find, to the nearest tenth of an ounce, the 98% confidence interval for the mean breaking strength of the thread.

7. A school principal compares the semester average and number of days absence of a sample of 100 students to establish a relationship between frequent absence and poor grades. The results are shown in the chart. Determine if the table indicates a relationship.

Absences	Average Grade		
	Above 85	70 – 85	Below 70
More Than 15	1	5	19
5 – 15	6	12	7
Less Than 5	16	22	12

> **Bonus** A set of data consists of 3, 17, 20, and two other numbers. If the mean of the five numbers is 12 and the standard deviation is 6, find the missing data points.

Answers to Suggested Test Items

1. a.

```
6 | 1 3
5 | 1 4 9
4 | 0 5 7
3 | 0 1 1 1 3 4 4 7 7
2 | 2 3 5 6 7 7 8 9
```

b. 37   **c.** 33   **d.**

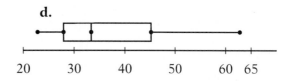

e. 31   **f.** 42.5   **g.** 64 percentile; 26 percentile   **h.**

i. $\mu = 35.7$, lower by about 1.3

j. median = 33.25, higher by 0.25

2. a. $\mu = 791.7$ **b.** median = 781.9   **c.** 160

3. 4%

4. No, set of wages is probably not a standard distribution.

5. a. 3.4   **b.** 0.663   **c.** 0.617
   d. $\mu = 3.4 \pm 0.13$ or from 3.27 to 3.53

6. $\mu = 12.5 \pm 1.2$ or from 11.3 to 13.7

7. $\chi^2 = 21.586$ Since the results would occur by chance less than 0.1% of the time, the table indicates a relationship.

x_i	f_i
60-64	2
55-59	1
50-54	2
45-49	2
40-44	1
35-39	2
30-34	6
25-29	6
20-24	2

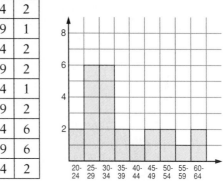

Bonus

x	$x - \mu$	$(x - \mu)^2$
3	–9	81
17	5	25
20	8	64
a	$a - 12$	$(a - 12)^2$
b	$b - 12$	$(b - 12)^2$
$\Sigma =$	$40 + a + b$	$170 + (a - 12)^2 + (b - 12)^2$

$$\frac{40 + a + b}{5} = 12$$

$$a + b = 20$$

$$\sqrt{\frac{170 + (a - 12)^2 + (b - 12)^2}{5}} = 6$$

$$170 + (a - 12)^2 + (b - 12)^2 = 180$$

$$(a - 12)^2 + (b - 12)^2 = 10$$

$$a^2 - 24a + 144 + b^2 - 24b + 144 = 10$$

$$a^2 - 24a + 144 + (20 - a)^2 - 24(20 - a) + 144 = 10$$

$$a^2 - 24a + 144 + 400 - 40a + a^2 - 480 + 24a + 144 = 10$$

$$2a^2 - 40a + 198 = 0$$

$$a^2 - 20a + 99 = 0$$

$$(a - 11)(a - 9) = 0$$

$$a - 11 = 0 \quad | \quad a - 9 = 0$$
$$a = 11 \quad | \quad a = 9$$
$$b = 9 \quad | \quad b = 11$$

Answer: $a = 11$ and $b = 9$ or $a = 9$ and $b = 11$

Chapter 13

Statistics

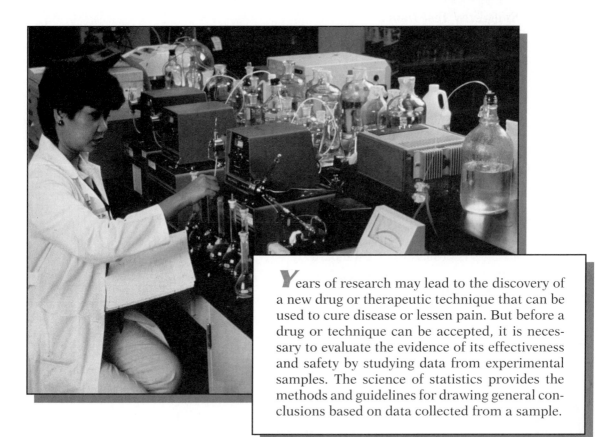

Years of research may lead to the discovery of a new drug or therapeutic technique that can be used to cure disease or lessen pain. But before a drug or technique can be accepted, it is necessary to evaluate the evidence of its effectiveness and safety by studying data from experimental samples. The science of statistics provides the methods and guidelines for drawing general conclusions based on data collected from a sample.

Chapter Table of Contents

13.1 *Organizing Data*

In our information-oriented world, it is important to condense and simplify facts and figures so that meanings and consequences can be understood. *Statistics* is the science that deals with the collection, organization, and interpretation of related pieces of numerical information called *data*. The procedures that are used to assemble, organize, display, and draw conclusions from a collection of data are known as *statistical methods*.

In statistical studies, information about a particular factor is collected from every member of a focus group called the *population*. Any group can act as a population. Often, the members have a common bond, for example, all students at a particular college, the employees of a business, or the users of a particular credit card. The population may be small, such as the eight members of a bridge club, or very large, such as all 18-year-old citizens of the United States.

Stem-and-Leaf Diagram

A *distribution* is the set of data about a particular factor from all members of a group. For example, the distribution shown consists of the scores of the members of a class of students on a recent test. To better understand these scores, it is desirable to place them in order. There are several statistical methods used to display a set of raw data in an ordered arrangement. One such way is called a *stem-and-leaf diagram*.

Test Scores

87, 54, 66, 72, 86,
77, 82, 75, 81, 78,
85, 69, 62, 73, 77,
88, 67, 93, 84, 75,
88, 82, 57, 90, 73.

To construct a stem-and-leaf diagram for the test scores shown, begin by choosing a part of the scores to serve as the stems. Since all the scores are two-digit numbers, the tens digits of the numbers are convenient stems. Then the units digits will be the leaves. Continue the construction as follows.

(1) List the stems under one another with a vertical line to the right.

(2) List each score by writing its leaf (the units digit) following the appropriate value on its stem (the tens digit). For example, 87 is listed by writing 7 to the right of the vertical line after its stem, 8.

(3) Continue to add the other scores to the diagram until all are entered.

(4) Arrange the entries in order, to further organize the data.

```
(1)        (2)        (3)
9|         9|         9| 3 0
8|         8| 7       8| 7 6 2 1 5 8 4 8 2
7|         7|         7| 2 7 5 8 3 7 5 3
6|         6|         6| 6 9 2 7
5|         5|         5| 4 7

                      (4)
                      9| 0 3
                      8| 1 2 2 4 5 6 7 8 8
                      7| 2 3 3 5 5 7 7 8
                      6| 2 6 7 9
                      5| 4 7
```

The finished stem-and-leaf diagram displays the data in order and shows, without loss of the original values, the relative number of students whose scores were in each of the decades. Thus, the diagram shows that most students scored in the 70's and 80's, that the highest score was 93, and the lowest score was 54.

Example 1

A movie-theater operator has added an afternoon showing to the schedule during the first two weeks of the summer. The number of persons who attended these showings each day is listed below.

112, 156, 93, 113, 121, 138, 127, 116, 121, 117, 102, 119, 131, 147.

Display the data in a stem-and-leaf diagram.

Solution

Use the first two digits of each number for the stem. Think of the lowest number, 93, as 093 and let the stems range from 09 to 15.

The stems are: The completed diagram is:

15	15 | 6
14	14 | 7
13	13 | 1 8
12	12 | 1 1 7
11	11 | 2 3 6 7 9
10	10 | 2
09	09 | 3

Frequency Table and Histogram

A *frequency table* is another method of organizing data. The *frequency* of a data value is the number of times that that value occurs in the distribution. The following example shows how to construct a frequency table and a corresponding frequency graph called a *histogram*.

Example 2

The owner of a small firm, who allows each employee to take a maximum of 10 paid absences due to illness each year, recorded the number of such days used by each employee in the past year as follows:

3, 6, 7, 8, 6, 10, 10, 3, 5, 2, 5, 2, 1, 9, 7, 10, 2, 0, 1, 0, 4, 4, 2, 2, 7, 4, 0, 10, 0, 2, 1, 0, 4, 10, 4, 6, 2, 1, 5, 1.

Display the data in a frequency table and in a histogram.

Solution

(1) List all possible values, in this case, 0 through 10.

(2) For each employee, place a tally mark in the row that corresponds to the number of paid absences. Group the tally marks in sets of 5 by drawing the 5th tally across the 4 preceding tallies.

(3) Count the tallies to obtain the frequency.

(4) Draw a histogram. For each value in the set of data, draw a vertical bar whose height represents the frequency of that value.

Frequency Table

# of days	Tally	Frequency
10	$\cancel{IIII}$	5
9	/	1
8	/	1
7	///	3
6	///	3
5	///	3
4	$\cancel{IIII}$	5
3	//	2
2	$\cancel{IIII}$ //	7
1	$\cancel{IIII}$	5
0	$\cancel{IIII}$	5

Histogram

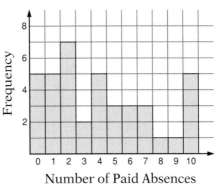

Number of Paid Absences

Graphing calculators and computer spreadsheets will display data in the form of a histogram. The statistical functions of a graphing calculator are accessed by entering [2nd] [STAT]. After these keys are pressed, press [▶] twice to highlight **DATA**. To clear any previously entered data, use [▼] to highlight **2:ClrStat** and then press [ENTER] twice. Now use the following sequence of keys to return to the **STAT** menu and choose **1:Edit** to enter the data from Example 2.

The display indicates that each entry consists of two values. Enter the number of days as the x-value and the frequency as the y-value. To begin recording the data, enter 0 as x_1 and 5 as y_1. As soon as the first of these values is entered, a new pair of variables, x_2 and y_2, appears on the screen. Enter 1 as x_2 and 5 as y_2. Continue until all 11 pairs have been entered.

To draw the histogram, first set the range of x values from 0 to 11 and the range of y values from 0 to 8. The xscl and yscl should be 1. Then press the following keys:

The calculator will display the histogram shown in Example 2. When the cursor keys are used to move to the top of each bar, the frequency is displayed as the *y*-value.

The raw data could have been entered into the calculator as 40 individual data values with the frequency of 1 for each. The calculator would have combined the frequencies for each time the same value was entered and displayed the correct histogram.

Grouped Data

When the *range* of a distribution, the difference between the largest and smallest numbers, is large, it is often useful to organize the data in a frequency table that shows intervals of scores rather than individual scores.

For example, the intervals that contain the scores of 250 students on a final examination in math are given in the table shown. Note that unlike the stem-and-leaf diagram, the table does not display the raw data from which the table is made, but displays the number of scores that occur in each interval. Thus, a large amount of data is shown in a relatively compact way.

Score	Frequency
96 – 100	8
91 – 95	12
86 – 90	38
81 – 85	62
76 – 80	43
71 – 75	36
66 – 70	33
61 – 65	12
56 – 60	2
51 – 55	3
46 – 50	1

When organizing data into intervals, choose enough intervals (at least 5) to allow the character of the distribution to be maintained, but a small enough number of intervals (no more than 15) to allow for a convenient representation. The intervals must be equal in size and include all of the values of the given data.

When data is obtained by measurement, the values are often rounded to a convenient degree of accuracy. For example, examination scores are given to the nearest integer. Thus, a score of 46 can be any score in the interval [45.5, 46.5). The interval 46 – 50 includes all values in the interval [45.5, 50.5) and the next interval, 51 – 55, includes all values in the interval [50.5, 55.5). Note that the upper boundary of 46 – 50 and the lower boundary of 51 – 55 is the midpoint between the endpoints 50 and 51.

Example 3

Given the weights of the 100 students in the senior class of Wayne Central High School, organize the data in a grouped frequency table.

130, 154, 143, 109, 155, 166, 183, 194, 145, 187,
203, 173, 109, 131, 179, 105, 153, 154, 127, 117,
120, 131, 97, 102, 126, 172, 158, 137, 103, 115,
112, 107, 163, 143, 131, 103, 125, 182, 127, 192,
198, 112, 184, 153, 107, 205, 162, 138, 144, 137,
133, 161, 163, 142, 144, 144, 116, 156, 145, 169,
116, 135, 117, 117, 128, 132, 174, 140, 182, 193,
131, 127, 221, 118, 196, 156, 121, 117, 167, 130,
122, 119, 177, 162, 154, 129, 114, 162, 150, 161,
106, 126, 142, 143, 121, 118, 175, 132, 148, 148.

Solution

The range of weights is from 97 to 221, or a difference of 124. Consider intervals of 10 starting with 95-104; there are 13 such intervals. Begin by making a tally mark for each weight, taking them in the order in which they are listed in the given data.

Weight	Tally	Frequency
215 – 224	/	1
205 – 214	/	1
195 – 204	///	3
185 – 194	////	4
175 – 184	ЖЖ //	7
165 – 174	ЖЖ /	6
155 – 164	ЖЖ ЖЖ /	11
145 – 154	ЖЖ ЖЖ	10
135 – 144	ЖЖ ЖЖ ///	13
125 – 134	ЖЖ ЖЖ ЖЖ //	17
115 – 124	ЖЖ ЖЖ ////	14
105 – 114	ЖЖ ////	9
95 – 104	////	4

Note that the use of 10 as the size of each interval made it easy to determine the endpoints of each interval since each can be found from that of the previous interval by increasing the tens digit by 1. However, this may not be the best way of organizing the data.

Often, when data is grouped, it is necessary to work with the midpoint of an interval, which is an integer when the size of the interval is an odd number, but contains a fraction when the size of the interval is an even number. Also, it may turn out that the data is better organized using a smaller number of intervals.

The same set of data might be organized into intervals of width 15 as follows:

Weight	Tally	Frequency
215 – 229	/	1
200 – 214	//	2
185 – 199	ЖЖ /	6
170 – 184	ЖЖ ЖЖ	10
155 – 169	ЖЖ ЖЖ ////	14
140 – 154	ЖЖ ЖЖ ЖЖ ////	19
125 – 139	ЖЖ ЖЖ ЖЖ ЖЖ /	21
110 – 124	ЖЖ ЖЖ ЖЖ //	17
95 – 109	ЖЖ ЖЖ	10

After data has been organized by intervals, certain information can be extracted. For example, from the frequency table on the previous page, the number of people with weights that are less than 110 pounds can be found by observing the frequency of the lowest group, 95 – 109, which is 10. The number of persons who weigh less than 125 pounds is the sum of the frequencies in the last two groups, 10 + 17 or 27.

In general, the number of people who weigh less than the lower endpoint of any group is the sum of the frequencies of all of the groups with smaller endpoints. These sums are called *cumulative frequencies*.

The table below shows the cumulative frequency for the weights given in Example 3 when organized into intervals of size 15.

Weight	Frequency	Cumulative Frequency
215 – 229	1	100
200 – 214	2	99
185 – 199	6	97
170 – 184	10	91
155 – 169	14	81
140 – 154	19	67
125 – 139	21	48
110 – 124	17	27
95 – 109	10	10

In this table, the interval with the lowest endpoints appears at the bottom of the listing. The table could also have been constructed with the intervals in increasing order, with the lowest endpoints at the top of the column. In either case, the calculation of cumulative frequency starts from the interval with the lowest endpoints. Note that the cumulative frequency for the interval with the highest endpoints is always the total number of scores in the distribution.

Exercises 13.1

In **1 – 8**, organize the set of data by using a stem-and-leaf diagram.

1. The cost, in dollars, of a gallon of 2% milk in 15 different stores:

2.54, 2.39, 2.32, 2.42, 2.55, 2.48, 2.54, 2.65, 2.56, 2.75, 2.60, 2.49, 2.45, 2.65, 2.69.

2. The grades on a history test:

96, 90, 85, 84, 74, 66, 78, 87, 53, 82, 73, 68, 91, 67, 89, 84, 77, 87, 85, 76, 87, 72, 75, 91, 77.

3. The weights of the members of a football squad:

189, 177, 212, 215, 197, 182, 227, 176, 201, 203, 196, 205, 196, 186, 247, 226, 207, 216, 193, 210, 227, 235.

4. The price, in cents, of a head of lettuce in the same store for 13 consecutive weeks:

98, 95, 85, 49, 75, 95, 90, 49, 72, 88, 87, 54, 87.

1. 2.7 | 5
2.6 | 0 5 5 9
2.5 | 4 4 5 6
2.4 | 2 5 8 9
2.3 | 2 9

2. 9 | 0 1 1 6
8 | 2 4 4 5 5 7 7 7 9
7 | 2 3 4 5 6 7 7 8
6 | 6 7 8
5 | 3

3. 24 | 7
23 | 5
22 | 6 7 7
21 | 0 2 5 6
20 | 1 3 5 7
19 | 3 6 6 7
18 | 2 6 9
17 | 6 7

4. 9 | 0 5 5 8
8 | 5 7 7 8
7 | 2 5
6 |
5 | 4
4 | 9 9

5. 7 | 10 25
6 | 24 42 51 90 97
5 | 05 22 30 61
4 | 10 17 43 76 86
3 | 18 33 69 92

6. 34 | 050
33 | 300 650
32 | 000 100 400 400 500 720 780

Exercises 13.1 *(continued)*

5. The scores, on a college aptitude test, of the members of the senior class of Euclid High School:

505, 642, 443, 697, 392, 410, 561, 318, 624, 417, 333, 476, 530, 690, 710, 651, 369, 486, 725, 522.

6. The salaries, in dollars, of teachers in an elementary school:

32,100; 32,000; 32,400; 32,720; 33,650; 32,400; 32,500; 33,300; 32,780; 34,050.

7. The average cost, in dollars, of a gallon of gasoline in 10 cities during a given month:

City	Cost	City	Cost
Los Angeles	1.05	Baton Rouge	1.12
Baltimore	1.08	Fairmont	1.40
Philadelphia	1.15	Dayton	1.17
St. Louis	1.20	Milwaukee	1.32
San Diego	1.09	Minneapolis	1.29

8. The earned run average (E.R.A.) of Dale Murray, a pitcher for the New York Yankees, for the 13 years of his career:

Year	E.R.A.	Year	E.R.A.	Year	E.R.A.
1972	2.42	1977	4.94	1981	1.85
1973	4.26	1978	3.78	1982	3.16
1974	1.47	1979	4.58	1983	4.48
1975	3.97	1980	1.64	1984	5.94
1976	3.27				

In **9 – 12**, organize the data into a frequency table and draw a histogram.

9. Number of family pets owned by students in a 5th grade:

1, 2, 2, 2, 1,
3, 1, 3, 0, 0,
0, 1, 5, 1, 0,
0, 0, 2, 2, 3,
1, 1, 3, 1, 4.

10. Number of books read by students during the summer vacation:

3, 5, 0, 5, 0, 0, 1, 4, 2, 8,
5, 8, 2, 7, 4, 1, 2, 10, 5, 1,
1, 2, 0, 9, 1, 7, 9, 8, 3, 3,
0, 5, 6, 10, 1, 2, 5, 1, 10, 1
7, 1, 3, 8, 5, 0, 1, 6, 9, 4,
12, 4, 10, 2, 2, 3, 0, 0, 3, 3.

7.

1.4	0
1.3	2
1.2	0 9
1.1	2 5 7
1.0	5 8 9

8.

5	.94
4	.26 .48 .58 .94
3	.16 .27 .78 .97
2	.42
1	.47 .64 .85

9.

x_i	f_i
5	1
4	1
3	4
2	5
1	8
0	6

10.

x_i	f_i
12	1
11	0
10	4
9	3
8	4
7	3
6	2
5	7
4	4
3	7
2	7
1	10
0	8

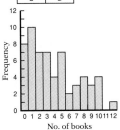

11.

x_i	f_i
149 – 151	1
146 – 148	1
143 – 145	3
140 – 142	11
137 – 139	14
134 – 136	8
131 – 133	6
128 – 130	4
125 – 127	2

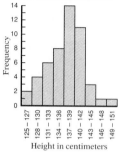

Height in centimeters

12.

x_i	f_i
41 – 45	4
36 – 40	5
31 – 35	1
26 – 30	4
21 – 25	2
16 – 20	3
11 – 15	5
6 – 10	5
1 – 5	1

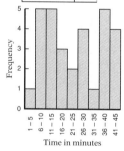

Time in minutes

Exercises 13.1 (continued)

11. Heights, in centimeters, of 50 ten-year-old children:

137, 134, 134, 134, 130, 126, 144, 137,
131, 135, 141, 149, 136, 136, 140, 141,
137, 139, 129, 139, 139, 139, 137, 133,
127, 129, 147, 142, 143, 133, 132, 135,
132, 142, 138, 137, 141, 143, 131, 135, 140,
140, 138, 142, 137, 138, 141, 142, 129, 139.

12. Length of time, in minutes, required by employees of Cimco to commute to work:

15, 38, 22, 10, 25, 44,
4, 40, 38, 42, 10, 12,
20, 8, 36, 27, 37, 27,
27, 33, 45, 17, 45, 12,
30, 18, 8, 12, 10, 14.

13. The table shows the frequency distribution of high temperatures for a city during the month of July. Find the cumulative frequency and answer the questions.

   a. What are the boundaries of the intervals used?

   b. On how many days was the temperature above 80?

   c. On how many days was the temperature below 71?

   d. On how many days was the high temperature between 71 and 80?

   e. In which interval is each of these high temperatures located?
   (1) 85.3°   **(2)** 80.5°

Temperature	Frequency
91 – 95	2
86 – 90	5
81 – 85	12
76 – 80	8
71 – 75	3
66 – 70	1

14. To determine the number of students born in a particular month, a sample of 100 students from a school of 1,200 students was chosen at random. The month that each was born was tallied in a frequency distribution as shown.

   a. What percent of the students was born in each month?

   b. If this data is used to estimate the number of students in the school born during each of the months, how many of the students would you estimate were born in each of these months?
   (1) January   **(2)** April   **(3)** September

Month of Birth	Frequency
January	7
February	7
March	5
April	8
May	6
June	8
July	9
August	11
September	12
October	6
November	9
December	12

---

13. a. Boundaries
   90.5 – 95.5
   85.5 – 90.5
   80.5 – 85.5
   75.5 – 80.5
   70.5 – 75.5
   65.5 – 70.5

b. 19
c. 1
d. 11
e. (1) 81 – 85
   (2) 81 – 85

14. a.

Jan.	7%	July	9%
Feb.	7%	Aug.	11%
Mar.	5%	Sept.	12%
Apr.	8%	Oct.	6%
May	6%	Nov.	9%
June	8%	Dec.	12%

b. (1) 84
   (2) 96
   (3) 144

Exercises 13.1 (continued)

15. Given the height, in inches, of 50 students in a local college, display the data in a grouped frequency table using intervals beginning with 48 – 52.

71, 64, 50, 62, 60,
73, 70, 69, 74, 80,
67, 83, 69, 70, 68,
53, 64, 71, 72, 73,
58, 62, 67, 69, 49,
65, 79, 70, 74, 71,
62, 78, 65, 69, 70,
61, 64, 71, 66, 68,
71, 68, 68, 70, 72,
73, 63, 57, 69, 68.

16. a. Reconstruct the group frequency table in Exercise 15 by using intervals of 4 inches beginning with 47 – 50.

  b. How many students are at least 5'7" tall but not taller than 5'10"?

  c. What percent of the students are at least 5'7" tall but not taller than 5'10"?

  d. Add a cumulative frequency column to the frequency table.

  e. According to the table, how many students are shorter than 5'3"?

17. In an American city, 560 adults were surveyed about the number of hours per week that they exercise. The results are summarized in the frequency table shown.

  a. Add a cumulative frequency column to the table.
  b. How many of the people in the group do exercise 5 or more hours per week?
  c. How many exercise less than 4 hours per week?
  d. Draw a histogram for the data.
  e. If you have access to a calculator that can draw a histogram, compare your graph to the one drawn by the calculator.
  f. Suppose each person in the group exercised one extra hour per week. How would the appearance of the histogram change?

Hours of Exercise	Frequency
0	50
1	65
2	32
3	41
4	65
5	75
6	80
7	35
8	22
9	28
10	30
11	22
12	15

In **18 – 23**, survey 25 persons to obtain the required data.
  a. Organize the data in a stem-and-leaf diagram.
  b. Organize the data in a frequency table.
  c. Draw a histogram.

18. Circumference of the right wrist, in cm.

19. Length of the little finger on the left hand, in cm.

20. Number of living brothers and sisters.

21. Number of hours spent watching TV during a 7-day period.

22. Number of meals eaten outside one's home in the past 7 days.

23. Estimate of the weight in grams of a nickel.

15.

x_i	f_i
83 – 87	1
78 – 82	3
73 – 77	5
68 – 72	22
63 – 67	9
58 – 62	6
53 – 57	2
48 – 52	2

16. a.

x_i	f_i	d.
83 – 86	1	50
79 – 82	2	49
75 – 78	1	47
71 – 74	12	46
67 – 70	17	34
63 – 66	7	17
59 – 62	5	10
55 – 58	2	5
51 – 54	1	3
47 – 50	2	2

  b. 17
  c. 34%
  e. 10

17. a. Cumulative Frequency
50
115
147
188
253
328
408
443
465
493
523
545
560

18 – 23. Answers will vary.

17. b. 307   **c.** 188   **d.**

  f. Histogram shifted to the right by the width of 1 bar.

13.2 Measures of Central Tendency

After data has been collected, it is useful to be able to represent all of the data by a single number that somehow reflects the central, or typical, piece of data. Such a number is called a *measure of central tendency*. The most common measures of central tendency are *mean*, *median*, *mode*, and *midrange*.

The Mean

The *mean* of a set of n numbers is the sum of the numbers divided by n.

If x_i, $i = 1$ to n, represents the set of data, then:

$$\text{mean } \mu = \frac{1}{n}\sum_{i=1}^{n} x_i = \frac{x_1 + x_2 + \cdots + x_n}{n}$$

The mean, or arithmetic mean, is commonly called the *average*, although this term can also be applied to the other measures of central tendency. Note that the Greek letter used to represent the mean, μ, is pronounced *mew*.

Example 1

Given the dollar amounts collected by a waitress in one evening:

1.25, 1.75, 5.00, 1.50, 3.85, 2.50, 3.00, 2.45, 2.70, 6.00, 3.15, 3.40, 3.10, 4.25, 2.75.

Find the average (mean) tip.

Solution

$$\text{mean} = \frac{\text{sum of tips}}{\text{number of tips}} = \frac{46.65}{15} = 3.11$$

Answer: The average tip is $3.11.

One way to think about the mean is to observe that if each tip had been $3.11, the total amount, $46.65, would be the same.

Example 2

This frequency table shows the number of spelling errors in the essays of 25 students. Find the mean number of spelling errors.

# of Errors	Frequency
10	2
9	1
8	2
7	4
6	9
5	1
4	2
3	1
2	1
1	0
0	2

Solution

To find the total number of errors, first multiply the number of errors by the frequency in each row. For example, since 2 papers each had 10 errors, these 2 papers contribute 2 × 10, or 20, errors to the total number of errors.

# of Errors (x)	Frequency (f)	$f \cdot x$
10	2	20
9	1	9
8	2	16
7	4	28
6	9	54
5	1	5
4	2	8
3	1	3
2	1	2
1	0	0
0	2	0
Totals	25	145

Answer: The mean is $\frac{145}{25}$ = 5.8 errors.

The mean is affected by the size of every entry in the set of data. For example, if the ages of the employees of a small convenience store are 21, 25, 27, 30, and 32, the mean age is $\frac{135}{5}$ = 27. If the person of age 30 quits and is replaced by someone of age 60, the mean age is $\frac{165}{5}$ = 33. Observe how the one data point that is an extreme value compared to the others in the distribution makes a significant difference in the mean.

The Median

When a set of data is in numerical order, the *median* is the middle score. If the number of scores is odd, the median score is an actual score. If the number of scores is even, the median score is the average of two actual scores.

For example, in a set of data consisting of 5 numbers, such as 3, 4, 7, 8, 12, the third number from each end, in the given set 7, is the median. When the set of data consists of 6 numbers, such as 3, 4, 7, 8, 12, 13, the average of the third and fourth numbers, in this set of six numbers, $\frac{7+8}{2}$ = 7.5, is the median. In general:

$$\text{the median} = x_{\frac{n+1}{2}} \text{ if } n \text{ is odd}$$

$$\text{the median} = \frac{x_{\frac{n}{2}} + x_{\frac{n}{2}+1}}{2} \text{ if } n \text{ is even}$$

In every set of data, half of the entries from the set of data will be less than or equal to the median and half will be greater than or equal to the median.

Example 3 _____

In Example 2, this frequency table, showing the number of spelling errors in the essays of 25 students, was used to determine the mean. Find the median number of spelling errors.

# of Errors	Frequency
10	2
9	1
8	2
7	4
6	9
5	1
4	2
3	1
2	1
1	0
0	2

Solution

Since there are 25 data points, the median is the $\frac{25+1}{2}$, or 13th, value. By adding frequencies from the top or bottom, the 13th value is seen to be one of the nine 6's.

Answer: The median is 6.

Unlike the mean value, the median value is not affected by extreme values that are large or small compared to other values in the distribution. For example, reconsider the ages of the employees mentioned earlier, namely, 21, 25, 27, 30, and 32. The median age is 27. If the person of age 30 quits and is replaced by someone of age 60, the ages are now 21, 25, 27, 32, 60, and the median is still 27.

The Mode

In a set of data, the number that occurs most frequently is the *mode*. If all numbers have the frequency 1, the distribution has no mode. If two numbers have the same frequency and the frequency of those numbers is greater than that of any other entry, then the distribution is *bimodal*. A distribution may have any number of modes.

Example 4 _____

This frequency table shows the number of correct answers on 35 papers for 2 different spelling tests of 10 words each. Find the mode of each distribution.

	Test 1		Test 2	
# Correct	Frequency	# Correct	Frequency
10	2	10	3
9	8	9	6
8	6	8	7
7	8	7	9
6	4	6	8
5	5	5	2
4	2	4	0

Solution

For Test 1, there are two modes, 9 and 7, since both 9 and 7 have a frequency of 8, which is greater than any other frequency. This distribution is bimodal.

For Test 2, the mode is 7, since 7 has the highest frequency in the distribution.

If each value in a distribution occurs once, the distribution has no mode. For example, the set of tips in Example 1 has no mode.

Mean and Median of Grouped Data

When data is given in a grouped frequency table, the mean and median can be approximated by assuming that the raw data is equally distributed throughout each interval. The mean is found by using the midpoint of the interval as the value of each entry in the interval.

Example 5

This table shows the heights, in centimeters, of the 5th-grade children at Perkins School. Find the mean height.

Height	Frequency
151 – 155	1
146 – 150	6
141 – 145	21
136 – 140	48
131 – 135	17
126 – 130	5
121 – 125	2

Solution

Use the value of the midpoint as the height of each student in that interval.

Height	Midpoint (x_i)	Frequency (f_i)	$f_i \cdot x_i$
151 – 155	153	1	153
146 – 150	148	6	888
141 – 145	143	21	3,003
136 – 140	138	48	6,624
131 – 135	133	17	2,261
126 – 130	128	5	640
121 – 125	123	2	246
Total		$\sum f_i = n = 100$	$\sum f_i x_i = 13{,}815$

$$\text{mean } \mu = \frac{1}{n}\sum_{i=1}^{n} f_i \cdot x_i = \frac{\sum f_i x_i}{\sum f_i} = \frac{13{,}815}{100} = 138.15$$

Answer: The mean height is 138.15 cm.

To find the median of a set of data that is displayed in a grouped frequency table, first find the interval in which the middle value lies. Then determine the median by using ratios, as shown in the following example.

Example 6

An insurance company asked 30 persons to record the time necessary to fill out a claim form. The results are summarized in the table shown. Find the median number of minutes required to fill out the form.

# of Minutes	Frequency	Cumulative Frequency
56 – 60	1	30
51 – 55	2	29
46 – 50	7	27
41 – 45	12	20
36 – 40	6	8
31 – 35	1	2
26 – 30	1	1

Solution

Since the total frequency is 30, the middle value is between the 15th and 16th entries from the bottom. That number is in the interval 41 – 45.

The 12 data points in the interval 41 – 45 correspond to cumulative entries 9 through 20. Think of the points as evenly spaced throughout the interval. Recall that the lower boundary of an interval is midway between its lower endpoint and the upper endpoint of the interval that precedes it.

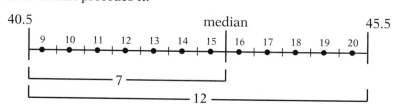

The midpoint between the 15th and 16th data points is $\frac{7}{12}$ of the distance between the endpoints of the interval. Therefore, the median is the lower endpoint plus $\frac{7}{12}$ of the interval size, that is:

$$40.5 + \frac{7}{12}(5) = 40.5 + 2.9 = 43.4$$

Answer: The median number of minutes is 43.4.

This method of finding the median of grouped data can be summarized in the following expression, where the median interval, M_c, is the class or interval in which the median lies and cf is the cumulative frequency.

$$\text{lower boundary of } M_c + \frac{\frac{n}{2} - cf \text{ for class below } M_c}{\text{frequency of } M_c} \cdot \text{interval size}$$

In Example 6: $\text{median} = 40.5 + \dfrac{\frac{30}{2} - 8}{12} \cdot 5 = 43.4$

The mode of any set of data can only be determined from the raw data, although for group data the interval with the highest frequency may be called the modal interval. In Example 6, the modal interval is 41 – 45. Note that the mode of the raw data may or may not lie in the modal interval.

Quartiles

The median of a set of data divides the data into two parts. The numbers that separate the data into four parts with an equal number of entries in each are called *quartiles*. The first, or lower, quartile is that number at or below which one-fourth of the entries lie and at or above which three-fourths of the entries lie. The lower quartile may be thought of as the median of the scores that are below the median of all the scores.

For example, consider the table below that lists in order the heights, in centimeters, of 26 children. Since the number of heights is even, the median is the average of the 13th and 14th heights.

| 54 | 58 | 60 | 65 | 65 | 66 | 68 | 70 | 72 | 72 | 74 | 75 | 80 |
| 81 | 82 | 82 | 83 | 85 | 86 | 86 | 87 | 88 | 88 | 90 | 93 | 97 |

$$\text{median} = \frac{80 + 81}{2} = 80.5$$

The lower quartile is the median of the first 13 heights, that is, 68, the 7th height.

| 54 | 58 | 60 | 65 | 65 | 66 | 68 | 70 | 72 | 72 | 74 | 75 | 80 |
| 81 | 82 | 82 | 83 | 85 | 86 | 86 | 87 | 88 | 88 | 90 | 93 | 97 |

The third, or upper, quartile is that number at or below which three-fourths of the scores lie and at or above which one-fourth of the scores lie. It is the median of the scores that lie above the median of all of the scores. In this example, the third quartile is 86, the 7th height above the median of all the scores.

| 54 | 58 | 60 | 65 | 65 | 66 | 68 | 70 | 72 | 72 | 74 | 75 | 80 |
| 81 | 82 | 82 | 83 | 85 | 86 | 86 | 87 | 88 | 88 | 90 | 93 | 97 |

The median is also called the second quartile. Thus, in this example, the quartiles, 68, 80.5, and 86 separate the data into four quarters. A score that is below 68 is in the first quartile, a score that is above 68 but below 80.5 is in the second quartile, a score above 80.5 but below 86 is in the third quartile, and a score that is above 86 is in the fourth quartile. Note that although the designation "third quartile" refers both to the value 86 and to the group of scores between 80.5 and 86, the correct meaning is clear from the context. The difference between the first and third quartile scores, 86 – 68 = 18, is the *interquartile range*.

A *whisker-box plot* is a statistical graph that uses the quartile scores to display the distribution of a set of scores. To construct a whisker-box plot, draw a scale that includes the numbers in the range of the distribution. Place a dot above the numbers that are the lowest score, the first, second and third quartiles, and the highest score.

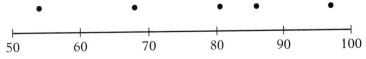

Then draw a box between the dots that represent the first and third quartiles, and a vertical line through the box at the median (second quartile).

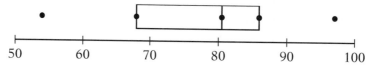

Draw the whiskers by joining the dots at the lowest score and the first quartile and the dots at the third quartile and the highest score.

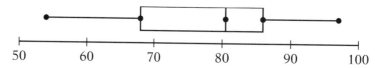

Information about the distribution of the data can be read from the display. In this case, the longer whisker at the left indicates a greater scattering of the lower marks as compared to the shorter right whisker. The larger part of the box below the median indicates a greater scattering of scores in the second quartile as compared to the third quartile.

Percentiles

The *percentile rank* of a score indicates its position relative to other scores in the distribution. For example, to say that the height of a particular girl, Sharon, is at the 45th percentile for girls of age 16 means that:

(1) all of the 16-year-old girls who are shorter than Sharon, and

(2) half of the 16-year-old girls who are the same height as Sharon

make up 45% of all 16-year-old girls.

To find the percentile rank of a score, think of the scores in three groups: those that are less than the given score, those that are equal to the given score, and those that are greater than the given score. Consider the middle of the group of equal scores and determine the percent of scores below that middle point, by adding the percent of the scores that are less than the given score to one-half of the percent of the scores that are equal to the given score. Thus:

The percentile rank of a score in a distribution is the sum of two percents:

(1) the percent of scores that are less than the given score, and

(2) half the percent of scores that are equal to the given score.

Example 7 —

This frequency table shows the number of words spelled correctly on a test of 12 words given to 25 students. Find the percentile rank of a student who spelled 9 words correctly.

# Correct	Frequency
12	1
11	0
10	2
9	5
8	7
7	6
6	3
5	1

Solution

Method 1: There are 17 tests with fewer than 9 words spelled correctly. This represents $\frac{17}{25} = \frac{68}{100} = 68\%$ of the tests. There are 5 tests that have 9 words spelled correctly. This represents $\frac{5}{25} = \frac{20}{100} = 20\%$ of the tests. The percentile rank of 9 words spelled correctly is $68 + \frac{1}{2}(20) = 78$.

Method 2: Add the number of papers that scored below 9 and half of the number of papers that scored 9, and find the ratio of this number to the total number of papers.

Number of papers with a score below 9 = 17

One-half the number of papers with a score of 9 = 2.5

Total = 19.5

$$\frac{19.5}{25} = 0.78 = 78\%$$

Answer: On this test, a student who correctly spelled 9 words is at the 78th percentile.

Midrange

The *midrange* of a set of data is the number that is the average of the largest and the smallest data points. If the set of data consists of n data points arranged in ascending or descending order, then:

$$\text{midrange} = \frac{x_1 + x_n}{2}$$

For example, if the number of correctly-spelled words on a spelling test ranges from 5 to 12, then the midrange of the distribution of correctly-spelled words is $\frac{5 + 12}{2} = \frac{17}{2} = 8.5$.

1. a. (1) 2.54
    (2) 2.54
    (3) 2.54, 2.65
    (4) 2.535
  b.

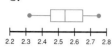

2.2 2.3 2.4 2.5 2.6 2.7 2.8

2. a. (1) 79.76
    (2) 82
    (3) 87
    (4) 74.5
  b.

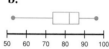

50   60   70   80   90   100

3. a. (1) 205.6
    (2) 204
    (3) 227, 196
    (4) 211.5
  b.

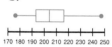

170 180 190 200 210 220 230 240 250

4. a. (1) 79
    (2) 87
    (3) 49, 87, 95
    (4) 73.5
  b.

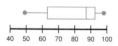

40   50   60   70   80   90   100

5. a. (1) 1.56
    (2) 1
    (3) 1
    (4) 2.5
  b.

0   1   2   3   4   5

Exercises 13.2

In **1 – 8**, using ungrouped data:

a. Find: **(1)** the mean **(2)** the median **(3)** the mode if one exists **(4)** the midrange

b. Draw a whisker-box plot for the data.

1. The cost, in dollars, of a gallon of skim milk in 15 different stores:

    2.54, 2.39, 2.32, 2.42, 2.55,
    2.48, 2.54, 2.65, 2.56, 2.75,
    2.60, 2.49, 2.45, 2.65, 2.69.

2. The grades on a history test:

    96, 90, 85, 84, 74, 66, 78, 87,
    53, 82, 73, 68, 91, 67, 89, 84,
    77, 87, 85, 76, 87, 72, 75, 91, 77.

3. The weights, in pounds, of the members of a football squad:

    189, 177, 212, 215, 197, 182, 227,
    176, 201, 203, 196, 205, 196, 186,
    247, 226, 207, 216, 193, 210, 227, 235.

4. For 13 consecutive weeks in the same store, the price, in cents, of a head of lettuce:

    98, 95, 85, 49, 75, 95, 90,
    49, 72, 88, 87, 54, 87.

5. The number of family pets owned by students in a 5th grade:

    1, 2, 2, 2, 1,
    3, 1, 3, 0, 0,
    0, 1, 5, 1, 0,
    0, 0, 2, 2, 3,
    1, 1, 3, 1, 4.

6. The length of time, in minutes, required by employees of Ascoa to commute:

    15, 38, 22, 10, 25, 44,
    4, 40, 38, 42, 10, 12,
    20, 8, 36, 27, 37, 27,
    27, 33, 45, 17, 45, 12,
    30, 18, 8, 12, 10, 14.

7. The number of books read by students in Class 4A during the summer:

    3, 5, 0, 5, 0, 0, 1, 4, 2, 8,
    5, 8, 2, 7, 4, 1, 2, 10, 5, 1,
    1, 2, 0, 9, 1, 7, 9, 8, 3, 3,
    0, 5, 6, 10, 1, 2, 5, 1, 10, 1,
    7, 1, 3, 8, 5, 0, 1, 6, 9, 4,
    12, 4, 10, 2, 2, 3, 0, 0, 3, 3.

8. The heights, in centimeters, of 50 ten-year-old children:

    137, 134, 134, 134, 130, 126, 144,
    137, 131, 135, 141, 149, 136, 136,
    140, 141, 137, 139, 129, 139, 139,
    139, 137, 133, 127, 129, 147, 142,
    143, 133, 132, 135, 132, 142, 138,
    137, 141, 143, 131, 135, 140, 140,
    138, 142, 137, 138, 141, 142, 129, 139.

9. a. Organize the data from Exercise 8 into groups, beginning with 125 – 129, and find the mean and median.

  b. Compare the results with those found using the ungrouped data.

---

6. a. (1) 24.2
    (2) 23.5
    (3) 10, 12, 27
    (4) 24.5
  b.

0 10 20 30 40 50

7. a. (1) 4
    (2) 3
    (3) 1
    (4) 6
  b.

0 2 4 6 8 10 12

8. a. (1) 137
    (2) 137
    (3) 137
    (4) 137.5
  b.

125 130 135 140 145 150

9. a. mean = 136.8
    median = 137.1

  b. mean differs by 0.2,
    median differs by 0.1

Exercises 13.2 *(continued)*

10. A survey was done comparing the ages of the employees in a firm of 30 workers in order to develop a group health insurance plan. The following data was collected.

28, 38, 41, 26, 21, 54, 19, 62, 49, 53, 51, 43, 44, 39, 27, 36, 32, 37, 41, 50, 29, 31, 35, 46, 56, 64, 43, 61, 47, 51.

a. Find the median.
b. Find the lower and upper quartiles.
c. Draw a whisker-box plot of the ages.

11. The monthly salaries of 100 employees of the ABC company are listed in the frequency table.

Monthly Salary in Dollars	Frequency
4,001 – 5,000	10
3,001 – 4,000	24
2,001 – 3,000	35
1,001 – 2,000	23
1 – 1,000	8

a. Find the median monthly salary to the nearest dollar.
b. Find the mean salary of the 100 employees.

12. The batting averages for the nine players on a baseball team are:

.190, .210, .220, .235, .248, .261, .274, .290, .324.

a. Calculate the median batting average.
b. Calculate the mean of the batting averages.
c. Is the mean for the batting averages the same as the team batting average? Explain. (Assume that only these nine players are on the team and each stays in every game.)

13. The following scores were obtained by 11 students who took a quiz with a maximum grade of 10:

6, 7, 6, 8, 10, 5, 4, 2, 8, 10, 10.

a. Compute the mode.
b. Compute the median.
c. Compute the midrange.
d. Which of these statistical measures is the best measure of central tendency for these scores?
e. Compute the first and third quartiles.
f. Draw a whisker-box plot.

In **14 – 21**, find the percentile rank of the data value for the data in the given exercise.

14. $2.45 in Exercise 1
15. 85 in Exercise 2
16. 196 lb. in Exercise 3
17. $.54 in Exercise 4
18. 1 pet in Exercise 5
19. 20 minutes in Exercise 6
20. 4 books in Exercise 7
21. 138 cm in Exercise 8

10. a. 42
   b. 32, 51
   c.

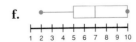

10 20 30 40 50 60 70

11. a. $2,543
   b. $2,550.50

12. a. 0.248
   b. 0.250
   c. May not be the same because each player may not have had the same number of times at bat. The team batting average would be the weighted average of the individual batting averages.

13. a. 10   **b.** 7
   c. 6   **d.** median
   e. 5 and 10
   f.

1 2 3 4 5 6 7 8 9 10

14. 23rd percentile
15. 64th percentile

16. 32nd percentile
17. 19th percentile
18. 40th percentile
19. 45th percentile
20. 57th percentile
21. 55th percentile

Exercises 13.2 *(continued)*

22. The 135 seniors at Irondequoit High School were asked the number of movies that they watched on television during the past week. The results were summarized in the table shown.

a. Find:

    (1) the mean

    (2) the median

    (3) the mode

b. Find the percentile rank:

    (1) of 1

    (2) of 4

    (3) of 7

# of Movies	Frequency
8	3
7	5
6	7
5	10
4	23
3	37
2	28
1	12
0	10

23. This frequency table shows the IQ's of students in Park School.

a. Find the mean.

b. Find the median.

IQ	Frequency
136 – 145	12
126 – 135	42
116 – 125	96
106 – 115	189
96 – 105	312
86 – 95	157
76 – 85	102
66 – 75	35

24. One of the factors used by a college to determine who will be admitted is the standardized math score of the applicant. The scores are placed in categories A – F, as shown. Only those students that are in the 65th percentile or above are considered for admission. The data was listed for 500 applicants.

a. Assuming that all of the students in a given category are equally qualified, find the percentile rank of a student in:

    (1) Category A

    (2) Category B

    (3) Category F

b. In what categories must their scores lie for students to be considered for admission?

Category	Math Score	Frequency
A	701 – 800	35
B	601 – 700	65
C	501 – 600	150
D	401 – 500	179
E	301 – 400	61
F	200 – 300	10

22. a. (1) 3.13

    (2) 3

    (3) 3

  b. (1) 12th percentile

    (2) 73rd percentile

    (3) 96th percentile

23. a. 101.44

  b. 101.22

24. a. (1) 96.5th percentile

    (2) 86.5th percentile

    (3) 1st percentile

  b. A, B, and C

13.3 Measures of Dispersion

Consider the following two distributions of 5 numbers:

(1) | 5 | 8 | 19 | 27 | 36 | Mean 19 | Median 19

(2) | 17 | 18 | 19 | 20 | 21 | Mean 19 | Median 19

Note that both distributions have the same mean and median, but the distributions are very different. Thus, it is important to have some ways of indicating how the individual numbers vary. One way to point out the difference in the natures of these distributions is to use the range, the difference between the largest and smallest values. Indeed, the two ranges are decidedly different: 36 – 5, or 31, for the first distribution and 21 – 17, or 4, for the second.

Another useful measure indicates the variation of the numbers within the range. The method involves measuring the difference between each element of a distribution and the mean for that distribution.

Consider a record of absences of a class of 40 students in a 3-month period. The first column, x_i, shows a number of absences, and the second column, f_i, shows the frequency, that is, the number of students absent that many times. The difference between the mean and an actual number of days is recorded as $x_i - \mu$.

Calculate the mean: $\mu = \dfrac{1}{n}\sum_{i=1}^{n} f_i \cdot x_i = \dfrac{160}{40} = 4$

x_i	f_i	$f_i \cdot x_i$	$x_i - \mu$	$f_i(x_i - \mu)$
10	5	50	6	30
9	1	9	5	5
8	1	8	4	4
7	3	21	3	9
6	2	12	2	4
5	3	15	1	3
4	5	20	0	0
3	2	6	–1	–2
2	7	14	–2	–14
1	5	5	–3	–15
0	6	0	–4	–24
Totals	40	160		0

Note that when the differences $x_i - \mu$ are multiplied by the frequency, the total of the 40 differences is 0. This is a property of the mean that is true for all distributions.

$$\sum_{i=1}^{n} f_i(x_i - \mu) = 0$$

The differences, called *deviations from the mean*, are used to find other statistical measures. One measure involves taking the absolute value of each difference, so that their sum is not 0. The average of all the absolute values is called the mean absolute deviation.

A measure that proves to be more useful is one that uses the square of the deviation from the mean. This gives greater weight to the numbers that are farther from the mean. The average of the squares of the deviations from the mean is called the *variance*, written σ^2. Note that this Greek letter is the lower case "sigma".

x_i	f_i	$x_i - \mu$	$(x_i - \mu)^2$	$f_i(x_i - \mu)^2$
10	5	6	36	180
9	1	5	25	25
8	1	4	16	16
7	3	3	9	27
6	2	2	4	8
5	3	1	1	3
4	5	0	0	0
3	2	−1	1	2
2	7	−2	4	28
1	5	−3	9	45
0	6	−4	16	96

$$\text{Variance } \sigma^2 = \frac{1}{n}\sum_{i=1}^{n} f_i(x_i - \mu)^2$$

$$= \frac{430}{40} = 10.75$$

Although the variance is a useful measure, it is in square units. In order to have a measure that is in the same unit as the given data, find the square root of the variance. The square root of the variance is called the *standard deviation*.

$$\text{Standard Deviation } \sigma = \sqrt{\frac{1}{n}\sum_{i=1}^{n} f_i(x_i - \mu)^2}$$

$$= \sqrt{10.75} \approx 3.2787$$

The standard deviation of a set of grouped data can be found by using the midpoint of each interval as the score for all entries in that interval, just as was done when finding the mean.

Example 1

Find the standard deviation of the weights of the seniors of Wayne Central High School that were given in Section 13.1, Example 3, and are shown in the first two columns of the table below.

Solution

First calculate the mean: $\mu = \frac{1}{n}\sum_{i=1}^{n} f_i \cdot x_i = \frac{14,400}{100} = 144$

Weight	f_i	Midpoint (x_i)	$f_i \cdot x_i$	$x_i - \mu$	$(x_i - \mu)^2$	$f_i(x_i - \mu)^2$
215 – 229	1	222	222	78	6,084	6,084
200 – 214	2	207	414	63	3,969	7,938
185 – 199	6	192	1,152	48	2,304	13,824
170 – 184	10	177	1,770	33	1,089	10,890
155 – 169	14	162	2,268	18	324	4,536
140 – 154	19	147	2,793	3	9	171
125 – 139	21	132	2,772	−12	144	3,024
110 – 124	17	117	1,989	−27	729	12,393
95 – 109	10	102	1,020	−42	1,764	17,640
Total	100		14,400			76,500

$$\sigma = \sqrt{\frac{1}{n}\sum_{i=1}^{n} f_i(x_i - \mu)^2} = \sqrt{\frac{76{,}500}{100}} = \sqrt{765} = 27.659$$

Answer: The standard deviation is 27.659 pounds.

Z-Scores

The standard deviation is useful in making comparisons between data obtained using different scales. One basis for comparison is how the data is distributed about the mean. A measure called a z-score gives the number of standard deviations of a score above or below the mean.

The z-score of a given data point is the quotient of the difference between the data point and the mean, divided by the standard deviation, that is

$$z = \frac{\text{data point} - \text{mean}}{\text{standard deviation}}$$

or in symbols:

$$z = \frac{x - \mu}{\sigma}$$

For example, in the distribution in Example 1, the z-score of a weight of 130 pounds is

$$\frac{130 - 144}{27.659} \approx -0.506$$

If a z-score is positive, the score is above the mean;
if a z-score is negative, the score is below the mean.

Example 2

Bert's teacher always records marks as z-scores. On a recent test, the mean grade for the class was 75 with a standard deviation of 8. Bert's z-score for that test was 1.5. What was his grade on the test?

Solution

$$z = \frac{x - \mu}{\sigma}$$
$$1.5 = \frac{x - 75}{8}$$
$$1.5(8) + 75 = x$$
$$87 = x$$

Answer: Bert's grade on the test was 87.

When the numbers in a set of data are converted to z-scores, the mean is 0 and the standard deviation is 1. In Section 13.4, z-scores will be used to determine the frequency with which a range of scores is expected to occur in a set of data.

1. **a.** 0.0130
   b. 0.1138
2. **a.** 94.42
   b. 9.72
3. **a.** 342
   b. 18.5
4. **a.** 286.8
   b. 16.93
5. **a.** 1.77
   b. 1.33
6. **a.** 164.3
   b. 12.8
7. **a.** 10.7
   b. 3.27
8. **a.** 24.72
   b. 4.97
9. 2
10. –2
11. 2.5
12. –2
13. $\frac{2}{3} = 0.\overline{6}$
14. $-\frac{6}{5} = -1.2$
15. **a.** 3.295
    b. 1.815
16. **a.** 213.187
    b. 14.601
    c. **(1)** 0.586
       (2) –0.784
       (3) 1.477
17. 55
18. 7.8
19. **a.** changes by the value of the constant
    b. no change
    c. no change
20. $\mu = 78;\ \sigma = 5$

Exercises 13.3

In **1 – 8**, use the data sets given in Exercises 1 – 8 on page 610 (Section 13.2) and:
  a. Find the variance.
  b. Find the standard deviation.

In **9 – 14**, find the z-score for the value of a data point x for the given mean and standard deviation.

9. $x = 40,\ \mu = 34,\ \sigma = 3$ **10.** $x = 4,\ \mu = 7,\ \sigma = 1.5$ **11.** $x = 125,\ \mu = 100,\ \sigma = 10$

12. $x = 75,\ \mu = 90,\ \sigma = 7.5$ **13.** $x = 208,\ \mu = 200,\ \sigma = 12$ **14.** $x = 19,\ \mu = 25,\ \sigma = 5$

15. Refer to the data of Exercise 22 on page 612 (Section 13.2) and:
  a. Find the variance.
  b. Find the standard deviation.

16. Refer to the data set of Exercise 23 on page 612 (Section 13.2) and:
  a. Find the variance.
  b. Find the standard deviation.
  c. Find the z-scores of the following IQ's.
     (1) 110 **(2)** 90 **(3)** 123

17. Find, to the nearest integer, the score that a student obtained on a standardized test if the z-score was 1.5, the mean was 46, and the standard deviation was 5.8.

18. In a class test, the mean was 76 and a student with a grade of 97 had a z-score of 2.7. Find, to the nearest tenth, the standard deviation of the data.

19. A constant is added to each number in a set of data.
  a. How does the mean change?
  b. How does the variance change?
  c. How does the standard deviation change?

20. Find the mean and standard deviation of a set of grades if a grade of 86 corresponds to a z-score of 1.6 and a grade of 72 corresponds to a z-score of –1.2.

21. Verify that: $\sigma = \sqrt{\dfrac{\sum\limits_{i=1}^{n} x_i^2}{n} - \mu^2}$

21.
$$\sigma^2 = \frac{\sum\limits_{i=1}^{n}(x_i - \mu)^2}{n} = \frac{1}{n}\sum_{i=1}^{n}(x_i^2 - 2x_i\mu + \mu^2)$$

$$= \frac{1}{n}\sum_{i=1}^{n}x_i^2 - 2\mu\left(\frac{1}{n}\sum_{i=1}^{n}x_i\right) + \frac{1}{n}\sum_{i=1}^{n}\mu^2$$

$$= \frac{1}{n}\sum_{i=1}^{n}x_i^2 - 2\mu(\mu) + \mu^2$$

$$\sigma = \sqrt{\frac{\sum\limits_{i=1}^{n}x_i^2}{n} - \mu^2}$$

13.4 The Normal Distribution

When a fair coin is tossed, the probability that the coin shows heads is $\frac{1}{2}$ and the probability that it shows tails is $\frac{1}{2}$. If 6 coins are tossed, the coins can show 0, 1, 2, 3, 4, 5, or 6 heads. When 6 coins are tossed 200 times, the approximate expected distribution of the number of heads is given by the frequency table and the histogram drawn.

# of heads	Frequency
0	3
1	19
2	47
3	62
4	47
5	19
6	3

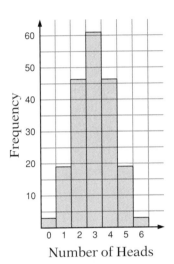

If 10 coins are tossed, the coins can show 0, 1, 2, 3, 4, 5, 6, 7, 8, 9, or 10 heads. If 10 coins are tossed 200 times the approximate expected frequency distribution for the number of heads and the histogram are shown below.

# of heads	Frequency
0	1
1	2
2	9
3	23
4	41
5	48
6	41
7	23
8	9
9	2
10	1

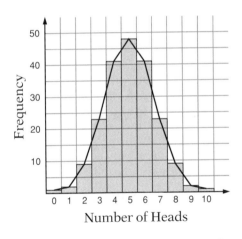

A *frequency distribution polygon* is the polygon formed by joining the midpoints of consecutive bars of the histogram by line segments. The graph is a symmetric bell-shaped figure. As the number of coins that are tossed and the number of times that they are tossed are increased, the polygon approaches a smooth curve called the *normal curve*.

The normal curve is an important standard, or norm, in statistics because sets of data can frequently be represented by this curve. For example, a set of data that gives the heights of all 10-year-old children, the weights of a particular breed of dog, the number of seconds needed to run a mile by the members of a track team, the number of hours that a particular brand of battery or light bulb will operate, are all examples of distributions that will have a frequency distribution similar to the frequency polygon shown above.

The equation of the normal curve is an exponential function dependent on the mean and the standard deviation of the set of normally distributed data that it represents. On the graph, the x-values are the data values for a given distribution, and the y-values are the frequencies. The x-coordinate of the midpoint of the curve represents the mean of the set of data. Because of the symmetry of the data represented by the curve, the mean is also the median for the distribution. If the mean is 0 and the standard deviation 1, the curve is called the *standard normal curve*.

This is the curve that represents a set of normally distributed data represented by z-scores. The area under the standard normal curve is 1, corresponding to the probability value 1, the sum of the probabilities of all the ways an event can occur.

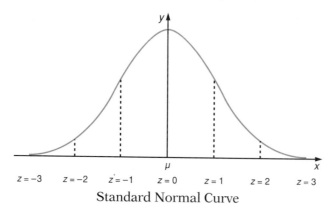

Standard Normal Curve

Table 6 in the Appendix shows the area under the standard normal curve between the mean and a point whose value is z. Since the curve is symmetric with respect to the mean, the same values apply to both positive and negative values of z.

The table can be used to approximate the expected frequencies of a range of scores in a normally distributed set of data. Read the table as you would a log table. Here, z-values in tenths are in the column at the left, and hundredths digits are across the top. For example, a z-score of 1.78 corresponds to a table entry of 0.4625.

Note that when $z = 1$, the area is 0.3413. Therefore, about 34% of the data in a normal distribution is between the mean and 1 standard deviation above the mean, or 68% is between 1 standard deviation below and 1 standard deviation above the mean.

When z is 2, the area is 0.4772. Therefore, about 2(0.4772) = 0.9544, or 95%, of the data is between 2 standard deviations below and 2 standard deviations above the mean. When $z = 3$, the area is 0.4987. Therefore, about 2(0.4987), or more than 99.7% of the data is between 3 standard deviations below and 3 standard deviations above the mean. A normal distribution is assumed for all sets of data in this section.

Example 1

The heights of all 10-year-old children are normally distributed, with a mean of 138 cm and a standard deviation of 5 cm.

a. Find the percent of the 10-year-old children that are between 138 and 140 cm tall.

b. Find the percent that are less than 140 cm tall.

Solution

a. Find z for $x = 140$. $z = \frac{140 - 138}{5} = \frac{2}{5} = 0.4$

In the table, the area from the mean to $z = 0.4$ is 0.1554.

Answer: 15.54% of the children are between 138 (the mean) and 140 cm tall.

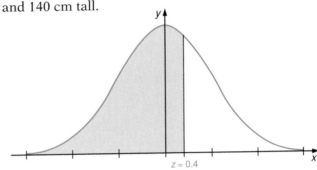

b. Since the heights of 50% of the children are less than the mean, $0.5000 + 0.1554 = 0.6554$.

Answer: 65.54% of the children are less than 140 cm tall.

Example 2

A school district administered IQ tests to all students in the district and found that the mean of the data was 102 and the standard deviation was 12.

a. What percent of the students in the district have an IQ that is above 125?

b. If there are 1,600 students in the district, how many can be expected to have an IQ over 125?

Solution

Find the z-score for 125. $z = \frac{125 - 102}{12} = \frac{23}{12} = 1.91\overline{6}$

The portion of the students with IQ's over 125 is represented by the area under the standard normal curve above 1.92. The table shows that for $z = 1.92$, the value is 0.4726. This means that 0.4726 of the scores are between the mean and 125. Since half, or 0.5, of the scores are above the mean, the portion of the scores above 125 is:

$0.5 - 0.4726 = 0.0274$

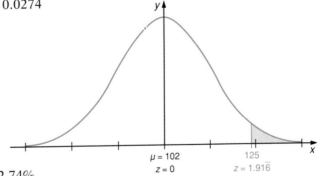

Answer: **a.** 2.74%

b. $1,600(0.0274) = 43.84$, or about 44 students

Answers 13.4

Answers 13.4

1. 34.13%

2. 15.87%

3. 34.13%

4. 15.87%

5. 72.58%

6. 73.87%

7. 5.99

8. 27.78

9. 3.64

10. 23 boxes

11. 2 days

12. 11

13. 6

Example 3

In a small town, the mean monthly rent for a one-bedroom apartment is $450. It was found that 80% of the apartments rented for between $350 and $550. Find the standard deviation of the monthly rent.

Solution

Since $350 is $100 below the mean and $550 is $100 above the mean, find the value of the standard deviation for a value of z that corresponds to one-half of 80%, or 0.40. The table shows that when the area is 0.3997, $z = 1.28$.

$$1.28 = \frac{100}{\sigma}$$

$$\sigma = \frac{100}{1.28} = 78.125$$

Answer: The standard deviation is $78.125.

Exercises 13.4

In **1 – 6**, the mean of a normal distribution is 12 and the standard deviation is 3. Find the percent of the data that lies within the given limits.

1. between 12 and 15
2. greater than 15
3. between 9 and 12
4. less than 9
5. between 10 and 18
6. between 5 and 14

In **7 – 9**, find the standard deviation of a normal distribution if the mean is 20 and the given percent of the data lies between the mean and 30.

7. 45.25%
8. 14%
9. 49.7%

10. The amount of cereal machine-packed into 16-oz. boxes varies. If the mean weight is 16.1 oz. and the standard deviation is 0.05 oz., determine the number of boxes that contain less than 16 oz. in each 1,000 boxes filled.

11. Charlotte knows that the length of time it takes her to walk to school varies, with a mean of 18 minutes and a standard deviation of 3 minutes. If she allows 25 minutes to get to school, how many times will she probably be late for school in one year if there are 180 days of school?

12. A librarian estimates that the average number of books checked out by a library patron is 5 with a standard deviation of 1.5. If 500 persons check out books today, about how many will have checked out more than 8 books?

13. The acceptable tolerance for the length of a part made for a computer is from 0.4 to 0.42 cm. The mean length for 500 of the parts is 0.41 and the standard deviation is 0.004 cm. Determine how many of the 500 parts are likely to be considered defective because they are not of the proper length.

Exercises 13.4 (continued)

14. The grades on a standardized test range from 20 to 80 inclusive. If the mean grade on the test is 45 and the standard deviation is 5:

    a. Find, to the nearest integer, the lowest grade a student can score to be in the top 10% of the group.

    b. What is the highest grade that a student can score and still be in the bottom fifth of the group that took the test?

    c. Find the percentage of the students scoring 60 on the test. *Hint:* Assume that the distribution of scores is continuous, and a grade of 60 includes grades from 59.5 to 60.5.

15. The mean grade for a classroom mathematics test was 76% and the standard deviation was 6. The instructor assigned a grade of A to students whose grades were in the top 4% of the group. What is the lowest grade that a student could obtain and still get an A?

16. Ms. Kingston compared her students' final grades to their scores on a standardized state examination given at the end of the term. The data for 5 students are listed below.

	Final Grade	State-Exam Grade
Alice	85	74
Bill	92	86
Carl	75	50
Daisy	80	75
Edward	90	83

The following information is known about the final grades and the state-exam grades obtained by all of the students in the class.

	Mean	Standard Deviation
Final Grade	78.4	5.6
State-Exam Grade	72.6	10.2

    a. Find the z-score of the final grades and the state-exam grades for each of the 5 students.

    b. In comparison to the rest of the class, did Alice do better on the state exam or her final grade?

    c. Which of the 5 students showed the least consistency of achievement between his/her final grade and the state-exam grade?

    d. Which of the 5 students was most consistent in his/her performance according to his/her achievement on the final grade and state-exam grade?

14. a. 51
    b. 40
    c. 0.09%

15. 87

16. a.
Alice	1.18	0.14
Bill	2.43	1.31
Carl	−0.61	−2.22
Daisy	0.29	0.24
Edward	2.07	1.02

    b. final
    c. Carl
    d. Daisy

13.5 *Sampling and Estimation*

Up to this point, discussions have focused on sets of data that represented information about a population, that is, the entire group under consideration. Often, it is not possible or convenient to gather data from every member of the population and, thus, information about the entire group is derived from information obtained from a smaller set, or *sample*. If the information obtained from the sample is to represent the population, an attempt must be made to obtain a truly representative sample.

A *random sample* is one in which each member of the population is equally likely to be included. If a random sample is selected by some plan, it is also a *systematic sample*. For example, from an alphabetical list of all graduates of a college, every 25th person is surveyed starting with some randomly-selected name chosen from the first 25 names. Such a sample is a random sample because every graduate is equally likely to be selected, and this sample is systematic because the selection is made by a definite plan.

Often the issue on which data is being collected is known to be influenced by some factor such as gender, race, religion, or education. The sample selected should reflect the makeup of the population with regard to the influential factor. Such a sample is called a *stratified sample*. For example, a survey that is intended to predict the outcome of an election should use a sample that has the same proportion of respondents from the major political parties and unaffiliated respondents as are present in the voting population.

Example 1

A group of parents who propose building a playground in a neighborhood park wants to determine the willingness of persons who live in the neighborhood to contribute time to the project. How can they select a sample of 20 families from the neighborhood to determine if there is sufficient interest?

Solution

Since parents with young children are probably more likely to show interest in the project, select a stratified sample. If 30% of the families in the neighborhood have young children, 30% of the sample should be families with young children.

Separate the population into two groups, those with young children and those without. Survey 3 families in the first group for every 7 surveyed in the second. Therefore, from the list of families with young children, select 6 families (30% of 20) at random by drawing names from a box or by numbering the families and generating 6 random numbers on a computer. Select 14 families (70% of 20) from the other group by the same method.

Sample Mean and Standard Deviation

Let us consider how the procedures for finding mean and standard deviation may change when a sample of a population, as opposed to the entire population, is being studied.

To draw some conclusions about the weights of peaches grown this year in a particular region, each of a sample of 30 peaches might be weighed. The mean weight of the sample, $\bar{x}$, would be calculated in the same manner as the mean weight for a population, μ.

$$\text{mean } \bar{x} = \frac{\sum\limits_{i=1}^{n} f_i x_i}{n} = \frac{117}{30} = 3.9$$

x_i	f_i	$x_i \cdot f_i$
4.4	1	4.4
4.3	0	0
4.2	1	4.2
4.1	3	12.3
4.0	6	24.0
3.9	8	31.2
3.8	5	19.0
3.7	4	14.8
3.6	1	3.6
3.5	1	3.5
Totals	30	117.0

When a small sample is taken from a population, it is very likely that the members of the sample will come from the center of the population distribution and that the range of the sample will be smaller than that of the population. As the size of the sample increases, it becomes more likely that values from the extremes of the population will be included. To adjust for this, the standard deviation s based on the sample, as an estimate of the standard deviation σ for the population, is found by the following formula:

$$s = \sqrt{\frac{\sum\limits_{i=1}^{n}(x_i - \bar{x})^2 \cdot f_i}{n-1}}$$

To find the standard deviation, first find the variance by finding the sum of the squares of the deviations from the mean. Divide this sum by 29, one less than the number of data points in the sample.

$$s^2 = \frac{\sum\limits_{i=1}^{n}(x_i - \bar{x})^2 \cdot f_i}{n-1} = \frac{0.98}{29} = 0.0338$$

$$s = \sqrt{0.0338} = 0.1838$$

Note that 0.1838 is not the standard deviation of the sample, but is an approximation for the standard deviation of the population from which the sample came.

x_i	f_i	$x_i - \bar{x}$	$(x_i - \bar{x})^2$	$f_i(x_i - \bar{x})^2$
4.4	1	0.5	0.25	0.25
4.3	0	0.4	0.16	0
4.2	1	0.3	0.09	0.09
4.1	3	0.2	0.04	0.12
4.0	6	0.1	0.01	0.06
3.9	8	0	0	0
3.8	5	−0.1	0.01	0.05
3.7	4	−0.2	0.04	0.16
3.6	1	−0.3	0.09	0.09
3.5	1	−0.4	0.16	0.16
Totals	30			0.98

A graphing calculator will display the statistical measures of mean and standard deviation. For the previous example, enter the weights of the peaches as the x's and the frequencies as the y's in the ▮ STAT ▮ menu. After the data has been entered, display the values of the mean and standard deviation by using the following sequence of keys:

▮ 2nd ▮ ▮ STAT ▮ ▮ ENTER ▮ ▮ ENTER ▮

The mean is given as $\bar{x}$. Two standard deviations, S_x and σ_x, are given. S_x is based on the data as a sample, and σ_x assumes the data to be the population.

Distribution of Sample Means

If all possible samples of the same size are taken from a population, the set of means of the samples is itself a set of data, $\overline{X}$.

For example, a physician doing research on calcium deficiencies in 18-year-old American women (the population for this study) may study the amount of calcium in the daily diet of the population by surveying samples of 100 women chosen at random from the population. The set of all means that could be obtained in this way is the distribution $\overline{X}$. The mean of each sample is one data point of $\overline{X}$. The mean $\mu_{\overline{x}}$ of $\overline{X}$ equals the mean μ of the population ($\mu_{\overline{x}} = \mu$).

Example 2

Let 18, 21, 32, and 56 be the ages of the 4 employees of a gift shop.
a. Find the mean of the population.
b. Find the mean of each sample of size 3 taken from the population.
c. Find the mean of the distribution $\overline{X}$ that consists of the sample means and compare it with the population mean.

Solution

a. $\mu = \dfrac{18 + 21 + 32 + 56}{4} = 31\frac{3}{4}$

b.

Sample of size 3	Sample Mean
18, 21, 32	$\dfrac{18 + 21 + 32}{3} = 23\frac{2}{3}$
18, 21, 56	$\dfrac{18 + 21 + 56}{3} = 31\frac{2}{3}$
18, 32, 56	$\dfrac{18 + 32 + 56}{3} = 35\frac{1}{3}$
21, 32, 56	$\dfrac{21 + 32 + 56}{3} = 36\frac{1}{3}$

c. $\mu_{\overline{x}} = \dfrac{23\frac{2}{3} + 31\frac{2}{3} + 35\frac{1}{3} + 36\frac{1}{3}}{4} = \dfrac{127}{4} = 31\frac{3}{4}$

Therefore, $\mu = \mu_{\overline{x}}$.

Any sample taken from a population must have a mean that is smaller than the largest data point of the population and larger than the smallest data point of the population. For example, in the four samples taken from the population in Example 2, each sample mean is smaller than 56 and larger than 18. Therefore, the sample means differ from μ by less than the original data and the standard deviation of the set of sample means must be smaller than the standard deviation of the population.

The size of the standard deviation of $\overline{X}$ is dependent on the number of data points in each sample. As the number of data points, n, in each sample increases, the standard deviation of $\overline{X}$ decreases. When the size of the population is large in comparison with the size of the samples, the standard deviation of the distribution of sample means is given by the equation: $\sigma_{\overline{x}} = \dfrac{\sigma}{\sqrt{n}}$

where σ is the standard deviation of the population and n is the sample size. The standard deviation of the distribution of sample means is called the *standard error of the mean*.

The distribution of sample means, $\overline{X}$, is always a set of data that is normally distributed, even when the population from which it is drawn is not. This important principle is given in the Central Limit Theorem, which states that if $\overline{x}$ is the mean of a random sample of size n from an infinite population with mean μ and standard deviation σ, then when

n is large, $z = \dfrac{\overline{x} - \mu}{\sigma_{\overline{x}}}$, or $z = \dfrac{\overline{x} - \mu}{\frac{\sigma}{\sqrt{n}}}$, has approximately the standard normal distribution.

Thus, the Central Limit Theorem allows us to consider the mean of each sample as a data point of a distribution that is normally distributed. The values of the standard normal distribution table can be used to determine the probability of obtaining samples with means that are within a given interval.

Example 3

The mean height of all 10-year-old children is 138 cm with a standard deviation of 5 cm.

a. What is the standard error of the mean for samples of 100 children?

b. What is the probability that the mean of a sample of 100 children will be between 137 and 139?

Solution

a. $\sigma_{\overline{x}} = \dfrac{5}{\sqrt{100}} = \dfrac{5}{10} = 0.5$ is the standard error of the mean, that is, the standard deviation of $\overline{X}$.

b. Since 137 is 2 standard deviations below the mean and 139 is 2 standard deviations above the mean, use the standard normal distribution table to find that the area from the mean to $z = 2$ is 0.4772. Therefore, from $z = -2$ to $z = 2$, the area is $2(0.4772) = 0.9544$ or approximately 95.4%.

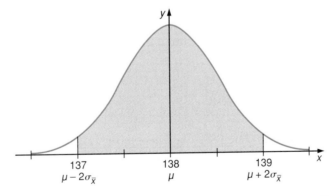

Although the theorem is stated for an infinite population, it applies to finite populations when the size of the sample, although large, is small when compared to the size of the population. It is difficult to assign actual values to the n for which the theorem is applicable, but 30 is usually regarded as sufficiently large. If the population itself has approximately the shape of a normal distribution, the size of the sample can be small.

Example 4

The population mean for IQ scores is 100 with a standard deviation of 15. Find the range within which the mean values of random samples of 200 persons can be expected to lie 99.5% of the time.

Solution

Find the range of z-scores for 99.5% of the cases by finding z when $A = \frac{1}{2}(0.995) = 0.4975$. That is, $z = \pm 2.81$.

$$z = \frac{\bar{x} - \mu}{\sigma_{\bar{x}}}$$

$$\pm 2.81 = \frac{\bar{x} - 100}{\frac{15}{\sqrt{200}}}$$

$$\bar{x} = 100 \pm 2.81 \left(\frac{15}{\sqrt{200}} \right)$$

$$= 100 \pm 2.98$$

Answer: $97 < \bar{x} < 103$

In Example 4, the range of expected mean values for samples of a given size was determined from the known population mean. Often the problem is just the opposite— to determine the expected limits of a population mean from sample means. The expected range of values is called a *confidence interval*.

Example 5

A random sample of 300 women were found to have a mean weight of 130 with a standard deviation of 15 pounds. Find the 98% confidence interval for the population mean.

Solution

The value of z for $A = \frac{1}{2}(0.98) = 0.49$ is 2.33.

$$z = \frac{\bar{x} - \mu}{\sigma_{\bar{x}}}$$

$$\pm 2.33 = \frac{130 - \mu}{\frac{15}{\sqrt{300}}}$$

$$\pm 2.33 \left(\frac{15}{\sqrt{300}} \right) = 130 - \mu$$

$$\mu = 130 \pm 2.02$$

Answer: The 98% confidence interval is between 128 and 132.

Note that, to find the endpoints of a confidence interval, the formula for the value of z can be solved for μ in terms of $\bar{x}$, z, and $\sigma_{\bar{x}}$:

$$\pm z = \frac{\bar{x} - \mu}{\sigma_{\bar{x}}}$$

$$\pm z \cdot \sigma_{\bar{x}} = \bar{x} - \mu$$

$$\mu = \bar{x} \pm z \cdot \sigma_{\bar{x}}$$

For example, for a 95% confidence interval: $z = \pm 1.96$, $\mu = \bar{x} \pm 1.96 \sigma_{\bar{x}}$

Exercises 13.5

In **1 – 6**, tell why the group of people selected is not a random sample from a population of all residents of the United States.

1. Names chosen from the telephone directory.
2. Passengers disembarking from a bus.
3. Subscribers to a fashion magazine.
4. Students at a business college.
5. People answering the phone in a private residence at 9:00 A.M.
6. Spectators at a baseball game.

In **7 – 10**, use the data from the sample to find the sample mean and the standard deviation.

7. The circumferences in inches of 20 apples selected from a day's harvest.

Circumference	Frequency
13	2
12	5
11	8
10	2
9	2
8	1

8. The ages of 40 people who attend evening college classes.

Age	Frequency
65 – 74	1
55 – 64	3
45 – 54	8
35 – 44	9
25 – 34	17
15 – 24	2

9. The weights of 50 adult golden retriever dogs.

Weight	Frequency
125 – 129	1
120 – 124	2
115 – 119	5
110 – 114	11
105 – 109	14
100 – 104	8
95 – 99	7
90 – 94	2

10. The number of absences from school in the last quarter for a sample of 35 students.

# of Absences	Frequency
8	1
7	2
6	4
5	5
4	7
3	8
2	2
1	2
0	4

1. Does not include persons who have no telephones.
2. Does not include persons who do not take a bus.
3. Does not include persons who do not subscribe to the magazine.
4. Does not include persons who are not students.
5. Does not include persons who are not home at 9AM.
6. Does not include persons who do not attend baseball games.

7. $\mu = 11$; $s = 1.298$
8. $\mu = 38.5$; $s = 11.72$
9. $\mu = 107.3$; $s = 7.786$
10. $\mu = 3.6857$; $s = 2.097$

Exercises 13.5 (continued)

11. From the set of data {4, 6, 7, 10, 13}, select all possible samples of 2 elements and verify that $\mu_{\bar{x}} = \mu$.

12. Given the set of data {12, 15, 16, 17}.
   a. Select all possible samples of 2 elements.
   b. Find the mean of each sample.
   c. Find the mean and standard deviation of the set of means.

13. The weights of the students of a large high school are normally distributed with a mean of 135 and a standard deviation of 15.
   a. If samples of 30 students each are studied, what would be the standard error of the mean for those samples?
   b. What is the probability that the mean of a particular sample of 30 students is between 130 and 140 pounds?
   c. If 100 samples were studied, in how many of the samples would you expect the mean to be above 150?

14. In a survey of 100 people, the average number of books of fiction read each year was 18. If the standard deviation of the population is 8, find the probability that the mean for the population is between 15.6 and 20.4.

15. A sample of 100 fish caught in Seneca Lake had a mean weight of 8.5 pounds. If the standard deviation of the weights of fish in Seneca Lake is estimated to be 2.8 pounds, find the 98% confidence interval for the mean of the weight of fish in Seneca Lake.

16. The mean age of a sample of 500 full-time college students is 25 and the estimated standard deviation is 2.3. Find the 98% confidence interval of the mean of all full-time college students.

17. Find the 98% confidence interval of the mean of a standardized test if a sample of 200 students had a mean of 50 and the standard deviation is 14.

18. Find the 99% confidence interval for the mean weight of 18-year-old women if a sample of 400 women had a mean of 128 pounds and the standard deviation is 12 pounds.

19. The mean height of the students of a college is 173 cm with a standard deviation of 9 cm. The mean height of the 50 students who are majoring in Physical Education is 180 cm. Can heights of the students in the Physical Education Department be considered a random sample of the heights of all students at this college? Explain.

20. An automobile manufacturer advertised the fuel consumption of a new model car as 35 m.p.g. with a standard deviation of 2.5 m.p.g. A car rental agency that bought 100 of this model found the cars averaged 31 m.p.g.
   a. What would be the 95% confidence interval for this model car based on the sample of 100 cars?
   b. What conclusion can you draw about the advertised claim based on this sample?

11. {4, 6}, {4, 7},{4, 10}, {4, 13}, {6, 7}, {6, 10}, {6, 13}, {7, 10}, {7, 13}, {10, 13}

$\mu = \dfrac{4 + 6 + 7 + 10 + 13}{5}$

$= \dfrac{40}{5} = 8$

$\mu_{\bar{x}} = \dfrac{5 + 5.5 + 7 + 8.5 + 6.5 + 8 + 9.5 + 8.5 + 10 + 11.5}{10}$

$= \dfrac{80}{10} = 8$

12. a. {12, 15}, {12, 16}, {12, 17}, {15, 16}, {15, 17}, {16, 17}
   b. 13.5, 14, 14.5, 15.5, 16, 16.5
   c. $\mu_{\bar{x}} = 15$; $\sigma_{\bar{x}} = 1.08$

13. a. 2.74 b. 93% c. none

14. 99.74%

15. 8.5 ± 0.65, that is, from 7.85 to 9.15

16. 25 ± 0.24, that is, from 24.76 to 25.24

17. 50 ± 2.3, that is, from 47.7 to 52.3

18. 128 ± 1.54, that is, from 126.46 to 129.54

19. probably not since the 99% confidence interval for a sample of 50 students is 173 ± 3.27, that is, from 169.73 to 176.27

20. a. 31 ± 0.49, that is, from 30.51 to 31.49
   b. claim is untrue or sample is not random

13.6 Chi-Square Distribution

When statistical data about the views of a group of persons are gathered, it is often important to determine whether the results are related to some other factor. For example, are views about a proposed tax increase influenced by the gender, political affiliation, or income bracket of the respondent? A test that is applied to do this, called the chi-square (χ^2) test, uses the differences between observed data and expected data, as shown in the following example.

Example 1

A restaurant owner wants to determine whether there is a significant difference between the number of customers who order dessert with an early-bird special dinner compared with those who order dessert with a regular dinner. A study of the past 100 customers provided the information shown. Find the chi-square value.

Dessert	Yes	No	Total
Early-bird	15	21	36
Regular	10	54	64
Total	25	75	100

Solution

Since $\frac{25}{100}$, or $\frac{1}{4}$, of the customers ordered dessert, we would expect $\frac{1}{4}$ of 36, or 9, persons who ordered early-bird and $\frac{1}{4}$ of 64, or 16, persons who ordered regular dinners to order dessert. Place these figures in the chart listing the given data with expected values shown in parentheses below the given data. The totals for each category remain unchanged.

Dessert	Yes	No	Total
Early-bird	15 (9)	21 (27)	36
Regular	10 (16)	54 (48)	64
Total	25	75	100

Let O represent the observed values and E represent the expected values. Chi-square is defined by the following formula:

$$\chi^2 = \sum \frac{(O - E)^2}{E}$$

For the given information:

$$\chi^2 = \frac{(15 - 9)^2}{9} + \frac{(21 - 27)^2}{27} + \frac{(10 - 16)^2}{16} + \frac{(54 - 48)^2}{48}$$

$$= \frac{36}{9} + \frac{36}{27} + \frac{36}{16} + \frac{36}{48}$$

$$= 4.00 + 1.33 + 2.25 + 0.75 = 8.33$$

Answer: The chi-square value is 8.33.

A χ^2-value has meaning only when it is compared with the values in the χ^2 Distribution Table (see Appendix, Table 7). These values depend on the number of *degrees of freedom*, df, of the observed data.

For example, the number of degrees of freedom for the data of Example 1 is determined as follows. Since the total of each row and column of the data table is known, we need to know only one other value to complete the table. That is, knowing that there was a total of 36 early-bird dinners, 15 with dessert, the number of early-bird dinners without dessert is determined, 36 – 15. Also, knowing that 25 people ordered dessert and, of these, 15 had ordered early-bird dinners determines the number of desserts ordered with a regular dinner, 25 – 15. Therefore, in addition to the given totals, since only 1 value, the number of people who order dessert with their early-bird dinners, needs to be known to determine all of the others, the number of degrees of freedom is 1.

A table organized into r rows and c columns, $r > 1$ and $c > 1$, is called a *contingency table*. When the total for each of r rows and c columns in the table is known, we need to know $r - 1$ values to complete each row and $c - 1$ values to complete each column. Therefore,

The number of degrees of freedom is $(r - 1)(c - 1)$.

To find the expected value for any cell of the table, multiply the row total by the column total and divide the product by the size of the sample. That is, the expected value for the i row and j column is:

$$E_{ij} = \frac{(\text{total of row } i)(\text{total of column } j)}{\text{size of the sample}}$$

In Example 1, the expected value for the regular dinner with dessert is the entry in the second row, first column.

$$E_{21} = \frac{64 \times 25}{100} = 16$$

To interpret the χ^2-value found in Example 1, use the first row of the table of χ^2-values for which $df = 1$. The highest value in the row, 6.635, corresponds to the heading 0.01, or 1%. The headings indicate the probability that the observed value would have occurred by chance. Note that, as the χ^2-values increase, the values in the heading decrease.

Since the value 8.33 is greater than 6.635 at the 0.01 level, the number of desserts ordered would occur by chance less than 1% of the time and, thus, a valid conclusion is that the number of desserts ordered is related to the type of dinner ordered.

Example 2

A soap company surveyed 100 men and 150 women to determine their preferences regarding scented soap. Each person was asked if he or she preferred soap with no scent (A), lightly-scented soap (B), or highly-scented soap (C). The table shows the responses. Does preference depend on gender?

	A	B	C
Men	27	64	9
Women	35	103	12
Total	62	167	21

Solution

Since 2 out of 5 respondents are men and 3 out of 5 are women, if there is no bias, the responses are expected to be in the same ratio. Therefore, of the 62 persons who prefer no scent, the expected values are $\frac{2}{5}(62)$, or 24.8 men, and $\frac{3}{5}(62)$, or 37.2, women. Expected values in the other categories can be found in a similar fashion.

	O	E	O – E	(O – E)²	$\frac{(O - E)^2}{E}$
Men (A)	27	24.8	2.2	4.84	0.20
Men (B)	64	66.8	–2.8	7.84	0.12
Men (C)	9	8.4	0.6	0.36	0.04
Women (A)	35	37.2	–2.2	4.84	0.13
Women (B)	103	100.2	2.8	7.84	0.08
Women (C)	12	12.6	–0.6	0.36	0.03

$$\chi^2 = 0.60$$

Since there are 2 rows and 3 columns in the given table, the number of degrees of freedom is $(2 - 1)(3 - 1) = 2$. For 2 degrees of freedom, the χ^2-value of 0.60 indicates that these values will occur in an unbiased sample between 70% and 80% of the time. It appears that preference does not depend on gender.

Goodness of Fit

Chi-square is often referred to as the *goodness-of-fit* statistic since it is used to test how observed results compare with probability expectations. When the probability of the observed results is very small, bias is indicated.

Example 3

When a coin was tossed 100 times, it showed heads 32 times. Is the coin a fair coin?

Solution

Since, for a fair coin, heads and tails are equally likely, the expected number of heads is 50.

Heads	32 (50)
Tails	68 (50)
Total	100

$$\chi^2 = \frac{(32 - 50)^2}{50} + \frac{(68 - 50)^2}{50}$$
$$= \frac{324}{50} + \frac{324}{50}$$
$$= 6.48 + 6.48$$
$$= 12.96$$

The data has 1 degree of freedom. Compare the value of χ^2, 12.96, with those given in the table for 1 degree of freedom. Since the value is greater than 6.635, a fair coin will show 32 heads in 100 tosses less than 1% of the times. The coin is probably not a fair coin.

1. $\chi^2 = 1.11$

Results will occur by chance about 30% of the time. Effectiveness is not established.

2. $\chi^2 = 7.01$

Results will occur by chance less than 5% of the time. There is probably a relationship.

3. $\chi^2 = 0.4074$

Results will occur by chance between 50% and 70% of the time. Effectiveness of the new method is not established.

4. $\chi^2 = 272.\overline{2}$

Claim is invalid.

5. $\chi^2 = 12.74$

Claim is invalid.

Exercises 13.6

1. A researcher tested a new drug by treating 60 diseased mice and leaving 20 untreated. Of the treated mice, 38 recovered, and of the untreated mice, 10 recovered. Can the treatment be considered to be effective?

2. A college teacher compared the grades of full-time and part-time students in one of his classes. The results are shown in the table at the right. Do part-time students get significantly higher grades?

Students	A & B	C & D	Failure
Full-Time	6	41	3
Part-Time	9	16	0

3. An educator proposed a new method of teaching speed-reading to adults. The results are shown in the table at the right. Is the new method significantly better?

Method	No Improvement	Improvement
New	40	80
Old	10	15

4. A TV commercial claims that 9 out of 10 hospitals use Tyad as a pain killer. In a survey of 200 hospitals, it was found that 110 used Tyad. Is the TV claim valid?

5. A tutoring service advertised that 95% of the students who attended their test-preparation classes raised their scores on college-entrance tests by 100 points or more. Is their claim justified if, from the last group of students, 42 improved their scores by 100 points or more and 8 did not?

6. A six-sided die with faces numbered 1 through 6 is tossed 102 times. The results are shown in the table at the right. Can the die be considered to be fair?

Outcome	1	2	3	4	5	6
Frequency	19	17	18	22	10	16

7. An industrial firm was accused of discrimination in their hiring practices. The company produced the record of people who applied for jobs, shown in the table at the right.

Applicants	Hired	Not Hired	Total Who Applied
White Male	14	13	27
White Female	2	13	15
Minority Male	8	13	21
Minority Female	1	11	12

a. The company refutes the charge, claiming that it refused employment to essentially the same number of people from each group who applied. Use the χ^2-test to prove or disprove that bias exists.

b. Reorganize the data into two categories, white and minority. Is there evidence of bias?

c. Reorganize the data into two categories, male and female. Is there evidence of bias?

---

6. $\chi^2 = 4.706$ 5 degrees of freedom
Results will occur by chance about 50% of the time. The die is probably fair.

7. a. $\chi^2 = 10.46$ 3 degrees of freedom
There is a less than 5% chance that the results would occur at random. Selection of employees can be considered to be biased.

7. b. $\chi^2 = 0.9740$ 1 degree of freedom
There is between 30% and 50% chance that the results would occur at random. Selection of employees cannot be considered to be biased based on racial background.

c. $\chi^2 = 9.375$ 1 degree of freedom
There is less than 1% chance that the results would occur at random. Selection of employees can be considered to be biased based on gender.

Exercises 13.6 *(continued)*

8. The birthdays of 700 people are recorded according to the day of the week that they occur in a given year. It is determined that 50 occur on Monday, 75 on Tuesday, 150 on Wednesday, 125 on Thursday, 110 on Friday, 90 on Saturday, and 100 on Sunday. Do the recorded values differ from the expected values significantly at the 0.01 level?

9. Two coins are tossed simultaneously 100 times with the outcomes shown.

Outcome	Frequency
2 Heads	20
1 Head & 1 Tail	50
2 Tails	30

The expected values can be obtained by the following expansion:

$$(h + t)^2 = h^2 + 2ht + t^2$$

indicating that the ratio of the outcomes are:

2 heads : 1 tail and 1 head : 2 tails = 1 : 2 : 1

a. Compute χ^2.

b. Determine if the observed results are consistent with the expected ratios at the

(1) 10% level

(2) 5% level

10. A Punnett square is used in biology to predict the numbers of offspring with certain traits when plants or animals reproduce. For example, if the trait being studied has to do with long-stemmed (S) and short-stemmed (s) plants and it is known that long-stemmed plants are dominant, the following Punnett square could be used.

	S	s
S	SS	Ss
s	sS	ss

The results suggest that $\frac{3}{4}$ of the resulting offspring are expected to be long-stemmed and $\frac{1}{4}$ are expected to be short-stemmed. 1,000 offspring of the plants resulted in the following traits:

Outcome	Frequency
SS	200
Ss or sS	300
ss	500

a. Are the observed results consistent with the expected values at the 5% level of significance?

b. Verify your answer by using a χ^2 goodness-of-fit test.

---

8. $\chi^2 = 64.5$ 6 degrees of freedom

    yes

9. a. $\chi^2 = 2$ 2 degrees of freedom

    b. (1) yes

       (2) yes

10. a. no   b. $\chi^2 = 333.\overline{3}$

Only characteristic long-stemmed (S) and short-stemmed (s) is being considered. All plants in the categories SS, Ss, and sS will be long-stemmed.

	O	E
long-stemmed	500	750
short-stemmed	500	250

$$\frac{(250)^2}{750} + \frac{(250)^2}{250} = 333.\overline{3}$$

11. a. (1) $\frac{9}{16}$

(2) $\frac{3}{16}$

(3) $\frac{3}{16}$

(4) $\frac{1}{16}$

b. 9 : 3 : 3 : 1

c. $\chi^2 = 335.56$

d. results are not consistent with expected values

12. a. 75 (70)
25 (30)
65 (70)
35 (30)

b. $df = 1$

c. $\chi^2 = 2.38$

d. no

13. yes

Exercises 13.6 (continued)

11. The Punnett square used to study two plant traits of garden peas is shown at the right. (S = long, s = short, W = round, and w = wrinkled, with long and round dominant)

	SW	Sw	sW	sw
SW	SSWW	SWSw	SWsW	SWsw
Sw	SwSW	SwSw	SwsW	Swsw
sW	sWSW	sWSw	sWsW	sWsw
sw	swSW	swSw	swsW	swsw

a. Analyze the table to determine how many of the outcomes result in plants that are:

(1) long-round   **(2)** long-wrinkled

(3) short-round   **(4)** short-wrinkled

b. Determine the expected ratio for

long round : long wrinkled : short round : short wrinkled

c. 500 of these plants were observed with the results shown in the table at right.

d. Use a χ^2-test to determine if these results are consistent with the expected results at the 1% level.

Outcome	Frequency
long-round	100
long-wrinkled	150
short-round	150
short-wrinkled	100

12. Two different brands of paint are tested as follows: 100 houses are painted on one side with Paint A and the other side is painted with Paint B. If either side of the houses shows peeling within 5 years, the paint used is considered unsatisfactory. The observed results are shown in the table.

Paint	Satisfactory	Not Satisfactory	Total
A	75	25	100
B	65	35	100
Total	140	60	200

a. Compute the expected value for each of the cells.

b. Determine the number of degrees of freedom.

c. Compute χ^2.

d. Are the paints significantly different at the 10% level?

13. A food company makes one type of breakfast cereal in two types of packaging. The company is interested in learning whether either package is preferred by men or women. A sample of 200 people are surveyed and asked which packaging they would select if they were buying the product. Determine from the data in the table whether there is a relationship between gender and the type of packaging preferred. Use the 10% significance level.

Gender	Package A	Package B	Total
Male	65	35	100
Female	40	60	100
Total	105	95	200

Chapter 13 Summary and Review

Organizing Data

- Statistics is the science that deals with the collection, organization, and interpretation of related pieces of numerical information called data.
- Statistical methods are used to assemble, organize, display, and draw conclusions from a collection of data.
- A distribution is the set of data about a particular factor from all members of a group called a population.
- Stem-and-leaf diagrams and frequency tables are used to organize data.
- A histogram is a vertical bar graph with no space between the bars, in which the height of each bar represents the frequency of the measure represented by the bar.
- When data is grouped into intervals, the lower boundary of an interval is the midpoint between its lower endpoint and the upper endpoint of the preceding interval.
- The cumulative frequency of an interval is the sum of the frequencies of that interval and all lower intervals.

Measures of Central Tendency

- The mean of a set of n numbers is the sum of the numbers divided by n.
  If x_i, where $i = 1$ to n, represents the set of data, then

$$\text{mean } \mu = \frac{1}{n}\sum_{i=1}^{n} x_i = \frac{x_1 + x_2 + \cdots + x_n}{n}$$

- The mean for grouped data is found by using the midpoint of the interval as the value of each entry in the interval.
- The median is the middle score of a set of data in numerical order.

$$\text{the median} = x_{\frac{n+1}{2}} \text{ if } n \text{ is odd}$$

$$\text{the median} = \frac{x_{\frac{n}{2}} + x_{\frac{n}{2}+1}}{2} \text{ if } n \text{ is even}$$

- The median of grouped data can be found by using the following expression, where the median interval, M_c, is the class or interval in which the median lies, and cf is the cumulative frequency.

$$\text{lower boundary of } M_c + \frac{\frac{n}{2} - cf \text{ for class below } M_c}{\text{frequency of } M_c} \cdot \text{interval size}$$

- The mode of a set of data is the number that occurs most frequently.
- Quartiles separate the data into four parts with an equal number of entries in each.
- The first, or lower, quartile is that number at or below which one-fourth of the entries lie and at or above which three-fourths of the entries lie. The lower quartile is the median of the scores that are below the median of all the scores.
- The third, or upper, quartile is that number at or below which three-fourths of the scores lie and at or above which one-fourth of the scores lie. It is the median of the scores that lie above the median of all of the scores.
- The interquartile range is the difference between the first and third quartile scores.
- A whisker-box plot is a diagram that uses the quartile scores to display the distribution of a set of scores.
- The percentile rank of a score in a distribution is the sum of the percent of the scores that are less than the given score and half of the percent of the scores that are equal to the given score.
- The midrange of a set of data is the number that is the average of the smallest and largest data points.

Measures of Dispersion

- The variance is the average of the squares of the deviations from the mean.
- The standard deviation is the square root of the variance.

Standard Deviation $\sigma = \sqrt{\dfrac{1}{n}\sum f(x - \mu)^2}$

- The standard deviation of a set of grouped data can be found by using the midpoint of each interval as the score for all entries in that interval.
- A z-score measures the number of standard deviations of a score above or below the mean.

$$z = \frac{x - \mu}{\sigma}$$

The Normal Distribution

- The expected frequency of the occurrence of r heads in n tosses of a coin follows a binomial distribution. As n increases, the histograms representing expected frequencies determine frequency polygons that approach a symmetric bell-shaped curve called the normal curve.
- The area under the standard normal curve can be used to predict the frequency of occurrence of data from a distribution that is normally distributed.
- In a normal distribution, about 68% of the scores lie within 1 standard deviation from the mean, about 95% lie within 2 standard deviations from the mean, and about 99.7% lie within 3 standard deviations from the mean.
- A table of areas under the standard normal curve can be used to determine the number of scores in a normal distribution that are expected to occur within a given range of values.

Samples

- A random sample is one in which each member of the population is equally likely to be included.

- A systematic sample is a random sample selected by some plan.

- A stratified sample is one selected to reflect the make-up of the population with regard to some factor such as race, gender, or political affiliation.

- The standard deviation based on a sample, as an estimate of the population standard deviation, is found by the following formula:

$$s = \sqrt{\frac{\sum\limits_{i=1}^{n}(x_i - \bar{x})^2 \cdot f_i}{n-1}}$$

- The means of the set of samples of size n taken from a population constitute a distribution $\overline{X}$. The mean of $\overline{X}$ is equal to the mean of the population ($\mu_{\bar{x}} = \mu$).

- The size of the standard deviation of $\overline{X}$ is dependent on the number of data points in each sample. As the number of data points, n, in each sample increases, the standard deviation of $\overline{X}$ decreases. When the size of the population is large in comparison with the size of the samples, the standard deviation of the distribution of the means of samples of size n is given by the equation

$$\sigma_{\bar{x}} = \frac{\sigma}{\sqrt{n}}$$

- The standard deviation of the distribution of sample means is called the standard error of the mean.

- The Central Limit Theorem: If $\bar{x}$ is the mean of a random sample of size n from an infinite population with mean μ and standard deviation σ, then, when n is large, $z = \dfrac{\bar{x} - \mu}{\frac{\sigma}{\sqrt{n}}}$ has approximately the standard normal distribution.

Chi-Square Distribution

- The chi-square test is used to determine whether observed results are consistent with expectations.

$$\chi^2 = \sum \frac{(O - E)^2}{E}$$

- When the totals for each of r rows and c columns in a table are known, $(r - 1)$ values are needed to complete each row and $(c - 1)$ values to complete each column, and the number of degrees of freedom is $(r - 1)(c - 1)$.

1.a. (*See below*)

Chapter 13   Review Exercises

1. The following list gives the number of years that each of the 30 employees of a retail store have worked for the company:

3, 14, 25, 6, 8, 2, 9, 12, 15, 31, 18, 9, 10, 7, 12, 20, 1, 11, 27, 2, 14, 17, 5, 16, 9, 2, 23, 10, 3, 4.

a. Use a stem-and-leaf diagram to organize the data.

b. Find the mean of the ungrouped data.

c. Find the median of the ungrouped data.

d. Find the first and third quartiles and draw a whisker-box plot.

e. What is the mode (or modes) of the data?

f. Find the midrange of the data.

g. What is the percentile rank of 16?   of 12?

h. Organize the data in intervals of 3, beginning with 1 – 3, and draw a histogram.

i. Find the mean of the grouped data and compare it with the mean found for the ungrouped data.

j. Find the median of the grouped data and compare it with the median found for the ungrouped data.

b. $\mu = 11.5$

c. 10

d.

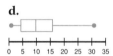

0 5 10 15 20 25 30 35

e. 2, 9

f. 16

g. 75th percentile; 60th percentile

h.

x_i	f_i
31 – 33	1
28 – 30	0
25 – 27	2
22 – 24	1
19 – 21	1
16 – 18	3
13 – 15	3
10 – 12	5
7 – 9	5
4 – 6	3
1 – 3	6

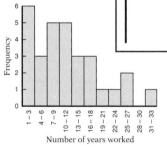

i. mean 11.4, smaller by 0.1

j. median 10.1, larger by 0.1

2. The data shows the weights, in pounds, of the fish entered in the annual fishing derby.

a. Find the mean weight of the fish entered in the derby.

b. Find the median weight of the fish entered in the derby.

Weight	Frequency
23 – 25	1
20 – 22	2
17 – 19	5
14 – 16	11
11 – 13	15
8 – 10	4

3. Before beginning an experiment in a science class, the students weighed the test tubes in which samples of material for the experiment were to be placed. The 15 students in the class obtained the following weights in grams:

37.5, 37.8, 36.9, 37.7, 38.0, 36.8, 37.2, 37.6, 37.4, 37.0, 37.3, 37.5, 37.4, 37.6, 37.3.

a. Find the mean weight.

b. Find the standard deviation.

c. Find the percent of the weights that are more than 1 standard deviation from the mean.

d. If the data were a normal distribution, what percent of the weights would be expected to be more than 1 standard deviation from the mean?

1. a.
```
3 | 1
2 | 0 3 5 7
1 | 0 0 1 2 2 4 4 5 6 7 8
0 | 1 2 2 2 3 3 4 5 6 7 8 9 9 9
```

4. A college determined that the mean score on a standardized math test for incoming freshmen was 550 with a standard deviation of 80. If the scores are approximately a normal distribution, what percent of the incoming freshmen can be expected to have earned a score above 700 on the test? Answer to the nearest whole percent.

5. The weights of a random sample of rabbits are approximately a normal distribution with a mean of 8 pounds. If 80% of the rabbits weigh less than 10 pounds, what is the standard deviation of the weights?

6. Find the mean and standard deviation of a set of data if a raw score of 15 corresponds to a z-score of 1.75 and a raw score of 4 corresponds to a z-score of –0.5.

7. A produce company packages carrots in 2-pound bags. A sample of 30 bags showed the following weights:

2.2, 2.5, 1.9, 2.1, 1.8, 2.0, 2.4, 2.3, 2.1, 2.0, 1.8, 2.4, 1.7, 2.1, 2.2, 2.0, 2.3, 1.9, 2.2, 1.9, 1.8, 2.1, 2.0, 2.3, 2.6, 2.1, 2.3, 2.0, 1.9, 2.1.

a. Find the mean of the sample.

b. Use the sample data to estimate the standard deviation.

8. A random sample of 150 people were found to spend a mean of $120.00 per year on magazines. If the standard deviation for all people in the United States is estimated to be $40.00, find the 95% confidence interval for the mean spent per year by all people in the United States.

9. A random sample of 100 grocery shoppers were asked the length of time needed to select and check out groceries. The sample mean was 50 minutes and the estimated standard deviation was 18 minutes. Find the 98% confidence interval for the population mean.

10. A coin is tossed 50 times and the coin shows a head 35 times. Is it reasonable to assume that the coin is a fair coin? Explain.

11. If the mean of $(a + 10)$, $(b + 20)$, $(c + 30)$, $(d + 40)$, $(e + 50)$ is 45, what is the mean of a, b, c, d, e?

12. A set of data consists of the numbers 5, 9, 12, 15, 20, and 23. All of the numbers except 20 have a frequency of 1. Find the frequency of 20 if the mean is 15.5.

13. Is it possible for the upper quartile of a set of data to be smaller than the mean? Explain.

14. In a normally distributed set of data, find, to the nearest tenth, the number of standard deviations of the first and third quartiles from the mean.

2. a. 14.13
   b. 13.5

3. a. 37.4
   b. 0.3204
   c. 33.3%
   d. 31.7%

4. 3%

5. 2.38

6. $\mu = 6.4$; $\sigma = 4.9$

7. a. 2.1
   b. 0.221

8. $\mu = \$120 \pm \$6.40$ or from $113.60 to $126.40

9. $\mu = 50 \pm 4$ or from 46 to 54 minutes

10. $\chi^2 = 8$

No, since these results would occur by chance less than 1% of the time.

11. 15

12. 3

13. Yes; example:
   1, 1, 1, 1, 1, 1, 10

14. 0.7

15. $\chi^2 = 32.464$

Yes, since these results would occur by chance less than 1% of the time.

16. $\chi^2 = 3.54$

No, since these results would occur by chance about 50% of the time.

15. A school board conducted a study of a high school population in grades 10, 11, and 12 to determine whether there is a correlation between the grade a student is in and the student's academic achievement. The table shows the academic averages of the 500 students.

Academic Average

Grade	Less Than 70	70 – 85	Greater Than 85
10th	25	150	25
11th	45	90	40
12th	30	60	35

Does the data indicate that the academic average is related to the grade the student is in?

16. Three different textbooks are used to teach first-year algebra. The final grades were recorded according to the textbook used, as shown in the table below. Determine whether there is statistical evidence that the grades are related to the textbook used.

Final Grade	Text A	Text B	Text C	Total
100 – 85	60	75	65	200
84 – 70	205	195	200	600
Below 70	75	60	65	200
Total	340	330	330	1,000

☑ Exercises for Challenge

1. The mean hourly wage for employees of a retail store is $6.30. The mean hourly wage for the part-time employees is $5.10 and for the full-time employees is $7.80. What is the ratio of the number of part-time employees to the number of full-time employees?

2. Find the mean of the numbers x_i, $i = 1$ to n, if $\frac{1}{n}\sum_{i=1}^{n}(80 - x_i) = 4.5$.

3. The mean of a set of 10 numbers is 84, of a set of 8 numbers is 72, and of a set of 6 numbers is 89. What is the mean of the 24 numbers?

4. A teacher raised each grade in a class of 25 students by 3 points. By how much did the standard deviation change?

5. By what factor will the variance of a set of data points change if the value of each data point is doubled?

6. The mean of a set of 10 numbers was doubled when one of the numbers was increased by 50 and the other numbers remained the same. What was the mean of the original set of numbers?

In **7 – 9**, write the answer in simplest form.

7. Express in terms of k the median of the numbers 2^k, 2^{k+1}, 2^{k+2}, 2^{k+3}.

8. Express in terms of k the mean of the numbers 2^k, 2^{k+1}, 2^{k+2}, 2^{k+3}.

9. A set of data consists of the integers 1 through n. The frequency of each number is equal to the number, as shown in the chart. Express the mean of the distribution in terms of n.

x	f
1	1
2	2
⋮	⋮
n	n

In questions **10** and **11**, the median of a set of integers was 10. After an additional integer was added to the set, the median was 10.5.

10. The number added must have been
(A) less than 10   (B) 10
(C) 10.5   (D) 11
(E) greater than 10

11. Before the additional integer was added to the distribution, which of the following must have been true?
I   There was an odd number of data points.
II   One of the data points was 10.
III   One of the data points was 11.
(A) I only   (B) I and II only
(C) I and III only (D) II and III only
(E) I, II, and III

12. The mean of the numbers log 4, log 5, and log 50 is
(A) 1   (B) $\frac{1}{3}$ log 59
(C) log $\frac{59}{3}$ (D) 3   (E) log $\frac{1{,}000}{3}$

Answers Exercises for Challenge

1. $\frac{5}{4}$

2. 75.5

3. 81.25

4. 0

5. 4

6. 5

7. $3 \cdot 2^k$

8. $15 \cdot 2^{k-2}$

9. $\frac{2n + 1}{3}$

10. E

11. B

12. A

13. E

14. B

15. $f_2 = 3$; $f_5 = 1$

16. C

17. B

18. C

19. A

20. D

21. C

☑ **Exercises for Challenge** (continued)

In **13** and **14**, the distribution consists of the integers from 1 to 20 such that the frequency of each integer k is 2^{k-1}.

13. The median of the distribution is
(A) 10   (B) 10.5   (C) 19
(D) 19.5   (E) 20

14. Which of the following is true?
(A) midrange = median = mean
(B) midrange < mean < median
(C) midrange < median < mean
(D) midrange = mean and
     mean < median
(E) midrange = median and
     median < mean

15. The table shows a set of data for which the frequency of two of the values is unknown. If

$$\sum_{i=1}^{6} f_i = 25 \text{ and } \mu = 3.28,$$

find the missing frequencies.

x_i	f_i
6	1
5	?
4	7
3	12
2	?
1	1

Questions 16 – 21 each consist of two quantities, one in Column A and one in Column B. You are to compare the two quantities and choose:

A   if the quantity in Column A is greater;
B   if the quantity in Column B is greater;
C   if the two quantities are equal;
D   if the relationship cannot be determined from the information given.

1. In certain questions, information concerning one or both of the quantities to be compared is centered above the two columns.

2. In a given question, a symbol that appears in both columns represents the same thing in Column A as it does in Column B.

3. x, n, and k, etc. stand for real numbers.

	Column A	Column B
16.	$\dfrac{1}{n}\sum_{i=1}^{n}(x_i - \mu)^2$	$\dfrac{1}{n}\sum_{i=1}^{n}x_i^2 - \mu^2$

	Column A	Column B
17.	Mean of the integers from 1 to $2n$	Median of the integers from 2 to $2n$

A distribution consists of the integers from 1 to 101.

	Column A	Column B
18.	midrange	mean

The mean of the set of data $\{x_1, x_2, ..., x_{10}\}$ is 8.5.

	Column A	Column B
19.	$\displaystyle\sum_{i=1}^{10}(x_i - 8)$	$\displaystyle\sum_{i=1}^{10}(x_i - 8.5)$

A distribution consists of the grades of 25 students in a statistics class.

	Column A	Column B
20.	The number of grades that are more than 2 standard deviations below the mean.	The number of grades that are more than 2 standard deviations above the mean.

	Column A	Column B
21.	Variance of the set of odd integers from 1 to 11	Variance of the set of even integers from 2 to 12

Appendix
Tables

Table of Contents

Table 1 Squares & Square Roots

N	N^2	$\sqrt{N}$	$\sqrt{10N}$	N	N^2	$\sqrt{N}$	$\sqrt{10N}$
1.0	1.00	1.000	3.162	5.5	30.25	2.345	7.416
1.1	1.21	1.049	3.317	5.6	31.36	2.366	7.483
1.2	1.44	1.095	3.464	5.7	32.49	2.387	7.550
1.3	1.69	1.140	3.606	5.8	33.64	2.408	7.616
1.4	1.96	1.183	3.742	5.9	34.81	2.429	7.681
1.5	2.25	1.225	3.873	6.0	36.00	2.449	7.746
1.6	2.56	1.265	4.000	6.1	37.21	2.470	7.810
1.7	2.89	1.304	4.123	6.2	38.44	2.490	7.874
1.8	3.24	1.342	4.243	6.3	39.69	2.510	7.937
1.9	3.61	1.378	4.359	6.4	40.96	2.530	8.000
2.0	4.00	1.414	4.472	6.5	42.25	2.550	8.062
2.1	4.41	1.449	4.583	6.6	43.56	2.569	8.124
2.2	4.84	1.483	4.690	6.7	44.89	2.588	8.185
2.3	5.29	1.517	4.796	6.8	46.24	2.608	8.246
2.4	5.76	1.549	4.899	6.9	47.61	2.627	8.307
2.5	6.25	1.581	5.000	7.0	49.00	2.646	8.367
2.6	6.76	1.612	5.099	7.1	50.41	2.665	8.426
2.7	7.29	1.643	5.196	7.2	51.84	2.683	8.485
2.8	7.84	1.673	5.292	7.3	53.29	2.702	8.544
2.9	8.41	1.703	5.385	7.4	54.76	2.720	8.602
3.0	9.00	1.732	5.477	7.5	56.25	2.739	8.660
3.1	9.61	1.761	5.568	7.6	57.76	2.757	8.718
3.2	10.24	1.789	5.657	7.7	59.29	2.775	8.775
3.3	10.89	1.817	5.745	7.8	60.84	2.793	8.832
3.4	11.56	1.844	5.831	7.9	62.41	2.811	8.888
3.5	12.25	1.871	5.916	8.0	64.00	2.828	8.944
3.6	12.96	1.897	6.000	8.1	65.61	2.846	9.000
3.7	13.69	1.924	6.083	8.2	67.24	2.864	9.055
3.8	14.44	1.949	6.164	8.3	68.89	2.881	9.110
3.9	15.21	1.975	6.245	8.4	70.56	2.898	9.165
4.0	16.00	2.000	6.325	8.5	72.25	2.915	9.220
4.1	16.81	2.025	6.403	8.6	73.96	2.933	9.274
4.2	17.64	2.049	6.481	8.7	75.69	2.950	9.327
4.3	18.49	2.074	6.557	8.8	77.44	2.966	9.381
4.4	19.36	2.098	6.633	8.9	79.21	2.983	9.434
4.5	20.25	2.121	6.708	9.0	81.00	3.000	9.487
4.6	21.16	2.145	6.782	9.1	82.81	3.017	9.539
4.7	22.09	2.168	6.856	9.2	84.64	3.033	9.592
4.8	23.04	2.191	6.928	9.3	86.49	3.050	9.644
4.9	24.01	2.214	7.000	9.4	88.36	3.066	9.695
5.0	25.00	2.236	7.071	9.5	90.25	3.082	9.747
5.1	26.01	2.258	7.141	9.6	92.16	3.098	9.798
5.2	27.04	2.280	7.211	9.7	94.09	3.114	9.849
5.3	28.09	2.302	7.280	9.8	96.04	3.130	9.899
5.4	29.16	2.324	7.348	9.9	98.01	3.146	9.950
5.5	30.25	2.345	7.416	10.0	100.00	3.162	10.000

T
A
B
L
E
S

Table 2 Common Logarithms

N	0	1	2	3	4	5	6	7	8	9
10	0000	0043	0086	0128	0170	0212	0253	0294	0334	0374
11	0414	0453	0492	0531	0569	0607	0645	0682	0719	0755
12	0792	0828	0864	0899	0934	0969	1004	1038	1072	1106
13	1139	1173	1206	1239	1271	1303	1335	1367	1399	1430
14	1461	1492	1523	1553	1584	1614	1644	I673	1703	1732
15	1761	1790	1818	1847	1875	1903	1931	1959	1987	2014
16	2041	2068	2095	2122	2148	2175	2201	2227	2253	2279
17	2304	2330	2355	2380	2405	2430	2455	2480	2504	2529
18	2553	2577	2601	2625	2648	2672	2695	2718	2742	2765
19	2788	2810	2833	2856	2878	2900	2923	2945	2967	2989
20	3010	3032	3054	3075	3096	3118	3139	3160	3181	3201
21	3222	3243	3263	3284	3304	3324	3345	3365	3385	3404
22	3424	3444	3464	3483	3502	3522	3541	3560	3579	3598
23	3617	3636	3655	3674	3692	3711	3729	3747	3766	3784
24	3802	3820	3838	3856	3874	3892	3909	3927	3945	3962
25	3979	3997	4014	4031	4048	4065	4082	4099	4116	4133
26	4150	4166	4183	4200	4216	4232	4249	4265	4281	4298
27	4314	4330	4346	4362	4378	4393	4409	4425	4440	4456
28	4472	4487	4502	4518	4533	4548	4564	4579	4594	4609
29	4624	4639	4654	4669	4683	4698	4713	4728	4742	4757
30	4771	4786	4800	4814	4829	4843	4857	4871	4886	4900
31	4914	4928	4942	4955	4969	4983	4997	5011	5024	5038
32	5051	5065	5079	5092	5105	5119	5132	5145	5159	5172
33	5185	5198	5211	5224	5237	5250	5263	5276	5289	5302
34	5315	5328	5340	5353	5366	5378	5391	5403	5416	5428
35	5441	5453	5465	5478	5490	5502	5514	5527	5539	5551
36	5563	5575	5587	5599	5611	5623	5635	5647	5658	5670
37	5682	5694	5705	5717	5729	5740	5752	5763	5775	5786
38	5798	5809	5821	5832	5843	5855	5866	5877	5888	5899
39	5911	5922	5933	5944	5955	5966	5977	5988	5999	6010
40	6021	6031	6042	6053	6064	6075	6085	6096	6107	6117
41	6128	6138	6149	6160	6170	6180	6191	6201	6212	6222
42	6232	6243	6253	6263	6274	6284	6294	6304	6314	6325
43	6335	6345	6355	6365	6375	6385	6395	6405	6415	6425
44	6435	6444	6454	6464	6474	6484	6493	6503	6513	6522
45	6532	6542	6551	6561	6571	6580	6590	6599	6609	6618
46	6628	6637	6646	6656	6665	6675	6684	6693	6702	6712
47	6721	6730	6739	6749	6758	6767	6776	6785	6794	6803
48	6812	6821	6830	6839	6848	6857	6866	6875	6884	6893
49	6902	6911	6920	6928	6937	6946	6955	6964	6972	6981
50	6990	6998	7007	7016	7024	7033	7042	7050	7059	7067
51	7076	7084	7093	7101	7110	7118	7126	7135	7143	7152
52	7160	7168	7177	7185	7193	7202	7210	7218	7226	7235
53	7243	7251	7259	7267	7275	7284	7292	7300	7308	7316
54	7324	7332	7340	7348	7356	7364	7372	7380	7388	7396

T A B L E S

Table 2 Common Logarithms

645

Table 2 Common Logarithms

N	0	1	2	3	4	5	6	7	8	9
55	7404	7412	7419	7427	7435	7443	7451	7459	7466	7474
56	7482	7490	7497	7505	7513	7520	7528	7536	7543	7551
57	7559	7566	7574	7582	7589	7597	7604	7612	7619	7627
58	7634	7642	7649	7657	7664	7672	7679	7686	7694	7701
59	7709	7716	7723	7731	7738	7745	7752	7760	7767	7774
60	7782	7789	7796	7803	7810	7818	7825	7832	7839	7846
61	7853	7860	7868	7875	7882	7889	7896	7903	7910	7917
62	7924	7931	7938	7945	7952	7959	7966	7973	7980	7987
63	7993	8000	8007	8014	8021	8028	8035	8041	8048	8055
64	8062	8069	8075	8082	8089	8096	8102	8109	8116	8122
65	8129	8136	8142	8149	8156	8162	8169	8176	8182	8189
66	8195	8202	8209	8215	8222	8228	8235	8241	8248	8254
67	8261	8267	8274	8280	8287	8293	8299	8306	8312	8319
68	8325	8331	8338	8344	8351	8357	8363	8370	8376	8382
69	8388	8395	8401	8407	8414	8420	8426	8432	8439	8445
70	8451	8457	8463	8470	8476	8482	8488	8494	8500	8506
71	8513	8519	8525	8531	8537	8543	8549	8555	8561	8567
72	8573	8579	8585	8591	8597	8603	8609	8615	8621	8627
73	8633	8639	8645	8651	8657	8663	8669	8675	8681	8686
74	8692	8698	8704	8710	8716	8722	8727	8733	8739	8745
75	8751	8756	8762	8768	8774	8779	8785	8791	8797	8802
76	8808	8814	8820	8825	8831	8837	8842	8848	8854	8859
77	8865	8871	8876	8882	8887	8893	8899	8904	8910	8915
78	8921	8927	8932	8938	8943	8949	8954	8960	8965	8971
79	8976	8982	8987	8993	8998	9004	9009	9015	9020	9025
80	9031	9036	9042	9047	9053	9058	9063	9069	9074	9079
81	9085	9090	9096	9101	9106	9112	9117	9122	9128	9133
82	9138	9143	9149	9154	9159	9165	9170	9175	9180	9186
83	9191	9196	9201	9206	9212	9217	9222	9227	9232	9238
84	9243	9248	9253	9258	9263	9269	9274	9279	9284	9289
85	9294	9299	9304	9309	9315	9320	9325	9330	9335	9340
86	9345	9350	9355	9360	9365	9370	9375	9380	9385	9390
87	9395	9400	9405	9410	9415	9420	9425	9430	9435	9440
88	9445	9450	9455	9460	9465	9469	9474	9479	9484	9489
89	9494	9499	9504	9509	9513	9518	9523	9528	9533	9538
90	9542	9547	9552	9557	9562	9566	9571	9576	9581	9586
91	9590	9595	9600	9605	9609	9614	9619	9624	9628	9633
92	9638	9643	9647	9652	9657	9661	9666	9671	9675	9680
93	9685	9689	9694	9699	9703	9708	9713	9717	9722	9727
94	9731	9736	9741	9745	9750	9754	9759	9763	9768	9773
95	9777	9782	9786	9791	9795	9800	9805	9809	9814	9818
96	9823	9827	9832	9836	9841	9845	9850	9854	9859	9863
97	9868	9872	9877	9881	9886	9890	9894	9899	9903	9908
98	9912	9917	9921	9926	9930	9934	9939	9943	9948	9952
99	9956	9961	9965	9969	9974	9978	9983	9987	9991	9996

T
A
B
L
E
S

Table 3 Values of e^x and e^{-x}

x	e^x	e^{-x}	x	e^x	e^{-x}
0.00	1.0000	1.0000	2.5	12.182	0.0821
0.05	1.0513	0.9512	2.6	13.464	0.0743
0.10	1.1052	0.9048	2.7	14.880	0.0672
0.15	1.1618	0.8607	2.8	16.445	0.0608
0.20	1.2214	0.8187	2.9	18.174	0.0550
0.25	1.2840	0.7788	3.0	20.086	0.0498
0.30	1.3499	0.7408	3.1	22.198	0.0450
0.35	1.4191	0.7047	3.2	24.533	0.0408
0.40	1.4918	0.6703	3.3	27.113	0.0369
0.45	1.5683	0.6376	3.4	29.964	0.0334
0.50	1.6487	0.6065	3.5	33.115	0.0302
0.55	1.7333	0.5769	3.6	36.598	0.0273
0.60	1.8221	0.5488	3.7	40.447	0.0247
0.65	1.9155	0.5220	3.8	44.701	0.0224
0.70	2.0138	0.4966	3.9	49.402	0.0202
0.75	2.1170	0.4724	4.0	54.598	0.0183
0.80	2.2255	0.4493	4.1	60.340	0.0166
0.85	2.3396	0.4274	4.2	66.686	0.0150
0.90	2.4596	0.4066	4.3	73.700	0.0136
0.95	2.5857	0.3867	4.4	81.451	0.0123
1.0	2.7183	0.3679	4.5	90.017	0.0111
1.1	3.0042	0.3329	4.6	99.484	0.0101
1.2	3.3201	0.3012	4.7	109.95	0.0091
1.3	3.6693	0.2725	4.8	121.51	0.0082
1.4	4.0552	0.2466	4.9	134.29	0.0074
1.5	4.4817	0.2231	5.0	148.41	0.0067
1.6	4.9530	0.2019	5.5	244.69	0.0041
1.7	5.4739	0.1827	6.0	403.43	0.0025
1.8	6.0496	0.1653	6.5	665.14	0.0015
1.9	6.6859	0.1496	7.0	1096.6	0.0009
2.0	7.3891	0.1353	7.5	1808.0	0.0006
2.1	8.1662	0.1225	8.0	2981.0	0.0003
2.2	9.0250	0.1108	8.5	4914.8	0.0002
2.3	9.9742	0.1003	9.0	8103.1	0.0001
2.4	11.023	0.0907	10.0	22026.0	0.00005

T
A
B
L
E
S

Table 3 Values of e^x and e^{-x}

647

Table 4 · *Natural Logarithms*

x	$\ln x$	x	$\ln x$	x	$\ln x$
0.0	$-\infty$	4.5	1.5041	9.0	2.1972
0.1	−2.3026	4.6	1.5261	9.1	2.2083
0.2	−1.6094	4.7	1.5476	9.2	2.2192
0.3	−1.2040	4.8	1.5686	9.3	2.2300
0.4	−0.9163	4.9	1.5892	9.4	2.2407
0.5	−0.6931	5.0	1.6094	9.5	2.2513
0.6	−0.5108	5.1	1.6292	9.6	2.2618
0.7	−0.3567	5.2	1.6487	9.7	2.2721
0.8	−0.2231	5.3	1.6677	9.8	2.2824
0.9	−0.1054	5.4	1.6864	9.9	2.2925
1.0	0.0000	5.5	1.7047	10	2.3026
1.1	0.0953	5.6	1.7228	11	2.3979
1.2	0.1823	5.7	1.7405	12	2.4849
1.3	0.2624	5.8	1.7579	13	2.5649
1.4	0.3365	5.9	1.7750	14	2.6391
1.5	0.4055	6.0	1.7918	15	2.7081
1.6	0.4700	6.1	1.8083	16	2.7726
1.7	0.5306	6.2	1.8245	17	2.8332
1.8	0.5878	6.3	1.8405	18	2.8904
1.9	0.6419	6.4	1.8563	19	2.9444
2.0	0.6931	6.5	1.8718	20	2.9957
2.1	0.7419	6.6	1.8871	25	3.2189
2.2	0.7885	6.7	1.9021	30	3.4012
2.3	0.8329	6.8	1.9169	35	3.5553
2.4	0.8755	6.9	1.9315	40	3.6889
2.5	0.9163	7.0	1.9459	45	3.8067
2.6	0.9555	7.1	1.9601	50	3.9120
2.7	0.9933	7.2	1.9741	55	4.0073
2.8	1.0296	7.3	1.9879	60	4.0943
2.9	1.0647	7.4	2.0015	65	4.1744
3.0	1.0986	7.5	2.0149	70	4.2485
3.1	1.1314	7.6	2.0281	75	4.3175
3.2	1.1632	7.7	2.0412	80	4.3820
3.3	1.1939	7.8	2.0541	85	4.4427
3.4	1.2238	7.9	2.0669	90	4.4998
3.5	1.2528	8.0	2.0794	100	4.6052
3.6	1.2809	8.1	2.0919	110	4.7005
3.7	1.3083	8.2	2.1041	120	4.7875
3.8	1.3350	8.3	2.1163	130	4.8676
3.9	1.3610	8.4	2.1282	140	4.9416
4.0	1.3863	8.5	2.1401	150	5.0106
4.1	1.4110	8.6	2.1518	160	5.0752
4.2	1.4351	8.7	2.1633	170	5.1358
4.3	1.4586	8.8	2.1748	180	5.1930
4.4	1.4816	8.9	2.1861	190	5.2470

T
A
B
L
E
S

Table 5 **Trigonometric Functions of** θ • **Degrees & Radians**

θ Degrees	θ Radians	sin θ	csc θ	tan θ	cot θ	sec θ	cos θ		
0° 00′	0.0000	0.0000	Undefined	0.0000	Undefined	1.000	1.0000	1.5708	90° 00′
10′	0.0029	0.0029	343.8	0.0029	343.8	1.000	1.0000	1.5679	50′
20′	0.0058	0.0058	171.9	0.0058	171.9	1.000	1.0000	1.5650	40′
30′	0.0087	0.0087	114.6	0.0087	114.6	1.000	1.0000	1.5621	30′
40′	0.0116	0.0116	85.95	0.0116	85.94	1.000	0.9999	1.5592	20′
50′	0.0145	0.0145	68.76	0.0145	68.75	1.000	0.9999	1.5563	10′
1° 00′	0.0175	0.0175	57.30	0.0175	57.29	1.000	0.9998	1.5533	89° 00′
10′	0.0204	0.0204	49.11	0.0204	49.10	1.000	0.9998	1.5504	50′
20′	0.0233	0.0233	42.98	0.0233	42.96	1.000	0.9997	1.5475	40′
30′	0.0262	0.0262	38.20	0.0262	38.19	1.000	0.9997	1.5446	30′
40′	0.0291	0.0291	34.38	0.0291	34.37	1.000	0.9996	1.5417	20′
50′	0.0320	0.0320	31.26	0.0320	31.24	1.001	0.9995	1.5388	10′
2° 00′	0.0349	0.0349	28.65	0.0349	28.64	1.001	0.9994	1.5359	88° 00′
10′	0.0378	0.0378	26.45	0.0378	26.43	1.001	0.9993	1.5330	50′
20′	0.0407	0.0407	24.56	0.0407	24.54	1.001	0.9992	1.5301	40′
30′	0.0436	0.0436	22.93	0.0437	22.90	1.001	0.9990	1.5272	30′
40′	0.0465	0.0465	21.49	0.0466	21.47	1.001	0.9989	1.5243	20′
50′	0.0495	0.0494	20.23	0.0495	20.21	1.001	0.9988	1.5213	10′
3° 00′	0.0524	0.0523	19.11	0.0524	19.08	1.001	0.9986	1.5184	87° 00′
10′	0.0553	0.0552	18.10	0.0553	18.07	1.002	0.9985	1.5155	50′
20′	0.0582	0.0581	17.20	0.0582	17.17	1.002	0.9983	1.5126	40′
30′	0.0611	0.0610	16.38	0.0612	16.35	1.002	0.9981	1.5097	30′
40′	0.0640	0.0640	15.64	0.0641	15.60	1.002	0.9980	1.5068	20′
50′	0.0669	0.0669	14.96	0.0670	14.92	1.002	0.9978	1.5039	10′
4° 00′	0.0698	0.0698	14.34	0.0699	14.30	1.002	0.9976	1.5010	86° 00′
10′	0.0727	0.0727	13.76	0.0729	13.73	1.003	0.9974	1.4981	50′
20′	0.0756	0.0756	13.23	0.0758	13.20	1.003	0.9971	1.4952	40′
30′	0.0785	0.0785	12.75	0.0787	12.71	1.003	0.9969	1.4923	30′
40′	0.0814	0.0814	12.29	0.0816	12.25	1.003	0.9967	1.4893	20′
50′	0.0844	0.0843	11.87	0.0846	11.83	1.004	0.9964	1.4864	10′
5° 00′	0.0873	0.0872	11.47	0.0875	11.43	1.004	0.9962	1.4835	85° 00′
10′	0.0902	0.0901	11.10	0.0904	11.06	1.004	0.9959	1.4806	50′
20′	0.0931	0.0929	10.76	0.0934	10.71	1.004	0.9957	1.4777	40′
30′	0.0960	0.0958	10.43	0.0963	10.39	1.005	0.9954	1.4748	30′
40′	0.0989	0.0987	10.13	0.0992	10.08	1.005	0.9951	1.4719	20′
50′	0.1018	0.1016	9.839	0.1022	9.788	1.005	0.9948	1.4690	10′
6° 00′	0.1047	0.1045	9.567	0.1051	9.514	1.006	0.9945	1.4661	84° 00′
10′	0.1076	0.1074	9.309	0.1080	9.255	1.006	0.9942	1.4632	50′
20′	0.1105	0.1103	9.065	0.1110	9.010	1.006	0.9939	1.4603	40′
30′	0.1134	0.1132	8.834	0.1139	8.777	1.006	0.9936	1.4573	30′
40′	0.1164	0.1161	8.614	0.1169	8.556	1.007	0.9932	1.4544	20′
50′	0.1193	0.1190	8.405	0.1198	8.345	1.007	0.9929	1.4515	10′
7° 00′	0.1222	0.1219	8.206	0.1228	8.144	1.008	0.9925	1.4486	83° 00′
10′	0.1251	0.1248	8.016	0.1257	7.953	1.008	0.9922	1.4457	50′
20′	0.1280	0.1276	7.834	0.1287	7.770	1.008	0.9918	1.4428	40′
30′	0.1309	0.1305	7.661	0.1317	7.596	1.009	0.9914	1.4399	30′
40′	0.1338	0.1334	7.496	0.1346	7.429	1.009	0.9911	1.4370	20′
50′	0.1367	0.1363	7.337	0.1376	7.269	1.009	0.9907	1.4341	10′
8° 00′	0.1396	0.1392	7.185	0.1405	7.115	1.010	0.9903	1.4312	82° 00′
10′	0.1425	0.1421	7.040	0.1435	6.968	1.010	0.9899	1.4283	50′
20′	0.1454	0.1449	6.900	0.1465	6.827	1.011	0.9894	1.4254	40′
30′	0.1484	0.1478	6.765	0.1495	6.691	1.011	0.9890	1.4224	30′
40′	0.1513	0.1507	6.636	0.1524	6.561	1.012	0.9886	1.4195	20′
50′	0.1542	0.1536	6.512	0.1554	6.435	1.012	0.9881	1.4166	10′
9° 00′	0.1571	0.i564	6.392	0.1584	6.314	1.012	0.9877	1.4137	81° 00′
		cos θ	sec θ	cot θ	tan θ	csc θ	sin θ	θ Radians	θ Degrees

T A B L E S

Table 5 **Trigonometric Functions of** θ • **Degrees & Radians**   **649**

Table 5 **Trigonometric Functions of** θ • *Degrees & Radians*

θ Degrees	θ Radians	sin θ	csc θ	tan θ	cot θ	sec θ	cos θ		
9 ° 00′	0.1571	0.1564	6.392	0.1584	6.314	1.012	0.9877	1.4137	**81° 00′**
10′	0.1600	0.1593	6.277	0.1614	6.197	1.013	0.9872	1.4108	50′
20′	0.1629	0.1622	6.166	0.1644	6.084	1.013	0.9868	1.4079	40′
30′	0.1658	0.1650	6.059	0.1673	5.976	1.014	0.9863	1.4050	30′
40′	0.1687	0.1679	5.955	0.1703	5.871	1.014	0.9858	1.4021	20′
50′	0.1716	0.1708	5.855	0.1733	5.769	1.015	0.9853	1.3992	10′
10° 00′	0.1745	0.1736	5.759	0.1763	5.671	1.015	0.9848	1.3963	**80° 00′**
10′	0.1774	0.1765	5.665	0.1793	5.576	1.016	0.9843	1.3934	50′
20′	0.1804	0.1794	5.575	0.1823	5.485	1.016	0.9838	1.3904	40′
30′	0.1833	0.1822	5.487	0.1853	5.396	1.017	0.9833	1.3875	30′
40′	0.1862	0.1851	5.403	0.1883	5.309	1.018	0.9827	1.3846	20′
50′	0.1891	0.1880	5.320	0.1914	5.226	1.018	0.9822	1.3817	10′
11° 00′	0.1920	0.1908	5.241	0.1944	5.145	1.019	0.9816	1.3788	**79° 00′**
10′	0.1949	0.1937	5.164	0.1974	5.066	1.019	0.9811	1.3759	50′
20′	0.1978	0.1965	5.089	0.2004	4.989	1.020	0.9805	1.3730	40′
30′	0.2007	0.1994	5.016	0.2035	4.915	1.020	0.9799	1.3701	30′
40′	0.2036	0.2022	4.945	0.2065	4.843	1.021	0.9793	1.3672	20′
50′	0.2065	0.2051	4.876	0.2095	4.773	1.022	0.9787	1.3643	10′
12° 00′	0.2094	0.2079	4.810	0.2126	4.705	1.022	0.9781	1.3614	**78° 00′**
10′	0.2123	0.2108	4.745	0.2156	4.638	1.023	0.9775	1.3584	50′
20′	0.2153	0.2136	4.682	0.2186	4.574	1.024	0.9769	1.3555	40′
30′	0.2182	0.2164	4.620	0.2217	4.511	1.024	0.9763	1.3526	30′
40′	0.2211	0.2193	4.560	0.2247	4.449	1.025	0.9757	1.3497	20′
50′	0.2240	0.2221	4.502	0.2278	4.390	1.026	0.9750	1.3468	10′
13° 00′	0.2269	0.2250	4.445	0.2309	4.331	1.026	0.9744	1.3439	**77° 00′**
10′	0.2298	0.2278	4.390	0.2339	4.275	1.027	0.9737	1.3410	50′
20′	0.2327	0.2306	4.336	0.2370	4.219	1.028	0.9730	1.3381	40′
30′	0.2356	0.2334	4.284	0.2401	4.165	1.028	0.9724	1.3352	30′
40′	0.2385	0.2363	4.232	0.2432	4.113	1.029	0.9717	1.3323	20′
50′	0.2414	0.2391	4.182	0.2462	4.061	1.030	0.9710	1.3294	10′
14° 00′	0.2443	0.2419	4.134	0.2493	4.011	1.031	0.9703	1.3265	**76° 00′**
10′	0.2473	0.2447	4.086	0.2524	3.962	1.031	0.9696	1.3235	50′
20′	0.2502	0.2476	4.039	0.2555	3.914	1.032	0.9689	1.3206	40′
30′	0.2531	0.2504	3.994	0.2586	3.867	1.033	0.9681	1.3177	30′
40′	0.2560	0.2532	3.950	0.2617	3.821	1.034	0.9674	1.3148	20′
50′	0.2589	0.2560	3.906	0.2648	3.776	1.034	0.9667	1.3119	10′
15° 00′	0.2618	0.2588	3.864	0.2679	3.732	1.035	0.9659	1.3090	**75° 00′**
10′	0.2647	0.2616	3.822	0.2711	3.689	1.036	0.9652	1.3061	50′
20′	0.2676	0.2644	3.782	0.2742	3.647	1.037	0.9644	1.3032	40′
30′	0.2705	0.2672	3.742	0.2773	3.606	1.038	0.9636	1.3003	30′
40′	0.2734	0.2700	3.703	0.2805	3.566	1.039	0.9628	1.2974	20′
50′	0.2763	0.2728	3.665	0.2836	3.526	1.039	0.9621	1.2945	10′
16° 00′	0.2793	0.2756	3.628	0.2867	3.487	1.040	0.9613	1.2915	**74° 00′**
10′	0.2822	0.2784	3.592	0.2899	3.450	1.041	0.9605	1.2886	50′
20′	0.2851	0.2812	3.556	0.2931	3.412	1.042	0.9596	1.2857	40′
30′	0.2880	0.2840	3.521	0.2962	3.376	1.043	0.9588	1.2828	30′
40′	0.2909	0.2868	3.487	0.2994	3.340	1.044	0.9580	1.2799	20′
50′	0.2938	0.2896	3.453	0.3026	3.305	1.045	0.9572	1.2770	10′
17° 00′	0.2967	0.2924	3.420	0.3057	3.271	1.046	0.9563	1.2741	**73° 00′**
10′	0.2996	0.2952	3.388	0.3089	3.237	1.047	0.9555	1.2712	50′
20′	0.3025	0.2979	3.357	0.3121	3.204	1.048	0.9546	1.2683	40′
30′	0.3054	0.3007	3.326	0.3153	3.172	1.049	0.9537	1.2654	30′
40′	0.3083	0.3035	3.295	0.3185	3.140	1.049	0.9528	1.2625	20′
50′	0.3113	0.3062	3.265	0.3217	3.108	1.050	0.9520	1.2595	10′
18° 00′	0.3142	0.3090	3.236	0.3249	3.078	1.051	0.9511	1.2566	**72° 00′**
		cos θ	sec θ	cot θ	tan θ	csc θ	sin θ	θ Radians	θ Degrees

T
A
B
L
E
S

Table 5 **Trigonometric Functions of** θ • *Degrees & Radians*

θ Degrees	θ Radians	sin θ	csc θ	tan θ	cot θ	sec θ	cos θ		
18° 00′	0.3142	0.3090	3.236	0.3249	3.078	1.051	0.9511	1.2566	**72° 00′**
10′	0.3171	0.3118	3.207	0.3281	3.047	1.052	0.9502	1.2537	50′
20′	0.3200	0.3145	3.179	0.3314	3.018	1.053	0.9492	1.2508	40′
30′	0.3229	0.3173	3.152	0.3346	2.989	1.054	0.9483	1.2479	30′
40′	0.3258	0.3201	3.124	0.3378	2.960	1.056	0.9474	1.2450	20′
50′	0.3287	0.3228	3.098	0.3411	2.932	1.057	0.9465	1.2421	10′
19° 00′	0.3316	0.3256	3.072	0.3443	2.904	1.058	0.9455	1.2392	**71° 00′**
10′	0.3345	0.3283	3.046	0.3476	2.877	1.059	0.9446	1.2363	50′
20′	0.3374	0.3311	3.021	0.3508	2.850	1.060	0.9436	1.2334	40′
30′	0.3403	0.3338	2.996	0.3541	2.824	1.061	0.9426	1.2305	30′
40′	0.3432	0.3365	2.971	0.3574	2.798	1.062	0.9417	1.2275	20′
50′	0.3462	0.3393	2.947	0.3607	2.773	1.063	0.9407	1.2246	10′
20° 00′	0.3491	0.3420	2.924	0.3640	2.747	1.064	0.9397	1.2217	**70° 00′**
10′	0.3520	0.3448	2.901	0.3673	2.723	1.065	0.9387	1.2188	50′
20′	0.3549	0.3475	2.878	0.3706	2.699	1.066	0.9377	1.2159	40′
30′	0.3578	0.3502	2.855	0.3739	2.675	1.068	0.9367	1.2130	30′
40′	0.3607	0.3529	2.833	0.3772	2.651	1.069	0.9356	1.2101	20′
50′	0.3636	0.3557	2.812	0.3805	2.628	1.070	0.9346	1.2072	10′
21° 00′	0.3665	0.3584	2.790	0.3839	2.605	1.071	0.9336	1.2043	**69° 00′**
10′	0.3694	0.3611	2.769	0.3872	2.583	1.072	0.9325	1.2014	50′
20′	0.3723	0.3638	2.749	0.3906	2.560	1.074	0.9315	1.1985	40′
30′	0.3752	0.3665	2.729	0.3939	2.539	1.075	0.9304	1.1956	30′
40′	0.3782	0.3692	2.709	0.3973	2.517	1.076	0.9293	1.1926	20′
50′	0.3811	0.3719	2.689	0.4006	2.496	1.077	0.9283	1.1897	10′
22° 00′	0.3840	0.3746	2.669	0.4040	2.475	1.079	0.9272	1.1868	**68° 00′**
10′	0.3869	0.3773	2.650	0.4074	2.455	1.080	0.9261	1.1839	50′
20′	0.3898	0.3800	2.632	0.4108	2.434	1.081	0.9250	1.1810	40′
30′	0.3927	0.3827	2.613	0.4142	2.414	1.082	0.9239	1.1781	30′
40′	0.3956	0.3854	2.595	0.4176	2.394	1.084	0.9228	1.1752	20′
50′	0.3985	0.3881	2.577	0.4210	2.375	1.085	0.9216	1.1723	10′
23° 00′	0.4014	0.3907	2.559	0.4245	2.356	1.086	0.9205	1.1694	**67° 00′**
10′	0.4043	0.3934	2.542	0.4279	2.337	1.088	0.9194	1.1665	50′
20′	0.4072	0.3961	2.525	0.4314	2.318	1.089	0.9182	1.1636	40′
30′	0.4102	0.3987	2.508	0.4348	2.300	1.090	0.9171	1.1606	30′
40′	0.4131	0.4014	2.491	0.4383	2.282	1.092	0.9159	1.1577	20′
50′	0.4160	0.4041	2.475	0.4417	2.264	1.093	0.9147	1.1548	10′
24° 00′	0.4189	0.4067	2.459	0.4452	2.246	1.095	0.9135	1.1519	**66° 00′**
10′	0.4218	0.4094	2.443	0.4487	2.229	1.096	0.9124	1.1490	50′
20′	0.4247	0.4120	2.427	0.4522	2.211	1.097	0.9112	1.1461	40′
30′	0.4276	0.4147	2.411	0.4557	2.194	1.099	0.9100	1.1432	30′
40′	0.4305	0.4173	2.396	0.4592	2.177	1.100	0.9088	1.1403	20′
50′	0.4334	0.4200	2.381	0.4628	2.161	1.102	0.9075	1.1374	10′
25° 00′	0.4363	0.4226	2.366	0.4663	2.145	1.103	0.9063	1.1345	**65° 00′**
10′	0.4392	0.4253	2.352	0.4699	2.128	1.105	0.9051	1.1316	50′
20′	0.4422	0.4279	2.337	0.4734	2.112	1.106	0.9038	1.1286	40′
30′	0.4451	0.4305	2.323	0.4770	2.097	1.108	0.9026	1.1257	30′
40′	0.4480	0.4331	2.309	0.4806	2.081	1.109	0.9013	1.1228	20′
50′	0.4509	0.4358	2.295	0.4841	2.066	1.111	0.9001	1.1199	10′
26° 00′	0.4538	0.4384	2.281	0.4877	2.050	1.113	0.8988	1.1170	**64° 00′**
10′	0.4567	0.4410	2.268	0.4913	2.035	1.114	0.8975	1.1141	50′
20′	0.4596	0.4436	2.254	0.4950	2.020	1.116	0.8962	1.1112	40′
30′	0.4625	0.4462	2.241	0.4986	2.006	1.117	0.8949	1.1083	30′
40′	0.4654	0.4488	2.228	0.5022	1.991	1.119	0.8936	1.1054	20′
50′	0.4683	0.4514	2.215	0.5059	1.977	1.121	0.8923	1.1025	10′
27° 00′	0.4712	0.4540	2.203	0.5095	1.963	1.122	0.8910	1.0996	**63° 00′**
		cos θ	sec θ	cot θ	tan θ	csc θ	sin θ	θ Radians	θ Degrees

T
A
B
L
E
S

Table 5 **Trigonometric Functions of** θ • *Degrees & Radians*   **651**

Table 5 Trigonometric Functions of θ • Degrees & Radians

θ Degrees	θ Radians	sin θ	csc θ	tan θ	cot θ	sec θ	cos θ		
27° 00′	0.4712	0.4540	2.203	0.5095	1.963	1.122	0.8910	1.0996	**63° 00′**
10′	0.4741	0.4566	2.190	0.5132	1.949	1.124	0.8897	1.0966	50′
20′	0.4771	0.4592	2.178	0.5169	1.935	1.126	0.8884	1.0937	40′
30′	0.4800	0.4617	2.166	0.5206	1.921	1.127	0.8870	1.0908	30′
40′	0.4829	0.4643	2.154	0.5243	1.907	1.129	0.8857	1.0879	20′
50′	0.4858	0.4669	2.142	0.5280	1.894	1.131	0.8843	1.0850	10′
28° 00′	0.4887	0.4695	2.130	0.5317	1.881	1.133	0.8829	1.0821	**62° 00′**
10′	0.4916	0.4720	2.118	0.5354	1.868	1.134	0.8816	1.0792	50′
20′	0.4945	0.4746	2.107	0.5392	1.855	1.136	0.8802	1.0763	40′
30′	0.4974	0.4772	2.096	0.5430	1.842	1.138	0.8788	1.0734	30′
40′	0.5003	0.4797	2.085	0.5467	1.829	1.140	0.8774	1.0705	20′
50′	0.5032	0.4823	2.074	0.5505	1.816	1.142	0.8760	1.0676	10′
29° 00′	0.5061	0.4848	2.063	0.5543	1.804	1.143	0.8746	1.0647	**61° 00′**
10′	0.5091	0.4874	2.052	0.5581	1.792	1.145	0.8732	1.0617	50′
20′	0.5120	0.4899	2.041	0.5619	1.780	1.147	0.8718	1.0588	40′
30′	0.5149	0.4924	2.031	0.5658	1.767	1.149	0.8704	1.0559	30′
40′	0.5178	0.4950	2.020	0.5696	1.756	1.151	0.8689	1.0530	20′
50′	0.5207	0.4975	2.010	0.5735	1.744	1.153	0.8675	1.0501	10′
30° 00′	0.5236	0.5000	2.000	0.5774	1.732	1.155	0.8660	1.0472	**60° 00′**
10′	0.5265	0.5025	1.990	0.5812	1.720	1.157	0.8646	1.0443	50′
20′	0.5294	0.5050	1.980	0.5851	1.709	1.159	0.8631	1.0414	40′
30′	0.5323	0.5075	1.970	0.5890	1.698	1.161	0.8616	1.0385	30′
40′	0.5352	0.5100	1.961	0.5930	1.686	1.163	0.8601	1.0356	20′
50′	0.5381	0.5125	1.951	0.5969	1.675	1.165	0.8587	1.0327	10′
31° 00′	0.5411	0.5150	1.942	0.6009	1.664	1.167	0.8572	1.0297	**59° 00′**
10′	0.5440	0.5175	1.932	0.6048	1.653	1.169	0.8557	1.0268	50′
20′	0.5469	0.5200	1.923	0.6088	1.643	1.171	0.8542	1.0239	40′
30′	0.5498	0.5225	1.914	0.6128	1.632	1.173	0.8526	1.0210	30′
40′	0.5527	0.5250	1.905	0.6168	1.621	1.175	0.8511	1.0181	20′
50′	0.5556	0.5275	1.896	0.6208	1.611	1.177	0.8496	1.0152	10′
32° 00′	0.5585	0.5299	1.887	0.6249	1.600	1.179	0.8480	1.0123	**58° 00′**
10′	0.5614	0.5324	1.878	0.6289	1.590	1.181	0.8465	1.0094	50′
20′	0.5643	0.5348	1.870	0.6330	1.580	1.184	0.8450	1.0065	40′
30′	0.5672	0.5373	1.861	0.6371	1.570	1.186	0.8434	1.0036	30′
40′	0.5701	0.5398	1.853	0.6412	1.560	1.188	0.8418	1.0007	20′
50′	0.5730	0.5422	1.844	0.6453	1.550	1.190	0.8403	0.9977	10′
33° 00′	0.5760	0.5446	1.836	0.6494	1.540	1.192	0.8387	0.9948	**57° 00′**
10′	0.5789	0.5471	1.828	0.6536	1.530	1.195	0.8371	0.9919	50′
20′	0.5818	0.5495	1.820	0.6577	1.520	1.197	0.8355	0.9890	40′
30′	0.5847	0.5519	1.812	0.6619	1.511	1.199	0.8339	0.9861	30′
40′	0.5876	0.5544	1.804	0.6661	1.501	1.202	0.8323	0.9832	20′
50′	0.5905	0.5568	1.796	0.6703	1.492	1.204	0.8307	0.9803	10′
34° 00′	0.5934	0.5592	1.788	0.6745	1.483	1.206	0.8290	0.9774	**56° 00′**
10′	0.5963	0.5616	1.781	0.6787	1.473	1.209	0.8274	0.9745	50′
20′	0.5992	0.5640	1.773	0.6830	1.464	1.211	0.8258	0.9716	40′
30′	0.6021	0.5664	1.766	0.6873	1.455	1.213	0.8241	0.9687	30′
40′	0.6050	0.5688	1.758	0.6916	1.446	1.216	0.8225	0.9657	20′
50′	0.6080	0.5712	1.751	0.6959	1.437	1.218	0.8208	0.9628	10′
35° 00′	0.6109	0.5736	1.743	0.7002	1.428	1.221	0.8192	0.9599	**55° 00′**
10′	0.6138	0.5760	1.736	0.7046	1.419	1.223	0.8175	0.9570	50′
20′	0.6167	0.5783	1.729	0.7089	1.411	1.226	0.8158	0.9541	40′
30′	0.6196	0.5807	1.722	0.7133	1.402	1.228	0.8141	0.9512	30′
40′	0.6225	0.5831	1.715	0.7177	1.393	1.231	0.8124	0.9483	20′
50′	0.6254	0.5854	1.708	0.7221	1.385	1.233	0.8107	0.9454	10′
36° 00′	0.6283	0.5878	1.701	0.7265	1.376	1.236	0.8090	0.9425	**54° 00′**
		cos θ	sec θ	cot θ	tan θ	csc θ	sin θ	θ Radians	θ Degrees

Table 5 Trigonometric Functions of θ • Degrees & Radians

θ Degrees	θ Radians	sin θ	csc θ	tan θ	cot θ	sec θ	cos θ		
36° 00′	0.6238	0.5878	1.701	0.7265	1.376	1.236	0.8090	0.9425	**54° 00′**
10′	0.6312	0.5901	1.695	0.7310	1.368	1.239	0.8073	0.9396	50′
20′	0.6341	0.5925	1.688	0.7355	1.360	1.241	0.8056	0.9367	40′
30′	0.6370	0.5948	1.681	0.7400	1.351	1.244	0.8039	0.9338	30′
40′	0.6400	0.5972	1.675	0.7445	1.343	1.247	0.8021	0.9308	20′
50′	0.6429	0.5995	1.668	0.7490	1.335	1.249	0.8004	0.9279	10′
37° 00′	0.6458	0.6018	1.662	0.7536	1.327	1.252	0.7986	0.9250	**53° 00′**
10′	0.6487	0.6041	1.655	0.7581	1.319	1.255	0.7969	0.9221	50′
20′	0.6516	0.6065	1.649	0.7627	1.311	1.258	0.7951	0.9192	40′
30′	0.6545	0.6088	1.643	0.7673	1.303	1.260	0.7934	0.9163	30′
40′	0.6574	0.6111	1.636	0.7720	1.295	1.263	0.7916	0.9134	20′
50′	0.6603	0.6134	1.630	0.7766	1.288	1.266	0.7898	0.9105	10′
38° 00′	0.6632	0.6157	1.624	0.7813	1.280	1.269	0.7880	0.9076	**52° 00′**
10′	0.6661	0.6180	1.618	0.7860	1.272	1.272	0.7862	0.9047	50′
20′	0.6690	0.6202	1.612	0.7907	1.265	1.275	0.7844	0.9018	40′
30′	0.6720	0.6225	1.606	0.7954	1.257	1.278	0.7826	0.8988	30′
40′	0.6749	0.6248	1.601	0.8002	1.250	1.281	0.7808	0.8959	20′
50′	0.6778	0.6271	1.595	0.8050	1.242	1.284	0.7790	0.8930	10′
39° 00′	0.6807	0.6293	1.589	0.8098	1.235	1.287	0.7771	0.8901	**51° 00′**
10′	0.6836	0.6316	1.583	0.8146	1.228	1.290	0.7753	0.8872	50′
20′	0.6865	0.6338	1.578	0.8195	1.220	1.293	0.7735	0.8843	40′
30′	0.6894	0.6361	1.572	0.8243	1.213	1.296	0.7716	0.8814	30′
40′	0.6923	0.6383	1.567	0.8292	1.206	1.299	0.7698	0.8785	20′
50′	0.6952	0.6406	1.561	0.8342	1.199	1.302	0.7679	0.8756	10′
40° 00′	0.6981	0.6428	1.556	0.8391	1.192	1.305	0.7660	0.8727	**50° 00′**
10′	0.7010	0.6450	1.550	0.8441	1.185	1.309	0.7642	0.8698	50′
20′	0.7039	0.6472	1.545	0.8491	1.178	1.312	0.7623	0.8668	40′
30′	0.7069	0.6494	1.540	0.8541	1.171	1.315	0.7604	0.8639	30′
40′	0.7098	0.6517	1.535	0.8591	1.164	1.318	0.7585	0.8610	20′
50′	0.7127	0.6539	1.529	0.8642	1.157	1.322	0.7566	0.8581	10′
41° 00′	0.7156	0.6561	1.524	0.8693	1.150	1.325	0.7547	0.8552	**49° 00′**
10′	0.7185	0.6583	1.519	0.8744	1.144	1.328	0.7528	0.8523	50′
20′	0.7214	0.6604	1.514	0.8796	1.137	1.332	0.7509	0.8494	40′
30′	0.7243	0.6626	1.509	0.8847	1.130	1.335	0.7490	0.8465	30′
40′	0.7272	0.6648	1.504	0.8899	1.124	1.339	0.7470	0.8436	20′
50′	0.7301	0.6670	1.499	0.8952	1.117	1.342	0.7451	0.8407	10′
42° 00′	0.7330	0.6691	1.494	0.9004	1.111	1.346	0.7431	0.8378	**48° 00′**
10′	0.7359	0.6713	1.490	0.9057	1.104	1.349	0.7412	0.8348	50′
20′	0.7389	0.6734	1.485	0.9110	1.098	1.353	0.7392	0.8319	40′
30′	0.7418	0.6756	1.480	0.9163	1.091	1.356	0.7373	0.8290	30′
40′	0.7447	0.6777	1.476	0.9217	1.085	1.360	0.7353	0.8261	20′
50′	0.7476	0.6799	1.471	0.9271	1.079	1.364	0.7333	0.8232	10′
43° 00′	0.7505	0.6820	1.466	0.9325	1.072	1.367	0.7314	0.8203	**47° 00′**
10′	0.7534	0.6841	1.462	0.9380	1.066	1.371	0.7294	0.8174	50′
20′	0.7563	0.6862	1.457	0.9435	1.060	1.375	0.7274	0.8145	40′
30′	0.7592	0.6884	1.453	0.9490	1.054	1.379	0.7254	0.8116	30′
40′	0.7621	0.6905	1.448	0.9545	1.048	1.382	0.7234	0.8087	20′
50′	0.7650	0.6926	1.444	0.9601	1.042	1.386	0.7214	0.8058	10′
44° 00′	0.7679	0.6947	1.440	0.9657	1.036	1.390	0.7193	0.8029	**46° 00′**
10′	0.7709	0.6967	1.435	0.9713	1.030	1.394	0.7173	0.7999	50′
20′	0.7738	0.6988	1.431	0.9770	1.024	1.398	0.7153	0.7970	40′
30′	0.7767	0.7009	1.427	0.9827	1.018	1.402	0.7133	0.7941	30′
40′	0.7796	0.7030	1.423	0.9884	1.012	1.406	0.7112	0.7912	20′
50′	0.7825	0.7050	1.418	0.9942	1.006	1.410	0.7092	0.7883	10′
45° 00′	0.7854	0.7071	1.414	1.0000	1.000	1.414	0.7071	0.7854	**45° 00′**
		cos θ	sec θ	cot θ	tan θ	csc θ	sin θ	θ Radians	θ Degrees

T A B L E S

Table 6 Area Under the Standard Normal Curve

z	0	1	2	3	4	5	6	7	8	9
0.0	0.0000	0.0040	0.0080	0.0120	0.0160	0.0199	0.0239	0.0279	0.0319	0.0359
0.1	0.0398	0.0438	0.0478	0.0517	0.0557	0.0596	0.0636	0.0675	0.0714	0.0754
0.2	0.0793	0.0832	0.0871	0.0910	0.0948	0.0987	0.1026	0.1064	0.1103	0.1141
0.3	0.1179	0.1217	0.1255	0.1293	0.1331	0.1368	0.1406	0.1443	0.1480	0.1517
0.4	0.1554	0.1591	0.1628	0.1664	0.1700	0.1736	0.1772	0.1808	0.1844	0.1879
0.5	0.1915	0.1950	0.1985	0.2019	0.2054	0.2088	0.2123	0.2157	0.2190	0.2224
0.6	0.2258	0.2291	0.2324	0.2357	0.2389	0.2422	0.2454	0.2486	0.2518	0.2549
0.7	0.2580	0.2612	0.2642	0.2673	0.2704	0.2734	0.2764	0.2794	0.2823	0.2852
0.8	0.2881	0.2910	0.2939	0.2967	0.2996	0.3023	0.3051	0.3078	0.3106	0.3133
0.9	0.3159	0.3186	0.3212	0.3238	0.3264	0.3289	0.3315	0.3340	0.3365	0.3389
1.0	0.3413	0.3438	0.3461	0.3485	0.3508	0.3531	0.3554	0.3577	0.3599	0.3621
1.1	0.3643	0.3665	0.3686	0.3708	0.3729	0.3749	0.3770	0.3790	0.3810	0.3830
1.2	0.3849	0.3869	0.3888	0.3907	0.3925	0.3944	0.3962	0.3980	0.3997	0.4015
1.3	0.4032	0.4049	0.4066	0.4082	0.4099	0.4115	0.4131	0.4147	0.4162	0.4177
1.4	0.4192	0.4207	0.4222	0.4236	0.4251	0.4265	0.4279	0.4292	0.4306	0.4319
1.5	0.4332	0.4345	0.4357	0.4370	0.4382	0.4394	0.4406	0.4418	0.4429	0.4441
1.6	0.4452	0.4463	0.4474	0.4484	0.4495	0.4505	0.4515	0.4525	0.4535	0.4545
1.7	0.4554	0.4564	0.4573	0.4582	0.4591	0.4599	0.4608	0.4616	0.4625	0.4633
1.8	0.4641	0.4649	0.4656	0.4664	0.4671	0.4678	0.4686	0.4693	0.4699	0.4706
1.9	0.4713	0.4719	0.4726	0.4732	0.4738	0.4744	0.4750	0.4756	0.4761	0.4767
2.0	0.4772	0.4778	0.4783	0.4788	0.4793	0.4798	0.4803	0.4808	0.4812	0.4817
2.1	0.4821	0.4826	0.4830	0.4834	0.4838	0.4842	0.4846	0.4850	0.4854	0.4857
2.2	0.4861	0.4864	0.4868	0.4871	0.4875	0.4878	0.4881	0.4884	0.4887	0.4890
2.3	0.4893	0.4896	0.4898	0.4901	0.4904	0.4906	0.4909	0.4911	0.4913	0.4916
2.4	0.4918	0.4920	0.4922	0.4925	0.4927	0.4929	0.4931	0.4932	0.4934	0.4936
2.5	0.4938	0.4940	0.4941	0.4943	0.4945	0.4946	0.4948	0.4949	0.4951	0.4952
2.6	0.4953	0.4955	0.4956	0.4957	0.4959	0.4960	0.4961	0.4962	0.4963	0.4964
2.7	0.4965	0.4966	0.4967	0.4968	0.4969	0.4970	0.4971	0.4972	0.4973	0.4974
2.8	0.4974	0.4975	0.4976	0.4977	0.4977	0.4978	0.4979	0.4979	0.4980	0.4981
2.9	0.4981	0.4982	0.4982	0.4983	0.4984	0.4984	0.4985	0.4985	0.4986	0.4986
3.0	0.4987	0.4987	0.4987	0.4988	0.4988	0.4989	0.4989	0.4989	0.4990	0.4990
3.1	0.4990	0.4991	0.4991	0.4991	0.4992	0.4992	0.4992	0.4992	0.4993	0.4993
3.2	0.4993	0.4993	0.4994	0.4994	0.4994	0.4994	0.4994	0.4995	0.4995	0.4995
3.3	0.4995	0.4995	0.4995	0.4996	0.4996	0.4996	0.4996	0.4996	0.4996	0.4997
3.4	0.4997	0.4997	0.4997	0.4997	0.4997	0.4997	0.4997	0.4997	0.4997	0.4998
3.5	0.4998	0.4998	0.4998	0.4998	0.4998	0.4998	0.4998	0.4998	0.4998	0.4998
3.6	0.4998	0.4998	0.4998	0.4998	0.4998	0.4998	0.4998	0.4998	0.4998	0.4998
3.7	0.4999	0.4999	0.4999	0.4999	0.4999	0.4999	0.4999	0.4999	0.4999	0.4999
3.8	0.4999	0.4999	0.4999	0.4999	0.4999	0.4999	0.4999	0.4999	0.4999	0.4999

T
A
B
L
E
S

Table 7 χ^2 *Distribution*

df	0.99	0.95	0.90	0.80	0.70	0.50	0.30	0.20	0.10	0.05	0.01
1	0.000157	0.0039	0.0158	0.0642	0.148	0.455	1.074	1.642	2.706	3.841	6.635
2	0.0201	0.103	0.211	0.446	0.713	1.386	2.408	3.219	4.605	5.991	9.210
3	0.115	0.352	0.584	1.005	1.424	2.366	3.665	4.642	6.251	7.815	11.341
4	0.297	0.711	1.064	1.649	2.195	3.357	4.878	5.989	7.779	9.488	13.277
5	0.554	1.145	1.610	2.343	3.000	4.351	6.064	7.289	9.236	11.070	15.086

T A B L E S

Table 7 χ^2 *Distribution* 655

I N D E X

I N D E X

INDEX

Prime factorization, 6 – 7
Prime numbers, 6 – 7
Principle of Mathematical
    Induction, 187 – 189
Probability
    and Chi-square values, 630 – 631
    and Monte Carlo method for
        finding area, TM 10-4
    and normal curve, 618
Product, *see* Multiplication
Proving identities, 417 – 419
Punnett square, 633 – 634
Pyramidal numbers, 188
Pythagorean relationships in
    trigonometry, 415

Q

Quadrants
    and reference angles, 381
    and signs of trigonometric
        functions, 362, 381
Quadratic equations, 62 – 71
    applications of, 67 – 68
    nature of roots, 66
    solving
        by completing the square, 63 – 64
        by factoring, 62 – 63
        by the quadratic formula, 64 – 65
        standard form of, 62
Quadratic form, converting higher-degree
    equations to, 74 – 75
Quadratic formula, 64 – 65
Quadratic functions, 114
Quadratic inequalities, 69 – 70
    graphs of, 485, 505 – 506
Quadric surfaces, 581 – 586
Quartiles, 607 – 608
Quotient, *see* Division

R

Radians, 357
    converting between degrees
        and, 357 – 358
    measuring angles using, 357 – 359
Radical equations, 72 – 75
Radicals
    operations with, 18 – 20

in simplest form, 19
Random sample, 622
Range
    of a distribution, 596
    of a function, 109
    interquartile, 607
Rational expressions, 33
    operations with, 38 – 39
Rational functions, 150 – 154
Rational numbers, 15 – 16
Rational Root Theorem, 86 – 91
Real numbers, 15 – 25
    field properties of, 22
    order properties of, 22 – 25
    see also Subsets of the real numbers
Reciprocal relationships in
    trigonometry, 415
Rectangular coordinates, *see* Coordinates
Recursive formula, 168
Reference angles, 380
Reflections
    line, 125 – 128
    matrices and, 323 – 325
    in origin, 128 – 129
    point, 128 – 129
    in x-axis, 141
    in y-axis, 140 – 141
Reflexive Property of Equality, 4
Remainder Theorem, 80 – 81, 85
Repeating decimals, 15 – 16
Resultant of vectors, 244
Revolution, surfaces of, 577 – 580
Right circular cones, 579
Right cylinders, 575
Right triangles
    applying trigonometry in, 372 – 379
    defining trigonometric
        functions in, 360 – 362
Roots, 49 – 107
    approximate, of polynomial
        equations, 93 – 101
    of complex numbers, 555 – 557
    finding by graphing, 78 – 82, 94, 99
    of an open sentence, 49
    of a quadratic equation, 63
Rose curves, 540 – 541
Rotations, 130
    matrices and, 326 – 327
    using polar coordinates, 536, TM 11-4

Row operations, 308

S

Sample(s), 622
   distribution of means of, 624 – 626
   mean and standard
     deviation of, 623 – 624
   random, 622
   stratified, 622
   systematic, 622
Sampling, 622 – 628
Scalar multiplication
   of matrices, 296 – 297
   of vectors, 246 – 247
Scalars, 246, 296
Scale change, 326
Scatter plot, 394
Secant function, 351
   graph of, 401
Secret codes, 335
Sentences, open, *see* Open sentences
Sequences
   arithmetic, 168 – 174
   common difference in, 168
   common ratio in, 175
   geometric, 175 – 179
   infinite, 163 – 164
   of numbers, 163 – 167
   sum of, 164 – 165
Series, 164
   geometric, 180 – 185
Shear, simple, 328 – 329
Sigma (Σ) notation, 164 – 165
Simple harmonic motion, 464 – 468
Simple shear, 328 – 329
Sine function, 350
   amplitude of, 390
   graph of, 387 – 398
   inverse of, 443
   period of, 387
   right-triangle definition of, 372
Singular matrices, 320 – 321
Slope-intercept form, 111
Snell's law of refraction of light, 371
Solution, 49
   algebraic, of a problem, 54
   extraneous, 72

Solving equations
   containing radicals, 72 – 74
   of degree greater than 2,
     78 – 79, 86 – 90
   first degree, 49 – 51
   with fractional exponents, 212
   literal, 52
   polynomial, 78 – 101
   quadratic, 62 – 65
   rational roots, 94 – 96
Solving inequalities
   first degree, 50 – 51
   quadratic, 69 – 70
Solving systems, *see* Systems of equations
Solving triangles, *see* Triangles
Sound waves, frequency of, 396
Space
   lines in, 268 – 273, 565 – 569
   planes in, 565 – 569
   vectors in, 252 – 256
Spheres, 570 – 572, 582
   surface areas of, 571 – 572
   translations of, 571
   volumes of, 571
Square matrices, 287
   determinants of, 288 – 290
Square numbers, 188
Square-root function, 116
Square roots, 18
Squares and square roots table, 645
Standard deviation, 614 – 615
   from a sample, 623 – 624
Standard error of mean, 624 – 625
Standard normal curve, 618 – 620
Statistical methods, 593
Statistics, 593 – 642
   defined, 593
Stem-and-leaf diagram, 593 – 594
Step function, 115
Stratified sample, 622
Subscripts, 163
Subsets of the real numbers
   integers, 3
   irrational numbers, 16
   natural numbers, 2
   rational numbers, 15
   whole numbers, 3
Subtraction
   of complex numbers, 27 – 28

I
N
D
E
X